FUNDAMENTALS OF COLLEGE
PHYSICS

SECOND EDITION

PETER J. NOLAN

State University of New York
at Farmingdale

WCB **Wm. C. Brown Publishers**

Dubuque, IA Bogota Boston Buenos Aires Caracas Chicago
Guilford, CT London Madrid Mexico City Sydney Toronto

Book Team

Editor *Megan Johnson*
Developmental Editor *Tom Riley*
Production Editor *Carol M. Besler*
Photo Editor *Lori Hancock*
Permissions Coordinator *Karen L. Storlie*
Art Processor *Joyce Watters*
Publishing Services Coordinator *Barbara J. Hodgson*

Wm. C. Brown Publishers
A Division of Wm. C. Brown Communications, Inc.

Vice President and General Manager *Beverly Kolz*
Vice President, Publisher *Jeffrey L. Hahn*
Vice President, Director of Sales and Marketing *Virginia S. Moffat*
Vice President, Director of Production *Colleen A. Yonda*
National Sales Manager *Douglas J. DiNardo*
Marketing Manager *Amy Halloran*
Advertising Manager *Janelle Keeffer*
Production Editorial Manager *Renée Menne*
Publishing Services Manager *Karen J. Slaght*
Royalty/Permissions Manager *Connie Allendorf*

Wm. C. Brown Communications, Inc.

President and Chief Executive Officer *G. Franklin Lewis*
Senior Vice President, Operations *James H. Higby*
Corporate Senior Vice President, President of WCB Manufacturing *Roger Meyer*
Corporate Senior Vice President and Chief Financial Officer *Robert Chesterman*

Copyedited by Patricia Steele

Cover/interior design by Rokusek Design

Cover/part opener images © Fundamental Photographs

The credits section for this book begins on page C–1 and is considered an extension of the copyright page.

A Times Mirror Company

Library of Congress Catalog Card Number: 94–70898

ISBN Fundamentals of College Physics   Casebound: 0–697–23138–0
ISBN Fundamentals of College Physics, Volume 1   Paperbound: 0–697–24392–3
ISBN Fundamentals of College Physics, Volume 2   Paperbound: 0–697–24393–1

Printed in the United States of America by Wm. C. Brown Communications, Inc., 2460 Kerper Boulevard, Dubuque, IA 52001

10 9 8 7 6 5 4 3 2 1

This book is dedicated to my sons—Thomas, James, John Michael, and Kevin—the joy of my life.

Contents In Brief

Contents

P A R T   T W O

Vibratory Motion, Wave Motion, and Fluids, 280

10 Elasticity, 283

11 Simple Harmonic Motion, 297

12 Wave Motion, 321

13 Fluids, 367

Contents

23 Electromagnetic Induction, 661

24 Alternating Current Circuits, 687

25 Maxwell's Equations and Electromagnetic Waves, 715

PART FIVE

Light and Optics, 744

26 The Law of Reflection, 747

27 The Law of Refraction, 765

28 Physical Optics, 807

PART SIX

Modern Physics, 840

29 Special Relativity, 843

30 Spacetime and General Relativity, 891

31 Quantum Physics, 921

Contents

Preface

This book is an outgrowth of more than twenty-five years of teaching college physics. The original manuscript has been used in the classroom for several of them, and the first edition is used in over one hundred colleges and universities throughout the country. Because the book is written for students, it contains a great many of the intermediate steps that are often left out of the derivations and illustrative problem solutions in many traditional college physics textbooks. Students new to physics often find it difficult to follow derivations when the intermediate steps are left out. In addition, the units of measurement are carried along, step by step, in the equations to make it easier for students to understand. This book does not require calculus; the only prerequisites are high school algebra and trigonometry. In fact, a short review of trigonometry is given in chapter 2, before the discussion of the components of a vector.

This text gives a good, fairly rigorous, traditional college physics coverage. Instructors are expected to choose those topics they deem most important for the particular course. Students can read on their own the detailed descriptions found in those chapters, or parts of chapters, omitted from the course. Unfortunately, many interesting and important topics in modern physics are never covered in college physics courses because of lack of time. These chapters, especially, are written in even more detail to enable students to read them on their own. Even years after taking the course, students can read these sections for their own edification and enjoyment. This is one of the reasons that students should never sell any of their college textbooks. They are an investment for a lifetime of reference, illumination, and relaxation.

The organization of the text follows the traditional sequence of mechanics, wave motion, heat, electricity and magnetism, optics, and modern physics. The emphasis throughout the book is on clarity. The book starts out at a very simple level, and advances as students' understanding grows.

Color has been used extensively throughout, especially in the diagrams, to help students visualize the material. To standardize the colors used, each main part of the text uses a specific color scheme. This color scheme is spelled out in the Color as a Study Aid sections at the beginning of each part. On a few occasions it was necessary to depart from the standards to avoid confusion. In addition, some standard colors, like red and blue, may mean one thing in one Part but something quite different in another Part. This is unavoidable because there are not enough different colors to account for every physical quantity. However, the colors are consistent within any one Part of the book.

There are a large number of diagrams and illustrative problems in the text to help students visualize physical ideas. Important equations are highlighted to help students find and recognize them. A summary of these important equations is given at the end of each chapter.

THE LEARNING SYSTEM

c. The image distance for the second lens, found from equation 27.26, is

$$\frac{1}{q_2} = \frac{1}{f_2} - \frac{1}{p_2} = \frac{1}{(-4.50\ \text{cm})} - \frac{1}{(-5.4\ \text{cm})}$$

$$q_2 = -27\ \text{cm}$$

Because q_2 is negative the final image is virtual.

Have you ever wondered how the human eye works? It is one of the greatest marvels of nature, but one that is usually taken for granted. The human eye is very much like a camera. It contains a lens, an iris diaphragm to let the light in, and its film is the retina at the back of the eye, figure 1. The lens focuses incident light onto the retina so a sharp image is observed. The iris diaphragm opens wide in subdued lighting and closes down in very bright light.

In the eye, the distance from the lens to the retina *q*, does not change as in a camera, instead the shape of the lens is changed by the muscles and ligaments of the eye. *Hence, the eye changes its focal length in order to bring objects at different object distances to a focus at the same image distance q, the distance from the lens to the retina. This changing of the focal length of the eye is called* **accommodation**. A schematic of the normal eye is shown in figure 2(a). The greatest amount of bending of light in the eye occurs at the cornea, because it is here where the greatest difference in the index of refraction occurs. The index of refraction of air is of course equal to one, whereas the index of refraction of the cornea is about 1.35. The index of refraction of the lens is about 1.44; that of the aqueous and vitreous humor is about 1.34.

Some eyes do not have the ability to change the shape of the lens as necessary and this is referred to as a defect in vision. One such defect is called **hyperopia, or farsightedness,** and is shown in figure 2(b). *A farsighted person can see distant objects*

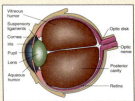

Figure 1
The human eye.

Vitreous humor / Suspensory ligaments / Cornea / Iris / Pupil / Lens / Aqueous humor / Optic disk / Optic nerve / Posterior cavity / Retina

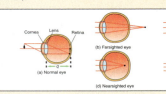

Figure 2
Defects of the human eye.

Cornea / Lens / Retina
(a) Normal eye
(b) Farsighted eye
(c) Farsighted—corrected
(d) Nearsighted eye
(e) Nearsighted—corrected

---

clearly, but objects a short distance away are blurred. The lens is not capable of bending the incoming light of near objects to a focus on the retina. Instead the light is focused behind the retina, leaving blurred vision. The defect is easily remedied by placing a converging lens in front of the eye, as shown in figure 2(c). The converging lens brings these incoming rays to a focus at the retina. Farsightedness is a defect that occurs to most people as they age. The lens hardens with age and the muscles weaken and are no longer capable of shaping the lens as needed.

Another defect is **myopia, or nearsightedness,** as illustrated in figure 2(d). *A nearsighted person can see nearby objects clearly, but distant objects are blurred.* The distant objects come to a focus in front of the retina leaving a blurred image on the retina. This defect can be eliminated by placing a diverging lens in front of the eye. The diverging lens causes the rays to converge more slowly and hence the image is formed farther back at the retina, as shown in figure 2(e).

The minimum distance from the eye at which an object can be seen distinctly is called the **near point of the eye.** *For the average person, the near point is about 25.0 cm.* That is, an object at a distance *p* of 25.0 cm is the minimum distance that an object can be placed to give a distinct image on the retina. If the distance is decreased below this value, the image will be blurred for most people. Remember, this is only an average value, as we get older, the minimum distance of distinct vision increases. For example, the near point can be as close as 6.00 cm for a child and can increase to over a 100 cm for an elderly person.

Another common defect of the eye is called *astigmatism*. Astigmatism occurs in some people, because their eye is not completely spherical. That is, the radius of curvature of the eye in the vertical direction is not the same as the radius of curvature in the horizontal direction. Hence, vertical rays do not converge to the same position as horizontal rays. This defect is usually corrected with a lens of cylindrical curvature.

Although the magnification of a lens was defined in equation 27.22 as $M = h_i/h_o$, when viewing an object with the eye it is sometimes more desirable to talk about the angular magnification of the lens. Figure 3(a) shows an object at three distances from the unaided eye. The size of the object, as determined by the unaided eye, depends on the angle subtended by the object at the eye. Thus, when the object is at position 1, it subtends an angle θ_1; when it is at position 2, it subtends the angle θ_2; and when it is at position 3, it subtends the angle θ_3. As seen from the diagram, $\theta_1 < \theta_2 < \theta_3$. Thus, the closer the object is to the eye, the greater the angle subtended and the easier it is for the eye to see the object. However, the object cannot be moved closer to the eye than the near point, because that is the closest point that the eye can still see the object distinctly. In figure 3(b), the object is placed at the object distance $p_1 = 25.0$ cm in front of a normal eye. The height of the object h_o subtends an angle α at the lens of the eye. The image of the object is focused on the retina of the eye. This is the largest image that the unaided eye can make of the object. If the object is placed within the focal point of a converging lens, the object can be brought even closer

Range of focal length for the eye. The distance from the cornea to the retina in the human eye is about 20.0 mm. What should the focal length of the human eye be in order to see clearly an object (a) at infinity and (b) at the near point of most distinct vision, 25 cm?

Solution

a. For the object at infinity, the image is located at the principal focus. Hence

$$f = q = 20.0\ \text{mm} = 2.00\ \text{cm}$$

is the focal length of the eye when it views an object at infinity.

b. For the object at $p = 25.0$ cm, the focal length would have to be

$$\frac{1}{f} = \frac{1}{p} + \frac{1}{q} = \frac{1}{25.0\ \text{cm}} + \frac{1}{2.00\ \text{cm}} = 0.540\ \text{cm}^{-1}$$

$$f = 1.85\ \text{cm}$$

Thus, the eye only has to change its focal length by (2.00 cm − 1.85 cm = 0.15 cm) 15 mm to see its entire range of vision.

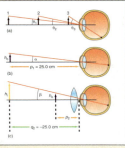

(a) / (b) / (c)
$p_1 = 25.00$ cm
$q_2 = -25.00$ cm

Figure 3
Angular magnification and the simple microscope.
A doctor examines a patient's eye.

From figure 3(c), we see that

$$\tan\beta = \frac{h_o}{p_2}$$

whereas from figure 3(b), we see that

$$\tan\alpha = \frac{h_o}{p_1} = \frac{h_o}{25.00\ \text{cm}}$$

Let us take the ratio

$$\frac{\tan\beta}{\tan\alpha} = \frac{h_o/p_2}{h_o/25.00\ \text{cm}}$$

$$= \frac{25.00\ \text{cm}}{p_2}$$

Because the angles considered are small, we can use the small-angle approximation, setting the tangent of an angle equal to the angle itself. Hence,

$$\frac{\beta}{\alpha} = \frac{25.00\ \text{cm}}{p_2}$$

But this ratio is the definition of the angular magnification, equation 27H.1, hence

$$M_A = \frac{25.00\ \text{cm}}{p_2} \qquad (27H.2)$$

The value of p_2, found from the lens equation, is

$$\frac{1}{p_2} = \frac{1}{f} - \frac{1}{q_2}$$

$$\frac{1}{p_2} = \frac{1}{f} - \frac{1}{-25.00\ \text{cm}} = \frac{25.00\ \text{cm} + f}{(25.00\ \text{cm})f} \qquad (27H.3)$$

And the object distance for the image to be at the near point is

$$p_2 = \frac{(25.00\ \text{cm})f}{25.00\ \text{cm} + f} \qquad (27H.4)$$

Substituting equation 27H.4 into equation 27H.2, gives

$$M_A = \frac{\beta}{\alpha} = \frac{25.00\ \text{cm}}{p_2} = \frac{25.00\ \text{cm}}{(25.00\ \text{cm})f/(25.00\ \text{cm} + f)}$$

$$= \frac{25.00\ \text{cm}}{f} + 1 \qquad (27H.5)$$

Equation 27H.5 gives the angular magnification of a converging lens when the image is located at the near point of the eye, which is assumed to be 25.0 cm for the normal eye.

$$= \frac{25.00\ \text{cm}}{10.00\ \text{cm}} + 1$$

$$= 3.5$$

The eye is more relaxed when viewing an object at infinity than at the near point. It is, therefore, sometimes more desirable to view an image at infinity than at the near point. For the image to be at infinity, the object must be placed at the principal focus. The lens equation, equation 27H.3, shows that if p is at infinity, then the object distance p_2 is equal to the focal length f. Hence substituting f for p_2 in equation 27H.2, gives

$$M_A = \frac{25.00\ \text{cm}}{f} \qquad (27H.6)$$

Equation 27H.6 gives the angular magnification of a simple microscope when the image is set for viewing at infinity.

The angular magnification of a lens when the image is at infinity. A magnifying glass of 10.0-cm focal length is used to view an object, such that the image is at infinity. Find the angular magnification of the lens.

Solution

The angular magnification, found from equation 27H.6, is

$$M_A = \frac{25.00\ \text{cm}}{f}$$

$$= \frac{25.00\ \text{cm}}{10.00\ \text{cm}} = 2.50$$

Notice that when the image is viewed at the near point the angular magnification is 3.5, whereas when viewed at infinity, it is 2.5. If viewing is for a long time, it might be desirable to view the image at infinity because it is easier on the eyes. If the magnification is the most important consideration, then the image should be viewed at the near point of the eye.

"Have You Ever Wondered?" boxes focus on the applied nature of physics by showing students how physics affects their everyday lives.

THE LEARNING SYSTEM

End-of-chapter *"Language of Physics"* features key concepts and terms that are defined and page-referenced.

The Language of Physics

Kinematics
The branch of mechanics that describes the motion of a body without regard to the cause of that motion (p. 39).

Average velocity
The average rate at which the displacement vector changes with time. Since a displacement is a vector, the velocity is also a vector (p. 39).

Average speed
The distance that a body moves per unit time. Speed is a scalar quantity (p. 41).

Constant velocity
A body moving in one direction in such a way that it always travels equal distances in equal times (p. 42).

Acceleration
The rate at which the velocity of a moving body changes with time (p. 43).

Instantaneous velocity
The velocity at a particular instant of time. It is defined as the limit of the ratio of the change in the displacement of the body to the change in time, as the time interval approaches zero. The magnitude of the instantaneous velocity is the instantaneous speed of the moving body (p. 45).

Kinematic equations of linear motion
A set of equations that gives the displacement and velocity of the moving body at any instant of time, and the velocity of the moving body at any displacement, if the acceleration of the body is a constant (p. 46).

Freely falling body
Any body that is moving under the influence of gravity only. Hence, any body that is dropped or thrown on the surface of the earth is a freely falling body (p. 51).

Acceleration due to gravity
If air friction is ignored, all objects that are dropped near the surface of the earth, are accelerated toward the center of the earth with an acceleration of 9.80 m/s² or 32 ft/s² (p. 52).

Projectile motion
The motion of a body thrown or fired with an initial velocity v_0 in a gravitational field (p. 55).

Trajectory
The path through space followed by a projectile (p. 56).

Range of a projectile
The horizontal distance from the point where the projectile is launched to the point where it returns to its launch height (p. 64).

Summary of Important Equations

"Summary of Important Equations" provides a quick reference of equations presented in each chapter.

Average velocity
$$v_{avg} = \frac{\Delta r}{\Delta t} = \frac{r_2 - r_1}{t_2 - t_1} \tag{3.32}$$

Acceleration
$$a = \frac{\Delta v}{\Delta t} = \frac{v - v_0}{t} \tag{3.33}$$

Instantaneous velocity in two or more directions, which is a generalization of the instantaneous velocity in one dimension
$$v = \lim_{\Delta t \to 0} \frac{\Delta r}{\Delta t}$$
$$v = \lim_{\Delta t \to 0} \frac{\Delta x}{\Delta t} \tag{3.8}$$

Velocity at any time
$$v = v_0 + at \tag{3.35}$$

Displacement at any time
$$r = v_0 t + \frac{1}{2}at^2 \tag{3.34}$$

Velocity at any displacement in the x-direction
$$v^2 = v_0^2 + 2ax \tag{3.16}$$

Velocity at any displacement in the y-direction
$$v^2 = v_0^2 + 2ay \tag{3.16}$$

For Projectile Motion

x-displacement
$$x = v_{0x} t \tag{3.38}$$

y-displacement
$$y = v_{0y} t - \frac{1}{2}gt^2 \tag{3.39}$$

x-component of velocity
$$v_x = v_{0x} \tag{3.40}$$

y-component of velocity
$$v_y = v_{0y} - gt \tag{3.41}$$

y-component of velocity at any height y
$$v_y^2 = v_{0y}^2 - 2gy \tag{3.48}$$

Range
$$R = \frac{v_0^2 \sin 2\theta}{g} \tag{3.47}$$

Questions for Chapter 3

"Questions" help students review key chapter concepts and prepare for *"Problems."*

1. Discuss the difference between distance and displacement.
2. Discuss the difference between speed and velocity.
3. Discuss the difference between average speed and instantaneous speed.
† 4. Although speed is the magnitude of the instantaneous velocity, is the average speed equal to the magnitude of the average velocity?
5. Why can the kinematic equations be used only for motion at constant acceleration?
6. In dealing with average velocities discuss the statement, "Straight line motion at 60 mph for 1 hr followed by motion in the same direction at 30 mph for 2 hr does not give an average of 45 mph but rather 40 mph."

7. What effect would air resistance have on the velocity of a body that is dropped near the surface of the earth?
8. What is the acceleration of a projectile when its instantaneous vertical velocity is zero at the top of its trajectory?
9. Can an object have zero velocity at the same time that it has an acceleration? Explain and give some examples.
10. Can the velocity of an object be in a different direction than the acceleration? Give some examples.
11. Can you devise a means of using two clocks to measure your reaction time?
†12. A person on a moving train throws a ball straight upward. Describe the

motion as seen by a person on the train and by a person on the station platform.
13. You are in free fall, and you let go of your watch. What is the relative velocity of the watch with respect to you?
†14. What kind of motion is indicated by a graph of displacement versus time, if the slope of the curve is (a) horizontal, (b) sloping upward to the right, and (c) sloping downward?
†15. What kind of motion is indicated by a graph of velocity versus time, if the slope of the curve is (a) horizontal, (b) sloping upward at a constant value, (c) sloping upward at a changing rate, (d) sloping downward at a constant value, and (e) sloping downward at a changing rate?

Chapter 3 Kinematics—The Study of Motion

THE LEARNING SYSTEM

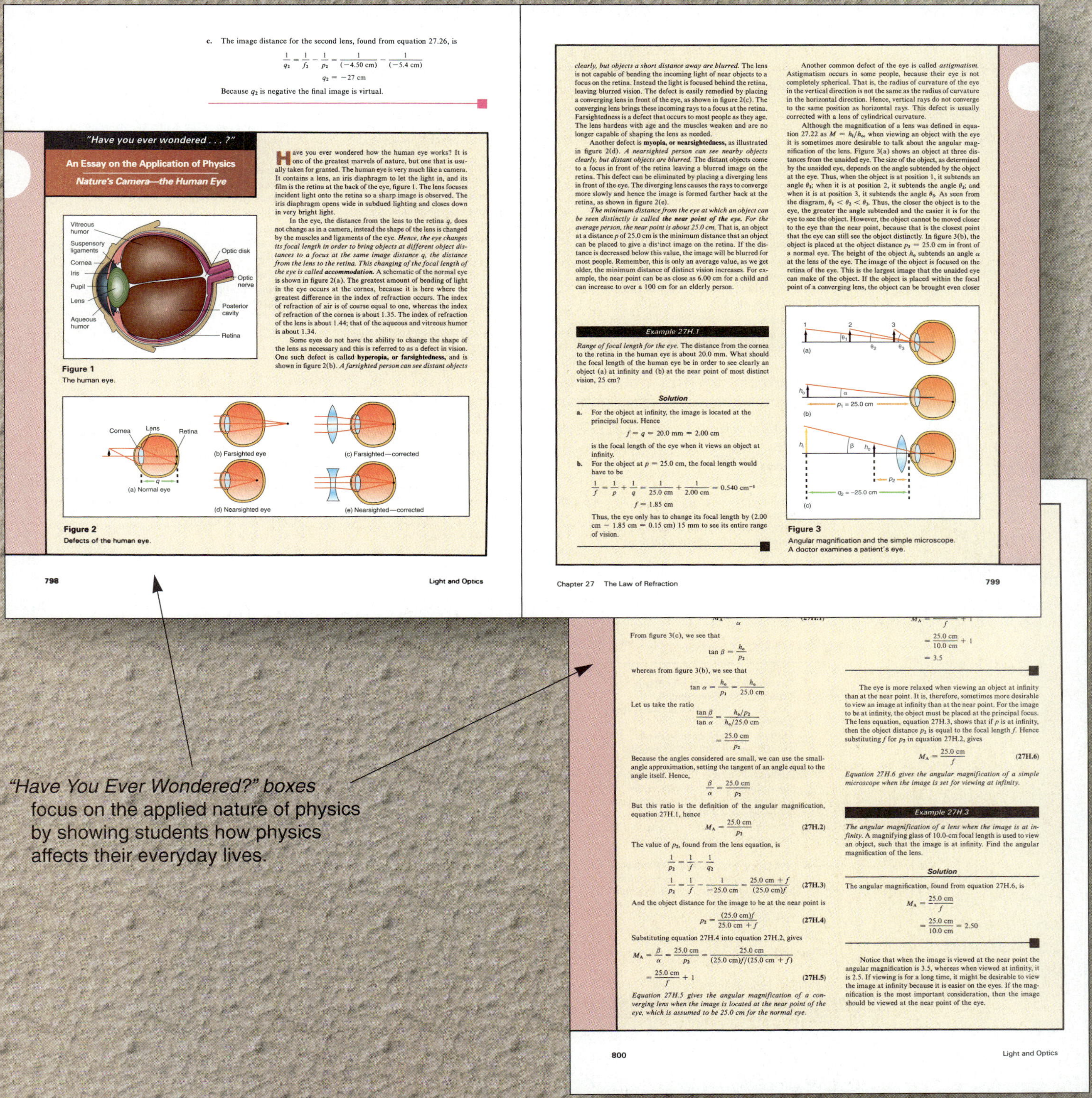

Page 798

c. The image distance for the second lens, found from equation 27.26, is

$$\frac{1}{q_2} = \frac{1}{f_2} - \frac{1}{p_2} = \frac{1}{(-4.50 \text{ cm})} - \frac{1}{(-5.4 \text{ cm})}$$

$$q_2 = -27 \text{ cm}$$

Because q_2 is negative the final image is virtual.

"Have you ever wondered . . . ?"

An Essay on the Application of Physics

Nature's Camera—the Human Eye

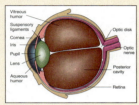

Have you ever wondered how the human eye works? It is one of the greatest marvels of nature, but one that is usually taken for granted. The human eye is very much like a camera. It contains a lens, an iris diaphragm to let the light in, and its film is the retina at the back of the eye, figure 1. The lens focuses incident light onto the retina so a sharp image is observed. The iris diaphragm opens wide in subdued lighting and closes down in very bright light.

In the eye, the distance from the lens to the retina q, does not change as in a camera, instead the shape of the lens is changed by the muscles and ligaments of the eye. Hence, *the eye changes its focal length in order to bring objects at different object distances to a focus at the same image distance q, the distance from the lens to the retina. This changing of the focal length of the eye is called accommodation.* A schematic of the normal eye is shown in figure 2(a). The greatest amount of bending of light in the eye occurs at the cornea, because it is here where the greatest difference in the index of refraction occurs. The index of refraction of air is of course equal to one, whereas the index of refraction of the cornea is about 1.35. The index of refraction of the lens is about 1.44; that of the aqueous and vitreous humor is about 1.34.

Some eyes do not have the ability to change the shape of the lens as necessary and this is referred to as a defect in vision. One such defect is called **hyperopia, or farsightedness,** and is shown in figure 2(b). *A farsighted person can see distant objects*

Figure 1
The human eye.

Figure 2
Defects of the human eye.

(a) Normal eye
(b) Farsighted eye
(c) Farsighted—corrected
(d) Nearsighted eye
(e) Nearsighted—corrected

Page 799

clearly, but objects a short distance away are blurred. The lens is not capable of bending the incoming light of near objects to a focus on the retina. Instead the light is focused behind the retina, leaving blurred vision. The defect is easily remedied by placing a converging lens in front of the eye, as shown in figure 2(c). The converging lens brings these incoming rays to a focus at the retina. Farsightedness is a defect that occurs to most people as they age. The lens hardens with age and the muscles weaken and are no longer capable of shaping the lens as needed.

Another defect is **myopia, or nearsightedness,** as illustrated in figure 2(d). *A nearsighted person can see nearby objects clearly, but distant objects are blurred.* The distant objects come to a focus in front of the retina leaving a blurred image on the retina. This defect can be eliminated by placing a diverging lens in front of the eye. The diverging lens causes the rays to converge more slowly and hence the image is formed farther back at the retina, as shown in figure 2(e).

The minimum distance from the eye at which an object can be seen distinctly is called the near point of the eye. For the average person, the near point is about 25.0 cm. That is, an object can be placed to give a distinct image on the retina. If the distance is decreased below this value, the image will be blurred for most people. Remember, this is only an average value, as we get older, the minimum distance of distinct vision increases. For example, the near point can be as close as 6.00 cm for a child and can increase to over a 100 cm for an elderly person.

Example 27H.1

Range of focal length for the eye. The distance from the cornea to the retina in the human eye is about 20.0 mm. What should the focal length of the human eye be in order to see clearly an object (a) at infinity and (b) at the near point of most distinct vision, 25 cm?

Solution

a. For the object at infinity, the image is located at the principal focus. Hence

$$f = q = 20.0 \text{ mm} = 2.00 \text{ cm}$$

is the focal length of the eye when it views an object at infinity.

b. For the object at $p = 25.0$ cm, the focal length would have to be

$$\frac{1}{f} = \frac{1}{p} + \frac{1}{q} = \frac{1}{25.0 \text{ cm}} + \frac{1}{2.00 \text{ cm}} = 0.540 \text{ cm}^{-1}$$

$$f = 1.85 \text{ cm}$$

Thus, the eye only has to change its focal length by (2.00 cm − 1.85 cm = 0.15 cm) 15 mm to see its entire range of vision.

Another common defect of the eye is called *astigmatism.* Astigmatism occurs in some people, because their eye is not completely spherical. That is, the radius of curvature of the eye in the vertical direction is not the same as the radius of curvature in the horizontal direction. Hence, vertical rays do not converge to the same position as horizontal rays. This defect is usually corrected with a lens of cylindrical curvature.

Although the magnification of a lens was defined in equation 27.22 as $M = h_i/h_o$, when viewing an object with the eye it is sometimes more desirable to talk about the angular magnification of the lens. Figure 3(a) shows an object at three distances from the unaided eye. The size of the object, as determined by the unaided eye, depends on the angle subtended by the object at the eye. Thus, when the object is at position 1, it subtends an angle θ_1; when it is at position 2, it subtends the angle θ_2; and when it is at position 3, it subtends the angle θ_3. As seen from the diagram, $\theta_1 < \theta_2 < \theta_3$. Thus, the closer the object is to the eye, the greater the angle subtended and the easier it is for the eye to see the object. However, the object cannot be moved closer to the eye than the near point, because that is the closest point that the eye can still see the object distinctly. In figure 3(b), the object is placed at the object distance $p_1 = 25.0$ cm in front of a normal eye. The height of the object h_o subtends an angle α at the lens of the eye. The image of the object is focused on the retina of the eye. This is the largest image that the unaided eye can make of the object. If the object is placed within the focal point of a converging lens, the object can be brought even closer

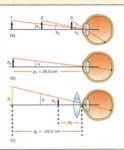

Figure 3
Angular magnification and the simple microscope.
A doctor examines a patient's eye.

Page 800

From figure 3(c), we see that

$$\tan \beta = \frac{h_o}{p_3}$$

whereas from figure 3(b), we see that

$$\tan \alpha = \frac{h_o}{p_1} = \frac{h_o}{25.0 \text{ cm}}$$

Let us take the ratio

$$\frac{\tan \beta}{\tan \alpha} = \frac{h_o/p_3}{h_o/25.0 \text{ cm}} = \frac{25.0 \text{ cm}}{p_3}$$

Because the angles considered are small, we can use the small-angle approximation, setting the tangent of an angle equal to the angle itself. Hence,

$$\frac{\beta}{\alpha} = \frac{25.0 \text{ cm}}{p_3}$$

But this ratio is the definition of the angular magnification, equation 27H.1, hence

$$M_A = \frac{25.0 \text{ cm}}{p_3} \tag{27H.2}$$

The value of p_3, found from the lens equation, is

$$\frac{1}{p_3} = \frac{1}{f} - \frac{1}{q_3}$$

$$\frac{1}{p_3} = \frac{1}{f} - \frac{1}{-25.0 \text{ cm}} = \frac{25.0 \text{ cm} + f}{(25.0 \text{ cm})f} \tag{27H.3}$$

And the object distance for the image to be at the near point is

$$p_3 = \frac{(25.0 \text{ cm})f}{25.0 \text{ cm} + f} \tag{27H.4}$$

Substituting equation 27H.4 into equation 27H.2, gives

$$M_A = \frac{\beta}{\alpha} = \frac{25.0 \text{ cm}}{p_3} = \frac{25.0 \text{ cm}}{(25.0 \text{ cm})f/(25.0 \text{ cm} + f)}$$

$$= \frac{25.0 \text{ cm}}{f} + 1 \tag{27H.5}$$

Equation 27H.5 gives the angular magnification of a converging lens when the image is located at the near point of the eye, which is assumed to be 25.0 cm for the normal eye.

$$= \frac{25.0 \text{ cm}}{10.0 \text{ cm}} + 1$$

$$= 3.5$$

The eye is more relaxed when viewing an object at infinity than at the near point. It is, therefore, sometimes more desirable to view an image at infinity than at the near point. For the image to be at infinity, the object must be placed at the principal focus. The lens equation, equation 27H.3, shows that if p is at infinity, then the object distance p_3 is equal to the focal length f. Hence substituting f for p_3 in equation 27H.2, gives

$$M_A = \frac{25.0 \text{ cm}}{f} \tag{27H.6}$$

Equation 27H.6 gives the angular magnification of a simple microscope when the image is set for viewing at infinity.

Example 27H.3

The angular magnification of a lens when the image is at infinity. A magnifying glass of 10.0-cm focal length is used to view an object, such that the image is at infinity. Find the angular magnification of the lens.

Solution

The angular magnification, found from equation 27H.6, is

$$M_A = \frac{25.0 \text{ cm}}{f}$$

$$= \frac{25.0 \text{ cm}}{10.0 \text{ cm}} = 2.50$$

Notice that when the image is viewed at the near point the angular magnification is 3.5, whereas when viewed at infinity, it is 2.5. If viewing is for a long time, it might be desirable to view the image at infinity because it is easier on the eyes. If the magnification is the most important consideration, then the image should be viewed at the near point of the eye.

"Have You Ever Wondered?" boxes focus on the applied nature of physics by showing students how physics affects their everyday lives.

THE LEARNING SYSTEM

End-of-chapter "Language of Physics" features key concepts and terms that are defined and page-referenced.

The Language of Physics

Kinematics
The branch of mechanics that describes the motion of a body without regard to the cause of that motion (p. 39).

Average velocity
The average rate at which the displacement vector changes with time. Since a displacement is a vector, the velocity is also a vector (p. 39).

Average speed
The distance that a body moves per unit time. Speed is a scalar quantity (p. 41).

Constant velocity
A body moving in one direction in such a way that it always travels equal distances in equal times (p. 42).

Acceleration
The rate at which the velocity of a moving body changes with time (p. 43).

Instantaneous velocity
The velocity at a particular instant of time. It is defined as the limit of the ratio of the change in the displacement of the body to the change in time, as the time interval approaches zero. The magnitude of the instantaneous velocity is the instantaneous speed of the moving body (p. 45).

Kinematic equations of linear motion
A set of equations that gives the displacement and velocity of the moving body at any instant of time, and the velocity of the moving body at any displacement, if the acceleration of the body is a constant (p. 46).

Freely falling body
Any body that is moving under the influence of gravity only. Hence, any body that is dropped or thrown on the surface of the earth is a freely falling body (p. 51).

Acceleration due to gravity
If air friction is ignored, all objects that are dropped near the surface of the earth, are accelerated toward the center of the earth with an acceleration of 9.80 m/s² or 32 ft/s² (p. 52).

Projectile motion
The motion of a body thrown or fired with an initial velocity v_0 in a gravitational field (p. 55).

Trajectory
The path through space followed by a projectile (p. 56).

Range of a projectile
The horizontal distance from the point where the projectile is launched to the point where it returns to its launch height (p. 64).

Summary of Important Equations

"Summary of Important Equations" provides a quick reference of equations presented in each chapter.

Average velocity
$$\mathbf{v}_{avg} = \frac{\Delta \mathbf{r}}{\Delta t} = \frac{\mathbf{r}_2 - \mathbf{r}_1}{t_2 - t_1} \tag{3.32}$$

Acceleration
$$\mathbf{a} = \frac{\Delta \mathbf{v}}{\Delta t} = \frac{\mathbf{v} - \mathbf{v}_0}{t} \tag{3.33}$$

Instantaneous velocity in two or more directions, which is a generalization of the instantaneous velocity in one dimension
$$\mathbf{v} = \lim_{\Delta t \to 0} \frac{\Delta \mathbf{r}}{\Delta t}$$
$$v = \lim_{\Delta t \to 0} \frac{\Delta x}{\Delta t} \tag{3.8}$$

Velocity at any time
$$\mathbf{v} = \mathbf{v}_0 + \mathbf{a}t \tag{3.35}$$

Displacement at any time
$$\mathbf{r} = \mathbf{v}_0 t + \tfrac{1}{2}\mathbf{a}t^2 \tag{3.34}$$

Velocity at any displacement in the x-direction
$$v^2 = v_0^2 + 2ax \tag{3.16}$$

Velocity at any displacement in the y-direction
$$v^2 = v_0^2 + 2ay \tag{3.16}$$

For Projectile Motion

x-displacement
$$x = v_{0x}t \tag{3.38}$$

y-displacement
$$y = v_{0y}t - \tfrac{1}{2}gt^2 \tag{3.39}$$

x-component of velocity
$$v_x = v_{0x} \tag{3.40}$$

y-component of velocity
$$v_y = v_{0y} - gt \tag{3.41}$$

y-component of velocity at any height y
$$v_y^2 = v_{0y}^2 - 2gy \tag{3.48}$$

Range
$$R = \frac{v_0^2 \sin 2\theta}{g} \tag{3.47}$$

Questions for Chapter 3

"Questions" help students review key chapter concepts and prepare for "Problems."

1. Discuss the difference between distance and displacement.
2. Discuss the difference between speed and velocity.
3. Discuss the difference between average speed and instantaneous speed.
† 4. Although speed is the magnitude of the instantaneous velocity, is the average speed equal to the magnitude of the average velocity?
5. Why can the kinematic equations be used only for motion at constant acceleration?
6. In dealing with average velocities discuss the statement, "Straight line motion at 60 mph for 1 hr followed by motion in the same direction at 30 mph for 2 hr does not give an average of 45 mph but rather 40 mph."

7. What effect would air resistance have on the velocity of a body that is dropped near the surface of the earth?
8. What is the acceleration of a projectile when its instantaneous vertical velocity is zero at the top of its trajectory?
9. Can an object have zero velocity at the same time that it has an acceleration? Explain and give some examples.
10. Can the velocity of an object be in a different direction than the acceleration? Give some examples.
11. Can you devise a means of using two clocks to measure your reaction time?
†12. A person on a moving train throws a ball straight upward. Describe the

motion as seen by a person on the train and by a person on the station platform.
13. You are in free fall, and you let go of your watch. What is the relative velocity of the watch with respect to you?
†14. What kind of motion is indicated by a graph of displacement versus time, if the slope of the curve is (a) horizontal, (b) sloping upward to the right, and (c) sloping downward?
†15. What kind of motion is indicated by a graph of velocity versus time, if the slope of the curve is (a) horizontal, (b) sloping upward at a constant value, (c) sloping upward at a changing rate, (d) sloping downward at a constant value, and (e) sloping downward at a changing rate?

THE LEARNING SYSTEM

†63. A 1.50-kg block moves along a smooth horizontal surface at 2.00 m/s. It then encounters a smooth inclined plane that makes an angle of 53.0° with the horizontal. How far up the incline will the block move before coming to rest?

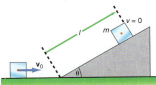

†64. Repeat problem 63, but in this case the inclined plane is rough and the coefficient of kinetic friction between the block and the plane is 0.400.

†65. In the diagram mass m_1 is located at the top of a rough inclined plane that has a length $l_1 = 0.500$ m. $m_1 = 0.500$ kg, $m_2 = 0.200$ kg, $\mu_{k_1} = 0.500$, $\mu_{k_2} = 0.300$, $\theta = 50.0°$, and $\phi = 50.0°$. (a) Find the total energy of the system in the position shown. (b) The system is released from rest. Find the work done for block 1 to overcome friction as it slides down the plane. (c) Find the work done for block 2 to overcome friction as it slides up the plane. (d) Find the potential energy of block 2 when it arrives at the top of the plane. (e) Find the velocity of block 1 as it reaches the bottom of the plane. (f) Find the kinetic energy of each block at the end of their travel.

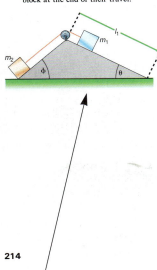

†66. If a constant force acting on a body is plotted against the displacement of the body from x_1 to x_2, as shown in the diagram, then the work done is given by

$$W = F(x_2 - x_1)$$
$$= \text{Area under the curve}$$

Show that this concept can be extended to cover the case of a variable force, and hence find the work done for the variable force, $F = kx$, where $k = 2.00$ N/m as the body is displaced from x_1 to x_2. Draw a graph showing your results.

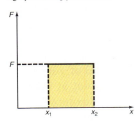

Interactive Tutorials

🖫 67. A projectile of mass $m = 100$ kg is fired vertically upward at a velocity $v_0 = 50.0$ m/s. Calculate its potential energy PE (relative to the ground), its kinetic energy KE, and its total energy E_{tot} for the first 10.0 s of flight.

🖫 68. Consider the general motion in an Atwood's machine such as the one shown in the diagram of problem 31; $m_A = 0.650$ kg and is at a height $h_A = 2.55$ m above the reference plane and mass $m_B = 0.420$ kg is at a height $h_B = 0.400$ m. If the system starts from rest, find (a) the initial potential energy of mass A, (b) the initial potential energy of mass B, and (c) the total energy of the system. When m_A has fallen a distance $y_A = 0.75$ m, find (d) the potential energy of mass A, (e) the potential energy of mass B, (f) the speed of each mass at that point, (g) the kinetic energy of mass A, and (h) the kinetic energy of mass B. (i) When mass A hits the ground, find the speed of each mass.

🖫 69. Consider the general motion in the combined system shown in the diagram of problem 42; $m_1 = 0.750$ kg and is at a height $h_1 = 1.85$ m above the reference plane and mass $m_2 = 0.285$ kg is at a height $h_2 = 2.25$ m, $\mu_k = 0.450$. If the system starts from rest, find (a) the initial potential energy of mass 1, (b) the initial potential energy of mass 2, and (c) the total energy of the system. When m_1 has fallen a distance $y_1 = 0.35$ m, find (d) the potential energy of mass 1, (e) the potential energy of mass 2, (f) the energy lost due to friction as mass 2 slides on the rough surface, (g) the speed of each mass at that point, (h) the kinetic energy of mass 1, and (i) the kinetic energy of mass 2. (j) When mass 1 hits the ground, find the speed of each mass.

🖫 70. Consider the general case of motion shown in the diagram with mass m_A initially located at the top of a rough inclined plane of length l_A, and mass m_B is at the bottom of the second plane; x_A is the distance from the mass A to the bottom of the plane. Let $m_A = 0.750$ kg, $m_B = 0.250$ kg, $l_A = 0.550$ m, $\theta = 40.0°$, $\phi = 30.0°$, $\mu_{k_A} = 0.400$, $\mu_{k_B} = 0.300$, and $x_A = 0.200$ m. When $x_A = 0.200$ m, find (a) the initial total energy of the system, (b) the distance block B has moved, (c) the potential energy of mass A, (d) the potential energy of mass B, (e) the energy lost due to friction for block A, (f) the energy lost due to friction for block B, (g) the velocity of each block, (h) the kinetic energy of mass A, and (i) the kinetic energy of mass B.

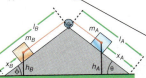

Mechanics

Getting Started on the Computer

The **Interactive Tutorials** was written on Lotus 1-2-3 Release 4 for windows. The Windows environment is used because it is the easiest way to use a computer. It relies on the basic concepts of point and click. An electronic mouse is used to control the location of the pointer on the computer screen. By moving the mouse, you can place the pointer at any location on the screen. If you place the pointer at a menu item and click the left mouse button you activate that particular function of the menu.

Procedure for using the Interactive Tutorials disk that comes with this textbook.

1. Turn on your PC. After a moment or two, the *Program Manager* window appears on the screen.
2. Move the pointer with the mouse until the pointer is directly over the *Lotus Applications* icon on the screen. Double click the left mouse button. A new window will open showing you another series of icons.
3. Place the mouse pointer directly over the icon labeled *Lotus 1-2-3 Release 4* and again double click the left mouse button. After another moment the Lotus 1-2-3 Spreadsheet program will appear on your computer monitor screen in the form of a blank spreadsheet.
4. At this point, insert your Interactive Tutorials disk that comes with your textbook into drive B of your computer.
5. Using the mouse, point at the upper left-hand corner of your screen at the *File* menu of the Lotus spreadsheet. Click on *File* with the left mouse button and a menu will open, falling down on the left side of the screen. Move the pointer down to the file option *Open...* and then click on *Open...* with the left mouse button.
6. A new *Open File* dialogue box will open. On the lower right-hand side is a small box labeled *Drives:*. Click on the *Drives:* box and a submenu will drop down showing the drives *a:, b:, c:,* and *d:*. Click on the *b:* drive, and the computer will show you, in a box on the left-hand side of the *Open File* dialogue box, all the files on the b: drive of your computer. Because there are more files on the disk than can be displayed in the box, you cannot see all the files in this box. However, notice that to the immediate right of the box is a narrow box with an arrow pointing upward at the top of the box and another arrow pointing downward at the bottom of the narrow box. If you place and hold the pointer on the down arrow, the screen will scroll downward, displaying the additional files on the disk. If you place and hold the pointer on the up arrow the screen will scroll upward. In this way you can observe all the files on the disk. The files are listed by chapter and problem number. As an example, the file Ch3P79.wk4 is problem 79 in chapter 3. The suffix .wk4 is the notation that is used to identify this file as a Lotus computer spreadsheet, release 4.
7. Double click on the file of your choice and the spreadsheet will open for you. The spreadsheet will have all the necessary data for the problem already in it. If you wish to change any data in the yellow-colored cells of the Initial Conditions, place the pointer at the particular cell in the spreadsheet and click the left mouse button. This will activate that particular cell of the spreadsheet and you can enter any numbers you wish into that cell.
8. If you wish to open a new spreadsheet problem repeat steps 5 through 7. Good luck to you, and have fun with these spreadsheets.

V is the sum of the voltages across R, L, and C in series and is given by

$$V = VR + VL + VC$$

and is shown in the graph below

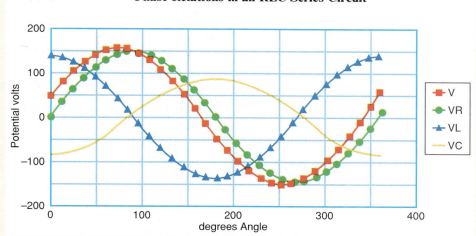

Phase Relations in an RLC Series Circuit

Since VR is in phase with the current i in the circuit, we can compare the phase relationship between the voltage V and the current i, by comparing the voltage V and the voltage VR. The phase angle can best be determined from this diagram by observing where the V curve intersects the horizontal axis, and where the VR curve intersects the horizontal axis. The difference between these two values represents the phase angle (phi), which was already calculated in part (i).

Special Cases
1) An RC series circuit. Let L = 0.
2) An RL series circuit. Let C = 0.
3) An LC series circuit. Let R = 0.

These Interactive Tutorials are a very helpful tool to aid in the learning of physics if they are used properly. The student should try to solve the particular problem in the traditional way using paper and an electronic calculator. Then she or he should open the spreadsheet, insert the appropriate data into the Initial Conditions cells, and see how the computer solves the problem. Go through each step on the computer and compare it to the steps you made on paper. Does your answer agree? If not, check through all the in-between steps on the computer and your paper and find where you made a mistake. Repeat the problem using different Initial Conditions on the computer and your paper. Again check your answers and all the in-between steps. Once you are sure that you know how to solve the problem, try some special cases. What would happen if you changed an angle? a weight? a force? and so forth. In this way you can get a great deal of insight into the physics of the problem and also learn a great deal of physics in the process.

You must be very careful not to just plug numbers into the Initial Conditions and look at the answers without understanding the in-between steps and the actual physics of the problem. You will only be deceiving yourself. Be careful. These spreadsheets can be extremely helpful if they are used properly.

(j) The resonant frequency fo is found from equation 24.21 as

$$fo = 1/[2 (pi) sqrt(L C)]$$

$$fo = (\boxed{1}) / (\boxed{6.28}) \times sqrt(\boxed{2} \ H) \times (\boxed{5.55E{-}06})$$

$$fo = \boxed{4.78E{+}01} \ Hz$$

The angular frequency (omega) of the alternating current is related to the frequency of the current by

$$(omega) = 2 (pi) f$$

$$(omega) = (\boxed{2}) \times (\boxed{3.14}) \times (\boxed{60} \ Hz)$$

$$(omega) = \boxed{3.77E{+}02} \ rad/s$$

The maximum value of the current, imax, is found from rearranging equation 24.5 as

$$imax = ieff / 0.707$$

$$imax = (\boxed{0.13} \ A) / (\boxed{0.71})$$

$$imax = \boxed{0.18} \ A$$

The equation for the alternating current is given by

$$i = imax \ sin[(omega) t]$$

$$i = (\boxed{0.18} \ A) \times sin[(\boxed{3.77E{+}02} \ rad/s) t]$$

The data to plot the current as a function of time is calculated from this equation and the results are shown in the graph below.

The data to plot the voltage drop across the resistor as a function of time is calculated from the equation

$$VR = i R$$

$$VR = imax \ R \ sin[(omega) t]$$

and the results are shown in the graph below.

The data to plot the voltage drop across the inductor as a function of time is calculated from the equation

$$VL = i XL$$

$$VL = imax \ XL \ sin[(omega) t +(pi)/2]$$

The term + (pi)/2 in the sine term is to show that VL leads VR by 90 degrees. The results are shown in the graph below.

The data to plot the voltage drop across the capacitor as a function of time is calculated from the equation

$$VC = i XC$$

$$VC = imax \ XC \ sin[(omega) t - (pi)/2]$$

The term − (pi)/2 in the sine term is to show that VC lags VR by 90 degrees. The results are shown in the graph below.

(continued)

(b) The capacitive reactance XC is found from equation 24.17 as

$$XC = 1/[2 \text{ (pi) f C}]$$

$$XC = (\quad 1 \quad) \ / \ (\quad 6.28 \quad) \times (\quad 60 \text{ Hz}) \times (\ 5.55E-06)$$

$$XC = \quad 4.78E+02 \text{ ohms}$$

(c) The impedance Z of the circuit is found from equation 24.18 as

$$Z = \text{Sqrt}[R\verb|^|2 + (XL - XC)\verb|^|2]$$

$$Z = \text{Sqrt}[(\quad 800 \text{ ohms})\verb|^|2 + ((\quad 753.98 \text{ ohms}) - (\quad 477.94 \text{ ohms}))\verb|^|2$$

$$Z = \quad 8.46E+02 \text{ ohms}$$

(d) The effective current in the circuit is found from Ohm's law, equation 24.19, as

$$i = V/Z$$

$$i = (\quad 110 \quad V) \ / \ (\ 8.46E+02 \text{ ohms})$$

$$i = \quad 1.30E-01 \text{ A}$$

(e) The voltage drop VR across the resistor is found from equation 24.13, as

$$VR = i R$$

$$VR = (\quad 1.30E-01 \quad A) \times (\ 8.00E+02 \text{ ohms})$$

$$VR = \quad 1.04E+02 \text{ V}$$

(f) The voltage drop VL across the inductor is found from equation 24.14, as

$$VL = i XL$$

$$VL = (\quad 1.30E-01 \quad A) \times (\ 7.54E+02 \text{ ohms})$$

$$VL = \quad 9.80E+01 \text{ V}$$

(g) The voltage drop VC across the capacitor is found from equation 24.16, as

$$VC = i XC$$

$$VC = (\quad 1.30E-01 \quad A) \times (\ 4.78E+02 \text{ ohms})$$

$$VC = \quad 6.21E+01 \text{ V}$$

(h) The total voltage across R, L, C in series is found from equation 24.11 as

$$V = \text{Sqrt}[VR\verb|^|2 + (VL - VC)\verb|^|2]$$

$$V = \text{Sqrt}[(\quad 103.98 \text{ ohms})\verb|^|2 + ((\quad 98 \text{ ohms}) - (\quad 62.12 \text{ ohms}))\verb|^|2$$

$$V = \quad 1.10E+02 \text{ V}$$

which is equal to the applied voltage as expected.

(i) The phase angle (phi) is found from equation 24.11 as

$$\text{(phi)} = \arctan[(VL - VC)/VR]$$

$$\text{(phi)} = \arctan[((\quad 98 \quad V) - (\quad 62.12 \quad V)) \ / \ (\ 103.98 \text{ V})]$$

$$\text{(phi)} = \quad 19.04 \text{ degrees}$$

Finally, we should note that the spreadsheets are "protected" by allowing you to enter data only in the designated light yellow-colored cells of the Initial Conditions area. Therefore, the student cannot damage the spreadsheets in any way, and they can be used over and over again.

Another example of an Interactive Tutorial, shown in figure 2, is an Interactive Tutorial for a problem on an *RLC* series *AC* circuit from chapter 24. The initial conditions in the yellow-colored cells set the values of the resistance R, the inductance L, the capacitance C, the frequency f, and the applied voltage V. The spreadsheet then completely solves the problem for these initial values. The student can follow through all the in-between steps of the problem in the light green-colored cells of the spreadsheet and all the answers in the light blue-green-colored cells. The spreadsheet then draws a graph showing the phase relations for the voltage across the resistor, V_R, the voltage across the inductance V_L, the voltage across the capacitance, V_C, and the resultant voltage V in the circuit depending upon the initial values in the yellow cells. As special cases, the student can then let $L = 0$ in the yellow cell area, and observe the characteristics of an *RC* series circuit; then replace a value of L in the initial conditions and then let $C = 0$ and observe the characteristics of an *RL* series circuit. Finally, inserting values for L and C and letting $R = 0$, the student can observe the characteristics of an *LC* series circuit. In all the cases the spreadsheet calculates all the quantities with all the in-between steps, and shows the phase relations of the voltages.

Chapter 24 Alternating Current Circuits

Computer Assisted Instruction

Interactive Tutorial

59. RLC series circuit. An RLC series circuit has a resistance R = 800 ohms, Inductance L = 2.00 H, capacitance C = 5.55×10^{-6} F, and they are connected to an applied voltage V = 110 V, operating at a frequency f = 60 Hz. Find (a) the inductive reactance XL, (b) the capacitive reactance, XC, (c) the impedance Z of the circuit, (d) the current i in the circuit, (e) the voltage drop VR across the resistor, (f) the voltage drop VL across the inductor, (g) the voltage drop VC across the capacitor, (h) the total voltage drop across RLC, (i) the phase angle (phi), and (j) the resonant frequency fo. (k) Plot the curves of the voltage across R, L, and C individually and the voltage across the combination. Show the phase relations for all these voltages.

Initial Conditions

R = 800 ohms   f = 60 Hz
L = 2.00E+00 H   V = 110 V
C = 5.55E−06 F

(a) The inductive reactance XL is found from equation 24.15 as

$$XL = 2 (pi) f L$$

XL = (2) × (3.14) × (60 Hz) × (2)

XL = 7.54E+02   (continued)

Figure 2.

An Interactive Tutorial for Chapter 24 Alternating Current Circuits.

Chapter 3 Kinematics

Computer Assisted Instruction

Interactive Tutorial

79. A golf ball is hit with an initial velocity vo = 53.0 m/s at an angle (theta) = 50.0 degrees above the horizontal. (a) How high will the ball go? (b) What is the total time the ball is in the air? (c) How far will the ball travel horizontally before it hits the ground?

Initial Conditions

The magnitude of the initial velocity vo =	53	m/s
The angle (theta) =	50	degrees
The acceleration of gravity g =	9.8	m/s^2

The x-component of the initial velocity is found as

vox = vo cos(theta)

vox = (53 m/s) × cos (50)

vox = 34.07 m/s

The y-component of the initial velocity is found as

voy = vo sin(theta)

voy = (53 m/s) × sin(50)

voy = 40.6 m/s

(a) The maximum height ymax of the golf ball above the launch point is found from the kinematic equation

vy^2 = voy^2 − 2 g y

When y = ymax, vy = 0. Therefore

0 = voy^2 − 2 g ymax

Solving for the maximum height ymax above the launch point, we get

ymax = (voy^2)/2g

ymax = (40.6 m/s)^2 / 2 (9.8 m/s^2)

ymax = 84.1 m

(b) The total time tt the ball is in the air is found from the kinematic equation

y = voy t − 0.5gt^2

When t is equal to the total time tt, the projectile is on the ground and y = 0, therefore

0 = voy tt −.5 g tt^2

or dividing each term by tt we get

0 = voy −.5 g tt

upon solving for the total time tt we get

tt = 2 voy/g

tt = 2 × (40.6 m/s) / (9.8 m/s^2)

tt = 8.29 s

(c) The maximum distance the ball travels in the x-direction before it hits the ground, is found from the kinematic equation for the displacement of the ball in the x-direction

x = vox t

The maximum distance xmax occurs for t = tt. Therefore

xmax = vox tt

xmax = (34.07 m/s) × (8.29 s)

xmax = 282.28 m

Figure 1.

A typical Interactive Tutorial.

Interactive Tutorials

For the last ten or more years, there has been an ongoing discussion in the physics community for the need to introduce the computer into the freshman physics class. However, to this day no one has been able to agree on how the computer should best be used. In those cases where a computer has been used, the traditional use has been to state the problem, write down a general equation, and give the result of the calculation, with all the in-between steps of the calculation missing. The student gets the result of the calculation, but does not see how it is attained. It is almost like magic. In this second edition of *The Fundamental of College Physics,* the computer is used in a highly effective and unique way. At the end of the problems section in each chapter, I have introduced a new section called **Interactive Tutorials.** The Interactive Tutorials are a series of physics problems, very much like the examples in the chapter and the problems at the end of the chapter. A computer disk is available to both the instructor and the student with each of these problems on the disk. These computer files are available for the IBM PC or compatible personal computer using the Lotus 1-2-3 computer spreadsheet or on a Macintosh computer using Microsoft's Excel computer spreadsheet.

What makes these Interactive Tutorials unique is the way the computer is used as an instructional tool. Figure 1 shows a typical tutorial for a problem in chapter 3 on Kinematics. When the student opens this particular spreadsheet, he or she sees the problem stated in the usual manner. That is, this problem is an example of a projectile fired at an initial velocity $v_0 = 53.0$ m/s at an angle $\theta = 50.0°$, and it is desired to find the maximum height of the projectile, the total time the projectile is in the air, and the range of the projectile. Directly below the stated problem is a group of yellow-colored cells labeled **Initial Conditions.** Into these yellow cells are placed the numerical values associated with the particular problem. For this problem the initial conditions consist of the initial velocity v_0, the initial angle θ, and the acceleration due to gravity g as shown in figure 1. The problem is now solved in the traditional way of a worked-out example in the book. Words are used to describe the physical principles and then the equations are written down. Then the in-between steps of the calculation are shown in light green-colored cells, and the final result of the calculation is shown in a light blue-green-colored cell. The entire problem is solved in this manner, as shown in figure 1. If the student wishes to change the problem by using a different initial velocity or a different launch angle, he or she then changes these values in the yellow-colored cells of the initial conditions. When the initial conditions are changed the computer spreadsheet recalculates all the new in-between steps in the problem and all the new final answers to the problem. In this way the problem is completely interactive. It changes for every new set of initial conditions. The Interactive Tutorials make the book a living book. The tutorials can be changed many times over to solve for all kinds of special cases.

One of the most troubling aspects of using a spreadsheet, however, is its inability to use subscripts or superscripts. Therefore, information contained on the spreadsheet can sometimes be confusing. As an example, when the textbook refers to an initial velocity, the notation v_0 is utilized. Since the spreadsheet cannot use subscripts, this will be written as vo. Other examples are $v_{ox} =$ vox; $y_{max} =$ ymax; $M_1 =$ M1; $M_2 =$ M2; and so on. The notation used for the subscript in the spreadsheet will, however, be the same notation used in the textbook. When in doubt about notation, refer to the stated problem in the textbook. For superscripts, the notation ^ is used. Hence, the quantity x^2 in your textbook will be written on the spreadsheet as x^2. A term containing both subscripts and superscripts, such as v_y^2, will be written on the spreadsheet as vy^2. Powers of ten that are used in scientific notation are written with the capital letter E. Hence, the number 5280, written in scientific notation as 5.280×10^3, will be written on the spreadsheet as 5.280E+3. Because the spreadsheet cannot mix fonts in a particular cell, the notation Δx will be written on the spreadsheet as (delta) x, and angles such as θ will be written as (theta).

A Special Note to the Student

"One thing I have learned in a long life: that all our science measured against reality, is primitive and childlike—and yet it is the most precious thing we have."

Albert Einstein
as quoted by Banesh Hoffmann in
Albert Einstein, Creator and Rebel

The language of physics is mathematics, so it is necessary to use mathematics in our study of nature. However, just as sometimes "you cannot see the forest for the trees," you must be careful or "you will not see the physics for the mathematics." Remember, mathematics is only a tool used to help describe the physical world. You must be careful to avoid getting lost in the mathematics and thereby losing sight of the physics. When solving problems, a sketch or diagram that represents the physics of the problem should be drawn first, then the mathematics should be added.

Physics is such a logical subject that when a student sees an illustrative problem worked out, either in the textbook or on the blackboard, it usually seems very simple. Unfortunately, for most students, it is simple only until they sit down and try to do a problem on their own. Then they often find themselves confused and frustrated because they do not know how to get started.

If this happens to you, do not feel discouraged. It is a normal phenomenon that happens to many students. The usual approach to overcoming this difficulty is going back to the illustrative problem in the text. When you do so, however, do not look at the solution of the problem first. Read the problem carefully, and then try to solve the problem on your own. At any point in the solution, when you cannot proceed to the next step on your own, peek at that step and only that step in the illustrative problem. The illustrative problem shows you what to do at that step. Then continue to solve the problem on your own. Every time you get stuck, look again at the appropriate solution step in the illustrative problem until you can finish the entire problem. The reason you had difficulty at a particular place in the problem is usually that you did not understand the physics at that point as well as you thought you did. It will help to reread the appropriate theory section. Getting stuck on a problem is not a bad thing, because each time you do, you have the opportunity to learn something. Getting stuck is the first step on the road to knowledge. I hope you will feel comforted to know that most of the students who have gone before you also had these difficulties. You are not alone. Just keep trying. Eventually, you will find that solving physics problems is not as difficult as you first thought; in fact, with time, you will find that they can even be fun to solve. The more problems that you solve, the easier they become, and the greater will be your enjoyment of the course.

Test Item File More than 4,000 classroom-tested test items are available to adopters of this textbook. The test items are available on the computer test generating program MicroTest for both the IBM and Macintosh platforms.

Physics Videodisc Over 600 lab demonstrations are available from the Wm. C. Brown Publishers videodisc library. These professionally produced experiments are considered the best available lab demonstrations on the market. Please call your local Wm. C. Brown Sales Representative for more information.

Experiments in Physics, second edition, by Nolan and Bigliani This laboratory text for the algebra-based introductory physics course is the first and only laboratory textbook that incorporates computer spreadsheets to calculate and analyze data obtained in the lab. This manual is completely customizable. Contact your Wm. C. Brown Sales Representative for more information.

For the Student

Student Solutions Manual Approximately twenty percent of the problems in the textbook can be found in the *Student Solutions Manual* to accompany *Fundamentals of College Physics.* Every solution is completely worked out for the student.

Physics Student Study Art Notebook A notebook of key physics art work is available packaged with the new edition of *Fundamentals of College Physics.* This notebook provides full-color line art from the text, with space for note taking by the student. This unique notebook helps students study for exams and take clearer, more accurate classroom notes.

Student Study Guide An excellent *Student Study Guide,* written by Lewis O'Kelly of Memphis State University, is available to accompany *Fundamentals of College Physics.* This study guide reinforces the concepts present in the main text and provides at-home quizzes and problem solving techniques.

Acknowledgments

I would like at this time to thank Eileen and my children for their love, understanding, patience, and encouragement throughout the long stages of writing and rewriting this book. I would also like to thank the following individuals for their encouragement: Dr. Lloyd Makarowitz, Chairman of the Physics Department at SUNY Farmingdale; Linda Rennie, our secretary; and all the members of the Physics Department at SUNY Farmingdale.

I would also like to thank all the very friendly and helpful people at Wm. C. Brown Publishers who helped in the production of this book. In particular, I would like to thank Executive Editor Jeffrey Hahn, Physics Editors Jane Ducham and Megan Johnson, Developmental Editor Tom Riley, and Production Editors Carol Besler and Julie Wilde. I would like to thank Jerry Hopper and all his people at Diphrent Strokes, Inc., who entered the realm of high technology and did all the beautiful artwork on computers. This will certainly be the way of the future. I would like to give a very special thank you to Pat Steele for the magnificent job of copy editing this book. Her help was indispensable. I would also like to thank the following individuals for their contribution to the overall text package: *Student Study Guide* by Lewis O'Kelly, Memphis State University; *Student's Solutions Manual and Instructor's Solutions Manual* by Ronald Cosby, Ball State University, Joel Levine, Orange Coast College, and Paul Morris, Abilene Christian University; *Test Item File* by Lloyd Makarowitz, SUNY-Farmingdale. Finally, I wish to acknowledge and thank the following reviewers for their helpful criticisms and suggestions: Joel M. Levine, Orange Coast College; Russell A. Roy, Santa Fe Community College; Paul Morris, Abilene Christian University; C. Sherman Frye, Northern Virginia Community College; Franklin Curtis Mason, Middle Tennessee State University; Franklin D. Trumpy, Des Moines Area Community College; Lewis B. O'Kelly, Memphis State University; Ronald M. Cosby, Ball State University; Paul Feldker, Florissant Valley Community College; J. W. Northrip, Southwest Missouri State University; Michael J. Matkovich, Oakton Community College; Michael K. Garrity, St. Cloud State University; and Phyllis A. Salmons, Embry-Riddle Aeronautical University.

Also new to this second edition is an additional 165 new physics problems, all in SI units, when added to the 60 new duplicated problems gives a total of 225 new physics problems, all in SI units. There are also 129 new computer Interactive Tutorials, giving a grand total of 354 new problems in this second edition.

The levels of college physics courses vary greatly, depending on the backgrounds of the students and the institution where the course is taught. Some instructors like to introduce the concepts of vector multiplication in the chapter on vectors and utilize these concepts throughout the course. Others feel that this introduces unnecessary mathematical difficulties. To satisfy both instruction preferences, vector multiplication is found in appendix F and can be used throughout the book if desired.

Scattered throughout the text, at the ends of chapters, are sections entitled "Have you ever wondered . . . ?" These are a series of essays on the application of physics to areas such as meteorology, astronomy, aviation, space travel, the health sciences, the environment, philosophy, traffic congestion, sports, and the like. Many students are unaware that physics has such far-reaching applications. These sections are intended to engage students' varied interests but can be omitted, if desired, without loss of continuity in the physics course.

The relation between theory and experiment is carried throughout the book, emphasizing that our models of nature are good only if they can be verified by experiment.

Concepts presented in the lecture and text can be well demonstrated in a laboratory setting. To this end, *Experiments In Physics, second edition* by Peter J. Nolan and Raymond E. Bigliani is also available from Wm. C. Brown Publishers. *Experiments in Physics* contains some 55 experiments covering all the traditional topics in a physics laboratory. New to the second edition of *Experiments in Physics* is a complete computer analysis of the data for every experiment. Your WCB Sales Representative can arrange to have any of your own department's lab exercises combined with those experiments you desire from the Nolan and Bigliani second edition and bound into your own personal laboratory manual.

A Bibliography, given at the end of the book, lists some of the large number of books that are accessible to students taking college physics. These books cover such topics in modern physics as relativity, quantum mechanics, and elementary particles. Although many of these books are of a popular nature, they do require some physics background. After finishing this book, students should be able to read any of them for pleasure without difficulty.

Finally, we should note that we are living in a rapidly changing world. Many of the changes in our world are sparked by advances in physics, engineering, and the high-technology industries. Since engineering and technology are the application of physics to the solution of practical problems, it behooves every individual to get as much background in physics as possible. *You can depend on the fact that there will be change in our society. You can be either the architect of that change or its victim, but there will be change.*

The Teaching Package

A very extensive teaching package supports the use of *Fundamentals of College Physics,* second edition. The author and the publisher have worked closely with your colleagues to develop this material.

For the Instructor

Instructor's Solutions Manual You can find worked-out solutions to every problem in the textbook in the *Instructor's Solutions Manual* offered with *Fundamentals of College Physics.*

Transparency Set A set of over 100 full-color acetate transparencies is available to adopters of this textbook.

Instructor's Resource Manual This package item includes selected transparency masters, documentation for the more than 125 Interactive Tutorials and spreadsheet diskettes that accompany the text. The tutorials are available on both the IBM Windows and Macintosh platforms.

Students sometimes have difficulty remembering the meanings of all the vocabulary associated with new physical ideas. Therefore, a section called The Language of Physics, found at the end of each chapter, contains the most important ideas and definitions discussed in that chapter.

To comprehend the physical ideas expressed in the theory class, students need to be able to solve physics problems for themselves. Problem sets at the end of each chapter are grouped according to the section where the topic is covered. Problems that are a mix of different sections are found in the Additional Problems section. If you have difficulty with a problem, refer to that section of the chapter for help. The problems begin with simple, plug-in problems to develop students' confidence and to give them a feel for the numerical magnitudes of some physical quantities. The problems then become progressively more difficult and end with some that are very challenging. The more difficult problems are indicated by a dagger (†). The starred problems are either conceptually more difficult or very long. However, just because a problem is starred is no reason to avoid attempting its solution. Many problems at the end of the chapter are very similar to the illustrative problems worked out in the text. When solving these problems, students can use the illustrative problems as a guide. However, students should be warned that physics cannot be learned by memorizing the exhaustive set of illustrative problems. These problems are only a guide to foster greater understanding. To facilitate setting up a problem, the Hints for Problem Solving section, which is found before the problem set in chapter 3, should be studied carefully.

New to the second edition of this book is a section called Interactive Tutorials, which uses computer spreadsheets in a unique and exciting way to solve physics problems. These Interactive Tutorials can be found at the end of the problems section in each chapter. They are also available on a computer disk for both the instructor and the student. More details on these Interactive Tutorials can be found in the section "A Special Note to the Student" at the end of the Preface.

A series of questions relating to the topics discussed in the chapter is also included at the end of each chapter. Students should try to answer these questions to see if they fully understand the ramifications of the theory discussed in the chapter. Just as with the problem sets, some of these questions are either conceptually more difficult or will entail some outside reading. These more difficult questions are also indicated by a dagger (†).

A word about units. Many recent college physics texts use only the International System of Units (SI units). Although working exclusively in SI units is a desirable goal, we do live in a world where other systems of units, particularly the British engineering system, are continuing to be used. Failure to show students how to work with these other units, or how to solve problems regardless of the units employed, does them a disservice. In addition, from the point of view of pedagogy, it is beneficial in the learning process to build on knowledge already possessed by students. For example, when we say a car is moving at 55 miles per hour, students immediately have a feel for this motion. With time, they will have that same feel for the car moving at 88.6 km/hr. Hence, both systems of units are found at the beginning of this book. As we progress through the book, however, the emphasis switches to the SI units, until, from about chapter 16 on, the SI units are used almost exclusively. This approach weans students from the British engineering system to the International System (SI) of units, and at the same time prepares students to convert to any unit or system of units when necessary.

However, for those instructors who prefer to work only in the International System of Units, for this second edition I have taken approximately 60 problems that are presently in the British Engineering System of Units and duplicated them so that they are now also available to be used in the International System of Units. This will give a greater choice of problems for those who wish to work only in SI units. These duplicated problems are not an exact duplicate in that the numbers used are not just a conversion from one system to another, but instead the numbers are totally different. Hence, an instructor wishing to use both systems can assign both problems if he or she desires.

THE LEARNING SYSTEM

Highlighted equations reinforce textual material.

In-text Examples are followed by Solutions and include intermediate steps.

Example 7.5

Power expended. A person pulls a block with a force of 15.0 N at an angle of 25.0° with the horizontal. If the block is moved 5.00 m in the horizontal direction in 5.00 s, how much power is expended?

Solution

The power expended, found from equations 7.3 and 7.2, is

$$P = \frac{W}{t} = \frac{Fx \cos \theta}{t}$$

$$= \frac{(15.0 \text{ N})(5.00 \text{ m}) \cos 25.0°}{5.00 \text{ s}} = 13.6 \frac{\text{N m}}{\text{s}} = 13.6 \text{ W}$$

■

When a constant force acts on a body in the direction of the body's motion, we can also express the power as

$$P = W = \frac{Fx}{t} = F \frac{x}{t}$$

but

$$\frac{x}{t} = v$$

the velocity of the moving body. Therefore,

$$P = Fv \tag{7.4}$$

is the power expended by a force F, acting on a body that is moving at the velocity v.

Example 7.6

Power to move your car. An applied force of 5500 N keeps a car moving at 95 km/hr. How much power is expended by the car?

Solution

The power expended by the car, found from equation 7.4, is

$$P = Fv = (5500 \text{ N}) \left(95 \frac{\text{km}}{\text{hr}} \right) \left(\frac{1 \text{ hr}}{3600 \text{ s}} \right) \left(\frac{1000 \text{ m}}{1 \text{ km}} \right)$$

$$= 1.45 \times 10^5 \text{ N m/s} = 1.45 \times 10^5 \text{ J/s}$$

$$= 1.45 \times 10^5 \text{ W}$$

■

7.4 Gravitational Potential Energy

Gravitational potential energy is defined as the energy that a body possesses by virtue of its position. If the block shown in figure 7.7, were lifted to a height h above the table, then that block would have potential energy in that raised position. That is, in the raised position, the block has the ability to do work whenever it is allowed to fall. The most obvious example of gravitational potential energy is a waterfall (figure 7.8). Water at the top of the falls has potential energy. When the water falls to the bottom, it can be used to turn turbines and thus do work. A similar example is a pile driver. A pile driver is basically a large weight that is raised above a pile

Figure 7.7
Gravitational potential energy.

Figure 7.8
Water at the top of the falls has potential energy.

194

Mechanics

---

that is to be driven into the ground. In the raised position, the driver has potential energy. When the weight is released, it falls and hits the pile and does work by driving the pile into the ground.

Therefore, whenever an object in the gravitational field of the earth is placed in a position above some reference plane, then that object will have potential energy because it has the ability to do work.

As in all the concepts studied in physics, we want to make this concept of potential energy quantitative. That is, how much potential energy does a body have in the raised position? How should potential energy be measured?

Because work must be done on a body to put the body into the position where it has potential energy, the work done is used as the measure of this potential energy. That is, the potential energy of a body is equal to the work done to put the body into the particular position. Thus, the potential energy (PE) is

$$\text{PE} = \text{Work done to put body into position} \tag{7.5}$$

We can now compute the potential energy of the block in figure 7.7 as

$$\text{PE} = \text{Work done}$$

$$\text{PE} = W = Fh = wh \tag{7.6}$$

The applied force F necessary to lift the weight is set equal to the weight w of the block. And since $w = mg$, the potential energy of the block becomes

$$\text{PE} = mgh \tag{7.7}$$

We should emphasize here that the potential energy of a body is referenced to a particular plane, as in figure 7.9. If we raise the block a height h_1 above the table, then with respect to the table it has a potential energy

$$\text{PE}_1 = mgh_1$$

While at the same position, it has the potential energy

$$\text{PE}_2 = mgh_2$$

with respect to the floor, and

$$\text{PE}_3 = mgh_3$$

with respect to the ground outside the room. All three potential energies are different because the block can do three different amounts of work depending on whether it falls to the table, the floor, or the ground. Therefore, it is very important that when the potential energy of a body is stated, it is stated with respect to a particular reference plane. We should also note that it is possible for the potential energy to be negative with respect to a reference plane. That is, if the body is not located above the plane but instead is found below it, it will have negative potential energy with respect to that plane. In such a position the body can not fall to the reference plane and do work, but instead work must be done on the body to move the body up to the reference plane.

Example 7.7

The potential energy. A mass of 1.00 kg is raised to a height of 1.00 m above the floor (figure 7.11). What is its potential energy with respect to the floor?

Solution

The potential energy, found from equation 7.7, is

$$\text{PE} = mgh = (1.00 \text{ kg})(9.80 \text{ m/s}^2)(1.00 \text{ m})$$

$$= 9.80 \text{ J}$$

Figure 7.10
Changing potential energy.

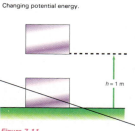

Figure 7.11
The potential energy of a block.

$h = 1 \text{ m}$

Chapter 7   Energy and Its Conservation

195

FUNDAMENTALS OF COLLEGE

PHYSICS

SECOND EDITION

Mechanics

Color as a Study Aid

As a pedagogical aid, color has been extensively used throughout the book. The purpose of the color is not just to make the book look good, but rather to help the student visualize the material. The colors have been standardized according to the following code.

Chapter 2

For multiple vectors in a diagram each vector has a different color:

■ First vector **a**

■ Second vector **b**

■ Third vector **c**

■ Fourth vector **d**

■ Resultant vector **R**

■ Negative of a vector

The components of a vector are always a lighter shade of the same color of the original vector:

■ Vector **a**

■ x- and y-components of **a**

■ Vector **R**

■ x- and y-components of **R**

The x, y, and z coordinates are always black.

Chapter 3

The color code for vectors:

■ Displacement vectors

■ Velocity vectors

■ Acceleration vectors

■ Trajectories

The components of a vector are always a lighter shade of the same color of the original vector:

■ Displacement vectors

■ Components of displacement vector

■ Velocity vectors

■ Components of the velocity vector

■ Acceleration vectors

■ Components of the acceleration vector

Coordinates are always black.

Chapter 4

Color code for force vectors:

■ Applied force vectors

■ Components of applied force vector

■ Tension force vectors

■ Components of tension force vector

■ Friction force vectors

■ Normal force vectors

■ Weight force vectors

■ Components of weight force vector

■ Reaction force vectors

■ Centripetal force vector

Note that a different shade of green is used for the centripetal force vector so that it is not the same dark green that is used for the applied force vector.

Also note that a different shade of dark blue is used for the weight force vector so that it is not the same blue that is used for the velocity vector.

Chapter 5

The same color code is used as in the previous chapters. In addition:

■ Lever arms

Chapter 6

The same color code is used as in the previous chapters. In addition:

■ Arc *s* of a circle

■ Circular orbits

■ Elliptical orbits

■ Gravitational force vector

Note that this is the same color as the dark blue of the weight vector.

Chapters 7 through 9

In all these chapters the same color code is used as in the previous chapters.

3

1 Introduction and Measurements

Earth rise from the moon.

In exploring nature, therefore, we must begin by trying to determine its first principles.

Aristotle

The method with which we shall follow in this treatise will be always to make what is said depend on what was said before.

Galileo Galilei

1.1 Historical Background

Physics has its birth in mankind's quest for knowledge and truth. In ancient times, people were hunters following the wild herds for their food supply. Since they had to move with the herds for their survival, they could not be tied down to one site with permanent houses for themselves and their families. Instead these early people lived in whatever caves they could find during their nomadic trips. Eventually these cavemen found that it was possible to domesticate such animals as sheep and cattle. They no longer needed to follow the wild herds. Once they stayed long enough in one place to take care of their herds, they found that seeds collected from various edible plants in one year could be planted the following year for a new crop. Thus, many of these ancient people became farmers, growing their own food supply. They, of course, found that they could grow a better crop in a warm climate near a readily available source of water. It is not surprising then that the earliest known civilizations sprang up on the banks of the great rivers: the Nile in Egypt and the Tigris and Euphrates in Mesopotamia. Once permanently located on their farms, these early people were able to build houses for themselves. Trades eventually developed and what would later be called civilization began.

To be successful farmers, these ancient people had to know when to plant the seeds and when to harvest the crop. If they planted the seeds too early, a frost could destroy the crop, causing starvation for their families. If they planted the seeds too late, there might not be sufficient growing time or adequate rain.

In those very dark nights, people could not help but notice the sky. It must have been a beautiful sight without the background street lights that are everywhere today. People began to study that sky and observed a regularity in the movements of the sun, moon, and stars. In ancient Egypt, for example, the Nile river would overflow when Sirius, the Dog Star, rose above the horizon just before dawn. People then developed a calendar based on the position of the stars. By their observation of the sky, they found that when certain known stars were in a particular position in the sky it was time to plant a new crop. With an abundant harvest it was now possible to store enough grain to feed the people for the entire year.

For the first time in the history of humanity, obtaining food for survival was not an all time-consuming job. These ancient people became affluent enough to afford the time to think and question. What is the cause of the regularity in the motion of the heavenly bodies? What makes the sun rise, move across the sky, and then set? What makes the stars and moon move in the night sky? What is the earth made of? What is man? And through this questioning of the world about them, **philosophy** was born—the search for knowledge or wisdom (*philos* in Greek means

Figure 1.1
The caveman steps out of his cave.

"love of" and *sophos* means "wisdom"). Philosophy, therefore, originated when these early people began to seek a rational explanation of the world about them, an explanation of the nature of the world without recourse to magic, myths, or revelation. Ancient philosophers studied ethics, morality, and the essence of beings as determined by the mind, but they also studied the natural world itself. This latter activity was called **natural philosophy**—the study of the phenomena of nature. Among early Greek natural philosophers were Thales of Miletus (ca. 624–547 B.C.), Democritus (ca. 460–370 B.C.), Aristarchus (ca. 320–250 B.C.), and Archimedes (ca. 287–212 B.C.), perhaps the greatest scientist and mathematician of ancient times.

For many centuries afterward, the study of nature continued to be called natural philosophy. In fact, one of the greatest scientific works ever written was by Sir Isaac Newton. When it was published in 1687, he entitled it *Philosophiae Naturalis Principia Mathematica* (*The Mathematical Principles of Natural Philosophy*).

*Natural philosophy, therefore, studied all of nature. The Greek word for "natural" is physikos. Therefore, the name **physics** came to mean the study of all of nature.* Physics became a separate entity from philosophy because it employed a different method to search for truth. Physics developed and employed an approach called the scientific method in its quest for knowledge.

The **scientific method** is the application of a logical process of reasoning to arrive at a model of nature that is consistent with experimental results. The scientific method consists of five steps:

1. Observation
2. Hypothesis
3. Experiment
4. Theory or law
5. Prediction

This process of scientific reasoning can be followed with the help of the flow diagram shown in figure 1.2.

1. *Observation.* The first step in the scientific method is to make an observation of nature, that is, to collect data about the world. The data may be drawn from a simple observation, or they may be the results of numerous experiments.

Figure 1.2

The scientific method.

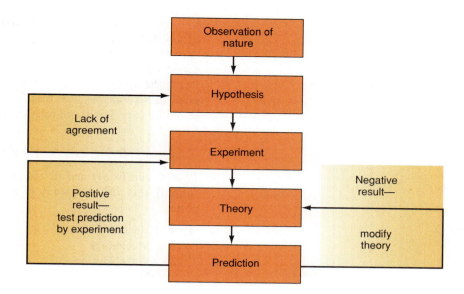

2. *Hypothesis.* From an analysis of these observations and experimental data, a model of nature is hypothesized. The dictionary defines a hypothesis as an assumption that is made in order to draw out and test its logical or empirical consequences; that is, an assumption is made that in a given situation nature will always work in a certain way. If this hypothesis is correct, we should be able to confirm it by testing. This testing of the hypothesis is called the experiment.

3. *Experiment.* An experiment is a controlled procedure carried out to discover, test, or demonstrate something. An experiment is performed to confirm that the hypothesis is valid. If the results of the experiment do not support the hypothesis, the experimental technique must be checked to make sure that the experiment was really measuring that aspect of nature that it was supposed to measure. If nothing wrong is found with the experimental technique, and the results still contradict the hypothesis, then the original hypothesis must be modified. Another experiment is then made to test the modified hypothesis. The hypothesis can be modified and experiments redesigned as often as necessary until the hypothesis is validated.

4. *Theory.* Finally, success: the experimental results confirm that the hypothesis is correct. The hypothesis now becomes a new theory about some specific aspect of nature, a scientifically acceptable general principle based on observed facts. After a careful analysis of the new theory, a prediction about some presently unknown aspect of nature can be made.

5. *Prediction.* Is the prediction correct? To answer that question, the prediction must be tested by performing a new experiment. If the new experiment does not agree with the prediction, then the theory is not as general as originally thought. Perhaps it is only a special case of some other more general model of nature. The theory must now be modified to conform to the negative results of the experiment. The modified theory is then analyzed to obtain a new prediction, which is then tested by a new experiment. If the new experiment confirms the prediction, then there is reasonable confidence that this theory of nature is correct. This process of prediction and experiment continues many times. As more and more predictions are confirmed by experiment, mounting evidence indicates that a good model of the way nature works has been developed. At this point, the theory can be called a *law of physics.*

This method of scientific reasoning demonstrates that the establishment of any theory is based on experiment. In fact, the success of physics lies in this agreement between theoretical models of the natural world and their experimental confirmation in the laboratory. A particular model of nature may be a great intellectual achievement but, if it does not agree with physical reality, then, from the point of view of physics, that hypothesis is useless. Only hypotheses that can be tested by experiment are relevant in the study of physics.

1.2 The Realm of Physics

Physics can be defined as the study of the entire natural or physical world. To simplify this task, the study of physics is usually divided into the following categories:

I. Classical Physics
1. Mechanics
2. Wave Motion
3. Heat
4. Electricity and Magnetism
5. Light
II. Modern Physics
1. Relativity
2. Quantum Mechanics
3. Atomic and Nuclear Physics
4. Condensed Matter Physics
5. Elementary Particle and High-Energy Physics

Although there are other sciences of nature besides physics, physics is the foundation of these other sciences. For example, astronomy is the application of physics to the study of all matter beyond the earth, including everything from within the solar system out to the remotest galaxies. Chemistry is the study of the properties of matter and the transformation of that matter. Geology is the application of physics to the study of the earth. Meteorology is the application of physics to the study of the atmosphere. Engineering is the application of physics to the solution of practical problems. The science of biology, which traditionally had been considered independent of physics, now uses many of the principles of physics in its study of molecular biology. The health sciences use so many new techniques and equipment based on physical principles that even there it has become necessary to have an understanding of physics.

This distinction between one science and another is usually not clear. In fact, there is often a great deal of overlap among them.

1.3 Physics Is a Science of Measurement

In order to study the entire physical world, we must first observe it. To be precise in the observation of nature, all the physical quantities that are observed should be measured and described by numbers. The importance of numerical measurements was stated by the Scottish physicist, William Thomson (1824–1907), who was made Baron Kelvin in 1892 and has since been referred to as Lord Kelvin:

> *I often say that when you can measure what you are speaking about, and express it in numbers, you know something about it; but when you cannot express it in numbers, your knowledge is of a meager and unsatisfactory kind.*

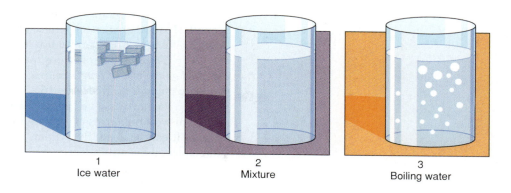

Figure 1.3

A thought experiment on temperature.

1 | 2 | 3
Ice water | Mixture | Boiling water

We can see the necessity for quantitative measurements from the following example. First, let us consider the following *thought experiment. (A thought experiment is an experiment that we can think through, rather than actually performing the experiment.)* Three beakers are placed on the table as shown in figure 1.3. In the first beaker, we place several ice cubes in water. We place boiling water in the third beaker. In the second beaker, we place a mixture of the ice water from beaker 1 and the boiling water from beaker 3. If you put your left hand into beaker 1, you will conclude that the ice water is *cold.* Now place your left hand into beaker 2, which contains the mixture. After coming from the ice water, your hand finds the second beaker to be hot by comparison. So you naturally conclude that the mixture is *hot.*

Now take your right hand and plunge it into the boiling water of beaker 3. (This is the reason that this is only a *thought experiment.* You can certainly appreciate what would happen in the experiment without actually risking bodily harm.) You would then conclude that the water in beaker 3 is certainly *hot.* Now place your right hand into beaker 2. After the boiling water, your hand finds the mixture cold by comparison, so you conclude that the mixture is *cold.* After this relatively "scientific" experiment, you find that you have contradictory conclusions. That is, you have found the middle mixture to be either hot or cold depending on the sequence of the measurements.

We can therefore conclude that in this particular observation of nature, describing something as hot or cold is not very accurate. Unless we can say numerically how hot or cold something is, our observation of nature is incomplete. In practice, of course, we would use a thermometer to measure the temperature of the contents of each beaker and read the hotness or coldness of each beaker as a number on the thermometer. For example, the thermometer might read, 0°C or 50°C or 100°C. We would now have assigned a number to our observation of nature and would thus have made a precise statement about that observation. This example points out the necessity of assigning a number to any observation of nature. The next logical question is, "What should we observe in nature?"

1.4 The Fundamental Quantities

If physics is the study of the entire natural world, where do we begin in our observations and measurements of it? It is desirable to describe the natural world in terms of the fewest possible number of quantities. This idea is not new; some of the ancient Greek philosophers thought that the entire world was composed of only four elements—earth, air, fire, and water. Although today we certainly would not accept these elements as building blocks of the world, *we do accept the basic principle that the world is describable in terms of a few fundamental quantities.*

When we look out at the world, we observe that the world occupies space, that within that space we find matter, and that space and matter exists within something we call time. So we will look for our observations of the world in terms of space, matter, and time. To measure space, we use the fundamental quantity of length. To measure matter, we use the fundamental quantities of mass and electrical charge. To measure time, we use the fundamental quantity of time itself.

*Therefore, to measure the entire physical world, we use the four **fundamental quantities** of length, mass, time, and charge.* We call all the other quantities that we observe derived quantities.

We have assigned ourselves an enormous task by trying to study the entire physical world in terms of only four quantities. The most remarkable part of it all is that it can be done. Everything in the world can be described in terms of these fundamental quantities. For example, consider a biological system, composed of very complex living tissue. But the tissue itself is made up of cells, and the cells are made of chemical molecules. The molecules are made of atoms, while the atoms consist of electrons, protons, and neutrons, which can be described in terms of the four fundamental quantities.

We might also ask of what electrons, protons, and neutrons are made. These particles are usually considered to be fundamental particles, however, the latest hypothesis in elementary particle physics is that protons and neutrons are made of even smaller particles called quarks. And although no one has yet actually found an isolated quark, and indeed some theories suggest that they are confined within the particles and will never be seen, the quark hypothesis has successfully predicted the existence of other particles, which *have* been found. The finding of these predicted particles gives a certain amount of credence to the existence of quarks. Of course if the quark is ever found then the next logical question will be, "Of what is the quark made?"

This progression from one logical question to the next in our effort to study the entire natural world is part of the adventure of physics. But to succeed on this adventure, we need to be precise in our observations, which brings us back to the subject at hand. If we intend to measure the world in terms of the four fundamental quantities of length, mass, time, and charge, we need to agree on some standard of measurement for each of these quantities.

1.5 The Standard of Length

The fundamental quantity of length is used to measure the location and the dimensions of any object in space. An object is located in space with reference to some coordinate system, as shown in figure 1.4. If the object is at the position P, then it can be located by moving a distance l_x in the x direction, then a distance l_y in the y direction, and finally a distance l_z in the z direction.

But before we can measure the distances l_x, l_y, and l_z, or for that matter, any distance, we need a standard of length that all observers can agree on. For example, suppose we wanted to measure the length of the room. We could use this text book as the standard of length. We would then place the text book on the floor and lay off the entire distance by placing the book end-over-end on the floor as often as necessary until the entire distance is covered. We might then say that the room is 25 books long. But this is not a very good standard of length because there are different sized books, and if you performed the measurement with another book, you would say that the floor has a different length.

We could even use the tile on the floor as a standard of length. To measure the length of the room all we would have to do is count the number of tiles. Indeed, if you worked at laying floor tiles, this would be a very good standard of length. The choice of a standard of length does seem somewhat arbitrary. In fact, just think of some of the units of measurement that you are familiar with:

The *foot*—historically the foot was used as a standard of length and it was literally the length of the king's foot. Every time you changed the king, you changed the measurement of the foot.

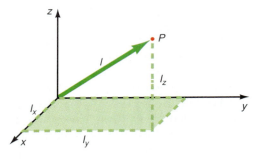

Figure 1.4

The location of an object in space.

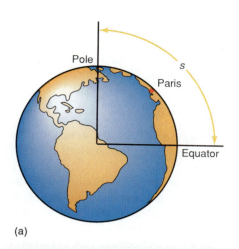

(a)

(b)

Figure 1.5

(a) The original definition of the meter.
(b) View of the earth from space.

The *yard*—the yard was the distance from the outstretched hand of the king to the back of his neck. Obviously, this standard of length also changed with each king.

The *inch*—the inch was the distance from the tip of the king's thumb to the thumb knuckle.

With these very arbitrary and constantly changing standards of length, it was obviously very difficult to make a measurement of length that all could agree on.

During the French Revolution, the French National Assembly initiated a proposal to the French Academy of Sciences to reform the systems of weights and measures. Under the guidance of such great physicists as Joseph L. Lagrange and Pierre S. de Laplace, the committee agreed on a measuring system based on the number 10 and its multiples. In this system, the unit of length chosen was one ten-millionth of the distance s from the North Pole to the equator along a meridian passing through Paris, France (figure 1.5). The entire distance from the pole to the equator was not actually measured. Instead a geodetic survey was undertaken for 10 degrees of latitude extending from Dunkirk, in northern France, to Barcelona, in Spain. From these data, the distance from the pole to the equator was found. The meter, the standard of length, was defined as one ten-millionth of this distance. A metal rod equal to this distance was made, and it was stored in Sèvres, just outside Paris. Copies of this rod were distributed to other nations to be used as their standard.

In time, with greater sophistication in measuring techniques, it turned out that the distance from the North Pole to the equator was in error, so the length of the meter could no longer represent one ten-millionth of that distance. But that really did not matter, as long as everyone agreed that this length of rod would be the standard of length. For years these rods were the accepted standard. However, they also had drawbacks. They were not readily accessible to all the nations of the world, and they could be destroyed by fire or war. A new standard had to be found. The standard remained the meter, but it was now defined in terms of something else. In 1960, the Eleventh General Conference of Weights and Measures defined the meter as a certain number of wavelengths of light from the krypton 86 atom.

Using a standard meter bar or a prescribed number of wavelengths indeed gives us a standard length. Such measurements are called *direct measurements*. But in addition to direct measuring procedures, an even more accurate determination of a quantity sometimes can be made by measuring something other than the desired quantity, and then obtaining the desired quantity by a calculation with the measured quantity. Such procedures, called *indirect measuring techniques*, have been used to obtain an even more precise definition of the meter.

We can measure the speed of light c, a derived quantity, to a very great accuracy. The speed of light has been measured at 299,792,458 meters/second, with an uncertainty of only four parts in 10^9, a very accurate value to be sure. Using this value of the speed of light, the standard meter can now be defined. On October 20, 1983, the Seventeenth General Conference on Weights and Measures redefined the meter as: *"The **meter** is the length of the path traveled by light in a vacuum during a time interval of 1/299,792,458 of a second."*

We will see in chapter 3 on kinematics that the distance an object moves at a constant speed is equal to the product of its speed and the time that it is in motion. Using this relation the meter is defined as

$$\text{distance} = (\text{speed of light})(\text{time})$$

$$= \left(\frac{299{,}792{,}458 \text{ meters}}{\text{second}}\right)\left(\frac{\text{second}}{299{,}792{,}458}\right) = 1 \text{ meter}$$

Hence the meter, the fundamental quantity of length, is now determined in terms of the speed of light and the fundamental quantity of time. The meter, thus defined, is a fixed standard of length accessible to everyone and is nonperishable. For everyone brought up to think of lengths in terms of the familiar inches, feet, or yards, the meter, abbreviated m, is equivalent to

$$1.000 \text{ m} = 39.37 \text{ in.} = 3.281 \text{ ft} = 1.094 \text{ yd}$$

For very precise work, the standard of length must be used in terms of its definition. For most work in a college physics course, however, the standard of length will be the simple meter stick.

The system of measurements based on the meter was originally called the metric system of measurements. Today it is called the **International System (SI) of units.** The letters are written SI rather than IS because the official international name follows French usage, "Le Système International d'Unités." This system of measurements is used by scientists throughout the world and commercially by almost all the countries of the world except the United States and one or two other countries. The United States is supposed to be changing over to this system also.

One of the advantages of using the meter as the standard of length is that the meter is divided into 100 parts called centimeters (abbreviated cm). The centimeter, in turn, is divided into ten smaller divisions called millimeters (mm). The kilometer (abbreviated km) is equal to a thousand meters, and is used to measure very large distances. Thus the units of length measurement become a decimal system, that is,

$$1 \text{ m} = 100 \text{ cm}$$
$$1 \text{ cm} = 10 \text{ mm}$$
$$1 \text{ km} = 1000 \text{ m}$$

A further breakdown of the units of length into powers of ten is facilitated by using the following prefixes:

$$\text{tera (T)} = 10^{12}$$
$$\text{giga (G)} = 10^{9}$$
$$\text{mega (M)} = 10^{6}$$
$$\text{milli (m)} = 10^{-3}$$
$$\text{micro } (\mu) = 10^{-6}$$
$$\text{nano (n)} = 10^{-9}$$
$$\text{pico (p)} = 10^{-12}$$
$$\text{femto (f)} = 10^{-15}$$

Students unfamiliar with the powers of ten notation, and scientific notation in general, should consult appendix B. Using these prefixes, the lengths of all observables can be measured as multiples or submultiples of the meter.

The decimal nature of the SI system makes it easier to use than the *British engineering system,* the system of units that is used in the United States. For example, compare the simplicity of the decimal metric system to the arbitrary units of the British engineering system:

$$12 \text{ inches} = 1 \text{ foot}$$
$$3 \text{ feet} = 1 \text{ yard}$$
$$5280 \text{ feet} = 1 \text{ mile}$$

In fact, these units are now officially defined in terms of the meter as

$$1 \text{ foot} = 0.3048 \text{ meters} = 30.48 \text{ centimeters}$$
$$1 \text{ yard} = 0.9144 \text{ meters} = 91.44 \text{ centimeters}$$
$$1 \text{ inch} = 0.0254 \text{ meters} = 2.54 \text{ centimeters}$$
$$1 \text{ mile} = 1.609 \text{ kilometers}$$

A complete list of equivalent measurements can be found in appendix A.

Figure 1.6

The standard kilogram mass.

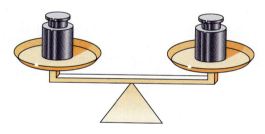

Figure 1.7

A simple balance.

1.6 The Standard of Mass

The simplest definition of **mass** is that *mass is a measure of the quantity of matter in a body.* This is not a particularly good definition, but it is one for which we have an intuitive grasp. Mass will be redefined more accurately in terms of its inertial and gravitational characteristics later. For now, let us think of the mass of a body as being the matter that is contained in the sum of all the atoms and molecules that make up that body. For example, the mass of this book is the matter of the billions upon billions of atoms that make up the pages and the print of the book itself.

The standard we use to measure mass can, like the standard of length, also be quite arbitrary. In 1795, the French Academy of Science initially defined the standard as the amount of matter in 1000 cm³ of water at 0°C and called this amount of mass, one kilogram. This definition was changed in 1799 to make the kilogram the amount of matter in 1000 cm³ of water at 4°C, the temperature of the maximum density of water. However, in 1889, *the new and current definition of the **kilogram** became the amount of matter in a specific platinum iridium cylinder 39 mm high and 39 mm in diameter.* The metal alloy of platinum and iridium was chosen because it was considered to be the most resistant to wear and tarnish. Copies of the cylinder (figure 1.6) are kept in the standards laboratories of most countries of the world.

The disadvantage of using this cylinder as the standard of mass is that it could be easily destroyed and it is not readily accessible to every country on earth. It seems likely that sometime in the future, when the necessary experimental techniques are developed, the kilogram will be redefined in terms of the mass of some specified number of atoms or molecules, thereby giving the standard of mass an atomic definition.

With the standard of mass, the kilogram, defined, any number of identical masses, multiple masses, or submultiple masses can be found by using a simple balance, as shown in figure 1.7. We place a standard kilogram on the left pan of the balance, and then place another piece of matter on the right pan. If the new piece of matter is exactly 1 kg, then the scale will balance and we have made another kilogram mass. If there is too much matter in the tested sample the scales will not balance. We then shave off a little matter from the sample until the scales do balance. On the other hand, if there is not enough matter in the sample, we add a little matter to the sample until the scales do balance. In this way, we can make as many one kilogram masses as we want.

Any multiple of the kilogram mass can now be made with the aid of the original one kilogram masses. That is, if we want to make a 5-kg mass, we place five 1-kg masses on the left pan of the balance and add mass to the right pan until the scale balances. When this is done, we will have made a 5-kg mass. Proceeding in this way, we can obtain any multiple of the kilogram.

To make submultiples of the kilogram mass, we cut a 1-kg mass in half, and place one half of the mass on each of the two pans of the balance. If we have cut the kilogram mass exactly in half, the scales will balance. If they do not, we shave off a little matter from one of the samples and add it to the other sample until the scales do balance. Two 1/2-kg masses thus result. Since the prefix kilo means a thousand, these half-kilogram masses each contain 500 grams (abbreviated g). If we now cut a 500-g mass in half and place each piece on one of the pans of the balance, making of course whatever corrections that are necessary, we have two 250-g masses. Continuing this process by taking various combinations of cuttings and placing them on the balance, eventually we can make any submultiple of the kilogram. The assembly of these multiples and submultiples of the kilogram is called a set of masses. (Quite often, this is erroneously referred to as a set of weights.)

We can now measure the unknown mass of any body by placing it on the left pan of the balance and adding any multiple, and/or submultiple, of the kilogram to the right pan until the scales balance. The sum of the combination of the masses placed on the right pan is the mass of the unknown body. So we can determine the mass of any body in terms of the standard kilogram.

The principle underlying the use of the balance is the gravitational force between masses. (The gravitational force will be discussed in detail in chapter 6.) The mass on the left pan is attracted toward the center of the earth and therefore pushes down on the left pan. The mass on the right pan is also attracted toward the earth and pushes down on the right pan. When the force down on the right pan is equal to the force down on the left pan, the scales are balanced and the mass on the right pan is equal to the mass on the left pan. Mass measured by a balance depends on the force of gravity acting on the mass. Hence, mass measured by a balance can be called *gravitational mass*. The balance will work on the moon or on any planet where there are gravitational forces. The equality of masses on the earth found by a balance will show the same equality on the moon or on any planet. But a balance at rest in outer space extremely far away from gravitational forces will not work at all.

1.7 The Standard of Time

What is time and how do we measure it? Time is such a fundamental concept that it is very difficult to define. We will try by *defining time as a duration between the passing of events.* (Do not ask me to define duration, because I would have to define it as the time during which something happens, and I would end up seemingly caught in circular reasoning. This is the way it is with fundamental quantities, they are so fundamental that we cannot define them in terms of something else. If we could, that something else would become the fundamental quantity.) As with all fundamental quantities, we must choose a standard and measure all durations in terms of that standard. To measure time we need something that will repeat itself at regular intervals. The number of intervals counted gives a quantitative measure of the duration. The simplest method of measuring a time interval is to use the rhythmic beating of your own heart as a time standard. Then, just as you measured a length by the number of times the standard length was used to mark off the unknown length, you can measure a time duration by the number of pulses from your heart that covers the particular unknown duration. Note that Galileo timed the swinging chandeliers in a church one morning by the use of his pulse, finding the time for one complete oscillation of the pendulum to be independent of the magnitude of that oscillation.

In this way, we can measure time durations by the number of heartbeats counted. However, if you start running or jumping up and down your heart will beat faster and the time interval recorded will be different than when you were at rest. Therefore, for any good timing device we need something that repeats itself over and over again, always with the same constant time interval. Obviously, the technique used to measure time intervals should be invariant, and the results obtained should be the same for different individuals. One such invariant, which occurs day after day, is the rotation of the earth.

It is not surprising, then, that the early technique used for measuring time was the rotation of the earth. One complete rotation of the earth was called a day, and the day was divided into 24 hours; each hour was divided into 60 minutes; and finally each minute was divided into 60 seconds. The standard of time became the second. It may seem strange that the day was divided into 24 hours, the hour into 60 minutes, and the minute into 60 seconds. But remember that the very earliest recorded studies of astronomy and mathematics began in ancient Mesopotamia and Babylonia, where the number system was based on the number 60, rather than on the number 10, which we base our number system on. Hence, a count of 60 of their base units was equal to 1 of their next larger units. When they got to a count of 120 base units, they set this equal to 2 of the larger units. Thus, a count of 60 seconds, their base unit, was equal to 1 unit of their next larger unit, the minute. When they got to 60 minutes, this was equal to their next larger unit, the hour.

Their time was also related to their angular measurements of the sky. Hence the year became 360 days, the approximate time for the earth to go once around the sun. They related the time for the earth to move once around the heavens, 360 days, to the angle moved through when moving once around a circle by also dividing the circle into 360 units, units that today are called *angular degrees*. They then divided their degree by their base number 60 to get their next smaller unit of angle, 1/60 of a degree, which they called a *minute of arc*. They then divided their minute by their base number 60 again to get an angle of 1 second, which is equal to 1/60 of a minute. The movement of the heavenly bodies across the sky became their calendar. Of course their minutes and seconds of arc are not the same as our minutes and seconds of time, but because of their base number 60 our measurements of arc and time are still based on the number 60.

What is even more interesting is that the same committee that originally introduced the meter and the kilogram proposed a clock that divided the day into 10 equal units, each called a *deciday*. They also divided a quadrant of a circle (90°) into a hundred parts each called the *grade*. They thus tried to place time and angle measurements into a decimal system also, but these units were never accepted by the people.

So the second, which is 1/86,400 part of a day, was kept as the measure of time. However, it was eventually found that the earth does not spin at a constant rate. It is very close to being a constant value, but it does vary ever so slightly. In 1967, the Thirteenth General Conference of Weights and Measures decided that the primary standard of time should be based on an atomic clock, figure 1.8. *The second is now defined as "the duration of 9,192,631,770 periods (or cycles) of the radiation corresponding to the transition between two hyperfine levels of the ground state of the cesium-133 atom."* The atomic clock is located at the National Bureau of Standards in Boulder, Colorado. The atomic clock is accurate to 1 second in a thousand years and can measure a time interval of one millionth of a second.

The atomic clock provides the reference time, from which certain specified radio stations (such as WWV in Fort Collins, Colorado) broadcast the correct time. This time is then transmitted to local radio and TV stations and telephone services, from which we usually obtain the time to set our watches.

For the accuracy required in a freshman college physics course, the unit of time, the second, is the time it takes for the second hand on a nondigital watch to move one interval.

Figure 1.8

The atomic clock.

1.8 The Standard of Electrical Charge

One of the fundamental characteristics of matter is that it has not only mass but also electrical charge. We now know that all matter is composed of atoms. These atoms in turn are composed of electrons, protons, and neutrons. Forces have been found that exist between these electrons and protons, forces caused by the electrical charge that these particles carry. The smallest charge ever found is the charge on the electron. By convention we call it a negative charge. The proton contains the same amount of charge, but it is a positive charge. Most matter contains equal numbers of electrons and protons, and hence is electrically neutral.

Although electrical charge is a fundamental property of matter, it is a quantity that is relatively difficult to measure directly, whereas the effects of electric current—the flow of charge per unit time—is much easier to measure. *Therefore, the fundamental unit of electricity is defined as the ampere, where "the ampere is that constant current that, if maintained in two straight parallel conductors of*

Table 1.1
Systems of Units

Physical Quantity	International System (SI)	British Engineering System
Length	meter (m)	foot (ft)
Mass	kilogram (kg)	slug
Time	second (s)	second (s)
Electric current	ampere (A)	
Electric charge	coulomb (C)	
Weight or force	newton (N)	pound (lb)

infinite length, of negligible circular cross section, and placed one meter apart in a vacuum, would produce between these conductors a force equal to 2 × 10⁻⁷ newtons per meter of length." This definition will be explained in more detail when electricity is studied in section 22.7. The ampere, the unit of current, is also defined as the passage of 1 **coulomb** of charge per second in a circuit. This represents a passage of 6.25×10^{18} electrons per second. Therefore, the charge on one electron is 1.60×10^{-19} coulombs.

1.9 Systems of Units

When the standards of the fundamental quantities are all assembled, they are called a *system of units*. The standards for the fundamental quantities, discussed in the previous sections, are part of a new system of units called the International System of units, abbreviated SI units. They were adopted by the Eleventh General Conference of Weights and Measures in 1960. This system of units refines and replaces the older metric system of units, and is very similar to it. Table 1.1 shows the two systems of units that will be considered.

Let us add another quantity to table 1.1, namely the quantity of weight or force. In SI units this is not a fundamental quantity, but rather a derived quantity. (A complete definition of the concepts of force and weight will be given in chapter 4.) For the present, let us add it to the table and say the weight and mass of an object are related but not identical quantities. As already indicated, mass is a measure of the quantity of matter in a body. The weight of a body here on earth is a measure of the gravitational force of attraction of the earth on that mass, pulling the mass of that body down toward the center of the earth. In the international system, the unit of weight or force is called the *newton,* named of course after Sir Isaac Newton.

An important distinction between mass and weight can easily be shown here. If you were to go to the moon, figure 1.9, you would find that the gravitational force on the moon is only 1/6 of the gravitational force found here on earth. Hence, on the moon you would only weigh 1/6 what you do on earth. That is, if you weigh 180 lb on earth, you would only weigh 30 lb on the surface of the moon. Yet your mass has not changed at all. The thing that you call you, all the complexity of atoms, molecules, cells, tissue, blood, bones, and the like, is still the same. Your weight would have changed, but not your mass. The difference between mass and weight will be explained in much more detail in a later chapter. The unit of weight or force, the newton, is only placed in the table now in order to compare it to the next system of units.

Figure 1.9
Your weight on the moon is very different from your weight on earth.

What has been said about systems of units so far is fine, but the system that most of you are familiar with has been only briefly mentioned. The system of units that you are accustomed to using is called the British engineering system of units (see table 1.1). In that system, the unit of length is the foot. (Recall that the unit of a foot is now defined in terms of the standard of length, the meter.) The unit of time is again the second.

In the British engineering system (BES), mass is not defined as a fundamental quantity; instead the weight of a body is described as fundamental, and its mass is derived from its weight. The fundamental unit of weight in the BES is defined as the pound with which we are all familiar. The unit of mass is derived from the unit of weight, and is called a *slug*. Whenever you hear or see the word pound it means a weight or a force, never a mass.

The history of some of the names of these units is quite interesting. The name slug came from the idea that when we try to move a large massive object and put it into motion, the more massive the object the more difficult it is to put that object into motion. The object is thus said to be sluggish. Therefore, the more massive the body, the more sluggish it is, and hence, it contains more slugs. This unit is often used in engineering, but not in our daily lives.

As you are probably aware, the United States is now in the process of changing from the British engineering system to the international system of units. The British engineering system will probably be obsolete within 10 to 20 years. (Even the British no longer use the British engineering system.) There may be a little confusion during the transition, but in the long run the international system is a better system because it is a much easier system to use.

However, problems have already arisen even before the SI units have been officially adopted. For example, if you go to the local supermarket and buy an average-sized can of beans, you will see printed on it "Net wt. 21 oz. (1 lb. 5 oz.) or 595 g." The business sector has already erroneously equated mass and weight by calling them the same name, grams or kilograms. What the businessman really means is that the can of beans has a mass of 595 grams. The weight of an object in SI units should be expressed in newtons. We will show how to deal with this new confusion later. In this book, however, whenever you see the word kilogram or gram it will refer to the mass of an object.

To simplify the use of units in equations, abbreviations will be used. All unit abbreviations in SI units are one or two letters long and the abbreviations do not require a period following them. The name of a unit based on a proper name is written in lower-case letters, while its abbreviation is capitalized. All other abbreviations are written in lower-case letters. The abbreviations are shown in table 1.1.

Most of the measurements used in this book will be in SI units, although many will be used in the British engineering system since this is the system you probably have the best feel for, because you grew up with it. Sometimes it will be necessary to convert from one of these systems of units to another. In order to do this, it is necessary to make use of a conversion factor.

1.10 Conversion Factors

A **conversion factor** is a factor by which a quantity expressed in one set of units must be multiplied in order to express that quantity in different units. The numbers for a conversion factor are usually expressed as an equation, relating the quantity in one system of units to the same quantity in different units. Appendix A, at the back of this book, contains a large number of conversion factors. An example of an equation leading to a conversion factor is

$$1 \text{ m} = 3.281 \text{ ft}$$

A conversion factor is also used to change a quantity expressed in one system of units to a value in different units of the same system or to a unit in another system of units. For example, if both sides of the above equality are divided by 3.281 ft we get

$$\frac{1 \text{ m}}{3.281 \text{ ft}} = \frac{3.281 \text{ ft}}{3.281 \text{ ft}} = 1$$

Thus,

$$\frac{1 \text{ m}}{3.281 \text{ ft}} = 1$$

is a conversion factor that is equal to unity. If a height is multiplied by a conversion factor, we do not physically change the height, because all we are doing is multiplying it by the number one. The effect, however, expresses the same height as a different number with a different unit.

Example 1.1

Converting feet to meters. The height of a building is 100.0 ft. Find the height in meters.

Solution

To express the height h in meters, multiply the height in feet by the conversion factor that converts feet to meters, that is,

$$h = 100.0 \text{ ft} \left(\frac{1 \text{ m}}{3.281 \text{ ft}} \right) = 30.48 \text{ m}$$

Notice that the units act like algebraic quantities. That is, the unit foot, which is in both the numerator and the denominator of the equation, divides out, leaving us with the single unit, meters.

The technique to remember in using a conversion factor is that the unit in the numerator that is to be eliminated, must be in the denominator of the conversion factor. Then, because units act like algebraic quantities, identical units can be divided out of the equation immediately.

Conversion factors should also be set up in a chain operation. This will make it easy to see which units cancel. For example, suppose we want to express the time T of one day in terms of seconds. This number can be found as follows:

$$T = 1 \text{ day} \left(\frac{24 \text{ hr}}{1 \text{ day}} \right) \left(\frac{60 \text{ min}}{1 \text{ hr}} \right) \left(\frac{60 \text{ s}}{1 \text{ min}} \right)$$

$$= 86,400 \text{ s}$$

By placing the conversion factors in this sequential fashion, the units that are not wanted divide out directly and the only unit left is the one we wanted, seconds. This technique is handy because if we make a mistake and use the wrong conversion factor, the error is immediately apparent. For example, let us go back to the first example and use the reciprocal of the conversion factor by mistake:

$$h = 100.0 \text{ ft} \left(\frac{3.281 \text{ ft}}{1 \text{ m}} \right) = 328.1 \frac{\text{ft}^2}{\text{m}}$$

It is immediately obvious that we have made a mistake because the final units do not represent the height in meters, the unit for which we were looking. Whenever something like this happens, we simply go back and use the correct conversion factor.

These examples are, of course, trivial, but the important thing to learn is the technique. Later when these ideas are applied to problems that are not trivial, if the technique is followed as shown, there should be no difficulty in obtaining the correct solutions.

1.11 **Derived Quantities**

Most of the quantities that are observed in the study of physics are derived in terms of the fundamental quantities. For example, the speed of a body is the ratio of the distance that an object moves to the time it takes to move that distance. This is expressed as

$$\text{speed} = \frac{\text{distance traveled}}{\text{time}}$$

$$v = \frac{\text{length}}{\text{time}}$$

That is, the speed v is the ratio of the fundamental quantity of length to the fundamental quantity of time. Thus, speed is derived from length and time. For example, the unit for speed in SI units is a meter per second (m/s).

Another example of a derived quantity is the volume V of a body. For a box, the volume is equal to the length times the width times the height. Thus,

$$V = (\text{length})(\text{width})(\text{height})$$

But because the length, width, and height of the box are measured by a distance, the volume is equal to the cube of the fundamental unit of length L. That is,

$$V = L^3$$

Hence, the SI unit for volume is m³.

As a final example of a derived quantity, the density of a body is defined as its mass per unit volume, that is,

$$\rho = \frac{m}{V}$$

$$= \frac{\text{mass}}{(\text{length})^3}$$

Hence, the density is defined as the ratio of the fundamental quantity of mass to the cube of the fundamental quantity of length. The SI unit for density is thus kg/m³. All the remaining quantities of physics are derived in this way, in terms of the four fundamental quantities of length, mass, charge, and time.

Note that the international system of units also recognizes temperature, luminous intensity, and "quantity of matter" (the mole) as fundamental. However, they are not fundamental in the same sense as mass, length, time, and charge. Later in the book we will see that temperature can be described as a measure of the mean kinetic energy of molecules, which is described in terms of length, mass, and time. Similarly, intensity can be derived in terms of energy, area, and time, which again are all describable in terms of length, mass, and time.

Learning physics at an early age.
PEANUTS reprinted by permission of UFS, Inc.

The Language of Physics

Philosophy
The search for knowledge or wisdom (p. 5).

Natural philosophy
The study of the natural or physical world (p. 6).

Physics
The Greek word for "natural" is *physikos*. Therefore, the word physics came to mean the study of the entire natural or physical world (p. 6).

Scientific method
The application of a logical process of reasoning to arrive at a model of nature that is consistent with experimental results. The scientific method consists of five steps: (1) observation, (2) hypothesis, (3) experiment, (4) theory or law, and (5) prediction (p. 6).

Fundamental quantities
The most basic quantities that can be used to describe the physical world. When we look out at the world, we observe that the world occupies space, and within that space we find matter, and that space and matter exists within something we call time. So the observation of the world can be made in terms of space, matter, and time. The fundamental quantity of length is used to describe space, the fundamental quantities of mass and electrical charge are used to describe matter, and the fundamental quantity of time is used to describe time. All other quantities, called derived quantities, can be described in terms of some combination of the fundamental quantities (p. 10).

International System (SI) of units
The internationally adopted system of units used by all the scientists and almost all the countries of the world (p. 12).

Meter
The standard of length. It is defined as the length of the path traveled by light in a vacuum during an interval of $1/299,792,458$ of a second (p. 11).

Mass
The measure of the quantity of matter in a body (p. 13).

Kilogram
The unit of mass. It is defined as the amount of matter in a specific platinum iridium cylinder 39 mm high and 39 mm in diameter (p. 13).

Second
The unit of time. It is defined as the duration of 9,192,631,770 periods of the radiation corresponding to the transition between two hyperfine levels of the ground state of the cesium-133 atom of the atomic clock (p. 15).

Coulomb
The unit of electrical charge. It is defined in terms of the unit of current, the ampere. The ampere is a flow of 1 coulomb of charge per second. The ampere is defined as that constant current that, if maintained in two straight parallel conductors of infinite length, of negligible circular cross section, and placed one meter apart in vacuum, would produce between these conductors a force equal to 2×10^{-7} newtons per meter of length (p. 16).

Conversion factor
A factor by which a quantity expressed in one set of units must be multiplied in order to express that quantity in different units (p. 17).

Questions for Chapter 1

1. Why should physics have separated from philosophy at all?
† 2. What were Aristotle's ideas on physics, and what was their effect on science in general, and on physics in particular?
3. Is the scientific method an oversimplification?
4. How does a law of physics compare with a civil law?
† 5. Is there a difference between saying that an experiment validates a law of nature and that an experiment verifies a law of nature? Where does the concept of truth fit in the study of physics?

6. How does physics relate to your field of study?
7. In the discussion of hot and cold in section 1.3, what would happen if you placed your right hand in the hot water and your left hand in the cold water, and then placed both of them in the mixture simultaneously?
8. Can you think of any more examples that show the need for quantitative measurements?
9. Compare the description of the world in terms of earth, air, fire, and water with the description in terms of length, mass, electrical charge, and time.

10. Discuss the pros and cons of dividing the day into decidays. Do you think this idea should be reintroduced into society? Using yes and no answers, have your classmates vote on a change to a deciday. Is the result surprising?
11. Discuss the difference between mass and weight.

Problems for Chapter 1

In all the examples and problems in this book we assume that whole numbers, such as 2 or 3, have as many significant figures as are necessary in the solution of the problem.

1. How many cubic centimeters are there in a cubic inch?

2. A liter contains 1000 cm³. How many liters are there in a cubic meter?
3. A speed of 60.0 miles per hour (mph) is equal to how many ft/s?
4. The density of 1 g/cm³ is equal to how many kg/liter?

5. How many seconds are there in a day? a month? a year?
6. Calculate your height in meters.
7. What is 90 km/hr expressed in mph?
8. How many square meters are there in 1 acre, if 1 acre is equal to 43,560 ft²?
9. How many feet are there in 1 km?

10. Express the age of the earth (approximately 4.6×10^9 years) in seconds.

11. The speed of sound in air is 331 m/s at 0°C. Express this speed in ft/s and mph.

12. The speedometer of a new car is calibrated in km/hr. If the speed limit is 55 mph, how fast can the car go in km/hr and still stay below the speed limit?

13. A floor has an area of 144 ft². What is this area expressed in m²?

14. A tank contains a volume of 50 ft³. Express this volume in cubic meters.

15. Assuming that an average person lives for 75 yrs, how many (a) seconds and (b) minutes are there in this lifetime? If the heart beats at an average of 70 pulses/min, how many beats does the average heart have?

16. A cube is 50 cm on each side. Find its surface area in m² and ft² and its volume in m³ and ft³.

17. The speed of light in a vacuum is approximately 186,000 miles/s. Express this speed in mph and m/s.

18. The distance from home plate to first base on a baseball field is 90 ft. What is this distance in meters?

19. In the game of football, a first down is 10 yd long. What is this distance in meters? If the field is 100 yd long, what is the length of the field in meters?

20. The diameter of a sphere is measured as 6.28 cm. What is its volume in cm³, m³, in.³, and ft³?

21. The Empire State Building is 1245 ft tall. Express this height in meters, miles, inches, and millimeters.

22. A drill is 1/4 in. in diameter. Express this in centimeters, and then millimeters.

23. The average diameter of the earth is 7927 miles. Express this in km.

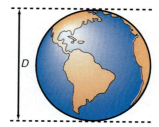

24. A 31-story building is 132 m tall. What is the average height of each story in feet?

25. Light of a certain color has a wavelength of 589 nm. Express this wavelength in (a) pm, (b) mm, (c) cm, (d) m. How many of these 589 nm waves are there in an inch?

26. Calculate the average distance to the moon in meters if the distance is 239,000 miles.

27. A basketball player is 7 ft tall. What is this height in meters?

28. The mass of a hydrogen atom is 1.67×10^{-24} g. Calculate the number of atoms in 1 g of hydrogen.

29. The Washington National Monument is 555 ft high. Express this height in meters.

30. The Statue of Liberty is 305 ft high. Express this height in meters.

31. Cells found in the human body have a volume generally in the range of 10^4 to 10^6 cubic microns. A micron is an older name of the unit that is now called a micrometer and is equal to 10^{-6} m. Express this volume in cubic meters and cubic inches.

32. The diameter of a deoxyribonucleic acid (DNA) molecule is about 20 angstroms. Express this diameter in picometers, nanometers, micrometers, millimeters, centimeters, meters, and inches. Note that the old unit angstrom is equal to 10^{-10} m.

33. A glucose molecule has a diameter of about 8.6 angstroms. Express this diameter in millimeters and inches.

34. Muscle fibers range in diameter from 10 microns to 100 microns. Express this range of diameters in centimeters and inches.

35. The axon of the neuron, the nerve cell of the human body, has a diameter of approximately 0.2 microns. Express this diameter in terms of (a) pm, (b) nm, (c) μm, (d) mm, and (e) cm.

36. The Sears Tower in Chicago, the world's tallest building, is 1454 ft high. Express this height in meters.

37. A baseball has a mass of 145 g. Express this mass in slugs.

38. One shipping ton is equal to 40 ft³. Express this volume in cubic meters.

39. A barrel of oil contains 42 U.S. gallons, each of 231 in.³. What is its volume in cubic meters?

40. The main span of the Verrazano Narrows Bridge in New York is 1298.4 m long. Express this distance in feet and miles.

41. The depth of the Mariana Trench in the Pacific Ocean is 10,911 m. Express this depth in feet.

42. Mount McKinley is 6194 m high. Express this height in feet.

43. The average radius of the earth is 6371 km. Find the area of the surface of the earth in m² and in ft². Find the volume of the earth in m³ and ft³. If the mass of the earth is 5.97×10^{24} kg, find the average density of the earth in kg/m³.

44. Cobalt-60 has a half-life of 5.27 yr. Express this time in (a) months, (b) days, (c) hours, (d) seconds, and (e) milliseconds.

45. On a certain European road in a quite residential area, the speed limit is posted as 40 km/hr. Express this speed limit in miles per hour.

46. In a recent storm, it rained 6.00 in. of rain in a period of 2.00 hr. If the size of your property is 100 ft by 100 ft, find the total volume of water that fell on your property. Express your answer in (a) cubic feet, (b) cubic meters, (c) liters, and (d) gallons.

47. A cheap wrist watch loses time at the rate of 8.5 seconds a day. How much time will the watch be off at the end of a month? A year?

48. A ream of paper contains 500 sheets of 8½ in. by 11 in. paper. If the package is 1 and ⅞ in. high, find (a) the thickness of each sheet of paper in inches and millimeters, (b) the dimensions of the page in millimeters, and (c) the area of a page in square meters and square millimeters.

Interactive Tutorials

🖫 49. Conversion Calculator. The Conversion Calculator will allow you to convert from a quantity in one system of units to that same quantity in another system of units and/or to convert to different units within the same system of units.

2

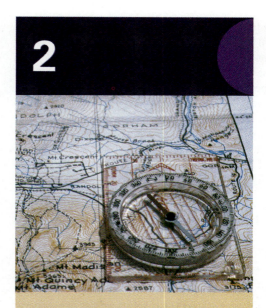

You need vectors to navigate.

Corollary I. A body, acted on by two forces simultaneously, will describe the diagonal of a parallelogram in the same time as it would describe the sides by those forces separately.

Isaac Newton

ectors

2.1 Introduction

Of all the varied quantities that are observed in nature, some have the characteristics of scalar quantities while others have the characteristics of vector quantities. *A **scalar** quantity is a quantity that can be completely described by a magnitude, that is, by a number and a unit.* Some examples of scalar quantities are mass, length, time, density, and temperature. The characteristic of scalar quantities is that they add up like ordinary numbers. That is, if we have a mass $M_1 = 3$ kg and another mass $M_2 = 4$ kg then the sum of the two masses is

$$M = M_1 + M_2 = 3 \text{ kg} + 4 \text{ kg} = 7 \text{ kg} \tag{2.1}$$

*A **vector** quantity, on the other hand, is a quantity that needs both a magnitude and a direction to completely describe it.* Some examples of vector quantities are force, displacement, velocity, and acceleration. The velocity of a car moving at 50 miles per hour (mph) due east can be represented by a vector. Velocity is a vector because it has a magnitude, 50 mph, and a direction, due east.

A vector is represented in this text book by boldface script, that is, **A**. Because we cannot write in boldface script on note paper or a blackboard, a vector is written there as the letter with an arrow over it. A vector can be represented on a diagram by an arrow. A picture of this vector can be obtained by drawing an arrow from the origin of a cartesian coordinate system. The length of the arrow represents the magnitude of the vector, while the direction of the arrow represents the direction of the vector. The direction is specified by the angle θ that the vector makes with an axis, usually the *x*-axis, and is shown in figure 2.1. The magnitude of vector **A** is written as the absolute value of **A** namely $|\mathbf{A}|$, or simply by the letter A without boldfacing. One of the defining characteristics of vector quantities is that they must be added in a way that takes their direction into account.

2.2 The Displacement

Probably the simplest vector that can be discussed is the displacement vector. Whenever a body moves from one position to another it undergoes a displacement. *The displacement can be represented as a vector that describes how far and in what direction the body has been displaced from its original position.* The tail of the displacement vector is located at the position where the displacement started, and its tip is located at the position at which the displacement ended. For example, if you walk 3 mi due east, this walk can be represented as a vector that is 3 units long

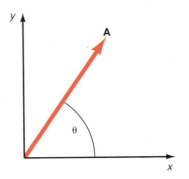

Figure 2.1

Representation of a vector.

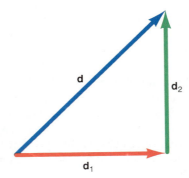

Figure 2.2

The displacement vector.

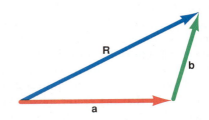

Figure 2.3

The addition of vectors.

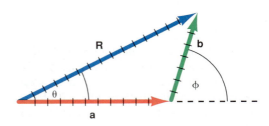

Figure 2.4

The graphical addition of vectors.

and points due east. It is shown as $\mathbf{d_1}$ in figure 2.2. This is an example of a displacement vector. Suppose you now walk 4 mi due north. This distance of 4 mi in a northerly direction can be represented as another displacement vector $\mathbf{d_2}$, which is also shown in figure 2.2. The result of these two displacements is a final displacement vector $\mathbf{d}$ that shows the total displacement from the original position.

We now ask how far did you walk? Well, you walked 3 mi east and 4 mi north and hence you have walked a total distance of 7 mi. But how far are you from where you started? Certainly not 7 mi, as we can easily see using a little high school geometry. In fact the final displacement $\mathbf{d}$ is a vector of magnitude d and that distance d can be immediately determined by simple geometry. Applying the Pythagorean theorem to the right triangle of figure 2.2 we get

$$d = \sqrt{d_1^2 + d_2^2} \qquad\qquad (2.2)$$
$$= \sqrt{(3 \text{ mi})^2 + (4 \text{ mi})^2} = \sqrt{25 \text{ mi}^2}$$

and thus,

$$d = 5 \text{ mi}$$

Even though you have walked a total distance of 7 mi, you are only 5 mi away from where you started. Hence, when these vector displacements are added

$$\mathbf{d} = \mathbf{d_1} + \mathbf{d_2} \qquad\qquad (2.3)$$

we do not get 7 mi for the magnitude of the final displacement, but 5 mi instead. *The displacement is thus a change in the position of a body from its initial position to its final position. Its magnitude is the distance between the initial position and the final position of the body.*

It should now be obvious that vectors do not add like ordinary scalar numbers. In fact, all the rules of algebra and arithmetic that you were taught in school are the rules of scalar algebra and scalar arithmetic, although the word scalar was probably never used at that time. To solve physical problems associated with vectors it is necessary to deal with vector algebra.

2.3 Vector Algebra—The Addition of Vectors

Let us now add any two arbitrary vectors $\mathbf{a}$ and $\mathbf{b}$. The result of adding the two vectors $\mathbf{a}$ and $\mathbf{b}$ forms a new **resultant** vector $\mathbf{R}$, which is the sum of $\mathbf{a}$ and $\mathbf{b}$. This can be shown graphically by laying off the first vector $\mathbf{a}$ in the horizontal direction and then placing the tail of the second vector $\mathbf{b}$ at the tip of vector $\mathbf{a}$, as shown in figure 2.3.

The resultant vector $\mathbf{R}$ is drawn from the origin of the first vector to the tip of the last vector. The resultant vector is written mathematically as

$$\mathbf{R} = \mathbf{a} + \mathbf{b} \qquad\qquad (2.4)$$

Remember that in this sum we do not mean scalar addition. The resultant vector is the vector sum of the individual vectors $\mathbf{a}$ and $\mathbf{b}$.

We can add these vectors graphically, with the aid of a ruler and a protractor. First, we need to choose a scale such that a unit distance on the graph paper corresponds to a unit of magnitude of the vector. Using this scale, we lay off the length that corresponds to the magnitude of vector $\mathbf{a}$ in the x-direction with a ruler. Then, at the tip of vector $\mathbf{a}$, place the center of the protractor and measure the angle ϕ that vector $\mathbf{b}$ makes with the x-axis. Mark that direction on the paper. Using the ruler, measure off the length of vector $\mathbf{b}$ in the marked direction, as shown in figure 2.4. Now draw a line from the tail of vector $\mathbf{a}$ to the tip of vector $\mathbf{b}$. This is the resultant vector $\mathbf{R}$. Take the ruler and measure the length of vector $\mathbf{R}$ from the diagram. This length R is the magnitude of vector $\mathbf{R}$. Using the protractor, measure

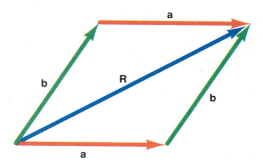

Figure 2.5

The addition of vectors by the parallelogram method.

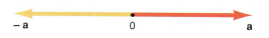

Figure 2.6

The negative of a vector.

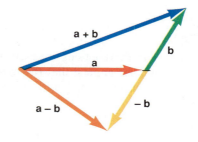

Figure 2.7

The subtraction of vectors.

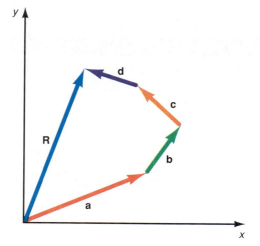

Figure 2.8

Addition of vectors by the polygon method.

the angle θ between **R** and the *x*-axis—this angle θ is the direction of the resultant vector **R.**

Although a vector is a quantity that has both magnitude and direction, it does not have a position. Consequently a vector may be moved parallel to itself without changing the characteristics of the vector. Because the magnitude of the moved vector is still the same, and its direction is still the same, the vector is the same.

Hence, when adding vectors **a** and **b**, we can move vector **a** parallel to itself until the tip of **a** touches the tip of **b**. Similarly, we can move vector **b** parallel to itself until the tip of **b** touches the tail end of the top vector **a**. In moving the vectors parallel to themselves we have formed a parallelogram, as shown in figure 2.5.

From what was said before about the resultant of **a** and **b**, we can see that the resultant of the two vectors is the main diagonal of the parallelogram formed by the vectors **a** and **b**, hence we call this process the **parallelogram method of vector addition.** It is sometimes stated as part of the definition of a vector, that vectors obey the parallelogram law of addition. Note from the diagram that

$$\mathbf{R} = \mathbf{a} + \mathbf{b} = \mathbf{b} + \mathbf{a} \tag{2.5}$$

that is, vectors can be added in any order. Mathematicians would say vector addition is commutative.

2.4 Vector Subtraction—The Negative of a Vector

If we are given a vector **a,** as shown in figure 2.6, then the vector minus **a** ($-\mathbf{a}$) is a vector of the same magnitude as **a** but in the opposite direction. That is, if vector **a** points to the right, then the vector $-\mathbf{a}$ points to the left. Vector $-\mathbf{a}$ is called the negative of the vector **a.** By defining the negative of a vector in this way, we can now determine the process of vector subtraction. The subtraction of vector **b** from vector **a** is defined as

$$\mathbf{a} - \mathbf{b} = \mathbf{a} + (-\mathbf{b}) \tag{2.6}$$

In other words, the subtraction of **b** from **a** is equivalent to adding vector **a** and the negative vector ($-\mathbf{b}$). This is shown graphically in figure 2.7.

2.5 Addition of Vectors by the Polygon Method

To find the sum of any number of vectors graphically, we use the polygon method. In the polygon method, we add each vector to the preceding vector by placing the tail of one vector to the head of the previous vector, as shown in figure 2.8. The resultant vector **R** is the sum of all these vectors. That is,

$$\mathbf{R} = \mathbf{a} + \mathbf{b} + \mathbf{c} + \mathbf{d} \tag{2.7}$$

We find **R** by drawing the vector from the origin of the coordinate system to the tip of the final vector, as shown in figure 2.8.

Vectors are usually added analytically or mathematically. In order to do that, we need to define the components of a vector. However, to discuss the components of a vector, we first need a brief review of trigonometry.

2.6 Review of Trigonometry

Although we assume that everybody reading this book has been exposed to the fundamentals of trigonometry, the essential ideas and definitions of trigonometry will now be reviewed.

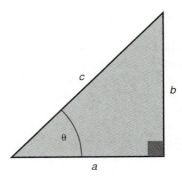

Figure 2.9

A simple right triangle.

Consider the right triangle shown in figure 2.9. It has sides a and b and hypotenuse c. Side a is called the side adjacent to the angle θ (theta), and the side b is called the side opposite to the angle θ. The trigonometric functions, defined with respect to this triangle, are

$$\text{sine } \theta = \frac{\text{opposite side}}{\text{hypotenuse}} \tag{2.8}$$

for the **sine function,** or

$$\sin \theta = \frac{b}{c} \tag{2.9}$$

and

$$\text{cosine } \theta = \frac{\text{adjacent side}}{\text{hypotenuse}} \tag{2.10}$$

for the **cosine function,** or

$$\cos \theta = \frac{a}{c} \tag{2.11}$$

and

$$\text{tangent } \theta = \frac{\text{opposite side}}{\text{adjacent side}} \tag{2.12}$$

for the **tangent function,** or

$$\tan \theta = \frac{b}{a} \tag{2.13}$$

Let us now review how these simple trigonometric functions are used. Assuming that the hypotenuse c of the right triangle and the angle θ that the hypotenuse makes with the x-axis are known, we want to determine the lengths of sides a and b of the triangle. From the definition,

$$\cos \theta = \frac{a}{c} \tag{2.11}$$

we can find the length of side a by multiplying both sides of equation 2.11 by c, that is,

$$a = c \cos \theta \tag{2.14}$$

Example 2.1

Using the cosine function to determine a side of the triangle. If the hypotenuse c is equal to 10.0 cm and the angle θ is equal to 60.0°, find the length of side a.

Solution

The length of side a is found from equation 2.14 as

$$a = c \cos \theta = 10.0 \text{ cm } \cos 60.0° = 10.0 \text{ cm } (0.500) = 5.00 \text{ cm}$$

(We assume here that anyone can compute the cos 60.0° with the aid of a handheld calculator.)

To find side b of the triangle we use the definition of the sine function:

$$\sin \theta = \frac{b}{c} \tag{2.9}$$

Multiplying both sides of equation 2.9 by c we obtain

$$b = c \sin \theta \qquad (2.15)$$

Example 2.2

Using the sine function to determine a side of the triangle. The hypotenuse c of a right triangle is 10.0 cm long, and the angle θ is equal to 60.0°. Find the length of side b.

Solution

The length of side b is found from equation 2.15 as

$$b = c \sin \theta = 10.0 \text{ cm} \sin 60.0° = 10.0 \text{ cm} (0.866) = 8.66 \text{ cm}$$

Therefore, if the hypotenuse and angle θ of a right triangle are given, the lengths of the sides a and b of that triangle can be determined by simple trigonometry.

Suppose that the lengths of sides a and b of a right triangle are given and we want to find the hypotenuse c and the angle θ of that triangle, as shown in figure 2.9. The hypotenuse is found by the **Pythagorean theorem** from elementary geometry as

$$c^2 = a^2 + b^2 \qquad (2.16)$$

Hence,

$$c = \sqrt{a^2 + b^2} \qquad (2.17)$$

The angle θ is found from the definition of the tangent function,

$$\tan \theta = \frac{b}{a} \qquad (2.13)$$

Using the inverse of the tangent function, sometimes called the arctangent, the angle θ becomes

$$\theta = \tan^{-1} \frac{b}{a} \qquad (2.18)$$

Example 2.3

Using the Pythagorean theorem and the inverse tangent. The lengths of two sides of a right triangle are $a = 5.00$ cm and $b = 8.66$ cm. Find the hypotenuse of the triangle and the angle θ.

Solution

The hypotenuse of the triangle is found from equation 2.17 as

$$c = \sqrt{a^2 + b^2} = \sqrt{(5.00 \text{ cm})^2 + (8.66 \text{ cm})^2} = 10.0 \text{ cm}$$

and the angle θ is found from equation 2.18 as

$$\theta = \tan^{-1} \frac{b}{a} = \tan^{-1} \frac{8.66 \text{ cm}}{5.00 \text{ cm}} = \tan^{-1} 1.73 = 60.0°$$

Therefore, if the lengths of the sides a and b of a right triangle are known we can easily calculate the hypotenuse and angle θ. We will repeatedly use these elementary concepts of trigonometry in the discussion of the components of a vector.

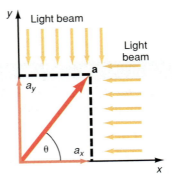

Figure 2.10

Defining the components of a vector.

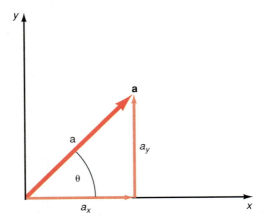

Figure 2.11

Finding the components of a vector mathematically.

2.7 Resolution of a Vector into Its Components

An arbitrary vector **a** is drawn onto an x,y-coordinate system, as in figure 2.10. Vector **a** makes an angle θ with the x-axis. To find the x-component a_x of vector **a**, we project vector **a** down onto the x-axis, that is, we drop a perpendicular from the tip of **a** to the x-axis. One way of visualizing this concept of a **component of a vector** is to place a light beam above vector **a** and parallel to the y-axis. The light hitting vector **a** will not make it to the x-axis, and will therefore leave a shadow on the x-axis. We call this shadow on the x-axis the x-component of vector **a** and denote it by a_x. The component is shown as the light red line on the x-axis in figure 2.10.

In the same way, we can determine the y-component of vector **a**, a_y, by projecting **a** onto the y-axis in figure 2.10. That is, we drop a perpendicular from the tip of **a** onto the y-axis. Again, we can visualize this by projecting light, which is parallel to the x-axis, onto vector **a**. The shadow of vector **a** on the y-axis is the y-component a_y, shown in figure 2.10 as the light red line on the y-axis.

The components of the vector are found mathematically by noting that the vector and its components constitute a triangle, as seen in figure 2.11. From trigonometry, we find the x-component of **a** from

$$\cos \theta = \frac{a_x}{a} \tag{2.19}$$

Solving for a_x, the x-component of vector **a** obtained is

$$a_x = a \cos \theta \tag{2.20}$$

We find the y-component of vector **a** from

$$\sin \theta = \frac{a_y}{a} \tag{2.21}$$

Hence, the y-component of vector **a** is

$$a_y = a \sin \theta \tag{2.22}$$

Example 2.4

Finding the components of a vector. The magnitude of vector **a** is 10.0 units and the vector makes an angle of 30.0° with the x-axis. Find the components of a.

Solution

The x-component of vector **a**, found from equation 2.20, is

$$a_x = a \cos \theta = 10.0 \text{ units } \cos 30.0° = 8.66 \text{ units}$$

The y-component of **a**, found from equation 2.22, is

$$a_y = a \sin \theta = 10.0 \text{ units } \sin 30.0° = 5.00 \text{ units}$$

What do these components of a vector represent physically? If vector **a** is a displacement, then a_x would be the distance that the object is east of its starting point and a_y would be the distance north of it. That is, if you walked a distance of 10.0 miles in a direction that is 30.0° north of east, you would be 8.66 miles east of where you started from and 5.00 miles north of where you started from. If, on the other hand, vector **a** were a force of 10.0 lb applied at an angle of 30.0° to the x-axis, then the x-component a_x is equivalent to a force of 8.66 lb in the x-direction, while the y-component a_y is equivalent to a force of 5.00 lb in the y-direction.

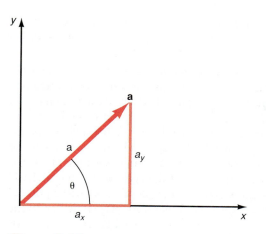

Figure 2.12

Determining a vector from its components.

2.8 Determination of a Vector from Its Components

If the components a_x and a_y of a vector are given, and we want to find the vector **a** itself, that is, its magnitude a and its direction θ, then the process is the inverse of the technique used in section 2.7. The components a_x and a_y of vector **a** are seen in figure 2.12. If we form the triangle with sides a_x and a_y, then the hypotenuse of that triangle is the magnitude a of the vector, and is determined by the Pythagorean theorem as

$$a^2 = a_x^2 + a_y^2 \tag{2.23}$$

Hence, the magnitude of vector **a** is

$$a = \sqrt{a_x^2 + a_y^2} \tag{2.24}$$

It is thus very simple to find the magnitude of a vector once its components are known.

To find the angle θ that vector **a** makes with the x-axis we use the definition of the tangent, namely

$$\text{tangent } \theta = \frac{\text{opposite side}}{\text{adjacent side}} \tag{2.12}$$

For the simple triangle of figure 2.12, the opposite side is a_y and the adjacent side is a_x. Therefore,

$$\tan \theta = \frac{a_y}{a_x} \tag{2.25}$$

We find the angle θ by using the inverse tangent, as

$$\theta = \tan^{-1} \frac{a_y}{a_x} \tag{2.26}$$

Example 2.5

Finding a vector from its components. The components of a certain vector are given as $a_x = 8.66$ and $a_y = 5.00$. Find the magnitude of the vector and the angle θ that it makes with the x-axis.

Solution

The magnitude of vector **a,** found from equation 2.24, is

$$a = \sqrt{a_x^2 + a_y^2} = \sqrt{(8.66)^2 + (5.00)^2}$$
$$= \sqrt{75.0 + 25.0}$$
$$= 10.0$$

The angle θ, found from equation 2.26, is

$$\theta = \tan^{-1} \frac{a_y}{a_x} = \tan^{-1} \frac{5.00}{8.66} = \tan^{-1} 0.577$$
$$= 30.0°$$

Therefore, the magnitude of vector **a** is 10.0 and the angle θ is 30.0°.

Note that the same values were used in this example as in example 2.4 to show that one process is indeed the inverse of the other. That is, we started with a vector with magnitude $a = 10.0$ and angle $\theta = 30.0°$, and found the components $a_x = 8.66$ and $a_y = 5.00$. Then we performed the inverse problem by giving

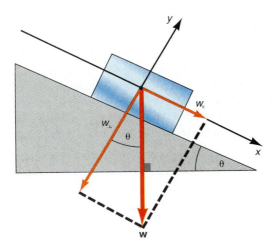

Figure 2.13

Components of the weight parallel and perpendicular to the inclined plane.

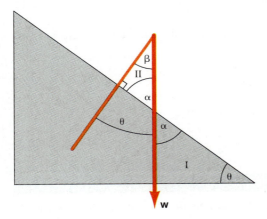

Figure 2.14

Comparison of two triangles.

the components $a_x = 8.66$ and $a_y = 5.00$ and finding the values of $a = 10.0$ and $\theta = 30.0°$. The techniques developed here will be very useful later for the addition of any number of vectors.

The components of a vector can also be found along axes other than the traditional horizontal and vertical ones. A coordinate system can be orientated any way we choose. For example, suppose a block is placed on an inclined plane that makes an angle θ with the horizontal, as shown in figure 2.13. Let us find the components of the weight of the block parallel and perpendicular to the inclined plane.

We draw in a set of axes that are parallel and perpendicular to the inclined plane, as shown in figure 2.13, with the positive x-axis pointing down the plane and the positive y-axis perpendicular to the plane. To find the components parallel and perpendicular to the plane, we draw the weight of the block as a vector pointed toward the center of the earth. The weight vector is therefore perpendicular to the base of the inclined plane. To find the component of **w** perpendicular to the plane, we drop a perpendicular line from the tip of vector **w** onto the negative y-axis. This length $w_\perp$ is the perpendicular component of vector **w**. Similarly, to find the parallel component of **w**, we drop a perpendicular line from the tip of **w** onto the positive x-axis. This length $w_\parallel$ is the parallel component of the vector **w**.

The angle between vector **w** and the perpendicular axis is also the inclined plane angle θ, as shown in the comparison of the two triangles in figure 2.14. (Figure 2.14 is an enlarged view of the two triangles of figure 2.13). In triangle I, the angles must add up to 180°. Thus,

$$\theta + \alpha + 90° = 180° \tag{2.27}$$

while for triangle II

$$\beta + \alpha + 90° = 180° \tag{2.28}$$

From equations 2.27 and 2.28 we see that

$$\beta = \theta \tag{2.29}$$

This is an important relation that we will use every time we use an inclined plane.

Example 2.6

Components of the weight perpendicular and parallel to the inclined plane. A 100-lb block is placed on an inclined plane with an angle $\theta = 50.0°$, as shown in figure 2.13. Find the components of the weight of the block parallel and perpendicular to the inclined plane.

Solution

We find the perpendicular component of **w** from figure 2.13 as

$$w_\perp = w \cos \theta \tag{2.30}$$
$$= 100 \text{ lb} \cos 50.0° = 64.2 \text{ lb}$$

The parallel component is

$$w_\parallel = w \sin \theta \tag{2.31}$$
$$= 100 \text{ lb} \sin 50.0° = 76.6 \text{ lb}$$

One of the interesting things about this inclined plane is that the component of the weight parallel to the inclined plane supplies the force responsible for making the block slide down the plane. Similarly, if you park your car on a hill with the gear in neutral and the emergency brake off, the car will roll down the hill. Why? You can now see that it is the component of the weight of the car that is parallel to the hill that essentially pulls the car down the hill. That force is just as real as if a person were pushing the car down the hill. That force, as can be seen from equation 2.31, is a function of the angle θ. If the angle of the plane is reduced to zero, then

$$w_{\parallel} = w \sin 0° = 0$$

Thus, we can reasonably conclude that when a car is not on a hill (i.e., when $\theta = 0°$) there is no force, due to the weight of the car, to cause the car to move. Also note that the steeper the hill, the greater the angle θ, and hence the greater the component of the force acting to move the car down the hill.

2.9 The Addition of Vectors by the Component Method

A very important technique for the addition of vectors is **the addition of vectors by the component method.** Let us assume that we are given two vectors, **a** and **b,** and we want to find their vector sum. The sum of the vectors is the resultant vector **R** given by

$$\mathbf{R} = \mathbf{a} + \mathbf{b} \tag{2.32}$$

and is shown in figure 2.15. We determine **R** as follows. First, we find the components a_x and a_y of vector **a** by making the projections onto the x- and y-axes, respectively. To find the components of the vector **b**, we again make a projection onto the x- and y-axes, but note that the tail of vector **b** is not at the origin of coordinates, but rather at the tip of **a.** So both the tip and the tail of **b** are projected onto the x-axis, as shown, to get b_x, the x-component of **b**. In the same way, we project **b** onto the y-axis to get b_y, the y-component of **b**. All these components are shown in figure 2.15(a).

The resultant vector **R** is given by equation 2.32, and because **R** is a vector it has components R_x and R_y, which are the projections of **R** onto the x- and y-axes, respectively. They are shown in figure 2.15(b). Now let us go back to the original diagram, figure 2.15(a), and project **R** onto the x-axis. Here R_x is shown a little distance below the x-axis, so as not to confuse R_x with the other components that are already there. Similarly, **R** is projected onto the y-axis to get R_y. Again R_y is slightly displaced from the y-axis, so as not to confuse R_y with the other components already there.

Look very carefully at figure 2.15(a). Note that the length of R_x is equal to the length of a_x plus the length of b_x. Because components are numbers and hence add like ordinary numbers, this addition can be written simply as

$$R_x = a_x + b_x \tag{2.33}$$

That is, *the x-component of the resultant vector is equal to the sum of the x-components of the individual vectors.*

In the same manner, look at the geometry on the y-axis of figure 2.15(a). The length R_y is equal to the sum of the lengths of a_y and b_y, and therefore

$$R_y = a_y + b_y \tag{2.34}$$

Thus, *the y-component of the resultant vector is equal to the sum of the y-components of the individual vectors.* We demonstrated the addition of vectors for only two vectors because it is easier to see the results in figure 2.15 for two vectors than it would be for many vectors. However, the technique is the same for the addition of any number of vectors. For the general case, where there are many vectors, equations 2.33 and 2.34 for R_x and R_y can be generalized to

$$R_x = a_x + b_x + c_x + d_x + \cdots \tag{2.35}$$

and

$$R_y = a_y + b_y + c_y + d_y + \cdots \tag{2.36}$$

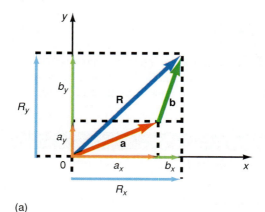

(a)

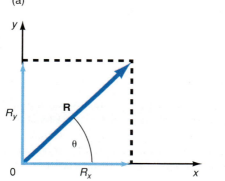

(b)

Figure 2.15

The addition of vectors by the component method.

The plus sign and the dots that appear at the far right in equations 2.35 and 2.36 indicate that additional components can be added for any additional vectors.

We now have R_x and R_y, the components of the resulting vector **R**. But if we know the components of **R**, we can find the magnitude of **R** by using the Pythagorean theorem, that is,

$$R = \sqrt{R_x^2 + R_y^2} \qquad (2.37)$$

The angle θ in figure 2.15(b), found from the geometry, is

$$\tan \theta = \frac{R_y}{R_x} \qquad (2.38)$$

Thus,

$$\theta = \tan^{-1}\frac{R_y}{R_x} \qquad (2.39)$$

where R_x and R_y are given by equations 2.35 and 2.36. Thus, we have found the magnitude R and the direction θ of the resultant vector **R**. Therefore, the sum of any number of vectors can be determined by the component method of vector addition.

Example 2.7

The addition of vectors by the component method. Find the resultant of the following four vectors:

$$A = 100, \theta_1 = 30.0°$$
$$B = 200, \theta_2 = 60.0°$$
$$C = 75.0, \theta_3 = 140°$$
$$D = 150, \theta_4 = 250°$$

Solution

The four vectors are drawn in figure 2.16. Because any vector can be moved parallel to itself, all the vectors have been moved so that they are drawn as emanating from the origin. Before actually solving the problem, let us first outline the solution. To find the resultant of these four vectors, we must first find the individual components of each vector, then we find the x- and y-components of the resulting vector from

$$R_x = A_x + B_x + C_x + D_x \qquad (2.35)$$
$$R_y = A_y + B_y + C_y + D_y \qquad (2.36)$$

We then find the resulting vector from

$$R = \sqrt{R_x^2 + R_y^2} \qquad (2.37)$$

and

$$\theta = \tan^{-1}\frac{R_y}{R_x} \qquad (2.39)$$

The actual solution of the problem is found as follows: we find the individual x-components as

$$
\begin{aligned}
A_x &= A \cos \theta_1 = 100 \cos 30.0° = 100(0.866) &= 86.6 \\
B_x &= B \cos \theta_2 = 200 \cos 60.0° = 200(0.500) &= 100.0 \\
C_x &= C \cos \theta_3 = 75 \cos 140° = 75(-0.766) &= -57.5 \\
D_x &= D \cos \theta_4 = 150 \cos 250° = 150(-0.342) &= -51.3 \\
\hline
R_x &= A_x + B_x + C_x + D_x = & 77.8
\end{aligned}
$$

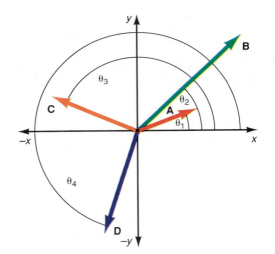

Figure 2.16
Addition of four vectors.

Mechanics

whereas the y-components are

$$
\begin{aligned}
A_y &= A \sin \theta_1 = 100 \sin 30.0° = 100(0.500) & &= & 50.0 \\
B_y &= B \sin \theta_2 = 200 \sin 60.0° = 200(0.866) & &= & 173.0 \\
C_y &= C \sin \theta_3 = 75 \sin 140° = 75(0.643) & &= & 48.2 \\
D_y &= D \sin \theta_4 = 150 \sin 250° = 150(-0.940) & &= & -141.0 \\
& & R_y = A_y + B_y + C_y + D_y &= & 130.2
\end{aligned}
$$

The x- and y-components of vector $\mathbf{R}$ are shown in figure 2.17. Because R_x and R_y are both positive, we find vector $\mathbf{R}$ in the first quadrant. If R_x were negative, $\mathbf{R}$ would have been in the second quadrant. It is a good idea to plot the components R_x and R_y for any addition so that the direction of $\mathbf{R}$ is immediately apparent. We find the magnitude of the resultant vector from equation 2.37 as

$$
\begin{aligned}
R &= \sqrt{R_x^2 + R_y^2} = \sqrt{(77.8)^2 + (130.2)^2} \\
&= \sqrt{23{,}004.8} \\
&= 152
\end{aligned}
$$

The angle θ that vector $\mathbf{R}$ makes with the x-axis is found as

$$
\begin{aligned}
\theta &= \tan^{-1}\frac{R_y}{R_x} = \tan^{-1}\frac{130.2}{77.8} = \tan^{-1} 1.674 \\
&= 59.1°
\end{aligned}
$$

as is seen in figure 2.17.

It is important to note here that the components C_x, D_x, and D_y are negative numbers. This is because C_x and D_x lie along the negative x-axis and D_y lies along the negative y-axis. We should note that in the solution of the components of the vector $\mathbf{C}$ in this problem, the angle of 140° was entered directly into the calculator to give the solution for the cosine and sine of that angle. The calculator automatically gives the correct sign for the components if we always measure the angle from the positive x-axis.[1]

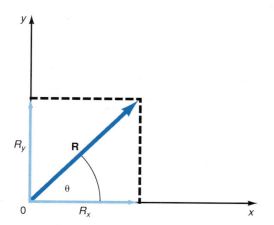

Figure 2.17

The resultant vector.

Example 2.8

The necessity of taking the wind velocity into account when flying an airplane. An airplane is flying due east from city A to city B with an airspeed of 150 mph. A wind is blowing from the northwest at 35 mph. Find the velocity of the airplane with respect to the ground.

Solution

The velocity of the plane with respect to the air is shown as the vector $\mathbf{v_{PA}}$ in figure 2.18. If there were no wind present, the plane would fly in a straight line from city A to city B. However, there is a wind blowing and it is shown as the vector $\mathbf{v_{AG}}$, the

---

1. We can also measure the angle that the vector makes with any axis other than the positive x-axis. For example, instead of using the angle of 140° with respect to the positive x-axis, an angle of 40° with respect to the negative x-axis can be used to describe the direction of vector $\mathbf{C}$. The x-component of vector $\mathbf{C}$ would then be given by $C_x = C \cos 40° = 75.0 \cos 40° = 57.5$. Note that this is the same numerical value we obtained before, however the answer given by the calculator is now positive. But as we can see in figure 2.16, C_x is a negative quantity because it lies along the negative x-axis. Hence, if you do not use the angle with respect to the positive x-axis, you must add the positive or negative sign that is associated with that component. In most of the problems that will be covered in this text, we will measure the angle from the positive x-axis because of the simplicity of the calculation. However, whenever it is more convenient to measure the angle from any other axis, we will do so.

(a)

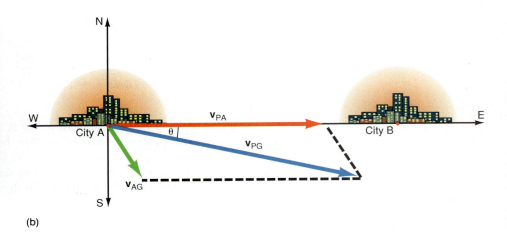

(b)

Figure 2.18

When flying an airplane, the velocity of the wind must be taken into account.

velocity of air with respect to the ground. This wind blows the plane away from the straight line motion from A to B. The total velocity of the plane with respect to the ground is the vector sum of $\mathbf{v_{PA}}$ and $\mathbf{v_{AG}}$. That is,

$$\mathbf{v_{PG}} = \mathbf{v_{PA}} + \mathbf{v_{AG}}$$

A wind from the northwest makes an angle of $-45°$ or $+315°$ with the positive x-axis. We find the x-component of the resulting velocity as

$$(v_{PA})_x = v_{PA} \cos \theta_2 = 150 \text{ mph} \cos 0° = 150 \text{ mph}$$
$$(v_{AG})_x = v_{AG} \cos \theta_1 = 35.0 \text{ mph} \cos 315° = \underline{24.7 \text{ mph}}$$
$$(v_{PG})_x = (v_{PA})_x + (v_{AG})_x = 174.7 \text{ mph}$$

While the y-component of the resulting velocity is

$$(v_{PA})_y = v_{PA} \sin \theta_2 = 150 \text{ mph} \sin 0° = 00.0 \text{ mph}$$
$$(v_{AG})_y = v_{AG} \sin \theta_1 = 35.0 \text{ mph} \sin 315° = \underline{-24.7 \text{ mph}}$$
$$(v_{PG})_y = (v_{PA})_y + (v_{AG})_y = -24.7 \text{ mph}$$

The magnitude of the resulting velocity of the plane with respect to the ground is

$$v_{PG} = \sqrt{[(v_{PG})_x]^2 + [(v_{PG})_y]^2}$$
$$= \sqrt{(174.7 \text{ mph})^2 + (-24.7 \text{ mph})^2}$$
$$= 176 \text{ mph}$$

The angle that the velocity vector $\mathbf{v_{PG}}$ makes with the positive x-axis is

$$\theta = \tan^{-1}\frac{(v_{PG})_y}{(v_{PG})_x}$$

$$= \tan^{-1}\frac{-24.7 \text{ mph}}{174.7 \text{ mph}}$$

$$= -8.05°$$

Thus the direction of the aircraft as it moves over the ground is 8.05° south of east. If the pilot does not make a correction, he or she will not arrive at city B as expected.

The concepts of the vector addition and vector subtraction have been discussed in this chapter. We will use these concepts throughout the book. The multiplication of vectors will be shown in appendix F.

The Language of Physics

Scalar
A scalar quantity is a quantity that can be completely described by a magnitude, that is, by a number and a unit (p. 23).

Vector
A vector quantity is a quantity that needs both a magnitude and direction to completely describe it (p. 23).

Resultant
The vector sum of any number of vectors is called the resultant vector (p. 24).

Parallelogram method of vector addition
The main diagonal of a parallelogram is equal to the magnitude of the sum of the vectors that make up the sides of the parallelogram (p. 25).

Sine function
The ratio of the length of the opposite side to the length of the hypotenuse in a right triangle (p. 26).

Cosine function
The ratio of the length of the adjacent side to the length of the hypotenuse in a right triangle (p. 26).

Tangent function
The ratio of the length of the opposite side of a right triangle to the length of the adjacent side (p. 26).

Pythagorean theorem
The sum of the squares of the lengths of two sides of a right triangle is equal to the square of the length of the hypotenuse (p. 27).

Component of a vector
The projection of a vector onto a specified axis. The length of the projection of the vector onto the x-axis is called the x-component of the vector. The length of the projection of the vector onto the y-axis is called the y-component of the vector (p. 28).

The addition of vectors by the component method
The x-component of the resultant vector R_x is equal to the sum of the x-components of the individual vectors, while the y-component of the resultant vector R_y is equal to the sum of the y-components of the individual vectors. The magnitude of the resultant vector is then found by the Pythagorean theorem applied to the right triangle with sides R_x and R_y. The direction of the resultant vector is found by trigonometry (p. 31).

Summary of Important Equations

Vector addition is commutative
$$\mathbf{R} = \mathbf{a} + \mathbf{b} = \mathbf{b} + \mathbf{a} \tag{2.5}$$

Subtraction of vectors
$$\mathbf{a} - \mathbf{b} = \mathbf{a} + (-\mathbf{b}) \tag{2.6}$$

Addition of vectors
$$\mathbf{R} = \mathbf{a} + \mathbf{b} + \mathbf{c} + \mathbf{d} \tag{2.7}$$

Definition of the sine
$$\text{sine } \theta = \frac{\text{opposite side}}{\text{hypotenuse}} \tag{2.8}$$

Definition of the cosine
$$\text{cosine } \theta = \frac{\text{adjacent side}}{\text{hypotenuse}} \tag{2.10}$$

Definition of the tangent
$$\text{tangent } \theta = \frac{\text{opposite side}}{\text{adjacent side}} \tag{2.12}$$

Pythagorean theorem
$$c = \sqrt{a^2 + b^2} \tag{2.17}$$

x-component of a vector
$$a_x = a \cos \theta \tag{2.20}$$

y-component of a vector
$$a_y = a \sin \theta \tag{2.22}$$

Magnitude of a vector
$$a = \sqrt{a_x^2 + a_y^2} \tag{2.24}$$

Direction of a vector
$$\theta = \tan^{-1}\frac{a_y}{a_x} \tag{2.26}$$

x-component of resultant vector
$$R_x = a_x + b_x + c_x + d_x \tag{2.35}$$

y-component of resultant vector
$$R_y = a_y + b_y + c_y + d_y \tag{2.36}$$

Magnitude of resultant vector
$$R = \sqrt{R_x^2 + R_y^2} \tag{2.37}$$

Direction of resultant vector
$$\theta = \tan^{-1}\frac{R_y}{R_x} \tag{2.39}$$

Questions for Chapter 2

1. Give an example of some quantities that are scalars and vectors other than those listed in section 2.1.
2. Can a vector ever be zero? What does a zero vector mean?
† 3. Since time seems to pass from the past to the present and then to the future, can you say that time has a direction and therefore could be represented as a vector quantity?
4. Does the subtraction of two vectors obey the commutative law?
5. What happens if you multiply a vector by a scalar?
6. What happens if you divide a vector by a scalar?
7. If a person walks around a block that is 800 ft on each side and ends up at the starting point, what is the person's displacement?
8. How can you add three vectors of equal magnitude in a plane such that their resultant is zero?
9. When are two vectors $\mathbf{a}$ and $\mathbf{b}$ equal?
†10. If a coordinate system is rotated, what does this do to the vector? to the components?
†11. Why are all the fundamental quantities scalars?
12. A vector equation is equivalent to how many component equations?
13. If the components of a vector $\mathbf{a}$ are a_x and a_y, what are the components of the vector $\mathbf{b} = -5\mathbf{a}$?
14. If $|\mathbf{a} + \mathbf{b}| = |\mathbf{a} - \mathbf{b}|$, what is the angle between $\mathbf{a}$ and $\mathbf{b}$?

Problems for Chapter 2

2.7 Resolution of a Vector into Its Components and
2.8 Determination of a Vector from Its Components

1. A strong child pulls a sled with a force of 100 lb at an angle of 35° above the horizontal. Find the vertical and horizontal components of this pull.

2. A 50-N force is directed at an angle of 50° above the horizontal. Resolve this force into vertical and horizontal components.

3. A girl wants to hold a 25-lb sled at rest on a snow covered hill. The hill makes an angle of 15° with the horizontal. What force must she exert parallel to the slope? What is the force perpendicular to the surface of the hill that presses the sled against the hill?

4. A boy wants to hold a 68.0-N sled at rest on a snow-covered hill. The hill makes an angle of 27.5° with the horizontal. (a) What force must he exert parallel to the slope? (b) What is the force perpendicular to the surface of the hill that presses the sled against the hill?

5. A displacement vector, at an angle of 35° with respect to a specified direction, has a y-component equal to 150 ft. What is the magnitude of the displacement vector?

6. A plane is traveling northeast at 200 km/hr. What is (a) the northward component of its velocity, and (b) the eastward component of its velocity?

7. While taking off, an airplane climbs at an 8° angle with respect to the ground. If the aircraft's speed is 200 km/hr, what are the vertical and horizontal components of its velocity?

8. A car that weighs 3200 lb is parked on a hill that makes an angle of 23° with the horizontal. Find the component of the car's weight parallel to the hill and perpendicular to the hill.

9. A car that weighs 8900 N is parked on a hill that makes an angle of 43° with the horizontal. Find the component of the car's weight parallel to the hill and perpendicular to the hill.

10. A girl pushes a lawn mower with a force of 90 N. The handle of the mower makes an angle of 40° with the ground. What are the vertical and horizontal components of this force and what are their physical significances? What effect does raising the handle to 50° have?

11. A missile is launched with a speed of 1000 m/s at an angle of 73° above the horizontal. What are the horizontal and vertical components of the missile's velocity?

12. When a ladder leans against a smooth wall, the wall exerts a horizontal force **F** on the ladder, as

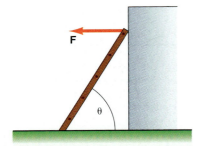

shown in the diagram. If F is equal to 50 N and θ is equal to 63°, find the component of the force perpendicular to the ladder and the component parallel to the ladder.

2.9 The Addition of Vectors by the Component Method

13. Find the resultant of the following three displacements; 3 mi due east, 6 mi east-northeast, and 7 mi northwest.

14. A girl drives 3 km north, then 12 km to the northwest, and finally 5 km south-southwest. How far has she traveled? What is her displacement?

15. An airplane flies due east at 200 mph straight from city A to city B. A northeast wind of 40 mph is blowing. (Note that all winds are defined in terms of the direction from which the wind blows. Hence, a northeast wind blows out of the northeast and blows toward the southwest.) What is the resultant velocity of the plane with respect to the ground?

16. An airplane flies due north at 380 km/hr straight from city A to city B. A southeast wind of 75 km/hr is blowing. (Note that all winds are defined in terms of the direction from which the wind blows. Hence, a southeast wind blows out of the southeast and blows toward the northwest.) What is the resultant velocity of the plane with respect to the ground?

17. Find the resultant of the following forces: (a) 30 N at an angle of 40° with respect to the x-axis, (b) 120 N at an angle of 135°, and (c) 60 N at an angle of 260°.

18. Find the resultant of the following set of forces. (a) **F₁** of 200 N at an angle of 53° with respect to the x-axis. (b) **F₂** of 300 N at an angle of 150° with respect to the x-axis. (c) **F₃** of 200 N at an angle of 270° with respect to the x-axis. (d) **F₄** of 350 N at an angle of 310° with respect to the x-axis.

Additional Problems

19. A heavy trunk weighing 150 lb is pulled along a smooth station platform by a 50 lb force making an angle of 37° above the horizontal. Find (a) the horizontal component of the force, (b) the vertical component of the force, and (c) the resultant downward force on the floor.

20. A heavy trunk weighing 800 N is pulled along a smooth station platform by a 210-N force making an angle of 53° above the horizontal. Find (a) the horizontal component of

the force, (b) the vertical component of the force, and (c) the resultant downward force on the floor.

21. Vector **A** has a magnitude of 50 ft and points in a direction of 50° north of east. What are the magnitudes and directions of the vectors, (a) 2**A**, (b) 0.5**A**, (c) −**A**, (d) −5**A**, (e) **A** + 4**A**, (f) **A** − 4**A**?

22. Given the two force vectors **F₁** = 20.0 N at an angle of 30.0° with the positive x-axis and **F₂** = 40.0 N at an angle of 150.0° with the positive x-axis, find the magnitude and direction of a third force that when added to **F₁** and **F₂** gives a zero resultant.

23. When vector **A**, of magnitude 5.00 m/s at an angle of 120° with respect to the positive x-axis, is added to a second vector **B**, the resultant vector has a magnitude $R = 8.00$ m/s and is at an angle of 85.0° with the positive x-axis. Find the vector **B**.

24. A car travels 100 mi due west and then 45 mi due north. How far is the car from its starting point? Solve graphically and analytically.

25. Find the resultant of the following forces graphically and analytically: 5 lb at an angle of 33° above the horizontal and 20 lb at an angle of 97° counterclockwise from the horizontal.

26. Find the resultant of the following forces graphically and analytically: 25 N at an angle of 53° above the horizontal and 100 N at an angle of 117° counterclockwise from the horizontal.

†27. The velocity of an aircraft is 200 km/hr due west. A northwest wind of 50 km/hr is blowing. (a) What is the velocity of the aircraft relative to the ground? (b) If the pilot's destination is due west, at what angle should he point his plane to get there? (c) If his destination is 400 km due west, how long will it take him to get there?

28. A plane flies east for 50.0 km, then at an angle of 30.0° north of east for 75.0 km. In what direction should it now fly and how far, such that it will be 200 km northwest of its original position?

†29. The current in a river flows north at 5 mph. A boat starts straight across the river at 8 mph relative to the water. (a) What is the speed of the boat relative to the land? (b) If the river is 2 mi wide, how long does it take the boat to cross the river? (c) If the boat sets out straight for the opposite side, how far north will it reach the opposite shore? (d) If we want to have the boat go straight across the river, at what angle should the boat be headed?

30. The current in a river flows south at 7 km/hr. A boat starts straight across the river at 19 km/hr relative to the water. (a) What is the speed of the boat relative to the land? (b) If the river is 1.5 km wide, how long does it take the boat to cross the river? (c) If the boat sets out straight for the opposite side, how far south will it reach the opposite shore? (d) If we want to have the boat go straight across the river, at what angle should the boat be headed?

†31. Show that if the angle between vectors **a** and **b** is an acute angle, then the sum **a** + **b** becomes the main diagonal of the parallelogram and the difference **a** − **b** becomes the minor diagonal of the parallelogram. Also show that if the angle is obtuse the results are reversed.

32. Find the resultant of the following three vectors. The magnitudes of the vectors are $a = 5.00$ km, $b = 10.0$ km, and $c = 20.0$ km.

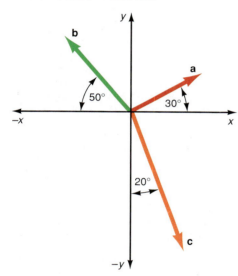

33. Find the resultant of the following three forces. The magnitudes of the forces are $F_1 = 2.00$ N, $F_2 = 8.00$ N, and $F_3 = 6.00$ N.

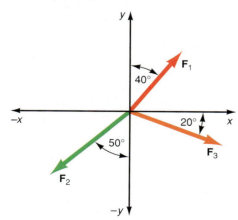

†34. Show that for three nonparallel vectors all in the same plane, any one of them can be represented as a linear sum of the other two.

†35. A unit vector is a vector that has a magnitude of one unit and is in a specified direction. If a unit vector **i** is defined to be in the x-direction, and a unit vector **j** is defined to be in the y-direction, show that any vector **a** can be written in the form

$$\mathbf{a} = a_x\mathbf{i} + a_y\mathbf{j}$$

†36. Prove that $|\mathbf{a} + \mathbf{b}| \le |\mathbf{a}| + |\mathbf{b}|$.

37. An airplane flies due east at 200 mph straight from city A to city B a distance of 200 mi. A wind of 40 mph from the northwest is blowing. If the pilot doesn't make any corrections, where will the plane be in 1 hr?

38. Given vectors **a** and **b**, where $a = 50$, $\theta_1 = 33°$, $b = 80$, and $\theta_2 = 128°$, find (a) **a** + **b**, (b) **a** − **b**, (c) **a** − 2**b**, (d) 3**a** + **b**, (e) 2**a** − **b**, and (f) 2**b** − **a**.

39. In the accompanying figure the tension T in the cable is 200 N. Find the vertical component T_y and the horizontal component T_x of this tension.

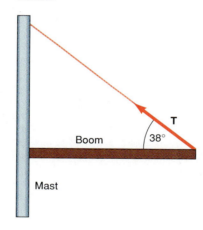

†40. In the accompanying diagram w_1 is 5 lb and w_2 is 3 lb. Find the angle θ such that the component of w_1 parallel to the incline is equal to w_2.

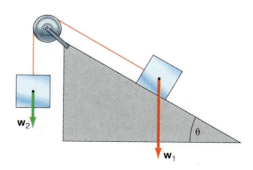

†41. In the accompanying diagram $w_1 = 2$ N, $w_2 = 5$ N, and $\theta = 65°$. Find the angle ϕ such that the components of the two forces parallel to the inclines are equal.

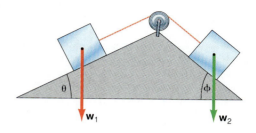

†42. In the accompanying diagram $w = 50$ lb, and $\theta = 10°$. What must be the value of F such that w will be held in place? What happens if the angle is doubled to 20°?

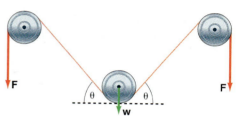

†43. In projectile motion in two dimensions the projectile is located by the displacement vector $\mathbf{r_1}$ at the time t_1 and by the displacement vector $\mathbf{r_2}$ at t_2, as shown in the diagram. If $r_1 = 20$ m, $\theta_1 = 60°$, $r_2 = 25$ m, and $\theta_2 = 25°$, find the magnitude and direction of the vector $\mathbf{r_2} - \mathbf{r_1}$.

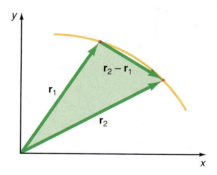

Interactive Tutorials

44. A 50.0-N force is directed at an angle of 50° above the horizontal. Resolve this force into vertical and horizontal components.

45. Find the resultant of any number of force vectors (up to five vectors).

3

Kinematics—The Study of Motion

Captured motion.

My purpose is to set forth a very new science dealing with a very ancient subject. There is, in nature, perhaps nothing older than motion, concerning which the books written by Philosophers are neither few nor small; nevertheless I have discovered by experiment some properties of it which are worth knowing and which have not hitherto been either observed or demonstrated . . . and what I consider more important, there has been opened up to this vast and most excellent science, of which my work is merely the beginning, ways and means by which other minds more acute than mine will explore its remote corners.

Galileo Galilei
Dialogues Concerning Two New Sciences

3.1 Introduction

Kinematics is defined as that branch of mechanics that studies the motion of a body without regard to the cause of that motion. In our everyday life we constantly observe objects in motion. For example, an object falls from the table, a car moves along the highway, or a plane flies through the air. In this process of motion, we observe that at one time the object is located at one position in space and then at a later time it has been displaced to some new position. Motion thus entails a movement from one position to another position. Therefore, to describe the motion of a body logically, we need to start by defining the position of a body. To do this we need a reference system. Thus, we introduce a coordinate system, as shown in figure 3.2. The body is located at the point 0 at the time $t = 0$. The point 0, the origin of the coordinate system, is the reference position. We measure the displacement of the moving body from there. After an elapse of time t_1 the object will have moved from 0 and will be found along the x-axis at position 1, a distance x_1 away from 0.

A little later in time, at $t = t_2$, the object will be located at point 2, a distance x_2 away from 0. (As an example, the moving body might be a car on the street. The reference point 0 might be a lamp post on the street, while points 1 and 2 might be telephone poles.) Let us now consider the motion between points 1 and 2.

The **average velocity** of the body in motion between the points 1 and 2 is defined as the displacement of the moving body divided by the time it takes for that displacement. That is,

$$v_{avg} = \frac{\text{displacement}}{\text{time for displacement}} \tag{3.1}$$

where v_{avg} is the notation used for the average velocity. For this description of one-dimensional motion, it is not necessary to use boldface vector notation. However, a positive value of v implies a velocity in the positive x- or y-direction, while a negative value of v implies a velocity in the negative x- or y-direction. A positive value of x implies a displacement in the positive x-direction, while a negative value of x implies a displacement in the negative x-direction. A positive value of y implies a displacement in the positive y-direction, while a negative value of y implies a displacement in the negative y-direction. The more general case, the velocity of a moving body in two dimensions, is treated in section 3.10.

Figure 3.1

Galileo Galilei.

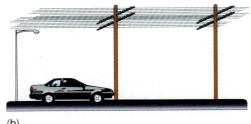

(a)

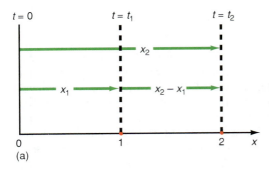

(b)

Figure 3.2

The position of an object at two different times.

From figure 3.2, we can see that during the time interval $t_2 - t_1$, the displacement or change in position of the body is simply $x_2 - x_1$. Therefore, the average velocity of the body in motion between points 1 and 2 is

$$v_{avg} = \frac{x_2 - x_1}{t_2 - t_1} \qquad (3.2)$$

Note here that in the example of the car and the telephone poles, t_1 is the time on a clock when the car passes the first telephone pole, position 1, and t_2 is the time on the same clock when the car passes the second telephone pole, position 2.

A convenient notation to describe this change in position with the change in time is the *delta notation*. Delta (the Greek letter Δ) is used as a symbolic way of writing "change in," that is,

$$\Delta x = (\text{change in } x) = x_2 - x_1 \qquad (3.3)$$

and

$$\Delta t = (\text{change in } t) = t_2 - t_1 \qquad (3.4)$$

Using this delta notation we can write the average velocity as

$$v_{avg} = \frac{x_2 - x_1}{t_2 - t_1} = \frac{\Delta x}{\Delta t} \qquad (3.5)$$

Example 3.1

Finding the average velocity using the Δ notation. A car passes telephone pole number 1, located 50.0 ft down the street from the corner lamp post, at a time $t_1 = 8.00$ s. It then passes telephone pole number 2, located 250 ft from the lamp post, at a time of $t_2 = 16.0$ s. What was the average velocity of the car between the positions 1 and 2?

Solution

The average velocity of the car, found from equation 3.5, is

$$v_{avg} = \frac{\Delta x}{\Delta t} = \frac{x_2 - x_1}{t_2 - t_1} = \frac{250 \text{ ft} - 50.0 \text{ ft}}{16.0 \text{ s} - 8.00 \text{ s}}$$

$$= \frac{200 \text{ ft}}{8.00 \text{ s}} = 25.0 \text{ ft/s}$$

(Note that according to the convention that we have adopted, the 25 ft/s represents a velocity because the magnitude of the velocity is 25 ft/s and the direction of the velocity vector is in the positive x-direction. If the answer were -25 ft/s the direction would have been in the negative x-direction.)

For convenience, the reference position 0 that is used to describe the motion is occasionally moved to position 1, then $x_1 = 0$, and the displacement is denoted by x, as shown in figure 3.3. The clock is started at this new reference position 1, so $t_1 = 0$ there. We now express the elapsed time for the displacement as t. In this simplified coordinate system the average velocity is

$$v_{avg} = \frac{x}{t} \qquad (3.6)$$

Remember, the average velocity is the same physically in both equations 3.5 and 3.6; the numerator is still the displacement of the moving body, and the denominator is still the elapsed time for this displacement. Because the reference point has been changed, the notation appears differently. We use both notations in the description of motion. The particular notation we use depends on the problem.

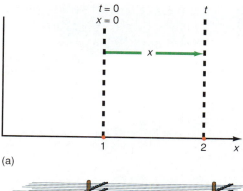

$t = 0$
$x = 0$

t

x

1
2
x

(a)

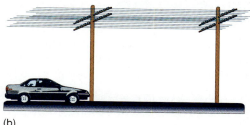

(b)

Figure 3.3

The position of an object determined from a new reference system.

Example 3.2

Changing the reference position. A car passes telephone pole number 1 at $t = 0$ on a watch. It passes a second telephone pole 200 ft down the block 8.00 seconds later. What is the car's average velocity?

Solution

The average velocity, found from equation 3.6, is

$$v_{\text{avg}} = \frac{x}{t} = \frac{200 \text{ ft}}{8.00 \text{ s}} = 25.0 \text{ ft/s}$$

Also note that this is the same problem solved in example 3.1; only the reference position for the measurement of the motion has been changed.

Before we leave this section, we should make a distinction between the average velocity of a body and the average speed of a body. *The average speed of a body is the distance that a body moves per unit time. The average velocity of a body is the displacement of a body per unit time.* Because the displacement of a body is a vector quantity, that is, it specifies the distance an object moves *in a specified direction,* its velocity is also a vector quantity. *Thus, velocity is a vector quantity while speed is a scalar quantity.* For example, if a girl runs 100 m in the x-direction and turns around and returns to the starting point in a total time of 90 s, her average velocity is zero because her displacement is zero. Her average speed, on the other hand, is the total distance she ran divided by the total time it took, or 200 m/90 s = 2.2 m/s. If she ran 100 m in 45 s in one direction only, let us say the positive x-direction, her average speed is 100 m/45 s = 2.2 m/s. Her average velocity is 2.2 m/s in the positive x-direction. In this case, the speed is the magnitude of the velocity vector. Speed is always a positive quantity, whereas velocity can be either positive or negative depending on whether the motion is in the positive x-direction or the negative x-direction, respectively.

Section 3.2 shows how the motion of a body can be studied in more detail in the laboratory.

3.2 Experimental Description of a Moving Body

Following Galileo's advice that motion should be studied by experiment, let us go into the laboratory and describe the motion of a moving body on an air track. An air track is a hollow aluminum track that has a right isosceles triangle for its cross section. Air is forced into the air track by a blower and flows out the sides of the track through many small holes. A glider, having an inverted Y cross section, is placed on the track. The air escaping from the holes in the track provides a cushion of air for the glider to move on, thereby substantially reducing the retarding force of friction on the glider. The setup of an air track in the laboratory is shown in figure 3.4.

We attach a piece of spark-timer tape to the air track to act as a permanent record of the position of the moving glider as a function of time. We connect a spark timer, a device that emits electrical pulses at certain prescribed times, to a wire on the air track. The pulse from the timer moves along this wire, which is parallel to the air track, and then jumps across an air gap as a spark to another wire attached to the moving glider. The resulting pulse proceeds along the wire of the glider down to the bottom of the track to the timer tape. A spark now jumps across an air gap between the glider wire and the air track, and in so doing it burns a hole in the timer tape. This burned hole on the tape, which appears as a dot, is a record of the position

Figure 3.4

Setup of an air track.

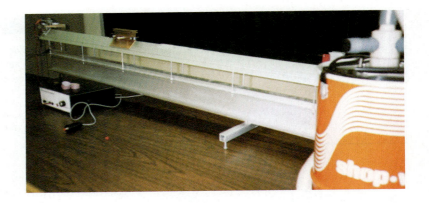

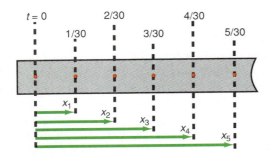

Figure 3.5

Spark-timer paper showing constant velocity.

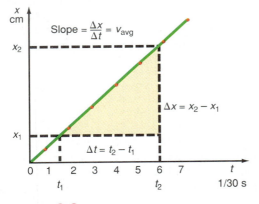

Figure 3.6

Graph of distance versus time for constant velocity.

of the glider at that instant of time. Thus, the combination of a glider, an air track, and a spark timer gives us a record of the position of a moving body at any instant of time. Let us now look at an experiment with a glider moving at constant velocity along the air track.

3.3 A Body Moving at Constant Velocity

To study a body moving at constant velocity we place a glider on a level air track and give it a slight push to initiate its motion along the track. The spark timer is turned on, leaving a permanent record of this motion on a piece of spark-timer tape. The distance traveled by the glider as a function of time is recorded on the spark-timer paper, and appears as in figure 3.5. The spark timer is set to give a spark every $1/30$ of a second. The first dot occurs at the time $t = 0$, and each succeeding dot occurs at a time interval of $1/30$ of a second later. We label the first dot as dot 0, the reference position, and then measure the total distance x from the first dot to each succeeding dot with a meter stick.

The measured data for the total distance traveled by the glider as a function of time are plotted in figure 3.6. Note that the plot is a straight line. If you measure the slope of this line you will observe that it is $\Delta x / \Delta t$, which is the average velocity defined in equation 3.5. Since all the points generate a straight line, which has a constant slope, the velocity of the glider is a constant equal to the slope of this graph. *Whenever a body moves in such a way that it always travels equal distances in equal times, that body is said to be moving with a **constant velocity.*** This can also be observed in figure 3.5 by noting that the dots are equally spaced.

The SI unit for velocity is m/s. The units cm/s and km/hr are also used. In the British engineering system we use ft/s, in./s, and mi/hr. Note that on a graph of the displacement of a moving body versus time, the slope $\Delta x / \Delta t$ always represents a velocity. If the slope is positive, the velocity is positive and the direction of the moving body is toward the right. If the slope is negative, the velocity is negative and the direction of the moving body is toward the left.

Example 3.3

The velocity of a glider on an air track. A glider goes from a position of 12.0 cm at a time of $t = 1/3$ s to a position of 75.0 cm at a time of $t = 5/3$ s. Find the average velocity of the glider during this interval.

Figure 3.7

The tilted air track.

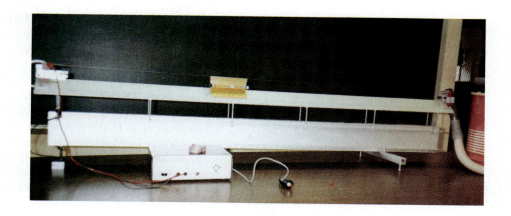

The average velocity of the glider, found from equation 3.5, is

$$v_{avg} = \frac{\Delta x}{\Delta t} = \frac{x_2 - x_1}{t_2 - t_1}$$

$$= \frac{75.0 \text{ cm} - 12.0 \text{ cm}}{5/3 \text{ s} - 1/3 \text{ s}} = \frac{63.0 \text{ cm}}{4/3 \text{ s}}$$

$$= 47.3 \text{ cm/s}$$

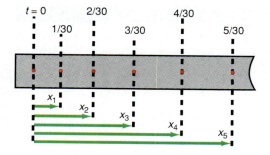

Figure 3.8

Spark-timer tape for accelerated motion.

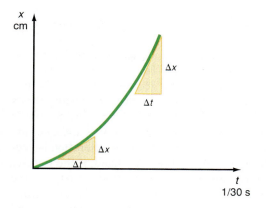

Figure 3.9

Graph of x versus t for constant acceleration.

3.4 A Body Moving at Constant Acceleration

If we tilt the air track at one end it effectively becomes a frictionless inclined plane. We place a glider at the top of the track and then release it from rest. Figure 3.7 is a picture of the glider in its motion on the inclined air track.

The spark timer is turned on, giving a record of the position of the moving glider as a function of time, as illustrated in figure 3.8. The most important feature to immediately note on this record of the motion, is that the dots, representing the positions of the glider, are no longer equally spaced as they were for motion at constant speed, but rather become farther and farther apart as the time increases. The total distance x that the glider moves is again measured as a function of time. If we plot this measured distance x against the time t, we obtain the graph shown in figure 3.9.

The first thing to note in this figure is that the graph of x versus t is not a straight line. However, as you may recall from section 3.3, the slope of the distance versus time graph, $\Delta x/\Delta t$, represents the velocity of the moving body. But in figure 3.9 there are many different slopes to this curve because it is continuously changing with time. Since the slope at any point represents the velocity at that point, we observe that the velocity of the moving body is changing with time. *The change of velocity with time is defined as the* **acceleration** *of the moving body, and the average acceleration is written as*

$$a_{avg} = \frac{\Delta v}{\Delta t} = \frac{v_2 - v_1}{t_2 - t_1} \tag{3.7}$$

Since the velocity is a vector quantity, acceleration, which is equal to the change in velocity with time, is also a vector quantity. More will be said about this shortly.

Because the velocity is changing continuously, the average velocity for every time interval can be computed from equation 3.5. Thus, subtracting each value of x from the next value of x gives us Δx, the distance the glider moves during one time interval. The average velocity during that interval can then be computed from $v_{avg} = \Delta x/\Delta t$. At the beginning of this interval the actual velocity is less than this

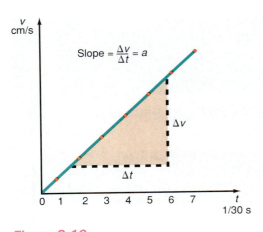

Figure 3.10

Graph of v versus t for constant acceleration.

value while at the end of the interval it is greater. Later we will see that *for constant acceleration, the velocity at the center of the time interval is equal to the average velocity for the entire time interval.*

If we plot the velocity at the center of the interval against the time, we obtain the graph in figure 3.10. We can immediately observe that the graph is a straight line. The slope of this line, $\Delta v/\Delta t$, is the experimental acceleration of the glider. Since this graph is a straight line, the slope is a constant; this implies that the acceleration is also a constant. Hence, the acceleration of a body moving down a frictionless inclined plane is a constant. In the case of more general motion, a body can also have its acceleration changing with time. However, most of the accelerated motion discussed in this book is at constant acceleration. The most notable exception is for simple harmonic motion, which we discuss in chapter 11. *Because in constantly accelerated motion the average acceleration is the same as the constant acceleration, the subscript avg will be deleted from the acceleration in all the equations dealing with this type of motion.*

Since acceleration is a change in velocity per unit time, the units for acceleration are velocity divided by the time. In SI units, the acceleration is

$$\frac{m/s}{s}$$

For convenience, this is usually written in the equivalent algebraic form as m/s^2. But we must not forget the physical meaning of a change in velocity of so many m/s every second. Other units used to express acceleration are ft/s^2, cm/s^2, $(km/hr)/s$, mph/s, and $in./s^2$.

Example 3.4

The acceleration of a glider on an air track. A glider's velocity on a tilted air track increases from 17.0 cm/s at the time $t = 1/3$ s to 55.0 cm/s at a time of $t = 7/3$ s. What is the acceleration of the glider?

Solution

The acceleration of the glider, found from equation 3.7, is

$$a = \frac{\Delta v}{\Delta t} = \frac{v_2 - v_1}{t_2 - t_1}$$

$$= \frac{55.0 \text{ cm/s} - 17.0 \text{ cm/s}}{7/3 \text{ s} - 1/3 \text{ s}} = \frac{38.0 \text{ cm/s}}{6/3 \text{ s}}$$

$$= 19.0 \text{ cm/s}^2$$

Before leaving this section we should note that since acceleration is a vector, if the acceleration is a positive quantity, the velocity is increasing with time, and the acceleration vector points toward the right. If the acceleration is a negative quantity, the velocity is decreasing with time, and the acceleration vector points toward the left. When the velocity is positive, indicating that the body is moving in the positive x-direction, and the acceleration is positive, the object is speeding up, or accelerating. However, when the velocity is positive, and the acceleration is negative, the object is slowing down, or decelerating. On the other hand, if the velocity is negative, indicating that the body is moving in the negative x-direction, and the acceleration is negative, the body is speeding up in the negative x-direction. However, when the velocity is negative and the acceleration is positive, the body is slowing down in the negative x-direction. If the acceleration lasts long enough, the body will eventually come to a stop and will then start moving in the positive x-direction. The velocity will then be positive and the body will be speeding up in the positive x-direction.

3.5 The Instantaneous Velocity of a Moving Body

In section 3.4 we observed that the velocity of the glider varies continuously as it "slides" down the frictionless inclined plane. We also stated that the average velocity could be computed from $v_{avg} = \Delta x/\Delta t$. At the beginning of the interval of motion the actual velocity is less than this value while at the end of the interval it is greater. If the interval is made smaller and smaller, the average velocity v_{avg} throughout the interval becomes closer to the actual velocity at the instant the body is at the center of the time interval. Finding the velocity at a particular instant of time leads us to the concept of instantaneous velocity. *Instantaneous velocity is defined as the limit of $\Delta x/\Delta t$ as Δt gets smaller and smaller, eventually approaching zero.* We write this concept mathematically as

$$v = \lim_{\Delta t \to 0} \frac{\Delta x}{\Delta t} \tag{3.8}$$

As in the case of average velocity in one-dimensional motion, if the limit of $\Delta x/\Delta t$ is a positive quantity, the velocity vector points toward the right. If the limit of $\Delta x/\Delta t$ is a negative quantity, the velocity vector points toward the left.

The concept of instantaneous velocity can be easily understood by performing the following experiment on an air track. First, we tilt the air track to again give an effectively frictionless inclined plane. Then we place a 20-cm length of metal, called a flag, at the top of the glider. A photocell gate, which is a device that can be used to automatically turn a clock on and off, is attached to a clock timer and is placed on the air track. We then allow the glider to slide down the track. When the flag of the glider interrupts the light beam to the photocell, the clock is turned on. When the flag has completely passed through the light beam, the photocell gate turns off the clock. The clock thus records the time for the 20-cm flag to pass through the photocell gate. We find the average velocity of the flag as it moves through the gate from equation 3.5 as $v = \Delta x/\Delta t$. The 20-cm length of the flag is Δx, and Δt is the time interval, as read from the clock.

We repeat the process for a 15-cm, 10-cm, and a 5-cm flag. For each case we measure the time Δt that it takes for the flag to move through the gate. The first thing that we observe is that the time for the flag to move through the gate, Δt, gets smaller for each smaller flag. You might first expect that if Δt approaches 0, the ratio of $\Delta x/\Delta t$ should approach infinity. However, since Δx, the length of the flag, is also getting smaller, the ratio of $\Delta x/\Delta t$ remains finite. If we plot $\Delta x/\Delta t$ as a function of Δt for each flag, we obtain the graph in figure 3.11.

Notice that as Δt approaches 0, ($\Delta t \to 0$), the plotted line intersects the $\Delta x/\Delta t$ axis. At this point, the distance interval Δx has been reduced from 20 cm to effectively 0 cm. The value of Δt has become progressively smaller so this point represents the limiting value of $\Delta x/\Delta t$ as Δt approaches 0. But this limit is the definition of the instantaneous velocity. Hence, the point where the line intersects the $\Delta x/\Delta t$ axis gives the value of the velocity of the glider at the instant of time that the glider is located at the position of the photocell gate. This limiting process allows us to describe the motion of a moving body in terms of the velocity of the body at any instant of time rather than in terms of the body's average velocity.

Usually we will be more interested in the instantaneous velocity of a moving body than its average velocity. The speedometer of a moving car is a physical example of instantaneous velocity. Whether the car's velocity is constant or changing with time, the instant that the speedometer is observed, the speedometer indicates the speed of the car at that particular instant of time. The instantaneous velocity of the car is that observed value of the speed in the direction that the car is traveling.

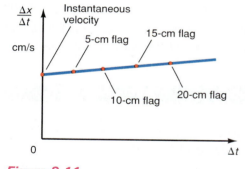

Figure 3.11

Graph of $\Delta x/\Delta t$ versus Δt to obtain the instantaneous velocity of the glider.

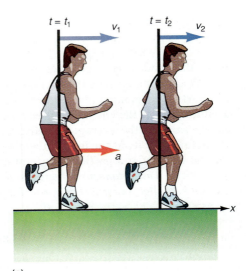

(a)

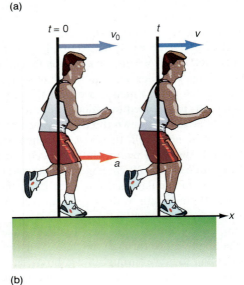

(b)

Figure 3.12
Change in reference system.

3.6 The Kinematic Equations in One Dimension

Because the previous experiments were based on motion at constant acceleration, we can only apply the results of those experiments to motion at a constant acceleration. Let us now compile those results into a set of equations, called the **kinematic equations of linear motion,** that will describe the motion of a moving body. For motion at constant acceleration, the average acceleration is equal to the constant acceleration. Hence, the subscript avg can be deleted from equation 3.7 and that equation now gives the constant acceleration of the moving body as

$$a = \frac{v_2 - v_1}{t_2 - t_1} \tag{3.7}$$

Equation 3.7 indicates that at the time t_1 the body is moving at the velocity v_1, while at the time t_2 the body is moving at the velocity v_2. This motion is represented in figure 3.12(a) for a runner.

Let us change the reference system by starting the clock at the time $t_1 = 0$, as shown in figure 3.12(b). We will now designate the velocity of the moving body at the time 0 as v_0 instead of the v_1 in the previous reference system of figure 3.12(a). Similarly, the time t_2 will correspond to any time t and the velocity v_2 will be denoted by v, the velocity at that time t. Thus, the velocity of the moving body will be v_0 when the time is equal to 0, and v when the time is equal to t. This change of reference system allows us to rewrite equation 3.7 as

$$a = \frac{v - v_0}{t} \tag{3.9}$$

Equation 3.9 is similar to equation 3.7 in that it gives the same definition for acceleration, namely a change in velocity with time, but in a slightly different but equivalent notation. Solving equation 3.9 for v gives the first of the very important kinematic equations, namely,

$$v = v_0 + at \tag{3.10}$$

Equation 3.10 says that the velocity v of the moving object can be found at any instant of time t once the acceleration a and the initial velocity v_0 of the moving body are known.

Example 3.5

Using the kinematic equation for the velocity as a function of time. A car passes a green traffic light while moving at a velocity of 6.00 m/s. It then accelerates at 0.300 m/s² for 15.0 s. What is the car's velocity at 15.0 s?

Solution

The velocity, found from equation 3.10, is

$$v = v_0 + at$$

$$= \left(6.00 \, \frac{m}{s}\right) + \left(0.300 \, \frac{m}{s^2}\right)(15.0 \, s)$$

$$= 10.5 \, m/s$$

The velocity of the car is 10.5 m/s. This means that the car is moving at a speed of 10.5 m/s in the positive x-direction.

In addition to the velocity of the moving body at any time t, we would also like to know the location of the body at that same time. That is, let us obtain an equation for the displacement of the moving body as a function of time. Solving equation 3.6 for the displacement x gives

$$x = v_{\text{avg}}t \tag{3.11}$$

Hence, the displacement of the moving body is equal to the average velocity of the body times the time it is in motion. For example, if you are driving your car at an average velocity of 50 miles per hour, and you drive for a period of time of two hours, then your displacement is

$$x = 50 \frac{\text{mi}}{\text{hr}} (2 \text{ hr})$$

$$= 100 \text{ mi}$$

You have traveled a total distance of 100 mi from where you started.

Equation 3.11 gives us the displacement of the moving body in terms of its average velocity. The actual velocity during the motion might be greater than or less than the average value. The average velocity does not tell us anything about the body's acceleration. We would like to express the displacement of the body in terms of its acceleration during a particular time interval, and in terms of its initial velocity at the beginning of that time interval.

For example, consider a car in motion along a road between the times $t = 0$ and $t = t$. At the beginning of the time interval the car has an initial velocity v_0, while at the end of the time interval it has the velocity v, as shown in figure 3.13. If the acceleration of the moving body is constant, then the average velocity throughout the entire time interval is

$$v_{\text{avg}} = \frac{v_0 + v}{2} \tag{3.12}$$

This averaging of velocities for bodies moving at constant acceleration is similar to determining a grade in a course. For example, if you have two test grades in the course, your course grade, the average of the two test grades, is the sum of the test grades divided by 2,

$$\text{Avg. Grade} = \frac{100 + 90}{2} = 95$$

If we substitute this value of the average velocity into equation 3.11, the displacement becomes

$$x = v_{\text{avg}}t = \left(\frac{v_0 + v}{2}\right)t \tag{3.13}$$

Figure 3.13

A car moving on a road.

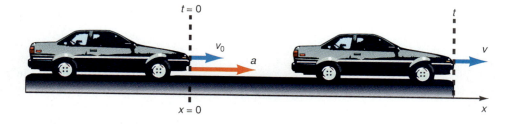

Note that v represents the final value of the velocity at the time t, the end of the time interval. But there already exists an equation for the value of v at the time t, namely equation 3.10. Therefore, substituting equation 3.10 into equation 3.13 gives

$$x = \left[\frac{v_0 + (v_0 + at)}{2} \right] t$$

Simplifying, we get

$$x = \left(\frac{2v_0 + at}{2} \right) t$$

$$= \frac{2v_0 t}{2} + \tfrac{1}{2} a t^2$$

$$x = v_0 t + \tfrac{1}{2} a t^2 \qquad (3.14)$$

Equation 3.14, the second of the kinematic equations, represents the displacement x of the moving body at any instant of time t. In other words, if the original velocity and the constant acceleration of the moving object are known, then we can determine the location of the moving object at any time t. This rather simple equation contains a tremendous amount of information.

Example 3.6

Using the kinematic equation for the displacement as a function of time. A car, initially traveling at 30.0 km/hr, accelerates at the constant rate of 1.50 m/s². How far will the car travel in 15.0 s?

Solution

To express the result in the proper units, km/hr is converted to m/s as

$$v_0 = 30.0 \, \frac{\text{km}}{\text{hr}} \left(\frac{1 \, \text{hr}}{3600 \, \text{s}} \right) \left(\frac{1000 \, \text{m}}{1 \, \text{km}} \right) = 8.33 \, \text{m/s}$$

The displacement of the car, found from equation 3.14, is

$$x = v_0 t + \tfrac{1}{2} a t^2$$

$$= \left(8.33 \, \frac{\text{m}}{\text{s}} \right) (15.0 \, \text{s}) + \frac{1}{2} \left(1.50 \, \frac{\text{m}}{\text{s}^2} \right) (15.0 \, \text{s})^2$$

$$= 125 \, \text{m} + 169 \, \text{m}$$

$$= 294 \, \text{m}$$

The first term in the answer, 125 m, represents the distance that the car would travel if there were no acceleration and the car continued to move at the velocity 8.33 m/s for 15.0 s. But there is an acceleration, and the second term shows how much farther the car moves because of that acceleration, namely 169 m. The total displacement of 294 m is the total distance that the car travels because of the two effects.

As a further example of the kinematics of a moving body, consider the car moving along a road at an initial velocity of 60.0 mph = 88.0 ft/s, as shown in figure 3.14. The driver sees a tree fall into the middle of the road 200 ft away. The driver immediately steps on the brakes, and the car starts to decelerate at the constant rate of $a = -18.0 \, \text{ft/s}^2$. (As mentioned previously, in one-dimensional motion a negative acceleration means that the acceleration vector is toward the left, in the opposite direction of the motion. If the velocity is positive, a negative value for the acceleration means that the body is slowing down or decelerating.) Will the car come to a stop before hitting the tree?

Mechanics

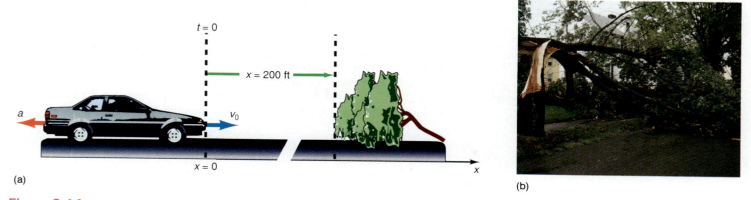

(a)

(b)

Figure 3.14

A tree falls on the road.

What we need for the solution of this problem is the actual distance the car travels before it can come to a stop while decelerating at the rate of 18.0 ft/s². Before we can find that distance, however, we must know the time it takes for the car to come to a stop. Then we substitute this stopping time into equation 3.14, and the equation tells us how far the car will travel before coming to a stop. (Note that most of the questions that might be asked about the motion of the car can be answered using the kinematic equations 3.10 and 3.14.)

Equation 3.10 tells us the velocity of the car at any instant of time. But when the car comes to rest its velocity is zero. Thus, at the time when the car comes to a stop t_{stop}, the velocity v will be equal to zero. Therefore, equation 3.10 becomes

$$0 = v_0 + at_{stop}$$

Solving for the time for the car to come to a stop, we have

$$t_{stop} = -\frac{v_0}{a} \tag{3.15}$$

the time interval from the moment the brakes are applied until the car comes to a complete stop. Substituting the values of the initial velocity v_0 and the constant acceleration a into equation 3.15, we have

$$t_{stop} = -\frac{v_0}{a} = \frac{-88.0 \text{ ft/s}}{-18.0 \text{ ft/s}^2} = 4.89 \text{ s}$$

It will take 4.89 s for the car to come to a stop if nothing gets in its way to change its rate of deceleration. Note how the units cancel in the equation until the final unit becomes seconds, that is,

$$\frac{v_0}{a} = \frac{\text{ft/s}}{\text{ft/s}^2} = \frac{1/s}{1/s^2} = \frac{1}{s}\frac{1}{1/s^2} = \frac{1}{1/s} = s$$

Thus, $(\text{ft/s})/(\text{ft/s}^2)$, comes out to have the unit seconds, which it must since it represents the time for the car to come to a stop.

Now that we know the time for the car to come to a stop, we can substitute that value back into equation 3.14 and find the distance the car will travel in the 4.89 s:

$$x = v_0t + \tfrac{1}{2}at^2$$

$$= 88.0\,\frac{\text{ft}}{\text{s}}\,(4.89 \text{ s}) + \frac{1}{2}\left(-18.0\,\frac{\text{ft}}{\text{s}^2}\right)(4.89 \text{ s})^2$$

$$= 430 \text{ ft} - 215 \text{ ft}$$

$$= 215 \text{ ft}$$

The car will come to a stop in 215 ft. Since the tree is only 200 ft in front of the car, it cannot come to a stop in time and will hit the tree.

In addition to the velocity and position of a moving body at any instant of time, we sometimes need to know the velocity of the moving body at a particular displacement x. In the example of the car hitting the tree, we might want to know the velocity of the car when it hits the tree. That is, what is the velocity of the car when the displacement x of the car is equal to 200 ft?

To find the velocity as a function of displacement x, we must eliminate time from our kinematic equations. To do this, we start with equation 3.13 for the displacement of the moving body in terms of the average velocity,

$$x = v_{avg}t = \left(\frac{v_0 + v}{2}\right)t \tag{3.13}$$

But v is the velocity of the moving body at any time t, given by

$$v = v_0 + at \tag{3.10}$$

Solving for t gives

$$t = \frac{v - v_0}{a}$$

Substituting this value into equation 3.13 gives

$$x = \left(\frac{v_0 + v}{2}\right)t = \left(\frac{v_0 + v}{2}\right)\left(\frac{v - v_0}{a}\right)$$

$$= \left(\frac{v_0 + v}{2a}\right)(v - v_0)$$

$$2ax = v_0 v + v^2 - v_0 v - v_0^2$$

$$= v^2 - v_0^2$$

Solving for v^2, we obtain *the third kinematic equation,*

$$v^2 = v_0^2 + 2ax \tag{3.16}$$

which is used to determine the velocity v of the moving body at any displacement x.

Let us now go back to the problem of the car moving down the road, with a tree lying in the road 200 ft in front of the car. We already know that the car will hit the tree, but at what velocity will it be going when it hits the tree? That is, what is the velocity of the car at the displacement of 200 feet? Using equation 3.16 with $x = 200$ ft, $v_0 = 60.0$ mph $= 88.0$ ft/s, and $a = -18.0$ ft/s^2, and solving for v gives

$$v^2 = v_0^2 + 2ax$$

$$= (88.0 \text{ ft/s})^2 + 2(-18.0 \text{ ft/s}^2)(200 \text{ ft})$$

$$= 7740 \text{ ft}^2/\text{s}^2 - 7200 \text{ ft}^2/\text{s}^2$$

$$= 540 \text{ ft}^2/\text{s}^2$$

$$v = \frac{23.2 \text{ ft}}{\text{s}}\left(\frac{60.0 \text{ mph}}{88.0 \text{ ft/s}}\right)$$

and finally,

$$v = 15.8 \text{ mph}$$

When the car hits the tree it will be moving at 15.8 mph, so the car may need a new bumper or fender. Equation 3.16 allows us to determine the velocity of the moving body at any displacement x.

A problem similar to that of the car and the tree involves the maximum velocity that a car can move and still have adequate time to stop before hitting something the driver sees on the road in front of the car. Let us again assume that the car decelerates at the same constant rate as before, $a = -18$ ft/s², and that the low beam headlights of the car are capable of illuminating a 200 ft distance of the road. Using equation 3.16, which gives the velocity of the car as a function of displacement, let us find the maximum value of v_0 such that v is equal to zero when the car has the displacement x. That is,

$$v^2 = v_0^2 + 2ax$$

$$0 = v_0^2 + 2ax$$

$$v_0 = \sqrt{-2ax}$$

$$= \sqrt{-2(-18.0 \text{ ft/s}^2)(200 \text{ ft})}$$

$$= \sqrt{7200 \text{ ft}^2/\text{s}^2}$$

$$= \frac{84.9 \text{ ft}}{\text{s}}\left(\frac{60.0 \text{ mph}}{88.0 \text{ ft/s}}\right)$$

$$= 57.9 \text{ mph}$$

If the car decelerates at the constant rate of 18.0 ft/s² and the low beam headlights are only capable of illuminating a distance of 200 ft, then the maximum safe velocity of the car at night without hitting something is 57.9 miles per hour. For velocities faster than this, the distance it takes to bring the car to a stop is greater than the distance the driver can see with low beam headlights. If you see it, you'll hit it! Of course these results are based on the assumption that the car decelerates at 18 ft/s². This number depends on the condition of the brakes and tires and road conditions, and will be different for each car.

In summary, the three kinematic equations,

$$x = v_0 t + \tfrac{1}{2}at^2 \tag{3.14}$$

$$v = v_0 + at \tag{3.10}$$

and

$$v^2 = v_0^2 + 2ax \tag{3.16}$$

are used to describe the motion of an object undergoing constant acceleration. The first equation gives the displacement of the object at any instant of time. The second equation gives the body's velocity at any instant of time. The third equation gives the velocity of the body at any displacement x.

These equations are used for either positive or negative accelerations. Remember the three kinematic equations hold only for constant acceleration. If the acceleration varies with time then more advanced techniques must be used to determine the position and velocity of the moving object.

3.7 **The Freely Falling Body**

Another example of the motion of a body in one dimension is the freely falling body. **A freely falling body** is defined as a body that is moving freely under the influence of gravity, where it is assumed that the effect of air resistance is negligible. The body can have an upward, downward, or even zero initial velocity. The simplest of the freely falling bodies we discuss is the body dropped in the vicinity of the surface of the earth. That is, the first case to be considered is the one with zero initial velocity, $v_0 = 0$. The motion of a body in the vicinity of the surface of the earth with either an upward or downward initial velocity will be considered in section 3.9.

In chapter 4 on Newton's second law of motion, we will see that whenever an unbalanced force acts on an object of mass m, it gives that object an acceleration, a. The gravitational force that the earth exerts on an object causes that object to have an acceleration. This acceleration is called the **acceleration due to gravity** and is denoted by the letter g. Therefore, any time a body is dropped near the surface of the earth, that body, ignoring air friction, experiences an acceleration g. From experiments in the laboratory we know that the value of g near the surface of the earth is constant and is given by

$$g = 9.80 \text{ m/s}^2 = 980 \text{ cm/s}^2 = 32.2 \text{ ft/s}^2$$

Any body that falls with the acceleration due to gravity, g, is called a freely falling body. To simplify the calculations in the British engineering system we will use the value $g = 32.0 \text{ ft/s}^2$.

Originally Aristotle said that a heavier body falls faster than a lighter body and on his authority this statement was accepted as truth for 1800 years. It was not disproved until the end of the sixteenth century when Simon Stevin (Stevinus) of Bruges (1548–1620) dropped balls of very different weights and found that they all fell at the same rate. That is, the balls were all dropped from the same height at the same time and all landed at the ground simultaneously. The argument still persisted that a ball certainly drops faster than a feather, but Galileo Galilei (1564–1642) explained the difference in the motion by saying that it is the air's resistance that slows up the feather. If the air were not present the ball and the feather would accelerate at the same rate.

A standard demonstration of the rate of fall is the penny and the feather demonstration. A long tube containing a penny and a feather is used, as shown in figure 3.15. If we turn the tube upside down, first we observe that the penny falls to the bottom of the tube before the feather. Then we connect the tube to a vacuum pump and evacuate most of the air from the tube. Again we turn the tube upside down, and now the penny and feather do indeed fall at the same rate and reach the bottom of the tube at the same time. Thus, it *is* the air friction that causes the feather to fall at the slower rate.

Another demonstration of a freely falling body, performed by the Apollo astronauts on the surface of the moon, was seen by millions of people on television. One of the astronauts dropped a feather and a hammer simultaneously and millions saw them fall at the same rate, figure 3.16. Remember, there is no atmosphere on the moon.

Therefore, neglecting air friction, all freely falling bodies accelerate downward at the same rate regardless of their mass. Recall that the acceleration of a body was defined as the change in its velocity with respect to time, that is,

$$a = \frac{\Delta v}{\Delta t} \tag{3.7}$$

Hence, a body that undergoes an acceleration due to gravity of 32.0 ft/s^2, has its velocity changing by 32.0 ft/s every second. If we neglect the effects of air friction, every body near the surface of the earth accelerates downward at that rate, whether the body is very large or very small. For all the problems considered in this book, we neglect the effects of air resistance.

Since the acceleration due to gravity is constant near the surface of the earth, we can determine the position and velocity of the freely falling body by using the kinematic equations 3.10, 3.14, and 3.16. However, because the motion is vertical, we designate the displacement by y in the kinematic equations:

$$v = v_0 + at \tag{3.10}$$

$$y = v_0 t + \tfrac{1}{2}at^2 \tag{3.14}$$

$$v^2 = v_0^2 + 2ay \tag{3.16}$$

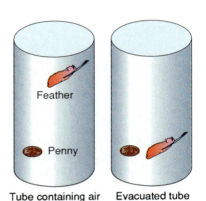

Figure 3.15

Free-fall of the penny and the feather.

Figure 3.16

Astronaut David R. Scott holds a geological hammer in his right hand and a feather in his left. The hammer and feather dropped to the lunar surface at the same instant.

Since the first case we consider is a body that is dropped, we will set the initial velocity v_0 equal to zero in the kinematic equations. Also the acceleration of the moving body is now the acceleration due to gravity, therefore we write the acceleration as

$$a = -g \tag{3.17}$$

The minus sign in equation 3.17 is consistent with our previous convention for one-dimensional motion. Motion in the direction of the positive axis is considered positive, while motion in the direction of the negative axis is considered negative. Hence, all quantities in the upward direction (positive y-direction) are considered positive, whether displacements, velocities, or accelerations. And all quantities in the downward direction (negative y-direction) are considered negative, whether displacements, velocities, or accelerations. The minus sign indicates that the direction of the acceleration is down, toward the center of the earth. This notation will be very useful later in describing the motion of projectiles. Therefore, the kinematic equations for a body dropped from rest near the surface of the earth are

$$y = -\tfrac{1}{2}gt^2 \tag{3.18}$$

$$v = -gt \tag{3.19}$$

$$v^2 = -2gy \tag{3.20}$$

Equation 3.18 gives the height or location of the freely falling body at any time, equation 3.19 gives its velocity at any time, and equation 3.20 gives the velocity of the freely falling body at any height y. This sign convention gives a negative value for the displacement y, which means that the zero position of the body is the position from which the body is dropped, and the body's location at any time t will always be below that point. The minus sign on the velocity indicates that the direction of the velocity is downward.

Equations 3.18, 3.19, and 3.20 completely describe the motion of the freely falling body that is dropped from rest. As an example, let us calculate the distance fallen and velocity of a freely falling body as a function of time for the first 5 s of its fall. Because most students are more familiar with the British engineering system of units, we will use this system for the calculations. However, the results of the computations are also written in parentheses in SI units in figure 3.17. At $t = 0$ the body is located at $y = 0$, (top of figure 3.17) and its velocity is zero. We then release the body. Where is it at $t = 1$ s? Using equation 3.18 and British engineering system units, y_1 is the displacement of the body (distance fallen) at the end of 1 s:

$$y_1 = -\tfrac{1}{2}gt^2 = -\tfrac{1}{2}(32 \text{ ft/s}^2)(1 \text{ s})^2 = -16 \text{ ft}$$

The minus sign indicates that the body is 16 ft *below* the starting point. To find the velocity at the end of 1 s, we use equation 3.19:

$$v_1 = -gt = (-32 \text{ ft/s}^2)(1 \text{ s}) = -32 \text{ ft/s}$$

The velocity is 32 ft/s downward at the end of 1 s. The position and velocity at the end of 1 s are shown in figure 3.17. For $t = 2$ s, the displacement and velocity are

$$y_2 = -\tfrac{1}{2}gt^2 = -\tfrac{1}{2}(32 \text{ ft/s}^2)(2 \text{ s})^2 = -64 \text{ ft}$$

$$v = -gt = -(32 \text{ ft/s}^2)(2 \text{ s}) = -64 \text{ ft/s}$$

At the end of 2 s the body has dropped a total distance downward of 64 ft and is moving at a velocity of 64 ft/s downward. For $t = 3$ s we obtain

$$y_3 = -\tfrac{1}{2}gt^2 = -\tfrac{1}{2}(32.0 \text{ ft/s}^2)(3 \text{ s})^2 = -144 \text{ ft}$$

$$v_3 = -gt = -(32.0 \text{ ft/s}^2)(3 \text{ s}) = -96 \text{ ft/s}$$

At the end of 3 s the body has fallen a distance of 144 ft and is moving downward at a velocity of 96 ft/s.

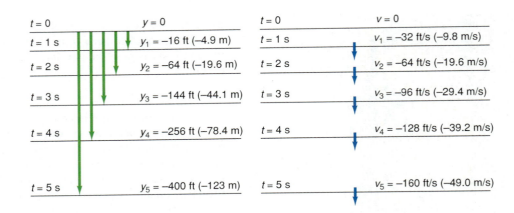

Figure 3.17

The distance and velocity for a freely falling body.

The distance and velocity for $t = 4$ s and $t = 5$ s are found similarly and are shown in figure 3.17. One of the first things to observe in figure 3.17 is that an object falls a relatively large distance in only a few seconds of time. Also note that the object does not fall equal distances in equal times, but rather the distance interval becomes greater for the same time interval as time increases. This is, of course, the result of the t^2 in equation 3.18 and is a characteristic of accelerated motion. Also note that the change in the velocity in any 1-s time interval is 32 ft/s, which is exactly what we meant by saying the acceleration due to gravity is 32 ft/s².

We stated previously that *the average velocity during a time interval is exactly equal to the instantaneous value of the velocity at the exact center of that time interval.* We can see that this is the case by inspecting figure 3.17. For example, if we take the time interval as between $t = 3$ s and $t = 5$ s, then the average velocity between the third and fifth second is

$$v_{35avg} = \frac{v_5 + v_3}{2} = \frac{-160 \text{ ft/s} + (-96 \text{ ft/s})}{2}$$

$$= \frac{-256 \text{ ft/s}}{2}$$

$$= -128 \text{ ft/s} = v_4$$

The average velocity between the time interval of 3 and 5 s, v_{35avg}, is exactly equal to v_4, the velocity at t equals 4 seconds, which is the exact center of the 3–5 time interval, as we can see in figure 3.17. The figure also shows the characteristic of an average velocity. At the beginning of the time interval the actual velocity is less than the average value, while at the end of the time interval the actual velocity is greater than the average value, but right at the center of the time interval the actual velocity is equal to the average velocity. Note that the average velocity occurs at the center of the time interval and not the center of the space interval.

In summary, we can see the enormous power inherent in the kinematic equations. An object was dropped from rest and the kinematic equations completely described the position and velocity of that object at any instant of time. All that information was contained in those equations.

3.8 Determination of Your Reaction Time by a Freely Falling Body

How long a period of time does it take for you to react to something? How can you measure this reaction time? It would be very difficult to use a clock to measure reaction time because it will take some reaction time to turn the clock on and off. However, a freely falling body can be used to measure reaction time. To see how this is accomplished, have one student hold a vertical meter stick near the top, as

(a)

(b)

Figure 3.18

Measurement of reaction time.

shown in figure 3.18(a). The second student then places his or her hand at the zero of the meter stick (the bottom of the stick) with thumb and forefinger extended. The thumb and forefinger should be open about 1 to 2 in. When the first student drops the meter stick, the second student catches it with the thumb and finger [figure 3.18(b)].

As the meter stick is released, it becomes a freely falling body and hence falls a distance y in a time t:

$$y = -\tfrac{1}{2}gt^2$$

The location of the fingers on the meter stick, where the meter stick was caught, represents the distance y that the meter stick has fallen. Solving for the time t we get

$$t = \sqrt{-\frac{2y}{g}} \tag{3.21}$$

Since we have measured y, the distance the meter stick has fallen, and we know the acceleration due to gravity g, we can do the simple calculation in equation 3.21 and determine your reaction time. (Remember that the value of y placed into equation 3.21 will be a negative number and hence we will take the square root of a positive quantity since the square root of a negative number is not defined.)

If you practice catching the meter stick, you will be able to catch it in less time. But eventually you reach a time that, no matter how much you practice, you cannot make smaller. This time is your *minimum reaction time*—the time it takes for your eye to first see the stick drop and then communicate this message to your brain. Your brain then communicates this information through nerves and muscles to your fingers and then you catch the stick. Your *normal reaction time* is most probably the time that you first caught the stick. A normal reaction time to catch the meter stick is about 0.2 to 0.3 seconds.

Note that this is not quite the same reaction time it would take to react to a red light while driving a car, because in that case, part of the communication from the brain would entail lifting your leg from the accelerator, placing it on the brake pedal, and then pressing. The motion of more muscles and mass would consequently take a longer period of time. A normal reaction time in a car is approximately 0.5 s. To obtain a more accurate value of the stopping distance for a car we also need to include the distance that the car moves while the driver reacts to the red light.

3.9 Projectile Motion in One Dimension

A case one step more general than the freely falling body dropped from rest, is the motion of a body that is thrown up or down with an initial velocity v_0 near the surface of the earth. This type of motion is called **projectile motion** in one dimension. Remember, however, that this type of motion still falls into the category of a freely falling body because the object experiences the acceleration g downward throughout its motion. The kinematic equations for projectile motion are

$$y = v_0 t - \tfrac{1}{2}gt^2 \tag{3.22}$$

$$v = v_0 - gt \tag{3.23}$$

$$v^2 = v_0^2 - 2gy \tag{3.24}$$

These three equations completely describe the motion of a projectile in one dimension. Note that these equations are more general than those for the body dropped from rest because they contain the initial velocity v_0. In fact, if v_0 is set equal to zero these equations reduce to the ones studied for the body dropped from rest.

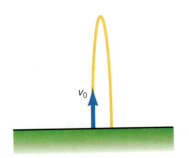

Figure 3.19

Trajectory of a projectile in one dimension.

In the previous cases of motion, we were concerned only with motion in one direction. Here there are two possible directions, up and down. According to our convention the upward direction is positive and the downward direction is negative. Hence, if the projectile is initially thrown upward, v_0 is positive; if the projectile is initially thrown downward, v_0 is negative. Also note that whether the projectile is thrown up or down, the acceleration of gravity always points downward. If it did not, then a ball thrown upward would continue to rise forever and would leave the earth, a result that is contrary to observation.

Let us now consider the motion of a projectile thrown upward. Figure 3.19 shows its path through space, which is called a **trajectory**. The projectile goes straight up, and then straight down. The downward portion of the motion is slightly displaced from the upward portion to clearly show the two different parts of the motion. For example, suppose the projectile is a baseball thrown straight upward with an initial velocity $v_0 = 100$ ft/s. We want to determine

1. The maximum height of the ball.
2. The time it takes for the ball to rise to the top of its trajectory.
3. The total time that the ball is in the air.
4. The velocity of the ball as it strikes the ground.
5. The position and velocity of the ball at any time t, for example, for $t = 4.00$ s.

We are asking for a great deal of information, especially considering that the only data given is the initial position and velocity of the ball. Yet all this information can be obtained using the three kinematic equations 3.22, 3.23, and 3.24. In fact, any time we try to solve any kinematic problem, the first thing is to write down the kinematic equations, because somehow, somewhere, the answers are in those equations. It is just a matter of manipulating them into the right form to obtain the information we want about the motion of the projectile.

Let us now solve the problem of projectile motion in one dimension.

Find the Maximum Height of the Ball

Equation 3.22 tells us the height of the ball at any instant of time. We could find the maximum height if we knew the time for the projectile to rise to the top of its trajectory. But at this point that time is unknown. (In fact, that is question 2.) Equation 3.24 tells us the velocity of the moving body at any height y. The velocity of the ball is positive on the way up, and negative on the way down, so therefore it must have gone through zero somewhere. In fact, the velocity of the ball is zero when the ball is at the very top of its trajectory. If it were greater than zero the ball would continue to rise, if it were less than zero the ball would be on its way down. Therefore, at the top of the trajectory, the position of maximum height, $v = 0$, and equation 3.24,

$$v^2 = v_0^2 - 2gy$$

becomes

$$0 = v_0^2 - 2gy_{max}$$

where y_{max} is the maximum height of the projectile. For any other height y, the velocity is either positive indicating that the ball is on its way up, or negative indicating that it is on the way down. Solving for y_{max}, the maximum height of the ball is

$$2gy_{max} = v_0^2$$

$$y_{max} = \frac{v_0^2}{2g}$$

(3.25)

Inserting numbers into equation 3.25, we get

$$y_{max} = \frac{v_0^2}{2g} = \frac{(100 \text{ ft/s})^2}{2(32.0 \text{ ft/s}^2)}$$

$$= 156 \text{ ft}$$

The ball will rise to a maximum height of 156 ft.

Find the Time for the Ball to Rise to the Top of the Trajectory

We have seen that when the projectile is at the top of its trajectory, $v = 0$. Therefore, equation 3.23,

$$v = v_0 - gt$$

becomes

$$0 = v_0 - gt_r$$

where t_r is the time for the projectile to rise to the top of its path. Only at this value of time does the velocity equal zero. At any other time the velocity is either positive or negative, depending on whether the ball is on its way up or down. Solving for t_r we get

$$t_r = \frac{v_0}{g} \tag{3.26}$$

the time for the ball to rise to the top of its trajectory. Inserting numbers into equation 3.26 we obtain

$$t_r = \frac{v_0}{g} = \frac{100 \text{ ft/s}}{32.0 \text{ ft/s}^2}$$

$$= 3.13 \text{ s}$$

It takes 3.13 s for the ball to rise to the top of the trajectory. Notice that the ball has the acceleration $-g$ at the top of the trajectory even though the velocity is zero at that instant. That is, in any kind of motion, we can have a nonzero acceleration even though the velocity is zero. The important thing for an acceleration is the *change in velocity,* not the velocity itself. At the top, the change in velocity is not zero, because the velocity is changing from positive values on the way up, to negative values on the way down.

The time t_r could also have been found using equation 3.22,

$$y = v_0 t - \tfrac{1}{2}gt^2$$

by substituting the maximum height of 156 ft for y. Even though this also gives the correct solution, the algebra and arithmetic are slightly more difficult because a quadratic equation for t would have to be solved.

Find the Total Time that the Object Is in the Air

When t is equal to the total time t_t, that the projectile is in the air, y is equal to zero. That is, during the time from $t = 0$ to $t = t_t$, the projectile goes from the ground to its maximum height and then falls back to the ground. Using equation 3.22, the height of the projectile at any time t,

$$y = v_0 t - \tfrac{1}{2}gt^2$$

with $t = t_t$ and $y = 0$, we get

$$0 = v_0 t_t - \tfrac{1}{2}gt_t^2 \tag{3.27}$$

Solving for t_t we obtain

$$t_t = \frac{2v_0}{g} \tag{3.28}$$

the total time that the projectile is in the air. Recall from equation 3.26 that the time for the ball to rise to the top of its trajectory is $t_r = v_0/g$. And the total time, equation 3.28, is just twice that value. Therefore, the total time that the projectile is in the air becomes

$$t_t = \frac{2v_0}{g} = 2t_r \tag{3.29}$$

The total time that the projectile is in the air is twice the time it takes the projectile to rise to the top of its trajectory. Stated in another way, the time for the ball to go up to the top of the trajectory is equal to the time for the ball to come down to the ground.

For this particular problem,

$$t_t = 2t_r = 2(3.13 \text{ s}) = 6.26 \text{ s}$$

The projectile will be in the air for a total of 6.26 s. Also note that equation 3.27 is really a quadratic equation with two roots. One of which we can see by inspection is $t = 0$, which is just the initial moment that the ball is launched.

Find the Velocity of the Ball as It Strikes the Ground

There are two ways to find the velocity of the ball at the ground. The simplest is to use equation 3.24,

$$v^2 = v_0^2 - 2gy$$

noting that the height is equal to zero ($y = 0$) when the ball is back on the ground. Therefore,

$$v_g^2 = v_0^2$$

and

$$v_g = \pm v_0 \tag{3.30}$$

The two roots represent the velocity at the two times that $y = 0$, namely, when the ball is first thrown up ($t = 0$), with an initial velocity $+v_0$, and when the ball lands ($t = t_t$) with a final velocity of $-v_0$ (the minus sign indicates that the ball is on its way down).

Another way to find the velocity at the ground is to use equation 3.23,

$$v = v_0 - gt$$

which represents the velocity of the projectile at any instant of time. If we let t be the total time that the projectile is in the air (i.e., $t = t_t$), then $v = v_g$, the velocity of the ball at the ground. Thus,

$$v_g = v_0 - gt_t \tag{3.31}$$

But we have already seen that

$$t_t = \frac{2v_0}{g} \tag{3.28}$$

Substituting equation 3.28 into equation 3.31 gives

$$v_g = v_0 - \frac{g(2v_0)}{g}$$

Hence,

$$v_g = -v_0$$

The velocity of the ball as it strikes the ground is equal to the negative of the original velocity with which the ball was thrown upward, that is,

$$v_g = -v_0 = -100 \text{ ft/s}$$

Find the Position and Velocity of the Ball at t = 4.00 s

The position of the ball at any time t is given by equation 3.22 as

$$y = v_0 t - \tfrac{1}{2}gt^2$$

Substituting in the values for $t = 4.00$ s gives

$$y_4 = (100 \text{ ft/s})(4.00 \text{ s}) - \tfrac{1}{2}(32.0 \text{ ft/s}^2)(4.00 \text{ s})^2$$
$$= 400 \text{ ft} - 256 \text{ ft}$$
$$= 144 \text{ ft}$$

At $t = 4.00$ s the ball is 144 ft above the ground.

The velocity of the ball at any time is given by equation 3.23 as

$$v = v_0 - gt$$

For $t = 4.00$ s, the velocity becomes

$$v_4 = 100 \text{ ft/s} - (32.0 \text{ ft/s}^2)(4.00 \text{ s})$$
$$= 100 \text{ ft/s} - 128 \text{ ft/s}$$
$$= -28.0 \text{ ft/s}$$

At the end of 4 s the velocity of the ball is -28.0 ft/s, where the negative sign indicates that the ball is on its way down. We could have used equation 3.22 for every value of time and plotted the entire trajectory, as shown in figure 3.20.

There is great beauty and power in these few simple equations, because with them we can completely predict the motion of the projectile for any time, simply by knowing its initial position and velocity. This is a characteristic of the field of physics. First we observe how nature works. Then we make a mathematical model of nature in terms of certain equations. We manipulate these equations until we can make a prediction, and this prediction yields information that we usually have no other way of knowing.

For example, how could you know that the velocity of the ball after 4.00 s would be -28.0 ft/s. In general, there is no way of knowing that. Yet we have actually captured a small piece of nature in our model and have seen how it works.

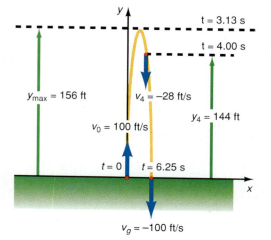

Figure 3.20

Results of projectile motion in one dimension.

Example 3.7

A projectile is fired straight up from the top of a building. A projectile is fired from the top of a building at an initial velocity of 35.0 m/s upward. The top of the building is 30.0 m above the ground. The motion is shown in figure 3.21. Find (a) the maximum height of the projectile, (b) the time for the projectile to reach its maximum height, (c) the velocity of the projectile as it strikes the ground, and (d) the total time that the projectile is in the air.

Solution

We will solve this problem using the techniques just developed.

a. To find the maximum height of the projectile we again note that at the top of the trajectory $v = 0$. Using equation 3.24,

$$v^2 = v_0^2 - 2gy$$

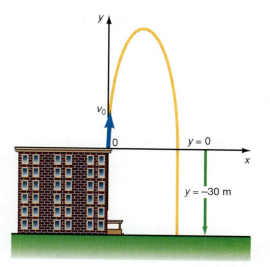

Figure 3.21

A projectile is fired vertically from the top of a building.

and setting $v = 0$ we obtain

$$0 = v_0^2 - 2gy_{max}$$

Solving for the maximum height,

$$y_{max} = \frac{v_0^2}{2g} = \frac{(35.0 \text{ m/s})^2}{2(9.80 \text{ m/s}^2)}$$

$$= 62.5 \text{ m}$$

The projectile's maximum height is 62.5 m above the roof of the building, or 92.5 m above the ground. Notice that since the initial data was given to us in SI units, we solve the problem in these units and use the value 9.80 m/s² for the acceleration due to gravity.

b. To find the time for the projectile to reach its maximum height we again note that at the maximum height $v = 0$. Substituting $v = 0$ into equation 3.23, we get

$$0 = v_0 - gt_r$$

Solving for the time to rise to the top of the trajectory, we get

$$t_r = \frac{v_0}{g} = \frac{35.0 \text{ m/s}}{9.80 \text{ m/s}^2}$$

$$= 3.57 \text{ s}$$

It takes 3.57 s for the ball to rise from the top of the roof to the top of its trajectory.

c. To find the velocity of the projectile when it strikes the ground, we use equation 3.24. When $y = -30.0$ m the projectile will be on the ground, and its velocity as it strikes the ground is

$$v^2 = v_0^2 - 2gy$$
$$(v_g)^2 = (35.0 \text{ m/s})^2 - 2(9.80 \text{ m/s}^2)(-30.0 \text{ m})$$
$$= 1225 \text{ m}^2/\text{s}^2 + 588 \text{ m}^2/\text{s}^2 = 1813 \text{ m}^2/\text{s}^2$$
$$= -42.6 \text{ m/s}$$

The projectile hits the ground at a velocity of -42.6 m/s. Note that this value is greater than the initial velocity v_0, because the projectile does not hit the roof on its way down, but rather hits the ground 30.0 m below the level of the roof. The acceleration has acted for a longer time, thereby imparting a greater velocity to the projectile.

d. To find the total time that the projectile is in the air we use equation 3.23,

$$v = v_0 - gt$$

But when t is equal to the total time that the projectile is in the air, the velocity is equal to the velocity at the ground (i.e., $v = v_g$). Therefore,

$$v_g = v_0 - gt_t$$

Solving for the total time, we get

$$t_t = \frac{v_0 - v_g}{g}$$

$$= \frac{35.0 \text{ m/s} - (-42.6 \text{ m/s})}{9.80 \text{ m/s}^2}$$

$$= \frac{(35.0 + 42.6)\text{m/s}}{9.80 \text{ m/s}^2}$$

$$= 7.92 \text{ s}$$

The total time that the projectile is in the air is 7.92 s. Note that it is not twice the time for the projectile to rise because the projectile did not return to the level where it started from, but rather to 30.0 m below that level.

3.10 The Kinematic Equations in Vector Form

Up to now we have discussed motion in one dimension only. And although the displacement, velocity, and acceleration of a body are vector quantities, we did not write them in the traditional boldface type, characteristic of vectors. We took into account their vector character by noting that when the displacement, velocity, and acceleration were in the positive x- or y-direction, we considered the quantities positive. When the displacement, velocity, and acceleration were in the negative x- or y-direction, we considered those quantities negative. For two-dimensional motion we must be more general and write the displacement, velocity, and acceleration in boldface type to show their full vector character. Let us now define the kinematic equations in terms of their vector characteristics.

The average velocity of a body is defined as the rate at which the displacement vector changes with time. That is,

$$\mathbf{v}_{avg} = \frac{\Delta \mathbf{r}}{\Delta t} = \frac{\mathbf{r}_2 - \mathbf{r}_1}{t_2 - t_1} \tag{3.32}$$

where the letter $\mathbf{r}$ is the displacement vector. The displacement vector $\mathbf{r}_1$ locates the position of the body at the time t_1, while the displacement vector $\mathbf{r}_2$ locates the position of the body at the time t_2. The displacement between the times t_1 and t_2 is just the difference between these vectors, $\mathbf{r}_2 - \mathbf{r}_1$, or $\Delta \mathbf{r}$, and is shown in figure 3.22.

We find the instantaneous velocity by taking the limit in equation 3.32 as Δt approaches zero, just as we did in equation 3.8. The magnitude of the instantaneous velocity vector is the instantaneous speed of the body, while the direction of the velocity vector is the direction that the body is moving, which is tangent to the trajectory at that point.

The average acceleration vector is defined as the rate at which the velocity vector changes with time:

$$\mathbf{a} = \frac{\Delta \mathbf{v}}{\Delta t} = \frac{\mathbf{v} - \mathbf{v}_0}{t} \tag{3.33}$$

Since the only cases that we will consider concern motion at constant acceleration, we will not use the subscript avg on $\mathbf{a}$. We find the kinematic equation for the displacement and velocity of the body at any instant of time as in section 3.6, only we write every term except t as a vector:

$$\mathbf{r} = \mathbf{v}_0 t + \tfrac{1}{2}\mathbf{a}t^2 \tag{3.34}$$

and

$$\mathbf{v} = \mathbf{v}_0 + \mathbf{a}t \tag{3.35}$$

Equation 3.34 represents the vector displacement of the moving body at any time t, while equation 3.35 represents the velocity of the moving body at any time. These vector equations are used to describe the motion of a moving body in two or three directions.

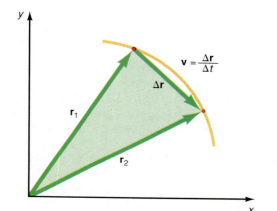

Figure 3.22

The change in the displacement vector.

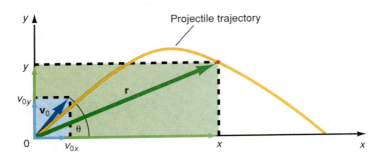

Figure 3.23

The trajectory of a projectile in two dimensions.

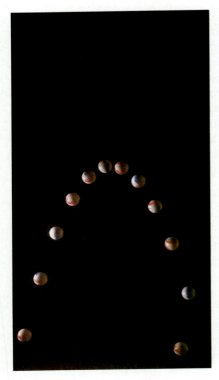

Stroboscopic photograph of a projectile in two dimensions.

3.11 Projectile Motion in Two Dimensions

In the study of kinematics we found that the displacement and velocity of a moving body can be determined if the original velocity $\mathbf{v_0}$ of the body and the acceleration $\mathbf{a}$ acting on it are known. The displacement of the body was given by

$$\mathbf{r} = \mathbf{v_0}t + \tfrac{1}{2}\mathbf{a}t^2 \tag{3.34}$$

while its velocity was given by

$$\mathbf{v} = \mathbf{v_0} + \mathbf{a}t \tag{3.35}$$

These two equations completely describe the resulting motion of the body. As an example of two-dimensional kinematics let us study the motion of a projectile in two dimensions. A projectile is thrown from the point 0 in figure 3.23 with an initial velocity $\mathbf{v_0}$. The trajectory of the projectile is shown in the figure. The initial velocity $\mathbf{v_0}$ has two components: v_{0x}, the x-component of the initial velocity, and v_{0y}, the y-component.

The location of the projectile at any instant of time is given by equation 3.34 and is shown as the displacement vector $\mathbf{r}$ in figure 3.23. We resolve the displacement vector $\mathbf{r}$ into two components: the distance the projectile has moved in the x-direction, we designate as x; the distance (or height) the projectile has moved in the y-direction we designate as y. We can now write the one vector equation 3.34 as two component equations, namely,

$$x = v_{0x}t + \tfrac{1}{2}a_xt^2 \tag{3.36}$$

$$y = v_{0y}t + \tfrac{1}{2}a_yt^2 \tag{3.37}$$

Figure 3.23 shows that v_{0x} is the x-component of the original velocity, given by $v_{0x} = v_0 \cos \theta$, while v_{0y} is the y-component of the original velocity, $v_{0y} = v_0 \sin \theta$. We have resolved the vector acceleration $\mathbf{a}$ into two components a_x and a_y.

In chapter 4 on Newton's laws of motion, we will see that whenever an unbalanced force F acts on a body of mass m, it gives that mass an acceleration a. Because there is no force acting on the projectile in the horizontal x-direction, the acceleration in the x-direction must be zero, that is, $a_x = 0$. Therefore, the x-component of the displacement $\mathbf{r}$ of the projectile, equation 3.36, takes the simple form

$$x = v_{0x}t \tag{3.38}$$

There is, however, a force acting on the projectile in the y-direction, the force of gravity that the earth exerts on any object, directed toward the center of the earth. We define the direction of this gravitational force to be in the negative y-direction. This gravitational force produces a constant acceleration called the acceleration due to gravity g. Hence, the y-component of the acceleration of the projectile is given by $-g$, that is, $a_y = -g$. The y-component of the displacement of the projectile therefore becomes

$$y = v_{0y}t - \tfrac{1}{2}gt^2 \tag{3.39}$$

Mechanics

Using the same arguments, we resolve the velocity **v** at any instant of time, equation 3.35, into the two scalar equations:

$$v_x = v_{0x} \tag{3.40}$$

$$v_y = v_{0y} - gt \tag{3.41}$$

Equation 3.40 does not contain the time t, and therefore the x-component of the velocity v_x is independent of time and is a constant. *Hence, the projectile motion consists of two motions: accelerated motion in the y-direction and motion at constant velocity in the x-direction.*

We can completely describe the motion of the projectile using the four equations, namely,

$$x = v_{0x}t \tag{3.38}$$

$$y = v_{0y}t - \tfrac{1}{2}gt^2 \tag{3.39}$$

$$v_x = v_{0x} \tag{3.40}$$

$$v_y = v_{0y} - gt \tag{3.41}$$

Now let us apply these equations to the projectile motion shown in figure 3.23. We essentially look for the same information that we found for projectile motion in one dimension. Because two-dimensional motion is a superposition of accelerated motion in the y-direction coupled to motion in the x-direction at constant velocity, we can use many of the techniques and much of the information we found in the one-dimensional case.

Let us find (1) the time for the projectile to rise to its maximum height, (2) the total time that the projectile is in the air, (3) the range (or maximum distance in the x-direction) of the projectile, (4) the maximum height of the projectile, (5) the velocity of the projectile as it strikes the ground, and (6) the location and velocity of the projectile at any time t.

To determine this information we use the kinematic equations 3.38 through 3.41.

The Time for the Projectile to Rise to Its Maximum Height

To determine the maximum height of the projectile we use the same reasoning used for the one-dimensional case. As the projectile is moving upward it has some positive vertical velocity v_y. When it is coming down it has some negative vertical velocity $-v_y$. At the very top of the trajectory, $v_y = 0$.

Therefore, at the top of the trajectory, equation 3.41 becomes

$$0 = v_{0y} - gt_r \tag{3.42}$$

Figure 3.24

A punted football is an example of a projectile in two dimensions.

Note that this is very similar to the equation for the one-dimensional case, except for the subscript y on v_0. This is an important distinction between the two motions, because the initial velocity upward v_{0y} is now less than the initial velocity upward v_0 in the one-dimensional case. Solving equation 3.42 for the time to rise to the top of the trajectory t_r, we get

$$t_r = \frac{v_{0y}}{g} \tag{3.43}$$

Since we know v_0 and hence v_{0y}, and because g is a constant, we can immediately compute t_r.

The Total Time the Projectile Is in the Air

To find the total time that the projectile is in the air, we use equation 3.39. When t is the total time t_t, the projectile is back on the ground and the height of the projectile is zero, $y = 0$. Therefore,

$$0 = v_{0y}t_t - \tfrac{1}{2}gt_t^2$$

Solving for the total time that the projectile is in the air, we get

$$t_t = \frac{2v_{0y}}{g} \tag{3.44}$$

But using equation 3.43 for the time to rise, $t_r = v_{0y}/g$, the total time that the projectile is in the air is exactly double this value,

$$t_t = \frac{2v_{0y}}{g} = 2t_r \tag{3.45}$$

which is the same as the one-dimensional case, as expected.

The Range of the Projectile

The **range of a projectile** is defined as the horizontal distance from the point where the projectile is launched to the point where it returns to its launch height. In this case, the range is the maximum distance that the projectile moves in the x-direction before it hits the ground. Because the maximum horizontal distance is the product of the horizontal velocity, which is a constant, and the total time of flight, the range, becomes

$$\text{range} = R = x_{\text{max}} = v_{0x}t_t \tag{3.46}$$

Sometimes it is convenient to express the range in another way. Since $v_{0x} = v_0 \cos \theta$, and the total time in the air is

$$t_t = \frac{2v_{0y}}{g} = \frac{2v_0 \sin \theta}{g}$$

we substitute these values into equation 3.46 to obtain

$$R = \frac{(v_0 \cos \theta)(2v_0 \sin \theta)}{g} = \frac{v_0^2 2 \sin \theta \cos \theta}{g}$$

However, using the well-known trigonometric identity,

$$2 \sin \theta \cos \theta = \sin 2\theta$$

the range of the projectile becomes

$$R = \frac{v_0^2 \sin 2\theta}{g} \tag{3.47}$$

We derived equation 3.47 based on the assumption that the initial and final elevations are the same, and we can use it only in problems where this assumption holds. This formulation of the range is particularly useful when we want to know at what angle a projectile should be fired in order to get the maximum possible range. From equation 3.47 we can see that for a given initial velocity v_0, the maximum range depends on the sine function. Because the sine function varies between -1 and $+1$, the maximum value occurs when $\sin 2\theta = 1$. But this happens when $2\theta = 90°$, hence the maximum range occurs when $\theta = 45°$. We obtain the maximum range of a projectile by firing it at an angle of $45°$.

The Maximum Height of the Projectile

We can find the maximum height of the projectile by substituting the time t_r into equation 3.39 and solving for the maximum height. However, since it is useful to have an equation for vertical velocity as a function of the height, we will use an alternate solution. Equation 3.39 represents the y-component of the displacement of the projectile at any instant of time and equation 3.41 is the y-component of the velocity at any instant of time. If the time is eliminated between these two equations (exactly as it was in section 3.6, for equation 3.16), we obtain the kinematic equation

$$v_y^2 = v_{0y}^2 - 2gy \tag{3.48}$$

which gives the y-component of the velocity of the moving body at any height y.

When the projectile has reached its maximum height, $v_y = 0$. Therefore, equation 3.48 becomes

$$0 = v_{0y}^2 - 2gy_{max}$$

Solving for y_{max} we obtain

$$y_{max} = \frac{v_{0y}^2}{2g} \tag{3.49}$$

the maximum height of the projectile.

The Velocity of the Projectile as It Strikes the Ground

The velocity of the projectile as it hits the ground $\mathbf{v_g}$ can be described in terms of its components, as shown in figure 3.25. The x-component of the velocity at the ground, found from equation 3.40, is

$$v_{xg} = v_x = v_{0x} \tag{3.50}$$

The y-component of the velocity at the ground, found from equation 3.41 with $t = t_t$ is

$$v_{yg} = v_{0y} - gt_t \tag{3.51}$$

$$= v_{0y} - \frac{g(2v_{0y})}{g}$$

$$v_{yg} = -v_{0y} \tag{3.52}$$

The y-component of the velocity of the projectile at the ground is equal to the negative of the y-component of the original velocity. The minus sign just indicates that the projectile is coming down. But this is exactly what we expected from the study of one-dimensional motion. The magnitude of the actual velocity at the ground, found from its two components, is

$$v_g = \sqrt{(v_{xg})^2 + (v_{yg})^2} \tag{3.53}$$

and using equations 3.50 and 3.52, becomes

$$v_g = \sqrt{(v_{0x})^2 + (-v_{0y})^2} = v_0 \tag{3.54}$$

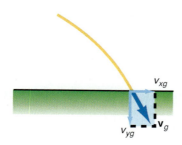

Figure 3.25

The velocity of the projectile at the ground.

The speed of the projectile as it strikes the ground is equal to the original speed of the projectile. The direction that the velocity vector makes with the ground is

$$\theta = \tan^{-1} \frac{v_{yg}}{v_{xg}} = \tan^{-1} -\frac{v_{0y}}{v_{0x}} = -\theta$$

The angle that the velocity vector makes as it hits the ground is the negative of the original angle. That is, if the projectile was fired at an original angle of 30° above the positive x-axis, it will make an angle of 30° below the positive x-axis when it hits the ground.

The Location and Velocity of the Projectile at Any Time t

We find the position and velocity of the projectile at any time t by substituting that value of t into equations 3.38, 3.39, 3.40, and 3.41. Let us look at some examples of projectile motion.

Example 3.8

Projectile motion in two dimensions. A ball is thrown with an initial velocity of 100 ft/s at an angle of 60.0° above the horizontal, as shown in figure 3.26. Find (a) the maximum height of the ball, (b) the time to rise to the top of the trajectory, (c) the total time the ball is in the air, (d) the range of the ball, (e) the velocity of the ball as it strikes the ground, and (f) the position and velocity of the ball at $t = 4$ s.

Solution

The x-component of the initial velocity is

$$v_{0x} = v_0 \cos \theta = (100 \text{ ft/s}) \cos 60° = 50.0 \text{ ft/s}$$

The y-component of the initial velocity is

$$v_{0y} = v_0 \sin \theta = (100 \text{ ft/s}) \sin 60.0° = 86.6 \text{ ft/s}$$

a. The maximum height of the ball, found from equation 3.49, is

$$y_{max} = \frac{v_{0y}^2}{2g} = \frac{(86.6 \text{ ft/s})^2}{2(32.0 \text{ ft/s}^2)} = 117 \text{ ft}$$

b. To find the time to rise to the top of the trajectory, we use equation 3.43,

$$t_r = \frac{v_{0y}}{g} = \frac{86.6 \text{ ft/s}}{32.0 \text{ ft/s}^2} = 2.71 \text{ s}$$

c. To find the total time that the ball is in the air, we use equation 3.45,

$$t_t = 2t_r = 2(2.71 \text{ s}) = 5.42 \text{ s}$$

d. The range of the ball, found from equation 3.47, is

$$R = \frac{v_0^2 \sin 2\theta}{g} = \frac{(100 \text{ ft/s})^2 \sin 120°}{32.0 \text{ ft/s}^2} = 271 \text{ ft}$$

Figure 3.26

Trajectory of thrown ball.

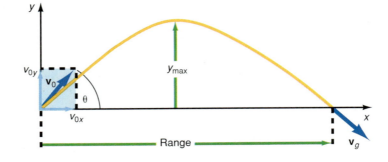

Mechanics

As a check, we can use equation 3.46 to get

$$R = x_{max} = v_{0x}t_t = (50.0 \text{ ft/s})(5.42 \text{ s}) = 271 \text{ ft}$$

e. To find the magnitude of the velocity of the ball at the ground, we use equation 3.53,

$$v_g = \sqrt{(v_{xg})^2 + (v_{yg})^2}$$

where

$$v_{xg} = v_{0x} = 50.0 \text{ ft/s}$$

and

$$v_{yg} = v_{0y} - gt_t = 86.6 \text{ ft/s} - (32 \text{ ft/s}^2)(5.42 \text{ s})$$
$$= 86.6 \text{ ft/s} - 173.4 \text{ ft/s} = -86.8 \text{ ft/s}$$

Hence,

$$v_g = \sqrt{(50.0 \text{ ft/s})^2 + (-86.8 \text{ ft/s})^2}$$
$$= 100 \text{ ft/s}$$

The direction that the velocity vector makes with the ground is

$$\theta = \tan^{-1}\frac{v_{yg}}{v_{xg}} = \tan^{-1}\frac{-86.8 \text{ ft/s}}{50.0 \text{ ft/s}} = -60.0°$$

f. To find the position and velocity of the ball at $t = 4$ s we use the kinematic equations 3.38 through 3.41.

1. $x = v_{0x}t = (50.0 \text{ ft/s})(4 \text{ s}) = 200 \text{ ft}$

2. $y = v_{0y}t - \frac{1}{2}gt^2$

$$= (86.6 \text{ ft/s})(4 \text{ s}) - \frac{1}{2}(32.0 \text{ ft/s}^2)(4 \text{ s})^2$$
$$= 90.4 \text{ ft}$$

The ball is 200 ft down range and is 90.4 ft high.
    The components of the velocity at 4 s are

3. $v_x = v_{0x} = 50.0 \text{ ft/s}$

4. $v_y = v_{0y} - gt$

$$= 86.6 \text{ ft/s} - (32.0 \text{ ft/s}^2)(4 \text{ s})$$
$$= -41.4 \text{ ft/s}$$

At the end of 4 s the x-component of the velocity is 50.0 ft/s and the y-component is -41.4 ft/s. To determine the magnitude of the velocity vector at 4 s we have

$$v = \sqrt{(v_x)^2 + (v_y)^2}$$
$$= \sqrt{(50.0 \text{ ft/s})^2 + (-41.4 \text{ ft/s})^2}$$
$$= 64.9 \text{ ft/s}$$

The direction of the velocity vector at 4 s is

$$\theta = \tan^{-1}\frac{v_y}{v_x} = \tan^{-1}\frac{-41.4 \text{ ft/s}}{50.0 \text{ ft/s}} = -39.6°$$

The velocity vector makes an angle of 39.6° below the horizontal at 4 s.

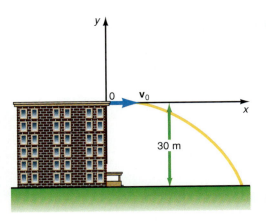

Figure 3.27

Trajectory of projectile thrown horizontally.

Example 3.9

A projectile is fired horizontally from the roof of a building. A projectile is fired horizontally from the roof of a building 30.0 m high at an initial velocity of 20.0 m/s, as shown in figure 3.27. Find (a) the total time the projectile is in the air, (b) where the projectile will hit the ground, and (c) the velocity of the projectile as it hits the ground.

Solution

The x- and y-components of the velocity are

$$v_{0x} = v_0 = 20.0 \text{ m/s}$$

$$v_{0y} = 0$$

a. To find the total time that the projectile is in the air, we use equation 3.39,

$$y = v_{0y}t - \tfrac{1}{2}gt^2$$

However, the initial conditions are that $v_{0y} = 0$. Therefore,

$$y = -\tfrac{1}{2}gt^2$$

Solving for t,

$$t = \sqrt{\frac{-2y}{g}}$$

However, when $t = t_t$, $y = -30.0$ m. Hence,

$$t_t = \sqrt{\frac{-2y}{g}} = \sqrt{\frac{-2(-30.0 \text{ m})}{9.80 \text{ m/s}^2}}$$

$$= 2.47 \text{ s}$$

b. To find where the projectile hits the ground, we use equation 3.38,

$$x = v_{0x}t$$

Now the projectile hits the ground when $t = t_t$, therefore,

$$x = v_{0x}t_t = (20.0 \text{ m/s})(2.47 \text{ s}) = 49.4 \text{ m}$$

The projectile hits the ground at the location $y = -30.0$ m and $x = 49.4$ m.

c. To find the velocity of the projectile at the ground we use equations 3.50, 3.51, and 3.53:

$$v_{xg} = v_{0x} = v_0 = 20.0 \text{ m/s}$$

$$v_{yg} = v_{0y} - gt_t = 0 - (9.80 \text{ m/s}^2)(2.47 \text{ s}) = -24.2 \text{ m/s}$$

$$v_g = \sqrt{v_{xg}^2 + v_{yg}^2}$$

$$= \sqrt{(20.0 \text{ m/s})^2 + (-24.2 \text{ m/s})^2}$$

$$= 31.4 \text{ m/s}$$

The direction that the velocity vector makes with the ground is

$$\theta = \tan^{-1}\frac{v_{yg}}{v_{xg}} = \tan^{-1}\frac{-24.2 \text{ m/s}}{20.0 \text{ m/s}} = -50.4°$$

The velocity vector makes an angle of 50.4° below the horizontal when the projectile hits the ground.

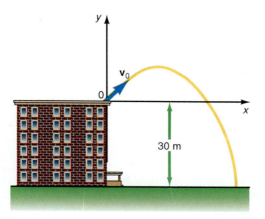

Figure 3.28

Trajectory of a projectile fired from the roof of a building.

Example 3.10

A projectile is fired at an angle from the roof of a building. A projectile is fired at an initial velocity of 35.0 m/s at an angle of 30.0° above the horizontal from the roof of a building 30.0 m high, as shown in figure 3.28. Find (a) the maximum height of the projectile, (b) the time to rise to the top of the trajectory, (c) the total time that the projectile is in the air, (d) the velocity of the projectile at the ground, and (e) the range of the projectile.

Solution

The x- and y-components of the original velocity are

$$v_{0x} = v_0 \cos \theta = (35.0 \text{ m/s}) \cos 30° = 30.3 \text{ m/s}$$

$$v_{0y} = v_0 \sin \theta = (35.0 \text{ m/s}) \sin 30° = 17.5 \text{ m/s}$$

a. To find the maximum height we use equation 3.49:

$$y_{max} = \frac{v_{0y}^2}{2g} = \frac{(17.5 \text{ m/s})^2}{2(9.80 \text{ m/s}^2)}$$

$$= 15.6 \text{ m}$$

above the building. Since the building is 30 m high, the maximum height with respect to the ground is 45.6 m.

b. To find the time to rise to the top of the trajectory we use equation 3.43:

$$t_r = \frac{v_{0y}}{g} = \frac{17.5 \text{ m/s}}{9.80 \text{ m/s}^2} = 1.79 \text{ s}$$

c. To find the total time the projectile is in the air we use equation 3.39:

$$y = v_{0y}t - \tfrac{1}{2}gt^2$$

When $t = t_t$, $y = -30.0$ m. Therefore,

$$-30.0 \text{ m} = (17.5 \text{ m/s})t_t - \tfrac{1}{2}(9.80 \text{ m/s}^2)t_t^2$$

Rearranging the equation, we get

$$4.90 \, t_t^2 - 17.5 \, t_t - 30.0 = 0$$

The units have been temporarily left out of the equation to simplify the following calculations. This is a quadratic equation of the form

$$ax^2 + bx + c = 0$$

with the solution

$$x = \frac{-b \pm \sqrt{b^2 - 4ac}}{2a}$$

In this problem, $x = t_t$, $a = 4.90$, $b = -17.5$, and $c = -30.0$. Therefore,

$$t_t = \frac{+17.5 \pm \sqrt{(17.5)^2 - 4(4.90)(-30.0)}}{2(4.90)}$$

$$= \frac{+17.5 \pm 29.9}{9.80} = 4.84 \text{ s}$$

The total time that the projectile is in the air is 4.84 s. If we had solved the equation for the negative root, we would have found a time of -1.27 s. This corresponds to a time when the height is -30.0 meters but it is a time before the projectile was thrown. If the projectile had been thrown from the ground it would have taken 1.27 seconds to reach the roof.

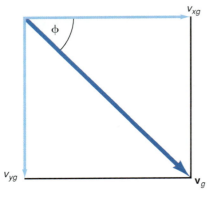

Figure 3.29

Angle of velocity vector as it strikes the ground.

d. To find the velocity of the projectile at the ground we use equations 3.50, 3.51, and 3.53:

$$v_{xg} = v_{0x} = 30.3 \text{ m/s}$$

$$v_{yg} = v_{0y} - gt_t = 17.5 \text{ m/s} - (9.80 \text{ m/s}^2)(4.84 \text{ s})$$

$$= -29.9 \text{ m/s}$$

$$v_g = \sqrt{(v_{xg})^2 + (v_{yg})^2}$$

$$= \sqrt{(30.3 \text{ m/s})^2 + (29.9 \text{ m/s})^2}$$

$$= 42.6 \text{ m/s}$$

The speed of the projectile as it strikes the ground is 42.6 m/s. The angle that the velocity vector makes with the ground, found from figure 3.29, is

$$\tan \phi = \frac{v_{yg}}{v_{xg}}$$

$$\phi = \tan^{-1}\frac{v_{yg}}{v_{xg}} = \tan^{-1}\left(\frac{-29.9}{30.3}\right)$$

$$= -44.6°$$

The velocity vector makes an angle of 44.6° below the horizontal when the projectile hits the ground.

e. To find the range of the projectile we use equation 3.46:

$$x_{max} = v_{0x}t_t = (30.3 \text{ m/s})(4.84 \text{ s})$$

$$= 147 \text{ m}$$

Figure 1

Does your expressway look like this?

Have you ever wondered, while sitting in heavy traffic on the expressway, as shown in figure 1, why there is so much traffic congestion? The local radio station tells you there are no accidents on the road, the traffic is heavy because of volume. What does that mean? Why can't cars move freely on the expressway? Why call it an expressway, if you have to move so slowly?

Let us apply some physics to the problem to help understand it. In particular, we will make a simplified model to help analyze the traffic congestion. In this model, we assume that the total length of the expressway L is 10,000 ft (approximately two miles), the length of the car x_c is 10 ft, and the speed of the car v_0 is 55 mph. How many cars of this size can safely fit on this expressway if they are all moving at 55 mph?

First, we need to determine the safe distance required for each car. If the car is moving at 55 mph (80.7 ft/s), and the car is capable of decelerating at -18.0 ft/s^2, the distance required to stop the car is found from equation 3.16,

$$v^2 = v_0^2 + 2ax$$

by noting that $v = 0$ when the car comes to a stop. Solving for the distance x_d that the car moves while decelerating to a stop we get

$$x_d = \frac{-v_0^2}{2a} = \frac{-(80.7 \text{ ft/s})^2}{2(-18 \text{ ft/s}^2)} = 181 \text{ ft}$$

Before the actual deceleration, the car will move, during the reaction time, a distance x_R given by

$$x_R = v_0 t_R = (80.7 \text{ ft/s})(0.500 \text{ s}) = 40.4 \text{ ft}$$

where we assume that it takes the driver 0.500 s to react. The total distance ΔL needed for each car on the expressway to safely come to rest is

$$\Delta L = x_c + x_R + x_d = 10 \text{ ft} + 40.4 \text{ ft} + 181 \text{ ft} = 231 \text{ ft}$$

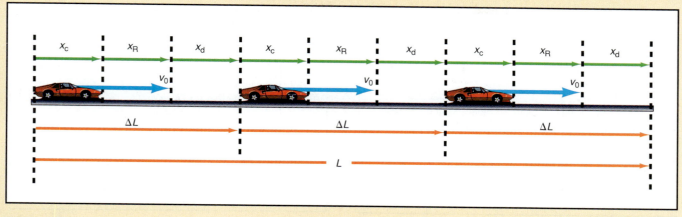

Figure 2

The number of cars on an expressway.

Because it takes a safe distance ΔL for one car to come to rest, N cars will take a distance of $N\Delta L$. The total length of the road L can then hold N cars, each requiring a distance ΔL to stop, as seen in figure 2. Stated mathematically this is

$$L = N\Delta L \qquad \text{(H3.1)}$$

Therefore, the number of cars N that can safely fit on this road is

$$N = \frac{L}{\Delta L} = \frac{10,000 \text{ ft}}{231 \text{ ft}} = 43 \text{ cars}$$

Thus for a road 10,000 ft long, only 43 cars can fit safely on it when each is moving at 55 mph. If the number of cars on the road doubles, then the safe distance per car ΔL must be halved because the product of N and ΔL must equal L the total length of the road, which is a constant. Rewriting equation H3.1 in the form

$$\Delta L = \frac{L}{N}$$

$$= x_c + x_R + x_d = \frac{L}{N}$$

$$x_c + v_0 t_R - \frac{v_0^2}{2a} = \frac{L}{N} \qquad \text{(H3.2)}$$

Notice in equation H3.2 that if the number of cars N increases, the only thing that can change on this fixed length road is the initial velocity v_0 of each car. That is, by increasing the number of cars on the road, the velocity of each car must decrease, in order for each car to move safely. Equation H3.2 can be written in the quadratic form

$$-\frac{v_0^2}{2a} + v_0 t_R + x_c - \frac{L}{N} = 0$$

which can be solved quadratically to yield

$$v_0 = at_R \mp \sqrt{(at_R)^2 + 2a(x_c - L/N)} \qquad \text{(H3.3)}$$

Equation H3.3 gives the maximum velocity that N cars can safely travel on a road L ft long. (Don't forget that a is a negative number.) Using the same numerical values of a, t_R, x_c, and L as

above, equation H3.3 is plotted in figure 3 to show the safe velocity (in miles per hour) for cars on an expressway as a function of the number of cars on that expressway. Notice from the form of the curve that as the number of cars increases, the safe velocity decreases. As the graph shows, increasing the number of cars on the road to 80, decreases the safe velocity to 38 mph. A further increase in the number of cars on the road to 200, decreases the safe velocity to 20 mph.

Hence, when that radio announcer says, "There is no accident on the road, the heavy traffic comes from volume," he means that by increasing the number of cars on the road, the safe velocity of each car must decrease.

You might wonder if there is some optimum number of cars that a road can handle safely. We can define the capacity C of a road as the number of cars that pass a particular place per unit time. Stated mathematically, this is

$$C = \frac{N}{t} \qquad \text{(H3.4)}$$

From the definition of velocity, the time for N cars to pass through a distance L, when moving at the velocity v_0, is

$$t = \frac{L}{v_0}$$

Substituting this into equation H3.4 gives

$$C = \frac{N}{t} = \frac{N}{L/v_0} = \frac{v_0}{L/N}$$

Substituting from equation H3.2 for L/N, the capacity of the road is

$$C = \frac{v_0}{x_c + v_0 t_R - v_0^2/2a} \qquad \text{(H3.5)}$$

Using the same values for x_c, t_R, and a as before, equation H3.5 is plotted in figure 4. The number of cars per hour that the road can hold is on the y-axis, and the speed of the cars in miles per hour is on the x-axis. Notice that at a speed of 60 mph, the road can handle 1200 cars per hour. By decreasing the speed of the cars, the number of cars per hour that the road can handle increases. As shown in the figure, if the speed decreases to 40 mph the road can handle about 1600 cars per hour. Notice that the curve peaks at a speed of about 13 mph, allowing about 2300

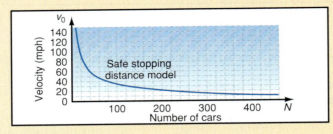

Figure 3

Plot of the velocity of cars (*y*-axis) as a function of the number of cars on the expressway (*x*-axis).

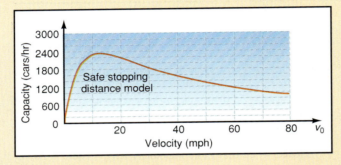

Figure 4

The capacity of a road as a function of the velocity of the cars.

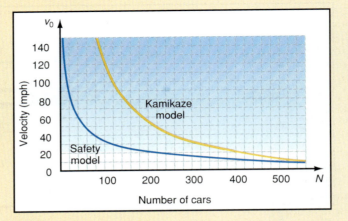

Figure 5

Comparison of traffic with the safe stopping distance model and the kamikaze model.

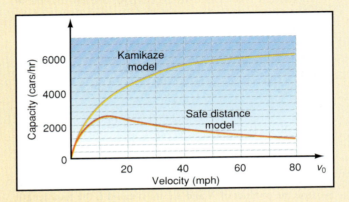

Figure 6

Comparison of the capacity versus velocity for the safe distance model and the kamikaze model.

cars/hour to flow on the expressway. Thus, according to this model, the optimum speed to pass the greatest number of cars per hour is only 13 mph. Hence, even though the road may be called an expressway, if the volume of cars increases significantly, the cars are not going to travel very rapidly. The solution to the problem is to build more lanes to handle the increased volume.

It should also be emphasized that this model is based on safe driving intervals between cars. If an object were to drop from the back of a truck you are following, you would need the safe distance to stop in time to avoid hitting the object. On the other hand, if the car in front of you, also traveling at 55 mph, has to stop, and if both drivers have the same reaction time and both cars decelerate at the same rate, then both cars will need 231 ft to come to a stop. Hence, when both cars come to a stop they will still be separated by the distance of 231 ft. For this reason, in areas of very heavy traffic, many people do not leave the safe distance between them and the car in front. Instead, they get closer and closer to the car in front of them until they are only separated by the reaction distance x_R. I call this the kamikaze model, for obvious reasons. The kamikaze model allows more cars to travel at a greater velocity than are allowed by the safe stopping distance model. The velocity of the cars as a function of the number of cars is found by solving equation H3.2 with the v_0^2 term, which is the term associated with the deceleration distance x_d set equal to zero. The result is shown in figure 5, which compares the safe stopping distance model with the kamikaze model. Notice that many more cars can now fit on the

road. For example, in the safe stopping model, only 40 cars, each traveling at 60 mph, can fit safely on this road. In the kamikaze model about 185 cars can fit on this road, but certainly not safely. There will be only 44 ft between each car, and if you have a slower reaction time than that of the car in front of you, you will almost certainly hit him when he steps on the brakes. This is the reason why there are so many rear-end collisions on expressways. The number of cars on a real expressway falls somewhere between the extremes of these two models. Note that even in the kamikaze model, the velocity of the cars must decrease with volume.

The capacity of the expressway for the kamikaze model is found by setting the v_0^2 term in equation H3.5 to zero. The result is shown in figure 6. Notice that in the kamikaze model the capacity increases with velocity, and there is no optimum speed for the maximum car flow. In practice, the actual capacity of an expressway lies somewhere between these two extremes.

In conclusion, if your expressway is not much of an expressway, it is time to petition your legislators to allocate more money for the widening of the expressway, or maybe it is time to move to a less populated part of the country.

The Language of Physics

Kinematics

The branch of mechanics that describes the motion of a body without regard to the cause of that motion (p. 39).

Average velocity

The average rate at which the displacement vector changes with time. Since a displacement is a vector, the velocity is also a vector (p. 39).

Average speed

The distance that a body moves per unit time. Speed is a scalar quantity (p. 41).

Constant velocity

A body moving in one direction in such a way that it always travels equal distances in equal times (p. 42).

Acceleration

The rate at which the velocity of a moving body changes with time (p. 43).

Instantaneous velocity

The velocity at a particular instant of time. It is defined as the limit of the ratio of the change in the displacement of the body to the change in time, as the time interval approaches zero. The magnitude of the instantaneous velocity is the instantaneous speed of the moving body (p. 45).

Kinematic equations of linear motion

A set of equations that gives the displacement and velocity of the moving body at any instant of time, and the velocity of the moving body at any displacement, if the acceleration of the body is a constant (p. 46).

Freely falling body

Any body that is moving under the influence of gravity only. Hence, any body that is dropped or thrown on the surface of the earth is a freely falling body (p. 51).

Acceleration due to gravity

If air friction is ignored, all objects that are dropped near the surface of the earth, are accelerated toward the center of the earth with an acceleration of 9.80 m/s^2 or 32 ft/s^2 (p. 52).

Projectile motion

The motion of a body thrown or fired with an initial velocity v_0 in a gravitational field (p. 55).

Trajectory

The path through space followed by a projectile (p. 56).

Range of a projectile

The horizontal distance from the point where the projectile is launched to the point where it returns to its launch height (p. 64).

Summary of Important Equations

Average velocity
$$\mathbf{v_{avg}} = \frac{\Delta \mathbf{r}}{\Delta t} = \frac{\mathbf{r_2} - \mathbf{r_1}}{t_2 - t_1} \tag{3.32}$$

Acceleration
$$\mathbf{a} = \frac{\Delta \mathbf{v}}{\Delta t} = \frac{\mathbf{v} - \mathbf{v_0}}{t} \tag{3.33}$$

Instantaneous velocity in two or more directions, which is a generalization of the instantaneous velocity in one dimension

$$\mathbf{v} = \lim_{\Delta t \to 0} \frac{\Delta \mathbf{r}}{\Delta t}$$
$$\qquad\qquad\qquad\qquad (3.8)$$
$$v = \lim_{\Delta t \to 0} \frac{\Delta x}{\Delta t}$$

Velocity at any time
$$\mathbf{v} = \mathbf{v_0} + \mathbf{a}t \tag{3.35}$$

Displacement at any time
$$\mathbf{r} = \mathbf{v_0}t + \frac{1}{2}\mathbf{a}t^2 \tag{3.34}$$

Velocity at any displacement in the x-direction
$$v^2 = v_0^2 + 2ax \tag{3.16}$$

Velocity at any displacement in the y-direction
$$v^2 = v_0^2 + 2ay \tag{3.16}$$

For Projectile Motion

x-displacement
$$x = v_{0x}t \tag{3.38}$$

y-displacement
$$y = v_{0y}t - \frac{1}{2}gt^2 \tag{3.39}$$

x-component of velocity
$$v_x = v_{0x} \tag{3.40}$$

y-component of velocity
$$v_y = v_{0y} - gt \tag{3.41}$$

y-component of velocity at any height y
$$v_y^2 = v_{0y}^2 - 2gy \tag{3.48}$$

Range
$$R = \frac{v_0^2 \sin 2\theta}{g} \tag{3.47}$$

Questions for Chapter 3

1. Discuss the difference between distance and displacement.
2. Discuss the difference between speed and velocity.
3. Discuss the difference between average speed and instantaneous speed.
† 4. Although speed is the magnitude of the instantaneous velocity, is the average speed equal to the magnitude of the average velocity?
5. Why can the kinematic equations be used only for motion at constant acceleration?
6. In dealing with average velocities discuss the statement, "Straight line motion at 60 mph for 1 hr followed by motion in the same direction at 30 mph for 2 hr does not give an average of 45 mph but rather 40 mph."

7. What effect would air resistance have on the velocity of a body that is dropped near the surface of the earth?
8. What is the acceleration of a projectile when its instantaneous vertical velocity is zero at the top of its trajectory?
9. Can an object have zero velocity at the same time that it has an acceleration? Explain and give some examples.
10. Can the velocity of an object be in a different direction than the acceleration? Give some examples.
11. Can you devise a means of using two clocks to measure your reaction time?
†12. A person on a moving train throws a ball straight upward. Describe the

motion as seen by a person on the train and by a person on the station platform.
13. You are in free fall, and you let go of your watch. What is the relative velocity of the watch with respect to you?
†14. What kind of motion is indicated by a graph of displacement versus time, if the slope of the curve is (a) horizontal, (b) sloping upward to the right, and (c) sloping downward?
†15. What kind of motion is indicated by a graph of velocity versus time, if the slope of the curve is (a) horizontal, (b) sloping upward at a constant value, (c) sloping upward at a changing rate, (d) sloping downward at a constant value, and (e) sloping downward at a changing rate?

Hints for Problem Solving

To be successful in a physics course it is necessary to be able to solve problems. The following procedure should prove helpful in solving the physics problems assigned. First, as a preliminary step, read the appropriate topic in the textbook. Do not attempt to solve the problems before doing this. Look at the appropriate illustrative problems to see how they are solved. With this background, now read the assigned problem. Now continue with the following procedure.

1. Draw a small picture showing the details of the problem. This is very useful so that you do not lose sight of the problem that you are trying to solve.
2. List all the information that you are given.
3. List all the answers you are expected to find.
4. From the summary of important equations or the text proper, list the equations that are appropriate to this topic.
5. Pick the equation that relates the variables that you are given.
6. Place a check mark ($\checkmark$) over each variable that is given and a question mark (?) over each variable that you are looking for.
7. Solve the equation for the unknown variable.
8. When the answer is obtained check to see if the answer is reasonable.

Let us apply this technique to the following example.

A car is traveling at 30 ft/s when it starts to accelerate at 10 ft/s². Find (a) the velocity and (b) the displacement of the car at the end of 5 s.

1. Draw a picture of the problem.

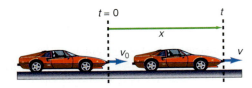

$t = 0$ t

x

v_0 v

2. Given: $v_0 = 30$ ft/s
$$a = 10 \text{ ft/s}^2$$
$$t = 5 \text{ s}$$

3. Find: $v = ?$
$$x = ?$$

4. The problem is one in kinematics and the kinematic equations apply. That is,

$$(1) \quad x = v_0 t + \tfrac{1}{2}at^2$$
$$(2) \quad v = v_0 + at$$
$$(3) \quad v^2 = v_0^2 + 2ax$$

5. Part a of the problem. To solve for the velocity v, we need an equation containing v. Equation 1 does not contain a velocity term v, and hence can not be used to solve for the velocity. Equations 2 and 3, on the other hand, both contain v. Thus, we can use one or possibly both of these equations to solve for the velocity.

6. Write down the equation and place a check mark over the known terms and a question mark over the unknown terms:

$$\overset{?}{(2)} \quad v = \overset{\checkmark}{v_0} + \overset{\checkmark\checkmark}{at}$$

The only unknown in equation 2 is the velocity v and we can now solve for it.

7. The velocity after 5 s, found from equation 2 is

$$v = v_0 + at$$
$$= 30 \text{ ft/s} + (10 \text{ ft/s}^2)(5 \text{ s})$$
$$= 30 \text{ ft/s} + 50 \text{ ft/s}$$
$$= 80 \text{ ft/s}$$

Notice what would happen if we tried to use equation 3 at this time:

$$\overset{?}{(3)} \quad v^2 = \overset{\checkmark}{v_0^2} + \overset{\checkmark\checkmark?}{2ax}$$

We can not solve for the velocity v from equation 3 because there are two unknowns, both v and x. However, if we had solved part b of the problem for x first, then we could have used this equation.

5. Part b of the problem. To solve for the displacement x, we need an equation containing x. Notice that equation 2 does not contain x, so we can not use it. Equations 1 and 3, on the other hand, do contain x, and we can use either to solve for x.

6. Looking at equation 1, we have

$$\overset{?}{(1)} \quad x = \overset{\checkmark\checkmark}{v_0 t} + \overset{\checkmark\checkmark\checkmark}{\tfrac{1}{2}at^2}$$

7. Solving for the only unknown in equation 1, x, we get

$$x = v_0 t + \tfrac{1}{2}at^2$$
$$= (30 \text{ ft/s})(5 \text{ s}) +$$
$$\tfrac{1}{2}(10 \text{ ft/s}^2)(5 \text{ s})^2$$
$$= 150 \text{ ft} + 125 \text{ ft}$$
$$= 275 \text{ ft}$$

Note that at this point we could also have used equation 3 to determine x, because we already found the velocity v in part a of the problem.

Problems for Chapter 3

3.1 Introduction

1. A driver travels 300 mi in 5 hr and 25 min. What is his average speed in (a) mph, (b) ft/s, (c) km/hr, and (d) m/s?
2. A car travels at 40 mph for 2 hr and 60 mph for 3 hr. What is its average speed?
3. A man hears the sound of thunder 5 s after he sees the lightning flash. If the speed of sound in air is 343 m/s, how far away is the lightning? Assume that the speed of light is so large that the lightning was seen essentially at the same time that it was created.

4. The earth-moon distance is 3.84×10^8 m. If it takes 3 days to get to the moon, what is the average speed?
5. Electronic transmission is broadcast at the speed of light, which is 3.00×10^8 m/s. How long would it take for a radio transmission from earth to an astronaut orbiting the planet Mars? Assume that at the time of transmission the distance from earth to Mars is 7.80×10^7 km.
6. In the game of baseball, some excellent fast-ball pitchers have managed to pitch a ball at approximately 100 mph. If the pitcher's mound is 60.5 ft from home

plate, how long does it take the ball to get to home plate? If the pitcher then throws a change-of-pace ball (a slow ball) at 60 mph, how long will it now take the ball to get to the plate?
7. Two students are having a race on a circular track. Student 1 is on the inside track, which has a radius of curvature $r_1 = 250$ m, and is moving at the speed $v_1 = 4.50$ m/s. With what speed must student 2 run to keep up with student 1 if student 2 is on the outside track of radius of curvature $r_2 = 255$ m?

8. A plot of the displacement of a car as a function of time is shown in the diagram. Find the velocity of the car along the paths (a) *O-A*, (b) *A-B*, (c) *B-C*, and (d) *C-D*.

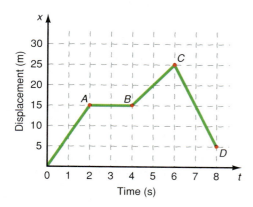

9. A plot of the velocity of a car as a function of time is shown in the diagram. Find the acceleration of the car along the paths (a) *O-A*, (b) *A-B*, (c) *B-C*, and (d) *C-D*.

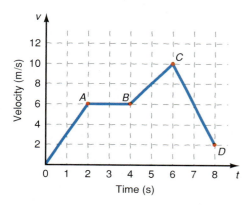

10. If an airplane is traveling at 110 knots, what is its velocity in (a) mph, (b) km/hr, (c) ft/s, and (d) m/s? A knot is a nautical mile per hour, and a nautical mile is equal to 6076 ft.

3.6 The Kinematic Equations in One Dimension

11. A girl who is initially running at 1.00 m/s increases her velocity to 2.50 m/s in 5.00 s. Find her acceleration.

12. A car is traveling at 95.0 km/hr. The driver steps on the brakes and the car comes to a stop in 60.0 m. What is the car's deceleration?

13. A train accelerates from an initial velocity of 20.0 mph to a final velocity of 35.0 mph in 11.8 s. Find its acceleration and the distance the train travels during this time.

14. A train accelerates from an initial velocity of 25.0 km/hr to a final velocity of 65.0 km/hr in 8.50 s. Find its acceleration and the distance the train travels during this time.

15. An airplane travels 1000 ft at a constant acceleration while taking off. If it starts from rest, and takes off in 25.0 s, what is its takeoff velocity?

16. An airplane travels 450 km at a constant acceleration while taking off. If it starts from rest, and takes off in 30.0 s, what is its takeoff velocity?

17. A car starts from rest and acquires a velocity of 20.0 mph in 10.0 s. Where is the car located and what is its velocity at 10.0, 15.0, 20.0, and 25.0 s?

18. A jet airplane goes from rest to a velocity of 250 ft/s in a distance of 400 ft. What is the airplane's average acceleration in ft/s²?

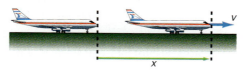

19. An electron in a vacuum tube acquires a velocity of 5.3×10^8 cm/s in a distance of 0.25 cm. Find the acceleration of the electron.

20. A driver traveling at 60.0 mph tries to stop the car and finds that the brakes have failed. The emergency brake is then pulled and the car comes to a stop in 456 ft. Find the car's deceleration.

21. An airplane has a touchdown velocity of 75.0 knots and comes to rest in 400 ft. What is the airplane's average deceleration? How long does it take the plane to stop?

22. A pitcher gives a baseball a horizontal velocity of 110 ft/s by moving his arm through a distance of approximately 3.00 ft. What is the average acceleration of the ball during this throwing process?

23. The speedometer of a car reads 60.0 mph when the brakes are applied. The car comes to rest in 4.55 s. How far does the car travel before coming to rest?

†24. A body with unknown initial velocity moves with constant acceleration. At the end of 8.00 s, it is moving at a velocity of 50.0 m/s and it is 200 m from where it started. Find the body's acceleration and its initial velocity.

†25. A driver traveling at 25.0 mph sees the light turn red at the intersection. If her reaction time is 0.600 s, and the car can decelerate at 18.0 ft/s², find the stopping distance of the car. What would the stopping distance be if the car were moving at 50.0 mph?

26. A driver traveling at 30.0 km/hr sees the light turn red at the intersection. If his reaction time is 0.600 s, and the car can decelerate at 4.50 m/s²,

find the stopping distance of the car. What would the stopping distance be if the car were moving at 90.0 km/hr?

†27. A uniformly accelerating train passes a green light signal at 25.0 km/hr. It passes a second light 125 m farther down the track, 12.0 s later. What is the train's acceleration? What is the train's velocity at the second light?

x = 125 m

28. A car accelerates from 50.0 mph to 80.0 mph in 26.9 s. Find its acceleration and the distance the car travels in this time.

†29. A motorcycle starts from rest and accelerates at 4.00 m/s² for 5.00 s. It then moves at constant velocity for 25.0 s, and then decelerates at 2.00 m/s² until it stops. Find the total distance that the motorcycle has moved.

†30. A car starts from rest and accelerates at a constant rate of 3.00 m/s² until it is moving at 18.0 m/s. The car then decreases its acceleration to 0.500 m/s² and continues moving for an additional distance of 250 m. Find the total time taken.

3.7 The Freely Falling Body

31. A passenger, in abandoning a sinking ship, steps over the side. The deck is 15.0 m above the water surface. With what velocity does the passenger hit the water?

32. How long does it take for a stone to fall from a bridge to the water 30.0 m below? With what velocity does the stone hit the water?

33. An automobile traveling at 60.0 mph hits a stone wall. From what height would the car have to fall to acquire the same velocity?

34. A rock is dropped from the top of a building and hits the ground 8.00 s later. How high is the building?

35. A stone is dropped from a bridge 100 ft high. How long will it take for the stone to hit the water below?

36. A ball is dropped from a building 50.0 meters high. How long will it take the ball to hit the ground below?

†37. A girl is standing in an elevator that is moving upward at a velocity of 12.0 ft/s when she drops her handbag. If she was originally holding the bag at a height of 4.00 ft above the elevator floor, how long will it take the bag to hit the floor?

3.9 Projectile Motion in One Dimension

38. A ball is thrown vertically upward with an initial velocity of 130 ft/s. Find its position and velocity at the end of 2, 4, 6, and 8 s and sketch these positions and velocities on a piece of graph paper.

39. A projectile is fired vertically upward with an initial velocity of 40.0 m/s. Find the position and velocity of the projectile at 1, 3, 5, and 7 s.

†40. A ball is thrown vertically upward from the top of a building 40.0 m high with an initial velocity of 25.0 m/s. What is the total time that the ball is in the air?

41. A stone is thrown vertically upward from a bridge 100 ft high at an initial velocity of 50.0 ft/s. How long will it take for the stone to hit the water below?

†42. A stone is thrown vertically downward from a bridge 100 ft high at an initial velocity of −50.0 ft/s. How long will it take for the stone to hit the water below?

43. A rock is thrown vertically downward from a building 40.0 m high at an initial velocity of −15.0 m/s. (a) What is the rock's velocity as it strikes the ground? (b) How long does it take for the rock to hit the ground?

44. A baseball batter fouls a ball vertically upward. The ball is caught right behind home plate at the same height that it was hit. How long was the baseball in flight if it rose a distance of 100 ft? What was the initial velocity of the baseball?

3.11 Projectile Motion in Two Dimensions

45. A projectile is thrown from the top of a building with a horizontal velocity of 15.0 m/s. The projectile lands on the street 85.0 m from the base of the building. How high is the building?

46. To find the velocity of water issuing from the nozzle of a garden hose, the nozzle is held horizontally and the stream is directed against a vertical wall. If the wall is 7.00 m from the nozzle and the water strikes the wall 0.650 m below the horizontal, what is the velocity of the water?

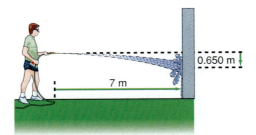

0.650 m

7 m

47. A bomb is dropped from an airplane in level flight at a velocity of 970 km/hr. The altitude of the aircraft is 2000 m. At what horizontal distance from the initial position of the aircraft will the bomb land?

†48. A cannon is placed on a hill 20.0 m above level ground. A shell is fired horizontally at a muzzle velocity of 300 m/s. At what horizontal distance from the cannon will the shell land? How long will this take? What will be the shell's velocity as it strikes its target?

49. A shell is fired from a cannon at a velocity of 300 m/s to hit a target 3000 m away. At what angle above the horizontal should the cannon be aimed?

50. In order to hit a target, a marksman finds he must aim 10.0 cm above the target, which is 300 m away. What is the initial speed of the bullet?

51. A golf ball is hit with an initial velocity of 175 ft/s at an angle of 50.0° above the horizontal. (a) How high will the ball go? (b) What is the total time the ball is in the air? (c) How far will the ball travel horizontally before it hits the ground?

52. A projectile is thrown from the ground with an initial velocity of 20.0 m/s at an angle of 40.0° above the horizontal. Find (a) the projectile's maximum height, (b) the time required to reach its maximum height, (c) its velocity at the top of the trajectory, (d) the range of the projectile, and (e) the total time of flight.

Additional Problems

53. A missile has a velocity of 10,000 mph at "burn-out," which occurs 2 min after ignition. Find the average acceleration in (a) ft/s², (b) m/s², and (c) in terms of g, the acceleration due to gravity at the surface of the earth.

54. A block slides down a smooth inclined plane that makes an angle of 25.0° with the horizontal. Find the acceleration of the block. If the plane is 10.0 meters long and the block starts from rest, what is its velocity at the bottom of the plane? How long does it take for the block to get to the bottom?

†55. At the instant that the traffic light turns green, a car starting from rest with an acceleration of 7.00 ft/s² is passed by a truck moving at a constant velocity of 30.0 mph. (a) How long will it take for the car

to overtake the truck? (b) How far from the starting point will the car overtake the truck? (c) At what velocity will the car be moving when it overtakes the truck?

56. At the instant that the traffic light turns green, a car starting from rest with an acceleration of 2.50 m/s² is passed by a truck moving at a constant velocity of 60.0 km/hr. (a) How long will it take for the car to overtake the truck? (b) How far from the starting point will the car overtake the truck? (c) At what velocity will the car be moving when it overtakes the truck?

†57. A boat passes a buoy while moving to the right at a velocity of 8.00 m/s. The boat has a constant acceleration to the left, and 10.0 s later the boat is found to be moving at a velocity of −3.00 m/s. Find (a) the acceleration of the boat, (b) the distance from the buoy when the boat reversed direction, (c) the time for the boat to return to the buoy, and (d) the velocity of the boat when it returns to the buoy.

†58. Two trains are initially at rest on parallel tracks with train 1 50.0 m ahead of train 2. Both trains accelerate simultaneously, train 1 at the rate of 2.00 m/s² and train 2 at the rate of 2.50 m/s². How long will it take train 2 to overtake train 1? How far will train 2 travel before it overtakes train 1?

†59. Repeat problem 58 but with train 1 initially moving at 5.00 m/s and train 2 initially moving at 7.00 m/s.

†60. A policeman driving at 55.0 mph observes a car 200 ft ahead of him speeding at 80.0 mph. If the county line is 1200 ft away from the police car, what must the acceleration of the police car be, in order to catch the speeder before he leaves the county?

61. A policewoman driving at 80.0 km/hr observes a car 50.0 m ahead of her speeding at 120 km/hr. If the county line is 400 m away from the police car, what must the acceleration of the police car be in order to catch the speeder before he leaves the county?

†62. Two trains are approaching each other along a straight and level track. The first train is heading east at 70.0 mph, while the second train is heading west at 45.0 mph. When they are 1.50 miles apart they see each other and start to decelerate. Train 1 decelerates at 5.00 ft/s², while train 2 decelerates at 3.00 ft/s². Will the trains be able to stop or will there be a collision?

63. Two trains are approaching each other along a straight and level track. The first train is heading south at 125 km/hr, while the second train is heading north at 80.0 km/hr. When they are 2.00 km apart, they see each other and start to decelerate. Train 1 decelerates at 2.00 m/s², while train 2 decelerates at 1.50 m/s². Will the trains be able to stop or will there be a collision?

†64. A boy in an elevator, which is descending at the constant velocity of −5.00 m/s, jumps to a height of 0.500 m above the elevator floor. How far will the elevator descend before the boy returns to the elevator floor?

65. The acceleration due to gravity on the moon is 1.62 m/s². If an astronaut on the moon throws a ball straight upward, with an initial velocity of 25.0 m/s, how high will the ball rise?

†66. A helicopter, at an altitude of 300 m, is rising vertically at 20.0 m/s when a wheel falls off. How high will the wheel go with respect to the ground? How long will it take for the wheel to hit the ground below? At what velocity will the wheel hit the ground?

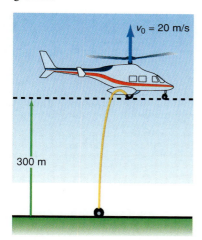

$v_0 = 20$ m/s

300 m

†67. A ball is dropped from the roof of a building 40.0 m high. Simultaneously, another ball is thrown upward from the ground and collides with the first ball at half the distance to the roof. What was the initial velocity of the ball that was thrown upward?

†68. A ball is dropped from the top of a 40.0-m high building. At what initial velocity must a second ball be thrown from the top of the building 2.00 s later, such that both balls arrive at the ground at the same time?

†69. Show that the range of a projectile is the same for either a projection angle of 45.0° + θ or an angle of 45.0° − θ.

70. A projectile hits a target 1.50 km away 10.5 s after it was fired. Find (a) the elevation angle of the gun and (b) the initial velocity of the projectile.

71. A football is kicked with an initial velocity of 70.0 ft/s at an angle of 65.0° above the horizontal. Find (a) how long the ball is in the air, (b) how far down field the ball lands, (c) how high the ball rises, and (d) the velocity of the ball when it strikes the ground.

†72. A baseball is hit at an initial velocity of 110 ft/s at an angle of 45.0° above the horizontal. Will the ball clear a 10.0 ft fence 300 ft from home plate for a home run? If so, by how much will it clear the fence?

†73. A ball is thrown from a bridge 100 m high at an initial velocity of 30.0 m/s at an angle of 50.0° above the horizontal. Find (a) how high the ball goes, (b) the total time the ball is in the air, (c) the maximum horizontal distance that the ball travels, and (d) the velocity of the ball as it strikes the ground.

74. A ball is thrown at an angle of 35.5° below the horizontal at a speed of 22.5 m/s from a building 20.0 m high. (a) How long will it take for the ball to hit the ground below? (b) How far from the building will the ball land?

†75. Using the kinematic equations for the x- and y-components of the displacement, find the equation of the trajectory for two-dimensional projectile motion. Compare this equation with the equation for a parabola expressed in its standard form.

†76. Using the kinematic equations, prove that if two balls are released simultaneously from a table, one with zero velocity and the other with a horizontal velocity v_{0x}, they will both reach the ground at the same time.

Interactive Tutorials

77. A train accelerates from an initial velocity of 20.0 m/s to a final velocity of 35.0 m/s in 11.8 s. Find its acceleration and the distance the train travels in this time.

78. A ball is dropped from a building 50.0 m high. How long will it take the ball to hit the ground below and with what final velocity?

79. A golf ball is hit with an initial velocity $v_0 = 53.0$ m/s at an angle $\theta = 50.0°$ above the horizontal. (a) How high will the ball go? (b) What is the total time the ball is in the air? (c) How far will the ball travel horizontally before it hits the ground?

80. Instantaneous velocity. If the equation for the displacement x of a body is known, the average velocity throughout an interval can be computed by the formula

$$v_{avg} = (\Delta x)/(\Delta t)$$

The instantaneous velocity is defined as the limit of the average velocity as Δt approaches zero. That is,

$$v = \lim (\Delta x)/(\Delta t)$$
$$\Delta t \to 0$$

For an acceleration with a displacement given by $x = 0.5\,at^2$, use different values of Δt to see how the average velocity approaches the instantaneous velocity. Compare this to the velocity determined by the equation $v = at$, and determine the percentage error. Plot the average velocity, $(\Delta x)/(\Delta t)$, versus Δt.

81. Free-fall and generalized one-dimensional projectile motion. A projectile is fired from a height y_0 above the ground with an initial velocity v_0 in a vertical direction. Find (a) the time t_r for the projectile to rise to its maximum height, (b) the total time t_t the ball is in the air, (c) the maximum height y_{max} of the projectile, (d) the velocity v_g of the projectile as it strikes the ground, and (e) the location and velocity of the projectile at any time t. (f) Plot a picture of the motion as a function of time.

82. Generalized two-dimensional projectile motion. A projectile is fired from a height y_0 above the horizontal with an initial velocity v_0 at an angle θ. Find (a) the time t_r for the projectile to rise to its maximum height; (b) the total time t_t the ball is in the air; (c) the maximum distance the ball travels in the x-direction, x_{max} before it hits the ground; (d) the maximum height y_{max} of the projectile; (e) the velocity v_g of the projectile as it strikes the ground; and (f) the location and velocity of the projectile at any time t. (g) Plot a picture of the trajectory.

4

Newton's Laws of Motion

Artwork to celebrate the 300th anniversary of the publication of Isaac Newton's great work, Philosophical Naturalis Principia Mathematica.

I do not know what I may appear to the world; but to myself I seem to have been only like a boy playing on the sea shore, and diverting myself in now and then finding a smoother pebble or a prettier shell than ordinary, while the great ocean of truth lay all undiscovered before me.

Sir Isaac Newton

4.1 Introduction

Chapter 3 dealt with kinematics, the study of motion. We saw that if the acceleration, initial position, and velocity of a body are known, then the future position and velocity of the moving body can be completely described. But one of the things left out of that discussion, was the cause of the body's acceleration. If a piece of chalk is dropped, it is immediately accelerated downward. The chalk falls because the earth exerts a force of gravity on the chalk pulling it down toward the center of the earth. We will see that any time there is an acceleration, there is always a force present to cause that acceleration. In fact, it is Newton's laws of motion that describe what happens to a body when forces are acting on it. That branch of mechanics concerned with the forces that change or produce the motions of bodies is called **dynamics.**

As an example, suppose you get into your car and accelerate from rest to 50 mph. What causes that acceleration? The acceleration is caused by a force that begins with the car engine. The engine supplies a force, through a series of shafts and gears to the tires, that pushes backward on the road. The road in turn exerts a force on the car to push it forward. Without that force you would never be able to accelerate your car. Similarly, when you step on the brakes, you exert a force through the brake linings, to the wheels and tires of the car to the road. The road exerts a force backward on the car that causes the car to decelerate. All motions are started or stopped by forces.

Before we start our discussion of Newton's laws of motion, let us spend a few moments discussing the life of Sir Isaac Newton, perhaps the greatest scientist who ever lived. Newton was born in the little hamlet of Woolsthorpe in Lincolnshire, England, on Christmas day, 1642. It was about the same time that Galileo Galilei died; it was as though the torch of knowledge had been passed from one generation to another. Newton was born prematurely and was not expected to live; somehow he managed to survive. His father had died three months previously. Isaac grew up with a great curiosity about the things around him. His chief delight was to sit under a tree reading a book. His uncle, a member of Trinity College at Cambridge University, urged that the young Newton be sent to college, and Newton went to Cambridge in June, 1661. He spent the first two years at college learning arithmetic, Euclidean geometry, and trigonometry. He also read and listened to lectures on the Copernican system of astronomy. After that he studied natural philosophy. In 1665 the bubonic plague hit London and Newton returned to his mother's farm at Woolsthorpe. It was there, while observing an apple fall from a tree, that Newton wondered that if the pull of the earth can act through space to pull an apple from

(a)

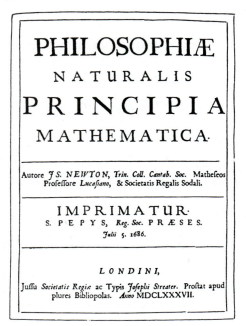

PHILOSOPHIÆ

NATURALIS

PRINCIPIA

MATHEMATICA.

Autore *JS. NEWTON,* Trin. Coll. Cantab. Soc. Mathefeos Profeffore *Lucafiano,* & Societatis Regalis Sodali.

IMPRIMATUR·
S. PEPYS, *Reg. Soc.* PRÆSES.
Julii 5. 1686.

LONDINI,

Juffu *Societatis Regiæ* ac Typis *Jofephi Streater.* Proftat apud plures Bibliopolas. *Anno* MDCLXXXVII.

(b)

Figure 4.1

(*a*) Sir Isaac Newton. (*b*) The first page of Newton's *Principia.*

a tree, could it not also reach out as far as the moon and pull the moon toward the earth? This reasoning became the basis for his law of universal gravitation.

Newton also invented the calculus (he called it fluxions) as a means of solving a problem in gravitation. (We should also note, however, that the German mathematician Gottfried Leibniz also invented the calculus independently of, and simultaneously with, Newton.) Newton's work on mechanics, gravity, and astronomy was published in 1687 as the *Mathematical Principles of Natural Philosophy.* It is commonly referred to as the *Principia,* from its Latin title. Because of its impact on science, it is perhaps one of the most important books ever written. A copy of the first page of the *Principia* is shown in figure 4.1. Newton died in London on March 20, 1727, at the age of 84.

4.2 Newton's First Law of Motion

Newton's first law of motion can be stated as: *A body at rest, will remain at rest and a body in motion at a constant velocity will continue in motion at that constant velocity, unless acted on by some unbalanced external force.* By a **force** we mean a push or a pull that acts on a body. A more sophisticated definition of force will be given after the discussion of Newton's second law.

There are really two statements in the first law. The first statement says that a body at rest will remain at rest unless acted on by some unbalanced force. As an example of this first statement, suppose you placed a book on the desk. That book would remain there forever, unless some unbalanced force moved it. That is, you might exert a force to pick up the book and move it someplace else. But if neither you nor anything else exerts a force on that book, that book will stay there forever. Books, and other inanimate objects, do not just jump up and fly around the room by themselves. A body at rest remains at rest and will stay in that position forever unless acted on by some unbalanced external force. This law is really a simple observation of nature. This is the first part of Newton's first law and it is so basic that it almost seems trivial and unnecessary.

The second part of the statement of Newton's first law is not quite so easy to see. This part states that a body in motion at a constant velocity will continue to move at that constant velocity unless acted on by some unbalanced external force. In fact, at first observation it actually seems to be wrong. For example, if you take this book and give it a shove along the desk, you immediately see that it does not keep on moving forever. In fact, it comes to a stop very quickly. So either Newton's law is wrong or there must be some force acting on the book while it is in motion along the desk. In fact there is a force acting on the book and this force is the force of friction, which tends to oppose the motion of one body sliding on another. (We will go into more details on friction later in this chapter.) But, if instead of trying to slide the book along the desk, we tried to slide it along a sheet of ice (say on a frozen lake), then the book would move a much greater distance before coming to rest. The frictional force acting on the book by the ice is much less than the frictional force that acted on the book by the desk. But there is still a force, regardless of how small, and the book eventually comes to rest. However, we can imagine that in the limiting case where these frictional forces are completely eliminated, an object moving at a constant velocity would continue to move at that same velocity forever, unless it were acted on by a nonzero net force. *The resistance of a body to a change in its motion is called* **inertia,** *and Newton's first law is also called the law of inertia.*

If you were in outer space and were to take an object and throw it away where no forces acted on it, it would continue to move at a constant velocity. Yet if you take your pen and try to throw it into space, it falls to the floor. Why? Because the force of gravity pulls on it and accelerates it to the ground. It is not free to move in straight line motion but instead follows a parabolic trajectory, as we have seen in the study of projectiles.

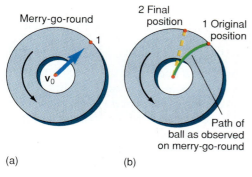

Figure 4.2

A noninertial coordinate system.

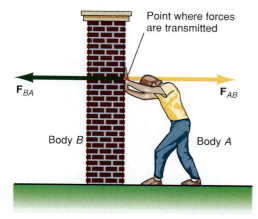

Figure 4.3

A merry-go-round is a noninertial coordinate system.

Figure 4.4

Forces involved when you lean against a wall.

The first part of Newton's first law—A body at rest, will remain at rest . . .—is really a special case of the second statement—a body in motion at some constant velocity. . . . A body at rest has zero velocity, and will therefore have that same zero velocity forever, unless acted on by some unbalanced external force.

Newton's first law of motion also defines what is called an inertial coordinate system. A coordinate system in which objects experiencing no unbalanced forces remain at rest or continue in uniform motion, is called an inertial coordinate system. *An **inertial coordinate system** (also called an inertial reference system) is a coordinate system that is either at rest or moving at a constant velocity with respect to another coordinate system that is either at rest or also moving at a constant velocity.* In such a coordinate system the first law of motion holds. A good way to understand an inertial coordinate system is to look at a noninertial coordinate system. A rotating coordinate system is an example of a noninertial coordinate system. Suppose you were to stand at rest at the center of a merry-go-round and throw a ball to another student who is on the outside of the rotating merry-go-round at the position 1 in figure 4.2(a). When the ball leaves your hand it is moving at a constant horizontal velocity, v_0. Remember that a velocity is a vector, that is, it has both magnitude and direction. The ball is moving at a constant horizontal speed in a constant direction. The y-component of the velocity changes because of gravity, but not the x-component. You, being at rest at the center, are in an inertial coordinate system. The person on the rotating merry-go-round is rotating and is in a noninertial coordinate system. As observed by you, at rest at the center of the merry-go-round, the ball moves through space at a constant horizontal velocity. But the person standing on the outside of the merry-go-round sees the ball start out toward her, but then it appears to be deflected to the right of its original path, as seen in figure 4.2(b). Thus, the person on the merry-go-round does not see the ball moving at a constant horizontal velocity, even though you, at the center, do, because she is rotating away from her original position. That student sees the ball changing its direction throughout its flight and the ball appears to be deflected to the right of its path. The person on the rotating merry-go-round is in a noninertial coordinate system and Newton's first law does not hold in such a coordinate system. That is, the ball in motion at a constant horizontal velocity does not appear to continue in motion at that same horizontal velocity. Thus, when Newton's first law is applied it must be done in an inertial coordinate system. In this book nearly all coordinate systems will be either inertial coordinate systems or ones that can be approximated by inertial coordinate systems, hence Newton's first law will be valid. The earth is technically not an inertial coordinate system because of its rotation about its axis and its revolution about the sun. The acceleration caused by the rotation about its axis is only about 1/300 of the acceleration caused by gravity, whereas the acceleration due to its orbital revolution is about 1/1650 of the acceleration due to gravity. Hence, as a first approximation, the earth can usually be used as an inertial coordinate system.

Before discussing the second law, let us first discuss Newton's third law because its discussion is somewhat shorter than the second.

4.3 Newton's Third Law of Motion

Newton stated his third law in the succinct form, "Every action has an equal but opposite reaction." Let us express **Newton's third law of motion** in the form, *if there are two bodies, A and B, and if body A exerts a force on body B, then body B will exert an equal but opposite force on body A.* The first thing to observe in Newton's third law is that two bodies are under consideration, body A and body B. This contrasts to the first (and second) law, which apply to a single body. As an example of the third law, consider the case of a person leaning against the wall, as shown in figure 4.4. The person is body A, the wall is body B. The person is exerting a force on the wall, and Newton's third law states that the wall is exerting an equal but opposite force on the person.

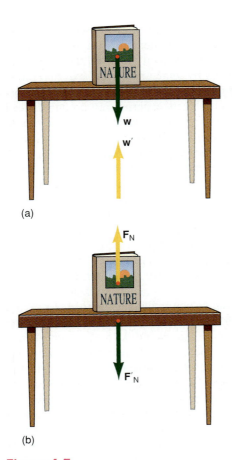

(a)

(b)

Figure 4.5

Newton's third law of motion.

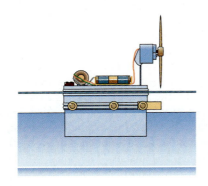

Figure 4.6

Glider and airplane motor.

The key to Newton's third law is that there are two different bodies exerting two equal but opposite forces on each other. Stated mathematically this becomes

$$\mathbf{F}_{AB} = -\mathbf{F}_{BA} \tag{4.1}$$

where $\mathbf{F}_{AB}$ is the force on body A exerted by body B and $\mathbf{F}_{BA}$ is the force on body B exerted by body A. Equation 4.1 says that all forces in nature exist in pairs. There is no such thing as a single isolated force. We call $\mathbf{F}_{BA}$ the *action force,* whereas we call $\mathbf{F}_{AB}$ the *reaction force* (although either force can be called the action or reaction force). Together these forces are an *action-reaction pair.*

Another example of the application of Newton's third law is a book resting on a table, as seen in figure 4.5. A gravitational force, directed toward the center of the earth, acts on that book. We call the gravitational force on the book its weight **w.** By Newton's third law there is an equal but opposite force **w'** acting on the earth. The forces **w** and **w'** are the action and reaction pair of Newton's third law, and note how they act on two different bodies, the book and the earth. The force **w** acting on the book should cause it to fall toward the earth. However, because the table is in the way, the force down on the book is applied to the table. Hence the book exerts a force down on the table. We label this force on the table, $\mathbf{F}'_{N}$. By Newton's third law the table exerts an equal but opposite force upward on the book. We call the equal but upward force acting on the book the *normal force,* and designate it as $\mathbf{F}_{N}$. *When used in this context, normal means perpendicular to the surface.*

If we are interested in the forces acting on the book, they are the gravitational force, which we call the weight **w,** and the normal force $\mathbf{F}_{N}$. Note however, that these two forces are not an action-reaction pair because they act on the same body, namely the book.

We will discuss Newton's third law in more detail when we consider the law of conservation of momentum in chapter 8.

4.4 Newton's Second Law of Motion

Newton's second law of motion is perhaps the most basic, if not the most important, law of all of physics. We begin our discussion of Newton's second law by noting that whenever an object is dropped, the object is accelerated down toward the earth. We know that there is a force acting on the body, a force called the force of gravity. The force of gravity appears to be the cause of the acceleration downward. We therefore ask the question, *Do all forces cause accelerations? And if so, what is the relation of the acceleration to the causal force?*

Experimental Determination of Newton's Second Law

To investigate the relation between forces and acceleration, we will go into the laboratory and perform an experiment with a propelled glider on an air track, as seen in figure 4.6.[1]

We turn a switch on the glider to apply 0.2 V to the airplane motor mounted on top of the glider. As the propeller turns, it exerts a force on the glider that pulls the glider down the track. We turn on a spark timer, giving a record of the position of the glider as a function of time. From the spark timer tape, we determine the acceleration of the glider as in the experiment on kinematics. We then connect a piece of Mylar tape to the back of the glider and pass it over an air pulley at the end of the track. Weights are hung from the Mylar tape until the force exerted by the weights is equal to the force exerted by the propeller. The glider will then be at rest. In this way, we determine the force exerted by the propeller. This procedure is repeated several times with battery voltages of 0.4, 0.6, 0.8, and 1.0 V. If we plot the acceleration of the glider against the force, we get the result shown in figure 4.7.

---

1. See Nolan and Bigliani, *Experiments in Physics,* 2d ed., Wm. C. Brown Publishers, Dubuque, Iowa.

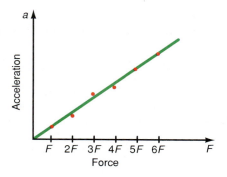

Figure 4.7

Plot of acceleration *a* versus the applied force *F* for a propelled glider.

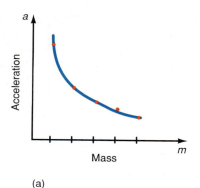

(a)

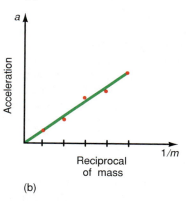

(b)

Figure 4.8

Plot of (*a*) the acceleration *a* versus the mass *m* and (*b*) the acceleration *a* versus the reciprocal of the mass (1/*m*) for the propelled gliders.

Whenever a graph of two variables is a straight line, as in figure 4.7, the dependent variable is directly proportional to the independent variable. (See appendix C for a discussion of proportions.) Therefore this graph tells us that the acceleration of the glider is directly proportional to the applied force, that is,

$$a \propto F \tag{4.2}$$

Thus, not only does a force cause an acceleration of a body but that acceleration is directly proportional to that force, and in the direction of that force. That is, if we double the force, we double the acceleration; if we triple the force, we triple the acceleration; and so forth.

Let us now ask, how is the acceleration affected by the mass of the object being moved? To answer this question we go back to the laboratory and our experiment. This time we connect together two gliders of known mass and place them on the air track. Hence, the mass of the body in motion is increased. We turn on the propeller and the gliders go down the air track with the spark timer again turned on. Then we analyze the spark timer tape to determine the acceleration of the two gliders. We repeat the experiment with three gliders and then with four gliders, all of known mass. We determine the acceleration for each increased mass and plot the acceleration of the gliders versus the mass of the gliders, as shown in figure 4.8(a). The relation between acceleration and mass is not particularly obvious from this graph except that as the mass gets larger, the acceleration gets smaller, which suggests that the acceleration may be related to the reciprocal of the mass. We then plot the acceleration against the reciprocal of the mass in figure 4.8(b), and obtain a straight line.

Again notice the linear relation. This time, however, the acceleration is directly proportional to the reciprocal of the mass. Or saying it another way, the acceleration is inversely proportional to the mass of the moving object. (See appendix C for discussion of inverse proportions.) That is,

$$a \propto \frac{1}{m} \tag{4.3}$$

Thus, the greater the mass of a body, the smaller will be its acceleration for a given force. *Hence, the mass of a body is a measure of the body's resistance to being put into accelerated motion.* Equations 4.2 and 4.3 can be combined into a single proportionality, namely

$$a \propto \frac{F}{m} \tag{4.4}$$

The result of this experiment shows that the acceleration of a body is directly proportional to the applied force and inversely proportional to the mass of the moving body. The proportionality in relation 4.4 can be rewritten as an equation if a constant of proportionality *k* is introduced (see the appendix on proportions). Thus,

$$F = kma \tag{4.5}$$

Let us now define the unit of force in such a way that *k* will be equal to the value one, thereby simplifying the equation. The unit of force in SI units, thus defined, is

$$1 \text{ newton} = 1 \text{ kg} \frac{\text{m}}{\text{s}^2}$$

The abbreviation for a newton is the capital letter N. *A newton is the net amount of force required to give a mass of 1 kg an acceleration of 1 m/s².* Hence, *force is now defined as more than a push or a pull, but rather a force is a quantity that causes a body of mass m to have an acceleration a.* Recall from chapter 1 that the mass of an object is a fundamental quantity. We now see that force is a derived quantity. It is derived from the fundamental quantities of mass in kilograms, length in meters, and time in seconds.

A check on dimensions shows that k is indeed equal to unity in this way of defining force, that is,

$$F = kma$$

$$\text{newton} = (k) \text{ kg m/s}^2$$

$$\text{kg m/s}^2 = (k) \text{ kg m/s}^2$$

$$k = 1$$

Equation 4.5 therefore becomes

$$F = ma \tag{4.6}$$

Equation 4.6 is the mathematical statement of Newton's second law of motion. This is perhaps the most fundamental of all the laws of classical physics. **Newton's second law of motion** can be stated in words as: *If an unbalanced external force F acts on a body of mass m, it will give that body an acceleration a. The acceleration is directly proportional to the applied force and inversely proportional to the mass of the body.* We must understand by Newton's second law that the force F is the resultant external force acting on the body. Sometimes, to be more explicit, Newton's second law is written in the form

$$\Sigma F = ma \tag{4.7}$$

where the greek letter sigma, Σ, means "the sum of." Thus, if there is more than one force acting on a body, it is the resultant unbalanced force that causes the body to be accelerated. For example, if a book is placed on a table as in figure 4.5, the forces acting on the book are the force of gravity pulling the book down toward the earth, while the table exerts a normal force upward on the book. These forces are equal and opposite, so that the resultant unbalanced force acting on the book is zero. Hence, even though forces act on the book, the resultant of these forces is zero and there is no acceleration of the book. It remains on the table at rest.

Newton's second law is the fundamental principle that relates forces to motions, and is the foundation of mechanics. Thus, if an unbalanced force acts on a body, it will give it an acceleration. In particular, the acceleration is found from equation 4.7 to be

$$a = \frac{\Sigma F}{m} \tag{4.8}$$

It is a matter of practice that Σ is usually left out of the equations but do not forget it; it is always implied because it is the resultant force that causes the acceleration.

Once the acceleration of the body is known, its future position and velocity at any time can be determined using the kinematic equations developed in chapter 3, namely,

$$x = v_0 t + \tfrac{1}{2}at^2 \tag{3.14}$$

$$v = v_0 + at \tag{3.10}$$

and

$$v^2 = v_0^2 + 2ax \tag{3.16}$$

provided, of course, that the force, and therefore the acceleration, are constant. When the force and acceleration are not constant, more advanced mathematical techniques are required.

Our determination of Newton's second law has been based on the experimental work performed on the air track. Since the air track is one dimensional, the equations have been written in their one dimensional form. However, recall that acceleration is a vector quantity and therefore force, which is equal to that acceleration times mass, must also be written as a vector quantity. Newton's second law should therefore be written in the more general vector form as

$$\mathbf{F} = m\mathbf{a} \tag{4.9}$$

The kinematic equations must also be used in their vector form.

Newton's First Law of Motion Is Consistent with His Second Law of Motion

Newton's first law of motion can be shown to be consistent with his second law of motion in the following manner. Let us start with Newton's second law

$$\mathbf{F} = m\mathbf{a} \tag{4.9}$$

However, the acceleration is defined as the change in velocity with time. Thus,

$$\mathbf{F} = m\mathbf{a} = m\frac{\Delta \mathbf{v}}{\Delta t}$$

If there is no resultant force acting on the body, then $\mathbf{F} = 0$. Hence,

$$0 = m\frac{\Delta \mathbf{v}}{\Delta t}$$

and therefore

$$\Delta \mathbf{v} = 0 \tag{4.10}$$

which says that there is no change in the velocity of a body if there is no resultant applied force. Another way to see this is to note that

$$\Delta \mathbf{v} = \mathbf{v_f} - \mathbf{v_0} = 0$$

Hence,

$$\mathbf{v_f} = \mathbf{v_0} \tag{4.11}$$

That is, if there is no applied force ($\mathbf{F} = 0$), then the final velocity $\mathbf{v_f}$ is always equal to the original velocity $\mathbf{v_0}$. But that in essence is the first law of motion—a body in motion at a constant velocity will continue in motion at that same constant velocity, unless acted on by some unbalanced external force.

Also note that the first part of the first law, a body at rest will remain at rest unless acted on by some unbalanced external force, is the special case of $\mathbf{v_0} = 0$. That is,

$$\mathbf{v_f} = \mathbf{v_0} = 0$$

indicates that if a body is initially at rest ($\mathbf{v_0} = 0$), then at any later time its final velocity is still zero ($\mathbf{v_f} = \mathbf{v_0} = 0$), and the body will remain at rest as long as $\mathbf{F}$ is equal to zero. Thus, the first law, in addition to defining an inertial coordinate system, is also consistent with Newton's second law.

The ancient Greeks knew that a body at rest under no forces would remain at rest. And they knew that by applying a force to the body they could set it into motion. However, they erroneously assumed that the force had to be exerted continuously in order to keep the body in motion. Galileo was the first to show that this is not true, and Newton showed in his second law that the net force is necessary only to start the body into motion, that is, to accelerate it from rest to a velocity $\mathbf{v}$. Once it is moving at the velocity $\mathbf{v}$, the net force can be removed and the body will continue in motion at that same velocity $\mathbf{v}$.

Figure 4.9

Motion of a block on a smooth horizontal surface.

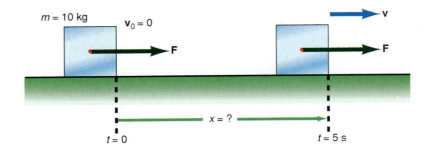

An Example of Newton's Second Law

Example 4.1

Motion of a block on a smooth horizontal surface. A 10.0-kg block is placed on a smooth horizontal table, as shown in figure 4.9. A horizontal force of 6.00 N is applied to the block. Find (a) the acceleration of the block, (b) the position of the block at $t = 5.00$ s, and (c) the velocity of the block at $t = 5.00$ s.

Solution

a. First we draw the forces acting on the block as in the diagram. The statement that the table is smooth implies that there is only a negligible frictional force between the block and the table and it can be ignored. The only unbalanced force[2] acting on the block is the force F, and the acceleration is immediately found from Newton's second law as

$$a = \frac{F}{m} = \frac{6.00 \text{ N}}{10.0 \text{ kg}} = 0.600 \, \frac{\text{kg m/s}^2}{\text{kg}}$$

$$= 0.600 \text{ m/s}^2$$

Note here that this acceleration takes place only as long as the force is applied. If the force is removed, for any reason, then the acceleration becomes zero, and the block continues to move with whatever velocity it had at the time that the force was removed.

b. Now that the acceleration of the block is known, its position at any time can be found using the kinematic equations developed in chapter 3, namely,

$$x = v_0 t + \tfrac{1}{2} a t^2 \tag{3.14}$$

But because the block is initially at rest $v_0 = 0$,

$$x = \tfrac{1}{2} a t^2 = \tfrac{1}{2} (0.600 \text{ m/s}^2)(5.00 \text{ s})^2$$

$$= 7.50 \text{ m}$$

c. The velocity at the end of 5.00 s, found from equation 3.10, is

$$v = v_0 + at$$

$$= 0 + (0.600 \text{ m/s}^2)(5.00 \text{ s})$$

$$= 3.00 \text{ m/s}$$

In summary, we see that Newton's second law tells us the acceleration imparted to a body because of the forces acting on it. Once this acceleration is known, the position and velocity of the body at any time can be determined by using the kinematic equations.

---

2. Note that there are two other forces acting on the block. One is the weight **w** of the block, which acts downward, and the other is the normal force $\mathbf{F_N}$ that the table exerts upward on the block. However, these forces are balanced and do not cause an acceleration of the block.

Special Case of Newton's Second Law—The Weight of a Body Near the Surface of the Earth

Newton's second law tells us that if an unbalanced force acts on a body of mass m, it will give it an acceleration a. Let the body be a pencil that you hold in your hand. Newton's second law says that if there is an unbalanced force acting on this pencil, it will receive an acceleration. If you let go of the pencil it immediately falls down to the surface of the earth. It is an object in free-fall and, as we have seen, an object in free-fall has an acceleration whose magnitude is g. That is, if Newton's second law is applied to the pencil

$$F = ma$$

But the acceleration a is the acceleration due to gravity, and its magnitude is g. Therefore, Newton's second law can be written as

$$F = mg$$

But this gravitational force pulling an object down toward the earth is called the weight of the body, and its magnitude is w. Hence,

$$F = w$$

and Newton's second law becomes

$$w = mg \qquad\qquad (4.12)$$

Equation 4.12 thus gives us a relationship between the mass of a body and the weight of a body.

Example 4.2

Finding the weight of a mass. Find the weight of a 1.00-kg mass.

Solution

The weight of a 1.00-kg mass, found from equation 4.12, is

$$w = mg = (1.00 \text{ kg})(9.80 \text{ m/s}^2) = 9.80 \text{ kg m/s}^2$$

$$= 9.80 \text{ N}$$

A mass of 1 kg has a weight of 9.80 N.

In pointing out the distinction between the weight of an object and the mass of an object in chapter 1, we said that a woman on the moon would weigh one-sixth of her weight on the earth. We can now see why. The acceleration due to gravity on the moon g_m is only about one-sixth of the acceleration due to gravity here on the surface of the earth g_E. That is,

$$g_m = \tfrac{1}{6} g_E$$

Hence, the weight of a woman on the moon would be

$$w_m = mg_m = m(\tfrac{1}{6} g_E) = \tfrac{1}{6}(mg_E) = \tfrac{1}{6} w_E$$

The weight of a woman on the moon would be one-sixth of her weight here on the earth. The mass of the woman would be the same on the earth as on the moon, but her weight would be different.

We can see from equations 4.6 and 4.12 that the weight of a body in SI units should be expressed in terms of newtons. And in the scientific community it is. However, the business community does not always follow science. The United

States is now switching over to SI units, but instead of expressing weights in newtons, as defined, the weights of objects are erroneously being expressed in terms of kilograms, a unit of mass.

As an example, if you go to the supermarket and buy a 16 oz can of beans (weight = 1 lb), you will see stamped on the can

NET WT 16 oz (1 lb) 0.453 kg

This is really a mistake, as we now know, because we know that there is a difference between the weight and the mass of a body. To get around this problem, a physics student should realize that in commercial and everyday use, the word "weight" nearly always means mass. So when you buy something that the businessman says weighs 1 kg, he means that it has the weight of a 1-kg mass. We have seen that the weight of a 1-kg mass is 9.80 N. *In this text the word kilogram will always mean mass, and only mass.* If however, you come across any item marked as a weight and expressed in kilograms in your everyday life, you can convert that mass to its proper weight in newtons by simply multiplying the mass by 9.80 m/s².

Example 4.3

Weight and mass at the supermarket. While at the supermarket you buy a bag of potatoes labeled, NET WT 5.00 kg. What is the *correct* weight expressed in newtons?

Solution

We find the weight in newtons by multiplying the mass in kg by 9.80 m/s². Hence,

$$w = (5.00 \text{ kg})(9.80 \text{ m/s}^2) = 49.0 \text{ N}$$

The Mass of an Object in the British Engineering System

In the British engineering system, weight is a fundamental quantity and mass a derived quantity. We can determine the mass from equation 4.12 by solving for m:

$$m = \frac{w}{g} \tag{4.13}$$

Thus, the mass of a body in the British engineering system is simply the weight of that body divided by the acceleration due to gravity. The unit of mass in the British engineering system is derived as

$$m = \frac{w}{g} = \frac{\text{lb}}{\text{ft/s}^2}$$

This unit of mass is defined as the slug, where

$$1 \text{ slug} = 1 \frac{\text{lb}}{\text{ft/s}^2}$$

The slug is defined as that amount of mass that receives an acceleration of 1 ft/s² when a force of 1 pound acts on it.

Example 4.4

Mass in the British engineering system. A man weighs 160 lb. What is his mass?

From equation 4.13 his mass is

$$m = \frac{w}{g} = \frac{160 \text{ lb}}{32.0 \text{ ft/s}^2} = 5.00 \text{ slugs}$$

If we want to use Newton's second law in the British engineering system we must express the mass of the body in slugs.

Example 4.5

Newton's second law in the British engineering system. A force of 1000 lb is used to move a car weighing 3200 lb. What is the acceleration of the car?

The acceleration is found from Newton's second law as

$$a = \frac{F}{m}$$

but the mass must first be found as

$$m = \frac{w}{g} = \frac{3200 \text{ lb}}{32.0 \text{ ft/s}^2} = 100 \text{ slugs}$$

Hence,

$$a = \frac{F}{m} = \frac{1000 \text{ lb}}{100 \text{ slugs}} = 10.0 \frac{\text{lb}}{\frac{\text{lb}}{\text{ft/s}^2}}$$

$$= 10.0 \text{ ft/s}^2$$

Because $m = w/g$ some engineering books even write Newton's second law in the form

$$F = \frac{w}{g} a \tag{4.14}$$

when dealing with the British engineering system of units. This amounts to the same thing but avoids having to use the unit of slugs directly.

4.5 Applications of Newton's Second Law

A Block on a Frictionless Inclined Plane

Let us find the acceleration of a block that is to slide down a frictionless inclined plane. (The statement that the plane is frictionless means that it is not necessary to take into account the effects of friction on the motion of the block.) The velocity and the displacement of the block at any time can then be found from the kinematic equations. (Note that this problem is equivalent to placing a glider on the tilted air track in the laboratory.) The first thing to do is to draw a diagram of all the forces acting on the block, as shown in figure 4.10. A diagram showing all the forces acting on a body is called a *force diagram* or a *free-body diagram*. Note that all the forces are drawn as if they were acting at the geometrical center of the body. (The reason

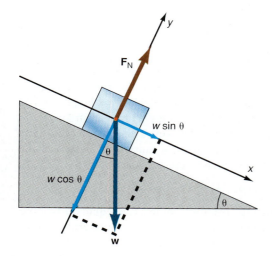

Figure 4.10

A block on a frictionless inclined plane.

for this will be discussed in more detail later when we study the center of mass of a body, but for now we will just say that the body moves as if all the forces were acting at the center of the body.)

The first force we consider is the weight of the body **w,** which acts down toward the center of the earth and is hence perpendicular to the base of the incline. The plane itself exerts a force upward on the block that we denote by the symbol F_N, and call the normal force. (Recall that a normal force is, by definition, a force that is always perpendicular to the surface.)

Let us now introduce a set of axes that are parallel and perpendicular to the plane, as shown in figure 4.10. Thus the parallel axis is the x-axis and lies in the direction of the motion, namely down the plane. The y-axis is perpendicular to the inclined plane, and points upward away from the plane. Take the weight of the block and resolve it into components, one parallel to the plane and one perpendicular to the plane. Recall from chapter 2, on the components of vectors, that if the plane makes an angle θ with the horizontal, then the acute angle between **w** and the perpendicular to the plane is also the angle θ. Hence, the component of **w** parallel to the plane $w_\parallel$ is

$$w_\parallel = w \sin \theta \qquad (4.15)$$

whereas the component perpendicular to the plane $w_\perp$ is

$$w_\perp = w \cos \theta \qquad (4.16)$$

as can be seen in figure 4.10. One component of the weight, namely $w \cos \theta$, holds the block against the plane, while the other component, $w \sin \theta$, is the force that acts on the block causing the block to accelerate down the plane. To find the acceleration of the block down the plane, we use Newton's second law,

$$F = ma \qquad (4.6)$$

The force acting on the block to cause the acceleration is given by equation 4.15. Hence,

$$w \sin \theta = ma \qquad (4.17)$$

But by equation 4.12

$$w = mg \qquad (4.12)$$

Substituting this into equation 4.17 gives

$$mg \sin \theta = ma$$

Because the mass is contained on both sides of the equation, it divides out, leaving

$$a = g \sin \theta \qquad (4.18)$$

as the acceleration of the block down a frictionless inclined plane. An interesting thing about this result is that equation 4.18 does not contain the mass m. That is, the acceleration down the plane is the same, whether the block has a large mass or a small mass. The acceleration is thus independent of mass. This is similar to the case of the freely falling body. There, a body fell at the same acceleration regardless of its mass. Hence, both accelerations are independent of mass. If the angle of the inclined plane is increased to 90°, then the acceleration becomes

$$a = g \sin \theta = g \sin 90° = g (1) = g.$$

Therefore, at $\theta = 90°$ the block goes into free-fall. When θ is equal to 0°, the acceleration is zero. We can use the inclined plane to obtain any acceleration from zero up to the acceleration due to gravity g, by simply changing the angle θ. Notice that the algebraic solution to a problem gives a formula rather than a number for the answer. One of the reasons why algebraic solutions to problems are superior to numerical ones is that we can examine what happens at the extremes (for example at 90° or 0°) to see if they make physical sense, and many times special cases can be considered.

Galileo used the inclined plane extensively to study motion. Since he did not have good devices available to him for measuring time, it was difficult for him to study the velocity and acceleration of a body. By using the inclined plane at relatively small angles of θ, however, he was able to slow down the motion so that he could more easily measure it.

Because we now know the acceleration of the block down the plane, we can determine its velocity and position at any time, or its velocity at any position, using the kinematic equations of chapter 3. However, now the acceleration a is determined from equation 4.18.

Note also in this discussion that if Newton's second law is applied to the perpendicular component we obtain

$$F_\perp = ma_\perp = 0$$

because there is no acceleration perpendicular to the plane. Hence,

$$F_\perp = F_N - w \cos \theta = 0$$

and

$$F_N = w \cos \theta \tag{4.19}$$

Example 4.6

A block sliding down a frictionless inclined plane. A 10.0-kg block is placed on a frictionless inclined plane, 5.00 m long, that makes an angle of 30.0° with the horizontal. If the block starts from rest at the top of the plane, what will its velocity be at the bottom of the incline?

Solution

The velocity of the block at the bottom of the plane is found from the kinematic equation

$$v^2 = v_0^2 + 2ax$$

Hence,

$$v = \sqrt{2ax}$$

Before solving for v, we must first determine the acceleration a. Using Newton's second law we obtain

$$a = \frac{F}{m} = \frac{w \sin \theta}{m} = \frac{mg \sin \theta}{m}$$

$$= g \sin \theta = (9.80 \text{ m/s}^2) \sin 30.0°$$

$$= 4.90 \text{ m/s}^2$$

Hence,

$$v = \sqrt{2ax} = \sqrt{2(4.90 \text{ m/s}^2)(5.00 \text{ m})}$$

$$= 7.00 \text{ m/s}$$

The velocity of the block at the bottom of the plane is 7.00 m/s in a direction pointing down the inclined plane.

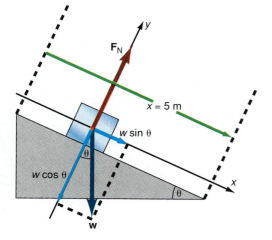

Figure 4.11

Diagram for example 4.6.

It is perhaps appropriate here to discuss the different concepts of mass. In chapter 1, we gave a very simplified definition of mass by saying that mass is a measure of the amount of matter in a body. We picked a certain amount of matter, called it a standard, and gave it the name kilogram. This amount of matter was not placed into motion. It was just the amount of matter in a platinum-iridium cylinder 39 mm in diameter and 39 mm high. The amount of matter in any other body was then compared to this standard kilogram mass. But this comparison was made by placing the different pieces of matter on a balance scale. As pointed out in chapter 1, the balance can be used to show an equality of the amount of matter in a body only because the gravitational force exerts a force downward on each pan of the balance. *Mass determined in this way is actually a measure of the gravitational force on that amount of matter, and hence mass measured on a balance is called gravitational mass.*

In the experimental determination of Newton's second law using the propeller glider, we added additional gliders to the air track to increase the mass that was in motion. The acceleration of the combined gliders was determined as a function of their mass and we observed that *the acceleration was inversely proportional to that mass. Thus, mass used in this way represents the resistance of matter to be placed into motion.* For a person, it would be more difficult to give the same acceleration to a very large mass of matter than to a very small mass of matter. *This characteristic of matter, whereby it resists motion is called inertia. The resistance of a body to be set into motion is called the* **inertial mass** *of that body.* Hence, in Newton's second law,

$$\mathbf{F} = m\mathbf{a} \tag{4.9}$$

the mass m stands for the inertial mass of the body. Just as we can determine the gravitational mass of any body in terms of the standard mass of 1 kg using a balance, we can determine the inertial mass of any body in terms of the standard mass of 1 kg using Newton's second law.

As an example, let us go back into the laboratory and use the propelled glider we used early in section 4.4. For a given battery voltage the glider has a constant force acting on the glider. For a glider of mass m_1, the force causes the glider to have an acceleration a_1, which can be represented by Newton's second law as

$$F = m_1 a_1 \tag{4.20}$$

If a new glider of mass m_2 is used with the same battery setting, and thus the same force F, the glider m_2 will experience the acceleration a_2. We can also represent this by Newton's second law as

$$F = m_2 a_2 \tag{4.21}$$

Because the force is the same in equations 4.20 and 4.21, the two equations can be set equal to each other giving

$$m_2 a_2 = m_1 a_1$$

Solving for m_2, we get

$$m_2 = \frac{a_1}{a_2} m_1 \tag{4.22}$$

Thus, the inertial mass of any body can be determined in terms of a mass m_1 and the ratio of the accelerations of the two masses. If the mass m_1 is taken to be the 1-kg mass of matter that we took as our standard, then the mass of any body can be determined inertially in this way. Equation 4.22 defines the inertial mass of a body.

Example 4.7

Finding the inertial mass of a body. A 1.00-kg mass experiences an acceleration of 3.00 m/s² when acted on by a certain force. A second mass experiences an acceleration of 8.00 m/s² when acted on by the same force. What is the value of the second mass?

Solution

The value of the second mass, found from equation 4.22, is

$$m_2 = \frac{a_1}{a_2} m_1$$

$$= \frac{3.00 \text{ m/s}^2}{8.00 \text{ m/s}^2} (1 \text{ kg})$$

$$= 0.375 \text{ kg}$$

Masses measured by the gravitational force can be denoted as m_g, while masses measured by their resistance to motion (i.e., inertial masses) can be represented as m_i. Then, for the motion of a block down the frictionless inclined plane, equation 4.17,

$$w \sin \theta = ma$$

should be changed as follows. The weight of the mass in equation 4.17 is determined in terms of a gravitational mass, and is written as

$$w = m_g g \qquad (4.23)$$

whereas the mass in Newton's second law is written in terms of the inertial mass m_i. Hence, equation 4.17 becomes

$$m_g g \sin \theta = m_i a \qquad (4.24)$$

It is, however, a fact of experiment that no differences have been found in the two masses even though they are determined differently. That is, experiments performed by Newton could detect no differences between gravitational and inertial masses. Experiments carried out by Roland von Eötvös (1848–1919) in 1890 showed that the relative difference between inertial and gravitational mass is at most 10^{-9}, and Robert H. Dicke found in 1961 the difference could be at most 10^{-11}. That is, the differences between the two masses are

$$m_i - m_g \leq 0.000000001 \text{ kg} \qquad \text{(Eötvös)},$$

$$m_i - m_g \leq 0.00000000001 \text{ kg} \qquad \text{(Dicke)}.$$

Hence, as best as can be determined,

$$m_i = m_g \qquad (4.25)$$

Because of this equivalence between the two different characteristics of mass, the masses on each side of equation 4.24 divide out, giving us the previously found relation, $a = g \sin \theta$. Since a freely falling body is the special case of a body on a 90° inclined plane, the equivalence of these two types of masses is the reason that all objects fall at the same acceleration g near the surface of the earth. This equivalence of gravitational and inertial mass led Einstein to propose it as a general principle called the equivalence principle of which more is said in chapter 30 when general relativity is discussed.

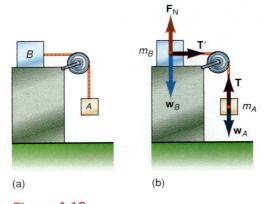

(a)　　　　　(b)

Figure 4.12
Combined motion.

Combined Motion of a Block on a Frictionless Horizontal Plane and a Block Falling Vertically

Let us now find the acceleration of a block, on a smooth horizontal table, that is connected by a cord that passes over a pulley to another block that is hanging over the end of the table, as shown in figure 4.12(a). By a smooth table, we mean there is a negligible frictional force between the block and the table so that the block will move freely over the table. We also assume that the mass of the connecting cord and pulley is negligible and can be ignored in this problem.

We want to find the motion of the blocks. In other words, what is the acceleration of the blocks, and their velocity and position at any time? The two blocks, taken together, are sometimes called a system. To determine the acceleration, we will use Newton's second law. However, before we can do so, we must draw a very careful free-body diagram showing all the forces that are acting on the two blocks, as is done in figure 4.12(b). The forces acting on block A are its weight $\mathbf{w}_A$, pulling it downward, and the tension $\mathbf{T}$ in the cord. It is this tension $\mathbf{T}$ in the cord that restrains block A from falling freely. The forces acting on body B are its weight $\mathbf{w}_B$, the normal force $\mathbf{F}_N$ that the table exerts on block B, and the tension $\mathbf{T}'$ in the cord that acts to pull block B toward the right. Newton's second law, applied to block A, gives

$$\mathbf{F} = m_A\mathbf{a}$$

Here $\mathbf{F}$ is the total resultant force acting on block A and therefore,

$$\mathbf{F} = \mathbf{T} + \mathbf{w}_A = m_A\mathbf{a} \tag{4.26}$$

Equation 4.26 is a vector equation. To simplify its solution, we use our previous convention with vectors in one dimension. That is, the upward direction $(+y)$ is taken as positive and the downward direction $(-y)$ as negative. Therefore, equation 4.26 can be simplified to

$$T - w_A = -m_A a \tag{4.27}$$

However, we can not yet solve equation 4.27 for the acceleration, because the tension T in the cord is unknown. We obviously need more information. We have one equation with two unknowns, the acceleration a and the tension T. *Whenever we want to solve a system of algebraic equations for some unknowns, we must always have as many equations as there are unknowns in order to obtain a solution.* Since there are two unknowns here, we need another equation. We obtain that second equation by applying Newton's second law to block B:

$$\mathbf{F} = m_B\mathbf{a}$$

Here $\mathbf{F}$ is the resultant force on block B and, from figure 4.12(b), we can see that

$$\mathbf{F}_N + \mathbf{w}_B + \mathbf{T}' = m_B\mathbf{a} \tag{4.28}$$

This vector equation is equivalent to the two component equations

$$F_N - w_B = 0 \tag{4.29}$$

and

$$T' = m_B a \tag{4.30}$$

The right-hand side of equation 4.29 is zero, because there is no acceleration of block B perpendicular to the table. It reduces to

$$F_N = w_B$$

That is, the normal force that the table exerts on block B is equal to the weight of block B. The magnitude of the acceleration of block B is also a because block B and block A are tied together by the string and therefore have the same motion.

Equation 4.30 is Newton's second law for the motion of block B to the right. Now we make the assumption that

$$T' = T$$

that is, the magnitude of the tension in the cord pulling on block B is the same as the magnitude of the tension in the cord restraining block A. This is a valid assumption providing the mass of the pulley is very small and friction in the pulley bearing is negligible. The only effect of the pulley is to change the direction of the string and hence the direction of the tension. (In chapter 9 we will again solve this problem, taking the rotational motion of the pulley into account without the assumption of equal tensions.) Therefore, equation 4.30 becomes

$$T = m_B a \qquad (4.31)$$

We now have enough information to solve for the acceleration of the system. That is, there are the two equations 4.27 and 4.31 and the two unknowns a and T. By subtracting equation 4.31 from equation 4.27, we eliminate the tension T from both equations:

$$
\begin{aligned}
T - w_A &= -m_A a && (4.27) \\
\text{Subtract} \qquad T &= m_B a && (4.31) \\
\hline
T - T - w_A &= -m_A a - m_B a \\
- w_A &= -m_A a - m_B a \\
w_A &= (m_A + m_B)a
\end{aligned}
$$

Solving for the acceleration a,

$$a = \frac{w_A}{m_A + m_B}$$

To simplify further we note that

$$w_A = m_A g$$

Therefore, the acceleration of the system of two blocks is

$$a = \frac{m_A}{m_A + m_B} g \qquad (4.32)$$

To determine the tension T in the cord, we use equations 4.31 and 4.32:

$$T = m_B a = \frac{m_B m_A}{m_A + m_B} g \qquad (4.33)$$

Since the acceleration of the system is a constant we can determine the position and velocity of block B in the x-direction at any time using the kinematic equations

$$x = v_0 t + \tfrac{1}{2} a t^2 \qquad (3.14)$$

$$v = v_0 + at \qquad (3.10)$$

and

$$v^2 = v_0^2 + 2ax \qquad (3.16)$$

with the acceleration now given by equation 4.32. We find the position of block A at any time using the same equations, but with x replaced by the displacement y.

Alternate Solution to the Problem There is another way to compute the acceleration of this system that in a sense is a lot easier. But it is an intuitive way of solving the problem. Some students can see the solution right away, others can not. Let us again start with Newton's second law and solve for the acceleration *a* of the system

$$a = \frac{F}{m} \tag{4.8}$$

Thus, *the acceleration of the system is equal to the total resultant force applied to the system divided by the total mass of the system that is in motion.* The total force that is accelerating the system is the weight w_A. The tension T in the string just transmits the total force from one block to another. The total mass that is in motion is the sum of the two masses, m_A and m_B. Therefore, the acceleration of the system, found from equation 4.8, is

$$a = \frac{w_A}{m_A + m_B}$$

or

$$a = \frac{m_A}{m_A + m_B} g$$

Notice that this is the same acceleration that we determined previously in equation 4.32. The only disadvantage of this second technique is that it does not tell the tension in the cord. Which technique should the student use in the solution of the problem? That depends on the student. If you can see the intuitive approach, and wish to use it, do so. If not, follow the first step-by-step approach.

Example 4.8

Combined motion of a block moving on a smooth horizontal surface and a mass falling vertically. A 6.00-kg block rests on a smooth table. It is connected by a string of negligible mass to a 2.00-kg block hanging over the end of the table, as shown in figure 4.13. Find (a) the acceleration of each block, (b) the tension in the connecting string, (c) the position of mass *A* after 0.400 s, and (d) the velocity of mass *A* at 0.400 s.

Solution

a. To solve the problem, we draw all the forces that are acting on the system and then apply Newton's second law. The magnitude of the acceleration, obtained from equation 4.32, is

$$a = \frac{m_A}{m_A + m_B} g = \frac{2.00 \text{ kg}}{2.00 \text{ kg} + 6.00 \text{ kg}} (9.80 \text{ m/s}^2)$$

$$= 2.45 \text{ m/s}^2$$

b. The tension, found from equation 4.31, is

$$T = m_B a = (6.00 \text{ kg})(2.45 \text{ m/s}^2) = 14.7 \text{ N}$$

c. The position of mass *A* after 0.400 s is found from the kinematic equation

$$y = v_0 t + \tfrac{1}{2}at^2$$

Because the block starts from rest, $v_0 = 0$, and the block falls the distance

$$y = \tfrac{1}{2}at^2 = \tfrac{1}{2}(-2.45 \text{ m/s}^2)(0.400 \text{ s})^2$$

$$= -0.196 \text{ m}$$

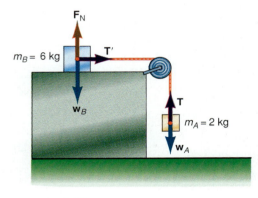

Figure 4.13

Diagram for example 4.8.

d. The velocity of block A is found from the kinematic equation

$$v = v_0 + at$$

$$= 0 + (-2.45 \text{ m/s}^2)(0.400 \text{ s})$$

$$= -0.980 \text{ m/s}$$

The negative sign is used for the acceleration of block A because it accelerated in the negative y-direction. Hence, $y = -0.196$ m indicates that the block is below its starting position. The negative sign on the velocity indicates that block A is moving in the negative y-direction. If we had done the same analysis for block B, the results would have been positive because block B is moving in the positive x-direction.

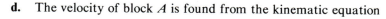

Atwood's Machine

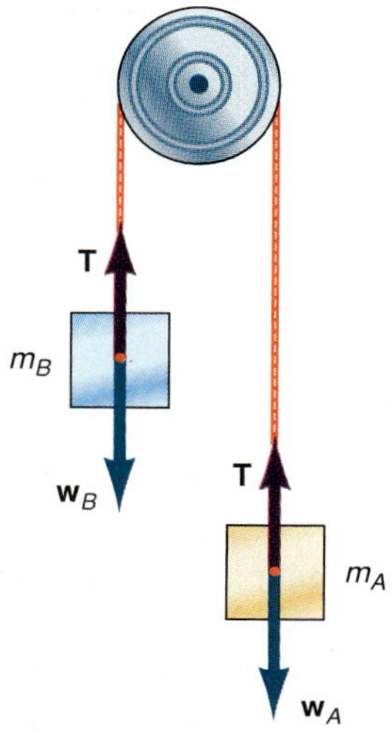

Atwood's machine is a system that consists of a pulley, with a mass m_A on one side, connected by a string of negligible mass to another mass m_B on the other side, as shown in figure 4.14.

We assume that m_A is larger than m_B. When the system is released, the mass m_A will fall downward, pulling the lighter mass m_B, on the other side, upward. We would like to determine the acceleration of the system of two masses. When we know the acceleration we can determine the position and velocity of each of the masses at any time from the kinematic equations.

Let us start by drawing all the forces acting on the masses in figure 4.14 and then apply Newton's second law to each mass. (The assumption that the tension T in the rope is the same for each mass is again utilized. We will solve this problem again in chapter 9, on rotational motion, where the rotating pulley is massive and hence the tensions on both sides of the pulley are not the same.)

For mass A, Newton's second law is

$$\mathbf{F}_A = m_A\mathbf{a}$$

or

$$\mathbf{T} + \mathbf{w}_A = m_A\mathbf{a} \tag{4.34}$$

We can simplify this equation by taking the upward direction as positive and the downward direction as negative, that is,

$$T - w_A = -m_A a \tag{4.35}$$

We cannot yet solve for the acceleration of the system, because the tension T in the string is unknown. Another equation is needed to eliminate T. We obtain this equation by applying Newton's second law to mass B:

$$\mathbf{F}_B = m_B\mathbf{a}$$

$$\mathbf{T} + \mathbf{w}_B = m_B\mathbf{a} \tag{4.36}$$

Simplifying again by taking the upward direction as positive and the downward direction as negative, we get

$$T - w_B = +m_B a \tag{4.37}$$

We thus have two equations, 4.35 and 4.37, in the two unknowns of acceleration a and tension T. The tension T is eliminated by subtracting equation 4.37 from equation 4.35. That is,

$$\begin{aligned} T - w_A &= -m_A a \qquad &(4.35) \\ \text{Subtract} \qquad T - w_B &= m_B a \qquad &(4.37) \\ \hline T - w_A - T + w_B &= -m_A a - m_B a \\ w_B - w_A &= -(m_A + m_B)a \end{aligned}$$

Figure 4.14

Atwood's machine.

Solving for a, we obtain

$$a = \frac{w_A - w_B}{m_A + m_B} = \frac{m_A g - m_B g}{m_A + m_B}$$

Hence, the acceleration of each mass of the system is

$$a = \left(\frac{m_A - m_B}{m_A + m_B}\right) g \qquad (4.38)$$

We find the tension T in the string by substituting equation 4.38 back into equation 4.35:

$$T = w_A - m_A a$$

$$= m_A g - m_A \left(\frac{m_A - m_B}{m_A + m_B}\right) g$$

$$= \left[1 + \left(\frac{m_B - m_A}{m_A + m_B}\right)\right] m_A g$$

$$= \left(\frac{m_A + m_B + m_B - m_A}{m_A + m_B}\right) m_A g \qquad (4.35)$$

Hence,

$$T = \left(\frac{2 m_A m_B}{m_A + m_B}\right) g \qquad (4.39)$$

is the tension in the string of the Atwood's machine.

Special Cases Any formulation in physics should reduce to some simple, recognizable form when certain restrictions are placed on the motion. As an example, suppose a 16-lb (256-oz) bowling ball is placed on one side of Atwood's machine and a small 0.1-oz marble on the other side. What kind of motion would we expect? The bowling ball is so large compared to the marble that the bowling ball should fall like a freely falling body. What does the formulation for the acceleration in equation 4.38 say?

If the bowling ball is m_A and the marble is m_B, then m_A is very much greater than m_B and can be written mathematically as

$$m_A \gg m_B$$

Then,

$$m_A + m_B \approx m_A$$

As an example,

$$\frac{256}{g} + \frac{0.1}{g} = \frac{256.1}{g} \approx \frac{256}{g} = m_A$$

Similarly,

$$m_A - m_B \approx m_A$$

As an example,

$$\frac{256}{g} - \frac{0.1}{g} = \frac{255.9}{g} \approx \frac{256}{g} = m_A$$

Therefore the acceleration of the system, equation 4.38, becomes

$$a = \left(\frac{m_A - m_B}{m_A + m_B}\right) g = \frac{m_A}{m_A} g = g$$

That is, the equation for the acceleration of the system reduces to the acceleration due to gravity, as we would expect if one mass is very much larger than the other.

Another special case is where both masses are equal. That is, if

$$m_A = m_B$$

then the acceleration of the system is

$$a = \left(\frac{m_A - m_B}{m_A + m_B}\right)g = \left(\frac{m_A - m_A}{2m_A}\right)g = 0$$

That is, if both masses are equal there is no acceleration of the system. The system is either at rest or moving at a constant velocity.

Alternate Solution to Atwood's Machine   A simpler solution to Atwood's machine can be obtained directly from Newton's second law by the intuitive approach. The acceleration of the system, found from Newton's second law, is

$$a = \frac{F}{m}$$

where F is the resultant force acting on the system and m is the total mass in motion. The resultant force acting on the system is the difference between the two weights, $w_A - w_B$, and the total mass of the system is the sum of the two masses that are in motion, namely $m_A + m_B$. Thus,

$$a = \frac{F}{m} = \frac{w_A - w_B}{m_A + m_B} = \left(\frac{m_A - m_B}{m_A + m_B}\right)g$$

the same result we found before in equation 4.38.

The Weight of a Person Riding in an Elevator

A scale is placed on the floor of an elevator. A 192-lb person enters the elevator when it is at rest and stands on the scale. What does the scale read when (a) the elevator is at rest, (b) the elevator is accelerating upward at 5.00 ft/s², (c) the acceleration becomes zero and the elevator moves at the constant velocity of 5.00 ft/s upward, (d) the elevator decelerates at 5.00 ft/s² before coming to rest, and (e) the cable breaks and the elevator is in free-fall?

A picture of the person in the elevator showing the forces that are acting is drawn in figure 4.15. The forces acting on the person are his weight **w**, acting down, and the reaction force of the elevator floor acting upward, which we call **F**$_N$. Applying Newton's second law we obtain

$$\mathbf{F_N + w} = m\mathbf{a} \tag{4.40}$$

a.   If the elevator is at rest then $\mathbf{a} = 0$ in equation 4.40. Therefore,

$$\mathbf{F_N + w} = 0$$

$$\mathbf{F_N} = -\mathbf{w}$$

which shows that the floor of the elevator is exerting a force upward, through the scale, on the person, that is equal and opposite to the force that the person is exerting on the floor. Hence,

$$F_N = w = 192 \text{ lb}$$

We usually think of the operation of a scale in terms of us pressing down on the scale, but we can just as easily think of the scale as pushing upward on us. Thus, the person would read 192 lb on the scale.

b.   The doors of the elevator are now closed and the elevator accelerates upward at a rate of 5.00 ft/s². Newton's second law is again given by equation 4.40. We can write this as a scalar equation if the usual convention of positive for up and negative for down is taken. Hence,

$$F_N - w = ma$$

(a)

(b)

Figure 4.15

Forces acting on a person in an elevator.

Solving for F_N, we get

$$F_N = w + ma \qquad (4.41)$$

Substituting the given values into equation 4.41 gives

$$F_N = 192 \text{ lb} + \frac{192 \text{ lb}}{32.0 \text{ ft/s}^2}(5.00 \text{ ft/s}^2)$$

$$= 192 \text{ lb} + 30.0 \text{ lb}$$

$$= 222 \text{ lb}$$

That is, the floor is exerting a force upward on the person of 222 lb. Therefore, the scale would now read 222 lb. Does the person now really weigh 222 lb? Of course not. What the scale is reading is the person's weight plus the additional force of 30.0 lb that is applied to the person, via the scales and floor of the elevator, to cause the person to be accelerated upward along with the elevator. I am sure that all of you have experienced this situation. When you step into an elevator and it accelerates upward you feel as though there is a force acting on you, pushing you down. Your knees feel like they might buckle. It is not that something is pushing you down, but rather that the floor is pushing you up. The floor is pushing upward on you with a force greater than your own weight in order to put you into accelerated motion. That extra force upward on you of 30.0 lb is exactly the force necessary to give you the acceleration of $+5.00$ ft/s^2.

c. The acceleration now stops and the elevator moves upward at the constant velocity of 5.00 ft/s. What does the scale read now?

Newton's second law is again given by equation 4.40, but since $a = 0$,

$$F_N = w = 192 \text{ lb}$$

Notice that this is the same value as when the elevator was at rest. This is a very interesting phenomenon. *The scale reads the same whether you are at rest or moving at a constant velocity. That is, if you are in motion at a constant velocity, and you have no external references to observe that motion, you cannot tell that you are in motion at all.*

I am sure you also have experienced this while riding an elevator. First you feel the acceleration and then you feel nothing. Your usual reaction is to ask "are we moving, or are we at rest?" You then look for a crack around the elevator door to see if you can see any signs of motion. *Without a visual reference, the only way you can sense a motion is if that motion is accelerated.*

d. The elevator now decelerates at 5.00 ft/s^2. What does the scale read? Newton's second law is again given by equation 4.40, and writing it in the simplified form, we have

$$F_N - w = -ma \qquad (4.42)$$

The minus sign on the right-hand side of equation 4.42 indicates that the acceleration vector is opposite to the direction of the motion because the elevator is decelerating. Solving equation 4.42 for F_N gives

$$F_N = w - ma$$

$$= 192 \text{ lb} - \frac{192 \text{ lb}}{32.0 \text{ ft/s}^2}(5.00 \text{ ft/s}^2)$$

$$= 192 \text{ lb} - 30.0 \text{ lb}$$

$$= 162 \text{ lb}$$

Hence, the force acting on the person is less than the person's weight. The effect is very noticeable when you walk into an elevator and accelerate downward (which is the same as decelerating when the elevator is going upward). You feel as if you are falling. Well, you are falling.

At rest the floor exerts a force upward on a 192-lb person of 192 lb, now it only exerts a force upward of 162 lb. The floor is not exerting enough force to hold the person up. Therefore, the person falls. It is a controlled fall of 5.00 ft/s², but a fall nonetheless. The scale in the elevator now reads 162 lb. The difference in that force and the person's weight is the force that accelerates the person downward.

e. Let us now assume that the cable breaks. What is the acceleration of the system now. Newton's second law is again given by equation 4.40, or in simplified form by

$$F_N - w = -ma \tag{4.42}$$

But if the cable breaks, the elevator becomes a freely falling body with an acceleration g. Therefore, equation 4.42 becomes

$$F_N - w = -mg$$

The force that the elevator exerts upward on the person becomes

$$F_N = w - mg$$

But the weight w is equal to mg. Thus,

$$F_N = w - w = 0$$

or

$$F_N = 0$$

Because the scale reads the force that the floor is pushing upward on the person, the scale now reads zero. This is why *it is sometimes said that in free-fall you are weightless, because in free-fall the scale that reads your weight now reads zero.* This is a somewhat misleading statement because you still have mass, and that mass is still attracted down toward the center of the earth. And in this sense you still have a weight pushing you downward. The difference here is that, while standing on the scale, the scale says that you are weightless, only because the scale itself is also in free-fall. As your feet try to press against the scale to read your weight, the scale falls away from them, and does not permit the pressure of your feet against the scale, and so the scale reads zero. From a reference system outside of the elevator, you would say that the falling person still has weight and that weight is causing that person to accelerate downward at the value g. *However, in the frame of reference of the elevator, not only the person seems weightless, but all weights and gravitational forces on anything around the person seem to have disappeared.* Normally, at the surface of the earth, if a person holds a pen and then lets go, the pen falls. But in the freely falling elevator, if a person lets go of the pen it will not fall to the floor, but will appear to be suspended in space in front of the person as if it were floating. According to the reference frame outside the elevator the pen is accelerating downward at the same rate as the person. But in the elevator, both are falling at the value g and therefore do not move with respect to one another. *In the freely falling reference system of the elevator, the force of gravity and its acceleration appear to have been eliminated.*

4.6 Friction

Whenever we try to slide one body over another body there is a force that opposes that motion. This opposing force is called the force of friction. For example, if this book is placed on the desk and a force is exerted on the book toward the right, there is a force of friction acting on the book toward the left opposing the applied force, as shown in figure 4.16.

Opposed frictional force

f

Applied force

F

Figure 4.16

The force of friction.

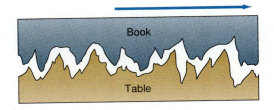

Figure 4.17

The "smooth" surfaces of contact that cause frictional forces.

Figure 4.18

You can walk because of friction.

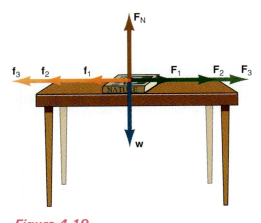

Figure 4.19

The force of static friction.

The basis of this frictional force stems from the fact that the surfaces that slide over each other are really not smooth at all. The top of the desk feels smooth to the hand, and so does the book, but that is because our hands themselves are not particularly smooth. In fact, if we magnified the surface of the book, or the desk, thousands of times, we would see a great irregularity in the supposedly "smooth" surface, as shown in figure 4.17.

As we try to slide the book along the desk these little microscopic chunks of the material get in each others way, and get stuck in the "mountains" and "valleys" of the other material, thereby opposing the tendency of motion. This is why it is difficult to slide one body over another. To get the body into motion we have to break off, or ride over, these microscopic chunks of matter. Because these chunks are microscopic, we do not immediately see the effect of this loss of material. Over a long period of time, however, the effect is very noticeable. As an example, if you observe any step of a stairway, which should be flat and level, you will notice that after a long period of time the middle of the stair is worn from the thousands of times a foot slid on the step in the process of walking up or down the stairs. This effect occurs whether the stairs are made of wood or even marble.

The same wearing process occurs on the soles and heels of shoes, and eventually they must be replaced. In fact the walking process can only take place because there is friction between the shoes and the ground. In the process of walking, in order to step forward, you must press your foot backward on the ground. But because there is friction between your shoe and the ground, there is a frictional force tending to oppose that motion of your shoe backward and therefore the ground pushes forward on your shoe, which allows you to walk forward, as shown in figure 4.18.

If there were no frictional force, your foot would slip backward and you would not be able to walk. This effect can be readily observed by trying to walk on ice. As you push your foot backward, it slips on the ice. You might be able to walk very slowly on the ice because there is some friction between your shoes and the ice. But try to run on the ice and see how difficult it is. If friction were entirely eliminated you could not walk at all.

Force of Static Friction

If this book is placed on the desk, as in figure 4.19, and a small force $\mathbf{F_1}$ is exerted to the right, we observe that the book does not move. There must be a frictional force $\mathbf{f_1}$ to the left that opposes the tendency of motion to the right. That is, $\mathbf{f_1} = -\mathbf{F_1}$.

If we increase the force to the right to $\mathbf{F_2}$, and again observe that the book does not move, the opposing frictional force must also have increased to some new value $\mathbf{f_2}$, where $\mathbf{f_2} = -\mathbf{F_2}$. If we now increase the force to the right to some value $\mathbf{F_3}$, the book just begins to move. The frictional force to the left has increased to some value $\mathbf{f_3}$, where $\mathbf{f_3}$ is infinitesimally less than $\mathbf{F_3}$. The force to the right is now greater than the frictional force to the left and the book starts to move to the right. When the object just begins to move, it has been found experimentally that the frictional force is

$$f_s = \mu_s F_N \tag{4.43}$$

where F_N is the normal or perpendicular force holding the two bodies in contact with each other. As we can see in figure 4.19, the forces acting on the book in the vertical are the weight of the body w, acting downward, and the normal force F_N of the desk, pushing upward on the book. In this case, since the acceleration of the book in the vertical is zero, the normal force F_N is exactly equal to the weight of the book w. (If the desk were tilted, F_N would still be the force holding the two objects together, but it would no longer be equal to w.)

The quantity μ_s in equation 4.43 is called the coefficient of static friction and depends on the materials of the two bodies which are in contact. Coefficients

Table 4.1

Approximate Coefficients of Static and Kinetic Friction
for Various Materials in Contact

Materials in Contact	μ_s	μ_k
Glass on glass	0.95	0.40
Steel on steel (lubricated)	0.15	0.09
Wood on wood	0.50	0.30
Wood on stone	0.50	0.40
Rubber tire on dry concrete	1.00	0.70
Rubber tire on wet concrete	0.70	0.50
Leather on wood	0.50	0.40
Teflon on steel	0.04	0.04
Copper on steel	0.53	0.36

of static friction for various materials are given in table 4.1. It should be noted that these values are approximate and will vary depending on the condition of the rubbing surfaces.

As we have seen, the force of static friction is not always equal to the product of μ_s and F_N, but can be less than that amount, depending on the value of the applied force tending to move the body. Therefore, the **force of static friction** should be written as

$$f_s \leq \mu_s F_N \tag{4.44}$$

where the symbol $\leq$ means "equal to, or less than." The only time that the equality holds is when the object is just about to go into motion.

Force of Kinetic Friction

Once the object is placed into motion, it is easier to keep it in motion. That is, the force that is necessary to keep the object in motion is much less than the force necessary to start the object into motion. In fact once the object is in motion, we no longer talk of the force of static friction, but rather we talk of the **force of kinetic friction** or sliding friction. For a moving object the frictional force is found experimentally as

$$f_k = \mu_k F_N \tag{4.45}$$

and is called the force of kinetic friction. The quantity μ_k is called the coefficient of kinetic friction and is also given for various materials in table 4.1. Note from the table that the coefficients of kinetic friction are less than the coefficients of static friction. This means that less force is needed to keep the object in motion, than it is to start it into motion.

We should note here, that these laws of friction are *empirical laws,* and are not exactly like the other laws of physics. For example, with Newton's second law, when we apply an unbalanced external force on a body of mass m, that body is accelerated by an amount given by $a = F/m$, and is always accelerated by that amount. Whereas the frictional forces are different, they are average results. That is, on the average equations 4.44 and 4.45 are correct. At any one given instant of time a force equal to $f_s = \mu_s F_N$, could be exerted on the book of figure 4.19, and yet the book might not move. At still another instance of time a force somewhat less than $f_s = \mu_s F_N$, is exerted and the book does move. Equation 4.43 represents an average result over very many trials. On the average, this equation is correct, but any one individual case may not conform to this result. Hence, this law is not quite as exact as the other laws of physics. In fact, if we return to figure 4.17, we see that it is not so surprising that the frictional laws are only averages, because at any one instant of time there are different interactions between the "mountains" and "valleys" of the two surfaces.

When two substances of the same material are slid over each other, as for example, copper on copper, we get the same kind of results. But if the two surfaces could be made "perfectly smooth," the frictional force would not decrease, but would rather increase. When we get down to the atomic level of each surface that is in contact, the atoms themselves have no way of knowing to which piece of copper they belong, that is, do the atoms belong to the top piece or to the bottom piece. The molecular forces between the atoms of copper would bind the two copper surfaces together.

In most applications of friction in technology, it is usually desirable to minimize the friction as much as possible. Since liquids and gases show much lower frictional effects (liquids and gases possess a quality called *viscosity*—a fluid friction), a layer of oil is usually placed between two metal surfaces as a lubricant, which reduces the friction enormously. The metal now no longer rubs on metal, but rather slides on the layer of the lubricant between the surfaces.

For example, when you put oil in your car, the oil is distributed to the moving parts of the engine. In particular, the oil coats the inside wall of the cylinders in the engine. As the piston moves up and down in the cylinder it slides on this coating of oil, and the friction between the piston and the cylinder is reduced.

Similarly when a glider is placed on an air track, the glider rests on a layer or cushion of air. The air acts as the lubricant, separating the two surfaces of glider and track. Hence, the frictional force between the glider and the air track is so small that in almost all cases it can be neglected in studying the motion of the glider.

When the skates of an ice skater press on the ice, the increased pressure causes a thin layer of the ice to melt. This liquid water acts as a lubricant to decrease the frictional force on the ice skater. Hence the ice skater seems to move effortlessly over the ice, figure 4.20.

Figure 4.20
An ice skater takes advantage of reduced friction.

Rolling Friction

To reduce friction still further, a wheel or ball of some type is introduced. When something can roll, the frictional force becomes very much less. Many machines in industry are designed with ball bearings, so that the moving object rolls on the ball bearings and friction is greatly reduced.

The whole idea of rolling friction is tied to the concept of the wheel. Some even consider the beginning of civilization as having started with the invention of the wheel, although many never even think of a wheel as something that was invented. The wheel goes so far back into the history of mankind that no one knows for certain when it was first used, but it was an invention. In fact, there were some societies that never discovered the wheel.

The frictional force of a wheel is very small compared with the force of sliding friction, because, theoretically, there is no relative motion between the rim of a wheel and the surface over which it rolls. Because the wheel touches the surface only at a point, there is no sliding friction. The small amount of rolling friction that does occur in practice is caused by the deformation of the wheel as it rolls over the surface, as shown in figure 4.21. Notice that the part of the tire in contact with the ground is actually flat, not circular.

In practice, that portion of the wheel that is deformed does have a tendency to slide along the surface and does produce a frictional force. So the smaller the

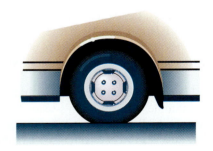

Figure 4.21
The deformation of a rolling wheel.

deformation, the smaller the frictional force. The harder the substance of the wheel, the less it deforms. For example, with steel on steel there is very little deformation and hence very little friction.

4.7 Applications of Newton's Second Law Taking Friction into Account

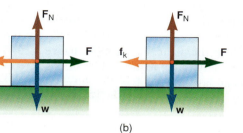

Figure 4.22

A box on a rough floor.

Example 4.9

A box on a rough floor. A 50.0-lb wooden box is at rest on a wooden floor, as shown in figure 4.22. (a) What horizontal force is necessary to start the box into motion? (b) If a force of 20.0 lb is continuously applied once the box is in motion, what will be its acceleration?

Solution

a. Whenever a problem says that a surface is rough, it means that we must take friction into account in the solution of the problem. The minimum force necessary to overcome static friction is found from equation 4.43. Hence, using table 4.1

$$F = f_s = \mu_s F_N$$

$$= \mu_s w = (0.50)(50.0 \text{ lb})$$

$$= 25 \text{ lb}$$

Note that whenever we say that $F = f_s$, we mean that F is an infinitesimal amount greater than f_s, and that it acts for an infinitesimal period of time. If the block is at rest, and $F = f_s$, then the net force acting on the block would be zero, its acceleration would be zero, and the block would therefore remain at rest forever. Thus, F must be an infinitesimal amount greater than f_s for the block to move. Now an infinitesimal quantity is, as the name implies, an extremely small quantity, so for all practical considerations we can assume that the force F plus the infinitesimal quantity, is just equal to the force F in all our equations. This is a standard technique that we will use throughout the study of physics. We will forget about the infinitesimal quantity and just say that the applied force is equal to the force to be overcome. But remember that there really must be that infinitesimal amount more, if the motion is to start.

b. Newton's second law applied to the box is

$$F - f_k = ma \tag{4.46}$$

The force of kinetic friction, found from equation 4.45 and table 4.1, is

$$f_k = \mu_k F_N = \mu_k w$$

$$= (0.30)(50.0 \text{ lb})$$

$$= 15 \text{ lb}$$

The acceleration of the block, found from equation 4.46, is

$$a = \frac{F - f_k}{m} = \frac{F - f_k}{w/g}$$

$$= \frac{20.0 \text{ lb} - 15 \text{ lb}}{50.0 \text{ lb}/32.0 \text{ ft/s}^2}$$

$$= 3.2 \text{ ft/s}^2$$

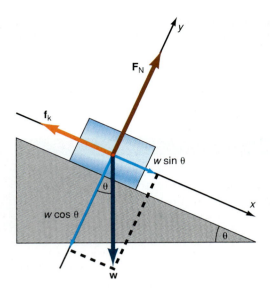

Figure 4.23

Block on an inclined plane with friction.

Example 4.10

A block on a rough inclined plane. **Find the acceleration of a block on an inclined plane, as shown in figure 4.23, taking friction into account.**

Solution

The problem is very similar to the one solved in figure 4.10, which was for a frictionless plane. We draw all the forces and their components as before, but now we introduce the frictional force. Because the frictional force always opposes the sliding motion, and $w \sin \theta$ acts to move the block down the plane, the frictional force f_k in opposing that motion must be pointed up the plane, as shown in figure 4.23. The block is given a slight push to overcome any force of static friction. To determine the acceleration, we use Newton's second law,

$$\mathbf{F} = m\mathbf{a} \qquad (4.9)$$

However, we can write this as two component equations, one parallel to the inclined plane and the other perpendicular to it.

Components Parallel to the Plane: Taking the direction down the plane as positive, Newton's second law becomes

$$w \sin \theta - f_k = ma \qquad (4.47)$$

Notice that this is very similar to the equation for the frictionless plane, except for the term f_k, the force of friction that is slowing down this motion.

Components Perpendicular to the Plane: Newton's second law for the perpendicular components is

$$F_N - w \cos \theta = 0 \qquad (4.48)$$

The right-hand side is zero because there is no acceleration perpendicular to the plane. That is, the block does not jump off the plane or crash through the plane so there is no acceleration perpendicular to the plane. The only acceleration is the one parallel to the plane, which was just found.

The frictional force f_k, given by equation 4.45, is

$$f_k = \mu_k F_N$$

where F_N is the normal force holding the block in contact with the plane. When the block was on a horizontal surface F_N was equal to the weight w. But now it is not. Now F_N, found from equation 4.48, is

$$F_N = w \cos \theta \qquad (4.49)$$

That is, because the plane is tilted, the force holding the block in contact with the plane is now $w \cos \theta$ rather than just w. Therefore, the frictional force becomes

$$f_k = \mu_k F_N = \mu_k w \cos \theta \qquad (4.50)$$

Substituting equation 4.50 back into Newton's second law, equation 4.47, we get

$$w \sin \theta - \mu_k w \cos \theta = ma$$

but since $w = mg$ this becomes

$$mg \sin \theta - \mu_k mg \cos \theta = ma$$

Since the mass m is in every term of the equation it can be divided out, and the acceleration of the block down the plane becomes

$$a = g \sin \theta - \mu_k g \cos \theta \qquad (4.51)$$

Note that the acceleration is independent of the mass m, since it canceled out of the equation. Also note that this equation reduces to the result for a frictionless plane, equation 4.18, when there is no friction, that is, when $\mu_k = 0$.

In this example, if $\mu_k = 0.300$ and $\theta = 30.0°$, the acceleration becomes

$$a = g \sin \theta - \mu_k g \cos \theta$$

$$= (9.80 \text{ m/s}^2)\sin 30.0° - (0.300)(9.80 \text{ m/s}^2)\cos 30.0°$$

$$= 4.90 \text{ m/s}^2 - 2.55 \text{ m/s}^2$$

$$= 2.35 \text{ m/s}^2$$

Notice the difference between the acceleration when there is no friction (4.90 m/s²) and when there is (2.35 m/s²). The block was certainly slowed down by friction.

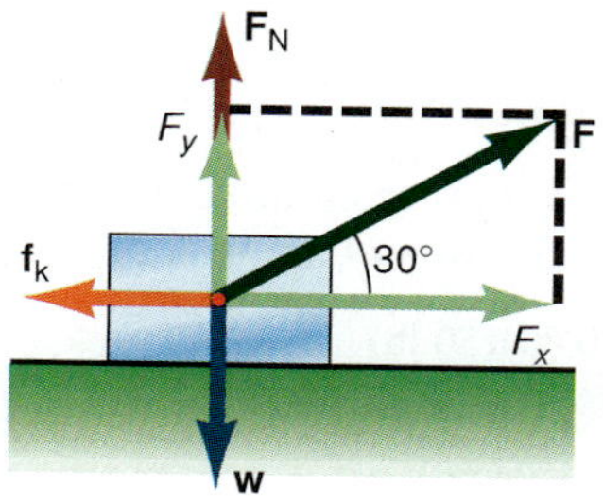

Figure 4.24

Pulling a block on a rough floor.

Example 4.11

Pulling a block on a rough floor. What force is necessary to pull a 50.0-lb wooden box at a constant speed over a wooden floor by a rope that makes an angle θ of 30° above the horizontal, as shown in figure 4.24?

Solution

Let us start by drawing all the forces that are acting on the box in figure 4.24. We break down the applied force into its components F_x and F_y. If Newton's second law is applied to the horizontal components, we obtain

$$F_x - f_k = ma_x \tag{4.52}$$

However, since the box is to move at constant speed, the acceleration a_x must be zero. Therefore,

$$F_x - f_k = 0$$

or

$$F \cos \theta - f_k = 0 \tag{4.53}$$

but

$$f_k = \mu_k F_N$$

where F_N is the normal force holding the box in contact with the floor. Before we can continue with our solution we must determine F_N.

If Newton's second law is applied to the vertical forces we have

$$F_y + F_N - w = ma_y \tag{4.54}$$

but because there is no acceleration in the vertical direction, a_y is equal to zero. Therefore,

$$F_y + F_N - w = 0$$

Solving for F_N we have

$$F_N = w - F_y$$

or

$$F_N = w - F \sin \theta \tag{4.55}$$

Note that F_N is not simply equal to w, as it was in example 4.9, but rather to $w - F \sin \theta$. The y-component of the applied force has the effect of lifting part of the weight from the floor. Hence, the force holding the box in contact with the floor is less than its weight. The frictional force therefore becomes

$$f_k = \mu_k F_N = \mu_k(w - F \sin \theta) \tag{4.56}$$

and substituting this back into equation 4.53, we obtain

$$F \cos \theta - \mu_k(w - F \sin \theta) = 0$$

or

$$F \cos \theta + \mu_k F \sin \theta - \mu_k w = 0$$

Factoring out the force F,

$$F(\cos \theta + \mu_k \sin \theta) = \mu_k w$$

and finally, solving for the force necessary to move the block at a constant speed, we get

$$F = \frac{\mu_k w}{\cos \theta + \mu_k \sin \theta} \tag{4.57}$$

Using the value of $\mu_k = 0.30$ (wood on wood) from table 4.1 and substituting the values for w, θ, and μ_k into equation 4.57, we obtain

$$F = \frac{\mu_k w}{\cos \theta + \mu_k \sin \theta} = \frac{(0.30)(50 \text{ lb})}{\cos 30° + 0.30 \sin 30°}$$

$$= 14.8 \text{ lb}$$

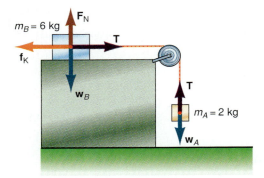

Figure 4.25

Combined motion of a block moving on a rough horizontal surface and a mass falling vertically.

Example 4.12

Combined motion of a block moving on a rough horizontal surface and a mass falling vertically. Find the acceleration of a block, on a "rough" table, connected by a cord passing over a pulley to a second block hanging over the table, as shown in figure 4.25.

Solution

This problem is similar to the problem solved in figure 4.12, only now the effects of friction are taken into account. We still assume that the mass of the string and the pulley are negligible. All the forces acting on the two blocks are drawn in figure 4.25. We apply Newton's second law to block A, obtaining

$$T - w_A = -m_A a \tag{4.58}$$

Applying it to block B, we obtain

$$T - f_k = m_B a \tag{4.59}$$

where the force of kinetic friction is

$$f_k = \mu_k F_N = \mu_k w_B \tag{4.60}$$

Substituting equation 4.60 into equation 4.59, we have

$$T - \mu_k w_B = m_B a \tag{4.61}$$

We eliminate the tension T in the equations by subtracting equation 4.58 from equation 4.61. Thus,

$$T - \mu_k w_B = m_B a \tag{4.61}$$

Subtract
$$T - w_A = -m_A a \tag{4.58}$$
$$\overline{T - \mu_k w_B - T + w_A = m_B a + m_A a}$$
$$w_A - \mu_k w_B = (m_B + m_A)a$$

Solving for the acceleration a, we have

$$a = \frac{w_A - \mu_k w_B}{m_A + m_B}$$

But since $w = mg$, this becomes

$$a = \left(\frac{m_A - \mu_k m_B}{m_A + m_B}\right) g \qquad (4.62)$$

the acceleration of the system. Note that if there is no friction, $\mu_k = 0$ and the equation reduces to equation 4.32, the acceleration without friction.

If $m_A = 2.00$ kg, $m_B = 6.00$ kg, and $\mu_k = 0.30$ (wood on wood), then the acceleration of the system is

$$a = \left(\frac{m_A - \mu_k m_B}{m_A + m_B}\right) g = \left[\frac{2.00 \text{ kg} - (0.30)6.00 \text{ kg}}{2.00 \text{ kg} + 6.00 \text{ kg}}\right] 9.80 \text{ m/s}^2$$

$$= 0.25 \text{ m/s}^2$$

This is only one-tenth of the acceleration obtained when there was no friction. It is interesting to see what happens if μ_k is equal to 0.40 instead of the value of 0.30 used in this problem. For this new value of μ_k, the acceleration becomes

$$a = \left(\frac{m_A - \mu_k m_B}{m_A + m_B}\right) g = \left[\frac{2.00 \text{ kg} - (0.40)6.00 \text{ kg}}{2.00 \text{ kg} + 6.00 \text{ kg}}\right] 9.80 \text{ m/s}^2$$

$$= -0.49 \text{ m/s}^2$$

The negative sign indicates that the acceleration is in the opposite direction of the applied force, which is of course absurd; that is, the block on the table m_B would be moving to the left while block m_A would be moving up. Something is very wrong here. In physics we try to analyze nature and the way it works. But, obviously nature just does not work this way. This is a very good example of trying to use a physics formula when it doesn't apply. Equation 4.62, like all equations, was derived using certain assumptions. If those assumptions hold in the application of the equation, then the equation is valid. If the assumptions do not hold, then the equation is no longer valid. Equation 4.62 was derived from Newton's second law on the basis that block m_B was moving to the right and therefore the force of kinetic friction that opposed that motion would be to the left. For $\mu_k = 0.40$ the force of friction is greater than the tension in the cord and the block does not move at all, that is, the acceleration of the system is zero. In fact if we look carefully at equation 4.62 we see that the acceleration will be zero if

$$m_A - \mu_k m_B = 0$$

which becomes

$$\mu_k m_B = m_A$$

and

$$\mu_k = \frac{m_A}{m_B} \qquad (4.63)$$

Whenever μ_k is equal to or greater than this ratio the acceleration is always zero. Even if we push the block to overcome static friction the kinetic friction is still too great and the block remains at rest. Whenever you solve a problem, always look at the numerical answer and see if it makes sense to you.

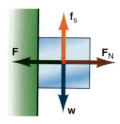

Figure 4.26

Pushing a block up a rough inclined plane.

Example 4.13

Pushing a block up a rough inclined plane. What force F is necessary to push a block up a rough inclined plane at a constant velocity?

Solution

The first thing to note is that if the block is to be pushed up the plane, then the frictional force, which always opposes the sliding motion, must act down the plane. The forces are shown in figure 4.26. Newton's second law for the parallel component becomes

$$-F + w \sin \theta + f_k = 0 \tag{4.64}$$

The right-hand side of equation 4.64 is 0 because the block is to be moved at constant velocity, that is, $a = 0$. The frictional force f_k is

$$f_k = \mu_k F_N = \mu_k w \cos \theta \tag{4.65}$$

Hence, equation 4.64 becomes

$$F = w \sin \theta + f_k = w \sin \theta + \mu_k w \cos \theta$$

Finally,

$$F = w(\sin \theta + \mu_k \cos \theta) \tag{4.66}$$

is the force necessary to push the block up the plane at a constant velocity.

It is appropriate to say something more about this force. If the block is initially at rest on the plane, then there is a force of static friction acting up the plane opposing the tendency of the block to slide down the plane. When the force is exerted to move the block up the plane, then the tendency for the sliding motion is up the plane. Now the force of static friction reverses and acts down the plane. When the applied force F is slightly greater than $w \sin \theta + f_s$, the block will just be put into motion up the plane. Now that the block is in motion, the frictional force to be overcome is the force of kinetic friction, which is less than the force of static friction. The force necessary to move the block up the plane at constant velocity is given by equation 4.66. Because the net force acting on the block is zero, the acceleration of the block is zero. If the block is at rest with a zero net force, then the block would have to remain at rest. However, the block was already set into motion by overcoming the static frictional forces, and since it is in motion, it will continue in that motion as long as the force given by equation 4.66 is applied.

Figure 4.27

A book pressed against a rough wall.

Example 4.14

A book pressed against a rough wall. A 1.00-lb book is held against a wall by pressing it against the wall with a force of 5.00 lb. What must be the minimum coefficient of friction between the book and the wall, such that the book does not slide down the wall? The forces acting on the book are shown in figure 4.27.

Solution

The book has a tendency to slide down the wall because of its weight. Because frictional forces always tend to oppose sliding motion, there is a force of static friction acting upward on the book. If the book is not to fall, then f_s must not be less than the weight of the book w. Therefore, let

$$f_s = w \tag{4.67}$$

but

$$f_s = \mu_s F_N = \mu_s F \tag{4.68}$$

Substituting equation 4.68 into equation 4.67, we obtain

$$\mu_s F = w$$

Solving for the coefficient of static friction, we obtain

$$\mu_s = \frac{w}{F} = \frac{1.00 \text{ lb}}{5.00 \text{ lb}} = 0.200$$

Therefore, the minimum coefficient of static friction to hold the book against the wall is $\mu_s = 0.200$.

4.8 Determination of the Coefficients of Friction

If the coefficient of friction for any two materials can not be found in a standardized table, it can always be found experimentally in the laboratory as follows.

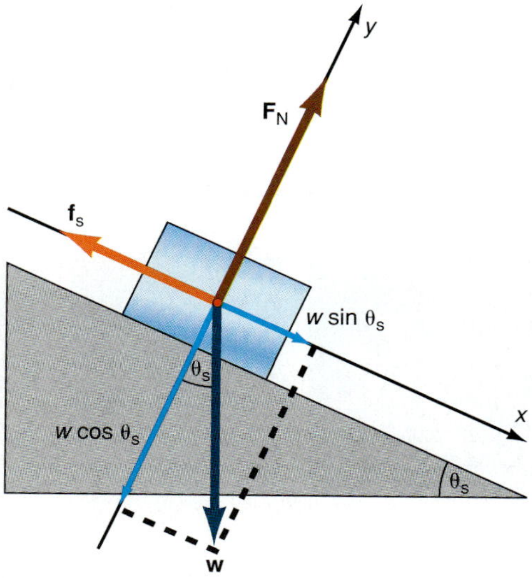

Coefficient of Static Friction

To determine the coefficient of static friction, we use an inclined plane whose surface is made up of one of the materials. As an example, let the plane be made of pine wood and the block that is placed on the plane will be made of oak wood. The forces acting on the block are shown in figure 4.28. We increase the angle θ of the plane until the block just begins to slide. We measure this angle where the block starts to slip and call it θ_s, the *angle of repose*.

We assume that the acceleration a of the block is still zero, because the block is just on the verge of slipping. Applying Newton's second law to the block gives

$$w \sin \theta_s - f_s = 0 \tag{4.69}$$

where

$$f_s = \mu_s F_N = \mu_s w \cos \theta_s \tag{4.70}$$

Substituting equation 4.70 back into equation 4.69 we have

$$w \sin \theta_s - \mu_s w \cos \theta_s = 0$$
$$w \sin \theta_s = \mu_s w \cos \theta_s$$
$$\mu_s = \frac{\sin \theta_s}{\cos \theta_s}$$

Therefore, the coefficient of static friction is

$$\mu_s = \tan \theta_s \tag{4.71}$$

That is, the coefficient of static friction μ_s is equal to the tangent of the angle θ_s, found experimentally. With this technique, the coefficient of static friction between any two materials can easily be found.

Coefficient of Kinetic Friction

The coefficient of kinetic friction is found in a similar way. We again place a block on the inclined plane and vary the angle, but now we give the block a slight push to overcome the force of static friction. The block then slides down the plane at a constant velocity. Experimentally, this is slightly more difficult to accomplish because it is difficult to tell when the block is moving at a constant velocity, rather than being accelerated. However, with a little practice we can determine when it is

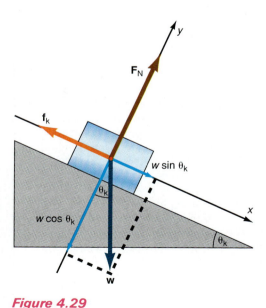

Figure 4.28

Determining the coefficient of static friction.

Figure 4.29

Determining the coefficient of kinetic friction.

moving at constant velocity. We measure the angle at which the block moves at constant velocity and call it θ_k. Since there is no acceleration, Newton's second law becomes

$$w \sin \theta_k - f_k = 0 \qquad (4.72)$$

but

$$f_k = \mu_k F_N = \mu_k w \cos \theta_k$$
$$w \sin \theta_k - \mu_k w \cos \theta_k = 0$$
$$w \sin \theta_k = \mu_k w \cos \theta_k$$
$$\mu_k = \frac{\sin \theta_k}{\cos \theta_k}$$

Therefore, the coefficient of kinetic friction for the two materials in contact is

$$\mu_k = \tan \theta_k \qquad (4.73)$$

"Have you ever wondered . . . ?"

An Essay on the Application of Physics

The Physics of Sports

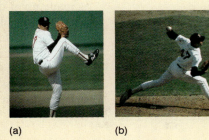

(a)   (b)   (c)

Figure 1

Look at that form.

Have you ever wondered, while watching a baseball game, why the pitcher goes through all those gyrations (figure 1) in order to throw the baseball to the batter? Why can't he throw the ball like all the rest of the players? No one else on the field goes through that big windup. Is there a reason for him to do that?

In order to understand why the pitcher goes through that big windup, let us first analyze the process of throwing a ball, figure 2. From what we already know about Newton's second law, we know you must exert a force on the ball to give it an acceleration. When you hold the ball initially in your hand, with your hand extended behind your head, the ball is at rest and hence has a zero initial velocity, that is, $v_0 = 0$. You now exert the force F on the ball as you move your arm through the distance x_1. The ball is now accelerated by your arm from a zero initial velocity to the final velocity v_1, as it leaves your hand. We find the velocity of the ball from the kinematic equation

$$v_1^2 = v_0^2 + 2ax_1 \qquad (H4.1)$$

But since v_0 is equal to zero, the velocity of the ball as it leaves your hand is

$$v_1^2 = 2ax_1$$
$$v_1 = \sqrt{2ax_1} \qquad (H4.2)$$

But the acceleration of the ball comes from Newton's second law as

$$a = \frac{F}{m}$$

Substituting this into the equation for the velocity we get

$$v_1 = \sqrt{2(F/m)x_1} \qquad (H4.3)$$

which tells us that the velocity of the ball depends on the mass m of the ball, the force F that your arm exerts on the ball, and the distance x_1 that you move the ball through while you are accelerating it. Since you cannot change the force F that your arm is capable of applying, nor the mass m of the ball, the only way to maximize the velocity v of the ball as it leaves your hand is to increase the value of x to as large a value as possible.

Maximizing the value of x is the reason for the pitcher's long windup. In figure 3, we see the pitcher moving his hand as far backward as possible. In order for the pitcher not to fall down as he leans that far backward, he lifts his left foot forward and upward to maintain his balance. As he lowers his left leg his right arm starts to move forward. As his left foot touches the ground, he lifts his right foot off the ground and swings his body around until his right foot is as far forward as he can make it, while bringing his right arm as far forward as he can, figure 3(b). By going through this long motion he has managed to increase the distance that he moves the ball through, to the value x_2. The velocity of the ball as it leaves his hand is v_2 and is given by

$$v_2 = \sqrt{2(F/m)x_2} \qquad (H4.4)$$

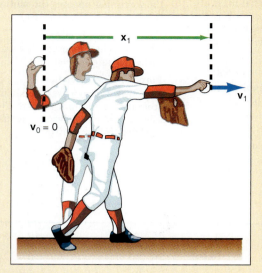

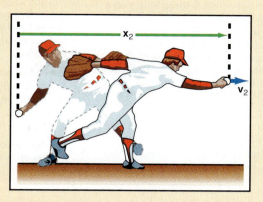

Figure 3

A pitcher throwing a baseball.

Figure 2

The process of throwing a ball.

Taking the ratio of these two velocities we obtain

$$\frac{v_2}{v_1} = \frac{\sqrt{2(F/m)x_2}}{\sqrt{2(F/m)x_1}}$$

which simplifies to

$$\frac{v_2}{v_1} = \sqrt{\frac{x_2}{x_1}}$$

The velocity v_2 becomes

$$v_2 = \sqrt{\frac{x_2}{x_1}}\, v_1 \qquad \textbf{(H4.5)}$$

Hence, by going through that long windup, the pitcher has increased the distance to x_2, thereby increasing the value of the

velocity that he can throw the baseball to v_2. For example, for an average person, x_1 is about 50 in., while x_2 is about 125 in. Therefore, the velocity becomes

$$v_2 = \sqrt{\frac{125}{50}}\, v_1$$
$$= 1.58\, v_1$$

Thus, if a pitcher is normally capable of throwing a baseball at a speed of 60 mph, by going through the long windup, the speed of the ball becomes

$$v_2 = 1.58(60 \text{ mph}) = 95 \text{ mph}$$

The long windup has allowed the pitcher to throw the baseball at 95 mph, much faster than the 60 mph that he could normally throw the ball. So this is why the pitcher goes through all those gyrations.

The Language of Physics

Dynamics

That branch of mechanics concerned with the forces that change or produce the motions of bodies. The foundation of dynamics is Newton's laws of motion (p. 79).

Newton's first law of motion

A body at rest will remain at rest, and a body in motion at a constant velocity will continue in motion at that constant velocity, unless acted on by some unbalanced external force. This is sometimes referred to as the law of inertia (p. 80).

Force

The simplest definition of a force is a push or a pull that acts on a body. Force can also be defined in a more general way by Newton's second law, that is, a force is that which causes a mass m to have an acceleration a (p. 80).

Inertia

The characteristic of matter that causes it to resist a change in motion is called inertia (p. 80).

Inertial coordinate system

A coordinate system that is either at rest or moving at a constant velocity with respect to another coordinate system that is either

at rest or also moving at some constant velocity. Newton's first law of motion defines an inertial coordinate system. That is, if a body is at rest or moving at a constant velocity in a coordinate system where there are no unbalanced forces acting on the body, the coordinate system is an inertial coordinate system. Newton's first law must be applied in an inertial coordinate system (p. 81).

Newton's third law of motion

If there are two bodies, A and B, and if body A exerts a force on body B, then body B exerts an equal but opposite force on body A (p. 81).

Newton's second law of motion

If an unbalanced external force **F** acts on a body of mass m, it will give that body an acceleration **a**. The acceleration is directly proportional to the applied force and inversely proportional to the mass of the body. Once the acceleration is determined by Newton's second law, the position and velocity of the body can be determined by the kinematic equations (p. 84).

Inertial mass

The measure of the resistance of a body to a change in its motion is called the inertial mass of the body. The mass of a body in Newton's second law is the inertial mass of the body. The best that can be determined at this time is that the inertial mass of a body is equal to the gravitational mass of the body (p. 92).

Atwood's machine

A simple pulley device that is used to study the acceleration of a system of bodies (p. 97).

Friction

The resistance offered to the relative motion of two bodies in contact. Whenever we try to slide one body over another body, the force that opposes the motion is called the force of friction (p. 101).

Force of static friction

The force that opposes a body at rest from being put into motion (p. 103).

Force of kinetic friction

The force that opposes a body in motion from continuing that motion. The force of kinetic friction is always less than the force of static friction (p. 103).

Summary of Important Equations

Newton's second law
$$\mathbf{F} = m\mathbf{a} \tag{4.9}$$

The weight of a body
$$w = mg \tag{4.12}$$

Definition of inertial mass
$$m_2 = \frac{a_1}{a_2} m_1 \tag{4.22}$$

Force of static friction
$$f_s \leq \mu_s F_N \tag{4.44}$$

Force of kinetic friction
$$f_k = \mu_k F_N \tag{4.45}$$

Coefficient of static friction
$$\mu_s = \tan \theta_s \tag{4.71}$$

Coefficient of kinetic friction
$$\mu_k = \tan \theta_k \tag{4.73}$$

Questions for Chapter 4

1. A force was originally defined as a push or a pull. Define the concept of force dynamically using Newton's laws of motion.

2. Discuss the difference between the ancient Greek philosophers' requirement of a constantly applied force as a condition for motion with Galileo's and Newton's concept of a force to initiate an acceleration.

3. Is a coordinate system that is accelerated in a straight line an inertial coordinate system? Describe the motion of a projectile in one dimension in a horizontally accelerated system.

4. If you drop an object near the surface of the earth it is accelerated downward to the earth. By Newton's third law, can you also assume that a force is exerted on the earth and the earth should be accelerated upward toward the object? Can you observe such an acceleration? Why or why not?

†5. Discuss an experiment that could be performed on a tilted air track whereby changing the angle of the track would allow you to prove that the acceleration of a body is proportional to the applied force.

Why could you not use this same experiment to show that the acceleration is inversely proportional to the mass?

†6. Discuss the concept of mass as a quantity of matter, a measure of the resistance of matter to being put into motion, and a measure of the gravitational force acting on the mass. Has the original platinum-iridium cylinder, which is stored in Paris, France, and defined as the standard of mass, ever been accelerated so that mass can be defined in terms of its inertial characteristics? Does it have to? Which is the most fundamental definition of mass?

7. From the point of view of the different concepts of mass, discuss why all bodies fall with the same acceleration near the surface of the earth.

8. Discuss why the normal force F_N is not always equal to the weight of the body that is in contact with a surface.

9. In the discussion of Atwood's machine, we assumed that the tension in the string is the same on both sides of the pulley. Can a pulley rotate if the tension is the same on both sides of the pulley?

†10. You are riding in an elevator and the cable breaks. The elevator goes into free fall. The instant before the elevator hits the ground, you jump upward about 3 ft. Will this do you any good? Discuss your motion with respect to the elevator and with respect to the ground. What will happen to you?

†11. Discuss the old saying: "If a horse pulls on a cart with a force F, then by Newton's third law the cart pulls backward on the horse with the same force F, therefore the horse can not move the cart."

12. A football is filled with mercury and taken into space where it is weightless. Will it hurt to kick this football since it is weightless?

†13. A 110-lb lady jumps out of a plane to go skydiving. She extends her body to obtain maximum frictional resistance from the air. After a while, she descends at a constant speed, called her terminal speed. At this time, what is the value of the frictional force of the air?

14. When a baseball player catches a ball he always pulls his glove backward. Why does he do this?

Problems for Chapter 4

In all problems assume that all objects are initially at rest, i.e., $v_0 = 0$, unless otherwise stated.

4.4 Newton's Second Law of Motion

1. What is the mass of a 200-lb person?
2. What is the weight of a 100-kg person at the surface of the earth? What would the person weigh on Mars where $g = 3.84$ m/s²?
3. What horizontal force must be applied to a 15.0-kg body in order to give it an acceleration of 5.00 m/s²?
4. A constant force accelerates a 3200-lb car from 0 to 60.0 mph in 12.0 s. Find (a) the acceleration of the car and (b) the force acting on the car that produces the acceleration.
5. A 3200-lb car is traveling along a highway at 60.0 mph. If the driver immediately applies his brakes and the car comes to rest in a distance of 250 ft, what average force acted on the car during the deceleration?
6. A 910-kg car is traveling along a highway at 88.0 km/hr. If the driver immediately applies his brakes and the car comes to rest in a distance of 70.0 m, what average force acted on the car during the deceleration?
7. A car is traveling at 60.0 mph when it collides with a stone wall. The car comes to rest after the first foot of the car is crushed. What was the average horizontal force acting on a 150-lb driver while the car came to rest? If five cardboard boxes, each 4 ft wide and filled with sand had been placed in front of the wall, and the car moved through all that sand before coming to rest, what would the average force acting on the driver have been then?
8. A rifle bullet of mass 12.0 g has a muzzle velocity of 750 m/s. What is the average force acting on the bullet when the rifle is fired, if the bullet is accelerated over the entire 1.00-m length of the rifle?
9. A car is to tow a 5000-lb truck with a rope. How strong should the rope be so that it will not break when accelerating the truck from rest to 10.0 ft/s in 12.0 s?
10. A force of 200 lb acts on a body that weighs 60.0 lb. (a) What is the mass of the body? (b) What is the acceleration of the body? (c) If the body starts from rest, how fast will it be going after it has moved 10.0 ft?

11. A cable supports an elevator that weighs 8000 N. (a) What is the tension T in the cable when the elevator accelerates upward at 1.50 m/s²? (b) What is the tension when the elevator accelerates downward at 1.50 m/s²?
12. A rope breaks when the tension exceeds 30.0 N. What is the minimum acceleration downward that a 60.0-N load can have without breaking the rope?
13. A 5.00-g bullet is fired at a speed of 100 m/s into a fixed block of wood and it comes to rest after penetrating 6.00 cm into the wood. What is the average force stopping the bullet?
14. A rope breaks when the tension exceeds 100 lb. What is the maximum vertical acceleration that can be given to an 80.0-lb load to lift it with this rope without breaking the rope?
15. What horizontal force must a locomotive exert on a 1000-ton train to increase its speed from 15.0 mph to 30.0 mph in moving 200 ft along a level track?
16. A steady force of 10.0 lb, exerted 30.0° above the horizontal, acts on a 50.0-lb sled on level snow. How far will the sled move in 10.0 s? (Neglect friction.)
17. A steady force of 70.0 N, exerted 43.5° above the horizontal, acts on a 30.0-kg sled on level snow. How far will the sled move in 8.50 s? (Neglect friction.)
18. A helicopter rescues a person at sea by pulling him upward with a cable. If the person weighs 180 lb and is accelerated upward at 1.00 ft/s², what is the tension in the cable?

4.5 Applications of Newton's Second Law

19. A force of 10.0 N acts horizontally on a 20.0-kg mass that is at rest on a smooth table. Find (a) the acceleration, (b) the velocity at 5.00 s, and (c) the position of the body at 5.00 s. (d) If the force is removed at 7.00 s, what is the body's velocity at 7.00, 8.00, 9.00, and 10.0 s?
20. A 50.0-lb box slides down a frictionless inclined plane that makes an angle of 37.0° with the horizontal. (a) What unbalanced force acts on the block? (b) What is the acceleration of the block?

21. A 20.0-kg block slides down a smooth inclined plane. The plane is 10.0 m long and is inclined at an angle of 30.0° with the horizontal. Find (a) the acceleration of the block, and (b) the velocity of the block at the bottom of the plane.
22. A 180-lb person stands on a scale in an elevator. What does the scale read when (a) the elevator is ascending with an acceleration of 4.00 ft/s², (b) it is ascending at a constant velocity of 8.00 ft/s, (c) it decelerates at 4.00 ft/s², (d) it descends at a constant velocity of 8.00 ft/s, and (e) the cable breaks and the elevator is in free-fall?
23. A 90.0-kg person stands on a scale in an elevator. What does the scale read when (a) the elevator is ascending with an acceleration of 1.50 m/s², (b) it is ascending at a constant velocity of 3.00 m/s, (c) it decelerates at 1.50 m/s², (d) it descends at a constant velocity of 3.00 m/s, and (e) the cable breaks and the elevator is in free-fall?
24. A spring scale is attached to the ceiling of an elevator. If a mass of 2.00 kg is placed in the pan of the scale, what will the scale read when (a) the elevator is accelerated upward at 1.50 m/s², (b) it is decelerated at 1.50 m/s², (c) it is moving at constant velocity, and (d) the cable breaks and the elevator is in free-fall?
†25. A block is propelled up a 48.0° frictionless inclined plane with an initial velocity $v_0 = 1.20$ m/s. (a) How far up the plane does the block go before coming to rest? (b) How long does it take to move to that position?
†26. In the diagram m_A is equal to 3.00 kg and m_B is equal to 1.50 kg. The angle of the inclined plane is 38.0°. (a) Find the acceleration of the system of two blocks. (b) Find the tension T_B in the connecting string.

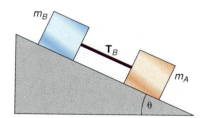

27. The two masses $m_A = 2.00$ kg and $m_B = 20.0$ kg are connected as shown. The table is frictionless. Find (a) the acceleration of the system, (b) the velocity of m_B at $t = 3.00$ s, and (c) the position of m_B at $t = 3.00$ s.

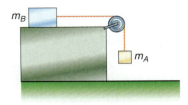

28. A 30.0-g mass and a 50.0-g mass are placed on an Atwood machine. Find (a) the acceleration of the system, (b) the velocity of the 50.0-g block at 4.00 s, (c) the position of the 50.0-g mass at the end of the fourth second, (d) the tension in the connecting string.

†29. Three blocks of mass $m_1 = 100$ g, $m_2 = 200$ g, and $m_3 = 300$ g are connected by strings as shown. (a) What force F is necessary to give the masses a horizontal acceleration of 4 m/s²? Find the tensions T_1 and T_2.

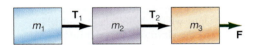

†30. A force of 20.0 lb acts as shown on the two blocks. If the blocks are on a frictionless surface, find the acceleration of each block and the horizontal force exerted on each block.

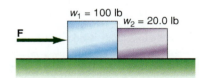

4.7 Applications of Newton's Second Law Taking Friction into Account

31. If the coefficient of friction between the tires of a car and the road is 0.300, what is the minimum stopping distance of a car traveling at 60.0 mph?

32. If the coefficient of friction between the tires of a car and the road is 0.300, what is the minimum stopping distance of a car traveling at 85.0 km/hr?

33. A 200-N container is to be pushed across a rough floor. The coefficient of static friction is 0.500 and the coefficient of kinetic friction is 0.400.

What force is necessary to start the container moving, and what force is necessary to keep it moving at a constant velocity?

34. A 2.00-kg toy accelerates from rest to 3.00 m/s in 8.00 s on a rough surface of $\mu_k = 0.300$. Find the applied force F.

35. A 50.0-lb box is to be moved along a rough floor at a constant velocity. The coefficient of friction is $\mu_k = 0.300$. (a) What force F_1 must you exert if you push downward on the box as shown? (b) What force F_2 must you exert if you pull upward on the box as shown? (c) Which is the better way to move the box?

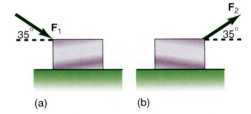

(a)   (b)

36. A 5.00-lb book is held against a rough vertical wall. If the coefficient of static friction between the book and the wall is 0.300, what force perpendicular to the wall is necessary to keep the book from sliding?

37. A block slides along a wooden table with an initial speed of 50.0 cm/s. If the block comes to rest in 150 cm, find the coefficient of kinetic friction between the block and the table.

38. What force must act horizontally on a 20.0-kg mass moving at a constant speed of 4.00 m/s on a rough table of coefficient of kinetic friction of 0.300? If the force is removed, when will the body come to rest? Where will it come to rest?

39. A 10.0-kg package slides down an inclined mail chute 15.0 m long. The top of the chute is 6.00 m above the floor. What is the speed of the package at the bottom of the chute if (a) the chute is frictionless and (b) the coefficient of kinetic friction is 0.300?

40. In order to place a 200-lb air conditioner in a window, a plank is laid between the window and the floor, making an angle of 40.0° with the horizontal. How much force is necessary to push the air conditioner up the plank at a constant speed if the coefficient of kinetic friction between the air conditioner and the plank is 0.300?

41. If a 4.00-kg container has a velocity of 3.00 m/s after sliding down a 2.00-m plane inclined at an angle of 30.0°, what is (a) the force of

friction acting on the container and (b) the coefficient of kinetic friction between the container and the plane?

†42. A 100-lb crate sits on the floor of a truck. If $\mu_s = 0.300$, what is the maximum acceleration of the truck before the crate starts to slip?

43. A skier starts from rest and slides a distance of 200 ft down the ski slope. The slope makes an angle of 35.0° with the horizontal. (a) If the coefficient of friction between the skis and the slope is 0.100, find the speed of the skier at the bottom of the slope. (b) At the bottom of the slope the skier continues to move on level snow. Where does the skier come to a stop?

44. A skier starts from rest and slides a distance of 85.0 m down the ski slope. The slope makes an angle of 23.0° with the horizontal. (a) If the coefficient of friction between the skis and the slope is 0.100, find the speed of the skier at the bottom of the slope. (b) At the bottom of the slope, the skier continues to move on level snow. Where does the skier come to a stop?

†45. A mass of 2.00 kg is pushed up an inclined plane that makes an angle of 50.0° with the horizontal. If the coefficient of kinetic friction between the mass and the plane is 0.400, and a force of 50.0 N is applied parallel to the plane, what is (a) the acceleration of the mass and (b) its velocity after moving 3.00 m up the plane?

46. The two masses $m_A = 20$ kg and $m_B = 20$ kg are connected as shown on a rough table. If the coefficient of friction between block B and the table is 0.45, find (a) the acceleration of each block and (b) the tension in the connecting string.

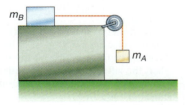

47. To determine the coefficient of static friction, the following system is set up. A mass, $m_B = 2.50$ kg, is placed on a rough horizontal table such as in the diagram for problem 46. When mass m_A is increased to the value of 1.50 kg the system just starts into motion. Determine the coefficient of static friction.

48. To determine the coefficient of kinetic friction, the following system is set up. A mass, $m_B = 2.50$ kg, is placed on a rough horizontal table such as in the diagram for problem 46. Mass m_A has the value of 1.85 kg, and the system goes into accelerated motion with a value a_1. While mass m_A falls to the floor, a distance $x_1 = 30.0$ cm below its starting point, mass m_B will also move through a distance x_1 and will have acquired a velocity v_1 at x_1. When m_A hits the floor, the acceleration a_1 becomes zero. From this point on, the only acceleration m_B experiences is the deceleration a_2 caused by the force of kinetic friction acting on m_B. Mass m_B moves on the rough surface until it comes to rest at the distance $x_2 = 20.0$ cm. From this information, determine the coefficient of kinetic friction.

Additional Problems

†49. Find the force F that is necessary for the system shown to move at constant velocity if $\mu_k = 0.300$ for all surfaces. The masses are $m_A = 6.00$ kg and $m_B = 2.00$ kg.

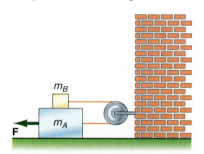

50. A pendulum is placed in a car at rest and hangs vertically. The car then accelerates forward and the pendulum bob is observed to move backward, the string making an angle of 15.0° with the vertical. Find the acceleration of the car.

51. Two gliders are tied together by a string after they are connected together by a compressed spring and placed on an air track. Glider A has a mass of 200 g and the mass of glider B is unknown. The string is now cut and the gliders fly apart. If glider B has an acceleration of 5.00 cm/s² to the right, and the acceleration of glider A to the left is 20.0 cm/s², find the mass of glider B.

52. A mass of 1.87 kg is pushed up a smooth inclined plane with an applied force of 3.50 N parallel to the plane. If the plane makes an angle of 35.8° with the horizontal,

find (a) the acceleration of the mass and (b) its velocity after moving 1.50 m up the plane.

†53. Two blocks $m_1 = 20.0$ kg and $m_2 = 10.0$ kg are connected as shown on a frictionless plane. The angle $\theta = 25.0°$ and $\phi = 35.0°$. Find the acceleration of each block and the tension in the connecting string.

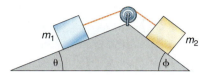

†54. What horizontal acceleration a_x must the inclined block M have in order for the smaller block m_A not to slide down the frictionless inclined plane? What force must be applied to the system to keep the block from sliding down the frictionless plane? $M = 10.0$ kg, $m_A = 1.50$ kg, and $\theta = 43°$.

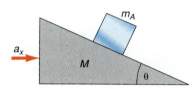

†55. If the acceleration of the system is 3.00 m/s² when it is lifted, and $m_A = 5.00$ kg, $m_B = 3.00$ kg, and $m_C = 2.00$ kg, find the tensions T_A, T_B, and T_C.

†56. Consider the double Atwood's machine as shown. If $m_1 = 50.0$ g, $m_2 = 20.0$ g, and $m_3 = 25.0$ g, what is the acceleration of m_3?

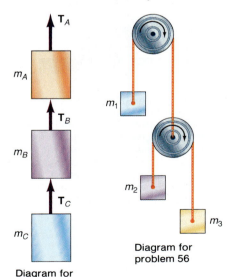

Diagram for problem 56

Diagram for problem 55

†57. Find the tension T_{23} in the string between mass m_2 and m_3, if $m_1 = 10.0$ kg, $m_2 = 2.00$ kg, and $m_3 = 1.00$ kg.

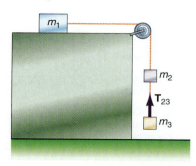

†58. If $m_A = 6.00$ kg, $m_B = 3.00$ kg, and $m_C = 2.00$ kg in the diagram, find the magnitude of the acceleration of the system and the tensions T_A, T_B, and T_C.

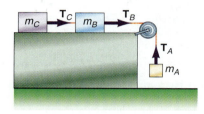

†59. Find (a) the acceleration of mass m_A in the diagram. All surfaces are frictionless. (b) Find the displacement of block A at $t = 0.500$ s. The value of the masses are $m_A = 3.00$ kg and $m_B = 5.00$ kg.

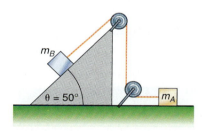

60. A force of 5.00 N acts on a body of mass $m = 2.00$ kg at an angle of 35.0° above the horizontal. If the coefficient of friction between the body and the surface upon which it is resting is 0.250, find the acceleration of the mass.

†61. Derive the formula for the magnitude of the acceleration of the system shown in the diagram.
   (a) What problem does this reduce to if $\phi = 90°$?
   (b) What problem does this reduce to if both θ and ϕ are equal to 90°?

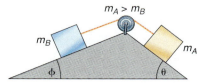

†62. What force is necessary to pull the two masses at constant speed if $m_1 = 2.00$ kg, $m_2 = 5.00$ kg, $\mu_{k_1} = 0.300$, and $\mu_{k_2} = 0.200$? What is the tension T_1 in the connecting string?

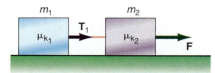

†63. If $m_A = 4.00$ kg, $m_B = 2.00$ kg, $\mu_{k_A} = 0.300$, and $\mu_{k_B} = 0.400$, find (a) the acceleration of the system down the plane and (b) the tension in the connecting string.

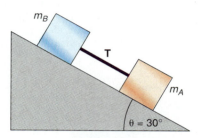

†64. A block $m = 0.500$ kg slides down a frictionless inclined plane 2.00 m long. It then slides on a rough horizontal table surface of $\mu_k = 0.300$ for 0.500 m. It then leaves the top of the table, which is 1.00 m high. How far from the base of the table does the block land?

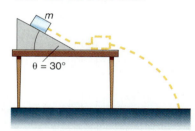

†65. In the diagram $m_A = 6.00$ kg, $m_B = 3.00$ kg, $m_C = 2.00$ kg, $\mu_{k_C} = 0.400$, and $\mu_{k_B} = 0.300$. Find the magnitude of the acceleration of the system and the tension in each string.

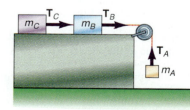

†66. In the diagram $m_A = 4.00$ kg, $m_B = 2.00$ kg, $m_C = 4.00$ kg, and $\theta = 58°$. If all the surfaces are frictionless, find the magnitude of the acceleration of the system.

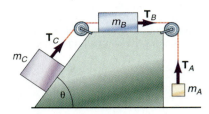

†67. If $m_A = 6.00$ kg, $m_B = 2.00$ kg, $m_C = 4.00$ kg, and the coefficient of kinetic friction for the surfaces are $\mu_{k_B} = 0.300$ and $\mu_{k_C} = 0.200$ find the magnitude of the acceleration of the system shown in the diagram and the tension in each string.

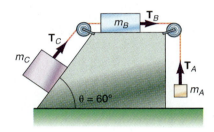

†68. Find (a) the magnitude of the acceleration of the system shown if $\mu_{k_B} = 0.300$, $\mu_{k_A} = 0.200$, $m_B = 3.00$ kg, and $m_A = 5.00$ kg, (b) the velocity of block A at 0.500 s.

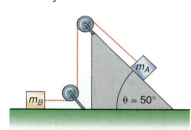

†69. In the diagram, block B rests on a frictionless surface but there is friction between blocks B and C. $m_A = 2.00$ kg, $m_B = 3.00$ kg, and $m_C = 1$ kg. Find (a) the magnitude of the acceleration of the system and (b) the minimum coefficient of friction between blocks C and B such that C will move with B.

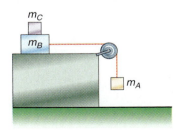

†70. When a body is moving through the air, the effect of air resistance can be taken into account. If the speed of the body is not too great, the force associated with the retarding force of air friction is proportional to the first power of the velocity of the moving body. This retarding force causes the velocity of a falling body at any time t to be

$$v = \frac{mg}{k}(1 - e^{-(k/m)t})$$

where m is the mass of the falling body and k is a constant that depends on the shape of the body. Show that this reduces to the case of a freely falling body if t and k are both small. (*Hint:* expand the term $e^{-(k/m)t}$ in a power series.)

71. Repeat problem 70, but now let the time t be very large (assume it is infinite). What does the velocity of the falling body become now? Discuss this result with Aristotle's statement that heavier objects fall faster than lighter objects. Clearly distinguish between the concepts of velocity and acceleration.

†72. If a body moves through the air at very large speeds the retarding force of friction is proportional to the square of the speed of the body, that is, $f = kv^2$, where k is a constant. Find the equation for the terminal velocity of such a falling body.

Interactive Tutorials

73. A 20.0-kg block slides down from the top of a smooth inclined plane that is 10.0 m long and is inclined at an angle θ of 30° with the horizontal. Find the acceleration a of the block and its velocity v at the bottom of the plane. Assume the initial velocity $v_0 = 0$.

74. Two masses $m_A = 40.0$ kg and $m_B = 30.0$ kg are connected by a massless string that hangs over a massless, frictionless pulley in an Atwood's machine arrangement as shown in figure 4.14. Calculate the acceleration a of the system and the tension T in the string.

75. A mass $m_A = 40.0$ kg hangs over a table connected by a massless string to a mass $m_B = 20.0$ kg that is on a rough horizontal table, with a coefficient of friction $\mu_k = 0.400$, that is similar to figure 4.25. Calculate the acceleration a of the system and the tension T in the string.

76. Generalization of problem 61 that also includes friction. Derive the formula for the magnitude of the acceleration of the system shown in the diagram for problem 61. As a general case, assume that the coefficient of kinetic friction between block A and the surface in μ_{k_A} and between block B and the surface is μ_{k_B}. Identify and solve for all the special cases that you can think of.

77. Free fall with friction—variable acceleration—terminal velocity. In the freely falling body studied in chapter 3, we assumed that the resistance of the air could be considered negligible. Let us now remove that constraint. Assume that there is frictional force caused by the motion through the air, and let us further assume that the frictional force is proportional to the square of the velocity of the moving body and is given by

$$f = kv^2$$

Find the displacement, velocity, and acceleration of the falling body and compare it to the displacement, velocity, and acceleration of a freely falling body without friction.

78. The mass of the connecting string is not negligible. In the problem of the combined motion of a block on a frictionless horizontal plane and a block falling vertically, as shown in figure 4.12, it was assumed that the mass of the connecting string was negligible and had no effect on the problem. Let us now remove that constraint. Assume that the string is a massive string. The string has a linear mass density of 0.050 kg/m and is 1.25 m long. Find the acceleration, velocity, and displacement y of the system as a function of time, and compare it to the acceleration, velocity, and displacement of the system with the string of negligible mass.

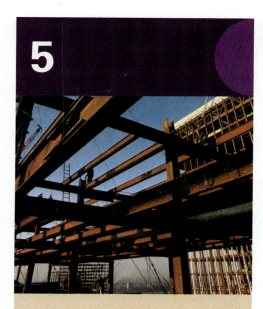

Statics in our lives.

Nature and Nature's laws lay hid in night:
God said, Let Newton be! and all was light.

Alexander Pope

5 Equilibrium

5.1 The First Condition of Equilibrium

The simplest way to define the **equilibrium** of a body is to say that *a body is in equilibrium if it has no acceleration.* That is, if the acceleration of a body is zero, then it is in equilibrium. Bodies in equilibrium under a system of forces are described as a special case of Newton's second law,

$$\mathbf{F} = m\mathbf{a} \qquad (4.9)$$

where $\mathbf{F}$ is the resultant force acting on the body. As pointed out in chapter 4, to emphasize the point that $\mathbf{F}$ is the resultant force, Newton's second law is sometimes written in the form

$$\Sigma\,\mathbf{F} = m\mathbf{a}$$

If there are forces acting on a body, but the body is not accelerated (i.e., $\mathbf{a} = 0$), then the body is in equilibrium under these forces and the condition for that body to be in equilibrium is simply

$$\Sigma\,\mathbf{F} = 0 \qquad (5.1)$$

Equation 5.1 is called the first condition of equilibrium. *The **first condition of equilibrium** states that for a body to be in equilibrium, the vector sum of all the forces acting on that body must be zero.* If the sum of the force vectors are added graphically they will form a closed figure because the resultant vector, which is equal to the sum of all the force vectors, is equal to zero.

Remember that if the acceleration is zero, then there is no change of the velocity with time. Most of the cases considered in this book deal with bodies that are at rest ($v = 0$) under the applications of forces. Occasionally we also consider a body that is moving at a constant velocity (also a case of zero acceleration). At first, we consider only examples where all the forces act through only one point of the body. Forces that act through only one point of the body are called *concurrent forces. That portion of the study of mechanics that deals with bodies in equilibrium is called **statics.*** When a body is at rest under a series of forces it is sometimes said to be in static equilibrium.

One of the simplest cases of a body in equilibrium is a book resting on the table, as shown in figure 5.1. The forces acting on the book are its weight $\mathbf{w}$, acting downward, and $\mathbf{F}_N$, the normal force that the table exerts upward on the book.

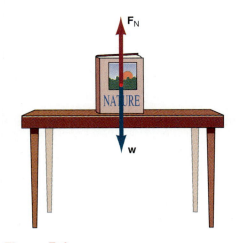

Figure 5.1

A body in equilibrium.

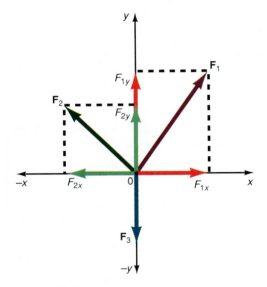

Figure 5.2

Three forces in equilibrium.

Because the book is resting on the table, it has zero acceleration. Hence, the sum of all the forces acting on the book must be zero and the book must be in equilibrium. The sum of all the forces are

$$\Sigma \mathbf{F} = \mathbf{F_N} + \mathbf{w} = 0$$

Taking the upward direction to be positive and the downward direction to be negative, this becomes

$$F_N - w = 0$$

Hence,

$$F_N = w$$

That is, the force that the table exerts upward on the book is exactly equal to the weight of the book acting downward. As we can easily see, this is nothing more than a special case of Newton's second law where the acceleration is zero. That is, forces can act on a body without it being accelerated if these forces balance each other out.

Let us consider another example of a body in equilibrium, as shown in figure 5.2. Suppose three forces $\mathbf{F_1}$, $\mathbf{F_2}$, and $\mathbf{F_3}$ are acting on the body that is located at the point 0, the origin of a Cartesian coordinate system. If the body is in equilibrium, then the vector sum of those forces must add up to zero and the body is not accelerating. Another way to observe that the body is in equilibrium is to look at the components of the forces, which are shown in figure 5.2. From the diagram we can see that if the sum of all the forces in the x-direction is zero, then there will be no acceleration in the x-direction. If the forces in the positive x-direction are taken as positive, and those in the negative x-direction as negative, then the sum of the forces in the x-direction is simply

$$F_{1x} - F_{2x} = 0 \tag{5.2}$$

Similarly, if the sum of all the forces in the y-direction is zero, there will be no acceleration in the y-direction. As seen in the diagram, this becomes

$$F_{1y} + F_{2y} - F_3 = 0 \tag{5.3}$$

A generalization of equations 5.2 and 5.3 is

$$\Sigma F_x = 0 \tag{5.4}$$

$$\Sigma F_y = 0 \tag{5.5}$$

which is another way of stating the first condition of equilibrium.

The first condition of equilibrium also states that the body is in equilibrium if the sum of all the forces in the x-direction is equal to zero and the sum of all the forces in the y-direction is equal to zero. Equations 5.4 and 5.5 are two component equations that are equivalent to the one vector equation 5.1.

Although only bodies in equilibrium in two dimensions will be treated in this book, if a third dimension were taken into account, an additional equation ($\Sigma F_z = 0$) would be necessary. Let us now consider some examples of bodies in equilibrium.

Example 5.1

A ball hanging from a vertical rope. A ball is hanging from a rope that is attached to the ceiling, as shown in figure 5.3. Find the tension in the rope. We assume that the mass of the rope is negligible and can be ignored in the problem.

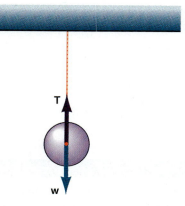

Figure 5.3

Ball hanging from a vertical rope.

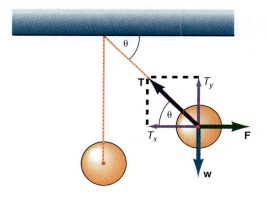

Figure 5.4

Ball pulled to one side.

The first thing that we should observe is that even though there are forces acting on the ball, the ball is at rest. That is, the ball is in *static equilibrium.* Therefore, the first condition of equilibrium must hold, that is,

$$\Sigma F_x = 0 \tag{5.4}$$

$$\Sigma F_y = 0 \tag{5.5}$$

The first step in solving the problem is to draw a diagram showing the forces that are acting on the ball. There is the weight **w,** acting downward in the negative y-direction, and the tension **T** in the rope, acting upward in the positive y-direction. Note that there are no forces in the x-direction so we do not use equation 5.4. The first condition of equilibrium for this problem is

$$\Sigma F_y = 0 \tag{5.5}$$

and, as we can see from the diagram in figure 5.3, this is equivalent to

$$T - w = 0$$

or

$$T = w$$

The tension in the rope is equal to the weight of the ball. If the ball weighs 5 N, then the tension in the rope is 5 N.

Example 5.2

The ball is pulled to one side. A ball hanging from a rope, is pulled to the right by a horizontal force **F** such that the rope makes an angle θ with the ceiling, as shown in figure 5.4. What is the tension in the rope?

Solution

The first thing that we should observe is that the system is at rest. Therefore, the ball is in static equilibrium and the first condition of equilibrium holds. But the tension **T** is neither in the x- nor y-direction. Before we can use equations 5.4 and 5.5, we must resolve the tension T into its components, T_x and T_y, as shown in figure 5.4. The first condition of equilibrium,

$$\Sigma F_x = 0 \tag{5.4}$$

is applied, which, as we see from figure 5.4 gives

$$\Sigma F_x = F - T_x = 0$$

or

$$F = T_x = T \cos \theta \tag{5.6}$$

Similarly,

$$\Sigma F_y = 0 \tag{5.5}$$

becomes

$$\Sigma F_y = T_y - w = 0$$
$$T_y = T \sin \theta = w \tag{5.7}$$

Note that there are four quantities T, θ, w, and F and only two equations, 5.6 and 5.7. Therefore, if any two of the four quantities are specified, the other two can be determined. *Recall that in order to solve a set of algebraic equations there must always be the same number of equations as unknowns.*

For example, if $w = 5.00$ N and $\theta = 40.0°$, what is the tension T and the force F. We use equation 5.7 to solve for the tension:

$$T = \frac{w}{\sin \theta} = \frac{5.00 \text{ N}}{\sin 40.0°} = 7.78 \text{ N}$$

We determine the force F, from equation 5.6, as

$$F = T \cos \theta = 7.78 \text{ N} \cos 40.0° = 5.96 \text{ N}$$

Example 5.3

Resting in your hammock. A person weighing 150 lb lies in a hammock, as shown in figure 5.5(a). The rope at the person's head makes an angle ϕ of 40.0° with the horizontal, while the rope at the person's feet makes an angle θ of 20.0°. Find the tension in the two ropes.

Solution

All the forces that are acting on the hammock are drawn in figure 5.5(b). The forces are resolved into their components, as shown in figure 5.5(b), where

$$T_{1x} = T_1 \cos \theta$$
$$T_{1y} = T_1 \sin \theta$$
$$T_{2x} = T_2 \cos \phi$$
$$T_{2y} = T_2 \sin \phi \quad (5.8)$$

The first thing we observe is that the hammock is at rest under the influence of several forces and is therefore in static equilibrium. Thus, the first condition of equilibrium must hold. Setting the forces in the x-direction to zero, equation 5.4,

$$\Sigma F_x = 0$$

gives

$$\Sigma F_x = T_{2x} - T_{1x} = 0$$

and

$$T_{2x} = T_{1x}$$

Figure 5.5
Lying in a hammock.

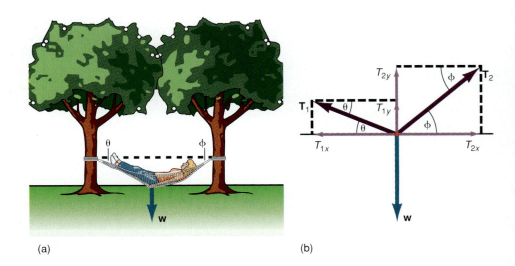

(a)   (b)

Using equations 5.8 for the components, this becomes

$$T_2 \cos \phi = T_1 \cos \theta \tag{5.9}$$

Taking all the forces in the y-direction and setting them equal to zero,

$$\Sigma F_y = 0 \tag{5.5}$$

gives

$$\Sigma F_y = T_{1y} + T_{2y} - w = 0$$

and

$$T_{1y} + T_{2y} = w$$

Using equations 5.8 for the components, this becomes

$$T_1 \sin \theta + T_2 \sin \phi = w \tag{5.10}$$

Equations 5.9 and 5.10 represent the first condition of equilibrium as it applies to this problem. Note that there are five quantities, T_1, T_2, w, θ, and ϕ and only two equations. Therefore, three of these quantities must be specified in order to solve the problem. In this case, θ, ϕ, and w are given and we will determine the tensions T_1 and T_2.

Let us start by solving equation 5.9 for T_2, thus,

$$T_2 = \frac{T_1 \cos \theta}{\cos \phi} \tag{5.11}$$

We cannot use equation 5.11 to solve for T_2 at this point, because T_1 is unknown. Equation 5.11 says that if T_1 is known, then T_2 can be determined. If we substitute this equation for T_2 into equation 5.10, thereby eliminating T_2 from the equations, we can solve for T_1. That is, equation 5.10 becomes

$$T_1 \sin \theta + \left(\frac{T_1 \cos \theta}{\cos \phi} \right) \sin \phi = w$$

Factoring out T_1 we get

$$T_1 \left(\sin \theta + \frac{\cos \theta \sin \phi}{\cos \phi} \right) = w \tag{5.12}$$

Finally, solving equation 5.12 for the tension T_1, we obtain

$$T_1 = \frac{w}{\sin \theta + \cos \theta \tan \phi} \tag{5.13}$$

Note that $\sin \phi / \cos \phi$ in equation 5.12 was replaced by $\tan \phi$, its equivalent, in equation 5.13. Substituting the values of $w = 150$ lb, $\theta = 20.0°$, and $\phi = 40.0°$ into equation 5.13, we find the tension T_1 as

$$T_1 = \frac{w}{\sin \theta + \cos \theta \tan \phi}$$

$$= \frac{150 \text{ lb}}{\sin 20.0° + \cos 20.0° \tan 40.0°}$$

$$= \frac{150 \text{ lb}}{0.342 + 0.940(0.839)} = \frac{150 \text{ lb}}{1.13}$$

$$= 132 \text{ lb} \tag{5.13}$$

Substituting this value of T_1 into equation 5.11, the tension T_2 in the second rope becomes

$$T_2 = T_1 \frac{\cos \theta}{\cos \phi} = 132 \text{ lb} \frac{\cos 20.0°}{\cos 40.0°}$$

$$= 162 \text{ lb}$$

Note that the tension in each rope is different, that is, T_1 is not equal to T_2. The ropes that are used for this hammock must be capable of withstanding these tensions or they will break.

An interesting special case arises when the angles θ and ϕ are equal. For this case equation 5.11 becomes

$$T_2 = T_1 \frac{\cos \theta}{\cos \phi} = T_1 \frac{\cos \theta}{\cos \theta} = T_1$$

that is,

$$T_2 = T_1$$

For this case, T_1, found from equation 5.13, is

$$T_1 = \frac{w}{\sin \theta + \cos \theta \, (\sin \theta / \cos \theta)}$$

$$= \frac{w}{2 \sin \theta} \tag{5.14}$$

Thus, when the angle θ is equal to the angle ϕ, the tension in each rope is the same and is given by equation 5.14. Note that if θ were equal to zero in equation 5.14, the tension in the ropes would become infinite. Since this is impossible, the rope must always sag by some amount.

Before leaving this section on the equilibrium of a body let us reiterate that although the problems considered here have been problems where the body is at rest under the action of forces, bodies moving at constant velocity are also in equilibrium. Some of these problems have already been dealt with in chapter 4, that is, examples 4.11 and 4.13 when a block was moving at a constant velocity under the action of several forces, it was a body in equilibrium.

5.2 The Concept of Torque

Let us now consider the familiar seesaw you played on in the local school yard during your childhood. Suppose a 60.0-lb child (w_1) is placed on the left side of a weightless seesaw and another 40.0-lb child (w_2) is placed on the right side, as shown in figure 5.6. The weights of the two children exert forces down on the seesaw, while the support in the middle exerts a force upward, which is exactly equal to the weight of the two children. According to the first condition of equilibrium,

$$\Sigma F_y = 0$$

the body should be in equilibrium. However, we know from experience that if a 60.0-lb child is at the left end, and a 40.0-lb child is at the right end, the 60.0-lb child will move downward, while the 40.0-lb child moves upward. That is, the seesaw rotates in a counterclockwise direction. Even though the first condition of equilibrium holds, the body is not in complete equilibrium because the seesaw has tilted. It is obvious that the first condition of equilibrium is not sufficient to describe equilibrium. The first condition takes care of the problem of *translational equilibrium* (i.e., the body will not accelerate either in the x-direction or the y-direction), but it says nothing about the problem of *rotational equilibrium*.

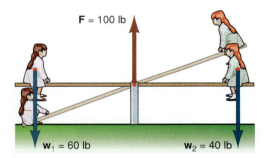
(a)

(b)

Figure 5.6
The seesaw.

Mechanics

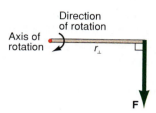

Figure 5.7

Torque defined.

In fact, up to this point in almost all our discussions we assumed that all the forces that act on a body all pass through the center of the body. With the seesaw, the forces do not all pass through the center of the body (figure 5.6), but rather act at different locations on the body. Forces acting on a body that do not all pass through one point of the body are called *nonconcurrent forces*. Hence, even though the forces acting on the body cause the body to be in translational equilibrium, the body is still capable of rotating. Therefore, we need to look into the problem of forces acting on a body at a point other than the center of the body; to determine how these off-center forces cause the rotation of the body; and finally to prevent this rotation so that the body will also be in rotational equilibrium. To do this, we need to introduce the concept of torque.

Torque is defined to be the product of the force times the lever arm. The *lever arm* is defined as the perpendicular distance from the axis of rotation to the line along which the force acts. The line along which the force acts is in the direction of the force vector **F**, and it is sometimes called the *line of action of the force.* The line of action of a force passes through the point of application of the force and is parallel to **F.** This is best seen in figure 5.7. The lever arm appears as $r_\perp$, and the force is denoted by **F.** Note that $r_\perp$ is perpendicular to **F.**

The magnitude of the torque τ (the Greek letter tau) is then defined mathematically as

$$\tau = r_\perp F \tag{5.15}$$

What does this mean physically? Let us consider a very simple example of a torque acting on a body. Let the body be the door to the room. The axes of rotation of the door pass through those hinges that you see at the edge of the door. The distance from the hinge to the door knob is the lever arm $r_\perp$, as shown in figure 5.8. If we exert a force on the door knob by pulling outward, perpendicular to the door, then we have created a torque that acts on the door and is given by equation 5.15. What happens to the door? It opens, just as we would expect. We have caused a rotational motion of the door by applying a torque. Therefore, *an unbalanced torque acting on a body at rest causes that body to be put into rotational motion.* Torque comes from the Latin word *torquere,* which means to twist. We will see in chapter 9, on

Figure 5.8

An example of a torque applied to a door.

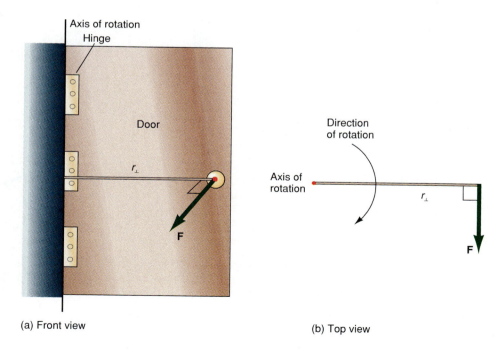

(a) Front view

(b) Top view

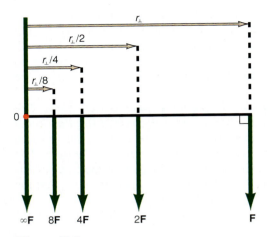

Figure 5.9

If the lever arm decreases, the force must be increased to give the same torque.

Figure 5.10

If the force is not perpendicular to r.

rotational motion, that torque is the rotational analogue of force. When an unbalanced force acts on a body, it gives that body a translational acceleration. When an unbalanced torque acts on a body, it gives that body a rotational acceleration.

It is not so much the applied force that opens a door, but rather the applied torque; the product of the force that we apply and the lever arm. A door knob is therefore placed as far away from the hinges as possible to give the maximum lever arm and hence the maximum torque for a given force.

Because the torque is the product of $r_\perp$ and the force F, for a given value of the force, if the distance $r_\perp$ is cut in half, the value of the torque will also be cut in half. If the torque is to remain the same when the lever arm is halved, the force must be doubled, as we easily see in equation 5.15. If a door knob was placed at the center of the door, then twice the original force would be necessary to give the door the same torque. It may even seem strange that some manufacturers of cabinets and furniture place door knobs in the center of cabinet doors because they may have a certain aesthetic value when placed there, but they cause greater exertion by the furniture owner in order to open those doors.

If the door knob was moved to a quarter of the original distance, then four times the original force would have to be exerted in order to supply the necessary torque to open the door. We can see this effect in the diagram of figure 5.9. If the lever arm was finally decreased to zero, then it would take an infinite force to open the door, which is of course impossible. In general, *if a force acts through the axis of rotation of a body, it has no lever arm (i.e., $r_\perp = 0$) and therefore cannot cause a torque to act on the body about that particular axis*, that is, from equation 5.15

$$\tau = r_\perp F = (0)F = 0$$

Instead of exerting a force perpendicular to the door, suppose we exert a force at some other angle θ, as shown in figure 5.10(a), where θ is the angle between the extension of r and the direction of F. Note that in this case r is not a lever arm since it is not perpendicular to F. The definition of a lever arm is the perpendicular distance from the axis of rotation to the line of action of the force. To obtain the lever arm, we extend a line in either the forward or backward direction of the force.

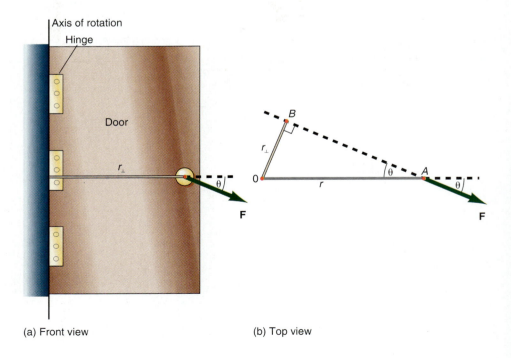

(a) Front view   (b) Top view

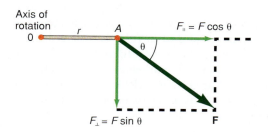

Figure 5.11

The parallel and perpendicular components of a force.

Then we drop a perpendicular to this line, as shown in figure 5.10(b). The line extended in the direction of the force vector, and through the point of application of the force, is the line of action of the force. The lever arm, obtained from the figure, is

$$r_\perp = r \sin \theta \qquad (5.16)$$

In general, if the force is not perpendicular to r, the torque equation 5.15 becomes

$$\tau = r_\perp F = rF \sin \theta \qquad (5.17)$$

Although this approach to using the lever arm to compute the torque is correct, it may seem somewhat artificial, since the force is really applied at the point A and not the point B in figure 5.10(b). Let us therefore look at the problem from a slightly different point of view, as shown in figure 5.11. Take r, exactly as it is given—the distance from the axis of rotation to the point of application of the force. Then take the force vector $\mathbf{F}$ and resolve it into two components: one, $F_\parallel$, lies along the direction of r (parallel to r), and the other, $F_\perp$, is perpendicular to r. The component $F_\parallel$ is a force component that goes right through 0, the axis of rotation. But as just shown, if the force goes through the axis of rotation it has no lever arm about that axis and therefore it cannot produce a torque about that axis. Hence, the component of the force parallel to r cannot create a torque about 0.

The component $F_\perp$, on the other hand, does produce a torque, because it is an application of a force that is perpendicular to a distance r. This perpendicular component produces a torque given by

$$\tau = rF_\perp \qquad (5.18)$$

But from figure 5.11 we see that

$$F_\perp = F \sin \theta \qquad (5.19)$$

Thus, the torque becomes

$$\tau = rF_\perp = rF \sin \theta \qquad (5.20)$$

Comparing equation 5.17 to equation 5.20, it is obvious that the results are identical and should be combined into one equation, namely

$$\tau = r_\perp F = rF_\perp = rF \sin \theta \qquad (5.21)$$

Therefore, *the torque acting on a body can be computed either by (a) the product of the force times the lever arm, (b) the product of the perpendicular component of the force times the distance r, or (c) simply the product of r and F times the sine of the angle between F and the extension of r.*

The unit of torque is given by the product of a distance times a force and in SI units, is a m N, (meter newton), while in the British engineering system the unit is a lb ft. According to our definition, we would expect that this unit should be ft lb. However, as we will see in chapter 7, the unit ft lb is used as a unit of energy. In order to avoid using the same unit for two different physical concepts, we will use the notation lb ft for torque in the British engineering system.

5.3 The Second Condition of Equilibrium

Let us now return to the problem of the two children on the seesaw in figure 5.6, which is redrawn schematically in figure 5.12. The entire length of the seesaw is 12.00 ft. From the discussion of torques, it is now obvious that each child produces a torque tending to rotate the seesaw plank. The first child produces a torque about the axis of rotation, sometimes called the *fulcrum*, given by

$$\tau_1 = F_1 r_1 = w_1 r_1 = (60.0 \text{ lb})(6.00 \text{ ft})$$
$$= 360 \text{ lb ft}$$

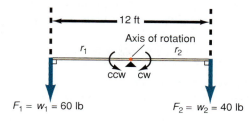

Figure 5.12

The seesaw revisited.

which has a tendency to rotate the seesaw counterclockwise (ccw). *A torque that produces a counterclockwise rotation is sometimes called a counterclockwise torque.* The second child produces a torque about the fulcrum given by

$$\tau_2 = F_2 r_2 = w_2 r_2 = (40.0 \text{ lb})(6.00 \text{ ft})$$
$$= 240 \text{ lb ft}$$

which has a tendency to rotate the seesaw clockwise (cw). *A torque that produces a clockwise rotation is sometimes called a clockwise torque.* These tendencies to rotate the seesaw are opposed to each other. That is, τ_1 tends to produce a counterclockwise rotation with a magnitude of 360 lb ft, while τ_2 has the tendency to produce a clockwise rotation with a magnitude of 240 lb ft. It is a longstanding convention among physicists to designate counterclockwise torques as positive, and clockwise torques as negative. This conforms to the mathematicians' practice of plotting positive angles on an xy plane as measured counterclockwise from the positive x-axis. Hence, τ_1 is a positive torque and τ_2 is a negative torque and the net torque will be the difference between the two, namely

$$\text{net } \tau = \tau_1 - \tau_2 = 360 \text{ lb ft} - 240 \text{ lb ft} = 120 \text{ lb ft}$$

or a net torque τ of 120 lb ft, which will rotate the seesaw counterclockwise.

It is now clear why the seesaw moved. Even though the forces acting on it were balanced, the torques were not. If the torques were balanced then there would be no tendency for the body to rotate, and the seesaw would also be in rotational equilibrium. That is, *the necessary condition for the body to be in rotational equilibrium is that the torques clockwise must be equal to the torques counterclockwise.* That is,

$$\tau_{cw} = \tau_{ccw} \tag{5.22}$$

For this case

$$w_1 r_1 = w_2 r_2 \tag{5.23}$$

We can now solve equation 5.23 for the position r_1 of the first child such that the torques are equal. That is,

$$r_1 = \frac{w_2 r_2}{w_1} = \left(\frac{40.0 \text{ lb}}{60.0 \text{ lb}}\right)(6.00 \text{ ft}) = 4.00 \text{ ft}$$

If the 60.0-lb child moves in toward the axis of rotation by 2.00 ft (4.00 ft from axis), then the torque counterclockwise becomes

$$\tau_1 = \tau_{ccw} = w_1 r_1 = (60.0 \text{ lb})(4.00 \text{ ft}) = 240 \text{ lb ft}$$

which is now equal to the torque τ_2 clockwise. Thus, the torque tending to rotate the seesaw counterclockwise (240 lb ft) is equal to the torque tending to rotate it clockwise (240 lb ft). Hence, the net torque is zero and the seesaw will not rotate. The seesaw is now said to be in rotational equilibrium. This equilibrium condition is shown in figure 5.13.

In general, *for any rigid body acted on by any number of planar torques, the condition for that body to be in rotational equilibrium is that the sum of all the torques clockwise must be equal to the sum of all the torques counterclockwise.* Stated mathematically, this becomes

$$\Sigma \tau_{cw} = \Sigma \tau_{ccw} \tag{5.24}$$

*This condition is called the **second condition of equilibrium**.*

If we subtract the term $\Sigma \tau_{cw}$ from both sides of the equation, we obtain

$$\Sigma \tau_{ccw} - \Sigma \tau_{cw} = 0$$

But the net torque is this difference between the counterclockwise and clockwise torques, *so that the second condition for equilibrium can also be written as: for a*

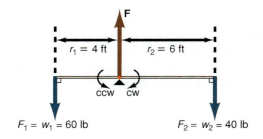

Figure 5.13

The seesaw in equilibrium.

rigid body acted on by any number of torques, the condition for that body to be in rotational equilibrium is that the sum of all the torques acting on that body must be zero, that is,

$$\Sigma \tau = 0 \qquad \text{(5.25)}$$

The torque is about an axis that is perpendicular to the plane of the paper. Since the plane of the paper is the x,y plane, the torque axis lies along the z-axis. Hence the torque can be represented as a vector that lies along the z-axis. Thus, we can also write equation 5.25 as

$$\Sigma \tau_z = 0$$

In general torques can also be exerted about the x-axis and the y-axis, and for such general cases we have

$$\Sigma \tau_x = 0$$
$$\Sigma \tau_y = 0$$

However, in this text we will restrict ourselves to forces in the x,y plane and torques along the z-axis.

5.4 Equilibrium of a Rigid Body

In general, for a body that is acted on by any number of planar forces, the conditions for that body to be in equilibrium are

$$\Sigma F_x = 0 \qquad \text{(5.4)}$$
$$\Sigma F_y = 0 \qquad \text{(5.5)}$$
$$\Sigma \tau_{cw} = \Sigma \tau_{ccw} \qquad \text{(5.24)}$$

  The first condition of equilibrium guarantees that the body will be in translational equilibrium, while the second condition guarantees that the body will be in rotational equilibrium. The solution of various problems of statics reduce to solving the three equations 5.4, 5.5, and 5.25. Section 5.5 is devoted to the solution of various problems of rigid bodies in equilibrium.

5.5 Examples of Rigid Bodies in Equilibrium

Parallel Forces

Two men are carrying a girl on a large plank that is 10.000 m long and weighs 200.0 N. If the girl weighs 445.0 N and sits 3.000 m from one end, how much weight must each man support?

  The diagram drawn in figure 5.14(a) shows all the forces that are acting on the plank. We assume that the plank is uniform and the weight of the plank can be located at its center.

  The first thing we note is that the body is in equilibrium and therefore the two conditions of equilibrium must hold. The first condition of equilibrium, equation 5.5, applied to figure 5.14 yields,

$$\Sigma F_y = 0$$
$$F_1 + F_2 - w_p - w_g = 0$$
$$F_1 + F_2 = w_p + w_g$$
$$= 200.0 \text{ N} + 445.0 \text{ N}$$
$$F_1 + F_2 = 645.0 \text{ N} \qquad \text{(5.26)}$$

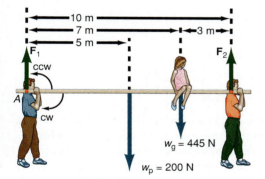

(a) Torques computed about point A

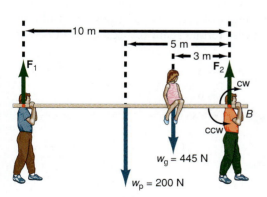

(b) Torques computed about point B

Figure 5.14

A plank in equilibrium under parallel forces.

Since there are no forces in the x-direction, we do not use equation 5.4. The second condition of equilibrium, given by equation 5.24, is

$$\Sigma \tau_{cw} = \Sigma \tau_{ccw}$$

However, before we can compute any torques, we must specify the axis about which the torques will be computed. (In a moment we will see that it does not matter what axis is taken.) For now, let us consider that the axis passes through the point A, where man 1 is holding the plank up with the force F_1. The torques tending to rotate the plank clockwise about axis A are caused by the weight of the plank and the weight of the girl, while the torque tending to rotate the plank counterclockwise about the same axis A is produced by the force F_2 of the second man. Therefore,

$$\Sigma \tau_{cw} = \Sigma \tau_{ccw}$$

$$w_p(5.000 \text{ m}) + w_g(7.000 \text{ m}) = F_2(10.000 \text{ m})$$

Solving for the force F_2 exerted by the second man,

$$F_2 = \frac{w_p(5.000 \text{ m}) + w_g(7.000 \text{ m})}{10.000 \text{ m}}$$

$$= \frac{(200.0 \text{ N})(5.000 \text{ m}) + (445.0 \text{ N})(7.000 \text{ m})}{10.000 \text{ m}}$$

$$= \frac{1000 \text{ m N} + 3115 \text{ m N}}{10.000 \text{ m}}$$

$$F_2 = 411.5 \text{ N} \tag{5.27}$$

Thus, the second man must exert a force upward of 411.5 N. The force that the first man must support, found from equations 5.26 and 5.27, is

$$F_1 + F_2 = 645.0 \text{ N}$$

$$F_1 = 645 \text{ N} - F_2 = 645.0 \text{ N} - 411.5 \text{ N}$$

$$F_1 = 233.5 \text{ N}$$

The first man must exert an upward force of 233.5 N while the second man carries the greater burden of 411.5 N. Note that the force exerted by each man is different. If the girl sat at the center of the plank, then each man would exert the same force.

Let us now see that the same results occur if the torques are computed about any other axis. Let us arbitrarily take the position of the axis to pass through the point B, the location of the force F_2. Since F_2 passes through the axis at point B it cannot produce any torque about that axis because it now has no lever arm. The force F_1 now produces a clockwise torque about the axis through B, while the forces w_p and w_g produce a counterclockwise torque about the axis through B. The solution is

$$\Sigma F_y = 0$$

$$F_1 + F_2 - w_p - w_g = 0$$

$$F_1 + F_2 = w_p + w_g = 645.0 \text{ N}$$

and

$$\Sigma \tau_{cw} = \Sigma \tau_{ccw}$$

$$F_1(10.000 \text{ m}) = w_p(5.000 \text{ m}) + w_g(3.000 \text{ m})$$

Solving for the force F_1,

$$F_1 = \frac{(200.0 \text{ N})(5.000 \text{ m}) + (445.0 \text{ N})(3.000 \text{ m})}{10.000 \text{ m}}$$

$$= \frac{1000 \text{ m N} + 1335 \text{ m N}}{10.000 \text{ m}}$$

$$= 233.5 \text{ N}$$

while the force F_2 is

$$F_2 = 645.0 \text{ N} - F_1$$

$$= 645.0 \text{ N} - 233.5 \text{ N}$$

$$= 411.5 \text{ N}$$

Notice that F_1 and F_2 have the same values as before. As an exercise, take the center of the plank as the point through which the axis passes. Compute the torques about this axis and show that the results are the same.

In general, whenever a rigid body is in equilibrium, every point of that body is in both translational equilibrium and rotational equilibrium, so any point of that body can serve as an axis to compute torques. Even a point outside the body can be used as an axis to compute torques if the body is in equilibrium.

As a general rule, in picking an axis for the computation of torques, try to pick the point that has the largest number of forces acting through it. These forces have no lever arm, and hence produce a zero torque about that axis. This makes the algebra of the problem easier to handle.

The Center of Gravity of a Body

A meter stick of negligible weight has a 10.0-N weight hung from each end. Where, and with what force, should the meter stick be picked up such that it remains horizontal while it moves upward at a constant velocity? This problem is illustrated in figure 5.15.

The meter stick and the two weights constitute a system. If the stick translates with a constant velocity, then the system is in equilibrium under the action of all the forces. The conditions of equilibrium must apply and hence the sum of the forces in the y-direction must equal zero,

$$\Sigma F_y = 0 \tag{5.5}$$

Applying equation 5.5 to this problem gives

$$F - w_1 - w_2 = 0$$

$$F = w_1 + w_2 = 10.0 \text{ N} + 10.0 \text{ N} = 20.0 \text{ N}$$

Therefore, a force of 20 N must be exerted in order to lift the stick. But where should this force be applied? In general, the exact position is unknown so we assume that it can be lifted at some point that is a distance x from the left end of the stick. If this is the correct position, then the body is also in rotational equilibrium and the second condition of equilibrium must also apply. Hence, the sum of the torques clockwise must be set equal to the sum of the torques counterclockwise,

$$\Sigma \tau_{cw} = \Sigma \tau_{ccw} \tag{5.24}$$

Taking the left end of the meter stick as the axis of rotation, the second condition, equation 5.24, becomes

$$w_2 l = Fx \tag{5.28}$$

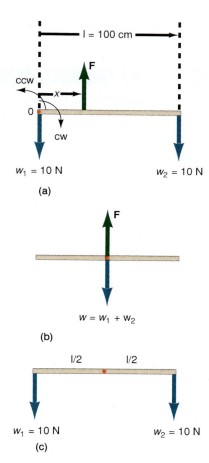

Figure 5.15

The center of gravity of a meter stick.

Since we already found F from the first condition, and w_2 and l are known, we can solve for x, the point where the stick should be lifted:

$$x = \frac{w_2 l}{F} = \frac{(10.0 \text{ N})(100 \text{ cm})}{20.0 \text{ N}}$$

$$= 50.0 \text{ cm}$$

The meter stick should be lifted at its exact geometrical center.

    The net effect of these forces can be seen in figure 5.15(b). The force up F is equal to the weight down W. The torque clockwise is balanced by the counterclockwise torque, and there is no tendency for rotation. The stick, with its equal weights at both ends, acts as though all the weights were concentrated at the geometrical center of the stick. *This point that behaves as if all the weight of the body acts through it, is called the **center of gravity (cg)** of the body.* Hence the center of gravity of the system, in this case a meter stick and two equal weights hanging at the ends, is located at the geometrical center of the meter stick.

    The center of gravity is located at the center of the stick because of the symmetry of the problem. The torque clockwise about the center of the stick is w_2 times $1/2$, while the torque counterclockwise about the center of the stick is w_1 times $1/2$, as seen in figure 5.15(c). Because the weights w_1 and w_2 are equal, and the lever arms ($1/2$) are equal, the torque clockwise is equal to the torque counterclockwise. Whenever such symmetry between the weights and the lever arms exists, the center of gravity is always located at the geometric center of the body or system of bodies.

The center of gravity when there is no symmetry. If weight w_2 in the preceding discussion is changed to 20.0 N, where will the center of gravity of the system be located?

Solution

The first condition of equilibrium yields

$$\Sigma F_y = 0$$

$$F - w_1 - w_2 = 0$$

$$F = w_1 + w_2 = 10.0 \text{ N} + 20.0 \text{ N}$$

$$= 30.0 \text{ N}$$

The second condition of equilibrium again yields equation 5.28,

$$w_2 l = Fx$$

The location of the center of gravity becomes

$$x = \frac{w_2 l}{F} = \frac{(20.0 \text{ N})(100 \text{ cm})}{30.0 \text{ N}}$$

$$= 66.7 \text{ cm}$$

Thus, when there is no longer the symmetry between weights and lever arms, the center of gravity is no longer located at the geometric center of the system.

General Definition of the Center of Gravity

In the previous section we assumed that the weight of the meter stick was negligible compared to the weights w_1 and w_2. Suppose the weights w_1 and w_2 are eliminated and we want to pick up the meter stick all by itself. The weight of the meter stick can no longer be ignored. But how can the weight of the meter stick be handled?

In the previous problem w_1 and w_2 were discrete weights. Here, the weight of the meter stick is distributed throughout the entire length of the stick. How can the center of gravity of a continuous mass distribution be determined instead of a discrete mass distribution? From the symmetry of the uniform meter stick, we expect that the center of gravity should be located at the geometric center of the 100-cm meter stick, that is, at the point $x = 50$ cm. At this center point, half the mass of the stick is to the left of center, while the other half of the mass is to the right of center. The half of the mass on the left side creates a torque counterclockwise about the center of the stick, while the half of the mass on the right side creates a torque clockwise. Thus, the uniform meter stick has the same symmetry as the stick with two equal weights acting at its ends, and thus must have its center of gravity located at the geometrical center of the meter stick, the 50-cm mark.

To find a general equation for the center of gravity of a body, let us find the equation for the center of gravity of the uniform meter stick shown in figure 5.16. The meter stick is divided up into 10 equal parts, each of length 10 cm. Because the meter stick is uniform, each 10-cm portion contains 1/10 of the total weight of the meter stick, W. Let us call each small weight w_i, where the i is a subscript that identifies which w is being considered.

Because of the symmetry of the uniform mass distribution, each small weight w_i acts at the center of each 10-cm portion. The center of each ith portion, denoted by x_i, is shown in the figure. If a force F is exerted upward at the center of gravity x_{cg}, the meter stick should be balanced. If we apply the first condition of equilibrium to the stick we obtain

$$\Sigma F_y = 0$$
$$F - w_1 - w_2 - w_3 - \cdots - w_{10} = 0$$
$$F = w_1 + w_2 + w_3 + \cdots + w_{10} \qquad (5.5)$$

A shorthand notation for this sum can be written as

$$w_1 + w_2 + w_3 + \cdots + w_{10} = \sum_{i=1}^{n} w_i$$

The Greek letter Σ again means "sum of," and when placed in front of w_i it means "the sum of each w_i." The notation $i = 1$ to n, means that we will sum up some n w_i's. In this case, $n = 10$. Using this notation, the first condition of equilibrium becomes

$$F = \sum_{i=1}^{n} w_i = W \qquad (5.29)$$

The sum of all these w_i's is equal to the total weight of the meter stick W.

The second condition of equilibrium,

$$\Sigma \tau_{cw} = \Sigma \tau_{ccw} \qquad (5.24)$$

Figure 5.16

The weight distribution of a uniform meter stick.

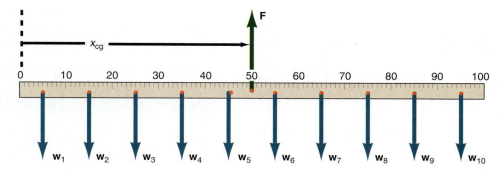

when applied to the meter stick, with the axis taken at the zero of the meter stick, yields

$$(w_1x_1 + w_2x_2 + w_3x_3 + \cdots + w_{10}x_{10}) = Fx_{cg}$$

In the shorthand notation this becomes

$$\sum_{i=1}^{n} w_ix_i = Fx_{cg}$$

Solving for x_{cg}, we have

$$x_{cg} = \frac{\sum\limits_{i=1}^{n} w_ix_i}{F} \tag{5.30}$$

Using equation 5.29, *the general expression for the x-coordinate of the center of gravity of a body is given by*

$$x_{cg} = \frac{\sum\limits_{i=1}^{n} w_ix_i}{W} \tag{5.31}$$

Applying equation 5.31 to the uniform meter stick we have

$$x_{cg} = \frac{\sum w_ix_i}{W} = \frac{w_1x_1 + w_2x_2 + \cdots + w_{10}x_{10}}{W}$$

but since $w_1 = w_2 = w_3 = w_4 = \cdots = w_{10} = W/10$, it can be factored out giving

$$x_{cg} = \frac{W/10}{W}(x_1 + x_2 + x_3 + \cdots + x_{10})$$

$$= 1/10\,(5 + 15 + 25 + 45 + \cdots + 95)$$

$$= 500/10$$

$$= 50 \text{ cm}$$

The center of gravity of the uniform meter stick is located at its geometrical center, just as expected from symmetry considerations. The assumption that the weight of a body can be located at its geometrical center, provided that its mass is uniformly distributed, has already been used throughout this book. Now we have seen that this was a correct assumption.

To find the center of gravity of a two-dimensional body, the *x*-coordinate of the cg is found from equation 5.31, while the *y*-coordinate, found in an analogous manner, is

$$y_{cg} = \frac{\sum\limits_{i=1}^{n} w_iy_i}{W} \tag{5.32}$$

For a nonuniform body or one with a nonsymmetrical shape, the problem becomes much more complicated with the sums in equations 5.31 and 5.32 becoming integrals and will not be treated in this book.

Examples Illustrating the Concept of the Center of Gravity

Example 5.5

The center of gravity of a weighted beam. A weight of 50.0 N is hung from one end of a uniform beam 12.0 m long. If the beam weighs 25.0 N, where and with what force should the beam be picked up so that it remains horizontal? The problem is illustrated in figure 5.17.

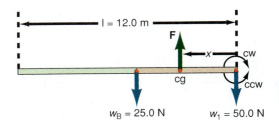

Figure 5.17

The center of gravity of a weighted beam.

Because the beam is uniform, the weight of the beam w_B is located at the geometric center of the beam. Let us assume that the center of gravity of the system of beam and weight is located at a distance x from the right side of the beam. The body is in equilibrium, and the equations of equilibrium become

$$\Sigma F_y = 0 \tag{5.5}$$

$$F - w_B - w_1 = 0$$

$$F = w_B + w_1$$

$$= 25.0 \text{ N} + 50.0 \text{ N} = 75.0 \text{ N}$$

Taking the right end of the beam as the axis about which the torques are computed, we have

$$\Sigma \tau_{cw} = \Sigma \tau_{ccw} \tag{5.24}$$

The force F will cause a torque clockwise about the right end, while the force w_B will cause a counterclockwise torque. Hence,

$$Fx = w_B \frac{l}{2}$$

Thus, the center of gravity of the system is located at

$$x_{cg} = \frac{w_B \, l/2}{F}$$

$$= \frac{(25.0 \text{ N})(6.0 \text{ m})}{75.0 \text{ N}} = 2.0 \text{ m}$$

Therefore, we should pick up the beam 2.0 m from the right hand side with a force of 75.0 N.

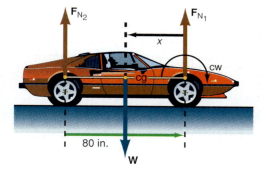

Figure 5.18

The center of gravity of an automobile.

Example 5.6

The center of gravity of an automobile. The front wheels of an automobile, when run onto a platform scale, are found to support 1800 lb, while the rear wheels can support 1500 lb. The auto has an 80.0-in. wheel base (distance from the front axle to the rear axle). Locate the center of gravity of the car. The car is shown in figure 5.18.

Solution

If the car pushes down on the scales with forces w_1 and w_2, then the scale exerts normal forces upward of F_{N_1} and F_{N_2}, respectively, on the car. The total weight of the car is W and can be located at the center of gravity of the car. Since the location of this cg is unknown, let us assume that it is at a distance x from the front wheels. Because the car is obviously in equilibrium, the conditions of equilibrium are applied. Thus,

$$\Sigma F_y = 0 \tag{5.5}$$

From figure 5.18, we see that this is

$$F_{N_1} + F_{N_2} - W = 0$$

$$F_{N_1} + F_{N_2} = W$$

Solving for W, the weight of the car, we get

$$W = 1800 \text{ lb} + 1500 \text{ lb} = 3300 \text{ lb}$$

The second condition of equilibrium, using the front axle of the car as the axis, gives

$$\Sigma \tau_{cw} = \Sigma \tau_{ccw} \tag{5.24}$$

The force F_{N_2} will cause a clockwise torque about the front axle, while W will cause a counterclockwise torque. Hence,

$$F_{N_2}(80.0 \text{ in.}) = Wx_{cg}$$

Solving for the center of gravity, we get

$$x_{cg} = F_{N_2} \frac{(80.0 \text{ in.})}{W}$$

$$= \frac{(1500 \text{ lb})(80.0 \text{ in.})}{3300 \text{ lb}}$$

$$= 36.4 \text{ in.}$$

That is, the cg of the car is located 36.4 in. behind the front axle of the car.

Center of Mass

*The **center of mass (cm)** of a body or system of bodies is defined as that point that moves in the same way that a single particle of the same mass would move when acted on by the same forces. Hence, the point reacts as if all the mass of the body were concentrated at that point.* All the external forces can be considered to act at the center of mass when the body undergoes any translational acceleration. The general motion of any rigid body can be resolved into the translational motion of the center of mass and the rotation about the center of mass. *On the surface of the earth, where g, the acceleration due to gravity, is relatively uniform, the center of mass (cm) of the body will coincide with the center of gravity (cg) of the body.* To see this, take equation 5.31 and note that

$$w_i = m_i g$$

Substituting this into equation 5.31 we get

$$x_{cg} = \frac{\Sigma w_i x_i}{\Sigma w_i} = \frac{\Sigma (m_i g) x_i}{\Sigma (m_i g)}$$

Factoring the g outside of the summations, we get

$$x_{cg} = \frac{g \Sigma m_i x_i}{g \Sigma m_i} \tag{5.33}$$

The right-hand side of equation 5.33 is the defining relation for the center of mass of a body, and we will write it as

$$x_{cm} = \frac{\Sigma m_i x_i}{\Sigma m_i} = \frac{\Sigma m_i x_i}{M} \tag{5.34}$$

where M is the total mass of the body. *Equation 5.34 represents the x-coordinate of the center of mass of the body.* We obtain a similar equation for the y-coordinate by replacing the letter x with the letter y in equation 5.34:

$$y_{cm} = \frac{\Sigma m_i y_i}{\Sigma m_i} = \frac{\Sigma m_i y_i}{M} \tag{5.35}$$

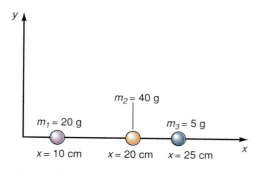

Figure 5.19

The center of mass.

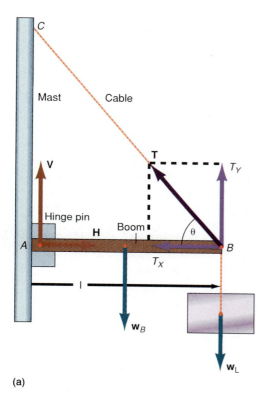

(a)

(b)

Figure 5.20

The crane boom.

Example 5.7

Finding the center of mass. Three masses, $m_1 = 20.0$ g, $m_2 = 40.0$ g, and $m_3 = 5.00$ g are located on the x-axis at 10.0, 20.0, and 25.0 cm, respectively, as shown in figure 5.19. Find the center of mass of the system of three masses.

Solution

The center of mass is found from equation 5.34 with $n = 3$. Thus,

$$x_{cm} = \frac{\Sigma m_i x_i}{\Sigma m_i} = \frac{m_1 x_1 + m_2 x_2 + m_3 x_3}{m_1 + m_2 + m_3}$$

$$= \frac{(20.0 \text{ g})(10.0 \text{ cm}) + (40.0 \text{ g})(20.0 \text{ cm}) + (5.00 \text{ g})(25.0 \text{ cm})}{20.0 \text{ g} + 40.0 \text{ g} + 5.00 \text{ g}}$$

$$= \frac{1125 \text{ g cm}}{65.0 \text{ g}}$$

$$= 17.3 \text{ cm}$$

The center of mass of the three masses is at 17.3 cm.

The Crane Boom

A large uniform boom is connected to the mast by a hinge pin at the point A in figure 5.20. A load w_L is to be supported at the other end B. A cable is also tied to B and connected to the mast at C to give additional support to the boom. We want to determine all the forces that are acting on the boom in order to make sure that the boom, hinge pin, and cable are capable of withstanding these forces when the boom is carrying the load w_L.

First, what are the forces acting on the boom? Because the boom is uniform, its weight $\mathbf{w_B}$ can be situated at its center of gravity, which coincides with its geometrical center. There is a tension $\mathbf{T}$ in the cable acting at an angle θ to the boom. At the hinge pin, there are two forces acting. The first, denoted by $\mathbf{V}$, is a vertical force acting on the end of the boom. If this force were not acting on the boom at this end point, this end of the boom would fall down. That is, the pin with this associated force $\mathbf{V}$ is holding the boom up.

Second, there is also a horizontal force $\mathbf{H}$ acting on the boom toward the right. The horizontal component of the tension $\mathbf{T}$ pushes the boom into the mast. The force $\mathbf{H}$ is the reaction force that the mast exerts on the boom. If there were no force $\mathbf{H}$, the boom would go right through the mast. The vector sum of these two forces, $\mathbf{V}$ and $\mathbf{H}$, is sometimes written as a single contact force at the location of the hinge pin. However, since we want to have the forces in the x- and y-directions, we will leave the forces in the vertical and horizontal directions. The tension $\mathbf{T}$ in the cable also has a vertical component T_y, which helps to hold up the load and the boom.

Let us now determine the forces V, H, and T acting on the system when $\theta = 30.0°$, $w_B = 270$ N, $w_L = 900$ N, and the length of the boom, $l = 6.00$ m. The first thing to do to solve this problem is to observe that the body, the boom, is at rest under the action of several different forces, and must therefore be in equilibrium. Hence, the first and second conditions of equilibrium must apply:

$$\Sigma F_y = 0 \tag{5.5}$$

$$\Sigma F_x = 0 \tag{5.4}$$

$$\Sigma \tau_{cw} = \Sigma \tau_{ccw} \tag{5.24}$$

Using figure 5.20, we observe which forces are acting in the y-direction. Equation 5.5 becomes

$$\Sigma F_y = V + T_y - w_B - w_L = 0$$

or

$$V + T_y = w_B + w_L \tag{5.36}$$

Note from figure 5.20 that $T_y = T \sin \theta$. The right-hand side of equation 5.36 is known, because w_B and w_L are known. But the left-hand side contains the two unknowns, V and T, so we can not proceed any further with this equation at this time.

Let us now consider the second of the equilibrium equations, namely equation 5.4. Using figure 5.20, notice that the force in the positive x-direction is H, while the force in the negative x-direction is T_x. Thus, the equilibrium equation 5.4 becomes

$$\Sigma F_x = H - T_x = 0$$

or

$$H = T_x = T \cos \theta \tag{5.37}$$

There are two unknowns in this equation, namely H and T. At this point, we have two equations with the three unknowns V, H, and T. We need another equation to determine the solution of the problem. This equation comes from the second condition of equilibrium, equation 5.24. In order to compute the torques, we must first pick an axis of rotation. Remember, any point can be picked for the axis to pass through. For convenience we pick the point A in figure 5.20, where the forces V and H are acting, for the axis of rotation to pass through. The forces w_B and w_L are the forces that produce the clockwise torques about the axis at A, while T_y produces the counterclockwise torque. Therefore, equation 5.24 becomes

$$w_B(l/2) + w_L(l) = T_y(l) = T \sin \theta \, (l) \tag{5.38}$$

After dividing term by term by the length l, we can solve equation 5.38 for T. Thus,

$$T \sin \theta = (w_B/2) + w_L$$

The tension in the cable is therefore

$$T = \frac{(w_B/2) + w_L}{\sin \theta} \tag{5.39}$$

Substituting the values of w_B, w_L, and θ, into equation 5.39 we get

$$T = \frac{(270 \text{ N}/2) + 900 \text{ N}}{\sin 30.0°}$$

or

$$T = 2070 \text{ N}$$

The tension in the cable is 2070 N. We can find the second unknown force H by substituting this value of T into equation 5.37:

$$H = T \cos \theta = (2070 \text{ N}) \cos 30.0°$$

and

$$H = 1790 \text{ N}$$

The horizontal force exerted on the boom by the hinge pin is 1790 N. We find the final unknown force V by substituting T into equation 5.36, and solving for V, we get

$$V = w_B + w_L - T \sin \theta \qquad (5.40)$$

$$= 270 \text{ N} + 900 \text{ N} - (2070 \text{ N})\sin 30.0°$$

$$= 135 \text{ N}$$

The hinge pin exerts a force of 135 N on the boom in the vertical direction. To summarize, the forces acting on the boom are $V = 135$ N, $H = 1790$ N, and $T = 2070$ N. The reason we are concerned with the value of these forces, is that the boom is designed to carry a particular load. If the boom system is not capable of withstanding these forces the boom will collapse. For example, we just found the tension in the cable to be 2070 N. Is the cable that will be used in the system capable of withstanding a tension of 2070 N? If it is not, the cable will break, the boom will collapse, and the load will fall down. On the other hand, is the hinge pin capable of taking a vertical stress of 135 N and a horizontal stress of 1790 N? If it is not designed to withstand these forces, the pin will be sheared and again the entire system will collapse. Also note that this is not a very well designed boom system in that the hinge pin must be able to withstand only 135 N in the vertical while the horizontal force is 1790 N. In designing a real system the cable could be moved to a much higher position on the mast thereby increasing the angle θ, reducing the component T_x, and hence decreasing the force component H.

There are many variations of the boom problem. Some have the boom placed at an angle to the horizontal. Others have the cable at any angle, and connected to almost any position on the boom. But the procedure for the solution is still the same. The boom is an object in equilibrium and equations 5.4, 5.5, and 5.24 must apply. Variations on the boom problem presented here are included in the problems at the end of the chapter.

The Ladder

A ladder of length L is placed against a wall, as shown in figure 5.21. A person, of weight $\mathbf{w_P}$, ascends the ladder until the person is located a distance d from the top of the ladder. We want to determine all the forces that are acting on the ladder. We assume that the ladder is uniform. Hence, the weight of the ladder $\mathbf{w_L}$ can be located at its geometrical center, that is, at $L/2$. There are two forces acting on the bottom of the ladder, $\mathbf{V}$ and $\mathbf{H}$. The vertical force $\mathbf{V}$ represents the reaction force that the ground exerts on the ladder. That is, since the ladder pushes against the ground, the ground must exert an equal but opposite force upward on the ladder.

With the ladder in this tilted position, there is a tendency for the ladder to slip to the left at the ground. If there is a tendency for the ladder to be in motion to the left, then there must be a frictional force tending to oppose that motion, and therefore that frictional force must act toward the right. We call this horizontal frictional force $\mathbf{H}$. At the top of the ladder there is a force $\mathbf{F}$ on the ladder that acts normal to the wall. This force is the force that the wall exerts on the ladder and is the reaction force to the force that the ladder exerts on the wall. There is also a tendency for the ladder to slide down the wall and therefore there should also be a frictional force on the ladder acting upward at the wall. To solve the general case where there is friction at the wall is extremely difficult. We simplify the problem by assuming that the wall is smooth and hence there is no frictional force acting on the top of the ladder. Thus, whatever results that are obtained in this problem are an approximation to reality.

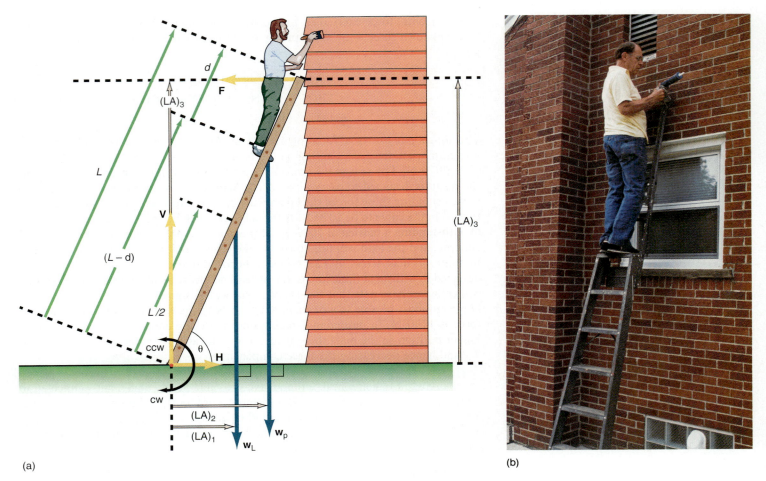

(a)

(b)

Figure 5.21

The ladder.

Since the ladder is at rest under the action of several forces it must be in static equilibrium. Hence, the first and second conditions of equilibrium must apply. Namely,

$$\Sigma F_y = 0 \tag{5.5}$$

$$\Sigma F_x = 0 \tag{5.4}$$

$$\Sigma \tau_{cw} = \Sigma \tau_{ccw} \tag{5.24}$$

Figure 5.21 shows that the force upward is V, while the forces downward are w_L and w_p. Substituting these values into equation 5.5 gives

$$\Sigma F_y = V - w_L - w_p = 0$$

or

$$V = w_L + w_p \tag{5.41}$$

The figure also shows that the force to the right is H, while the force to the left is F. Equation 5.4 therefore becomes

$$\Sigma F_x = H - F = 0$$

or

$$H = F \tag{5.42}$$

It is important that you see how equations 5.41 and 5.42 are obtained from figure 5.21. This is the part that really deals with the physics of the problem. Once all the equations are obtained, their solution is really a matter of simple mathematics.

Before we can compute any of the torques for the second condition of equilibrium, we must pick an axis of rotation. As already pointed out, we can pick any axis to compute the torques. We pick the base of the ladder as the axis of rotation. The forces V and H go through this axis and, therefore, V and H produce no torques about this axis, because they have no lever arms. Observe from the figure that the weights w_L and w_p are the forces that produce clockwise torques, while F is the force that produces the counterclockwise torque. Recall, that torque is the product of the force times the lever arm, where the lever arm is the perpendicular distance from the axis of rotation to the direction or line of action of the force. Note from figure 5.21 that the distance from the axis of rotation to the center of gravity of the ladder does not make a 90° angle with the force w_L, and therefore $L/2$ cannot be a lever arm. If we drop a perpendicular from the axis of rotation to the line showing the direction of the vector $\mathbf{w_L}$, we obtain the lever arm (LA) given by

$$(LA)_1 = (L/2) \cos \theta$$

Thus, the torque clockwise produced by w_L is

$$\tau_{1cw} = w_L(L/2) \cos \theta \tag{5.43}$$

Similarly, the lever arm associated with the weight of the person is

$$(LA)_2 = (L - d) \cos \theta$$

Hence, the second torque clockwise is

$$\tau_{2cw} = w_p(L - d) \cos \theta \tag{5.44}$$

The counterclockwise torque is caused by the force F. However, the ladder does not make an angle of 90° with the force F, and the length L from the axis of rotation to the wall, is not a lever arm. We obtain the lever arm associated with the force F by dropping a perpendicular from the axis of rotation to the direction of the force vector $\mathbf{F}$, as shown in figure 5.21. Note that in order for the force vector to intersect the lever arm, the line from the force had to be extended until it did intersect the lever arm. We call this extended line the *line of action of the force*. This lever arm $(LA)_3$ is equal to the height on the wall where the ladder touches the wall, and is found by the trigonometry of the figure as

$$(LA)_3 = L \sin \theta$$

Hence, the counterclockwise torque produced by F is

$$\tau_{ccw} = FL \sin \theta \tag{5.45}$$

Substituting equations 5.43, 5.44, and 5.45 into equation 5.24 for the second condition of equilibrium, yields

$$w_L(L/2)\cos \theta + w_p(L - d)\cos \theta = FL \sin \theta \tag{5.46}$$

The physics of the problem is now complete. It only remains to solve the three equations 5.41, 5.42, and 5.46 mathematically. There are three equations with the three unknowns V, H, and F.

As a typical problem, let us assume that the following data are given: $\theta = 60.0°$, $w_L = 40.0$ lb, $w_p = 160$ lb, $L = 20.0$ ft, and $d = 5.00$ ft. Equation 5.46, solved for the force F, gives

$$F = \frac{w_L(L/2)\cos \theta + w_p(L - d)\cos \theta}{L \sin \theta} \tag{5.47}$$

Substituting the values just given, we have

$$F = \frac{40.0 \text{ lb}(10.0 \text{ ft})\cos 60.0° + 160 \text{ lb}(15.0 \text{ ft}) \cos 60.0°}{20.00 \text{ ft} \sin 60.0°}$$

$$= \frac{200 \text{ lb ft} + 1200 \text{ lb ft}}{17.32 \text{ ft}}$$

$$= 80.8 \text{ lb}$$

However, since $H = F$ from equation 5.42, we have

$$H = 80.8 \text{ lb}$$

Solving for V from equation 5.41 we obtain

$$V = w_L + w_p = 40.0 \text{ lb} + 160 \text{ lb}$$

$$= 200 \text{ lb}$$

Thus, we have found the three forces F, V, and H acting on the ladder.

As a variation of this problem, we might ask, "What is the minimum value of the coefficient of friction between the ladder and the ground, such that the ladder will not slip out at the ground?" Recall from setting up this problem, that H is indeed a frictional force, opposing the tendency of the bottom of the ladder to slip out, and as such is given by

$$H = f_s = \mu_s F_N \tag{5.48}$$

But the normal force F_N that the ground exerts on the ladder, seen from figure 5.21, is the vertical force V. Hence,

$$H = \mu_s V$$

The coefficient of friction between the ground and the ladder is therefore

$$\mu_s = \frac{H}{V}$$

For this particular example, the minimum coefficient of friction is

$$\mu_s = \frac{80.8 \text{ lb}}{200 \text{ lb}}$$

$$\mu_s = 0.404$$

If μ_s is not equal to, or greater than 0.404, then the necessary frictional force H is absent and the ladder will slide out at the ground.

Applications of the Theory of Equilibrium to the Health Sciences

Example 5.8

A weight lifter's dumbbell curls. A weight lifter is lifting a dumbbell that weighs 75.0 lb, as shown in figure 5.22(a) The biceps muscle exerts a force F_M upward on the forearm at a point approximately 2.00 in. from the elbow joint. The forearm weighs approximately 15.0 lb and its center of gravity is located approximately 7.30 in. from the elbow joint. The upper arm exerts a force at the elbow joint that we denote by F_J. The dumbbell is located approximately 14.5 in. from the elbow. What force must be exerted by the biceps muscle in order to lift the dumbbell?

Figure 5.22

The arm lifting a weight.

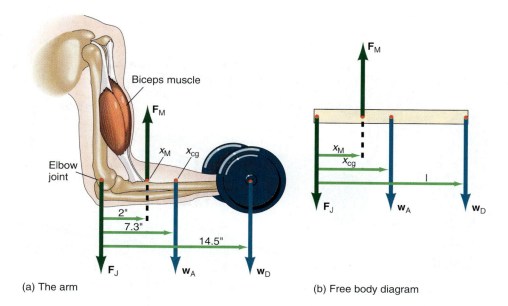

(a) The arm   (b) Free body diagram

Solution

The free body diagram for the arm is shown in figure 5.22(b). The first condition of equilibrium gives

$$\Sigma F_y = F_M - F_J - w_A - w_D = 0$$

$$F_M = F_J + w_A + w_D$$

$$F_M = F_J + 15.0 \text{ lb} + 75.0 \text{ lb} \tag{5.49}$$

$$F_M = F_J + 90.0 \text{ lb} \tag{5.50}$$

Taking the elbow joint as the axis, the second condition of equilibrium gives

$$\Sigma \tau_{cw} = \Sigma \tau_{ccw}$$

$$w_A x_{cg} + w_D l = F_M x_M \tag{5.51}$$

The force exerted by the biceps muscle becomes

$$F_M = \frac{w_A x_{cg} + w_D l}{x_M} \tag{5.52}$$

$$= \frac{(15.0 \text{ lb})(7.30 \text{ in.}) + (75.0 \text{ lb})(14.5 \text{ in.})}{2.00 \text{ in.}}$$

$$= 599 \text{ lb}$$

Thus, the biceps muscle exerts the relatively large force of 599 lb in lifting the 75.0 lb dumbbell. We can now find the force at the joint, from equation 5.50, as

$$F_J = F_M - 90.0 \text{ lb}$$

$$= 599 \text{ lb} - 90 \text{ lb} = 509 \text{ lb}$$

Example 5.9

A weight lifter's bend over rowing. A weight lifter bends over at an angle of 50.0° to the horizontal, as shown in figure 5.23(a). He holds a barbell that weighs 150 lb, w_B. The spina muscle in his back supplies the force F_M to hold the spine of his back in this position. The length of the man's spine is approximately 27.0 in. The spina muscle acts approximately 18.0 in. from the base of the spine and makes an angle

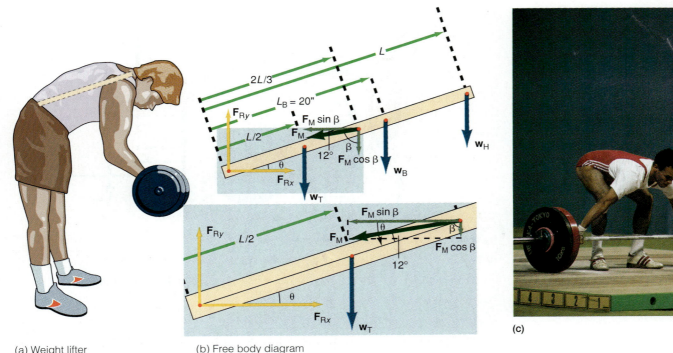

(a) Weight lifter

(b) Free body diagram

(c)

Figure 5.23

Forces on the spinal column.

of 12.0° with the spine, as shown. The man's head weighs about 14.0 lb, w_H, and this force acts at the top of the spinal column, as shown. The torso of the man weighs about 80 lb and this is denoted by w_T, and is located at the center of gravity of the torso, which is taken as 13.5 in. At the base of the spinal column is the fifth lumbar vertebra, which acts as the axis about which the body bends. A reaction force F_R acts on this fifth lumbar vertebra, as shown in the figure. Determine the reaction force F_R and the muscular force F_M on the spine.

Solution

A free body diagram of all the forces is shown in figure 5.23(b). Note that the angle β is

$$\beta = 90° - \theta + 12° = 90° - 50° + 12° = 52°$$

The first condition of equilibrium yields

$$\Sigma F_y = 0$$

$$F_R \sin \theta - w_T - w_B - w_H - F_M \cos \beta = 0$$

or

$$F_R \sin \theta = w_T + w_B + w_H + F_M \cos \beta \qquad (5.53)$$

and

$$\Sigma F_x = 0$$

$$F_R \cos \theta - F_M \sin \beta = 0$$

or

$$F_R \cos \theta = F_M \sin \beta \qquad (5.54)$$

The second condition of equilibrium gives

$$\Sigma \tau_{cw} = \Sigma \tau_{ccw}$$

$$w_T(L/2)\cos \theta + F_M\cos \beta(2L/3)\cos \theta + w_BL_B\cos \theta$$

$$+ w_HL \cos \theta = F_M \sin \beta(2L/3)\sin \theta \qquad (5.55)$$

Solving for F_M, the force exerted by the muscles, gives

$$F_M = \frac{w_T(L/2)\cos \theta + w_BL_B\cos \theta + w_HL \cos \theta}{\sin \beta(2L/3)\sin \theta - \cos \beta(2L/3)\cos \theta} \qquad (5.56)$$

$$= \frac{(80 \text{ lb})(13.5 \text{ in.})(\cos 50°) + (150 \text{ lb})(20 \text{ in.})(\cos 50°) + (14 \text{ lb})(27 \text{ in.})(\cos 50°)}{(\sin 52°)(18 \text{ in.})(\sin 50°) - (\cos 52°)(18 \text{ in.})(\cos 50°)}$$

$$= 764 \text{ lb}$$

The reaction force F_R on the base of the spine, found from equation 5.54, is

$$F_R = \frac{F_M \sin \beta}{\cos \theta}$$

$$= \frac{(764 \text{ lb})\sin 52°}{\cos 50°} = 937 \text{ lb}$$

Thus in lifting a 150 lb barbell there is a force on the spinal disk at the base of the spine of 937 lb.

"Have you ever wondered . . . ?"

An Essay on the Application of Physics

Traction

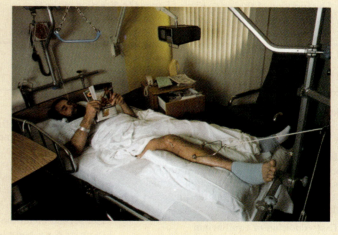

Have you ever wondered, while visiting Uncle Johnny in the hospital, what they were doing to that poor man in the other bed (figure 1)? As you can see in figure 2, they have him connected to all kinds of pulleys, ropes, and weights. It looks like some kind of medieval torture rack, where they are stretching the man until he tells all he knows. Or perhaps the man is a little short for his weight and they are just trying to stretch him to normal size.

Of course it is none of these things, but the idea of stretching is correct. Actually the man in the other bed is in traction. Traction is essentially a process of exerting a force on a skeletal structure in order to hold a bone in a prescribed position. Traction is used in the treatment of fractures and is a direct application of a body in equilibrium under a number of forces. The object of traction is to exert sufficient force to keep the two sections of the fractured bone in alignment and just touching while they heal. The traction process thus prevents muscle contraction that might cause misalignment at the fracture. The traction force can be exerted through a splint or by a steel pin passed directly through the bone.

Figure 1

A man in traction.

An example of one type of traction, shown in figure 2, is known as Russell traction and is used in the treatment of a fracture of the femur.

Let us analyze the problem from the point of view of equilibrium. First note that almost all of the forces on the bone are transmitted by the ropes that pass around the pulleys. The characteristic of all the systems with pulleys and ropes that are used

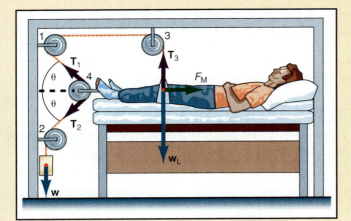

Figure 2
Russell traction.

in traction is that the tension in the taut connecting rope is everywhere the same. Thus, the forces exerted on the bone are the tensions T_1, T_2, T_3, the weight of the leg w_L, and the force exerted by the muscles F_M. The first condition of equilibrium applied to the leg yields

$$\Sigma F_y = 0$$

$$= T_1 \sin \theta + T_3 - T_2 \sin \theta - w_L = 0 \quad \textbf{(5H.1)}$$

The function of the pulleys is to change the direction of the force, but the tension in the rope is everywhere the same. But the tension T is supplied by the weight w that is hung from the end of the bed and is thus equal to the weight w. Hence,

$$T_1 = T_2 = T_3 = w \quad \textbf{(5H.2)}$$

Equation 5H.1 now becomes

$$w \sin \theta + w - w \sin \theta - w_L = 0$$

or

$$w = w_L \quad \textbf{(5H.3)}$$

Thus the weight w hung from the bottom of the bed must be equal to the weight of the leg w_L.

The second equation of the first condition of equilibrium is

$$\Sigma F_x = 0$$

$$F_M - T_1 \cos \theta - T_2 \cos \theta = 0 \quad \textbf{(5H.4)}$$

Using equation 5H.2 this becomes

$$F_M - w \cos \theta - w \cos \theta = 0$$

$$F_M = w \cos \theta + w \cos \theta$$

Thus,

$$F_M = 2w \cos \theta \quad \textbf{(5H.5)}$$

which says that by varying the angle θ, the force to overcome muscle contraction can be varied to any value desired. In this analysis, the force exerted to overcome the muscle contraction lies along the axis of the bone. Variations of this technique can be used if we want to have the traction force exerted at any angle because of the nature of the medical problem.

The Language of Physics

Statics
That portion of the study of mechanics that deals with bodies in equilibrium (p. 121).

Equilibrium
A body is said to be in equilibrium under the action of several forces if the body has zero translational acceleration and no rotational motion (p. 121).

The first condition of equilibrium
For a body to be in equilibrium the vector sum of all the forces acting on the body must be zero. This can also be stated as: a body is in equilibrium if the sum of all the forces in the x-direction is equal to zero and the sum of all the forces in the y-direction is equal to zero (p. 121).

Torque
Torque is defined as the product of the force times the lever arm. Whenever an unbalanced torque acts on a body at rest, it will put that body into rotational motion (p. 127).

Lever arm
The lever arm is defined as the perpendicular distance from the axis of rotation to the direction or line of action of the force. If the force acts through the axis of rotation of the body, it has a zero lever arm and cannot cause a torque to act on the body (p. 127).

The second condition of equilibrium
In order for a body to be in rotational equilibrium, the sum of the torques acting on the body must be equal to zero. This can also be stated as: the necessary condition for a body to be in rotational equilibrium is that the sum of all the torques clockwise must be equal to the sum of all the torques counterclockwise (p. 130).

Center of gravity (cg)
The point that behaves as though the entire weight of the body is located at that point. For a body with a uniform mass distribution located in a uniform gravitational field, the center of gravity is located at the geometrical center of the body (p. 134).

Center of mass (cm)
The point of a body at which all the mass of the body is assumed to be concentrated. For a body with a uniform mass distribution, the center of mass coincides with the geometrical center of the body. When external forces act on a body to put the body into translational motion, all the forces can be considered to act at the center of mass of the body. For a body in a uniform gravitational field, the center of gravity coincides with the center of mass of the body (p. 138).

Summary of Important Equations

First condition of equilibrium
$\Sigma \mathbf{F} = 0$ **(5.1)**

First condition of equilibrium
$\Sigma F_x = 0$ **(5.4)**
$\Sigma F_y = 0$ **(5.5)**

Torque
$\tau = r_\perp F = rF_\perp$ **(5.21)**
$\quad = rF \sin \theta$

Second condition
of equilibrium
$\Sigma \tau = 0$ **(5.25)**

Second condition
of equilibrium
$\Sigma \tau_{cw} = \Sigma \tau_{ccw}$ **(5.24)**

Center of gravity
$x_{cg} = \dfrac{\Sigma w_i x_i}{W}$ **(5.31)**

$y_{cg} = \dfrac{\Sigma w_i y_i}{W}$ **(5.32)**

Center of mass
$x_{cm} = \dfrac{\Sigma m_i x_i}{\Sigma m_i} = \dfrac{\Sigma m_i x_i}{M}$ **(5.34)**

$y_{cm} = \dfrac{\Sigma m_i y_i}{\Sigma m_i} = \dfrac{\Sigma m_i y_i}{M}$ **(5.35)**

Questions for Chapter 5

1. Why can a body moving at constant velocity be considered as a body in equilibrium?
2. Why cannot an accelerated body be considered as in equilibrium?
3. Why can a point outside the body in equilibrium be considered as an axis to compute torques?
4. What is the difference between the center of mass of a body and its center of gravity?
5. A ladder is resting against a wall and a person climbs up the ladder. Is the ladder more likely to slip out at the bottom as the person climbs closer to the top of the ladder? Explain.
6. When flying an airplane a pilot frequently changes from the fuel tank in the right wing to the fuel tank in the left wing. Why does he do this?

7. Where would you expect the center of gravity of a sphere to be located? A cylinder?
† 8. When lifting heavy objects why is it said that you should bend your knees and lift with your legs instead of your back? Explain.
9. A short box and a tall box are sitting on the floor of a truck. If the truck makes a sudden stop, which box is more likely to tumble over? Why?
†10. A person is sitting at the end of a row boat that is at rest in the middle of the lake. If the person gets up and walks toward the front of the boat, what will happen to the boat? Explain in terms of the center of mass of the system.

11. Is it possible for the center of gravity of a body to lie outside of the body? (*Hint:* consider a doughnut.)
†12. Why does an obese person have more trouble with lower back problems than a thin person?
13. Describe how a lever works in terms of the concept of torque.
†14. Describe how you could determine the center of gravity of an irregular body such as a plate, experimentally.
†15. Engineers often talk about the moment of a force acting on a body. Is there any difference between the concept of a torque acting on a body and the moment of a force acting on a body?

Problems for Chapter 5

5.1 The First Condition of Equilibrium

1. In a laboratory experiment on a force table, three forces are in equilibrium. One force of 0.300 N acts at an angle of 40.0°. A second force of 0.800 N acts at an angle of 120°. What is the magnitude and direction of the force that causes equilibrium?
2. Two ropes each 10.0 ft long are attached to the ceiling at two points located 15.0 ft apart. The ropes are tied together in a knot at their lower end and a load of 70.0 lb is hung on the knot. What is the tension in each rope?

3. What force must be applied parallel to the plane to make the block move up the frictionless plane at constant speed?

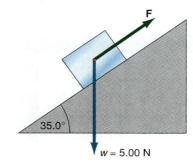

4. Two ropes are attached to the ceiling as shown, making angles of 40.0° and 20.0°. A weight of 100 N is hung from the knot. What is the tension in each rope?

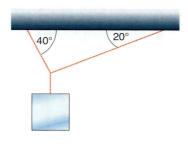

5. Find the force F, parallel to the frictionless plane, that will allow the system to move at constant speed.

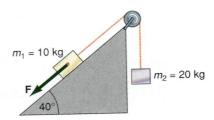

6. A weightless rope is stretched horizontally between two poles 25.0 ft apart. Spiderman, who weighs 160 lb, balances himself at the center of the rope, and the rope is observed to sag 0.500 ft at the center. Find the tension in each part of the rope.

7. A weightless rope is stretched horizontally between two poles 25.0 ft apart. Spiderman, who weighs 160 lb, balances himself 5.00 ft from one end, and the rope is observed to sag 0.300 ft there. What is the tension in each part of the rope?

8. A force of 15.0 N is applied to a 15.0-N block on a rough inclined plane that makes an angle of 52.0° with the horizontal. The force is parallel to the plane. The block moves up the plane at constant velocity. Find the coefficient of kinetic friction between the block and the plane.

9. With what force must a 5.00-N eraser be pressed against a blackboard for it to be in static equilibrium? The coefficient of static friction between the board and the eraser is 0.250.

10. A traffic light, weighing 150 lb is hung from the center of a cable of negligible weight that is stretched horizontally between two poles that are 60.0 ft apart. The cable is observed to sag 2.00 ft. What is the tension in the cable?

11. A traffic light that weighs 600 N is hung from the cable as shown. What is the tension in each cable? Assume the cable to be massless.

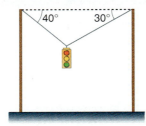

12. Your car is stuck in a snow drift. You attach one end of a 50.0-ft rope to the front of the car and attach the other end to a nearby tree, as shown in the figure. If you can exert a force of 150 lb on the center of the rope, thereby displacing it 3.00 ft to the side, what will be the force exerted on the car?

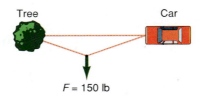

$F = 150$ lb

†13. What force is indicated on the scale in part a and part b of the diagram if $m_1 = m_2 = 20.0$ kg?

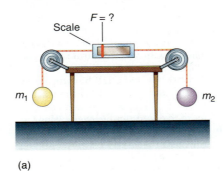

(a)

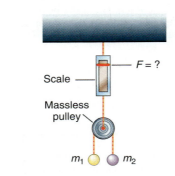

(b)

†14. Find the tension in each cord of the figure, if the block weighs 100 N.

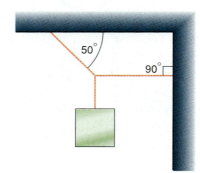

5.2 The Concept of Torque

15. A force of 10.0 lb is applied to a door knob perpendicular to a 30.0-in. door. What torque is produced to open the door?

16. A horizontal force of 10.0 lb is applied at an angle of 40.0° to the door knob of a 30.0-in. door. What torque is produced to open the door?

17. A horizontal force of 50.0 N is applied at an angle of 28.5° to a door knob of a 75.0-cm door. What torque is produced to open the door?

18. A door knob is placed in the center of a 30.0-in. door. If a force of 10.0 lb is exerted perpendicular to the door at the knob, what torque is produced to open the door?

19. Compute the net torque acting on the pulley in the diagram if the radius of the pulley is 0.250 m and the tensions are $T_1 = 50.0$ N and $T_2 = 30.0$ N.

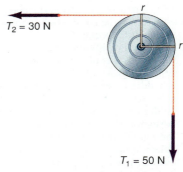

20. Find the torque produced by the bicycle pedal in the diagram if the force $F = 25.0$ lb, the radius of the crank $r = 7.00$ inches, and angle $\theta = 37.0°$.

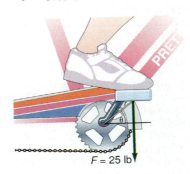

$F = 25$ lb

5.5 Examples of Rigid Bodies in Equilibrium

Parallel Forces

21. Two men are carrying a 25.0-ft telephone pole that weighs 200 lb. If the center of gravity of the pole is 10.0 ft from the right end, and the men lift the pole at the ends, how much weight must each man support?

22. Two men are carrying a 9.00-m telephone pole that has a mass of 115 kg. If the center of gravity of the pole is 3.00 m from the right end, and the men lift the pole at the ends, how much weight must each man support?

23. A uniform board that is 5.00 m long and weighs 450 N is supported by two wooden horses, 0.500 m from each end. If a 800-N person stands on the board 2.00 m from the right end, what force will be exerted on each wooden horse?

24. A 300-N boy and a 250-N girl sit at opposite ends of a 4.00-m seesaw. Where should another 250-N girl sit in order to balance the seesaw?

25. A uniform beam 10.0 ft long and weighing 15.0 lb carries a load of 20.0 lb at one end and 45.0 lb at the other end. It is held horizontal, while resting on a wooden horse 4.00 ft from the heavier load. What torque must be applied to keep it at rest in this position?

26. A uniform beam 3.50 m long and weighing 90.0 N carries a load of 110 N at one end and 225 N at the other end. It is held horizontal, while resting on a wooden horse 1.50 m from the heavier load. What torque must be applied to keep it at rest in this position?

27. A uniform pole 5.00 m long and weighing 100 N is to be carried at its ends by a man and his son. Where should a 250-N load be hung on the pole, such that the father will carry twice the load of his son?

28. A meter stick is hung from two scales that are located at the 20.0- and 70.0-cm marks of the meter stick. Weights of 2.00 N are placed at the 10.0- and 40.0-cm marks, while a weight of 1.00 N is placed at the 90.0-cm mark. The weight of the uniform meter stick is 1.50 N. Determine the scale readings at A and B in the diagram.

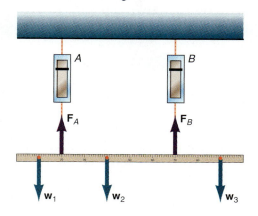

Center of Gravity of a Body

29. A tapered pole 10.0 ft long weighs 25.0 lb. The pole balances at its midpoint when a 5.00-lb weight hangs from the slimmer end. Where is the center of gravity of the pole?

†30. A loaded wheelbarrow that weighs 75.0 lb has its center of gravity 2.00 ft from the front wheel axis. If the distance from the wheel axis to the end of the handles is 6.00 ft, how much of the weight of the wheelbarrow is supported by each arm?

†31. Find the center of gravity of the carpenters square shown in the diagram.

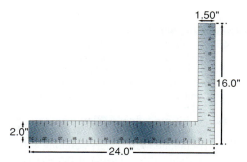

32. The front and rear axles of a 3000-lb car are 8.00 ft apart. If the center of gravity of the car is located 3.00 ft behind the front axle, find the load supported by the front and rear wheels of the car.

33. The front and rear axles of a 1110-kg car are 2.50 m apart. If the center of gravity of the car is located 1.15 m behind the front axle, find the load supported by the front and rear wheels of the car.

34. A very bright but lonesome child decides to make a seesaw for one. The child has a large plank, and a wooden horse to act as a fulcrum. Where should the child place the fulcrum, such that the plank will balance, when the child is sitting on the end? The child weighs 60.0 lb and the plank weighs 40.0 lb and is 10.0 ft long. (*Hint:* find the center of gravity of the system.)

Center of Mass

35. Four masses of 20.0, 40.0, 60.0, and 80.0 g are located at the respective distances of 10.0, 20.0, 30.0, and 40.0 cm from an origin. Find the center of mass of the system.

36. Three masses of 15.0, 45.0, and 25.0 g are located on the *x*-axis at 10.0, 25.0, and 45.0 cm. Two masses of 25.0 and 33.0 g are located on the *y*-axis at 35.0 and 50.0 cm, respectively. Find the center of mass of the system.

†37. A 1.00-kg circular metal plate of radius 0.500 m has attached to it a smaller circular plate of the same material of 0.100 m radius, as shown in the diagram. Find the center of mass of the combination with respect to the center of the large plate.

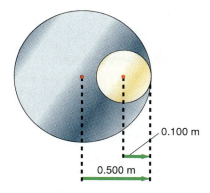

†38. This is the same problem as 37 except the smaller circle of material is removed from the larger plate. Where is the center of mass now?

Crane Boom Problems

39. A horizontal uniform boom that weighs 200 N and is 5.00 m long supports a load w_L of 1000 N, as shown in the figure. Find all the forces acting on the boom.

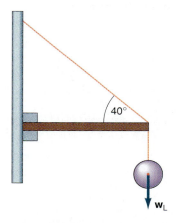

40. A horizontal, uniform boom 4.00 m long that weighs 200 N supports a load w_L of 1000 N. A guy wire that helps to support the boom, is attached 1.00 m in from the end of the boom. Find all the forces acting on the boom.

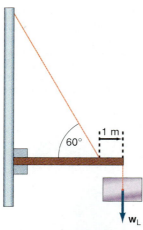

Diagram for problem 40.

41. A horizontal, uniform boom 4.50 m long that weighs 250 N supports a 3500 N load w_L. A guy wire that helps to support the boom is attached 1.0 m in from the end of the boom, as in the diagram for problem 40. If the maximum tension that the cable can withstand is 3500 N, how far out on the boom can a 95.0-kg repairman walk without the cable breaking?

42. A uniform beam 4.00 m long that weighs 200 N is supported, as shown in the figure. The boom lifts a load w_L of 1000 N. Find all the forces acting on the boom.

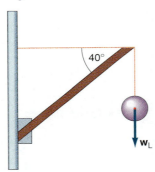

†43. A uniform beam 4.00 m long that weighs 200 N is supported, as shown in the figure. The boom lifts a load w_L of 1000 N. Find all the forces acting on the boom.

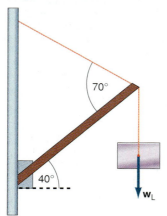

44. An 80.0-lb sign is hung on a uniform steel pole that weighs 25.0 lb, as shown in the figure. Find all the forces acting on the boom.

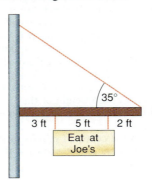

Ladder Problems

45. A uniform ladder 6.00 m long weighing 120 N leans against a frictionless wall. The base of the ladder is 1.00 m away from the wall. Find all the forces acting on the ladder.

46. A uniform ladder 6.00 m long weighing 120 N leans against a frictionless wall. A girl weighing 400 N climbs three-fourths of the way up the ladder. If the base of the ladder makes an angle of 75.0° with the ground, find all the forces acting on the ladder. Compute all torques about the base of the ladder.

47. Repeat problem 46, but compute all torques about the top of the ladder. Is there any difference in the results of the problem?

48. A uniform ladder 15.0 ft long weighing 25.0 lb leans against a frictionless wall. If the base of the ladder makes an angle of 40.0° with the ground, what is the minimum coefficient of friction between the ladder and the ground such that the ladder will not slip out?

†49. A uniform ladder 20.0 ft long weighing 35.0 lb leans against a frictionless wall. The base of the ladder makes an angle of 60.0° with the ground. If the coefficient of friction between the ladder and the ground is 0.300, how high can a 160-lb man climb the ladder before the ladder starts to slip?

50. A uniform ladder 5.50 m long with a mass of 12.5 kg leans against a frictionless wall. The base of the ladder makes an angle of 48.0° with the ground. If the coefficient of friction between the ladder and the ground is 0.300, how high can a 82.3-kg man climb the ladder before the ladder starts to slip?

Applications to the Health Sciences

51. A weight lifter is lifting a dumbbell as in the example shown in figure 5.22 only now the forearm makes an angle of 30.0° with the horizontal. Using the same data as for that problem find the force F_M exerted by the biceps muscle and the reaction force at the elbow joint F_J. Assume that the force F_M remains perpendicular to the arm.

52. Consider the weight lifter in the example shown in figure 5.23. Determine the forces F_M and F_R if the angle $\theta = 00.0°$.

†53. The weight of the upper body of the person in the accompanying diagram acts downward about 8.00 cm in front of the fifth lumbar vertebra. This weight produces a torque about the fifth lumbar vertebra. To counterbalance this torque the muscles in the lower back exert a force F_M that produces a counter torque. These muscles exert their force about 5.00 cm behind the fifth lumbar vertebra. If the person weighs 180 lb find the force exerted by the lower back muscles F_M and the reaction force F_R that the sacrum exerts upward on the fifth lumbar vertebra. The weight of the upper portion of the body is about 65% of the total body weight.

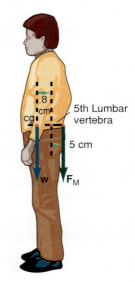

†54. Consider the same situation as in problem 53 except that the person is overweight. The center of gravity with the additional weight is now located 15.0 cm in front of the fifth lumbar vertebra instead of the previous 8.00 cm. Hence a greater torque will be exerted by this additional weight. The distance of the lower back muscles is only slightly greater at 6.00 cm. If the person weighs 240 lb find the force F_R on the fifth lumbar vertebra and the force F_M exerted by the lower back muscles.

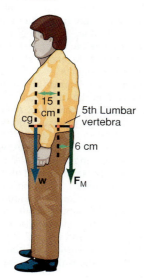

†55. A 668-N person stands evenly on the balls of both feet. The Achilles tendon, which is located at the back of the ankle, provides a tension T_A to help balance the weight of the body as seen in the diagram. The distance from the ball of the foot to the Achilles tendon is approximately 18.0 cm. The tibia leg bone pushes down on the foot with a force F_T. The distance from the tibia to the ball of the foot is about 14.0 cm. The ground exerts a reaction force F_N upward on the ball of the foot that is equal to half of the body weight. Draw a free body diagram of the forces acting and determine the force exerted by the Achilles tendon and the tibia.

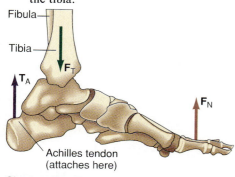

Additional Problems

†56. If w weighs 100 N, find (a) the tension in ropes 1, 2, and 3 and (b) the tension in ropes 4, 5, and 6. The angle $\theta = 52.0°$ and the angle $\phi = 33.0°$.

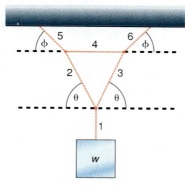

†57. Block A rests on a table and is connected to another block B by a rope that is also connected to a wall. If $M_A = 15.0$ kg and $\mu_s = 0.200$, what must be the value of M_B to start the system into motion?

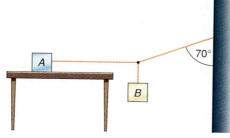

58. In the pulley system shown, what force F is necessary to keep the system in equilibrium?

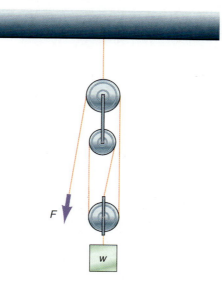

†59. A sling is used to support a leg as shown in the diagram. The leg is elevated at an angle of 20.0°. The bed exerts a reaction force **R** on the thigh as shown. The weight of the thigh, leg, and ankle are given by w_T, w_L, and w_A, respectively, and the locations of these weights are as shown. The sling is located 27.0 in. from the point O in the diagram. A free body diagram is shown in part b of the diagram. Find the weight **w** that is necessary to put the leg into equilibrium.

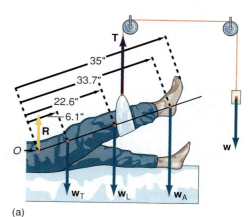

(a)

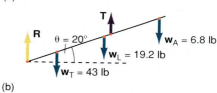

(b)

†60. Find the tensions T_1, T, and T_2 in the figure if $w_1 = 500$ N and $w_2 = 300$ N. The angle $\theta = 35.0°$ and the angle $\phi = 25.0°$.

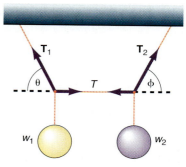

†61. The steering wheel of an auto has a diameter of 18.0 in. The axle that it is connected to has a diameter of 2.00 in. If a force of 25.0 lb is exerted on the rim of the wheel, (a) what is the torque exerted on the steering wheel, (b) what is the torque exerted on the axle, and (c) what force is exerted on the rim of the axle?

62. One type of simple machine is called a wheel and axle. A wheel of radius 35.0 cm is connected to an axle of 2.00 cm radius. A force of $F_{in} = 10.0$ N is applied tangentially to the wheel. What force F_{out} is exerted on the axle? The ratio of the output force F_{out} to the input force F_{in} is called the ideal mechanical advantage (IMA) of the system. Find the IMA of this system.

†63. A box 1.00 m on a side rests on a floor next to a small piece of wood that is fixed to the floor. The box weighs 500 N. At what height h should a force of 400 N be applied so as to just tip the box?

†64. A 200-N door, 0.760 m wide and 2.00 m long, is hung by two hinges. The top hinge is located 0.230 m down from the top, while the bottom hinge is located 0.330 m up from the bottom. Assume that the center of gravity of the door is at its geometrical center. Find the horizontal force exerted by each hinge on the door.

†65. A uniform ladder 6.00 m long weighing 100 N leans against a frictionless wall. If the coefficient of friction between the ladder and the ground is 0.400, what is the smallest angle θ that the ladder can make with the ground before the ladder starts to slip?

†66. If an 800-N man wants to climb a distance of 5.00 m up the ladder of problem 65, what angle θ should the ladder make with the ground such that the ladder will not slip?

†67. A uniform ladder 20.0 ft long weighing 30.0 lb leans against a rough wall, that is, a wall where there is a frictional force between the top of the ladder and the wall. The coefficient of static friction is 0.400. If the base of the ladder makes an angle θ of 40.0° with the ground when the ladder begins to slip down the wall, find all the forces acting on the ladder. (*Hint:* With a rough wall there will be a vertical force f_s acting upward at the top of the ladder. In general, this force is unknown but we do know that it must be less than $\mu_s F_N$. At the moment the ladder starts to slip, this frictional force is known and is given by the equation of static friction, namely, $f_s = \mu_s F_N = \mu_s F$. Although there are now four unknowns, there are also four equations to solve for them.)

†68. A 225-lb person stands three-quarters of the way up a stepladder. The step side weighs 20.0 lb, is 6.00 ft long, and is uniform. The rear side weighs 10.0 lb, is also uniform, and is also 6.00 ft long. A hinge connects the front and back of the ladder at the top. A weightless tie rod, 1.50 ft in length, is connected 2.00 ft from the top of the ladder. Find the forces exerted by the floor on the ladder and the tension in the tie rod.

Interactive Tutorials

69. Two ropes are attached to the ceiling, making angles $\theta = 20.0°$ and $\phi = 40.0°$, suspending a mass $m = 50.0$ kg. Calculate the tensions T_1 and T_2 in each rope.

70. A uniform beam of length $L = 10.0$ m and mass $m = 5.00$ kg is held up at each end by a force F_A (at 0.00 m) and force F_B (at 10.0 m). If a weight $W = 400$ N is placed at the position $x = 8.00$ m, calculate forces F_A and F_B.

71. The crane boom. A uniform boom of weight $w_B = 250$ N and length $l = 8.00$ m is connected to the mast by a hinge pin at the point A in figure 5.20. A load $w_L = 1200$ N is supported at the other end. A cable is connected at the end of the boom making an angle $\theta = 55.0°$, as shown in the diagram. Find the tension T in the cable and the vertical V and horizontal H forces that the hinge pin exerts on the boom.

72. A uniform ladder of weight $w_l = 100$ N and length $L = 20.0$ m leans against a frictionless wall at a base angle $\theta = 60.0°$. A person weighing $w_p = 150$ N climbs the ladder a distance $d = 6.00$ m from the base of the ladder. Calculate the horizontal H and vertical V forces acting on the ladder, and the force F exerted by the wall on the top of the ladder.

6 Uniform Circular Motion, Gravitation, and Satellites

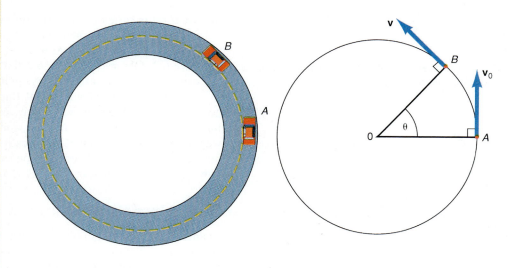

On the moon.

That's one small step for man, one giant leap for mankind.

Neil Armstrong, as he stepped on the surface of the moon July 20, 1969

6.1 Uniform Circular Motion

Uniform circular motion is defined as motion in a circle at constant speed. Motion in a circle with changing speeds will be discussed in chapter 9. A car moving in a circle at the constant speed of 20 km/hr is an example of a body in uniform circular motion. At every point on that circle the car would be moving at 20 km/hr. This type of motion is shown in figure 6.1. At the time t_0, the car is located at the point A and is moving with the velocity $\mathbf{v_0}$, which is tangent to the circle at that point. At a later time t, the car will have moved through the angle θ, and will be located at the point B. At the point B the car has a velocity $\mathbf{v}$, which is tangent to the circle at B. (The velocity is always tangent to the circle, because at any instant the tangent specifies the direction of motion.) The lengths of the two vectors, $\mathbf{v}$ and $\mathbf{v_0}$, are the same because the magnitude of any vector is represented as the length of that vector. The magnitude of the velocity is the speed, which is a constant for uniform circular motion.

(a)   (b)

Figure 6.1
Uniform circular motion.

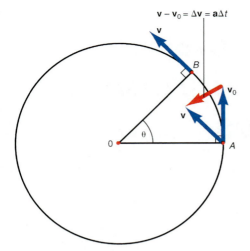

Figure 6.2

The direction of the centripetal acceleration.

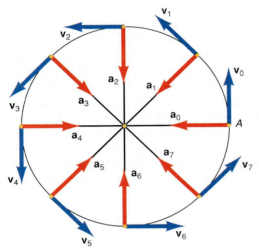

Figure 6.3

The centripetal acceleration always points toward the center of the circle.

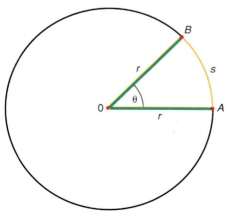

Figure 6.4

Definition of an angle expressed in radians.

The first thing we observe in figure 6.1 is that the direction of the velocity vector has changed in going from the point A to the point B. Recall from chapter 3, on kinematics, that the acceleration is defined as the change in velocity with time, that is,

$$\mathbf{a} = \frac{\Delta \mathbf{v}}{\Delta t} \tag{6.1}$$

Even though the speed is a constant in uniform circular motion, the direction is always changing with time. Hence, the velocity is changing with time, and there must be an acceleration. *Thus, motion in a circle at constant speed is accelerated motion.* We must now determine the direction of this acceleration and its magnitude.

6.2 Centripetal Acceleration and its Direction

To determine the direction of the centripetal acceleration, let us start by moving the vector $\mathbf{v}$, located at the point B in figure 6.1, parallel to itself to the point A, as shown in figure 6.2. The difference between the two velocity vectors is $\mathbf{v} - \mathbf{v_0}$ and points approximately toward the center of the circle in the direction shown. But this difference between the velocity vectors is the change in the velocity vector $\Delta \mathbf{v}$, that is,

$$\Delta \mathbf{v} = \mathbf{v} - \mathbf{v_0} \tag{6.2}$$

But from equation 6.1

$$\Delta \mathbf{v} = \mathbf{a} \Delta t \tag{6.3}$$

This is a vector equation, and whatever direction the left-hand side of the equation has, the right-hand side must have the same direction. Therefore, the vector $\Delta \mathbf{v}$ points in the same direction as the acceleration vector $\mathbf{a}$. Observe from figure 6.2 that $\Delta \mathbf{v}$ points approximately toward the center of the circle. (If the angle θ, between the points A and B, were made very small, then $\Delta \mathbf{v}$ would point exactly at the center of the circle.) Thus, since $\Delta \mathbf{v}$ points toward the center, the acceleration vector must also point toward the center of the circle. *This is the characteristic of uniform circular motion. Even though the body is moving at constant speed, there is an acceleration and the acceleration vector points toward the center of the circle. This acceleration is called the* **centripetal acceleration.**

The word centripetal means "center seeking" or seeking the center. If this circular motion were shown at intervals of 45°, we would obtain the picture shown in figure 6.3. Observe in figure 6.3 that no matter where the body is on the circle, the centripetal acceleration always points toward the center of the circle.

What is the magnitude of this acceleration? The problem of calculating accelerations of objects moving in circles at constant speed was first solved by Christian Huygens (1629–1695) in 1673, and his solution is effectively the same one that we use today. The argument is basically a geometric one. However, before the magnitude of the centripetal acceleration can be determined, we need first to determine how an angle is defined in terms of radian measure.

6.3 Angles Measured in Radians

In addition to the usual unit of degrees used to measure an angle, an angle can also be measured in another unit called a **radian.** As the body moves along the arc s of the circle from point A to point B in figure 6.4, it sweeps out an angle θ in the time t. This angle θ, measured in radians (abbreviated rad), is defined as the ratio of the arc length s traversed to the radius of the circle r. That is,

$$\theta = \frac{s}{r} = \frac{\text{arc length}}{\text{radius}} \tag{6.4}$$

Thus an angle of 1 radian is an angle swept out such that the distance s, traversed along the arc, is equal to the radius of the circle:

$$\theta = \frac{s}{r} = \frac{r}{r} = 1 \text{ rad}$$

Notice that a radian is a dimensionless quantity. If s is measured in meters and r is measured in meters, then the ratio yields units of meters over meters and the units will thus cancel.

For an entire rotation around the circle, that is, for one revolution, the arc subtended is the circumference of the circle, $2\pi r$. Therefore, an angle of one revolution, measured in radians, becomes

$$\theta = \frac{s}{r} = \frac{2\pi r}{r} = 2\pi \text{ rad}$$

That is, one revolution is equal to 2π rad. The relationship between an angle measured in degrees, and one measured in radians can be found from the fact that one revolution is also equal to 360 degrees. Thus,

$$1 \text{ rev} = 2\pi \text{ rad} = 360°$$

and solving for a radian, we get

$$1 \text{ rad} = \frac{360°}{2\pi} = 57.296°$$

Similarly,

$$1 \text{ degree} = 0.01745 \text{ rad}$$

For ease in calculations on an electronic calculator it is helpful to use the conversion factor

$$\pi \text{ rad} = 180°$$

In almost all problems in circular motion the angles will be measured in radians.

The relationship between the arc length s and the angle θ, measured in radians, for circular motion, found from equation 6.4, is

$$s = r\theta \tag{6.5}$$

6.4 The Magnitude of the Centripetal Acceleration

Having determined the relation between the arc length s and the angle θ swept out, we can now determine the magnitude of the centripetal acceleration. In moving at the constant speed v, along the arc of the circle from A to B in figure 6.2, the body has traveled the distance

$$s = vt \tag{6.6}$$

But in this same time t, the angle θ has been swept out in moving the distance s along the arc. If the distance s moved along the arc from equations 6.5 and 6.6 are equated, we have

$$r\theta = vt$$

Solving for θ, we obtain

$$\theta = \frac{vt}{r} \tag{6.7}$$

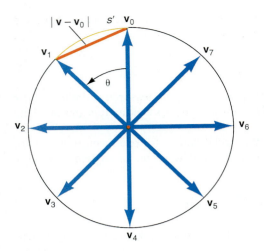

Figure 6.5

The velocity circle.

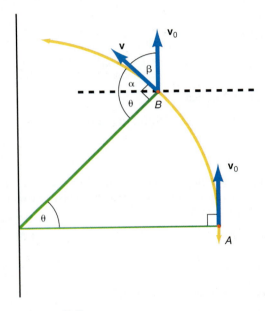

Figure 6.6

The angle between the velocity vectors **v** and **v₀** is the same as the angle θ swept out in moving from point *A* to point *B*.

This is the angle θ swept out in the uniform circular motion, in terms of the speed v, time t, and the radius r of the circle. We will return to equation 6.7 in a moment, but first let us look at the way that these velocity vectors are changing with time.

As we see in figure 6.3, the velocity vector **v** points in a different direction at every instant of time. Let us slide each velocity vector in figure 6.3 parallel to itself to a common point. If we draw a curve connecting the tips of each velocity vector, we obtain the circle shown in figure 6.5. That is, since the magnitude of the velocity vector is a constant, a circle of radius v is generated. As the object moves from *A* to *B* and sweeps out the angle θ in figure 6.2, the velocity vector also moves through the same angle θ, figure 6.5. To prove this, notice that the velocity vectors **v₀** at *A* and **v** at *B* are each tangent to the circle there, figure 6.6. In moving through the angle θ in going from *A* to *B*, the velocity vector turns through this same angle θ. This is easily seen in figure 6.6. The angle α is

$$\alpha = \frac{\pi}{2} - \theta \qquad (6.8)$$

while the angle β is

$$\beta = \frac{\pi}{2} - \alpha \qquad (6.9)$$

Substituting equation 6.8 into equation 6.9 gives

$$\beta = \frac{\pi}{2} - \left(\frac{\pi}{2} - \theta \right)$$

Hence, β, the angle between **v** and **v₀** in figure 6.6, is

$$\beta = \theta$$

Thus, the angle between the velocity vectors **v** and **v₀** is the same as the angle θ swept out in moving from point *A* to point *B*.

Therefore, in moving along the velocity circle in figure 6.5, an amount of arc s' is swept out with the angle θ. This velocity circle has a radius of v, the constant speed in the circle. Using equation 6.4, as it applies to the velocity circle, we have

$$\theta = \frac{\text{arc length}}{\text{radius}} = \frac{s'}{v} \qquad (6.10)$$

If the angle θ is relatively small, then the arc of the circle s' is approximately equal to the chord of the circle $|\mathbf{v} - \mathbf{v_0}|$ in figure 6.5.[1] That is,

$$\text{arc} \approx \text{chord}$$
$$s' = |\mathbf{v} - \mathbf{v_0}|$$

But

$$\mathbf{v} - \mathbf{v_0} = \mathbf{a}t$$

hence,

$$s' = at$$

Substituting this result into equation 6.10 gives

$$\theta = \frac{at}{v} \qquad (6.11)$$

Thus we have obtained a second relation for the angle θ swept out, expressed now in terms of acceleration, speed, and time. Returning to equation 6.7, which gave us

---

1. Note that $|\mathbf{v} - \mathbf{v_0}|$ is the magnitude of the difference in the velocity vectors and is the straight line between the tip of the velocity vector **v₀** and the tip of the velocity vector **v**, and as such, is equal to the chord of the circle in figure 6.5.

Mechanics

the angle θ swept out as the moving body went from point A to point B along the circular path, and comparing it to equation 6.11, which gives the angle θ swept out in the velocity circle. Because both angles θ are equal, equation 6.11 can now be equated to equation 6.7, giving

$$\theta = \theta$$

$$\frac{at}{v} = \frac{vt}{r}$$

Solving for the acceleration we obtain

$$a = \frac{v^2}{r}$$

Placing a subscript c on the acceleration to remind us that this is the centripetal acceleration, we then have

$$a_c = \frac{v^2}{r} \tag{6.12}$$

Therefore, *for the uniform circular motion of an object moving at constant speed v in a circle of radius r, the object undergoes an acceleration a_c, pointed toward the center of the circle, and having the magnitude given by equation 6.12.*

Example 6.1

Find the centripetal acceleration. An object moves in a circle of 10.0-m radius, at a constant speed of 5.00 m/s. What is its centripetal acceleration?

Solution

The centripetal acceleration, found from equation 6.12, is

$$a_c = \frac{v^2}{r} \tag{6.12}$$

$$= \frac{(5.00 \text{ m/s})^2}{10.0 \text{ m}}$$

$$= 2.50 \text{ m/s}^2$$

Example 6.2

The special case of the centripetal acceleration equal to the gravitational acceleration. At what uniform speed should a body move in a circular path of 10.0 m radius such that the acceleration experienced will be the same as the acceleration of gravity?

Solution

We find the velocity of the moving body in terms of the centripetal acceleration by solving equation 6.12 for v:

$$v = \sqrt{ra_c}$$

To have the body experience the same acceleration as the acceleration of gravity, we set $a_c = g$ and get

$$v = \sqrt{ra_c} = \sqrt{rg}$$

$$= \sqrt{(10.0 \text{ m})(9.80 \text{ m/s}^2)}$$

$$= 9.90 \text{ m/s}$$

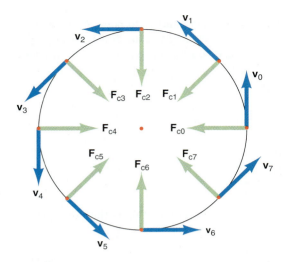

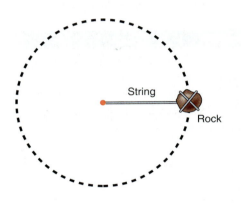

Figure 6.7

The centripetal force always points toward the center of the circle.

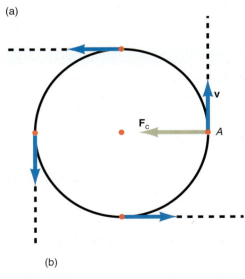

(a)

(b)

Figure 6.8

The string supplies the centripetal force on a rock moving in a circle.

6.5 The Centripetal Force

We have just seen that an object in uniform circular motion experiences a centripetal acceleration. However, because of Newton's second law of motion, there must be a force acting on the object to give it the necessary centripetal acceleration. Applying Newton's second law to the body in uniform circular motion we have

$$F = ma = ma_c = \frac{mv^2}{r} \qquad (6.13)$$

A subscript c is placed on the force to remind us that this is the centripetal force, and equation 6.13 becomes

$$F_c = \frac{mv^2}{r} \qquad (6.14)$$

*The force, given by equation 6.14, that causes an object to move in a circle at constant speed is called the **centripetal force**.* Because the centripetal acceleration is pointed toward the center of the circle, then from Newton's second law in vector form, we see that

$$\mathbf{F_c} = m\mathbf{a_c} \qquad (6.15)$$

Hence, the centripetal force must also point toward the center of the circle. Therefore, *when an object moves in uniform circular motion there must always be a centripetal force acting on the object toward the center of the circle as seen in figure 6.7.*

We should note here that we need to physically supply the force to cause the body to go into uniform circular motion. The centripetal force is the amount of force necessary to put the body into uniform circular motion, but it is not a real physical force in itself that is applied to the body. It is the amount of force necessary, but something must supply that force, such as a tension, a weight, gravity, and the like. As an example, consider the motion of a rock, tied to a string of negligible mass, and whirled in a horizontal circle, at constant speed v. At every instant of time there must be a centripetal force acting on the rock to pull it toward the center of the circle, if the rock is to move in the circle. This force is supplied by your hand, and transmitted to the rock, by the string. It is evident that such a force must be acting by the following consideration. Consider the object at point A in figure 6.8 moving with a velocity $\mathbf{v}$ at a time t. By Newton's first law, a body in motion at a constant velocity will continue in motion at that same constant velocity, unless acted on by some unbalanced external force. Therefore, if there were no centripetal force acting on the object, the object would continue to move at its same constant velocity and would fly off in a direction tangent to the circle. In fact, if you were to cut the string, while the rock is in motion, you would indeed observe the rock flying off tangentially to the original circle. (Cutting the string removes the centripetal force.)

Example 6.3

Finding the centripetal force. A 500-g rock attached to a string is whirled in a horizontal circle at the constant speed of 10.0 m/s. The length of the string is 1.00 m. Neglecting the effects of gravity, find (a) the centripetal acceleration of the rock and (b) the centripetal force acting on the rock.

Solution

a. The centripetal acceleration, found from equation 6.12, is

$$a_c = \frac{v^2}{r} = \frac{(10.0 \text{ m/s})^2}{1.00 \text{ m}} = 100 \frac{\text{m}^2/\text{s}^2}{\text{m}}$$

$$= 100 \text{ m/s}^2$$

b. The centripetal force, which is supplied by the tension in the string, found from equation 6.14, is

$$F_c = \frac{mv^2}{r} = \frac{(0.500 \text{ kg})(10.0 \text{ m/s})^2}{1.00 \text{ m}} = 50.0 \frac{\text{kg m}^2}{\text{m s}^2}$$

$$= 50.0 \text{ N}$$

Notice how the units combine so that the final unit is a newton, the unit of force.

6.6 The Centrifugal Force

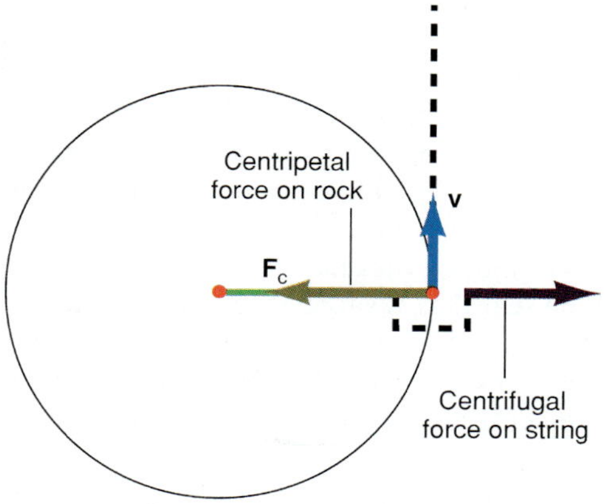

In the preceding example of the rock revolving in a horizontal circle, there was a centripetal force acting on the rock by the string. But by Newton's third law, if body *A* exerts a force on body *B*, then body *B* exerts an equal but opposite force on body *A*. Thus, if the string (body *A*) exerts a force on the rock (body *B*), then the rock (body *B*) must exert an equal but opposite force on the string (body *A*). *This reaction force to the centripetal force is called the* **centrifugal force.** Note that the centrifugal force does not act on the same body as does the centripetal force. The centripetal force acts on the rock, the centrifugal force acts on the string. The centrifugal force is shown in figure 6.9 as the dashed line that goes around the rock to emphasize that the force does not act on the rock but on the string.

If we wish to describe the motion of the rock, then we must use the centripetal force, because it is the centripetal force that acts on the rock and is necessary for the rock to move in a circle. The reaction force is the centrifugal force. But *the centrifugal force does not act on the rock, which is the object in motion.*

The word centrifugal means to fly from the center, and hence the centrifugal force acts away from the center. This has been the cause of a great deal of confusion. Many people mistakenly believe that the centrifugal force acts outward on the rock, keeping it out on the end of the string. We can show that this reasoning is incorrect by merely cutting the string. If there really were a centrifugal force acting outward on the rock, then the moment the string is cut the rock should fly radially away from the center of the circle, as in figure 6.10(a). It is a matter of observation that the rock does not fly away radially but rather flies away tangentially as predicted by Newton's first law.

A similar example is furnished by a car wheel when it goes through a puddle of water, as in figure 6.10(b). Water droplets adhere to the wheel. The water droplet is held to the wheel by the adhesive forces between the water molecules and the tire. As the wheel turns, the drop of water wants to move in a straight line as it is governed by Newton's first law but the adhesive force keeps the drop attached to the

Figure 6.9

The centrifugal force is the reaction force on the string.

Figure 6.10

There is no radial force outward acting on the rock.

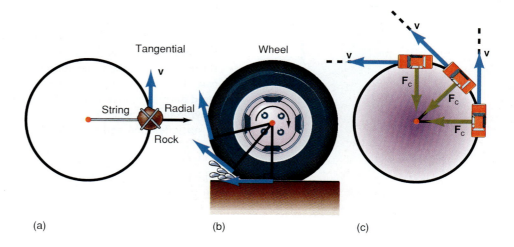

(a)   (b)   (c)

wheel. That is, the adhesive force is supplying the necessary centripetal force. As the wheel spins faster, v increases and the centripetal force necessary to keep the droplet attached to the wheel also increases ($F_c = mv^2/r$). If the wheel spins fast enough, the adhesive force is no longer large enough to supply the necessary centripetal force and the water droplet on the rotating wheel flies away tangentially from the wheel according to Newton's first law.

Another example illustrating the difference between centripetal force and centrifugal force is supplied by a car when it goes around a turn, as in figure 6.10(c). Suppose you are in the passenger seat as the driver makes a left turn. Your first impression as you go through the turn is that you feel a force pushing you outward against the right side of the car. We might assume that there is a centrifugal force acting on you and you can feel that centrifugal force pushing you outward toward the right. This however is not a correct assumption. Instead what is really happening is that at the instant the driver turns the wheels, a frictional force between the wheels and the pavement acts on the car to deviate it from its straight line motion, and deflects it toward the left. You were originally moving in a straight line at an initial velocity v. By Newton's first law, you want to continue in that same straight ahead motion. But now the car has turned and starts to push inward on you to change your motion from the straight ahead motion, to a motion that curves toward the left. It is the right side of the car, the floor, and the seat that is supplying, through friction, the necessary centripetal force on you to turn your straight ahead motion into circular motion. There is no centrifugal force pushing you toward the right, but rather the car, through friction, is supplying the centripetal force on you to push you to the left.

Other mistaken beliefs about the centrifugal force will be mentioned as we proceed. However, in almost all of the physical problems that you will encounter, you can forget entirely about the centrifugal force, because it will not be acting on the body in motion. Only in a noninertial coordinate system, such as a rotating coordinate system, do "fictitious" forces such as the centrifugal force need to be introduced. However, in this book we will limit ourselves to inertial coordinate systems.

6.7 Examples of Centripetal Force

The Rotating Disk in the Amusement Park

Amusement parks furnish many examples of the application of centripetal force and circular motion. In one such park there is a large, horizontal, highly polished wooden disk, very close to a highly polished wooden floor. While the disk is at rest, children come and sit down on it. Then the disk starts to rotate faster and faster until the children slide off the disk onto the floor.

Let us analyze this circular motion. In particular let us determine the maximum velocity that the child can move and still continue to move in the circular path. At any instant of time, the child has some tangential velocity v, as seen in figure 6.11. By Newton's first law, the child has the tendency to continue moving in that tangential direction at the velocity v. However, if the child is to move in a circle, there must be some force acting on the child toward the center of the circle. In this case that force is supplied by the static friction between the seat of the pants of the child and the wooden disk. If that frictional force is present, the child will continue moving in the circle. That is, the necessary centripetal force is supplied by the force of static friction and therefore

$$F_c = f_s \qquad (6.16)$$

The frictional force, obtained from equation 4.44, is

$$f_s \leq \mu_s F_N$$

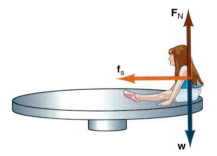

(a) Side view

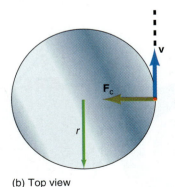

(b) Top view

Figure 6.11

The rotating disk.

Recall that the frictional force is usually less than the product $\mu_s F_N$, and is only equal at the moment that the body is about to slip. In this example, we are finding the maximum velocity of the child and that occurs when the child is about to slip off the disk. Hence, we will use the equality sign for the frictional force in equation 4.44. Using the centripetal force from equation 6.14 and the frictional force from equation 4.44, we obtain

$$\frac{mv^2}{r} = \mu_s F_N \qquad (6.17)$$

As seen from figure 6.11, $F_N = w = mg$. Therefore, equation 6.17 becomes

$$\frac{mv^2}{r} = \mu_s mg \qquad (6.18)$$

The first thing that we observe in equation 6.18 is that the mass m of the child is on both sides of the equation and divides out. Thus, whatever happens to the child, it will happen to a big massive child or a very small one. When equation 6.18 is solved for v, we get

$$v = \sqrt{\mu_s rg} \qquad (6.19)$$

This is the maximum speed that the child can move and still stay in the circular path. For a speed greater than this, the frictional force will not be great enough to supply the necessary centripetal force. Depending on the nature of the children's clothing, μ_s will, in general, be different for each child, and therefore each child will have a different maximum value of v allowable. If the disk's speed is slowly increased until v is greater than that given by equation 6.19, there is not enough frictional force to supply the necessary centripetal force, and the children gleefully slide tangentially from the disk in all directions across the highly polished floor.

Example 6.4

The rotating disk. A child is sitting 1.50 m from the center of a highly polished, wooden, rotating disk. The coefficient of static friction between the disk and the child is 0.30. What is the maximum tangential speed that the child can have before slipping off the disk?

Solution

The maximum speed, obtained from equation 6.19, is

$$v = \sqrt{\mu_s rg}$$
$$= \sqrt{(0.30)(1.50 \text{ m})(9.80 \text{ m/s}^2)}$$
$$= 2.1 \text{ m/s}$$

The Rotating Circular Room in the Amusement Park

In another amusement park there is a ride that consists of a large circular room. (It looks as if you were on the inside of a very large barrel.) Everyone enters the room and stands against the wall. The door closes, and the entire room starts to rotate. As the speed increases each person feels as if he or she is being pressed up against the wall. Eventually, as everyone is pinned against the wall, the floor of the room drops out about 4 or 5 ft, leaving all the children apparently hanging on the wall. After several minutes of motion, the rotation slows down and the children eventually slide down the wall to the lowered floor and the ride ends.

Let us analyze the motion, in particular let us find the value of μ_s, the minimum value of the coefficient of static friction such that the child will not slide down the wall. The room is shown in figure 6.12. As the room reaches its operational

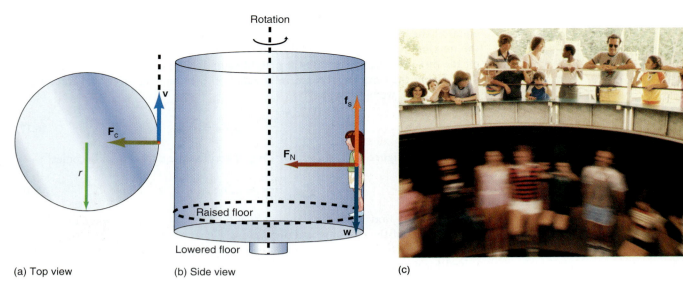

(a) Top view   (b) Side view   (c)

Figure 6.12

Circular room in an amusement park.

speed, the child, at any instant, has a velocity **v** that is tangential to the room, as in figure 6.12(a). By Newton's first law, the child should continue in this straight line motion, but the wall of the room exerts a normal force on the child toward the center of the room, causing the child to deviate from the straight line motion and into the circular motion of the wall of the room. This normal force of the wall on the child, toward the center of the room, supplies the necessary centripetal force. When the floor drops out, the weight **w** of the child is acting downward and would cause the child to slide down the wall. But the frictional force f_s between the wall and the child's clothing opposes the weight, as seen in figure 6.12(b). The frictional force f_s is again given by equation 4.44. We are looking for the minimum value of μ_s that will just keep the child pinned against the wall. That is, the child will be just on the verge of slipping down the wall. Hence, we use the equality sign in equation 4.44. Thus, the frictional force is

$$f_s = \mu_s F_N \tag{6.20}$$

where F_N is the normal force holding the two objects in contact. The centripetal force F_c is supplied by the normal force F_N, that is,

$$F_c = F_N = \frac{mv^2}{r} \tag{6.21}$$

Therefore, the greater the value of the normal force, the greater will be the frictional force.

The child does not slide down the wall because the frictional force f_s is equal to the weight of the child:

$$f_s = w \tag{6.22}$$

Substituting equation 6.20 into 6.22 gives

$$\mu_s F_N = w \tag{6.23}$$

Substituting the normal force F_N from equation 6.21 into equation 6.23 gives

$$\mu_s \frac{mv^2}{r} = w$$

But the weight w of the child is equal to mg. Thus,

$$\mu_s \frac{mv^2}{r} = mg \tag{6.24}$$

Notice that the mass m is contained on both sides of equation 6.24 and can be cancelled. Hence,

$$\mu_s \frac{v^2}{r} = g \tag{6.25}$$

We can solve equation 6.25 for μ_s, the minimum value of the coefficient of static friction such that the child will not slide down the wall:

$$\mu_s = \frac{rg}{v^2} \tag{6.26}$$

Example 6.5

The rotating room. The radius r of the rotating room is 15 ft, and the speed v of the child is 40 ft/s. Find the minimum value of μ_s to keep the child pinned against the wall.

Solution

The minimum value of μ_s, found from equation 6.26, is

$$\mu_s = \frac{rg}{v^2} = \frac{(15 \text{ ft})(32 \text{ ft/s}^2)}{(40 \text{ ft/s})^2} = 0.30$$

As long as μ_s is greater than 0.30, the child will be held against the wall.

As the ride comes to an end, the speed v decreases, thereby decreasing the centripetal force, which is supplied by the normal force F_N. The frictional force, $f_s = \mu_s F_N$, also decreases and is no longer capable of holding up the weight w of the child, and the child slides slowly down the wall.

Again, we should note that there is no centrifugal force acting on the child pushing the child against the wall. It is the wall that is pushing against the child with the centripetal force that is supplied by the normal force. There are many variations of this ride in different amusement parks, where you are held tight against a rotating wall. The analysis will be similar.

A Car Making a Turn on a Level Road

Consider a car making a turn at a corner. The portion of the turn can be approximated by an arc of a circle of radius r, as shown in figure 6.13. If the car makes the turn at a constant speed v, then during that turn, the car is going through uniform circular motion and there must be some centripetal force acting on the car. The necessary centripetal force is supplied by the frictional force between the tires of the car and the pavement.

Let us analyze the motion, in particular let us find the minimum coefficient of static friction that must be present between the tires of the car and the pavement in order for the car to make the turn without skidding. As the steering wheel of the car is turned, the tires turn into the direction of the turn. But the tire also wants to continue in straight line motion by Newton's first law. Because all real tires are slightly deformed, part of the tire in contact with the road is actually flat. Hence, the portion of the tire in contact with the ground has a tendency to slip and there is therefore a frictional force that opposes this motion. Hence, the force that allows the car to go into that circular path is the static frictional force f_s between the flat portion of the tire and the road. The problem is therefore very similar to the rotating disk discussed previously. The frictional force f_s is again given by equation 4.44. We are looking for the minimum value of μ_s that will just keep the car moving in the

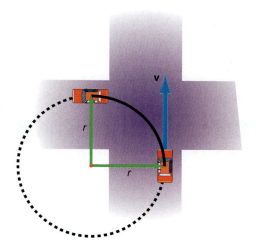

(a) Top view

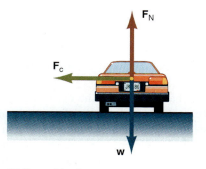

(b) Rear side view

Figure 6.13

A car making a turn on a level road.

circle. That is, the car will be just on the verge of slipping. Hence, we use the equality sign in equation 4.44. Because the centripetal force is supplied by the frictional force, we equate them as

$$F_c = f_s \tag{6.27}$$

We obtain the centripetal force from equation 6.14 and the frictional force from equation 6.20. Hence,

$$\frac{mv^2}{r} = \mu_s F_N$$

But as we can see from figure 6.13, the normal force F_N is equal to the weight w, thus,

$$\frac{mv^2}{r} = \mu_s w$$

The weight of the car $w = mg$, therefore,

$$\frac{mv^2}{r} = \mu_s mg \tag{6.28}$$

Notice that the mass m is on each side of equation 6.28 and can be divided out. Solving equation 6.28 for the minimum coefficient of static friction that must be present between the tires of the car and the pavement, gives

$$\mu_s = \frac{v^2}{rg} \tag{6.29}$$

Because equation 6.29 is independent of the mass of the car, the effect will be the same for a large massive car or a small one.

Example 6.6

Making a turn on a level road. A car is traveling at 20.0 mph in a circle of radius $r = 200$ ft. Find the minimum value of μ_s for the car to make the turn without skidding.

Solution

The minimum coefficient of friction, found from equation 6.29, is

$$\mu_s = \frac{v^2}{rg}$$

$$= \frac{[(20.0 \text{ mph})(88.0 \text{ ft/s})/(60.0 \text{ mph})]^2}{(200 \text{ ft})(32.0 \text{ ft/s}^2)}$$

$$= 0.134$$

The minimum value of the coefficient of static friction between the tires and the road must be 0.134.

For all values of μ_s, equal to or greater than this value, the car can make the turn without skidding. From table 4.1, the coefficient of friction for a tire on concrete is much greater than this, and there will be no problem in making the turn. However, if there is snow or freezing rain, then the coefficient of static friction between the tires and the snow or ice will be much lower. If it is lower than the value of 0.134 just determined, then the car will skid out in the turn. That is, there will no longer be enough frictional force to supply the necessary centripetal force.

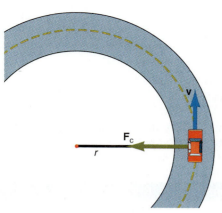

(a) Top view

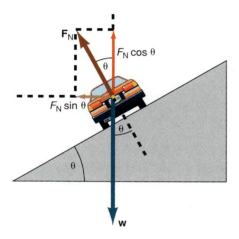

(b) Rear view

(c)

Figure 6.14

A car making a turn on a banked road.

If you ever go into a skid what should you do? The standard procedure is to turn the wheels of the car into the direction of the skid. You are then no longer trying to make the turn, and therefore you no longer need the centripetal force. You will stop skidding and proceed in the direction that was originally tangent to the circle. By tapping the brakes, you then slow down so that you can again try to make the turn. At a slower speed you may now be able to make the turn. As an example, if the speed of the car in example 6.6 is reduced from 20 mph to 10 mph, that is, in half, then from equation 6.29 the minimum value of μ_s would be cut by a fourth. Therefore, $\mu_s = 0.034$. The car should then be able to make the turn.

Even on a hot sunny day with excellent road conditions there could be a problem in making the original turn, if the car is going too fast.

Example 6.7

Making a level turn while driving too fast. If the car in example 6.6 tried to make the turn at a speed of 60 mph, that is, three times faster than before, what would the value of μ_s have to be?

Solution

The minimum coefficient of friction, found from equation 6.29, is

$$\mu_s = \frac{(3v_0)^2}{rg} = 9\frac{(v_0^2)}{rg} = 9\,\mu_{s0} = 1.21$$

That is, by increasing the speed by a factor of three, the necessary value of μ_s has been increased by a factor of 9 to the value of 1.21.

From the possible values of μ_s given in table 4.1, we cannot get such high values of μ_s. Therefore, the car will definitely go into a skid. When the original road was designed, it could have been made into a more gentle curve with a much larger value of the radius of curvature r, thereby reducing the minimum value of μ_s needed. This would certainly help, but there are practical constraints on how large we can make r.

A Car Making a Turn on a Banked Road

On large highways that handle cars at high speeds, the roads are usually banked to make the turns easier. By banking the road, a component of the reaction force of the road points into the center of curvature of the road, and that component will supply the necessary centripetal force to move the car in the circle. The car on the banked road is shown in figure 6.14. The road is banked at an angle θ. The forces acting on the car are the weight **w,** acting downward, and the reaction force of the road $\mathbf{F_N}$, acting upward on the car, perpendicular to the road. We resolve the force $\mathbf{F_N}$ into vertical and horizontal components. Using the value of θ as shown, the vertical component is $F_N \cos\theta$, while the horizontal component is $F_N \sin\theta$. As we can see from the figure, the horizontal component points toward the center of the circle. Hence, the necessary centripetal force is supplied by the component $F_N \sin\theta$. That is,

$$F_c = F_N \sin\theta \tag{6.30}$$

The vertical component is equal to the weight of the car, that is,

$$w = F_N \cos\theta \tag{6.31}$$

The problem can be simplified by eliminating F_N by dividing equation 6.30 by equation 6.31:

$$\frac{F_N \sin\theta}{F_N \cos\theta} = \frac{F_c}{w} = \frac{mv^2/r}{mg}$$

and finally, using the fact that $\sin\theta/\cos\theta = \tan\theta$, we have

$$\tan\theta = \frac{v^2}{rg} \tag{6.32}$$

Solving for θ, the angle of bank, we get

$$\theta = \tan^{-1}\frac{v^2}{rg} \tag{6.33}$$

which says that if the road is banked by this angle θ, then the necessary centripetal force for any car to go into the circular path will be supplied by the horizontal component of the reaction force of the road.

Example 6.8

Making a turn on a banked road. The car from example 6.7 is to manipulate a turn with a radius of curvature of 200 ft at a speed of 60 mph = 88 ft/s. At what angle should the road be banked for the car to make the turn?

Solution

To have the necessary centripetal force, the road should be banked at the angle θ given by equation 6.33 as

$$\theta = \tan^{-1}\frac{v^2}{rg} = \tan^{-1}\left[\frac{(88\ \text{ft/s})^2}{(200\ \text{ft})(32\ \text{ft/s}^2)}\right]$$

$$= 50°$$

This angle is a little large for practical purposes. A reasonable compromise might be to increase the radius of curvature r, to a higher value, say $r = 600$ ft, then,

$$\theta = \tan^{-1}\frac{v^2}{rg} = \tan^{-1}\left[\frac{(88\ \text{ft/s})^2}{(600\ \text{ft})(32\ \text{ft/s}^2)}\right]$$

$$= 22°$$

a more reasonable angle of bank.

The design of the road becomes a trade-off between the angle of bank and the radius of curvature, but the necessary centripetal force is supplied by the horizontal component of the reaction force of the road on the car.

An Airplane Making a Turn

During straight and level flight, the following forces act on the aircraft shown in figure 6.15: **T** is the thrust on the aircraft pulling it forward into the air, **w** is the weight of the aircraft acting downward, **L** is the lift on the aircraft that causes the

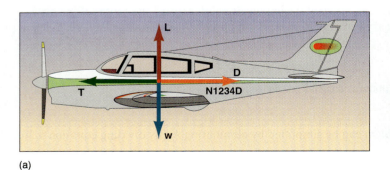

(a)　　　　　　　　　　　　　　　　　(b)

Figure 6.15

Forces acting on an aircraft in flight.

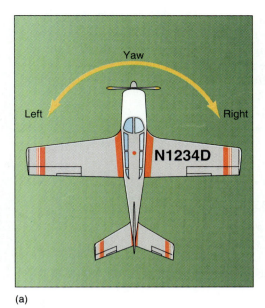

(a)

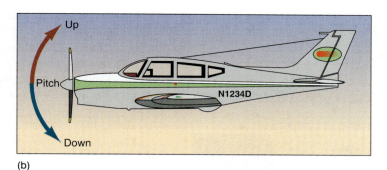

(b)

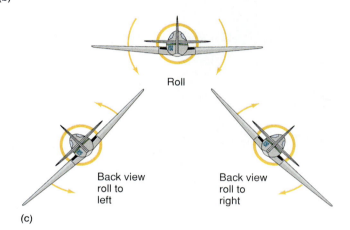

Roll

Back view roll to left

Back view roll to right

(c)

Figure 6.16

(*a*) Yaw of an aircraft. (*b*) Pitch of an aircraft. (*c*) Roll of an aircraft.

(a)

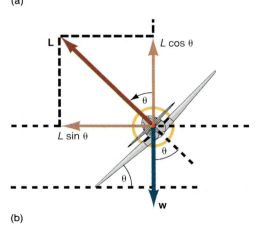

(b)

Figure 6.17

Forces acting on the aircraft during a turn.

plane to ascend, and **D** is the drag on the aircraft that tends to slow down the aircraft. The drag is opposite to the thrust. Lift and drag are just the vertical and horizontal components of the fluid force of the air on the aircraft. In normal straight and level flight, the aircraft is in equilibrium under all these forces. The lift overcomes the weight and holds the plane up; the thrust overcomes the frictional drag forces, allowing the plane to fly at a constant speed. An aircraft has three ways of changing the direction of its motion.

Yaw Control: By applying a force on the rudder pedals with his feet, the pilot can make the plane turn to the right or left, as shown in figure 6.16(a).

Pitch Control: By pushing the stick forward, the pilot can make the plane dive; by pulling the stick backward, the pilot can make the plane climb, as shown in figure 6.16(b).

Roll Control: By pushing the stick to the right, the pilot makes the plane roll to the right; by pushing the stick to the left, the pilot makes the plane roll to the left, as shown in figure 6.16(c).

To make a turn to the right or left, therefore, the pilot could simply use the rudder pedals and yaw the aircraft to the right or left. However, this is not an efficient way to turn an aircraft. As the aircraft yaws, it exposes a larger portion of its fuselage to the air, causing a great deal of friction. This increased drag causes the plane to slow down. To make the most efficient turn, a pilot performs a coordinated turn. In a coordinated turn, the pilot yaws, rolls, and pitches the aircraft simultaneously. The attitude of the aircraft is as shown in figure 6.17. In level flight the forces acting on the plane in the vertical are the lift **L** and the weight **w.** If the aircraft was originally in equilibrium in level flight, then $L = w$. Because of the turn, however, only a component of the lift is in the vertical, that is,

$$L \cos \theta = w$$

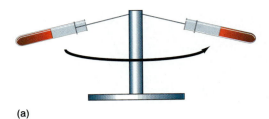

(a)

(b)

Figure 6.18

The centrifuge.

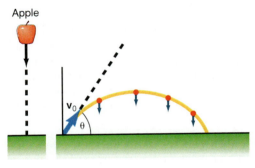

Figure 6.19

Motion of an apple or a projectile.

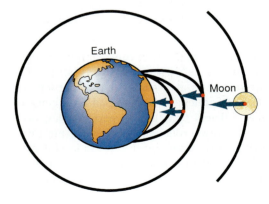

Figure 6.20

The same force acting on a projectile acts on the moon.

Therefore, the aircraft loses altitude in a turn, unless the pilot pulls back on the stick, pitching the nose of the aircraft upward. This new attitude of the aircraft increases the angle of attack of the wings, thereby increasing the lift L of the aircraft. In this way the turn can be made at a constant altitude.

The second component of the lift, $L \sin \theta$, supplies the necessary centripetal force for the aircraft to make its turn. That is,

$$L \sin \theta = F_c = \frac{mv^2}{r} \tag{6.34}$$

while

$$L \cos \theta = w = mg \tag{6.35}$$

Dividing equation 6.34 by equation 6.35 gives

$$\frac{L \sin \theta}{L \cos \theta} = \frac{mv^2/r}{mg}$$

$$\tan \theta = \frac{v^2}{rg} \tag{6.36}$$

That is, for an aircraft traveling at a speed v, and trying to make a turn of radius of curvature r, the pilot must bank or roll the aircraft to the angle θ given by equation 6.36. Note that this is the same equation found for the banking of a road. A similar analysis would show that when a bicycle makes a turn on a level road, the rider leans into the turn by the same angle θ given by equation 6.36, to obtain the necessary centripetal force to make the turn.

The Centrifuge

The **centrifuge** is a device for separating particles of different densities in a liquid. The liquid is placed in a test tube and the test tube in the centrifuge, as shown in figure 6.18. The centrifuge spins at a high speed. The more massive particles in the mixture separate to the bottom of the test tube while the particles of smaller mass separate to the top. There is no centrifugal force acting on these particles to separate them as is often stated in chemistry, biology, and medical books. Instead, each particle at any instant has a tangential velocity v and wants to continue at that same velocity by Newton's first law. The centripetal force necessary to move the particles in a circle is given by equation 6.14 ($F_c = mv^2/r$). The normal force of the bottom of the glass tube on the particles supplies the necessary centripetal force on the particles to cause them to go into circular motion. The same normal force on a small mass causes it to go into circular motion more easily than on a large massive particle. The result is that the more massive particles are found at the bottom of the test tube, while the particles of smaller mass are found at the top of the test tube.

6.8 Newton's Law of Universal Gravitation

Newton observed that an object, an apple, released near the surface of the earth, was accelerated toward the earth. Since the cause of an acceleration is an unbalanced force, there must, therefore, be a force pulling objects toward the earth. If you throw a projectile at an initial velocity v_0, as seen in figure 6.19, then instead of that object moving off into space in a straight line as Newton's first law dictates, it is continually acted on by a force pulling it back to earth. If you were strong enough to throw the projectile with greater and greater initial velocities, then the projectile paths would be as shown in figure 6.20. The distance down range would become greater and greater until at some initial velocity, the projectile would not hit the earth at all, but would go right around it in an orbit. But at any point along force, as it is a force which is directed towards some centre; and as it regards more particularly a body in that centre, we call it circumsolar, circumterrestrial, circumjovial; and so in respect of other central bodies.

[3.] The action of centripetal forces.

That by means of centripetal forces the planets may be retained in certain orbits, we may easily understand, if we consider the motions of projectiles (pp. 2–4); for a stone that is projected is by the pressure of its own weight forced out of the rectilinear path, which by the initial projection alone it should have pursued, and made to describe a curved line in the air; and

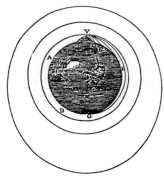

through that crooked way is at last brought down to the ground; and the greater the velocity is with which it is projected, the farther it goes before it falls to the earth. We may therefore suppose the velocity to be so increased, that it would describe an arc of 1, 2, 5, 10, 100, 1000 miles before it arrived at the earth, till at last, exceeding the limits of the earth, it should pass into space without touching it.

Figure 6.21

A page from Newton's *Principia*.

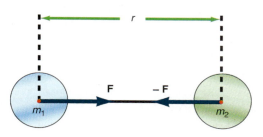

Figure 6.22

Newton's law of universal gravitation.

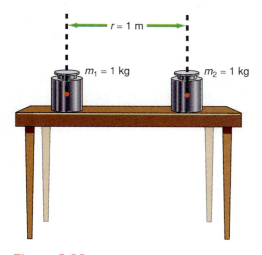

Figure 6.23

Two 1-kg masses sitting on a table.

its path the projectile would still have a force acting on it pulling it down toward the surface of the earth just as it had in figure 6.19. Figure 6.21 shows a page from the translated version of Newton's *Principia* showing these ideas.

Newton was led to the conclusion that the same force that causes the apple to fall to the earth also causes the moon to be pulled to the earth. Thus, the moon moves in its orbit about the earth because it is pulled toward the earth. It is falling toward the earth. But if there is a force between the moon and the earth, why not a force between the sun and the earth? Or for that matter why not a force between the sun and all the planets? Newton proposed that the same gravitational force that acts on objects near the surface of the earth also acts on all the heavenly bodies. This was a revolutionary hypothesis at that time, for no one knew why the planets revolved around the sun. Following this line of reasoning to its natural conclusion, Newton proposed that there was a force of gravitation between each and every mass in the universe.

Newton's law of universal gravitation was stated as follows: between every two masses in the universe there is a force of attraction between them that is directly proportional to the product of their masses, and inversely proportional to the square of the distance separating them. If the two masses are as shown in figure 6.22 with r the distance between the centers of the two masses, then the force of attraction is

$$F = \frac{Gm_1m_2}{r^2} \qquad (6.37)$$

where G is a constant, called the universal gravitational constant, given by

$$G = 6.67 \times 10^{-11} \frac{\text{N m}^2}{\text{kg}^2}$$

We assume here that the radii of the masses are relatively small compared to the distance separating them so that the distance separating the masses is drawn to the center of the masses. Such masses are sometimes treated as *point masses* or *particles*. Spherical masses are usually treated like particles.

Note here that the numerical value of the constant G was determined by a celebrated experiment by Henry Cavendish (1731–1810) over 100 years after Newton's statement of the law of gravitation. Cavendish used a torsion balance with known masses. The force between the masses was measured and G was then calculated.

6.9 Gravitational Force between Two 1-Kg Masses

Newton's law of universal gravitation says that there is a force between any two masses in the universe. Let us set up a little experiment to test this law. Let us take two standard 1-kg masses and place them on the desk, so that they are 1 m apart, as shown in figure 6.23. According to Newton's theory of gravitation, there is a force between these masses, and according to Newton's second law, they should be accelerated toward each other. However, we observe that the two masses stay right where they are. They do not move together! Is Newton's law of universal gravitation correct or isn't it?

Let us compute the gravitational force between these two 1-kg masses. We assume that the gravitational force acts at the center of each of the 1-kg masses. By equation 6.37, we have

$$F = \frac{Gm_1m_2}{r^2} = \frac{6.67 \times 10^{-11} \text{ N m}^2}{\text{kg}^2} \frac{(1 \text{ kg})(1 \text{ kg})}{(1 \text{ m})^2}$$

and therefore the force acting between these two 1-kg masses is

$$F = 6.67 \times 10^{-11} \text{ N}$$

This is, of course, a very small force. In fact, if this is written in ordinary decimal notation we have

$$F = 0.0000000000667 \text{ N}$$

A very, very small force indeed. (Sometimes it is worth while for the beginning student to write numbers in this ordinary notation to get a better "feel" for the meaning of the numbers that are expressed in scientific notation.)

If we redraw figure 6.23 showing all the forces acting on the masses, we get figure 6.24. The gravitational force on mass m_2 is trying to pull it toward the left. But if the body tends to slide toward the left, there is a force of static friction that acts to oppose that tendency and acts toward the right. The frictional force that must be overcome is

$$f_s = \mu_s F_{N2} = \mu_s w_2 = \mu_s m_2 g$$

Assuming a reasonable value of $\mu_s = 0.50$ we obtain for this frictional force,

$$f_s = \mu_s m_2 g = (0.50)(1.00 \text{ kg})(9.80 \text{ m/s}^2) = 4.90 \text{ N}$$

Hence, to initiate the movement of the 1-kg mass across the table, a force greater than 4.90 N is needed. As you can see, the gravitational force (6.67×10^{-11} N) is nowhere near this value, and is thus not great enough to overcome the force of static friction. Hence you do not, in general, observe different masses attracting each other. That is, two chairs do not slide across the room and collide due to the gravitational force between them.

If these small 1-kg masses were taken somewhere out in space, where there is no frictional force opposing the gravitational force, the two masses would be pulled together. It will take a relatively long time for the masses to come together because the force, and hence the acceleration is small, but they will come together within a few days.

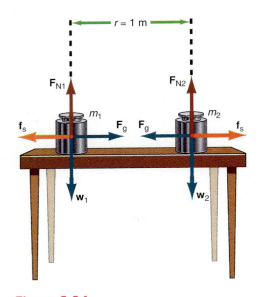

Figure 6.24

Gravitational force on two 1-kg masses.

6.10 Gravitational Force between a 1-Kg Mass and the Earth

The reason why the computed gravitational force between the two 1-kg masses was so small is because G, the universal gravitational constant, is very small compared to the masses involved. If, instead of considering two 1-kg masses, we consider one mass to be the 1-kg mass and the second mass to be the earth, then the force between them is very noticeable. If you let go of the 1-kg mass, the gravitational force acting on it immediately pulls it toward the surface of the earth. The cause of the greater force in this case is the larger mass of the earth. In fact, let us determine the gravitational force on a 1-kg mass near the surface of the earth. Figure 6.25 shows a mass m_1 of 1 kg a distance h above the surface of the earth. The radius of the earth r_e is $r_e = 6.371 \times 10^6$ m, and its mass m_e is $m_e = 5.977 \times 10^{24}$ kg. The separation distance between m_1 and m_e is

$$r = r_e + h \approx r_e \qquad (6.38)$$

since $r_e \gg h$. The gravitational force acting on that 1-kg mass is

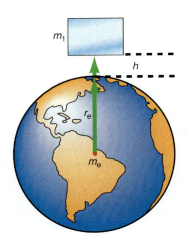

Figure 6.25

Gravitational force on a 1-kg mass near the surface of the earth.

$$F_g = \frac{G m_e m_1}{r_e^2} \qquad (6.39)$$

$$= \left(\frac{6.67 \times 10^{-11} \text{ N m}^2}{\text{kg}^2}\right) \frac{(5.977 \times 10^{24} \text{ kg})(1.00 \text{ kg})}{(6.371 \times 10^6 \text{ m})^2}$$

$$= 9.82 \text{ N}$$

This number should be rather familiar. Recall that the weight of a 1.00-kg mass was determined from Newton's second law as

$$w = mg = (1.00 \text{ kg})(9.80 \text{ m/s}^2) = 9.80 \text{ N}$$

(The standard value of $g = 9.80$ m/s^2 has been used. We will see shortly that g can actually vary between 9.78 m/s^2 at the equator to 9.83 m/s^2 at the pole. Also the radius of the earth r_e used in equation 6.39 is the mean value of r_e. The actual value of r_e varies slightly with latitude.)

The point to notice here is that the weight of a body is in fact the gravitational force acting on that body by the earth and pulling it down toward the center of the earth. Thus, the weight of a body is actually determined by Newton's law of universal gravitation. This points up even further the difference between the mass and the weight of a body.

6.11 The Acceleration Due to Gravity and Newton's Law of Universal Gravitation

Newton's second law states that when an external unbalanced force acts on an object, it will give that object an acceleration a, that is,

$$F = ma$$

But if the force acting on a body near the surface of the earth is the gravitational force, then that body experiences the acceleration g of a freely falling body, as shown in section 3.7. That is, the force acting on the object is called its weight, and it experiences the acceleration g. Thus, Newton's second law becomes

$$w = mg \tag{6.40}$$

But as just shown, the weight of a body is equal to the gravitational force acting on that body and therefore,

$$w = F_g \tag{6.41}$$

Using equations 6.40 and 6.39 we get

$$mg = \frac{Gm_e m}{r_e^2} \tag{6.42}$$

Solving for g, we obtain

$$g = \frac{Gm_e}{r_e^2} \tag{6.43}$$

That is, *the acceleration due to gravity, g, which in chapter 3 was accepted as an experimental fact, can be deduced from theoretical considerations of Newton's second law and his law of universal gravitation.*

Example 6.9

Determine the acceleration of gravity g at the surface of the earth.

Solution

The value of g, determined from equation 6.43, is

$$g = \frac{Gm_e}{r_e^2} = \left(\frac{6.67 \times 10^{-11} \text{ N m}^2}{\text{kg}^2} \right) \frac{(5.977 \times 10^{24} \text{ kg})}{(6.371 \times 10^6 \text{ m})^2}$$

$$= 9.82 \text{ m/s}^2$$

Newton introduced his law of universal gravitation, and a by-product of it is a theoretical explanation of the acceleration due to gravity g. This is an example of the beauty and simplicity of physics. There is no way that we could have predicted the relation of equation 6.43 from purely experimental grounds. Yet Newton's second law and his law of universal gravitation, in combination, have made that prediction.

6.12 Variation of the Acceleration Due to Gravity

We can see from equation 6.43 why the acceleration due to gravity g is very nearly a constant. G is a constant and m_e is a constant, but r_e is not exactly a constant. The earth is not, in fact, a perfect sphere. It is, rather, an oblate spheroid. That is, the radius of the earth at the equator r_{ee} is slightly greater than the radius of the earth at the poles r_{ep}, as seen in figure 6.26. The diagram is, of course, exaggerated to show this difference. The actual values of r_{ee} and r_{ep} are

$$r_{ee} = 6.378 \times 10^6 \text{ m}$$
$$r_{ep} = 6.356 \times 10^6 \text{ m}$$

with the mean radius

$$r_e = 6.371 \times 10^6 \text{ m}$$

The variation in the radius of the earth is thus quite small. However, the variation, although small, does contribute to the observed variation in the acceleration due to gravity on the earth from a low of 9.78 m/s² at the equator to a high of 9.83 m/s² at the North Pole, as seen in table 6.1. This analysis also assumes that the earth is not rotating. A more sophisticated analysis takes into account the variation in g caused by the centripetal acceleration, which varies with latitude on the surface of the earth. The standard value of g, adopted for most calculations in physics, is

$$g = 9.80 \text{ m/s}^2$$

the value at 45° north latitude at the surface of the earth.

At greater heights, g also varies slightly from that given in equation 6.43 because of the approximation

$$r_e \approx r_e + h$$

that was made for that equation. Although this approximation is, in general, quite good for most locations, if you are on the top of a mountain, such as Pikes Peak, this higher altitude (large value of h) will give you a slightly smaller value of g, as we can see in table 6.1.

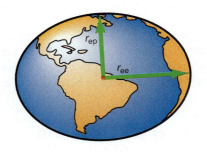

Figure 6.26
The earth is an oblate spheroid.

Table 6.1 Different Values of g on the Earth	
Location	**Value of g in m/s²**
Equator at sea level	9.78
New York City	9.80
45° N latitude (standard)	9.80
North Pole	9.83
Pikes Peak—elevation 4293 m	9.79
Denver, Colorado—elevation 1638 m	9.80

Mechanics

Again it is quite remarkable that these slight variations in the observed experimental values of g on the surface of this earth can be explained and predicted by Newton's law of universal gravitation, with slight corrections for the radius of the earth, the centripetal acceleration (which is a function of latitude), and the height of the location above mean sea level. There are also slight local variations in g due to the nonhomogeneous nature of the mass distribution of the earth. These variations in g due to different mass distributions are used in geophysical explorations. One of the many scientific experiments performed on the moon was a mapping of the acceleration due to gravity on the moon to disclose the possible locations of different mineral deposits.

6.13 Acceleration Due to Gravity on the Moon and on Other Planets

Equation 6.43 was derived on the basis of the gravitational force of the earth acting on a mass near the surface of the earth. The result is perfectly general however. If, for example, an object were placed close to the surface of the moon, as shown in figure 6.27, the force on that mass would be its lunar weight, which is just the gravitational force of the moon acting on it. Therefore the weight of an object on the moon is

$$w_{\text{m}} = F_{\text{g}} \qquad (6.44)$$

This becomes

$$mg_{\text{m}} = \frac{Gm_{\text{m}}m}{r_{\text{m}}^2} \qquad (6.45)$$

where g_{m} is the acceleration due to gravity on the moon and m_{m} and r_{m} are the mass and radius of the moon, respectively. Hence, the acceleration due to gravity on the moon is

$$g_{\text{m}} = \frac{Gm_{\text{m}}}{r_{\text{m}}^2} \qquad (6.46)$$

Equation 6.46 is identical to equation 6.43 except for the subscripts. Therefore, we can use equation 6.43 to determine the acceleration due to gravity on any of the planets, simply by using the mass of that planet and the radius of that planet in equation 6.43.

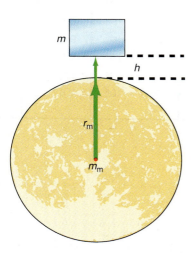

Figure 6.27

A mass placed close to the surface of the moon.

Example 6.10

Determine the acceleration due to gravity on the moon.

Solution

The acceleration due to gravity on the moon, found by solving equation 6.46, is

$$g_{\text{m}} = \frac{Gm_{\text{m}}}{r_{\text{m}}^2} = \left(\frac{6.67 \times 10^{-11} \ \text{N m}^2}{\text{kg}^2} \right) \frac{(7.34 \times 10^{22} \ \text{kg})}{(1.738 \times 10^6 \ \text{m})^2}$$

$$= 1.62 \ \text{m/s}^2 = 0.165 \ g_{\text{e}} \approx \tfrac{1}{6} g_{\text{e}}$$

The acceleration due to gravity on the moon is approximately 1/6 the acceleration due to gravity on the earth.

Table 6.2

Some Characteristics of the Planets and the Moon

Planet	Mean Orbital Radius (m)	Mass (kg)	Mean Radius of Planet (m)	g at Equator (m/s^2)	(g_e)
Mercury	5.80×10^{10}	3.24×10^{23}	2.340×10^6	3.95	0.4
Venus	1.08×10^{11}	4.86×10^{24}	6.10×10^6	8.72	0.89
Earth	1.50×10^{11}	5.97×10^{24}	6.371×10^6	9.78	1.00
Mars	2.28×10^{11}	6.40×10^{23}	3.32×10^6	3.84	0.39
Jupiter	7.80×10^{11}	1.89×10^{27}	69.8×10^6	23.16	2.37
Saturn	1.43×10^{12}	5.67×10^{26}	58.2×10^6	8.77	0.90
Uranus	2.88×10^{12}	8.67×10^{25}	23.8×10^6	9.46	0.97
Neptune	4.52×10^{12}	1.05×10^{26}	22.4×10^6	13.66	1.40
Pluto	5.91×10^{12}	6.6×10^{23}	2.9×10^6	5.23	0.53
Earth's Moon	3.84×10^8	7.34×10^{22}	1.738×10^6	1.62	1/6

The Earth and its satellite.

Because the weight of an object is

$$w = mg$$

the weight of an object on the moon is

$$w_m = mg_m = m(1/6 \, g_e) = 1/6 \, (mg_e) = 1/6 \, w_e$$

which is 1/6 of the weight that it had on the earth. That is, if you weigh 180 lb on earth, you will only weigh 30 lb on the moon.

Table 6.2 is a list of the masses, radii, and values of g on the various planets. Note that the most massive planet is Jupiter, and it has an acceleration due to gravity of

$$g_J = 2.37 \, g_e$$

Therefore, the weight of an object on Jupiter will be

$$w_J = 2.37 \, w_e$$

If you weighed 180 lb on earth, you would weigh 427 lb on Jupiter.

6.14 Satellite Motion

Consider the motion of a satellite around its parent body. This could be the motion of the earth around the sun, the motion of any planet around the sun, the motion of the moon around the earth, the motion of any other moon around its planet, or the motion of an artificial satellite around the earth, the moon, another planet, and so forth.

Let us start with the analysis of the motion of an artificial satellite in a circular orbit around the earth. Perhaps the first person to ever conceive of the possibility of an artificial earth satellite was Sir Isaac Newton, when he wrote in 1686 in his *Principia:*

But if we now imagine bodies to be projected in the directions of lines parallel to the horizon from greater heights, as of 5, 10, 100, 1000, or more miles or rather as many semi-diameters of the earth, those bodies according

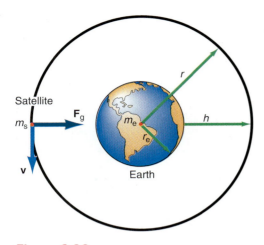

Figure 6.28

Satellite motion.

to their different velocity, and the different force of gravity in different heights, will describe arcs either concentric with the earth, or variously eccentric, and go on revolving through the heavens in those orbits just as the planets do in their orbits.[2]

For the satellite to be in motion in a circular orbit, there must be a centripetal force acting on the satellite to force it into the circular motion. This centripetal force acting on the satellite, is supplied by the force of gravity of the earth. Let us assume that the satellite is in orbit a distance *h* above the surface of the earth, as shown in figure 6.28. Because the centripetal force is supplied by the gravitational force, we have

$$F_c = F_g \tag{6.47}$$

or

$$\frac{m_s v^2}{r} = \frac{G m_e m_s}{r^2} \tag{6.48}$$

Solving for the speed of the satellite in the circular orbit, we obtain

$$v = \sqrt{\frac{G m_e}{r}} \tag{6.49}$$

The first thing that we note is that m_s, the mass of the satellite, divided out of equation 6.48. This means that the speed of the satellite is independent of its mass. That is, the speed is the same, whether it is a very large massive satellite or a very small one.

Equation 6.49 represents the speed that a satellite must have if it is to remain in a circular orbit, at a distance *r* from the center of the earth. Because the satellite is at a height *h* above the surface of the earth, the orbital radius *r* is

$$r = r_e + h \tag{6.50}$$

Equation 6.49 also says that the speed depends only on the radius of the orbit *r*. For large values of *r*, the required speed will be relatively small; whereas for small values of *r*, the speed *v* must be much larger. If the actual speed of a satellite at a distance *r* is less than the value of *v*, given by equation 6.49, then the satellite will move closer toward the earth. If it gets close enough to the earth's atmosphere, the air friction will slow the satellite down even further, eventually causing it to spiral into the earth. The increased air friction will then cause it to burn up and crash. If the actual speed at the distance *r* is greater than that given by equation 6.49, then the satellite will go farther out into space, eventually going into either an elliptical, parabolic, or hyperbolic orbit, depending on the speed *v*.

How do we get the satellite into this circular orbit? The satellite is placed in the orbit by a rocket. The rocket is launched vertically from the earth, and at a predetermined altitude it pitches over, so as to approach the desired circular orbit tangentially. The engines are usually turned off and the rocket coasts on a projectile trajectory to the orbital intercept point *I* in figure 6.29. Let the velocity of the rocket on the ascent trajectory at the point of intercept be v_A. The velocity necessary for the satellite to be in circular orbit at the height *h* is **v** and its speed is given by equation 6.49. Therefore, the rocket must undergo a change in velocity Δv to match its ascent velocity to the velocity necessary for the circular orbit. That is,

$$\Delta v = v - v_A \tag{6.51}$$

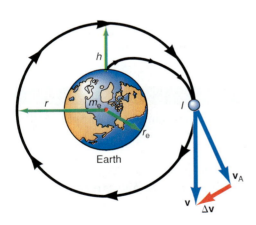

Figure 6.29

Placing a satellite in a circular orbit.

2. Quoted from Sir Isaac Newton's *Mathematical Principles of Natural Philosophy*, p. 552. Translated by A. Motte. University of California Press, 1960.

This change in velocity is of course produced by the thrust of the rocket engines. How long should these engines be turned on to get this necessary change in speed Δv? As a first approximation we take Newton's second law in the form

$$F = ma = m\frac{\Delta v}{\Delta t}$$

Solving Newton's second law for Δt, gives

$$\Delta t = \frac{m}{F}\Delta v \qquad (6.52)$$

where F is the thrust of the rocket engine, Δv is the necessary speed change, determined from equation 6.51; and m is the mass of the space craft at this instant of time. Therefore, equation 6.52 tells the astronaut the length of time to "burn" his engines. At the end of this time the engines are shut off, and the spacecraft has the necessary orbital speed to stay in its circular orbit.

This is, of course, a greatly simplified version of orbital insertion, for we need not only the magnitude of Δv but also its direction. An attitude control system is necessary to determine the proper direction for the $\Delta \mathbf{v}$. Also it is important to note that using equation 6.52 is only an approximation, because as the spacecraft burns its rocket propellant, its mass m is continually changing. This example points out a deficiency in using Newton's second law in the form $F = ma$, because this form assumes that the mass under consideration is a constant. In chapter 8, on momentum, we will write Newton's second law in another form that allows for the case of variable mass.

We should also note here that the orbits of all the planets around the sun are ellipses rather than circles. But, in general, the amount of ellipticity is relatively small, and as a first approximation it is quite often assumed that their orbits are circular. For this approximation, we can use equation 6.49, with the appropriate change in subscripts, to determine the approximate speed of any of the planets. For precise astronomical work, the elliptical orbit must be used. Extremely precise experimental determinations of the orbits of the planets were made by the Danish astronomer Tycho Brahe (1546–1609). Johannes Kepler (1571–1630) analyzed this work and expressed the result in what are now called **Kepler's laws of planetary motion.** Kepler's laws are

1. The orbit of each planet is an ellipse with the sun at one focus.
2. The speed of the planet varies in such a way that the line joining the planet and the sun sweeps out equal areas in equal times.
3. The cube of the semimajor axes of the elliptical orbit is proportional to the square of the time for the planet to make a complete revolution about the sun.

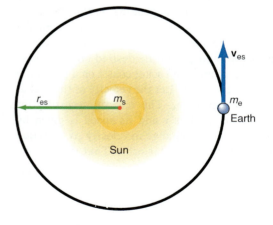

Figure 6.30

The speed of the earth in its orbit about the sun.

Example 6.11

Determine the speed of the earth in its orbit about the sun, shown in figure 6.30.

Solution

The mass of the sun is $m_{sun} = 1.99 \times 10^{30}$ kg. The mean orbital radius of the earth around the sun, found in table 6.2, is $r_{es} = 1.50 \times 10^{11}$ m. The speed of the earth around the sun v_{es}, found from equation 6.49, is

$$v_{es} = \sqrt{\frac{Gm_s}{r_{es}}}$$

$$= \sqrt{\left(\frac{6.67 \times 10^{-11}\ \text{N m}^2}{\text{kg}^2}\right)\frac{1.99 \times 10^{30}\ \text{kg}}{1.50 \times 10^{11}\ \text{m}}}$$

$$= 2.97 \times 10^4\ \text{m/s} = 29.7\ \text{km/s} = 66,600\ \text{mi/hr}$$

That is, as you sit and read this book, you are speeding through space at 66,600 mph. This is a little over 18 miles each second. The mean orbital speed of any of the planets or satellites can be determined in the same way.

6.15 The Geosynchronous Satellite

An interesting example of satellite motion is the geosynchronous satellite. The **geosynchronous satellite** is a satellite whose orbital motion is synchronized with the rotation of the earth. In this way the geosynchronous satellite is always over the same point on the equator as the earth turns. The geosynchronous satellite is obviously very useful for world communication, weather observations, and military use.

What should the orbital radius of such a satellite be, in order to stay over the same point on the earth's surface? The speed necessary for the circular orbit, given by equation 6.49, is

$$v = \sqrt{\frac{Gm_e}{r}}$$

But this speed must be equal to the average speed of the satellite in one day, namely

$$v = \frac{s}{t} = \frac{2\pi r}{\tau} \tag{6.53}$$

where τ is the period of revolution of the satellite that is equal to one day. That is, the satellite must move in one complete orbit in a time of exactly one day. Because the earth rotates in one day and the satellite will revolve around the earth in one day, the satellite at A' will always stay over the same point on the earth A, as in figure 6.31(a). That is, the satellite is at A', which is directly above the point A on the earth. As the earth rotates, A' is always directly above A. Setting equation 6.53 equal to equation 6.49 for the speed of the satellite, we have

$$\frac{2\pi r}{\tau} = \sqrt{\frac{Gm_e}{r}} \tag{6.54}$$

Squaring both sides of equation 6.54 gives

$$\frac{4\pi^2 r^2}{\tau^2} = \frac{Gm_e}{r}$$

or

$$r^3 = \frac{Gm_e \tau^2}{4\pi^2}$$

Solving for r, gives the required orbital radius of

$$r = \left(\frac{Gm_e \tau^2}{4\pi^2}\right)^{1/3} \tag{6.55}$$

Substituting the values for the earth into equation 6.55 gives

$$r = \left\{\frac{(6.67 \times 10^{-11}\ \text{Nm}^2/\text{kg}^2)(5.97 \times 10^{24}\ \text{kg})[24\ \text{hr}(3600\ \text{s/hr})]^2}{4(3.14159)^2}\right\}^{1/3}$$

$$r = 4.22 \times 10^7\ \text{m} = 4.22 \times 10^4\ \text{km} = 26{,}200\ \text{miles}$$

the orbital radius, measured from the center of the earth, for a geosynchronous satellite. A satellite at this height will always stay directly above a particular point on the surface of the earth.

A satellite communication system can be set up by placing several geosynchronous satellites in orbits over different points on the surface of the earth. As an

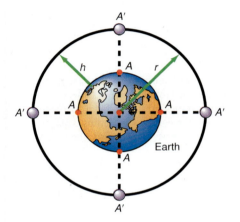

(a)

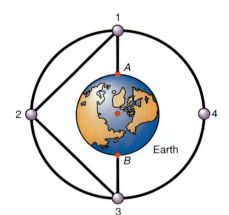

(b)

Figure 6.31

The geosynchronous satellite.

example, suppose four geosynchronous satellites were placed in orbit, as shown in figure 6.31(b). Let us say that we want to communicate, by radio or television, between the points *A* and *B*, which are on opposite sides of the earth. The communication would first be sent from point *A* to geosynchronous satellite 1, which would retransmit the communication to geosynchronous satellite 2. This satellite would then transmit to geosynchronous satellite 3, which would then transmit to the point *B* on the opposite side of the earth. Since these geosynchronous satellites appear to hover over one place on earth, continuous communication with any place on the surface of the earth can be attained.

"Have you ever wondered . . . ?"

An Essay on the Application of Physics

Space Travel

Earth is the cradle of man, but man was never made to stay in a cradle forever.

K. Tsiolkovsky

Have you ever wondered what it would be like to go to the moon or perhaps to another planet or to travel anywhere in outer space? But how can you get there? How can you travel into space?

Man has long had a fascination with the possibility of space travel. Jules Verne's novel, *From the Earth to the Moon,* was first published in 1868. In it he describes a trip to the moon inside a gigantic cannon shell. It is interesting to note that he says

Now as the Moon is never in the zenith, or directly overhead, in countries further than 28° from the equator, to decide on the exact spot for casting the Columbiad became a question that required some nice consultation. [And then a little further on] The 28th parallel of north latitude, as every school boy knows, strikes the American continent a little below Cape Canaveral. (pp. 66 and 68)

As I am sure we all know, Cape Canaveral is the site of the present Kennedy Space Center, the launch site for the Apollo mission to the moon. The first astronauts landed on the moon on July 20, 1969, just over a hundred years after the publication of Verne's novel. (Actually Jules Verne did not pick Cape Canaveral as the launch site, but rather Tampa, Florida, a relatively short distance away, because of its "position and easiness of approach, both by sea and land.")[3]

The idea of space travel left the realm of science fiction by the work of three men, Konstantin Tsiolkovsky, a Russian; Robert Goddard, an American; and Hermann Oberth, a German. Tsiolkovsky's first paper, "Free Space," was published in 1883. In his *Dreams of Earth and Sky,* 1895, he wrote of an artificial earth satellite. Goddard's first paper, "A Method of Reaching Extreme Altitudes," was written in 1919. The extreme altitudes he was referring to was the moon. Goddard launched the first liquid-

[3] *From the Earth to the Moon* in *The Space Novels of Jules Verne,* p. 69, Dover Publications, N.Y.

fueled rocket in history on March 16, 1926. Meanwhile, Oberth published his work, *The Rocket into Inter Planetary Space,* in 1923, which culminated with the German V-2 rocket in World War II. Another analysis of the problems associated with space flight was published in 1925 by Walter Hohmann in *Die Erreichbarkeit der Himmelskorper (The Attainability of the Heavenly Bodies).* In the preface, Hohmann says,

The present work will contribute to the recognition that space travel is to be taken seriously and that the final successful solution of the problem cannot be doubted, if existing technical possibilities are purposefully perfected as shown by conservative mathematical treatment.

Hohmann's original work had been written 10 years previous to its publication. In this work, Hohmann shows how to get to the Moon, Venus, and Mars. His simple approach to reach these heavenly bodies is by use of the cotangential ellipse. Before this approach is described, let us first say a word about the elliptical orbit.

Just as the speed of a satellite in a circular orbit is determined by equation 6.49, we can determine the speed of a satellite in an elliptical orbit. The mathematical derivation is slightly more complicated and will not be given here, but the result is quite simple. The speed of a satellite in an elliptical orbit is given by

$$v = \sqrt{GM\left(\frac{2}{r} - \frac{1}{a}\right)} \qquad \text{(6H.1)}$$

where *a* is a constant of the orbit called the semimajor axis of the ellipse and is shown in figure 1.

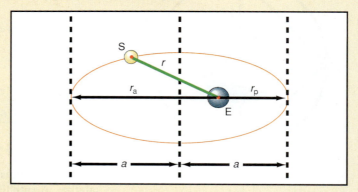

Figure 1

An elliptical orbit.

Let us assume that this is an elliptical satellite orbit about the earth. The earth is located at the focus of the ellipse, labeled E in figure 1 and r is the distance from the center of the earth to the satellite S, at any instant of time. The first thing we observe about elliptical motion is that the speed v is not a constant as it is for circular orbital motion. The speed varies with the location r in the orbit as we see in equation 6H.1. When the satellite is at its closest approach to the earth, $r = r_p$, the satellite is said to be at its perigee position. From equation 6H.1 we see that because this is the smallest value of r in the orbit, this corresponds to the greatest speed of the satellite. Hence, the satellite moves at its greatest speed when it is closest to the earth. When the satellite is at its farthest position from the earth, $r = r_a$, the satellite is said to be at its apogee position. Because this is the largest value of r in the orbit, it is the largest value of r in equation 6H.1. Since r is in the denominator of equation 6H.1, the largest value of r corresponds to the smallest value of v. Hence, the satellite moves at its slowest speed when it is the farthest distance from the earth. Thus, the motion in the orbit is not uniform, it speeds up as the satellite approaches the earth and slows down as the satellite recedes away from the earth.

We can express the semimajor axis of the ellipse in terms of the perigee and apogee distances by observing from figure 1 that

$$2a = r_a + r_p$$

or

$$a = \frac{r_a + r_p}{2} \qquad \text{(6H.2)}$$

For the special case where the ellipse degenerates into a circle, $r_a = r_p = r$, the radius of the circular orbit and then

$$a = \frac{r_a + r_p}{2} = \frac{r + r}{2} = \frac{2r}{2} = r$$

The equation for the speed, equation 6H.1 then becomes

$$v = \sqrt{GM\left(\frac{2}{r} - \frac{1}{r}\right)}$$
$$= \sqrt{\frac{GM}{r}} \qquad \text{(6H.3)}$$

But equation 6H.3 is the equation for the speed of a satellite in a circular orbit, equation 6.49. Hence, the elliptical orbit is the more general orbit, with the circular orbit as a special case.

Example 6H.1

The earth is at its closest position to the sun, its perihelion, on about January 3 when it is approximately 1.47×10^{11} m away from the sun. The earth reaches its aphelion distance, its greatest distance, on July 4, when it is about 1.53×10^{11} m away from the sun. Find the speed of the earth at its perihelion and aphelion position in its orbit.

Solution

The semimajor axis of the elliptical orbit, found from equation 6H.2, is

$$a = \frac{r_a + r_p}{2}$$
$$= \frac{1.53 \times 10^{11} \text{ m} + 1.47 \times 10^{11} \text{ m}}{2}$$
$$= 1.50 \times 10^{11} \text{ m} \qquad \text{(6H.2)}$$

The speed of the earth at perihelion is found from equation 6H.1 with $r = r_p$, the perihelion distance,

$$v = \sqrt{GM_s\left(\frac{2}{r} - \frac{1}{a}\right)}$$

Hence,

$$v_p = \sqrt{GM_s\left(\frac{2}{r_p} - \frac{1}{a}\right)}$$
$$= \sqrt{(6.67 \times 10^{-11} \text{ Nm}^2/\text{kg}^2)(1.99 \times 10^{30} \text{ kg})(2/1.47 \times 10^{11} \text{ m} - 1/1.50 \times 10^{11} \text{ m})}$$
$$= 3.03 \times 10^4 \text{ m/s}$$

The speed of the earth at aphelion is found from equation 6H.1 with $r = r_a = 1.53 \times 10^{11}$ m,

$$v = \sqrt{GM_s\left(\frac{2}{r} - \frac{1}{a}\right)}$$
$$v_a = \sqrt{GM_s\left(\frac{2}{r_a} - \frac{1}{a}\right)}$$
$$= \sqrt{(6.67 \times 10^{-11} \text{ Nm}^2/\text{kg}^2)(1.99 \times 10^{30} \text{ kg})(2/1.53 \times 10^{11} \text{ m} - 1/1.50 \times 10^{11} \text{ m})}$$
$$= 2.92 \times 10^4 \text{ m/s}$$

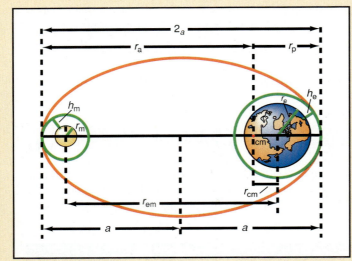

Figure 2

The Hohmann transfer orbit.

It is thus easy to see why the earth, in its orbit about the sun, is sometimes approximated as a circular orbit. The aphelion distance, perihelion distance, and the mean distance are very close, that is, 1.53×10^{11} m, 1.47×10^{11} m, and 1.50×10^{11} m, respectively. Also the speed of the earth at aphelion, perihelion, and in a circular orbit is 2.92×10^4 m/s, 3.03×10^4 m/s, and 2.97×10^4 m/s, respectively, which are also very close. The error in using the circular approximation rather than the elliptical analysis is no more than about 2%.

The simplest approach to space flight to the moon or to a planet is by use of the *Hohmann transfer ellipse*. Let us assume that the spacecraft is launched from the surface of the earth on an ascent trajectory. It is then desired to place the spacecraft in a circular parking orbit about the earth. If the circular parking orbit is to be at a height h_e above the surface of the earth then the necessary speed for the spacecraft, given by equation 6.49, is

$$v_{oe} = \sqrt{\frac{GM_e}{r_e + h_e}} \qquad \text{(6H.4)}$$

Knowing the speed of the spacecraft on the ascent trajectory from an on-board inertial navigational system, equation 6.51 is then used to determine the necessary "delta v," Δv, to get into this orbit. The engines are then turned on for the value of Δt, determined by equation 6.52, and the spacecraft is thus inserted into the circular parking orbit about the earth.

Before descending to the surface of the moon, it would be desirable to first go into a circular lunar parking orbit. To get to this circular lunar parking orbit, a "cotangential ellipse," the Hohmann transfer ellipse, is placed onto the two parking orbits, such that the focus of the ellipse is placed at the center of mass of the earth-moon system, and the ellipse is tangential to each parking orbit, as seen in figure 2. All positions in the orbit are measured from the center of mass of the earth-moon system. The semimajor axis a, of this ellipse, found from figure 2, is

$$a = \frac{r_{em} + r_m + h_m + r_e + h_e}{2} \qquad \text{(6H.5)}$$

where

r_{em} is the distance from the center of the earth to the center of the moon.

r_m is the radius of the moon.

h_m is the height of the spacecraft above the surface of the moon.

r_e is the radius of the earth.

h_m is the height of the spacecraft above the surface of the earth when it is in its circular parking orbit.

The insertion of the spacecraft into the transfer ellipse occurs at the perigee position of the elliptical orbit, which from figure 2 is

$$r_p = r_{cm} + r_e + h_e \qquad \text{(6H.6)}$$

where r_{cm} is the distance from the center of earth to the center of mass of the earth-moon system. The necessary speed to get into this cotangential ellipse at the perigee position, found from equation 6H.1, is

$$v_{TEp} = \sqrt{GM_e\left(\frac{2}{r_p} - \frac{1}{a}\right)} \qquad \text{(6H.7)}$$

where a and r_p are found from equations 6H.5 and 6H.6, respectively. The notation v_{TEp} stands for the speed in the transfer ellipse at perigee.

Because the speed of the spacecraft in the earth parking orbit is known, equation 6H.4, and the necessary speed for the transfer orbit is known, equation 6H.7, the necessary change in speed (Δv_I) of the spacecraft is just the difference between these speeds. Hence, the required Δv for insertion into the transfer ellipse is given by

$$\Delta v_I = v_{TEp} - v_{oe} \qquad \text{(6H.8)}$$

The spacecraft engines must be turned on to supply this necessary change in speed (Δv). When this Δv_I is achieved, the spacecraft engines are turned off and the spacecraft coasts toward the moon. If the engines are not turned on again, then the spacecraft would coast to the moon, reach it, and would then continue back toward the earth on the second half of the transfer ellipse. Thus, if there were some type of malfunction on the spacecraft, it would automatically return to earth.

Assuming there is no failure, the astronauts on board the spacecraft would like to change from their transfer orbit into the circular lunar parking orbit. The speed of the spacecraft on the transfer ellipse is given by equation 6H.1, with $r = r_a$ the apogee distance, as

$$v_{TEa} = \sqrt{GM_e\left(\frac{2}{r_a} - \frac{1}{a}\right)} \qquad \text{(6H.9)}$$

The apogee distance r_a, found from figure 2, is

$$r_a = r_{em} + r_m + h_m - r_{cm} \qquad \text{(6H.10)}$$

The necessary speed that the spacecraft must have to enter a circular lunar parking orbit v_{om} is found from modifying equation 6.49 to

$$v_{om} = \sqrt{\frac{GM_m}{r_m + h_m}} \qquad \text{(6H.11)}$$

where M_m is the mass of the moon, r_m is the radius of the moon, and h_m is the height of the spacecraft above the surface of the moon in its circular lunar parking orbit. The necessary change in speed to transfer from the Hohmann ellipse to the circular lunar parking orbit is obtained by subtracting equation 6H.11 from equation 6H.9. Thus the necessary Δv is

$$\Delta v_{II} = v_{TEa} - v_{om} \qquad \text{(6H.12)}$$

The spacecraft engines are turned on to obtain this necessary change in speed. When the engines are shut off the spacecraft will have the speed v_{om}, and will stay in the circular lunar parking orbit until the astronauts are ready to descend to the lunar surface. The process is repeated for the return to earth.

The Hohmann transfer is the simplest of the transfer orbits and is also the orbit of minimum energy. However, it has the disadvantage of having a large flight time. In the very early stages of the Apollo program, the Hohmann transfer ellipse was considered for the lunar transfer orbit. However, because of its long flight time, it was discarded for a hyperbolic transfer orbit that had been perfected by the Jet Propulsion Laboratories in California on its *Ranger, Surveyor,* and *Lunar Orbiter* unmanned spacecrafts to the moon. The hyperbolic orbit requires a great deal more energy, but its flight time is relatively small. The procedure for a trip on a hyperbolic orbit is similar to the elliptical orbit, only another equation is necessary for the speed of the spacecraft in the hyperbolic orbit. The principle however is the

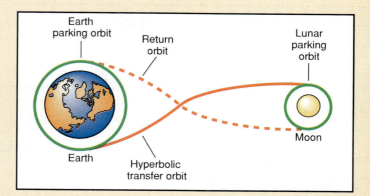

Figure 3

A hyperbolic transfer orbit.

same. Determine the current speed in the particular orbit, then determine the speed that is necessary for the other orbit. The difference between the two of them is the necessary Δv. The spacecraft engines are turned on until this value of Δv is obtained. A typical orbital picture for this type of transfer is shown in figure 3.

Unmanned satellites have since traveled to Mars, Venus, Saturn, Jupiter, Uranus, and Neptune. And what about manned trips to these planets? On July 20, 1989, the twentieth anniversary of the first landing on the moon, the president of the United States, George Bush, announced to the world that the United States will begin planning a manned trip to the planet Mars and eventually to an exploration of our entire solar system. Man is thus getting ready to leave his cradle.

The Language of Physics

Uniform circular motion
Motion in a circle at constant speed. Because the velocity vector changes in direction with time, this type of motion is accelerated motion (p. 155).

Centripetal acceleration
When a body moves in uniform circular motion, the acceleration is called centripetal acceleration. The direction of the centripetal acceleration is toward the center of the circle (p. 156).

Radian
A unit that is used to measure an angle. It is defined as the ratio of the arc length subtended to the radius of the circle, where 2π radians equals 360° (p. 156).

Centripetal force
The force that is necessary to cause an object to move in a circle at constant speed. The centripetal force acts toward the center of the circle (p. 160).

Centrifugal force
The reaction force to the centripetal force. The reaction force does not act on the same body as the centripetal force. That is, if a string were tied to a rock and the rock were swung in a horizontal circle at constant speed, the centripetal force would act on the rock while the centrifugal force would act on the string (p. 161).

Centrifuge
A device for separating particles of different densities in a liquid. The centrifuge spins at a high speed. The more massive particles in the mixture will separate to the bottom of the test tube while the particles of smaller mass will separate to the top (p. 170).

Newton's law of universal gravitation
Between every two masses in the universe there is a force of attraction that is directly proportional to the product of their masses

and inversely proportional to the square of the distance separating them (p. 171).

Kepler's laws of planetary motion
(1) The orbit of each planet is an ellipse with the sun at one focus. (2) The speed of the planet varies in such a way that the line joining the planet and the sun sweeps out equal areas in equal times. (3) The cube of the semimajor axes of the elliptical orbit is proportional to the square of the time for the planet to make a complete revolution about the sun (p. 178).

Geosynchronous satellite
A satellite whose orbital motion is synchronized with the rotation of the earth. In this way the satellite is always over the same point on the equator as the earth turns (p. 179).

Summary of Important Equations

Definition of angle in radians

$$\theta = \frac{s}{r} \tag{6.4}$$

Arc length

$$s = r\,\theta \tag{6.5}$$

Centripetal acceleration

$$a_c = \frac{v^2}{r} \tag{6.12}$$

Centripetal force

$$F_c = ma_c = \frac{mv^2}{r} \tag{6.14}$$

Angle of bank for circular turn

$$\theta = \tan^{-1}\frac{v^2}{rg} \tag{6.33}$$

Newton's law of universal gravitation

$$F = \frac{Gm_1m_2}{r^2} \tag{6.37}$$

The acceleration due to gravity on earth

$$g_e = \frac{Gm_e}{r^2} \tag{6.43}$$

The acceleration due to gravity on the moon

$$g_m = \frac{Gm_m}{r_m^2} \tag{6.46}$$

Speed of a satellite in a circular orbit

$$v = \sqrt{Gm_e/r} \tag{6.49}$$

Questions for Chapter 6

1. If a car is moving in uniform circular motion at a speed of 5.00 m/s and has a centripetal acceleration of 2.50 m/s², will the speed of the car increase at 2.50 m/s every second?
2. Does it make any sense to say that a car in uniform circular motion is moving with a velocity that is tangent to a circle and yet the acceleration is perpendicular to the tangent? Should not the acceleration be tangential because that is the direction that the car is moving?
3. If a car is moving in uniform circular motion, and the acceleration is toward the center of that circle, why does the car not move into the center of the circle?
4. Answer the student's question, "If an object moving in uniform circular motion is accelerated motion, why doesn't the speed change with time?"
5. Reply to the student's statement, "I know there is a centrifugal force acting on me when I move in circular motion in my car because I can feel the force pushing me against the side of the car."
† 6. Is it possible to change to a noninertial coordinate system, say a coordinate system that is fixed to the rotating body, to study uniform circular motion? In this rotating coordinate system is there a centrifugal force?

7. If you take a pail of water and turn it upside down all the water will spill out. But if you take the pail of water, attach a rope to the handle, and turn it rapidly in a vertical circle the water will not spill out when it is upside down at the top of the path. Why is this?
† 8. In high-performance jet aircraft the pilot must wear a pressure suit that exerts pressure on the abdomen and upper thighs of the pilot when the pilot pulls out of a steep dive. Why is this necessary?
9. If the force of gravity acting on a body is directly proportional to its mass, why does a massive body fall at the same rate as a less massive body?
10. Why does the earth bulge at the equator and not at the poles?
11. If the acceleration of gravity varies from place to place on the surface of the earth, how does this affect records made in the Olympics in such sports as shot put, javelin throwing, high jump, and the like?
12. What is wrong with applying Newton's second law in the form $F = ma$ to satellite motion? Does this same problem occur in the motion of an airplane?
†13. How can you use Kepler's second law to explain that the earth moves faster in its motion about the sun when it is closer to the sun?

14. Could you place a synchronous satellite in a polar orbit about the earth? At 45° latitude?
15. Explain how you can use a Hohmann transfer orbit to allow one satellite in an earth orbit to rendezvous with another satellite in a different earth orbit.
†16. A satellite is in a circular orbit. Explain what happens to the orbit if the engines are momentarily turned on to exert a thrust (a) in the direction of the velocity, (b) opposite to the velocity, (c) toward the earth, and (d) away from the earth.
17. A projectile fired close to the earth falls toward the earth and eventually crashes to the earth. The moon in its orbit about the earth is also falling toward the earth. Why doesn't it crash into the earth?
†18. The gravitational force on the earth caused by the sun is greater than the gravitational force on the earth caused by the moon. Why then does the moon have a greater effect on the tides than the sun?
†19. How was the universal gravitational constant G determined experimentally?
20. A string is tied to a rock and then the rock is put into motion in a vertical circle. Is this an example of uniform circular motion?

Problems for Chapter 6

6.3 Angles Measured in Radians

1. Express the following angles in radians: (a) 360°, (b) 270°, (c) 180°, (d) 90°, (e) 60°, (f) 30°, and (g) 1 rev.
2. Express the following angles in degrees: (a) 2π rad, (b) π rad, (c) 1 rad, and (d) 0.500 rad.
3. A record player turns at 33-1/3 rpm. What distance along the arc has a point on the edge moved in 1.00 min if the record has a diameter of 10.0 in.?

6.4 and 6.5 The Centripetal Acceleration and the Centripetal Force

4. A 4.00-kg stone is whirled at the end of a 2.00-m rope in a horizontal circle at a speed of 15.0 m/s. Ignoring the gravitational effects (a) calculate the centripetal acceleration and (b) calculate the centripetal force.
5. An automatic washing machine, in the spin cycle, is spinning wet clothes at the outer edge at 8.00 m/s. The diameter of the drum is 0.450 m. Find the acceleration of a piece of clothing in this spin cycle.
6. A 3200-lb car moving at 60.0 mph goes around a curve of 1000-ft radius. What is the centripetal acceleration? What is the centripetal force on the car?
7. A 1500-kg car moving at 86.0 km/hr goes around a curve of 325-m radius. What is the centripetal acceleration? What is the centripetal force on the car?
8. An electron is moving at a speed of 2.00×10^6 m/s in a circle of radius 0.0500 m. What is the force on the electron?
9. Find the centripetal force on a 70.0-lb girl on a merry-go-round that turns through one revolution in 40.0 s. The radius of the merry-go-round is 10.0 ft.

6.7 Examples of Centripetal Force

10. A boy sits on the edge of a polished wooden disk. The disk has a radius of 3.00 m and the coefficient of friction between his pants and the disk is 0.300. What is the maximum speed of the disk at the moment the boy slides off?

11. A 3200-lb car begins to skid when traveling at 60.0 mph around a level curve of 500-ft radius. Find the centripetal acceleration and the coefficient of friction between the tires and the road.
12. A 1200-kg car begins to skid when traveling at 80.0 km/hr around a level curve of 125-m radius. Find the centripetal acceleration and the coefficient of friction between the tires and the road.
13. At what angle should a bobsled turn be banked if the sled, moving at 26.0 m/s, is to round a turn of radius 100 m?
14. A motorcyclist goes around a curve of 400-ft radius at a speed of 60.0 mph, without leaning into the turn. (a) What must the coefficient of friction between the tires and the road be in order to supply the necessary centripetal force? (b) If the road is iced and the motorcyclist can not depend on friction, at what angle from the vertical should the motorcyclist lean to supply the necessary centripetal force?
15. A motorcyclist goes around a curve of 100-m radius at a speed of 95.0 km/hr, without leaning into the turn. (a) What must the coefficient of friction between the tires and the road be in order to supply the necessary centripetal force? (b) If the road is iced and the motorcyclist can not depend on friction, at what angle from the vertical should the motorcyclist lean to supply the necessary centripetal force?
16. At what angle should a highway be banked for cars traveling at a speed of 60.0 mph, if the radius of the road is 1000 ft and no frictional forces are involved?
17. At what angle should a highway be banked for cars traveling at a speed of 100 km/hr, if the radius of the road is 400 m and no frictional forces are involved?
18. An airplane is flying in a circle with a speed of 200 knots. (A knot is a nautical mile per hour.) The aircraft weighs 2000 lb and is banked at an angle of 30.0°. Find the radius of the turn in feet.
19. An airplane is flying in a circle with a speed of 650 km/hr. At what angle with the horizon should a pilot make a turn of radius of 8.00 km such that a component of the lift of the aircraft supplies the necessary centripetal force for the turn?

6.8 Newton's Law of Universal Gravitation

20. Two large metal spheres are separated by a distance of 2.00 m from center to center. If each sphere has a mass of 5000 kg, what is the gravitational force between them?
21. A 5.00-kg mass is 1.00 m from a 10.0-kg mass. (a) What is the gravitational force that the 5.00-kg mass exerts on the 10.0-kg mass? (b) What is the gravitational force that the 10.0-kg mass exerts on the 5.00-kg mass? (c) If both masses are free to move, what will their initial acceleration be?
22. Three point masses of 10.0 kg, 20.0 kg, and 30.0 kg are located on a line at 10.0 cm, 50.0 cm, and 80.0 cm, respectively. Find the resultant gravitational force on (a) the 10.0-kg mass, (b) the 20.0-kg mass, and (c) the 30.0-kg mass.
23. Pete meets Eileen for the first time and is immediately attracted to her. If Pete has a mass of 75.0 kg and Eileen has a mass of 50.0 kg and they are separated by a distance of 3.00 m, is their attraction purely physical?

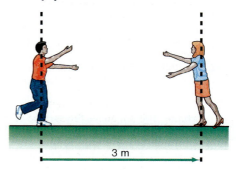

3 m

24. What is the gravitational force between a proton and an electron in a hydrogen atom if they are separated by a distance of 5.29×10^{-11} m?

6.11–6.13 The Acceleration Due to Gravity

25. What is the value of g at a distance from the center of the earth of (a) 1 earth radius, (b) 2 earth radii, (c) 10 earth radii, and (d) at the distance of the moon?

26. What is the weight of a body, in terms of its weight at the surface of the earth, at a distance from the center of the earth of (a) 1 earth radius, (b) 2 earth radii, (c) 10 earth radii, and (d) at the distance of the moon? How can an object in a satellite, at say 2 earth radii, be considered to be weightless?

27. Calculate the acceleration of gravity on the surface of Mars. What would a man who weighs 180 lb on earth weigh on Mars?

†28. It is the year 2020 and a base has been established on Mars. An enterprising businessman decides to buy coffee on earth at $5.00/lb and sell it on Mars for $10.00/lb. How much does he make or lose per lb when he sells it on Mars? Ignore the cost of transportation from earth to Mars.

29. The sun's radius is 110 times that of the earth, and its mass is 333,000 times as large. What would be the weight of a 1.00-kg object at the surface of the sun, assuming that it does not melt or evaporate there?

6.14 Satellite Motion

30. What is the velocity of the moon around the earth in a circular orbit? What is the time for one revolution?

31. Calculate the velocity of the earth in an approximate circular orbit about the sun. Calculate the time for one revolution.

32. A satellite orbits the earth 700 mi above the surface. Find its speed and its period of revolution.

33. Calculate the speed of a satellite orbiting 100 km above the surface of Mars. What is its period?

†34. An Apollo space capsule orbited the moon in a circular orbit at a height of 112 km above the surface. The time for one complete orbit, the period T, was 120 min. Find the mass of the moon.

†35. A satellite orbits the earth in a circular orbit in 130 min. What is the distance of the satellite to the center of the earth? What is its height above the surface? What is its speed?

†36. A rock attached to a string hangs from the roof of a moving train. If the train is traveling at 50.0 mph around a level curve of 500-ft radius, find the angle that the string makes with the vertical.

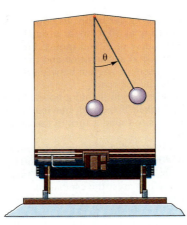

37. Find the centripetal force due to the rotation of the earth acting on a 100 kg person at (a) the equator, (b) 45.0° north latitude, and (c) the north pole.

38. Find the resultant vector acceleration caused by the acceleration due to gravity and the centripetal acceleration for a person located at (a) the equator, (b) 45.0° north latitude, and (c) the north pole.

†39. A pilot who weighs 180 lb pulls out of a vertical dive at 400 mph along an arc of a circle of 5280-ft radius. Find the centripetal acceleration, centripetal force, and the net normal force on the pilot at the bottom of the dive.

40. A 90-kg pilot pulls out of a vertical dive at 685 km/hr along an arc of a circle of 1500-m radius. Find the centripetal acceleration, centripetal force, and the net force on the pilot at the bottom of the dive.

†41. What is the minimum speed of an airplane in making a vertical loop such that an object in the plane will not fall during the peak of the loop? The radius of the loop is 1000 ft.

†42. A rope is attached to a pail of water and the pail is then rotated in a vertical circle of 80.0-cm radius. What must the minimum speed of the pail of water be such that the water will not spill out?

43. A mass is attached to a string and is swung in a vertical circle. At a particular instant the mass is moving at a speed v, and its velocity vector makes an angle θ with the horizontal. Show that the normal component of the acceleration is given by

$$T + w \sin\theta = mv^2/r$$

and the tangential component of the acceleration is given by

$$a_T = -g \cos\theta$$

Hence show why this motion in a vertical circle is not uniform circular motion.

†44. A 10.0-N ball attached to a string 1.00 m long moves in a horizontal circle. The string makes an angle of 60.0° with the vertical. (a) Find the tension in the string. (b) Find the component of the tension that supplies the necessary centripetal force. (c) Find the speed of the ball.

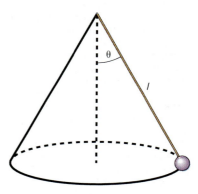

45. A mass $m_A = 35.0$ g is on a smooth horizontal table. It is connected by a string that passes through the center of the table to a mass $m_B = 25.0$ g. At what uniform speed should m_A move in a circle of radius $r = 40.0$ cm such that mass m_B remains motionless?

†46. Three point masses of 30.0 kg, 50.0 kg, and 70.0 kg are located at the vertices of an equilateral triangle 1.00 m on a side. Find the resultant gravitational force on each mass.

†47. Four metal spheres are located at the corners of a square of sides of 0.300 m. If each sphere has a mass of 10.0 kg, find the force on the sphere in the lower right-hand corner.

48. What is the gravitational force between the earth and the moon? If a steel cable can withstand a force of 7.50×10^4 N/cm², what must the diameter of a steel cable be to sustain the equivalent force?

†49. At what speed would the earth have to rotate such that the centripetal force at the equator would be equal to the weight of a body there? If the earth rotated at this velocity, how long would a day be? If a 200-lb man stood on a weighing scale there, what would the scales read?

†50. What would the mass of the earth have to be in order that the gravitational force is inadequate to supply the necessary centripetal force to keep a person on the surface of the earth at the equator? What density would this correspond to? Compare this to the actual density of the earth.

†51. Compute the gravitational force of the sun on the earth. Then compute the gravitational force of the moon on the earth. Which do you think would have a greater effect on the tides, the sun or the moon? Which has the greatest effect?

†52. Find the force exerted on 1.00 kg of water by the moon when (a) the 1.00 kg is on the side nearest the moon and (b) when the 1.00 kg is on the side farthest from the moon. Would this account for tides?

†53. By how much does (a) the sun and (b) the moon change the value of g at the surface of the earth?

54. How much greater would the range of a projectile be on the moon than on the earth?

†55. Find the point between the earth and the moon where the gravitational forces of earth and moon are equal. Would this be a good place to put a satellite?

†56. An earth satellite is in a circular orbit 110 mi above the earth. The period, the time for one orbit, is 88.0 min. Determine the velocity of the satellite and the acceleration of gravity in the satellite at the satellite altitude.

†57. Show that Kepler's third law, which shows the relationship between the period of motion and the radius of the orbit, can be found for circular orbits by equating the centripetal force to the gravitational force, and obtaining

$$T^2 = \frac{4\pi^2 r^3}{Gm}$$

†58. Using Kepler's third law from problem 57, find the mass of the sun. If the radius of the sun is 7.00×10^8 m, find its density.

†59. The speed of the earth around the sun was found, using dynamical principles in the example 6.11 of section 6.14, to be 66,600 mph. Show that this result is consistent with a purely kinematical calculation of the speed of the earth about the sun.

†60. A better approximation for equation 6.52, the "burn time" for the rocket engines, can be obtained if the rate at which the rocket fuel burns, is a known constant. The rate at which the fuel burns is then given by $\Delta m/\Delta t = K$. Hence, the mass at any time during the burn will be given by $(m_0 - K\Delta t)$, where m_0 is the initial mass of the rocket ship before the engines are turned on. Show that for this approximation the time of burn becomes

$$\Delta t = \frac{m_0 \Delta v}{F + K \Delta v}$$

†61. If a spacecraft is to transfer from a 200 nautical mile earth parking orbit to an 80.0 nautical mile lunar parking orbit by a Hohmann transfer ellipse, find (a) the location of the center of mass of the earth-moon system, (b) the perigee distance of the transfer ellipse, (c) the apogee distance, (d) the semimajor axis of the ellipse, (e) the speed of the spacecraft in the earth circular

parking orbit, (f) the speed necessary for insertion into the Hohmann transfer ellipse, (g) the necessary Δv for this insertion, (h) the speed of the spacecraft in a circular lunar parking orbit, (i) the speed of the spacecraft on the Hohmann transfer at time of lunar insertion, and (j) the necessary Δv for insertion into the lunar parking orbit.

Interactive Tutorials

▥ 62. Two masses $m_1 = 5.10 \times 10^{21}$ kg and $m_2 = 3.00 \times 10^{14}$ kg are separated by a distance $r = 4.30 \times 10^5$ m. Calculate their gravitational force of attraction.

▥ 63. Planet X has mass $m_p = 3.10 \times 10^{25}$ kg and a radius $r_p = 5.40 \times 10^7$ m. Calculate the acceleration due to gravity g at distances of 1–10 planet radii from the planet's surface.

▥ 64. Find the angle of bank for a car making a turn on a banked road.

▥ 65. Find the speed of a satellite in a circular orbit about its parent body.

▥ 66. You are to plan a trip to the planet Mars using the Hohmann transfer ellipse described in the "Have you ever wondered . . . ?" section. The spacecraft is to transfer from a 925-km earth circular parking orbit to a 185-km circular parking orbit around Mars. Find (a) the center of mass of the Earth-Sun-Mars system, (b) the perigee distance of the transfer ellipse, (c) the apogee distance of the transfer ellipse, (d) the semimajor axis of the ellipse, (e) the speed of the spacecraft in the earth parking orbit, (f) the speed necessary for insertion into the Hohmann transfer ellipse, (g) the necessary Δv for insertion into the transfer ellipse, (h) the necessary speed in the Mars circular parking orbit, (i) the speed of the spacecraft in the transfer ellipse at Mars, and (j) the necessary Δv for insertion into the Mars parking orbit.

Energy and Its Conservation

An ellipsoid roller coaster where potential energy is converted to kinetic energy.

The fundamental principle of natural philosophy is to attempt to reduce the apparently complex physical phenomena to some simple fundamental ideas and relations.

Einstein and Infeld

7.1 Energy

The fundamental concept that connects all of the apparently diverse areas of natural phenomena such as mechanics, heat, sound, light, electricity, magnetism, chemistry, and others, is the concept of energy. Energy can be subdivided into well-defined forms, such as (1) mechanical energy, (2) heat energy, (3) electrical energy, (4) chemical energy, and (5) atomic energy. In any process that occurs in nature, energy may be transformed from one form to another. The history of technology is one of a continuing process of transforming one type of energy into another. Some examples include the light bulb, generator, motor, microphone, and loudspeakers.

In its simplest form, *energy can be defined as the ability of a body or system of bodies to perform work. A system is an aggregate of two or more particles that is treated as an individual unit.* In order to describe the energy of a body or a system, we must first define the concept of work.

7.2 Work

Almost everyone has an intuitive grasp for the concept of work. However, we need a precise definition of the concept of work so let us define it as follows. Let us exert a force **F** on the block in figure 7.1, causing it to be displaced a distance x along the table. The **work** *W done in displacing the body a distance x along the table is defined as the product of the force acting on the body, in the direction of the displacement, times the displacement x of the body.* Mathematically this is

$$W = Fx \tag{7.1}$$

We will always use a capital W to designate the work done, in order to distinguish it from the weight of a body, for which we use the lower case w. The important thing to observe here is that there must be a displacement x if work is to be done. If you push as hard as you can against the wall with your hands, then from the point of view of physics, you do no work on the wall as long as the wall has not moved through a displacement x. This may not appeal to you intuitively because after pushing against that wall for a while, you will become tired and will feel that you certainly did do work. But again, from the point of view of physics, no work on the wall is accomplished because there is no displacement of the wall. In order to do work on an object, you must exert a force F on that object and move that object from one place to another. If that object is not moved, no work is done.

From the point of view of expending energy in pushing against the immovable wall, your body used chemical energy in its tissues and muscles to hold

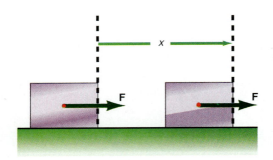

Figure 7.1

The concept of work.

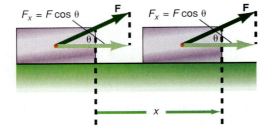

Figure 7.2

Work done when the force is not in the direction of the displacement.

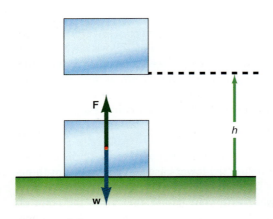

Figure 7.3

Work done in lifting a box.

your hands against the wall. As the body uses this energy, it becomes tired and that energy must eventually be replaced by eating. We will consider the energy used by the body in sustaining the force chemical energy. But, in terms of mechanical energy, no work is done in pressing your hands against an immovable wall. Hence, work as it is used here, is *mechanical work*.

In order to be consistent with the definition of work stated above, if the force acting on the body is not parallel to the displacement, as in figure 7.2, then the work done is the product of the force in the direction of the displacement, times the displacement. That is, the x-component of the force,

$$F_x = F \cos \theta$$

is the component of the force in the direction of the displacement. Therefore, the work done on the body is

$$W = (F \cos \theta)x$$

which is usually written as

$$W = Fx \cos \theta \tag{7.2}$$

This is the general equation used to find the work done on a body. If the force is in the same direction as the displacement, then the angle θ equals zero. But $\cos 0° = 1$, and equation 7.2 reduces to equation 7.1, where the force was in the direction of the displacement.

Units of Work

Since the unit of force in SI units is a newton, and the unit of length is a meter, the SI unit of work is defined as 1 newton meter, which we call 1 joule, that is,

$$1 \text{ joule} = 1 \text{ newton meter}$$

Abbreviated, this is

$$1 \text{ J} = 1 \text{ N m}$$

One joule of work is done when a force of one newton acts on a body, moving it through a distance of one meter. The unit joule is named after James Prescott Joule (1818–1889), a British physicist. *Since energy is the ability to do work, the units of work will also be the units of energy.* In the British engineering system, the force is expressed in pounds and the distance in feet. Hence, the unit of work is defined as

$$1 \text{ unit of work} = 1 \text{ ft lb}$$

One foot-pound is the work done when a force of one pound acts on a body moving it through a distance of one foot. Unlike SI units, the unit of work in the British engineering system is not given a special name. (Remember that the unit lb ft was used for torque to distinguish it from the unit of ft lb for energy. The concepts are very distinct, in that torque was equal to a force times a perpendicular distance, while work is equal to a force times a parallel distance.)

Example 7.1

Work done in lifting a box. What is the minimum amount of work that is necessary to lift a 1.00-lb box to a height of 5.00 ft (figure 7.3)?

Solution

We find the work done by noting that F is the force that is necessary to lift the block and h is the distance that the block is lifted. Since the force is in the same direction as the displacement, θ is equal to zero in equation 7.2. Thus,

$$W = Fx \cos \theta = Fh \cos 0°$$

$$= Fh = (1.00 \text{ lb})(5.00 \text{ ft})$$

$$= 5.00 \text{ ft lb}$$

Note here that if a force of only 1.00 lb is exerted to lift the block, then the block will be in equilibrium and will not be lifted from the table at all. If, however, a force that is just infinitesimally greater than w is exerted for just an infinitesimal period of time, then this will be enough to set w into motion. Once the block is moving, then a force F, equal to w, will keep it moving upward at a constant velocity, regardless of how small that velocity may be. In all such cases where forces are exerted to lift objects, such that $F = w$, we will tacitly assume that some additional force was applied for an infinitesimal period of time, to start the motion.

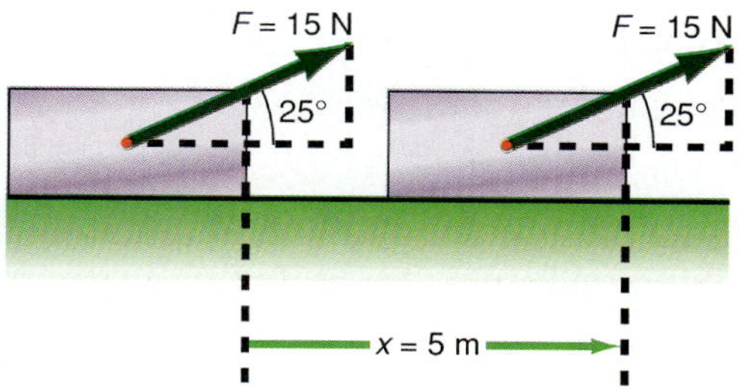

Figure 7.4

Work done when pulling a box.

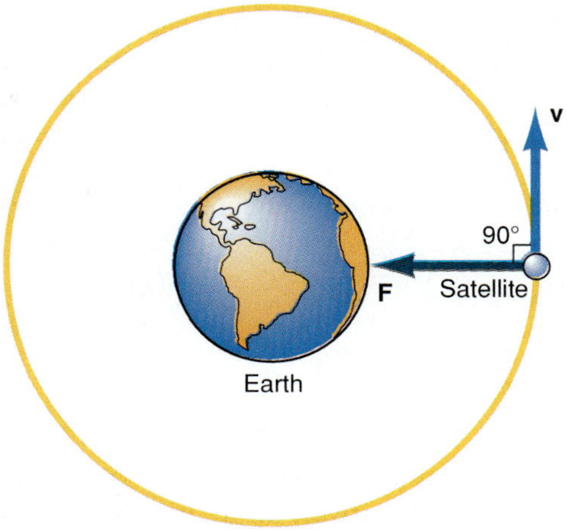

Figure 7.5

The work done to keep a satellite in orbit.

Example 7.2

When the force is not in the same direction as the displacement. A force of 15.0 N acting at an angle of 25.0° to the horizontal is used to pull a box a distance of 5.00 m across a floor (figure 7.4). How much work is done?

Solution

The work done, found by using equation 7.2, is

$$W = Fx \cos \theta = (15.0 \text{ N})(5.00 \text{ m})(\cos 25.0°)$$

$$= 68.0 \text{ N m} = 68.0 \text{ J}$$

Example 7.3

Work done keeping a satellite in orbit. Find the work done to keep a satellite in a circular orbit about the earth.

Solution

A satellite in a circular orbit about the earth has a gravitational force acting on it that is perpendicular to the orbit, as seen in figure 7.5. The displacement of the satellite in its orbit is perpendicular to that gravitational force. Note that if the displacement is perpendicular to the direction of the applied force, then θ is equal to 90°, and cos 90° = 0. Hence, the work done on the satellite by gravity, found from equation 7.2, is

$$W = Fx \cos \theta = Fx \cos 90° = 0$$

Therefore, no work is done by gravity on the satellite as it moves in its orbit. Work had to be done to get the satellite into the orbit, but once there, no additional work is required to keep it moving in that orbit. *In general, whenever the applied force is perpendicular to the displacement, no work is done by that applied force.*

Figure 7.6

Work done in stopping a car.

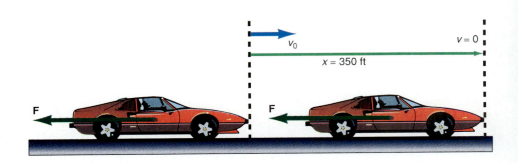

Example 7.4

Work done in stopping a car. A force of 867 lb is applied to a car to bring it to rest in a distance of 350 ft, as shown in figure 7.6. How much work is done in stopping the car?

Solution

To determine the work done in bringing the car to rest, note that the applied force is opposite to the displacement of the car. Therefore, θ is equal to 180° in equation 7.2. Hence, the work done, found from equation 7.2, is

$$W = Fx \cos \theta = (867 \text{ lb})(350 \text{ ft}) \cos 180°$$

$$= -3.03 \times 10^5 \text{ ft lb}$$

Notice that cos 180° = −1, and hence, the work done is negative. In general, whenever the force is opposite to the displacement, the work will always be negative.

7.3 Power

When you walk up a flight of stairs, you do work because you are lifting your body up those stairs. You know, however, that there is quite a difference between walking up those stairs slowly and running up them very rapidly. The work that is done is the same in either case because the net result is that you lifted up the same weight w to the same height h. But you know that if you ran up the stairs you would be more tired than if you walked up them slowly. There is, therefore, a difference in the rate at which work is done.

Power is defined as the time rate of doing work. We express this mathematically as

$$\text{Power} = \frac{\text{work done}}{\text{time}}$$

$$P = \frac{W}{t} \tag{7.3}$$

When you ran up the stairs rapidly, the time t was small, and therefore the power P, which is the work divided by that small time, was relatively large. Whereas, when you walked up the stairs slowly, t was much larger, and therefore the power P was smaller than before. Hence, when you go up the stairs rapidly you expend more power than when you go slowly.

Units of Power

In SI units, the unit of power is defined as a watt, that is,

$$1 \text{ watt} = 1 \frac{\text{joule}}{\text{second}}$$

which we abbreviate as

$$1 \text{ W} = 1 \frac{\text{J}}{\text{s}}$$

One watt of power is expended when one joule of work is done each second. The watt is named in honor of James Watt (1736–1819), a Scottish engineer who perfected the steam engine. The kilowatt, a unit with which you may already be more familiar, is a thousand watts:

$$1 \text{ kw} = 1000 \text{ W}$$

Another unit with which you may also be familiar is the kilowatt-hour (kwh), but this is not a unit of power, but energy, as can be seen from equation 7.3. Since

$$P = \frac{W}{t}$$

then

$$W = Pt = (\text{kilowatt})(\text{hour})$$

Your monthly electric bill is usually expressed in kilowatt-hours, which is the amount of electric energy you have used for that month. It is the number of kilowatts of power that you used times the number of hours that you used them. To convert kilowatt-hours to joules note

$$1 \text{ kwh} = (1000 \text{ J/s})(1 \text{ hr})(3600 \text{ s/hr}) = 3.6 \times 10^6 \text{ J}$$

To determine the units of power in the British engineering system, we have

$$P = \frac{W}{t} = \frac{\text{ft lb}}{\text{s}}$$

and although this would be the logical unit to express power in the British engineering system, it is not the unit used. Instead, the unit of power in the British engineering system is the horsepower. The horsepower is defined as

$$1 \text{ horsepower} = 1 \text{ hp} = 550 \frac{\text{ft lb}}{\text{s}}$$

The story goes that when Watt had perfected the steam engine, he approached the coal mine owners and leading industrialists of the time, and tried to convince them that they should use his steam engine to take the coal out of the mines instead of the workhorses being used. Supposedly, they wanted to know how good his engine really was. For instance, how many horses could one of his engines replace? At that time, Watt did not know the answer, so he went down into the mines and measured the average amount of work that each horse could do per minute. Each workhorse could do, on the average, 33,000 ft lb of work per minute or 550 ft lb per second. Watt then figured out how many ft lb per second his steam engine could do, and then rated his engine in terms of horsepower. He was then in a position to tell the mine owners how many horses could be replaced by his engine. Even today, in the United States, most engine power is still rated in terms of horsepower. As an example, your car engine may be rated at 200 hp, or the electric motor in your washing machine may be rated at 3/4 hp. A handy conversion factor from watts to horsepower is

$$1 \text{ hp} = 745.7 \text{ W}$$

It seems strange in this day and age of high technology to continue rating engines in terms of horses. It is another reason for abandoning the British engineering system of units.

Example 7.5

Power expended. A person pulls a block with a force of 15.0 N at an angle of 25.0° with the horizontal. If the block is moved 5.00 m in the horizontal direction in 5.00 s, how much power is expended?

Solution

The power expended, found from equations 7.3 and 7.2, is

$$P = \frac{W}{t} = \frac{Fx \cos \theta}{t}$$

$$= \frac{(15.0 \text{ N})(5.00 \text{ m})\cos 25.0°}{5.00 \text{ s}} = 13.6 \frac{\text{N m}}{\text{s}} = 13.6 \text{ W}$$

When a constant force acts on a body in the direction of the body's motion, we can also express the power as

$$P = W = \frac{Fx}{t} = F\frac{x}{t}$$

but

$$\frac{x}{t} = v$$

the velocity of the moving body. Therefore,

$$P = Fv \tag{7.4}$$

is the power expended by a force *F*, acting on a body that is moving at the velocity *v*.

Example 7.6

Power to move your car. An applied force of 5500 N keeps a car moving at 95 km/hr. How much power is expended by the car?

Solution

The power expended by the car, found from equation 7.4, is

$$P = Fv = (5500 \text{ N})\left(95 \frac{\text{km}}{\text{hr}}\right)\left(\frac{1 \text{ hr}}{3600 \text{ s}}\right)\left(\frac{1000 \text{ m}}{1 \text{ km}}\right)$$

$$= 1.45 \times 10^5 \text{ N m/s} = 1.45 \times 10^5 \text{ J/s}$$

$$= 1.45 \times 10^5 \text{ W}$$

7.4 Gravitational Potential Energy

Gravitational potential energy is defined as the energy that a body possesses by virtue of its position. If the block shown in figure 7.7, were lifted to a height *h* above the table, then that block would have potential energy in that raised position. That is, in the raised position, the block has the ability to do work whenever it is allowed to fall. The most obvious example of gravitational potential energy is a waterfall (figure 7.8). Water at the top of the falls has potential energy. When the water falls to the bottom, it can be used to turn turbines and thus do work. A similar example is a pile driver. A pile driver is basically a large weight that is raised above a pile

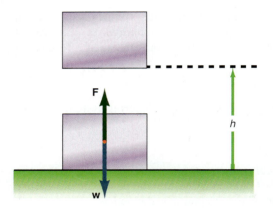

Figure 7.7

Gravitational potential energy.

Figure 7.8

Water at the top of the falls has potential energy.

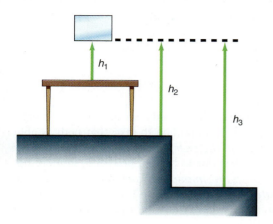

Figure 7.9

Reference plane for potential energy.

Figure 7.10

Changing potential energy.

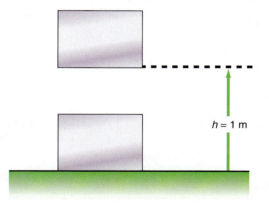

Figure 7.11

The potential energy of a block.

that is to be driven into the ground. In the raised position, the driver has potential energy. When the weight is released, it falls and hits the pile and does work by driving the pile into the ground.

Therefore, whenever an object in the gravitational field of the earth is placed in a position above some reference plane, then that object will have potential energy because it has the ability to do work.

As in all the concepts studied in physics, we want to make this concept of potential energy quantitative. That is, how much potential energy does a body have in the raised position? How should potential energy be measured?

Because work must be done on a body to put the body into the position where it has potential energy, the work done is used as the measure of this potential energy. That is, the potential energy of a body is equal to the work done to put the body into the particular position. Thus, the potential energy (PE) is

$$PE = \text{Work done to put body into position} \qquad (7.5)$$

We can now compute the potential energy of the block in figure 7.7 as

$$PE = \text{Work done}$$

$$PE = W = Fh = wh \qquad (7.6)$$

The applied force F necessary to lift the weight is set equal to the weight w of the block. And since $w = mg$, the potential energy of the block becomes

$$PE = mgh \qquad (7.7)$$

We should emphasize here that the potential energy of a body is referenced to a particular plane, as in figure 7.9. If we raise the block a height h_1 above the table, then with respect to the table it has a potential energy

$$PE_1 = mgh_1$$

While at the same position, it has the potential energy

$$PE_2 = mgh_2$$

with respect to the floor, and

$$PE_3 = mgh_3$$

with respect to the ground outside the room. All three potential energies are different because the block can do three different amounts of work depending on whether it falls to the table, the floor, or the ground. Therefore, it is very important that when the potential energy of a body is stated, it is stated with respect to a particular reference plane. We should also note that it is possible for the potential energy to be negative with respect to a reference plane. That is, if the body is not located above the plane but instead is found below it, it will have negative potential energy with respect to that plane. In such a position the body can not fall to the reference plane and do work, but instead work must be done on the body to move the body up to the reference plane.

Example 7.7

The potential energy. A mass of 1.00 kg is raised to a height of 1.00 m above the floor (figure 7.11). What is its potential energy with respect to the floor?

Solution

The potential energy, found from equation 7.7, is

$$PE = mgh = (1.00 \text{ kg})(9.80 \text{ m/s}^2)(1.00 \text{ m})$$

$$= 9.80 \text{ J}$$

In addition to gravitational potential energy, a body can have elastic potential energy and electrical potential energy. An example of elastic potential energy is a compressed spring. When the spring is compressed, the spring has potential energy because when it is released, it has the ability to do work as it expands to its normal position. Its potential energy is equal to the work that is done to compress it. We will discuss the spring and its potential energy in much greater detail in chapter 11 on simple harmonic motion. We will discuss electric potential energy in chapter 19 on electric fields.

7.5 Kinetic Energy

In addition to having energy by virtue of its position, a body can also possess energy by virtue of its motion. When we bring a body in motion to rest, that body is able to do work. *The kinetic energy of a body is the energy that a body possesses by virtue of its motion.* Because work had to be done to place a body into motion, *the kinetic energy of a moving body is equal to the amount of work that must be done to bring a body from rest into that state of motion. Conversely, the amount of work that you must do in order to bring a moving body to rest is equal to the negative of the kinetic energy of the body.* That is,

$$\text{Kinetic energy (KE)} = \text{Work done to put body into motion}$$

$$= -\text{Work done to bring body to a stop} \qquad (7.8)$$

The work done to put a body at rest into motion is positive and hence the kinetic energy is positive, and the body has gained energy. The work done to bring a body in motion to a stop is negative, and hence the change in its kinetic energy is negative. This means that the body has lost energy as it goes from a velocity v to a zero velocity.

Consider a block at rest on the frictionless table as shown in figure 7.12. A constant net force F is applied to the block to put it into motion. When it is a distance x away, it is moving at a speed v. What is its kinetic energy at this point? The kinetic energy, found from equation 7.8, is

$$\text{KE} = \text{Work done} = W = Fx \qquad (7.9)$$

But by Newton's second law, the force acting on the body gives the body an acceleration. That is, $F = ma$, and substituting this into equation 7.9 we have

$$\text{KE} = Fx = max \qquad (7.10)$$

But for a body moving at constant acceleration, the kinematic equation 3.16 was

$$v^2 = v_0^2 + 2ax$$

Since the block started from rest, $v_0 = 0$, giving us

$$v^2 = 2ax$$

Solving for the term ax,

$$ax = \frac{v^2}{2} \qquad (7.11)$$

Substituting equation 7.11 back into equation 7.10, we have

$$\text{KE} = m(ax) = \frac{mv^2}{2}$$

or

$$\text{KE} = \tfrac{1}{2}mv^2 \qquad (7.12)$$

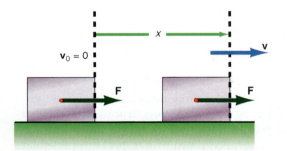

Figure 7.12

The kinetic energy of a body.

$v_0 = 0$

x

v

F F

Equation 7.12 is the classical expression for the kinetic energy of a body in motion at speed v.

Example 7.8

Kinetic energy. Let the block of figure 7.12 have a mass $m = 2.00$ kg and let it be moving at a speed of 5.00 m/s when $x = 5.00$ m. What is its kinetic energy at $x = 5.00$ m?

Solution

Using equation 7.12 for the kinetic energy we obtain

$$KE = \tfrac{1}{2}mv^2 = \tfrac{1}{2}(2.00 \text{ kg})(5.00 \text{ m/s})^2$$

$$= 25.0 \text{ kg m}^2/\text{s}^2 = 25.0(\text{kg m/s}^2)\text{m} = 25.0 \text{ N m}$$

$$= 25.0 \text{ J}$$

Example 7.9

Kinetic energy in the British system of units. A block of mass 10.0 slug is moving at a speed of 5.00 ft/s. Find its kinetic energy.

Solution

The kinetic energy, found from equation 7.12, is

$$KE = \tfrac{1}{2}mv^2 = \tfrac{1}{2}(10.0 \text{ slug})(5.00 \text{ ft/s})^2$$

$$= \left(125 \frac{\text{slug ft}^2}{\text{s}^2}\right)\left(\frac{1}{\text{slug}}\right)\left(\frac{\text{lb}}{\text{ft/s}^2}\right)$$

$$= 125 \text{ ft lb}$$

Example 7.10

The effect of doubling the speed on the kinetic energy. If a car doubles its speed, what happens to its kinetic energy?

Solution

Let us assume that the car of mass m is originally moving at a speed v_0. Its original kinetic energy is

$$(KE)_0 = \tfrac{1}{2}mv_0^2$$

If the speed is doubled, then $v = 2v_0$ and its kinetic energy is

$$KE = \tfrac{1}{2}mv^2 = \tfrac{1}{2}m(2v_0)^2 = \tfrac{1}{2}m4v_0^2$$

$$= 4(\tfrac{1}{2}mv_0^2) = 4KE_0$$

That is, doubling the speed results in quadrupling the kinetic energy. Increasing the speed by a factor of 4 increases the kinetic energy by a factor of 16. This is why automobile accidents at high speeds cause so much damage.

Before we leave this section, we should note that in our derivation of the kinetic energy, work was done to bring an object from rest into motion. The work done on the body to place it into motion was equal to the acquired kinetic energy of the body. If an object is already in motion when the constant force is applied to it, the work done is equal to the change in kinetic energy of the body. That is, equation 7.9 can be written as

$$\text{Work done} = W = Fx$$

$$W = Fx = max$$

but if the block is already in motion at an initial velocity v_0 when the force was applied,

$$v^2 = v_0^2 + 2ax$$

$$ax = \frac{v^2 - v_0^2}{2}$$

Hence,

$$W = Fx = max = m\left(\frac{v^2 - v_0^2}{2}\right)$$

$$= m\frac{v^2}{2} - m\frac{v_0^2}{2}$$

$$= KE_f - KE_i = \Delta KE$$

7.6 The Conservation of Energy

When we say that something is conserved, we mean that that quantity is a constant and does not change with time. It is a somewhat surprising aspect of nature that when a body is in motion, its position is changing with time, its velocity is changing with time, yet certain characteristics of that motion still remain constant. One of the quantities that remain constant during motion is the total energy of the body. The analysis of systems whose energy is conserved leads us to the law of conservation of energy.

*In any **closed system,** that is, an isolated system, the total energy of the system remains a constant. This is the **law of conservation of energy.*** There may be a transfer of energy from one form to another, but the total energy remains the same.

As an example of the conservation of energy applied to a mechanical system without friction, let us go back and look at the motion of a projectile in one dimension. Assume that a ball is thrown straight upward with an initial velocity v_0. The ball rises to some maximum height and then descends to the ground, as shown in figure 7.13. At the point 1, a height h_1 above the ground, the ball has a potential energy given by

$$PE_1 = mgh_1 \tag{7.13}$$

At this same point it is moving at a velocity v_1 and thus has a kinetic energy given by

$$KE_1 = \tfrac{1}{2}mv_1^2 \tag{7.14}$$

The total energy of the ball at point 1 is the sum of its potential energy and its kinetic energy. Hence, using equations 7.13 and 7.14, we get

$$E_1 = PE_1 + KE_1 \tag{7.15}$$

$$E_1 = mgh_1 + \tfrac{1}{2}mv_1^2 \tag{7.16}$$

When the ball reaches point 2 it has a new potential energy because it is higher up, at the height h_2. Hence, its potential energy is

$$PE_2 = mgh_2$$

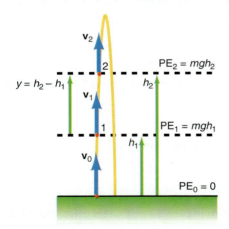

Figure 7.13

The conservation of energy and projectile motion.

As the ball rises, it slows down. Hence, it has a smaller velocity v_2 at point 2 than it had at point 1. Its kinetic energy is now

$$KE_2 = \frac{1}{2}mv_2^2$$

The total energy of the ball at position 2 is the sum of its potential energy and its kinetic energy:

$$E_2 = PE_2 + KE_2 \qquad (7.17)$$

$$E_2 = mgh_2 + \frac{1}{2}mv_2^2 \qquad (7.18)$$

Let us now look at the difference in the total energy of the ball between when it is at position 2 and when it is at position 1. The change in the total energy of the ball between position 2 and position 1 is

$$\Delta E = E_2 - E_1 \qquad (7.19)$$

Using equations 7.16 and 7.18, this becomes

$$\Delta E = mgh_2 + \frac{1}{2}mv_2^2 - mgh_1 - \frac{1}{2}mv_1^2$$

Simplifying,

$$\Delta E = mg(h_2 - h_1) + \frac{1}{2}m(v_2^2 - v_1^2) \qquad (7.20)$$

Let us return, for the moment, to the third of the kinematic equations for projectile motion developed in chapter 3, namely

$$v^2 = v_0^2 - 2gy \qquad (3.24)$$

Recall that v was the velocity of the ball at a height y above the ground, and v_0 was the initial velocity at the ground. We can apply equation 3.24 to the present situation by noting that v_2 is the velocity of the ball at a height $h_2 - h_1 = y$, above the level where the velocity was v_1. Hence, we can rewrite equation 3.24 as

$$v_2^2 = v_1^2 - 2gy$$

Rearranging terms, this becomes

$$v_2^2 - v_1^2 = -2gy \qquad (7.21)$$

If we substitute equation 7.21 into equation 7.20, we get

$$\Delta E = mg(h_2 - h_1) + \frac{1}{2}m(-2gy)$$

But, as we can see from figure 7.13, $h_2 - h_1 = y$. Hence,

$$\Delta E = mgy - mgy$$

or

$$\Delta E = 0 \qquad (7.22)$$

which tells us that there is no change in the total energy of the ball between the arbitrary levels 1 and 2. But, since $\Delta E = E_2 - E_1$ from equation 7.19, equation 7.22 is also equivalent to

$$\Delta E = E_2 - E_1 = 0 \qquad (7.23)$$

Therefore,

$$E_2 = E_1 = \text{constant} \qquad (7.24)$$

That is, the total energy of the ball at position 2 is equal to the total energy of the ball at position 1. *Equations 7.22, 7.23, and 7.24 are equivalent statements of the law of conservation of energy. There is no change in the total energy of the ball throughout its entire flight. Or similarly, the total energy of the ball remains the same throughout its entire flight, that is, it is a constant.*

We can glean even more information from these equations by combining equations 7.15, 7.17, and 7.23 into

$$\Delta E = E_2 - E_1 = PE_2 + KE_2 - PE_1 - KE_1 = 0$$

$$PE_2 - PE_1 + KE_2 - KE_1 = 0 \qquad (7.25)$$

But,

$$PE_2 - PE_1 = \Delta PE \qquad (7.26)$$

is the change in the potential energy of the ball, and

$$KE_2 - KE_1 = \Delta KE \qquad (7.27)$$

is the change in the kinetic energy of the ball. Substituting equations 7.26 and 7.27 back into equation 7.25 gives

$$\Delta PE + \Delta KE = 0 \qquad (7.28)$$

or

$$\Delta PE = -\Delta KE \qquad (7.29)$$

Equation 7.29 says that the change in potential energy of the ball will always be equal to the change in the kinetic energy of the ball. Hence, if the velocity decreases between level 1 and level 2, ΔKE will be negative. When this is multiplied by the minus sign in equation 7.29, we obtain a positive number. Hence, there is a positive increase in the potential energy ΔPE. *Thus, the amount of kinetic energy of the ball lost between levels 1 and 2 will be equal to the gain in potential energy of the ball between the same two levels. Thus, energy can be transformed between kinetic energy and potential energy but, the total energy will always remain a constant.* The energy described here is mechanical energy. But the law of conservation of energy is, in fact, more general and applies to all forms of energy, not only mechanical energy. We will say more about this later.

This transformation of energy between kinetic and potential is illustrated in figure 7.14. When the ball is launched at the ground with an initial velocity v_0, all the energy is kinetic, as seen on the bar graph. When the ball reaches position 1, it is at a height h_1 above the ground and hence has a potential energy associated with that height. But since the ball has slowed down to v_1, its kinetic energy has decreased. But the sum of the kinetic energy and the potential energy is still the same constant energy, E_{tot}. The ball has lost kinetic energy but its potential energy has increased by the same amount lost. That is, energy was transformed from kinetic energy to potential energy. At position 2 the kinetic energy has decreased even further but the potential energy has increased correspondingly. At position 3, the ball is at the top of its trajectory. Its velocity is zero, hence its kinetic energy at the top is also zero. The total energy of the ball is all potential. At position 4, the ball

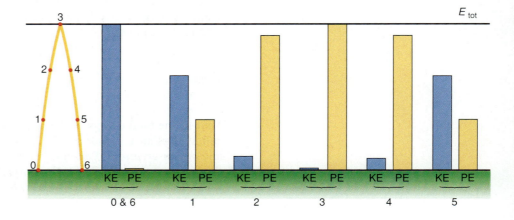

Figure 7.14

Bar graph of energy during projectile motion.

Mechanics

has started down. Its kinetic energy is small but nonzero, and its potential energy is starting to decrease. At position 5, the ball is moving much faster and the kinetic energy has increased accordingly. The potential energy has decreased to account for the increase in the kinetic energy. At position 6, the ball is back on the ground, and hence has no potential energy. All of the energy has been converted back into kinetic energy. As we can observe from the bar graph, the total energy remained constant throughout the flight.

Example 7.11

Conservation of energy and projectile motion. A 0.140-kg ball is thrown upward with an initial velocity of 35.0 m/s. Find (a) the total energy of the ball, (b) the maximum height of the ball, and (c) the kinetic energy and velocity of the ball at 30.0 m.

Solution

a. The total energy of the ball is equal to the initial kinetic energy of the ball, that is,

$$E_{tot} = KE_i = \tfrac{1}{2}mv^2$$

$$= \tfrac{1}{2}(0.140 \text{ kg})(35.0 \text{ m/s})^2$$

$$= 85.8 \text{ J}$$

b. At the top of the trajectory the velocity of the ball is equal to zero and hence its kinetic energy is also zero there. Thus, the total energy at the top of the trajectory is all in the form of potential energy. Therefore,

$$E_{tot} = PE = mgh$$

and the maximum height is

$$h = \frac{E_{tot}}{mg}$$

$$= \frac{85.8 \text{ J}}{(0.140 \text{ kg})(9.80 \text{ m/s}^2)}$$

$$= 62.5 \text{ m}$$

c. The total energy of the ball at 30 m is equal to the total energy of the ball initially. That is,

$$E_{30} = PE_{30} + KE_{30} = E_{tot}$$

The kinetic energy of the ball at 30.0 m is

$$KE_{30} = E_{tot} - PE_{30} = E_{tot} - mgh_{30}$$

$$= 85.8 \text{ J} - (0.140 \text{ kg})(9.80 \text{ m/s}^2)(30.0 \text{ m})$$

$$= 44.6 \text{ J}$$

The velocity of the ball at 30 m is found from

$$\tfrac{1}{2}mv^2 = KE_{30}$$

$$v = \sqrt{\frac{2KE_{30}}{m}}$$

$$= \sqrt{\frac{2(44.6 \text{ J})}{0.140 \text{ kg}}}$$

$$= 25.2 \text{ m/s}$$

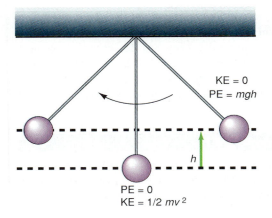

KE = 0
PE = mgh

PE = 0
KE = 1/2 mv^2

h

Figure 7.15

The simple pendulum.

Another example of this transformation of energy back and forth between kinetic and potential is given by the pendulum. The simple pendulum, as shown in figure 7.15, is a string, one end of which is attached to the ceiling, the other to a bob. The pendulum is pulled to the right so that it is a height h above its starting point. All its energy is in the form of potential energy. When it is released, it falls toward the center. As its height h decreases, it loses potential energy, but its velocity increases, increasing its kinetic energy. At the center position h is zero, hence its potential energy is zero. All its energy is now kinetic, and the bob is moving at its greatest velocity. Because of the inertia of the bob it keeps moving toward the left. As it does, it starts to rise, gaining potential energy. This gain in potential energy is of course accompanied by a corresponding loss in kinetic energy, until the bob is all the way to the left. At that time its velocity and hence kinetic energy is zero and, since it is again at the height h, all its energy is potential and equal to the potential energy at the start.

We can find the maximum velocity, which occurs at the bottom of the swing, by equating the total energy at the bottom of the swing to the total energy at the top of the swing:

$$E_{bottom} = E_{top}$$
$$KE_{bottom} = PE_{top} \tag{7.30}$$

$$\tfrac{1}{2}mv^2 = mgh$$

$$v = \sqrt{2gh} \tag{7.31}$$

Thus, the velocity at the bottom of the swing is independent of the mass of the bob and depends only on the height.

Let us now consider the following important example showing the relationship between work, potential energy, and kinetic energy.

Example 7.12

When the work done is not equal to the potential energy. A 5.00-kg block is lifted vertically through a height of 5.00 m by a force of 60.0 N. Find (a) the work done in lifting the block, (b) the potential energy of the block at 5.00 m, (c) the kinetic energy of the block at 5.00 m, (d) the velocity of the block at 5.00 m.

Solution

a. The work done in lifting the block, found from equation 7.1, is

$$W = Fy = (60.0 \text{ N})(5.00 \text{ m}) = 300 \text{ J}$$

b. The potential energy of the block at 5.00 m, found from equation 7.7, is

$$PE = mgh = (5.00 \text{ kg})(9.80 \text{ m/s}^2)(5.00 \text{ m}) = 245 \text{ J}$$

It is important to notice something here. We defined the potential energy as the work done to move the body into its particular position. Yet in this problem the work done to lift the block is 300 J, while the PE is only 245 J. The numbers are not the same. It seems as though something is wrong. Looking at the problem more carefully, however, we see that everything is okay. In the defining relation for the potential energy, we assumed that the work done to raise the block to the height h is done at a constant velocity, approximately a zero velocity. (Remember the force up F was just equal to the weight of the block). In this problem, the weight of the block is

$$w = mg = (5.00 \text{ kg})(9.80 \text{ m/s}^2) = 49.0 \text{ N}$$

Since the force exerted upward of 60.0 N is greater than the weight of the block, 49.0 N, the block is accelerated upward and arrives at the height of

5.00 m with a nonzero velocity and hence kinetic energy. Thus, the work done has raised the mass and changed its velocity so that the block arrives at the 5.00-m height with both a potential energy and a kinetic energy.

c. The kinetic energy is found by the law of conservation of energy, equation 7.15,

$$E_{tot} = KE + PE$$

Hence, the kinetic energy is

$$KE = E_{tot} - PE$$

The total energy of the block is equal to the total amount of work done on the block, namely 300 J, and as shown, the potential energy of the block is 245 J. Hence, the kinetic energy of the block at a height of 5.00 m is

$$KE = E_{tot} - PE = 300 \text{ J} - 245 \text{ J} = 55 \text{ J}$$

d. The velocity of the block at 5.00 m, found from equation 7.12 for the kinetic energy of the block, is

$$KE = \tfrac{1}{2}mv^2$$
$$v = \sqrt{2KE/m} = \sqrt{2(55 \text{ J}/5.00 \text{ kg})}$$
$$= 4.69 \text{ m/s}$$

7.7 Further Analysis of the Conservation of Energy

There are many rather difficult problems in physics that are greatly simplified and easily solved by the principle of conservation of energy. In fact, in advanced physics courses, most of the analysis is done by energy methods. Let us consider the following simple example. A block starts from rest at the top of the frictionless plane, as seen in figure 7.16. What is the speed of the block at the bottom of the plane?

Let us first solve this problem by Newton's second law. The force acting on the block down the plane is $w \sin \theta$, which is a constant. Newton's second law gives

$$F = ma$$
$$w \sin \theta = ma$$
$$mg \sin \theta = ma$$

Hence, the acceleration down the plane is

$$a = g \sin \theta \qquad (7.32)$$

which is a constant. The speed of the block at the bottom of the plane is found from the kinematic formula,

$$v^2 = v_0^2 + 2ax$$
$$v = \sqrt{2ax}$$

or, since $a = g \sin \theta$,

$$v = \sqrt{2g \sin \theta \, x} \qquad (7.33)$$

but

$$x \sin \theta = h$$

Therefore,

$$v = \sqrt{2gh} \qquad (7.34)$$

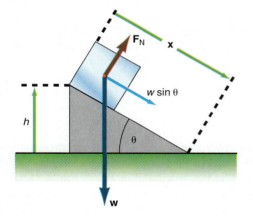

Figure 7.16

A block on an inclined plane.

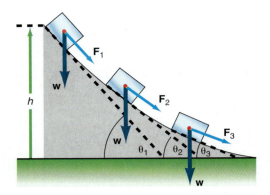

Figure 7.17

A block on a curved surface.

The problem is, of course, quite simple because the force acting on the block is a constant and hence the acceleration is a constant. The kinematic equations were derived on the basis of a constant acceleration and can be used only when the acceleration is a constant. What happens if the forces and accelerations are not constant? As an example, consider the motion of a block that starts from rest at the top of a frictionless *curved* surface, as shown in figure 7.17. The weight w acting downward is always the same, but at each position, the angle the block makes with the horizontal is different. Therefore, the force is different at every position on the surface, and hence the acceleration is different at every point. Thus, the simple techniques developed so far can not be used. (The calculus would be needed for the solution of this case of variable acceleration.)

Let us now look at the same problem from the point of view of energy. The law of conservation of energy says that the total energy of the system is a constant. Therefore, the total energy at the top must equal the total energy at the bottom, that is,

$$E_{top} = E_{bot}$$

The total energy at the top is all potential because the block starts from rest ($v_0 = 0$, hence KE = 0), while at the bottom all the energy is kinetic because at the bottom $h = 0$ and hence PE = 0. Therefore,

$$PE_{top} = KE_{bot}$$
$$mgh = \tfrac{1}{2}mv^2$$
$$v = \sqrt{2gh} \qquad (7.35)$$

the speed of the block at the bottom of the plane. We have just solved a very difficult problem, but by using the law of conservation of energy, its solution is very simple.

A very interesting thing to observe here is that the speed of the block down a frictionless inclined plane of height *h,* equation 7.34, is the same as the speed of a block down the frictionless curved surface of height *h,* equation 7.35. In fact, if the block were dropped over the top of the inclined plane (or curved surface) so that it fell freely to the ground, its speed at the bottom would be found as

$$E_{top} = E_{bot}$$
$$mgh = \tfrac{1}{2}mv^2$$
$$v = \sqrt{2gh}$$

which is the same speed obtained for the other two cases. This is a characteristic of the law of conservation of energy. The speed of the moving object at the bottom is the same regardless of the path followed by the moving object to get to the final position. This is a consequence of the fact that the same amount of energy was used to place the block at the top of the plane for all three cases, and therefore that same amount of energy is obtained when the block returns to the bottom of the plane.

The energy that the block has at the top of the plane is equal to the work done on the block to place the block at the top of the plane. If the block in figure 7.18 is lifted vertically to the top of the plane, the work done is

$$W = Fh = wh = mgh \qquad (7.36)$$

If the block is pushed up the frictionless plane at a constant speed, then the work done is

$$W = Fx = w \sin \theta \, x$$
$$W = mgx \sin \theta \qquad (7.37)$$

but

$$x \sin \theta = h$$

Figure 7.18

A conservative system.

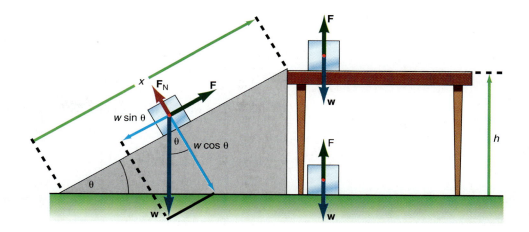

and hence, the work done in pushing the block up the plane is

$$W = mgh \qquad (7.38)$$

which is the identical amount of work just found in lifting the block vertically into the same position. Therefore, the energy at the top is independent of the path taken to get to the top. *Systems for which the energy is the same regardless of the path taken to get to that position are called conservative systems.* **Conservative systems** are systems for which the energy is conserved, that is, the energy remains constant throughout the motion. A conservative system is a system in which the difference in energy is the same regardless of the path taken between two different positions. In a conservative system the total mechanical energy is conserved.

For a better understanding of a conservative system it is worthwhile to consider a nonconservative system. The nonconservative system that we will examine is an inclined plane on which friction is present, as shown in figure 7.19. Let us compute the work done in moving the block up the plane at a constant speed. The force F, exerted up the plane, is

$$F = w \sin \theta + f_k \qquad (7.39)$$

where

$$f_k = \mu_k F_N = \mu_k w \cos \theta \qquad (7.40)$$

Figure 7.19

A nonconservative system.

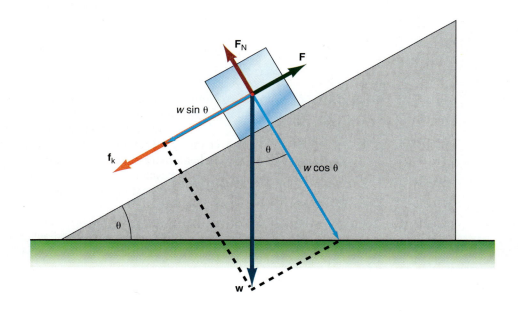

or

$$F = w \sin \theta + \mu_k w \cos \theta$$

or

$$F = mg \sin \theta + \mu_k mg \cos \theta \qquad (7.41)$$

The work done in sliding the block up the plane is

$$W_s = Fx = (mg \sin \theta + \mu_k mg \cos \theta)x$$
$$= mgx \sin \theta + \mu_k mgx \cos \theta \qquad (7.42)$$

but

$$x \sin \theta = h$$

Therefore,

$$W_s = mgh + \mu_k mgx \cos \theta \qquad (7.43)$$

That is, the work done in sliding the block up the plane against friction is greater than the amount of work necessary to lift the block to the top of the plane. The work done in lifting it is

$$W_L = mgh$$

But there appears to be a contradiction here. Since both blocks end up at the same height h above the ground, they should have the same energy mgh. This seems to be a violation of the law of conservation of energy. The problem is that an inclined plane with friction is not a conservative system. Energy is expended by the person exerting the force, to overcome the friction of the inclined plane. The amount of energy lost is found from equation 7.43 as

$$E_{lost} = \mu_k mgx \cos \theta \qquad (7.44)$$

This energy that is lost in overcoming friction shows up as heat energy in the block and the plane. At the top of the plane, both blocks will have the same potential energy. But we must do more work to slide the block up the frictional plane than in lifting it straight upward to the top.

If we now let the block slide down the plane, the same amount of energy, equation 7.44, is lost in overcoming friction as it slides down. Therefore, the total energy of the block at the bottom of the plane is less than in the frictionless case and therefore its speed is also less. That is, the total energy at the bottom is now

$$\tfrac{1}{2}mv^2 = mgh - \mu_k mgx \cos \theta \qquad (7.45)$$

and the speed at the bottom is now

$$v = \sqrt{2gh - 2\mu_k gx \cos \theta} \qquad (7.46)$$

Notice that the speed of the block down the rough plane, equation 7.46, is less than the speed of the block down a smooth plane, equation 7.34.

Any time a body moves against friction, there is always an amount of mechanical energy lost in overcoming this friction. This lost energy always shows up as heat energy. The law of conservation of energy, therefore, holds for a nonconservative system, if we account for the lost mechanical energy of the system as an increase in heat energy of the system, that is,

$$E_{tot} = KE + PE + Q \qquad (7.47)$$

where Q is the heat energy gained or lost during the process. We will say more about this when we discuss the first law of thermodynamics in chapter 17.

Example 7.13

Losing kinetic energy to friction. A 1.50-kg block slides along a smooth horizontal surface at 2.00 m/s. It then encounters a rough horizontal surface. The coefficient of kinetic friction between the block and the rough surface is $\mu_k = 0.400$. How far will the block move along the rough surface before coming to rest?

Solution

When the block slides along the smooth surface it has a total energy that is equal to its kinetic energy. When the block slides over the rough surface it slows down and loses its kinetic energy. Its kinetic energy is equal to the work done on the block by friction as it is slowed to a stop. Therefore,

$$\mathrm{KE} = W_f$$

$$\tfrac{1}{2}mv^2 = f_k x = \mu_k F_N x = \mu_k w x = \mu_k mgx$$

Solving for *x*, the distance the block moves as it comes to a stop, we get

$$\mu_k mgx = \tfrac{1}{2}mv^2$$

$$x = \frac{\tfrac{1}{2}v^2}{\mu_k g}$$

$$= \frac{\tfrac{1}{2}(2 \text{ m/s})^2}{(0.400)(9.80 \text{ m/s}^2)}$$

$$= 0.510 \text{ m}$$

"Have you ever wondered . . . ?"

An Essay on the Application of Physics

The Great Pyramids

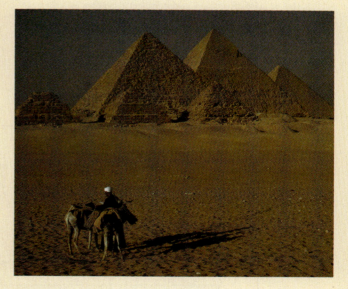

Figure 1

The great pyramid of Cheops.[1]

Have you ever wondered how the great pyramids of Egypt were built? The largest, Cheops, located 10 mi outside of the city of Cairo, figure 1, is about 400 ft high and contains more than 2½ million blocks of limestone and granite weighing between 2 and 70 ton, apiece. Yet these pyramids were built over 4000 years ago. How did these ancient people ever raise these large stones to such great heights with the very limited equipment available to them?

It is usually supposed that the pyramids were built using the principle of the mechanical advantage obtained by the inclined plane. The first level of stones for the pyramid were assembled on the flat surface, as in figure 2(a). Then an incline was built out of sand and pressed against the pyramid, as in figure 2(b). Another level of stones were then put into place. As each succeeding level was made, more sand was added to the incline in order to reach the next level. The process continued with additional sand added to the incline for each new level of stones. When the final stones were at the top, the sand was removed leaving the pyramids as seen today.

The advantage gained by using the inclined plane can be explained as follows. An ideal frictionless inclined plane is shown in figure 3. A stone that has the weight w_s is to be lifted from the ground to the height h. If it is lifted straight up, the work that must be done to lift the stone to the height h, is

$$W_1 = F_A h = w_s h \qquad \text{(H7.1)}$$

where F_A is the applied force to lift the stone and w_s is the weight of the stone.

If the same stone is on an inclined plane, then the component of the weight of the stone, $w_s \sin \theta$, acts down the plane and hence a force, $F = w_s \sin \theta$, must be exerted on the stone in order to push the stone up the plane. The work done pushing the stone a distance L up the plane is

$$W_2 = FL \qquad \text{(H7.2)}$$

Whether the stone is lifted to the top of the plane directly, or pushed up the inclined plane to the top, the stone ends up at the top and the work done in pushing the stone up the plane is equal to the work done in lifting the stone to the height h. Therefore,

$$W_2 = W_1 \qquad \text{(H7.3)}$$

$$FL = w_s h \qquad \text{(H7.4)}$$

Hence, the force F that must be exerted to push the block up the inclined plane is

$$F = \frac{h}{L} w_s \qquad \text{(H7.5)}$$

If the length of the incline L is twice as large as the height h (i.e., $L = 2h$), then the force necessary to push the stone up the incline is

$$F = \frac{h}{L} w_s = \frac{h}{2h} w_s = \frac{w_s}{2}$$

Therefore, if the length of the incline is twice the length of the height, the force necessary to push the stone up the incline is only half the weight of the stone. If the length of the incline is increased to $L = 10h$, then the force F is

$$F = \frac{h}{L} w_s = \frac{h}{10h} w_s = \frac{w_s}{10}$$

That is, by increasing the length of the incline to ten times the height, the force that we must exert to push the stone up the incline is only 1/10 of the weight of the stone. Thus by making L very large, the force that we must exert to push the stone up the inclined plane is made relatively small. If $L = 100h$, then the force necessary would only be one-hundredth of the weight of the stone.

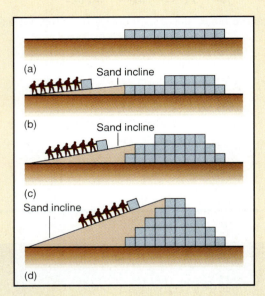

Figure 2

The construction of the pyramids.

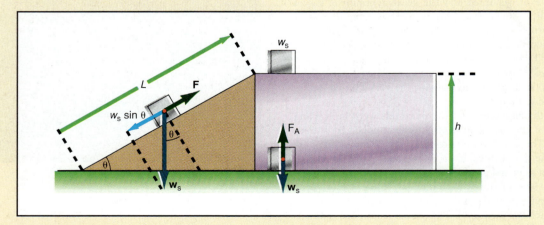

Figure 3

The inclined plane.

Figure 4

Aerial view of the pyramid of Dashur.

The inclined plane is called a simple machine. With it, we have amplified our ability to move a very heavy stone to the top of the hill. This amplification is called the *ideal mechanical advantage* (IMA) of the inclined plane and is defined as

$$\text{Ideal mechanical advantage} = \frac{\text{Force out}}{\text{Force in}} \qquad \textbf{(H7.6)}$$

or

$$\text{IMA} = \frac{F_{\text{out}}}{F_{\text{in}}} \qquad \textbf{(H7.7)}$$

The force that we get out of the machine, in this example, is the weight of the stone w_s, which ends up at the top of the incline, while the force into the machine is equal to the force F that is exerted on the stone in pushing it up the incline. Thus, the ideal mechanical advantage is

$$\text{IMA} = \frac{w_s}{F} \qquad \textbf{(H7.8)}$$

Using equation H7.4 this becomes

$$\text{IMA} = \frac{w_s}{F} = \frac{L}{h} \qquad \textbf{(H7.9)}$$

Hence if $L = 10h$, the IMA is

$$\text{IMA} = 10\frac{h}{h} = 10$$

and the amplification of the force is 10.

The angle θ of the inclined plane, found from the geometry of figure 3, is

$$\sin\theta = \frac{h}{L} \qquad \textbf{(H7.10)}$$

Thus, by making θ very small, a slight incline, a very small force could be applied to move the very massive stones of the pyramid into position. The inclined plane does not give us something for nothing, however. The work done in lifting the stone or pushing the stone is the same. Hence, the smaller force F must be exerted for a very large distance L to do the same work as lifting the very massive stone to the relatively short height h. However, if we are limited by the force F that we can exert, as were the ancient Egyptians, then the inclined plane gives us a decided advantage. An aerial view of the pyramid of Dashur is shown in figure 4. Notice the ramp under the sands leading to the pyramid.[1]

---

1. This picture is taken from *Secrets of the Great Pyramids* by Peter Tompkins, Harper Colophon Books, 1978.

The Language of Physics

Energy
The ability of a body or system of bodies to perform work (p. 189).

System
An aggregate of two or more particles that is treated as an individual unit (p. 189).

Work
The product of the force acting on a body in the direction of the displacement, times the displacement of the body (p. 189).

Power
The time rate of doing work (p. 192).

Gravitational potential energy
The energy that a body possesses by virtue of its position in a gravitational field. The potential energy is equal to the work that must be done to put the body into that particular position (p. 194).

Kinetic energy
The energy that a body possesses by virtue of its motion. The kinetic energy is equal to the work that must be done to bring the body from rest into that state of motion (p. 196).

Closed system
An isolated system that is not affected by any external influences (p. 198).

Law of conservation of energy
In any closed system, the total energy of the system remains a constant. To say that energy is conserved means that the energy is a constant (p. 198).

Conservative system
A system in which the difference in energy is the same regardless of the path taken between two different positions. In a conservative system the total mechanical energy is conserved (p. 205).

Summary of Important Equations

Work done $W = Fx$	(7.1)	Power of moving system $P = Fv$	(7.4)	Total mechanical energy $E_{tot} = KE + PE$		
Work done in general $W = Fx \cos \theta$	(7.2)	Gravitational potential energy $PE = mgh$	(7.7)	Conservation of mechanical energy $\Delta E = E_2 - E_1 = 0$	(7.23)	
Power $P = W/t$	(7.3)	Kinetic energy $KE = \frac{1}{2}mv^2$	(7.12)	$E_2 = E_1 = $ constant	(7.24)	

Questions for Chapter 7

1. If the force acting on a body is perpendicular to the displacement, how much work is done in moving the body?
2. A person is carrying a heavy suitcase while walking along a horizontal corridor. Does the person do work (a) against gravity (b) against friction?
3. A car is moving at 90 km/hr when it is braked to a stop. Where does all the kinetic energy of the moving car go?

†4. A rowboat moves in a northerly direction upstream at 3 mph relative to the water. If the current moves south at 3 mph relative to the bank, is any work being done?
†5. For a person to lose weight, is it more effective to exercise or to cut down on the intake of food?
6. If you lift a body to a height h with a force that is greater than the weight of a body, where does the extra energy go?
7. Potential energy is energy that a body possesses by virtue of its position, while kinetic energy is

energy that a body possesses by virtue of its speed. Could there be an energy that a body possesses by virtue of its acceleration? Discuss.
8. For a conservative system, what is $\Delta E/\Delta t$?
9. Describe the transformation of energy in a pendulum as it moves back and forth.
10. If positive work is done putting a body into motion, is the work done in bringing a moving body to rest negative work? Explain.

Problems for Chapter 7

7.2 Work

1. A 500-lb box is raised through a height of 15.0 ft. How much work is done in lifting the box at a constant velocity?
2. How much work is done if (a) a force of 150 N is used to lift a 10.0-kg mass to a height of 5.00 m and (b) a force of 150 N, parallel to the surface, is used to pull a 10.0-kg mass, 5.00 m on a horizontal surface?
3. A force of 8.00 N is used to pull a sled through a distance of 100 m. If the force makes an angle of 40.0° with the horizontal, how much work is done?
4. A person pushes a lawn mower with a force of 50.0 N at an angle of 35.0° below the horizontal. If the mower is moved through a distance of 25.0 m, how much work is done?
5. A consumer's gas bill indicates that they have used a total of 37 therms of gas for a 30-day period. Express this energy in joules. A therm is a unit of energy equal to 100,000 Btu and a Btu (British thermal unit) is a unit of energy equal to 778 ft lb.

6. A 670-kg man lifts a 200-kg mass to a height of 1.00 m above the floor and then carries it through a horizontal distance of 10.0 m. How much work is done (a) against gravity in lifting the mass, (b) against gravity in carrying it through the horizontal distance, and (c) against friction in carrying it through the horizontal distance?
7. Calculate the work done in (a) pushing a 4.00-kg block up a frictionless inclined plane 10.0 m long that makes an angle of 30.0° with the horizontal and (b) lifting the block vertically from the ground to the top of the plane, 5.00 m high. (c) Compare the force used in parts a and b.

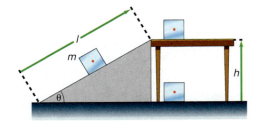

8. A football player weighs 200 lb and does a chin-up by pulling himself up by his arms to an additional height of 30.0 in. above the floor. If he does a total of 20 chin-ups, how much work does he do?
9. A 110-kg football player does a chin-up by pulling himself up by his arms an additional height of 50.0 cm above the floor. If he does a total of 25 chin-ups, how much work does he do?

7.3 Power

10. A consumer's electric bill indicates that they have used a total of 793 kwh of electricity for a 30-day period. Express this energy in (a) joules and (b) ft lb. (c) What is the average power used per hour?
11. A 150-lb person climbs a rope at a constant velocity of 2.00 ft/s in a period of time of 10.0 s. (a) How much power does the person expend? (b) How much work is done?
12. You are designing an elevator that must be capable of lifting a load (elevator plus passengers) of 4000 lb to a height of 12 floors (120 ft) in

1 min. What horsepower motor should you require if half of the power is used to overcome friction?

13. A locomotive pulls a train at a velocity of 40.0 mph with a force of 15,000 lb. What power is exerted by the locomotive?

14. A locomotive pulls a train at a velocity of 88.0 km/hr with a force of 55,000 N. What power is exerted by the locomotive?

7.4 Gravitational Potential Energy

15. Find the potential energy of a 7.00-kg mass that is raised 2.00 m above the desk. If the desk is 1.00 m high, what is the potential energy of the mass with respect to the floor?

16. A 5.00-kg block is at the top of an inclined plane that is 4.00 m long and makes an angle of 35.0° with the horizontal. Find the potential energy of the block.

17. A 15.0-kg sledge hammer is 2.00 m high. How much work can it do when it falls to the ground?

18. A pile driver lifts a 500-lb hammer 10.0 ft before dropping it on a pile. If the pile is driven 4.00 in. into the ground when hit by the hammer, what is the average force exerted on the pile?

7.5 Kinetic Energy

19. What is the kinetic energy of the earth as it travels at a velocity of 30.0 km/s in its orbit about the sun?

20. Compute the kinetic energy of a 3200-lb auto traveling at (a) 15.0 mph, (b) 30.0 mph, and (c) 60.0 mph.

21. Compare the kinetic energy of a 1200-kg auto traveling at (a) 30.0 km/hr, (b) 60.0 km/hr, and (c) 120 km/hr.

22. If an electron in a hydrogen atom has a velocity of 2.19×10^6 m/s, what is its kinetic energy?

23. A 1000-lb airplane traveling at 150 mph is 3000 ft above the terrain. What is its kinetic energy and its potential energy?

24. A 700-kg airplane traveling at 320 km/hr is 1500 m above the terrain. What is its kinetic energy and its potential energy?

25. A 10.0-g bullet, traveling at a velocity of 900 m/s hits and is embedded 2.00 cm into a large piece of oak wood that is fixed at rest. What is the kinetic energy of the bullet? What is the average force stopping the bullet?

26. A little league baseball player throws a baseball (0.15 kg) at a speed of 8.94 m/s. (a) How much work must be done to catch this baseball? (b) If the catcher moves his glove backward by 2.00 cm while catching the ball, what is the average force exerted on his glove by the ball? (c) What is the average force if the distance is 20.0 cm? Is there an advantage in moving the glove backward?

7.6 The Conservation of Energy

27. A 2.00-kg block is pushed along a horizontal frictionless table a distance of 3.00 m, by a horizontal force of 12.0 N. Find (a) how much work is done by the force, (b) the final kinetic energy of the block, and (c) the final velocity of the block. (d) Using Newton's second law, find the acceleration and then the final velocity.

28. A 2.75-kg block is placed at the top of a 40.0° frictionless inclined plane that is 40.0 cm high. Find (a) the work done in lifting the block to the top of the plane, (b) the potential energy at the top of the plane, (c) the kinetic energy when the block slides down to the bottom of the plane, (d) the velocity of the block at the bottom of the plane, and (e) the work done in sliding down the plane.

29. A projectile is fired vertically with an initial velocity of 200 ft/s. Using the law of conservation of energy, find how high the projectile rises.

30. A 3.00-kg block is lifted vertically through a height of 6.00 m by a force of 40.0 N. Find (a) the work done in lifting the block, (b) the potential energy of the block at 6.00 m, (c) the kinetic energy of the block at 6.00 m, and (d) the velocity of the block at 6.00 m.

31. Apply the law of conservation of energy to an Atwood's machine and find the velocity of block A as it hits the ground. $m_B = 40.0$ g, $m_A = 50.0$ g, $h_B = 0.500$ m, and $h_A = 1.00$ m.

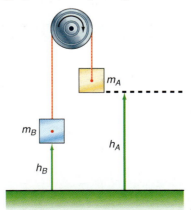

†32. Determine the velocity of block 2 when the height of block 1 is equal to $h_0/4$. $m_2 = 35.0$ g, $m_1 = 20.0$ g, $h_0 = 1.50$ m, and $h_2 = 2.00$ m.

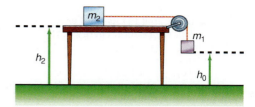

33. A 250-g bob is attached to a string 1.00 m long to make a pendulum. If the pendulum bob is pulled to the right, such that the string makes an angle of 15.0° with the vertical, what is (a) the maximum potential energy, (b) the maximum kinetic energy, and (c) the maximum velocity of the bob and where does it occur?

34. A 45.0-kg girl is on a swing that is 2.00 m long. If the swing is pulled to the right, such that the rope makes an angle of 30.0° with the vertical, what is (a) the maximum potential energy of the girl, (b) her maximum kinetic energy, and (c) the maximum velocity of the swing and where does it occur?

7.7 Further Analysis of the Conservation of Energy

35. A 3.56-kg mass moving at a speed of 3.25 m/s enters a region where the coefficient of kinetic friction is 0.500. How far will the block move before it comes to rest?

36. A 5.00-kg mass is placed at the top of a 35.0° rough inclined plane that is 30.0 cm high. The coefficient of kinetic friction between the mass and the plane is 0.400. Find (a) the potential energy at the top of the plane, (b) the work done against friction as it slides down the plane, (c) the kinetic energy of the mass at the bottom of the plane, and (d) the velocity of the mass at the bottom of the plane.

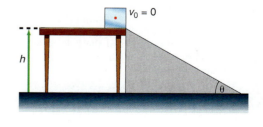

37. A 100-g block is pushed down a rough inclined plane with an initial velocity of 1.50 m/s. The plane is 2.00 m long and makes an angle of 35.0° with the horizontal. If the block comes to rest at the bottom of the plane, find (a) its total energy at the top, (b) its total energy at the bottom, (c) the total energy lost due to friction, (d) the frictional force, and (e) the coefficient of friction.

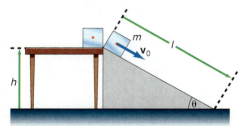

38. A 1.00-kg block is pushed along a rough horizontal floor with a horizontal force of 5.00 N for a distance of 5.00 m. If the block is moving at a constant velocity of 4.00 m/s, find (a) the work done on the block by the force, (b) the kinetic energy of the block, and (c) the energy lost to friction.

39. A 500-lb box is pushed along a rough floor by a horizontal force. The block moves at constant velocity for a distance of 15.0 ft. If the coefficient of friction between the box and the floor is 0.30, how much work is done in moving the box?

40. A 10.0-lb package slides from rest down a portion of a circular mail chute that is 20.0 ft above the ground. Its velocity at the bottom is 20.0 ft/s. How much energy is lost due to friction?

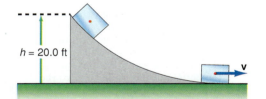

41. A 6.68-kg package slides from rest down a portion of a circular mail chute that is 4.58 m above the ground. Its velocity at the bottom is 7.63 m/s. How much energy is lost due to friction?

42. In the diagram $m_2 = 3.00$ kg, $m_1 = 5.00$ kg, $h_2 = 1.00$ m, $h_1 = 0.750$ m, and $\mu_k = 0.400$. Find (a) the initial total energy of the system, (b) the work done against friction as m_2 slides on the rough surface, (c) the

velocity v_1 of mass m_1 as it hits the ground, and (d) the kinetic energy of m_1 as it hits the ground.

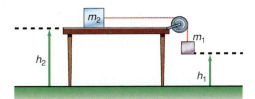

†43. A 5.00-kg body is placed at the top of the track, position A, 2.00 m above the base of the track, as shown in the diagram. (a) Find the total energy of the block. (b) The block is allowed to slide from rest down the frictionless track to the position B. Find the velocity of the body at B. (c) The block then moves over the level rough surface of $\mu_k = 0.300$. How far will the block move before coming to rest?

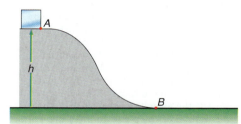

44. A 0.500-kg ball is dropped from a height of 3.00 m. Upon hitting the ground it rebounds to a height of 1.50 m. (a) How much mechanical energy is lost in the rebound, and what happens to this energy? (b) What is the velocity just before and just after hitting the ground?

Additional Problems

†45. The concept of work can be used to describe the action of a lever. Using the principle of work in equals work out, show that

$$F_{out} = \frac{r_{in}}{r_{out}} F_{in}$$

Show how this can be expressed in terms of a mechanical advantage.

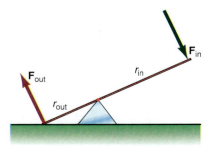

†46. Show how the inclined plane can be considered as a simple machine by comparing the work done in sliding an object up the plane with the work done in lifting the block to the top of the plane. How does the inclined plane supply a mechanical advantage?

47. A force acting on a 300-g mass causes it to move at a constant speed over a rough surface. The coefficient of kinetic friction is 0.350. Find the work required to move the mass a distance of 2.00 m.

48. A 5.00-kg projectile is fired at an angle of 58.0° above the horizontal with the initial velocity of 30.0 m/s. Find (a) the total energy of the projectile, (b) the total energy in the vertical direction, (c) the total energy in the horizontal direction, (d) the total energy at the top of the trajectory, (e) the potential energy at the top of the trajectory, (f) the maximum height of the projectile, (g) the kinetic energy at the top of the trajectory, and (h) the velocity of the projectile as it hits the ground.

49. It takes 20,000 W to keep a 1600-kg car moving at a constant speed of 60.0 km/hr on a level road. How much power is required to keep the car moving at the same speed up a hill inclined at an angle of 22.0° with the horizontal?

50. John consumes 5000 kcal/day. His metabolic efficiency is 70.0%. If his normal activity utilizes 2000 kcal/day, how many hours will John have to exercise to work off the excess calories by (a) walking, which uses 3.80 kcal/hr; (b) swimming, which uses 8.00 kcal/hr; and (c) running, which uses 11.0 kcal/hr?

51. A 2.50-kg mass is at rest at the bottom of a 5.00-m-long rough inclined plane that makes an angle of 25.0° with the horizontal. When a constant force is applied up the plane and parallel to it, it causes the mass to arrive at the top of the incline at a speed of 0.855 m/s. Find (a) the total energy of the mass when it is at the top of the incline, (b) the work done against friction, and (c) the magnitude of the applied force. The coefficient of friction between the mass and the plane is 0.350.

†52. A 2.00-kg block is placed at the position A on the track that is 3.00 m above the ground. Paths A-B and C-D of the track are frictionless, while section B-C is rough with a coefficient of kinetic friction of 0.350 and a length of 1.50 m. Find (a) the total energy of the block at *A*, (b) the velocity of the block at *B*, (c) the energy lost along path *B-C*, and (d) how high the block rises along path *C-D*.

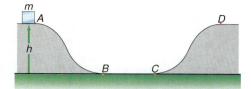

53. A mass *m* = 3.50 kg is launched with an initial velocity v_0 = 1.50 m/s from the position *A* at a height *h* = 3.80 m above the reference plane in the diagram for problem 52. Paths *A-B* and *C-D* of the track are frictionless, while path *B-C* is rough with a coefficient of kinetic friction of 0.300 and a length of 3.00 m. Find (a) the number of oscillations the block makes before coming to rest along the path *B-C* and (b) where the block comes to rest on path *B-C*.

54. A ball starts from rest at position *A* at the top of the track. Find (a) the total energy at *A*, (b) the total energy at *B*, (c) the velocity of the ball at *B*, and (d) the velocity of the ball at *C*.

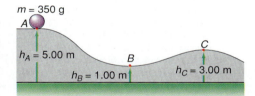

55. A 20.0-kg mass is at rest on a rough horizontal surface. It is then accelerated by a net constant force of 8.6 N. After the mass has moved 1.5 m from rest, the force is removed and the mass comes to rest in 2.00 m. Using energy methods find the coefficient of kinetic friction.

56. In an Atwood's machine m_B = 30.0 g, m_A = 50.0 g, h_B = 0.400 m, and h_A = 0.800 m. The machine starts from rest and mass m_A acquires a

velocity of 1.25 m/s as it strikes the ground. Find the energy lost due to friction in the bearings of the pulley.

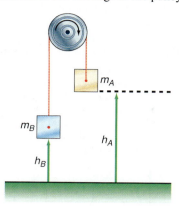

†57. What is the total energy of the Atwood's machine in the position shown in the diagram? If the blocks are released and m_1 falls through a distance of 1.00 m, what is the kinetic and potential energy of each block, and what are their velocities?

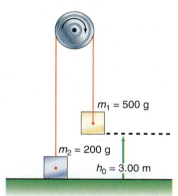

†58. The gravitational potential energy of a mass *m* with respect to infinity is given by

$$PE = -\frac{Gm_Em}{r}$$

where *G* is the universal gravitational constant, m_E is the mass of the earth, and *r* is the distance from the center of the earth to the mass *m*. Find the escape velocity of a spaceship from the earth. (The escape velocity is the necessary velocity to remove a body from the gravitational attraction of the earth.)

†59. Modify problem 58 and find the escape velocity for (a) the moon, (b) Mars, and (c) Jupiter.

†60. The entire Atwood's machine shown is allowed to go into free-fall. Find the velocity of m_1 and m_2 when the entire system has fallen 1.00 m.

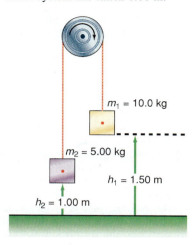

†61. A 1.50-kg block moves along a smooth horizontal surface at 2.00 m/s. The horizontal surface is at a height h_0 above the ground. The block then slides down a rough hill, 20.0 m long, that makes an angle of 30.0° with the horizontal. The coefficient of kinetic friction between the block and the hill is 0.600. How far down the hill will the block move before coming to rest?

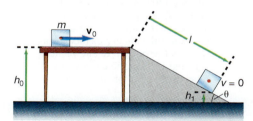

†62. At what point above the ground must a car be released such that when it rolls down the track and into the circular loop it will be going fast enough to make it completely around the loop? The radius of the circular loop is *R*.

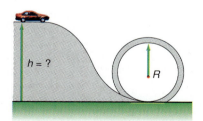

†63. A 1.50-kg block moves along a smooth horizontal surface at 2.00 m/s. It then encounters a smooth inclined plane that makes an angle of 53.0° with the horizontal. How far up the incline will the block move before coming to rest?

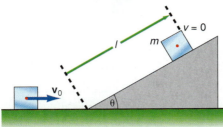

†64. Repeat problem 63, but in this case the inclined plane is rough and the coefficient of kinetic friction between the block and the plane is 0.400.

†65. In the diagram mass m_1 is located at the top of a rough inclined plane that has a length $l_1 = 0.500$ m. $m_1 = 0.500$ kg, $m_2 = 0.200$ kg, $\mu_{k_1} = 0.500$, $\mu_{k_2} = 0.300$, $\theta = 50.0°$, and $\phi = 50.0°$. (a) Find the total energy of the system in the position shown. (b) The system is released from rest. Find the work done for block 1 to overcome friction as it slides down the plane. (c) Find the work done for block 2 to overcome friction as it slides up the plane. (d) Find the potential energy of block 2 when it arrives at the top of the plane. (e) Find the velocity of block 1 as it reaches the bottom of the plane. (f) Find the kinetic energy of each block at the end of their travel.

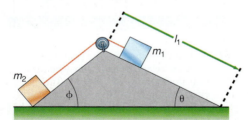

†66. If a constant force acting on a body is plotted against the displacement of the body from x_1 to x_2, as shown in the diagram, then the work done is given by

$$W = F(x_2 - x_1)$$
$$= \text{Area under the curve}$$

Show that this concept can be extended to cover the case of a variable force, and hence find the work done for the variable force, $F = kx$, where $k = 2.00$ N/m as the body is displaced from x_1 to x_2. Draw a graph showing your results.

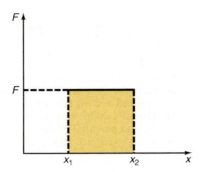

Interactive Tutorials

🔲 67. A projectile of mass $m = 100$ kg is fired vertically upward at a velocity $v_0 = 50.0$ m/s. Calculate its potential energy PE (relative to the ground), its kinetic energy KE, and its total energy E_{tot} for the first 10.0 s of flight.

🔲 68. Consider the general motion in an Atwood's machine such as the one shown in the diagram of problem 31; $m_A = 0.650$ kg and is at a height $h_A = 2.55$ m above the reference plane and mass $m_B = 0.420$ kg is at a height $h_B = 0.400$ m. If the system starts from rest, find (a) the initial potential energy of mass A, (b) the initial potential energy of mass B, and (c) the total energy of the system. When m_A has fallen a distance $y_A = 0.75$ m, find (d) the potential energy of mass A, (e) the potential energy of mass B, (f) the speed of each mass at that point, (g) the kinetic energy of mass A, and (h) the kinetic energy of mass B. (i) When mass A hits the ground, find the speed of each mass.

🔲 69. Consider the general motion in the combined system shown in the diagram of problem 42; $m_1 = 0.750$ kg and is at a height $h_1 = 1.85$ m above the reference plane and mass $m_2 = 0.285$ kg is at a height $h_2 = 2.25$ m, $\mu_k = 0.450$. If the system starts from rest, find (a) the initial potential energy of mass 1, (b) the initial potential energy of mass 2, and (c) the total energy of the system. When m_1 has fallen a distance $y_1 = 0.35$ m, find (d) the potential energy of mass 1, (e) the potential energy of mass 2, (f) the energy lost due to friction as mass 2 slides on the rough surface, (g) the speed of each mass at that point, (h) the kinetic energy of mass 1, and (i) the kinetic energy of mass 2. (j) When mass 1 hits the ground, find the speed of each mass.

🔲 70. Consider the general case of motion shown in the diagram with mass m_A initially located at the top of a rough inclined plane of length l_A, and mass m_B is at the bottom of the second plane; x_A is the distance from the mass A to the bottom of the plane. Let $m_A = 0.750$ kg, $m_B = 0.250$ kg, $l_A = 0.550$ m, $\theta = 40.0°$, $\phi = 30.0°$, $\mu_{k_A} = 0.400$, $\mu_{k_B} = 0.300$, and $x_A = 0.200$ m. When $x_A = 0.200$ m, find (a) the initial total energy of the system, (b) the distance block B has moved, (c) the potential energy of mass A, (d) the potential energy of mass B, (e) the energy lost due to friction for block A, (f) the energy lost due to friction for block B, (g) the velocity of each block, (h) the kinetic energy of mass A, and (i) the kinetic energy of mass B.

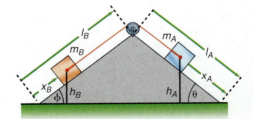

8

Momentum and Its Conservation

Conservation of momentum in a collision.

The quantity of motion is the measure of the same, arising from the velocity and quantity conjointly.

Isaac Newton, *Principia*

8.1 Momentum

In dealing with some problems in mechanics, we find that in many cases, it is exceedingly difficult, if not impossible, to determine the forces that are acting on a body, and/or for how long the forces are acting. These difficulties can be overcome, however, by using the concept of momentum.

    The **linear momentum** *of a body is defined as the product of the mass of the body in motion times its velocity.* That is,

$$\mathbf{p} = m\mathbf{v} \tag{8.1}$$

Because velocity is a vector, linear momentum is also a vector, and points in the same direction as the velocity vector. We use the word linear here to indicate that the momentum of the body is along a line, in order to distinguish it from the concept of angular momentum. Angular momentum applies to bodies in rotational motion and will be discussed in chapter 9. In this book, whenever the word momentum is used by itself it will mean linear momentum.

    This definition of momentum may at first seem rather arbitrary. Why not define it in terms of v^2, or v^3? We will see that this definition is not arbitrary at all. Let us consider Newton's second law

$$\mathbf{F} = m\mathbf{a} = m\frac{\Delta\mathbf{v}}{\Delta t}$$

However, since $\Delta\mathbf{v} = \mathbf{v_f} - \mathbf{v_i}$, we can write this as

$$\mathbf{F} = m\left(\frac{\mathbf{v_f} - \mathbf{v_i}}{\Delta t}\right)$$

$$\mathbf{F} = \frac{m\mathbf{v_f} - m\mathbf{v_i}}{\Delta t} \tag{8.2}$$

But $m\mathbf{v_f} = \mathbf{p_f}$, the final value of the momentum, and $m\mathbf{v_i} = \mathbf{p_i}$, the initial value of the momentum. Substituting this into equation 8.2, we get

$$\mathbf{F} = \frac{\mathbf{p_f} - \mathbf{p_i}}{\Delta t} \tag{8.3}$$

However, the final value of any quantity, minus the initial value of that quantity, is equal to the change of that quantity and is denoted by the delta Δ symbol. Hence,

$$\mathbf{p_f} - \mathbf{p_i} = \Delta\mathbf{p} \tag{8.4}$$

the change in the momentum. Therefore, Newton's second law becomes

$$F = \frac{\Delta p}{\Delta t} \qquad (8.5)$$

Newton's second law in terms of momentum can be stated as: *When a resultant applied force F acts on a body, it causes the linear momentum of that body to change with time.*

The interesting thing we note here is that this is essentially the form in which Newton expressed his second law. Newton did not use the word momentum, however, but rather the expression, "quantity of motion," which is what today would be called momentum. Thus, defining momentum as $p = mv$ is not arbitrary at all. In fact, Newton's second law in terms of the time rate of change of momentum is more basic than the form $F = ma$. In the form $F = ma$, we assume that the mass of the body remains constant. But suppose the mass does not remain constant? As an example, consider an airplane in flight. As it burns fuel its mass decreases with time. At any one instant, Newton's second law in the form $F = ma$, certainly holds and the aircraft's acceleration is

$$a = \frac{F}{m}$$

But only a short time later the mass of the aircraft is no longer m, and therefore the acceleration changes. Another example of a changing mass system is a rocket. Newton's second law in the form $F = ma$ does not properly describe the motion because the mass is constantly changing. Also when objects move at speeds approaching the speed of light, the theory of relativity predicts that the mass of the body does not remain a constant, but rather it increases. In all these variable mass systems, Newton's second law in the form $F = \Delta p / \Delta t$ is still valid, even though $F = ma$ is not.

8.2 The Law of Conservation of Momentum

A very interesting result, and one of extreme importance, is found by considering the behavior of mechanical systems containing two or more particles. Recall from chapter 7 that a system is an aggregate of two or more particles that is treated as an individual unit. Newton's second law, in the form of equation 8.5, can be applied to the entire system if F is the total force acting on the system and p is the total momentum of the system. Forces acting on a system can be divided into two categories: external forces and internal forces. *External forces are forces that originate outside the system and act on the system.* *Internal forces are forces that originate within the system and act on the particles within the system.* The net force acting on and within the system is equal to the sum of the external forces and the internal forces. If the total external force F acting on the system is zero then, since

$$F = \frac{\Delta p}{\Delta t} \qquad (8.5)$$

this implies that

$$\frac{\Delta p}{\Delta t} = 0$$

or

$$\Delta p = 0 \qquad (8.6)$$

But

$$\Delta p = p_f - p_i$$

Figure 8.1

Collision of billiard balls in an example of conservation of momentum.

Therefore,

$$\mathbf{p_f} - \mathbf{p_i} = 0$$

and

$$\mathbf{p_f} = \mathbf{p_i} \qquad (8.7)$$

Equation 8.7 is called the law of conservation of linear momentum. It says that if the total external force acting on a system is equal to zero, then the final value of the total momentum of the system is equal to the initial value of the total momentum of the system. That is, the total momentum is a constant, or as usually stated, the total momentum is conserved.

As an example of the law of conservation of momentum let us consider the head-on collision of two billiard balls. The collision is shown in a stroboscopic picture in figure 8.1 and schematically in figure 8.2. Initially the ball of mass m_1 is moving to the right with an initial velocity $\mathbf{v_{1i}}$, while the second ball of mass m_2 is moving to the left with an initial velocity $\mathbf{v_{2i}}$.

At impact, the two balls collide, and ball 1 exerts a force $\mathbf{F_{21}}$ on ball 2, toward the right. But by Newton's third law, ball 2 exerts an equal but opposite force on ball 1, namely $\mathbf{F_{12}}$. (The notation, $\mathbf{F}_{ij}$, means that this is the force on ball i, caused by ball j.) If the system is defined as consisting of the two balls that are enclosed within the green region of figure 8.2, then the net force on the system of the two balls is equal to the forces on ball 1 plus the forces on ball 2, plus any external forces acting on these balls. The forces $\mathbf{F_{12}}$ and $\mathbf{F_{21}}$ are internal forces in that they act completely within the system.

It is assumed in this problem that there are no external horizontal forces acting on either of the balls. Hence, the net force on the system is

$$\text{Net } \mathbf{F} = \mathbf{F_{12}} + \mathbf{F_{21}}$$

But by Newton's third law

$$\mathbf{F_{21}} = -\mathbf{F_{12}}$$

Therefore, the net force becomes

$$\text{Net } \mathbf{F} = \mathbf{F_{12}} + (-\mathbf{F_{12}}) = 0 \qquad (8.8)$$

That is, the net force acting on the system of the two balls during impact is zero, and equation 8.7, the law of conservation of momentum, must hold. The

Figure 8.2

Example of conservation of momentum.

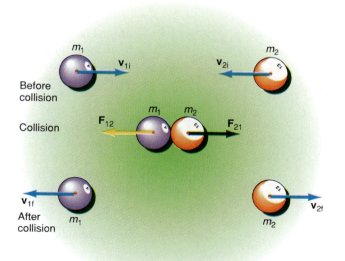

total momentum of the system after the collision must be equal to the total momentum of the system before the collision. Although the momentum of the individual bodies within the system may change, the total momentum will not. After the collision, ball m_1 moves to the left with a final velocity $\mathbf{v_{1f}}$, and ball m_2 moves off to the right with a final velocity $\mathbf{v_{2f}}$.

We will go into more detail on collisions in section 8.5. The important thing to observe here, is what takes place during impact. First, we are no longer considering the motion of a single body, but rather the motion of two bodies. The two bodies are the system. Even though there is a force on ball 1 and ball 2, these forces are internal forces, and the internal forces can not exert a net force on the system, only an external force can do that. *Whenever a system exists without external forces—a system that we call a closed system—the net force on the system is always zero and the law of conservation of momentum always holds.*

The law of conservation of momentum is a consequence of Newton's third law. Recall that because of the third law, all forces in nature exist in pairs; there is no such thing as a single isolated force. Because all internal forces act in pairs, the net force on an isolated system must always be zero, and the system's momentum must always be conserved. Therefore, all systems to which the law of conservation of momentum apply, must consist of at least two bodies and could consist of even millions or more, such as the number of atoms in a gas. If the entire universe is considered as a closed system, then it follows that the total momentum of the universe is also a constant.

The law of conservation of momentum, like the law of conservation of energy, is independent of the type of interaction between the interacting bodies, that is, it applies to colliding billiard balls as well as to gravitational, electrical, magnetic, and other similar interactions. It applies on the atomic and nuclear level as well as on the astronomical level. It even applies in cases where Newtonian mechanics fails. *Like the conservation of energy, the conservation of momentum is one of the fundamental laws of physics.*

8.3 **Examples of the Law of Conservation of Momentum**

Firing a Gun or a Cannon

Let us consider the case of firing a bullet from a gun. The bullet and the gun are the system to be analyzed and they are initially at rest in our frame of reference. We also assume that there are no external forces acting on the system. Because there is no motion of the bullet with respect to the gun at this point, the initial total momentum of the system of bullet and gun $\mathbf{p_i}$ is zero, as shown in figure 8.3(a).

At the moment the trigger of the gun is pulled, a controlled chemical explosion takes place within the gun, figure 8.3(b). A force $\mathbf{F_{BG}}$ is exerted on the bullet by the gun through the gases caused by the exploding gun powder. But by Newton's third law, an equal but opposite force $\mathbf{F_{GB}}$ is exerted on the gun by the bullet. Since there are no external forces, the net force on the system of bullet and gun is

$$\text{Net Force} = \mathbf{F_{BG}} + \mathbf{F_{GB}} \tag{8.9}$$

But by Newton's third law

$$\mathbf{F_{BG}} = -\mathbf{F_{GB}}$$

Therefore, in the absence of external forces, the net force on the system of bullet and gun is equal to zero:

$$\text{Net Force} = \mathbf{F_{BG}} - \mathbf{F_{GB}} = 0 \tag{8.10}$$

Thus, momentum is conserved and

$$\mathbf{p_f} = \mathbf{p_i} \tag{8.11}$$

Figure 8.3

Conservation of momentum in firing a gun.

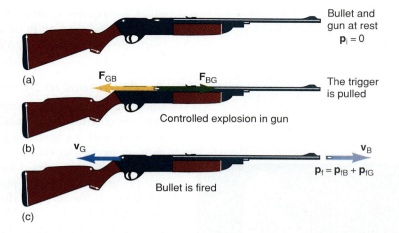

Bullet and gun at rest
$\mathbf{p}_i = 0$

(a)

$\mathbf{F}_{GB}$ $\mathbf{F}_{BG}$

The trigger is pulled

Controlled explosion in gun

(b) $\mathbf{v}_G$

$\mathbf{v}_B$

$\mathbf{p}_f = \mathbf{p}_{fB} + \mathbf{p}_{fG}$

Bullet is fired

(c)

However, because the initial total momentum was zero,

$$\mathbf{p}_i = 0 \qquad (8.12)$$

the total final momentum must also be zero. But because the bullet is moving with a velocity $\mathbf{v}_B$ to the right, and therefore has momentum to the right, the gun must move to the left with the same amount of momentum in order for the final total momentum to be zero, figure 8.3(c). That is, calling $\mathbf{p}_{fB}$ the final momentum of the bullet, and $\mathbf{p}_{fG}$ the final momentum of the gun, the total final momentum is

$$\mathbf{p}_f = \mathbf{p}_{fB} + \mathbf{p}_{fG} = 0$$

$$m_B\mathbf{v}_B + m_G\mathbf{v}_G = 0$$

Solving for the velocity $\mathbf{v}_G$ of the gun, we get

$$\mathbf{v}_G = -\frac{m_B}{m_G}\mathbf{v}_B \qquad (8.13)$$

Because $\mathbf{v}_B$ is the velocity of the bullet to the right, we see that because of the minus sign in equation 8.13, the velocity of the gun must be in the opposite direction, namely to the left. We call $\mathbf{v}_G$ the recoil velocity and its magnitude is

$$v_G = \frac{m_B}{m_G}v_B \qquad (8.14)$$

Even though v_B, the speed of the bullet, is quite large, v_G, the recoil speed of the gun, is relatively small because v_B is multiplied by the ratio of the mass of the bullet m_B to the mass of the gun m_G. Because m_B is relatively small, while m_G is relatively large, the ratio is a small number.

Example 8.1

Recoil of a gun. If the mass of the bullet is 5.00 g, and the mass of the gun is 10.0 kg, and the velocity of the bullet is 300 m/s, find the recoil speed of the gun.

Solution

The recoil speed of the gun, found from equation 8.14, is

$$v_G = \frac{m_B}{m_G}v_B$$

$$= \frac{5.00 \times 10^{-3}\text{ kg}}{10.0\text{ kg}}\,300\text{ m/s}$$

$$= 0.150\text{ m/s} = 15.0\text{ cm/s}$$

Figure 8.4

Recoil of a cannon.

which is relatively small compared to the speed of the bullet. Because it is necessary for this recoil velocity to be relatively small, the mass of the gun must always be relatively large compared to the mass of the bullet.

An Astronaut in Space Throws an Object Away

Consider the case of an astronaut repairing the outside of his spaceship while on an untethered extravehicular activity. While trying to repair the radar antenna he bangs his finger with a wrench. In pain and frustration he throws the wrench away. What happens to the astronaut?

Let us consider the system as an isolated system consisting of the wrench and the astronaut. Let us place a coordinate system, a frame of reference, on the spaceship. In the analysis that follows, we will measure all motion with respect to this reference system. In this frame of reference there is no relative motion of the wrench and the astronaut initially and hence their total initial momentum is zero, as shown in figure 8.5(a).

During the throwing process, the astronaut exerts a force $\mathbf{F}_{wA}$ on the wrench. But by Newton's third law, the wrench exerts an equal but opposite force $\mathbf{F}_{Aw}$ on the astronaut, figure 8.5(b). The net force on this isolated system is therefore zero and the law of conservation of momentum must hold. Thus, the final total momentum must equal the initial total momentum, that is,

$$\mathbf{p}_f = \mathbf{p}_i$$

But initially, $\mathbf{p}_i = 0$ in our frame of reference. Also, the final total momentum is the sum of the final momentum of the wrench and the astronaut, figure 8.5(c). Therefore,

$$\mathbf{p}_f = \mathbf{p}_{fw} + \mathbf{p}_{fA} = 0$$
$$m_w\mathbf{v}_{fw} + m_A\mathbf{v}_{fA} = 0$$

Solving for the final velocity of the astronaut, we get

$$\mathbf{v}_{fA} = \frac{-m_w}{m_A}\mathbf{v}_{fw} \tag{8.15}$$

Thus, as the wrench moves toward the left, the astronaut must recoil toward the right. The magnitude of the final velocity of the astronaut is

$$v_{fA} = \frac{m_w}{m_A}v_{fw} \tag{8.16}$$

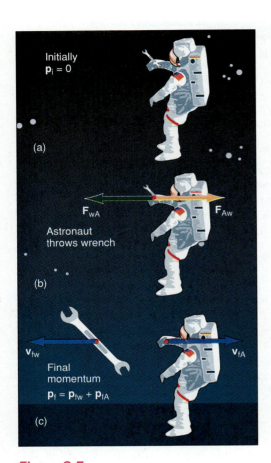

Figure 8.5

Conservation of momentum and an astronaut.

Initially
$p_i = 0$
(a)

F_{wA} F_{Aw}
Astronaut throws wrench
(b)

v_{fw} v_{fA}
Final momentum
$p_f = p_{fw} + p_{iA}$
(c)

Example 8.2

The hazards of being an astronaut. An 80.0-kg astronaut throws a 0.250-kg wrench away at a speed of 3.00 m/s. Find (a) the speed of the astronaut as he recoils away from his space station and (b) how far will he be from the space ship in 1 hr?

Solution

a. The recoil speed of the astronaut, found from equation 8.16, is

$$v_{fA} = \frac{m_w}{m_A} v_{fw}$$

$$= \frac{(0.250 \text{ kg})(3.00 \text{ m/s})}{80.0 \text{ kg}}$$

$$= 9.38 \times 10^{-3} \text{ m/s}$$

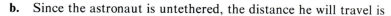

b. Since the astronaut is untethered, the distance he will travel is

$$x_A = v_{fA}t = (9.38 \times 10^{-3} \text{ m/s})(3600 \text{ s})$$
$$= 33.8 \text{ m}$$

The astronaut will have moved a distance of 33.8 m away from his space ship in 1 hr.

A Person on the Surface of the Earth Throws a Rock Away

The result of the previous subsection may at first seem somewhat difficult to believe. An astronaut throws an object away in space and as a consequence of it, the astronaut moves off in the opposite direction. This seems to defy our ordinary experiences, for if a person on the surface of the earth throws an object away, the person does not move backward. What is the difference?

Let an 80.0-kg person throw a 0.250-kg rock away, as shown in figure 8.6. As the person holds the rock, its initial velocity is zero. The person then applies a force to the rock accelerating it from zero velocity to a final velocity v_f. While the rock is leaving the person's hand, the force $\mathbf{F_{Rp}}$ is exerted on the rock by the person. But by Newton's third law, the rock is exerting an equal but opposite force $\mathbf{F_{pR}}$ on the person. But the system that is now being analyzed is not an isolated system, consisting only of the person and the rock. Instead, the system also contains the surface of the earth, because the person is connected to it by friction. The force $\mathbf{F_{pR}}$, acting on the person, is now opposed by the frictional force between the person and the earth and prevents any motion of the person.

As an example, let us assume that in throwing the rock the person's hand moves through a distance x of 1.00 m, as shown in figure 8.6(a), and it leaves the person's hand at a velocity of 3.00 m/s. The acceleration of the rock can be found from the kinematic equation

$$v^2 = v_0^2 + 2a_R x$$

by solving for a_R. Thus,

$$a_R = \frac{v^2}{2x} = \frac{(3.00 \text{ m/s})^2}{2(1.00 \text{ m})} = 4.50 \text{ m/s}^2$$

The force acting on the rock F_{Rp}, found by Newton's second law, is

$$F_{Rp} = m_R a_R = (0.250 \text{ kg})(4.50 \text{ m/s}^2)$$
$$= 1.13 \text{ N}$$

But by Newton's third law this must also be the force exerted on the person by the rock, F_{pR}. That is, there is a force of 1.13 N acting on the person, tending to push that person to the left. But since the person is standing on the surface of the earth there is a frictional force that tends to oppose that motion and is shown in figure 8.6(b). The maximum value of that frictional force is

$$f_s = \mu_s F_N = \mu_s w_p$$

The weight of the person w_p is

$$w_p = mg = (80.0 \text{ kg})(9.80 \text{ m/s}^2) = 784 \text{ N}$$

Assuming a reasonable value of $\mu_s = 0.500$ (leather on wood), we have

$$f_s = \mu_s w_p = (0.500)(784 \text{ N})$$
$$= 392 \text{ N}$$

That is, before the person will recoil from the process of throwing the rock, the recoil force F_{pR}, acting on the person, must be greater than the maximum frictional force of 392 N. We found the actual reaction force on the person to be only

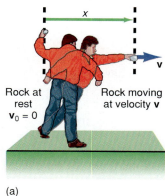

Rock at rest $v_0 = 0$ Rock moving at velocity **v**

(a)

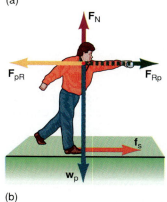

(b)

Figure 8.6

A person throwing a rock on the surface of the earth.

1.13 N, which is no where near the amount necessary to overcome friction. Hence, when a person on the surface of the earth throws an object, the person does not recoil like an astronaut in space.

If friction could be minimized, then the throwing of the object would result in a recoil velocity. For example, if a person threw a rock to the right, while standing in a boat on water, then because the frictional force between the boat and the water is relatively small, the person and the boat would recoil to the left.

In a similar way, if a person is standing at the back of a boat, which is at rest, and then walks toward the front of the boat, the boat will recoil backward to compensate for his forward momentum.

8.4 Impulse

Let us consider Newton's second law in the form of change in momentum as found in equation 8.5,

$$\mathbf{F} = \frac{\Delta \mathbf{p}}{\Delta t}$$

If both sides of equation 8.5 are multiplied by Δt, we have

$$\mathbf{F}\Delta t = \Delta \mathbf{p} \tag{8.17}$$

The quantity $\mathbf{F}\Delta t$, is called the **impulse**[1] of the force and is given by

$$\mathbf{J} = \mathbf{F}\Delta t \tag{8.18}$$

The impulse $\mathbf{J}$ is a measure of the force that is acting, times the time that force is acting. Equation 8.17 then becomes

$$\mathbf{J} = \Delta \mathbf{p} \tag{8.19}$$

That is, the impulse acting on a body changes the momentum of that body. Since $\Delta \mathbf{p} = \mathbf{p_f} - \mathbf{p_i}$, equation 8.19 also can be written as

$$\mathbf{J} = \mathbf{p_f} - \mathbf{p_i} \tag{8.20}$$

In many cases, the force $\mathbf{F}$ that is exerted is not a constant during the collision process. In that case an average force $\mathbf{F_{avg}}$ can be used in the computation of the impulse. That is,

$$\mathbf{F_{avg}}\Delta t = \Delta \mathbf{p} \tag{8.21}$$

Examples of the use of the concept of impulse can be found in such sports as baseball, golf, tennis, and the like, see figure 8.7. If you participated in such sports, you were most likely told that the "follow through" is extremely important. For example, consider the process of hitting a golf ball. The ball is initially at rest on the tee. As the club hits the ball, the club exerts an average force $\mathbf{F_{avg}}$ on the ball. By "following through" with the golf club, as shown in figure 8.8, we mean that the longer the time interval Δt that the club is exerting its force on the ball, the greater is the impulse imparted to the ball and hence the greater will be the change in momentum of the ball. The greater change in momentum implies a greater change in the velocity of the ball and hence the ball will travel a greater distance.

The principle is the same in baseball, tennis, and other similar sports. The better the follow through, the longer the bat or racket is in contact with the ball and the greater the change in momentum the ball will have. Those interested in the application of physics to sports can read the excellent book, *Sport Science* by Peter Brancazio (Simon and Schuster, 1984).

(a)

(b)

Figure 8.7

Physics in sports. When hitting (a) a baseball or (b) a tennis ball, the "follow-through" is very important.

---

1. In some books the letter I is used to denote the impulse. In order to not confuse it with the moment of inertia of a body, also designated by the letter I and treated in detail in chapter 9, we will use the letter J for impulse.

Mechanics

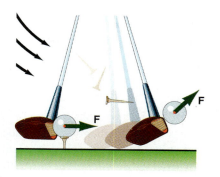

(a)

(b)

Figure 8.8

The effect of ''follow through'' in hitting a golf ball.

Figure 8.9

A perfectly elastic collision.

8.5 Collisions in One Dimension

We saw in section 8.2 that momentum is always conserved in a collision if the net external force on the system is zero. In physics three different kinds of collisions are usually studied. Momentum is conserved in all of them, but kinetic energy is conserved in only one. These different types of collisions are

1. *A perfectly elastic collision*—a collision in which no kinetic energy is lost, that is, kinetic energy is conserved.
2. *An inelastic collision*—a collision in which some kinetic energy is lost. All real collisions belong to this category.
3. *A perfectly inelastic collision*—a collision in which the two objects stick together during the collision. A great deal of kinetic energy is usually lost in this collision.

In all real collisions in the macroscopic world, some kinetic energy is lost. As an example, consider a collision between two billiard balls. As the balls collide they are temporarily deformed. Some of the kinetic energy of the balls goes into the potential energy of deformation. Ideally, as each ball returns to its original shape, all the potential energy stored by the ball is converted back into the kinetic energy of the ball. In reality, some kinetic energy is lost in the form of heat and sound during the deformation process. The mere fact that we can hear the collision indicates that some of the mechanical energy has been transformed into sound energy. But in many cases, the amount of kinetic energy that is lost is so small that, as a first approximation, it can be neglected. For such cases we assume that no energy is lost during the collision, and the collision is treated as a perfectly elastic collision. The reason why we like to solve perfectly elastic collisions is simply that they are much easier to analyze than inelastic collisions.

Perfectly Elastic Collisions

Between Unequal Masses Consider the collision shown in figure 8.9 between two different masses, m_1 and m_2, having initial velocities v_{1i} and v_{2i}, respectively. We assume that v_{1i} is greater than v_{2i}, so that a collision will occur. We can write the law of conservation of momentum as

$$\mathbf{p_i} = \mathbf{p_f}$$

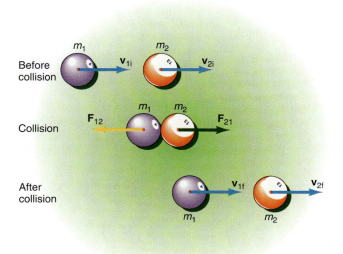

That is,

$$\text{Total momentum before collision} = \text{Total momentum after collision}$$

$$\mathbf{p}_{1i} + \mathbf{p}_{2i} = \mathbf{p}_{1f} + \mathbf{p}_{2f}$$

or

$$m_1\mathbf{v}_{1i} + m_2\mathbf{v}_{2i} = m_1\mathbf{v}_{1f} + m_2\mathbf{v}_{2f} \tag{8.22}$$

where the subscript i stands for the initial values of the momentum and velocity (before the collision) while f stands for the final values (after the collision). This is a vector equation. *If the collision is in one dimension only, and motion to the right is considered positive, then we can rewrite equation 8.22 as the scalar equation*

$$m_1 v_{1i} + m_2 v_{2i} = m_1 v_{1f} + m_2 v_{2f} \tag{8.23}$$

Usually we know v_{1i} and v_{2i} and need to find v_{1f} and v_{2f}. In order to solve for these final velocities, we need another equation.

The second equation comes from the law of conservation of energy. Since the collision occurs on a flat surface, which we take as our reference level and assign the height zero, there is no change in potential energy to consider during the collision. Thus, we need only consider the conservation of kinetic energy. The law of conservation of energy, therefore, becomes

$$KE_{BC} = KE_{AC} \tag{8.24}$$

That is,

$$\text{Kinetic energy before collision} = \text{Kinetic energy after collision} \tag{8.25}$$

which becomes

$$\tfrac{1}{2}m_1 v_{1i}^2 + \tfrac{1}{2}m_2 v_{2i}^2 = \tfrac{1}{2}m_1 v_{1f}^2 + \tfrac{1}{2}m_2 v_{2f}^2 \tag{8.26}$$

If the initial values of the speed of the two bodies are known, then we find the final values of the speed by solving equations 8.23 and 8.26 simultaneously. The algebra involved can be quite messy for a direct simultaneous solution. (A simplified solution is given below. However, even the simplified solution is a little long. Those students not interested in the derivation can skip directly to the solution in equation 8.30.)

To simplify the solution, we rewrite equation 8.23, the conservation of momentum, in the form

$$m_1(v_{1i} - v_{1f}) = m_2(v_{2f} - v_{2i}) \tag{8.27}$$

where the masses have been factored out. Similarly, we factor the masses out in equation 8.26, the conservation of energy, and rewrite it in the form

$$m_1(v_{1i}^2 - v_{1f}^2) = m_2(v_{2f}^2 - v_{2i}^2) \tag{8.28}$$

We divide equation 8.28 by equation 8.27 to eliminate the mass terms:

$$\frac{m_1(v_{1i}^2 - v_{1f}^2)}{m_1(v_{1i} - v_{1f})} = \frac{m_2(v_{2f}^2 - v_{2i}^2)}{m_2(v_{2f} - v_{2i})}$$

Note that we can rewrite the numerators as products of factors:

$$\frac{(v_{1i} + v_{1f})(v_{1i} - v_{1f})}{v_{1i} - v_{1f}} = \frac{(v_{2i} + v_{2f})(v_{2f} - v_{2i})}{v_{2f} - v_{2i}}$$

which simplifies to

$$v_{1i} + v_{1f} = v_{2i} + v_{2f} \tag{8.29}$$

Solving for v_{2f} in equation 8.29, we get

$$v_{2f} = v_{1i} + v_{1f} - v_{2i}$$

Substituting this into equation 8.27, we have

$$m_1(v_{1i} - v_{1f}) = m_2[(v_{1i} + v_{1f} - v_{2i}) - v_{2i}]$$

$$m_1 v_{1i} - m_1 v_{1f} = m_2 v_{1i} + m_2 v_{1f} - m_2 v_{2i} - m_2 v_{2i}$$

Collecting terms of v_{1f}, we have

$$-m_1 v_{1f} - m_2 v_{1f} = -2m_2 v_{2i} + m_2 v_{1i} - m_1 v_{1i}$$

Multiplying both sides of the equation by -1, we get

$$+m_1 v_{1f} + m_2 v_{1f} = +2m_2 v_{2i} - m_2 v_{1i} + m_1 v_{1i}$$

Simplifying,

$$(m_1 + m_2)v_{1f} = (m_1 - m_2)v_{1i} + 2m_2 v_{2i}$$

Solving for the final speed of ball 1, we have

$$v_{1f} = \left(\frac{m_1 - m_2}{m_1 + m_2}\right)v_{1i} + \left(\frac{2m_2}{m_1 + m_2}\right)v_{2i} \qquad (8.30)$$

In a similar way, we can solve equation 8.29 for v_{1f}, which we then substitute into equation 8.27. After the same algebraic treatment (which is left as an exercise), the final speed of the second ball becomes

$$v_{2f} = \left(\frac{2\,m_1}{m_1 + m_2}\right)v_{1i} - \left(\frac{m_1 - m_2}{m_1 + m_2}\right)v_{2i} \qquad (8.31)$$

Equations 8.30 and 8.31 were derived on the assumption that balls 1 and 2 were originally moving with a positive velocity to the right before the collision, and both balls had a positive velocity to the right after the collision. If v_{1f} comes out to be a negative number, ball 1 will have a negative velocity after the collision and will rebound to the left.

If the collision looks like the one depicted in figure 8.2, we can still use equations 8.30 and 8.31. However, ball 2 will be moving to the left, initially, and will thus have a negative velocity v_{2i}. This means that v_{2i} has to be a negative number when placed in these equations. If v_{1f} comes out to be a negative number in the calculations, that means that ball 1 has a negative final velocity and will be moving to the left.

<hr>

Example 8.3

Perfectly elastic collision, ball 1 catches up with ball 2. Consider the perfectly elastic collision between masses $m_1 = 100$ g and $m_2 = 200$ g. Ball 1 is moving with a velocity v_{1i} of 30.0 cm/s to the right, and ball 2 has a velocity $v_{2i} = 20.0$ cm/s, also to the right, as shown in figure 8.9. Find the final velocities of the two balls.

Solution

The final velocity of the first ball, found from equation 8.30, is

$$v_{1f} = \left(\frac{m_1 - m_2}{m_1 + m_2}\right)v_{1i} + \left(\frac{2m_2}{m_1 + m_2}\right)v_{2i}$$

$$= \left(\frac{100\text{ g} - 200\text{ g}}{100\text{ g} + 200\text{ g}}\right)(30.0\text{ cm/s}) + \frac{2(200\text{ g})}{100\text{ g} + 200\text{ g}}(20.0\text{ cm/s})$$

$$= 16.7\text{ cm/s}$$

Since v_{1f} is a positive quantity, the final velocity of ball 1 is toward the right. The final velocity of the second ball, obtained from equation 8.31, is

$$v_{2f} = \left(\frac{2m_1}{m_1 + m_2}\right) v_{1i} - \left(\frac{m_1 - m_2}{m_1 + m_2}\right) v_{2i}$$

$$= \left[\frac{2(100 \text{ g})}{100 \text{ g} + 200 \text{ g}}\right](30.0 \text{ cm/s}) - \left(\frac{100 \text{ g} - 200 \text{ g}}{100 \text{ g} + 200 \text{ g}}\right)(20.0 \text{ cm/s})$$

$$= 26.7 \text{ cm/s}$$

Since v_{2f} is a positive quantity, the second ball has a positive velocity and is moving toward the right.

Example 8.4

Perfectly elastic collision with masses approaching each other. Consider the perfectly elastic collision between masses $m_1 = 100$ g, $m_2 = 200$ g, with velocity $v_{1i} = 20.0$ cm/s to the right, and velocity $v_{2i} = -30.0$ cm/s to the left, as shown in figure 8.2. Find the final velocities of the two balls.

Solution

The final velocity of ball 1, found from equation 8.30, is

$$v_{1f} = \left(\frac{m_1 - m_2}{m_1 + m_2}\right) v_{1i} + \left(\frac{2 m_2}{m_1 + m_2}\right) v_{2i}$$

$$= \left(\frac{100 \text{ g} - 200 \text{ g}}{100 \text{ g} + 200 \text{ g}}\right)(20.0 \text{ cm/s}) + \left[\frac{2(200 \text{ g})}{100 \text{ g} + 200 \text{ g}}\right](-30.0 \text{ cm/s})$$

$$= -46.7 \text{ cm/s}$$

Since v_{1f} is a negative quantity, the final velocity of the first ball is negative, indicating that the first ball moves to the left after the collision. The final velocity of the second ball, found from equation 8.31, is

$$v_{2f} = \left(\frac{2 m_1}{m_1 + m_2}\right) v_{1i} - \left(\frac{m_1 - m_2}{m_1 + m_2}\right) v_{2i}$$

$$= \left[\frac{2(100 \text{ g})}{100 \text{ g} + 200 \text{ g}}\right](20.0 \text{ cm/s}) - \left(\frac{100 \text{ g} - 200 \text{ g}}{100 \text{ g} + 200 \text{ g}}\right)(-30.0 \text{ cm/s})$$

$$= 3.33 \text{ cm/s}$$

Since v_{2f} is a positive quantity, the final velocity of ball 2 is positive, and the ball will move toward the right.

Let us now look at a few special types of collisions.

Between Equal Masses If the elastic collision occurs between two equal masses, then the final velocities after the collision are again given by equations 8.30 and 8.31, only with mass m_1 set equal to m_2. That is,

$$v_{1f} = \left(\frac{m_2 - m_2}{m_2 + m_2}\right) v_{1i} + \left(\frac{2m_2}{m_2 + m_2}\right) v_{2i}$$

$$= 0 + \frac{2m_2}{2m_2} v_{2i}$$

$$v_{1f} = v_{2i} \tag{8.32}$$

and

$$v_{2f} = \left(\frac{2 m_2}{m_2 + m_2}\right) v_{1i} - \left(\frac{m_2 - m_2}{m_2 + m_2}\right) v_{2i}$$

$$= \frac{2m_2}{2m_2} v_{1i} + 0$$

$$v_{2f} = v_{1i} \qquad (8.33)$$

Equations 8.32 and 8.33 tell us that the bodies exchange their velocities during the collision.

Both Masses Equal, One Initially at Rest This is the same case, except that one mass is initially at rest, that is, $v_{2i} = 0$. From equation 8.32 we get

$$v_{1f} = v_{2i} = 0 \qquad (8.34)$$

while equation 8.33 remains the same

$$v_{2f} = v_{1i}$$

as before. This is an example of the first body being "stopped cold" while the second one "takes off" with the original velocity of the first ball.

A Ball Thrown against a Wall When you throw a ball against a wall, figure 8.10, you have another example of a collision. Assuming the collision to be elastic, equations 8.30 and 8.31 apply. The wall is initially at rest, so $v_{2i} = 0$. Because the wall is very massive compared to the ball we can say that

$$m_2 \gg m_1$$

which implies that

$$m_1 - m_2 \approx -m_2$$

and

$$m_1 + m_2 \approx m_2$$

Solving equation 8.30 for v_{1f}, we have

$$v_{1f} = \left(\frac{m_1 - m_2}{m_1 + m_2}\right) v_{1i} + \left(\frac{2m_2}{m_1 + m_2}\right) v_{2i}$$

$$= -\frac{m_2}{m_2} v_{1i} + 0$$

Therefore, the final velocity of the ball is

$$v_{1f} = -v_{1i} \qquad (8.35)$$

The negative sign indicates that the final velocity of the ball is negative, so the ball rebounds from the wall and is now moving toward the left with the original speed. The velocity of the wall, found from equation 8.31, is

$$v_{2f} = \left(\frac{2m_1}{m_1 + m_2}\right) v_{1i} - \left(\frac{m_1 - m_2}{m_1 + m_2}\right) v_{2i}$$

$$= \frac{2m_1}{m_2} v_{1i} + 0$$

However, since

$$m_2 \gg m_1$$

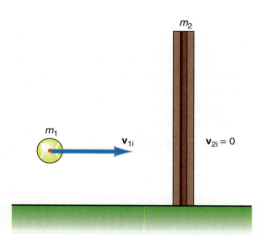

Figure 8.10

A ball bouncing off a wall.

then

$$\frac{m_1}{m_2} \approx 0$$

Therefore,

$$v_{2f} = 0 \qquad\qquad (8.36)$$

The ball rebounds from the wall with the same speed that it hit the wall, and the wall, because it is so massive, remains at rest.

Inelastic Collisions

Let us consider for a moment equation 8.29, which we developed earlier in the section, namely

$$v_{1i} + v_{1f} = v_{2f} + v_{2i}$$

If we rearrange this equation by placing all the initial velocities on one side of the equation and all the final velocities on the other, we have

$$v_{1i} - v_{2i} = v_{2f} - v_{1f} \qquad\qquad (8.37)$$

However, as we can observe from figure 8.9,

$$v_{1i} - v_{2i} = V_A \qquad\qquad (8.38)$$

that is, the difference in the velocities of the two balls is equal to the *velocity of approach* V_A of the two billiard balls. (The velocity of approach is also called the *relative velocity* between the two balls.) As an example, if ball 1 is moving to the right initially at 10.00 cm/s and ball 2 is moving to the right initially at 5.00 cm/s, then the velocity at which they approach each other is

$$V_A = v_{1i} - v_{2i} = 10.00 \text{ cm/s} - 5.00 \text{ cm/s}$$
$$= 5.00 \text{ cm/s}$$

Similarly,

$$v_{2f} - v_{1f} = V_S \qquad\qquad (8.39)$$

is the velocity at which the two balls separate. That is, if the final velocity of ball 1 is toward the left at the velocity $v_{1f} = -10.0$ cm/s, and ball 2 is moving to the right at the velocity $v_{2f} = 5.00$ cm/s, then the velocity at which they move away from each other, the *velocity of separation,* is

$$V_S = v_{2f} - v_{1f} = 5.00 \text{ cm/s} - (-10.0 \text{ cm/s})$$
$$= 15.0 \text{ cm/s}$$

Therefore, we can write equation 8.37 as

$$V_A = V_S \qquad\qquad (8.40)$$

That is, *in a perfectly elastic collision, the velocity of approach of the two bodies is equal to the velocity of separation.*

In an inelastic collision, the velocity of separation is not equal to the velocity of approach, and a new parameter, the **coefficient of restitution,** is defined as a measure of the inelastic collision. That is, we define the coefficient of restitution e as

$$e = \frac{V_S}{V_A} \qquad\qquad (8.41)$$

and the velocity of separation becomes

$$V_S = eV_A \qquad\qquad (8.42)$$

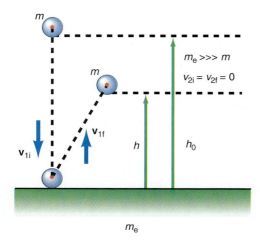

(a)

(b)

Figure 8.11

Imperfectly elastic collision of a ball with the earth.

For a perfectly elastic collision $e = 1$. For a perfectly inelastic collision $e = 0$, which implies $V_S = 0$. Thus, the objects stick together and do not separate at all. *For the inelastic collision*

$$0 < e < 1 \qquad (8.43)$$

Determination of the Coefficient of Restitution If the inelastic collision is between a ball and the earth, as shown in figure 8.11, then, because the earth is so massive, $v_{2i} = v_{2f} = 0$. Equation 8.42 reduces to

$$v_{1f} = e v_{1i} \qquad (8.44)$$

The ball attained its speed v_{1i} by falling from the height h_0, where it had the potential energy

$$PE_0 = mgh_0$$

Immediately before impact its kinetic energy is

$$KE_i = \tfrac{1}{2} m v_{1i}^2$$

And, by the law of the conservation of energy,

$$KE_i = PE_0$$

or

$$\tfrac{1}{2} m v_{1i}^2 = mgh_0$$

Thus, the initial speed before impact with the earth is

$$v_{1i} = \sqrt{2gh_0} \qquad (8.45)$$

After impact, the ball rebounds with a speed v_{1f}, and has a kinetic energy of

$$KE_f = \tfrac{1}{2} m v_{1f}^2$$

which will be less than KE_i because some energy is lost in the collision. After the collision the ball rises to a new height h, as seen in the figure. The final potential energy of the ball is

$$PE_f = mgh$$

However, by the law of conservation of energy

$$KE_f = PE_f$$
$$\tfrac{1}{2} m v_{1f}^2 = mgh$$

Hence, the final speed after the collision is

$$v_{1f} = \sqrt{2gh} \qquad (8.46)$$

We can now find the coefficient of restitution from equations 8.44, 8.45, and 8.46, as

$$e = \frac{v_{1f}}{v_{1i}} = \frac{\sqrt{2gh}}{\sqrt{2gh_0}} = \sqrt{\frac{h}{h_0}} \qquad (8.47)$$

Thus, by measuring the final and initial heights of the ball and taking their ratio, we can find the coefficient of restitution.

The loss of energy in an inelastic collision can easily be found using equation 8.42,

$$V_S = e V_A$$

The kinetic energy after separation is

$$KE_S = \tfrac{1}{2} m V_S^2 \qquad (8.48)$$

Substituting for V_S from equation 8.42 gives,

$$KE_S = \tfrac{1}{2}m(eV_A)^2$$

$$KE_S = \tfrac{1}{2}me^2V_A^2$$

$$KE_S = e^2(\tfrac{1}{2}mV_A^2)$$

But $\tfrac{1}{2}mV_A^2$ is the kinetic energy of approach. Therefore the relation between the kinetic energy after separation and the initial kinetic energy is given by

$$KE_S = e^2KE_A \tag{8.49}$$

The total amount of energy lost in the collision can now be found as

$$\Delta E_{lost} = KE_A - KE_S$$

$$= KE_A - e^2KE_A \tag{8.50}$$

$$\Delta E_{lost} = (1 - e^2)KE_A \tag{8.51}$$

Example 8.5

An imperfectly elastic collision. A 20.0-g racquet ball is dropped from a height of 1.00 m and impacts a tile floor. If the ball rebounds to a height of 76.0 cm, (a) what is the coefficient of restitution, (b) what percentage of the initial energy is lost in the collision, and (c) what is the actual energy lost in the collision?

Solution

a. The coefficient of restitution, found from equation 8.47, is

$$e = \sqrt{\frac{h}{h_0}} = \sqrt{\frac{76.0 \text{ cm}}{100 \text{ cm}}} = 0.872$$

b. The percentage energy lost, found from equation 8.51, is

$$\Delta E_{lost} = (1 - e^2)KE_A$$

$$= (1 - (0.872)^2)KE_A$$

$$= 0.240 \text{ KE}_A$$

$$= 24.0\% \text{ of the initial KE}$$

c. The actual energy lost in the collision with the floor is

$$\Delta E = PE_0 - PE_f$$

$$= mgh_0 - mgh$$

$$= (0.020 \text{ kg})(9.80 \text{ m/s}^2)(1.00 \text{ m}) - (0.020 \text{ kg})(9.80 \text{ m/s}^2)(0.76 \text{ m})$$

$$= 0.047 \text{ J lost}$$

Perfectly Inelastic Collision

Between Unequal Masses In the perfectly inelastic collision, figure 8.12, the two bodies join together during the collision process and move off together as one body after the collision. We assume that v_{1i} is greater than v_{2i}, so a collision will occur. The law of conservation of momentum, when applied to figure 8.12, becomes

$$m_1\mathbf{v}_{1i} + m_2\mathbf{v}_{2i} = (m_1 + m_2)\mathbf{V}_f \tag{8.52}$$

Taking motion to the right as positive, we write this in the scalar form,

$$m_1v_{1i} + m_2v_{2i} = (m_1 + m_2)V_f \tag{8.53}$$

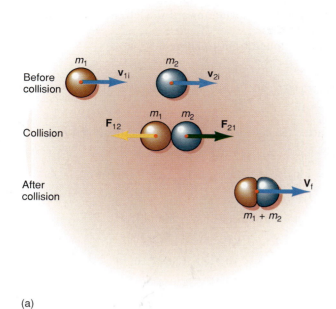

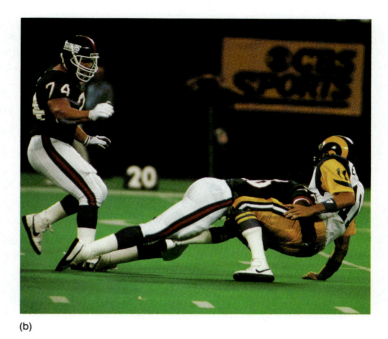

(a)

(b)

Figure 8.12

(a) Perfectly inelastic collision. (b) A football player being tackled is also an example of a perfectly inelastic collision.

Solving for the final speed V_f of the combined masses, we get

$$V_f = \left(\frac{m_1}{m_1 + m_2}\right)v_{1i} + \left(\frac{m_2}{m_1 + m_2}\right)v_{2i} \qquad (8.54)$$

It is interesting to determine the initial and final values of the kinetic energy of the colliding bodies.

$$\text{KE}_i = \tfrac{1}{2}m_1 v_{1i}^2 + \tfrac{1}{2}m_2 v_{2i}^2 \qquad (8.55)$$

$$\text{KE}_f = \tfrac{1}{2}(m_1 + m_2)V_f^2 \qquad (8.56)$$

Is kinetic energy conserved for this collision?

Example 8.6

A perfectly inelastic collision. A 50.0-g piece of clay moving at a velocity of 5.00 cm/s to the right has a head-on collision with a 100-g piece of clay moving at a velocity of -10.0 cm/s to the left. The two pieces of clay stick together during the impact. Find (a) the final velocity of the clay, (b) the initial kinetic energy, (c) the final kinetic energy, and (d) the amount of energy lost in the collision.

Solution

a. The initial velocity of the first piece of clay is positive, because it is in motion toward the right. The initial velocity of the second piece of clay is negative, because it is in motion toward the left. The final velocity of the clay, given by equation 8.54, is

$$V_f = \left(\frac{m_1}{m_1 + m_2}\right)v_{1i} + \left(\frac{m_2}{m_1 + m_2}\right)v_{2i}$$

$$= \frac{(50.0 \text{ g})(5.00 \text{ cm/s})}{50 \text{ g} + 100 \text{ g}} + \frac{(100 \text{ g})(-10.0 \text{ cm/s})}{50 \text{ g} + 100 \text{ g}}$$

$$= -5.00 \text{ cm/s} = -5.00 \times 10^{-2} \text{ m/s}$$

The minus sign means that the velocity of the combined pieces of clay is negative and they are therefore moving toward the left, not toward the right as we assumed in figure 8.12.

b. The initial kinetic energy, found from equation 8.55, is

$$KE_i = \tfrac{1}{2}m_1v_{1i}^2 + \tfrac{1}{2}m_2v_{2i}^2$$
$$= \tfrac{1}{2}(0.050 \text{ kg})(5.00 \times 10^{-2} \text{ m/s})^2 + \tfrac{1}{2}(0.100 \text{ kg})(-10.0 \times 10^{-2} \text{ m/s})^2$$
$$= 5.63 \times 10^{-4} \text{ J}$$

c. The kinetic energy after the collision, found from equation 8.56, is

$$KE_f = \tfrac{1}{2}(m_1 + m_2)V_f^2$$
$$= \tfrac{1}{2}(0.050 \text{ kg} + 0.100 \text{ kg})(-5.00 \times 10^{-2} \text{ m/s})^2$$
$$= 1.88 \times 10^{-4} \text{ J}$$

d. The mechanical energy lost in the collision is found from

$$\Delta E = KE_i - KE_f$$
$$= 5.63 \times 10^{-4} \text{ J} - 1.88 \times 10^{-4} \text{ J}$$
$$= 3.75 \times 10^{-4} \text{ J}$$

Hence, 3.75×10^{-4} J of energy are lost in the deformation caused by the collision.

†8.6 Collisions in Two Dimensions— Glancing Collisions

In the collisions treated so far, the collisions were head-on collisions, and the forces exerted on the two colliding bodies were on a line in the direction of motion of the two bodies. As an example, consider the collision to be between two billiard balls. For a head-on collision, as in figure 8.13(a), the force on ball 2 caused by ball 1,

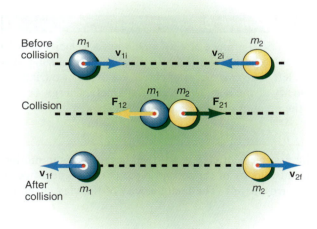

(a) One-dimensional collision (on centers)

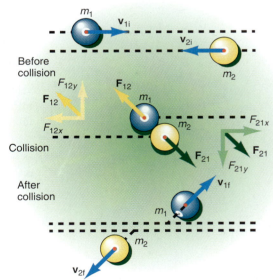

(b) Two-dimensional collision (off centers)

Figure 8.13

Comparison of one-dimensional and two-dimensional collisions.

Mechanics

$\mathbf{F}_{21}$, is in the positive x-direction, while $\mathbf{F}_{12}$, the force on ball 1 caused by ball 2, is in the negative x-direction. After the collision, the two balls move along the original line of action. In a glancing collision, on the other hand, the motion of the centers of mass of each of the two balls do not lie along the same line of action, figure 8.13(b). Hence, when the balls collide, the force exerted on each ball does not lie along the original line of action but is instead a force that is exerted along the line connecting the center of mass of each ball, as shown in the diagram. Thus the force on ball 2 caused by ball 1, $\mathbf{F}_{21}$, is a two-dimensional vector, and so is $\mathbf{F}_{12}$, the force on ball 1 caused by ball 2. As we can see in the diagram, these forces can be decomposed into x- and y-components. Hence, a y-component of force has been exerted on each ball causing it to move out of its original direction of motion. Therefore, after the collision, the two balls move off in the directions indicated. All glancing collisions must be treated as two-dimensional problems. Since the general solution of the two-dimensional collision problem is even more complicated than the one-dimensional problem solved in the last section, we will solve only some special cases of the two-dimensional problem.

Consider the glancing collision between two billiard balls shown in figure 8.14. Ball 1 is moving to the right at the velocity $\mathbf{v}_{1i}$ and ball 2 is at rest ($\mathbf{v}_{2i} = 0$). After the collision, ball 1 is found to be moving at an angle $\theta = 45.0°$ above the horizontal and ball 2 is moving at an angle $\phi = 45.0°$ below the horizontal. Let us find the velocities of both balls after the collision. As in all collisions, the law of conservation of momentum holds, that is,

$$\mathbf{p}_f = \mathbf{p}_i$$

$$m_1\mathbf{v}_{1f} + m_2\mathbf{v}_{2f} = m_1\mathbf{v}_{1i}$$

The last single vector equation is equivalent to the two scalar equations

$$m_1 v_{1f} \cos \theta + m_2 v_{2f} \cos \phi = m_1 v_{1i} \tag{8.57}$$

$$m_1 v_{1f} \sin \theta - m_2 v_{2f} \sin \phi = 0 \tag{8.58}$$

Solving equation 8.58 for v_{2f} with $\theta = \phi = 45.0°$, we get

$$m_1 v_{1f} \sin 45.0° = m_2 v_{2f} \sin 45.0°$$

$$v_{2f} = \frac{m_1}{m_2} v_{1f} \tag{8.59}$$

Figure 8.14

A glancing collision.

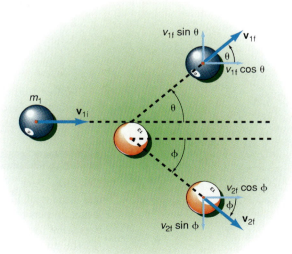

Inserting equation 8.59 into equation 8.57 we can solve for v_{1f} as

$$m_1 v_{1f} \cos 45.0° + m_2 \left(\frac{m_1}{m_2} v_{1f} \right) \cos 45.0° = m_1 v_{1i}$$

$$2m_1 v_{1f} \cos 45.0° = m_1 v_{1i}$$

$$v_{1f} = \frac{v_{1i}}{2 \cos 45.0°} \qquad \textbf{(8.60)}$$

Example 8.7

A glancing collision. Billiard ball 1 is moving at a speed of $v_{1i} = 10.0$ cm/s, when it has a glancing collision with an identical billiard ball that is at rest. After the collision, $\theta = \phi = 45.0°$. The mass of the billiard ball is 0.170 kg. (a) Find the speed of ball 1 and 2 after the collision. (b) Is energy conserved in this collision?

Solution

a. The speed of ball 1, found from equation 8.60, is

$$v_{1f} = \frac{v_{1i}}{2 \cos 45.0°}$$

$$= \frac{10.0 \text{ cm/s}}{2 \cos 45.0°}$$

$$= 7.07 \text{ cm/s}$$

and the speed of ball 2, found from equation 8.59, is

$$v_{2f} = \frac{m_1}{m_2} v_{1f}$$

$$= \frac{m_1}{m_1} v_{1f}$$

$$= v_{1f} = 7.07 \text{ cm/s}$$

b. The kinetic energy before the collision is

$$\text{KE}_i = \tfrac{1}{2} m_1 v_{1i}^2 = \tfrac{1}{2}(0.170 \text{ kg})(0.100 \text{ m/s})^2$$

$$= 8.50 \times 10^{-4} \text{ J}$$

while the kinetic energy after the collision is

$$\text{KE}_f = \tfrac{1}{2} m_1 v_{1f}^2 + \tfrac{1}{2} m_2 v_{2f}^2$$

$$= \tfrac{1}{2}(0.170 \text{ kg})(0.0707 \text{ m/s})^2 + \tfrac{1}{2}(0.170 \text{ kg})(0.0707 \text{ m/s})^2$$

$$= 8.50 \times 10^{-4} \text{ J}$$

Notice that the kinetic energy after the collision is equal to the kinetic energy before the collision. Therefore the collision is perfectly elastic.

Example 8.8

Colliding cars. Two cars collide at an intersection as shown in figure 8.15. Car 1 has a mass of 1200 kg and is moving at a velocity of 95.0 km/hr due east and car 2 has a mass of 1400 kg and is moving at a velocity of 100 km/hr due north. The cars stick together and move off as one at an angle θ as shown in the diagram. Find (a) the angle θ and (b) the final velocity of the combined cars.

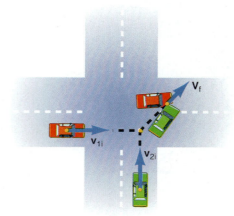

Figure 8.15

A perfectly inelastic glancing collision.

a. This is an example of a perfectly inelastic collision in two dimensions. The law of conservation of momentum yields

$$\mathbf{p_f} = \mathbf{p_i}$$

$$(m_1 + m_2)\mathbf{V_f} = m_1\mathbf{v_{1i}} + m_2\mathbf{v_{2i}} \tag{8.61}$$

Resolving this equation into its x- and y-component equations, we get for the x-component:

$$(m_1 + m_2)V_f \cos\theta = m_1 v_{1i} \tag{8.62}$$

and for the y-component:

$$(m_1 + m_2)V_f \sin\theta = m_2 v_{2i} \tag{8.63}$$

Dividing the y-component equation by the x-component equation we get

$$\frac{(m_1 + m_2)V_f \sin\theta}{(m_1 + m_2)V_f \cos\theta} = \frac{m_2 v_{2i}}{m_1 v_{1i}}$$

$$\frac{\sin\theta}{\cos\theta} = \frac{m_2 v_{2i}}{m_1 v_{1i}}$$

$$\tan\theta = \frac{m_2 v_{2i}}{m_1 v_{1i}}$$

$$\tan\theta = \frac{(1400 \text{ kg})(100 \text{ km/hr})}{(1200 \text{ kg})(95.0 \text{ km/hr})}$$

$$\theta = 50.8°$$

b. The combined final speed, found by solving for V_f in equation 8.62, is

$$V_f = \frac{m_1 v_{1i}}{(m_1 + m_2)\cos\theta}$$

$$= \frac{(1200 \text{ kg})(95.0 \text{ km/hr})}{(1200 \text{ kg} + 1400 \text{ kg})\cos 50.8°}$$

$$= 69.4 \text{ km/hr}$$

The Language of Physics

Linear momentum
The product of the mass of the body in motion times its velocity (p. 215).

Newton's second law in terms of linear momentum
When a resultant applied force acts on a body, it causes the linear momentum of that body to change with time (p. 216).

External forces
Forces that originate outside the system and act on the system (p. 216).

Internal forces
Forces that originate within the system and act on the particles within the system (p. 216).

Law of conservation of linear momentum
If the total external force acting on a system is equal to zero, then the final value of the total momentum of the system is equal to the initial value of the total momentum of the system. Thus, the total momentum is a constant, or as usually stated, the total momentum is conserved. The law of conservation of momentum is a consequence of Newton's third law (p. 217).

Impulse
The product of the force that is acting and the time that the force is acting. The impulse acting on a body is equal to the change in momentum of the body (p. 222).

Perfectly elastic collision
A collision in which no kinetic energy is lost, that is, the kinetic energy is conserved. Momentum is conserved in all collisions for which there are no external forces. In this type of collision, the velocity of separation of the two bodies is equal to the velocity of approach (p. 223).

Inelastic collision
A collision in which some kinetic energy is lost. The velocity of separation of the two bodies in this type of collision is not equal to the velocity of approach. The coefficient of restitution is a measure of the inelastic collision (p. 223).

Perfectly inelastic collision
A collision in which the two objects stick together during the collision. A great deal of kinetic energy is usually lost in this type of collision (p. 223).

Coefficient of restitution
The measure of the amount of the inelastic collision. It is equal to the ratio of the velocity of separation of the two bodies to the velocity of approach (p. 228).

Summary of Important Equations

Definition of momentum
$$\mathbf{p} = m\mathbf{v} \tag{8.1}$$

Newton's second law in terms of momentum
$$\mathbf{F} = \frac{\Delta \mathbf{p}}{\Delta t} \tag{8.5}$$

Law of conservation of momentum for $\mathbf{F}_{net} = 0$
$$\mathbf{p}_f = \mathbf{p}_i \tag{8.7}$$

Recoil speed of a gun
$$v_G = \frac{m_B}{m_G} v_B \tag{8.14}$$

Impulse
$$\mathbf{J} = \mathbf{F}\Delta t \tag{8.18}$$

Impulse is equal to the change in momentum
$$\mathbf{J} = \Delta \mathbf{p} \tag{8.19}$$

Conservation of momentum in a collision
$$m_1\mathbf{v}_{1i} + m_2\mathbf{v}_{2i} = m_1\mathbf{v}_{1f} + m_2\mathbf{v}_{2f} \tag{8.22}$$

Conservation of momentum in scalar form, both bodies in motion in same direction, and $v_{1i} > v_{2i}$.
$$m_1v_{1i} + m_2v_{2i} = m_1v_{1f} + m_2v_{2f} \tag{8.23}$$

Conservation of energy in a perfectly elastic collision
$$\tfrac{1}{2}m_1v_{1i}^2 + \tfrac{1}{2}m_2v_{2i}^2$$
$$= \tfrac{1}{2}m_1v_{1f}^2 + \tfrac{1}{2}m_2v_{2f}^2 \tag{8.26}$$

Final velocity of ball 1 in a perfectly elastic collision
$$v_{1f} = \left(\frac{m_1 - m_2}{m_1 + m_2}\right)v_{1i}$$
$$+ \left(\frac{2m_2}{m_1 + m_2}\right)v_{2i} \tag{8.30}$$

Final velocity of ball 2 in a perfectly elastic collision
$$v_{2f} = \left(\frac{2m_1}{m_1 + m_2}\right)v_{1i}$$
$$- \left(\frac{m_1 - m_2}{m_1 + m_2}\right)v_{2i} \tag{8.31}$$

The velocity of approach
$$v_{1i} - v_{2i} = V_A \tag{8.38}$$

The velocity of separation
$$v_{2f} - v_{1f} = V_S \tag{8.39}$$

For any collision
$$V_S = eV_A \tag{8.42}$$

For a perfectly elastic collision
$$e = 1$$

For an inelastic collision
$$0 < e < 1 \tag{8.43}$$

For a perfectly inelastic collision
$$e = 0$$

Perfectly inelastic collision
$$V_f = \left(\frac{m_1}{m_1 + m_2}\right)v_{1i}$$
$$+ \left(\frac{m_2}{m_1 + m_2}\right)v_{2i} \tag{8.54}$$

Questions for Chapter 8

1. If the velocity of a moving body is doubled, what does this do to the kinetic energy and the momentum of the body?
2. Why is Newton's second law in terms of momentum more appropriate than the form $F = ma$?
3. State and discuss the law of conservation of momentum and show its relation to Newton's third law of motion.
4. Discuss what is meant by an isolated system and how it is related to the law of conservation of momentum.
5. Is it possible to have a collision in which all the kinetic energy is lost? Describe such a collision.

6. An airplane is initially flying at a constant velocity in plane and level flight. If the throttle setting is not changed, explain what happens to the plane as it continues to burn its fuel?
†7. In the early days of rocketry it was assumed by many people that a rocket would not work in outer space because there was no air for the exhaust gases to push against. Explain why the rocket does work in outer space.
8. Discuss the possibility of a fourth type of collision, a super elastic collision, in which the particles have more kinetic energy after the collision than before. As an example,

consider a car colliding with a truck loaded with dynamite.
9. If the net force acting on a body is equal to zero, what happens to the center of mass of the body?
†10. A bird is sitting on a swing in an enclosed bird cage that is resting on a mass balance. If the bird leaves the swing and flies around the cage without touching anything, does the balance show any change in its reading?
11. From the point of view of impulse, explain why an egg thrown against a wall will break, while an egg thrown against a loose vertical sheet will not.

Problems for Chapter 8

8.1 Momentum

1. What is the momentum of a 3200-lb car traveling at 55.0 mph?
2. A 3200-lb car traveling at 55.0 mph collides with a tree and comes to a stop in 0.100 s. What is the change in momentum of the car? What average force acted on the car during impact? What is the impulse?

3. Answer the same questions in problem 2 if the car hit a sand barrier in front of the tree and came to rest in 0.300 s.
4. A 0.150-kg ball is thrown straight upward at an initial velocity of 30.0 m/s. Two seconds later the ball has a velocity of 10.4 m/s. Find (a) the initial momentum of the ball, (b) the momentum of the ball at 2 s, (c) the force acting on the ball, and (d) the weight of the ball.

5. How long must a force of 5.00 N act on a block of 3.00-kg mass in order to give it a velocity of 4.00 m/s?
6. A force of 25.0 N acts on a 10.0-kg mass in the positive x-direction, while another force of 13.5 N acts in the negative x-direction. If the mass is initially at rest, find (a) the time rate of change of momentum, (b) the change in momentum after 1.85 s, and (c) the velocity of the mass at the end of 1.85 s.

8.2 and 8.3 Conservation of Momentum

7. A 10.0-g bullet is fired from a 5.00-kg rifle with a velocity of 300 m/s. What is the recoil velocity of the rifle?

8. In an ice skating show, a 200-lb man at rest pushes a 100-lb woman away from him at a speed of 4.00 ft/s. What happens to the man?

9. A 5000-kg cannon fires a shell of 3.00-kg mass with a velocity of 250 m/s. What is the recoil velocity of the cannon?

10. A cannon of 3.50×10^3 kg fires a shell of 2.50 kg with a muzzle speed of 300 m/s. What is the recoil velocity of the cannon?

11. A 150-lb boy at rest on roller skates throws a 2.00-lb ball horizontally with a speed of 25.0 ft/s. With what speed does the boy recoil?

12. An 80.0-kg astronaut pushes herself away from a 1200-kg space capsule at a velocity of 3.00 m/s. Find the recoil velocity of the space capsule.

13. A 200-lb man is standing in a 500-lb boat. The man walks forward at 3.00 ft/s relative to the water. What is the final velocity of the boat? Neglect any resistive force of the water on the boat.

14. A 78.5-kg man is standing in a 275-kg boat. The man walks forward at 1.25 m/s relative to the water. What is the final velocity of the boat? Neglect any resistive force of the water on the boat.

15. A water hose sprays 2 kg of water against the side of a building in 1 s. If the velocity of the water is 15 m/s, what force is exerted on the wall by the water? (Assume that the water does not bounce off the wall of the building.)

8.4 Impulse

16. A boy kicks a football with an average force of 20.0 lb for a time of 0.200 s. (a) What is the impulse? (b) What is the change in momentum of the football? (c) If the football has a mass of 250 g, what is the velocity of the football as it leaves the kicker's foot?

17. A boy kicks a football with an average force of 66.8 N for a time of 0.185 s. (a) What is the impulse? (b) What is the change in momentum of the football? (c) If the football has mass of 250 g, what is the velocity of the football as it leaves the kicker's foot?

18. A baseball traveling at 150 km/hr is struck by a bat and goes straight back to the pitcher at the same speed. If the baseball has a mass of 200 g, find (a) the change in momentum of the baseball, (b) the impulse imparted to the ball, and (c) the average force acting if the bat was in contact with the ball for 0.100 s.

19. A 10.0-kg hammer strikes a nail at a velocity of 12.5 m/s and comes to rest in a time interval of 0.004 s. Find (a) the impulse imparted to the nail and (b) the average force imparted to the nail.

20. If a gas molecule of mass 5.30×10^{-26} kg and an average speed of 425 m/s collides perpendicularly with a wall of a room and rebounds at the same speed, what is its change of momentum? What impulse is imparted to the wall?

8.5 Collisions in One Dimension

21. Two gliders moving toward each other, one of mass 200 g and the other of 250 g, collide on a frictionless air track. If the first glider has an initial velocity of 25.0 cm/s toward the right and the second of -35.0 cm/s toward the left, find the velocities after the collision if the collision is perfectly elastic.

22. A 250-g glider overtakes and collides with a 200-g glider on an air track. If the 250-g glider is moving at 35.0 cm/s and the second glider at 25.0 cm/s, find the velocities after the collision if the collision is perfectly elastic.

†23. A 200-g ball makes a perfectly elastic collision with an unknown mass that is at rest. If the first ball rebounds with a final speed of $v_{1f} = \frac{1}{2}v_{1i}$, (a) what is the unknown mass, and (b) what is the final velocity of the unknown mass?

24. A 30.0-g ball, m_1, collides perfectly elastically with a 20.0-g ball, m_2. If the initial velocities are $v_{1i} = 50.0$ cm/s to the right and $v_{2i} = -30.0$ cm/s to the left, find the final velocities v_{1f} and v_{2f}. Compute the initial and final momenta. Compute the initial and final kinetic energies.

25. A 150-g ball moving at a velocity of 25.0 cm/s to the right collides with a 250-g ball moving at a velocity of 18.5 cm/s to the left. The collision is imperfectly elastic with a coefficient of restitution of 0.65. Find (a) the velocity of each ball after the collision, (b) the kinetic energy before the collision, (c) the kinetic

energy after the collision, and (d) the percentage of energy lost in the collision.

26. A 2000-lb car traveling at 60.0 mph collides "head-on" with a 20,000-lb truck traveling toward the car at 30.0 mph. The car becomes stuck to the truck during the collision. What is the final velocity of the car and truck?

27. A 1150-kg car traveling at 110 km/hr collides "head-on" with a 9500-kg truck traveling toward the car at 40.0 km/hr. The car becomes stuck to the truck during the collision. What is the final velocity of the car and truck?

28. A 3.00-g bullet is fired at 200 m/s into a wooden block of 10-kg mass that is at rest. If the bullet becomes embedded in the wooden block, find the velocity of the block and bullet after impact.

29. A 20,000-lb freight car traveling at 3.00 ft/s collides with a stationary freight car that weighs 15,000 lb. If the cars couple together find the resultant velocity of the cars after the collision.

30. A 9500-kg freight car traveling at 5.50 km/hr collides with an 8000-kg stationary freight car. If the cars couple together, find the resultant velocity of the cars after the collision.

31. Two gliders are moving toward each other on a frictionless air track. Glider 1 has a mass of 200 g and glider 2 of 250 g. The first glider has an initial speed of 25.0 cm/s while the second has a speed of 35.0 cm/s. If the collision is perfectly inelastic, find (a) the final velocity of the gliders, (b) the kinetic energy before the collision, and (c) the kinetic energy after the collision. (d) How much energy is lost, and where did it go?

8.6 Collisions in Two Dimensions— Glancing Collisions

32. A 230-lb linebacker moving due east at 25 mph tackles a 175-lb halfback moving south at 40 mph. The two stick together during the collision. What is the resultant velocity of the two of them?

33. A 10,000-kg truck enters an intersection heading north at 45 km/hr when it makes a perfectly inelastic collision with a 1000-kg car traveling at 90 km/hr due east. What is the final velocity of the car and truck?

†34. Billiard ball 2 is at rest when it is hit with a glancing collision by ball 1 moving at a velocity of 50.0 cm/s toward the right. After the collision ball 1 moves off at an angle of 35.0° from the original direction while ball 2 moves at an angle of 40.0°, as shown in the diagram. The mass of each billiard ball is 0.017 kg. Find the final velocity of each ball after the collision. Find the kinetic energy before and after the collision. Is the collision elastic?

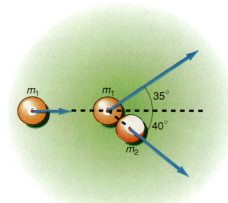

35. A 0.150-kg ball, moving at a speed of 25.0 m/s, makes an elastic collision with a wall at an angle of 40.0°, and rebounds at an angle of 40.0°. Find (a) the change in momentum of the ball and (b) the magnitude and direction of the momentum imparted to the wall. The diagram is a view from the top.

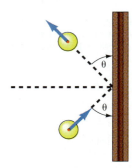

Additional Problems

†36. A 0.250-kg ball is dropped from a height of 1.00 m. It rebounds to a height of 0.750 m. If the ground exerts a force of 300 N on the ball, find the time the ball is in contact with the ground.

37. A 200-g ball is dropped from the top of a building. If the speed of the ball before impact is 40.0 m/s, and right after impact it is 25.0 cm/s, find (a) the momentum of the ball before impact, (b) the momentum of the ball after impact, (c) the kinetic energy of the ball before impact, (d) the kinetic energy of the ball after impact, and (e) the coefficient of restitution of the ball.

†38. A ball is dropped from a height of 1.00 m and rebounds to a height of 0.920 m. Approximately how many bounces will the ball make before losing 90% of its energy?

39. A 60.0-g tennis ball is dropped from a height of 1.00 m. If it rebounds to a height of 0.560 m, (a) what is the coefficient of restitution of the tennis ball and the floor, and (b) how much energy is lost in the collision?

†40. A 25.0-g bullet strikes a 5.00-kg ballistic pendulum that is initially at rest. The pendulum rises to a height of 14.0 cm. What is the initial speed of the bullet?

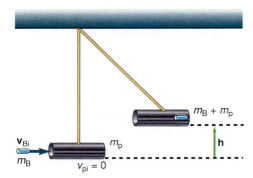

41. A 25.0-g bullet with an initial speed of 400 m/s strikes a 5-kg ballistic pendulum that is initially at rest. (a) What is the speed of the combined bullet-pendulum after the collision? (b) How high will the pendulum rise?

42. An 80-kg caveman, standing on a branch of a tree 5 m high, swings on a vine and catches a 60-kg cavegirl at the bottom of the swing. How high will both of them rise?

†43. A hunter fires an automatic rifle at an attacking lion that weighs 300 lb. If the lion is moving toward the hunter at 10.0 ft/s, and the rifle bullets weigh 2.00 oz each and have a muzzle velocity of 2500 ft/s, how many bullets must the man fire at the lion in order to stop the lion in his tracks?

†44. Two gliders on an air track are connected by a compressed spring and a piece of thread as shown; m_1 = 300 g and m_2 is unknown. If the connecting string is cut, the gliders separate. Glider 1 experiences the velocity v_1 = 10.0 cm/s, and glider 2 experiences the velocity v_2 = 20.0 cm/s, what is the unknown mass?

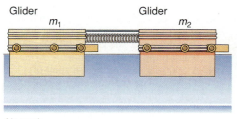

Air track

†45. Two gliders on an air track are connected by a compressed spring and a piece of thread as shown. The masses of the gliders are m_1 = 300 g and m_2 = 250 g. The connecting string is cut and the compressed string causes the two gliders to separate from each other. If glider 1 has moved 35.0 cm from its starting point, where is glider 2 located?

†46. Two balls, m_1 = 100 g and m_2 = 200 g, are suspended near each other as shown. The two balls are initially in contact. Ball 2 is then pulled away so that it makes a 45.0° angle with the vertical and is then released. (a) Find the velocity of ball 2 just before impact and the velocity of each ball after the perfectly elastic impact. (b) How high will each ball rise?

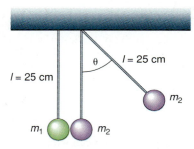

†47. Two swimmers simultaneously dive off opposite ends of a 200-lb boat. If the first swimmer has a weight w_1 = 200 lb and a velocity of 5.00 ft/s toward the right, while the second swimmer has a weight w_2 = 150 lb and a velocity of −3.00 ft/s toward the left, what is the final velocity of the boat?

48. Two swimmers simultaneously dive off opposite ends of a 110-kg boat. If the first swimmer has a mass m_1 = 66.7 kg and a velocity of 1.98 m/s toward the right, while the second swimmer has a mass m_2 = 77.8 kg and a velocity of −7.63 m/s toward the left, what is the final velocity of the boat?

†49. Show that the kinetic energy of a moving body can be expressed in terms of the linear momentum as KE $= p^2/2m$.

†50. A machine gun is mounted on a small train car and fires 100 bullets per minute backward. If the mass of each bullet is 10.0 g and the speed of each bullet as it leaves the gun is 900 m/s, find the average force exerted on the gun. If the mass of the car and machine gun is 225 kg, what is the acceleration of the train car while the gun is firing?

†51. An open toy railroad car of mass 250 g is moving at a constant speed of 30 cm/s when a wooden block of 50 g is dropped into the open car. What is the final speed of the car and block?

†52. Masses m_1 and m_2 are located on the top of the two frictionless inclined planes as shown in the diagram. It is given that $m_1 = 30.0$ g, $m_2 = 50.0$ g, $l_1 = 50.0$ cm, $l_2 = 20.0$ cm, $l = 100$ cm, $\theta_1 = 50.0°$, and $\theta_2 = 25.0°$. Find (a) the speeds v_1 and v_2 at the bottom of each inclined plane, note that ball 1 reaches the bottom of the plane before ball 2; (b) the position between the planes where the masses will collide elastically; (c) the speeds of the two masses after the collision; and (d) the final locations l_1' and l_2' where the two masses will rise up the plane after the collision.

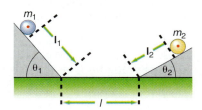

†53. The mass $m_1 = 40.0$ g is initially located at a height $h_1 = 1.00$ m on the frictionless surface shown in the diagram. It is then released from rest and collides with the mass $m_2 = 70.0$ g, which is at rest at the bottom of the surface. After the collision, will the mass m_2 make it over the top of the hill at position B, which is at a height of 0.500 m?

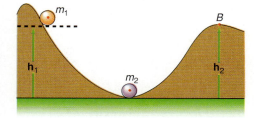

†54. Two balls of mass m_1 and m_2 are placed on a frictionless surface as shown in the diagram. Mass $m_1 = 30.0$ g is at a height $h_1 = 50.0$ cm above the bottom of the bowl, while mass $m_2 = 60.0$ g is at a height of $3/4$ h_1. The distance $l = 100$ cm. Assuming that both balls reach the bottom at the same time, find (a) the speed of each ball at the bottom of each surface, (b) the position where the two balls collide, (c) the speed of each ball after the collision, and (d) the height that each ball will rise to after the collision.

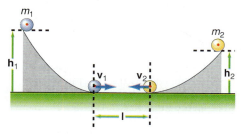

†55. A person is in a small train car that weighs 500 lb and contains 500 lb of rocks. The train is initially at rest. The person starts to throw large rocks, each weighing 100 lb, from the rear of the train at a speed of 5.00 ft/s. (a) If the person throws out 1 rock what will the recoil velocity of the train be? The person then throws out another rock at the same speed. (b) What is the recoil velocity now? (c) The person continues to throw out the rest of the rocks one at a time. What is the final velocity of the train when all the rocks have been thrown out?

†56. A bullet of mass 20.0 g is fired into a block of mass 5.00 kg that is initially at rest. The combined block and bullet moves a distance of 5.00 m over a rough surface of coefficient of kinetic friction of 0.500, before coming to rest. Find the initial velocity of the bullet.

†57. A bullet of mass 20.0 g is fired at an initial velocity of 200 m/s into a 15.0-kg block that is initially at rest. The combined bullet and block move over a rough surface of coefficient of kinetic friction of 0.500. How far will the combined bullet and block move before coming to rest?

58. A 15.0-kg bullet moving at a speed of 250 m/s hits a 2.00-kg block of wood, which is initially at rest. The bullet emerges from the block of wood at 150 m/s. Find (a) the final velocity of the block of wood and (b) the amount of energy lost in the collision.

†59. A 5-kg pendulum bob, at a height of 0.750 m above the floor, swings down to the ground where it hits a 2.15 kg block that is initially at rest. The block then slides up a 30.0° incline. Find how far up the incline the block will slide if (a) the plane is frictionless and (b) if the plane is rough with a value of $\mu_k = 0.450$.

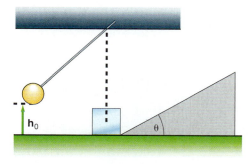

†60. A 0.15-kg baseball is thrown upward at an initial velocity of 35.0 m/s. Two seconds later, a 20.0-g bullet is fired at 250 m/s into the rising baseball. How high will the combined bullet and baseball rise?

†61. A 25-g ball slides down a smooth inclined plane, 0.850 m high, that makes an angle of 35.0° with the horizontal. The ball slides into an open box of 200-g mass and the ball and box slide on a rough surface of $\mu_k = 0.450$. How far will the combined ball and box move before coming to rest?

†62. A 25-g ball slides down a smooth inclined plane, 0.850 m high, that makes an angle of 35.0° with the horizontal. The ball slides into an open box of 200-g mass and the ball and box slide off the end of a table 1.00 m high. How far from the base of the table will the combined ball and box hit the ground?

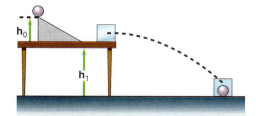

†63. A 1300-kg car collides with a 15,000-kg truck at an intersection and they couple together and move off as one leaving a skid mark 5 m long that makes an angle of 30.0° with the original direction of the car. If $\mu_k = 0.700$, find the initial velocities of the car and truck before the collision.

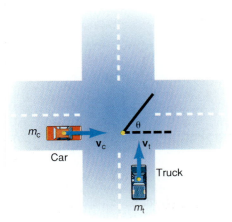

Car

Truck

64. A bomb of mass $M = 2.50$ kg, moving in the x-direction at a speed of 10.5 m/s, explodes into three pieces. One fragment, $m_1 = 0.850$ kg, flies off at a velocity of 3.5 m/s at an angle of 30.0° above the x-axis. Fragment $m_2 = 0.750$ kg, flies off at an angle of 43.5° below the positive x-axis, and the third fragment flies off at an angle of 150° with respect to the positive x-axis. Find the velocities of m_2 and m_3.

Interactive Tutorials

65. Recoil velocity of a gun. A bullet of mass $m_b = 10.0$ g is fired at a velocity $v_b = 300$ m/s from a rifle of mass $m_r = 5.00$ kg. Calculate the recoil velocity v_r of the rifle. If the bullet is in the barrel of the rifle for $t = 0.004$ s, what is the bullet's acceleration and what force acted on the bullet? Assume the force is a constant.

66. An inelastic collision. A car of mass $m_1 = 1000$ kg is moving at a velocity $v_1 = 50.0$ m/s and collides inelastically with a car of mass $m_2 = 750$ kg moving in the same direction at a velocity of $v_2 = 20.0$ m/s. Calculate (a) the final velocity v_f of both vehicles; (b) the initial momentum p_i; (c) the final momentum p_f; (d) the initial kinetic energy KE_i; (e) the final kinetic energy KE_f of the system; (f) the energy lost in the collision ΔE; and (g) the percentage of the original energy lost in the collision, %E_{lost}.

67. A perfectly elastic collision. A mass, $m_1 = 3.57$ kg, moving at a velocity, $v_1 = 2.55$ m/s, overtakes and collides with a second mass, $m_2 = 1.95$ kg, moving at a velocity $v_2 = 1.35$ m/s. If the collision is perfectly elastic, find (a) the velocities after the collision, (b) the momentum before the collision, (c) the momentum after the collision, (d) the kinetic energy before the collision, and (e) the kinetic energy after the collision.

68. An imperfectly elastic collision. A mass, $m = 2.84$ kg, is dropped from a height $h_0 = 3.42$ m and hits a wooden floor. The mass rebounds to a height $h = 2.34$ m. If the collision is imperfectly elastic, find (a) the velocity of the mass as it hits the floor, v_{1i}; (b) the velocity of the mass after it rebounds from the floor, v_{1i}; (c) the coefficient of restitution, e; (d) the kinetic energy, KE_A, just as the mass approached the floor; (e) the kinetic energy, KE_S, after the separation of the mass from the floor; (f) the actual energy lost in the collision; (g) the percentage of energy lost in the collision; (h) the momentum before the collision; and (i) the momentum after the collision.

69. An imperfectly elastic collision—the bouncing ball. A ball of mass, $m = 1.53$ kg, is dropped from a height $h_0 = 1.50$ m and hits a wooden floor. The collision with the floor is imperfectly elastic and the ball only rebounds to a height $h = 1.12$ m for the first bounce. Find (a) the initial velocity of the ball, v_i, as it hits the floor on its first bounce; (b) the velocity of the ball v_f, after it rebounds from the floor on its first bounce; (c) the coefficient of restitution, e; (d) the initial kinetic energy, KE_i, just as the ball approaches the floor; (e) the final kinetic energy, KE_f, of the ball after the bounce from the floor; (f) the actual energy lost in the bounce, E_{lost}/bounce; and (g) the percentage of the initial kinetic energy lost by the ball in the bounce, %E_{lost}/bounce. The ball continues to bounce until it loses all its energy. (h) Find the cumulative total percentage energy lost, % Energy lost, for all the bounces. (i) Plot a graph of the % of Total Energy lost as a function of the number of bounces.

70. A variable mass system. A train car of mass $m_T = 1500$ kg, contains 35 rocks each of mass $m_r = 30$ kg. The train is initially at rest. A man throws out each rock from the rear of the train at a speed $v_r = 8.50$ m/s. (a) When the man throws out one rock, what will the recoil velocity, V_T, of the train be? (b) What is the recoil velocity when the man throws out the second rock? (c) What is the recoil velocity of the train when the nth rock is thrown out? (d) If the man throws out each rock at the rate $R = 1.5$ rocks/s, find the change in the velocity of the train and its acceleration. (e) Draw a graph of the velocity of the train as a function of the number of rocks thrown out of the train. (f) Draw a graph of the acceleration of the train as a function of time.

9

Rotational Motion

9.1 Introduction

Up to now, the main emphasis in the description of the motion of a body dealt with the translational motion of that body. But in addition to translating, a body can also rotate about some axis, called the *axis of rotation*. Therefore, for a complete description of the motion of a body we also need to describe any rotational motion the body might have. As a matter of fact, the most general motion of a rigid body is composed of the translation of the center of mass of the body plus its rotation about the center of mass.

In the analysis of rotational motion, we will see a great similarity to translational motion. In fact, this chapter can serve as a review of all the mechanics discussed so far.

9.2 Rotational Kinematics

In the study of translational kinematics the first concept we defined was the position of an object. The position of the body was defined as the displacement x from a reference point. In a similar way, the position of a point on a rotating body is defined by the **angular displacement** θ from some reference line that connects the point to the axis of rotation, as shown in figure 9.1. That is, if the point was originally at P, and a little later it is at the point P', then the body has rotated through the angular displacement θ. The angular displacement can be represented as a vector that is perpendicular to the plane of the motion. If the angular displacement is a positive quantity, the rotation of the body is counterclockwise and the angular displacement vector points upward. If the angular displacement is a negative quantity, the rotation of the body is clockwise and the angular displacement vector points downward. The magnitude of the angular displacement is the angle θ itself. We measure the angle θ in radians, which we defined in section 6.3. The linear distance between the points P and P' is given by the arc length s, and is related to the angular displacement by

$$s = r\theta \tag{6.5}$$

The average velocity of a translating body was defined as the displacement of the body divided by the time for that displacement:

$$v_{\text{avg}} = \frac{\Delta x}{\Delta t}$$

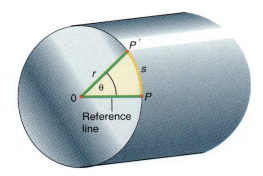

Figure 9.1

The angular displacement.

(a)

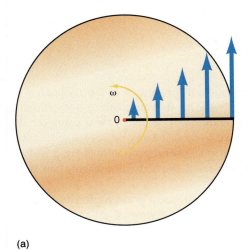

(b)

Figure 9.2

The linear velocity varies with the distance r from the center of the rotating body.

In the same way, *the average **angular velocity** of a rotating body is defined as the angular displacement of the body about the axis of rotation divided by the time for that displacement:*

$$\omega_{\text{avg}} = \frac{\Delta\theta}{\Delta t} \tag{9.1}$$

The units for angular velocity are radians/second, abbreviated as rad/s. It is important to remember that the radian is a dimensionless quantity, and can be added or deleted from an equation whenever it is convenient. The angular velocity, like the angular displacement, can also be represented as a vector quantity that is also perpendicular to the plane of the motion. It is positive and points upward for counterclockwise rotations and is negative and points downward for clockwise rotations.

The similarities between translational and rotational motion can be seen in table 9.1. The relation between the linear velocity of a point on the rotating body and the angular velocity of the body is found by dividing both sides of equation 6.5 by t, that is,

$$\frac{s}{t} = \frac{r\theta}{t}$$

but $s/t = v$ and $\theta/t = \omega$. Therefore,

$$v = r\omega \tag{9.2}$$

Equation 9.2 says that for a body rotating at an angular velocity ω, the farther the distance r that the body is from the axis of rotation, the greater is its linear velocity. This can be seen in figure 9.2(a). You may recall when you were a child and went on the merry-go-round, you usually wanted to ride on the outside horses because they moved the fastest. You can now see why. They are at the greatest distance from the axis of rotation and hence have the greatest linear velocity. The linear velocity of a point on the rotating body can also be called the *tangential velocity* because the point is moving along the tangential direction at any instant.

Table 9.1

Comparison of Translational and Rotational Motion

Translational Motion	Rotational Motion
$v_{\text{avg}} = \dfrac{\Delta x}{\Delta t} = \dfrac{x}{t}$	$\omega_{\text{avg}} = \dfrac{\Delta\theta}{\Delta t} = \dfrac{\theta}{t}$
$a = \dfrac{\Delta v}{\Delta t} = \dfrac{v - v_0}{t}$	$\alpha = \dfrac{\Delta\omega}{\Delta t} = \dfrac{\omega - \omega_0}{t}$
$v = v_0 + at$	$\omega = \omega_0 + \alpha t$
$x = v_0 t + \frac{1}{2}at^2$	$\theta = \omega_0 t + \frac{1}{2}\alpha t^2$
$v^2 = v_0^2 + 2ax$	$\omega^2 = \omega_0^2 + 2\alpha\theta$
$KE = \frac{1}{2}mv^2$	$KE = \frac{1}{2}I\omega^2$
$F = ma$	$\tau = I\alpha$
$p = mv$	$L = I\omega$
$F = \dfrac{\Delta p}{\Delta t}$	$\tau = \dfrac{\Delta L}{\Delta t}$
$p_f = p_i$	$L_f = L_i$

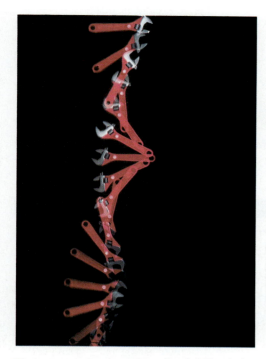

The most general motion of a rigid body is composed of the translation of the center of mass of the body plus its rotation about the center of mass.

Another example of the relation between the tangential velocity and the angular velocity is seen in the old fashioned "whip" that you formed by holding hands while you were ice skating or roller skating, figure 9.2(b). The person at the inside end of the "whip" hardly moved at all ($r = 0$), but the person at the far end of the whip (maximum r) moved at very high speeds.

Example 9.1

The angular velocity of a wheel. A wheel of radius 15.0 cm starts from rest and turns through 2.00 rev in 3.00 s. (a) What is its average angular velocity? (b) What is the tangential velocity of a point on the rim of the wheel?

Solution

a. The average angular velocity, found from equation 9.1, is

$$\omega_{avg} = \frac{\theta}{t}$$

$$= \frac{(2.00 \text{ rev})(2\pi \text{ rad})}{3.00 \text{ s} (1 \text{ rev})}$$

$$= 4.19 \text{ rad/s}$$

Note that we accomplished the conversion from revolutions to radians using the identity that one revolution is equal to 2π radians.

b. The tangential velocity of a point on the rim of the wheel, found from equation 9.2, is

$$v = r\omega$$

$$= (0.150 \text{ m})(4.19 \text{ rad/s})$$

$$= 0.628 \text{ m/s}$$

In the study of kinematics we defined the average translational acceleration of a body in equations 3.7 and 3.9 as the change in the velocity of the body with time, that is

$$a = \frac{\Delta v}{\Delta t} = \frac{v - v_0}{t}$$

Because we considered only problems where motion was at constant acceleration, the average acceleration was equal to the instantaneous acceleration. In the same way, *we now define the average **angular acceleration** α of the rotating body as the change in the angular velocity of the body with time,* that is,

$$\alpha = \frac{\Delta \omega}{\Delta t} = \frac{\omega - \omega_0}{t} \tag{9.3}$$

Again, since the only problems that we will consider concern motion at constant angular acceleration, the average angular acceleration is equal to the angular acceleration at any instant of time. We should note that angular acceleration, like angular velocity, can also be represented as a vector that lies along the axis of rotation of the rotating body. If the angular velocity vector is increasing with time, α is positive, and points upward from the plane of the rotation. If the angular velocity is decreasing with time, α is negative, and points downward into the plane of the rotation. The units for angular acceleration are radians/second per second, abbreviated as rad/s².

From the definition of the acceleration, equation 3.9, the first of the kinematic equations became

$$v = v_0 + at \qquad (3.10)$$

the velocity of the moving body at any instant of time. Similarly, if equation 9.3 is solved for ω, we have

$$\omega = \omega_0 + \alpha t \qquad (9.4)$$

the first of the **kinematical equations for rotational motion.** *Equation 9.4 gives the angular velocity of the rotating body at any instant of time for a constant acceleration, α.*

Example 9.2

The angular acceleration of a cylinder. A cylinder rotating at 10.0 rad/s is accelerated to 50.0 rad/s in 10.0 s. What is the angular acceleration of the cylinder?

Solution

The angular acceleration, found from equation 9.4, is

$$\alpha = \frac{\omega - \omega_0}{t} = \frac{50.0 \text{ rad/s} - 10.0 \text{ rad/s}}{10.0 \text{ s}}$$

$$= 4.00 \text{ rad/s}^2$$

Example 9.3

The angular velocity of a crankshaft. A crankshaft rotating at 10.0 rad/s undergoes an angular acceleration of 0.500 rad/s². What is the angular velocity of the shaft after 10.0 s?

Solution

The angular velocity, found from equation 9.4, is

$$\omega = \omega_0 + \alpha t$$

$$= 10.0 \text{ rad/s} + (0.500 \text{ rad/s}^2)(10.0 \text{ s})$$

$$= 15.0 \text{ rad/s}$$

The relationship between the magnitude of the tangential acceleration of a point on the rim of the rotating body and the angular acceleration is found by dividing both sides of equation 9.2 by t, that is,

$$v = r\omega \qquad (9.2)$$

$$\frac{v}{t} = \frac{r\omega}{t}$$

but $v/t = a$ and $\omega/t = \alpha$, therefore,

$$a = r\alpha \qquad (9.5)$$

Equation 9.5 gives the relationship between the magnitude of the tangential acceleration and the angular acceleration.

The next kinematic derivation was the equation giving the position of the moving body as a function of time. Recall that the average velocity was substituted in the equation $x = v_{avg}t$ to yield the kinematic equation for the position of the moving body as a function of time as

$$x = v_0 t + \tfrac{1}{2}at^2$$

Similarly, to find the angular displacement of a rotating body at any instant of time, we use equation 9.1 in the form

$$\theta = \omega_{avg}t \qquad (9.6)$$

But for a body rotating at constant angular acceleration, the average angular velocity is

$$\omega_{avg} = \frac{\omega + \omega_0}{2} \qquad (9.7)$$

where ω_0 is the initial angular velocity and ω is the final angular velocity at some time t. Substituting 9.7 into 9.6 gives

$$\theta = \left(\frac{\omega + \omega_0}{2}\right)t \qquad (9.8)$$

Substituting equation 9.4 for the angular velocity ω into equation 9.8 gives

$$\theta = \left[\frac{(\omega_0 + \alpha t) + \omega_0}{2}\right]t$$

Rearranging terms we get

$$\theta = \omega_0 t + \tfrac{1}{2}\alpha t^2 \qquad (9.9)$$

Equation 9.9 gives the angular displacement of the rotating body as a function of time for constant angular acceleration. This is the second kinematic equation for rotational motion.

Example 9.4

The angular displacement of a wheel. A wheel rotating at 15.0 rad/s undergoes an angular acceleration of 10.0 rad/s². Through what angle has the wheel turned when $t = 5.00$ s?

Solution

The angular displacement, found from equation 9.9, is

$$\theta = \omega_0 t + \tfrac{1}{2}\alpha t^2$$
$$= (15.0 \text{ rad/s})(5.00 \text{ s}) + \tfrac{1}{2}(10.0 \text{ rad/s}^2)(5.00 \text{ s})^2$$
$$= 200 \text{ rad}$$

We obtained the third translational kinematic equation,

$$v^2 = v_0^2 + 2ax \qquad (3.16)$$

from the first two translational kinematic equations by eliminating the time t between them. We can find a similar equation for the angular velocity as a function of the angular displacement by eliminating the time between equations 9.4 and 9.9 and we suggest that the student do this as an exercise. We will obtain the third kinematic equation for rotational motion in a slightly different manner, however. Let us start with

$$v^2 = v_0^2 + 2ax \qquad (3.16)$$

But we know that a relationship exists between the translational variables and the rotational variables. Those relationships are

$$s = r\theta \qquad (6.5)$$

$$v = r\omega \qquad (9.2)$$

$$a = r\alpha \qquad (9.5)$$

For the rotating body, we replace the linear distance x by the distance s along the arc of the circle. If we substitute the above equations into equation 3.16, we get

$$v^2 = v_0^2 + 2as$$
$$(r\omega)^2 = (r\omega_0)^2 + 2(r\alpha)(r\theta)$$
$$r^2\omega^2 = r^2\omega_0^2 + 2r^2\alpha\theta$$

Dividing each term by r^2, we obtain

$$\omega^2 = \omega_0^2 + 2\alpha\theta \qquad (9.10)$$

Equation 9.10 represents the angular velocity of the rotating body at any angular displacement θ for constant angular acceleration α.

Example 9.5

The angular velocity at a particular angular displacement. A wheel, initially rotating at 10.0 rad/s, undergoes an angular acceleration of 5.00 rad/s^2. What is the angular velocity when the wheel has turned through an angle of 50.0 rad?

Solution

The angular velocity, found from equation 9.10, is

$$\omega^2 = \omega_0^2 + 2\alpha\theta$$
$$= (10.0 \text{ rad/s})^2 + 2(5.00 \text{ rad/s}^2)(50.0 \text{ rad})$$
$$= 100 \text{ rad}^2/\text{s}^2 + 500 \text{ rad}^2/\text{s}^2 = 600 \text{ rad}^2/\text{s}^2$$
$$\omega = 24.5 \text{ rad/s}$$

Note in table 9.1 the similarity in the translational and rotational equations. Everywhere there is an x in the translational equations, there is a θ in the rotational equations. Everywhere there is a v in the translational equations, there is an ω in the rotational equations. And finally, everywhere there is an a in the translational equations, there is an α in the rotational equations. We will see additional analogues as we proceed in the discussion of rotational motion.

Another way to express the magnitude of the centripetal acceleration discussed in chapter 6,

$$a_c = \frac{v^2}{r} \qquad (6.12)$$

is to use

$$v = r\omega \qquad (9.2)$$

to obtain

$$a_c = \frac{\omega^2 r^2}{r}$$

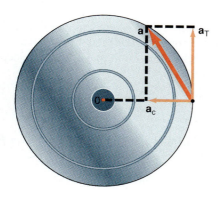

Figure 9.3

The total acceleration of a point on a rotating body is equal to the vector sum of the tangential acceleration and the centripetal acceleration.

Hence, we can represent the magnitude of the centripetal acceleration in terms of the angular velocity as

$$a_c = \omega^2 r \qquad (9.11)$$

For nonuniform circular motion, the resultant acceleration of a point on a rim of a rotating body becomes the vector sum of the tangential acceleration and the centripetal acceleration, as seen in figure 9.3.

Example 9.6

The total acceleration of a point on a rotating body. A cylinder 35.0 cm in diameter is at rest initially. It is then given an angular acceleration of 0.0400 rad/s². Find (a) the angular velocity at 7.00 s, (b) the centripetal acceleration of a point at the edge of the cylinder at 7.00 s, (c) the tangential acceleration at the edge of the cylinder at 7.00 s, and (d) the resultant acceleration of a point at the edge of the cylinder at 7.00 s.

Solution

a. The angular velocity at 7.00 s, found from equation 9.4, is

$$\omega = \omega_0 + \alpha t$$
$$= 0 + (0.0400 \text{ rad/s}^2)(7.00 \text{ s})$$
$$= 0.280 \text{ rad/s}$$

b. The centripetal acceleration, found from equation 9.11, is

$$a_c = \omega^2 r$$
$$= (0.280 \text{ rad/s})^2(17.5 \text{ cm})$$
$$= 1.37 \text{ cm/s}^2$$

c. The tangential acceleration, found from equation 9.5, is

$$a_T = r\alpha = (17.5 \text{ cm})(0.0400 \text{ rad/s}^2)$$
$$= 0.700 \text{ cm/s}^2$$

d. The resultant acceleration at 7.00 s, found from figure 9.3, is

$$a = \sqrt{(a_c)^2 + (a_T)^2}$$
$$= \sqrt{(1.37 \text{ cm/s}^2)^2 + (0.700 \text{ cm/s}^2)^2}$$
$$= 1.54 \text{ cm/s}^2$$

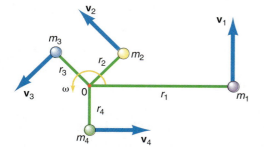

Figure 9.4

Rotational kinetic energy.

9.3 The Kinetic Energy of Rotation

Let us consider the motion of four point masses m_1, m_2, m_3, and m_4 located at distances r_1, r_2, r_3, and r_4, respectively, rotating at an angular speed ω about an axis through the point 0, as shown in figure 9.4.

Let us assume that the masses are connected to the center of rotation by rigid, massless rods. (A massless rod is one whose mass is so small compared to the mass at the end of the rod that we can neglect it in the analysis.) Let us determine the total kinetic energy of these rotating masses. The total energy is equal to the sum of the kinetic energy of each mass. That is,

$$KE_{total} = KE_1 + KE_2 + KE_3 + KE_4 + \cdots$$

The plus sign and dots after the last term indicate that if there are more than the four masses considered, another term is added for each additional mass. Because each mass is rotating with the same angular velocity ω, each has a linear velocity v, as shown. Since the kinetic energy of each mass is

$$KE = \tfrac{1}{2}mv^2$$

the total kinetic energy is

$$KE_{tot} = \tfrac{1}{2}m_1v_1^2 + \tfrac{1}{2}m_2v_2^2 + \tfrac{1}{2}m_3v_3^2 + \tfrac{1}{2}m_4v_4^2 + \cdots \qquad (9.12)$$

but from equation 9.2,

$$v = r\omega$$

hence, for each mass

$$\left.\begin{aligned} v_1 &= r_1\omega \\ v_2 &= r_2\omega \\ v_3 &= r_3\omega \\ v_4 &= r_4\omega \end{aligned}\right\} \qquad (9.13)$$

Substituting equations 9.13 back into equation 9.12, gives

$$KE_{tot} = \tfrac{1}{2}m_1(r_1\omega)^2 + \tfrac{1}{2}m_2(r_2\omega)^2 + \tfrac{1}{2}m_3(r_3\omega)^2 + \tfrac{1}{2}m_4(r_4\omega)^2 + \cdots$$
$$= \tfrac{1}{2}m_1r_1^2\omega^2 + \tfrac{1}{2}m_2r_2^2\omega^2 + \tfrac{1}{2}m_3r_3^2\omega^2 + \tfrac{1}{2}m_4r_4^2\omega^2 + \cdots$$

Note that there is a $1/2$ and an ω^2 in every term, so let us factor them out:

$$KE_{tot} = \tfrac{1}{2}(m_1r_1^2 + m_2r_2^2 + m_3r_3^2 + m_4r_4^2 + \cdots)\omega^2$$

Looking at the form of the equation for the translational kinetic energy $(\tfrac{1}{2}mv^2)$, and remembering all the symmetry in the translational-rotational equations, it is reasonable to expect that the equation for the rotational kinetic energy might have an analogous form. That symmetry is maintained by defining the term in parentheses as the moment of inertia, the rotational analogue of the mass m. That is, the moment of inertia about the axis of rotation for these four masses is

$$I = m_1r_1^2 + m_2r_2^2 + m_3r_3^2 + m_4r_4^2 \qquad (9.14)$$

We will discuss the concept of the moment of inertia in more detail in section 9.4. For now, we see that the equation for the total energy of the four rotating masses is

$$KE_{tot} = \tfrac{1}{2}I\omega^2$$

And finally let us note that the total kinetic energy of the rotating masses can simply be called the **kinetic energy of rotation.** Therefore, the kinetic energy of rotation about a specified axis is

$$KE_{rot} = \tfrac{1}{2}I\omega^2 \qquad (9.15)$$

9.4 The Moment of Inertia

The concept of mass m was introduced to give a measure of the inertia of a body, that is, its resistance to a change in its translational motion. *Now we introduce the moment of inertia to give a measurement of the resistance of the body to a change in its rotational motion.* For example, the larger the moment of inertia of a body, the more difficult it is to put that body into rotational motion. Conversely, the larger the moment of inertia of a body, the more difficult it is to stop its rotational motion.

For the particular configuration studied in section 9.3, the moment of inertia about the axis of rotation was defined as

$$I = m_1 r_1^2 + m_2 r_2^2 + m_3 r_3^2 + m_4 r_4^2 \tag{9.14}$$

For any number of masses, this definition can be generalized to

$$I = \sum_{i=1}^{n} m_i r_i^2 \tag{9.16}$$

where the Greek letter sigma, Σ, means the "sum of," as used before. The subscript i on m and r means that the index i takes on the values from 1 up to the number n. So when $n = 4$ the identical result found in equation 9.14 is obtained.

For the very special case of *the moment of inertia of a single mass m rotating about an axis,* equation 9.16 reduces to ($i = n = 1$),

$$I = mr^2 \tag{9.17}$$

Thus, the significant feature for rotational motion is not the mass of the rotating body, but rather the square of the distance of that body from the axis of rotation. A small mass m, at a great distance r from the axis of rotation, has a greater moment of inertia than a large mass, very close to the axis of rotation.

For continuous mass distributions, the moments of inertia are given in figure 9.5. More extensive tables of moments of inertia are found in various handbooks, such as the *Handbook of Chemistry and Physics* (published by the Chemical Rubber Co. Press, Cleveland, Ohio), if the need for them arises.

It is important to note here that when we ask for the moment of inertia of a body, we must specify about what axis the rotation will occur. Because r is different for each axis, and since I varies as r^2, I is also different for each axis. As an example,

Figure 9.5

Moments of inertia for various mass distributions.

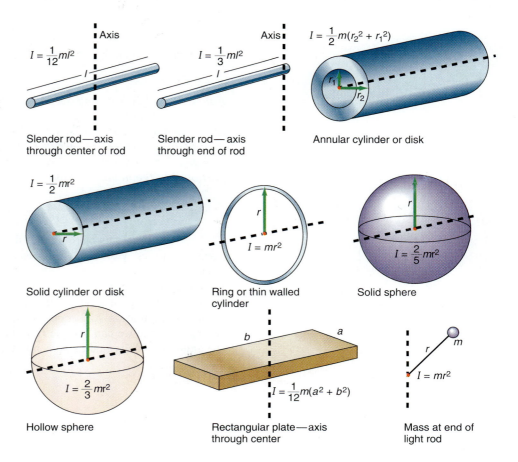

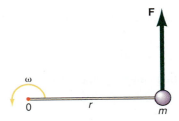

Figure 9.6

Torque causes a body to rotate.

consider the slender rod in figure 9.5. When the axis is taken through the center of the rod, as shown, $I = \frac{1}{12} ml^2$, while if the axis of rotation is at the end of the rod, then $I = \frac{1}{3} ml^2$. The unit for the moment of inertia is kg m² and has no special name.

9.5 Newton's Laws for Rotational Motion

Let us consider a single mass m connected by a rigid rod, of negligible mass, to an axis passing through the point 0, as shown in figure 9.6. Let us apply a tangential force F, in the plane of the page, to the body of mass m. The force acting on the constrained body causes a torque, given by

$$\tau = rF \tag{9.18}$$

This torque causes the body to rotate about the axis through 0. The force F acting on the mass m causes a tangential acceleration given by Newton's second law as

$$F = ma \tag{9.19}$$

If we substitute equation 9.19 into 9.18, we have

$$\tau = rma \tag{9.20}$$

But the tangential acceleration a is related to the angular acceleration by

$$a = r\alpha \tag{9.5}$$

Substituting this into equation 9.20, gives

$$\tau = rm(r\alpha) = mr^2\alpha \tag{9.21}$$

But, as already seen, the moment of inertia of a single mass rotating about an axis is

$$I = mr^2 \tag{9.17}$$

Therefore, equation 9.21 becomes

$$\tau = I\alpha \tag{9.22}$$

Equation 9.22 is Newton's second law for rotational motion. Although this equation was derived for a single mass, it is true in general, and **Newton's second law for rotational motion** can be stated as: *When an unbalanced external torque acts on a body of moment of inertia I, it gives that body an angular acceleration, α. The angular acceleration is directly proportional to the torque and inversely proportional to the moment of inertia,* that is,

$$\alpha = \frac{\tau}{I} \tag{9.23}$$

The problems of rotational dynamics are very similar to those in translational dynamics. We will consider rotational motion only in the x-y plane. The angular displacement vector, angular velocity vector, angular acceleration vector, and the torque vector are all perpendicular to the plane of the rotation. By determining the torque acting on a body, we can find the angular acceleration from Newton's second law, equation 9.23. For constant torque, the angular acceleration is a constant and hence we can use the rotational kinematic equations. Therefore, we find the angular velocity and displacement at any time from the kinematic equations

$$\omega = \omega_0 + \alpha t \tag{9.4}$$

and

$$\theta = \omega_0 t + \frac{1}{2}\alpha t^2 \tag{9.9}$$

Mechanics

To determine Newton's first law for rotational motion, we note that

$$\tau = I\alpha = I\frac{\Delta\omega}{\Delta t}$$

and if there is no external torque (i.e., if $\tau = 0$), then

$$\Delta\omega = 0$$
$$\omega_f - \omega_i = 0$$

or

$$\omega_f = \omega_i \qquad\qquad \text{(9.24)}$$

That is, equation 9.24 says that if there is no external torque acting on a body, then a body rotating at an initial angular velocity ω_i will continue to rotate at that same angular velocity forever.

Stated in more formal terms, **Newton's first law for rotational motion** is *A body in motion at a constant angular velocity will continue in motion at that same angular velocity, unless acted on by some unbalanced external torque.*

One of the most obvious examples of Newton's first law for rotational motion is the earth itself. Somehow, someway in its creation, the earth was given an initial angular velocity ω_i of 7.27×10^{-5} rad/s. Since there is no external torque acting on the earth it continues to rotate at this same angular velocity.

For completeness, we can state **Newton's third law of rotational motion** as *If body A and body B have the same common axis of rotation, and if body A exerts a torque on body B, then body B exerts an equal but opposite torque on body A.* That is, if body A exerts a torque on body B that tends to rotate body B in a clockwise direction, then body B will exert a torque on body A that will tend to rotate body A in a counterclockwise direction. An application of this principle is found in a helicopter (see figure 9.7). As the main rotor blades above the helicopter turn counterclockwise, the helicopter itself would start to turn clockwise. To prevent this rotation of the helicopter, a second but smaller set of rotor blades are located at the side and end of the helicopter to furnish a countertorque to prevent the helicopter from turning.

9.6 Rotational Dynamics

Now let us look at some examples of the use of Newton's laws in solving problems in rotational motion.

Example 9.7

Rotational dynamics of a cylinder. Consider a solid cylinder of mass $m = 3.00$ kg and radius $r = 0.500$ m, which is free to rotate about an axis through its center, as shown in figure 9.8. The cylinder is initially at rest when a constant force of 8.00 N is applied tangentially to the cylinder. Find (a) the moment of inertia of the cylinder, (b) the torque acting on the cylinder, (c) the angular acceleration of the cylinder, (d) its angular velocity after 10.0 s, and (e) its angular displacement after 10.0 s.

Solution

a. The equation for the moment of inertia of a cylinder about its main axis, found in figure 9.5, is

$$I = \tfrac{1}{2}mr^2$$
$$= \tfrac{1}{2}(3.00 \text{ kg})(0.500 \text{ m})^2$$
$$= 0.375 \text{ kg m}^2$$

Figure 9.7

Newton's third law for rotational motion and the helicopter.

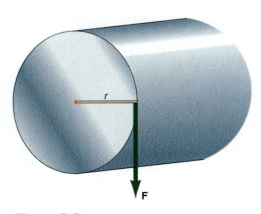

Figure 9.8

Rotational motion of a cylinder.

b. The torque acting on the cylinder is the product of the force times the lever arm. From figure 9.8, we see that the lever arm is just the radius of the cylinder. Therefore,

$$\tau = rF \tag{9.18}$$

$$= (0.500 \text{ m})(8.00 \text{ N})$$

$$= 4.00 \text{ m N}$$

c. The angular acceleration of the cylinder, determined by Newton's second law, is

$$\alpha = \frac{\tau}{I} \tag{9.23}$$

$$= \frac{4.00 \text{ m N}}{0.375 \text{ kg m}^2} = \frac{10.7 \text{ m kg m/s}^2}{\text{kg m}^2}$$

$$= 10.7 \text{ rad/s}^2$$

Note that in the solution all the units cancel out except the s² in the denominator. We then introduced the unit radian in the numerator to give us the desired unit for angular acceleration, namely, rad/s². Recall that the radian is a unit that can be multiplied by or divided into an equation at will, because the radian is a dimensionless quantity. It was defined as the ratio of the arc length to the radius of the circle,

$$\theta = \frac{s}{r} = \frac{\text{meter}}{\text{meter}} = 1 = \text{radian}$$

d. To determine the angular velocity of the rotating cylinder we use the kinematic equation for the angular velocity, namely

$$\omega = \omega_0 + \alpha t \tag{9.4}$$

$$= 0 + \left(10.7 \frac{\text{rad}}{\text{s}^2}\right)(10.0 \text{ s})$$

$$= 107 \text{ rad/s}$$

e. The angular displacement, found by the kinematic equation, is

$$\theta = \omega_0 t + \tfrac{1}{2}\alpha t^2 \tag{9.9}$$

$$= 0 + \tfrac{1}{2}(10.7 \text{ rad/s}^2)(10.0 \text{ s})^2$$

$$= 535 \text{ rad}$$

Example 9.8

Combined translational and rotational motion of a sphere rolling down an inclined plane. A solid sphere of 1.00 kg mass rolls down an inclined plane, as shown in figure 9.9. Find (a) the acceleration of the sphere, (b) its velocity at the bottom of the 1.00 m long plane, and (c) the frictional force acting on the sphere.

Solution

a. First, we draw all the forces acting on the sphere. The component of the weight acting down the plane, $w \sin \theta$, is shown acting through the center of mass of the sphere. Because of this force there is a tendency for the sphere to slide down the plane. A force of static friction opposes this motion and is directed up the plane, as shown. This frictional force can not be shown as acting at the center of the body as was done in problems with "blocks" sliding on the inclined plane. It is this frictional force acting at the point of

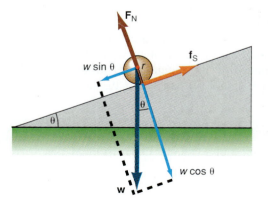

Figure 9.9

A sphere rolling down an inclined plane.

contact of the sphere that creates the necessary torque to rotate the sphere so that it rolls down the plane. The motion is therefore composed of two motions, the translation of the center of mass of the sphere, and the rotation about the center of mass of the sphere. Applying Newton's second law for the translational motion of the center of mass of the sphere gives

$$F = ma$$

$$w \sin \theta - f_s = ma \qquad (9.25)$$

Applying the second law for the rotation of the sphere about its center of mass gives

$$\tau = I\alpha \qquad (9.22)$$

But the torque is the product of the frictional force f_s and the radius of the sphere. Therefore,

$$f_s r = I\alpha \qquad (9.26)$$

Now we eliminate the frictional force f_s between the two equations 9.25 and 9.26. That is, from 9.26,

$$f_s = \frac{I\alpha}{r}$$

Substituting this into equation 9.25, we get

$$w \sin \theta - \frac{I\alpha}{r} = ma$$

The moment of inertia of a solid sphere, found from figure 9.5, is

$$I = \tfrac{2}{5}mr^2 \qquad (9.27)$$

Therefore,

$$ma = w \sin \theta - (\tfrac{2}{5}mr^2)\frac{\alpha}{r}$$

$$= w \sin \theta - \tfrac{2}{5}mr\alpha$$

But recall that

$$a = r\alpha \qquad (9.5)$$

Therefore,

$$ma = w \sin \theta - \tfrac{2}{5}ma$$

$$ma + \tfrac{2}{5}ma = mg \sin \theta$$

$$\tfrac{7}{5}a = g \sin \theta$$

Solving for the acceleration of the sphere, we get

$$a = \tfrac{5}{7}g \sin \theta \qquad (9.28)$$

$$a = \tfrac{5}{7}(9.80 \text{ m/s}^2)\sin 30.0°$$

$$= 3.50 \text{ m/s}^2$$

b. The velocity of the center of mass of the sphere at the bottom of the plane is found from the kinematic equation

$$v^2 = v_0^2 + 2ax \qquad (3.16)$$

Because the sphere starts from rest, $v_0 = 0$. Therefore,

$$\begin{aligned}
v &= \sqrt{2ax} \\
&= \sqrt{(2)(3.50 \text{ m/s}^2)(1.00 \text{ m})} \\
&= 2.65 \text{ m/s}
\end{aligned}$$

c. The frictional force can be determined from equation 9.25, that is,

$$\begin{aligned}
w \sin \theta - f_s &= ma \\
f_s &= w \sin \theta - ma \\
&= mg \sin \theta - m(\tfrac{5}{7}g \sin \theta) \\
&= \tfrac{2}{7}mg \sin \theta \\
&= \tfrac{2}{7}(1.00 \text{ kg})(9.80 \text{ m/s}^2)\sin 30.0° \\
&= 1.40 \text{ N}
\end{aligned}$$

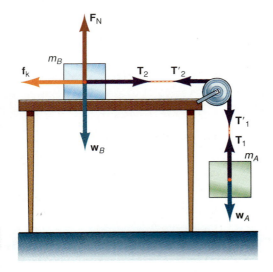

Figure 9.10

Combined motion taking the rotational motion of the pulley into account.

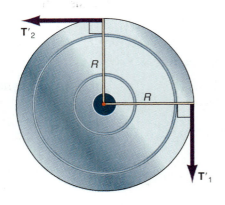

Figure 9.11

Forces acting on the pulley.

As we can see the general motion of a rigid body can become quite complicated. We will see in section 9.8 how these problems can be simplified by the use of the law of conservation of energy.

†Combined Translational and Rotational Motion Treated by Newton's Second Law

It is appropriate here to return to some of the problems discussed in chapter 4, in which we assumed that the tension in the rope on both sides of a pulley are equal. Let us analyze these problems taking the rotational motion of the pulley into account. Consider the problem of a block moving on a rough horizontal surface, as shown in figure 9.10. What is the acceleration of each block in the system?

Applying Newton's second law to block A, we obtain

$$T_1 - w_A = -m_A a \qquad (9.29)$$

Applying the second law to block B, we get

$$T_2 - f_k = m_B a \qquad (9.30)$$

We find the frictional force f_k from

$$f_k = \mu_k F_N = \mu_k w_B$$

Substituting this into equation 9.30, gives

$$T_2 - \mu_k w_B = m_B a \qquad (9.31)$$

It was at this point in chapter 4 that we made the assumption that the tension $T_1 = T_2$, and then determined the acceleration of each block of the system. Let us now look a little more closely at the assumption of the equality of tensions. The string exerts a force T_1 upward on weight w_A, but by Newton's third law the weight w_A exerts a force down on the string, call it T_1'. Figure 9.11 shows the pulley with the appropriate tensions in the string. The force T_1' acting on the pulley causes a torque

$$\tau_1 = T_1' R$$

which tends to rotate the pulley clockwise. The radius of the pulley is R.

Similarly, the string exerts a tension force T_2 on mass m_B. But by Newton's third law, block B exerts a force on the string, which we call T_2'. The force T_2' causes a counterclockwise torque about the axis of the pulley, given by

$$\tau_2 = T_2' R$$

Because the motion of the system causes the pulley to rotate in a clockwise direction, the net torque on the pulley is equal to the difference in these two torques, namely,

$$\tau = \tau_1 - \tau_2$$

$$\tau = T_1' R - T_2' R \qquad (9.32)$$

But by Newton's second law for rotational motion,

$$\tau = I\alpha \qquad (9.22)$$

Substituting equation 9.22 into equation 9.32, gives

$$I\alpha = T_1' R - T_2' R$$

$$I\alpha = (T_1' - T_2')R \qquad (9.33)$$

From figure 9.10 and Newton's third law, we have

$$\left. \begin{array}{c} T_1' = T_1 \\ T_2' = T_2 \end{array} \right\} \qquad (9.34)$$

Substituting equations 9.34 into equation 9.33, gives

$$I\alpha = (T_1 - T_2)R \qquad (9.35)$$

However, the angular acceleration $\alpha = a/R$. Therefore, equation 9.35 becomes

$$I\frac{a}{R} = (T_1 - T_2)R \qquad (9.36)$$

The pulley resembles a disk, whose moment of inertia, found from figure 9.5, is $I_{Disk} = \frac{1}{2}MR^2$, where M is the mass of the pulley and R is the radius of the pulley. Substituting this result into equation 9.36, gives

$$(\tfrac{1}{2}MR^2)\frac{a}{R} = (T_1 - T_2)R$$

Simplifying,

$$\tfrac{1}{2}Ma = (T_1 - T_2) \qquad (9.37)$$

There are now three equations 9.29, 9.31, and 9.37 in terms of the three unknowns a, T_1, and T_2. Solving equation 9.29 for T_1, gives

$$T_1 = w_A - m_A a \qquad (9.38)$$

Solving equation 9.31 for T_2, gives

$$T_2 = \mu_k w_B + m_B a \qquad (9.39)$$

Subtracting equation 9.39 from equation 9.38, we get

$$T_1 - T_2 = w_A - m_A a - \mu_k w_B - m_B a$$

Substituting for $T_1 - T_2$ from equation 9.37, gives

$$\tfrac{1}{2}Ma = w_A - m_A a - \mu_k w_B - m_B a$$

Gathering the terms with a in them to the left-hand side of the equation, we get

$$\tfrac{1}{2}Ma + m_A a + m_B a = w_A - \mu_k w_B$$

Factoring out the a, and writing each weight w as mg, we get

$$a(\tfrac{1}{2}M + m_A + m_B) = m_A g - \mu_k m_B g$$

Solving for the acceleration of the system, we have

$$a = \frac{(m_A - \mu_k m_B)g}{m_A + m_B + M/2} \tag{9.40}$$

It is immediately apparent in equation 9.40 that the acceleration of the system depends on the mass M of the pulley. If this mass is very small compared to the masses m_A and m_B (i.e., $M \approx 0$), then equation 9.40 would reduce to the simpler problem already found in equation 4.62.

Example 9.9

Combined translational and rotational motion. If $m_A = 2.00$ kg, $m_B = 6.00$ kg, $\mu_k = 0.300$, and $M = 8.00$ kg in figure 9.10, find the acceleration of each block of the system.

Solution

The acceleration of each block in the system, found from equation 9.40, is

$$a = \frac{(m_A - \mu_k m_B)g}{m_A + m_B + M/2}$$

$$= \frac{[2.00 \text{ kg} - (0.300)(6.00 \text{ kg})](9.80 \text{ m/s}^2)}{2.00 \text{ kg} + 6.00 \text{ kg} + 8.00 \text{ kg}/2}$$

$$= 0.163 \text{ m/s}^2$$

If we compare this example with example 4.12 in chapter 4, we see a relatively large difference in the acceleration of the system by assuming M to be negligible.

Example 9.10

The effect of a smaller pulley. Let us repeat example 9.9, but now let us use a much smaller plastic pulley, with $M = 25$ g. Find the acceleration of each block of the system.

Solution

The acceleration of each block, again found from equation 9.40, is

$$a = \frac{(m_A - \mu_k m_B)g}{m_A + m_B + M/2}$$

$$= \frac{[2.00 \text{ kg} - (0.300)(6.00 \text{ kg})](9.80 \text{ m/s}^2)}{2.00 \text{ kg} + 6.00 \text{ kg} + 0.025 \text{ kg}/2}$$

$$= 0.244 \text{ m/s}^2$$

which agrees very closely to the value found in example 4.12, of chapter 4, when the effect of the pulley was assumed to be negligible.

Example 9.11

The tension in the strings. If the radius of the pulley is 5.00 cm, find the tension in the strings of examples 9.9 and 9.10.

For example 9.9, the tension T_1, found from equation 9.38, is

$$T_1 = w_A - m_A a = m_A g - m_A a = m_A(g - a)$$
$$= (2.00 \text{ kg})(9.80 \text{ m/s}^2 - 0.163 \text{ m/s}^2)$$
$$= 19.3 \text{ N}$$

Tension T_2, found from equation 9.39, is

$$T_2 = \mu_k w_B + m_B a$$
$$= \mu_k m_B g + m_B a$$
$$= (0.300)(6.00 \text{ kg})(9.80 \text{ m/s}^2) + (6.00 \text{ kg})(0.163 \text{ m/s}^2)$$
$$= 17.6 \text{ N} + 0.978 \text{ N}$$
$$= 18.6 \text{ N}$$

Thus the tensions in the strings on both sides of the pulley are unequal. It is this difference in the tensions that causes the torque,

$$\tau = R(T_1 - T_2)$$
$$= (0.05 \text{ m})(19.3 \text{ N} - 18.6 \text{ N})$$
$$= 3.50 \times 10^{-2} \text{ m N}$$

on the pulley. This torque gives the pulley its angular acceleration.

For example 9.10, the tension T_1 is again found from equation 9.38, only now the acceleration of the system is 0.244 m/s². Thus,

$$T_1 = w_A - m_A a = m_A g - m_A a = m_A(g - a)$$
$$= (2.00 \text{ kg})(9.80 \text{ m/s}^2 - 0.244 \text{ m/s}^2)$$
$$= 19.1 \text{ N}$$

Tension T_2, found from equation 9.39, is

$$T_2 = \mu_k w_B + m_B a$$
$$= \mu_k m_B g + m_B a$$
$$= (0.300)(6.00 \text{ kg})(9.80 \text{ m/s}^2) + (6.00 \text{ kg})(0.244 \text{ m/s}^2)$$
$$= 17.6 \text{ N} + 1.46 \text{ N}$$
$$= 19.1 \text{ N}$$

Hence in this case, where the pulley has a small mass, the tensions are equal, at least to three significant figures, and there is no resultant torque to cause the pulley to rotate. The two tensions must be different to cause a net torque to rotate the pulley.

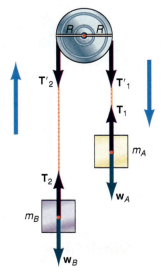

Figure 9.12

Atwood's machine with the rotational motion of the pulley taken into account.

†Atwood's Machine

Let us reconsider the Atwood's machine problem solved in chapter 4, only this time we no longer assume the tensions on each side of the pulley to be equal, figure 9.12. We apply Newton's second law to mass m_A to obtain

$$T_1 - w_A = -m_A a \qquad (9.41)$$

Applying the second law to mass m_B, we obtain

$$T_2 - w_B = m_B a \tag{9.42}$$

Let us now consider the pulley. The tension T_1' causes a clockwise torque,

$$\tau_1 = T_1' R$$

whereas the tension T_2' causes the counterclockwise torque,

$$\tau_2 = T_2' R$$

The net torque acting on the pulley is

$$\tau = \tau_1 - \tau_2 = T_1' R - T_2' R$$

But by Newton's second law for rotational motion

$$\tau = I\alpha$$

Therefore,

$$I\alpha = (T_1' - T_2')R \tag{9.43}$$

However, by Newton's third law of motion

$$T_2' = T_2$$
$$T_1' = T_1$$

Hence, the second law, equation 9.43, becomes

$$I\alpha = (T_1 - T_2)R \tag{9.44}$$

The moment of inertia of the pulley, found in figure 9.5, is

$$I_{\text{disk}} = \tfrac{1}{2}MR^2$$

and the angular acceleration is given by $\alpha = a/R$. Substituting these two values into equation 9.44, gives

$$\tfrac{1}{2}MR^2(a/R) = (T_1 - T_2)R$$

or

$$\tfrac{1}{2}Ma = (T_1 - T_2) \tag{9.45}$$

There are now three equations, 9.41, 9.42, and 9.45, for the three unknowns, a, T_1, and T_2. Subtracting equation 9.41 from equation 9.42, we get

$$T_2 - w_B - T_1 + w_A = m_B a + m_A a$$

or

$$T_2 - T_1 = w_B - w_A + m_B a + m_A a \tag{9.46}$$

Substituting the value for $T_2 - T_1$ from equation 9.45 into equation 9.46, we get

$$-\tfrac{1}{2}Ma = w_B - w_A + m_B a + m_A a$$

Gathering the terms with the acceleration a onto one side of the equation,

$$\tfrac{1}{2}Ma + m_B a + m_A a = w_A - w_B$$

Factoring out the a, we get,

$$a(\tfrac{1}{2}M + m_B + m_A) = w_A - w_B$$

Expressing the weights as $w = mg$, and solving for the acceleration of each mass of the system, we get

$$a = \frac{(m_A - m_B)g}{m_A + m_B + M/2} \tag{9.47}$$

Equation 9.47 is the acceleration of each mass in Atwood's machine, when the rotational motion of the pulley is taken into account. If the mass of the pulley M is very small, then equation 9.47 reduces to the simplified solution in equation 4.38.

Example 9.12

Combined motion in an Atwood's machine. In an Atwood's machine, $m_B = 30.0$ g, $m_A = 50.0$ g, and the mass M of the pulley is 2.00 kg. Find the acceleration of each mass.

Solution

The acceleration, found from equation 9.47, is

$$a = \frac{(m_A - m_B)g}{m_A + m_B + M/2}$$

$$= \frac{(50.0\ \text{g} - 30.0\ \text{g})(9.80\ \text{m/s}^2)}{[50.0\ \text{g} + 30.0\ \text{g} + (2000\ \text{g})/2]}$$

$$= 0.181\ \text{m/s}^2$$

If the pulley were made of light plastic and, therefore, had a negligible mass, then the acceleration of the system would have been, $a = 2.45$ m/s², which is a very significant difference.

Example 9.13

Velocity in an Atwood's machine. If block m_A of example 9.12, located a distance $h_A = 2.00$ m above the floor, falls from rest, find its velocity as it hits the floor.

Solution

Because m_A falls at the constant acceleration given by equation 9.47, the kinematic equation can be used to find its velocity at the floor. Thus,

$$v^2 = v_0^2 + 2ay$$

$$v = \sqrt{2ay} = \sqrt{2ah_A}$$

$$= \sqrt{2(0.181\ \text{m/s}^2)(2.00\ \text{m})}$$

$$= 0.850\ \text{m/s}$$

9.7 Angular Momentum and Its Conservation

Just as the linear momentum of a body was defined as the product of its mass and its linear velocity, $p = mv$, the **angular momentum** of a rotating body is defined as *the product of its moment of inertia and its angular velocity.* That is, the angular momentum L, with respect to a given axis, is defined as

$$L = I\omega \tag{9.48}$$

As the concept of momentum led to an alternative form of Newton's second law for the translational case, angular momentum also leads to an alternative form for the rotational case, as shown below.

Translational Case	Rotational Case
$F = ma = m\dfrac{\Delta v}{\Delta t}$	$\tau = I\alpha = I\dfrac{\Delta \omega}{\Delta t}$
$F = m\left(\dfrac{v_f - v_i}{\Delta t}\right)$	$\tau = I\left(\dfrac{\omega_f - \omega_i}{\Delta t}\right)$
$F = \dfrac{mv_f - mv_i}{\Delta t} = \dfrac{p_f - p_i}{\Delta t}$	$\tau = \dfrac{I\omega_f - I\omega_i}{\Delta t} = \dfrac{L_f - L_i}{\Delta t}$
$F = \dfrac{\Delta p}{\Delta t}$	$\tau = \dfrac{\Delta L}{\Delta t}$

Thus, we can write Newton's second law in terms of angular momentum as

$$\tau = \frac{\Delta L}{\Delta t} \tag{9.49}$$

If we apply equation 9.49 to a system of bodies, the total torque τ arises from two sources, external torques and internal torques. Because of Newton's third law for rotational motion, the internal torques will add to zero and equation 9.49 becomes

$$\tau_{\text{ext}} = \frac{\Delta L}{\Delta t} \tag{9.50}$$

If the total external torque acting on the system is zero, then

$$0 = \frac{\Delta L}{\Delta t}$$

$$\Delta L = 0 \tag{9.51}$$

$$L_f - L_i = 0$$

Therefore,

$$L_f = L_i \tag{9.52}$$

*Equations 9.51 and 9.52 are a statement of the **law of conservation of angular momentum**. They say: if the total external torque acting on a system is zero, then there is no change in the angular momentum of the system, and the final angular momentum is equal to the initial angular momentum.*

Let us now consider some examples of the conservation of angular momentum.

The Rotating Earth

Because there is no external torque acting on the earth, $\tau = 0$, and there is conservation of angular momentum. Hence,

$$L_f = L_i \tag{9.52}$$

But since the angular momentum is the product of the moment of inertia and the angular velocity, this becomes

$$I_f\omega_f = I_i\omega_i \tag{9.53}$$

Figure 9.13

Because there is no torque acting on the earth, its angular momentum is conserved, and it will continue to spin with the same angular velocity forever.

However, the moment of inertia of the earth does not change with time and thus, $I_f = I_i$. Therefore,

$$\omega_f = \omega_i$$

That is, the angular velocity of the earth is a constant and will continue to spin forever with the same angular velocity unless it is acted on by some external torque. We also assume that the moment of inertia of the earth does not change.

The Spinning Ice Skater

The familiar picture of the spinning ice skater, as shown in figure 9.14, gives another example of the conservation of angular momentum. As the skater (body A) pushes against the ice (body B), thereby creating a torque, the ice (body B) pushes back on the skater (body A), creating a torque on her. The net torque on the skater and the ice is therefore zero and angular momentum is conserved.

Because the earth is so massive there will be no measurable change in the angular momentum of the earth and we need consider only the skater. The skater first starts spinning relatively slowly with her hands outstretched. We assume that any friction between the skater and the ice is negligible. As the skater draws her arms to her sides, she starts to spin very rapidly. Let us analyze the motion by the law of conservation of angular momentum. The conservation of angular momentum gives

$$L_f = L_i \tag{9.52}$$

or

$$I_f \omega_f = I_i \omega_i \tag{9.53}$$

For simplicity of calculation, let us assume that the skater is holding a set of dumbbells in her hands so that her moment of inertia can be considered to come only from the dumbbells. (That is, we assume that the moment of inertia of the girl's hands and arms can be considered negligible compared to the dumbbells in order to simplify the calculation.) The skater's initial moment of inertia is

$$I_i = mr_i^2$$

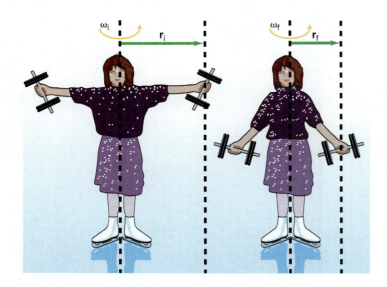

Figure 9.14
The spinning ice skater.

where m is the mass of the dumbbells and r_i the distance from the center of the body (the axis of rotation) to the outstretched dumbbells. When the skater pulls her hand down to her side the new moment of inertia is

$$I_f = mr_f^2$$

where r_f is now the distance from the axis of rotation to the dumbbell, as seen in figure 9.14(b). As we can immediately see from the figure, r_f is less than r_i, therefore I_f is less than I_i. But if the moment of inertia is changing, what happens to the skater as a consequence of the conservation of angular momentum? The angular momentum must remain the same, as given by equation 9.53. The final angular momentum must be equal to the initial angular momentum, which is equal to the product of $I_i\omega_i$, which remains a constant. Thus, the final angular momentum $I_f\omega_f$ must equal that same constant. But if I_f has decreased, the only way to maintain the equality is to have the final value of the angular velocity ω_f increase. And this is, in fact, exactly what happens. As the girl's arms are dropped to her side, the spinning increases. When the skater wishes to come out of the spin, she merely raises her arms to the original outstretched position, her moment of inertia increases and her angular velocity decreases.

A Man Diving from a Diving Board

When a man pushes down on a diving board, the board reacts by pushing back on him, as in figure 9.15. As the man leans forward at the start of the dive, the reaction force on him causes a torque to set him into rotational motion, about an axis through his center of mass, with a relatively small angular velocity ω_i. As the man leaves the board there is no longer a torque acting on him, and his angular momentum must be conserved. His initial moment of inertia is I_i, and he is spinning at an an-

Figure 9.15

A man diving from a diving board.

Mechanics

gular velocity ω_i. If he now bends his knees and pulls his legs and arms up toward himself to form a ball, his moment of inertia decreases to a value I_f. But by the conservation of angular momentum

$$I_f\omega_f = I_i\omega_i \qquad (9.53)$$

Since I_f has decreased, his angular velocity ω_f must increase to maintain the equality of the conservation of momentum. The man now rotates relatively rapidly for one or two turns. He then stretches his body out to its original configuration with the larger value of the moment of inertia. His angular velocity then decreases to the relatively low value ω_i that he started with. If he has timed his dive properly, his outstretched body will enter the water head first at the end of his dive. The force of gravity acts on the man throughout the motion and causes the center of mass of the man to follow the parabolic trajectory associated with any projectile. Thus, the center of mass of the man is moving under the force of gravity while the man rotates around his center of mass. A trapeze artist uses the same general techniques when she rotates her body while moving through the air from one trapeze to another.

Rotational Collision (an Idealized Clutch)

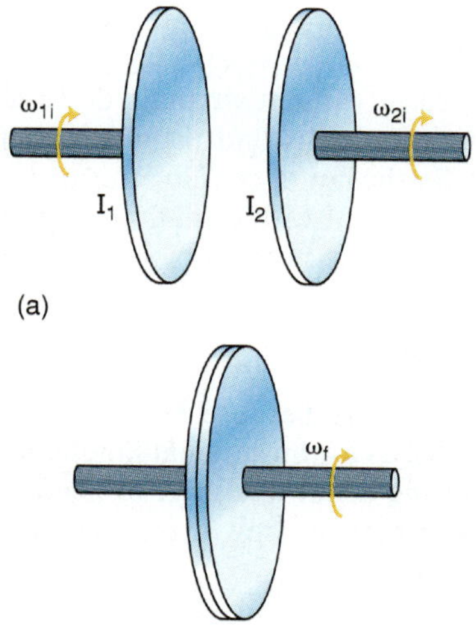

Consider two disks rotating independently, as shown in figure 9.16(a). The original angular momentum of the two rotating disks is the sum of the angular momentum of each disk, that is,

$$L_i = L_{1i} + L_{2i} \qquad (9.54)$$

The initial angular momentum of disk 1 is

$$L_{1i} = I_1\omega_{1i}$$

and the initial angular momentum of disk 2 is

$$L_{2i} = I_2\omega_{2i}$$

Hence, the total initial angular momentum is

$$L_i = I_1\omega_{1i} + I_2\omega_{2i} \qquad (9.55)$$

The two disks are now forced together along their axes. Initially there may be some slipping of the disks but very quickly the two disks are coupled together by friction and spin as one, with one final angular velocity ω_f, as shown in figure 9.16(b). During the coupling process disk 1 exerted a torque on disk 2, while by Newton's third law, disk 2 exerted an equal but opposite torque on disk 1. Therefore, the net torque is zero and angular momentum must be conserved; that is, the final value of the angular momentum must equal the initial value:

$$L_f = L_i \qquad (9.52)$$

The final value of the angular momentum is the sum of the angular momentum of each disk:

$$L_f = L_{1f} + L_{2f}$$

The final value of the angular momentum of disk 1 is

$$L_{1f} = I_1\omega_f$$

while for disk 2, we have

$$L_{2f} = I_2\omega_f$$

Note that both disks have the same final angular velocity, since they are coupled together. The final momentum is therefore

$$L_f = I_1\omega_f + I_2\omega_f = (I_1 + I_2)\omega_f \qquad (9.56)$$

Figure 9.16

Rotational collision—the clutch.

Substituting equations 9.55 and 9.56 into the conservation of angular momentum, equation 9.52, we get

$$(I_1 + I_2)\omega_f = I_1\omega_{1i} + I_2\omega_{2i} \qquad (9.57)$$

Solving for the final angular velocity of the coupled disks, we have

$$\omega_f = \frac{I_1\omega_{1i} + I_2\omega_{2i}}{I_1 + I_2} \qquad (9.58)$$

This idealized device is the basis of a clutch. For a real clutch, the first spinning disk could be attached to the shaft of a motor, while the second disk could be connected through a set of gears to the wheels of the vehicle. When disk 2 is coupled to disk 1, the wheels of the vehicle turn. When the disks are separated, the wheels are disengaged.

9.8 Combined Translational and Rotational Motion Treated by the Law of Conservation of Energy

Let us now consider the motion of a ball that rolls, without slipping, down an inclined plane, as shown in figure 9.17. In particular, let us find the velocity of the ball at the bottom of the one meter long incline. By the law of conservation of energy, the total energy at the top of the plane must be equal to the total energy at the bottom of the plane. Because the ball is initially at rest at the top of the plane, all the energy at the top is potential energy:

$$E_{top} = PE_{top} = mgh$$

At the bottom of the plane the potential energy is zero because $h = 0$. Since the body is translating at the bottom of the incline, it has a translational kinetic energy of its center of mass of $\frac{1}{2}mv^2$. But it is also rotating about its center of mass at the bottom of the plane, and therefore it also has a kinetic energy of rotation of $\frac{1}{2}I\omega^2$. Therefore the total energy at the bottom of the plane is

$$E_{bot} = KE_{trans} + KE_{rot}$$
$$E_{bot} = \frac{1}{2}mv^2 + \frac{1}{2}I\omega^2 \qquad (9.59)$$

Equating the total energy at the bottom to the total energy at the top, we have

$$E_{bot} = E_{top}$$
$$\frac{1}{2}mv^2 + \frac{1}{2}I\omega^2 = mgh \qquad (9.60)$$

The moment of inertia for the ball is the same as a solid sphere,

$$I = \frac{2}{5}mr^2 \qquad (9.61)$$

The angular velocity ω of the rotating ball is related to the linear velocity of a point on the surface of the ball by

$$\omega = \frac{v}{r} \qquad (9.62)$$

The distance that a point on the edge of the ball moves along the incline is the same as the distance that the center of mass of the ball moves along the incline. Hence,

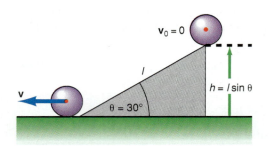

Figure 9.17

Combined translational and rotational motion.

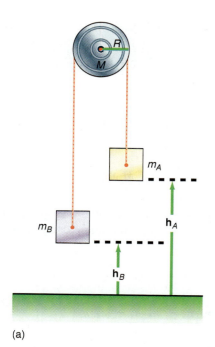

(a)

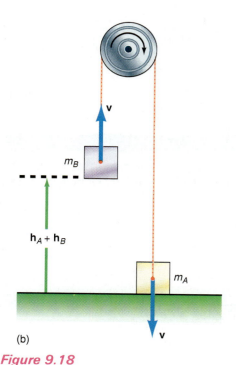

(b)

Figure 9.18
Atwood's machine revisited.

the velocity of the edge of the ball is equal to the velocity of the center of mass of the ball. Substituting equations 9.61 and 9.62 into equation 9.60, we have

$$\frac{1}{2}mv^2 + \frac{1}{2}\left(\frac{2}{5} mr^2\right)\left(\frac{v}{r}\right)^2 = mgh$$

$$\frac{1}{2} mv^2 + \frac{1}{2}\frac{2}{5} mr^2\frac{v^2}{r^2} = mgh$$

Simplifying,

$$\frac{v^2}{2} + \frac{v^2}{5} = gh$$

$$\frac{5v^2 + 2v^2}{10} = gh$$

$$\frac{7v^2}{10} = gh$$

$$v = \sqrt{\left(\frac{10}{7}\right)gh} \qquad (9.63)$$

the velocity of the ball at the bottom of the plane. The height h, found from the trigonometry of the triangle in figure 9.17, is

$$h = l \sin \theta = (1 \text{ m})\sin 30.0° = 0.500 \text{ m}$$

Therefore the velocity at the bottom of the plane is

$$v = \sqrt{(10/7)(9.80 \text{ m/s}^2)(0.500 \text{ m})}$$
$$= 2.65 \text{ m/s}$$

Note that this is the same result obtained in example 9.8 in section 9.6. The energy approach is obviously much easier.

As another example of the combined translational and rotational motion of a rigid body, let us consider the Atwood's machine shown in figure 9.18(a). Using the law of conservation of energy let us find the velocity of the mass m_A as it hits the ground.

The total energy of the system in the configuration shown consists only of the potential energy of the two masses m_A and m_B, that is,

$$E_{tot} = m_Agh_A + m_Bgh_B \qquad (9.64)$$

When the system is released, m_A loses potential energy as it falls but gains kinetic energy due to its motion. Mass, m_B gains potential energy as it rises and also acquires a kinetic energy. The pulley, when set into rotational motion, also has kinetic energy of rotation. The total energy of the system as m_A strikes the ground, found from figure 9.18(b), is

$$E_{tot} = PE_B + KE_A + KE_B + KE_{pulley}$$

$$E_{tot} = m_Bg(h_A + h_B) + \tfrac{1}{2}m_Av^2 + \tfrac{1}{2}m_Bv^2 + \tfrac{1}{2}I\omega^2 \qquad (9.65)$$

The speed of masses A and B are equal because they are tied together by the string. The moment of inertia of the pulley (disk), found from figure 9.5, is

$$I_{disk} = \tfrac{1}{2}MR^2 \qquad (9.66)$$

Also, the angular velocity ω of the disk is related to the tangential velocity of the string as it passes over the pulley by

$$\omega = \frac{v}{R} \qquad (9.67)$$

Substituting equations 9.66 and 9.67 into equation 9.65, gives

$$E_{tot} = m_Bg(h_A + h_B) + \frac{1}{2}m_Av^2 + \frac{1}{2}m_Bv^2 + \frac{1}{2}\left(\frac{1}{2}MR^2\right)\left(\frac{v}{R}\right)^2$$

Simplifying,

$$E_{tot} = m_Bg(h_A + h_B) + \frac{1}{2}(m_A + m_B)v^2 + \frac{1}{4}Mv^2$$

or

$$E_{tot} = m_Bg(h_A + h_B) + \frac{1}{2}\left(m_A + m_B + \frac{M}{2}\right)v^2 \tag{9.68}$$

By the law of conservation of energy, we equate the total energy in the initial configuration, equation 9.64, to the total energy in the final configuration, equation 9.68, obtaining

$$m_Agh_A + m_Bgh_B = m_Bg(h_A + h_B) + \frac{1}{2}\left(m_A + m_B + \frac{M}{2}\right)v^2$$

$$\frac{1}{2}\left(m_A + m_B + \frac{M}{2}\right)v^2 = m_Agh_A + m_Bgh_B - m_Bgh_A - m_Bgh_B,$$

$$\frac{1}{2}\left(m_A + m_B + \frac{M}{2}\right)v^2 = (m_A - m_B)gh_A$$

$$v^2 = \frac{(m_A - m_B)gh_A}{\frac{1}{2}(m_A + m_B + M/2)}$$

Solving for v, we get

$$v = \sqrt{\frac{(m_A - m_B)gh_A}{\frac{1}{2}(m_A + m_B + M/2)}} \tag{9.69}$$

Example 9.14

Conservation of energy and combined translational and rotational motion. If $m_B = 30.0$ g, $m_A = 50.0$ g, and the mass of the pulley M is 2.00 kg in figure 9.18, find the velocity of mass m_A as it falls through the distance $h_A = 2.00$ m.

Solution

The velocity of block B, found from equation 9.69, is

$$v = \sqrt{\frac{(m_A - m_B)gh_A}{\frac{1}{2}(m_A + m_B + M/2)}}$$

$$= \sqrt{\frac{(0.0500 \text{ kg} - 0.0300 \text{ kg})(9.80 \text{ m/s}^2)(2.00 \text{ m})}{\frac{1}{2}(0.0300 \text{ kg} + 0.0500 \text{ kg} + 2.00 \text{ kg}/2)}}$$

$$= 0.850 \text{ m/s}$$

Note that this is the same result obtained by treating the Atwood's machine by Newton's laws of motion rather than the energy technique.

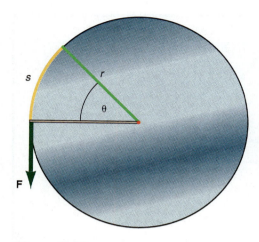

Figure 9.19

Work in rotational motion.

9.9 Work in Rotational Motion

The work done in translating a body from one position to another was found in chapter 7 as

$$W = Fx \qquad (7.1)$$

where F is the force in the direction of the displacement and x is the magnitude of the displacement. We can find the work done in causing a body to rotate from equation 7.1 and figure 9.19. In figure 9.19, a string is wrapped around the disk and pulled with a constant force F, causing the disk to rotate through the angle θ. The rim of the disk moves through the distance s. The work done by the force is

$$W = Fx$$

But $x = s$ and $s = r\theta$. Therefore,

$$W = Fr\theta \qquad (9.70)$$

But F times r is equal to the torque τ acting on the disk, that is

$$Fr = \tau \qquad (9.71)$$

Substituting equation 9.71 into equation 9.70 gives the work done to rotate the disk as

$$W = \tau\theta \qquad (9.72)$$

The power expended in rotating the disk for a time t is

$$P = \frac{W}{t} = \tau\frac{\theta}{t} \qquad (9.73)$$

but $\theta/t = \omega$, the angular velocity. Therefore,

$$P = \tau\omega \qquad (9.74)$$

Example 9.15

Work done in rotational motion. A constant force of 5.00 N is applied to a string that is wrapped around a disk of 0.500-m radius. If the wheel rotates through an angle of 2.00 rev, how much work is done?

Solution

The work done, given by equation 9.72, is

$$W = \tau\theta = rF\theta$$

$$= (0.500 \text{ m})(5.00 \text{ N})(2.00 \text{ rev})\left(\frac{2\pi \text{ rad}}{\text{rev}}\right)$$

$$= 31.4 \text{ J}$$

An Essay on the Application of Physics

Attitude Control of Airplanes and Spaceships

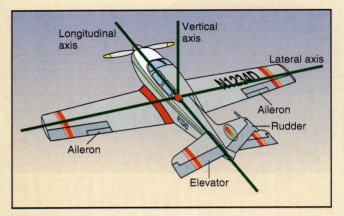

Figure 1
Control surfaces.

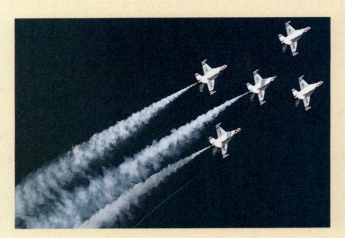

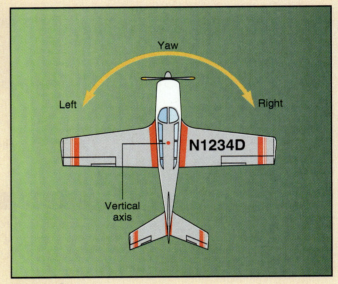

Figure 2
Aircraft yaw.

Have you ever wondered how an airplane or space vehicle is able to change its direction of flight? A plane or spacecraft can turn, climb, and dive. But how does it do this?

Attitude Control of Aircraft

An aircraft changes its attitude by the use of control surfaces, figure 1. As we saw in section 6.7, an airplane has three ways of changing the direction of its motion. They are yaw, pitch, and roll. Yaw is a rotation about the vertical axis of the aircraft. The control surface to yaw the aircraft is the rudder, which is located at the rear of the vertical stabilizer. Pitch is a rotation about the lateral axis of the aircraft. The control surface to pitch the aircraft is the elevator, which is located at the rear of the horizontal stabilizer. Roll is a rotation about the longitudinal axis of the aircraft. The control surfaces to roll the aircraft are the ailerons, which are located on the trailing edge of the wings.

1. **Yaw Control:** Yaw is a rotation of the aircraft about a vertical axis that passes through the center of gravity of the aircraft, as shown in figure 2. The aircraft can yaw to the right or left, as seen from the position of the pilot in the aircraft.

    Before the pilot presses either rudder pedal in the cockpit, the rudder is aligned with the vertical stabilizer and the air streams past the rudder exerting no unbalanced forces on it. When the pilot presses the right rudder pedal the rudder moves toward the right, as seen from above and behind the aircraft, figure 3(a). In this position the air stream exerts a normal force **F** on the rudder surface, as shown in the figure. If we draw the line r from the center of gravity of the aircraft to the point of application of the force, we see that this force produces a

torque about the vertical axis. We find the lever arm for this torque by dropping a perpendicular from the axis of rotation to the line of action of the force. As seen in the figure, the lever arm is $r \sin \psi$. Hence, the torque is

$$\tau = Fr \sin \psi \qquad (9H.1)$$

This torque produces a clockwise torque about the center of gravity causing the aircraft to rotate (yaw) to the right. The greater the angle ψ, the greater the torque acting on the aircraft, and hence the greater will be its yaw. When the pilot moves the rudder pedals back to the neutral position, the force acting on the rudder is reduced to zero, the torque $\tau = 0$, and there is no further yaw of the aircraft.

When the pilot presses the left rudder pedal, the rudder moves toward the left, as seen from above and

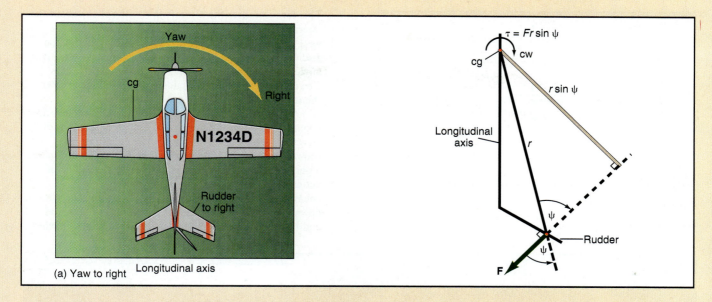

(a) Yaw to right

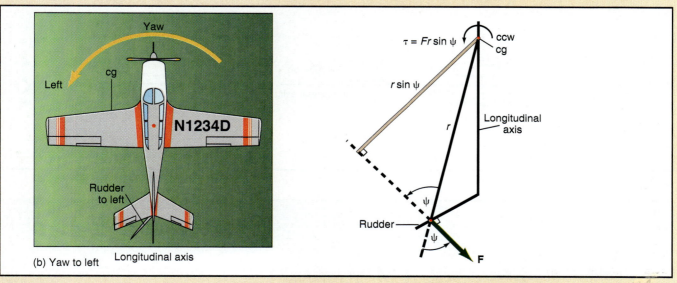

(b) Yaw to left

Figure 3

Dynamics of aircraft yaw.

behind the aircraft, figure 3(b). For this case the force of the air on the rudder produces a counterclockwise torque that causes the aircraft to rotate to the left, as seen in the diagram. Thus the rudder is a control surface that produces a torque on the aircraft that causes it to rotate either clockwise or counterclockwise about the vertical axis.

2. **Pitch Control:** Pitch is a rotation of the aircraft about a lateral axis that passes through the center of gravity of the aircraft, figure 4. In straight and level flight, the thrust vector of the aircraft lies along the longitudinal axis of the aircraft and thus the aircraft moves straight ahead.

When the pilot pulls the "stick" backward, the elevator is pushed upward, figure 5(a). The air that hits the elevator exerts a normal force $\mathbf{F}$ on the elevator, as seen in the diagram. If we draw the line r from the center of gravity of the aircraft to the point of application of the force, we see that this force produces a clockwise torque about the lateral axis of the aircraft. We find the lever arm for this torque by dropping a perpendicular from the axis of rotation to the line of action of the force. As seen in the figure, the lever arm is $r \sin \theta$. Hence the torque acting on the aircraft is given by

$$\tau = Fr \sin \theta \qquad (9H.2)$$

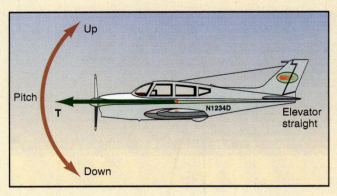

Figure 4
Aircraft pitch.

This torque causes the aircraft to rotate (pitch) about the lateral axis, such that the tail goes downward and the nose goes upward, figure 5(a). The thrust vector of the aircraft is no longer horizontal but now makes a positive angle with the horizontal, and hence the plane climbs. The farther back the pilot pulls on the stick the greater the torque and hence the steeper the climb.

When the pilot pushes the stick forward, the elevator is pushed downward, figure 5(b). The air that hits the elevator exerts a normal force **F** on the elevator, as shown. We find the lever arm for this torque by dropping a perpendicular from the axis of rotation to the line of action of the force, as shown. The resulting counterclockwise torque pushes the tail up and the nose down. The thrust vector now falls below the horizontal and the plane dives. The farther forward the pilot pushes

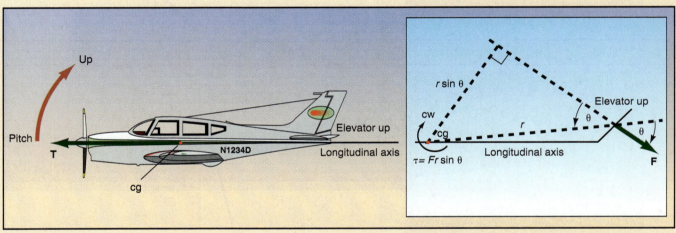

(a) Pitch up

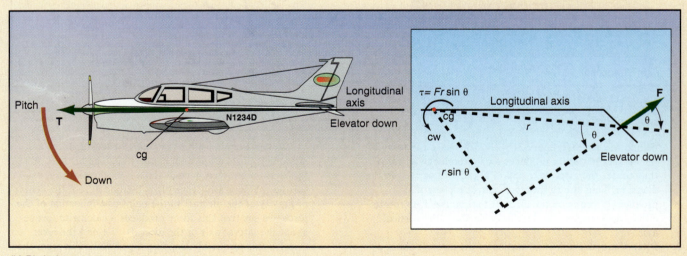

(b) Pitch down

Figure 5
Dynamics of aircraft pitch.

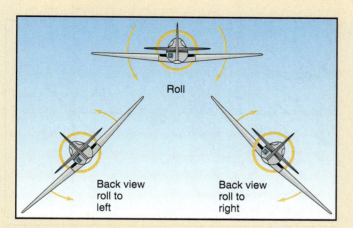

Figure 6

Aircraft roll.

the stick, the greater the torque and hence the steeper the dive. In this way the pilot can make the aircraft climb or dive.

3. **Roll Control:** Roll is a rotation of the aircraft about the longitudinal axis of the aircraft. When the pilot pushes the stick to the left, the plane will roll to the left; when he pushes the stick to the right, the plane will roll to the right, figure 6.

When the pilot pushes the stick to the left, the right aileron is pushed downward and the left aileron is pushed upward, figure 7(c). The wind blowing over the wings exerts a force on the ailerons as shown in figure 7(a,b). The force acting on the raised left aileron pushes the left wing downward, while the force acting on the lowered

right aileron pushes the right wing upward. The ailerons act similar to the elevator in that they produce a torque about the lateral axis of the aircraft. However, with one aileron up and one down the torques they produce to pitch the aircraft are equal and opposite and hence have no effect on pitching the aircraft. However, the force up on the right wing and the force down on the left wing cause a counterclockwise torque about the longitudinal axis, as viewed from the rear of the aircraft (the view that is seen by the pilot). Therefore the aircraft rolls to the left, figure 7(c). When the aircraft has rolled to the required bank angle, the pilot places the stick back to the neutral position and the aircraft stays at this angle of bank. To bring the aircraft back to level flight the pilot must push the stick to the right. The aircraft now rolls to the right until the aircraft is level. Then the pilot places the stick in the neutral position.

To roll the aircraft to the right the pilot pushes the stick to the right. The right aileron now goes up and the left aileron now goes down, figure 7(d). The force down on the right wing and the force up on the left wing causes a clockwise torque about the longitudinal axis. Thus, the aircraft rotates (rolls) to the right.

The force exerted on a control surface by the air creates the necessary torque to rotate the aircraft in any specified direction.

Attitude Control of Space Vehicles

An aircraft will not work in space because there is no air to exert the necessary lift on the wings of the aircraft. Nor can rudders, elevators, or ailerons work in space because there is no air to exert forces on the control surfaces to change the attitude of the vehicle.

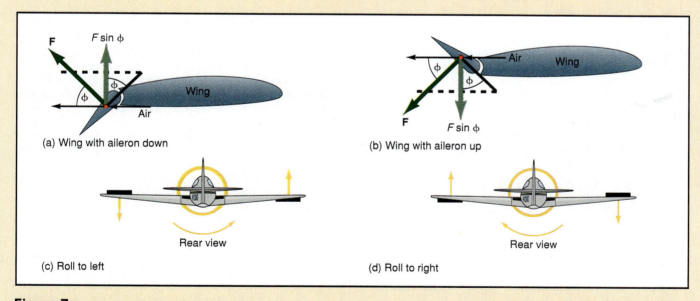

(a) Wing with aileron down

(b) Wing with aileron up

(c) Roll to left

(d) Roll to right

Figure 7

Dynamics of aircraft roll.

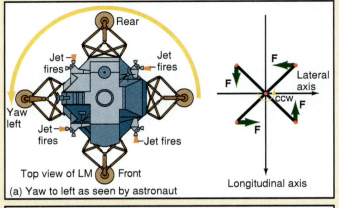

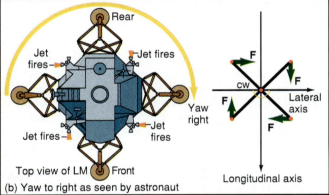

(a) Yaw to left as seen by astronaut

(b) Yaw to right as seen by astronaut

Figure 9

Dynamics of Lunar Module yaw.

Source (left): NASA.

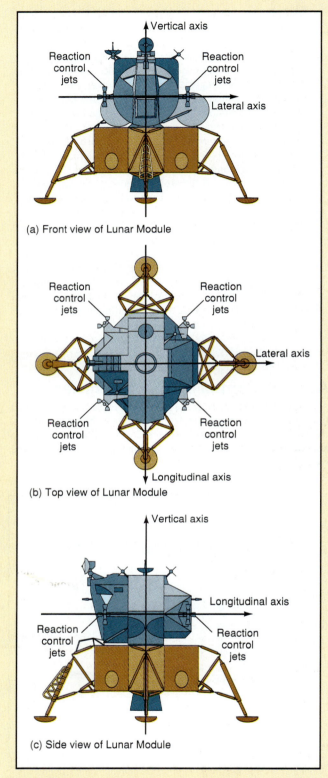

(a) Front view of Lunar Module

(b) Top view of Lunar Module

(c) Side view of Lunar Module

Figure 8

The Lunar Module.

Source: NASA.

To control the attitude of a space vehicle, reaction control jets are used. Figure 8 is a line drawing of the Lunar Module (LM) that landed on the moon. Notice the reaction control jets located on the sides of the Lunar Module. The reaction control system consists of 16 small rocket thrusters placed around the vehicle to control the translation and rotation of the Lunar Module. Also notice that the axes of the spacecraft are the same as the axes of an aircraft. Thus a rotation about the vertical axis of the spacecraft is called yaw, rotation about the lateral axis is called pitch, and rotation about the longitudinal axis is called roll.

Figure 8(b) is a top view of the Lunar Module. Notice that there are four thruster assemblies, each containing four rocket jets, located on the Lunar Module.

1. **Yaw Control:** For the spacecraft to yaw to the left the four reaction jets shown in figure 9(a) are fired to create a torque counterclockwise about the vertical axis of the Lunar Module. Each jet exerts a force **F** on the Lunar Module, which in turn creates a torque about the vertical axis. The total torque is the sum of the four torques. For the spacecraft to yaw to the right the four reaction jets shown in figure 9(b) are fired to create a torque clockwise

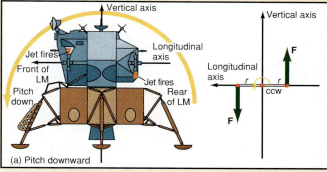

(a) Pitch downward

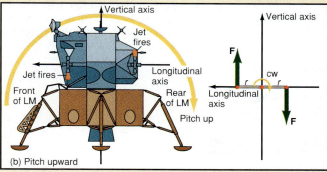

(b) Pitch upward

Figure 10

Dynamics of Lunar Module pitch.

Source (left): NASA.

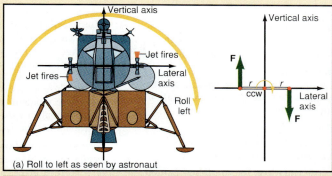

(a) Roll to left as seen by astronaut

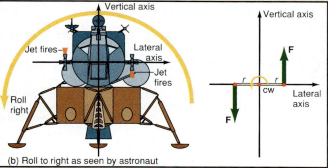

(b) Roll to right as seen by astronaut

Figure 11

Dynamics of Lunar Module roll.

Source (left): NASA.

about the vertical axis of the Lunar Module. Notice that a different set of jets are used to yaw to the right than to yaw to the left.

2. **Pitch Control:** For the Lunar Module to pitch downward, the two reaction jets on each side of the Lunar Module (total of 4 jets) shown in figure 10(a) are fired to create a torque counterclockwise about the lateral axis of the Lunar Module. For the spacecraft to pitch upward the two reaction jets on each side of the Lunar Module (total of 4 jets) shown in figure 10(b) are fired to create a torque clockwise about the lateral axis of the Lunar Module.

3. **Roll Control:** For the spacecraft to roll to the left the two reaction jets on each side of the Lunar Module (total of 4

jets) shown in figure 11(a) are fired to create a torque counterclockwise about the longitudinal axis of the Lunar Module. (Don't forget that left and right are defined from the position of the pilot. Figure 11 shows the Lunar Module from a front view, and hence left and right appear to be reversed.)

A roll to the right is accomplished by firing the two reaction jets on each side of the Lunar Module (total of 4 jets) shown in figure 11(b) to create a torque clockwise about the longitudinal axis of the Lunar Module.

Thus the Lunar Module, and any spacecraft for that matter, can control its attitude by supplying torques for its rotation by the suitable firing of the different reaction control jets.

The Language of Physics

Angular displacement
The angle that a body rotates through while in rotational motion (p. 241).

Angular velocity
The change in the angular displacement of a rotating body about the axis of rotation with time (p. 242).

Angular acceleration
The change in the angular velocity of a rotating body with time (p. 243).

Kinematic equations for rotational motion
A set of equations that give the angular displacement and angular velocity of a rotating body at any instant of time, and the angular velocity at a particular angular displacement, if the angular acceleration of the body is constant (p. 244).

Kinetic energy of rotation
The energy that a body possesses by virtue of its rotational motion (p. 248).

Moment of inertia
The measure of the resistance of a body to a change in its rotational motion. It is the rotational analogue of mass, which is a measure of the resistance of a body to a change in its translational motion. The larger the moment of inertia of a body the more difficult it is to put that body into rotational motion (p. 248).

Newton's second law for rotational motion
When an unbalanced external torque acts on a body, it gives that body an angular acceleration. The angular acceleration is directly proportional to the torque and inversely proportional to the moment of inertia (p. 250).

Newton's first law for rotational motion
A body in motion at a constant angular velocity will continue in motion at that same constant angular velocity unless acted upon by some unbalanced external torque (p. 251).

Newton's third law of rotational motion
If body A and body B have the same axis of rotation, and if body A exerts a torque on body B, then body B exerts an equal but opposite torque on body A (p. 251).

Angular momentum
The product of the moment of inertia of a rotating body and its angular velocity (p. 259).

Law of conservation of angular momentum
If the total external torque acting on a system is zero, then there is no change in the angular momentum of the system, and the final angular momentum is equal to the initial angular momentum (p. 260).

Summary of Important Equations

Angular velocity
$$\omega = \frac{\Delta\theta}{\Delta t} = \frac{\theta}{t} \tag{9.1}$$

Angular acceleration
$$\alpha = \frac{\Delta\omega}{\Delta t} = \frac{\omega - \omega_0}{t} \tag{9.3}$$

Kinematic equations
$$\omega = \omega_0 + \alpha t \tag{9.4}$$
$$\theta = \omega_0 t + \tfrac{1}{2}\alpha t^2 \tag{9.9}$$
$$\omega^2 = \omega_0^2 + 2\alpha\theta \tag{9.10}$$

Relations between translational and rotational variables
$$s = r\theta \tag{6.5}$$
$$v = r\omega \tag{9.2}$$
$$a = r\alpha \tag{9.5}$$

Centripetal acceleration
$$a_c = \omega^2 r \tag{9.11}$$

Kinetic energy of rotation
$$KE_{rot} = \tfrac{1}{2}I\omega^2 \tag{9.15}$$

Moment of inertia
$$I = \sum_{i=1}^{n} m_i r_i^2 \tag{9.16}$$

Moment of inertia for a single mass
$$I = mr^2 \tag{9.17}$$

Newton's second law for rotational motion
$$\tau = I\alpha \tag{9.22}$$

Angular momentum
$$L = I\omega \tag{9.48}$$

Newton's second law in terms of momentum
$$\tau = \frac{\Delta L}{\Delta t} \tag{9.49}$$

Law of conservation of angular momentum (no external torques)
$$L_f = L_i \tag{9.52}$$

Work done in rotational motion
$$W = \tau\theta \tag{9.72}$$

Power expended in rotational motion
$$P = \tau\omega \tag{9.74}$$

Questions for Chapter 9

1. Discuss the similarity between the equations for translational motion and the equations for rotational motion.
2. When moving in circular motion at a constant angular velocity, why does the body at the greatest distance from the axis of rotation move faster than the body closest to the axis of rotation?
3. It is easy to observe the angular velocity of the second hand of a clock. Why is it more difficult to observe the angular velocity of the minute and hour hands of the clock?
†4. If a cylinder, a ball, and a ring are placed at the top of an inclined plane and then allowed to roll down the plane, in what order will they arrive at the bottom of the plane? Why?
†5. How would you go about approximating the rotational kinetic energy of our galaxy?
6. Which would be more difficult to put into rotational motion, a large sphere or a small sphere? Why?
7. Why must the axis of rotation be specified when giving the moment of inertia of an object?

†8. If two balls collide such that the force transmitted lies along a line connecting the center of mass of each body, can either ball be put into rotational motion? If the balls collide in a glancing collision in which there is also friction between the two surfaces as they collide, can either ball be put into rotational motion? Draw a diagram of the collision in both cases and discuss both possibilities.

†9. As long as there are no external torques acting on the earth, the earth will continue to spin forever at its present angular velocity. Discuss the possibility of small perturbative torques that might act on the earth and what effect they might have.

†10. Can the angular displacement, angular velocity, and angular acceleration be treated as vectors? Consider a rotation of your book through an angular displacement of 90° about the x-axis, then a rotation through an angular displacement of 90° about the y-axis, and finally a rotation through an angular displacement of 90° about the z-axis. Would you get the same result if you changed the order of the rotations to the y-, x-, and then z-axis? What happens if the rotations are infinitesimal?

†11. If the instantaneous angular velocity can be considered as a vector, should the angular momentum also be considered as a vector? If so, what direction would it have? What would the change in the direction of the angular momentum look like?

†12. It is said that if you throw a cat, upside down, into the air, it will always land on its feet. Discuss this possibility from the point of view of the cat moving his legs and tail and thus changing his moment of inertia and hence his angular velocity.

Problems for Chapter 9

9.2 Rotational Kinematics

1. Express the following angular velocities of a phonograph turntable in terms of rad/s. (a) 33 1/3 rpm (revolutions per minute), (b) 45 rpm, and (c) 78 rpm.

2. Determine the angular velocity of the following hands of a clock: (a) the second hand, (b) the minute hand, and (c) the hour hand.

3. A cylinder 15.0 cm in diameter rotates at 1000 rpm. (a) What is its angular velocity in rad/s? (b) What is the tangential velocity of a point on the rim of the cylinder?

4. A circular saw blade rotating at 3600 rpm is reduced to 3450 rpm in 2.00 s. What is the angular acceleration of the blade?

5. A circular saw blade rotating at 3600 rpm is braked to a stop in 6 s. What is the angular acceleration? How many revolutions did the blade make before coming to a stop?

6. A wheel 50.0 cm in diameter is rotating at an initial angular velocity of 6.00 rad/s. It is given an acceleration of 2.00 rad/s². Find (a) the angular velocity at 5.00 s, (b) the angular displacement at 5.00 s, (c) the tangential velocity of a point on the rim at 5.00 s, (d) the tangential acceleration of a point on the rim, (e) the centripetal acceleration of a point on the rim, and (f) the resulting acceleration of a point on the rim.

9.3 The Kinetic Energy of Rotation

7. Find the kinetic energy of a 2.00-kg cylinder, 25.0 cm in diameter, if it is rotating about its longitudinal axis at an angular velocity of 0.550 rad/s.

8. A 3.00-kg ball, 15.0 cm in diameter, rotates at an angular velocity of 3.45 rad/s. Find its kinetic energy.

9.4 The Moment of Inertia

9. Calculate the moment of inertia of a 0.500-kg meterstick about an axis through its center, and perpendicular to its length.

10. Compute the moment of inertia through its center of a 16.0-lb bowling ball of radius 4.00 in.

11. Find the moment of inertia for the system of point masses shown for (a) rotation about the y-axis and (b) for rotation about the x-axis. Given are $m_1 = 2.00$ kg, $m_2 = 3.50$ kg, $r_1 = 0.750$ m, and $r_2 = 0.873$ m.

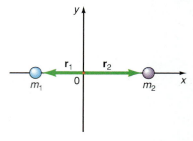

†12. Find the moment of inertia for the system shown for rotation about (a) the y-axis, (b) the x-axis, and (c) an axis going through masses m_2 and m_4. Assume $m_1 = 0.532$ kg, $m_2 = 0.425$ kg, $m_3 = 0.879$ kg, and $m_4 = 0.235$ kg.

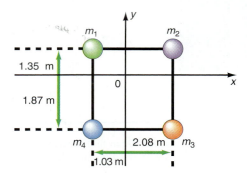

9.5 Newton's Laws for Rotational Motion and 9.6 Rotational Dynamics

13. A solid wheel of mass 5.00 kg and radius 0.350 m is set in motion by a constant force of 6.00 N applied tangentially. Determine the angular acceleration of the wheel.

14. A torque of 5.00 m N is applied to a body. Of this torque, 2.00 m N of it is used to overcome friction in the bearings. The body has a resultant angular acceleration of 5.00 rad/s². (a) When the applied torque is removed, what is the angular acceleration of the body? (b) If the angular velocity of the body was 100 rad/s when the applied torque was removed how long will it take the body to come to rest?

15. A mass of 200 g is attached to a wheel by a string wrapped around the wheel. The wheel has a mass of 1.00 kg. Find the acceleration of the mass. Assume that the moment of inertia of the wheel is the same as a disk.

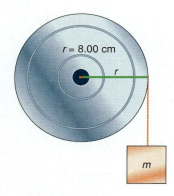

16. A mass m_A of 10.0 kg is attached to another mass m_B of 4.00 kg by a string that passes over a pulley of mass $M = 1.00$ kg. The coefficient of kinetic friction between block B and the table is 0.400. Find (a) the acceleration of each block of the system, (b) the tensions in the cords, and (c) the velocity of block A as it hits the floor 0.800 m below its starting point.

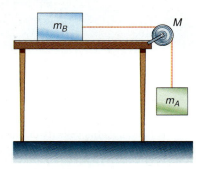

17. A mass $m_A = 100$ g, and another mass $m_B = 200$ g are attached to an Atwood's machine that has a pulley mass $M = 1.00$ kg. (a) Find the acceleration of each block of the system. (b) Find the velocity of mass B as it hits the floor 1.50 meters below its starting point.

9.7 Angular Momentum and Its Conservation

†18. A 75-kg student stands at the edge of a large disk of 150-kg mass that is rotating freely at an angular velocity of 0.800 rad/s. The disk has a radius of $R = 3.00$ m. (a) Find the initial moment of inertia of the disk and student and its kinetic energy. The student now walks toward the center of the disk. Find the moment of inertia, the angular velocity, and the kinetic energy when the student is at (b) $3R/4$, (c) $R/2$, and (d) $R/4$.

19. Two disks are to be made into an idealized clutch. Disk 1 has a mass of 3.00 kg and a radius of 20.0 cm, while disk 2 has a mass of 1.00 kg and a radius of 20.0 cm. If disk 2 is originally at rest and disk 1 is rotating at 2000 rpm, what is the final angular velocity of the coupled disks?

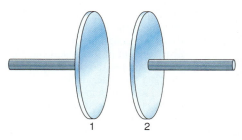

†20. Two beads are fixed on a thin long wire on the x-axis at $r_1 = 0.700$ m and $r_2 = 0.800$ m, as shown in the diagram. Assume $m_1 = 85.0$ g and $m_2 = 63.0$ g. The combination is spinning about the y-axis at an angular velocity of 4.00 rad/s. A catch is then released allowing the beads to move freely to the stops at the end of the wire, which is 1.00 m from the origin. Find (a) the initial moment of inertia of the system, (b) the initial angular momentum of the system, (c) the initial kinetic energy of the system, (d) the final angular momentum of the system, (e) the final angular velocity of the system, and (f) the final kinetic energy of the system.

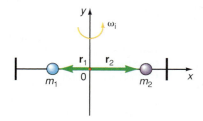

9.8 Combined Translational and Rotational Motion Treated by the Law of Conservation of Energy

21. Find the velocity of (a) a cylinder and (b) a ring at the bottom of an inclined plane that is 2.00 m high. The cylinder and ring start from rest and roll down the plane.

22. Compute the velocity of a cylinder at the bottom of a plane 1.5 m high if (a) it slides without rotating on a frictionless plane and (b) it rotates on a rough plane.

23. A 1.50-kg ball, 10.0 cm in radius, is rolling on a table at a velocity of 0.500 m/s. (a) What is its angular velocity about its center of mass? (b) What is the translational kinetic energy of its center of mass? (c) What is its rotational kinetic energy about its center of mass? (d) What is its total kinetic energy?

24. Using the law of conservation of energy for the Atwood's machine shown, find the velocity of m_A at the ground, if $m_A = 20.0$ g, $m_B = 10.0$ g, $M = 1.00$ kg, and $r = 15.0$ cm.

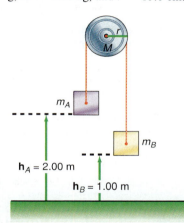

9.9 Work in Rotational Motion

25. A constant force of 2.50 N acts tangentially on a cylinder of 12.5-cm radius and the cylinder rotates through an angle of 5.00 rev. How much work is done in rotating the cylinder?

26. An engine operating at 1800 rpm develops 200 hp, what is the torque developed?

Additional Problems

27. Determine (a) the angular velocity of the earth, (b) its moment of inertia, and (c) its kinetic energy of rotation. (d) Compare this with its kinetic energy of translation. (e) Find the angular momentum of the earth.

28. The earth rotates once in a day. If the earth could collapse into a smaller sphere, what would be the radius of that sphere that would give a point on the equator a linear velocity equal to the velocity of light $c = 3.00 \times 10^8$ m/s? Use the initial angular velocity of the earth and the moment of inertia determined in problem 27.

29. A disk of 10.0-cm radius, having a mass of 100 g, is set into motion by a constant tangential force of 2.00 N. Determine (a) the moment of inertia of the disk, (b) the torque applied to the disk, (c) the angular acceleration of the disk, (d) the angular velocity at 2.00 s, (e) the angular displacement at 2.00 s, (f) the kinetic energy at 2.00 s, and (g) the angular momentum at 2.00 s.

30. A 3.50-kg solid disk of 25.5 cm diameter has a cylindrical hole of 3.00-cm radius cut into it. The hole is 1.00 cm in from the edge of the solid disk. Find (a) the initial moment of inertia of the disk about an axis perpendicular to the disk before the hole was cut into it and (b) the moment of inertia of the solid disk with the hole in it. State the assumptions you use in solving the problem.

†31. Due to slight effects caused by tidal friction between the water and the land and the nonsphericity of the sun, there is a slight angular deceleration of the earth. The length of a day will increase by approximately 1.5×10^{-3} s in a century. (a) What will be the angular velocity of the earth after one century? (b) What will be the change in the angular velocity of the earth per century? (c) As a first approximation, is it reasonable to assume that there are no external torques acting on the earth and the angular velocity of the earth is a constant?

32. A string of length 1.50 m with a small bob at one end is connected to a horizontal disk of negligible radius at the other end. The disk is put into rotational motion and is now rotating at an angular velocity $\omega = 5.00$ rad/s. Find the angle that the string makes with the vertical.

33. A constant force of 5.00 N acts on a disk of 3.00-kg mass and diameter of 50.0 cm for 10.0 s. Determine (a) the angular acceleration, (b) the angular velocity after 10.0 s, and (c) the kinetic energy after 10.0 s. (d) Compute the work done to cause the disk to rotate and compare with your answer to part c.

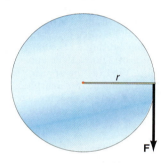

†34. A 5.00-kg block is at rest at the top of the inclined plane shown in the diagram. The plane makes an angle of 32.5° with the horizontal. A string is attached to the block and tied around the disk, which has a mass of 2.00 kg and a radius of 8.00 cm. Find the acceleration of the block down the plane if (a) the plane is frictionless, and (b) the plane is rough with a value of $\mu_k = 0.54$.

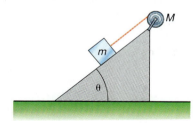

35. A large cylinder has a radius of 12.5 cm and it is pressed against a smaller cylinder of radius 4.50 cm such that the two axes of the cylinders are parallel. When the larger cylinder rotates about its axis, it causes the smaller cylinder to rotate about its axis. The larger cylinder accelerates from rest to a constant angular velocity of 20 rad/s. Find (a) the tangential velocity of a point on the surface of the large cylinder, (b) the tangential velocity of a point on the surface of the smaller cylinder, and (c) the angular velocity of the smaller cylinder. Can you think of this setup as a kind of mechanical advantage?

†36. A small disk of $r_1 = 5.00$-cm radius is attached to a larger disk of $r_2 = 15.00$-cm radius such that they have a common axis of rotation, as shown in the diagram. The small disk has a mass $M_1 = 0.250$ kg and the large disk has a mass $M_2 = 0.850$ kg. A string is wrapped around the small disk and a force is applied to the string causing a constant tangential force of 2.00 N to be applied to the disk. Find (a) the applied torque, (b) the moment of inertia of the system, (c) the angular acceleration of the system, and (d) the angular velocity at 4.00 s.

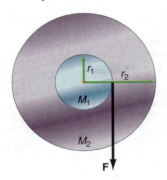

†37. Repeat problem 36 with the string wrapped around the large disk instead of the small disk.

†38. A small disk of mass $M_1 = 50.0$ g is connected to a larger disk of mass $M_2 = 200.0$ g such that they have a common axis of rotation. The small disk has a radius $r_1 = 10.0$ cm, while the large disk has a radius of $r_2 = 30.0$ cm. A mass $m_1 = 25.0$ g is connected to a string that is wrapped around the small disk, while a mass $m_2 = 35.0$ g is connected to a string and wrapped around the large disk, as shown in the diagram. Find (a) the moment of inertia of each disk, (b) the moment of inertia of the combined disks, (c) the net torque acting on the disks, (d) the angular acceleration of the disks, (e) the angular velocity of the disks at 4.00 s, (f) the kinetic energy of the disks at 4.00 s, and (g) the angular momentum of the disks at 4.00 s.

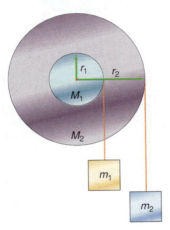

†39. One end of a string is wrapped around a pulley and the other end is connected to the ceiling, which is 3.00 m above the floor. The mass of the pulley is 200 g and has a radius of 10.0 cm. The pulley is released from rest and is allowed to fall. Find (a) the initial total energy of the system, and (b) the velocity of the pulley just before it hits the floor.

40. This is essentially the same problem as problem 39 but is to be treated by Newton's second law for rotational motion. Find the angular acceleration of the cylinder and the tension in the string.

†41. A 1.5-kg disk of 0.500-m radius is rotating freely at an angular velocity of 2.00 rad/s. Small 5-g balls of clay are dropped onto the disk at 3/4 of the radius at a rate of 4 per second. Find the angular velocity of the disk at 10.0 s.

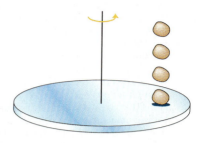

†42. A space station is to be built in orbit in the shape of a large wheel of outside radius 100.0 m and inside radius of 97.0 m. The satellite is to rotate such that it will have a centripetal acceleration exactly equal to the acceleration of gravity g on earth. The astronauts will then be able to walk about and work on the rim of the wheel in an environment similar to earth. (a) At what angular velocity must the wheel rotate to simulate the earth's gravity? (b) If the mass of the spaceship is 40,000 kg, what is its approximate moment of inertia? (c) How much energy will be necessary to rotate the space station? (d) If it takes 20.0 rev to bring the space station up to its operating angular velocity, what torque must be applied in the form of gas jets attached to the outer rim of the wheel?

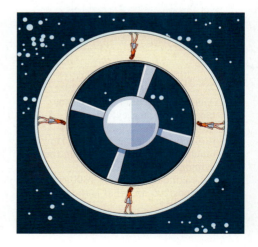

†43. At the instant that ball 1 is released from rest at the top of a rough inclined plane a second ball (2) moves past it on the horizontal surface below at a constant velocity of 2.30 m/s. The plane makes an angle $\theta = 35.0°$ with the horizontal and the height of the plane is 0.500 m. Using Newton's second law for combined translational and rotational motion find (a) the acceleration a of ball 1 down the plane, (b) the velocity of ball 1 at the base of the incline, (c) the time it takes for ball 1 to reach the bottom of the plane, (d) the distance that ball 2 has moved in this time, and (e) at what horizontal distance from the base of the incline will ball 1 overtake ball 2.

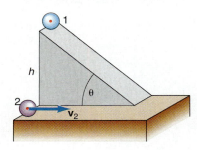

Interactive Tutorials

🖳 44. A cylinder of mass $m = 4.00$ kg and radius $r = 2.00$ m is rotating at an angular velocity $\omega = 3600$ rpm. Calculate (a) its angular velocity ω in rad/s, (b) its moment of inertia I, (c) its rotational kinetic energy KE_{rot}, and (d) its angular momentum L.

🖳 45. A mass $m = 2.00$ kg is attached by a string that is wrapped around a frictionless solid cylinder of mass $M = 8.00$ kg and radius $R = 0.700$ m that is free to rotate. Calculate (a) the acceleration a of the mass m and (b) the tension T in the string.

🖳 46. A cylinder of mass $m = 2.35$ kg and radius $r = 0.345$ m is initially rotating at an angular velocity $\omega_0 = 1.55$ rad/s when a constant force $F = 9.25$ N is applied tangentially to the cylinder as in figure 9.8. Find

(a) the moment of inertia I of the cylinder, (b) the torque τ acting on the cylinder, (c) the angular acceleration α of the cylinder, (d) the angular velocity ω of the cylinder at $t = 4.55$ s, and (e) the angular displacement θ at $t = 4.55$ s.

🖳 47. The moment of inertia of a continuous mass distribution. A meterstick, $m = 0.149$ kg, lies on the x-axis with the zero of the meterstick at the origin of the coordinate system. Determine the moment of inertia of the meterstick about an axis that passes through the zero of the meterstick and perpendicular to it. Assume that the meterstick can be divided into $N = 10$ equal parts.

🖳 48. This is a generalization of Interactive Tutorial problem 76 of chapter 4 but it also takes the rotational motion of the pulley into account. Derive the formula for the magnitude of the acceleration of the system shown in the diagram for problem 61 of chapter 4. The pulley has a mass M and the radius R. As a general case assume that the coefficient of kinetic friction between block A and the surface is μ_{kA} and between block B and the surface is μ_{kB}. Solve for all the special cases that you can think of. In all the cases, consider different values for the mass M of the pulley and see the effect it has on the results of the problem.

🖳 49. Consider the general motion in an Atwood's machine such as the one shown in figure 9.18. Mass $m_A = 0.650$ kg and is at a height $h_A = 2.55$ m above the reference plane and mass $m_B = 0.420$ kg is at a height $h_B = 0.400$ m. The pulley has a mass of $M = 2.00$ kg and a radius $R = 0.100$ m. If the system starts from rest, find (a) the initial potential energy of mass A, (b) the initial potential energy of mass B, and (c) the total energy of the system. When mass m_A has fallen a distance $y_A = 0.750$ m, find (d) the potential energy of mass A, (e) the potential energy of mass B, (f) the speed of

each mass at that point, (g) the kinetic energy of mass A, (h) the kinetic energy of mass B, (i) the moment of inertia of the pulley (assume it to be a disk), (j) the angular velocity ω of the pulley, and (k) the rotational kinetic energy of the pulley. (l) When mass A hits the ground, find the speed of each mass and the angular velocity of the pulley.

🖳 50. Consider the general motion in the combined system shown in the diagram of problem 16. Mass $m_A = 0.750$ kg and is at a height $h_A = 1.85$ m above the reference plane and mass $m_B = 0.285$ kg is at a height $h_B = 2.25$ m, $\mu_k = 0.450$. The pulley has a mass $M = 1.85$ kg and a radius $R = 0.0800$ m. If the system starts from rest, find (a) the initial potential energy of mass A, (b) the initial potential energy of mass B, and (c) the total energy of the system. When m_A has fallen a distance $y_A = 0.35$ m, find (d) the potential energy of mass A, (e) the potential energy of mass B, (f) the energy lost due to friction as mass B slides on the rough surface, (g) the speed of each mass at that point, (h) the kinetic energy of mass A, (i) the kinetic energy of mass B, (j) the moment of inertia of the pulley (assumed to be a disk), (k) the angular velocity ω of the pulley, and (l) the rotational kinetic energy of the pulley. (m) When mass A hits the ground, find the speed of each mass.

🖳 51. A disk of mass $M = 3.55$ kg and a radius $R = 1.25$ m is rotating freely at an initial angular velocity $\omega_i = 1.45$ rad/s. Small balls of clay of mass $m_b = 0.025$ kg are dropped onto the rotating disk at the radius $r = 0.85$ m at the rate of $n = 5$ ball/s. Find (a) the initial moment of inertia of the disk, (b) the initial angular momentum of the disk, and (c) the angular velocity ω at $t = 6.00$s. (d) Plot the angular velocity ω as a function of the number of balls dropped.

Vibratory Motion, Wave Motion, and Fluids

P A R T   T W O

Color as a Study Aid
Chapters 10–13

The same color code used in previous chapters is used in chapters 10 through 13. In general, the x, y, and z coordinates are always black.

The color code used in chapters 10 through 13:

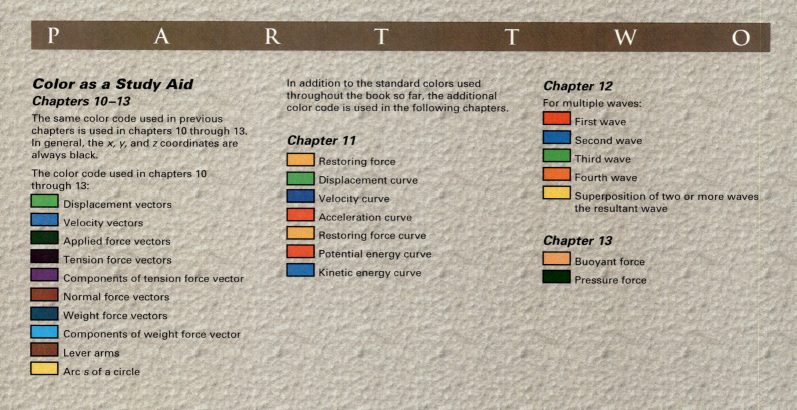

- Displacement vectors
- Velocity vectors
- Applied force vectors
- Tension force vectors
- Components of tension force vector
- Normal force vectors
- Weight force vectors
- Components of weight force vector
- Lever arms
- Arc s of a circle

In addition to the standard colors used throughout the book so far, the additional color code is used in the following chapters.

Chapter 11

- Restoring force
- Displacement curve
- Velocity curve
- Acceleration curve
- Restoring force curve
- Potential energy curve
- Kinetic energy curve

Chapter 12

For multiple waves:

- First wave
- Second wave
- Third wave
- Fourth wave
- Superposition of two or more waves the resultant wave

Chapter 13

- Buoyant force
- Pressure force

Elasticity

10.1 The Atomic Nature of Elasticity

Elasticity is that property of a body by which it experiences a change in size or shape whenever a deforming force acts on the body. When the force is removed the body returns to its original size and shape. Most people are familiar with the stretching of a rubber band. All materials, however, have this same elastic property, but in most materials it is not so pronounced.

The explanation of the elastic property of solids is found in an atomic description of a solid. Most solids are composed of a very large number of atoms or molecules arranged in a fixed pattern called the **lattice structure of a solid** and shown schematically in figure 10.1(a). These atoms or molecules are held in their positions by electrical forces. The electrical force between the molecules is attractive and tends to pull the molecules together. Thus, the solid resists being pulled apart. Any one molecule in figure 10.1(a) has an attractive force pulling it to the right and an equal attractive force pulling it to the left. There are also equal attractive forces pulling the molecule up and down, and in and out. A repulsive force between the molecules also tends to repel the molecules if they get too close together. This is why solids are difficult to compress. To explain this repulsive force we would need to invoke the Pauli exclusion principle of quantum mechanics (which we discuss in section 32.8). Here we simply refer to all these forces as molecular forces.

The net result of all these molecular forces is that each molecule is in a position of equilibrium. If we try to pull one side of a solid material to the right, let us say, then we are in effect pulling all these molecules slightly away from their equilibrium position. The displacement of any one molecule from its equilibrium position is quite small, but since there are billions of molecules, the total molecular displacements are directly measurable as a change in length of the material. When the applied force is removed, the attractive molecular forces pull all the molecules back to their original positions, and the material returns to its original length.

If we now exert a force on the material in order to compress it, we cause the molecules to be again displaced from their equilibrium position, but this time they are pushed closer together. The repulsive molecular force prevents them from getting too close together, but the total molecular displacement is directly measurable as a reduction in size of the original material. When the compressive force is removed, the repulsive molecular force causes the atoms to return to their equilibrium position and the solid returns to its original size. *Hence, the elastic properties of matter are a manifestation of the molecular forces that hold solids together.* Figure 10.1(b) shows a typical lattice structure of atoms in a solar cell analyzed with a scanning tunneling microscope.

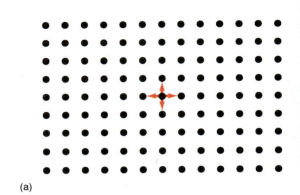

(a)

(b)

Figure 10.1

(a) Lattice structure of a solid. (b) Actual pictures of atoms in a solar cell.

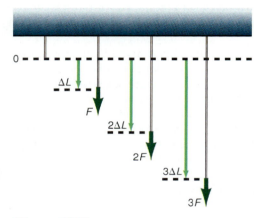

Figure 10.2

Stretching an object.

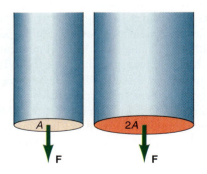

Figure 10.3

The deformation is inversely proportional to the cross-sectional area of the wire.

10.2 Hooke's Law—Stress and Strain

If we apply a force to a rubber band, we find that the rubber band stretches. Similarly, if we attach a wire to a support, as shown in figure 10.2, and sequentially apply forces of magnitude F, $2F$, and $3F$ to the wire, we find that the wire stretches by an amount ΔL, $2\Delta L$, and $3\Delta L$, respectively. (Note that the amount of stretching is greatly exaggerated in the diagram for illustrative purposes.) The deformation, ΔL, is directly proportional to the magnitude of the applied force F and is written mathematically as

$$\Delta L \propto F \tag{10.1}$$

This aspect of elasticity is true for all solids. It would be tempting to use equation 10.1 as it stands to formulate a theory of elasticity, but with a little thought it becomes obvious that although it is correct in its description, it is incomplete.

Let us consider two wires, one of cross-sectional area A, and another with twice that area, namely $2A$, as shown in figure 10.3. When we apply a force $\mathbf{F}$ to the first wire, that force is distributed over all the atoms in that cross-sectional area A. If we subject the second wire to the same applied force $\mathbf{F}$, then this same force is distributed over twice as many atoms in the area $2A$ as it was in the area A. Equivalently we can say that each atom receives only half the force in the area $2A$ that it received in the area A. Hence, the total stretching of the $2A$ wire is only $1/2$ of what it was in wire A. Thus, the elongation of the wire ΔL is inversely proportional to the cross-sectional area A of the wire, and this is written

$$\Delta L \propto \frac{1}{A} \tag{10.2}$$

Note also that the original length of the wire must have something to do with the amount of stretch of the wire. For if a force of magnitude F is applied to two wires of the same cross-sectional area, but one has length L_0 and the other has length $2L_0$, the same force is transmitted to every molecule in the length of the wire. But because there are twice as many molecules to stretch apart in the wire having length $2L_0$, there is twice the deformation, or $2\Delta L$, as shown in figure 10.4. We write this as the proportion

$$\Delta L \propto L_0 \tag{10.3}$$

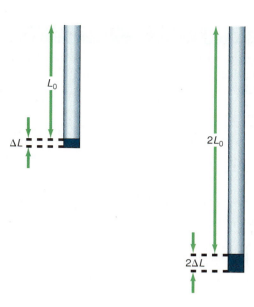

Figure 10.4

The deformation is directly proportional to the original length of the wire.

The results of equations 10.1, 10.2 and 10.3 are, of course, also deduced experimentally. *The deformation ΔL of the wire is thus directly proportional to the magnitude of the applied force F (equation 10.1), inversely proportional to the cross-sectional area A (equation 10.2), and directly proportional to the original length of the wire L_0 (equation 10.3).* These results can be incorporated into the one proportionality:

$$\Delta L \propto \frac{FL_0}{A}$$

which we rewrite in the form

$$\frac{F}{A} \propto \frac{\Delta L}{L_0} \tag{10.4}$$

The ratio of the magnitude of the applied force to the cross-sectional area of the wire is called the **stress** *acting on the wire, while the ratio of the change in length to the original length of the wire is called the* **strain** *of the wire.* Equation 10.4 is a statement of **Hooke's law** *of elasticity, which says that in an elastic body the stress is directly proportional to the strain,* that is,

$$\text{stress} \propto \text{strain} \tag{10.5}$$

The stress is what is applied to the body, while the resulting effect is called the strain.

To make an equality out of this proportion, we must introduce a constant of proportionality (see appendix C on proportionalities). This constant depends on the type of material used, since the molecules, and hence the molecular forces of each material, are different. This constant, called **Young's modulus of elasticity** is denoted by the letter Y. Equation 10.4 thus becomes

$$\frac{F}{A} = Y\frac{\Delta L}{L_0} \tag{10.6}$$

The value of Y for various materials is given in table 10.1.

Table 10.1

Some Elastic Constants

Substance	Young's Modulus		Shear Modulus		Bulk Modulus		Elastic Limit		Ultimate Tensile Stress	
	$N/m^2 \times 10^{10}$	$lb/in.^2 \times 10^6$	$N/m^2 \times 10^{10}$	$lb/in.^2 \times 10^6$	$N/m^2 \times 10^{10}$	$lb/in.^2 \times 10^6$	$N/m^2 \times 10^{10}$	$lb/in.^2 \times 10^6$	$N/m^2 \times 10^8$	$lb/in.^2 \times 10^4$
Aluminum	7.0	10	3	3.8	7	10	1.4	1.9	1.4	3.1
Bone	1.5		8.0						1.30	
Brass	9.1	14	3.6	5.1	6	8.5	3.5	5.5	4.5	6.5
Copper	11.0	17	4.2	6.0	14	17	1.6	2.2	4.1	4.9
Iron	9.1	13	7.0		10	15	1.7	2.5	3.2	4.6
Lead	1.6	2.3	0.56		0.77	1.1			0.2	0.29
Steel	21	30	8.4	12	16	24	2.4	3.7	4.8	7.0

Example 10.1

Stretching a wire. A steel wire 1.00 m long with a diameter $d = 1.00$ mm has a 10.0-kg mass hung from it. (a) How much will the wire stretch? (b) What is the stress on the wire? (c) What is the strain?

Solution

a. The cross-sectional area of the wire is given by

$$A = \frac{\pi d^2}{4} = \frac{\pi (1.00 \times 10^{-3} \text{ m})^2}{4} = 7.85 \times 10^{-7} \text{ m}^2$$

We assume that the cross-sectional area of the wire does not change during the stretching process. The force stretching the wire is the weight of the 10.0-kg mass, that is,

$$F = mg = (10.0 \text{ kg})(9.80 \text{ m/s}^2) = 98.0 \text{ N}$$

Young's modulus for steel is found in table 10.1 as $Y = 21 \times 10^{10}$ N/m². The elongation of the wire, found from modifying equation 10.6, is

$$\Delta L = \frac{FL_0}{AY}$$

$$= \frac{(98.0 \text{ N})(1.00 \text{ m})}{(7.85 \times 10^{-7} \text{ m}^2)(21.0 \times 10^{10} \text{ N/m}^2)}$$

$$= 0.594 \times 10^{-3} \text{ m} = 0.594 \text{ mm}$$

b. The stress acting on the wire is

$$\frac{F}{A} = \frac{98.0 \text{ N}}{7.85 \times 10^{-7} \text{ m}^2} = 1.25 \times 10^8 \text{ N/m}^2$$

c. The strain of the wire is

$$\frac{\Delta L}{L_0} = \frac{0.594 \times 10^{-3} \text{ m}}{1.00 \text{ m}} = 0.594 \times 10^{-3}$$

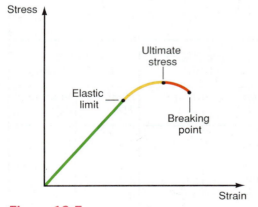

Figure 10.5

Stress-strain relationship.

The applied stress on the wire cannot be increased indefinitely if the wire is to remain elastic. Eventually a point is reached where the stress becomes so great that the atoms are pulled permanently away from their equilibrium position in the lattice structure. This point is called the **elastic limit** of the material and is shown in figure 10.5. When the stress exceeds the elastic limit the material does not return to its original size or shape when the stress is removed. The entire lattice structure of the material has been altered.

If the stress is increased beyond the elastic limit, eventually the ultimate stress point is reached. This is the highest point on the stress-strain curve and represents the greatest stress that the material can bear. Brittle materials break suddenly at this point, while some ductile materials can be stretched a little more due to a decrease in the cross-sectional area of the material. But they too break shortly thereafter at the breaking point. *Hooke's law is only valid below the elastic limit,* and it is only that region that will concern us.

Although we have been discussing the stretching of an elastic body, a body is also elastic under compression. If a large load is placed on a column, then the column is compressed, that is, it shrinks by an amount ΔL. When the load is removed the column returns to its original length.

Example 10.2

Compressing a steel column. A 100,000-lb load is placed on top of a steel column 10.0 ft long and 4.00 in. in diameter. By how much is the column compressed?

The cross-sectional area of the column is

$$A = \frac{\pi d^2}{4} = \frac{\pi (4.00 \text{ in.})^2}{4} = 12.6 \text{ in.}^2$$

The change in length of the column, found from equation 10.6, is

$$\Delta L = \frac{FL_0}{AY}$$

$$= \frac{(100,000 \text{ lb})(10.0 \text{ ft})}{(12.6 \text{ in.}^2)(30 \times 10^6 \text{ lb/in.}^2)} \left(\frac{12 \text{ in.}}{1 \text{ ft}} \right)$$

$$= 3.18 \times 10^{-2} \text{ in.} = 0.032 \text{ in.}$$

Note that the compression is quite small (0.032 in.) considering the very large load (100,000 lb). This is indicative of the very strong molecular forces in the lattice structure of the solid.

Example 10.3

Exceeding the ultimate compressive strength. A human bone is subjected to a compressive force of 5.00×10^5 N/m^2. The bone is 25.0 cm long and has an approximate area of 4.00 cm^2. If the ultimate compressive strength for a bone is 1.70×10^8 N/m^2, will the bone be compressed or will it break under this force?

The stress acting on the bone is found from

$$\frac{F}{A} = \frac{5.00 \times 10^5 \text{ N}}{4.00 \times 10^{-4} \text{ m}^2} = 12.5 \times 10^8 \text{ N/m}^2$$

Since this stress exceeds the ultimate compressive stress of a bone, 1.70×10^8 N/m^2, the bone will break.

10.3 Hooke's Law for a Spring

A simpler formulation of Hooke's law is sometimes useful and can be found from equation 10.6 by a slight rearrangement of terms. That is, solving equation 10.6 for F gives

$$F = \frac{AY}{L_0} \Delta L$$

Because A, Y, and L_0 are all constants, the term AY/L_0 can be set equal to a new constant k, namely

$$k = \frac{AY}{L_0} \tag{10.7}$$

We call k a force constant or a spring constant. Then,

$$F = k\Delta L \tag{10.8}$$

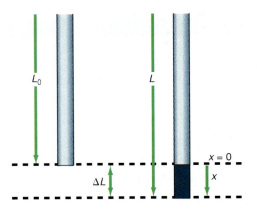

Figure 10.6

Changing the reference system.

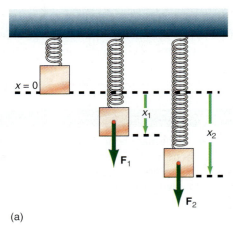

(a)

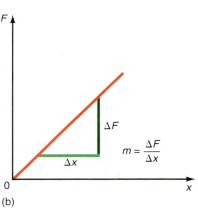

(b)

Figure 10.7

Experimental determination of a spring constant.

The change in length ΔL of the material is simply the final length L minus the original length L_0. We can introduce a new reference system to measure the elongation, by calling the location of the end of the material in its unstretched position, $x = 0$. Then we measure the stretch by the value of the displacement x from the unstretched position, as seen in figure 10.6. Thus, $\Delta L = x$, in the new reference system, and we can write equation 10.8 as

$$F = kx \qquad (10.9)$$

Equation 10.9 is a simplified form of Hooke's law that we use in vibratory motion containing springs. For a helical spring, we can not obtain the spring constant from equation 10.7 because the geometry of a spring is not the same as a simple straight wire. However, we can find k experimentally by adding various weights to a spring and measuring the associated elongation x, as seen in figure 10.7(a). A plot of the magnitude of the applied force F versus the elongation x gives a straight line that goes through the origin, as in figure 10.7(b). Because Hooke's law for the spring, equation 10.9, is an equation of the form of a straight line passing through the origin, that is,

$$y = mx$$

the slope m of the straight line is the spring constant k. In this way, we can determine experimentally the spring constant for any spring.

Example 10.4

The elongation of a spring. A spring with a force constant of 50.0 N/m is loaded with a 0.500-kg mass. Find the elongation of the spring.

Solution

The elongation of the spring, found from Hooke's law, equation 10.9, is

$$x = \frac{F}{k} = \frac{mg}{k}$$

$$= \frac{(0.500 \text{ kg})(9.80 \text{ m/s}^2)}{50.0 \text{ N/m}}$$

$$= 0.098 \text{ m}$$

10.4 Elasticity of Shape—Shear

In addition to being stretched or compressed, a body can be deformed by changing the shape of the body. If the body returns to its original shape when the distorting stress is removed, the body exhibits the property of elasticity of shape, sometimes called **shear**.

As an example, consider the cube fixed to the surface in figure 10.8(a). A tangential force $\mathbf{F_t}$ is applied at the top of the cube, a distance h above the bottom. The magnitude of this force F_t times the height h of the cube would normally cause a torque to act on the cube to rotate it. However, since the cube is not free to rotate, the body instead becomes deformed and changes its shape, as shown in figure 10.8(b). The normal lattice structure is shown in figure 10.8(c), and the deformed lattice structure in figure 10.8(d). The tangential force applied to the body causes the layers of atoms to be displaced sideways; one layer of the lattice structure slides over another.

Figure 10.8

Elasticity of shear.

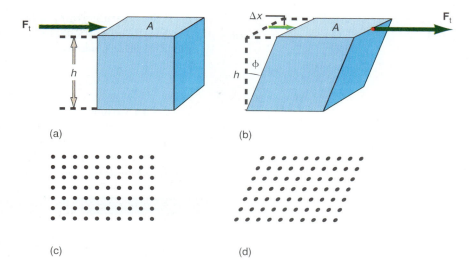

(a) (b)

(c) (d)

The tangential force thus causes a change in the shape of the body that is measured by the angle ϕ, called the *angle of shear*. We can also relate ϕ to the linear change from the original position of the body by noting from figure 10.8(b) that

$$\tan \phi = \frac{\Delta x}{h}$$

Because the deformations are usually quite small, as a first approximation the $\tan \phi$ can be replaced by the angle ϕ itself, expressed in radians. Thus,

$$\phi = \frac{\Delta x}{h} \tag{10.10}$$

*Equation 10.10 represents the **shearing strain** of the body.*

The tangential force $\mathbf{F_t}$ causes a deformation ϕ of the body and we find experimentally that

$$\phi \propto F_t \tag{10.11}$$

That is, the angle of shear is directly proportional to the magnitude of the applied tangential force F_t. We also find the deformation of the cube experimentally to be inversely proportional to the area of the top of the cube. With a larger area, the distorting force is spread over more molecules and hence the corresponding deformation is less. Thus,

$$\phi \propto \frac{1}{A} \tag{10.12}$$

Equations 10.11 and 10.12 can be combined into the single equation

$$\phi \propto \frac{F_t}{A} \tag{10.13}$$

Note that F_t/A has the dimensions of a stress and it is now defined as the **shearing stress:**

$$\text{Shearing stress} = \frac{F_t}{A} \tag{10.14}$$

Since ϕ is the shearing strain, equation 10.13 shows the familiar proportionality that stress is directly proportional to the strain. Introducing a constant of proportionality S, called the **shear modulus,** Hooke's law for the elasticity of shear is given by

$$\frac{F_t}{A} = S\phi \tag{10.15}$$

Values of S for various materials are given in table 10.1. The larger the value of S, the greater the resistance to shear. Note that the shear modulus is smaller than Young's modulus Y. This implies that it is easier to slide layers of molecules over each other than it is to compress or stretch them. The shear modulus is also known as the *torsion modulus* and the *modulus of rigidity.*

Example 10.5

Elasticity of shear. A sheet of copper 0.750 m long, 1.00 m high, and 0.500 cm thick is acted on by a tangential force of 50,000 N, as shown in figure 10.9. The value of S for copper is 4.20×10^{10} N/m². Find (a) the shearing stress, (b) the shearing strain, and (c) the linear displacement Δx.

Solution

a. The area that the tangential force is acting over is

$$A = bt = (0.750 \text{ m})(5.00 \times 10^{-3} \text{ m})$$

$$= 3.75 \times 10^{-3} \text{ m}^2$$

where b is the length of the base and t is the thickness of the copper wire shown in figure 10.9. The shearing stress is

$$\frac{F_t}{A} = \frac{50{,}000 \text{ N}}{3.75 \times 10^{-3} \text{ m}^2} = 1.33 \times 10^7 \text{ N/m}^2$$

b. The shearing strain, found from equation 10.15, is

$$\phi = \frac{F_t}{AS} = \frac{1.33 \times 10^7 \text{ N/m}^2}{4.20 \times 10^{10} \text{ N/m}^2}$$

$$= 3.17 \times 10^{-4} \text{ rad}$$

c. The linear displacement Δx, found from equation 10.10, is

$$\Delta x = h\phi = (1.00 \text{ m})(3.17 \times 10^{-4} \text{ rad})$$

$$= 3.17 \times 10^{-4} \text{ m} = 0.317 \text{ mm}$$

Figure 10.9

An example of shear.

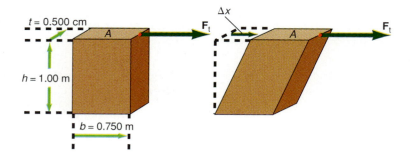

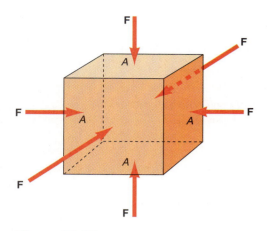

Figure 10.10
Volume elasticity.

10.5 Elasticity of Volume

If a uniform force is exerted on all sides of an object, as in figure 10.10, such as a block under water, each side of the block is compressed. Thus, the entire volume of the block decreases. The compressional stress is defined as

$$\text{stress} = \frac{F}{A} \qquad (10.16)$$

where F is the magnitude of the normal force acting on the cross-sectional area A of the block. The strain is measured by the change in volume per unit volume, that is,

$$\text{strain} = \frac{\Delta V}{V_0} \qquad (10.17)$$

Since the stress is directly proportional to the strain, by Hooke's law, we have

$$\frac{F}{A} \propto \frac{\Delta V}{V_0} \qquad (10.18)$$

To obtain an equality, we introduce a constant of proportionality B, called the **bulk modulus,** and Hooke's law for **elasticity of volume** becomes

$$\frac{F}{A} = -B\frac{\Delta V}{V_0} \qquad (10.19)$$

The minus sign is introduced in equation 10.19 because an increase in the stress (F/A) causes a decrease in the volume, leaving ΔV negative. The bulk modulus is a measure of how difficult it is to compress a substance. The reciprocal of the bulk modulus B, called the *compressibility k,* is a measure of how easy it is to compress the substance. The bulk modulus B is used for solids, while the compressibility k is usually used for liquids.

Quite often the body to be compressed is immersed in a liquid. In dealing with liquids and gases it is convenient to deal with the pressure exerted by the liquid or gas. We will see in detail in chapter 13 that pressure is defined as the force that is acting over a unit area of the body, that is,

$$p = \frac{F}{A}$$

For the case of volume elasticity, the stress F/A, acting on the body by the fluid, can be replaced by the pressure of the fluid itself. Thus, Hooke's law for volume elasticity can also be written as

$$p = -B\frac{\Delta V}{V_0} \qquad (10.20)$$

Example 10.6

Elasticity of volume. A solid copper sphere of 0.500-m³ volume is placed 100 ft below the ocean surface where the pressure is 3.00×10^5 N/m². What is the change in volume of the sphere? The bulk modulus for copper is 14×10^{10} N/m².

The change in volume, found from equation 10.20, is

$$\Delta V = -\frac{V_0}{B} p$$

$$= -\frac{(0.500 \text{ m}^3)(3.00 \times 10^5 \text{ N/m}^2)}{14 \times 10^{10} \text{ N/m}^2}$$

$$= -1.1 \times 10^{-6} \text{ m}^3$$

The minus sign indicates that the volume has decreased.

The Language of Physics

Elasticity
That property of a body by which it experiences a change in size or shape whenever a deforming force acts on the body. The elastic properties of matter are a manifestation of the molecular forces that hold solids together (p. 283).

Lattice structure of a solid
A regular, periodically repeated, three-dimensional array of the atoms or molecules comprising the solid (p. 283).

Stress
For a body that can be either stretched or compressed, the stress is the ratio of the applied force acting on a body to the cross-sectional area of the body (p. 285).

Strain
For a body that can be either stretched or compressed, the ratio of the change in length to the original length of the body is called the strain (p. 285).

Hooke's law
In an elastic body, the stress is directly proportional to the strain (p. 285).

Young's modulus of elasticity
The proportionality constant in Hooke's law. It is equal to the ratio of the stress to the strain (p. 285).

Elastic limit
The point where the stress on a body becomes so great that the atoms of the body are pulled permanently away from their equilibrium position in the lattice structure. When the stress exceeds the elastic limit, the material will not return to its original size or shape when the stress is removed. Hooke's law is no longer valid above the elastic limit (p. 286).

Shear
That elastic property of a body that causes the shape of the body to be changed when a stress is applied. When the stress is removed the body returns to its original shape (p. 288).

Shearing strain
The angle of shear, which is a measure of how much the body's shape has been deformed (p. 289).

Shearing stress
The ratio of the tangential force acting on the body to the area of the body over which the tangential force acts (p. 289).

Shear modulus
The constant of proportionality in Hooke's law for shear. It is equal to the ratio of the shearing stress to the shearing strain (p. 290).

Bulk modulus
The constant of proportionality in Hooke's law for volume elasticity. It is equal to the ratio of the compressional stress to the strain. The strain for this case is equal to the change in volume per unit volume (p. 291).

Elasticity of volume
When a uniform force is exerted on all sides of an object, each side of the object becomes compressed. Hence, the entire volume of the body decreases. When the force is removed the body returns to its original volume (p. 291).

Summary of Important Equations

Hooke's law in general
stress ∝ strain **(10.5)**

Hooke's law for stretching
or compression
$$\frac{F}{A} = Y\frac{\Delta L}{L_0} \quad \textbf{(10.6)}$$

Hooke's law for a spring
$$F = kx \quad \textbf{(10.9)}$$

Hooke's law for shear
$$\frac{F_t}{A} = S\phi \quad \textbf{(10.15)}$$

Hooke's law for volume
elasticity
$$\frac{F}{A} = -B\frac{\Delta V}{V_0} \quad \textbf{(10.19)}$$

Hooke's law for volume
elasticity
$$p = -B\frac{\Delta V}{V_0} \quad \textbf{(10.20)}$$

Questions for Chapter 10

1. Why is concrete often reinforced with steel?
†2. An amorphous solid such as glass does not have the simple lattice structure shown in figure 10.1. What effect does this have on the elastic properties of glass?
3. Discuss the assumption that the diameter of a wire does not change when under stress.
4. Compare the elastic constants of a human bone with the elastic constants of other materials listed in table 10.1. From this standpoint discuss the bone as a structural element.

5. Why are there no Young's moduli for liquids or gases?
6. Describe the elastic properties of a cube of jello.
7. If you doubled the diameter of a human bone, what would happen to the maximum compressive force that the bone could withstand without breaking?
†8. In the profession of Orthodontics, a dentist uses braces to realign teeth. Discuss this process from the point of view of stress and strain.

†9. Discuss Hooke's law as it applies to the bending of a beam that is fixed at one end and has a load placed at the other end.

(a)

(b)

†10. How do the elastic properties of a material affect the vibration of that material?

Problems for Chapter 10

10.2 Hooke's Law—Stress and Strain

1. An aluminum wire has a diameter of 0.850 mm and is subjected to a force of 1000 N. Find the stress acting on the wire.
2. A copper wire experiences a stress of 5.00×10^3 N/m². If the diameter of the wire is 0.750 mm, find the force acting on the wire.
3. A brass wire 0.750 cm long is stretched by 0.001 cm. Find the strain of the wire.
4. A steel wire, 1.00 m long, has a diameter of 1.50 mm. If a mass of 3.00 kg is hung from the wire, by how much will it stretch?
5. A load of 50,000 lb is placed on an aluminum column 4.00 in. in diameter. If the column was originally 4.00 ft high find the amount that the column has shrunk.
6. A mass of 25,000 kg is placed on a steel column, 3.00 m high and 15.0 cm in diameter. Find the decrease in length of the column under this compression.

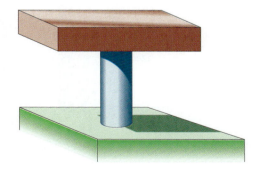

7. An aluminum wire, 1.50 m long, has a diameter of 0.750 mm. If a force of 60.0 N is suspended from the wire, find (a) the stress on the wire, (b) the elongation of the wire, and (c) the strain of the wire.
8. A copper wire, 1.00 m long, has a diameter of 0.750 mm. When an unknown weight is suspended from the wire it stretches 0.200 mm. What was the load placed on the wire?
9. A steel wire is 1.00 m long and has a diameter of 0.75 mm. Find the maximum value of a mass that can be suspended from the wire before exceeding the elastic limit of the wire.
10. A steel wire is 1.00 m long and has a 10.0-kg mass suspended from it. What is the minimum diameter of the wire such that the load will not exceed the elastic limit of the wire?
11. Find the maximum load that can be applied to a brass wire, 0.750 mm in diameter, without exceeding the elastic limit of the wire.
12. Find the maximum change in length of a 1.00-m brass wire, of 0.800 mm diameter, such that the elastic limit of the wire is not exceeded.
13. If the thigh bone is about 25.0 cm in length and about 4.00 cm in diameter determine the maximum compression of the bone before it will break. The ultimate compressive strength of bone is 1.70×10^8 N/m².
14. If the ultimate tensile strength of glass is 7.00×10^7 N/m², find the maximum weight that can be placed on a glass cylinder of 0.100 m² area, 25.0 cm long, if the glass is not to break.

15. A human bone is 2.00 cm in diameter. Find the maximum compression force the bone can withstand without fracture. The ultimate compressive strength of bone is 1.70×10^8 N/m².
16. A copper rod, 0.400 cm in diameter, supports a load of 150 kg suspended from one end. Will the rod return to its initial length when the load is removed or has this load exceeded the elastic limit of the rod?

10.3 Hooke's Law for a Spring

17. A coil spring stretches 4.00 cm when a mass of 500 g is suspended from it. What is the force constant of the spring?
18. A coil spring stretches by 2.00 cm when an unknown load is placed on the spring. If the spring has a force constant of 3.5 N/m, find the value of the unknown force.
19. A coil spring stretches by 2.50 cm when a mass of 750 g is suspended from it. (a) Find the force constant of the spring. (b) How much will the spring stretch if 800 g is suspended from it?
20. A horizontal spring stretches 20.0 cm when a force of 10.0 N is applied to the spring. By how much will it stretch if a 30.0-N force is now applied to the spring? If the same spring is placed in the vertical and a weight of 10.0 N is hung from the spring, will the results change?
21. A coil spring stretches by 4.50 cm when a mass of 250 g is suspended from it. What force is necessary to stretch the spring an additional 2.50 cm?

10.4 Elasticity of Shape—Shear

22. A brass cube, 5.00 cm on a side, is subjected to a tangential force. If the angle of shear is measured in radians to be 0.010 rad, what is the magnitude of the tangential force?

23. A copper block, 7.50 cm on a side, is subjected to a tangential force of 3.5 $\times$ 10³ N. Find the angle of shear.

24. A copper cylinder, 7.50 cm high, and 7.50 cm in diameter, is subjected to a tangential force of 3.5 $\times$ 10³ N. Find the angle of shear. Compare this result with problem 23.

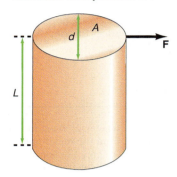

25. An annular copper cylinder, 7.50 cm high, inner radius of 2.00 cm and outer radius of 3.75 cm, is subjected to a tangential force of 3.5 $\times$ 10³ N. Find the angle of shear. Compare this result with problems 23 and 24.

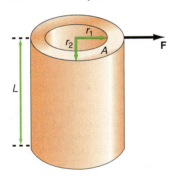

10.5 Elasticity of Volume

26. A cube of lead 15.0 cm on a side is subjected to a uniform pressure of 5.00 $\times$ 10⁵ N/m². By how much does the volume of the cube change?

27. A liter of glycerine contracts 0.21 cm³ when subject to a pressure of 9.8 $\times$ 10⁵ N/m². Calculate the bulk modulus of glycerine.

28. A pressure of 1.013 $\times$ 10⁷ N/m² is applied to a volume of 15.0 m³ of water. If the bulk modulus of water is 0.020 $\times$ 10¹⁰ N/m², by how much will the water be compressed?

29. Repeat problem 28, only this time use glycerine that has a bulk modulus of 0.45 $\times$ 10¹⁰ N/m².

30. Normal atmospheric pressure is 1.013 $\times$ 10⁵ N/m². How many atmospheres of pressure must be applied to a volume of water to compress it to 1.00% of its original volume? The bulk modulus of water is 0.020 $\times$ 10¹⁰ N/m².

31. Find the ratio of the density of water at the bottom of a 50.0-m lake to the density of water at the surface of the lake. The pressure at the bottom of the lake is 4.90 $\times$ 10⁵ N/m². (*Hint:* the volume of the water will be decreased by the pressure of the water above it.) The bulk modulus for water is 0.21 $\times$ 10¹⁰ N/m².

Additional Problems

32. A lead block 50.0 cm long, 10.0 cm wide, and 10.0 cm thick, has a force of 200,000 N placed on it. Find the stress, the strain, and the change in length if (a) the block is standing upright, and (b) the block is lying flat.

33. An aluminum cylinder must support a load of 450,000 N. The cylinder is 5.00 m high. If the maximum allowable stress is 1.4 $\times$ 10⁸, what must be the minimum radius of the cylinder in order for the cylinder to support the load? What will be the length of the cylinder when under load?

34. This is essentially the same problem as 33, but now the cylinder is made of steel. Find the minimum radius of the steel cylinder that is necessary to support the load and compare it to the radius of the aluminum cylinder. The maximum allowable stress for steel is 2.4 $\times$ 10¹⁰ N/m².

35. How many 1.00-kg masses may be hung from a 1.00-m steel wire, 0.750 mm in diameter, without exceeding the elastic limit of the wire?

36. A solid copper cylinder 1.50 m long and 10.0 cm in diameter, has a mass of 5000 kg placed on its top. Find the compression of the cylinder.

37. This is the same problem as 36, except that the cylinder is an annular cylinder with an inner radius of 3.50 cm and outer radius of 5.00 cm. Find the compression of the cylinder and compare with problem 36.

38. This is the same problem as problem 36 except the body is an I-beam with the dimensions shown in the diagram. Find the compression of the I-beam and compare to problems 36 and 37. The crossbar width is 2.00 cm.

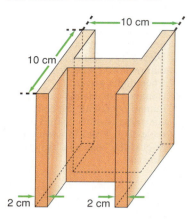

†39. Two pieces of metal rod, 2.00 cm thick, are to be connected together by riveting a steel plate to them as shown in the diagram. Two rivets, each 1.00 cm in diameter, are used. What is the maximum force that can be applied to the metal rod without exceeding a shearing stress of 8.4 $\times$ 10⁸ N/m².

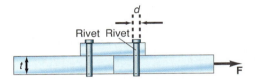

†40. A copper and steel wire are welded together at their ends as shown. The original length of each wire is 50.0 cm and each has a diameter of 0.780 mm. A mass of 10.0 kg is suspended from the combined wire. By how much will the combined wire stretch?

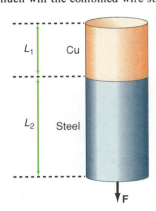

†41. A copper and steel wire each 50.0 cm in length and 0.780 mm in diameter are connected in parallel to a load of 98.0 N, as shown in the diagram. If the strain is the same for each wire, find (a) the force on wire 1, (b) the force on wire 2, and (c) the total displacement of the load.

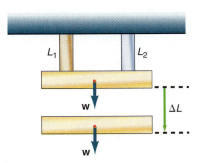

†42. Repeat problem 41 with the diameter of wire 1 equal to 1.00 mm and the diameter of wire 2 equal to 1.50 mm.

†43. Two steel wires of diameters 1.50 mm and 1.00 mm, and each 50.0 cm long, are welded together in series as shown in the diagram. If a weight of 98.0 N is suspended from the bottom of the combined wire, by how much will the combined wire stretch?

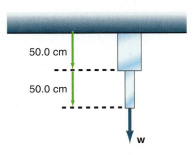

†44. Two springs are connected in parallel as shown in the diagram. The spring constants are $k_1 = 5.00$ N/m and $k_2 = 3.00$ N/m. A force of 10.0 N is applied as shown. If the strain is the same in each spring, find (a) the displacement of mass m, (b) the force on spring 1, and (c) the force on spring 2.

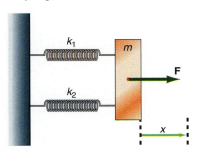

†45. Two springs are connected in series as shown in the diagram. The spring constants are $k_1 = 5.00$ N/m and $k_2 = 3.00$ N/m. A force of 10.0 N is applied as shown. Find (a) the displacement of mass m, (b) the displacement of spring 1, and (c) the displacement of spring 2.

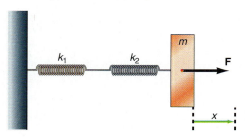

Interactive Tutorials

📟 46. Young's modulus for a wire is $Y = 2.10 \times 10^{11}$ N/m². The wire has an initial length of $L_0 = 0.700$ m and a diameter $d = 0.310$ mm. A force $F = 1.00$ N is applied in steps from 1.00 to 10.0 N. Calculate the wire's change in length ΔL with increasing load F, and graph the result.

A simple pendulum.

We are to admit no more causes of natural things than such as are both true and sufficient to explain their appearances.

Isaac Newton

11 Simple Harmonic Motion

11.1 Introduction to Periodic Motion

Periodic motion is any motion that repeats itself in equal intervals of time. The uniformly rotating earth represents a periodic motion that repeats itself every 24 hours. The motion of the earth around the sun is periodic, repeating itself every 12 months. A vibrating spring and a pendulum also exhibit periodic motion. The period of the motion is defined as the time for the motion to repeat itself. A special type of periodic motion is simple harmonic motion and we now proceed to investigate it.

11.2 Simple Harmonic Motion

An example of simple harmonic motion is the vibration of a mass m, attached to a spring of negligible mass, as the mass slides on a frictionless surface, as shown in figure 11.1. We say that the mass, in the unstretched position, figure 11.1(a), is in its equilibrium position. If an applied force $\mathbf{F_A}$ acts on the mass, the mass will be displaced to the right of its equilibrium position a distance x, figure 11.1(b). The distance that the spring stretches, obtained from Hooke's law, is

$$F_A = kx$$

The **displacement** x *is defined as the distance the body moves from its equilibrium position.* Because F_A is a force that pulls the mass to the right, it is also the force that pulls the spring to the right. By Newton's third law there is an equal but opposite elastic force exerted by the spring on the mass pulling the mass toward the left. Since this force tends to restore the mass to its original position, we call it the restoring force F_R. Because the restoring force is opposite to the applied force, it is given by

$$F_R = -F_A = -kx \tag{11.1}$$

When the applied force F_A is removed, the elastic restoring force F_R is then the only force acting on the mass m, figure 11.1(c), and it tries to restore m to its equilibrium position. We can then find the acceleration of the mass from Newton's second law as

$$ma = F_R$$
$$= -kx$$

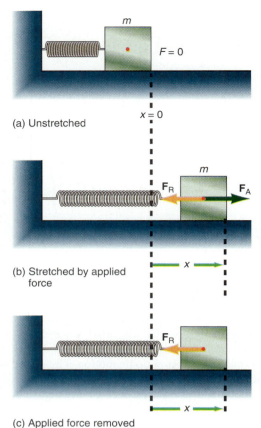

(a) Unstretched

(b) Stretched by applied force

(c) Applied force removed

Figure 11.1

The vibrating spring.

Thus,

$$a = -\frac{k}{m}x \qquad (11.2)$$

Equation 11.2 is the defining equation for simple harmonic motion. **Simple harmonic motion** *is motion in which the acceleration of a body is directly proportional to its displacement from the equilibrium position but in the opposite direction.* A vibrating system that executes simple harmonic motion is sometimes called a *harmonic oscillator. Because the acceleration is directly proportional to the displacement x in simple harmonic motion, the acceleration of the system is not constant but varies with x.* At large displacements, the acceleration is large, at small displacements the acceleration is small. Describing the vibratory motion of the mass *m* requires some new techniques because the kinematic equations derived in chapter 3 were based on the assumption that the acceleration of the system was a constant. As we can see from equation 11.2, this assumption is no longer valid. We need to derive a new set of kinematic equations to describe simple harmonic motion, and we will do so in section 11.3. However, let us first look at the motion from a physical point of view. The mass *m* in figure 11.2(a) is pulled a distance $x = A$ to the right, and is then released. The maximum restoring force on *m* acts to the left at this position because

$$F_{Rmax} = -kx_{max} = -kA$$

*The maximum displacement A is called the **amplitude** of the motion.* At this position the mass experiences its maximum acceleration to the left. From equation 11.2 we obtain

$$a = -\frac{k}{m}A$$

The mass continues to move toward the left while the acceleration continuously decreases. At the equilibrium position, figure 11.2(b), $x = 0$ and hence, from equation 11.2, the acceleration is also zero. However, because the mass has inertia it moves past the equilibrium position to negative values of *x,* thereby compressing the spring. The restoring force F_R now points to the right, since for negative values of *x,*

$$F_R = -k(-x) = kx$$

The force acting toward the right causes the mass to slow down, eventually coming to rest at $x = -A$. At this point, figure 11.2(c), there is a maximum restoring force pointing toward the right

$$F_{Rmax} = -k(-A)_{max} = kA$$

and hence a maximum acceleration

$$a_{max} = -\frac{k}{m}(-A) = \frac{k}{m}A$$

also to the right. The mass moves to the right while the force F_R and the acceleration *a* decreases with *x* until *x* is again equal to zero, figure 11.2(d). Then F_R and *a* are also zero. Because of the inertia of the mass, it moves past the equilibrium position to positive values of *x.* The restoring force again acts toward the left, slowing down the mass. When the displacement *x* is equal to *A,* figure 11.2(e), the mass momentarily comes to rest and then the motion repeats itself. *One complete motion from $x = A$ and back to $x = A$ is called a **cycle** or an **oscillation.** The **period** T is the*

Figure 11.2

Detailed motion of the vibrating spring.

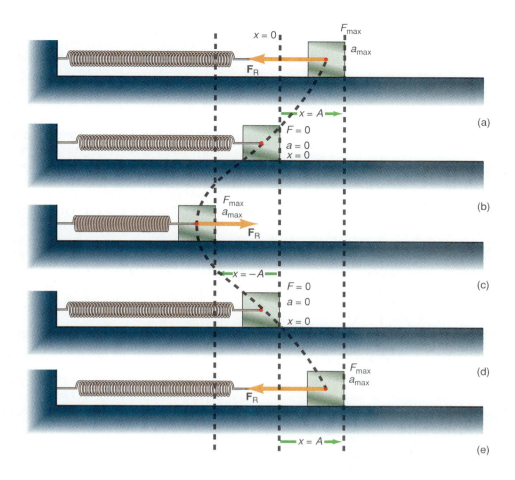

time for one complete oscillation, and the **frequency** *f* is the number of complete oscillations or cycles made in unit time. The period and the frequency are reciprocal to each other, that is,

$$f = \frac{1}{T} \tag{11.3}$$

The unit for a period is the second, while the unit for frequency, called a hertz, is one cycle per second. The hertz is abbreviated, Hz. Also note that a cycle is a number not a dimensional quantity and can be dropped from the computations whenever doing so is useful.

11.3 Analysis of Simple Harmonic Motion—The Reference Circle

As pointed out in section 11.2, the acceleration of the mass in the vibrating spring system is not a constant, but rather varies with the displacement *x*. Hence, the kinematic equations of chapter 3 can not be used to describe the motion. (We derived those equations on the assumption that the acceleration was constant.) Thus, a new set of equations must be derived to describe simple harmonic motion.

Simple harmonic motion is related to the uniform circular motion studied in chapter 6. An analysis of uniform circular motion gives us a set of equations to

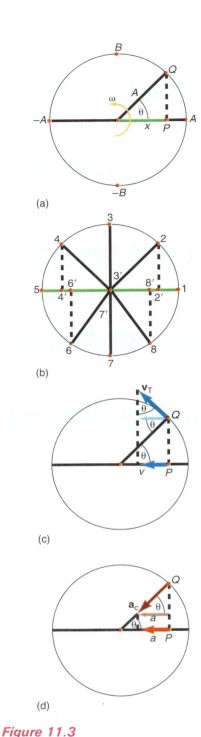

Figure 11.3

Simple harmonic motion and the reference circle.

describe simple harmonic motion. As an example, consider a point Q moving in uniform circular motion with an angular velocity ω, as shown in figure 11.3(a). At a particular instant of time t, the angle θ that Q has turned through is

$$\theta = \omega t \qquad (11.4)$$

The projection of point Q onto the x-axis gives the point P. As Q rotates in the circle, P oscillates back and forth along the x-axis, figure 11.3(b). That is, when Q is at position 1, P is at 1. As Q moves to position 2 on the circle, P moves to the left along the x-axis to position 2′. As Q moves to position 3, P moves on the x-axis to position 3′, which is of course the value of $x = 0$. As Q moves to position 4 on the circle, P moves along the negative x-axis to position 4′. When Q arrives at position 5, P is also there. As Q moves to position 6 on the circle, P moves to position 6′ on the x-axis. Then finally, as Q moves through positions 7, 8, and 1, P moves through 7′, 8′, and 1, respectively. The oscillatory motion of point P on the x-axis corresponds to the simple harmonic motion of a body m moving under the influence of an elastic restoring force, as shown in figure 11.2.

The position of P on the x-axis and hence the position of the mass m is described in terms of the point Q and the angle θ found in figure 11.3(a) as

$$x = A \cos \theta \qquad (11.5)$$

Here A is the amplitude of the vibratory motion and using the value of θ from equation 11.4 we have

$$x = A \cos \omega t \qquad (11.6)$$

Equation 11.6 is the first kinematic equation for simple harmonic motion; it gives the displacement of the vibrating body at any instant of time t. The angular velocity ω of point Q in the **reference circle** is related to the frequency of the simple harmonic motion. Because the angular velocity was defined as

$$\omega = \frac{\theta}{t} \qquad (11.7)$$

then, for a complete rotation of point Q, θ rotates through an angle of 2π rad. But this occurs in exactly the time for P to execute one complete vibration. We call this time for one complete vibration the period T. Hence, we can also write the angular velocity, equation 11.7, as

$$\omega = \frac{\theta}{t} = \frac{2\pi}{T} \qquad (11.8)$$

Since the frequency f is the reciprocal of the period T (equation 11.3) we can write equation 11.8 as

$$\omega = 2\pi f \qquad (11.9)$$

*Thus, the **angular velocity** of the uniform circular motion in the reference circle is related to the frequency of the vibrating system.* Because of this relation between the angular velocity and the frequency of the system, we usually call the angular velocity ω the *angular frequency of the vibrating system.* We can substitute equation 11.9 into equation 11.6 to give *another form for the first kinematic equation of simple harmonic motion,* namely

$$x = A \cos(2\pi f t) \qquad (11.10)$$

We can find the velocity of the mass m attached to the end of the spring in figure 11.2 with the help of the reference circle in figure 11.3(c). The point Q moves with the tangential velocity $\mathbf{V_T}$. The x-component of this velocity is the velocity of the point P and hence the velocity of the mass m. From figure 11.3(c) we can see that

$$v = -V_T \sin \theta \tag{11.11}$$

The minus sign indicates that the velocity of P is toward the left at this position. The linear velocity V_T of the point Q is related to the angular velocity ω by equation 9.2 of chapter 9, that is

$$v = r\omega$$

For the reference circle, $v = V_T$ and r is the amplitude A. Hence, the tangential velocity V_T is given by

$$V_T = \omega A \tag{11.12}$$

Using equations 11.11, 11.12, and 11.4, the velocity of point P becomes

$$v = -\omega A \sin \omega t \tag{11.13}$$

Equation 11.13 is the second of the kinematic equations for simple harmonic motion and it gives the speed of the vibrating mass at any time t.
A third kinematic equation for simple harmonic motion giving the speed of the vibrating body as a function of displacement can be found from equation 11.13 by using the trigonometric identity

$$\sin^2\theta + \cos^2\theta = 1$$

or

$$\sin \theta = \pm \sqrt{1 - \cos^2\theta}$$

From figure 11.3(a) or equation 11.5, we have

$$\cos \theta = \frac{x}{A}$$

Hence,

$$\sin \theta = \pm \sqrt{1 - \frac{x^2}{A^2}} \tag{11.14}$$

Substituting equation 11.14 back into equation 11.13, we get

$$v = \pm \omega A \sqrt{1 - \frac{x^2}{A^2}}$$

or

$$v = \pm \omega \sqrt{A^2 - x^2} \tag{11.15}$$

Equation 11.15 is the third of the kinematic equations for simple harmonic motion and it gives the velocity of the moving body at any displacement x. The $\pm$ sign in equation 11.15 indicates the direction of the vibrating body. If the body is moving to the right, then the positive sign $(+)$ is used. If the body is moving to the left, then the negative sign $(-)$ is used.
Finally, we can find the acceleration of the vibrating body using the reference circle in figure 11.3(d). The point Q in uniform circular motion experiences a centripetal acceleration $\mathbf{a_c}$ pointing toward the center of the circle in figure 11.3(d). The x-component of the centripetal acceleration is the acceleration of the vibrating body at the point P. That is,

$$a = -a_c \cos \theta \tag{11.16}$$

The minus sign again indicates that the acceleration is toward the left. But recall from chapter 6 that the magnitude of the centripetal acceleration is

$$a_c = \frac{v^2}{r} \tag{6.12}$$

where v represents the tangential speed of the rotating object, which in the present case is V_T, and r is the radius of the circle, which in the present case is the radius of the reference circle A. Thus,

$$a_c = \frac{V_T^2}{A}$$

But we saw in equation 11.12 that $V_T = \omega A$, therefore

$$a_c = \omega^2 A$$

The acceleration of the mass m, equation 11.16, thus becomes

$$a = -\omega^2 A \cos \omega t \tag{11.17}$$

Equation 11.17 is the fourth of the kinematic equations for simple harmonic motion. It gives the acceleration of the vibrating body at any time t. This equation has no counterpart in chapter 3, because there the acceleration was always a constant. Also, since $F = ma$ by Newton's second law, the force acting on the mass m, becomes

$$F = -m\omega^2 A \cos \omega t \tag{11.18}$$

Thus, the force acting on the mass m is a variable force.

Equations 11.6 and 11.17 can be combined into the simple equation

$$a = -\omega^2 x \tag{11.19}$$

If equation 11.19 is compared with equation 11.2,

$$a = -\frac{k}{m} x$$

we see that the acceleration of the mass at P, equation 11.19, is directly proportional to the displacement x and in the opposite direction. But this is the definition of simple harmonic motion as stated in equation 11.2. Hence, the projection of a point at Q, in uniform circular motion, onto the x-axis does indeed represent simple harmonic motion. Thus, the kinematic equations developed to describe the motion of the point P, also describe the motion of a mass attached to a vibrating spring.

An important relation between the characteristics of the spring and the vibratory motion can be easily deduced from equations 11.2 and 11.19. That is, because both equations represent the acceleration of the vibrating body they can be equated to each other, giving

$$\omega^2 = \frac{k}{m}$$

or

$$\omega = \sqrt{\frac{k}{m}} \tag{11.20}$$

The value of ω in the kinematic equations is expressed in terms of the force constant k of the spring and the mass m attached to the spring. The physics of simple harmonic motion is thus connected to the angular frequency ω of the vibration.

In summary, the kinematic equations for simple harmonic motion are

$$x = A \cos \omega t \tag{11.6}$$

$$v = -\omega A \sin \omega t \tag{11.13}$$

$$v = \pm \omega \sqrt{A^2 - x^2} \tag{11.15}$$

$$a = -\omega^2 A \cos \omega t \tag{11.17}$$

$$F = -m\omega^2 A \cos \omega t \tag{11.18}$$

where, from equations 11.9 and 11.20, we have

$$\omega = 2\pi f = \sqrt{\frac{k}{m}}$$

A plot of the displacement, velocity, and acceleration of the vibrating body as a function of time are shown in figure 11.4. We can see that the mathematical description follows the physical description in figure 11.2. When $x = A$, the maximum displacement, the velocity v is zero, while the acceleration is at its maximum value of $-\omega^2 A$. The minus sign indicates that the acceleration is toward the left. The force is at its maximum value of $-m\omega^2 A$, where the minus sign shows that the restoring force is pulling the mass back toward its equilibrium position. At the equilibrium position $x = 0$, $a = 0$, and $F = 0$, but v has its maximum velocity of $-\omega A$ toward the left. As x goes to negative values, the force and the acceleration become positive, slowing down the motion to the left, and hence v starts to decrease. At $x = -A$ the velocity is zero and the force and acceleration take on their maximum values toward the right, tending to restore the mass to its equilibrium position. As

Figure 11.4

Displacement, velocity, and acceleration in simple harmonic motion.

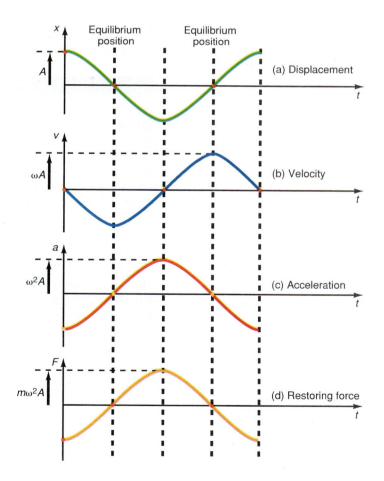

(a) Displacement

(b) Velocity

(c) Acceleration

(d) Restoring force

x becomes less negative, the velocity to the right increases, until it picks up its maximum value of ωA at $x = 0$, the equilibrium position, where F and a are both zero. Because of this large velocity, the mass passes the equilibrium position in its motion toward the right. However, as soon as x becomes positive, the force and the acceleration become negative thereby slowing down the mass until its velocity becomes zero at the maximum displacement A. One entire cycle has been completed, and the motion starts over again. (We should emphasize here that in this vibratory motion there are two places where the velocity is instantaneously zero, $x = A$ and $x = -A$, even though the instantaneous acceleration is nonzero there.)

Sometimes the vibratory motion is so rapid that the actual displacement, velocity, and acceleration at every instant of time are not as important as the gross motion, which can be described in terms of the frequency or period of the motion. We can find the frequency of the vibrating mass in terms of the spring constant k and the vibrating mass m by setting equation 11.9 equal to equation 11.20. Thus,

$$\omega = 2\pi f = \sqrt{\frac{k}{m}}$$

Solving for the frequency f, we obtain

$$f = \frac{1}{2\pi} \sqrt{\frac{k}{m}} \tag{11.21}$$

Equation 11.21 gives the frequency of the vibration. Because the period of the vibrating motion is the reciprocal of the frequency, we get for the period

$$T = 2\pi \sqrt{\frac{m}{k}} \tag{11.22}$$

Equation 11.22 gives the period of the simple harmonic motion in terms of the mass m in motion and the spring constant k. Notice that for a particular value of m and k, the period of the motion remains a constant throughout the motion.

Example 11.1

An example of simple harmonic motion. A mass of 0.300 kg is placed on a vertical spring and the spring stretches by 10.0 cm. It is then pulled down an additional 5.00 cm and then released. Find (a) the spring constant k, (b) the angular frequency ω, (c) the frequency f, (d) the period T, (e) the maximum velocity of the vibrating mass, (f) the maximum acceleration of the mass, (g) the maximum restoring force, (h) the velocity of the mass at $x = 2.00$ cm, and (i) the equation of the displacement, velocity, and acceleration at any time t.

Solution

Although the original analysis dealt with a mass on a horizontal frictionless surface, the results also apply to a mass attached to a spring that is allowed to vibrate in the vertical direction. The constant force of gravity on the 0.300-kg mass displaces the equilibrium position to $x = 10.0$ cm. When the additional force is applied to displace the mass another 5.00 cm, the mass oscillates about the equilibrium position, located at the 10.0-cm mark. Thus, the force of gravity only displaces the equilibrium position, but does not otherwise influence the result of the dynamic motion.

a.   The spring constant, found from Hooke's law, is

$$k = \frac{F_A}{x} = \frac{mg}{x}$$

$$= \frac{(0.300 \text{ kg})(9.80 \text{ m/s}^2)}{0.100 \text{ m}}$$

$$= 29.4 \text{ N/m}$$

b. The angular frequency ω, found from equation 11.20, is

$$\omega = \sqrt{\frac{k}{m}}$$

$$= \sqrt{\frac{29.4 \text{ N/m}}{0.300 \text{ kg}}}$$

$$= 9.90 \text{ rad/s}$$

c. The frequency of the motion, found from equation 11.9, is

$$f = \frac{\omega}{2\pi}$$

$$= \frac{9.90 \text{ rad/s}}{2\pi \text{ rad}}$$

$$= 1.58 \frac{\text{cycles}}{\text{s}} = 1.58 \text{ Hz}$$

d. We could find the period from equation 11.22 but since we already know the frequency f, it is easier to compute T from equation 11.3. Thus,

$$T = \frac{1}{f} = \frac{1}{1.58 \text{ cycles/s}} = 0.633 \text{ s}$$

e. The maximum velocity, found from equation 11.13, is

$$v_{max} = \omega A = (9.90 \text{ rad/s})(5.00 \times 10^{-2} \text{ m})$$

$$= 0.495 \text{ m/s}$$

f. The maximum acceleration, found from equation 11.17, is

$$a_{max} = \omega^2 A = (9.90 \text{ rad/s})^2(5.00 \times 10^{-2} \text{ m})$$

$$= 4.90 \text{ m/s}^2$$

g. The maximum restoring force, found from Hooke's law, is

$$F_{max} = kx_{max} = kA$$

$$= (29.4 \text{ N/m})(5.00 \times 10^{-2} \text{ m})$$

$$= 1.47 \text{ N}$$

h. The velocity of the mass at $x = 2.00$ cm, found from equation 11.15, is

$$v = \pm\omega\sqrt{A^2 - x^2}$$

$$= \pm(9.90 \text{ rad/s})\sqrt{(5.00 \times 10^{-2} \text{ m})^2 - (2.00 \times 10^{-2} \text{ m})^2}$$

$$= \pm 0.454 \text{ m/s}$$

where v is positive when moving to the right and negative when moving to the left.

i. The equation of the displacement at any instant of time, found from equation 11.6, is

$$x = A \cos \omega t$$

$$= (5.00 \times 10^{-2} \text{ m}) \cos(9.90 \text{ rad/s})t$$

The equation of the velocity at any instant of time, found from equation 11.13, is

$$v = -\omega A \sin \omega t$$

$$= -(9.90 \text{ rad/s})(5.00 \times 10^{-2} \text{ m})\sin(9.90 \text{ rad/s})t$$

$$= -(0.495 \text{ m/s})\sin(9.90 \text{ rad/s})t$$

The equation of the acceleration at any time, found from equation 11.17, is

$$a = -\omega^2 A \cos \omega t$$
$$= -(9.90 \text{ rad/s})^2 (5.00 \times 10^{-2} \text{ m})\cos(9.90 \text{ rad/s})t$$
$$= -(4.90 \text{ m/s}^2)\cos(9.90 \text{ rad/s})t$$

11.4 The Potential Energy of a Spring

In chapter 7 we defined the gravitational potential energy of a body as the energy that a body possesses by virtue of its position in a gravitational field. A body can also have elastic potential energy. For example, *a compressed spring has potential energy because it has the ability to do work as it expands to its equilibrium configuration. Similarly, a stretched spring must also contain potential energy because it has the ability to do work as it returns to its equilibrium position.* Because work must be done on a body to put the body into the configuration where it has the elastic potential energy, this work is used as the measure of the potential energy. Thus, the **potential energy of a spring** is equal to the work that you, the external agent, must do to compress (or stretch) the spring to its present configuration. We defined the potential energy as

$$PE = W = Fx \tag{11.23}$$

However, we can not use equation 11.23 in its present form to determine the potential energy of a spring. Recall that the work defined in this way, in chapter 7, was for a constant force. We have seen in this chapter that the force necessary to compress or stretch a spring is not a constant but is rather a variable force depending on the value of x, $(F = -kx)$. We can still solve the problem, however, by using the average value of the force between the value at the equilibrium position and the value at the position x. That is, because the restoring force is directly proportional to the displacement, the average force exerted in moving the mass m from $x = 0$ to the value x in figure 11.5(a) is

$$F_{\text{avg}} = \frac{F_0 + F}{2}$$

Thus, we find the potential energy in this configuration by using the average force, that is,

$$PE = W = F_{\text{avg}}x$$
$$= W = \left(\frac{F_0 + F}{2}\right)x$$
$$= \left(\frac{0 + kx}{2}\right)x$$

Hence,

$$PE = \tfrac{1}{2}kx^2 \tag{11.24}$$

Because of the x^2 in equation 11.24, the potential energy of a spring is always positive, whether x is positive or negative. The zero of potential energy is defined at the equilibrium position, $x = 0$.

Note that equation 11.24 could also be derived by plotting the force F acting on the spring versus the displacement x of the spring, as shown in figure 11.5(b). Because the work is equal to the product of the force F and the displacement x, the work is also equal to the area under the curve in figure 11.5(b). The area of that triangle is $\tfrac{1}{2}(x)(F) = \tfrac{1}{2}(x)(kx) = \tfrac{1}{2}kx^2$. (For the more general problem where the

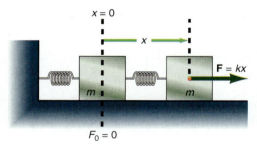

(a)

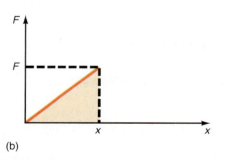

(b)

Figure 11.5
The potential energy of a spring.

force is not a linear function of the displacement x, if the force is plotted versus the displacement x, the work done, and hence the potential energy, will still be equal to the area under the curve.)

Example 11.2

The potential energy of a spring. A spring, with a spring constant of 29.4 N/m, is stretched 5.00 cm. How much potential energy does the spring possess?

Solution

The potential energy of the spring, found from equation 11.24, is

$$PE = \tfrac{1}{2}kx^2$$
$$= \tfrac{1}{2}(29.4 \text{ N/m})(5.00 \times 10^{-2} \text{ m})^2$$
$$= 3.68 \times 10^{-2} \text{ J}$$

11.5 Conservation of Energy and the Vibrating Spring

The vibrating spring system of figure 11.2 can also be described in terms of the law of conservation of energy. When the spring is stretched to its maximum displacement A, work is done on the spring, and hence the spring contains potential energy. The mass m attached to the spring also has that potential energy. The total energy of the system is equal to the potential energy at the maximum displacement because at that point, $v = 0$, and therefore the kinetic energy is equal to zero, that is,

$$E_{\text{tot}} = PE = \tfrac{1}{2}kA^2 \tag{11.25}$$

When the spring is released, the mass moves to a smaller displacement x, and is moving at a speed v. At this arbitrary position x, the mass will have both potential energy and kinetic energy. The law of conservation of energy then yields

$$E_{\text{tot}} = PE + KE$$
$$E_{\text{tot}} = \tfrac{1}{2}kx^2 + \tfrac{1}{2}mv^2 \tag{11.26}$$

But the total energy imparted to the mass m is given by equation 11.25. Hence, the law of conservation of energy gives

$$E_{\text{tot}} = E_{\text{tot}}$$
$$\tfrac{1}{2}kA^2 = \tfrac{1}{2}kx^2 + \tfrac{1}{2}mv^2 \tag{11.27}$$

We can also use equation 11.27 to find the velocity of the moving body at any displacement x. Thus,

$$\frac{1}{2}mv^2 = \frac{1}{2}kA^2 - \frac{1}{2}kx^2$$

$$v^2 = \frac{k}{m}(A^2 - x^2)$$

$$v = \pm\sqrt{\frac{k}{m}(A^2 - x^2)} \tag{11.28}$$

We should note that this is the same equation for the velocity as derived earlier (equation 11.15). It is informative to replace the values of x and v from their respective equations 11.6 and 11.13 into equation 11.26. Thus,

$$E_{\text{tot}} = \tfrac{1}{2}k(A\cos\omega t)^2 + \tfrac{1}{2}m(-\omega A\sin\omega t)^2$$

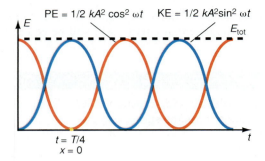

$PE = 1/2\ kA^2 \cos^2 \omega t$ $KE = 1/2\ kA^2\sin^2 \omega t$

E_{tot}

$t = T/4$
$x = 0$

Figure 11.6

Conservation of energy and simple harmonic motion.

or

$$E_{tot} = \tfrac{1}{2}kA^2 \cos^2\omega t + \tfrac{1}{2}m\omega^2 A^2 \sin^2\omega t$$

but since

$$\omega^2 = \frac{k}{m}$$

$$E_{tot} = \frac{1}{2}kA^2 \cos^2\omega t + \frac{1}{2}m\frac{k}{m}A^2 \sin^2\omega t$$

$$= \frac{1}{2}kA^2 \cos^2\omega t + \frac{1}{2}kA^2 \sin^2\omega t \qquad (11.29)$$

These terms are plotted in figure 11.6.

The total energy of the vibrating system is a constant and this is shown as the horizontal line, E_{tot}. At $t = 0$ the total energy of the system is potential energy (v is zero, hence the kinetic energy is zero). As the time increases the potential energy decreases and the kinetic energy increases, as shown. However, the total energy remains the same. From equation 11.24 and figure 11.6, we see that at $x = 0$ the potential energy is zero and hence all the energy is kinetic. This occurs when $t = T/4$. The maximum velocity of the mass m occurs here and is easily found by equating the maximum kinetic energy to the total energy, that is,

$$\frac{1}{2}mv^2_{max} = \frac{1}{2}kA^2$$

$$v_{max} = \sqrt{\frac{k}{m}}\,A = \omega A \qquad (11.30)$$

When the oscillating mass reaches $x = A$, the kinetic energy becomes zero since

$$\tfrac{1}{2}kA^2 = \tfrac{1}{2}kA^2 + \tfrac{1}{2}mv^2$$

$$\tfrac{1}{2}mv^2 = \tfrac{1}{2}kA^2 - \tfrac{1}{2}kA^2 = 0$$

$$= KE = 0$$

As the oscillation continues there is a constant interchange of energy between potential energy and kinetic energy but the total energy of the system remains a constant.

Example 11.3

Conservation of energy applied to a spring. A horizontal spring has a spring constant of 29.4 N/m. A mass of 300 g is attached to the spring and displaced 5.00 cm. The mass is then released. Find (a) the total energy of the system, (b) the maximum velocity of the system, and (c) the potential energy and kinetic energy for $x = 2.00$ cm.

Solution

a. The total energy of the system is

$$E_{tot} = \tfrac{1}{2}kA^2$$

$$= \tfrac{1}{2}(29.4\ \text{N/m})(5.00 \times 10^{-2}\ \text{m})^2$$

$$= 3.68 \times 10^{-2}\ \text{J}$$

b. The maximum velocity occurs when $x = 0$ and the potential energy is zero. Therefore, using equation 11.30,

$$v_{max} = \sqrt{\frac{k}{m}}\, A$$

$$= \sqrt{\frac{29.4 \text{ N/m}}{3.00 \times 10^{-1} \text{ kg}}}\, (5.00 \times 10^{-2} \text{ m})$$

$$= 0.495 \text{ m/s}$$

c. The potential energy at 2.00 cm is

$$PE = \tfrac{1}{2}kx^2 = \tfrac{1}{2}(29.4 \text{ N/m})(2.00 \times 10^{-2} \text{ m})^2$$

$$= 5.88 \times 10^{-3} \text{ J}$$

The kinetic energy at 2.00 cm is

$$KE = \frac{1}{2}mv^2 = \frac{1}{2}m\frac{k}{m}(A^2 - x^2)$$

$$= \frac{1}{2}(29.4 \text{ N/m})[(5.00 \times 10^{-2} \text{ m})^2 - (2.00 \times 10^{-2} \text{ m})^2]$$

$$= 3.09 \times 10^{-2} \text{ J}$$

Note that the sum of the potential energy and the kinetic energy is equal to the same value for the total energy found in part a.

11.6 The Simple Pendulum

Another example of periodic motion is a pendulum. A **simple pendulum** is a bob that is attached to a string and allowed to oscillate, as shown in figure 11.7(a). The bob oscillates because there is a restoring force, given by

$$\text{Restoring force} = -mg \sin \theta \tag{11.31}$$

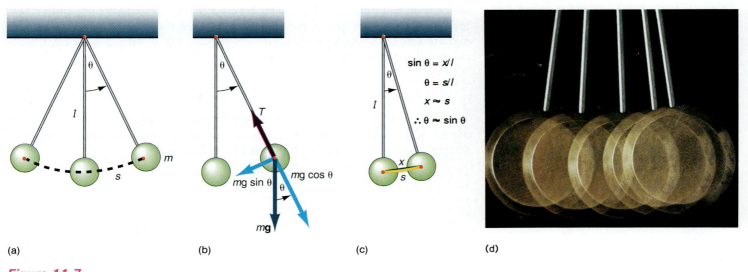

(a)   (b)   (c)   (d)

Figure 11.7
The simple pendulum.

This restoring force is just the component of the weight of the bob that is perpendicular to the string, as shown in figure 11.7(b). If Newton's second law, $F = ma$, is applied to the motion of the pendulum bob, we get

$$-mg \sin \theta = ma$$

The tangential acceleration of the pendulum bob is thus

$$a = -g \sin \theta \tag{11.32}$$

Note that although this pendulum motion is periodic, it is not, in general, simple harmonic motion because the acceleration is not directly proportional to the displacement of the pendulum bob from its equilibrium position. However, if the angle θ of the simple pendulum is small, then the sine of θ can be replaced by the angle θ itself, expressed in radians. (The discrepancy in using θ rather than the $\sin \theta$ is less than 0.2% for angles less than 10 degrees.) That is, for small angles

$$\sin \theta \approx \theta$$

The acceleration of the bob is then

$$a = -g\theta \tag{11.33}$$

From figure 11.7 and the definition of an angle in radians ($\theta = $ arc length/radius), we have

$$\theta = \frac{s}{l}$$

where s is the actual path length followed by the bob. Thus

$$a = -g\frac{s}{l} \tag{11.34}$$

The path length s is curved, but if the angle θ is small, the arc length s is approximately equal to the chord x, figure 11.7(c). Hence,

$$a = -\frac{g}{l}x \tag{11.35}$$

which is an equation having the same form as that of the equation for simple harmonic motion. Therefore, if the angle of oscillation θ is small, the pendulum will execute simple harmonic motion. For simple harmonic motion of a spring, the acceleration was found to be

$$a = -\frac{k}{m}x \tag{11.2}$$

We can now use the equations developed for the vibrating spring to describe the motion of the pendulum. We find an equivalent spring constant of the pendulum by setting equation 11.2 equal to equation 11.35. That is

$$\frac{k}{m} = \frac{g}{l}$$

or

$$k_P = \frac{mg}{l} \tag{11.36}$$

Equation 11.36 states that the motion of a pendulum can be described by the equations developed for the vibrating spring by using the equivalent spring constant of the pendulum k_p. Thus, the period of motion of the pendulum, found from equation 11.22, is

$$T_p = 2\pi \sqrt{\frac{m}{k_p}}$$

$$= 2\pi \sqrt{\frac{m}{mg/l}}$$

$$T_p = 2\pi \sqrt{\frac{l}{g}} \qquad (11.37)$$

The period of motion of the pendulum is independent of the mass m of the bob but is directly proportional to the square root of the length of the string. If the angle θ is equal to 15° on either side of the central position, then the true period differs from that given by equation 11.37 by less than 0.5%.

The pendulum can be used as a very simple device to measure the acceleration of gravity at a particular location. We measure the length l of the pendulum and then set the pendulum into motion. We measure the period by a clock and obtain the acceleration of gravity from equation 11.37 as

$$g = \frac{4\pi^2}{T_p^2} l \qquad (11.38)$$

Example 11.4

The period of a pendulum. Find the period of a simple pendulum 1.50 m long.

Solution

The period, found from equation 11.37, is

$$T_p = 2\pi \sqrt{\frac{l}{g}}$$

$$= 2\pi \sqrt{\frac{1.50 \text{ m}}{9.80 \text{ m/s}^2}}$$

$$= 2.46 \text{ s}$$

Example 11.5

The length of a pendulum. Find the length of a simple pendulum whose period is 1.00 s.

Solution

The length of the pendulum, found from equation 11.37, is

$$l = \frac{T_p^2}{4\pi^2} g$$

$$= \frac{(1.00 \text{ s})^2}{4\pi^2} (9.80 \text{ m/s}^2)$$

$$= 0.248 \text{ m}$$

Example 11.6

The pendulum and the acceleration of gravity. A pendulum 1.50 m long is observed to have a period of 2.47 s at a certain location. Find the acceleration of gravity there.

The acceleration of gravity, found from equation 11.38, is

$$g = \frac{4\pi^2}{T_p^2} l$$

$$= \frac{4\pi^2}{(2.47 \text{ s})^2}(1.50 \text{ m})$$

$$= 9.71 \text{ m/s}^2$$

We can also use a pendulum to measure an acceleration. If a pendulum is placed on board a rocket ship in interstellar space and the rocket ship is accelerated at 9.80 m/s², the pendulum oscillates with the same period as it would at rest on the surface of the earth. An enclosed person or thing on the rocket ship could not distinguish between the acceleration of the rocket ship at 9.80 m/s² and the acceleration of gravity of 9.80 m/s² on the earth. (This is an example of Einstein's principle of equivalence in general relativity.) An oscillating pendulum of measured length *l* can be placed in an elevator and the period *T* measured. We can use equation 11.38 to measure the resultant acceleration experienced by the pendulum in the elevator.

11.7 Springs in Parallel and in Series

Sometimes more than one spring is used in a vibrating system. The motion of the system will depend on the way the springs are connected. As an example, suppose there are three massless springs with spring constants k_1, k_2, and k_3. These springs can be connected in parallel, as shown in figure 11.8(a), or in series, as in figure 11.8(b). The period of motion of either configuration can be found by using an equivalent spring constant k_E.

Springs in Parallel

If the total force pulling the mass *m* a distance *x* to the right is F_{tot}, this force will distribute itself among the three springs such that there will be a force F_1 on spring 1, a force F_2 on spring 2, and a force F_3 on spring 3. If the displacement of each spring is equal to *x*, then the springs are said to be in parallel. Then we can write the total force as

$$F_{tot} = F_1 + F_2 + F_3 \tag{11.39}$$

However, since we assumed that each spring was displaced the same distance *x*, Hooke's law for each spring is

$$F_1 = k_1 x$$
$$F_2 = k_2 x$$
$$F_3 = k_3 x \tag{11.40}$$

Substituting equation 11.40 into equation 11.39 gives

$$F_{tot} = k_1 x + k_2 x + k_3 x$$
$$= (k_1 + k_2 + k_3)x$$

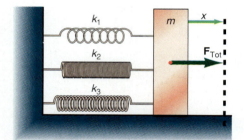

(a) Springs in parallel

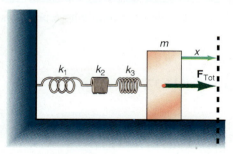

(b) Springs in series

Figure 11.8

Springs in parallel and in series.

We now define an equivalent spring constant k_E for springs connected in parallel as

$$k_E = k_1 + k_2 + k_3 \qquad (11.41)$$

Hooke's law for the combination of springs is given by

$$F_{tot} = k_E x \qquad (11.42)$$

The springs in parallel will execute a simple harmonic motion whose period, found from equation 11.22, is

$$T = 2\pi \sqrt{\frac{m}{k_E}} = 2\pi \sqrt{\frac{m}{k_1 + k_2 + k_3}} \qquad (11.43)$$

Springs in Series

If the same springs are connected in series, as in figure 11.8(b), the total force F_{tot} displaces the mass m a distance x to the right. But in this configuration, each spring stretches a different amount. Thus, the total displacement x is the sum of the displacements of each spring, that is,

$$x = x_1 + x_2 + x_3 \qquad (11.44)$$

The displacement of each spring, found from Hooke's law, is

$$x_1 = \frac{F_1}{k_1}$$

$$x_2 = \frac{F_2}{k_2}$$

$$x_3 = \frac{F_3}{k_3} \qquad (11.45)$$

Substituting these values of the displacement into equation 11.44, yields

$$x = \frac{F_1}{k_1} + \frac{F_2}{k_2} + \frac{F_3}{k_3} \qquad (11.46)$$

But because the springs are in series the total applied force is transmitted equally from spring to spring. Hence,

$$F_{tot} = F_1 = F_2 = F_3 \qquad (11.47)$$

Substituting equation 11.47 into equation 11.46, gives

$$x = \frac{F_{tot}}{k_1} + \frac{F_{tot}}{k_2} + \frac{F_{tot}}{k_3}$$

and

$$x = \left(\frac{1}{k_1} + \frac{1}{k_2} + \frac{1}{k_3} \right) F_{tot} \qquad (11.48)$$

We now define the equivalent spring constant for springs connected in series as

$$\frac{1}{k_E} = \frac{1}{k_1} + \frac{1}{k_2} + \frac{1}{k_3} \qquad (11.49)$$

Thus, the total displacement, equation 11.48, becomes

$$x = \frac{F_{tot}}{k_E} \qquad (11.50)$$

and Hooke's law becomes

$$F_{tot} = k_E x \qquad (11.51)$$

where k_E is given by equation 11.49. Hence, the combination of springs in series executes simple harmonic motion and the period of that motion, given by equation 11.22, is

$$T = 2\pi \sqrt{\frac{m}{k_E}} = 2\pi \sqrt{m\left(\frac{1}{k_1} + \frac{1}{k_2} + \frac{1}{k_3}\right)}$$

(11.52)

Example 11.7

Springs in parallel. Three springs with force constants $k_1 = 10.0$ N/m, $k_2 = 12.5$ N/m, and $k_3 = 15.0$ N/m are connected in parallel to a mass of 0.500 kg. The mass is then pulled to the right and released. Find the period of the motion.

Solution

The period of the motion, found from equation 11.43, is

$$T = 2\pi \sqrt{\frac{m}{k_1 + k_2 + k_3}}$$

$$= 2\pi \sqrt{\frac{0.500 \text{ kg}}{10.0 \text{ N/m} + 12.5 \text{ N/m} + 15.0 \text{ N/m}}}$$

$$= 0.726 \text{ s}$$

Example 11.8

Springs in series. The same three springs as in example 11.7 are now connected in series. Find the period of the motion.

Solution

The period, found from equation 11.52, is

$$T = 2\pi \sqrt{m\left(\frac{1}{k_1} + \frac{1}{k_2} + \frac{1}{k_3}\right)}$$

$$= 2\pi \sqrt{(0.500 \text{ kg})\left(\frac{1}{10.0 \text{ N/m}} + \frac{1}{12.5 \text{ N/m}} + \frac{1}{15.0 \text{ N/m}}\right)}$$

$$= 1.56 \text{ s}$$

The Language of Physics

Periodic motion
Motion that repeats itself in equal intervals of time (p. 297).

Displacement
The distance a vibrating body moves from its equilibrium position (p. 297).

Simple harmonic motion
Periodic motion in which the acceleration of a body is directly proportional to its displacement from the equilibrium position but in the opposite direction. Because the acceleration is directly proportional to the

displacement, the acceleration of the body is not constant. The kinematic equations developed in chapter 3 are no longer valid to describe this type of motion (p. 298).

Amplitude
The maximum displacement of the vibrating body (p. 298).

Cycle
One complete oscillation or vibratory motion (p. 298).

Period
The time for the vibrating body to complete one cycle (p. 298).

Frequency
The number of complete cycles or oscillations in unit time. The frequency is the reciprocal of the period (p. 299).

Reference circle
A body executing uniform circular motion does so in a circle. The projection of the position of the rotating body onto the *x*- or *y*-axis is equivalent to simple harmonic motion along that axis. Thus, vibratory motion is related to motion in a circle, the reference circle (p. 300).

Vibratory Motion, Wave Motion, and Fluids

Angular velocity
The angular velocity of the uniform circular motion in the reference circle is related to the frequency of the vibrating system. Hence, the angular velocity is called the angular frequency of the vibrating system (p. 300).

Potential energy of a spring
The energy that a body possesses by virtue of its configuration. A compressed spring has potential energy because it has the ability to do work as it expands to its equilibrium configuration. A stretched spring can also do work as it returns to its equilibrium configuration (p. 306).

Simple pendulum
A bob that is attached to a string and allowed to oscillate to and fro under the action of gravity. If the angle of the pendulum is small the pendulum will oscillate in simple harmonic motion (p. 309).

Summary of Important Equations

Restoring force in a spring
$$F_R = -kx \tag{11.1}$$

Defining relation for simple harmonic motion
$$a = -\frac{k}{m}x \tag{11.2}$$

Frequency
$$f = \frac{1}{T} \tag{11.3}$$

Displacement in simple harmonic motion
$$x = A \cos \omega t \tag{11.6}$$

Angular frequency
$$\omega = 2\pi f \tag{11.9}$$

Velocity as a function of time in simple harmonic motion
$$v = -\omega A \sin \omega t \tag{11.13}$$

Velocity as a function of displacement
$$v = \pm\omega\sqrt{A^2 - x^2} \tag{11.15}$$

Acceleration as a function of time
$$a = -\omega^2 A \cos \omega t \tag{11.17}$$

Angular frequency of a spring
$$\omega = \sqrt{\frac{k}{m}} \tag{11.20}$$

Frequency in simple harmonic motion
$$f = \frac{1}{2\pi}\sqrt{\frac{k}{m}} \tag{11.21}$$

Period in simple harmonic motion
$$T = 2\pi\sqrt{\frac{m}{k}} \tag{11.22}$$

Potential energy of a spring
$$PE = \tfrac{1}{2}kx^2 \tag{11.24}$$

Conservation of energy for a vibrating spring
$$\tfrac{1}{2}kA^2 = \tfrac{1}{2}kx^2 + \tfrac{1}{2}mv^2 \tag{11.27}$$

Period of motion of a simple pendulum
$$T_p = 2\pi\sqrt{\frac{l}{g}} \tag{11.37}$$

Equivalent spring constant for springs in parallel
$$k_E = k_1 + k_2 + k_3 \tag{11.41}$$

Period of motion for springs in parallel
$$T = 2\pi\sqrt{\frac{m}{k_1 + k_2 + k_3}} \tag{11.43}$$

Equivalent spring constant for springs in series
$$\frac{1}{k_E} = \frac{1}{k_1} + \frac{1}{k_2} + \frac{1}{k_3} \tag{11.49}$$

Period of motion for springs in series
$$T = 2\pi\sqrt{m\left(\frac{1}{k_1} + \frac{1}{k_2} + \frac{1}{k_3}\right)} \tag{11.52}$$

Questions for Chapter 11

1. Can the periodic motion of the earth be considered to be an example of simple harmonic motion?
2. Can the kinematic equations derived in chapter 3 be used to describe simple harmonic motion?
3. In the simple harmonic motion of a mass attached to a spring, the velocity of the mass is equal to zero when the acceleration has its maximum value. How is this possible? Can you think of other examples in which a body has zero velocity with a nonzero acceleration?
4. What is the characteristic of the restoring force that makes simple harmonic motion possible?
5. Discuss the significance of the reference circle in the analysis of simple harmonic motion.

6. How can a mass that is undergoing a one-dimensional translational simple harmonic motion have anything to do with an angular velocity or an angular frequency, which is a characteristic of two or more dimensions?
7. How is the angular frequency related to the physical characteristics of the spring and the vibrating mass in simple harmonic motion?
†8. In the entire derivation of the equations for simple harmonic motion we have assumed that the springs are massless and friction can be neglected. Discuss these assumptions. Describe qualitatively what you would expect to happen to the motion if the springs are not small enough to be considered massless.

†9. Describe how a geological survey for iron might be undertaken on the moon using a simple pendulum.
†10. How could a simple pendulum be used to make an accelerometer?
†11. Discuss the assumption that the displacement of each spring is the same when the springs are in parallel. Under what conditions is this assumption valid and when would it be invalid?

Problems for Chapter 11

11.2 Simple Harmonic Motion and 11.3 Analysis of Simple Harmonic Motion

1. A mass of 0.200 kg is attached to a spring of spring constant 30.0 N/m. If the mass executes simple harmonic motion, what will be its frequency?
2. A 30.0-g mass is attached to a vertical spring and it stretches 10.0 cm. It is then stretched an additional 5.00 cm and released. Find its period of motion and its frequency.
3. A 0.200-kg mass on a spring executes simple harmonic motion at a frequency f. What mass is necessary for the system to vibrate at a frequency of $2f$?
4. A simple harmonic oscillator has a frequency of 2.00 Hz and an amplitude of 10.0 cm. What is its maximum acceleration? What is its acceleration at $t = 0.25$ s?
5. A ball attached to a string travels in uniform circular motion in a horizontal circle of 50.0 cm radius in 1.00 s. Sunlight shining on the ball throws its shadow on a wall. Find the velocity of the shadow at (a) the end of its path and (b) the center of its path.
6. A 50.0-g mass is attached to a spring of force constant 10.0 N/m. The spring is stretched 20.0 cm and then released. Find the displacement, velocity, and acceleration of the mass at 0.200 s.
7. A 25.0-g mass is attached to a vertical spring and it stretches 15.0 cm. It is then stretched an additional 10.0 cm and then released. What is the maximum velocity of the mass? What is its maximum acceleration?
8. The displacement of a body in simple harmonic motion is given by $x = (0.15 \text{ m})\cos[(5.00 \text{ rad/s})t]$. Find (a) the amplitude of the motion, (b) the angular frequency, (c) the frequency, (d) the period, and (e) the displacement at 3.00 s.
9. A 500-g mass is hung from a coiled spring and it stretches 10.0 cm. It is then stretched an additional 15.0 cm and released. Find (a) the frequency of vibration, (b) the period, and (c) the velocity and acceleration at a displacement of 10.0 cm.

10. A mass of 0.200 kg is placed on a vertical spring and the spring stretches by 15.0 cm. It is then pulled down an additional 10.0 cm and then released. Find (a) the spring constant, (b) the angular frequency, (c) the frequency, (d) the period, (e) the maximum velocity of the mass, (f) the maximum acceleration of the mass, (g) the maximum restoring force, and (h) the equation of the displacement, velocity, and acceleration at any time t.

11.5 Conservation of Energy and the Vibrating Spring

11. A simple harmonic oscillator has a spring constant of 5.00 N/m. If the amplitude of the motion is 15.0 cm, find the total energy of the oscillator.
12. A body is executing simple harmonic motion. At what displacement is the potential energy equal to the kinetic energy?
13. A 20.0-g mass is attached to a horizontal spring on a smooth table. The spring constant is 3.00 N/m. The spring is then stretched 15.0 cm and then released. What is the total energy of the motion? What is the potential and kinetic energy when $x = 5.00$ cm?
14. A body is executing simple harmonic motion. At what displacement is the speed v equal to one-half the maximum speed?

11.6 The Simple Pendulum

15. Find the period and frequency of a simple pendulum 0.75 m long.
16. If a pendulum has a length L and a period T, what will be the period when (a) L is doubled and (b) L is halved?
17. Find the frequency of a child's swing whose ropes have a length of 3.25 m.
18. What is the period of a 0.500-m pendulum on the moon where $g_m = (1/6)g_e$?
19. What is the period of a pendulum 0.750 m long on a spaceship (a) accelerating at 4.90 m/s² and (b) moving at constant velocity?
20. What is the period of a pendulum in free-fall?
21. A pendulum has a period of 0.750 s at the equator at sea level. The pendulum is carried to another place on the earth and the period is now found to be 0.748 s. Find the acceleration due to gravity at this location.

11.7 Springs in Parallel and in Series

†22. Springs with spring constants of 5.00 N/m and 10.0 N/m are connected in parallel to a 5.00-kg mass. Find (a) the equivalent spring constant and (b) the period of the motion.
†23. Springs with spring constants 5.00 N/m and 10.0 N/m are connected in series to a 5.00-kg mass. Find (a) the equivalent spring constant and (b) the period of the motion.

Additional Problems

24. A 500-g mass is attached to a vertical spring of spring constant 30.0 N/m. How far should the spring be stretched in order to give the mass an upward acceleration of 3.00 m/s²?
25. A ball is caused to move in a horizontal circle of 40.0-cm radius in uniform circular motion at a speed of 25.0 cm/s. Its projection on the wall moves in simple harmonic motion. Find the velocity and acceleration of the shadow of the ball at (a) the end of its motion, (b) the center of its motion, and (c) halfway between the center and the end of the motion.
†26. The motion of the piston in the engine of an automobile is approximately simple harmonic. If the stroke of the piston (twice the amplitude) is equal to 8.00 in. and the engine turns at 1800 rpm, find (a) the acceleration at $x = A$ and (b) the speed of the piston at the midpoint of the stroke.
†27. A 535-g mass is dropped from a height of 25.0 cm above an uncompressed spring of $k = 20.0$ N/m. By how much will the spring be compressed?
28. A simple pendulum is used to operate an electrical device. When the pendulum bob sweeps through the midpoint of its swing, it causes an electrical spark to be given off. Find the length of the pendulum that will give a spark rate of 30.0 sparks per minute.
†29. The general solution for the period of a simple pendulum, without making the assumption of small angles of swing, is given by

$$T = 2\pi \sqrt{\frac{l}{g}} \left[1 + \frac{(\frac{1}{2})^2 \sin^2\theta}{2} + \frac{(\frac{1}{2})^2(\frac{3}{4})^2 \sin^4\theta}{2} + \cdots \right]$$

Find the period of a 1.00-m pendulum for $\theta = 10.0°$, 30.0°, and 50.0° and compare with the period obtained with the small angle approximation. Determine the percentage error in each case by using the small angle approximation.

30. A pendulum clock on the earth has a period of 1.00 s. Will this clock run slow or fast, and by how much if taken to (a) Mars, (b) Moon, and (c) Venus?

†31. A pendulum bob, 355 g, is raised to a height of 12.5 cm before it is released. At the bottom of its path it makes a perfectly elastic collision with a 500-g mass that is connected to a horizontal spring of spring constant 15.8 N/m, that is at rest on a smooth surface. How far will the spring be compressed?

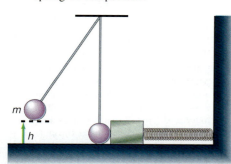

†32. A 500-g block is in simple harmonic motion as shown in the diagram. A mass m' is added to the top of the block when the block is at its maximum extension. How much mass should be added to change the frequency by 25%?

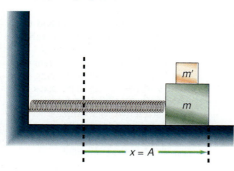

†33. A pendulum clock keeps correct time at a location at sea level where the acceleration of gravity is equal to 9.80 m/s². The clock is then taken up to the top of a mountain and the clock loses 3.00 s per day. How high is the mountain?

†34. Three people, who together weigh 422 lb, get into a car and the car is observed to move 2.00 in. closer to the ground. What is the spring constant of the car springs?

†35. In the accompanying diagram, the mass m is pulled down a distance of 9.50 cm from its equilibrium position and is then released. The mass then executes simple harmonic motion. Find (a) the total potential energy of the mass with respect to the ground when the mass is located at positions 1, 2, and 3; (b) the total energy of the mass at positions 1, 2, and 3; and (c) the speed of the mass at position 2. Assume $m = 55.6$ g, $k = 25.0$ N/m, $h_0 = 50.0$ cm.

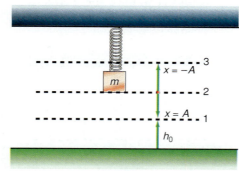

†36. A 20.0-g ball rests on top of a vertical spring gun whose spring constant is 20 N/m. The spring is compressed 10.0 cm and the gun is then fired. Find how high the ball rises in its vertical trajectory.

†37. A toy spring gun is used to fire a ball as a projectile. Find the value of the spring constant, such that when the spring is compressed 10.0 cm, and the gun is fired at an angle of 62.5°, the range of the projectile will be 1.50 m. The mass of the ball is 25.2 g.

†38. In the simple pendulum shown in the diagram, find the tension in the string when the height of the pendulum is (a) h, (b) $h/2$, and (c) $h = 0$. The mass $m = 500$ g, the initial height $h = 15.0$ cm, and the length of the pendulum $l = 1.00$ m.

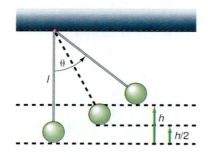

†39. A mass is attached to a horizontal spring. The mass is given an initial amplitude of 10.0 cm on a rough surface and is then released to oscillate in simple harmonic motion. If 10.0% of the energy is lost per cycle due to the friction of the mass moving over the rough surface, find the maximum displacement of the mass after 1, 2, 4, 6, and 8 complete oscillations.

†40. Find the maximum amplitude of vibration after 2 periods for a 85.0-g mass executing simple harmonic motion on a rough horizontal surface of $\mu_k = 0.350$. The spring constant is 24.0 N/m and the initial amplitude is 20.0 cm.

41. A 240-g mass slides down a circular chute without friction and collides with a horizontal spring, as shown in the diagram. If the original position of the mass is 25.0 cm above the table top and the spring has a spring constant of 18 N/m, find the maximum distance that the spring will be compressed.

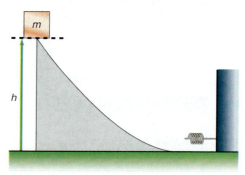

†42. A 235-g block slides down a frictionless inclined plane, 1.25 m long, that makes an angle of 34.0° with the horizontal. At the bottom of the plane the block slides along a rough horizontal surface 1.50 m long. The coefficient of kinetic friction between the block and the rough horizontal surface is 0.45. The block then collides with a horizontal spring of $k = 20.0$ N/m. Find the maximum compression of the spring.

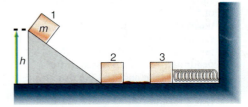

†43. A 335-g disk that is free to rotate about its axis is connected to a spring that is stretched 35.0 cm. The spring constant is 10.0 N/m. When the disk is released it rolls without slipping as it moves toward the equilibrium position. Find the speed of the disk when the displacement of the spring is equal to −17.5 cm.

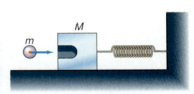

†44. A 25.0-g ball moving at a velocity of 200 cm/s to the right makes an inelastic collision with a 200-g block that is initially at rest on a frictionless surface. There is a hole in the large block for the small ball to fit into. If $k = 10$ N/m, find the maximum compression of the spring.

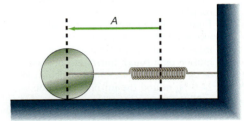

†45. Show that the period of simple harmonic motion for the mass shown is equivalent to the period for two springs in parallel.

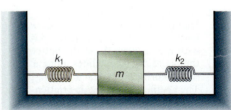

†46. A nail is placed in the wall at a distance of $l/2$ from the top, as shown in the diagram. A pendulum of length 85.0 cm is released from position 1. (a) Find the time it takes for the pendulum bob to reach position 2. When the bob of the pendulum reaches position 2, the string hits the nail. (b) Find the time it takes for the pendulum bob to reach position 3.

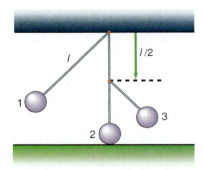

†47. A spring is attached to the top of an Atwood's machine as shown. The spring is stretched to $A = 10$ cm before being released. Find the velocity of m_2 when $x = -A/2$. Assume $m_1 = 28.0$ g, $m_2 = 43.0$ g, and $k = 10.0$ N/m.

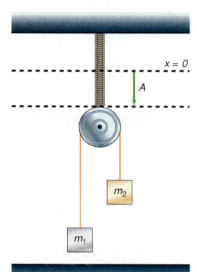

†48. A 280-g block is connected to a spring on a rough inclined plane that makes an angle of 35.5° with the horizontal. The block is pulled down the plane a distance $A = 20.0$ cm, and is then released. The spring constant is 40.0 N/m and the coefficient of kinetic friction is 0.100. Find the speed of the block when the displacement $x = -A/2$.

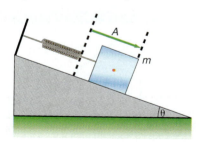

49. The rotational analog of simple harmonic motion, is angular simple harmonic motion, wherein a body rotates periodically clockwise and then counterclockwise. Hooke's law for rotational motion is given by

$$\tau = -C\theta$$

where τ is the torque acting on the body, θ is the angular displacement, and C is a constant, like the spring constant. Use Newton's second law for rotational motion to show

$$\alpha = \frac{C}{I}\theta$$

Use the analogy between the linear result, $a = -\omega^2 x$, to show that the frequency of vibration of an object executing angular simple harmonic motion is given by

$$f = \frac{1}{2\pi}\sqrt{\frac{C}{I}}$$

Interactive Tutorials

50. Calculate the period T of a simple pendulum located on a planet having a gravitational acceleration of $g = 9.80$ m/s², if its length $l = 1.00$ m is increased from 1 to 10 m in steps of 1.00 m. Plot the results as the period T versus the length l.

51. The displacement x of a body undergoing simple harmonic motion is given by the formula $x = A\cos\omega t$, where A is the amplitude of the vibration, ω is the angular frequency in rad/s, and t is the time in seconds. Plot the displacement x as a function of t for an amplitude $A = 0.150$ m and an angular frequency $\omega = 5.00$ rad/s as t increases from 0 to 2 s in 0.10 s increments.

52. A mass $m = 0.500$ kg is attached to a spring on a smooth horizontal table. An applied force $F_A = 4.00$ N is used to stretch the spring a distance $x_0 = 0.150$ m. (a) Find the spring constant k of the spring. The mass is returned to its equilibrium position and then stretched to a value $A = 0.15$ m and then released. The mass then executes simple harmonic motion. Find (b) the angular frequency ω, (c) the frequency f, (d) the period T, (e) the maximum velocity v_{max} of the vibrating mass, (f) the maximum acceleration a_{max} of the vibrating mass, (g) the maximum restoring force F_{Rmax}, and (h) the velocity of the mass at the displacement $x = 0.08$ m. (i) Plot the displacement x, velocity v, acceleration a, and the restoring force F_R at any time t.

53. A mass $m = 0.350$ kg is attached to a horizontal spring. The mass is then pulled a distance $x = A = 0.200$ m from its equilibrium position and when released the mass executes simple harmonic motion. Find (a) the total energy E_{tot} of the mass when it is at its maximum displacement A from its equilibrium position. When the mass is at the displacement $x = 0.120$ m find, (b) its potential energy PE, (c) its kinetic energy KE, and (d) its speed v. (e) Plot the total energy, potential energy, and kinetic energy of the mass as a function of the displacement x. The spring constant $k = 35.5$ N/m.

54. A mass $m = 0.350$ kg is attached to a vertical spring. The mass is at a height $h_0 = 1.50$ m from the floor. The mass is then pulled down a distance $A = 0.220$ m from its equilibrium position and when released executes simple harmonic motion. Find (a) the total energy of the mass when it is at its maximum displacement A below its equilibrium position, (b) the gravitational potential energy when it is at the displacement $x = 0.120$ m, (c) the elastic potential energy when it is at the same displacement x, (d) the kinetic energy at the displacement x, and (e) the speed of the mass when it is at the displacement x. The spring constant $k = 35.5$ N/m.

12

Standing transverse waves in a vibrating string.

Query 17. If a stone be thrown into stagnating water, the waves excited thereby continue some time to arise in the place where the stone fell into the water, and are propagated from thence in concentric circles upon the surface of the water to great distances.

Isaac Newton

Wave Motion

12.1 Introduction

Everyone has observed that when a rock is thrown into a pond of water, waves are produced that move out from the point of the disturbance in a series of concentric circles. *The wave is a propagation of the disturbance through the medium without any net displacement of the medium.* In this case the rock hitting the water initiates the disturbance and the water is the medium through which the wave travels. Of the many possible kinds of waves, the simplest to understand, and the one that we will analyze, is the wave that is generated by an object executing simple harmonic motion. As an example, consider the mass m executing simple harmonic motion in figure 12.1. Attached to the right of m is a very long spring. The spring is so long that it is not necessary to consider what happens to the spring at its far end at this time. When the mass m is pushed out to the position $x = A$, the portion of the spring immediately to the right of A is compressed. This *compression* exerts a force on the portion of the spring immediately to its right, thereby compressing it. It in turn compresses part of the spring to its immediate right. The process continues with the compression moving along the spring, as shown in figure 12.1. As the mass m moves in simple harmonic motion to the displacement $x = -A$, the spring immediately to its right becomes elongated. We call the elongation of the spring a *rarefaction;* it is the converse of a compression. As the mass m returns to its equilibrium position, the rarefaction moves down the length of the spring. The combination of a compression and rarefaction comprise part of a longitudinal wave. *A longitudinal wave is a wave in which the particles of the medium oscillate in simple harmonic motion parallel to the direction of the wave propagation.* The compressions and rarefactions propagate down the spring, as shown in figure 12.1(f). The mass m in simple harmonic motion generated the wave and the wave moves to the right with a velocity v. Every portion of the medium, in this case the spring, executes simple harmonic motion around its equilibrium position. The medium oscillates back and forth with motion parallel to the wave velocity. Sound is an example of a longitudinal wave.

Another type of wave, and one easier to visualize, is a transverse wave. *A transverse wave is a wave in which the particles of the medium execute simple harmonic motion in a direction perpendicular to its direction of propagation.* A transverse wave can be generated by a mass having simple harmonic motion in the vertical direction, as shown in figure 12.2. A horizontal string is connected to the mass as shown. As the mass executes simple harmonic motion in the vertical direction, the end of the string does likewise. As the end moves up and down, it causes the particle next to it to follow suit. It, in turn, causes the next particle to move. Each particle

Figure 12.1

Generation of a longitudinal wave.

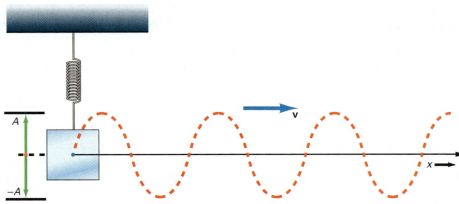

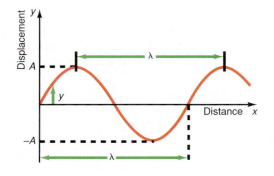

Figure 12.2

A transverse wave.

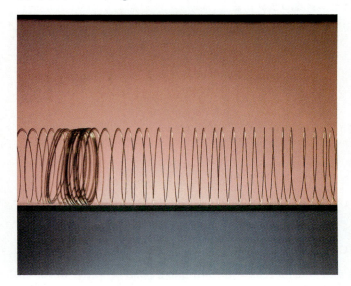

Figure 12.3

Characteristics of a simple wave.

transmits the motion to the next particle along the entire length of the string. The resulting wave propagates in the horizontal direction with a velocity v, while any one particle of the string executes simple harmonic motion in the vertical direction. The particle of the string is moving perpendicular to the direction of wave propagation, and is not moving in the direction of the wave.

Using figure 12.3, let us now define the characteristics of a transverse wave moving in a horizontal direction.

*The **displacement** of any particle of the wave is the displacement of that particle from its equilibrium position* and is measured by the vertical distance y.

*The **amplitude** of the wave is the maximum value of the displacement* and is denoted by A in figure 12.3.

Vibratory Motion, Wave Motion, and Fluids

*The **wavelength** of a wave is the distance, in the direction of propagation, in which the wave repeats itself and is denoted by λ.*

*The **period** T of a wave is the time it takes for one complete wave to pass a particular point.*

*The **frequency** f of a wave is defined as the number of waves passing a particular point per second.*

It is obvious from the definitions that the frequency is the reciprocal of the period, that is,

$$f = \frac{1}{T} \tag{12.1}$$

The speed of propagation of the wave is the distance the wave travels in unit time. Because a wave of one wavelength passes a point in a time of one period, its speed of propagation is

$$v = \frac{\text{distance traveled}}{\text{time}} = \frac{\lambda}{T} \tag{12.2}$$

Using equation 12.1, this becomes

$$v = \lambda f \tag{12.3}$$

Equation 12.3 is the fundamental equation of wave propagation. It relates the speed of the wave to its wavelength and frequency.

Example 12.1

Wavelength of sound. The human ear can hear sounds from a low of 20.0 Hz up to a maximum frequency of about 20,000 Hz. If the speed of sound in air at a temperature of 0 °C is 331 m/s, find the wavelengths associated with these frequencies.

Solution

The wavelength of a sound wave, determined from equation 12.3, is

$$\lambda = \frac{v}{f}$$

$$= \frac{331 \text{ m/s}}{20.0 \text{ cycles/s}}$$

$$= 16.6 \text{ m}$$

and

$$\lambda = \frac{v}{f} = \frac{331 \text{ m/s}}{20,000 \text{ cycles/s}}$$

$$= 0.0166 \text{ m}$$

The type of waves we consider in this chapter are called mechanical waves. *The wave causes a transfer of energy from one point in the medium to another point in the medium without the actual transfer of matter between these points.* Another type of wave, called an electromagnetic wave, is capable of traveling through empty space without the benefit of a medium. This type of wave is extremely unusual in this respect and we will treat it in more detail in chapters 25 and 29.

12.2 Mathematical Representation of a Wave

The simple wave shown in figure 12.3 is a picture of a transverse wave in a string at a particular time, let us say at $t = 0$. The wave can be described as a sine wave and can be expressed mathematically as

$$y = A \sin x \qquad (12.4)$$

The value of y represents the displacement of the string at every position x along the string, and A is the maximum displacement, and is called the amplitude of the wave. Equation 12.4 is plotted in figure 12.4. We see that the wave repeats itself for $x = 360° = 2\pi$ rad. Also plotted in figure 12.4 is $y = A \sin 2x$ and $y = A \sin 3x$. Notice from the figure that $y = A \sin 2x$ repeats itself twice in the same interval of 2π that $y = A \sin x$ repeats itself only once. Also note that $y = A \sin 3x$ repeats itself three times in that same interval of 2π. The wave $y = A \sin kx$ would repeat itself k times in the interval of 2π. We call the space interval in which $y = A \sin x$ repeats itself its wavelength, denoted by λ_1. Thus, when $x = \lambda_1 = 2\pi$, the wave starts to repeat itself. The wave represented by $y = A \sin 2x$ repeats itself for $2x = 2\pi$, and hence its wavelength is

$$\lambda_2 = x = \frac{2\pi}{2} = \pi$$

The wave $y = A \sin 3x$ repeats itself when $3x = 2\pi$, hence its wavelength is

$$\lambda_3 = x = \frac{2\pi}{3}$$

Using this notation any wave can be represented as

$$y = A \sin kx \qquad (12.5)$$

where k is a number, called the *wave number*. The wave repeats itself whenever

$$kx = 2\pi \qquad (12.6)$$

Because the value of x for a wave to repeat itself is its wavelength λ, equation 12.6 can be written as

$$k\lambda = 2\pi \qquad (12.7)$$

Figure 12.4

Plot of $A \sin x$, $A \sin 2x$, and $A \sin 3x$.

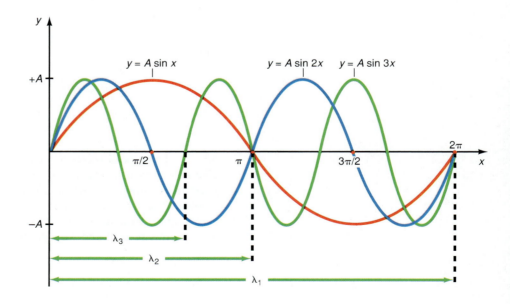

We can obtain the wavelength λ from equation 12.7 as

$$\lambda = \frac{2\pi}{k} \qquad (12.8)$$

Note that equation 12.8 gives the wavelengths in figure 12.4 by letting k have the values 1, 2, 3, and so forth, that is,

$$\lambda_1 = \frac{2\pi}{1}$$

$$\lambda_2 = \frac{2\pi}{2}$$

$$\lambda_3 = \frac{2\pi}{3}$$

$$\lambda_4 = \frac{2\pi}{4}$$

We observe that the wave number k is the number of waves contained in the interval of 2π. We can express the wave number k in terms of the wavelength λ by rearranging equation 12.8 into the form

$$k = \frac{2\pi}{\lambda} \qquad (12.9)$$

Note that in order for the units to be consistent, the wave number must have units of m^{-1}. The quantity x in equation 12.5 represents the location of any point on the string and is measured in meters. The quantity kx in equation 12.5 has the units ($m^{-1}m = 1$) and is thus a dimensionless quantity and represents an angle measured in radians. Also note that the wave number k is a different quantity than the spring constant k, discussed in chapter 11.

Equation 12.5 represents a snapshot of the wave at $t = 0$. That is, it gives the displacement of every particle of the string at time $t = 0$. As time passes, this wave, and every point on it, moves. Since each particle of the string executes simple harmonic motion in the vertical, we can look at the particle located at the point $x = 0$ and see how that particle moves up and down with time. Because the particle executes simple harmonic motion in the vertical, it is reasonable to represent the displacement of the particle of the string at any time t as

$$y = A \sin \omega t \qquad (12.10)$$

just as a simple harmonic motion on the x-axis was represented as $x = A \cos \omega t$ in chapter 11. The quantity t is the time and is measured in seconds, whereas the quantity ω is an angular velocity or an angular frequency and is measured in radians per second. Hence the quantity ωt represents an angle measured in radians. The displacement y repeats itself when $t = T$, the period of the wave. Since the sine function repeats itself when the argument is equal to 2π, we have

$$\omega T = 2\pi \qquad (12.11)$$

The period of the wave is thus

$$T = \frac{2\pi}{\omega}$$

but the period of the wave is the reciprocal of the frequency. Therefore,

$$T = \frac{1}{f} = \frac{2\pi}{\omega}$$

Solving for the angular frequency ω, in terms of the frequency f, we get

$$\omega = 2\pi f \qquad (12.12)$$

Notice that the wave is periodic in both space and time. The space period is represented by the wavelength λ, and the time by the period T.

Equation 12.5 represents every point on the string at $t = 0$, while equation 12.10 represents the point $x = 0$ for every time t. Obviously the general equation for a wave must represent every point x of the wave at every time t. We can arrange this by combining equations 12.5 and 12.10 into the one equation for a wave given by

$$y = A \sin(kx - \omega t) \tag{12.13}$$

The reason for the minus sign for ωt is explained below. We can find the relation between the wave number k and the angular frequency ω by combining equations 12.7 and 12.11 as

$$k\lambda = 2\pi \tag{12.7}$$

$$\omega T = 2\pi \tag{12.11}$$

Thus,

$$\omega T = k\lambda$$

and

$$\omega = \frac{k\lambda}{T}$$

However, the wavelength λ, divided by the period T is equal to the velocity of propagation of the wave v, equation 12.2. Therefore, the angular frequency becomes

$$\omega = kv \tag{12.14}$$

Now we can write equation 12.13 as

$$y = A \sin(kx - kvt) \tag{12.15}$$

or

$$y = A \sin k(x - vt) \tag{12.16}$$

The minus sign before the velocity v determines the direction of propagation of the wave. As an example, consider the wave

$$y_1 = A \sin k(x - vt) \tag{12.17}$$

We will now see that this is the equation of a wave traveling to the right with a speed v at any time t. A little later in time, Δt, the wave has moved a distance Δx to the right such that the same point of the wave now has the coordinates $x + \Delta x$ and $t + \Delta t$, figure 12.5(a). Then we represent the wave as

$$y_2 = A \sin k[(x + \Delta x) - v(t + \Delta t)]$$

or

$$y_2 = A \sin k[(x - vt) + \Delta x - v\Delta t] \tag{12.18}$$

If this equation for y_2 is to represent the same wave as y_1, then y_2 must be equal to y_1. It is clear from equations 12.18 and 12.17 that if

$$v = \frac{\Delta x}{\Delta t} \tag{12.19}$$

the velocity of the wave to the right, then

$$\Delta x - v\Delta t = \Delta x - \frac{\Delta x}{\Delta t}\Delta t = 0$$

Figure 12.5

A traveling wave.

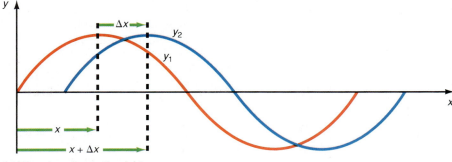

(a) Wave traveling to the right

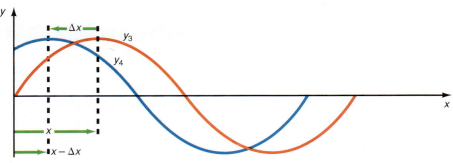

(b) Wave traveling to the left

and y_2 is equal to y_1. Because the term $\Delta x - v\Delta t$ is indeed equal to zero, y_2 is the same wave as y_1 only displaced a distance Δx to the right in the time Δt. Thus, equation 12.17 represents a wave traveling to the right with a velocity of propagation v.

A wave traveling to the left is depicted in figure 12.5(b) and we will begin by representing it as

$$y_3 = A \sin k(x - vt) \qquad (12.20)$$

In a time Δt, the wave y_3 moves a distance $-\Delta x$ to the left. The coordinates (x,t) of a point on y_3 now has the coordinates $x - \Delta x$ and $t + \Delta t$ for the same point on y_4. We can now write the new wave as

$$y_4 = A \sin k[(x - \Delta x) - v(t + \Delta t)]$$

or

$$y_4 = A \sin k[(x - vt) + (-\Delta x - v\Delta t)] \qquad (12.21)$$

The wave y_4 represents the same wave as y_3, providing $-\Delta x - v\Delta t = 0$ in equation 12.21. If $v = -\Delta x/\Delta t$, the velocity of the wave to the left, then

$$-\Delta x - v\Delta t = -\Delta x - \left(\frac{-\Delta x}{\Delta t}\right)\Delta t = 0$$

Thus, $-\Delta x - v\Delta t$ is indeed equal to zero, and wave y_4 represents the same wave as y_3 only it is displaced a distance $-\Delta x$ to the left in the time Δt. Instead of writing the equation 12.20 as a wave to the left with v a negative number, it is easier to write the equation for the wave to the left as

$$y = A \sin k(x + vt) \qquad (12.22)$$

where v is now a positive number. Therefore, equation 12.22 represents a wave traveling to the left, with a speed v. In summary, a wave traveling to the right can be represented either as

$$y = A \sin k(x - vt) \tag{12.23}$$

or

$$y = A \sin(kx - \omega t) \tag{12.24}$$

and a wave traveling to the left can be represented as either

$$y = A \sin k(x + vt) \tag{12.25}$$

or

$$y = A \sin(kx + \omega t) \tag{12.26}$$

Example 12.2

Characteristics of a wave. A particular wave is given by

$$y = (0.200 \text{ m}) \sin[(0.500 \text{ m}^{-1})x - (8.20 \text{ rad/s})t]$$

Find (a) the amplitude of the wave, (b) the wave number, (c) the wavelength, (d) the angular frequency, (e) the frequency, (f) the period, (g) the velocity of the wave (i.e., its speed and direction), and (h) the displacement of the wave at $x = 10.0$ m and $t = 0.500$ s.

Solution

The characteristics of the wave are determined by writing the wave in the standard form

$$y = A \sin(kx - \omega t)$$

a. The amplitude A is determined by inspection of both equations as $A = 0.200$ m.
b. The wave number k is found from inspection to be $k = 0.500$ m^{-1} or a half a wave in an interval of 2π.
c. The wavelength λ, found from equation 12.8, is

$$\lambda = \frac{2\pi}{k} = \frac{2\pi}{0.500 \text{ m}^{-1}}$$

$$= 12.6 \text{ m}$$

d. The angular frequency ω, found by inspection, is

$$\omega = 8.20 \text{ rad/s}$$

e. The frequency f of the wave, found from equation 12.12, is

$$f = \frac{\omega}{2\pi} = \frac{8.20 \text{ rad/s}}{2\pi \text{ rad}} = 1.31 \text{ cycles/s} = 1.31 \text{ Hz}$$

f. The period of the wave is the reciprocal of the frequency, thus

$$T = \frac{1}{f} = \frac{1}{1.31 \text{ Hz}} = 0.766 \text{ s}$$

g. The speed of the wave, found from equation 12.14, is

$$v = \frac{\omega}{k} = \frac{8.20 \text{ rad/s}}{0.500 \text{ m}^{-1}} = 16.4 \text{ m/s}$$

Vibratory Motion, Wave Motion, and Fluids

We could also have determined this by

$$v = f\lambda = \left(1.31 \frac{1}{s}\right)(12.6 \text{ m}) = 16.4 \text{ m/s}$$

The direction of the wave is to the right because the sign in front of ω is negative.

h. The displacement of the wave at $x = 10.0$ m and $t = 0.500$ s is

$$y = (0.200 \text{ m})\sin[(0.500 \text{ m}^{-1})(10.0 \text{ m}) - (8.20 \text{ rad/s})(0.500 \text{ s})]$$

$$= (0.200 \text{ m})\sin[0.900 \text{ rad}] = (0.200 \text{ m})(0.783)$$

$$= 0.157 \text{ m}$$

12.3 The Speed of a Transverse Wave on a String

Let us consider the motion of a transverse wave on a string, as shown in figure 12.6. The wave is moving to the right with a velocity v. Let us observe the wave by moving with the wave at the same velocity v. In this reference frame, the wave appears stationary, while the particles composing the string appear to be moving through the wave to the left. One such particle is shown at the top of the wave of figure 12.6 moving to the left at the velocity v. If we consider only a small portion of the top of the wave, we can approximate it by an arc of a circle of radius R, as shown. If the angle θ is small, the length of the string considered is small and the mass m of this small portion of the string can be approximated by a mass moving in uniform circular motion. Hence, there must be a centripetal force acting on this small portion of the string and its magnitude is given by

$$F_c = \frac{mv^2}{R} \tag{12.27}$$

This centripetal force is supplied by the tension in the string. In figure 12.6, the tensions on the right and left side of m are resolved into components. There is a force $T \cos \theta$ acting to the right of m and a force $T \cos \theta$ acting to the left. These components are equal and opposite and cancel each other out, thus exerting a zero

Figure 12.6

Velocity of a transverse wave.

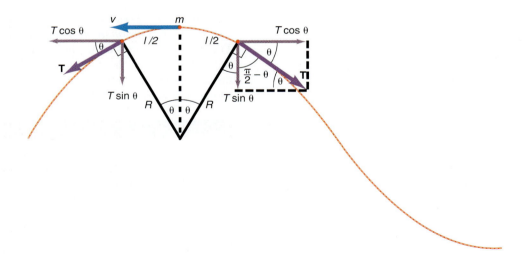

net force in the horizontal direction. The components $T \sin \theta$ on the right and left side of m act downward on m and thus supply the necessary centripetal force for m to be in uniform circular motion. Thus,

$$F_c = 2T \sin \theta$$

Since we assume that θ is small, the $\sin \theta$ can be replaced by the angle θ itself, expressed in radians. Thus,

$$F_c = 2T \theta \tag{12.28}$$

The small portion of the string l approximates an arc of a circle and the arc of a circle is given by $s = R\theta$. Therefore,

$$s = \frac{l}{2} + \frac{l}{2} = R(\theta + \theta) = 2\theta R$$

and

$$\theta = \frac{l}{2R}$$

The centripetal force, equation 12.28, becomes

$$F_c = 2T \frac{l}{2R}$$

and

$$F_c = \frac{Tl}{R} \tag{12.29}$$

Equating the centripetal force in equations 12.27 and 12.29 we get

$$\frac{mv^2}{R} = \frac{Tl}{R}$$

$$v^2 = \frac{Tl}{m} = \frac{T}{m/l}$$

Solving for the speed of the wave we get

$$v = \sqrt{\frac{T}{m/l}} \tag{12.30}$$

Therefore, the speed of a transverse wave in a string is given by equation 12.30, where T is the tension in the string and m/l is the mass per unit length of the string. The greater the tension in the string, the greater the speed of propagation of the wave. The greater the mass per unit length of the string, the smaller the speed of the wave. We will discuss equation 12.30 in more detail when dealing with traveling waves on a vibrating string in section 12.6.

Example 12.3

Play that guitar. Find the tension in a 60.0-cm guitar string that has a mass of 1.40 g if it is to play the note G with a frequency of 396 Hz. Assume that the wavelength of the note will be two times the length of the string, or $\lambda = 120$ cm (this assumption will be justified in section 12.6).

Solution

The speed of the wave, found from equation 12.3, is

$$v = \lambda f = (1.20 \text{ m})(396 \text{ cycles/s})$$

$$= 475 \text{ m/s}$$

The mass density of the string is

$$\frac{m}{l} = \frac{1.40 \times 10^{-3}\ \text{kg}}{0.600\ \text{m}} = 2.33 \times 10^{-3}\ \text{kg/m}$$

The tension that the guitar string must have in order to play this note, found from equation 12.30, is

$$T = v^2\frac{m}{l} = (475\ \text{m/s})^2(2.33 \times 10^{-3}\ \text{kg/m})$$

$$= 526\ \text{N}$$

Example 12.4

Sounds flat to me. If the tension in the guitar string of example 12.3 was 450 N, would the guitar play that note flat or sharp?

Solution

The mass density of the string is 2.33×10^{-3} kg/m. With a tension of 450 N, the speed of the wave is

$$v = \sqrt{\frac{T}{m/l}} = \sqrt{\frac{450\ \text{N}}{2.33 \times 10^{-3}\ \text{kg/m}}}$$

$$= 439\ \text{m/s}$$

The frequency of the wave is then

$$f = \frac{v}{\lambda} = \frac{439\ \text{m/s}}{1.20\ \text{m}} = 366\ \text{Hz}$$

The string now plays a note at too low a frequency and the note is flat by 396 Hz − 366 Hz = 30 Hz.

12.4 Reflection of a Wave at a Boundary

In the analysis of the vibrating string we assumed that the string was infinitely long so that it was not necessary to consider what happens when the wave gets to the end of the string. Now we need to rectify this omission by considering the **reflection of a wave at a boundary.** To simplify the discussion let us deal with a single pulse rather than the continuous waves dealt with in the preceding sections.

First, let us consider how a pulse is generated. Take a piece of string fixed at one end and hold the other end in your hand, as shown in figure 12.7(a). The string and hand are at rest. If the hand is moved up rapidly, the string near the hand will also be pulled up. This is shown in figure 12.7(b) with arrows pointing upward representing the force upward on the particles of the string. Each particle that moves upward exerts a force on the particle immediately to its right by the tension in the string. In this way, the force upward is passed from particle to particle along the string. In figure 12.7(c), the hand is quickly moved downward pulling the end of the string down with it. The force acting on the string downward is shown by the arrows pointing downward in figure 12.7(c). Note that the arrows pointing upward caused by the force upward in figure 12.7(b) are still upward and moving toward the right. In figure 12.7(d), the hand has returned to the equilibrium position and is at rest. However, the motion of the hand upward and downward has created a pulse that is moving along the string with a velocity of propagation v. The arrows upward represent the force pulling the string upward in advance of the center of the pulse, while the arrows downward represent the force pulling the string down-

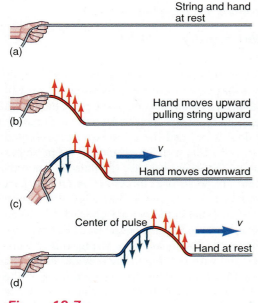

String and hand at rest

Hand moves upward pulling string upward

Hand moves downward

Center of pulse

Hand at rest

Figure 12.7

Creation of a pulse on a string.

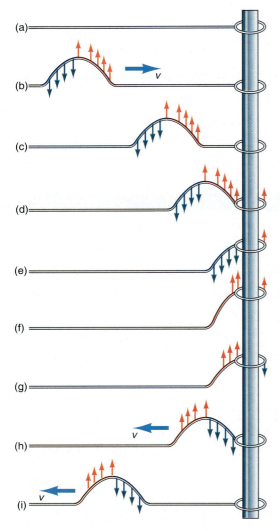

Figure 12.8

Reflection of a pulse on a string that is free to move in the transverse direction.

ward, behind the center of the pulse, back to its rest position. As the pulse propagates so will these forces. Let us now consider what happens to this pulse as it comes to a boundary, in this case, the end of the string.

The End of the String Is Not Fixed Rigidly but Is Allowed to Move

Let us first consider the case of a pulse propagating along a string that is free to move at its end point. This is shown in figure 12.8. The end of the stationary string, figure 12.8(a), is attached to a ring that is free to move in the vertical direction on a frictionless pole. A pulse is now sent down the string in figure 12.8(b). Let us consider the forces on the string that come from the pulse, that is, we will ignore the gravitational forces on the string. The arrows up and down on the pulse represent the forces upward and downward, respectively, on the particles of the string. The pulse propagates to the right in figure 12.8(c) and reaches the ring in figure 12.8(d). The force upward that has been propagating to the right causes the ring on the end of the string to move upward. The ring now rises to the height of the pulse as the center of the pulse arrives at the ring, figure 12.8(e). Although there is no additional force upward on the ring, the ring continues to move upward because of its momentum.

We can also consider this from an energy standpoint. At the location of the top of the pulse, the ring has a kinetic energy upward. The ring continues upward until this kinetic energy is lost. As the ring moves upward it now pulls the string up with it, eventually overcoming the forces downward at the rear of the pulse, until the string to the left of the ring has a net force upward acting on it, figure 12.8(f). This upward force is now propagated along the string to the left by pulling each adjacent particle to its left upward. Because the ring pulled upward on the string, by Newton's third law the string also pulls downward on the ring, and the ring eventually starts downward, figure 12.8(g). As the ring moves downward it exerts a force downward on the string, as shown by the arrows in figure 12.8(h). The forces upward and downward propagate to the left as the pulse shown in figure 12.8(i).

The net result of the interaction of the pulse to the right with the movable ring is a reflected pulse of the same size and shape that now moves to the left with the same speed of propagation. The incoming pulse was right side up, and the reflected pulse is also right side up. The movable ring at the end of the string acts like the hand, moving up and down to create the pulse in figure 12.7.

The End of the String Is Fixed Rigidly and Not Allowed to Move

A pulse is sent down a string that has the end fixed to a wall, as in figure 12.9(a). Let us consider the forces on the string that come from the pulse, that is, we will ignore the gravitational forces on the string. The front portion of the pulse has forces that are acting upward and are represented by arrows pointing upward. The back portion of the pulse has forces acting downward, and these forces are represented by arrows pointing downward. Any portion of the string in advance of the pulse has no force in the vertical direction acting on the string because the pulse has not arrived yet. Hence, in figure 12.9(a), there are no vertical forces acting on that part of the string that is tied to the wall. In figure 12.9(b), the leading edge of the pulse has just arrived at the wall. This leading edge has forces acting upward, and when they make contact with the wall they exert a force upward on the wall. But because of the enormous mass of the wall compared to the mass of the string, this upward force can not move the wall upward as it did with the ring in the previous case. The end of the string remains fixed. But by Newton's third law this upward force on the wall causes a reaction force downward on the string pulling the string down below the equilibrium position of the string, figure 12.9(c). This initiates the beginning of a pulse that moves to the left. At this point the picture becomes rather complicated,

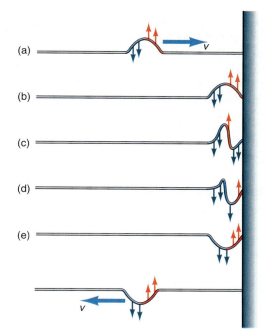

(a)

(b)

(c)

(d)

(e)

Figure 12.9

A reflected pulse on a string with a fixed end.

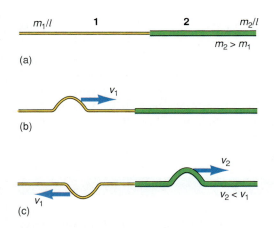

(a)

(b)

(c)

Figure 12.10

A pulse goes from a less dense medium to a more dense medium.

because while the back portion of the original pulse is still moving toward the right, the front portion has become reflected and is moving toward the left, figure 12.9(d). The resulting pulse becomes a superposition of these two pulses, one traveling toward the right, the other traveling toward the left. The forces acting on each particle of the string become the sum of the forces caused by each pulse. When the back portion of the original pulse reaches the wall it exerts a force downward on the wall. By Newton's third law the wall now exerts a reaction force upward on the string that pulls the string of the rear of the pulse upward to its equilibrium position, figure 12.9(e). The pulse has now been completely reflected by the wall and moves to the left with the same speed v, figure 12.9(f). Note, however, that in reflecting the pulse, the reaction force of the wall on the string has caused the reflected pulse to be inverted or turned upside down. *Hence, a wave or pulse that is reflected from a fixed end is inverted.* The reflected pulse is said to be 180° out of phase with the incident pulse. This was not the case for the string whose end was free to move.

Reflection and Transmission of a Wave at the Boundary of Two Different Media

When an incident wave impinges upon a boundary separating two different media, part of the incident wave is transmitted into the second medium while part is reflected back into the first medium. We can easily see this effect by connecting together two strings of different mass densities. Let us consider two different cases of the **reflection and transmission of a wave at the boundary of two different media.**

Case 1: *The Wave Goes from the Less Dense Medium to the More Dense Medium*

Consider the string in figure 12.10(a). The left-hand side of the combined string is a light string of mass density m_1/l, while the right-hand side is a heavier string of mass density m_2/l, where m_2 is greater than m_1. The combined string is pulled tight so that both strings have the same tension T. A pulse is now sent down the lighter string at a velocity v_1 to the right in figure 12.10(b). As the pulse hits the boundary between the two strings, the upward force in the leading edge of the pulse on string 1 exerts an upward force on string 2. Because string 2 is much more massive than string 1, the boundary acts like the fixed end in figure 12.9, and the reaction force of the massive string causes an inverted reflected pulse to travel back along string 1, as shown in figure 12.10(c). Because the massive string is not infinite, like a rigid wall, the forces of the incident pulse pass through to the massive string, thus also transmitting a pulse along string 2, as shown in figure 12.10(c). Since string 2 is more massive than string 1, the transmitted force can not displace the massive string elements of string 2 as much as in string 1. Hence the amplitude of the transmitted pulse is less than the amplitude of the incident pulse.

Because the tension in each string is the same, the speed of the pulses in medium 1 and 2 are

$$v_1 = \sqrt{\frac{T}{m_1/l}} \tag{12.31}$$

$$v_2 = \sqrt{\frac{T}{m_2/l}} \tag{12.32}$$

Because the tension T in each string is the same, they can be equated to find the speed of the pulse in medium 2 as

$$v_2 = \sqrt{\frac{m_1/l}{m_2/l}}\, v_1 \tag{12.33}$$

However, because m_2 is greater than m_1, equation 12.33 implies that v_2 will be less than v_1. That is, the speed v_2 of the transmitted pulse will be less than v_1, the speed of the incident and reflected pulses. *Thus, the pulse slows down in going from the* less dense medium to the more dense medium. If a sinusoidal wave were propagated along the string instead of the pulse, part of the wave would be reflected and part would be transmitted. However, because of the boundary, the wavelength of the transmitted wave would be different from the incident wave. To see this, note that the frequency of the wave must be the same on both sides of the boundary. (Since the frequency is the number of waves per second, and the same number pass from medium 1 into medium 2, we have $f_1 = f_2$.) The wavelength of the incident wave, found from equation 12.3, is

$$\lambda_1 = \frac{v_1}{f}$$

whereas the wavelength of the transmitted wave λ_2 is

$$\lambda_2 = \frac{v_2}{f}$$

Because the frequency is the same,

$$f = \frac{v_1}{\lambda_1}$$

and

$$f = \frac{v_2}{\lambda_2}$$

they can be equated giving

$$\frac{v_1}{\lambda_1} = \frac{v_2}{\lambda_2}$$

Thus, the wavelength of the transmitted wave λ_2 is

$$\lambda_2 = \frac{v_2}{v_1} \lambda_1 \qquad\qquad (12.34)$$

Since v_2 is less than v_1, equation 12.34 tells us that λ_2 is less than λ_1. *Hence, when a wave goes from a less dense medium to a more dense medium, the wavelength of the transmitted wave is less than the wavelength of the incident wave.*

Although these results were derived from waves on a string, they are quite general. In chapter 27 we will see that when a light wave goes from a region of low density such as a vacuum or air, into a more dense region, such as glass, the speed of the light wave decreases and its wavelength also decreases.

Example 12.5

A wave going from a less dense to a more dense medium. One end of a 60.0-cm steel wire of mass 1.40 g is welded to the end of a 60.0-cm steel wire of 6.00 g mass. The combined wires are placed under uniform tension. (a) If a wave propagates down the lighter wire at a speed of 475 m/s, at what speed will it be transmitted along the heavier wire? (b) If the wavelength on the first wire is 1.20 m, what is the wavelength on the second wire?

Solution

a.   The mass per unit length of each wire is

$$\frac{m_1}{l} = \frac{1.40 \times 10^{-3}\ \text{kg}}{0.600\ \text{m}} = 2.20 \times 10^{-3}\ \text{kg/m}$$

$$\frac{m_2}{l} = \frac{6.00 \times 10^{-3}\ \text{kg}}{0.600\ \text{m}} = 1.00 \times 10^{-2}\ \text{kg/m}$$

Vibratory Motion, Wave Motion, and Fluids

The speed of the transmitted wave, found from equation 12.33, is

$$v_2 = \sqrt{\frac{m_1/l}{m_2/l}}\, v_1$$

$$= \sqrt{\frac{2.20 \times 10^{-3}\ \mathrm{kg/m}}{1.00 \times 10^{-2}\ \mathrm{kg/m}}}\ (475\ \mathrm{m/s})$$

$$= 223\ \mathrm{m/s}$$

b. The wavelength of the transmitted wave, found from equation 12.34, is

$$\lambda_2 = \frac{v_2}{v_1}\, \lambda_1$$

$$= \left(\frac{223\ \mathrm{m/s}}{475\ \mathrm{m/s}}\right)(1.20\ \mathrm{m})$$

$$= 0.563\ \mathrm{m}$$

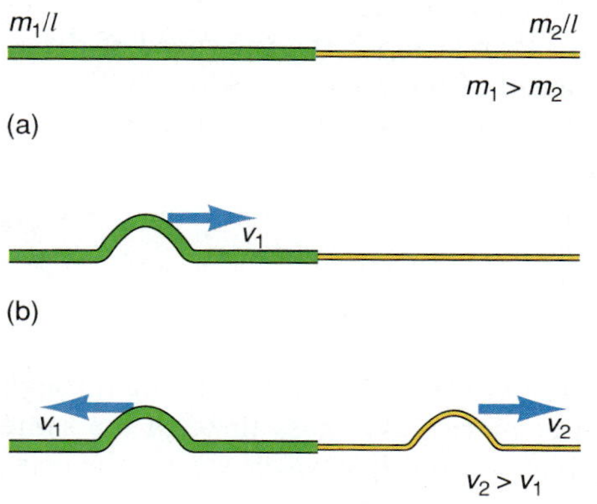

m_1/l　　　　　　　　m_2/l

$m_1 > m_2$

(a)

(b)

v_1

v_1　　　　　　　v_2

$v_2 > v_1$

(c)

Figure 12.11

A pulse goes from a more dense medium to a less dense medium.

Case 2: *A Wave Goes from a More Dense Medium to a Less Dense Medium*

Consider the string in figure 12.11(a). The left-hand side of the combined string is a heavy string of mass density m_1/l, whereas the right-hand side is a light string of mass density m_2/l, where m_2 is now less than m_1. A pulse is sent down the string in figure 12.11(b). When the pulse hits the boundary the boundary acts like the free end of the string in figure 12.8 because of the low mass of the second string. A pulse is reflected along the string that is erect or right side up, figure 12.11(c). However, the forces of the incident pulse are transmitted very easily to the lighter second string and a transmitted pulse also appears in string 2, figure 12.11(c). Because the tension is the same in both strings, a similar analysis to case 1 shows that *when a wave goes from a more dense medium to a less dense medium, the transmitted wave moves faster than the incident wave and has a longer wavelength.*

Example 12.6

A wave going from a more dense medium to a less dense medium. The first half of a combined string has a linear mass density of 0.100 kg/m, whereas the second half has a linear mass density of 0.0500 kg/m. A sinusoidal wave of wavelength 1.20 m is sent along string 1. If the combined string is under a tension of 10.0 N, find (a) the speed of the incident wave in string 1, (b) the speed of the transmitted wave in string 2, (c) the wavelength of the transmitted wave, and (d) the speed and wavelength of the reflected wave.

Solution

a. The speed of the incident wave in string 1, found from equation 12.31, is

$$v_1 = \sqrt{\frac{T}{m_1/l}}$$

$$= \sqrt{\frac{10.0\ \mathrm{N}}{0.100\ \mathrm{kg/m}}}$$

$$= 10.0\ \mathrm{m/s}$$

b. The speed of the transmitted wave in string 2, found from equation 12.32, is

$$v_2 = \sqrt{\frac{T}{m_2/l}}$$

$$= \sqrt{\frac{10.0 \text{ N}}{0.0500 \text{ kg/m}}}$$

$$= 14.1 \text{ m/s}$$

c. The wavelength of the transmitted wave, found from equation 12.34, is

$$\lambda_2 = \frac{v_2}{v_1}\lambda_1$$

$$= \left(\frac{14.1 \text{ m/s}}{10.0 \text{ m/s}}\right)(1.20 \text{ m})$$

$$= 1.69 \text{ m}$$

d. The speed and wavelength of the reflected wave are the same as the incident wave because the reflected wave is in the same medium as the incident wave. Note that the mass of string 1 is greater than string 2 and the speed of the wave in medium 2 is greater than the speed of the wave in medium 1 (i.e., $v_2 > v_1$). Also note that the wavelength of the transmitted wave is greater than the wavelength of the incident wave (i.e., $\lambda_2 > \lambda_1$).

The superposition of two waves in water.

12.5 The Principle of Superposition

Up to this point in our discussion we have considered only one wave passing through a medium at a time. What happens if two or more waves pass through the same medium at the same time? To solve the problem of multiple waves we use the principle of superposition. This principle is based on the vector addition of the displacement associated with each wave. *The **principle of superposition** states that whenever two or more wave disturbances pass a particular point in a medium, the resultant displacement of the point of the medium is the sum of the displacements of each individual wave disturbance.* For example, if the two waves

$$y_1 = A_1 \sin(k_1 x - \omega_1 t)$$
$$y_2 = A_2 \sin(k_2 x - \omega_2 t)$$

are acting in a medium at the same time, the resultant wave is given by

$$y = y_1 + y_2 \tag{12.35}$$

or

$$y_1 = A_1 \sin(k_1 x - \omega_1 t) + A_2 \sin(k_2 x - \omega_2 t) \tag{12.36}$$

The superposition principle holds as long as the resultant displacement of the medium does not exceed its elastic limit. Sometimes the two waves are said to interfere with each other, or cause *interference*.

Example 12.7

Superposition. The following two waves interfere with each other:

$$y_1 = (5.00 \text{ m})\sin[(0.800 \text{ m}^{-1})x - (6.00 \text{ rad/s})t]$$
$$y_2 = (10.00 \text{ m})\sin[(0.900 \text{ m}^{-1})x - (3.00 \text{ rad/s})t]$$

Find the resultant displacement when $x = 5.00$ m and $t = 1.10$ s.

The resultant displacement found by the superposition principle, equation 12.35, is

$$y = y_1 + y_2$$

where

$$y_1 = (5.00 \text{ m})\sin[(0.800 \text{ m}^{-1})(5.00 \text{ m}) - (6.00 \text{ rad/s})(1.10 \text{ s})]$$
$$= (5.00 \text{ m})\sin(4.00 \text{ rad} - 6.6 \text{ rad})$$
$$= (5.00 \text{ m})\sin(-2.60 \text{ rad})$$
$$= (5.00 \text{ m})(-0.5155) = -2.58 \text{ m}$$

and

$$y_2 = (10.00 \text{ m})\sin[(0.900 \text{ m}^{-1})(5.00 \text{ m}) - (3.00 \text{ rad/s})(1.10 \text{ s})]$$
$$= (10.00 \text{ m})\sin(4.50 \text{ rad} - 3.30 \text{ rad})$$
$$= (10.00 \text{ m})\sin(1.20 \text{ rad})$$
$$= (10.00 \text{ m})(0.932) = 9.32 \text{ m}$$

Hence, the resultant displacement is

$$y = y_1 + y_2 = -2.58 \text{ m} + 9.32 \text{ m}$$
$$= 6.74 \text{ m}$$

Note that this is the resultant displacement only for the values of $x = 5.00$ m and $t = 1.10$ s. We can find the entire resultant wave for any value of the time by substituting a series of values of x into the equation for that value of t. Then we determine the resultant displacement y for each value of x. A graph of the resultant displacement y versus x gives a snapshot of the resultant wave at that value of time t. The process can be repeated for various values of t, and the sequence of the graphs will show how that resultant wave travels with time.

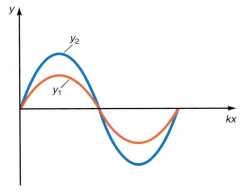

(a) Waves in phase

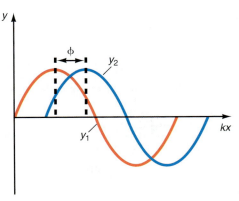

(b) Waves out of phase

Figure 12.12

Phase of a wave.

It is possible that when dealing with two or more waves the waves may not be in phase with each other. *Two waves are in phase if they reach their maximum amplitudes at the same time, are zero at the same time, and have their minimum amplitudes at the same time.* An example of two waves in phase is shown in figure 12.12(a). An example of two waves that are out of phase with each other is shown in figure 12.12(b). Note that the second wave does not have its maximum, zero, and minimum displacements at the same place as the first wave. Instead these positions are translated to the right of their position in wave y_1. We say that wave 2 is out of phase with wave 1 by an angle ϕ, where ϕ is measured in radians. The equation for the first wave is

$$y_1 = A_1 \sin(kx - \omega t) \tag{12.37}$$

whereas the equation for the wave displaced to the right is

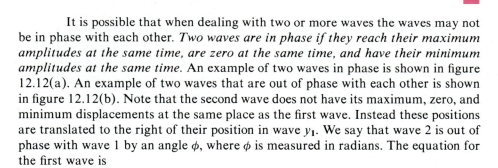

$$y_2 = A_2 \sin(kx - \omega t - \phi) \tag{12.38}$$

*The angle ϕ is called the **phase angle** and is a measure of how far wave 2 is displaced in the horizontal from wave 1. Just as the minus sign on $-\omega t$ indicated a wave traveling to the right, the minus sign on ϕ indicates a wave displaced to the right. The second wave lags the first wave by the phase angle ϕ. That is, wave 2 has its maximum, zero, and minimum displacements after wave 1 does, and the amount of lag is given by the phase angle ϕ.* If the wave was displaced to the left, the equation for the wave would be

$$y_2 = A_2 \sin(kx - \omega t + \phi) \tag{12.39}$$

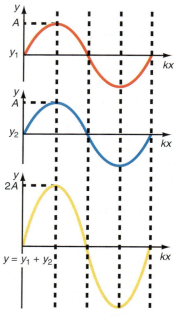

(a) Constructive interference

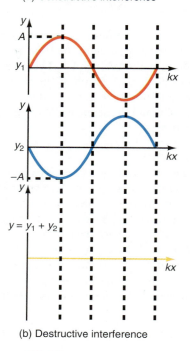

(b) Destructive interference

Figure 12.13

Interference of waves.

An important special case of the addition of waves is shown in figure 12.13. *When two waves are in phase with each other, $\phi = 0$ in equation 12.38, and the waves are said to exhibit **constructive interference**, figure 12.13(a).* That is,

$$y_1 = A \sin(kx - \omega t)$$
$$y_2 = A \sin(kx - \omega t)$$

and the resultant wave is

$$y = y_1 + y_2 = 2A \sin(kx - \omega t) \qquad (12.40)$$

That is, the resultant amplitude has doubled. If the two waves are 180°, or π rad, out of phase with each other, then y_2 is

$$y_2 = A_2 \sin(kx - \omega t - \pi)$$

Setting $kx - \omega t = B$ and $\pi = C$, we can use the formula for the sine of the difference between two angles, which is found in appendix B. That is,

$$\sin(B - C) = \sin B \cos C - \cos B \sin C \qquad (12.41)$$

Thus,

$$\sin[(kx - \omega t) - \pi] = \sin(kx - \omega t)\cos \pi - \cos(kx - \omega t)\sin \pi \qquad (12.42)$$

But $\sin \pi = 0$, and the last term drops out. And because the $\cos \pi = -1$, equation 12.42 becomes

$$\sin[(kx - \omega t) - \pi] = -\sin(kx - \omega t)$$

Therefore we can write the second wave as

$$y_2 = -A \sin(kx - \omega t) \qquad (12.43)$$

The superposition principle now yields

$$y = y_1 + y_2 = A \sin(kx - \omega t) - A \sin(kx - \omega t) = 0 \qquad (12.44)$$

*Thus, if the waves are 180° out of phase the resultant wave is zero everywhere. This is shown in figure 12.13(b) and is called **destructive interference**.* Wave 2 has completely canceled out the effects of wave 1.

A more general solution for the interference of two waves of the same frequency, same wave number, same amplitude, and in the same direction but out of phase with each other by a phase angle ϕ, can be easily determined by the superposition principle. Let the two waves be

$$y_1 = A \sin(kx - \omega t) \qquad (12.45)$$
$$y_2 = A \sin(kx - \omega t - \phi) \qquad (12.46)$$

The resultant wave is

$$y = y_1 + y_2 = A \sin(kx - \omega t) + A \sin(kx - \omega t - \phi) \qquad (12.47)$$

To simplify this result, we use the trigonometric identity found in appendix B for the sum of two sine functions, namely

$$\sin B + \sin C = 2 \sin\left(\frac{B + C}{2}\right)\cos\left(\frac{B - C}{2}\right) \qquad (12.48)$$

For this problem

$$B = kx - \omega t$$

and

$$C = kx - \omega t - \phi$$

Thus,

$$\sin(kx - \omega t) + \sin(kx - \omega t - \phi) =$$

$$2 \sin\left(\frac{kx - \omega t + kx - \omega t - \phi}{2}\right) \cos\left(\frac{kx - \omega t - kx + \omega t + \phi}{2}\right)$$

$$= 2 \sin\left(kx - \omega t - \frac{\phi}{2}\right)\cos\left(\frac{\phi}{2}\right) \qquad \textbf{(12.49)}$$

Substituting equation 12.49 into equation 12.47 we obtain for the resultant wave

$$y = 2A \cos\left(\frac{\phi}{2}\right)\sin\left(kx - \omega t - \frac{\phi}{2}\right) \qquad \textbf{(12.50)}$$

Equation 12.50 is a more general result than found before and contains constructive and destructive interference as special cases. For example, if $\phi = 0$ the two waves are in phase and since $\cos 0 = 1$, equation 12.50 becomes

$$y = 2A \sin(kx - \omega t)$$

which is identical to equation 12.40 for constructive interference. Also for the special case of $\phi = 180° = \pi$ rad, the $\cos(\pi/2) = \cos 90° = 0$, and equation 12.50 becomes $y = 0$, the special case of destructive interference, equation 12.44.

Example 12.8

Interference. The following two waves interfere:

$$y_1 = (5.00 \text{ m})\sin[(0.200 \text{ m}^{-1})x - (5.00 \text{ rad/s})t]$$

$$y_2 = (5.00 \text{ m})\sin[(0.200 \text{ m}^{-1})x - (5.00 \text{ rad/s})t - 0.500 \text{ rad}]$$

Find the equation for the resultant wave.

Solution

The resultant wave, found from equation 12.50, is

$$y = 2A \cos\left(\frac{\phi}{2}\right)\sin\left(kx - \omega t - \frac{\phi}{2}\right)$$

$$= 2(5.00 \text{ m})\cos\left(\frac{0.500 \text{ rad}}{2}\right)\sin\left[(0.200 \text{ m}^{-1})x - (5.00 \text{ rad/s})t - \left(\frac{0.500 \text{ rad}}{2}\right)\right]$$

$$= (10.00 \text{ m})(0.9689)\sin[(0.200 \text{ m}^{-1})x - (5.00 \text{ rad/s})t - 0.250 \text{ rad}]$$

$$= (9.69 \text{ m})\sin[(0.200 \text{ m}^{-1})x - (5.00 \text{ rad/s})t - 0.250 \text{ rad}]$$

We can now plot an actual picture of the resultant wave for a particular value of t for a range of values of x.

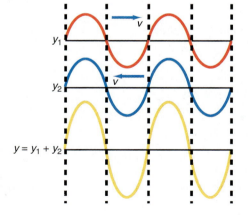

Figure 12.14

Formation of standing waves.

12.6 Standing Waves—The Vibrating String

If a string is fixed at both ends and a wave train is sent down the string, then, as shown before, the wave is reflected from the fixed ends. Hence, there are two wave trains on the string at the same time. One is traveling to the right, while the reflected wave is traveling toward the left, figure 12.14. We can find the resultant wave by the superposition principle. That is, if wave 1 is a wave to the right, we can express it

$$y_1 = A \sin(kx - \omega t) \qquad \textbf{(12.51)}$$

whereas we can express the wave to the left as

$$y_2 = A \sin(kx + \omega t) \tag{12.52}$$

The resultant wave is the sum of these two waves or

$$y = y_1 + y_2 = A \sin(kx - \omega t) + A \sin(kx + \omega t) \tag{12.53}$$

To add these two sine functions, we use the trigonometric identity in equation 12.48. That is,

$$\sin B + \sin C = 2 \sin\left(\frac{B + C}{2}\right)\cos\left(\frac{B - C}{2}\right)$$

where $B = kx - \omega t$ and $C = kx + \omega t$. Thus,

$$y = 2A \sin\left[\frac{(kx - \omega t) + (kx + \omega t)}{2}\right]\cos\left[\frac{(kx - \omega t) - (kx + \omega t)}{2}\right]$$

$$= 2A \sin\left(\frac{2kx}{2}\right)\cos\left(\frac{-2\omega t}{2}\right)$$

and

$$y = 2A \sin(kx)\cos(-\omega t)$$

Using the fact that

$$\cos(-\theta) = \cos(\theta)$$

the resultant wave is

$$y = 2A \sin(kx)\cos(\omega t) \tag{12.54}$$

For reasons that will appear shortly, *this is the equation of a standing wave or a stationary wave.*

The amplitude of the resultant standing wave is $2A \sin(kx)$, and note that it varies with x. To find the positions along x where this new amplitude has its minimum values, note that $\sin(kx) = 0$ whenever

$$kx = n\pi$$

for values of $n = 1, 2, 3, \ldots$. That is, the sine function is zero whenever the argument kx is a multiple of π. Thus, solving for x,

$$x = \frac{n\pi}{k} \tag{12.55}$$

But the wave number k was defined in equation 12.9 as

$$k = \frac{2\pi}{\lambda}$$

Substituting this value into equation 12.55, we get

$$x = \frac{n\pi}{k} = \frac{n\pi}{2\pi/\lambda}$$

and

$$x = \frac{n\lambda}{2} \tag{12.56}$$

Equation 12.56 gives us the location of the zero values of the amplitude. Thus we see that they occur for values of x of $\lambda/2$, $2\lambda/2 = \lambda$, $3\lambda/2$, $4\lambda/2 = 2\lambda$, and so on, as measured from either end. *These points, where the amplitude of the standing wave is zero, are called nodes.* Stated another way, *a **node** is the position of zero amplitude.* These nodes are independent of time, that is, the amplitude at these points is always zero.

The maximum values of the amplitude occur whenever $\sin(kx) = 1$, which happens whenever kx is an odd multiple of $\pi/2$. That is, $\sin(kx) = 1$ when

$$kx = (2n - 1)\frac{\pi}{2}$$

for $n = 1, 2, 3, \ldots$.

The term $2n - 1$ always gives an odd number for any value of n. (As an example, when $n = 2$, $2n - 1 = 3$, etc.) The location of the maximum amplitudes is therefore at

$$x = \left(\frac{2n - 1}{2k}\right)\pi$$

But since $k = 2\pi/\lambda$, this becomes

$$x = \frac{(2n - 1)\pi}{2(2\pi/\lambda)}$$

and

$$x = \frac{(2n - 1)\lambda}{4} = (2n - 1)\frac{\lambda}{4} \tag{12.57}$$

The maximum amplitudes are thus located at $x = \lambda/4$, $3\lambda/4$, $5\lambda/4$, and so forth. *The position of maximum amplitude is called an **antinode.*** Note that at this position the displacement of the resultant wave is a function of time. The original two traveling waves and the resultant standing wave are shown in figure 12.15 for values

Figure 12.15

Standing wave on a string.

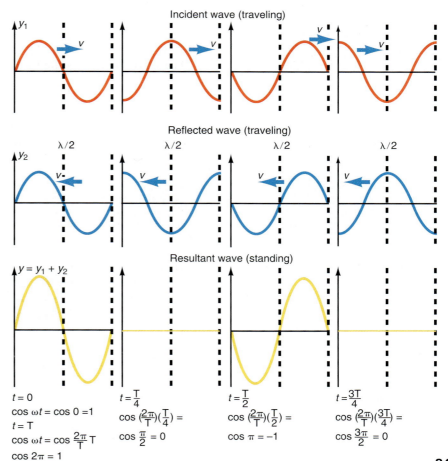

Chapter 12   Wave Motion

of time of 0, $T/4$, $T/2$, $3T/4$, and T, where T is the period of the wave. Recall that $\omega = 2\pi/T$. Therefore, $\cos \omega t = \cos(2\pi t/T)$. *Note that the waves are moving to the left and the right, but the resultant wave does not travel at all, it is a **standing wave on a string**. The node of the standing wave at $x = \lambda/2$ remains a node for all times. Thus, the string can not move up and down at that point, and can not therefore transmit any energy past that point. Thus, the resultant wave does not move along the string but is stationary or standing.*

How many different types of standing waves can be produced on this string? The only restriction on the number or types of different waves is that the ends of the string must be tied down or fixed. That is, there must be a node at the ends of the string, which implies that the displacement y must always equal zero for $x = 0$, and for $x = L$, the length of the string. When x is equal to zero the displacement is

$$y = 2A \sin[k(0)]\cos(\omega t) = 0 \tag{12.58}$$

When $x = L$, the displacement of the standing wave is

$$y = 2A \sin(kL)\cos(\omega t) \tag{12.59}$$

Equation 12.59 is not in general always equal to zero. Because it must always be zero in order to satisfy the boundary condition of $y = 0$ for $x = L$, it is necessary that

$$\sin(kL) = 0$$

which is true whenever kL is a multiple of π, that is,

$$kL = n\pi$$

for $n = 1, 2, 3, \ldots$. This places a restriction on the number of waves that can be placed on the string. The only possible wave numbers the wave can have are therefore

$$k = \frac{n\pi}{L} \tag{12.60}$$

Therefore, we must write the displacement of the standing wave as

$$y = 2A \sin\left(\frac{n\pi x}{L}\right)\cos(\omega t) \tag{12.61}$$

Because $k = 2\pi/\lambda$, a restriction on the possible wave numbers k is also a restriction on the possible wavelengths λ that can be found on the string. Thus,

$$k = \frac{2\pi}{\lambda} = \frac{n\pi}{L}$$

or

$$\lambda = \frac{2L}{n} \tag{12.62}$$

That is, the only wavelengths that are allowed on the string are $\lambda = 2L, L, 2L/3$, and so forth. In other words, not all wavelengths are possible; only those that satisfy equation 12.62 will have fixed end points. Only a discrete set of wavelengths is possible. Figure 12.16 shows some of the possible modes of vibration.

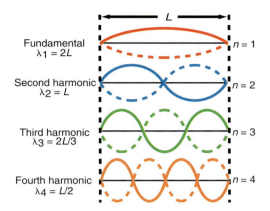

Fundamental
$\lambda_1 = 2L$ $n = 1$

Second harmonic
$\lambda_2 = L$ $n = 2$

Third harmonic
$\lambda_3 = 2L/3$ $n = 3$

Fourth harmonic
$\lambda_4 = L/2$ $n = 4$

Figure 12.16

The normal modes of vibration of a string.

We can find the frequency of any wave on the string with the aid of equations 12.3 and 12.30 as

$$f = \frac{v}{\lambda}$$

$$v = \sqrt{\frac{T}{m/l}}$$

Thus,

$$f = \frac{1}{\lambda} \sqrt{\frac{T}{m/l}} \qquad (12.63)$$

However, since the only wavelengths possible are those for $\lambda = 2L/n$, equation 12.62, the frequencies of vibration are

$$f = \frac{n}{2L} \sqrt{\frac{T}{m/l}} \qquad (12.64)$$

with $n = 1, 2, 3, \ldots$.

Equation 12.64 points out that there are only a discrete number of frequencies possible for the vibrating string, depending on the value of n. The simplest mode of vibration, for $n = 1$, is called the *fundamental mode of vibration*. As we can see from figure 12.16, a half of a wavelength fits within the length L of the string (i.e., $L = \lambda/2$ or $\lambda = 2L$). Thus, the fundamental mode of vibration has a wavelength of $2L$. We obtain the **fundamental frequency** f_1 from equation 12.64 by setting $n = 1$. Thus,

$$f_1 = \frac{1}{2L} \sqrt{\frac{T}{m/l}} \qquad (12.65)$$

For $n = 2$ we have what is called the *first overtone* or *second harmonic*. From figure 12.16, we see that one entire wavelength fits within one length L of the string (i.e., $L = \lambda$). We obtain the frequency of the second harmonic from equation 12.64 by letting $n = 2$. Hence,

$$f_2 = \frac{2}{2L} \sqrt{\frac{T}{m/l}} = 2f_1 \qquad (12.66)$$

Figure 12.17

Forced vibration of a string.

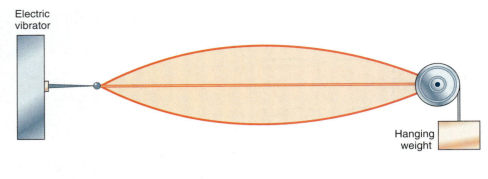

Electric
vibrator

Hanging
weight

In general, we find the frequency of any higher mode of vibration from

$$f_n = nf_1 \qquad (12.67)$$

A string that vibrates at a frequency given by equation 12.64 or 12.67 is said to be vibrating at one of its *natural frequencies.*

The possible waves for $n = 3$ and $n = 4$ are also shown in figure 12.16. Note that the nth harmonic contains n half wavelengths within the distance L. We can also observe that the location of the nodes and antinodes agrees with equations 12.56 and 12.57. Also note from equation 12.64 that the larger the tension T in the string, the higher the frequency of vibration. If we were considering a violin string, we would hear this higher frequency as a higher pitch. The smaller the tension in the string the lower the frequency or pitch. The string of any stringed instrument, such as a guitar, violin, cello, and the like, is tuned by changing the tension of the string. Also note from equation 12.64 that the larger the mass density m/l of the string, the smaller the frequency of vibration, whereas the smaller the mass density, the higher the frequency of the vibration. Thus, the mass density of each string of a stringed instrument is different in order to give a larger range of possible frequencies. Moving the finger of the left hand, which is in contact with the vibrating string, changes the point of contact of the string and thus changes the value of L, the effective length of the vibrating string. This in turn changes the possible wavelengths and frequencies that can be obtained from that string.

When we pluck a string, one or more of the natural frequencies of the string is excited. In a real string, internal frictions soon cause these vibrations to die out. However, if we apply a periodic force to the string at any one of these natural frequencies, the mode of vibration continues as long as the driving force is continued. We call this type of vibration a *forced vibration,* and we can easily set up a demonstration of forced vibration in the laboratory, as shown in figure 12.17. We connect one end of a string to an electrical vibrator of a fixed frequency and pass the other end over a pulley that hangs over the end of the table. We place weights on this end of the string to produce the tension in the string. We add weights until the string vibrates in its fundamental mode. When the tension is adjusted so that the natural frequency of the string is the same as that of the electrical vibrator, the amplitude of vibration increases rapidly. *This condition where the driving frequency is equal to the natural frequency of the system is called* **resonance.**

The tension in the string can be changed by changing the weights that are added to the end of the string, until all the harmonics are produced one at a time. The string vibrates so rapidly that the eye perceives only a blur whose shape is that of the envelope of the vibration, as shown in figure 12.18.

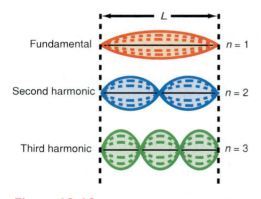

Fundamental $n = 1$

Second harmonic $n = 2$

Third harmonic $n = 3$

Figure 12.18

The envelope of the vibration.

Example 12.9

The tension in a guitar string. A guitar string 60.0 cm long has a linear mass density of 6.50×10^{-3} kg/m. If this string is to play a fundamental frequency of 220 Hz, what must the tension be in the string?

Solution

We find the tension necessary in the string by solving equation 12.64 for T, that is

$$f = \frac{n}{2L} \sqrt{\frac{T}{m/l}}$$

$$f^2 = \frac{n^2 T}{4L^2(m/l)}$$

and

$$T = \frac{4L^2 f^2 (m/l)}{n^2}$$

$$= \frac{4(0.600 \text{ m})^2 (220 \text{ Hz})^2 (6.5 \times 10^{-3} \text{ kg/m})}{1^2}$$

$$= 4.53 \times 10^2 \text{ N}$$

Example 12.10

The frequencies and wavelengths of a guitar string. Find (a) the frequencies and (b) the wavelengths of the fundamental, second, third, and fourth harmonics of example 12.9.

Solution

a. The fundamental frequency is given in example 12.9 as 220 Hz. The frequency of the next three harmonics, found from equation 12.67, are

$$f_n = n f_1$$
$$f_2 = 2f_1 = 2(220 \text{ Hz}) = 440 \text{ Hz}$$
$$f_3 = 3f_1 = 3(220 \text{ Hz}) = 660 \text{ Hz}$$
$$f_4 = 4f_1 = 4(220 \text{ Hz}) = 880 \text{ Hz}$$

b. The wavelength of the fundamental, found from equation 12.62, is

$$\lambda_n = \frac{2L}{n}$$

$$\lambda_1 = \frac{2(60.0 \text{ cm})}{1} = 120 \text{ cm}$$

The wavelengths of the harmonics are

$$\lambda_2 = \frac{2L}{2} = \frac{2(60.0 \text{ cm})}{2} = 60.0 \text{ cm}$$

$$\lambda_3 = \frac{2L}{3} = \frac{2(60.0 \text{ cm})}{3} = 40.0 \text{ cm}$$

$$\lambda_4 = \frac{2L}{4} = \frac{2(60.0 \text{ cm})}{4} = 30.0 \text{ cm}$$

Example 12.11

The displacement of the third harmonic. Find the value of the displacement for the third harmonic of example 12.10 if $x = 30.0$ cm and $t = 0$.

Solution

This displacement, found from equation 12.61, is

$$y = 2A \sin\left(\frac{n\pi x}{L}\right)\cos(\omega t)$$

$$= 2A \sin\left[\frac{3\pi(30.0 \text{ cm})}{60.0 \text{ cm}}\right]\cos[\omega(0)]$$

$$= 2A \sin\left(\frac{3\pi}{2}\right) = -2A$$

12.7 Sound Waves

A **sound wave** is a longitudinal wave, that is, a particle of the medium executes simple harmonic motion in a direction that is parallel to the velocity of propagation. A sound wave can be propagated in a solid, liquid, or a gas. The speed of sound in the medium depends on the density of the medium and on its elastic properties. We will state without proof that the speed of sound in a solid is

$$v = \sqrt{\frac{Y}{\rho}} \tag{12.68}$$

where Y is Young's modulus and ρ is the density of the medium. The speed of sound in a fluid is

$$v = \sqrt{\frac{B}{\rho}} \tag{12.69}$$

where B is the bulk modulus and ρ is the density. The speed of sound in a gas is

$$v = \sqrt{\frac{\gamma p}{\rho}} \tag{12.70}$$

where γ is a constant called the ratio of the specific heats of the gas and is equal to 1.40 for air (we discuss the specific heats of gases and their ratio in detail in chapter 17); p is the pressure of the gas; and ρ is the density of the gas. Note that the pressure and the density of a gas varies with the temperature of the gas and hence the speed of sound in a gas depends on the gas temperature. It can be shown, with the help of the ideal gas equation derived in chapter 15, that the speed of sound in air is

$$v = (331 + 0.606t) \text{ m/s} \tag{12.71}$$

where t is the temperature of the air in degrees Celsius.

Example 12.12

The speed of sound. Find the speed of sound in (a) iron, (b) water, and (c) air.

Solution

a. We find the speed of sound in iron from equation 12.68, where Y for iron is 9.1×10^{10} N/m² (from table 10.1). The density of iron is 7.8×10^3 kg/m³. Hence,

$$v = \sqrt{\frac{Y}{\rho}} \qquad (12.68)$$

$$= \sqrt{\frac{9.1 \times 10^{10} \text{ N/m}^2}{7.8 \times 10^3 \text{ kg/m}^3}}$$

$$= 3400 \text{ m/s}$$

b. We find the speed of sound in water from equation 12.69, where $B = 2.30 \times 10^9$ N/m² and $\rho = 1.00 \times 10^3$ kg/m³. Thus,

$$v = \sqrt{\frac{B}{\rho}} \qquad (12.69)$$

$$= \sqrt{\frac{2.30 \times 10^9 \text{ N/m}^2}{1.00 \times 10^3 \text{ kg/m}^3}}$$

$$= 1520 \text{ m/s}$$

c. We find the speed of sound in air from equation 12.70 with normal atmospheric pressure $p = 1.013 \times 10^5$ N/m² and $\rho = 1.29$ kg/m³. Hence,

$$v = \sqrt{\frac{\gamma p}{\rho}} \qquad (12.70)$$

$$= \sqrt{\frac{(1.40)(1.013 \times 10^5 \text{ N/m}^2)}{1.29 \text{ kg/m}^3}}$$

$$= 331 \text{ m/s}$$

Example 12.13

The speed of sound as a function of temperature. Find the speed of sound in air at a room temperature of 20.0 °C.

Solution

The speed of sound at 20.0 °C, found from equation 12.71, is

$$v = (331 + 0.606t) \text{ m/s}$$

$$= [331 + 0.606(20.0)] \text{ m/s}$$

$$= 343 \text{ m/s}$$

Example 12.14

Range of wavelengths. The human ear can detect sound only in the frequency spectrum of about 20.0 to 20,000 Hz. Find the wavelengths corresponding to these frequencies at room temperature.

(a) Fundamental

$$L = \lambda/4$$
or
$$\lambda = 4L$$

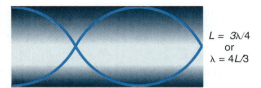

(b) Third harmonic

$$L = 3\lambda/4$$
or
$$\lambda = 4L/3$$

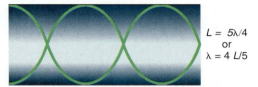

(c) Fifth harmonic

$$L = 5\lambda/4$$
or
$$\lambda = 4\,L/5$$

(d)

Figure 12.19

Standing waves in a closed organ pipe.

The corresponding wavelengths, found from equation 12.3, with $v = 343$ m/s as found in example 12.13, are

$$\lambda = \frac{v}{f} = \frac{343 \text{ m/s}}{20.0 \text{ Hz}} = 17.2 \text{ m}$$

and

$$\lambda = \frac{v}{f} = \frac{343 \text{ m/s}}{20,000 \text{ Hz}} = 0.0172 \text{ m}$$

Just as standing transverse waves can be set up on a vibrating string, standing longitudinal waves can be set up in closed and open organ pipes. Let us first consider the closed organ pipe. A traveling sound wave is sent down the closed organ pipe and is reflected at the closed end. Thus, there are two traveling waves in the pipe and they superimpose to form a standing wave in the pipe. An analysis of this standing longitudinal wave would lead to equation 12.54 for the resultant standing wave found from the superposition of a wave moving to the right and one moving to the left. The boundary conditions that must be satisfied are that the pressure wave must have a node at the closed end of the organ pipe and an antinode at the open end. The simplest wave is shown in figure 12.19(a). Although sound waves are longitudinal, the standing wave in the pipe is shown as a transverse standing wave to more easily show the nodes and antinodes. At the node, the longitudinal wave has zero amplitude, whereas at the antinode, the longitudinal wave has its maximum amplitude. It is obvious from the figure that only a quarter of a wavelength can fit in the length L of the pipe, hence the wavelength of the fundamental λ is equal to four times the length of the pipe. We must make a distinction here between an overtone and a harmonic. *An overtone is a frequency higher than the fundamental frequency. A harmonic is an overtone that is a multiple of the fundamental frequency. Hence, the nth harmonic is n times the fundamental, or first harmonic.* A harmonic is an overtone but an overtone is not necessarily a harmonic. We call the second possible standing wave in figure 12.19 the second overtone; it contains three quarter wavelengths in the distance L, whereas the third overtone has five. We can generalize the wavelength of all possible standing waves in the closed pipe to

$$\lambda_n = \frac{4L}{2n - 1} \qquad (12.72)$$

for $n = 1, 2, 3, \ldots$, where the value of $2n - 1$ always gives an odd number for any value of n. We can obtain the frequency of each standing wave from equations 12.3 and 12.72 as

$$f_n = \frac{v}{\lambda_n} = \frac{v}{4L/(2n - 1)}$$

$$f_n = \frac{(2n - 1)v}{4L} \qquad (12.73)$$

When n is equal to 1, the frequency is $f_1 = v/4L$; when n is equal to 2, the frequency is $f = 3(v/4L)$, which is three times the fundamental frequency and is thus the third harmonic. Because $n = 2$ is the first frequency above the fundamental it is called the first overtone, but it is also the third harmonic; $n = 3$ gives the second overtone, which is equal to the fifth harmonic. Thus, for an organ pipe closed at one end and open at the other, the $(n - 1)$th overtone is equal to the $(2n - 1)$th harmonic. Note that the allowable frequencies are all odd harmonics of the fundamental frequency. That is, $n = 1$ gives the fundamental frequency (zeroth overtone),

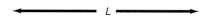

$L = \lambda/2$
or
$\lambda = 2L$

(a) Fundamental

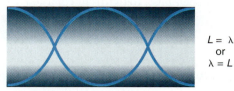

$L = \lambda$
or
$\lambda = L$

(b) Second harmonic

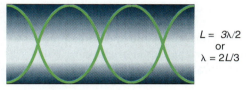

$L = 3\lambda/2$
or
$\lambda = 2L/3$

(c) Third harmonic

Figure 12.20

Standing waves in an open organ pipe.

which is the first harmonic; $n = 2$ gives the first overtone, which is the third harmonic; $n = 3$ gives the second overtone, which is the fifth harmonic; and so forth. Because the even harmonics are missing the distinction between overtones and harmonics must be made. Figure 12.19(d) shows a typical pipe organ.

Standing waves can also be set up in an open organ pipe, but now the boundary conditions necessitate an antinode at both ends of the pipe, as shown in figure 12.20. For simplicity, the longitudinal standing wave is again depicted as a transverse standing wave in the figure. The node is the position of zero amplitude and the antinodes at the ends of the open pipe are the position of maximum amplitude of the longitudinal standing wave. From inspection of the figure we can see that the wavelength of the nth harmonic is

$$\lambda_n = \frac{2L}{n} \tag{12.74}$$

and its frequency is

$$f_n = \frac{v}{\lambda_n} = \frac{v}{2L/n}$$

$$f_n = \frac{nv}{2L} \tag{12.75}$$

Equation 12.75 gives the frequency of the nth harmonic for an organ pipe open at both ends. For the open organ pipe, even and odd multiples of the fundamental frequency are possible. The $(n - 1)$th overtone is equal to the nth harmonic.

Example 12.15

The length of a closed organ pipe. Find the length of a closed organ pipe that can produce the musical note $A = 440$ Hz. Assume that the speed of sound in air is 343 m/s.

Solution

We find the length of the closed organ pipe from equation 12.73 by solving for L. Thus,

$$L = \frac{(2n - 1)v}{4f_n}$$

We obtain the fundamental note for $n = 1$:

$$L = \frac{(2n - 1)(343 \text{ m/s})}{4(440 \text{ cycles/s})}$$

$$= 0.195 \text{ m}$$

Example 12.16

Open and closed organ pipes. If the length of an organ pipe is 4.00 m, find the frequency of the fundamental for (a) a closed pipe and (b) an open pipe. Assume that the speed of sound is 343 m/s.

a. The frequency of the fundamental for the closed pipe, found from equation 12.73 with $n = 1$, is

$$f_n = \frac{(2n - 1)v}{4L}$$

$$f_1 = \frac{[2(1) - 1](343 \text{ m/s})}{4(4.00 \text{ m})}$$

$$= 21.4 \text{ Hz}$$

b. The frequency of the fundamental for an open pipe, found from equation 12.75, is

$$f_n = \frac{nv}{2L}$$

$$f_1 = \frac{(1)(343 \text{ m/s})}{2(4.00 \text{ m})}$$

$$= 42.8 \text{ Hz}$$

Note that the frequency of the fundamental of the open organ pipe is exactly twice the frequency of the fundamental in the closed organ pipe. Table 12.1 gives a summary of the harmonics for the vibrating string and the organ pipe.

Table 12.1

Summary of Some Different Harmonics for the Musical Note A, Which Has the Fundamental Frequency of 440 Hz

Vibrating String

n	Harmonic	Overtone	Frequency (Hz)
1	first	fundamental	440
2	second	first	880
3	third	second	1320

Organ Pipe Opened at One End

n	Harmonic	Overtone	Frequency (Hz)
1	first	fundamental	440
2	third	first	1320
3	fifth	second	2200

12.8 The Doppler Effect

Almost everyone has observed the change in frequency of a train whistle or a car horn as it approaches an observer and as it recedes from the observer. The change in frequency of the sound due to the motion of the sound source is an example of the **Doppler effect.** In general, this change in frequency of the sound wave can be caused by the motion of the sound source, the motion of the observer, or both. Let us consider the different possibilities.

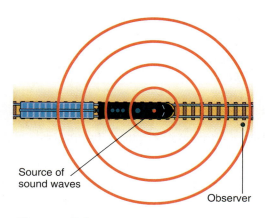

Figure 12.21

Observer and source stationary.

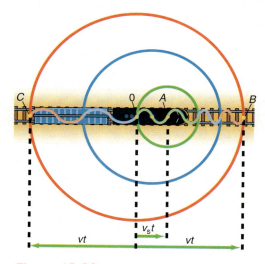

Figure 12.22

Doppler effect with the source moving and the observer stationary.

Case 1: *The Observer and the Sound Source Are Stationary*

This case is the normal case where there is no relative motion between the source and the observer and is shown in figure 12.21. When the source emits a sound of frequency f_s, the sound waves propagate out from the source in a series of concentric circles. The distance between each circle is the wavelength of the sound. The sound propagates at a speed v, and the frequency heard by the observer f_o is simply

$$f_o = \frac{v}{\lambda} = f_s \tag{12.76}$$

That is, the stationary observer hears the same frequency as the one emitted from the stationary source.

Case 2: *The Observer Is Stationary But the Source Is Moving*

When the source of sound moves with a velocity, v_s, to the right, the emitted waves are no longer concentric circles but rather appear as in figure 12.22. Each wave is symmetrical about the point of emission, but since the point of emission is moving to the right, the circular wave associated with each emission is also moving to the right. Hence the waves bunch up in advance of the moving source and spread out behind the source. The frequency that an observer hears is just the speed of propagation of the wave divided by its wavelength, that is,

$$f = \frac{v}{\lambda} \tag{12.77}$$

We can use equation 12.77 to describe qualitatively what the observer hears. As the wave approaches the observer, the waves bunch up and hence the effective wavelength λ appears smaller in the front of the wave. From equation 12.77 we can see that if λ decreases, the frequency f must increase. *Thus, when a moving source approaches a stationary observer the observed frequency is higher than the emitted frequency of the source.* When the source reaches the observer, the observer hears the frequency emitted. As the source passes and recedes from the observer the effective wavelength λ appears longer. Hence, from equation 12.77, if λ increases, the frequency f, heard by the observer, is lower than the frequency emitted by the source. *Thus, when a moving source recedes from a stationary observer the observed frequency is lower than the emitted frequency of the source.* To get a quantitative description of the observed frequency we proceed as follows.

a) *Moving Source Approaches a Stationary Observer*

The effective wavelength measured by the stationary observer in front of the moving source is simply the total distance *AB*, in figure 12.22, divided by the total number of waves in that distance, that is,

$$\lambda_f = \frac{\text{distance } AB}{\# \text{ of waves}} \tag{12.78}$$

In a time t, the moving source has moved a distance *0A*, in figure 12.22, which is given by the speed of the source v_s times the time t. The distance *0B* is given by the speed of the wave v times the time t. Hence, the distance *AB* in figure 12.22 is

$$\text{distance } AB = vt - v_st \tag{12.79}$$

whereas the number of waves between *A* and *B* is just the number of waves emitted per unit time, f_s, the frequency of the source, times the time t. Thus,

$$\# \text{ of waves in } AB = \left(\frac{(\# \text{ of waves emitted})t}{\text{time}} \right) = f_st \tag{12.80}$$

Substituting equations 12.79 and 12.80 into equation 12.78, the effective wavelength in front of the source is

$$\lambda_f = \frac{vt - v_st}{f_st}$$

$$\lambda_f = \frac{v - v_s}{f_s} \tag{12.81}$$

However, from equation 12.77, the observed frequency in front of the approaching source f_{of} is

$$f_{of} = \frac{v}{\lambda_f} = \frac{v}{(v - v_s)/f_s}$$

or

$$f_{of} = \frac{v}{v - v_s}f_s \tag{12.82}$$

Equation 12.82 gives the frequency that is observed by a stationary observer who is in front of an approaching source.

b) A Moving Source Recedes from a Stationary Observer

The effective wavelength λ_b heard by the stationary observer behind the moving source is equal to the total distance CA divided by the number of waves between C and A, that is,

$$\lambda_b = \frac{\text{distance } CA}{\# \text{ of waves}} \tag{12.83}$$

However, from figure 12.22, we see that the distance CA is

$$\text{distance } CA = vt + v_st \tag{12.84}$$

whereas the number of waves between C and A is the number of waves emitted per unit time, times the time t, that is,

$$\# \text{ of waves in } CA = \left(\frac{\# \text{ of waves emitted}}{\text{time}}\right)t = f_st \tag{12.85}$$

Substituting equations 12.84 and 12.85 into equation 12.83 yields the effective wavelength behind the receding source

$$\lambda_b = \frac{vt + v_st}{f_st}$$

or

$$\lambda_b = \frac{v + v_s}{f_s} \tag{12.86}$$

Substituting this effective wavelength into equation 12.77, we obtain

$$f_{ob} = \frac{v}{\lambda_b} = \frac{v}{(v + v_s)/f_s}$$

or

$$f_{ob} = \frac{v}{v + v_s}f_s \tag{12.87}$$

Equation 12.87 gives the frequency observed by a stationary observer who is behind the receding source.

Example 12.17

Doppler effect—moving source. A train moving at 25.00 m/s emits a whistle of frequency 200.0 Hz. If the speed of sound in air is 343.0 m/s, find the frequency observed by a stationary observer (a) in advance of the moving source and (b) behind the moving source.

Solution

a. The observed frequency in advance of the approaching source, found from equation 12.82, is

$$f_{of} = \frac{v}{v - v_s} f_s$$

$$= \left(\frac{343.0 \text{ m/s}}{343.0 \text{ m/s} - 25.00 \text{ m/s}} \right)(200.0 \text{ Hz})$$

$$= 215.7 \text{ Hz}$$

Note that the observed frequency in front of the approaching source is higher than the frequency emitted by the source.

b. The observed frequency behind the receding source, found from equation 12.87, is

$$f_{ob} = \frac{v}{v + v_s} f_s$$

$$= \left(\frac{343.0 \text{ m/s}}{343.0 \text{ m/s} + 25.00 \text{ m/s}} \right)(200.0 \text{ Hz})$$

$$= 186.4 \text{ Hz}$$

Note that the observed frequency behind the receding source is lower than the frequency emitted by the source. Also note that the change in the frequency is not symmetrical. That is, the change in frequency when the train is approaching is equal to 15.7 Hz, whereas the change in frequency when the train is receding is equal to 13.6 Hz.

Case 3: *The Source Is Stationary But the Observer Is Moving*

For a stationary source the sound waves are emitted as concentric circles, as shown in figure 12.23.

Figure 12.23

Doppler effect, the source is stationary but the observer is moving.

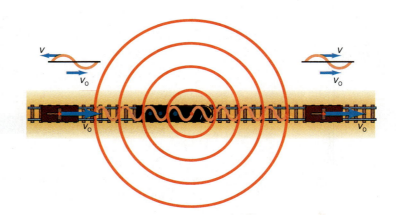

a) The Observer Is Moving toward the Source at a Velocity v_o

When the observer approaches the stationary source at a velocity v_o, the relative velocity between the observer and the wave is

$$v_{rel} = v_o + v$$

This relative velocity has the effect of having the emitted wave pass the observer at a greater velocity than emitted. The observed frequency heard while approaching the source is

$$f_{oA} = \frac{v_{rel}}{\lambda} = \frac{v_o + v}{\lambda}$$

But the wavelength emitted by the source does not change, and is simply

$$\lambda = \frac{v}{f_s} \tag{12.88}$$

Hence,

$$f_{oA} = \frac{v_o + v}{v/f_s}$$

Hence, *the frequency observed by the moving observer as it approaches the stationary source is*

$$f_{oA} = \frac{v_o + v}{v} f_s \tag{12.89}$$

b) The Observer Is Moving Away from the Source at a Velocity v_o

When the observer moves away from the source, the relative velocity between the wave and the observer is

$$v_{rel} = v - v_o$$

This reduced relative velocity has the effect of having the sound waves move past the receding observer at a slower rate. Thus, the observed frequency of the receding observer f_{oR} is

$$f_{oR} = \frac{v_{rel}}{\lambda} = \frac{v - v_o}{\lambda}$$

The wavelength λ of the emitted sound is still given by equation 12.88, and the observed frequency becomes

$$f_{oR} = \frac{v - v_o}{v/f_s}$$

Thus, *the frequency observed by an observer who is receding from a stationary source is*

$$f_{oR} = \frac{v - v_o}{v} f_s \tag{12.90}$$

Example 12.18

Doppler effect—moving observer. A stationary source emits a whistle at a frequency of 200.0 Hz. If the velocity of propagation of the sound wave is 343.0 m/s, find the observed frequency if (a) the observer is approaching the source at 25.00 m/s and (b) the observer is receding from the source at 25.00 m/s.

a. The frequency observed by the approaching observer, found from equation 12.89, is

$$f_{oA} = \frac{v_o + v}{v} f_s$$

$$= \left(\frac{25.00 \text{ m/s} + 343.0 \text{ m/s}}{343.0 \text{ m/s}} \right) (200.0 \text{ Hz})$$

$$= 214.6 \text{ Hz}$$

Note that the frequency observed by the approaching observer is greater than the emitted frequency of 200.0 Hz. However, observe that it is not the same numerical value found when the source was moving (215.7 Hz). The reason for the difference in the observed frequency is that the physical problems are not the same.

b. The frequency observed by the receding observer, found from equation 12.90, is

$$f_{oR} = \frac{v - v_o}{v} f_s$$

$$= \left(\frac{343.0 \text{ m/s} - 25.00 \text{ m/s}}{343.0 \text{ m/s}} \right) (200.0 \text{ Hz})$$

$$= 185.4 \text{ Hz}$$

Note that the frequency observed by the receding observer is less than the frequency emitted by the source. Also note that the frequency observed by the receding observer for a stationary source, $f_{oR} = 185.4$ Hz, is not the same frequency as observed by a stationary observer behind the receding source $f_{ob} = 186.4$ Hz. Finally, notice that when the source is stationary, the change in the frequency of approach, 14.6 Hz, is equal to the change in the frequency of recession, 14.6 Hz.

Case 4: *Both the Source and the Observer Are Moving*

If both the source and the observer are moving we can combine equations 12.82, 12.87, 12.89, and 12.90 into the one single equation

$$f_o = \frac{v \pm v_o}{v \mp v_s} f_s \qquad (12.91)$$

with the convention that

$+ v_o$ corresponds to the observer approaching
$- v_o$ corresponds to the observer receding
$- v_s$ corresponds to the source approaching
$+ v_s$ corresponds to the source receding

Example 12.19

Doppler effect—both source and observer move. A sound source emits a frequency of 200.0 Hz at a velocity of 343.0 m/s. If both the source and the observer move at a velocity of 12.50 m/s, find the observed frequency if (a) the source and the observer are moving toward each other and (b) the source and the observer are moving away from each other.

a. If the source and observer are approaching each other, the observed frequency, found from equation 12.91 with v_o positive and v_s negative, is

$$f_o = \frac{v + v_o}{v - v_s} f_s$$

$$= \left(\frac{343.0 \text{ m/s} + 12.50 \text{ m/s}}{343.0 \text{ m/s} - 12.50 \text{ m/s}} \right) (200.0 \text{ Hz})$$

$$= 215.1 \text{ Hz}$$

Note that the frequency observed is higher than the frequency emitted, and although the relative motion between the source and the observer is still 25.00 m/s, the observed frequency is different from that found in both examples 12.17 and 12.18 (215.7 Hz and 214.6 Hz).

b. If the source and observer are moving away from each other, then the observed frequency, found from equation 12.91 with v_o negative and v_s positive, is

$$f_o = \frac{v - v_o}{v + v_s} f_s$$

$$= \left(\frac{343.0 \text{ m/s} - 12.50 \text{ m/s}}{343.0 \text{ m/s} + 12.50 \text{ m/s}} \right) (200.0 \text{ Hz})$$

$$= 185.9 \text{ Hz}$$

Note that the observed frequency is lower than the emitted frequency, and although the relative velocity between observer and source is still 25.00 m/s, the observed frequency is different from that found in examples 12.17 and 12.18 (186.4 Hz and 185.4 Hz).

12.9 The Transmission of Energy in a Wave and the Intensity of a Wave

We have defined a wave as the propagation of a disturbance through a medium. The disturbance causes the particles of the medium to be set into motion. As we have seen, if the wave is a transverse wave, the particles oscillate in a direction perpendicular to the direction of the wave propagation. The oscillating particles possess energy, and this energy is passed from particle to particle of the medium. Thus, the wave transmits energy as it propagates. Let us now determine the amount of energy transmitted by a wave.

Let us consider a transverse wave on a string, whose particles are executing simple harmonic motion. If there is no energy loss due to friction, the total energy transmitted by the wave is equal to the total energy of the vibrating particle, that is,

$$E_{\text{transmitted}} = (E_{\text{tot}})_{\text{particle}}$$

The total energy possessed by a single particle in simple harmonic motion, given by a variation of equation 11.25, is

$$E_{\text{tot}} = \tfrac{1}{2} k R^2 \qquad \textbf{(12.92)}$$

where the letter R is now used for the amplitude of the vibration and hence the wave. Recall that k, in this equation, is the spring constant that was shown in chapter 11 to be related to the angular frequency by

$$\omega^2 = \frac{k}{m}$$

where m was the mass of the particle in motion. Solving for the spring constant k, we get

$$k = \omega^2 m \qquad (12.93)$$

Also recall that the angular frequency was related to the frequency of vibration by equation 12.12 as

$$\omega = 2\pi f$$

Substituting equation 12.12 into equation 12.93 yields

$$k = (2\pi f)^2 m \qquad (12.94)$$

Substituting equation 12.94 into equation 12.92 gives

$$E_{tot} = \tfrac{1}{2}(2\pi f)^2 m R^2$$

The energy transmitted by the wave is therefore

$$E_{transmitted} = 2\pi^2 m f^2 R^2 \qquad (12.95)$$

Notice that the energy transmitted by the wave is directly proportional to the square of the frequency of the wave and directly proportional to the square of the amplitude of the wave.

Example 12.20

Energy transmitted by a wave. The frequency and amplitude of a transverse wave on a string are doubled. What effect does this have on the amount of energy transmitted?

Solution

The energy of the original wave, given by equation 12.95, is

$$E_o = 2\pi^2 m f_o^2 R_o^2$$

The frequency of the new wave is $f = 2f_o$, and the amplitude of the new wave is $R = 2R_o$. The energy transmitted by the new wave is

$$E = 2\pi^2 m f^2 R^2 = 2\pi^2 m (2f_o)^2 (2R_o)^2$$
$$= 16(2\pi^2 m f_o^2 R_o^2) = 16E_o$$

We derived equation 12.95 for the transmission of energy by a transverse wave on a string. It is, however, completely general and can be used for any mechanical wave.

It is sometimes more convenient to describe the wave in terms of its intensity. *The **intensity of a wave** is defined as the energy of the wave that passes a unit area in a unit time.* That is, we define the intensity mathematically as

$$I = \frac{E}{At} \qquad (12.96)$$

The unit of intensity is the watt per square meter, W/m^2. Substituting the energy of a wave from equation 12.95 into equation 12.96 gives

$$I = \frac{2\pi^2 m f^2 R^2}{At} \qquad (12.97)$$

Because the density of a medium is defined as the mass per unit volume, the mass of the particle in simple harmonic motion can be replaced by

$$m = \rho V$$

And the volume of the medium can be expressed as the cross-sectional area A of the medium that the wave is moving through times a distance l the wave moves through (i.e., $V = Al$). The mass m then becomes

$$m = \rho A l \tag{12.98}$$

Substituting equation 12.98 into equation 12.97 yields

$$I = \frac{2\pi^2 \rho A l f^2 R^2}{At}$$

Notice that the cross-sectional area term in both numerator and denominator cancel, and l/t is the velocity of the wave v. Hence,

$$I = 2\pi^2 \rho v f^2 R^2 \tag{12.99}$$

Equation 12.99 gives the intensity of a mechanical wave of frequency f and amplitude R, moving at a velocity v in a medium of density ρ.

Example 12.21

The intensity of a sound wave. A trumpet player plays the note A at a frequency of 440 Hz, with an amplitude of 7.80×10^{-3} mm. If the density of air is 1.29 kg/m³ and the speed of sound is 331 m/s, find the intensity of the sound wave.

Solution

The intensity of the sound wave, found from equation 12.99, is

$$
\begin{aligned}
I &= 2\pi^2 \rho v f^2 R^2 \\
&= 2\pi^2(1.29 \text{ kg/m}^3)(331 \text{ m/s})(440 \text{ Hz})^2(7.80 \times 10^{-6} \text{ m})^2 \\
&= 9.93 \times 10^{-2} \text{ W/m}^2
\end{aligned}
$$

"Have you ever wondered . . . ?"

An Essay on the Application of Physics

The Production and Reception of Human Sound

Humans use sound waves to communicate with each other. Sound is produced in the larynx, sometimes called the voice box, which is a cartilaginous organ of the throat that contains the vocal cords, figure 1(a). It is the vocal cords that are responsible for producing human sound. The cords are horizontal folds in the mucous membrane lining of the larynx, figure 1(b).

The vocal cords contain elastic fibers. As air is exhaled from the lungs, it passes over these elastic fibers and sets them into vibration. The cords can be visualized as the vibrating strings studied in this chapter. The frequency of the produced sound can be varied by changing the tension in the vocal cords similar to the vibrating string. The greater the tension on the cords, the higher the frequency, or pitch, of the emitted sound. The lower the tension on the cords, the lower the pitch.

When you hum, you set up a standing wave of a particular frequency on your vocal cords. The exhaled air that passes over these cords picks up the vibration of the cords. As the air is expelled from your mouth, it is observed as a longitudinal wave at the frequency of the vibrating vocal cords. When you speak, the expelled air and vibrating vocal cords initiate the sound, but your mouth, lips, and tongue modify it to produce the vowels and consonants that make up the words of speech.

An interesting observation in the production of sound can be demonstrated by humming with your mouth closed. If you now pinch your nose closed, the hufnming will stop because the air will no longer flow over the vocal cords. If you are fortunate enough to have survived a case of choking on either food or drink, you will recall that when the choking begins you usually panic and try to yell to anyone to tell them that you are choking. Unfortunately, as you try to speak you find that you are unable to do so. Since the windpipe has been closed, no air can pass over the vocal cords to initiate the vibration that starts the speaking process. Many people die from choking simply because they are unable to communicate their condition to someone who can help. The usual procedure to communicate your choking condition is to get someone's attention. Then, point to your throat and cross your throat with your finger as though you were cutting your throat. If the other person is aware of the sign and realizes that

Figure 1

The vocal cords.

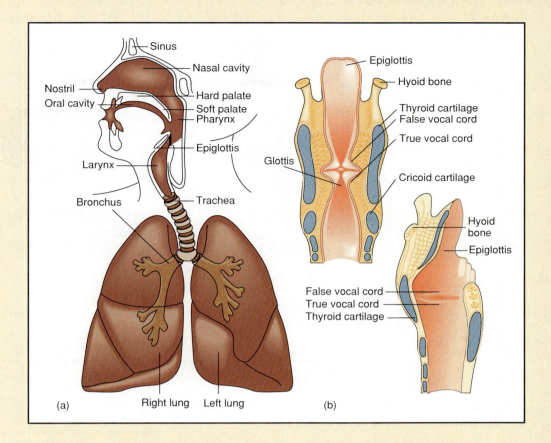

Sinus
Nasal cavity
Nostril
Oral cavity
Hard palate
Soft palate
Pharynx
Epiglottis
Larynx
Bronchus
Trachea
(a)
Right lung   Left lung

Epiglottis
Hyoid bone
Thyroid cartilage
False vocal cord
True vocal cord
Glottis
Cricoid cartilage
Hyoid bone
Epiglottis
False vocal cord
True vocal cord
Thyroid cartilage
(b)

you are choking, he or she can save you by initiating the Heimlich maneuver. This consists of holding you from behind and wrapping his or her arms around you. Then the person presses against your diaphragm with his or her arms. By pressing in and upward, a force is exerted on your lungs that tends to compress the lungs. This in turn increases the pressure of the air in your lungs until it is great enough to force the closed valve open, thereby expelling the food that was causing you to choke. This then permits you to breathe normally and you observe that you now have your voice back to communicate with anyone.

Your ears are used to detect sound. The human ear can be divided into three parts: the outer ear, the middle ear, and the inner ear, figure 2. The outer ear acts as a funnel to channel the sound wave to the ear drum. These sound vibrations set the ear drum into vibration. These vibrations are then passed through the middle ear by three bones called the malleus (hammer), incus (anvil), and stapes (stirrup). These bones effectively amplify the amplitude of the vibration and then pass it on to the inner ear. The inner ear is a system of cavities. One of these cavities is the cochlea, a bony labyrinth in the shape of a spiral. The cochlea contains a fluid, through which the amplified vibration is passed to the auditory nerve on its way to the brain. The brain then interprets this sound as either speech, music, noise, and so forth.

The loudness of a sound as heard by the human ear is not directly proportional to the intensity of the sound, but rather is proportional to the logarithm of the intensity. The human ear can hear sounds of intensities as low as $I_0 = 1.00 \times 10^{-12}$ W/m², which is called the threshold of hearing, to higher than 1.00 W/m², which is called the threshold of pain. Because of the enormous variation in intensity that the human ear can hear, a logarithmic scale is usually used to measure sound. *The intensity level β of a sound wave, measured in decibels (dB), is defined as*

$$\beta = 10 \log\left(\frac{I}{I_0}\right) \tag{12H.1}$$

where I_0 is the reference level, taken to be the threshold of hearing. The decibel is 1/10 of a bel, which was named to honor Alexander Graham Bell.

Example 12H.1

The intensity of sound in decibels. Find the intensity level of sound waves that are the following multiples of the threshold of hearing: (a) $I = I_0$, (b) $I = 2I_0$, (c) $I = 5I_0$, (d) $I = 10I_0$, and (e) $I = 100I_0$.

Figure 2

The human ear.

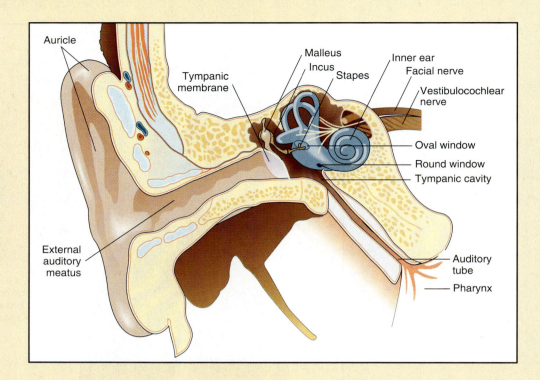

Auricle

Malleus
Incus
Stapes

Inner ear
Facial nerve

Tympanic
membrane

Vestibulocochlear
nerve

Oval window
Round window
Tympanic cavity

Auditory
tube

Pharynx

External
auditory
meatus

Solution

a.   The intensity level, found from equation 12H.1, is

$$\beta = 10 \log\left(\frac{I}{I_0}\right) \qquad \textbf{(12H.1)}$$

$$= 10 \log\left(\frac{I_0}{I_0}\right) = 10 \log 1 = 0 \text{ dB}$$

Because the log of 1 is equal to zero the intensity level at the threshold of hearing is 0 dB.

b.   For $I = 2I_0$, the intensity level is

$$\beta = 10 \log\left(\frac{2I_0}{I_0}\right) = 10 \log 2 = 3.01 \text{ dB}$$

c.   For $I = 5I_0$ the intensity level is

$$\beta = 10 \log\left(\frac{5I_0}{I_0}\right) = 10 \log 5 = 6.99 \text{ dB}$$

d.   For $I = 10I_0$ the intensity level is

$$\beta = 10 \log\left(\frac{10I_0}{I_0}\right) = 10 \log 10 = 10.00 \text{ dB}$$

e.   For $I = 100I_0$ the intensity level is

$$\beta = 10 \log\left(\frac{100I_0}{I_0}\right) = 10 \log 100 = 20.00 \text{ dB}$$

Thus, doubling the intensity level of a sound from 10 to 20 dB, a factor of 2, actually corresponds to an intensity increase from $10I_0$ to $100I_0$, or by a factor of 10. Similarly, an increase in the intensity level from 10 to 30 dB, a factor of 3, would correspond to an increase in the intensity from $10I_0$ to $1000I_0$, or a factor of 100.

Example 12H.2

Heavy traffic noise. A busy street with heavy traffic has an intensity level of 70 dB. Find the intensity of the sound.

Solution

We find the intensity by solving equation 12H.1 for I. Hence,

$$\frac{\beta}{10} = \log\left(\frac{I}{I_0}\right)$$

But the definition of the common logarithm is

$$\text{if } y = \log x \quad \text{then } x = 10^y$$

For our case this becomes

$$\text{if } \frac{\beta}{10} = \log\left(\frac{I}{I_0}\right) \quad \text{then } \frac{I}{I_0} = 10^{\beta/10}$$

Hence,

$$I = I_0 10^{\beta/10}$$

And

$$I = I_0 10^{70/10} = (1.00 \times 10^{-12} \text{ W/m}^2)(10^7)$$
$$= 1.00 \times 10^{-5} \text{ W/m}^2$$

Figure 3

Graph of intensity level of sound versus the frequency of the sound for a human ear.

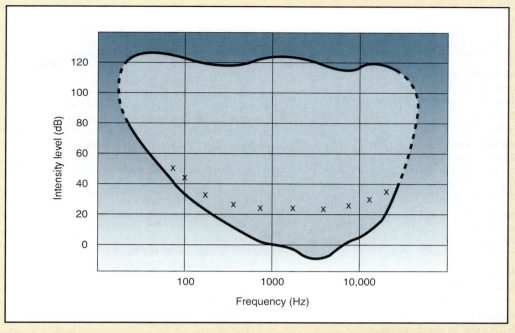

The human ear is responsive to a large range of frequencies and intensities. A typical response curve is shown in figure 3. The intensity level of sound is plotted against the frequency of the sound. The continuous curved line at the bottom represents the response curve of a normal ear. The lowest region on the curve occurs from about 1000 Hz to about 4000 Hz. These frequencies can be heard by the normal ear at very low intensity levels. On the other hand, to hear a frequency of 100 Hz the intensity level would have to be increased to about 35 dB. And for a normal ear to hear a frequency of about 20,000 Hz the intensity level would have to be increased to about 40 dB. At an intensity of 20 dB a frequency of 1000 Hz can easily be heard, but a frequency of 100 Hz could not be heard at all.

With age, the frequencies that the human ear can hear decreases. Many people resort to a hearing aid to overcome this hearing deficiency. A test is made of the person's ability to hear a sound of a known intensity level and frequency. The person is placed in a soundproof booth and earphones are placed over his or her ears. The examiner then plays pure sounds at a known frequency. He or she starts at a low intensity level and increases the intensity in small steps until the individual hears that particular frequency. When the individual hears the sound, he or she presses a button to let the examiner know that he or she has heard the sound. The examiner then marks an x on the graph of the frequency and intensity level of the normal ear shown in figure

3. The x's represent the actual frequencies heard at a particular intensity level. By knowing the frequencies that the person can no longer hear very well, a hearing aid, which is essentially a miniature electronic amplifier, is designed to amplify those frequencies, and thus bring the sound of that frequency up to a normal intensity level for that individual. For example the x's in figure 3 indicate that the individual's hearing response has deteriorated. In particular, the hearing response in the midrange frequency is much worse than at the low end or the high end of the spectrum. (The x's in the midrange are farther away from the normal curve.) Thus a hearing aid that amplifies the frequencies in the middle range of the audio spectrum would be useful for the individual. We would certainly not want to amplify the entire audio spectrum, for then we would be amplifying some of the frequencies that the person already hears reasonably well.

It is interesting to note that not only humans use sounds to communicate but animals do also. Some animals communicate at a higher frequency than can be heard by humans. These sounds are called ultrasonic and occur at frequencies above 20,000 Hz. Birds and dogs can hear these ultrasounds and bats even use them for navigation in a kind of sound radar. Ultrasound is used in sonar systems to detect submarines. It is also used in a variety of medical applications, including diagnosis and treatment. For example, chiropractors and physical therapists routinely use ultrasound for relief of lower back pain.

The Language of Physics

Wave
A wave is a propagation of a disturbance through a medium (p. 321).

Longitudinal wave
A wave in which the particles of the medium oscillate in simple harmonic motion parallel to the direction of the wave propagation (p. 321).

Transverse wave
A wave in which the particles of the medium execute simple harmonic motion in a direction perpendicular to its direction of propagation (p. 321).

Displacement
The distance that a particle of the medium is displaced from its equilibrium position as the wave passes by (p. 322).

Amplitude
The maximum value of the displacement (p. 322).

Wavelength
The distance, in the direction of propagation, in which the wave repeats itself (p. 323).

Period
The time it takes for one complete wave to pass a particular point. Hence, it is the time for a wave to repeat itself (p. 323).

Frequency
The number of waves passing a particular point per second (p. 323).

Reflection of a wave at a boundary
If a wave on a string traveling to the right is reflected from a nonfixed end, the reflected wave moves to the left with the same size and shape as the incident wave. If a wave on a string is traveling to the right and is reflected from a fixed end, the reflected wave is the same size and shape but is now inverted (p. 331).

Reflection and transmission of a wave at the boundary of two different media
(1) *Boundary between a less dense medium and a more dense medium.* The boundary acts as a fixed end and the reflected wave is inverted. The transmitted wave slows down on entering the more dense medium and the wavelength of the transmitted wave is less than the wavelength of the incident wave (p. 333).
(2) *Boundary between a more dense medium and a less dense medium.* The boundary acts as a nonfixed end and the reflected wave is not inverted, but is rather right side up. The transmitted wave speeds up on entering the less dense medium and the wavelength of the transmitted wave is greater than the wavelength of the incident wave (p. 335).

Principle of superposition
Whenever two or more wave disturbances pass a particular point in a medium, the resultant displacement of the point of the medium is the sum of the displacements of each individual wave disturbance (p. 336).

Phase angle
The measure of how far one wave is displaced in the direction of propagation from another wave (p. 337).

Constructive interference
When two interfering waves are in phase with each other (phase angle = 0) the amplitude of the combined wave is a maximum (p. 338).

Destructive interference
When two interfering waves are 180° out of phase with each other the amplitude of the combined wave is zero (p. 338).

Node
The point where the amplitude of a standing wave is zero (p. 341).

Antinode
The point where the amplitude of a standing wave is a maximum (p. 341).

Standing wave on a string
For a string fixed at both ends, a wave train is sent down the string. The wave is reflected from the fixed ends. Hence, there are two wave trains on the string, one traveling to the right and one traveling to the left. The resultant wave is the superposition of the two traveling waves. It is called a standing wave or a stationary wave because the resultant standing wave does not travel at all. The node of the standing wave remains a node for all times.

Thus, the string can not move up and down at that point, and can not transmit any energy past that point. Because the string is fixed at both ends, only certain wavelengths and frequencies are possible. When the string vibrates at these specified wavelengths, the string is said to be vibrating at one of its normal modes of vibration, and the string is vibrating at one of its natural frequencies (p. 342).

Fundamental frequency
The lowest of the natural frequencies of a vibrating system (p. 343).

Resonance
When a force is applied, whose frequency is equal to the natural frequency of the system, the system vibrates at maximum amplitude (p. 344).

Sound wave
A sound wave is a longitudinal wave that can be propagated in a solid, liquid, or gas (p. 346).

Overtone
An overtone is a frequency higher than the fundamental frequency (p. 348).

Harmonic
A harmonic is an overtone that is a multiple of the fundamental frequency. Hence, the nth harmonic is n times the fundamental frequency, or first harmonic (p. 348).

Doppler effect
The change in the wavelength and hence the frequency of a sound caused by the relative motion between the source and the observer. When a moving source approaches a stationary observer the observed frequency is higher than the emitted frequency of the source. When a moving source recedes from a stationary observer, the observed frequency is lower than the emitted frequency of the source (p. 350).

Intensity of a wave
The energy of a wave that passes a unit area in a unit time (p. 357).

Summary of Important Equations

Frequency of a wave
$$f = \frac{1}{T} \tag{12.1}$$

Fundamental equation of wave propagation
$$v = \lambda f \tag{12.3}$$

Wave number
$$k = \frac{2\pi}{\lambda} \tag{12.9}$$

Equation of a wave traveling to the right
$$y = A\sin(kx - \omega t) \tag{12.13}$$

Equation of a wave traveling to the left
$$y = A\sin(kx + \omega t) \tag{12.26}$$

Angular frequency
$$\omega = 2\pi f \tag{12.12}$$

Angular frequency
$$\omega = kv \tag{12.14}$$

Velocity of transverse wave on a string
$$v = \sqrt{\frac{T}{m/l}} \tag{12.30}$$

Change in wavelength in second medium
$$\lambda_2 = \frac{v_2}{v_1}\lambda_1 \tag{12.34}$$

Principle of superposition
$$y = y_1 + y_2 + y_3 + \dots \tag{12.35}$$

Equation of wave displaced to the right by phase angle ϕ
$$y = A\sin(kx - \omega t - \phi) \tag{12.38}$$

Interference of two waves out of phase by angle ϕ
$$y = 2A\cos\left(\frac{\phi}{2}\right)\sin\left(kx - \omega t - \frac{\phi}{2}\right) \tag{12.50}$$

The equation of the displacement of a standing wave on a string
$$y = 2A\sin\left(\frac{n\pi x}{L}\right)\cos(\omega t) \tag{12.61}$$

Location of nodes of standing wave
$$x = \frac{n\lambda}{2} \tag{12.56}$$

Location of antinodes
$$x = (2n - 1)\frac{\lambda}{4} \tag{12.57}$$

Possible wavelengths on vibrating string
$$\lambda = \frac{2L}{n} \tag{12.62}$$

Frequency of vibrating string
$$f = \frac{n}{2L}\sqrt{\frac{T}{m/l}} \tag{12.64}$$

Frequency of higher modes of vibration
$$f_n = nf_1 \tag{12.67}$$

Speed of sound in a solid
$$v = \sqrt{\frac{Y}{\rho}} \tag{12.68}$$

Speed of sound in a fluid
$$v = \sqrt{\frac{B}{\rho}} \tag{12.69}$$

Speed of sound in a gas
$$v = \sqrt{\frac{\gamma p}{\rho}} \tag{12.70}$$

Doppler frequency shift
$$f_o = \frac{v \pm v_o}{v \mp v_s}f_s \tag{12.91}$$

Energy transmitted by wave
$$E_{\text{transmitted}} = 2\pi^2 mf^2 R^2 \tag{12.95}$$

Intensity of a wave
$$I = 2\pi^2 \rho v f^2 R^2 \tag{12.99}$$

Intensity of a sound wave in decibels
$$\beta = 10\log\left(\frac{I}{I_0}\right) \tag{12H.1}$$

Questions for Chapter 12

1. Discuss the relation between simple harmonic motion and wave motion. Is it possible to create waves in a medium where the particles do not execute simple harmonic motion?
2. State the differences between transverse waves and longitudinal waves.
3. Describe how sound is made and heard by a human.
4. Discuss the statement "When a person is young enough to hear all the frequencies of a good stereo system, he can not afford to buy it. And when he can afford to buy it, he can not hear all the frequencies."
5. Discuss the statement that a wave is periodic in both space and time.

6. Why are there four different strings on a violin? Describe what a violin player does when she "tunes" the violin.
7. Discuss what happens to a pulse that is reflected from a fixed end and a free end.
†8. A wave is reflected from, and transmitted through, a more dense medium. What criteria would you use to estimate how much energy is reflected and how much is transmitted?
9. If the wavelength of a wave decreases as it enters a medium, what does this tell you about the medium?
10. When does the superposition principle fail in the analysis of combined wave motions?
11. Discuss what is meant by a standing wave and give some examples.

12. Discuss the difference between overtones and harmonics for a vibrating string, an open organ pipe, and a closed organ pipe.
†13. Discuss the possible uses of ultrasound in medicine.
†14. Discuss the Doppler effect on sound waves. Could the Doppler effect be applied to light waves? What would be the medium for the propagation?
†15. How could the Doppler effect be used to determine if the universe is expanding or contracting?
16. If two sounds of very nearly the same frequency are played together, the two waves will interfere with each other. The slight difference in frequency will cause an alternate rising and lowering of the intensity of the combined sound. This phenomenon is called beats. How can this technique be used to tune a piano?

Problems for Chapter 12

12.1 Introduction

1. Find the period of a sound wave of (a) 20.0 Hz and (b) 20,000 Hz.
2. A sound wave has a wavelength of 2.25×10^{-2} m and a frequency of 15,000 Hz. Find its speed.
3. Find the wavelength of a sound wave of 60.0 Hz at 20.0 °C.

12.2 Mathematical Representation of a Wave

4. At a time $t = 0$, a certain wave is given by $y = 10 \sin 5x$. Find the (a) amplitude of the wave and (b) its wavelength.
5. You want to generate a wave that has a wavelength of 20.0 cm and moves with a speed of 80.0 m/s. Find (a) the frequency of such a wave, (b) its wave number, and (c) its angular frequency.
6. A particular wave is given by $y = (8.50$ m$)\sin[(0.800$ m$^{-1})x - (5.40$ rad/s$)t]$. Find (a) the amplitude of the wave, (b) the wave number, (c) the wavelength, (d) the angular frequency, (e) the frequency, (f) the period, (g) the velocity of the wave, and (h) the displacement of the wave at $x = 5.87$ m and $t = 2.59$ s.
7. A certain wave has a wavelength of 25.0 cm, a frequency of 230 Hz, and an amplitude of 1.85 cm. Find (a) the wave number k and (b) the angular frequency ω. (c) Write the equation for this wave in the standard form $y = A \sin(kx - \omega t)$.

12.3 The Speed of a Transverse Wave on a String

8. A 60.0-cm guitar string has a mass of 1.40 g. If it is to play the note A at a frequency of 440 Hz, what must the tension be in the string? Assume that the wavelength of the note is twice the length of the string.
9. A 1.50-m length of wire with a mass of 0.035 kg is stretched between two points. Find the necessary tension in the wire such that the wave may travel from one end to another in a time of 0.0900 s.
10. A guitar string that has a mass per unit length of 2.33×10^{-3} kg/m is tightened to a tension of 655 N. What frequency will be heard if the string is 60.0 cm long? Is this a standard note or is it sharp or flat? (Remember that the wavelength of the note played is twice the length of the string.)

12.4 Reflection of a Wave at a Boundary

11. One end of a 100-cm wire of 3.45 g is welded to a 90.0-cm wire of 9.43 g. (a) If a wave moves along the first wire at a speed of 528 m/s, find its speed along the second wire. (b) If the wavelength on the first wire is 1.76 cm, find the wavelength of the wave on the second wire.

12. The first end of a combined string has a linear mass density of 4.20×10^{-3} kg/m, whereas the second end has a mass density of 10.5 kg/m. (a) If a 60.0-cm wave is to be sent along the first string at a speed of 8.56 m/s, what must the tension in the string be? (b) What is the wavelength of the reflected and transmitted wave?
13. The first end of a combined string has a linear mass density of 8.00 kg/m, whereas the second string has a density of 2.00 kg/m. If the speed of the wave in the first string is 10.0 m/s, find (a) the speed of the wave in the second string and (b) the tension in the string. (c) If a wave of length 60.0 cm is observed in the first string, find the wavelength and frequency of the wave in the second string.
14. The first end of a combined string has a linear mass density of 6.00 kg/m, whereas the second string has a density of 2.55 kg/m. The tension in the string is 350 N. If a vibration with a frequency of 20 vibrations is imparted to the first string, find the frequency, velocity, and wavelength of (a) the incident wave, (b) the reflected wave, and (c) the transmitted wave.

12.5 The Principle of Superposition

15. The following two waves interfere with each other:
$$y_1 = (10.8 \text{ m})\sin[(0.654 \text{ m}^{-1})x - (2.45 \text{ rad/s})t]$$
$$y_2 = (6.73 \text{ m})\sin[(0.893 \text{ m}^{-1})x - (6.82 \text{ rad/s})t]$$
Find the resultant displacement when $x = 0.782$ m and $t = 5.42$ s.

16. The following two waves combine:
$$y_1 = (10.8 \text{ m})\sin[(0.654 \text{ m}^{-1})x - (2.45 \text{ rad/s})t]$$
$$y_2 = (10.8 \text{ m})\sin[(0.654 \text{ m}^{-1})x - (2.45 \text{ rad/s})t - 0.834 \text{ rad}]$$
(a) Find the equation of the resultant wave. (b) Find the displacement of the resultant wave when $x = 0.895$ m and $t = 6.94$ s.

12.6 Standing Waves—The Vibrating String

17. The E string of a violin is vibrating at a fundamental frequency of 659 Hz. Find the wavelength and frequency of the third, fifth, and seventh harmonics. Let the length of the string be 60.0 cm.
18. A steel wire that is 1.45 m long and has a mass of 45 g is placed under a tension of 865 N. What is the frequency of its fifth harmonic?
19. A violin string plays a note at 440 Hz. What would the frequency of the wave on the string be if the tension in the string is (a) increased by 20.0% and (b) decreased by 20.0%?
20. A note is played on a guitar string 60.0 cm long at a frequency of 432 Hz. By how much should the string be shortened by pressing on it to play a note of 440 Hz?
21. A cello string, 75.0 cm long with a linear mass density of 7.25×10^{-3} kg/m, is to produce a fundamental frequency of 440 Hz. (a) What must be the tension in the string? (b) Find the frequency of the next three higher harmonics. (c) Find the wavelength of the fundamental and the next three higher harmonics.

12.7 Sound Waves

22. A sound wave in air has a velocity of 335 m/s. Find the temperature of the air.
23. A lightning flash is observed and 12 s later the associated thunder is heard. How far away is the lightning if the air temperature is 15.0 °C?

24. A soldier sees the flash from a cannon that is fired in the distance and 10 s later he hears the roar of the cannon. If the air temperature is 33 °C, how far away is the cannon?

25. A sound wave is sent to the bottom of the ocean by a ship in order to determine the depth of the ocean at that point. The sound wave returns to the boat in a time of 1.45 s. Find the depth of the ocean at this point. Use the bulk modulus of water to be 2.30 $\times$ 10⁹ N/m² and the density of seawater to be 1.03 $\times$ 10³ kg/m³.

26. Find the speed of sound in aluminum, copper, and lead.

27. You are trying to design three pipes for a closed organ pipe system that will give the following notes with their corresponding fundamental frequencies, C = 261.7 Hz, D = 293.7 Hz, E = 329.7 Hz. Find the length of each pipe. Assume that the speed of sound in air is 343 m/s.

28. Repeat problem 27 for an open organ pipe.

12.8 The Doppler Effect

29. A train is moving at a speed of 90.0 m/s and emits a whistle of frequency 400.0 Hz. If the speed of sound is 343 m/s, find the frequency observed by an observer who is at rest (a) in advance of the moving source and (b) behind the moving source.

30. A stationary police car turns on a siren at a frequency of 300 Hz. If the speed of sound in air is 343 m/s find the observed frequency if (a) the observer is approaching the police car at 35.0 m/s and (b) the observer is receding from the police car at 35.0 m/s.

31. A police car traveling at 90.0 m/s, turns on a siren at a frequency of 350 Hz as it tries to overtake a gangster's car moving away from the police car at a speed of 85 m/s. If the speed of sound in air is 343 m/s find the frequency heard by the gangster.

32. Two trains are approaching each other, each at a speed of 100 m/s. They each emit a whistle at a frequency of 225 Hz. If the speed of sound in air is 343 m/s, find the frequency that each train engineer hears.

33. A train moving east at a velocity of 20 m/s emits a whistle at a frequency of 348 Hz. Another train, farther up the track and moving east at a velocity of 30 m/s, hears the whistle from the first train. If the speed of sound in air is 343 m/s, what is the frequency of the sound heard by the second train engineer?

Additional Problems

34. One end of a violin string is connected to an electrical vibrator of 120 Hz, whereas the other end passes over a pulley and supports a mass of 10.0 kg, as shown in figure 12.17. The string is 60.0 cm long and has a mass of 12.5 g. What is the wavelength and speed of the wave produced?

†35. Three pipes, the first of lead, the second of brass, and the third of aluminum, each 10.0 m long, are welded together. If the first pipe is struck with a hammer at its end, how long will it take for the sound to pass through the pipes?

†36. A sound wave of 200 Hz in a steel pipe is transmitted into water and then into air. Find the wavelength of the sound in each medium.

†37. A railroad worker hits a steel track with a hammer. The sound wave through the steel track reaches an observer and 3.00 s later the sound wave through the air also reaches the observer. If the air temperature is 22.0 °C, how far away is the worker?

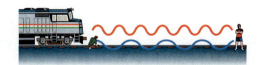

38. A tuning fork of 512 Hz is set into vibration above a long vertical tube containing water. A standing wave is observed as a resonance between the original wave and the reflected wave. If the speed of sound in air is 343 m/s, how far below the top of the tube is the water level?

39. The intensity of an ordinary conversation is about 3 $\times$ 10⁻⁶ W/m². Find the intensity level of the sound.

40. An indoor rock concert has an intensity level of 70 dB. Find the intensity of the sound.

41. The intensity level of a 500 Hz sound from a television program is about 40 dB. If the speed of sound is 343 m/s, find the amplitude of the sound wave.

†42. The speed of high-performance aircraft is sometimes given in terms of Mach numbers, where a Mach number is the ratio of the speed of the aircraft to the speed of sound at that level. Thus, a plane traveling at a speed of 343 m/s at sea level where the temperature is 20.0 °C would be traveling at Mach 1. If the temperature of the atmosphere increased to 30.0 °C, and the aircraft is still moving at 343 m/s, what is its Mach number?

Interactive Tutorials

⌨43. A tuning fork of frequency f = 512 Hz is set into vibration above a long vertical cylinder filled with water. As the water in the tube is lowered, resonance occurs between the initial wave traveling down the cylinder and the second wave that is reflected from the water surface below. Calculate (a) the wavelength λ of the sound wave in air and (b) the three resonance positions as measured from the top of the tube. The velocity of sound in air is v = 343 m/s.

⌨44. The superposition of any two waves. Given the following two waves:

$$y_1 = A_1 \sin(k_1 x - \omega_1 t)$$
$$y_2 = A_2 \sin(k_2 x - \omega_2 t - \phi)$$

For each wave find (a) the wavelength λ, (b) the frequency f, (c) the period T, and (d) the velocity v. Since each wave is periodic in both space and time, (e) for the value x = 2.00 m plot each wave and the sum of the two waves as a function of time t. (f) For the time t = 0.500 s, plot each wave and the sum of the two waves as a function of the distance x. For the initial conditions take A_1 = 3.50 m, k_1 = 0.55 m⁻¹, ω_1 = 4.25 rad/s, A_2 = 4.85 m, k_2 = 0.85 m⁻¹, ω_2 = 2.58 rad/s, and ϕ = 0. Then consider all the special cases listed in the tutorial itself.

⌨45. A 60.0-cm string with a mass of 1.40 g is to produce a fundamental frequency of 440 Hz. Find (a) the tension in the string, (b) the frequency of the next four higher harmonics, and (c) the wavelength of the fundamental and the next four higher harmonics.

⌨46. General purpose Doppler Effect Calculator. The Doppler Effect Calculator will calculate the observed frequency of a sound wave for the motion of the source or the observer, whether either or both are approaching or receding.

Fluids

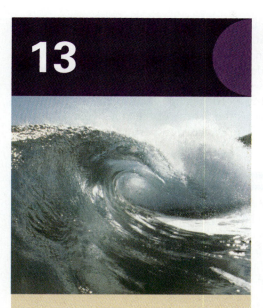

A fluid in motion.

When did science begin? Where did it begin? It began whenever and wherever men tried to solve the innumerable problems of life. The first solutions were mere expedients, but that must do for a beginning. Gradually the expedients would be compared, generalized, rationalized, simplified, interrelated, integrated; the texture of science would be slowly woven.

George Sarton

13.1 Introduction

Matter is usually said to exist in three phases: solid, liquid, and gas. Solids are hard bodies that resist deformations, whereas liquids and gases have the characteristic of being able to flow. A liquid flows and takes the shape of whatever container in which it is placed. A gas also flows into a container and spreads out until it occupies the entire volume of the container. *A fluid is defined as any substance that can flow, and hence liquids and gases are both considered to be **fluids.***

Liquids and gases are made up of billions upon billions of molecules in motion and to properly describe their behavior, Newton's second law should be applied to each of these molecules. However, this would be a formidable task, if not outright impossible, even with the use of modern high-speed computers. Also, the actual motion of a particular molecule is sometimes not as important as the overall effect of all those molecules when they are combined into the substance that is called the fluid. Hence, instead of using the microscopic approach of dealing with each molecule, we will treat the fluid from a macroscopic approach. That is, we will analyze the fluid in terms of its large-scale characteristics, such as its mass, density, pressure, and its distribution in space.

The study of fluids will be treated from two different approaches. First, we will consider only fluids that are at rest. This portion of the study of fluids is called **fluid statics or hydrostatics.** Second, we will study the behavior of fluids when they are in motion. This part of the study is called **fluid dynamics or hydrodynamics.** Let us start the study of fluids by defining and analyzing the macroscopic variables.

13.2 Density

*The **density** of a substance is defined as the amount of mass in a unit volume of that substance.* We use the symbol ρ (the lower case Greek letter rho) to designate the density and write it as

$$\rho = \frac{m}{V} \qquad\qquad (13.1)$$

A substance that has a large density has a great deal of mass in a unit volume, whereas a substance of low density has a small amount of mass in a unit volume. Density is expressed in SI units as kg/m^3, and occasionally in the laboratory as g/cm^3. Densities of solids and most liquids are very nearly constant but the densities of gases vary greatly with temperature and pressure. Table 13.1 is a list of densities for various materials. We observe from the table that in interstellar space

Table 13.1
Densities of Various Materials

Substance	kg/m³
Air (0 °C, 1 atm pressure)	1.29
Aluminum	2,700
Benzene	879
Blood	1.05×10^3
Bone	1.7×10^3
Brass	8,600
Copper	8,920
Critical density for universe to collapse under gravitation	5×10^{-27}
Planet Earth	5,520
Ethyl alcohol	810
Glycerine	1,260
Gold	19,300
Hydrogen atom	2,680
Ice	920
Interstellar space	$10^{-18} - 10^{-21}$
Iron	7,860
Lead	11,340
Mercury	13,630
Nucleus	10^{17}
Proton	1.5×10^{17}
Silver	10,500
Sun (avg)	1,400
Water (pure)	1,000
(sea)	1,030
Wood (maple)	620–750

the densities are extremely small, of the order of 10^{-18} to 10^{-21} kg/m³. That is, interstellar space is almost empty space. The density of the proton and neutron is of the order of 10^{17} kg/m³, which is an extremely large density. Hence, the nucleus of a chemical element is extremely dense. Because an atom of hydrogen has an approximate density of 2680 kg/m³, whereas the proton in the nucleus of that hydrogen atom has a density of about 1.5×10^{17} kg/m³, we see that the density of the nucleus is about 10^{13} times as great as the density of the atom. Hence, an atom consists almost entirely of empty space with the greatest portion of its mass residing in a very small nucleus.

Example 13.1

The density of an irregularly shaped object. In order to find the density of an irregularly shaped object, the object is placed in a beaker of water that is filled completely to the top. Since no two objects can occupy the same space at the same time, 25.0 cm³ of the water, which is equal to the volume of the unknown object, overflows into an attached calibrated beaker. The object is placed on a balance scale and is found to have a mass of 262.5 g. Find the density of the material.

Solution

The density, found from equation 13.1, is

$$\rho = \frac{m}{V} = \frac{262.5 \text{ g}}{25.0 \text{ cm}^3} = 10.5 \frac{\text{g}}{\text{cm}^3} = 10,500 \frac{\text{kg}}{\text{m}^3}$$

Example 13.2

Your own water bed. A person would like to design a water bed for the home. If the size of the bed is to be 2.20 m long, 1.80 m wide, and 0.300 m deep, what mass of water is necessary to fill the bed?

Solution

The mass of the water, found from equation 13.1, is

$$m = \rho V \tag{13.2}$$

The density is found from table 13.1. Hence, the mass of water required is

$$m = \rho V = \left(1000 \frac{\text{kg}}{\text{m}^3}\right)(2.20 \text{ m})(1.80 \text{ m})(0.300 \text{ m})$$

$$= 1190 \text{ kg}$$

As a matter of curiosity let us compute the weight of this water. The weight of the water is given by

$$w = mg = (1190 \text{ kg})(9.80 \text{ m/s}^2) = 11,600 \text{ N}$$

To give you a "feel" for this weight of water, it is equivalent to 2620 lb. In some cases, it will be necessary to reinforce the floor underneath this water bed or the bed might end up in the basement below.

13.3 Pressure

Pressure *is defined as the magnitude of the normal force acting per unit surface area.* The pressure is thus a scalar quantity. We write this mathematically as

$$p = \frac{F}{A} \tag{13.3}$$

The SI unit for pressure is newton/meter², which is given the special name pascal, in honor of the French mathematician, physicist, and philosopher, Blaise Pascal (1623–1662) and is abbreviated Pa. Hence, 1 Pa = 1 N/m². In the British engineering system the units are lb/in.², which is sometimes denoted by psi. Pressures are not limited to fluids, as the following examples show.

Example 13.3

Pressure exerted by a man. A man weighs 200 lb. At one particular moment when he walks, his right heel is the only part of his body that touches the ground. If the heel of his shoe measures 3.50 in. by 3.25 in., what pressure does the man exert on the ground?

Solution

The pressure that the man exerts on the ground, given by equation 13.3, is

$$p = \frac{F}{A}$$

$$= \frac{w}{A} = \frac{200 \text{ lb}}{(3.50 \text{ in.})(3.25 \text{ in.})}$$

$$= 17.6 \, \frac{\text{lb}}{\text{in.}^2}$$

Example 13.4

Pressure exerted by a woman. A woman who weighs 100 lb is wearing "high-heel" shoes. The cross section of her high-heel shoe measures 1/2 in. by 5/8 in. At a particular moment when she is walking, only one heel of her shoe makes contact with the ground. What is the pressure exerted on the ground by the woman?

Solution

The pressure exerted on the ground, found from equation 13.3, is

$$p = \frac{F}{A}$$

$$= \frac{w}{A} = \frac{100 \text{ lb}}{(0.500 \text{ in.})(6.25 \text{ in.})}$$

$$= 320 \, \frac{\text{lb}}{\text{in.}^2}$$

Thus, the 100-lb woman exerts a pressure through her high heel of 320 lb/in.², whereas the man, who weighs twice as much, exerts a pressure of only 17.6 lb/in.². That is, the woman exerts about 18 times more pressure than the man. The key to the great difference lies in the definition of pressure. Pressure is the force exerted per unit area. Because the area of the woman's high heel is so very small, the pressure becomes very large. The area of the man's heel is relatively large, hence the pressure he exerts is relatively small. When they are wearing high heels, women usually do not like to walk on soft ground because the large pressure causes the shoe to sink into the ground.

A further example of the effect of the surface area on pressure is found in the application of snowshoes. Here, a person's weight is distributed over such a large

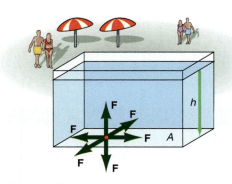

Figure 13.1

Pressure in a pool of water.

area that the pressure exerted on the snow is very small. Hence, the person is capable of walking in deep snow, while another person, wearing ordinary shoes, would find walking almost impossible.

Pressure exerted by a fluid is easily determined with the aid of figure 13.1, which represents a pool of water. We want to determine the pressure p at the bottom of the pool caused by the water in the pool. By our definition, equation 13.3, the pressure at the bottom of the pool is the magnitude of the force acting on a unit area of the bottom of the pool. But the force acting on the bottom of the pool is caused by the weight of all the water above it. Thus,

$$p = \frac{F}{A} = \frac{\text{weight of water}}{\text{area}} \qquad (13.4)$$

$$p = \frac{w}{A} = \frac{mg}{A} \qquad (13.5)$$

We have set the weight w of the water equal to mg in equation 13.5. The mass of the water in the pool, given by equation 13.2, is

$$m = \rho V$$

The volume of all the water in the pool is just equal to the area A of the bottom of the pool times the depth h of the water in the pool, that is,

$$V = Ah \qquad (13.6)$$

Substituting equations 13.2 and 13.6 into equation 13.5 gives for the pressure at the bottom of the pool:

$$p = \frac{mg}{A} = \frac{\rho V g}{A} = \frac{\rho A h g}{A}$$

Thus,

$$p = \rho g h \qquad (13.7)$$

(Although we derived equation 13.7 to determine the water pressure at the bottom of a pool of water, it is completely general and gives the water pressure at any depth h in the pool.) Equation 13.7 says that the water pressure at any depth h in any pool is given by the product of the density of the water in the pool, the acceleration due to gravity g, and the depth h in the pool. *Equation 13.7 is sometimes called **the hydrostatic equation.***

Example 13.5

Pressure in a swimming pool. Find the water pressure at a depth of 3.00 m in a swimming pool.

Solution

The density of water, found in table 13.1, is 1000 kg/m³, and the water pressure, found from equation 13.7, is

$$p = \rho g h$$
$$= (1000 \text{ kg/m}^3)(9.80 \text{ m/s}^2)(3.00 \text{ m})$$
$$= 2.94 \times 10^4 \text{ N/m}^2 = 2.94 \times 10^4 \text{ Pa}$$

The pressure at the depth of 3 m in the pool in figure 13.1 is the same everywhere. Hence, the force exerted by the fluid is the same in all directions. That is, the force is the same in up-down, right-left, or in-out directions. If the force due to the fluid were not the same in all directions, then the fluid would flow in the direction away from the greatest pressure and would not be a fluid at rest. A fluid at rest is a fluid in equilibrium. Thus, in example 13.5, the pressure is 2.94×10^4 Pa at every point at a depth of 3 m in the pool and exerts the same force in every direction at that depth. You experience this pressure when swimming at a depth of 3.00 m as a pressure on your ears. As you swim up to the surface, the pressure on your ears decreases because h is decreasing. Or to look at it another way, the closer you swim up toward the surface, the smaller is the amount of water that is above you. Because the pressure is caused by the weight of that water above you, the smaller the amount of water, the smaller will be the pressure.

Just as there is a water pressure at the bottom of a swimming pool caused by the weight of all the water above the bottom, there is also an air pressure exerted on every object at the surface of the earth caused by the weight of all the air that is above us in the atmosphere. That is, there is an atmospheric pressure exerted on us, given by equation 13.3 as

$$p = \frac{F}{A} = \frac{\text{weight of air}}{\text{area}} \tag{13.8}$$

However we can not use the same result obtained for the pressure in the pool of water, the hydrostatic equation 13.7, because air is compressible and hence its density ρ is not constant with height throughout the vertical portion of the atmosphere. The pressure of air at any height in the atmosphere can be found by the use of calculus and the density variation in the atmosphere. However, since calculus is beyond the scope of this course, we will revert to the use of experimentation to determine the pressure of the atmosphere.

The pressure of the air in the atmosphere was first measured by Evangelista Torricelli (1608–1647), a student of Galileo, by the use of a mercury **barometer**. A long narrow tube is filled to the top with mercury, chemical symbol Hg. It is then placed upside down into a reservoir filled with mercury, as shown in figure 13.2. The mercury in the tube starts to flow out into the reservoir, but it comes to a stop when the top of the mercury column is at a height h above the top of the mercury reservoir, as also shown in figure 13.2. The mercury does not empty completely because the normal pressure of the atmosphere p_0 pushes downward on the mercury reservoir. Because the force caused by the pressure of a fluid is the same in all directions, there is also a force acting upward inside the tube at the height of the mercury reservoir, and hence there is also a pressure p_0 acting upward as shown in figure 13.2. This force upward is capable of holding the weight of the mercury in the tube up to a height h. Thus, the pressure exerted by the mercury in the tube is exactly balanced by the normal atmospheric pressure on the reservoir, that is,

$$p_0 = p_{Hg} \tag{13.9}$$

But the pressure of the mercury in the tube p_{Hg}, given by equation 13.7, is

$$p_{Hg} = \rho_{Hg}gh \tag{13.10}$$

Substituting equation 13.10 back into equation 13.9, gives

$$p_0 = \rho_{Hg}gh \tag{13.11}$$

Equation 13.11 says that normal atmospheric pressure can be determined by measuring the height h of the column of mercury in the tube. It is found experimentally, that on the average, normal atmospheric pressure can support a column of mercury 76.0 cm high, or 760 mm high. The unit of 1.00 mm of Hg is sometimes called a torr in honor of Torricelli. Hence, normal atmospheric pressure can also be given as 760 torr. Using the value of the density of mercury of 1.360×10^4 kg/m³, found in table 13.1, normal atmospheric pressure, determined from equation 13.11, is

$$p_0 = \rho_{Hg}gh = \left(1.360 \times 10^4 \frac{\text{kg}}{\text{m}^3}\right)\left(9.80 \frac{\text{m}}{\text{s}^2}\right)(0.760 \text{ m})$$

$$= 1.013 \times 10^5 \text{ N/m}^2 = 1.013 \times 10^5 \text{ Pa}$$

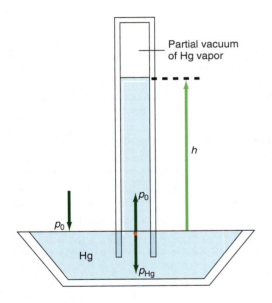

Figure 13.2

A mercury barometer.

Thus, the average or normal atmospheric pressure acting on us at the surface of the earth is 1.013×10^5 Pa, which is a rather large number as we will see presently. In the study of meteorology, the science of the weather, a different unit of pressure is usually employed, namely the millibar, abbreviated mb. The conversion factor between millibars and Pa (see appendix A) is

$$1 \text{ Pa} = 10^{-2} \text{ mb}$$

Using this conversion factor, normal atmospheric pressure can also be expressed as

$$p_0 = (1.013 \times 10^5 \text{ Pa})\left(\frac{10^{-2} \text{ mb}}{1 \text{ Pa}}\right)$$

$$= 1013 \text{ mb}$$

On all surface weather maps in a weather station, pressures are always expressed in terms of millibars.

To express normal atmospheric pressure in the British engineering system, the conversion factor

$$1 \text{ Pa} = 1.45 \times 10^{-4} \text{ lb/in.}^2$$

found in appendix A, is used. Hence, normal atmospheric pressure can also be expressed as

$$p_0 = (1.013 \times 10^5 \text{ Pa})\left(\frac{1.45 \times 10^{-4} \text{ lb/in.}^2}{1 \text{ Pa}}\right)$$

$$= 14.7 \text{ lb/in.}^2$$

The mercury barometer is thus a very accurate means of determining air pressure. The value of 76.0 cm or 1013 mb are only normal or average values. When the barometer is kept at the same location and the height of the mercury column is recorded daily, the value of h is found to vary slightly. When the value of h becomes greater than 76.0 cm of Hg, the pressure of the atmosphere has increased to a higher pressure. It is then said that a high-pressure area has moved into your region. When the value of h becomes less than 76.0 cm of Hg, the pressure of the atmosphere has decreased to a lower pressure and a low-pressure area has moved in. The barometer is extremely important in weather observation and prediction because, as a general rule of thumb, high atmospheric pressures usually are associated with clear skies and good weather. Low-pressure areas, on the other hand, are usually associated with cloudy skies, precipitation, and in general bad weather. (For further detail on the weather see the "Have You Ever Wondered" section at the end of chapter 17.)

The mercury barometer, after certain corrections for instrument height above sea level and ambient temperature, is an extremely accurate device to measure atmospheric pressure and can be found in every weather station throughout the world. Its chief limitation is its size. It must always remain vertical, and the glass tube and reservoir are somewhat fragile. Hence, another type of barometer is also used to measure atmospheric pressure. It is called an *aneroid barometer,* and is shown in figure 13.3. It is based on the principle of a partially evacuated, waferlike, metal cylinder called a Sylphon cell. When the atmospheric pressure increases, the cell decreases in size. A combination of linkages and springs are connected to the cell and to a pointer needle that moves over a calibrated scale that indicates the pressure. The aneroid barometer is a more portable device that is rugged and easily used, although it is originally calibrated with a mercury barometer. The word *aneroid* means not containing fluid. The aneroid barometer is calibrated in both centimeters of Hg and inches of Hg. Using a conversion factor, we can easily see that a height of 29.92 in. of Hg also corresponds to normal atmospheric pressure. Hence, as seen in figure 13.3, the pressure can be measured in terms of inches of

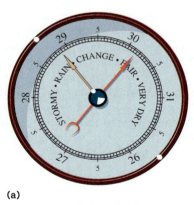

(a)

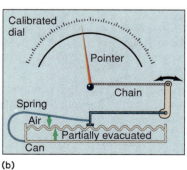

(b)

Figure 13.3

An aneroid barometer.

mercury. Also note that regions of high pressure (30 in. of Hg) are labeled to indicate fair weather, while regions of low pressure (29 in. of Hg) are labeled to indicate rain or poor weather.

As we go up into the atmosphere the pressure decreases, because there is less air above us. The aneroid barometer will read smaller and smaller pressures with altitude. Instead of calibrating the aneroid barometer in terms of centimeters of mercury or inches of mercury, we can also calibrate it in terms of feet or meters above the surface of the earth where this air pressure is found. An aneroid barometer so calibrated is called an *altimeter,* a device to measure the altitude or height of an airplane. The height of the plane is not really measured, the pressure is. But in the standard atmosphere, a particular pressure is found at a particular height above the ground. Hence, when the aneroid barometer measures this pressure, it corresponds to a fixed altitude above the ground. The pilot can read this height directly from the newly calibrated aneroid barometer, the altimeter.

Let us now look at some examples associated with atmospheric pressure.

Example 13.6

Why you get tired by the end of the day. The top of a student's head is approximately circular with a radius of 3.50 inches. What force is exerted on the top of the student's head by normal atmospheric pressure?

Solution

The area of the top of the student's head is found from

$$A = \pi r^2 = \pi(3.50 \text{ in.})^2 = 38.5 \text{ in.}^2$$

We find the magnitude of the force exerted on the top of the student's head by rearranging equation 13.3 into the form

$$F = pA \qquad (13.12)$$

Hence,

$$F = \left(14.7 \frac{\text{lb}}{\text{in.}^2}\right)(38.5 \text{ in.}^2)$$

$$= 566 \text{ lb}$$

This is a rather large force to have exerted on our heads all day long. However, we do not notice this enormous force because when we breathe air into our nose or mouth that air is exerting the same force upward inside our head. Thus, the difference in force between the top of the head and the inside of the head is zero.

Example 13.7

Atmospheric pressure on the walls of your house. Find the force on the outside wall of a ranch house, 10.0 ft high and 35.0 ft long, caused by normal atmospheric pressure.

Solution

The area of the wall of the house is given by

$$A = (\text{length})(\text{height})$$

$$= (35.0 \text{ ft})(10.0 \text{ ft})\left(\frac{144 \text{ in.}^2}{1 \text{ ft}^2}\right)$$

$$= 50,400 \text{ in.}^2$$

The force on the wall, given by equation 13.12, is

$$F = pA = \left(14.7 \frac{\text{lb}}{\text{in.}^2}\right)(50{,}400 \text{ in.}^2)$$

$$= 740{,}000 \text{ lb}$$

The force on the outside wall of the house in example 13.7 is thus 740,000 lb. This is truly an enormous force. Why doesn't the wall collapse under this great force? The wall does not collapse because that same atmospheric air is also inside the house. Remember that air is a fluid and flows. Hence, in addition to being outside the house, the air also flows to the inside of the house. Because the force exerted by the pressure in the fluid is the same in all directions, the air inside the house exerts the same force of 740,000 lb against the inside wall of the house, as shown in figure 13.4(a). The net force on the wall is therefore

$$\text{Net force} = (\text{force})_{\text{in}} - (\text{force})_{\text{out}}$$

$$= 740{,}000 \text{ lb} - 740{,}000 \text{ lb}$$

$$= 0$$

A very interesting case occurs when this net force is not zero. Suppose a tornado, an extremely violent storm, were to move over your house, as shown in figure 13.4(b). The pressure inside the tornado is very low. No one knows for sure how low, because it is slightly difficult to run into a tornado with a barometer to measure it. In the very few cases on record where tornadoes actually went over a weather station, there was never anything left of the weather station, to say nothing of the barometer that was in that station. That is, neither the barometer nor the weather station were ever found again. The pressure can be estimated, however, from the very high winds associated with the tornado. A good estimate is that the pressure inside the tornado is at least 10% below the actual atmospheric pressure. Let us assume that the actual pressure is the normal atmospheric pressure of 1013 mb, then 10% of that is 101 mb. Thus, the pressure in the tornado is approximately

$$1013 \text{ mb} - 101 \text{ mb} = (912 \text{ mb})\left(\frac{14.7 \text{ lb/in.}^2}{1013 \text{ mb}}\right) = 13.2 \text{ lb/in.}^2$$

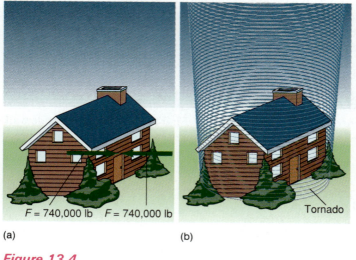

$F = 740{,}000 \text{ lb} \quad F = 740{,}000 \text{ lb}$

(a)

Tornado

(b)

(c)

Figure 13.4

Pressure on the walls in a house.

When the tornado goes over the house, the force on the outside wall is given by

$$F = pA = \left(13.2 \, \frac{\text{lb}}{\text{in.}^2}\right)(50{,}400 \text{ in.}^2)$$

$$= 665{,}000 \text{ lb}$$

But the original air inside the house is still there and is still exerting a force of 740,000 lb outward on the walls. The net force on the house is now

$$\text{Net force} = 740{,}000 \text{ lb} - 665{,}000 \text{ lb}$$

$$= 75{,}000 \text{ lb}$$

There is now a net force acting outward on the wall of 75,000 lb, enough to literally explode the walls of the house outward. This pressure differential, with its accompanying winds, accounts for the enormous destruction associated with a tornado. Thus, the force exerted by atmospheric pressure can be extremely significant.

It has always been customary to open the doors and windows in a house whenever a tornado is in the vicinity in the hope that a great deal of the air inside the house will flow out through these open windows and doors. Hence, the pressure differential between the inside and the outside walls of the house will be minimized. Unfortunately many victims of tornadoes do not follow this procedure, because tornadoes are spawned out of severe thunderstorms, which are usually accompanied by torrential rain. Usually the first thing one does in a house is to close the windows once the rain starts. A picture of a typical tornado is shown in figure 13.4(c).

Now that we have discussed atmospheric pressure, it is obvious that the total pressure exerted at a depth h in a pool of water must be greater than the value determined previously, because the air above the pool is exerting an atmospheric pressure on the top of the pool. This additional pressure is transmitted undiminished throughout the pool. Hence, the total or *absolute pressure* observed at the depth h in the pool is the sum of the atmospheric pressure plus the pressure of the water itself, that is,

$$p_{abs} = p_0 + p_w \tag{13.13}$$

Using equation 13.7, this becomes

$$p_{abs} = p_0 + \rho g h \tag{13.14}$$

Example 13.8

Absolute pressure. What is the absolute pressure at a depth of 3.00 m in a swimming pool?

Solution

The water pressure at a depth of 3.00 m has already been found to be $p_w = 2.94 \times 10^4$ Pa, the absolute pressure, found by equation 13.13, is

$$p_{abs} = p_0 + p_w$$

$$= 1.013 \times 10^5 \text{ Pa} + 2.94 \times 10^4 \text{ Pa}$$

$$= 1.31 \times 10^5 \text{ Pa}$$

When the pressure of the air in an automobile tire is measured, the actual pressure being measured is called the **gauge pressure,** that is, the pressure as indicated on the measuring device that is called a gauge. This measuring device, the gauge, reads zero when it is actually under normal atmospheric pressure. Thus, the total pressure or absolute pressure of the air inside the tire is the sum of the pressure recorded on the gauge plus normal atmospheric pressure. We can write this mathematically as

$$p_{abs} = p_{gauge} + p_0 \qquad (13.15)$$

Example 13.9

Gauge pressure and absolute pressure. A gauge placed on an automobile tire reads a pressure of 34.0 lb/in.². What is the absolute pressure of the air in the tire?

Solution

The absolute pressure of the air in the tire, found from equation 13.15, is

$$p_{abs} = p_{gauge} + p_0$$

$$= 34.0 \, \frac{lb}{in.^2} + 14.7 \, \frac{lb}{in.^2}$$

$$= 48.7 \, lb/in.^2$$

13.4 Pascal's Principle

The pressure exerted on the bottom of a pool of water by the water itself is given by $\rho g h$. However, there is also an atmosphere over the pool, and, as we saw in section 13.3, there is thus an additional pressure, normal atmospheric pressure p_0, exerted on the top of the pool. This pressure on the top of the pool is transmitted through the pool waters so that the total pressure at the bottom of the pool is the sum of the pressure of the water plus the pressure of the atmosphere, equations 13.13 and 13.14. The addition of both pressures is a special case of a principle, called *Pascal's principle* and it states that *if the pressure at any point in an enclosed fluid at rest is changed (Δp), the pressure changes by an equal amount (Δp), at all points in the fluid.* As an example of the use of Pascal's principle, let us consider the hydraulic lift shown in figure 13.5. A noncompressible fluid fills both cylinders and the connecting pipe. The smaller cylinder has a piston of cross-sectional area a, whereas the larger cylinder has a cross-sectional area A. As we can see in the figure, the cross-sectional area A of the larger cylinder is greater than the cross-sectional area a of the smaller cylinder. If a small force f is applied to the piston of the small cylinder, this creates a change in the pressure of the fluid given by

$$\Delta p = \frac{f}{a} \qquad (13.16)$$

But by Pascal's principle, this pressure change occurs at all points in the fluid, and in particular at the large piston on the right. This same pressure change applied to the right piston gives

$$\Delta p = \frac{F}{A} \qquad (13.17)$$

where F is the force that the fluid now exerts on the large piston of area A. Because these two pressure changes are equal by Pascal's principle, we can set equation 13.17 equal to equation 13.16. Thus,

$$\Delta p = \Delta p$$

$$\frac{F}{A} = \frac{f}{a}$$

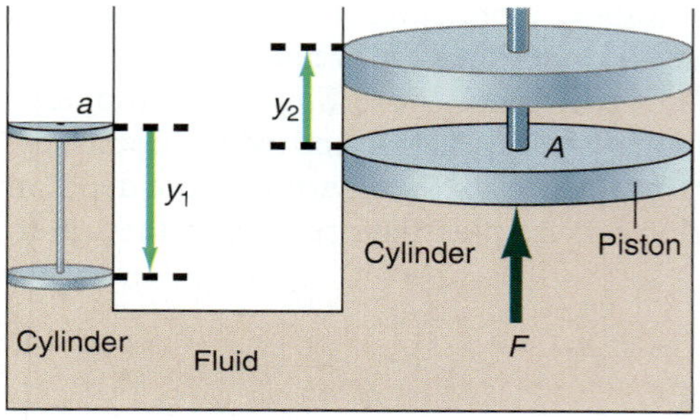

Figure 13.5
The hydraulic lift.

The force F on the large piston is therefore

$$F = \frac{A}{a} f \qquad \text{(13.18)}$$

Since the area A is greater than the area a, the force F will be greater than f. Thus, the hydraulic lift is a device that is capable of multiplying forces.

Example 13.10

Amplifying a force. The radius of the small piston in figure 13.5 is 5.00 cm, whereas the radius of the large piston is 30.0 cm. If a force of 2.00 N is applied to the small piston, what force will occur at the large piston?

Solution

The area of the small piston is

$$a = \pi r_1^2 = \pi (5.00 \text{ cm})^2 = 78.5 \text{ cm}^2$$

while the area of the large piston is

$$A = \pi r_2^2 = \pi (30.0 \text{ cm})^2 = 2830 \text{ cm}^2$$

The force exerted by the fluid on the large piston, found from equation 13.18, is

$$F = \frac{A}{a} f$$

$$= \left(\frac{2830 \text{ cm}^2}{78.5 \text{ cm}^2} \right) (2.00 \text{ N})$$

$$= 72.1 \text{ N}$$

Thus, the relatively small force of 2.00 N applied to the small piston produces the rather large force of 72.1 N at the large piston. The force has been magnified by a factor of 36.

The amplifying force discussed in Example 13.10 is what allows this hydraulic lift to raise this automobile.

It is interesting to compute the work that is done when the force f is applied to the small piston in figure 13.5. When the force f is applied, the piston moves through a displacement y_1, such that the work done is given by

$$W_1 = f y_1$$

But from equation 13.16

$$f = a \Delta p$$

Hence, the work done is

$$W_1 = a (\Delta p) y_1 \qquad \text{(13.19)}$$

When the change in pressure is transmitted through the fluid, the force F is exerted against the large piston and the work done by the fluid on the large piston is

$$W_2 = F y_2$$

where y_2 is the distance that the large piston moves and is shown in figure 13.5. But the force F, found from equation 13.17, is

$$F = A \Delta p$$

The work done on the large piston by the fluid becomes

$$W_2 = A (\Delta p) y_2 \qquad \text{(13.20)}$$

Applying the law of conservation of energy to a frictionless hydraulic lift, the work done to the fluid at the small piston must equal the work done by the fluid at the large piston, hence

$$W_1 = W_2 \qquad (13.21)$$

Substituting equations 13.19 and 13.20 into equation 13.21, gives

$$a(\Delta p)y_1 = A(\Delta p)y_2 \qquad (13.22)$$

Because the pressure change Δp is the same throughout the fluid, it cancels out of equation 13.22, leaving

$$ay_1 = Ay_2$$

Solving for the distance y_1 that the small piston moves

$$y_1 = \frac{A}{a} y_2 \qquad (13.23)$$

Since A is much greater than a, it follows that y_1 must be much greater than y_2.

Example 13.11

You can never get something for nothing. The large piston of example 13.10 moves through a distance of 0.200 cm. By how much must the small piston be moved?

Solution

The areas of the pistons are given from example 13.10 as $A = 2830$ cm² and $a = 78.5$ cm², hence the distance that the small piston must move, given by equation 13.23, is

$$y_1 = \frac{A}{a} y_2$$

$$= \left(\frac{2830 \text{ cm}^2}{78.5 \text{ cm}^2} \right)(0.200 \text{ cm})$$

$$= 7.21 \text{ cm}$$

Although a very large force is obtained at the large piston, the large piston is displaced by only a very small amount. Whereas the input force f, on the small piston is relatively small, the small piston must move through a relatively large displacement (36 times greater than the large piston). Usually there are a series of valves in the connecting pipe and the small cylinder is connected to a fluid reservoir also by valves. Hence, many displacements of the small piston can be made, each time adding additional fluid to the right cylinder. In this way the final displacement y_2 can be made as large as desired.

13.5 Archimedes' Principle

The variation of pressure with depth has a surprising consequence, it allows the fluid to exert buoyant forces on bodies immersed in the fluid. If this buoyant force is equal to the weight of the body, the body floats in the fluid. This result was first annunciated by Archimedes (287–212 BC) and is now called Archimedes' principle.

Archimedes' principle states that a body immersed in a fluid is buoyed up by a force that is equal to the weight of the fluid displaced. This principle can be verified with the help of figure 13.6.

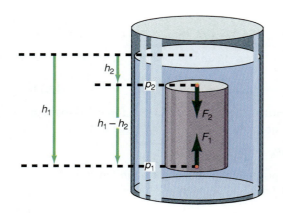

Figure 13.6

Archimedes' principle.

If we submerge a cylindrical body into a fluid, such as water, then the bottom of the body is at some depth h_1 below the surface of the water and experiences a water pressure p_1 given by

$$p_1 = \rho g h_1 \qquad (13.24)$$

where ρ is the density of the water. Because the force due to the pressure acts equally in all directions, there is an upward force on the bottom of the body. The force upward on the body is given by

$$F_1 = p_1 A \qquad (13.25)$$

where A is the cross-sectional area of the cylinder. Similarly, the top of the body is at a depth h_2 below the surface of the water, and experiences the water pressure p_2 given by

$$p_2 = \rho g h_2 \qquad (13.26)$$

However, in this case the force due to the water pressure is acting downward on the body causing a force downward given by

$$F_2 = p_2 A \qquad (13.27)$$

Because of the difference in pressure at the two depths, h_1 and h_2, there is a different force on the bottom of the body than on the top of the body. Since the bottom of the submerged body is at the greater depth, it experiences the greater force. Hence, there is a net force upward on the submerged body given by

$$\text{Net force upward} = F_1 - F_2$$

Replacing the forces F_1 and F_2 by their values in equations 13.25 and 13.27, this becomes

$$\text{Net force upward} = p_1 A - p_2 A$$

Replacing the pressures p_1 and p_2 from equations 13.24 and 13.26, this becomes

$$\text{Net force upward} = \rho g h_1 A - \rho g h_2 A$$
$$= \rho g A (h_1 - h_2) \qquad (13.28)$$

But

$$A(h_1 - h_2) = V$$

the volume of the cylindrical body, and hence the volume of the water displaced. Equation 13.28 thus becomes

$$\text{Net force upward} = \rho g V \qquad (13.29)$$

But ρ is the density of the water and from the definition of the density

$$\rho = \frac{m}{V} \qquad (13.1)$$

Substituting equation 13.1 back into equation 13.29 gives

$$\text{Net force upward} = \frac{m}{V} g V$$
$$= mg$$

But $mg = w$, the weight of the water displaced. Hence,

$$\text{Net force upward} = \text{Weight of water displaced} \qquad (13.30)$$

The net force upward on the body is called the *buoyant force* (BF). When the buoyant force on the body is equal to the weight of the body, the body does not sink in the

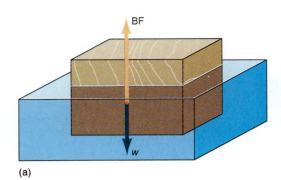

(a)

(b)

Figure 13.7

A body floats when the buoyant force is equal to the weight of the body.

water but rather floats, figure 13.7(b). Since the buoyant force is equal to the weight of the water displaced, *a body floats when the weight of the body is equal to the weight of the fluid displaced.*

Example 13.12

Wood floats. A block of oak wood 5.00 cm high, 5.00 cm wide, and 10.0 cm long is placed into a tub of water, figure 13.7(a). The density of the wood is 7.20×10^2 kg/m³. How far will the block of wood sink before it floats?

Solution

The block of wood will float when the buoyant force (BF), which is the weight of the fluid displaced by the volume of the body submerged, is equal to the weight of the body. The weight of the block of wood is found from

$$w = mg = \rho V g$$

The volume of the wooden block is $V = Ah$. Thus, the weight of the wooden block is

$$w = (7.20 \times 10^2 \text{ kg/m}^3)(0.0500 \text{ m})(0.0500 \text{ m})(0.100 \text{ m})(9.80 \text{ m/s}^2)$$

$$= 1.76 \text{ N}$$

The buoyant force is equal to the weight of the water displaced, and for the body to float this buoyant force must also equal the weight of the block. Hence,

$$\text{BF} = w_{\text{water}} = w_{\text{wood}}$$

$$w_{\text{water}} = m_{\text{water}}g = \rho_{\text{water}}Vg = \rho_{\text{water}}Ahg \tag{13.31}$$

Thus,

$$\rho_{\text{water}}Ahg = w_{\text{wood}}$$

$$h = \frac{w_{\text{wood}}}{\rho_{\text{water}}Ag} \tag{13.32}$$

$$= \frac{1.76 \text{ N}}{(1.00 \times 10^3 \text{ kg/m}^3)(0.0500 \text{ m})(0.100 \text{ m})(9.80 \text{ m/s}^2)}$$

$$= 0.0359 \text{ m} = 3.59 \text{ cm}$$

Thus, the block sinks to a depth of 3.59 cm. At this point the buoyant force becomes equal to the weight of the wooden block and the wooden block floats.

Example 13.13

Iron sinks. Repeat example 13.12 for a block of iron of the same dimensions.

Solution

The density of iron, found from table 13.1, is 7860 kg/m³. The weight of the iron block is given by

$$w_{\text{iron}} = mg = \rho V g$$

$$= (7860 \text{ kg/m}^3)(0.0500 \text{ m})(0.0500 \text{ m})(0.100 \text{ m})(9.80 \text{ m/s}^2)$$

$$= 19.3 \text{ N}$$

The depth that the iron block would have to sink in order to displace its own weight, again found from equation 13.32, is

$$h = \frac{w_{\text{iron}}}{\rho_{\text{water}} A g}$$

$$= \frac{19.3 \text{ N}}{(1.00 \times 10^3 \text{ kg/m}^3)(0.0500 \text{ m})(0.100 \text{ m})(9.80 \text{ m/s}^2)}$$

$$= 39.4 \text{ cm}$$

But the block is only 10 cm high. Hence, the buoyant force is not great enough to lift an iron block of this size, and the iron block sinks to the bottom.

Another way to look at this problem is to calculate the buoyant force on this piece of iron. The buoyant force on the iron, given by equation 13.29, is

Net force upward $= \rho g V$

$$= (1 \times 10^3 \text{ kg/m}^3)(9.80 \text{ m/s}^2)(0.0500 \text{ m})(0.500 \text{ m})(0.100 \text{ m})$$

$$= 2.45 \text{ N}$$

Thus, the net force upward on a block of iron of this size is 2.45 N. But the block weighs 19.3 N. Hence, the weight of the iron is greater than the buoyant force and the iron block sinks to the bottom.

Even an enormous ship is able to remain afloat.

But ships are made of iron and they do not sink. Why should the block sink and not the ship? If this same weight of iron is made into thin slabs, these thin slabs could be welded together into a boat structure of some kind. By increasing the size and hence the volume of this iron boat, a greater volume of water can be displaced. An increase in the volume of water displaced increases the buoyant force. If this can be made equal to the weight of the iron boat, then the boat floats.

Example 13.14

An iron boat. The iron block of example 13.13 is cut into 16 slices, each 5.00 cm by 10.0 cm by 5/16 cm. They are now welded together to form a box 20.0 cm wide by 10.0 cm long by 10.0 cm high, as shown in figure 13.8. Will this iron body now float or will it sink?

Solution

In this new configuration the iron displaces a much greater volume of water, and since the buoyant force is equal to the weight of the water displaced it is possible that this new configuration will float. We assume that no mass of iron is lost in cutting the blocks into the 16 slabs, and that the weight of the welding material is negligible. Thus, the weight of the box is also equal to 19.3 N. This example is analyzed in the same way as the previous example. Let us solve for the depth that the iron box must sink in order that the buoyant force be equal to the weight of the box. Thus, the depth that the box sinks, again found from the modified equation 13.32, is

$$h = \frac{w_{\text{box}}}{\rho_{\text{water}} A g}$$

$$= \frac{19.3 \text{ N}}{(1.00 \times 10^3 \text{ kg})(0.200 \text{ m})(0.100 \text{ m})(9.80 \text{ m/s}^2)}$$

$$= 9.84 \times 10^{-2} \text{ m} = 9.84 \text{ cm}$$

Because the iron box is 10 cm high, it sinks to a depth of 9.84 cm and it then floats. Note that this is the same mass of iron that sank in example 13.13. That same mass

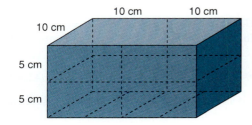

Figure 13.8

Iron can float.

can now float because the new distribution of that mass results in a displacement of a much larger volume of water. Since the buoyant force is equal to the weight of the water displaced, by increasing the volume taken up by the iron and the enclosed space, the amount of the water displaced has increased and so has the buoyant force.

Examples 13.12–13.14 dealt with bodies submerged in water, but remember that Archimedes' principle applies to all fluids.

13.6 The Equation of Continuity

Up to now, we have studied only fluids at rest. *Let us now study fluids in motion, the subject matter of hydrodynamics.* The study of fluids in motion is relatively complicated, but the analysis can be simplified by making a few assumptions. Let us assume that the fluid is incompressible and flows freely without any turbulence or friction between the various parts of the fluid itself and any boundary containing the fluid, such as the walls of a pipe. A fluid in which friction can be neglected is called a *nonviscous fluid.* A fluid, flowing steadily without turbulence, is usually referred to as being in *streamline flow.* The rather complicated analysis is further simplified by the use of two great conservation principles: the conservation of mass, and the conservation of energy. *The law of conservation of mass results in a mathematical equation, usually called the equation of continuity. The law of conservation of energy is the basis of Bernoulli's theorem,* the subject matter of section 13.7.

Let us consider an incompressible fluid flowing in the pipe of figure 13.9. At a particular instant of time the small mass of fluid Δm, shown in the left-hand portion of the pipe will be considered. This mass is given by a slight modification of equation 13.2, as

$$\Delta m = \rho \Delta V \tag{13.33}$$

Because the pipe is cylindrical, the small portion of volume of fluid is given by the product of the cross-sectional area A_1 times the length of the pipe Δx_1 containing the mass Δm, that is,

$$\Delta V = A_1 \Delta x_1 \tag{13.34}$$

The length Δx_1 of the fluid in the pipe is related to the velocity v_1 of the fluid in the left-hand pipe. Because the fluid in Δx_1 moves a distance Δx_1 in time Δt, $\Delta x_1 = v_1 \Delta t$. Thus,

$$\Delta x_1 = v_1 \Delta t \tag{13.35}$$

Substituting equation 13.35 into equation 13.34, we get for the volume of fluid,

$$\Delta V = A_1 v_1 \Delta t \tag{13.36}$$

Figure 13.9

The law of conservation of mass and the equation of continuity.

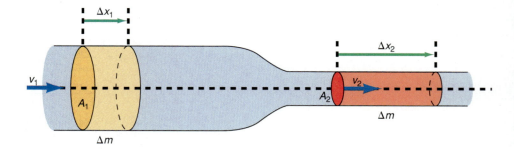

Vibratory Motion, Wave Motion, and Fluids

Substituting equation 13.36 into equation 13.33 yields the mass of the fluid as

$$\Delta m = \rho A_1 v_1 \Delta t \tag{13.37}$$

We can also express this as the rate at which the mass is flowing in the left-hand portion of the pipe by dividing both sides of equation 13.37 by Δt, thus

$$\frac{\Delta m}{\Delta t} = \rho A_1 v_1 \tag{13.38}$$

Example 13.15

Flow rate. What is the mass flow rate of water in a pipe whose diameter d is 10.0 cm when the water is moving at a velocity of 0.322 m/s.

Solution

The cross-sectional area of the pipe is

$$A_1 = \frac{\pi d_1^2}{4} = \frac{\pi (0.100 \text{ m})^2}{4}$$

$$= 7.85 \times 10^{-3} \text{ m}^2$$

The flow rate, found from equation 13.38, is

$$\frac{\Delta m}{\Delta t} = \rho A_1 v_1$$

$$= (1.00 \times 10^3 \text{ kg/m}^3)(7.85 \times 10^{-3} \text{ m}^2)(0.322 \text{ m/s})$$

$$= 2.53 \text{ kg/s}$$

Thus 2.53 kg of water flow through the pipe per second.

When this fluid reaches the narrow constricted portion of the pipe to the right in figure 13.9, the same amount of mass Δm is given by

$$\Delta m = \rho \Delta V \tag{13.39}$$

But since ρ is a constant, the same mass Δm must occupy the same volume ΔV. However, the right-hand pipe is constricted to the narrow cross-sectional area A_2. Thus, the length of the pipe holding this same volume must increase to a larger value Δx_2, as shown in figure 13.9. Hence, the volume of fluid is given by

$$\Delta V = A_2 \Delta x_2 \tag{13.40}$$

The length of pipe Δx_2 occupied by the fluid is related to the velocity of the fluid by

$$\Delta x_2 = v_2 \Delta t \tag{13.41}$$

Substituting equation 13.41 back into equation 13.40, we get for the volume of fluid,

$$\Delta V = A_2 v_2 \Delta t \tag{13.42}$$

It is immediately obvious that since A_2 has decreased, v_2 must have increased for the same volume of fluid to flow. Substituting equation 13.42 back into equation 13.39, the mass of the fluid flowing in the right-hand portion of the pipe becomes

$$\Delta m = \rho A_2 v_2 \Delta t \tag{13.43}$$

Dividing both sides of equation 13.43 by Δt yields the rate at which the mass of fluid flows through the right-hand side of the pipe, that is,

$$\frac{\Delta m}{\Delta t} = \rho A_2 v_2 \tag{13.44}$$

But *the **law of conservation of mass** states that mass is neither created nor destroyed in any ordinary mechanical or chemical process.* Hence, the law of conservation of mass can be written as

Mass flowing into the pipe = mass flowing out of the pipe

or

$$\frac{\Delta m}{\Delta t} = \frac{\Delta m}{\Delta t} \tag{13.45}$$

Thus, setting equation 13.38 equal to equation 13.44 yields

$$\rho A_1 v_1 = \rho A_2 v_2 \tag{13.46}$$

Equation 13.46 is called **the equation of continuity** and is an indirect statement of the law of conservation of mass. Since we have assumed an incompressible fluid, the densities on both sides of equation 13.46 are equal and can be canceled out leaving

$$A_1 v_1 = A_2 v_2 \tag{13.47}$$

Equation 13.47 is a special form of the equation of continuity for incompressible fluids (i.e., liquids).

Applying equation 13.47 to figure 13.9, we see that the velocity of the fluid v_2 in the narrow pipe to the right is given by

$$v_2 = \frac{A_1}{A_2} v_1 \tag{13.48}$$

Because the cross-sectional area A_1 is greater than the cross-sectional area A_2, the ratio A_1/A_2 is greater than one and thus the velocity v_2 must be greater than v_1.

Example 13.16

Applying the equation of continuity. In example 13.15 the cross-sectional area A_1 was 7.85×10^{-3} m² and the velocity v_1 was 0.322 m/s. If the diameter of the pipe to the right in figure 13.9 is 4.00 cm, find the velocity of the fluid in the right-hand pipe.

Solution

The cross-sectional area of the right-hand side of the pipe is

$$A_2 = \frac{\pi d_2^2}{4} = \frac{\pi (0.0400 \text{ m})^2}{4}$$

$$= 1.26 \times 10^{-3} \text{ m}^2$$

The velocity of the fluid on the right-hand side v_2, found from equation 13.48, is

$$v_2 = \frac{A_1}{A_2} v_1 = \left(\frac{7.85 \times 10^{-3} \text{ m}^2}{1.26 \times 10^{-3} \text{ m}^2} \right) (0.322 \text{ m/s})$$

$$= 2.01 \text{ m/s}$$

The fluid velocity increased more than six times when it flowed through the constricted pipe.

Therefore, as a general rule, the equation of continuity for liquids, equation 13.47, says that when the cross-sectional area of a pipe gets smaller, the velocity of the fluid must become greater in order that the same amount of mass *passes a given point in a given time. Conversely, when the cross-sectional area increases, the velocity of the fluid must decrease.* Equation 13.47, the equation of continuity, is sometimes written in the equivalent form

$$Av = \text{constant} \qquad (13.49)$$

Example 13.17

Flow rate revisited. What is the flow of mass per unit time for the example 13.16?

Solution

The rate of mass flow for the right-hand side of the pipe, given by equation 13.44, is

$$\frac{\Delta m}{\Delta t} = \rho A_2 v_2$$

$$= (1.0 \times 10^3 \text{ kg/m}^3)(1.26 \times 10^{-3} \text{ m}^2)(2.01 \text{ m/s})$$

$$= 2.53 \text{ kg/s}$$

Note that this is the same rate of flow found earlier for the left-hand side of the pipe, as it must be by the law of conservation of mass.

A compressible fluid (i.e., a gas) can have a variable density, and requires an additional equation to specify the flow velocity.

13.7 Bernoulli's Theorem

Bernoulli's theorem is a fundamental theory of hydrodynamics that describes a fluid in motion. It is really the application of the law of conservation of energy to fluid flow. Let us consider the fluid flowing in the pipe of figure 13.10. The left-hand side of the pipe has a uniform cross-sectional area A_1, which eventually tapers to the uniform cross-sectional area A_2 of the right-hand side of the pipe. The pipe is filled with a nonviscous, incompressible fluid. A uniform pressure p_1 is applied, such as from a piston, to a small element of mass of the fluid Δm and causes this mass to move through a distance Δx_1 of the pipe. Because the fluid is incompressible, the fluid moves throughout the rest of the pipe. The same small mass Δm, at the right-

Figure 13.10

Bernoulli's theorem.

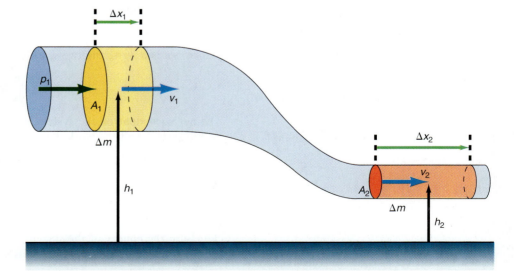

hand side of the pipe, moves through a distance Δx_2. The work done on the system by moving the small mass through the distance Δx_1 is given by the definition of work as

$$W_1 = F_1 \Delta x_1$$

Using equation 13.12, we can express the force F_1 moving the mass to the right in terms of the pressure exerted on the fluid as

$$F_1 = p_1 A_1$$

Hence,

$$W_1 = p_1 A_1 \Delta x_1$$

But

$$A_1 \Delta x_1 = \Delta V$$

the volume of the fluid moved through the pipe. Thus, we can write the work done on the system as

$$W_1 = p_1 \Delta V_1 \qquad (13.50)$$

As this fluid moves through the system, the fluid itself does work by exerting a force F_2 on the mass Δm on the right side, moving it through the distance Δx_2. Hence, the work done by the fluid system is

$$W_2 = F_2 \Delta x_2$$

But we can express the force F_2 in terms of the pressure p_2 on the right side by

$$F_2 = p_2 A_2$$

Therefore, the work done by the system is

$$W_2 = p_2 A_2 \Delta x_2$$

But

$$A_2 \Delta x_2 = \Delta V_2$$

the volume moved through the right side of the pipe. Thus, the work done by the system becomes

$$W_2 = p_2 \Delta V_2 \qquad (13.51)$$

But since the fluid is incompressible,

$$\Delta V_1 = \Delta V_2 = \Delta V$$

Hence, we can write the two work terms, equations 13.50 and 13.51, as

$$W_1 = p_1 \Delta V$$
$$W_2 = p_2 \Delta V$$

The net work done on the system is equal to the difference between the work done *on* the system and the work done *by* the system. Hence,

$$\text{Net work done on the system} = W_{\text{on}} - W_{\text{by}}$$
$$= W_1 - W_2 = p_1 \Delta V - p_2 \Delta V$$

$$\text{Net work done on the system} = (p_1 - p_2)\Delta V \qquad (13.52)$$

By the law of conservation of energy, the net work done on the system produces a change in the energy of the system. The fluid at position 1 is at a height h_1 above the reference level and therefore possesses a potential energy given by

$$PE_1 = (\Delta m)gh_1 \qquad (13.53)$$

Because this same fluid is in motion at a velocity v_1, it possesses a kinetic energy given by

$$KE_1 = \tfrac{1}{2}(\Delta m)v_1^2 \qquad (13.54)$$

Similarly at position 2, the fluid possesses the potential energy

$$PE_2 = (\Delta m)gh_2 \qquad (13.55)$$

and the kinetic energy

$$KE_2 = \tfrac{1}{2}(\Delta m)v_2^2 \qquad (13.56)$$

Therefore, we can now write the law of conservation of energy as

$$\text{Net work done on the system} = \text{Change in energy of the system} \qquad (13.57)$$

$$\text{Net work done on the system} = (E_{tot})_2 - (E_{tot})_1 \qquad (13.58)$$

$$\text{Net work done on the system} = (PE_2 + KE_2) - (PE_1 + KE_1) \qquad (13.59)$$

Substituting equations 13.52 through 13.56 into equation 13.59 we get

$$(p_1 - p_2)\Delta V = [(\Delta m)gh_2 + \tfrac{1}{2}(\Delta m)v_2^2] - [(\Delta m)gh_1 + \tfrac{1}{2}(\Delta m)v_1^2] \qquad (13.60)$$

But the total mass of fluid moved Δm is given by

$$\Delta m = \rho \Delta V \qquad (13.61)$$

Substituting equation 13.61 back into equation 13.60, gives

$$(p_1 - p_2)\Delta V = \rho(\Delta V)gh_2 + \tfrac{1}{2}\rho(\Delta V)v_2^2 - \rho(\Delta V)gh_1 - \tfrac{1}{2}\rho(\Delta V)v_1^2$$

Dividing each term by ΔV gives

$$(p_1 - p_2) = \rho gh_2 + \tfrac{1}{2}\rho v_2^2 - \rho gh_1 - \tfrac{1}{2}\rho v_1^2 \qquad (13.62)$$

If we place all the terms associated with the fluid at position 1 on the left-hand side of the equation and all the terms associated with the fluid at position 2 on the right-hand side, we obtain

$$p_1 + \rho gh_1 + \tfrac{1}{2}\rho v_1^2 = p_2 + \rho gh_2 + \tfrac{1}{2}\rho v_2^2 \qquad (13.63)$$

Equation 13.63 is the mathematical statement of

> **Bernoulli's theorem.** It says that the sum of the pressure, the potential energy per unit volume, and the kinetic energy per unit volume at any one location of the fluid is equal to the sum of the pressure, the potential energy per unit volume, and the kinetic energy per unit volume at any other location in the fluid, for a nonviscous, incompressible fluid in streamlined flow.

Since this sum is the same at any arbitrary point in the fluid, the sum itself must therefore be a constant. Thus, we sometimes write Bernoulli's equation in the equivalent form

$$p + \rho gh + \tfrac{1}{2}\rho v^2 = \text{constant} \qquad (13.64)$$

Example 13.18

Applying Bernoulli's theorem. In figure 13.10, the pressure $p_1 = 2.94 \times 10^3$ N/m², whereas the velocity of the water is $v_1 = 0.322$ m/s. The diameter of the pipe at location 1 is 10.0 cm and it is 5.00 m above the ground. If the diameter of the pipe at location 2 is 4.00 cm, and the pipe is 2.00 m above the ground, find the velocity of the water v_2 at position 2, and the pressure p_2 of the water at position 2.

The area A_1 is

$$A_1 = \frac{\pi d_1^2}{4} = \frac{\pi (0.100 \text{ m})^2}{4} = 7.85 \times 10^{-3} \text{ m}^2$$

whereas the area A_2 is

$$A_2 = \frac{\pi d_2^2}{4} = \frac{\pi (0.0400 \text{ m})^2}{4} = 1.26 \times 10^{-3} \text{ m}^2$$

The velocity at location 2 is found from the equation of continuity, equation 13.48, as

$$v_2 = \frac{A_1}{A_2} v_1 = \left(\frac{7.85 \times 10^{-3} \text{ m}^2}{1.26 \times 10^{-3} \text{ m}^2} \right)(0.322 \text{ m/s})$$

$$= 2.01 \text{ m/s}$$

The pressure at location 2 is found from rearranging Bernoulli's equation 13.63 as

$$p_2 = p_1 + \rho g h_1 + \tfrac{1}{2}\rho v_1^2 - \rho g h_2 - \tfrac{1}{2}\rho v_2^2$$

$$= 2.94 \times 10^3 \frac{\text{N}}{\text{m}^2} + \left(1 \times 10^3 \frac{\text{kg}}{\text{m}^3} \right) \left(9.80 \frac{\text{m}}{\text{s}^2} \right)(5.00 \text{ m})$$

$$+ \frac{1}{2} \left(1 \times 10^3 \frac{\text{kg}}{\text{m}^3} \right)(0.322 \text{ m/s})^2 - \left(1 \times 10^3 \frac{\text{kg}}{\text{m}^3} \right) \left(9.80 \frac{\text{m}}{\text{s}^2} \right)(2.00 \text{ m})$$

$$- \frac{1}{2} \left(1 \times 10^3 \frac{\text{kg}}{\text{m}^3} \right)(2.01 \text{ m/s})^2$$

$$= 2.94 \times 10^3 \text{ N/m}^2 + 4.9 \times 10^4 \text{ N/m}^2 + 5.18 \times 10^1 \text{ N/m}^2$$

$$- 1.96 \times 10^4 \text{ N/m}^2 - 2.02 \times 10^3 \text{ N/m}^2$$

$$= 3.04 \times 10^4 \text{ N/m}^2$$

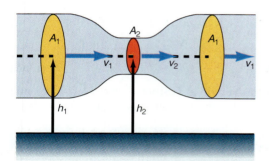

(a)

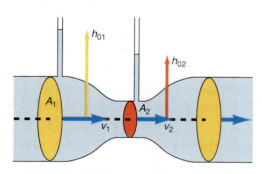

(b)

Figure 13.11

A Venturi meter.

13.8 Application of Bernoulli's Theorem

Let us now consider some special cases of Bernoulli's theorem.

The Venturi Meter

Let us first consider the constricted tube studied in figure 13.9 and slightly modified and redrawn in figure 13.11(a). Since the tube is completely horizontal $h_1 = h_2$ and there is no difference in potential energy between the locations 1 and 2. Bernoulli's equation therefore reduces to

$$p_1 + \tfrac{1}{2}\rho v_1^2 = p_2 + \tfrac{1}{2}\rho v_2^2 \qquad \text{(13.65)}$$

But by the equation of continuity,

$$v_2 = \frac{A_1}{A_2} v_1 \qquad \text{(13.48)}$$

Since A_1 is greater than A_2, v_2 must be greater than v_1, as shown before. Let us rewrite equation 13.65 as

$$p_2 = p_1 + \tfrac{1}{2}\rho v_1^2 - \tfrac{1}{2}\rho v_2^2$$

or

$$p_2 = p_1 + \tfrac{1}{2}\rho(v_1^2 - v_2^2) \qquad \text{(13.66)}$$

But since v_2 is greater than v_1, the quantity $(1/2)\rho(v_1^2 - v_2^2)$ is a negative quantity and when we subtract it from p_1, p_2 must be less than p_1. Thus, not only does the fluid speed up in the constricted tube, but the pressure in the constricted tube also decreases.

Example 13.19

When the velocity increases, the pressure decreases. In example 13.16, associated with figure 13.9, the velocity v_1 in area A_1 was 0.322 m/s and the velocity v_2 in area A_2 was found to be 2.01 m/s. If the pressure in the left pipe is 2.94×10^3 Pa, what is the pressure p_2 in the constricted pipe?

Solution

The pressure p_2, found from equation 13.66, is

$$p_2 = p_1 + \tfrac{1}{2}\rho(v_1^2 - v_2^2)$$

$$= 2.94 \times 10^3 \text{ Pa} + (\tfrac{1}{2})(1 \times 10^3 \text{ kg/m}^3)[(0.322 \text{ m/s})^2 - (2.01 \text{ m/s})^2]$$

$$= 2.94 \times 10^3 \text{ N/m}^2 - 1.97 \times 10^3 \text{ N/m}^2 = 9.7 \times 10^2 \text{ Pa}$$

Thus, the pressure of the water in the constricted portion of the tube has decreased to 9.7×10^2 Pa. Note that in example 13.18 of section 13.7 the pressure in the constricted area of the pipe was greater than in the larger area of the pipe. This is because in that example the pipe was not all at the same level (i.e., $h_1 \neq h_2$). An additional pressure arose on the right side because of the differences in the heights of the two pipes.

The effect of the decrease in pressure with the increase in speed of the fluid in a horizontal pipe is called the **Venturi effect,** and a simple device called a **Venturi meter,** based on this Venturi effect, is used to measure the velocity of fluids in pipes. A Venturi meter is shown schematically in figure 13.11(b). The device is basically the same as the pipe in 13.11(a) except for the two vertical pipes connected to the main pipe as shown. These open vertical pipes allow some of the water in the pipe to flow upward into the vertical pipes. The height that the water rises in the vertical pipes is a function of the pressure in the horizontal pipe. As just seen, the pressure in pipe 1 is greater than in pipe 2 and thus the height of the vertical column of water in pipe 1 will be greater than the height in pipe 2. By actually measuring the height of the fluid in the vertical columns the pressure in the horizontal pipe can be determined by the hydrostatic equation 13.7. Thus, the pressure in pipe 1 is

$$p_1 = \rho g h_{01}$$

and the pressure in pipe 2 is

$$p_2 = \rho g h_{02}$$

where h_{01} and h_{02} are the heights shown in figure 13.11(b). We can now write Bernoulli's equation 13.65 as

$$\rho g h_{01} + \tfrac{1}{2}\rho v_1^2 = \rho g h_{02} + \tfrac{1}{2}\rho v_2^2$$

Replacing v_2 by its value from the continuity equation 13.65, we get

$$\rho g h_{01} + \frac{1}{2}\rho v_1^2 = \rho g h_{02} + \frac{1}{2}\rho\left[\left(\frac{A_1}{A_2}\right)v_1\right]^2$$

$$\rho g h_{01} - \rho g h_{02} = \frac{1}{2}\rho\frac{A_1^2}{A_2^2}v_1^2 - \frac{1}{2}\rho v_1^2$$

$$\rho g(h_{01} - h_{02}) = \frac{1}{2}\rho\left(\frac{A_1^2}{A_2^2} - 1\right)v_1^2$$

Solving for v_1^2, we have

$$v_1^2 = \frac{\rho g(h_{01} - h_{02})}{\frac{1}{2}\rho[(A_1^2/A_2^2) - 1]}$$

Solving for v_1, we get

$$v_1 = \sqrt{\frac{2g(h_{01} - h_{02})}{(A_1^2/A_2^2) - 1}} \tag{13.67}$$

Equation 13.67 now gives us a simple means of determining the velocity of fluid flow in a pipe. The main pipe containing the fluid is opened and the Venturi meter is connected between the opened pipes. When the fluid starts to move, the heights h_{01} and h_{02} are measured. Since the cross-sectional areas are easily determined by measuring the diameters of the pipes, the velocity of the fluid flow is easily calculated from equation 13.67.

Example 13.20

A Venturi meter. A Venturi meter reads heights of $h_{01} = 30.0$ cm and $h_{02} = 10.0$ cm. Find the velocity of flow v_1 in the pipe. The area $A_1 = 7.85 \times 10^{-3}$ m^2 and the area of $A_2 = 1.26 \times 10^{-3}$ m^2.

Solution

The velocity of flow v_1 in the main pipe, found from equation 13.67, is

$$v_1 = \sqrt{\frac{2g(h_{01} - h_{02})}{(A_1^2/A_2^2) - 1}}$$

$$= \sqrt{\frac{2(9.80 \text{ m/s}^2)(0.300 \text{ m} - 0.100 \text{ m})}{\dfrac{(7.85 \times 10^{-3} \text{ m}^2)^2}{(1.26 \times 10^{-3} \text{ m}^2)^2} - 1}}$$

$$= 0.322 \text{ m/s}$$

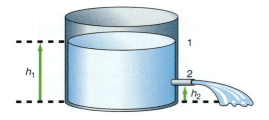

Figure 13.12
Flow from an orifice.

The Flow of a Liquid Through an Orifice

Let us consider the large tank of water shown in figure 13.12. Let the top of the fluid be location 1 and the orifice be location 2. Bernoulli's theorem applied to the tank, taken from equation 13.63, is

$$p_1 + \rho g h_1 + \tfrac{1}{2}\rho v_1^2 = p_2 + \rho g h_2 + \tfrac{1}{2}\rho v_2^2$$

But the pressure at the top of the tank and the outside pressure at the orifice are both p_0, the normal atmospheric pressure. Also, because of the very large volume of fluid, the small loss through the orifice causes an insignificant vertical motion of the top of the fluid. Thus, $v_1 \approx 0$. Bernoulli's equation becomes

$$p_0 + \rho g h_1 = p_0 + \rho g h_2 + \tfrac{1}{2}\rho v_2^2$$

The pressure term p_0 on both sides of the equation cancels out. Also h_2 is very small compared to h_1 and it can be neglected, leaving

$$\rho g h_1 = \tfrac{1}{2}\rho v_2^2$$

Solving for the velocity of efflux, we get

$$v_2 = \sqrt{2gh_1} \tag{13.68}$$

Notice that the velocity of efflux is equal to the velocity that an object would acquire when dropped from the height h_1.

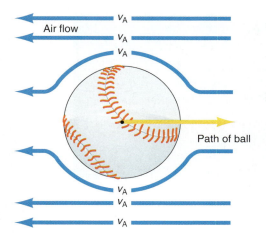

Air flow

(a) Nonspinning baseball

Path of ball

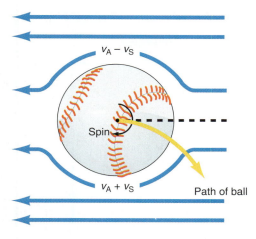

(b) Spinning baseball

Spin

$v_A - v_S$

$v_A + v_S$

Path of ball

Figure 13.13
The curving baseball.

Example 13.21

The velocity of efflux. A large water tank, 10.0 m high, springs a leak at the bottom of the tank. Find the velocity of the escaping water.

Solution

The velocity of efflux, found from equation 13.68, is

$$v_2 = \sqrt{2gh_1} = \sqrt{2(9.80 \text{ m/s}^2)(10.0 \text{ m})}$$
$$= 14.0 \text{ m/s}$$

The Curving Baseball

When a nonspinning ball is thrown through the air it follows the straight line path shown in figure 13.13(a). The air moves over the top and bottom of the ball with a speed v_A. If the ball is now released with a downward spin, as shown in figure 13.13(b), then the spinning ball drags some air around with it. At the top of the ball, there is a velocity of the air v_A to the left, and a velocity of the dragged air on the spinning baseball v_S to the right. Thus, the relative velocity of the air with respect to the ball is $v_A - v_S$ at the top of the ball. At the bottom of the ball the dragged air caused by the spin of the baseball v_S is in the same direction as the velocity of the air v_A moving past the ball. Thus, the relative velocity of the air with respect to the bottom of the ball is $v_A + v_S$. Hence, the velocity of the air at the top of the ball, $v_A - v_S$, is less than the velocity of the air at the bottom of the ball, $v_A + v_S$. By the Venturi principle, the pressure of the fluid is smaller where the velocity is greater. Thus, the pressure on the bottom of the ball is less than the pressure on the top, that is,

$$p_{top} > p_{bottom}$$

But the pressure is related to the force by $p = F/A$. Hence, the force acting on the top of the ball is greater than the force acting on the bottom of the ball, that is,

$$F_{top} > F_{bottom}$$

Therefore, the ball curves downward, or sinks, as it approaches the batter. By spinning the ball to the right (i.e., clockwise) as viewed from above, the ball curves toward the right. By spinning the ball to the left (i.e., counterclockwise) as viewed from above the ball, the ball curves toward the left. Spins about various axes through the ball can cause the ball to curve to the left and downward, to the left and upward, and so on.

Lift on an Airplane Wing

Another example of the Venturi effect can be seen with an aircraft wing, as shown in figure 13.14. The air flowing over the top of the wing has a greater distance to travel than the air flowing under the bottom of the wing. In order for the flow to be streamlined and for the air at the leading edge of the wing to arrive at the trailing edge at the same time, whether it goes above or below the wing, the velocity of the air over the top of the wing must be greater than the velocity of the air at the bottom of the wing. But by the Venturi principle, if the velocity is greater at the top of the wing, the pressure must be less there than at the bottom of the wing. Thus, p_2 is greater than p_1 and therefore $F_2 > F_1$. That is, there is a net positive force $F_2 - F_1$ acting upward on the wing, producing lift on the airplane wing.

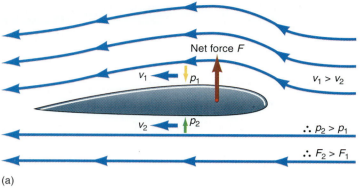

Net force F

v_1 p_1

$v_1 > v_2$

v_2 p_2

$\therefore p_2 > p_1$

$\therefore F_2 > F_1$

(a)

(b)

Figure 13.14

An airfoil.

An Essay on the Application of Physics

The Flow of Blood in the Human Body

Human blood consists of a plasma, the fluid, and red and white corpuscles that are immersed in the plasma. Because blood is a fluid, the laws of physics can be applied to the flow of blood throughout the body. A schematic diagram of the circulatory system, which transports blood and oxygen around the body, is shown in figure 1. It consists of (1) the heart, which is the pump that is responsible for supplying the pressure to move the blood; (2) the lungs, which are the source of oxygen for all the cells of the body; (3) the arteries, which are connecting blood vessels that pass the blood from the heart to various parts of the body; (4) the capillaries, which are extremely small blood vessels that bring the oxygenated blood down to the layer of human cells; and (5) the veins, which are blood vessels that return deoxygenated blood to the heart to complete the circulatory system.

The heart is the pump that circulates the blood throughout the body and a diagram of it is shown in figure 2. Blood, containing carbon dioxide, returns to the heart by the veins and enters the right auricle. It is then pumped from the right ventricle to the pulmonary artery to the lungs where it dumps the waste carbon dioxide and picks up a new supply of oxygen. It then returns to the left auricle of the heart. The left ventricle then pumps this oxygen rich blood to the aorta, the main artery of the body, for distribution to the rest of the body.

For a person at rest, the heart pumps approximately 5.00 liters of blood per minute (8.33×10^{-5} m³/s) at a rate of about 70 beats per minute. For a person engaged in very strenuous exercise the heart can pump up to 25.0 liters of blood per minute

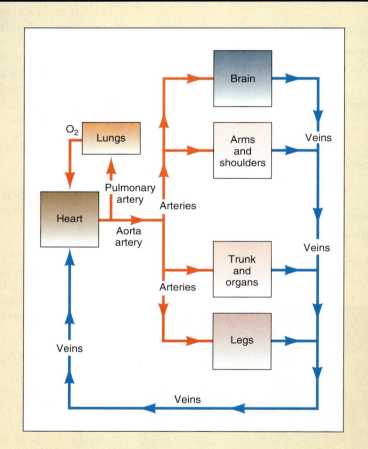

Figure 1

The circulatory system.

Figure 2

The human heart.

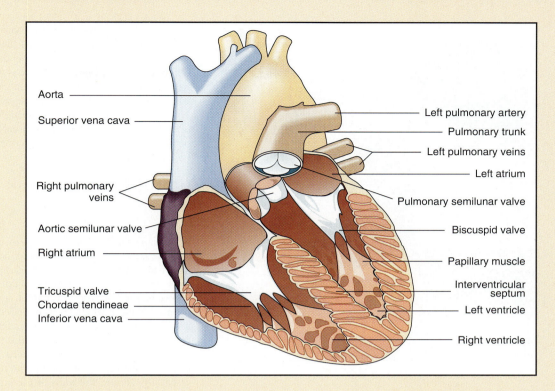

Aorta

Superior vena cava

Right pulmonary veins

Aortic semilunar valve

Right atrium

Tricuspid valve

Chordae tendineae

Inferior vena cava

Left pulmonary artery

Pulmonary trunk

Left pulmonary veins

Left atrium

Pulmonary semilunar valve

Biscuspid valve

Papillary muscle

Interventricular septum

Left ventricle

Right ventricle

$(41.7 \times 10^{-5} \text{ m}^3/\text{s})$ at a rate of about 180 beats per minute. We can determine the speed of the blood as it enters the aorta by a generalization of equation 13.36, as

$$\frac{\Delta V}{\Delta t} = A_A v_A \qquad \textbf{(13H.1)}$$

where $\Delta V/\Delta t$ is the rate at which the blood is flowing from the heart into the aorta, A_A is the cross-sectional area of the aorta, and v_A is the speed of the blood in the aorta. The diameter of the aorta is about 2.00 cm giving an area of

$$A = \pi r^2$$
$$= \pi(0.01 \text{ m})^2 = 3.14 \times 10^{-4} \text{ m}^2$$

The speed of the blood in the aorta is therefore

$$v_A = \frac{\Delta V/\Delta t}{A_A}$$
$$= \frac{8.33 \times 10^{-5} \text{ m}^3/\text{s}}{3.14 \times 10^{-4} \text{ m}^2}$$
$$= 0.265 \text{ m/s} = 26.5 \text{ cm/s} \qquad \textbf{(13H.2)}$$

We can determine the speed of the blood in the capillaries by the continuity equation 13.47, as

$$A_A v_A = A_C v_C \qquad \textbf{(13H.3)}$$

where A_A is the cross-sectional area of the aorta, which was just determined as 3.14×10^{-4} m²; v_A is the speed of the blood in the aorta, which was just found to be 26.5 cm/s; and A_C is the cross-sectional area of a capillary tube, which is quite small.

However, because there are literally billions of these capillaries the effective cross-sectional area of all these capillaries combined is approximately 2500×10^{-4} m². The speed of the blood in the capillary becomes

$$v_C = \frac{A_A}{A_C} v_A$$
$$= \left(\frac{3.14 \times 10^{-4} \text{ m}^2}{2500 \times 10^{-4} \text{ m}^2} \right)(26.5 \text{ cm/s})$$
$$= 0.0333 \text{ cm/s}$$

Thus, the blood moves relatively slowly at the level of the capillaries.

Finally, we should note that the body controls the flow of blood through the arteries by muscles that surround the arteries. When the muscles contract, the diameter of the artery is reduced. From the equation of continuity, $Av =$ constant. By decreasing the diameter of the artery, the cross-sectional area of the artery decreases and hence the speed of blood must increase through the artery. Alternatively, when the muscles are relaxed, the diameter of the artery increases to its former size, the cross-sectional area increases, and the speed of the blood decreases. With advancing age the arterial muscles lose some of this ability to contract, a situation called hardening of the arteries, and the control of blood flow is somewhat diminished.

A good indication of how well the heart is functioning is obtained by measuring the pressure that the heart exerts when pumping blood, and when at rest. The device used to measure blood pressure is called a sphygmomanometer. (The word is derived from the Greek word *sphygmos*, meaning pulse, and the

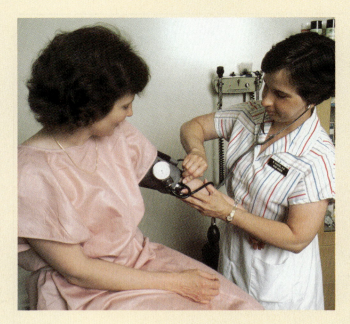

Figure 3

A nurse measures the blood pressure of a patient.

word *manometer,* which is a pressure measuring device. Hence, a sphygmomanometer is a device for measuring pulse pressure, or blood pressure.) The device consists of an air bag, called a cuff, that is wrapped around the upper arm of the patient at the level of the heart. A hand pump is used to inflate the cuff, and the pressure exerted by the cuff on the arm is measured by the mercury manometer. The pressure exerted by the cuff is increased until the pressure is great enough to collapse the brachial artery in the arm, cutting off the blood supply to the rest of the arm. A stethoscope is placed over the brachial artery and the pressure in the cuff is slowly decreased. When the pressure in the cuff becomes low enough, the pressure exerted by the heart is large enough to force the artery open and some blood squirts through. This blood flowing through the narrow restriction becomes turbulent and makes a noise as it enters the open portion of the artery. The physician hears this noise through the stethoscope, and simultaneously observes the pressure indicated on the manometer, expressed in terms of mm of Hg. At this point the pressure exerted by the heart, called the systolic pressure, is equal to the pressure exerted by the cuff. A normal systolic pressure is around 120 mm of Hg.

As the pressure in the cuff is decreased the turbulent flow noise is still heard in the stethoscope until the lowest pressure exerted by the heart, the diastolic pressure, is equal to the pressure exerted by the cuff. At this point the artery is completely open and the blood is no longer in turbulent flow and the characteristic noise disappears. The pressure is read from the mercury manometer at this point. This pressure is the pressure that the heart exerts when it is at rest. The normal diastolic pressure is around 80 mm of Hg. The combined systolic and diastolic pressures are usually indicated in the form 120/80. If the sys-

tolic pressure becomes too high, above about 150 mm of Hg, the patient has high blood pressure. If the systolic pressure becomes too large for a long period of time, damage can be done to the different organs of the body. If the systolic pressure becomes extremely large, arteries in the brain can rupture and the person will have a stroke. If the diastolic pressure exceeds 90 mm of Hg, the person is also said to have high blood pressure. This type of high blood pressure causes eventual damage to the heart itself, because it is operating under high pressures even while it is supposed to be resting.

For the type of streamlined flow considered in this chapter the flow of fluid per unit time was shown to be

$$\frac{\Delta V}{\Delta t} = Av \qquad (13.36)$$

which is essentially the equation of continuity. In this type of flow the speed v was the same throughout the cross-sectional area A considered. However, some fluids have a significant frictional force between the layers of the fluid, and this frictional effect, known as the *viscosity* of the fluid, must then be taken into account. A fluid in which frictional effects are significant is called a *viscous fluid* and the fluid flow is referred to as *laminar flow,* flow in layers. For such viscous fluids the speed v is not the same throughout the cross-sectional area A. The maximum speed occurs at the center of the pipe or tube, whereas the speed is essentially zero at the walls of the pipe. Experimental work by J. L. Poiseuille (1799–1869), a French scientist, and subsequently confirmed by theory, showed that the flow rate for viscous fluids is given by

$$\frac{\Delta V}{\Delta t} = \frac{(\Delta p)\pi R^4}{8\eta L} \qquad (13H.4)$$

where Δp is the pressure difference between both ends of the pipe, R is the radius of the pipe, L is the length of the pipe, and η is the coefficient of viscosity of the fluid. Equation 13H.4 is called *Poiseuille's equation.* Note that the flow rate is inversely proportional to the coefficient of viscosity of the fluid. Thus, a very viscous fluid (high value of η) flows very slowly compared to a fluid of low viscosity. That is, everything else being equal, molasses flows at a slower rate than water. Human blood is a viscous fluid, the greater the number of red corpuscles in the blood the greater the viscosity. The viscosity of human blood varies from about 1.50×10^{-3} (N/m²)s for plasma, to about 4.00×10^{-3} (N/m²)s for whole blood. Also note that the flow rate depends on the fourth power of the radius of the pipe. If the radius is doubled, the flow rate is multiplied by a factor of 16. This relation is important in the selection of the size of hypodermic needles.

Example 13H.1

A blood transfusion. A person is receiving a blood transfusion. The bottle containing the blood is elevated 75.0 cm above the arm of the person. The needle is 4.00 cm long and has a diameter of 0.500 mm. Find the rate at which the blood flows through the needle.

Solution

The rate of flow of blood is found from equation 13H.4, where η, the viscosity of blood, is 4.00×10^{-3} Ns/m². Let us assume that the total pressure differential is obtained by the effects of gravity from the hydrostatic equation, equation 13.7. The density of blood is about 1050 kg/m³. Thus,

$$\Delta p = \rho g h$$

$$= (1050 \text{ kg/m}^3)(9.80 \text{ m/s}^2)(0.750 \text{ m})$$

$$= 7.72 \times 10^3 \text{ Pa}$$

The blood flow rate now obtained is

$$\frac{\Delta V}{\Delta t} = \frac{(\Delta p)\pi R^4}{8\eta L} \qquad \text{(13H.4)}$$

$$= \frac{(7.72 \times 10^3 \text{ N/m}^2)(\pi)(0.250 \times 10^{-3} \text{ m})^4}{8(4.00 \times 10^{-3} \text{ Ns/m}^2)(0.0400 \text{ m})}$$

$$= 7.40 \times 10^{-8} \text{ m}^3/\text{s}$$

The Language of Physics

Fluids
A fluid is any substance that can flow. Hence, liquids and gases are both considered to be fluids (p. 367).

Fluid statics or hydrostatics
The study of fluids at rest (p. 367).

Fluid dynamics or hydrodynamics
The study of fluids in motion (p. 367).

Density
The amount of mass in a unit volume of a substance (p. 367).

Pressure
The magnitude of the normal force acting per unit surface area (p. 368).

The hydrostatic equation
An equation that gives the pressure of a fluid at a particular depth (p. 370).

Barometer
An instrument that measures atmospheric pressure (p. 371).

Gauge pressure
The pressure indicated on a pressure measuring gauge. It is equal to the absolute pressure minus normal atmospheric pressure (p. 375).

Pascal's principle
If the pressure at any point in an enclosed fluid at rest is changed, the pressure changes by an equal amount at all points in the fluid (p. 376).

Archimedes' principle
A body immersed in a fluid is buoyed up by a force that is equal to the weight of the fluid displaced. A body floats when the weight of the body is equal to the weight of the fluid displaced (p. 378).

Law of conservation of mass
In any ordinary mechanical or chemical process, mass is neither created nor destroyed (p. 384).

The equation of continuity
An equation based on the law of conservation of mass, that indicates that

when the cross-sectional area of a pipe gets smaller, the velocity of the fluid must become greater. Conversely, when the cross-sectional area increases, the velocity of the fluid must decrease (p. 384).

Bernoulli's theorem
The sum of the pressure, the potential energy per unit volume, and the kinetic energy per unit volume at any one location of the fluid is equal to the sum of the pressure, the potential energy per unit volume, and the kinetic energy per unit volume at any other location in the fluid, for a nonviscous, incompressible fluid in streamlined flow (p. 387).

Venturi effect
The effect of the decrease in pressure with the increase in speed of the fluid in a horizontal pipe (p. 389).

Venturi meter
A device that uses the Venturi effect to measure the velocity of fluids in pipes (p. 389).

Summary of Important Equations

Density
$$\rho = \frac{m}{V} \tag{13.1}$$

Mass
$$m = \rho V \tag{13.2}$$

Pressure
$$p = \frac{F}{A} \tag{13.3}$$

Hydrostatic equation
$$p = \rho g h \tag{13.7}$$

Force
$$F = pA \tag{13.12}$$

Absolute and gauge pressure
$$p_{abs} = p_{gauge} + p_0 \tag{13.15}$$

Hydraulic lift
$$F = \frac{A}{a} f \tag{13.18}$$

$$y_1 = \frac{A}{a} y_2 \tag{13.23}$$

Archimedes' principle
$$\text{Buoyant force} = \text{Weight of water displaced} \tag{13.30}$$

Mass flow rate
$$\frac{\Delta m}{\Delta t} = \rho A v \tag{13.38}$$

Equation of continuity
$$A_1 v_1 = A_2 v_2 \tag{13.47}$$
$$Av = \text{constant} \tag{13.49}$$

Work done in moving a fluid
$$W = p \Delta V \tag{13.50}$$

Bernoulli's theorem
$$p_1 + \rho g h_1 + \tfrac{1}{2}\rho v_1^2 = p_2 \\ + \rho g h_2 + \tfrac{1}{2}\rho v_2^2 \tag{13.63}$$
and
$$p + \rho g h \\ + \tfrac{1}{2}\rho v^2 = \text{constant} \tag{13.64}$$

Questions for Chapter 13

1. Discuss the differences between solids, liquids, and gases.
†2. Hieron II, King of Syracuse in ancient Greece, asked his relative Archimedes to determine if the gold crown made for him by the local goldsmith, was solid gold or a mixture of gold and silver. How did Archimedes, or how could you, determine whether or not the crown was pure gold?

3. When you fly in an airplane you find that your ears keep "popping" when the plane is ascending or descending. Explain why.
4. Using a barometer and the direction of the wind, describe how you could make a reasonable weather forecast.
†5. A pilot uses an aneroid barometer as an altimeter that is calibrated to a standard atmosphere. What happens to the aircraft if the temperature of the atmosphere does not coincide with the standard atmosphere?
†6. Does a sphygmomanometer measure gauge pressure or absolute pressure?

7. How would you define a mechanical advantage for the hydraulic lift?
8. In example 13.13, could the iron block sink to a depth of 39.4 cm in a pool of water 100 cm deep and then float at that point? Why or why not?
†9. How does eating foods very high in cholesterol have an effect on the arteries and hence the flow of blood in the body?
†10. Why is an intravenous bottle placed at a height h above the arm of a patient?

Problems for Chapter 13

13.2 Density

1. A cylinder 3.00 cm in diameter and 3.00 cm high has a mass of 15.0 g. What is its density?
2. Find the mass of a cube of iron 10.0 cm on a side.
3. A gold ingot is 50.0 cm by 20.0 cm by 10.0 cm. Find (a) its mass and (b) its weight.
4. Find the mass of the air in a room 6.00 m by 8.00 m by 3.00 m.
5. Assume that the earth is a sphere. Compute the average density of the earth.
6. Find the weight of 1.00 liter of air.
7. A crown, supposedly made of gold, has a mass of 8.00 kg. When it is placed in a full container of water, 691 cm³ of water overflows. Is the crown made of pure gold or is it mixed with some other materials?

8. A solid brass cylinder 10.0 cm in diameter and 25.0 cm long is soldered to a solid iron cylinder 10.0 cm in diameter and 50.0 cm long. Find the weight of the combined cylinder.
9. An annular cylinder of 2.50-cm inside radius and 4.55-cm outside radius is 10.5 cm high. If the cylinder has a mass of 5.35 kg, find its density.

13.3 Pressure

10. As mentioned in the text, a non-SI unit of pressure is the torr, named after Torricelli, which is equal to the pressure exerted by a column of mercury 1 mm high. Express a pressure of 2.53×10^5 Pa in torrs.

†11. From the knowledge of normal atmospheric pressure at the surface of the earth, compute the approximate mass of the atmosphere.
12. A barometer reads a height of 72.0 cm of Hg. Express this atmospheric pressure in terms of (a) in. of Hg, (b) mb, (c) lb/in.², and (d) Pa.
13. (a) A "high" pressure area of 1030 mb moves into an area. What is this pressure expressed in lb/in.²? (b) A "low" pressure area of 980 mb moves into an area. What is this pressure expressed in lb/in.²?
14. Normal systolic blood pressure is approximately 120 mm of Hg and normal diastolic pressure is 80 mm of Hg. Express these pressures in terms of Pa and lb/in.².

Vibratory Motion, Wave Motion, and Fluids

15. The point of a 10-penny nail has a diameter of 1.00 mm. If the nail is driven into a piece of wood with a force of 150 N, find the pressure that the tip of the nail exerts on the wood.

16. The gauge pressure in the tires of your car is 30.0 lb/in.². What is the absolute pressure of the air in the tires?

17. The gauge pressure in the tires of your car is 2.42×10^5 N/m². What is the absolute pressure of the air in the tires?

18. What is the water pressure and the absolute pressure in a swimming pool at depths of (a) 1.00 m, (b) 2.00 m, (c) 3.00 m, and (d) 4.00 m?

19. Find the force exerted by normal atmospheric pressure on the top of a table 1.00 m high, 1.00 m long, 0.75 m wide, and 0.10 m thick. What is the force on the underside of the table top exerted by normal atmospheric pressure?

20. What force is exerted on the top of the roof by normal atmospheric pressure?

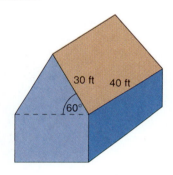

21. If normal atmospheric pressure can support a column of Hg 76.0 cm high, how high a column will it support of (a) water, (b) benzene, (c) alcohol, and (d) glycerine?

22. What is the minimum pressure of water entering a building if the pressure at the second floor faucet, 16.0 ft above the ground, is to be 5.00 lb/in.²?

23. The water main pressure entering a house is 31.0 N/cm². What is the pressure at the second floor faucet, 6.00 m above the ground? What is the maximum height of any faucet such that water will still flow from it?

24. A barometer reads 76.0 cm of Hg at the base of a tall building. The barometer is carried to the roof of the building and now reads 75.6 cm of Hg. If the average density of the air is 1.28 kg/m³, what is the height of the building?

25. The hatch of a submarine is 100 cm by 50.0 cm. What force is exerted on this hatch by the water when the submarine is 50.0 m below the surface?

13.4 Pascal's Principle

26. In the hydraulic lift of figure 13.5, the diameter $d_1 = 10.0$ cm and $d_2 = 50.0$ cm. If a force of 10.0 N is applied at the small piston, (a) what force will appear at the large piston? (b) If the large piston is to move through a height of 2.00 m, what must the total displacement of the small piston be?

27. In a hydraulic lift, the large piston exerts a force of 25.0 N when a force of 3.50 N is applied to the smaller piston. If the smaller piston has a radius of 12.5 cm, and the lift is 65.0% efficient, what must be the radius of the larger piston?

28. The theoretical mechanical advantage (TMA) of a hydraulic lift is equal to the ratio of the force that you get out of the lift to the force that you must put into the lift. Show that the theoretical mechanical advantage of the hydraulic lift is given by

$$\text{TMA} = \frac{F_{\text{out}}}{F_{\text{in}}} = \frac{A_{\text{out}}}{A_{\text{in}}} = \frac{y_{\text{in}}}{y_{\text{out}}}$$

where A_{out} is the area of the output piston, A_{in} is the area of the input piston, y_{in} is the distance that the input piston moves, and y_{out} is the distance that the output piston moves.

13.5 Archimedes' Principle

29. Find the weight of a cubic block of iron 5.00 cm on a side. This block is now hung from a spring scale such that the block is totally submerged in water. What would the scale indicate for the weight (called the apparent weight) of the block?

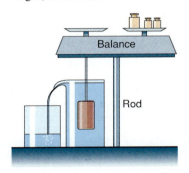

30. A copper cylinder 5.00 cm high and 3.00 cm in diameter is hung from a spring scale such that the cylinder is totally submerged in ethyl alcohol. Find the apparent weight of the block.

31. Find the buoyant force on a brass block 10.5 cm long by 12.3 cm wide by 15.0 cm high when placed in (a) water, (b) glycerine, and (c) mercury.

32. If the iron block in example 13.13 were placed in a pool of mercury instead of the water would it float or sink? If it floats, to what depth does it sink before it floats?

†33. A block of wood sinks 8.00 cm in pure water. How far will it sink in salt water?

34. A weather balloon contains 33.5 m³ of helium at the surface of the earth. Find the largest load this balloon is capable of lifting. The density of helium is 0.1785 kg/m³.

13.6 The Equation of Continuity

35. A 3/4-in. pipe is connected to a 1/2-in. pipe. If the velocity of the fluid in the 3/4-in. pipe is 2.00 ft/s, what is the velocity in the 1/2-in. pipe? How much water flows per second from the 1/2-in. pipe?

36. A 2.50-cm pipe is connected to a 0.900-cm pipe. If the velocity of the fluid in the 2.50-cm pipe is 1.50 m/s, what is the velocity in the 0.900-cm pipe? How much water flows per second from the 0.900-cm pipe?

37. A duct for a home air conditioning unit is 10.0 in. in diameter. If the duct is to remove the air in a room 20.0 ft by 24.0 ft by 8.00 ft high every 20.0 min, what must the velocity of the air in the duct be?

38. A duct for a home air-conditioning unit is 35.0 cm in diameter. If the duct is to remove the air in a room 9.00 m by 6.00 m by 3.00 m high every 15.0 min, what must the velocity of the air in the duct be?

13.7 Bernoulli's Theorem

39. Water enters the house from a main at a pressure of 1.5×10^5 Pa at a speed of 40.0 cm/s in a pipe 4.00 cm in diameter. What will be the pressure in a 2.00-cm pipe located on the second floor 6.00 m high when no water is flowing from the upstairs pipe? When the water starts flowing, at what velocity will it emerge from the upstairs pipe?

40. A can of water 30.0 cm high sits on a table 80.0 cm high. If the can develops a leak 5.00 cm from the bottom, how far away from the table will the water hit the floor?

41. Water rises to a height $h_{01} = 35.0$ cm, and $h_{02} = 10.0$ cm, in a Venturi meter, figure 13.11(b). The diameter of the first pipe is 4.00 cm, whereas the diameter of the second pipe is 2.00 cm. What is the velocity of the water in the first and second pipe? What is the mass flow rate and the volume flow rate?

Additional Problems

42. A car weighs 2890 lb and the gauge pressure of the air in each tire is 30 lb/in.². Assuming that the weight of the car is evenly distributed over the four tires, (a) find the area of each tire that is flat on the ground and (b) if the width of the tire is 5.00 inches, find the length of the tire that is in contact with the ground.

43. A certain portion of a rectangular, concrete flood wall is 12.0 m high and 30.0 m long. During severe flooding of the river, the water level rises to a height of 10.0 m. Find (a) the water pressure at the base of the flood wall, (b) the average water pressure exerted on the flood wall, and (c) the average force exerted on the flood wall by the water.

44. The Vehicle Assembly Building at the Kennedy Space Center is 160 m high. Assuming the density of air to be a constant, find the difference in atmospheric pressure between the ground floor and the ceiling of the building.

45. If the height of a water tower is 20.0 m, what is the pressure of the water as it comes out of a pipe at the ground?

†46. A 20.0-g block of wood floats in water to a depth of 5.00 cm. A 10.0-g block is now placed on top of the first block, but it does not touch the water. How far does the combination sink?

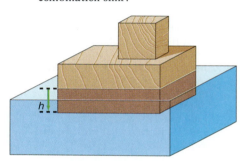

†47. An iron ball, 4.00 cm in diameter, is dropped into a tank of water. Assuming that the only forces acting on the ball are gravity and the buoyant force, determine the acceleration of the ball. Discuss the assumption made in this problem.

†48. If 80% of a floating cylinder is beneath the water, what is the density of the cylinder?

†49. From knowing that the density of an ice cube is 920 kg/m³ can you determine what percentage of the ice cube will be submerged when in a glass of water?

†50. Find the equation for the length of the side of a cube of material that will give the same buoyant force as (a) a sphere of radius r and (b) a cylinder of radius r and height h, if both objects are completely submerged.

†51. Find the radius of a solid cylinder that will experience the same buoyant force as an annular cylinder of radii $r_2 = 4.00$ cm and $r_1 = 3.00$ cm. Both cylinders have the same height h.

†52. A cone of maximum radius r_0 and height h_0, is placed in a fluid, as shown in the diagram. The volume of a right circular cone is given by

$$V_{\text{cone}} = \tfrac{1}{3}\pi r^2 h$$

(a) Find the equation for the weight of the cone. (b) If the cone sinks so that a height h_1 remains out of the fluid, find the equation for the volume of the cone that is immersed in the fluid. (c) Find the equation for the buoyant force acting on the cone. (d) Show that the height h_1 that remains out of the fluid is given by

$$h_1 = \sqrt[3]{1 - (\rho_c/\rho_f)}\, h_0$$

where ρ_c is the density of the cone and ρ_f is the density of the fluid. (e) If we approximate an iceberg by a cone, find the percentage height of the iceberg that sticks out of the salt water, and the percentage volume of the iceberg that is below the water.

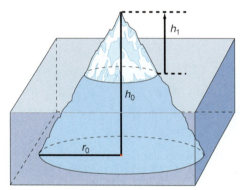

†53. A can 30.0 cm high is filled to the top with water. Where should a hole be made in the side of the can such that the escaping water reaches the maximum distance x in the horizontal direction? (*Hint:* calculate the distance x for values of h from 0 to 30.0 cm in steps of 5.00 cm.)

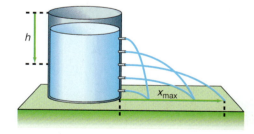

54. In the flow of fluid from an orifice in figure 13.12, it was assumed that the vertical motion of the water at the top of the tank was very small, and hence v_1 was set equal to zero. Show that if this assumption does not hold, the velocity of the fluid from the orifice v_2 can be given by

$$v_2 = \sqrt{\frac{2gh}{1 - (d_2^4/d_1^4)}}$$

where d_1 is the diameter of the tank and d_2 is the diameter of the orifice.

†55. A wind blows over the roof of your house at 100 mph. What is the difference in pressure acting on the roof because of this velocity? (*Hint:* the air inside the attic is still, that is, $v = 0$ inside the house.)

56. A wind blows over the roof of a house at 136 km/hr. What is the difference in pressure acting on the roof because of this velocity? (*Hint:* the air inside the attic is still, that is, $v = 0$ inside the house.)

†57. If air moves over the top of an airplane wing at 150 m/s and 120 m/s across the bottom of the wing, find the difference in pressure between the top of the wing and the bottom of the wing. If the area of the wing is 15.0 m², find the force acting upward on the wing.

Interactive Tutorials

⌨58. Find the buoyant force BF and apparent weight AW of a solid sphere of radius $r = 0.500$ m and density $\rho = 7.86 \times 10^3$ kg/m³, when immersed in a fluid whose density is $\rho_f = 1.00 \times 10^3$ kg/m³.

⌨59. Archimedes' principle. A solid block of wood of length $L = 15.0$ cm, width $W = 20.0$ cm, and height $h_0 = 10.0$ cm, is placed into a pool of water. The density of the block is 680 kg/m³. (a) Will the block sink or float? (b) If it floats, how deep will the block be submerged when it floats? (c) What percentage of the original volume is submerged?

⌨60. The equation of continuity and flow rate. Water flows in a pipe of diameter $d_1 = 4.00$ cm at a velocity of 35.0 cm/s, as shown in figure 13.9. The diameter of the tapered part of the pipe is $d_2 = 2.55$ cm. Find (a) the velocity of the fluid in the tapered part of the pipe, (b) the mass flow rate, and (c) the volume flow rate of the fluid.

⌨61. Bernoulli's theorem. Water flows in an elevated, tapered pipe, as shown in figure 13.10. The first part of the pipe is at a height $h_1 = 3.58$ m above the ground and the water is at a pressure $p_1 = 5000$ N/m², the diameter $d_1 = 25.0$ cm, and the velocity of the water is $v_1 = 0.553$ m/s. If the diameter of the tapered part of the pipe is $d_2 = 10.0$ cm and the height of the pipe above the ground is $h_2 = 1.25$ m, find (a) the velocity v_2 of the fluid in the tapered part of the pipe and (b) the pressure p_2 of the water in the tapered part of the pipe.

Thermodynamics

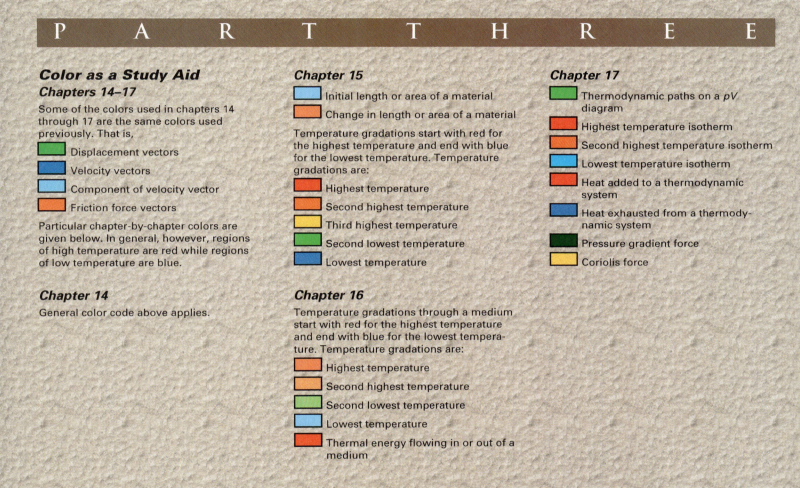

Color as a Study Aid
Chapters 14–17

Some of the colors used in chapters 14 through 17 are the same colors used previously. That is,

- Displacement vectors
- Velocity vectors
- Component of velocity vector
- Friction force vectors

Particular chapter-by-chapter colors are given below. In general, however, regions of high temperature are red while regions of low temperature are blue.

Chapter 14

General color code above applies.

Chapter 15

- Initial length or area of a material
- Change in length or area of a material

Temperature gradations start with red for the highest temperature and end with blue for the lowest temperature. Temperature gradations are:

- Highest temperature
- Second highest temperature
- Third highest temperature
- Second lowest temperature
- Lowest temperature

Chapter 16

Temperature gradations through a medium start with red for the highest temperature and end with blue for the lowest temperature. Temperature gradations are:

- Highest temperature
- Second highest temperature
- Second lowest temperature
- Lowest temperature
- Thermal energy flowing in or out of a medium

Chapter 17

- Thermodynamic paths on a pV diagram
- Highest temperature isotherm
- Second highest temperature isotherm
- Lowest temperature isotherm
- Heat added to a thermodynamic system
- Heat exhausted from a thermodynamic system
- Pressure gradient force
- Coriolis force

Temperature and Heat

The determination of temperature has long been recognized as a problem of the greatest importance in physical science. It has accordingly been made a subject of most careful attention, and, especially in late years, of very elaborate and refined experimental researches: and we are thus at present in possession of as complete a practical solution of the problem as can be desired even for the most accurate investigation.

William Thompson, Lord Kelvin

14.1 Temperature

The simplest and most intuitive definition of temperature is that **temperature is a measure of the hotness or coldness of a body.** That is, if a body is hot it has a high temperature, if it is cold it has a low temperature. This is not a very good definition, as we will see in a moment, but it is one that most people have a "feel" for, because we all know what hot and cold is. Or do we?

Let us reconsider the "thought experiment" treated in chapter 1. We place three beakers on the table, as shown in figure 14.1. Several ice cubes are placed into the first beaker of water, whereas boiling water is poured into the third beaker. We place equal amounts of the ice water from beaker one and the boiling water from beaker three into the second beaker to form a mixture. I now take my right hand and plunge it into beaker one, and conclude that it is *cold.* After drying off my right hand, I place it into the middle mixture. After coming from the ice water, the mixture in the second beaker feels hot by comparison. So I conclude that the mixture is *hot.*

I now take my left hand and plunge it into the boiling water of beaker three. (This is of course the reason why this is only a "thought experiment.") I conclude that the water in beaker three is certainly *hot.* Drying off my hand again I then place it into beaker two. After the boiling water, the mixture feels cold by comparison, so I conclude that the mixture is *cold.* After this relatively scientific experiment, my conclusion is contradictory. That is, I found the middle mixture to be either hot or cold depending on the sequence of the measurement. Thus, the hotness or coldness of a body is not a good concept to use to define the temperature of a body. Although we may have an intuitive feel for hotness or coldness, we can not use our intuition for any precise scientific work.

The Thermometer

In order to make a measurement of the temperature of a body, a new technique, other than estimating hotness or coldness, must be found. Let us look for some characteristic of matter that changes as it is heated. The simplest such characteristic is that most materials expand when they are heated. Using this characteristic of matter we take a glass tube and fill it with a liquid, as shown in figure 14.2. When the liquid is heated it expands and rises up the tube. The height of the liquid in the tube can be used to measure the hotness or coldness of a body. The device will become a thermometer.

Figure 14.1

A "thought experiment" on temperature.

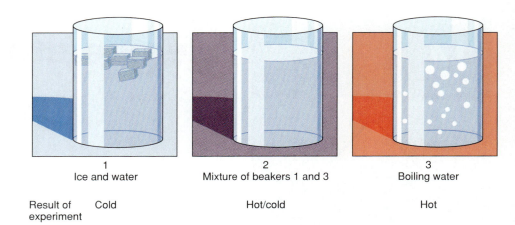

1	2	3
Ice and water	Mixture of beakers 1 and 3	Boiling water

Result of experiment   Cold   Hot/cold   Hot

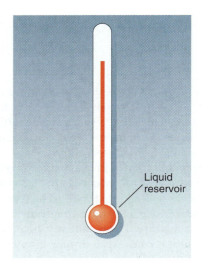

Liquid reservoir

Figure 14.2

A thermometer.

In order to quantify the process, we need to place numerical values on the glass tube, thus assigning a number that can be associated with the hotness or coldness of a body. This is the process of *calibrating* the thermometer.

First, we place the thermometer into the mixture of ice and water of beaker 1 in figure 14.1. The liquid lowers to a certain height in the glass tube. We scratch a mark on the glass at that height, and arbitrarily call it 0 degrees. Since it is the point where ice is melting in the water, we call 0° the melting point of ice. (Or similarly, the freezing point of water.)

Then we place the glass tube into beaker three, which contains the boiling water. (We assume that heat is continuously applied to beaker three to keep the water boiling.) The liquid in the glass tube is thus heated and expands to a new height. We mark this new height on the glass tube and arbitrarily call it 100°. Since the water is boiling at this point, we call it the boiling point of water.

Because the liquid in the tube expands linearly, to a first approximation, the distance between 0° and 100° can be divided into 100 equal parts. Any one of these divisions can be further divided into fractions of a degree. Thus, we obtain a complete scale of temperatures ranging from 0 to 100 degrees. Then we place this thermometer into the mixture of beaker two. The liquid in the glass rises to some number, and that number, whatever it may be, is the temperature of the mixture. That number is a numerical measure of the hotness or coldness of the body. We call this device a **thermometer,** and in particular *this scale of temperature that has 0° for the melting point of ice and 100° for the boiling point of water is called the Celsius temperature scale* and is shown in figure 14.3(a). This scale is named after the Swedish astronomer, Anders Celsius, who proposed it in 1742.

Figure 14.3

The temperature scales.

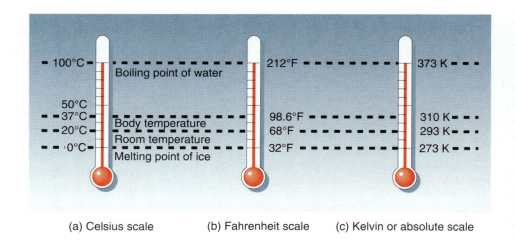

(a) Celsius scale   (b) Fahrenheit scale   (c) Kelvin or absolute scale

Figure 14.4

Simple lattice structure.

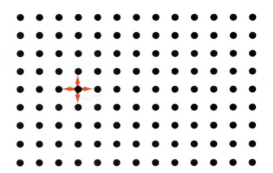

Another, perhaps more familiar, temperature scale is the **Fahrenheit temperature scale** shown in figure 14.3(b). The melting point of ice on this scale is 32 °F and the boiling point of water is 212 °F. At first glance it might seem rather strange to pick 32° for the freezing point and 212° for the boiling point of water. As a matter of fact Gabriel Fahrenheit, the German physicist, was not trying to use pure water as his calibration points. When the scale was first made, 0 °F corresponded to the lowest temperature then known, the temperature of freezing brine (a salt water mixture), and 100 °F was meant to be the temperature of the human body. Fahrenheit proposed his scale in 1714.

In addition to the Celsius and Fahrenheit scales there are other temperature scales, the most important of which is the Kelvin or *absolute* scale, as shown in figure 14.3(c). The melting point of ice on this scale is 273 K and the boiling point of water is 373 K. The **Kelvin temperature scale** does not use the degree symbol for a temperature. To use the terminology correctly, we should say that, "zero degrees Celsius corresponds to a temperature of 273 Kelvin." The Kelvin scale is extremely important in dealing with the behavior of gases. In fact, it was in the study of gases that Lord Kelvin first proposed the absolute scale in 1848. We will discuss this more natural introduction to the Kelvin scale in the study of gases in chapter 15. For the present, however, the implications of the Kelvin scale can still be appreciated by looking at the molecular structure of a solid.

The simplest picture of a solid, if it could be magnified trillions of times, is a large array of atoms or molecules in what is called a lattice structure, as shown in figure 14.4. Each dot in the figure represents an atom or molecule, depending on the nature of the substance. Each molecule is in equilibrium with all the molecules around it. The molecule above exerts a force upward on the molecule, whereas the molecule below exerts a force downward. Similarly, there are balanced forces from right and left and in and out. The molecule is therefore in equilibrium. In fact every molecule of the solid is in equilibrium. When heat is applied to a solid body, the added energy causes a molecule to vibrate around its equilibrium position. As any one molecule vibrates, it interacts with its nearest neighbors causing them to vibrate, which in turn causes its nearest neighbors to vibrate, and so on. Hence, the heat energy applied to the solid shows up as vibrational energy of the molecules of the solid. The higher the temperature of the solid, the larger is the vibrational motion of its molecules. The lower the temperature, the smaller is the vibrational motion of its molecules. Thus, the temperature of a body is really a measure of the mean or average kinetic energy of the vibrating molecules of the body.

It is therefore conceivable that if you could lower and lower the temperature of the body, the motion of the molecules would become less and less until at some very low temperature, the vibrational motion of the molecules would cease altogether. They would be frozen in one position. This point is called *absolute zero,* and is 0 on the Kelvin temperature scale. From work in quantum mechanics, however, it is found that even at absolute zero, the molecules contain a certain amount of energy called the *zero point energy.*

Figure 14.5

Converting one temperature scale to another.

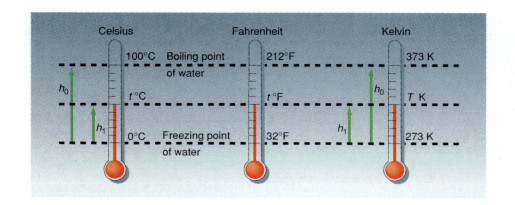

Even though temperature is really a measure of the mean kinetic energy of the molecules of a substance, from an experimental point of view it is difficult to make a standard of temperature in this way. Therefore, the International System of units considers temperature to be a fourth fundamental quantity and it is added to the three fundamental quantities of length, mass, and time. *The SI unit of temperature is the kelvin, and is defined as 1/273.16 of the temperature of the triple point of water.* The triple point of water is that point on a pressure-temperature diagram where the three phases of water, the solid, the liquid, and the gas, can coexist in equilibrium at the same pressure and temperature.

Temperature Conversions

The Celsius temperature scale is the recognized temperature scale in most scientific work and in most countries of the world. The Fahrenheit scale will eventually become obsolete along with the entire British engineering system of units. For the present, however, it is still necessary to convert from one temperature scale to another. That is, if a temperature is given in degrees Fahrenheit, how can it be expressed in degrees Celsius, and vice versa? It is easy to see how this conversion can be made.

The principle of the thermometer is based on the linear expansion of the liquid in the tube. For two identical glass tubes containing the same liquid, the expansion of the liquid is the same in both tubes. Therefore, the height of the liquid columns is the same for each thermometer, as shown in figure 14.5. The ratio of these heights in each thermometer is also equal. Therefore,

$$\left(\frac{h_1}{h_0}\right)_{\text{Celsius}} = \left(\frac{h_1}{h_0}\right)_{\text{Fahrenheit}}$$

These ratios, found from figure 14.5, are

$$\frac{t\,°C - 0°}{100° - 0°} = \frac{t\,°F - 32°}{212° - 32°}$$

$$\frac{t\,°C}{100°} = \frac{t\,°F - 32°}{180°}$$

Solving for the temperature in degrees Celsius

$$t\,°C = \frac{100°}{180°}(t\,°F - 32°)$$

Simplifying,

$$t\,°C = \tfrac{5}{9}(t\,°F - 32°) \tag{14.1}$$

Equation 14.1 allows us to convert a temperature in degrees Fahrenheit to degrees Celsius.

Example 14.1

Fahrenheit to Celsius. If room temperature is 68 °F, what is this temperature in Celsius degrees?

The temperature in Celsius degrees, found from equation 14.1, is

$$t\,°C = \tfrac{5}{9}(t\,°F - 32°) = \tfrac{5}{9}(68° - 32°) = \tfrac{5}{9}(36)$$
$$= 20\,°C$$

To convert a temperature in degrees Celsius to one in Fahrenheit, we solve equation 14.1 for $t\,°F$ to obtain

$$t\,°F = \tfrac{9}{5}t\,°C + 32° \qquad (14.2)$$

Example 14.2

Celsius to Fahrenheit. A temperature of $-5.00\,°C$ is equivalent to what Fahrenheit temperature?

The temperature in degrees Fahrenheit, found from equation 14.2, is

$$t\,°F = \tfrac{9}{5}t\,°C + 32° = \tfrac{9}{5}(-5.00°) + 32° = -9 + 32°$$
$$= 23\,°F$$

We can also find a conversion of absolute temperature to Celsius temperatures from figure 14.5, as

$$\left(\frac{h_1}{h_0}\right)_{\text{Celsius}} = \left(\frac{h_1}{h_0}\right)_{\text{Kelvin}}$$

$$\frac{t\,°C - 0°}{100° - 0°} = \frac{T\,K - 273}{373 - 273}$$

$$\frac{t\,°C}{100} = \frac{T\,K - 273}{100}$$

Therefore, the conversion of Kelvin temperature to Celsius temperatures is given by

$$t\,°C = T\,K - 273 \qquad (14.3)$$

And the reverse conversion by

$$T\,K = t\,°C + 273 \qquad (14.4)$$

For very precise work, 0 °C is actually equal to 273.16 K. In such cases, equations 14.3 and 14.4 should be modified accordingly.

Example 14.3

Celsius to Kelvin. Normal room temperature is considered to be 20.0 °C, find the value of this temperature on the Kelvin scale.

The absolute temperature, found from equation 14.4, is

$$T\,K = t\,°C + 273 = 20.0 + 273 = 293\,K$$

Note, in this book we will try to use the following convention: temperatures in Celsius and Fahrenheit will be represented by the lower case t, whereas Kelvin or absolute temperatures will be represented by a capital T. However, in some cases where time and temperature are found in the same equation, the lower case t will be used for time, and the upper case T will be used for temperature regardless of the unit used for temperature.

14.2 Heat

A solid body is composed of trillions upon trillions of atoms or molecules arranged in a lattice structure, as shown in figure 14.4. Each of these molecules possess an electrical potential energy and a vibrational kinetic energy. *The sum of the potential energy and kinetic energy of all these molecules is called the* **internal energy** *of the body.* When that internal energy is transferred between two bodies as a result of the difference in temperatures between the two bodies it is called heat.

Heat is thus the amount of internal energy flowing from a body at a higher temperature to a body at a lower temperature. Hence, a body does not contain heat, it contains internal energy. When the body cools, its internal energy is decreased; when it is heated, its internal energy is increased. A useful analogy is to compare the internal energy of a body to the money you have in a savings bank, whereas heat is analogous to the deposits or withdrawals of money.

Whenever two bodies at different temperatures are brought into contact, thermal energy always flows from the hotter body to the cooler body until they are both at the same temperature. When this occurs we say the two bodies are in **thermal equilibrium.** This is essentially the principle behind the thermometer. The thermometer is placed in contact with the body whose temperature is desired. Thermal energy flows from the hotter body to the cooler body until thermal equilibrium is reached. At that point, the thermometer is at the same temperature as the body. Hence, the thermometer is capable of measuring the temperature of a body.

The traditional unit of heat that we will use is the *kilocalorie, which is defined as the quantity of heat required to raise the temperature of 1 kg of water 1 °C, from 14.5 °C to 15.5 °C.*

The unit of heat in the British engineering system is the **British thermal unit,** abbreviated Btu. *One Btu is the heat required to raise the temperature of 1 lb of water 1 °F, from 58.5 °F to 59.5 °F.* The relation between the kilocalorie (kcal) and the Btu is

$$1 \text{ Btu} = 0.252 \text{ kcal}$$

It may seem strange to use the unit of kilocalorie or Btu for heat since heat is a flow of energy, and the unit of energy is a joule or a foot-pound. Historically it was not known that heat was a form of energy, but rather it was assumed that heat was a material quantity contained in bodies and was called Caloric. It was not until Benjamin Thompson's (1753–1814) experiments on the boring of cannons in 1798, that it became known that heat was a form of energy. Later James Prescott Joule (1818–1889) performed experiments to show *the exact equivalence between mechanical energy and heat energy. That equivalence is called the* **mechanical equivalent of heat** and is

$$1 \text{ kcal} = 1000 \text{ calories} = 4186 \text{ J}$$
$$1 \text{ Btu} = 778 \text{ ft lb} = 1055 \text{ J}$$

We will continue to follow past custom by expressing heat in units of kilocalories and Btu's, but we will also express it in terms of joules, the SI unit of energy.

We should also mention that the kilocalorie is sometimes called the large calorie and is identical to the unit used by dieticians. Thus when dieticians specify a diet as consisting of 1500 calories a day, they really mean that it is 1500 kcal per day.

Figure 14.6

Tyndall's demonstration.

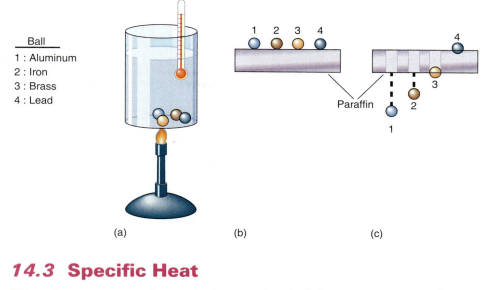

Ball

1 : Aluminum
2 : Iron
3 : Brass
4 : Lead

(a)   (b)   (c)

14.3 Specific Heat

When the temperature of several substances is raised the same amount, each substance does not absorb the same amount of thermal energy. This can be shown by Tyndall's demonstration in figure 14.6.

Four balls made of aluminum, iron, brass, and lead, all of the same mass, are placed in a beaker of boiling water, as shown in figure 14.6(a). After about 10 or 15 min, these balls will reach thermal equilibrium with the water and will all be at the same temperature as the boiling water. The four balls are then placed on a piece of paraffin, as shown in figure 14.6(b). Almost immediately, the aluminum ball melts the wax and falls through the paraffin, as shown in figure 14.6(c). A little later in time the iron ball melts its way through the wax. The brass ball melts part of the wax and sinks into it deeply. However, it does not melt enough wax to fall through. The lead ball barely melts the wax and sits on the top of the sheet of paraffin.

How can this strange behavior of the four different balls be explained? Since each ball was initially in the boiling water, each absorbed energy from the boiling water. When the balls were placed on the sheet of paraffin, each ball gave up that energy to the wax, thereby melting the wax. But since each ball melted a different amount of wax in a given time, each ball must have given up a different amount of energy to the wax. Therefore each ball must have absorbed a different quantity of energy while it was in the boiling water. Hence, *different bodies absorb a different quantity of thermal energy even when subjected to the same temperature change.*

To handle the problem of different bodies absorbing different quantities of thermal energy when subjected to the same temperature change, *the **specific heat** c of a body is defined as the amount of thermal energy Q required to raise the temperature of a unit mass of the material 1 °C. In terms of the traditional unit of kilocalories, the specific heat c of a body is defined as the number of kilocalories Q required to raise the temperature of 1 kg of the material 1 °C.* Thus,

$$c = \frac{Q}{m\Delta t} \tag{14.5}$$

We observe from this definition that the specific heat of water is 1 kcal/kg °C, since 1 kcal raises the temperature of 1 kg of water 1 °C. All other materials have a different value for the specific heat. Some specific heats are shown in table 14.1. Note that water has the largest specific heat. If the thermal energy is expressed in the SI unit of joules, *the specific heat c of a body is defined as the number of joules Q required to raise the temperature of 1 kg of the material 1 °C.* In SI units the specific heat of water is 4186 J/(kg °C).

Table 14.1

Specific Heats of Various Materials

Material	$\dfrac{\text{kcal}}{\text{kg °C}}$	$\dfrac{\text{J}}{\text{kg °C}}$
Air	0.241	1009
Aluminum	0.215	900.0
Brass	0.094	393.5
Copper	0.092	385.1
Glass	0.20	837.2
Gold	0.031	129.8
Iron	0.108	452.1
Lead	0.031	129.8
Platinum	0.032	134.0
Silver	0.057	238.6
Steel	0.108	452.1
Tin	0.054	226.0
Tungsten	0.032	134.0
Zinc	0.093	389.3
Water	1.000	4186
Ice	0.500	2093
Steam	0.481	2013

Having defined the specific heat by equation 14.5, we can rearrange that equation into the form

$$Q = mc\Delta t \tag{14.6}$$

Equation 14.6 represents the amount of thermal energy Q that will be absorbed or liberated in any process.

Using equation 14.6 it is now easier to explain the Tyndall demonstration. The thermal energy absorbed by each ball while in the boiling water is

$$Q_{Al} = mc_{Al}\Delta t$$
$$Q_{iron} = mc_{iron}\Delta t$$
$$Q_{brass} = mc_{brass}\Delta t$$
$$Q_{lead} = mc_{lead}\Delta t$$

Because all the balls went from room temperature to 100 °C, the boiling point of water, they all experienced the same temperature change Δt. Because all the masses were equal, the thermal energy absorbed by each ball is directly proportional to its specific heat. We can observe from table 14.1 that

$$c_{Al} = 0.215 \text{ kcal/(kg °C)}$$
$$c_{iron} = 0.108 \text{ kcal/(kg °C)}$$
$$c_{brass} = 0.094 \text{ kcal/(kg °C)}$$
$$c_{lead} = 0.031 \text{ kcal/(kg °C)}$$

Because the specific heat of aluminum is the largest of the four materials, the aluminum ball absorbs the greatest amount of thermal energy while in the water. Hence, it also liberates the greatest amount of thermal energy to melt the wax and should be the first ball to melt through the wax. Iron, brass, and lead absorb less thermal energy respectively because of their lower specific heats and consequently liberate thermal energy to melt the wax in this same sequence.

If the masses are not the same, then the amount of thermal energy absorbed depends on the product of the mass m and the specific heat c. The ball with the largest value of mc absorbs the most heat energy.

Example 14.4

Absorption of thermal energy. A steel ball at room temperature is placed in a pan of boiling water. If the mass of the ball is 200 g, how much thermal energy is absorbed by the ball?

Solution

The thermal energy absorbed by the ball, given by equation 14.6, is

$$Q = mc\Delta t$$

$$= (0.200 \text{ kg})\left(0.108 \ \frac{\text{kcal}}{\text{kg °C}}\right)(100 \text{ °C} - 20.0 \text{ °C})$$

$$= 1.73 \text{ kcal}$$

An interesting thing to note is that once the ball reaches the 100 °C mark, it is at the same temperature as the water and hence, there is no longer a transfer of thermal energy into the ball no matter how long the ball is left in the boiling water. All the thermal energy supplied to the pot containing the ball and the water will then go into boiling away the water.

Example 14.5

Absorption in SI units. A 500-g aluminum block at 10.0 °C is placed in an oven at 200 °C. How many joules of thermal energy are absorbed by the block?

Solution

The specific heat in SI units, found from table 14.1, is 900 J/(kg °C). The thermal energy absorbed is

$$Q = mc\Delta t$$

$$= (0.500 \text{ kg})\left(900 \ \frac{J}{\text{kg} \ °C}\right)(200 \ °C - 10.0 \ °C)$$

$$= 85,500 \text{ J}$$

14.4 Calorimetry

Calorimetry is defined as the measurement of heat. These measurements are performed in a device called a **calorimeter.** The simplest of all calorimeters consists of a metal container placed on a plastic insulating ring inside a larger highly polished metallic container, as shown in figure 14.7. The space between the two containers is filled with air to minimize the thermal energy lost from the inner calorimeter cup to the environment. The highly polished outer container reflects any external radiated energy that might otherwise make its way to the inner cup. A plastic cover is placed on the top of the calorimeter to prevent any additional loss of thermal energy to the environment. The inner cup is thus insulated from the environment, and all measurements of thermal energy absorption or liberation are made here. A thermometer is placed through a hole in the cover so that the temperature inside the calorimeter can be measured. The calorimeter is used to measure the specific heat of various substances, and the latent heat of fusion and vaporization of water.

   The basic principle underlying the calorimeter is the conservation of energy. The thermal energy lost by those bodies that lose thermal energy is equal to the thermal energy gained by those bodies that gain thermal energy. We write this conservation principle mathematically as

Thermal energy lost = Thermal energy gained	**(14.7)**

   As an example of the use of the calorimeter, let us determine the specific heat of a sample of iron. We place the iron sample in a pot of boiling water until the iron sample eventually reaches the temperature of boiling water, namely 100 °C. Meanwhile we place the inner calorimeter cup on a scale and determine its mass m_c. Then we place water within the cup and again place it on the scale to determine its mass. The difference between these two scale readings is the mass of water m_w in the cup. We place the inner cup into the calorimeter and place a thermometer through a hole in the cover of the calorimeter so that the initial temperature of the water t_{iw} is measured.

   After the iron sample reaches 100 °C, we place it within the inner calorimeter cup, and close the cover quickly. As time progresses, the temperature of the water, as recorded by the thermometer, starts to rise. It eventually stops at a final equilibrium temperature t_{fw} of the water, the sample, and the calorimeter can. The iron sample was the hot body and it lost thermal energy, whereas the water and the

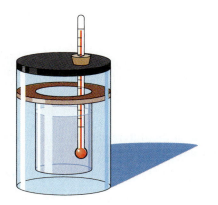

Figure 14.7
A calorimeter.

can, which is in contact with the water, absorb this thermal energy as is seen by the increased temperature of the mixture. We analyze the problem by the conservation of energy, equation 14.7, as

$$\text{Thermal energy lost} = \text{Thermal energy gained}$$

$$Q_s = Q_w + Q_c \qquad (14.8)$$

That is, the thermal energy lost by the sample Q_s is equal to the thermal energy gained by the water Q_w plus the thermal energy gained by the calorimeter cup Q_c. However, the thermal energy absorbed or liberated in any process, given by equation 14.6, is

$$Q = mc\Delta t$$

Using equation 14.6 in equation 14.8, gives

$$m_s c_s \Delta t_s = m_w c_w \Delta t_w + m_c c_c \Delta t_c \qquad (14.9)$$

where

m_s is the mass of the sample
m_w is the mass of the water
m_c is the mass of the calorimeter cup
c_s is the specific heat of the unknown sample
c_w is the specific heat of the water
c_c is the specific heat of the calorimeter cup

The change in the temperature of the sample is the difference between its initial temperature of 100 °C and its final equilibrium temperature t_{fw}. That is,

$$\Delta t_s = 100\ °C - t_{fw} \qquad (14.10)$$

The change in temperature of the water and calorimeter cup are equal since the water is in contact with the cup and thus has the same temperature. Therefore,

$$\Delta t_w = \Delta t_c = t_{fw} - t_{iw} \qquad (14.11)$$

Substituting equations 14.10 and 14.11 into equation 14.9, yields

$$m_s c_s(100 - t_{fw}) = m_w c_w(t_{fw} - t_{iw}) + m_c c_c(t_{fw} - t_{iw}) \qquad (14.12)$$

All the quantities in equation 14.12 are known except for the specific heat of the unknown sample, c_s. Solving for the specific heat yields

$$c_s = \frac{m_w c_w(t_{fw} - t_{iw}) + m_c c_c(t_{fw} - t_{iw})}{m_s(100 - t_{fw})} \qquad (14.13)$$

Example 14.6

Find the specific heat. A 0.0700-kg iron specimen is used to determine the specific heat of iron. The following laboratory data were found:

$$m_s = 0.0700\ \text{kg} \qquad\qquad t_{iw} = 20.0\ °C$$
$$m_c = 0.0600\ \text{kg} \qquad\qquad t_{fw} = 23.5\ °C$$
$$c_c = 0.215\ \text{kcal/kg }°C \qquad m_w = 0.150\ \text{kg}$$
$$t_s = 100\ °C$$

Find the specific heat of the specimen.

The specific heat of the iron specimen, found from equation 14.13, is

$$c_s = \frac{m_w c_w(t_{fw} - t_{iw}) + m_c c_c(t_{fw} - t_{iw})}{m_s(100 - t_{fw})}$$

$$= \frac{\begin{array}{l}(0.150\ \text{kg})(1.00\ \text{kcal/kg}\ °\text{C})(23.5\ °\text{C} - 20.0\ °\text{C}) \\ + (0.0600\ \text{kg})(0.215\ \text{kcal/kg}\ °\text{C})(23.5\ °\text{C} - 20\ °\text{C})\end{array}}{(0.0700\ \text{kg})(100\ °\text{C} - 23.5\ °\text{C})}$$

$$= 0.106\ \text{kcal/kg}\ °\text{C}$$

which is in good agreement with the accepted value of the specific heat of iron of 0.108 kcal/kg °C.

14.5 Change of Phase

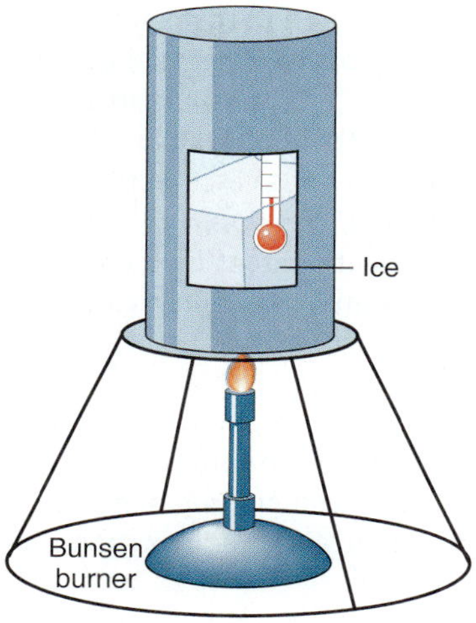

Matter exists in three states called the **phases of matter.** They are the solid phase, the liquid phase, and the gaseous phase. Let us see how one phase of matter is changed into another.

Let us examine the behavior of matter when it is heated over a relatively large range of temperatures. In particular, let us start with a piece of ice at −20.0 °C and heat it to a temperature of 120 °C. We place the ice inside a strong, tightly sealed, windowed enclosure containing a thermometer. Then we apply heat, as shown in figure 14.8. We observe the temperature as a function of time and plot it, as in figure 14.9.

As the heat is applied to the solid ice, the temperature of the block increases with time until 0 °C is reached. At this point the temperature remains constant, even though heat is being continuously applied. Looking at the block of ice, through the window in the container, we observe small drops of liquid water forming on the block of ice. The ice is starting to melt. We observe that the temperature remains constant until every bit of the solid ice is converted into the liquid water. We are observing a **change of phase.** That is, the ice is changing from the solid phase into the liquid phase. As soon as all the ice is melted, we again observe an increase in the temperature of the liquid water. The temperature increases up to 100 °C, and then levels off. Thermal energy is being applied, but the temperature is not changing. Looking through the window into the container, we see that there are bubbles forming throughout the liquid. The water is boiling. The liquid water is being converted to steam, the gaseous state of water. The temperature remains at this constant value of 100 °C until every drop of the liquid water has been converted to the gaseous steam. After that, as we continuously supply heat, we observe an increase in the temperature of the steam. Superheated steam is being made. (Note, you should not try to do this experiment on your own, because enormous pressures can be built up by the steam, causing the closed container to explode.)

Let us go back and analyze this experiment more carefully. As the thermal energy was supplied to the below freezing ice, its temperature increased to 0 °C. At this point the temperature remained constant even though heat was being continuously applied. Where did this thermal energy go if the temperature never changed? The thermal energy went into the melting of the ice, changing its phase from the solid to the liquid phase. If we observe the solid in terms of its lattice structure, figure 14.4, we can see that each molecule is vibrating about its equilibrium position. As heat is applied, the vibration increases, until at 0 °C, the vibrations of the molecules become so intense that the molecules literally pull apart from one another changing the entire structure of the material. This is the melting process. The amount of heat necessary to tear these molecules apart is a constant and

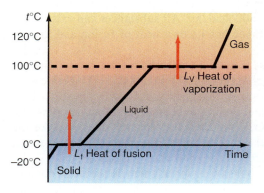

Figure 14.8

Converting ice to water to steam.

Figure 14.9

Changes of phase.

is called the latent heat of fusion of that material. *The **latent heat of fusion** is the amount of heat necessary to convert 1 kg of the solid to 1 kg of the liquid.* For water, it is found experimentally that it takes 79.7 kcal or 334,000 J of thermal energy to melt 1 kg of ice. To simplify our calculations, we will take the latent heat of fusion of water as

$$L_f = 80.0 \text{ kcal/kg} = 3.34 \times 10^5 \text{ J/kg}$$

If we must supply 80.0 kcal/kg to melt ice, then we must take away 80.0 kcal/kg to freeze water. Thus, the heat of fusion is equal to the heat of melting. The word *latent* means hidden or invisible, and not detectable as a temperature change. Heat supplied that does change the temperature is called *sensible heat.*

In the liquid state there are still molecular forces holding the molecules together, but because of the energy and motion of the molecules, these forces can not hold the molecules in the relatively rigid position they had in the solid state. This is why the liquid is able to flow and take the shape of any container in which it is placed.

As the water at 0 °C is further heated, the molecules absorb more and more energy, increasing their mean velocity within the liquid. This appears as a rise in temperature of the liquid. At 100 °C, so much energy has been imparted to the water molecules, that the molecular speeds have increased to the point that the molecules are ready to pull away from the molecular forces holding the liquid together. As further thermal energy is applied, the molecules fly away into space as steam. The temperature of the water does not rise above 100 °C because all the applied heat is supplying the molecules with the necessary energy to escape from the liquid.

*The heat that is necessary to convert 1 kg of the liquid to 1 kg of the gas is called the **latent heat of vaporization.*** For water, the latent heat of vaporization is 539 kcal/kg = 2,260,000 J/kg. However, for simplicity in our calculations we will use the standard value of

$$L_v = 540 \text{ kcal/kg} = 2.26 \times 10^6 \text{ J/kg}$$

Because this amount of thermal energy must be given to water to convert it to steam, this same quantity of thermal energy is given up to the environment when steam condenses back into the liquid state. Therefore, the heat of vaporization is equal to the heat of condensation.

Liquid water can also be converted to the gaseous state at any temperature, a process called *evaporation.* Thus, water left in an open saucer overnight will be gone by morning. Even though the temperature of the water remained at the room temperature, the liquid was converted to a gas. It evaporated into the air. The gaseous state of water is then usually referred to as *water vapor* rather than steam. At 0 °C the latent heat of vaporization is 600 kcal/kg. All substances can exist in the three states of matter, and each substance has its own heat of fusion and heat of vaporization.

Note also that another process is possible whereby a solid can go directly to a gas and vice versa without ever going through the liquid state. This process is called **sublimation.** Many students have seen this phenomenon with dry ice (which is carbon dioxide in the solid state). The ice seems to be smoking. Actually, however, the solid carbon dioxide is going directly into the gaseous state. The gas, like the dry ice, is so cold that it causes water vapor in the surrounding air to condense, which is seen as the "smoky" clouds around the solid carbon dioxide.

A more common phenomena, but not as spectacular, is the conversion of water vapor, a gas, directly into ice crystals, a solid, in the sublimation process commonly known as frost. On wintry mornings when you first get up and go outside your home, you sometimes see ice all over the tips of the grass in the yard and over the windshield and other parts of your car. The water vapor in the air did not first condense to water droplets and then the water droplets froze. Instead, the grass and the car surfaces were so cold that the water vapor in the air went directly from the gaseous state into the solid state without ever going through the liquid state.

The reverse process whereby the solid goes directly into the gas also occurs in nature, but it is not as noticeable as frost. There are times in the winter when a light covering of snow is observed on the ground. The temperature may remain below freezing, and an overcast sky may prevent any sun from heating up or melting that snow. Yet, in a day or so, some of that snow will have disappeared. It did not melt, because the temperature always remained below freezing. Some of the snow crystals went directly into the gaseous state as water vapor.

Just as there is a latent heat of fusion L_f and latent heat of vaporization L_v there is also a latent heat of sublimation L_s. Its value is given by

$$L_s = 677 \text{ kcal/kg} = 2.83 \times 10^6 \text{ J/kg}$$

Thus, *the heat that is necessary to convert 1.00 kg of the solid ice into 1.00 kg of the gaseous water vapor is called the **latent heat of sublimation.***

It is interesting to note here that there is no essential difference in the water molecule when it is either a solid, a liquid, or a gas. The molecule consists of the same two hydrogen atoms bonded to one oxygen atom. The difference in the state is related to the different energy, and hence speed of the molecule in the different states.

Notice that it takes much more energy to convert 1 kg of water to 1 kg of steam, than it does to convert 1 kg of ice to 1 kg of liquid water, almost seven times as much. This is also why a steam burn can be so serious, since the steam contains so much energy. Let us now consider some more examples.

Example 14.7

Converting ice to steam. Let us compute the thermal energy that is necessary to convert 5.00 kg of ice at $-20.0\ °C$ to superheated steam at $120\ °C$.

Solution

The necessary thermal energy is given by

$$Q = Q_i + Q_f + Q_w + Q_v + Q_s \tag{14.14}$$

where

Q_i is the energy needed to heat the ice up to $0\ °C$
Q_f is the energy needed to melt the ice
Q_w is the energy needed to heat the water to $100\ °C$
Q_v is the energy needed to boil the water
Q_s is the energy needed to heat the steam to $120\ °C$

The necessary thermal energy to warm up the ice from $-20.0\ °C$ to $0\ °C$ is found from

$$Q_i = m_i c_i [0° - (-20.0\ °C)]$$

The latent heat of fusion is the amount of heat needed per kilogram to melt the ice. The total amount of heat needed to melt all the ice is the heat of fusion times the number of kilograms of ice present. Hence, the thermal energy needed to melt the ice is

$$Q_f = m_i L_f \tag{14.15}$$

The thermal energy needed to warm the water from $0\ °C$ to $100\ °C$ is

$$Q_w = m_w c_w (100\ °C - 0\ °C)$$

The latent heat of vaporization is the amount of heat needed per kilogram to boil the water. The total amount of heat needed to boil all the water is the heat of vaporization times the number of kilograms of water present. Hence, the thermal energy needed to convert the liquid water at 100 °C to steam at 100 °C is

$$Q_v = m_w L_v \qquad (14.16)$$

and

$$Q_s = m_s c_s (120\ °C - 100\ °C)$$

is the thermal energy needed to convert the steam at 100 °C to superheated steam at 120 °C. Substituting all these equations into equation 14.14 gives

$$Q = m_i c_i[0\ °C - (-20\ °C)] + m_i L_f + m_w c_w(100\ °C - 0\ °C)$$
$$+ m_w L_v + m_s c_s(120\ °C - 100\ °C) \qquad (14.17)$$

Using the values of the specific heat from table 14.1, we get

$$Q = (5.00\ \text{kg})\left(0.500\ \frac{\text{kcal}}{\text{kg °C}}\right)(20\ °C) + (5.00\ \text{kg})\left(80.0\ \frac{\text{kcal}}{\text{kg}}\right)$$
$$+ (5.00\ \text{kg})\left(1.00\ \frac{\text{kcal}}{\text{kg °C}}\right)(100\ °C) + (5.00\ \text{kg})\left(540\ \frac{\text{kcal}}{\text{kg}}\right)$$
$$+ (5.00\ \text{kg})\left(0.481\ \frac{\text{kcal}}{\text{kg °C}}\right)(20.0\ °C)$$

$$= 50.0\ \text{kcal} + 400\ \text{kcal} + 500\ \text{kcal} + 2700\ \text{kcal} + 48.1\ \text{kcal}$$

$$= 3700\ \text{kcal}$$

Therefore, we need 3700 kcal of thermal energy to convert 5.00 kg of ice at −20.0 °C to superheated steam at 120 °C. Note the relative size of each term's contribution to the total thermal energy.

Example 14.8

Latent heat of fusion. The heat of fusion of water L_f can be found in the laboratory using a calorimeter. If 31.0 g of ice m_i at 0 °C are placed in a 60.0-g calorimeter cup m_c that contains 170 g of water m_w at an initial temperature t_{iw} of 20.0 °C, after the ice melts, the final temperature of the water t_{fw} is found to be 5.57 °C. Find the heat of fusion of water from this data. The specific heat of the calorimeter is 0.215 cal/g °C.

Solution

From the fundamental principle of calorimetry

$$\text{Thermal energy gained} = \text{Thermal energy lost}$$

$$Q_f + Q_{iw} = Q_w + Q_c \qquad (14.18)$$

where Q_f is the thermal energy necessary to melt the ice through the fusion process and Q_{iw} is the thermal energy necessary to warm the water that came from the melted ice. We call this water ice water to distinguish it from the original water in the container. This liquid water is formed at 0 °C and will be warmed to the final equilibrium temperature of the mixture t_{fw}. The thermal energy lost by the original water in the calorimeter is Q_w, and Q_c is the thermal energy lost by the calorimeter itself. Equation 14.18 therefore becomes

$$m_i L_f + m_{iw} c_w(t_{fw} - 0\ °C) = m_w c_w(t_{iw} - t_{fw}) + m_c c_c(t_{iw} - t_{fw})$$

We find the heat of fusion by solving for L_f, as

$$L_f = \frac{(m_w c_w + m_c c_c)(t_{iw} - t_{fw}) - m_{iw} c_w(t_{fw} - 0\ °C)}{m_i} \qquad (14.19)$$

Thermodynamics

Since the laboratory data were taken in grams we can do the calculation in those units by noting that the specific heat of any material has the same numerical value when given in either

$$\frac{kcal}{kg \, °C} \quad or \quad \frac{cal}{g \, °C}$$

because the prefix kilo is found in both the numerator and denominator of the units and can thus cancel out. The same thing applies to the heat of fusion and the heat of vaporization. Therefore, using the laboratory data supplied, the heat of fusion is found as

$$L_f = \frac{[(170 \text{ g})(1.00 \text{ cal/g } °C) + (60.0 \text{ g})(0.215 \text{ cal/g } °C)](20.0 \, °C - 5.57 \, °C)}{31.0 \text{ g}}$$
$$\frac{- (31.0 \text{ g})(1.00 \text{ cal/g } °C)(5.57 \, °C - 0 \, °C)}{31.0 \text{ g}}$$

$$L_f = 79.6 \text{ cal/g}$$

Note that this is in very good agreement with the standard value of 80.0 cal/g.

Example 14.9

Latent heat of vaporization. The heat of vaporization L_v of water can be found in the laboratory by passing steam at 100 °C into a calorimeter containing water. As the steam condenses and cools it gives up thermal energy to the water and the calorimeter. In the experiment the following data were taken:

mass of calorimeter cup	$m_c = 60.0$ gm
mass of water	$m_w = 170$ gm
mass of condensed steam	$m_s = 3.00$ gm
initial temperature of water	$t_{iw} = 19.9 \, °C$
final temperature of water	$t_{fw} = 30.0 \, °C$
specific heat of calorimeter	$c_c = 0.215 \text{ cal/g } °C$

Find the heat of vaporization from this data.

Solution

To determine the heat of vaporization let us start with the fundamental principle of calorimetry

$$\text{Thermal energy lost} = \text{Thermal energy gained}$$

$$Q_v + Q_{sw} = Q_w + Q_c \qquad (14.20)$$

where Q_v is the thermal energy necessary to condense the steam and Q_{sw} is the thermal energy necessary to cool the water that came from the condensed steam. We use the subscript sw to remind us that this is the water that came from the steam in order to distinguish it from the original water in the container. This liquid water is formed at 100 °C and will be cooled to the final equilibrium temperature of the mixture t_{fw}. Here Q_w is the thermal energy gained by the original water in the calorimeter and Q_c is the thermal energy gained by the calorimeter itself. Equation 14.20 therefore becomes

$$m_s L_v + m_{sw} c_w (100 \, °C - t_{fw}) = m_w c_w (t_{fw} - t_{iw}) + m_c c_c (t_{fw} - t_{iw})$$

Solving for the heat of vaporization,

$$L_v = \frac{m_w c_w (t_{fw} - t_{iw}) + m_c c_c (t_{fw} - t_{iw}) - m_{sw} c_w (100 \, °C - t_{fw})}{m_s} \qquad (14.21)$$

Therefore,

$$L_v = \frac{\begin{array}{l}(170 \text{ g})(1.00 \text{ cal/g }°\text{C})(30.0 °\text{C} - 19.9 °\text{C}) \\ + (60.0 \text{ g})(0.215 \text{ cal/g }°\text{C})(30.0° - 19.9 °\text{C}) \\ - (3.00 \text{ g})(1.00 \text{ cal/g }°\text{C})(100 °\text{C} - 30.0°)\end{array}}{3.00 \text{ g}}$$

Thus, we find from the experimental data that the heat of vaporization is

$$L_v = 546 \text{ cal/g}$$

which is in good agreement with the standard value of 540 cal/g.

Example 14.10

Mixing ice and water. If 10.0 g of ice, at 0 °C, are mixed with 50.0 g of water at 80.0 °C, what is the final temperature of the mixture?

Solution

When the ice is mixed with the water it will gain heat from the water. The law of conservation of thermal energy becomes

$$\text{Thermal energy gained} = \text{Thermal energy lost}$$

$$Q_f + Q_{iw} = Q_w$$

where Q_f is the heat gained by the ice as it goes through the melting process. When the ice melts, it becomes water at 0 °C. Let us call this water ice water to distinguish it from the original water in the container. Thus, Q_{iw} is the heat gained by the ice water as it warms up from 0 °C to the final equilibrium temperature t_{fw}. Finally, Q_w is the heat lost by the original water, which is at the initial temperature t_{iw}. Thus,

$$m_i L_f + m_{iw} c_w(t_{fw} - 0 °\text{C}) = m_w c_w(t_{iw} - t_{fw})$$

where m_i is the mass of the ice, m_{iw} is the mass of the ice water, and m_w is the mass of the original water. Solving for the final temperature of the water we get

$$m_i L_f + m_{iw} c_w t_{fw} = m_w c_w t_{iw} - m_w c_w t_{fw}$$
$$m_{iw} c_w t_{fw} + m_w c_w t_{fw} = m_w c_w t_{iw} - m_i L_f$$
$$(m_{iw} c_w + m_w c_w)t_{fw} = m_w c_w t_{iw} - m_i L_f$$
$$t_{fw} = \frac{m_w c_w t_{iw} - m_i L_f}{m_{iw} c_w + m_w c_w}$$

The final equilibrium temperature of the water becomes

$$t_{fw} = \frac{(50.0 \text{ g})(1 \text{ cal/g }°\text{C})(80.0 °\text{C}) - (10.0 \text{ g})(80.0 \text{ cal/g})}{(10.0 \text{ g})(1 \text{ cal/g }°\text{C}) + (50.0 \text{ g})(1 \text{ cal/g }°\text{C})}$$

$$= \frac{4000 \text{ cal} - 800 \text{ cal}}{60.0 \text{ cal/}°\text{C}}$$

$$= 53.3 °\text{C}$$

Example 14.11

Something is wrong here. Repeat example 14.10 with the initial temperature of the water at 10.0 °C.

Using the same equation as for the final water temperature in example 14.10, we get

$$t_{fw} = \frac{m_w c_w t_{iw} - m_i L_f}{m_{iw} c_w + m_w c_w}$$

Thus,

$$t_{fw} = \frac{(50.0 \text{ g})(1 \text{ cal/g } °C)(10.0 °C) - (10.0 \text{ g})(80.0 \text{ cal/g})}{(10.0 \text{ g})(1 \text{ cal/g } °C) + (50.0 \text{ g})(1 \text{ cal/g } °C)}$$

$$= \frac{500 \text{ cal} - 800 \text{ cal}}{60.0 \text{ cal/}°C}$$

$$= -5.00 °C$$

There is something very wrong here! Our answer says that the final temperature is 5° below zero. But this is impossible. The temperature of the water can not be below 0 °C and still be water, and the ice that was placed in the water can not convert all the water to ice and cause all the ice to be at a temperature of 5° below zero. Something is wrong. Let us check our equation. The equation worked for the last example, why not now? The equation was derived with the assumption that all the ice that was placed in the water melted. Is this a correct assumption? The energy necessary to melt all the ice is found from

$$Q_f = m_i L_f = (10.0 \text{ g})(80.0 \text{ cal/g}) = 800 \text{ cal}$$

The energy available to melt the ice comes from the water. The maximum thermal energy available occurs when all the water is cooled to 0 °C. Therefore, the maximum available energy is

$$Q_w = m_w c_w (t_{iw} - 0 °C) = (50.0 \text{ g})(1.00 \text{ cal/g } °C)(10.0 °C)$$

$$= 500 \text{ cal}$$

The amount of energy available to melt all the ice is 500 cal and it would take 800 cal to melt all the ice present. Therefore, there is not enough energy to melt the ice. Hence, our initial assumption that all the ice melted is incorrect. Thus, our equation is no longer valid. There is an important lesson to be learned here. All through our study of physics we make assumptions in order to derive equations. If the assumptions are correct, the equations are valid and can be used to predict some physical phenomenon. If the assumptions are not correct, the final equations are useless. In this problem there is still ice left and hence the final temperature of the mixture is 0 °C. The amount of ice that actually melted can be found by using the relation

$$f Q_f = Q_w$$

where f is the fraction of the ice that melts. Thus,

$$f = \frac{Q_w}{Q_f} = \frac{500 \text{ cal}}{800 \text{ cal}}$$

$$= 0.625$$

Therefore, only 62.5% of the ice melted and the final temperature of the mixture is 0 °C.

The Language of Physics

Temperature
The simplest definition of temperature is that temperature is a measure of the hotness or coldness of a body. A better definition is that temperature is a measure of the mean kinetic energy of the molecules of the body (p. 403).

Thermometer
A device for measuring the temperature of a body (p. 404).

Celsius temperature scale
A temperature scale that uses 0° for the melting point of ice and 100° for the boiling point of water (p. 404).

Fahrenheit temperature scale
A temperature scale that uses 32° for the melting point of ice and 212° for the boiling point of water (p. 405).

Kelvin temperature scale
The absolute temperature scale. The lowest temperature attainable is absolute zero, the 0 K of this scale. The temperature for the melting point of ice is 273 K and 373 K for the boiling point of water (p. 405).

Internal energy
The sum of the potential and kinetic energy of all the molecules of a body (p. 408).

Heat
The flow of thermal energy from a body at a higher temperature to a body at a lower temperature. When a body cools, its internal energy is decreased; when it is heated, its internal energy is increased (p. 408).

Thermal equilibrium
Whenever two bodies at different temperatures are touched together, thermal energy always flows from the hotter body to the cooler body until they are both at the same temperature. When this occurs the two bodies are said to be in thermal equilibrium (p. 408).

Kilocalorie
A unit of heat. It is defined as the amount of thermal energy required to raise the temperature of 1 kg of water 1 °C (p. 408).

British thermal unit (Btu)
The amount of thermal energy required to raise the temperature of 1 lb of water 1 °F (p. 408).

Mechanical equivalent of heat
The equivalence between mechanical energy and thermal energy. One kcal is equal to 4186 J (p. 408).

Specific heat
A characteristic of a material. It is defined as the number of kilocalories required to raise the temperature of 1 kg of the material 1 °C (p. 409).

Calorimetry
The measurement of heat (p. 411).

Calorimeter
An instrument that is used to make measurements of heat. The basic principle underlying the calorimeter is the conservation of energy. The thermal energy lost by those bodies that lose thermal energy is equal to the thermal energy gained by those bodies that gain thermal energy (p. 411).

Phases of matter
Matter exists in three phases, the solid phase, the liquid phase, and the gaseous phase (p. 413).

Change of phase
The change in a body from one phase of matter to another. As an example, melting is a change from the solid state of a body to the liquid state. Boiling is a change in state from the liquid state to the gaseous state (p. 413).

Latent heat of fusion
The amount of heat necessary to convert 1 kg of the solid to 1 kg of the liquid (p. 414).

Latent heat of vaporization
The amount of heat necessary to convert 1 kg of the liquid to 1 kg of the gas (p. 414).

Summary of Important Equations

Convert Fahrenheit temperature to Celsius
$$t\ ^\circ C = \tfrac{5}{9}(t\ ^\circ F - 32^\circ) \tag{14.1}$$

Convert Celsius temperature to Fahrenheit
$$t\ ^\circ F = \tfrac{9}{5}t\ ^\circ C + 32^\circ \tag{14.2}$$

Convert Celsius temperature to Kelvin
$$T\ K = t\ ^\circ C + 273 \tag{14.4}$$

Thermal energy absorbed or liberated
$$Q = mc\Delta t \tag{14.6}$$

Principle of calorimetry
Thermal = Thermal
energy lost energy gained   (14.7)

Fusion
$$Q_f = m_i L_f \tag{14.15}$$

Vaporization
$$Q_v = m_w L_v \tag{14.16}$$

Questions for Chapter 14

1. What is the difference between temperature and heat?
2. Explain how a bathtub of water at 5 °C can contain more thermal energy than a cup of coffee at 95 °C.
3. Discuss how the human body uses the latent heat of vaporization to cool itself through the process of evaporation.
†4. Relative humidity is defined as the percentage of the amount of water vapor in the air to the maximum amount of water vapor that the air can hold at that temperature. Discuss how the relative humidity affects the process of evaporation in general and how it affects the human body in particular.
†5. It is possible for a gas to go directly to the solid state without going through the liquid state, and vice versa. The process is called *sublimation.* An example of such a process is the formation of frost. Discuss the entire process of sublimation, the latent heat involved, and give some more examples of the process.
†6. Why does ice melt when an object is placed upon it? Describe the process of ice skating from the pressure of the skate on the ice.
7. Why are the numerical values of the specific heat the same in SI units and British engineering units?

Problems for Chapter 14

14.1 Temperature

1. Convert the following normal body temperatures to degrees Celsius: (a) oral temperature of 98.6 °F, (b) rectal temperature of 99.6 °F, and (c) axial (armpit) temperature of 97.6 °F.
2. Find the value of absolute zero on the Fahrenheit scale.
3. For what value is the Fahrenheit temperature equal to the Celsius temperature?
4. Convert the following temperatures to Fahrenheit: (a) 38.0 °C, (b) 68.0 °C, (c) 250 °C, (d) −10.0 °C, and (e) −20.0 °C.
5. Convert the following Fahrenheit temperatures to Celsius: (a) −23.0 °F, (b) 12.5 °F, (c) 55.0 °F, (d) 90.0 °F, and (e) 180 °F.
6. A temperature change of 5 °F corresponds to what temperature change in Celsius degrees?
†7. Derive an equation to convert the temperature in Fahrenheit degrees to its corresponding Kelvin temperature.
†8. Derive an equation to convert the change in temperature in Celsius degrees to a change in temperature in Fahrenheit degrees.

14.3 Specific Heat

9. A 450-g ball of copper at 20.0 °C is placed in a pot of boiling water until equilibrium is reached. How much thermal energy is absorbed by the ball?
10. A 250-g glass marble is taken from a freezer at −23.0 °C and placed into a beaker of boiling water. How much thermal energy is absorbed by the marble?
11. How much thermal energy must be supplied by an electric immersion heater if you wish to raise the temperature of 5.00 kg of water from 20.0 °C to 100 °C?

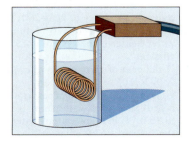

12. A 2.00-kg mass of copper falls from a height of 3.00 m to an insulated floor. What is the maximum possible temperature increase of the copper?
13. An iron block slides down an iron inclined plane at a constant speed. The plane is 10.0 m long and is inclined at an angle of 35.0° with the horizontal. Assuming that half the energy lost to friction goes into the block, what is the difference in temperature of the block from the top of the plane to the bottom of the plane?

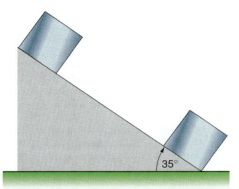

14. A 2000-kg car is traveling at 96.6 km/hr when it is braked to a stop. What is the maximum possible thermal energy generated in the brakes?
15. How much thermal energy is absorbed by an aluminum ball 20.0 cm in diameter, initially at a temperature of 20.0 °C, if it is placed in boiling water?

14.4 Calorimetry

16. If 30.0 g of water at 5.00 °C are mixed with 50.0 g of water at 70.0 °C and 25.0 g of water at 100 °C, find the resultant temperature of the mixture.
17. If 80.0 g of lead shot at 100 °C is placed into 100 g of water at 20.0 °C in an aluminum calorimeter of 60.0-g mass, what is the final temperature?
18. A 100-g mass of an unknown material at 100 °C is placed in an aluminum calorimeter of 60.0 g that contains 150 g of water at an initial temperature of 20.0 °C. The final temperature is observed to be 21.5 °C. What is the specific heat of the substance and what substance do you think it is?

19. A 100-g mass of an unknown material at 100 °C, is placed in an aluminum calorimeter of 60.0-g mass that contains 150 g of water at an initial temperature of 15.0 °C. At equilibrium the final temperature is 19.5 °C. What is the specific heat of the material and what material is it?
20. How much water at 50.0 °C must be added to 60.0 kg of water at 10.0 °C to bring the final mixture to 20.0 °C?
†21. A 100-g aluminum calorimeter contains 200 g of water at 15.0 °C. If 100.0 g of lead at 50.0 °C and 60.0 g of copper at 60.0 °C are placed in the calorimeter, what is the final temperature in the calorimeter?
22. A 200-g piece of platinum is placed inside a furnace until it is in thermal equilibrium. The platinum is then placed in a 100-g aluminum calorimeter containing 400 g of water at 5.00 °C. If the final equilibrium temperature of the water is 10.0 °C, find the temperature of the furnace.

14.5 Change of Phase

23. How many calories are needed to change 50.0 g of ice at −10.0 °C to water at 20.0 °C?
24. If 50.0 g of ice at 0.0 °C are mixed with 50.0 g of water at 80.0 °C what is the final temperature of the mixture?
25. How much ice at 0 °C must be mixed with 50.0 g of water at 75.0 °C to give a final water temperature of 20 °C?
26. If 50.0 g of ice at 0.0 °C are mixed with 50.0 g of water at 20.0 °C, what is the final temperature of the mixture? How much ice is left in the mixture?
27. How much heat is required to convert 10.0 g of ice at −15.0 °C to steam at 105 °C?
28. In the laboratory, 31.0 g of ice at 0 °C is placed into an 85.0-g copper calorimeter cup that contains 155 g

of water at an initial temperature of 23.0 °C. After the ice melts, the final temperature of the water is found to be 6.25 °C. From this laboratory data, find the heat of fusion of water and the percentage error between the standard value and this experimental value.

29. A 100-g iron ball is heated to 100 °C and then placed in a hole in a cake of ice at 0.00 °C. How much ice will melt?

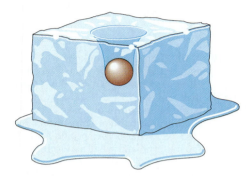

30. How much steam at 100 °C must be mixed with 300 g of water at 20.0 °C to obtain a final water temperature of 80.0 °C?

31. How much steam at 100 °C must be mixed with 1 kg of ice at 0.00 °C to produce water at 20.0 °C?

32. In the laboratory, 6.00 g of steam at 100 °C is placed into an 85.0-g copper calorimeter cup that contains 155 g of water at an initial temperature of 18.5 °C. After the steam condenses, the final temperature of the water is found to be 41.0 °C. From this laboratory data, find the heat of vaporization of water and the percentage error between the standard value and this experimental value.

†33. An electric stove is rated at 1 kW of power. If a pan containing 1.00 kg of water at 20.0 °C is placed on this stove, how long will it take to boil away all the water?

34. An electric immersion heater is rated at 0.200 kW of power. How long will it take to boil 100 cm³ of water at an initial temperature of 20.0 °C?

Additional Problems

35. Using the specific heat of copper in British engineering units determine how many Btu's are required to raise the temperature of 30 lb of copper from 70.0 °F to 300 °F?

36. A 890-N man consumes 3000 kcal of food per day. If this same energy were used to heat the same weight of water, by how much would the temperature of the water change?

37. An electric space heater is rated at 1.50 kW of power. How many kcal of thermal energy does it produce per second? How many Btu's of thermal energy per hour does it produce?

†38. A 0.055-kg mass of lead at an initial temperature of 135 °C, a 0.075-kg mass of brass at an initial temperature of 185 °C, and a 0.0445-kg of ice at an initial temperature of −5.25 °C is placed into a calorimeter containing 0.250 kg of water at an initial temperature of 23.0 °C. The aluminum calorimeter has a mass of 0.085 kg. Find the final temperature of the mixture.

39. A 100-g lead bullet is fired into a fixed block of wood at a speed of 350 m/s. If the bullet comes to rest in the block, what is the maximum change in temperature of the bullet?

†40. A 35-g lead bullet is fired into a 6.5-kg block of a ballistic pendulum that is initially at rest. The combined bullet-pendulum rises to a height of 0.125 m. Find (a) the speed of the combined bullet-pendulum after the collision, (b) the original speed of the bullet, (c) the original kinetic energy of the bullet, (d) the kinetic energy of the combined bullet-pendulum after the collision, and (e) how much of the initial mechanical energy was converted to thermal energy in the collision. If 50% of the energy lost shows up as thermal energy in the bullet, what is the change in energy of the bullet?

†41. After 50.0 g of ice at 0 °C is mixed with 200 g of water, also at 0 °C, in an insulated cup of 15.0-cm radius, a paddle wheel, 15.0 cm in radius, is placed inside the cup and set into rotational motion. What force, applied at the end of the paddle wheel, is necessary to rotate the paddle wheel at 60 rpm, for 10.0 minutes such that the final temperature of the mixture will be 15.0 °C?

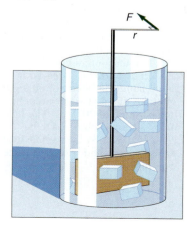

†42. A 75.0-kg patient is running a fever of 105 °F and is given an alcohol rub down to lower the body temperature. If the specific heat of the human body is approximately 0.830 kcal/ (kg °C), and the heat of vaporization of alcohol is 203 kcal/kg, find (a) the amount of heat that must be removed to lower the temperature to 102 °F and (b) the volume of alcohol required.

43. How much thermal energy is required to heat the air in a house from 15.0 °C to 20.0 °C if the house is 14.0 m long, 9.00 m wide, and 3.00 m high?

44. A classroom is at an initial temperature of 20 °C. If 35 students enter the class and each liberates heat to the air at the rate of 100 W, find the final temperature of the air in the room 50 min later, assuming all the heat from the students goes into heating the air. The classroom is 10.0 m long, 9.00 m wide, and 4.00 m high.

45. How much fuel oil is needed to heat a 100-gal tank of water from 40.0 °F to 180 °F, if oil is capable of supplying 140,000 Btu of thermal energy per gallon of oil?

46. How much fuel oil is needed to heat a 570-liter tank of water from 10.0 °C to 80.0 °C if oil is capable of supplying 3.88 × 10⁷ J of thermal energy per liter of oil?

47. How much heat is necessary to melt 100 kg of aluminum initially at a temperature of 20 °C? The melting point of aluminum is 660 °C and its heat of fusion is 90 kcal/kg.

†48. If the heat of combustion of natural gas is 1000 Btu/ft³, how many ft³ are needed to heat 26 ft³ of water from 60.0 °F to 180 °F in a hot water heater if the system is 75% efficient?

49. If the heat of combustion of natural gas is 3.71 × 10⁷ J/m³, how many cubic meters are needed to heat 0.580 m³ of water from 10.0 °C to 75.0 °C in a hot water heater if the system is 63% efficient?

†50. If the heat of combustion of coal is 12,000 Btu/lb, how many pounds of coal are necessary to heat 26 ft³ of water from 60 °F to 180 °F in a hot water heater if the system is 75% efficient?

51. If the heat of combustion of coal is 2.78 × 10⁷ J/kg, how many kilograms of coal are necessary to heat 0.580 m³ of water from 10.0 °C to 75.0 °C in a hot water heater if the system is 63% efficient?

†52. The *solar constant* is the amount of energy from the sun falling on the earth per second, per unit area and is given as SC = 1350 J/(s m²). If an average roof of a house is 60.0 m², how much energy impinges on the house in an 8-hr period? Express the answer in joules, kWhr, Btu, and kcal. Assuming you could convert all of this heat at 100% efficiency, how much fuel could you save if #2 fuel oil supplies 140,000 Btu/gal; natural gas supplies 1000 Btu/ft³; electricity supplies 3415 Btu/kWhr?

53. How much thermal energy can you store in a 2000-gal tank of water if the water has been subjected to a temperature change of 70.0 °F in a solar collector?

54. How much thermal energy can you store in a 5680-liter tank of water if the water has been subjected to a temperature change of 35.0 °C in a solar collector?

55. The refrigerating capacity of some older air conditioners were rated in terms of tons. A 1-ton air conditioner extracted enough thermal energy to freeze 1 ton of water at 32 °F per day. How much thermal energy is necessary to freeze 1 ton of water? If it takes 24 hr to freeze this water, how much power is expended in Btu/hr and watts? Compare this value to the rating of modern air conditioners.

56. A 5.94-kg lead ball rolls without slipping down a rough inclined plane 1.32 m long that makes an angle of 40.0° with the horizontal. The ball has an initial velocity $v_0 = 0.25$ m/s. The ball is not perfectly spherical and some energy is lost due to friction as it rolls down the plane. The ball arrives at the bottom of the plane with a velocity $v = 3.00$ m/s, and 80.0% of the energy lost shows up as a rise in the temperature of the ball. Find (a) the height of the incline, (b) the initial potential energy of the ball, (c) the initial kinetic energy of translation, (d) the initial kinetic energy of rotation, (e) the initial total energy of the ball, (f) the final kinetic energy of translation, (g) the final kinetic energy of rotation, (h) the final total mechanical energy of the ball at the bottom of the plane, (i) the energy lost by the ball due to friction, and (j) the increase in the temperature of the ball.

†57. The energy that fuels thunderstorms and hurricanes comes from the heat of condensation released when saturated water vapor condenses to form the droplets of water that become the clouds that we see in the sky. Consider the amount of air contained in an imaginary box 5.00 km long, 10.0 km wide, and 30.0 m high that covers the ground at the surface of the earth at a particular time. The air temperature is 20 °C and is saturated with all the water vapor it can contain at that temperature, which is 17.3×10^{-3} kg of water vapor per m³. The air in this imaginary box is now lifted into the atmosphere where it is cooled to 0 °C. Since the air is saturated, condensation occurs throughout the cooling process. The maximum water vapor the air can contain at 0 °C is 4.847×10^{-3} kg of water vapor per m³. (The heat of vaporization of water varies with temperatures from 600 kcal/kg at 0 °C to 540 kcal/kg at 100 °C. We will assume an average temperature of 10.0 °C for the cooling process.) Find (a) the volume of saturated air in the imaginary box, (b) the mass of water vapor in this volume at 20.0 °C, (c) the mass of water vapor in this volume at 0 °C, (d) the heat of vaporization of water at 10.0 °C, and (e) the thermal energy given off in the condensation process. (f) Discuss this quantity of energy in terms of the energy that powers thunderstorms and hurricanes.

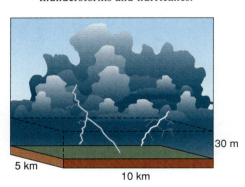

5 km   10 km   30 m

Interactive Tutorials

🔖 58. Find the total amount of thermal energy in joules necessary to convert ice of mass $m_i = 2.00$ kg at an initial temperature $t_{ii} = -20.0$ °C to water at a final water temperature of $t_{fw} = 88.3$ °C. The specific heat of ice is $c_i = 2093$ J/kg °C, water is $c_w = 4186$ J/kg °C, and the latent heat of fusion of water is $L_f = 3.34 \times 10^5$ J/kg.

🔖 59. If a sample of lead shot of mass $m_s = 0.080$ kg and initial temperature $t_{is} = 100$ °C is placed into a mass of water $m_w = 0.100$ kg in an aluminum calorimeter of mass $m_c = 0.060$ kg at an initial temperature $t_{iw} = 20.0$ °C, what is the final equilibrium temperature of the water, calorimeter, and lead shot? The specific heats are water $c_w = 4186$ J/kg °C, calorimeter $c_c = 900$ J/kg °C, and lead sample $c_s = 129.8$ J/kg °C.

🔖 60. Temperature Conversion Calculator. The Temperature Conversion Calculator will permit you to convert temperatures in one unit to a temperature in another unit.

🔖 61. Specific heat. A specimen of lead, $m_s = 0.250$ kg, is placed into an oven where it acquires an initial temperature $t_{is} = 200$ °C. It is then removed and placed into a calorimeter of mass $m_c = 0.060$ kg and specific heat $c_c = 900$ J/kg °C that contains water, $m_w = 0.200$ kg, at an initial temperature $t_{iw} = 10.0$ °C. The specific heat of water is $c_w = 4186$ J/kg °C. The final equilibrium temperature of the water in the calorimeter is observed to be $t_{fw} = 16.7$ °C. Find the specific heat c_s of this sample.

🔖 62. Find the total amount of thermal energy in joules necessary to convert ice of mass $m_i = 12.5$ kg at an initial temperature $t_{ii} = -25.0$ °C to superheated steam at a temperature $t_{ss} = 125$ °C. The specific heat of ice is $c_i = 2093$ J/kg °C, water is $c_w = 4186$ J/kg °C, and steam is $c_s = 2013$ J/kg °C. The latent heat of fusion of water is $L_f = 3.34 \times 10^5$ J/kg, and the latent heat of vaporization is $L_v = 2.26 \times 10^6$ J/kg.

🔖 63. A mixture. How much ice at an initial temperature of $t_{ii} = -15.0$ °C must be added to a mixture of three specimens contained in a calorimeter in order to make the final equilibrium temperature of the water $t_{fw} = 12.5$ °C? The three specimens and their characteristics are sample 1: zinc; $m_{s1} = 0.350$ kg, $c_{s1} = 389$ J/kg °C, initial temperature $t_{is1} = 150$ °C; sample 2: copper; $m_{s2} = 0.180$ kg, $c_{s2} = 385$ J/kg °C, initial temperature $t_{is2} = 100$ °C; and sample 3: tin; $m_{s3} = 0.350$ kg, $c_{s3} = 226$ J/kg °C, initial temperature $t_{is3} = 180$ °C. The calorimeter has a mass $m_c = 0.060$ kg and specific heat $c_c = 900$ J/kg °C and contains water, $m_w = 0.200$ kg, at an initial temperature $t_{iw} = 19.5$ °C. The specific heat of water is $c_w = 4186$ J/kg °C.

15

Thermal Expansion and the Gas Laws

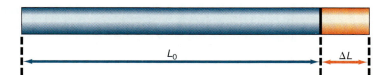

Thermal expansion of air. By heating the air with a flame it expands and inflates the balloon.

So many of the properties of matter, especially when in the gaseous form, can be deduced from the hypothesis that their minute parts are in rapid motion, the velocity increasing with the temperature, that the precise nature of this motion becomes a subject of rational curiosity... The relations between pressure, temperature and density in a perfect gas can be explained by supposing the particles to move with uniform velocity in straight lines, striking against the sides of the containing vessel and thus producing pressure.

James Clerk Maxwell

15.1 Linear Expansion of Solids

It is a well-known fact that most materials expand when heated. This expansion is called **thermal expansion.** (Recall that the phenomenon of thermal expansion was used in chapter 14 to devise the thermometer.) If a long thin rod of length L_0, at an initial temperature t_i, is heated to a final temperature t_f, then the rod expands by a small length ΔL, as shown in figure 15.1.

It is found by experiment that the change in length ΔL depends on the temperature change, $\Delta t = t_f - t_i$; the initial length of the rod L_0; and a constant that is characteristic of the material being heated. The experimentally observed linearity between ΔL and $L_0 \Delta t$ can be represented by the equation

$$\Delta L = \alpha L_0 \Delta t \qquad (15.1)$$

We call the constant α the *coefficient of linear expansion;* table 15.1 gives this value for various materials. The change in length is rather small, but it is, nonetheless, very significant.

Example 15.1

Expansion of a railroad track. A steel railroad track was 100 ft long when it was initially laid at a temperature of 20 °F. What is the change in length of the track when the temperature rises to 95 °F?

Solution

The coefficient of linear expansion for steel, found from table 15.1, is $\alpha_{\text{steel}} = 0.67 \times 10^{-5}/°F$. The change in length becomes

$$\Delta L = \alpha L_0 \Delta t$$
$$= (0.67 \times 10^{-5}/°F)(100 \text{ ft})(95 °F - 20 °F)$$
$$= (0.05 \text{ ft})\left(\frac{12 \text{ in.}}{\text{ft}}\right) = 0.60 \text{ in.}$$

Even though the change in length is relatively small, 0.60 inches in a distance of 100 ft, it is easily measurable. Associated with this small change in length is a very

Figure 15.1

Linear expansion.

Table 15.1

Coefficients of Thermal Expansion

Material	α Coefficient of Linear Expansion		β Coefficient of Volume Expansion	
	$\times 10^{-5}/°C$	$\times 10^{-5}/°F$	$\times 10^{-4}/°C$	$\times 10^{-4}/°F$
Aluminum	2.4	1.3		
Brass	1.8	1.0		
Copper	1.7	0.94		
Iron	1.2	0.67		
Lead	3.0	1.7		
Steel	1.2	0.67		
Zinc	2.6			
Glass (ordinary)	0.9			
Glass (Pyrex)	0.32			
Ethyl alcohol			11.0	6.1
Water			2.1	1.2
Mercury			1.8	1.0
Glass (Pyrex)			0.096	0.05
All noncondensing gases at constant pressure and 0 °C.			36.6	

large force. We can determine the force associated with this expansion by computing the force that is necessary to compress the rail back to its former length. Recall from chapter 10 that the amount that a body is stretched or compressed is given by Hooke's law as

$$\frac{F}{A} = Y\frac{\Delta L}{L_0} \qquad (10.6)$$

We can solve this equation for the force that is associated with a compression. Taking the compression of the rail as 0.05 ft, Young's modulus Y for steel as 2.9×10^7 lb/in.^{2}, and assuming that the cross-sectional area of the rail is 20 in.^{2}, the force necessary to compress the rail is

$$F = AY\frac{\Delta L}{L_0}$$

$$= (20 \text{ in.}^2)\left(2.9 \times 10^7 \frac{\text{lb}}{\text{in.}^2}\right)\left(\frac{0.05 \text{ ft}}{100 \text{ ft}}\right)$$

$$= 290,000 \text{ lb}$$

This force of 290,000 lb that is necessary to compress the rail by 0.05 ft, is also the force that is necessary to prevent the rail from expanding. It is obviously an extremely large force. It is this large force associated with the thermal expansion that makes thermal expansion so important. It is no wonder that we see and hear of cases where rails and roads have buckled during periods of very high temperatures.

Expansion joints allow bridges to expand and contract without damage.

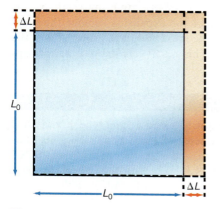

Figure 15.2

Expansion in area.

The expansion of the solid can be explained by looking at the molecular structure of the solid. The molecules of the substance are in a lattice structure. Any one molecule is in equilibrium with its neighbors, but vibrates about that equilibrium position. As the temperature of the solid is increased, the vibration of the molecule increases. However, the vibration is not symmetrical about the original equilibrium position. As the temperature increases the equilibrium position is displaced from the original equilibrium position. Hence, the mean displacement of the molecule from the original equilibrium position also increases, thereby spacing all the molecules farther apart than they were at the lower temperature. The fact that all the molecules are farther apart manifests itself as an increase in length of the material. Hence, linear expansion can be explained as a molecular phenomenon. The large force associated with the expansion comes from the large molecular forces between the molecules.

15.2 Area Expansion of Solids

For the long thin rod of section 15.1, only the length change was significant and that was all that we considered. But solids expand in all directions. If a square of thin material of length L_0 and width L_0, at an initial temperature of t_i, is heated to a new temperature t_f, the square of material expands, as shown in figure 15.2. The original area of the square is given by

$$A_0 = L_0^2$$

But each side expands by ΔL, forming a new square with sides $(L_0 + \Delta L)$. Thus, the final area becomes

$$A = (L_0 + \Delta L)^2$$
$$= L_0^2 + 2L_0\Delta L + (\Delta L)^2$$

The change in length ΔL is quite small to begin with, and its square $(\Delta L)^2$ is even smaller, and can be neglected in comparison to the magnitudes of the other terms. That is, we will set the quantity $(\Delta L)^2$ equal to zero in our analysis. Using this assumption, the final area becomes

$$A = L_0^2 + 2L_0\Delta L$$

The change in area, caused by the thermal expansion, is

$$\Delta A = \text{Final area} - \text{Original area}$$
$$= A - A_0$$
$$= L_0^2 + 2L_0\Delta L - L_0^2$$

Therefore

$$\Delta A = 2L_0\Delta L \qquad (15.2)$$

However, we have already seen that

$$\Delta L = \alpha L_0\Delta t \qquad (15.1)$$

Substituting equation 15.1 into 15.2 gives

$$\Delta A = 2L_0\alpha L_0\Delta t$$

and

$$\Delta A = 2\alpha L_0^2\Delta t$$

However, $L_0^2 = A_0$, the original area. Therefore

$$\Delta A = 2\alpha A_0\Delta t \qquad (15.3)$$

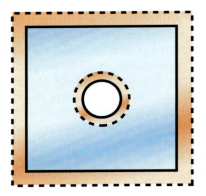

Figure 15.3

The empty hole expands at the same rate as if there were material in the hole.

Equation 15.3 gives us the area expansion of a material of original area A_0 when subjected to a temperature change Δt. Note that the coefficient of area expansion is twice the coefficient of linear expansion. Although we have derived this result for a square it is perfectly general and applies to any area. For example, if the material was circular in shape, the original area A_0 would be computed from the area of a circle of radius r_0 as

$$A_0 = \pi r_0^2$$

We would then find the change in area from equation 15.3.

All parts of the material expand at the same rate. For example, if there was a circular hole in the material, the empty hole would expand at the same rate as if material were actually present in the hole. We can see this in figure 15.3. The solid line represents the original material, whereas the dotted lines represent the expanded material. Many students feel that the material should expand into the hole, thereby causing the hole to shrink. The best way to show that the hole does indeed expand is to fill the hole with a plug made of the same material. As the material expands, so does the plug. At the end of the expansion remove the plug, leaving the hole. Since the plug expanded, the hole must also have grown. Thus, the hole expands as though it contained material. This result has many practical applications.

Example 15.2

Fitting a small wheel on a large shaft. We want to place a steel wheel on a steel shaft with a good tight fit. The shaft has a diameter of 10.010 cm. The wheel has a hole in the middle, with a diameter of 10.000 cm, and is at a temperature of 20 °C. If the wheel is heated to a temperature of 132 °C, will the wheel fit over the shaft? The coefficient of linear expansion for steel is found in table 15.1 as $\alpha = 1.20 \times 10^{-5}/°C$.

Solution

With the present dimensions the wheel can not fit over the shaft. If we place the wheel in an oven at 132 °C, the wheel expands. The diameter of the hole increases by

$$\Delta L_H = \alpha L_0 \Delta t$$
$$= (1.20 \times 10^{-5}/°C)(10.000 \text{ cm})(132 °C - 20 °C)$$
$$= 1.34 \times 10^{-2} \text{ cm}$$

The new hole in the wheel has the diameter

$$L = L_0 + \Delta L = 10.000 \text{ cm} + 0.013 \text{ cm}$$
$$= 10.013 \text{ cm}$$

Because the diameter of the hole in the wheel is now greater than the diameter of the shaft, the wheel now fits over the shaft. When the combined wheel and shaft is allowed to cool back to the original temperature of 20 °C, the hole in the wheel tries to contract to its original size, but is not able to do so, because of the presence of the shaft. Therefore, enormous forces are exerted on the shaft by the wheel, holding the wheel permanently on the shaft.

This problem could also have been treated by using the change in area by equation 15.3, but, as just seen, it is easier to find the change in the diameter.

Thermodynamics

15.3 Volume Expansion of Solids and Liquids

All materials have three dimensions, length, width, and height. When a body is heated, all three dimensions should expand and hence its volume should increase. Let us consider a cube of length L_0 on each side, at an initial temperature t_i. Its initial volume is

$$V_0 = L_0^3$$

If the material is heated to a new temperature t_f, then each side L_0 of the cube undergoes an expansion ΔL. The final volume of the cube is

$$V = (L_0 + \Delta L)^3$$
$$= L_0^3 + 3L_0^2\Delta L + 3L_0(\Delta L)^2 + (\Delta L)^3$$

Because ΔL is itself a very small quantity, the terms in $(\Delta L)^2$ and $(\Delta L)^3$ can be neglected. Therefore,

$$V = L_0^3 + 3L_0^2\Delta L$$

The change in volume due to the expansion becomes

$$\Delta V = V - V_0$$
$$= L_0^3 + 3L_0^2\Delta L - L_0^3$$
$$\Delta V = 3L_0^2\Delta L \tag{15.4}$$

However, the linear expansion ΔL was given by

$$\Delta L = \alpha L_0 \Delta t \tag{15.1}$$

Substituting this into equation 15.4 gives

$$\Delta V = 3L_0^2 \alpha L_0 \Delta t$$
$$= 3\alpha L_0^3 \Delta t$$

Since L_0^3 is equal to V_0, this becomes

$$\Delta V = 3\alpha V_0 \Delta t \tag{15.5}$$

We now define a new coefficient, called the *coefficient of volume expansion* β, for solids as

$$\beta = 3\alpha \tag{15.6}$$

Therefore, the change in volume of a substance when subjected to a change in temperature is

$$\Delta V = \beta V_0 \Delta t \tag{15.7}$$

Although we derived equation 15.7 for a solid cube, it is perfectly general and applies to any volume of a solid and even for any volume of a liquid. However, since α has no meaning for a liquid, we must determine β experimentally for the liquid. Just as a hole in a surface area expands with the surface area, a hole in a volume also expands with the volume of the solid. Hence, when a hollow glass tube expands, the empty volume inside the tube expands as though there were solid glass present.

Example 15.3

How much mercury overflows? An open glass tube is filled to the top with 25.0 cm³ of mercury at an initial temperature of 20.0 °C. If the mercury and the tube are heated to 100 °C, how much mercury will overflow from the tube?

The change in volume of the mercury, found from equation 15.7 with $\beta_{Hg} = 1.80 \times 10^{-4}/°C$ found from table 15.1, is

$$\Delta V_{Hg} = \beta_{Hg} V_0 \Delta t$$
$$= (1.80 \times 10^{-4}/°C)(25.0 \text{ cm}^3)(100 °C - 20 °C)$$
$$= 0.360 \text{ cm}^3$$

If the glass tube did not expand, this would be the amount of mercury that overflows. But the glass tube does expand and is therefore capable of holding a larger volume. The increased volume of the glass tube is found from equation 15.7 but this time with $\beta_g = 0.27 \times 10^{-4}/°C$

$$\Delta V_g = \beta_g V_0 \Delta t$$
$$= (0.27 \times 10^{-4}/°C)(25.0 \text{ cm}^3)(100 °C - 20.0 °C)$$
$$= 0.054 \text{ cm}^3$$

That is, the tube is now capable of holding an additional 0.054 cm³ of mercury. The amount of mercury that overflows is equal to the difference in the two volume expansions. That is,

$$\text{Overflow} = \Delta V_{Hg} - \Delta V_g$$
$$= 0.360 \text{ cm}^3 - 0.054 \text{ cm}^3$$
$$= 0.306 \text{ cm}^3$$

15.4 Volume Expansion of Gases: Charles' Law

Consider a gas placed in a tank, as shown in figure 15.4. The weight of the piston exerts a constant pressure on the gas. When the tank is heated, the pressure of the gas first increases. But the increased pressure in the tank pushes against the freely moving piston, and the piston moves until the pressure inside the tank is the same as the pressure exerted by the weight of the piston. Therefore the pressure in the tank remains a constant throughout the entire heating process. The volume of the gas increases during the heating process, as we can see by the new volume occupied by the gas in the top cylinder. In fact, we find the increased volume by multiplying the area of the cylinder by the distance the piston moves in the cylinder. If the volume of the gas is plotted against the temperature of the gas, in Celsius degrees, we obtain the straight line graph in figure 15.5. If the equation for this straight line is written in the point-slope form[1]

$$y - y_1 = m(x - x_1)$$

we get

$$V - V_0 = m(t - t_0)$$

---

1. The point-slope form of a straight line is obtained by the definition of the slope of a straight line, namely

$$m = \frac{\Delta y}{\Delta x}$$

or

$$\Delta y = m\Delta x$$

Using the meaning of Δy and Δx, we get

$$y - y_1 = m(x - x_1)$$

Figure 15.4

Volume expansion of a gas.

Freely moving piston

Thermometer

Source of heat

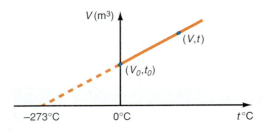

Figure 15.5

Plot of V versus t for a gas at constant pressure.

where V is the volume of the gas at the temperature t, V_0 is the volume of the gas at $t_0 = 0\ °C$, and m is the slope of the line. We can also write this equation in the form

$$\Delta V = m\Delta t \tag{15.8}$$

Note that equation 15.8, which shows the change in volume of a gas, looks like the volume expansion formula 15.7, for the change in volume of solids and liquids, that is,

$$\Delta V = \beta V_0 \Delta t \tag{15.7}$$

Let us assume, therefore, that the form of the equation for volume expansion is the same for gases as it is for solids and liquids. If we use this assumption, then

$$\beta V_0 = m$$

Hence the coefficient of volume expansion for the gas is found experimentally as

$$\beta = \frac{m}{V_0}$$

where m is the measured slope of the line. If we repeat this experiment many times for many different gases we find that

$$\beta = \frac{1}{273\ °C} = 3.66 \times 10^{-3}/°C$$

for all noncondensing gases at constant pressure. This result was first found by the French physicist, J. Charles (1746–1823). This is a rather interesting result, since the value of β is different for different solids and liquids, and yet it is a constant for all gases.

Equation 15.7 can now be rewritten as

$$V - V_0 = \beta V_0(t - t_0)$$

Because $t_0 = 0\ °C$, we can simplify this to

$$V - V_0 = \beta V_0 t$$

and

$$V = V_0 + \beta V_0 t$$

or

$$V = V_0(1 + \beta t) \tag{15.9}$$

Note that if the temperature $t = -273\ °C$, then

$$V = V_0\left(1 + -\frac{273}{273}\right) = V_0(1 - 1) = 0$$

That is, the plot of V versus t intersects the t-axis at $-273\ °C$, as shown in figure 15.5. Also observe that there is a linear relation between the volume of a gas and its temperature in degrees Celsius. Since $\beta = 1/273\ °C$, equation 15.9 can be simplified further into

$$V = V_0\left(1 + \frac{t}{273\ °C}\right) = V_0\left(\frac{273\ °C + t}{273\ °C}\right)$$

It was the form of this equation that led to the definition of the Kelvin or absolute temperature scale in the form

$$T\ K = t\ °C + 273 \tag{15.10}$$

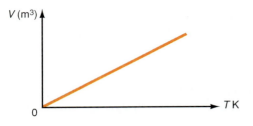

Figure 15.6

The volume V of a gas is directly proportional to its absolute temperature T.

With this definition of temperature, the volume of the gas is directly proportional to the absolute temperature of the gas, that is,

$$V = \left(\frac{V_0}{273}\right) T \qquad (15.11)$$

Changing the temperature scale is equivalent to moving the vertical coordinate of the graph, the volume, from the 0 °C mark in figure 15.5, to the −273 °C mark, and this is shown in figure 15.6. Thus, *the volume of a gas at constant pressure is directly proportional to the absolute temperature of the gas. This result is known as* **Charles' law.**

In general, if the state of the gas is considered at two different temperatures, we have

$$V_1 = \left(\frac{V_0}{273}\right) T_1$$

and

$$V_2 = \left(\frac{V_0}{273}\right) T_2$$

Hence,

$$\frac{V_1}{T_1} = \frac{V_0}{273} = \frac{V_2}{T_2}$$

Therefore,

$$\frac{V_1}{T_1} = \frac{V_2}{T_2} \qquad p = \text{constant} \qquad (15.12)$$

which is another form of Charles' law.

Figures 15.5 and 15.6 are slightly misleading in that they show the variation of the volume V with the temperature T of a gas down to −273 °C or 0 K. However, the gas will have condensed to a liquid and eventually to a solid way before this point is reached. A plot of V versus T for all real gases is shown in figure 15.7. Note that when each line is extrapolated, they all intersect at −273 °C or 0 K. Although they all have different slopes m, the coefficient of volume expansion ($\beta = m/V_0$) is the same for all the gases.

Figure 15.7

Plot of volume versus temperature for real gases.

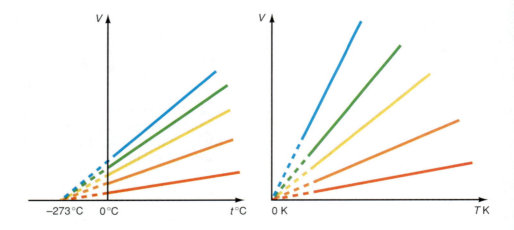

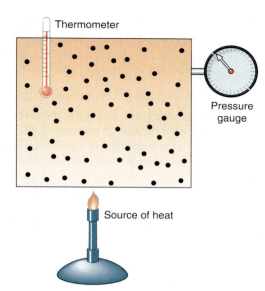

Thermometer

Pressure
gauge

Source of heat

Figure 15.8

Changing the pressure of a gas.

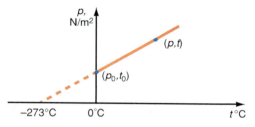

Figure 15.9

A plot of pressure versus temperature for
a gas.

15.5 Gay-Lussac's Law

Consider a gas contained in a tank, as shown in figure 15.8. The tank is made of steel and there is a negligible change in the volume of the tank, and hence the gas, as it is heated. A pressure gauge attached directly to the tank, is calibrated to read the absolute pressure of the gas in the tank. A thermometer reads the temperature of the gas in degrees Celsius. The tank is heated, thereby increasing the temperature and the pressure of the gas, which are then recorded. If we plot the pressure of the gas versus the temperature, we obtain the graph of figure 15.9. The equation of the resulting straight line is

$$p - p_0 = m'(t - t_0)$$

where p is the pressure of the gas at the temperature t, p_0 is the pressure at the temperature t_0, and m' is the slope of the line. The prime is placed on the slope to distinguish it from the slope determined in section 15.4. Because $t_0 = 0\ °C$, this simplifies to

$$p - p_0 = m't$$

or

$$p = m't + p_0 \qquad (15.13)$$

It is found experimentally that the slope is

$$m' = p_0\beta$$

where p_0 is the absolute pressure of the gas and β is the coefficient of volume expansion for a gas. Therefore equation 15.13 becomes

$$p = p_0\beta t + p_0$$

and

$$p = p_0(\beta t + 1) \qquad (15.14)$$

Thus, the pressure of the gas is a linear function of the temperature, as in the case of Charles' law. Since $\beta = 1/273\ °C$ this can be written as

$$p = p_0\left(\frac{t}{273\ °C} + 1\right) = p_0\left(\frac{t + 273\ °C}{273\ °C}\right) \qquad (15.15)$$

But the absolute or Kelvin scale has already been defined as

$$T\ K = t\ °C + 273$$

Therefore, equation 15.15 becomes

$$p = \left(\frac{p_0}{273}\right)T \qquad (15.16)$$

which shows that *the absolute pressure of a gas at constant volume is directly proportional to the absolute temperature of the gas, a result known as **Gay-Lussac's law,*** in honor of the French chemist Joseph Gay-Lussac (1778–1850). For a gas in different states at two different temperatures, we have

$$p_1 = \left(\frac{p_0}{273}\right)T_1 \quad \text{and} \quad p_2 = \left(\frac{p_0}{273}\right)T_2$$

or

$$\frac{p_1}{T_1} = \frac{p_2}{T_2} \qquad\qquad V = \text{constant} \qquad (15.17)$$

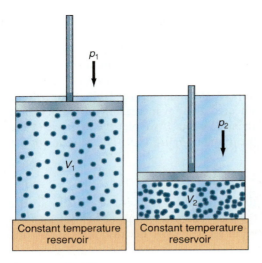

Figure 15.10

The change in pressure and volume of a gas at constant temperature.

Figure 15.11

Plot of the pressure p versus the reciprocal of the volume $1/V$ for a gas.

Equation 15.17 is another form of Gay-Lussac's law. (Sometimes this law is also called Charles' law, since Charles and Gay-Lussac developed these laws independently of each other.)

15.6 Boyle's Law

Consider a gas contained in a cylinder at a constant temperature, as shown in figure 15.10. By pushing the piston down into the cylinder, we increase the pressure of the gas and decrease the volume of the gas. If the pressure is increased in small increments, the gas remains in thermal equilibrium with the temperature reservoir, and the temperature of the gas remains a constant. We measure the volume of the gas for each increase in pressure and then plot the pressure of the gas as a function of the reciprocal of the volume of the gas. The result is shown in figure 15.11. Notice that the pressure is inversely proportional to the volume of the gas at constant temperature. We can write this as

$$p \propto \frac{1}{V}$$

or

$$pV = \text{constant} \tag{15.18}$$

That is, *the product of the pressure and volume of a gas at constant temperature is equal to a constant, a result known as **Boyle's law**,* in honor of the British physicist and chemist Robert Boyle (1627–1691). For a gas in two different equilibrium states at the same temperature, we write this as

$$p_1 V_1 = \text{constant} \quad \text{and} \quad p_2 V_2 = \text{constant}$$

Therefore,

$$p_1 V_1 = p_2 V_2 \qquad\qquad T = \text{constant} \tag{15.19}$$

another form of Boyle's law.

15.7 The Ideal Gas Law

The three gas laws,

$$\frac{V_1}{T_1} = \frac{V_2}{T_2} \qquad\qquad p = \text{constant} \tag{15.12}$$

$$\frac{p_1}{T_1} = \frac{p_2}{T_2} \qquad\qquad V = \text{constant} \tag{15.17}$$

$$p_1 V_1 = p_2 V_2 \qquad\qquad T = \text{constant} \tag{15.19}$$

can be combined into one equation, namely,

$$\frac{p_1 V_1}{T_1} = \frac{p_2 V_2}{T_2} \tag{15.20}$$

Equation 15.20 is a special case of a relation known as the **ideal gas law.** Hence, we see that the three previous laws, which were developed experimentally, are special cases of this ideal gas law, when either the pressure, volume, or temperature is held constant. The ideal gas law is a more general equation in that none of the variables must be held constant. Equation 15.20 expresses the relation between the pressure, volume, and temperature of the gas at one time, with the pressure, volume, and temperature at any other time. For this equality to hold for any time, it is necessary that

$$\frac{pV}{T} = \text{constant} \tag{15.21}$$

The cork is blown off this champagne bottle by the large build-up of CO_2 gas pressure.

This constant must depend on the quantity or mass of the gas. A convenient unit to describe the amount of the gas is the mole. *One mole of any gas is that amount of the gas that has a mass in grams equal to the atomic or molecular mass (M) of the gas.* The terms atomic mass and molecular mass are often erroneously called atomic weight and molecular weight in chemistry.

As an example of the use of the mole, consider the gas oxygen. One molecule of oxygen gas consists of two atoms of oxygen, and is denoted by O_2. The atomic mass of oxygen is found in the Periodic Table of the Elements in appendix E, as 16.00. The molecular mass of one mole of oxygen gas is therefore

$$M_{O_2} = 2(16) = 32 \text{ g/mole}$$

Thus, one mole of oxygen has a mass of 32 g. The mole is a convenient quantity to express the mass of a gas because *one mole of any gas at a temperature of 0 °C and a pressure of 1 atmosphere, has a volume of 22.4 liters. Also Avogadro's law states that every mole of a gas contains the same number of molecules. This number is called Avogadro's number N_A and is equal to 6.022×10^{23} molecules/mole.*

The mass of any gas will now be represented in terms of the number of moles, n. We can write the constant in equation 15.21 as n times a new constant, which shall be called R, that is,

$$\frac{pV}{T} = nR \tag{15.22}$$

To determine this constant R let us evaluate it for 1 mole of gas at a pressure of 1 atm and a temperature of 0 °C, or 273 K, and a volume of 22.4 L. That is,

$$R = \frac{pV}{nT} = \frac{(1 \text{ atm})(22.4 \text{ L})}{(1 \text{ mole})(273 \text{ K})}$$

$$R = 0.08205 \frac{\text{atm L}}{\text{mole K}}$$

Converted to SI units, this constant is

$$R = \left(0.08205 \frac{\text{L atm}}{\text{mole K}}\right)\left(1.013 \times 10^5 \frac{\text{N/m}^2}{\text{atm}}\right)\left(\frac{10^{-3} \text{ m}^3}{1 \text{ L}}\right)$$

$$R = 8.314 \frac{\text{J}}{\text{mole K}}$$

Another convenient unit for R is obtained by using the conversion factor, 4.186 J = 1 cal. Therefore,

$$R = 1.987 \frac{\text{cal}}{\text{mole K}}$$

We call the constant R the *universal gas constant,* and it is the same for all gases. We can now write equation 15.22 as

$$pV = nRT \tag{15.23}$$

Equation 15.23 is called the **ideal gas equation.** An ideal gas is one that is described by the ideal gas equation. Real gases can be described by the ideal gas equation as long as their density is low and the temperature is well above the condensation point (boiling point) of the gas. *Remember that the temperature T must always be expressed in Kelvin units.* Let us now look at some examples of the use of the ideal gas equation.

Example 15.4

Find the temperature of the gas. The pressure of an ideal gas is kept constant while 3.00 m³ of the gas, at an initial temperature of 50.0 °C, is expanded to 6.00 m³. What is the final temperature of the gas?

The temperature must be expressed in Kelvin units. Hence the initial temperature becomes

$$T = t \ °C + 273 = 50.0 + 273 = 323 \text{ K}$$

We find the final temperature of the gas by using the ideal gas equation in the form of equation 15.20, namely,

$$\frac{p_1 V_1}{T_1} = \frac{p_2 V_2}{T_2}$$

However, since the pressure is kept constant, $p_1 = p_2$, and cancels out of the equation. Therefore,

$$\frac{V_1}{T_1} = \frac{V_2}{T_2}$$

and the final temperature of the gas becomes

$$T_2 = \frac{V_2}{V_1} T_1$$

$$= \left(\frac{6.00 \text{ m}^3}{3.00 \text{ m}^3} \right)(323 \text{ K})$$

$$= 646 \text{ K}$$

Example 15.5

Find the volume of the gas. A balloon is filled with helium at a pressure of 2 atm, a temperature of 35.0 °C, and occupies a volume of 3.00 m³. The balloon rises in the atmosphere. When it reaches a height where the pressure is 0.500 atm, and the temperature is −20.0 °C, what is its volume?

First we convert the two temperatures to absolute temperature units as

$$T_1 = 35.0 \ °C + 273 = 308 \text{ K}$$

and

$$T_2 = -20.0 \ °C + 273 = 253 \text{ K}$$

We use the ideal gas law in the form

$$\frac{p_1 V_1}{T_1} = \frac{p_2 V_2}{T_2}$$

Solving for V_2 gives, for the final volume,

$$V_2 = \frac{p_1 T_2}{p_2 T_1} V_1$$

$$= \left[\frac{(2.00 \text{ atm})(253 \text{ K})}{(0.500 \text{ atm})(308 \text{ K})} \right](3.00 \text{ m}^3)$$

$$= 9.86 \text{ m}^3$$

Example 15.6

Find the pressure of the gas. What is the pressure produced by 2.00 moles of a gas at 35.0 °C contained in a volume of 5.00 L?

Solution

We convert the temperature of 35.0 °C to Kelvin by

$$T = 35.0 \text{ °C} + 273 = 308 \text{ K}$$

We use the ideal gas law in the form

$$pV = nRT \tag{15.23}$$

Solving for p,

$$p = \frac{nRT}{V} = \frac{(2.00 \text{ moles})(0.0821 \text{ atm L/mole K})(308 \text{ K})}{5.00 \text{ L}}$$

$$= 10.1 \text{ atm}$$

Example 15.7

Find the number of molecules in the gas. Compute the number of molecules in a gas contained in a volume of 10.0 cm³ at a pressure of 1.013×10^5 N/m², and a temperature of 30 K.

Solution

The number of molecules in a mole of a gas is given by Avogadro's number N_A, and hence the total number of molecules N in the gas is given by

$$N = nN_A \tag{15.24}$$

Therefore we first need to determine the number of moles of gas that are present. From the ideal gas law,

$$pV = nRT$$

$$n = \frac{pV}{RT} = \frac{(1.013 \times 10^5 \text{ N/m}^2)(10.0 \text{ cm}^3)}{(8.314 \text{ J/mole K})(30 \text{ K})} \left(\frac{1.00 \text{ m}^3}{10^6 \text{ cm}^3} \right)$$

$$= 4.06 \times 10^{-4} \text{ moles}$$

The number of molecules is now found as

$$N = nN_A = (4.06 \times 10^{-4} \text{ mole}) \left(6.022 \times 10^{23} \frac{\text{molecules}}{\text{mole}} \right)$$

$$= 2.44 \times 10^{20} \text{ molecules}$$

Example 15.8

Find the gauge pressure of the gas. An automobile tire has a volume of 5000 in.^{3} and contains air at a gauge pressure of 30.0 lb/in.^{2} when the temperature is 0.00 °C. What is the gauge pressure when the temperature rises to 30.0 °C?

Solution

When a gauge is used to measure pressure, it reads zero when it is under normal atmospheric pressure of 1.013×10^5 N/m^2 or 14.7 lb/in.^{2}. The pressure used in the ideal gas equation must be the absolute pressure, that is, the total pressure, which is the pressure read by the gauge plus atmospheric pressure. Therefore,

$$p_{\text{absolute}} = p_{\text{gauge}} + p_{\text{atm}} \tag{15.25}$$

Thus, the initial pressure of the gas is

$$p_1 = p_{\text{gauge}} + p_{\text{atm}} = 30.0 \text{ lb/in.}^2 + 14.7 \text{ lb/in.}^2$$
$$= 44.7 \text{ lb/in.}^2$$

The initial volume of the tire is $V_1 = 5000$ in.^{3} and the change in that volume is small enough to be neglected, so $V_2 = 5000$ in.^{3}. The initial temperature is

$$T_1 = 0.00 \text{ °C} + 273 = 273 \text{ K}$$

and the final temperature is

$$T_2 = 30.0 \text{ °C} + 273 = 303 \text{ K}$$

Solving the ideal gas equation for the final pressure, we get

$$p_2 = \frac{V_1 T_2}{V_2 T_1} p_1$$
$$= \left[\frac{(5000 \text{ in.}^3)(303 \text{ K})}{(5000 \text{ in.}^3)(273 \text{ K})} \right] (44.7 \text{ lb/in.}^2)$$
$$= 49.6 \text{ lb/in.}^2 \text{ absolute pressure}$$

Expressing this pressure in terms of gauge pressure we get

$$p_{2\text{gauge}} = p_{2\text{absolute}} - p_{\text{atm}}$$
$$= 49.6 \text{ lb/in.}^2 - 14.7 \text{ lb/in.}^2$$
$$= 34.9 \text{ lb/in.}^2$$

15.8 The Kinetic Theory of Gases

Up to now the description of a gas has been on the macroscopic level, a large-scale level, where the characteristics of a gas, such as its pressure, volume, and temperature, are measured without regard to the internal structure of the gas itself. In reality, a gas is composed of a large number of molecules in random motion. The large-scale characteristics of gases should be explainable in terms of the motion of these molecules. *The analysis of a gas at this microscopic level (the molecular level) is called the* **kinetic theory of gases.**

In the analysis of a gas at the microscopic level we make the following assumptions:

1. A gas is composed of a very large number of molecules that are in random motion.
2. The volume of the individual molecules is very small compared to the total volume of the gas.

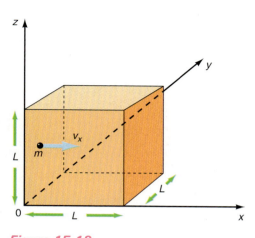

Figure 15.12

The kinetic theory of a gas.

3. The collisions of the molecules with the walls and other molecules are elastic and hence there is no energy lost during a collision.

4. The forces between molecules are negligible except during a collision. Hence, there is no potential energy associated with any molecule.

5. Finally, we assume that the molecules obey Newton's laws of motion.

Let us consider one of the very many molecules contained in the box shown in figure 15.12. For simplicity we assume that the box is a cube of length L. The gas molecule has a mass m and is moving at a velocity v. The x-component of its velocity is v_x. For the moment we only consider the motion in the x-direction. The pressure that the gas exerts on the walls of the box is caused by the collision of the gas molecule with the walls. The pressure is defined as the force acting per unit area, that is,

$$p = \frac{F}{A} \tag{15.26}$$

where A is the area of the wall where the collision occurs, and is simply

$$A = L^2$$

and F is the force exerted on the wall as the molecule collides with the wall and can be found by Newton's second law in the form

$$F = \frac{\Delta P}{\Delta t} \tag{15.27}$$

So as not to confuse the symbols for pressure and momentum, we will use the lower case p for pressure, and we will use the upper case P for momentum. Because momentum is conserved in a collision, the change in momentum of the molecule ΔP, is the difference between the momentum after the collision P_{AC} and the momentum before the collision P_{BC}. Also, since the collision is elastic the velocity of the molecule after the collision is $-v_x$. Therefore, the change in momentum of the molecule is

$$\Delta P = P_{AC} - P_{BC} = -mv_x - mv_x$$

$$= -2mv_x \quad \text{change in momentum of the molecule}$$

But the change in the momentum imparted to the wall is the negative of this, or

$$\Delta P = 2mv_x \quad \text{momentum imparted to wall}$$

Therefore, using Newton's second law, the force imparted to the wall becomes

$$F = \frac{\Delta P}{\Delta t} = \frac{2mv_x}{\Delta t} \tag{15.28}$$

The quantity Δt should be the time that the molecule is in contact with the wall. But this time is unknown. The impulse that the gas particle gives to the wall by the collision is given by

$$\text{Impulse} = F\Delta t = \Delta P \tag{15.29}$$

and is shown as the area under the force-time graph of figure 15.13. Because the time Δt for the collision is unknown, a larger time interval t_{bc}, the time between collisions, can be used with an average force F_{avg}, such that the product of $F_{avg}t_{bc}$ is equal to the same impulse as $F\Delta t$. We can see this in figure 15.13. We see that the impulse, which is the area under the curve, is the same in both cases.

At first this may seem strange, but if you think about it, it does make sense. The actual force in the collision is large, but acts for a very short time. After the collision, the gas particle rebounds from the first wall, travels back to the far wall, rebounds from it, and then travels to the first wall again, where a new collision

Figure 15.13

Since the impulse (the area under the curve) is the same, the change in momentum is the same.

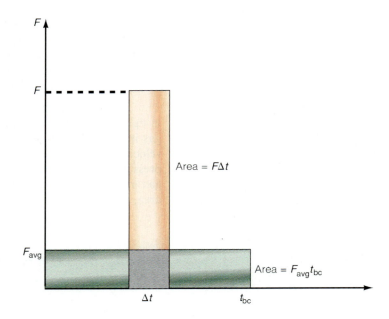

occurs. For the entire traveling time of the particle the actual force on the wall is zero. Because we think of the pressure on a wall as being present at all times, it is reasonable to talk about a smaller average force that is acting continuously for the entire time t_{bc}. As long as the impulse is the same in both cases, the momentum imparted to the wall is the same in both cases. Equation 15.29 becomes

$$\text{Impulse} = F\Delta t = F_{avg}t_{bc} = \Delta P \tag{15.30}$$

The force imparted to the wall, equation 15.28, becomes

$$F_{avg} = \frac{\Delta P}{t_{bc}} = \frac{2mv_x}{t_{bc}} \tag{15.31}$$

We find the time between the collision t_{bc} by noting that the particle moves a distance $2L$ between the collisions. Since the speed v_x is the distance traveled per unit time, we have

$$v_x = \frac{2L}{t_{bc}}$$

Hence, the time between collisions is

$$t_{bc} = \frac{2L}{v_x} \tag{15.32}$$

Therefore, the force imparted to the wall by this single collision becomes

$$F_{avg} = \frac{2mv_x}{2L/v_x} = \frac{mv_x^2}{L} \tag{15.33}$$

The total change in momentum per second, and hence the total force on the wall caused by all the molecules is the sum of the forces caused by all of the molecules, that is,

$$F_{avg} = F_{1avg} + F_{2avg} + F_{3avg} + \cdots + F_{Navg} \tag{15.34}$$

where N is the total number of molecules. Substituting equation 15.33 for each gas molecule, we have

$$F_{avg} = \frac{mv_{x1}^2}{L} + \frac{mv_{x2}^2}{L} + \frac{mv_{x3}^2}{L} + \cdots + \frac{mv_{xN}^2}{L}$$

$$F_{avg} = \frac{m}{L}(v_{x1}^2 + v_{x2}^2 + v_{x3}^2 + \cdots + v_{xN}^2) \tag{15.35}$$

Let us multiply and divide equation 15.35 by the total number of molecules N, that is,

$$F_{avg} = \frac{mN}{L}\left(\frac{v_{x1}^2 + v_{x2}^2 + v_{x3}^2 + \cdots + v_{xN}^2}{N}\right) \tag{15.36}$$

But the term in parentheses is the definition of an average value. That is,

$$v_{xavg}^2 = \left(\frac{v_{x1}^2 + v_{x2}^2 + v_{x3}^2 + \cdots + v_{xN}^2}{N}\right) \tag{15.37}$$

As an example, if you have four exams in the semester, your average grade is the sum of the four exams divided by 4. Here, the sum of the squares of the x-component of the velocity of each molecule, divided by the total number of molecules, is equal to the average of the square of the x-component of velocity. Therefore equation 15.36 becomes

$$F_{avg} = \frac{mN}{L}v_{xavg}^2$$

But since the pressure is defined as $p = F/A$, from equation 15.26, we have

$$p = \frac{F_{avg}}{A} = \frac{F_{avg}}{L^2} = \frac{mN}{L^3}v_{xavg}^2 = \frac{mN}{V}v_{xavg}^2 \tag{15.38}$$

or

$$pV = Nmv_{xavg}^2 \tag{15.39}$$

The square of the actual three-dimensional speed is

$$v^2 = v_x^2 + v_y^2 + v_z^2$$

and averaging over all molecules

$$v_{avg}^2 = v_{xavg}^2 + v_{yavg}^2 + v_{zavg}^2$$

But because the motion of any gas molecule is random,

$$v_{xavg}^2 = v_{yavg}^2 = v_{zavg}^2$$

That is, there is no reason why the velocity in one direction should be any different than in any other direction, hence their average speeds should be the same. Therefore,

$$v_{avg}^2 = 3v_{xavg}^2$$

or

$$v_{xavg}^2 = \frac{v_{avg}^2}{3} \tag{15.40}$$

Substituting equation 15.40 into equation 15.39, we get

$$pV = \frac{Nm}{3}v_{avg}^2$$

Multiplying and dividing the right-hand side by 2, gives

$$pV = \frac{2}{3}N\left(\frac{mv_{avg}^2}{2}\right) \tag{15.41}$$

The total number of molecules of the gas is equal to the number of moles of gas times Avogadro's number—the number of molecules in one mole of gas—that is,

$$N = nN_A \tag{15.24}$$

Substituting equation 15.24 into equation 15.41, gives

$$pV = \frac{2}{3}nN_A\left(\frac{mv_{avg}^2}{2}\right) \tag{15.42}$$

Recall that the ideal gas equation was derived from experimental data as

$$pV = nRT \tag{15.23}$$

The left-hand side of equation 15.23 contains the pressure and volume of the gas, all macroscopic quantities, and all determined experimentally. The left-hand side of equation 15.42, on the other hand, contains the pressure and volume of the gas as determined theoretically by Newton's second law. If the theoretical formulation is to agree with the experimental results, then these two equations must be equal. Therefore equating equation 15.23 to equation 15.42, we have

$$nRT = \frac{2}{3}nN_A\left(\frac{mv_{avg}^2}{2}\right)$$

or

$$\frac{3}{2}\left(\frac{R}{N_A}\right)T = \frac{mv_{avg}^2}{2} \tag{15.43}$$

where R/N_A is the gas constant per molecule. It appears so often that it is given the special name *the Boltzmann constant* and is designated by the letter k. Thus,

$$k = \frac{R}{N_A} = 1.38 \times 10^{-23} \text{ J/K} \tag{15.44}$$

Therefore, equation 15.43 becomes

$$\frac{3}{2}kT = \frac{1}{2}mv_{avg}^2 \tag{15.45}$$

Equation 15.45 relates the macroscopic view of a gas to the microscopic view. Notice that *the absolute temperature T of the gas (a macroscopic variable) is a measure of the mean translational kinetic energy of the molecules of the gas (a microscopic variable).* The higher the temperature of the gas, the greater the average kinetic energy of the gas, the lower the temperature, the smaller the average kinetic energy. Observe from equation 15.45 that if the absolute temperature of a gas is 0 K, then the mean kinetic energy of the molecule would be zero and its speed would also be zero. This was the original concept of absolute zero, a point where all molecular motion would cease. This concept of absolute zero can not really be derived from equation 15.45 because all gases condense to a liquid and usually a solid before they reach absolute zero. So the assumptions used to derive equation 15.45 do not hold and hence the equation can not hold down to absolute zero. Also, in more advanced studies of quantum mechanics it is found that even at absolute zero a molecule has energy, called its zero point energy. Equation 15.45 is, of course, perfectly valid as long as the gas remains a gas.

Example 15.9

The kinetic energy of a gas molecule. What is the average kinetic energy of the oxygen and nitrogen molecules in a room at room temperature?

Solution

Room temperature is considered to be 20 °C or 293 K. Therefore the mean kinetic energy, found from equation 15.45, is

$$KE_{avg} = \frac{1}{2}mv^2_{avg} = \frac{3}{2}kT$$

$$= \frac{3}{2}\left(1.38 \times 10^{-23} \frac{J}{K}\right)(293 \text{ K})$$

$$= 6.07 \times 10^{-21} \text{ J}$$

Notice that the average kinetic energy of any one molecule is quite small. This is because the mass of any molecule is quite small. The energy of the gas does become significant, however, because there are usually so many molecules in the gas. Because the average kinetic energy is given by $\frac{3}{2}kT$, we see that oxygen and nitrogen and any other molecule of gas at the same temperature all have the same average kinetic energy. Their speeds, however, are not all the same because the different molecules have different masses.

The average speed of a gas molecule can be determined by solving equation 15.45 for v_{avg}. That is,

$$\frac{1}{2}mv^2_{avg} = \frac{3}{2}kT$$

$$v^2_{avg} = \frac{3kT}{m}$$

and

$$v_{rms} = \sqrt{\frac{3kT}{m}} \qquad (15.46)$$

This particular average value of the speed, v_{rms}, is usually called the root-mean-square value, or rms value for short, of the speed v. It is called the rms speed, because it is the square *root* of the *mean* of the *square* of the speed. Occasionally the rms speed of a gas molecule is called the *thermal speed*. To determine the rms speed from equation 15.46, we must know the mass m of one molecule. The mass m of any molecule is found from

$$m = \frac{M}{N_A} \qquad (15.47)$$

That is, *the mass m of one molecule is equal to the molecular mass M of that gas divided by Avogadro's number N_A.*

Example 15.10

The rms speed of a gas molecule. Find the rms speed of an oxygen and nitrogen molecule at room temperature.

The molecular mass of O_2 is 32 g/mole. Therefore the mass of one molecule of O_2 is

$$m_{O_2} = \frac{M}{N_A} = \frac{32 \text{ g/mole}}{6.022 \times 10^{23} \text{ molecules/mole}}$$

$$= 5.31 \times 10^{-23} \text{ g/molecule}$$

The rms speed, found from 15.46, is

$$v_{rms} = \sqrt{\frac{3kT}{m}} = \sqrt{\frac{(3)(1.38 \times 10^{-23} \text{ J/K})(293 \text{ K})}{5.31 \times 10^{-23} \text{ gm}}}$$

$$= \sqrt{\frac{(228 \text{ kg m}^2)(10^3 \text{ gm})}{(\text{gm s}^2)(1 \text{ kg})}}$$

$$= 478 \text{ m/s}$$

Notice that the rms speed of an oxygen molecule is 478 m/s at room temperature, whereas the speed of sound at this temperature is about 343 m/s.

The mass of a nitrogen molecule is found from

$$m_{N_2} = \frac{M}{N_A}$$

The atomic mass of nitrogen is 14, and since there are two atoms of nitrogen in one molecule of nitrogen gas N_2, the molecular mass of nitrogen is

$$M = 2(14) = 28 \text{ g/mole}$$

Therefore

$$m_{N_2} = \frac{M}{N_A} = \frac{28 \text{ g/mole}}{6.022 \times 10^{23} \text{ molecules/mole}}$$

$$= 4.65 \times 10^{-23} \text{ g/molecule} = 4.65 \times 10^{-26} \text{ kg/molecule}$$

The rms speed of a nitrogen molecule is therefore

$$v_{rms} = \sqrt{\frac{3kT}{m}} = \sqrt{\frac{(3)(1.38 \times 10^{-23} \text{ J/K})(293 \text{ K})}{4.65 \times 10^{-26} \text{ kg}}}$$

$$= 511 \text{ m/s}$$

Note from the example that both speeds are quite high. The average speed of nitrogen is greater than the average speed of oxygen because the mass of the nitrogen molecule is less than the mass of the oxygen molecule.

"Have you ever wondered . . . ?"

An Essay on the Application of Physics

Relative Humidity and the Cooling of the Human Body

Figure 1

One of those dog days of summer when you never stop perspiring.

Have you ever wondered why you feel so uncomfortable on those dog days of August when the weatherman says that it is very hot and humid (figure 1)? What has humidity got to do with your being comfortable? What is humidity in the first place?

To understand the concept of humidity, we must first understand the concept of evaporation. Consider the two bowls shown in figure 2. Both are filled with water. Bowl 1 is open to the environment, whereas a glass plate is placed over bowl 2. If we leave the two bowls overnight, on returning the next day we would find bowl 1 empty while bowl 2 would still be filled with water. What happened to the water in bowl 1? The water in bowl 1 has evaporated into the air and is gone. *Evaporation is the process by which water goes from the liquid state to the gaseous state at any temperature.* Boiling, as you recall, is the process by which water goes from the liquid state to the gaseous state at the boiling point of 100 °C. That is, it is possible for liquid water to go to the gaseous state at any temperature.

Just as there is a latent heat of vaporization for boiling water ($L_v = 540$ kcal/kg), the latent heat of vaporization of water at 0 °C is $L_v = 600$ kcal/kg. The latent heat at any in-between temperature can be found by interpolation. Thus, in order to evaporate 1 kg of water into the air at 0 °C, you would have to supply 600 kcal of thermal energy to the water.

The molecules in the water in bowl 1 are moving about in a random order. But their attractive molecular forces still keep them together. These molecules can now absorb heat from the surroundings. This absorbed energy shows up as an increase in the kinetic energy of the molecule, and hence an increase in the velocity of the molecule. When the liquid molecule has absorbed enough energy it moves right out of the liquid water into the air above as a molecule of water vapor. (Remember the water molecule is the same whether it is a solid, liquid, or gas, namely H_2O, two atoms of hydrogen and one atom of oxygen. The difference is only in the energy of the molecule.)

Since the most energetic of the water molecules escape from the liquid, the molecules left behind have lower energy, hence the temperature of the remaining liquid decreases. *Hence, evaporation is a cooling process.* The water molecule that evaporated took the thermal energy with it, and the water left behind is just that much cooler.

The remaining water in bowl 1 now absorbs energy from the environment, thereby increasing the temperature of the water in the bowl. This increased thermal energy is used by more liquid water molecules to escape into the air as more water vapor. The process continues until all the water in bowl 1 is evaporated.

Now when we look at bowl 2, the water is still there. Why didn't all that water evaporate into the air? To explain what hap-

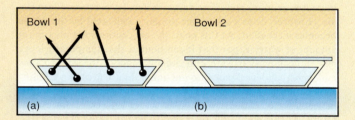

Figure 2
Evaporation.

pens in bowl 2 let us do the following experiment. We place water in a container and place a plate over the water. Then we allow dry air, air that does not contain water vapor, to fill the top portion of the closed container, figure 3(a). Using a thermometer, we measure the temperature of the air as $t = 20$ °C, and using a pressure gauge we measure the pressure of the air p_0, in the container. Now we remove the plate separating the dry air from the water by sliding it out of the closed container. As time goes by, we observe that the pressure recorded by the pressure gauge increases, figure 3(b). This occurs because some of the liquid water molecules evaporate into the air as water vapor. Water vapor is a gas like any other gas and it exerts a pressure. It is this water vapor pressure that is being recorded as the increased pressure on the gauge. The gauge is reading the air pressure of the dry air plus the actual water vapor pressure of the gas, $p_0 + p_{awv}$. Subtracting p_0 from $p_0 + p_{awv}$, gives the actual water

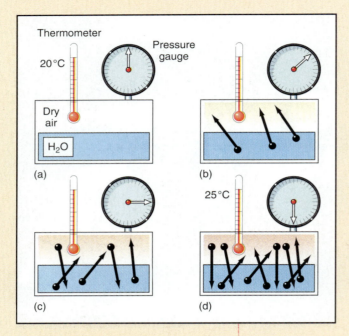

Figure 3

Water vapor in the air.

vapor pressure, p_{awv}. As time goes on, the water vapor pressure increases as more and more water molecules evaporate into the air. However, after a while, the pressure indicated by the gauge becomes a constant. At this point the air contains the maximum amount of water vapor that it can hold at that temperature. As new molecules evaporate into the air, some of the water vapor molecules condense back into the liquid, figure 3(c). An equilibrium condition is established, whereby just as many water vapor molecules are condensing as liquid water molecules are evaporating. At this point, the air is said to be *saturated*. That is, the air contains the maximum amount of water vapor that it can hold at that temperature. The vapor pressure read by the gauge is now called the saturation water vapor pressure, p_{swv}.

The amount of water vapor in the air is called humidity. *A measure of the amount of water vapor in the air is given by the relative humidity, RH, and is defined as the ratio of the amount of water vapor actually present in the air to the amount of water vapor that the air can hold at a given temperature and pressure, times 100%.* The amount of water vapor in the air is directly proportional to the water vapor pressure. Therefore, we can determine the relative humidity, RH, of the air as

$$RH = \left(\frac{\text{actual vapor pressure}}{\text{saturation vapor pressure}} \right) 100\% \quad \textbf{(15H.1)}$$

$$RH = \left(\frac{p_{awv}}{p_{swv}} \right) 100\% \quad \textbf{(15H.2)}$$

When the air is saturated, the actual vapor pressure recorded by the gauge is equal to the saturation vapor pressure and hence, the relative humidity is 100%.

If the air in the container is heated, we notice that the pressure indicated by the pressure gauge increases, figure 3(d). Part of the increased pressure is caused by the increase of the pressure of the air. This increase can be calculated by the ideal gas equation and subtracted from the gauge reading, so that we can determine any increase in pressure that would come from an increase in the actual water vapor pressure. We notice that by increasing the air temperature to 25 °C, the water vapor pressure also increases. After a while, however, the water vapor pressure again becomes a constant. The air is again saturated. We see from this experiment that *the maximum amount of water vapor that the air can hold is a function of temperature*. At low temperatures the air can hold only a little water vapor, while at high temperatures the air can hold much more water vapor.

We can now see why the water in bowl 2 in figure 2 did not disappear. Water evaporated from the liquid into the air above, increasing the relative humidity of the air. However, once the air became saturated, the relative humidity was equal to 100%, and no more water vapor could evaporate into it. This is why you can still see the water in bowl 2, there is no place for it to go.

Because of the temperature dependence of water vapor in the air, when the temperature of the air is increased, the capacity of the air to hold water increases. Therefore, if no additional water is added to the air, the relative humidity will decrease because the capacity of the air to hold water vapor has increased. Conversely, when the air temperature is decreased, its capacity to hold water vapor decreases, and therefore the relative humidity of the air increases. This temperature dependence causes a decrease in the relative humidity during the day light hours, and an increase in the relative humidity during the night time hours, with the maximum relative humidity occurring in the early morning hours just before sunrise.

The amount of evaporation depends on the following factors:

1. The vapor pressure. Whenever the actual vapor pressure is less than the maximum vapor pressure allowable at that temperature, the saturation vapor pressure, then evaporation will readily occur. Greater evaporation occurs whenever the air is dry, that is, at low relative humidities. Less evaporation occurs when the air is moist, that is, at high relative humidities.

2. Wind movement and turbulence. Air movement and turbulence replaces air near the water surface with less moist air and increases the rate of evaporation.

Now that we have discussed the concepts of relative humidity we can understand how the body cools itself. Through the process of perspiration, the body secretes microscopic droplets of water onto the surface of the skin of the body. As these tiny droplets of water evaporate into the air, they cool the body. As long as the relative humidity of the air is low, evaporation occurs readily, and the body cools itself. *However whenever the relative humidity becomes high, it is more difficult for the microscopic droplets of water to evaporate into the air. The body can not cool itself, and the person feels very uncomfortable.*

We are all aware of the discomfort caused by the hot and humid days of August. The high relative humidity prevents the normal evaporation and cooling of the body. As some evaporation occurs from the body, the air next to the skin becomes saturated, and no further cooling can occur. If a fan is used, we feel more comfortable because the fan blows the saturated air next to our skin away and replaces it with air that is slightly less saturated. Hence, the evaporation process can continue while the fan is in operation and the body cools itself. Another way to cool the human body in the summer is to use an air conditioner. The air conditioner not only cools the air to a lower temperature, but it also removes a great deal of water vapor from the air, thereby decreasing the relative humidity of the air and permitting the normal evaporation of moisture from the skin. (Note that if the air conditioner did not remove water vapor from the air, cooling the air would increase the relative humidity making us even more uncomfortable.)

In the hot summertime, people enjoy swimming as a cooling experience. Not only the immersion of the body in the cool water is so satisfying, but when the person comes out of the water, evaporation of the sea or pool water from the person adds to the cooling. It is also customary to wear loose clothing in the summertime. The reason for this is to facilitate the flow of air over the body and hence assist in the evaporation process. Tight fitting clothing prevents this evaporation process and the person feels hotter. If you happen to live in a dry climate (low relative humidity), then you can feel quite comfortable at 85 °F, while a person living in a moist climate (high relative humidity) is very uncomfortable at the same 85 °F.

What many people do not realize is that you can also feel quite uncomfortable even in the wintertime, because of the humidity of the air. If the relative humidity is very low in your home then evaporation occurs very rapidly, cooling the body perhaps more than is desirable. As an example, the air temperature might be 70 °F but if the relative humidity is low, say 30%, then evaporation readily occurs from the skin of the body, and the person feels cold even though the air temperature is 70 °F. In this case the person can feel more comfortable if he or she uses a humidifier. A humidifier is a device that adds water vapor to the air. By increasing the water vapor in the air, and hence increasing the relative humidity, the rate of evaporation from the body decreases. The person no longer feels cold at 70 °F, but feels quite comfortable. If too much water vapor is added to the air, increasing the relative humidity to near a 100%, then evaporation from the body is hampered, the body is not able to cool itself, and the person feels too hot even though the temperature is only 70 °F. Thus too high or too low a relative humidity makes the human body uncomfortable.

We should also note that the evaporation process is also used to cool the human body for medical purposes. If a person is running a high fever, then an alcohol rub down helps cool the body down to normal temperature. The principle of evaporation as a cooling device is the same, only alcohol is very volatile and evaporates very rapidly. This is because the saturation vapor pressure of alcohol at 20 °C is much higher than the saturation vapor pressure of water. At 20 °C, water has a saturation vapor pressure of 17.4 mm of Hg, whereas ethyl alcohol has a saturation vapor pressure of 44 mm of Hg. The larger the saturation vapor pressure of a liquid, the greater is the amount of its vapor that the air can hold and hence the greater is the rate of vaporization. Because the alcohol evaporates much more rapidly than water, much greater cooling occurs than when water evaporates. Ethyl ether and ethyl chloride have saturation vapor pressures of 442 mm and 988 mm of Hg, respectively. Ethyl chloride with its very high saturation vapor pressure, evaporates so rapidly that it freezes the skin, and is often used as a local anesthetic for minor surgery.

The Language of Physics

Thermal expansion
Most materials expand when heated (p. 425).

Charles' law
The volume of a gas at constant pressure is directly proportional to the absolute temperature of the gas (p. 432).

Gay-Lussac's law
The absolute pressure of a gas at constant volume is directly proportional to the absolute temperature of the gas (p. 433).

Boyle's law
The product of the pressure and volume of a gas at constant temperature is equal to a constant (p. 434).

The ideal gas law
The general gas law that contains Charles', Gay-Lussac's, and Boyle's law as special cases. It states that the product of the pressure and volume of a gas divided by the absolute temperature of the gas is a constant (p. 434).

Mole
One mole of any gas is that amount of the gas that has a mass in grams equal to the atomic or molecular mass of the gas. One mole of any gas at a temperature of 0 °C and a pressure of one atmosphere, has a volume of 22.4 liters (p. 435).

Avogadro's number
Every mole of a gas contains the same number of molecules, namely, 6.022×10^{23} molecules. The mass of one molecule is equal to the molecular mass of that gas divided by Avogadro's number (p. 435).

Kinetic theory of gases
The analysis of a gas at the microscopic level, treated by Newton's laws of motion. The kinetic theory shows that the absolute temperature of a gas is a measure of the mean translational kinetic energy of the molecules of the gas (p. 438).

Summary of Important Equations

Linear expansion
$$\Delta L = \alpha L_0 \Delta t \quad \text{(15.1)}$$

Area expansion
$$\Delta A = 2\alpha A_0 \Delta t \quad \text{(15.3)}$$

Volume expansion
$$\Delta V = 3\alpha V_0 \Delta t \quad \text{(15.5)}$$

Coefficient of volume expansion
for solids
$$\beta = 3\alpha \quad \text{(15.6)}$$

Volume expansion
$$\Delta V = \beta V_0 \Delta t \quad \text{(15.7)}$$

Ideal gas law
$$\frac{p_1 V_1}{T_1} = \frac{p_2 V_2}{T_2} \quad \text{(15.20)}$$
$$pV = nRT \quad \text{(15.23)}$$

Number of molecules
$$N = nN_A \quad \text{(15.24)}$$

Absolute pressure
$$p_{abs} = p_{gauge} + p_{atm} \quad \text{(15.25)}$$

Temperature and mean
kinetic energy
$$\tfrac{3}{2}kT = \tfrac{1}{2}mv_{avg}^2 \quad \text{(15.45)}$$

rms speed of a molecule
$$v_{rms} = \sqrt{\frac{3kT}{m}} \quad \text{(15.46)}$$

Mass of a molecule
$$m = \frac{M}{N_A} \quad \text{(15.47)}$$

Total mass of the gas
$$m_{total} = nM$$

Questions for Chapter 15

1. Describe the process of expansion from a microscopic point of view.
2. Explain why it is necessary to make a temperature correction when measuring atmospheric pressure with a barometer.
†3. In the very upper portions of the atmosphere there are extremely few molecules present. Discuss the concept of temperature as it would be applied in this portion of the atmosphere.

4. Explain the introduction of the Kelvin temperature scale in the application of Charles' law.
5. Describe the meaning and application of gauge pressure.
†6. Would you expect the ideal gas equation to be applicable to a volume that is of the same order of magnitude as the size of a molecule?
7. If a gas is at an extremely high density, what effect would this have on the assumptions underlying the kinetic theory of gases?

8. From the point of view of the time between collisions of a gas molecule and the walls of the container, what happens if the container is reduced to half its original size?
9. From the point of view of the kinetic theory of gases, explain why there is no atmosphere on the moon.
10. When an astronomer observes the stars at night in an observatory, the observatory is not heated but remains at the same temperature as the outside air. Why should the astronomer do this?

Problems for Chapter 15

15.1 Linear Expansion of Solids

1. An aluminum rod measures 2.00 m at 10.0 °C. Find its length when the temperature rises to 135 °C.
2. A brass ring has a diameter of 20.0 cm when placed in melting ice at 0 °C. What will its diameter be if it is placed in boiling water?
3. A steel ring of 2.00-in. diameter at 0 °C is to be heated and slipped over a steel shaft whose diameter is 2.003 in. at 0 °C. To what temperature should the ring be heated? If the ring is not heated, to what temperature should the shaft be cooled such that the ring will fit over the shaft?

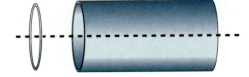

4. An aluminum ring, 7.00 cm in diameter at 5.00 °C, is to be heated and slipped over an aluminum shaft whose diameter is 7.003 cm at 5.00 °C. To what temperature should the ring be heated? If the ring is not heated, to what temperature should the shaft be cooled such that the ring will fit over the shaft?
5. The iron rim of a wagon wheel has an internal diameter of 80.0 cm when the temperature is 100 °C. What is its diameter when it cools to 0.00 °C?
6. A steel measuring tape, correct at 0.00 °C measures a distance L when the temperature is 30.0 °C. What is the error in the measurement due to the expansion of the tape?
7. Steel rails 50.0 ft long are laid when the temperature is 0.00 °C. What separation should be left between the rails to allow for thermal expansion when the temperature rises to 40.0 °C? If the cross-sectional area of a rail is 30.0 in.², what force is associated with this expansion?

8. Steel rails 20.0 m long are laid when the temperature is 5.00 °C. What separation should be left between the rails to allow for thermal expansion when the temperature rises to 38.5 °C? If the cross-sectional area of a rail is 230 cm², what force is associated with this expansion?

15.2 Area Expansion of Solids

9. A sheet of aluminum measures 10.0 ft by 5.00 ft at 0.00 °C. What is the area of the sheet at 150 °C?
10. A sheet of brass measures 4.00 m by 3.00 m at 5.00 °C. What is the area of the sheet at 175 °C?
11. If the radius of a copper circle is 20.0 cm at 0.00 °C, what will its area be at 100 °C?
12. A piece of aluminum has a hole 0.850 cm in diameter at 20.0 °C. To what temperature should the sheet be heated so that an aluminum bolt 0.865 cm in diameter will just fit into the hole?

15.3 Volume Expansion of Solids and Liquids

13. A chemistry student fills a Pyrex glass flask to the top with 100 cm³ of Hg at 0.00 °C. How much mercury will spill out of the tube, and have to be cleaned up by the student, if the temperature rises to 35.0 °C?

14. A tube is filled to a height of 20.0 cm with mercury at 0.00 °C. If the tube has a cross-sectional area of 25.0 mm², how high will the mercury rise in the tube when the temperature is 30.0 °C? Neglect the expansion of the tube.

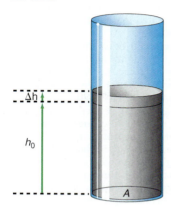

15. Since the volume of a material changes with a change in temperature, show that the density ρ at any temperature is given by

$$\rho = \frac{\rho_0}{1 + \beta \Delta t}$$

where ρ_0 is the density at the lower temperature.

15.7 The Ideal Gas Law

16. If 2.00 g of oxygen gas are contained in a tank of 500 cm³ at a pressure of 20.0 psi, what is the temperature of the gas?

17. What is the pressure produced by 2 moles of gas at 20.0 °C contained in a volume of 0.500 liters?

18. One mole of hydrogen is at a pressure of 2.00 atm and a volume of 0.25 m³. What is its temperature?

19. Compute the number of molecules in a gas contained in a volume of 50.0 cm³ at a pressure of 2.00 atm and a temperature of 300 K.

20. An automobile tire has a volume of 4500 in.³ and contains air at a gauge pressure of 32.0 lb/in.² when the temperature is 0.00 °C. What is the gauge pressure when the temperature rises to 35.0 °C?

21. An automobile tire has a volume of 0.0800 m³ and contains air at a gauge pressure of 2.48 × 10⁵ N/m² when the temperature is 3.50 °C. What is the gauge pressure when the temperature rises to 37.0 °C?

22. (a) How many moles of gas are contained in 0.300 kg of H_2 gas? (b) How many molecules of H_2 are there in this mass?

23. Nitrogen gas, at a pressure of 150 N/m², occupies a volume of 20.0 m³ at a temperature of 30.0 °C. Find the mass of this nitrogen gas in kilograms.

24. One mole of nitrogen gas at a pressure of 1.00 atm and a temperature of 300 K expands isothermally to double its volume. What is its new pressure? (Isothermal means at constant temperature.)

25. An ideal gas occupies a volume of 4.00 liters at a pressure of 1.00 atm and a temperature of 273 K. The gas is then compressed isothermally to one half of its original volume. Determine the final pressure of the gas.

26. The pressure of a gas is kept constant while 3.00 m³ of the gas at an initial temperature of 50.0 °C is expanded to 6.00 m³. What is the final temperature of the gas?

27. The volume of O_2 gas at a temperature of 20.0 °C is 4.00 liters. The temperature of the gas is raised to 100 °C while the pressure remains constant. What is the new volume of the gas?

28. A balloon is filled with helium at a pressure of 1.50 atm, a temperature of 25.0 °C, and occupies a volume of 3.00 m³. The balloon rises in the atmosphere. When it reaches a height where the pressure is 0.500 atm and the temperature is −20.0 °C, what is its volume?

†29. An air bubble of 32.0 cm³ volume is at the bottom of a lake 10.0 m deep where the temperature is 5.00 °C. The bubble rises to the surface where the temperature is 20.0 °C. Find the volume of the bubble just before it reaches the surface.

30. One mole of helium is at a temperature of 300 K and a volume of 10.0 liters. What is its pressure? The gas is warmed at constant volume to 600 K. What is its new pressure? How many molecules are there?

15.8 The Kinetic Theory of Gases

31. Find the rms speed of a helium atom at a temperature of 10.0 K.

32. Find the kinetic energy of a single molecule when it is at a temperature of (a) 0.00 °C, (b) 20.0 °C, (c) 100 °C, (d) 1000 °C, and (e) 5000 °C.

33. Find the mass of a carbon dioxide molecule (CO_2).

34. Find the rms speed of a helium atom on the surface of the sun, if the sun's surface temperature is approximately 6000 K.

35. At what temperature will the rms speed of an oxygen molecule be twice its speed at room temperature?

36. The rms speed of a gas molecule is v at a temperature of 300 K. What is the speed if the temperature is increased to 900 K?

†37. Find the total kinetic energy of all the nitrogen molecules in the air in a room 7.00 m by 10.0 m by 4.00 m, if the air is at a temperature of 22.0 °C and 1 atm of pressure.

38. If the rms speed of a monatomic gas is 445 m/s at 350 K, what is the atomic mass of the atom? What gas do you think it is?

Additional Problems

39. A barometer reads normal atmospheric pressure when the mercury column in the tube is at 76.0 cm of Hg at 0.00 °C. If the pressure of the atmosphere does not change, but the air temperature rises to 35.0 °C, what pressure will the barometer indicate? The tube has a diameter of 5.00 mm. Neglect the expansion of the tube.

40. Find the stress necessary to give the same strain that occurs when a steel rod undergoes a temperature change of 120 °C.

†41. The symbol π is defined as the ratio of the circumference of a circle to its diameter. If a circular sheet of metal expands by heating, show that the ratio of the expanded circumference to the expanded diameter is still equal to π.

42. A 15.0-cm strip of steel is welded to the left side of a 15.0-cm strip of aluminum. When the strip undergoes a temperature change Δt, will the combined strip bend to the right or to the left?

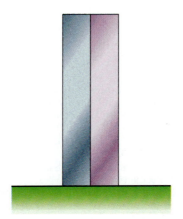

†43. A 350-g mass is connected by a thin brass rod 25.0 cm long to a rotating shaft that is rotating at an initial angular speed of 5.00 rad/s. If the temperature changes by 35 °C, (a) find the change in the moment of inertia of the system and (b) using the law of conservation of angular momentum, find the change in the rotational energy of the system.

44. The focal length of a polished aluminum spherical mirror is given by $f = R/2$, where R is the radius of curvature of the mirror, and is 23.5 cm. Find the new focal length of the mirror if the temperature changes by 45.0 °C.

†45. A 50.0-g silver ring, 12.0 cm in diameter, is spinning about an axis through its center at a constant speed of 11.4 rad/s. If the temperature changes by 185 °C, what is the change in the angular momentum of the ring? The coefficient of linear expansion for silver is $1.90 \times 10^{-5}/°C$.

†46. An aluminum rod is at room temperature. To what temperature would this rod have to be heated such that the thermal expansion is enough to exceed the elastic limit of aluminum? Compare this temperature with the melting point of aluminum. What conclusion can you draw?

47. A steel pendulum is 60.0 cm long, at 20.0 °C. By how much does the period of the pendulum change when the temperature is 35.0 °C?

48. Find the number of air molecules in a classroom 10.0 m long, 10.0 m wide, and 3.5 m high, if the air is at normal atmospheric pressure and a temperature of 20.0 °C.

49. A brass cylinder 5.00 cm in diameter and 8.00 cm long is at an initial temperature of 380 °C. It is placed in a calorimeter containing 0.120 kg of water at an initial temperature of 5.00 °C. The aluminum calorimeter has a mass of 0.060 kg. Find (a) the final temperature of the water and (b) the change in volume of the cylinder.

†50. Dalton's law of partial pressure says that when two or more gases are mixed together, the resultant pressure is the sum of the individual pressures of each gas. That is,

$$p = p_1 + p_2 + p_3 + p_4 + \cdots$$

If one mole of oxygen at 20.0 °C and occupying a volume of 2.00 m³ is added to two moles of nitrogen also at 20.0 °C and occupying a volume of 10.0 m³ and the final volume is 10.0 m³, find the resultant pressure of the mixture.

†51. The escape velocity from the earth is $v_E = 1.12 \times 10^4$ m/s. At what temperature is the rms speed equal to this for: (a) hydrogen (H_2), (b) helium (He), (c) nitrogen (N_2), (d) oxygen (O_2), (e) carbon dioxide (CO_2), and (f) water vapor (H_2O)? From these results, what can you infer about the earth's atmosphere?

†52. The escape velocity from the moon is $v_M = 0.24 \times 10^4$ m/s. At what temperature is the rms speed equal to this for (a) hydrogen (H_2), (b) helium (He), (c) nitrogen (N_2), (d) oxygen (O_2), (e) carbon dioxide (CO_2), and (f) water vapor (H_2O)? From these results, what can you infer about the possibility of an atmosphere on the moon?

†53. Show that the velocity of a gas molecule at one temperature is related to the velocity of the molecule at a second temperature by

$$v_2 = \sqrt{\frac{T_2}{T_1}} v_1$$

†54. A room is filled with nitrogen gas at a temperature of 293 K. (a) What is the average kinetic energy of a nitrogen molecule? (b) What is the rms speed of the molecule? (c) What is the rms value of the momentum of this molecule? (d) If the room is 4.00 m wide what is the average force exerted on the wall by this molecule? (e) If the wall is 4.00 m by 3.00 m, what is the pressure exerted on the wall by this molecule? (f) How many molecules moving at this speed are necessary to cause a pressure of 1.00 atm?

†55. Two isotopes of a gaseous substance can be separated by diffusion if each has a different velocity. Show that the rms speed of an isotope can be given by

$$v_2 = \sqrt{\frac{m_1}{m_2}} v_1$$

where the subscript 1 refers to isotope 1 and the subscript 2 refers to isotope 2.

Interactive Tutorials

56. A copper tube has the length $L_0 = 1.58$ m at the initial temperature $t_i = 20.0$ °C. Find its length L when it is heated to a final temperature $t_f = 100$ °C.

57. A circular brass sheet has an area $A_0 = 2.56$ m² at the initial temperature $t_i = 0$ °C. Find its new area A when it is heated to a final temperature $t_f = 90$ °C.

58. A glass tube is filled to a height $h_0 = 0.762$ m of mercury at the initial temperature $t_i = 0$ °C. The diameter of the tube is 0.085 m. How high will the mercury rise when the final temperature $t_f = 50$ °C? Neglect the expansion of the glass.

59. A gas has a pressure $p_1 = 1$ atm, a volume $V_1 = 4.58$ m³, and a temperature $t_i = 20.0$ °C. It is then compressed to a volume $V_2 = 1.78$ m³ and a pressure $p_2 = 3.57$ atm. Find the final temperature of the gas t_2.

60. Find the number of moles and the number of molecules in a gas under a pressure $p = 1$ atm and a temperature $t = 20.0$ °C. The room has a length $L = 15.0$ m, a width $W = 10.0$ m, and a height $h = 4.00$ m.

61. Kinetic theory. Oxygen gas is in a room under a pressure $p = 1$ atm and a temperature of $t = 20.0$ °C. The room has a length $L = 18.5$ m, a width $W = 12.5$ m, and a height $h = 5.50$ m. For the oxygen gas, find (a) the kinetic energy of a single molecule, (b) the total kinetic energy of all the oxygen molecules, (c) the mass of an oxygen molecule, and (d) the speed of the oxygen molecule. The molecular mass of oxygen is $M_{O_2} = 32.0$ g/mole.

62. Ideal Gas Equation Calculator.

Thermodynamics

16 Heat Transfer

Thermogram showing the distribution of heat on a man's face, wearing glasses.

There can be no doubt now, in the mind of the physicist who has associated himself with inductive methods, that matter is constituted of atoms, heat is movement of molecules and conduction of heat, like all other irreversible phenomena, obeys, not dynamical, but statistical laws, namely, the laws of probability.

Max Planck

16.1 Heat Transfer

In chapter 14 we saw that an amount of thermal energy Q, given by

$$Q = mc\Delta T \tag{14.6}$$

is absorbed or liberated in a sensible heating process. But how is this thermal energy transferred to, or from, the body so that it can be absorbed, or liberated? To answer that question, we need to discuss the mechanism of thermal energy transfer. The transfer of thermal energy has historically been called heat transfer.

Thermal energy can be transferred from one body to another by any or all of the following mechanisms:

1. Convection
2. Conduction
3. Radiation

Convection *is the transfer of thermal energy by the actual motion of the medium itself.* The medium in motion is usually a gas or a liquid. Convection is the most important heat transfer process for liquids and gases.

Conduction *is the transfer of thermal energy by molecular action, without any motion of the medium.* Conduction can occur in solids, liquids, and gases, but it is usually most important in solids.

Radiation *is a transfer of thermal energy by electromagnetic waves.*

We will discuss the details of electromagnetic waves in chapter 25. For now we will say that it is not necessary to have a medium for the transfer of energy by radiation. For example, energy is radiated from the sun as an electromagnetic wave, and this wave travels through the vacuum of space, until it impinges on the earth, thereby heating the earth.

Let us now go into more detail about each of these mechanisms of heat transfer.

16.2 Convection

Consider the large mass m of air at the surface of the earth that is shown in figure 16.1. The lines labeled T_0, T_1, T_2, and so on are called isotherms and represent the temperature distribution of the air at the time t. *An **isotherm** is a line along which the temperature is constant.* Thus everywhere along the line T_0 the air temperature is T_0, and everywhere along the line T_1 the air temperature is T_1, and so forth.

Figure 16.1

Horizontal convection.

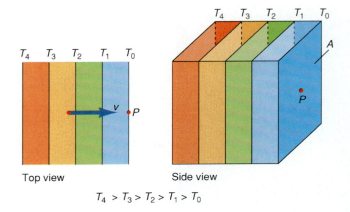

Top view   Side view

$$T_4 > T_3 > T_2 > T_1 > T_0$$

Consider a point P on the surface of the earth that is at a temperature T_0 at the time t. How can thermal energy be transferred to this point P thereby changing its temperature? That is, how does the thermal energy at that point change with time? If we assume that there is no local infusion of thermal energy into the air at P, such as heating from the sun and the like, then the only way that thermal energy can be transferred to P is by moving the hotter air, presently to the left of point P, to point P itself. That is, if energy can be transferred to the point P by convection, then the air temperature at the point P increases. This is equivalent to moving an isotherm that is to the left of P to the point P itself. The transfer of thermal energy per unit time to the point P is given by $\Delta Q/\Delta t$. By multiplying and dividing by the distance Δx, we can write this as

$$\frac{\Delta Q}{\Delta t} = \frac{\Delta Q}{\Delta x}\frac{\Delta x}{\Delta t} \qquad (16.1)$$

But

$$\frac{\Delta x}{\Delta t} = v$$

the velocity of the air moving toward P. Therefore, equation 16.1 becomes

$$\frac{\Delta Q}{\Delta t} = v\frac{\Delta Q}{\Delta x} \qquad (16.2)$$

But ΔQ, on the right-hand side of equation 16.2, can be replaced with

$$\Delta Q = mc\Delta T \qquad (14.6)$$

(We will need to depart from our custom of using the lower case t for temperatures in Celsius or Fahrenheit degrees, because we will use t to represent time. Thus, the upper case T is now used for temperature in either Celsius or Fahrenheit degrees.) Therefore,

$$\frac{\Delta Q}{\Delta t} = vmc\frac{\Delta T}{\Delta x} \qquad (16.3)$$

(Equation 16.3 is a good example of the difference between heat and thermal energy. Here Q is the thermal energy, and $\Delta Q/\Delta t$ is the thermal energy per unit time that is transferred, and this is what is called heat.) Hence, the thermal energy transferred to the point P by convection becomes

$$\Delta Q = vmc\frac{\Delta T}{\Delta x}\Delta t \qquad (16.4)$$

The term $\Delta T / \Delta x$ is called the **temperature gradient,** and tells how the temperature changes as we move in the x-direction. We will assume in our analysis that the temperature gradient remains a constant.

Example 16.1

Energy transfer per unit mass. If the temperature gradient is 2.00 °C per 100 km and if the specific heat of air is 0.250 kcal/(kg °C), how much thermal energy per unit mass is convected to the point P in 12.0 hr if the air is moving at a speed of 10.0 km/hr?

Solution

The heat transferred per unit mass, found from equation 16.4, is

$$\frac{\Delta Q}{m} = vc\frac{\Delta T}{\Delta x}\Delta t$$

$$= \left(10.0\,\frac{km}{hr}\right)\left(0.250\,\frac{kcal}{kg\,°C}\right)\left(\frac{2.00\,°C}{100\,km}\right)(12.0\,hr)$$

$$= 0.600\ kcal/kg$$

If the mass m of the air that is in motion is unknown, the density of the fluid can be used to represent the mass. Because the density $\rho = m/V$, where V is the volume of the air, we can write the mass as

$$m = \rho V \tag{16.5}$$

Therefore, the thermal energy transferred by convection to the point P becomes

$$\Delta Q = v\rho Vc\frac{\Delta T}{\Delta x}\Delta t \tag{16.6}$$

Sometimes it is more convenient to find the thermal energy transferred per unit volume. In this case, we can use equation 16.6 as

$$\frac{\Delta Q}{V} = v\rho c\frac{\Delta T}{\Delta x}\Delta t$$

Example 16.2

Energy transfer per unit volume. Using the data from example 16.1, find the thermal energy per unit volume transferred by convection to the point P. Assume that the density of air is $\rho_{air} = 1.293\ kg/m^3$.

Solution

The thermal energy transferred per unit volume is found as

$$\frac{\Delta Q}{V} = v\rho c\frac{\Delta T}{\Delta x}\Delta t$$

$$= \left(10.0\,\frac{km}{hr}\right)\left(1.293\,\frac{kg}{m^3}\right)\left(0.250\,\frac{kcal}{kg\,°C}\right)\left(\frac{2.00\,°C}{100\,km}\right)(12.0\,hr)$$

$$= 0.776\ kcal/m^3$$

Note that although the number 0.776 kcal/m³ may seem small, there are thousands upon thousands of cubic meters of air in motion in the atmosphere. Thus, the thermal energy transfer by convection can be quite significant.

Convection is the main mechanism of thermal energy transfer in the atmosphere. On a global basis, the nonuniform temperature distribution on the surface of the earth causes convection cycles that result in the prevailing winds. If the earth were not rotating, a huge convection cell would be established as shown in figure 16.2(a). The equator is the hottest portion of the earth because it gets the

Figure 16.2

Convection in the atmosphere. Lutgens/Tarbuck, *The Atmosphere: An Introduction to Meteorology,* 4/E, © 1989, pp. 186–187. © Prentice-Hall, Inc., Englewood Cliffs, NJ.

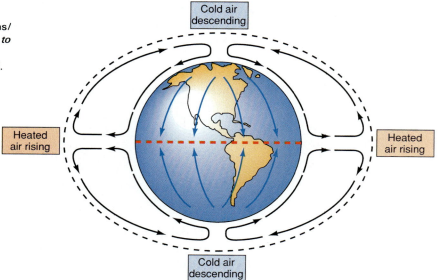

(a) Nonrotating earth

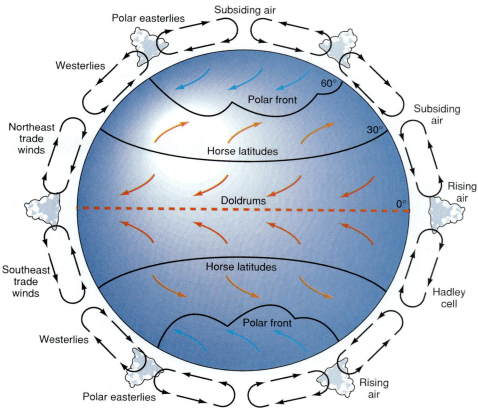

(b) Rotating earth

maximum radiation from the sun. Hot air at the equator expands and rises into the atmosphere. Cooler air at the surface flows toward the equator to replace the rising air. Colder air at the poles travels toward the equator. Air aloft over the poles descends to replace the air at the surface that just moved toward the equator. The initial rising air at the equator flows toward the pole, completing the convection cycle. The net result of the cycle is to bring hot air at the surface of the equator, aloft, then north to the poles, returning cold air at the polar surface back to the equator.

This simplified picture of convection on the surface of the earth is not quite correct, because the effect produced by the rotating earth, called the Coriolis effect, has been neglected. The **Coriolis effect** is caused by the rotation of the earth and can best be described by an example. If a projectile, aimed at New York, were fired from the North Pole, its path through space would be in a fixed vertical plane that has the North Pole as the starting point of the trajectory and New York as the ending point at the moment that the projectile is fired. However, by the time that the projectile arrived at the end point of its trajectory, New York would no longer be there, because while the projectile was in motion, the earth was rotating, and New York will have rotated away from the initial position it was in when the projectile was fired. A person fixed to the rotating earth would see the projectile veer away to the right of its initial path, and would assume that a force was acting on the projectile toward the right of its trajectory. This fictitious force is called the Coriolis force and this seemingly strange behavior occurs because the rotating earth is not an inertial coordinate system.

The Coriolis effect can be applied to the global circulation of air in the atmosphere, causing winds in the northern hemisphere to be deflected to the right of their original path. The global convection cycle described above still occurs, but instead of one huge convection cell, there are three smaller ones, as shown in figure 16.2(b). The winds from the North Pole flowing south at the surface of the earth are deflected to the right of their path and become the polar easterlies, as shown in figure 16.2(b). As the air aloft at the equator flows north it is deflected to the right of its path and eventually flows in a easterly direction at approximately 30° north latitude. The piling up of air at this latitude causes the air aloft to sink to the surface where it emerges from a semipermanent high-pressure area called the subtropical high. The air at the surface that flows north from this high-pressure area is deflected to the right of its path producing the mid-latitude westerlies. The air at the surface that flows south from this high-pressure area is also deflected to the right of its path and produces the northeast trade winds, also shown in figure 16.2(b). *Thus, it is the nonuniform temperature distribution on the surface of the earth that is responsible for the global winds.*

Transfer of thermal energy by convection is also very important in the process called the *sea breeze,* which is shown in figure 16.3. Water has a higher specific heat than land and for the same radiation from the sun, the temperature of the water does not rise as high as the temperature of the land. Therefore, the land mass becomes hotter than the neighboring water. The hot air over the land rises and a cool breeze blows off the ocean to replace the rising hot air. Air aloft descends to replace this cooler sea air and the complete cycle is as shown in figure 16.3. The net result

Figure 16.3
The sea breeze.

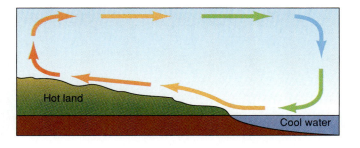
Hot land

Cool water

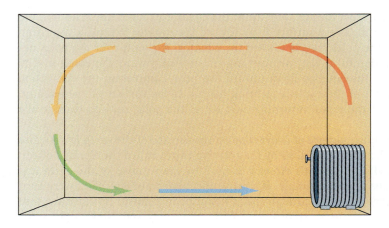

Figure 16.4

Natural convection in a room.

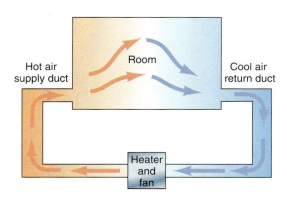

Figure 16.5

Forced convection.

of the process is to replace hot air over the land surface by cool air from the sea. This is one of the reasons why so many people flock to the ocean beaches during the hot summer months. The process reverses at night when the land cools faster than the water. The air then flows from the land to the sea and is called a *land breeze*.

This same process of thermal energy transfer takes place on a smaller scale in any room in your home or office. Let us assume there is a radiator situated at one wall of the room, as shown in figure 16.4. The air in contact with the heater is warmed, and then rises. Cooler air moves in to replace the rising air and a convection cycle is started. The net result of the cycle is to transfer thermal energy from the heater to the rest of the room. All these cases are examples of what is called *natural convection*.

To help the transfer of thermal energy by convection, fans can be used to blow the hot air into the room. Such a hot air heating system, shown in figure 16.5, is called a *forced convection system*. A metal plate is heated to a high temperature in the furnace. A fan blows air over the hot metal plate, then through some ducts, to a low-level vent in the room to be heated. The hot air emerges from the vent and rises into the room. A cold air return duct is located near the floor on the other side of the room, returning cool air to the furnace to start the convection cycle over again. The final result of the process is the transfer of thermal energy from the hot furnace to the cool room.

To analyze the transfer of thermal energy by this forced convection we will assume that a certain amount of mass of air Δm is moved from the furnace to the room. The thermal energy transferred by the convection of this amount of mass Δm is written as

$$\Delta Q = (\Delta m) c \Delta T \qquad (16.7)$$

where $\Delta T = T_h - T_c$, T_h is the temperature of the air at the hot plate of the furnace, and T_c is the temperature of the colder air as it leaves the room. The transfer of thermal energy per unit time becomes

$$\frac{\Delta Q}{\Delta t} = \frac{\Delta m}{\Delta t} c(T_h - T_c)$$

However,

$$m = \rho V$$

therefore

$$\Delta m = \rho \Delta V$$

Therefore, the thermal energy transfer becomes

$$\frac{\Delta Q}{\Delta t} = \rho c \frac{\Delta V}{\Delta t}(T_h - T_c) \qquad (16.8)$$

where $\Delta V/\Delta t$ is the volume flow rate, usually expressed as m³/min in SI units and cubic feet per minute (CFM) in the British engineering system of units.

Example 16.3

Forced convection. A hot air heating system is rated at 80,000 Btu/hr. If the heated air in the furnace reaches a temperature of 250 °F, the room temperature is 60.0 °F, and the fan can deliver 250 ft³/min, what is the thermal energy transfer per hour from the furnace to the room, and the efficiency of this system? The specific heat of air at constant pressure in the British engineering system is $c_{air} = 0.250$ Btu/lb °F and the weight density of air is $d_w = 0.0806$ lb/ft³.

Solution

We find the thermal energy transfer per hour in the British engineering system from a modification of equation 16.8. Instead of the mass density ρ, used in the SI system, the weight density d_w is used in the British engineering system. Therefore, the energy transfer is found as

$$\frac{\Delta Q}{\Delta t} = d_w c \frac{\Delta V}{\Delta t}(T_h - T_c)$$

$$= \left(0.0806 \frac{lb}{ft^3}\right)\left(0.250 \frac{Btu}{lb\,°F}\right)\left(250 \frac{ft^3}{min}\right)(250\,°F - 60.0\,°F)\left(60 \frac{min}{hr}\right)$$

$$= 57,500 \text{ Btu/hr}$$

We determine the efficiency, or rated value, of the heater as the ratio of the thermal energy out of the system to the thermal energy in, times 100%. Therefore,

$$\text{Eff} = \left(\frac{57,500 \text{ Btu/hr}}{80,000 \text{ Btu/hr}}\right)(100\%)$$

$$= 72\%$$

16.3 Conduction

Conduction is the transfer of thermal energy by molecular action, without any motion of the medium. Conduction occurs in solids, liquids, and gases, but the effect is most pronounced in solids. If one end of an iron bar is placed in a fire, in a relatively short time, the other end becomes hot. Thermal energy is conducted from the hot end of the bar to the cold end. The atoms or molecules in the hotter part of the body vibrate around their equilibrium position with greater amplitude than normal. This greater vibration causes the molecules to interact with their nearest neighbors, causing them to vibrate more also. These in turn interact with their nearest neighbors passing on this energy as *kinetic energy of vibration.* The thermal energy is thus passed from molecule to molecule along the entire length of the bar. The net result of these molecular vibrations is a transfer of thermal energy through the solid.

Heat Flow Through a Slab of Material

We can determine the amount of thermal energy conducted through a solid with the aid of figure 16.6. A slab of material of cross-sectional area A and thickness d is subjected to a high temperature T_h on the hot side and a colder temperature T_c on the other side.

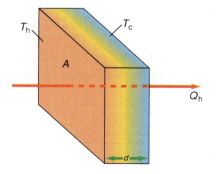

Figure 16.6

Heat conduction through a slab.

It is found experimentally that the thermal energy conducted through this slab is directly proportional to (1) the area A of the slab—the larger the area, the more thermal energy transmitted; (2) the time t—the longer the period of time, the more thermal energy transmitted; and finally (3) the temperature difference, $T_h - T_c$, between the faces of the slab. If there is a large temperature difference, a large amount of thermal energy flows. We can express these observations as the direct proportion

$$Q \propto A(T_h - T_c)t$$

The thermal energy transmitted is also found to be inversely proportional to the thickness of the slab, that is,

$$Q \propto \frac{1}{d}$$

This is very reasonable because the thicker the slab the greater the distance that the thermal energy must pass through. Thus, a thick slab implies a small amount of energy transfer, whereas a thin slab implies a larger amount of energy transfer.

These two proportions can be combined into one as

$$Q \propto \frac{A(T_h - T_c)t}{d} \tag{16.9}$$

To make an equality out of this proportion we must introduce a constant of proportionality. The constant must also depend on the material that the slab is made of, since it is a known fact that different materials transfer different quantities of thermal energy. We will call this constant *the coefficient of thermal conductivity*, and will denote it by k. Equation 16.9 becomes

$$Q = \frac{kA(T_h - T_c)t}{d} \tag{16.10}$$

Equation 16.10 gives the amount of thermal energy transferred by conduction. Table 16.1 gives the thermal conductivity k for various materials. If k is large, then a large amount of thermal energy will flow through the slab, and the material is called a good **conductor** of heat. If k is small then only a small amount of thermal energy will flow through the slab, and the material is called a poor conductor or a good **insulator**. Note from table 16.1 that most metals are good conductors while most nonmetals are good insulators. The ratio $(T_h - T_c)/d$ is the temperature gradient, $\Delta T / \Delta x$. Let us look at some examples of heat conduction.

This thermogram image, which was produced with an infrared scanner, shows heat escaping a house during winter.

Example 16.4

Heat transfer by conduction. Find the amount of thermal energy that flows per day through a solid oak wall 4.00 in. thick, 10.0 ft long, and 8.00 ft high, if the temperature of the inside wall is 70.0 °F while the temperature of the outside wall is 20.0 °F.

Solution

The thermal energy conducted through the wall, found from equation 16.10, is

$$Q = \frac{kA(T_h - T_c)t}{d}$$

$$= \frac{(1.02 \text{ Btu in./hr ft}^2 \, ^\circ\text{F})(80.0 \text{ ft}^2)(70.0 \, ^\circ\text{F} - 20.0 \, ^\circ\text{F})(24 \text{ hr})}{(4.00 \text{ in.})}$$

$$= 24{,}500 \text{ Btu}$$

Note that T_h and T_c are the temperatures of the wall and in general will be different from the air temperature inside and outside the room. The value T_h is usually lower

Table 16.1
Coefficient of Thermal Conductivity for Various Materials

	$\dfrac{\text{kcal}}{\text{m s }^\circ\text{C}}$	$\dfrac{\text{J}}{\text{m s }^\circ\text{C}}$	$\dfrac{\text{Btu in.}}{\text{hr ft}^2\ ^\circ\text{F}}$
Aluminum	0.056	2.34×10^2	1.42×10^3
Brass	0.026	1.09×10^2	0.754×10^3
Copper	0.096	4.02×10^2	2.79×10^3
Gold	0.076	3.13×10^2	2.21×10^3
Iron	0.021	8.79×10^1	0.609×10^3
Lead	0.0085	3.56×10^1	0.247×10^3
Nickel	0.022	9.21×10^1	0.639×10^3
Platinum	0.017	7.12×10^1	0.493×10^3
Silver	0.102	4.27×10^2	2.96×10^3
Zinc	0.028	1.17×10^2	0.813×10^3
Glass	1.89×10^{-4}	7.91×10^{-1}	5.5
Concrete	3.10×10^{-4}	1.30	9.00
Brick	1.55×10^{-4}	6.49×10^{-1}	4.50
Plaster	1.12×10^{-4}	4.69×10^{-1}	3.25
White pine	2.69×10^{-5}	1.13×10^{-1}	0.78
Oak	3.51×10^{-5}	1.47×10^{-1}	1.02
Cork board	8.6×10^{-6}	3.60×10^{-2}	0.25
Sawdust	0.141×10^{-4}	5.90×10^{-2}	0.41
Glass wool	0.099×10^{-4}	4.14×10^{-2}	0.29
Rock wool	0.093×10^{-4}	3.89×10^{-2}	0.27
Nitrogen	6.21×10^{-6}	2.60×10^{-2}	0.180
Helium	3.58×10^{-5}	1.50×10^{-1}	1.04
Air	5.5×10^{-6}	2.30×10^{-2}	0.160

than the room air temperature, whereas T_c is usually higher than the outside air temperature. This thermal energy loss through the wall must be replaced by the home heating unit in order to maintain a comfortable room temperature.

Equivalent Thickness of Various Walls

The walls in most modern homes are insulated with 4 in. of glass wool that is placed within the wall itself. But suppose the walls do not have this glass wool insulation. What should the equivalent thickness of another wall be, in order to give the same amount of insulation as a glass wool wall if the wall is made of (a) concrete, (b) brick, (c) glass, (d) oak wood, or (e) aluminum?

The amount of thermal energy that flows through the wall containing the glass wool, found from equation 16.10, is

$$Q_{gw} = \frac{k_{gw}A(T_h - T_c)t}{d_{gw}}$$

The thermal energy flowing through a concrete wall is given by

$$Q_c = \frac{k_c A (T_h - T_c) t}{d_c}$$

We assume in both equations that the walls have the same area, A; the same temperature difference $(T_h - T_c)$ is applied across each wall; and the thermal energy flows for the same time t. The subscript gw has been used for the wall containing the glass wool and the subscript c for the concrete wall. If both walls provide the same insulation then the thermal energy flow through each must be equal, that is,

$$Q_c = Q_{gw} \tag{16.11}$$

$$\frac{k_c A (T_h - T_c) t}{d_c} = \frac{k_{gw} A (T_h - T_c) t}{d_{gw}} \tag{16.12}$$

Therefore,

$$\frac{k_c}{d_c} = \frac{k_{gw}}{d_{gw}} \tag{16.13}$$

The equivalent thickness of the concrete wall to give the same insulation as the glass wool wall is

$$d_c = \frac{k_c}{k_{gw}} d_{gw} \tag{16.14}$$

Using the values of thermal conductivity from table 16.1 gives for the thickness of the concrete wall

$$d_c = \frac{k_c}{k_{gw}} d_{gw} = \left(\frac{9.00 \text{ Btu in./hr ft}^2 \, {}^\circ\text{F}}{0.29 \text{ Btu in./hr ft}^2 \, {}^\circ\text{F}} \right) (4.00 \text{ in.})$$

$$d_c = 124 \text{ in.} = 10.3 \text{ ft}$$

Therefore it would take a concrete wall 10 ft thick to give the same insulating ability as a 4-in. wall containing glass wool. Concrete is effectively a thermal sieve. Thermal energy flows through it, almost as fast as if there were no wall present at all. This is why uninsulated basements in most homes are difficult to keep warm.

To determine the equivalent thickness of the brick, glass, oak wood, and aluminum walls, we equate the thermal energy flow through each wall to the thermal energy flow through the wall containing the glass wool as in equations 16.11 and 16.12. We obtain a generalization of equation 16.13 as

$$\frac{k_b}{d_b} = \frac{k_g}{d_g} = \frac{k_{ow}}{d_{ow}} = \frac{k_{Al}}{d_{Al}} = \frac{k_{gw}}{d_{gw}} \tag{16.15}$$

with the results

brick

$$d_b = \frac{k_b}{k_{gw}} d_{gw} = \left(\frac{4.50}{0.29} \right) (4 \text{ in.}) = 62 \text{ in.} = 5.2 \text{ ft}$$

glass

$$d_g = \frac{k_g}{k_{gw}} d_{gw} = \left(\frac{5.5}{0.29} \right) (4 \text{ in.}) = 75.9 \text{ in.} = 6.3 \text{ ft}$$

oak wood

$$d_{ow} = \frac{k_{ow}}{k_{gw}} d_{gw} = \left(\frac{1.02}{0.29} \right) (4 \text{ in.}) = 14.1 \text{ in.} = 1.2 \text{ ft}$$

aluminum

$$d_{Al} = \frac{k_{Al}}{k_{gw}} d_{gw} = \left(\frac{1.42 \times 10^3}{0.29}\right)(4 \text{ in.}) = 19,600 \text{ in.} = 1630 \text{ ft}$$

We see from these results that concrete, brick, glass, wood, and aluminum are not very efficient as insulated walls. A standard wood frame, studded wall with four inches of glass wool placed between the studs is far more efficient.

A few years ago, aluminum siding for the home was very popular. There were countless home improvement advertisements that said, "You can insulate your home with beautiful maintenance free aluminum siding." As you can see from the preceding calculations, such statements were extremely misleading if not outright fraudulent. As just calculated, the aluminum wall would have to be 1630 ft thick, just to give the same insulation as the 4 in. of glass wool. Admittedly, some aluminum siding had a 1/8- to 1/4-in. backing of cork board, which is a good insulating material, but since the heat flow is inversely proportional to the thickness of the wall, four to eight times the amount of thermal energy will flow through the cork board as through the glass wool wall. Aluminum siding may provide a beautiful, maintenance free home, but it will not insulate it. Today most siding for the home is made of vinyl rather than aluminum because vinyl is a good insulator. Most cooking utensils, pots and pans, are made of aluminum because the aluminum will readily conduct the thermal energy from the fire to the food to be cooked.

Another interesting result from these calculations is the realization that a glass window would have to be 6 ft thick to give the same insulation as the 4 in. of glass wool in the normal wall. Since glass windows are usually only about 1/8 in. or less thick, relatively large thermal energy losses are experienced through the windows of the home.

Convection Cycle in the Walls of a Home

All these results are based on the fact that different materials have different thermal conductivities. The smaller the value of k, the better the insulator. If we look carefully at table 16.1, we notice that the smallest value of k is for the air itself, that is, $k = 0.160$ Btu in./(hr ft^2 °F). This would seem to imply that if the space between the studs of a wall were left completely empty, that is, if no insulating material were placed in the wall, the air in that space would be the best insulator. Something seems to be wrong, since anyone who has an uninsulated wall in a home knows that there is a tremendous thermal energy loss through it. The reason is that air is a good insulator only if it is not in motion. But the difficulty is that the air in an empty wall is not at rest, as we can see from figure 16.7. Air molecules in contact with the hot wall T_h are heated by this hot wall absorbing a quantity of thermal energy Q. This heated air, being less dense than the surrounding air, rises to the top. The air that was originally at the top now moves down along the cold outside wall. This air is warmer than the cold wall and transmits some of its thermal energy to the cold wall where it is conducted to the outside. The air now sinks down along the outside wall and moves inward to the hot inside wall where it is again warmed and rises. A convection cycle has been established within the wall, whose final result is the absorption of thermal energy Q at the hot wall and its liberation at the cold wall, thereby producing a heat transfer through the wall. A great deal of thermal energy can be lost through the air in the wall, not by conduction, but by convection. If the air could be prevented from moving, that is, by stopping the convection current, then air would be a good insulator. This is basically what is done in using glass wool for insulation. The glass wool consists of millions of fibers of glass that create millions of tiny air pockets. These air pockets cannot move and hence there is no convection. The air between the fibers is still or dead air and acts as a good insulator. It is the dead air that is doing the insulating, not the glass fibers, because as we have just seen glass is not a good insulator.

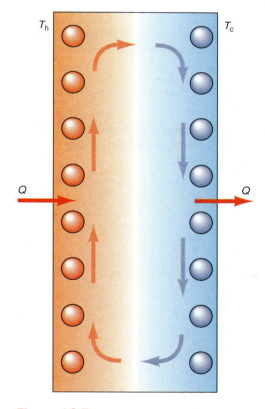

Figure 16.7

Convection currents in an empty wall.

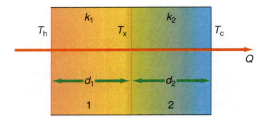

Figure 16.8

The compound wall.

As already mentioned, glass windows are a source of large thermal energy losses in a house. The use of storm windows or thermal windows cuts down on the thermal energy loss significantly. However, even storm windows or thermal windows are not as effective as a normally insulated wall because of the convection currents that occur between the panes of the glass windows.

The Compound Wall

Up to now a wall has been treated as if it consisted of only one material. In general this is not the case. Walls are made up of many different materials of different thicknesses. We solve this more general problem by considering the compound wall in figure 16.8. We assume, for the present, that the wall is made up of only two materials. This assumption will be extended to cover the case of any number of materials later. (The analysis, although simple is a little long. Those students weak in algebra and only interested in the results for the heat conduction through a compound wall can skip ahead to equation 16.18.)

Let us assume that the inside wall is the hot wall and it is at a temperature T_h, whereas the outside wall is the cold wall and it is at a temperature T_c. The temperature at the interface of the two materials is unknown at this time and will be designated by T_x. The first wall has a thickness d_1, and a thermal conductivity k_1, whereas wall 2 has a thickness d_2, and a thermal conductivity k_2. The thermal energy flow through the first wall, given by equation 16.10, is

$$Q_1 = \frac{k_1 A(T_h - T_x)t}{d_1} \qquad (16.16)$$

The thermal energy flow through the second wall is given by

$$Q_2 = \frac{k_2 A(T_x - T_c)t}{d_2}$$

Under a steady-state condition, the thermal energy flowing through the first wall is the same as the thermal energy flowing through the second wall. That is,

$$Q_1 = Q_2$$
$$\frac{k_1 A(T_h - T_x)t}{d_1} = \frac{k_2 A(T_x - T_c)t}{d_2}$$

Because the cross-sectional area of the wall A is the same for each wall and the time for the thermal energy flow t is the same, they can be canceled out, giving

$$\frac{k_1(T_h - T_x)}{d_1} = \frac{k_2(T_x - T_c)}{d_2}$$

or

$$\frac{k_1 T_h}{d_1} - \frac{k_1 T_x}{d_1} = \frac{k_2 T_x}{d_2} - \frac{k_2 T_c}{d_2}$$

Placing the terms containing T_x on one side of the equation, we get

$$-\frac{k_1 T_x}{d_1} - \frac{k_2 T_x}{d_2} = -\frac{k_1 T_h}{d_1} - \frac{k_2 T_c}{d_2}$$

$$\left(\frac{k_1}{d_1} + \frac{k_2}{d_2}\right) T_x = \frac{k_1}{d_1} T_h + \frac{k_2}{d_2} T_c$$

Solving for T_x, we get

$$T_x = \frac{(k_1/d_1)T_h + (k_2/d_2)T_c}{k_1/d_1 + k_2/d_2} \qquad (16.17)$$

If T_x, in equation 16.16, is replaced by T_x, from equation 16.17, we get

$$Q_1 = \frac{k_1 A\{T_h - [(k_1/d_1)T_h + (k_2/d_2)T_c]/(k_1/d_1 + k_2/d_2)\}t}{d_1}$$

$$= \frac{k_1 A}{d_1} \frac{[T_h(k_1/d_1 + k_2/d_2) - (k_1/d_1)T_h - (k_2/d_2)T_c]t}{k_1/d_1 + k_2/d_2}$$

$$= \frac{k_1 A}{d_1} \frac{[(k_1/d_1)T_h + (k_2/d_2)T_h - (k_1/d_1)T_h - (k_2/d_2)T_c]t}{k_1/d_1 + k_2/d_2}$$

$$= \frac{k_1 A k_2 (T_h - T_c)t}{d_1 d_2 (k_1/d_1 + k_2/d_2)}$$

$$= \frac{A(T_h - T_c)t}{(d_1 d_2/k_1 k_2)(k_1/d_1 + k_2/d_2)}$$

$$= \frac{A(T_h - T_c)t}{d_2/k_2 + d_1/k_1}$$

The thermal energy flow Q_1 through the first wall is equal to the thermal energy flow Q_2 through the second wall, which is just the thermal energy flow Q going through the compound wall. Therefore, the thermal energy flow through the compound wall is given by

$$Q = \frac{A(T_h - T_c)t}{d_1/k_1 + d_2/k_2} \tag{16.18}$$

If the compound wall had been made up of more materials, then there would be additional terms, d_i/k_i, in the denominator of equation 16.18 for each additional material. That is,

$$Q = \frac{A(T_h - T_c)t}{\sum\limits_{i=1}^{n} d_i/k_i} \tag{16.19}$$

The problem is usually simplified further by defining a new quantity called the **thermal resistance R,** or the **R value of the insulation,** as

$$R = \frac{d}{k} \tag{16.20}$$

The thermal resistance R acts to impede the flow of thermal energy through the material. The larger the value of R, the smaller the quantity of thermal energy conducted through the wall. For a compound wall, the total thermal resistance to thermal energy flow is simply

$$R_{total} = \frac{d_1}{k_1} + \frac{d_2}{k_2} + \frac{d_3}{k_3} + \frac{d_4}{k_4} + \cdots \tag{16.21}$$

or

$$R_{total} = R_1 + R_2 + R_3 + R_4 + \cdots \tag{16.22}$$

And the thermal energy flow through a compound wall is given by

$$Q = \frac{A(T_h - T_c)t}{\sum\limits_{i=1}^{n} R_i} \tag{16.23}$$

Example 16.5

Heat flow through a compound wall. A wall 10.0 ft by 8.00 ft is made up of a thickness of 4.00 in. of brick, 4.00 in. of glass wool, 1/2 in. of plaster, and 1/4 in. of oak wood paneling. If the inside temperature of the wall is $T_h = 65.0\ °F$ and the outside temperature is 20.0 °F, how much thermal energy flows through this wall per day?

The R value of each material, found with the aid of table 16.1, is

$$R_{\text{brick}} = \frac{d_{\text{brick}}}{k_{\text{brick}}} = \frac{4.00 \text{ in.}}{4.50 \text{ Btu in./hr ft}^2 \, °\text{F}} = 0.88 \, \frac{\text{hr ft}^2 \, °\text{F}}{\text{Btu}}$$

$$R_{\text{glass wool}} = \frac{d_{\text{gw}}}{k_{\text{gw}}} = \frac{4.00 \text{ in.}}{0.29 \text{ Btu in./hr ft}^2 \, °\text{F}} = 13.8 \, \frac{\text{hr ft}^2 \, °\text{F}}{\text{Btu}}$$

$$R_{\text{plaster}} = \frac{d_{\text{p}}}{k_{\text{p}}} = \frac{0.500 \text{ in.}}{3.25 \text{ Btu in./hr ft}^2 \, °\text{F}} = 0.15 \, \frac{\text{hr ft}^2 \, °\text{F}}{\text{Btu}}$$

$$R_{\text{wood}} = \frac{d_{\text{w}}}{k_{\text{w}}} = \frac{0.250 \text{ in.}}{1.02 \text{ Btu in./hr ft}^2 \, °\text{F}} = 0.25 \, \frac{\text{hr ft}^2 \, °\text{F}}{\text{Btu}}$$

The R value of the total compound wall, found from equation 16.22, is

$$R = R_1 + R_2 + R_3 + R_4 = 0.88 + 13.8 + 0.15 + 0.25$$

$$= 15.1 \, \frac{\text{hr ft}^2 \, °\text{F}}{\text{Btu}}$$

Note that the greatest portion of the thermal resistance comes from the glass wool. The total thermal energy conducted through the wall, found from equation 16.23, is

$$Q = \frac{A(T_{\text{h}} - T_{\text{c}})t}{\displaystyle\sum_{i=1}^{n} R_i}$$

$$= \frac{(10.0 \text{ ft})(8.00 \text{ ft})(65.0 \, °\text{F} - 20.0 \, °\text{F})(24 \text{ hr})}{15.1 \text{ hr ft}^2 \, °\text{F/Btu}}$$

$$= 5720 \text{ Btu}$$

Note that if there were no glass wool in the wall, the R value would be $R = 1.28$, and the thermal energy conducted through the wall would be 67,500 Btu, almost 12 times as much as the insulated wall. Remember, all these heat losses must be replaced by the home furnace in order to keep the temperature inside the home reasonably comfortable, and will require the use of fuel for this purpose. Finally, we should note that there is also a great heat loss in the winter through the roof of the house. To eliminate this energy loss there should be at least 6 in. of insulation in the roof of the house, and in some locations 12 in. is preferable.

You should note that when you buy insulation for your home in your local lumberyard or home materials store, you will see ratings such as an R value of 12 for a nominal 4 in. of glass wool insulation, or an R value of 19 for a nominal 6 in. of glass wool insulation. The units associated with these numbers are as given here for the British engineering system of units, namely

$$\frac{\text{hr ft}^2 \, °\text{F}}{\text{Btu}}$$

which is in the standard form used in the American construction industry today. So when using these products you must use the British engineering system of units for your calculations. You can still use the definition of $R = d/k$ in problems in SI units, but then the units for R will be

$$\frac{\text{s m}^2 \, °\text{C}}{\text{J}}$$

and the numerical values will not correspond to the R values listed on the insulation itself.

Everything that has been said about insulating our homes to prevent the loss of thermal energy in the winter, also applies in the summer. Only then the problem is reversed. The hot air is outside the house and the cool air is inside the house. The insulation will decrease the conduction of thermal energy through the walls into the room, keeping the room cool and cutting down or eliminating the use of air conditioning to cool the home.

16.4 Radiation

Radiation is the transfer of thermal energy by electromagnetic waves. As pointed out in chapter 12 on wave motion, any wave is characterized by its wavelength λ and frequency[1] ν. The electromagnetic waves in the visible portion of the spectrum are called *light waves.* These light waves have wavelengths that vary from about 0.38×10^{-6} m for violet light to about 0.72×10^{-6} m for red light. Above visible red light there is an invisible, infrared portion of the electromagnetic spectrum. The wavelengths range from 0.72×10^{-6} m to 1.5×10^{-6} m for the near infrared, from 1.5×10^{-6} m to 5.6×10^{-6} m for the middle infrared, and from 5.6×10^{-6} m up to 1×10^{-3} m for the far infrared. Most, but not all, of the radiation from a hot body falls in the infrared region of the electromagnetic spectrum. Every thing around you is radiating electromagnetic energy, but the radiation is in the infrared portion of the spectrum, which your eyes are not capable of detecting. Therefore, you are usually not aware of this radiation.

The Stefan-Boltzmann Law

Joseph Stefan (1835–1893) found experimentally, and Ludwig Boltzmann (1844–1906) found theoretically, that every body at an absolute temperature T radiates energy that is proportional to the fourth power of the absolute temperature. The result, which is called the **Stefan-Boltzmann law** is given by

$$Q = e\sigma AT^4 t \tag{16.24}$$

where Q is the thermal energy emitted; e is the emissivity of the body, which varies from 0 to 1; σ is a constant, called the Stefan-Boltzmann constant and is given by

$$\sigma = 5.67 \times 10^{-8} \frac{J}{s\ m^2\ K^4}$$

A is the area of the emitting body; T is the absolute temperature of the body; and t is the time.

Radiation from a Blackbody

The amount of radiation depends on the radiating surface. Polished surfaces are usually poor radiators, while blackened surfaces are usually good radiators. Good radiators of heat are also good absorbers of radiation, while poor radiators are also poor absorbers. *A body that absorbs all the radiation incident upon it is called a blackbody.* The name blackbody is really a misnomer, since the sun acts as a blackbody and it is certainly not black. A blackbody is a perfect absorber and a perfect emitter. The substance lampblack, a finely powdered black soot, makes a very good approximation to a blackbody. A box, whose insides are lined with a black material like lampblack, can act as a blackbody. If a tiny hole is made in the side of the box and then a light wave is made to enter the box through the hole, the light wave will be absorbed and re-emitted from the walls of the box, over and over. Such a device is called a *cavity resonator.* For a blackbody, the emissivity e in equation 16.24 is equal to 1. *The amount of heat absorbed or emitted from a blackbody is*

$$Q = \sigma AT^4 t \tag{16.25}$$

---

1. When dealing with electromagnetic waves, the symbol ν (Greek letter nu) is used to designate the frequency instead of the letter f used for conventional waves.

Example 16.6

Energy radiated from the sun. If the surface temperature of the sun is approximately 5800 K, how much thermal energy is radiated from the sun per unit time? Assume that the sun can be treated as a blackbody.

Solution

We can find the energy radiated from the sun per unit time from equation 16.25. The radius of the sun is about 6.96×10^8 m. Its area is therefore

$$A = 4\pi r^2 = 4\pi (6.96 \times 10^8 \text{ m})^2$$
$$= 6.09 \times 10^{18} \text{ m}^2$$

The heat radiated from the sun is therefore

$$\frac{Q}{t} = \sigma A T^4$$

$$= \left(5.67 \times 10^{-8} \frac{\text{J}}{\text{s m}^2 \text{ K}^4}\right)(6.09 \times 10^{18} \text{ m}^2)(5800 \text{ K})^4$$

$$= 3.91 \times 10^{26} \text{ J/s} = 9.33 \times 10^{22} \text{ kcal/s}$$

Example 16.7

The solar constant. How much energy from the sun impinges on the top of the earth's atmosphere per unit time per unit area?

Solution

The energy per unit time emitted by the sun is power and was found in example 16.6 to be 3.91×10^{26} J/s. This total power emitted by the sun does not all fall on the earth because that power is distributed throughout space, in all directions, figure 16.9. Hence, only a small portion of it is emitted in the direction of the earth. To find the amount of that power that reaches the earth, we first find the distribution of that power over a sphere, whose radius is the radius of the earth's orbit, $r = 1.5 \times 10^{11}$ m. This gives us the power, or energy per unit time, falling on a unit area at the distance of the earth from the sun. The area of this sphere is

$$A = 4\pi r^2 = 4\pi (1.5 \times 10^{11} \text{ m})^2$$
$$= 2.83 \times 10^{23} \text{ m}^2$$

The energy per unit area per unit time impinging on the earth is therefore

$$\frac{Q}{At} = \frac{3.91 \times 10^{26} \text{ J/s}}{2.83 \times 10^{23} \text{ m}^2} = 1.38 \times 10^3 \frac{\text{W}}{\text{m}^2}$$

This value, 1.38×10^3 W/m², *the energy per unit area per unit time impinging on the edge of the atmosphere, is called the* **solar constant,** *and is designated as S_0.*

Example 16.8

Solar energy reaching the earth. Find the total energy from the sun impinging on the top of the atmosphere during a 24-hr period.

Solution

The actual power impinging on the earth at the top of the atmosphere can be found by multiplying the solar constant S_0 by the effective area A subtended by the earth. The area subtended by the earth is found from the area of a disk whose radius is equal to the mean radius of the earth, $R_E = 6.37 \times 10^6$ m. That is,

$$A = \pi R_E^2 = \pi (6.37 \times 10^6 \text{ m})^2 = 1.27 \times 10^{14} \text{ m}^2$$

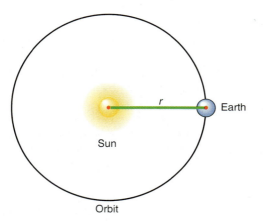

Figure 16.9

Radiation received on the earth from the sun.

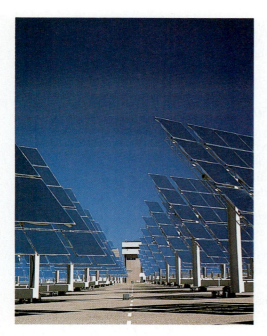

An array of solar collectors collect energy from the sun.

Power impinging on earth = (Solar constant)(Area)

$$P = \left(1.38 \times 10^3 \, \frac{\text{W}}{\text{m}^2}\right)(1.27 \times 10^{14} \, \text{m}^2) = 1.76 \times 10^{17} \, \text{W}$$

The energy impinging on the earth in a 24-hr period is found from

$$Q = Pt = (1.76 \times 10^{17} \, \text{W})(24 \, \text{hr})(3600 \, \text{s/hr})$$

$$= 1.52 \times 10^{22} \, \text{J}$$

This is an enormous quantity of energy. Obviously, solar energy, as a source of available energy for the world needs to be tapped.

All the solar energy incident on the upper atmosphere does not make it down to the surface of the earth because of reflection from clouds; scattering by dust particles in the atmosphere; and some absorption by water vapor, carbon dioxide, and ozone in the atmosphere. What is even more interesting is that this enormous energy received by the sun is reradiated back into space. If the earth did not re-emit this energy the mean temperature of the earth would constantly rise until the earth burned up.

A body placed in any environment absorbs energy from the environment. The net energy absorbed by the body Q is equal to the difference between the energy absorbed by the body from the environment Q_A and the energy radiated by the body to the environment Q_R, that is,

$$Q = Q_A - Q_R \tag{16.26}$$

If T_B is the absolute temperature of the radiating body and T_E is the absolute temperature of the environment, then the net heat absorbed by the body is

$$Q = Q_A - Q_R = e_E \sigma A T_E^4 t - e_B \sigma A T_B^4 t$$

$$Q = \sigma A (e_E T_E^4 - e_B T_B^4)t \tag{16.27}$$

where e_E is the emissivity of the environment and e_B is the emissivity of the body. In general these values, which are characteristic of the particular body and environment, must be determined experimentally. If the body and the environment can be approximated as blackbodies, then $e_B = e_E = 1$, and equation 16.27 reduces to the simpler form

$$Q = \sigma A (T_E^4 - T_B^4)t \tag{16.28}$$

If the value of Q comes out negative, it represents a net loss of energy from the body.

Example 16.9

Look at that person radiating. A person, at normal body temperature of 98.6 °F (37 °C) stands near a wall of a room whose temperature is 50.0 °F (10 °C). If the person's surface area is approximately 2.00 m², how much heat is lost from the person per minute?

Solution

The absolute temperature of the person is 310 K while the absolute temperature of the wall is 283 K. Let us assume that we can treat the person and the wall as blackbodies, then the heat lost by the person, given by equation 16.28, is

$$Q = \sigma A (T_E^4 - T_B^4)t$$

$$= \left(5.67 \times 10^{-8} \, \frac{\text{J}}{\text{s m}^2}\right)\left(\frac{2.00 \, \text{m}^2}{\text{K}^4}\right)[(283 \, \text{K})^4 - (310 \, \text{K})^4](60.0 \, \text{s})$$

$$= -320 \, \text{J} = -0.076 \, \text{kcal}$$

This thermal energy lost must be replaced by food energy. This result is of course only approximate, since the person is not a blackbody and no consideration was taken into account for the shape of the body and the insulation effect of the person's clothes.

†Blackbody Radiation as a Function of Wavelength

The Stefan-Boltzmann law tells us only about the total energy emitted and nothing about the wavelengths of the radiation. Because all this radiation consists of electromagnetic waves, the energy is actually distributed among many different wavelengths. The energy distribution per unit area per unit time per unit frequency $\Delta \nu$ is given by a relation known as **Planck's radiation law** as

$$\frac{Q}{At\Delta \nu} = \frac{2\pi h\nu^3}{c^2}\left(\frac{1}{e^{h\nu/kT} - 1}\right) \tag{16.29}$$

where c is the speed of light and is equal to 3×10^8 m/s, ν is the frequency of the electromagnetic wave, e is a constant equal to 2.71828 and is the base e used in natural logarithms, k is the Boltzmann constant given in chapter 15, and h is a new constant, called Planck's constant, given by

$$h = 6.625 \times 10^{-34} \text{ J s}$$

This analysis of blackbody radiation by Max Planck (1858–1947) was revolutionary in its time (December 1900) because Planck assumed that energy was quantized into little bundles of energy equal to $h\nu$. This was the beginning of what has come to be known as quantum mechanics, which will be discussed later in chapter 31. Equation 16.29 can also be expressed in terms of the wavelength λ as

$$\frac{Q}{At\Delta \lambda} = \frac{2\pi hc^2}{\lambda^5}\left(\frac{1}{e^{hc/\lambda kT} - 1}\right) \tag{16.30}$$

A plot of equation 16.30 is shown in figure 16.10 for various temperatures. Note that $T_4 > T_3 > T_2 > T_1$. The first thing to observe in this graph is that the intensity of the radiation for a given temperature varies with the wavelength from zero up to a maximum value and then decreases. That is, for any one temperature, there is one wavelength λ_{max} for which the intensity is a maximum. Second, as the temperature increases, the wavelength λ_{max} where the maximum or peak intensity occurs shifts to shorter wavelengths. This was recognized earlier by the German physicist Wilhelm Wien (1864–1928) and was written in the form

$$\lambda_{max}T = \text{constant} = 2.898 \times 10^{-3} \text{ m K} \tag{16.31}$$

and was called the **Wien displacement law.** Third, the visible portion of the electromagnetic spectrum (shown in the hatched area) is only a small portion of the spec-

Figure 16.10

The intensity of blackbody radiation as a function of wavelength and temperature.

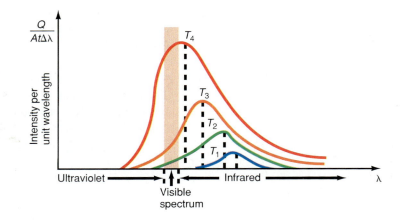

Thermodynamics

trum, and most of the radiation from a blackbody falls in the infrared range of the electromagnetic spectrum. Because our eyes are not sensitive to these wavelengths, the infrared radiation coming from a hot body is invisible. But as the temperature of the blackbody rises, the peak intensity shifts to lower wavelengths, until, when the temperature is high enough, some of the blackbody radiation is emitted in the visible red portion of the spectrum and the heated body takes on a red glow. If the temperature continues to rise, the red glow becomes a bright red, then an orange, then yellow-white, and finally blue-white as the blackbody emits more and more radiation in the visible range. When the blackbody emits all wavelengths in the visible portion of the spectrum, it appears white. (The visible range of the electromagnetic spectrum, starting from the infrared end, has the colors red, orange, yellow, green, blue, and violet before the ultraviolet portion of the spectrum begins.)

Example 16.10

The wavelength of the maximum intensity of radiation from the sun. Find the wavelength of the maximum intensity of radiation from the sun, assuming the sun to be a blackbody at 5800 K.

Solution

The wavelength of the maximum intensity of radiation from the sun is found from the Wien displacement law, equation 16.31, as

$$\lambda_{\text{max}} = \frac{2.898 \times 10^{-3}}{T} \text{ m K}$$

$$= \frac{2.898 \times 10^{-3}}{5800 \text{ K}} \text{ m K}$$

$$= 0.499 \times 10^{-6} \text{ m} = 0.499 \ \mu\text{m}$$

That is, the wavelength of the maximum intensity from the sun lies at 0.499 μm, which is in the blue-green portion of the visible spectrum. It is interesting to note that some other stars, which are extremely hot, radiate mostly in the ultraviolet region.

"Have you ever wondered . . . ?"

An Essay on the Application of Physics

The Greenhouse Effect and Global Warming

Have you ever wondered what the newscaster was talking about when she said that the earth is getting warmer because of the Greenhouse Effect? What is the Greenhouse Effect and what does it have to do with the heating of the earth?

The name Greenhouse Effect comes from the way the earth and its atmosphere is heated. The ultimate cause of heating of the earth's atmosphere is the sun. But if this is so, then why is the top of the atmosphere (closer to the sun) colder than the lower atmosphere (farther from the sun)? You may have noticed

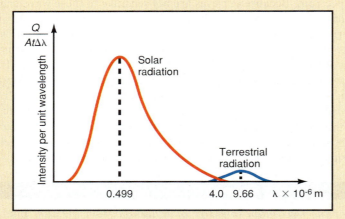

Figure 1

Comparison of radiation from the sun and the earth.

snow and ice on the colder mountain tops while the valleys below are relatively warm. We can explain this paradox in terms of the radiation of the sun, the radiation of the earth, and the constituents of the atmosphere. The sun radiates approximately as a blackbody at 5800 K with a peak intensity occurring at 0.499 × 10⁻⁶ m, as shown in figure 1.

Example 16H.1

The wavelength of the maximum intensity of radiation from the earth. Assuming that the earth has a mean temperature of about 300 K use the Wien displacement law to estimate the wavelength of the peak radiation from the earth.

Solution

The wavelength of the peak radiation from the earth, found from equation 16.31, is

$$\lambda_{max} = \frac{2.898 \times 10^{-3}}{T} \text{ m K} = \frac{2.898 \times 10^{-3} \text{ m K}}{300 \text{ K}}$$

$$= 9.66 \times 10^{-6} \text{ m}$$

which is also shown in figure 1. Notice that the maximum radiation from the earth lies well in the longer wave infrared region, whereas the maximum solar radiation lies in much shorter wavelengths. (Ninety-nine percent of the solar radiation is in wavelengths shorter than 4.0 μm, and almost all terrestrial radiation is at wavelengths greater than 4.0 μm.) Therefore, *solar radiation is usually referred to as short-wave radiation, while terrestrial radiation is usually referred to as long-wave radiation.*

Of all the gases in the atmosphere only oxygen, ozone, water vapor, and carbon dioxide are significant absorbers of radiation. Moreover these gases are selective absorbers, that is, they absorb strongly in some wavelengths and hardly at all in others. The absorption spectrum for oxygen and ozone is shown in figure 2(b).

The absorption of radiation is plotted against the wavelength of the radiation. An absorptivity of 1 means total absorption at that wavelength, whereas an absorptivity of 0 means that the gas does not absorb any radiation at that wavelength. Thus, when the absorptivity is 0, the gas is totally transparent to that wavelength of radiation. Observe from figure 2(b) that oxygen and ozone absorb almost all the ultraviolet radiation from the sun in wavelengths below 0.3 μm. A slight amount of ultraviolet light from the sun reaches the earth in the range 0.3 μm to the beginning of visible light in the violet at 0.38 μm. Also notice that oxygen and ozone are almost transparent to radiation in the visible and infrared region of the electromagnetic spectrum.

Figure 2(d) shows the absorption spectrum for water vapor (H_2O). Notice that there is no absorption in the ultraviolet or visible region of the electromagnetic spectrum for water vapor. However, there are a significant number of regions in the infrared where water vapor does absorb radiation.

Figure 2(c) shows the absorption spectrum for carbon dioxide (CO_2). Notice that there is no absorption in the ultraviolet or visible region of the electromagnetic spectrum for carbon dioxide. However, there are a significant number of regions in the infrared where carbon dioxide does absorb radiation. The bands are not quite as wide as for water vapor, but they are very significant as we will see shortly. Also note in figure 2(a) that nitrous oxide (N_2O) also absorbs some energy in the infrared portion of the spectrum.

Figure 2(e) shows the combined absorption spectrum for the atmosphere. We can see that the atmosphere is effectively transparent in the visible portion of the spectrum. Because the peak of the sun's radiation falls in this region, the atmosphere is effectively transparent to most of the sun's rays, and hence most of the sun's radiation passes through the atmosphere as if there were no atmosphere at all. The atmosphere is like an open window to let in all the sun's rays. Hence, the sun's rays pass directly through the atmosphere where they are then absorbed by the surface of the earth. The earth then reradiates as a blackbody, but since its average temperature is so low (250–300 K), its radiation is all in the infrared region as was shown in figure 1. But the water vapor, H_2O, and carbon dioxide, CO_2, in the atmosphere absorb almost all the energy in the infrared region. Thus, *the earth's atmosphere is mainly heated by the absorption of the infrared radiation from the earth.* Therefore, the air closest to the ground becomes warmer than air at much higher altitudes, and therefore the temperature of the atmosphere decreases with height. The warm air at the surface rises by convection, distributing the thermal energy throughout the rest of the atmosphere.

This process of heating the earth's atmosphere by terrestrial radiation is called the Greenhouse Effect. The reason for the name is that it was once thought that this was the way a greenhouse was heated. That is, short-wavelength radiation from the sun passed through the glass into the greenhouse. The plants and ground in the greenhouse absorbed this short-wave radiation and reradiated in the infrared. The glass in the greenhouse was essentially opaque to this infrared radiation and reflected this radiation back into the greenhouse thus keeping the greenhouse warm. Because the mechanism for heating the atmosphere was

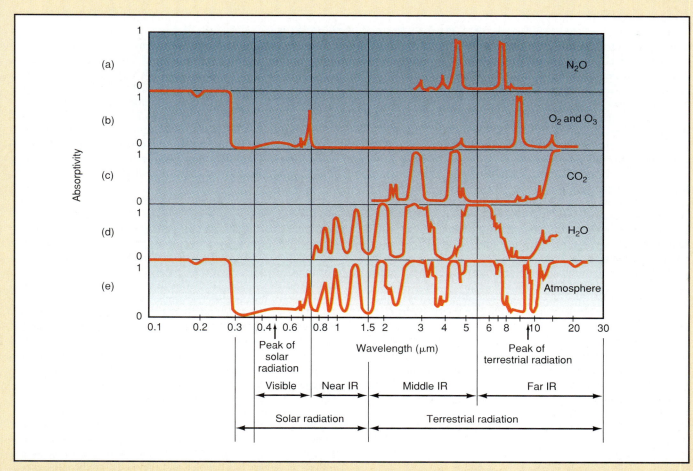

Figure 2

Absorption of radiation at various wavelengths for atmospheric constituents. Lutgens/Tarbuck, *The Atmosphere,* 3/E, p. 44. © Prentice-Hall, Inc., Englewood Cliffs, NJ.

thought to be similar to the mechanism for heating the greenhouse, the heating of the atmosphere came to be called the Greenhouse Effect. (It has since been shown that the dominant reason for keeping the greenhouse warm is the prevention of the convection of the hot air out of the greenhouse by the glass. However, the name Greenhouse Effect continues to be used.)

Because carbon dioxide is an absorber of the earth's infrared radiation, it has led to a concern over the possible warming of the atmosphere caused by excessive amounts of carbon dioxide that comes from the burning of fossil fuels, such as coal and oil, and the deforestation of large areas of trees, whose leaves normally absorb some of the excess carbon dioxide in the atmosphere. "For example, since 1958 concentrations of CO_2 have increased from 315 to 352 parts per million, an increase of approximately 15%."[1] Also, "During the last 100–200 years carbon

dioxide has increased by 25%."[2] And "Everyday 100 square miles of rain forest go up in smoke, pumping one billion tons of carbon dioxide into the atmosphere."[3]

Almost everyone agrees that the increase in carbon dioxide in the atmosphere is not beneficial, but this is where the agreement ends. There is wide disagreement on the consequences of this increased carbon dioxide level. Let us first describe the two most extreme views.

One scenario says that the increased level of CO_2 will cause the mean temperature of the atmosphere to increase. This increased temperature will cause the polar ice caps to melt and increase the height of the mean sea level throughout the world. This in turn will cause great flooding in the low-lying regions of the world. The increased temperature is also assumed to cause the destruction of much of the world's crops and hence its food supply.

A second scenario says that the increased temperatures from the excessive carbon dioxide will cause greater evaporation from the oceans and hence greater cloud cover over the entire globe.

1. "Computer Simulation of the Greenhouse Effect," Washington, Warren M. and Bettge, Thomas W., *Computers in Physics,* May/June 1990.

2. "Climate and the Earth's Radiation Budget," Ramanathan, V.; Barkstrom, Bruce R.; and Harrison, Edwin F., *Physics Today,* May 1989.

3. NOVA TV series, "The Infinite Voyage, Crisis in the Atmosphere."

It is then assumed that this greater cloud cover will reflect more of the incident solar radiation into space. This reflected radiation never makes it to the surface of the earth to heat up the surface. Less radiation comes from the earth to be absorbed by the atmosphere and hence there is a decrease in the mean temperature of the earth. This lower temperature will then initiate the beginning of a new ice age.

Thus one scenario has the earth burning up, the other has it freezing down. It is obvious from these two scenarios that much greater information on the effect of the increase in carbon dioxide in the atmosphere is necessary.

Another way to look at the Greenhouse Effect is to consider the earth as a planet in space that is in equilibrium between the incoming solar radiation and the outgoing terrestrial radiation. As we saw in example 16.7, the amount of energy per unit area per unit time falling on the earth from the sun is given by the solar constant, $S_0 = 1.38 \times 10^3$ J/(s m^2). The actual energy per unit time impinging on the earth at the top of the atmosphere can be found by multiplying the solar constant S_0 by the effective area A_d subtended by the earth. That is, $Q/t = S_0A_d$. The area subtended by the earth A_d is found from the area of a disk whose radius is equal to the mean radius of the earth. That is, $A_d = \pi R_E^2$.

The solar radiation reaching the surface of the earth is equal to the solar radiation impinging on the top of the atmosphere S_0A_d minus the amount of solar radiation reflected from the atmosphere, mostly from clouds. The albedo of the earth a, the ratio of the amount of radiation reflected to the total incident radiation, has been measured by satellites to be $a = 0.300$. Hence the amount of solar energy reaching the earth per second is given by

$$\frac{Q}{t} = S_0A_d - aS_0A_d = S_0A_d(1 - a)$$

Assuming that the earth radiates as a blackbody it will emit the radiation

$$\frac{Q}{t} = \sigma A_s T^4$$

The radiating area of the earth, $A_s = 4\pi R_E^2$, is the spherical area of the earth because the earth is radiating everywhere, not only in the region where it is receiving radiation from the sun. Because the earth must be in thermal equilibrium in its position in space, the radiation in must equal the radiation out, or

$$\frac{Q}{t} = S_0A_d(1 - a) = \sigma A_s T^4$$

Solving for the temperature T of the earth, we get

$$T^4 = \frac{S_0A_d(1 - a)}{\sigma A_s} = \frac{S_0\pi R_E^2(1 - a)}{\sigma 4\pi R_E^2}$$

$$= \frac{S_0(1 - a)}{4\sigma}$$

$$= \frac{[1.38 \times 10^3 \text{ J/(s m}^2)](1 - 0.300)}{4[5.67 \times 10^{-8} \text{ J/(s m}^2\text{ K}^4)]}$$

$$T = 255 \text{ K}$$

That is the radiative equilibrium temperature of the earth should be 255 K. This mean radiative temperature of 255 K is sometimes called the *planetary temperature* and/or the *effective temperature* of the earth. It is observed, however, that the mean temperature of the surface of the earth, averaged over time and place, is actually 288 K, some 33 K higher than this temperature.[4] This difference in the mean temperature of the earth is attributed to the Greenhouse Effect. That is, the energy absorbed by the water vapor and carbon dioxide in the atmosphere causes the surface of the earth to be much warmer than if there were no atmosphere. It is for this reason that environmentalists are so concerned with the abundance of carbon dioxide in the atmosphere.

As a contrast let us consider the planet Venus, whose main constituent in the atmosphere is carbon dioxide. Performing the same calculation for the solar constant in example 16.7, only using the orbital radius of Venus of 1.08×10^{11} m, gives a solar constant of 2668 W/m^2, roughly twice that of the earth. The mean albedo of Venus is about 0.80 because of the large amount of clouds covering the planet. Performing the same calculation for the planetary temperature of Venus gives 220 K. Even though the solar constant is roughly double that of the earth, because of the very high albedo, the planetary temperature is some 30 K colder than the earth. However, the surface temperature of Venus has been found to be 750 K due to the very large amount of carbon dioxide in the atmosphere. Hence the Greenhouse Effect on Venus has caused the mean surface temperature to be 891 °F. There is apparently no limitation to the warming that can result from the Greenhouse Effect.

More detailed computer studies of the earth's atmosphere, using general circulation models (GCM), have been made. In these models, it is assumed that the amount of carbon dioxide in the atmosphere has doubled and the model predicts the general condition of the atmosphere over a period of twenty years. The model indicates a global warming of about 4.0 to 4.5 °C. (A temperature of 4 or 5 °C may not seem like much, but when you recall that the mean temperature of the earth during an ice age was only 3 °C cooler than presently, the variation can be quite significant.) The effect of the warming was to cause greater extremes of temperature. That is, hot areas were hotter than normal, while cold areas were colder than normal.

Stephen H. Schneider[5] has said, "Sometime between 15,000 and 5,000 years ago the planet warmed up 5 °C. Sea levels rose 300 feet and forests moved. Literally that change in 5 °C revamped the ecological face of this planet. Species went extinct,

4. We should note that the radiative temperature of the earth is 255 K. This is a mean temperature located somewhere in the middle of the atmosphere. The surface temperature is much higher and temperatures in the very upper atmosphere are much lower, giving the mean of 255 K.

5. Stephen H. Schneider, *Global Warming*, Sierra Club Books, San Francisco, 1989.

Thermodynamics

The Language of Physics

Convection
The transfer of thermal energy by the actual motion of the medium itself (p. 451).

Conduction
The transfer of thermal energy by molecular action. Conduction occurs in solids, liquids, and gases, but the effect is most pronounced in solids (p. 451).

Radiation
The transfer of thermal energy by electromagnetic waves (p. 451).

Isotherm
A line along which the temperature is a constant (p. 451).

Temperature gradient
The rate at which the temperature changes with distance (p. 453).

Coriolis effect
On a rotating coordinate system, such as the earth, objects in straight line motion appear to be deflected to the right of their straight line path. Their actual motion in space is straight, but the earth rotates out from under them. The direction of the prevailing winds is a manifestation of the Coriolis effect (p. 455).

Conductor
A material that easily transmits heat by conduction. A conductor has a large value of thermal conductivity (p. 458).

Insulator
A material that is a poor conductor of heat. An insulator has a small value of thermal conductivity (p. 458).

Thermal resistance, or R value of an insulator
The ratio of the thickness of a piece of insulating material to its thermal conductivity (p. 463).

Stefan-Boltzmann law
Every body radiates energy that is proportional to the fourth power of the absolute temperature of the body (p. 465).

Blackbody
A body that absorbs all the radiation incident upon it. A blackbody is a perfect absorber and a perfect emitter. The substance lampblack, a finely powdered black soot, makes a very good approximation to a blackbody. The name is a misnomer, since many bodies, such as the sun, act like blackbodies and are not black (p. 465).

Solar constant
The power per unit area impinging on the edge of the earth's atmosphere. It is equal to 1.38×10^3 W/m^2 (p. 466).

Planck's radiation law
An equation that shows how the energy of a radiating body is distributed over the emitted wavelengths. Planck assumed that the radiated energy was quantized into little bundles of energy, eventually called quanta (p. 468).

Wien displacement law
The product of the wavelength that gives maximum radiation times the absolute temperature is a constant (p. 468).

Summary of Important Equations

Heat transferred by convection
$$\Delta Q = vmc\frac{\Delta T}{\Delta x}\Delta t \qquad (16.4)$$

$$\Delta Q = v\rho Vc\frac{\Delta T}{\Delta x}\Delta t \qquad (16.6)$$

$$\frac{\Delta Q}{\Delta t} = \rho c\frac{\Delta V}{\Delta t}(T_h - T_c) \qquad (16.8)$$

Heat transferred by conduction
$$Q = \frac{kA(T_h - T_c)t}{d} \qquad (16.10)$$

Heat transferred by conduction through a compound wall
$$Q = \frac{A(T_h - T_c)t}{\sum\limits_{i=1}^{n} d_i/k_i} \qquad (16.19)$$

$$Q = \frac{A(T_h - T_c)t}{\sum\limits_{i=1}^{n} R_i} \qquad (16.23)$$

R value of insulation
$$R = \frac{d}{k} \qquad (16.20)$$

Stefan-Boltzmann law, heat transferred by radiation
$$Q = e\sigma AT^4 t \qquad (16.24)$$

Radiation from a blackbody
$$Q = \sigma AT^4 t \qquad (16.25)$$

Energy absorbed by radiation from environment
$$Q = \sigma A(e_E T_E^4 - e_B T_B^4)t \qquad (16.27)$$

Planck's radiation law
$$\frac{Q}{At\Delta\nu} = \frac{2\pi h\nu^3}{c^2}\left(\frac{1}{e^{h\nu/kT} - 1}\right) \qquad (16.29)$$

$$\frac{Q}{At\Delta\lambda} = \frac{2\pi hc^2}{\lambda^5}\left(\frac{1}{e^{hc/\lambda kT} - 1}\right) \qquad (16.30)$$

Wien displacement law
$$\lambda_{max}T = \text{constant} \qquad (16.31)$$

Questions for Chapter 16

1. Explain the differences and similarities between convection, conduction, and radiation.
†2. Explain how the process of convection of ocean water is responsible for relatively mild winters in Ireland and the United Kingdom even though they are as far north as Hudson's Bay in Canada.
†3. Explain from the process of convection why the temperature of the Pacific Ocean off the west coast of the United States is colder than the temperature of the Atlantic Ocean off the east coast of the United States.
†4. Explain from the process of convection why it gets colder after the passing of a cold front and warmer at the approach and passing of a warm front.
5. Explain the process of heat conduction in a gas and a liquid.

6. Considering the process of heat conduction through the walls of your home, explain why there is a greater loss of thermal energy through the walls on a very windy day.
7. In the winter time, why does a metal door knob feel colder than the wooden door even though both are at the same temperature?
8. Explain the use of venetian blinds for the windows of the home as a temperature controlling device. What advantage do they have over shades?
9. Why are thermal lined drapes used to cover the windows of a home at night?
10. Why is it desirable to wear light colored clothing in very hot climates rather than dark colored clothing?
11. Explain how you can still feel cold while sitting in a room whose air temperature is 70 °F, if the temperature of the walls is very much lower.

12. From what you now know about the processes of heat transfer, discuss the insulation of a calorimeter.
†13. On a very clear night, radiation fog can develop if there is sufficient moisture in the air. Explain.
14. If the maximum radiation from the sun falls in the blue-green portion of the visible spectrum, why doesn't the sun appear blue-green?
†15. From the point of view of radiation, discuss the process of thermography, whereby a specialized camera takes pictures of an object in the infrared portion of the spectrum. Explain how this could be used in medicine to detect tumors in the human body. (The tumors are usually several degrees hotter than normal body tissue.)

Problems for Chapter 16

16.2 Convection

1. How much thermal energy per unit mass is transferred by convection in 6.00 hr if air at the surface of the earth is moving at 15.0 mph? The temperature gradient is measured as 4.00 °C per 100 miles.
2. Air is moving over the surface of the earth at 30.0 km/hr. The temperature gradient is 2.50 °C per 100 km. How much thermal energy per unit mass is transferred by convection in an 8.00-hour period?
3. An air conditioner can cool 300 ft³ of air per minute from 95.0 °F to 60.0 °F. How much thermal energy is removed from the room per hour?
4. An air conditioner can cool 10.5 m³ of air per minute from 30.0 °C to 18.5 °C. How much thermal energy per hour is removed from the room in one hour?
5. In a hot air heating system, air at the furnace is heated to 200 °F. A window is open in the house and the house temperature remains at 55 °F. If the furnace can deliver 200 ft³/min of air, how much thermal energy per hour is transferred from the furnace to the room?

6. A hot air heating system rated at 84,000 Btu/hr has an efficiency of 65.0%. The fan can move 225 ft³ of air per minute. If air enters the furnace at 60.0 °F, what is the temperature of the outlet air?
7. A hot air heating system rated at 6.3 × 10⁷ J/hr has an efficiency of 58.0%. The fan is capable of moving 5.30 m³ of air per minute. If air enters the furnace at 17.0 °C, what is the temperature of the outlet air?

16.3 Conduction

8. How much thermal energy flows through a glass window 1/8 in. thick, 4.00 ft high, and 3.00 ft wide in 12.0 hr if the temperature on the outside of the window is 20.0 °F and the temperature on the inside of the window is 70.0 °F?
9. How much thermal energy flows through a glass window 0.350 cm thick, 1.20 m high, and 0.80 m wide in 12.0 hr if the temperature on the outside of the window is −8.00 °C and the temperature on the inside of the window is 20.0 °C?
10. Repeat problem 8, but now assume that there are strong gusty winds whose air temperature is 5.00 °F.

11. Find the amount of thermal energy that will flow through a concrete wall 10.0 m long, 2.80 m high, and 22.0 cm wide, in a period of 24.0 hr, if the inside temperature of the wall is 20.0 °C and the outside temperature of the wall is 5.00 °C.
12. Find the amount of thermal energy transferred through a pine wood door in 6.00 hr if the door is 0.91 m wide, 1.73 m high, and 5.00 cm thick. The inside temperature of the door is 20.0 °C and the outside temperature of the door is −5.00 °C.
13. How much thermal energy will flow per hour through a copper rod, 5.00 cm in diameter and 1.50 m long, if one end of the rod is maintained at a temperature of 225 °C and the other end at 20.0 °C?
14. One end of a copper rod has a temperature of 100 °C, whereas the other end has a temperature of 20.0 °C. The rod is 1.25 m long and 3.00 cm in diameter. Find the amount of thermal energy that flows through the rod in 5.00 min. Find the temperature of the rod at 45.0 cm from the hot end.

15. On a hot summer day the outside temperature is 98.0 °F. A home air conditioner is trying to maintain a temperature of 70.0 °F. If there are 10 windows in the house, each 1/8 in. thick and 12.0 ft² in area, how much thermal energy must the air conditioner remove per hour to eliminate the thermal energy transferred through the windows?

16. On a hot summer day the outside temperature is 35.0 °C. A home air conditioner is trying to maintain a temperature of 22.0 °C. If there are 12 windows in the house, each 0.350 cm thick and 0.960 m² in area, how much thermal energy must the air conditioner remove per hour to eliminate the thermal energy transferred through the windows?

†17. A styrofoam cooler ($k = 4.8 \times 10^{-5}$ kcal/m s °C) is filled with ice at 0 °C for a summertime party. The cooler is 40.0 cm high, 50.0 cm long, 40.0 cm wide, and 3.00 cm thick. The air temperature is 95.0 °F. Find (a) the mass of ice in the cooler, (b) how much thermal energy is needed to melt all the ice, and (c) how long it will take for all the ice to melt. Assume that the energy to melt the ice is only conducted through the four sides of the cooler. Also take the thickness of the cooler walls into account when computing the size of the walls of the container.

18. An aluminum rod 50.0 cm long and 3.00 cm in diameter has one end in a steam bath at 100 °C and the other end in an ice bath at 0.00 °C. How much ice melts per hour?

19. If the home thermostat is turned from 70.0 °F down to 60.0 °F for an 8-hr period at night when the outside temperature is 20.0 °F, what percentage saving in fuel can the home owner realize?

20. If the internal temperature of the human body is 37.0 °C, the surface temperature is 35.0 °C, and there is a separation of 4.00 cm of tissue between, how much thermal energy is conducted to the skin of the body each second? Take the thermal conductivity of human tissue to be 0.5×10^{-4} kcal/s m °C and the area of the human skin to be 1.9 m².

21. What is the R value of (a) 4.00 in. of glass wool and (b) 6.00 in. of glass wool?

22. How thick should a layer of plaster be in order to provide the same R value as a 5.00 cm of concrete?

23. A basement wall consists of 8.00 in. of concrete, 1 1/2 in. of glass wool, 3/8 in. of sheetrock (plaster), and 3/4 in. of knotty pine paneling. The wall is 7.00 ft high and 30.0 ft long. The outside temperature is 40.0 °F and we want to maintain the inside temperature at 70.0 °F. How much thermal energy will be lost through four such walls in a 24-hr period?

24. A basement wall consists of 20.0 cm of concrete, 3.00 cm of glass wool, 0.800 cm of sheetrock (plaster), and 2.00 cm of knotty pine paneling. The wall is 2.50 m high and 10.0 m long. The outside temperature is 1.00 °C, and we want to maintain the inside temperature of 22.0 °C. How much thermal energy will be lost through four such walls in a 24-hr period?

25. On a summer day the attic temperature of a house is 160 °F. The ceiling of the house is 8.00 m wide by 13.0 m long and 3/8-in. thick plasterboard. The house is cooled by an air conditioner and maintains a 70.0 °F temperature in the house. (a) Find the amount of thermal energy transferred from the attic to the house in 2.00 hr. (b) If 6.00 in. of glass wool is now placed in the attic floor, find the amount of thermal energy transferred into the house.

26. How much thermal energy is conducted through a thermopane window, in 8.00 hr if the window is 32.0 in. wide by 45.0 in. high, and consists of two sheets of glass 1/8 in. thick separated by an air gap of 1/2 in.? The temperature of the inside window is 68.0 °F and the temperature of the outside window is 20.0 °F. Treat the thermopane window as a compound wall.

27. How much thermal energy is conducted through a thermopane window in 8.00 hr if the window is 80.0 cm wide by 120 cm high, and it consists of two sheets of glass 0.350 cm thick separated by an air gap of 1.50 cm? The temperature of the inside window is 22.0 °C and the temperature of the outside window is −5.00 °C. Treat the thermopane window as a compound wall.

28. How much thermal energy is conducted through a combined glass window and storm window in 8.00 hr if the window is 32.0 in. wide by 45.0 in. high and 1/8 in. thick? The storm window is the same size but is separated from the inside window by an air gap of 2.00 in. The temperature of the inside window is 68.0 °F and the temperature of the outside window is 20.0 °F. Treat the combination as a compound wall.

16.4 Radiation

29. How much thermal energy from the sun falls on the surface of the earth during an 8-hr period? (Ignore reflected solar radiation from clouds that does not make it to the surface of the earth.)

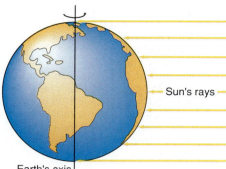

Earth's axis

Sun's rays

30. If the mean temperature of the surface of the earth is 288 K, how much thermal energy is radiated into space per second?

31. Assuming the human body has an emissivity, $e = 1$, and an area of approximately 2.23 m², find the amount of thermal energy radiated by the body in 8 hr if the surface temperature is 95.0 °F.

32. If the surface temperature of the human body is 35.0 °C, find the wavelength of the maximum intensity of radiation from the human body. Compare this wavelength to the wavelengths of visible light.

33. How much energy is radiated per second by an aluminum sphere 5.00 cm in radius, at a temperature of (a) 20.0 °C, and (b) 200 °C? Assume that the sphere emits as a blackbody.

34. How much energy is radiated per second by an iron cylinder 5.00 cm in radius and 10.0 cm long, at a temperature of (a) 20.0 °C and (b) 200 °C? Assume blackbody radiation.

35. How much energy is radiated per second from a wall 2.50 m high and 3.00 m wide, at a temperature of 20.0 °C? What is the wavelength of the maximum intensity of radiation?

36. A blackbody initially at 100 °C is heated to 300 °C. How much more power is radiated at the higher temperature?

37. A blackbody is at a temperature of 200 °C. Find the wavelength of the maximum intensity of radiation.

38. A blackbody is radiating at a temperature of 300 K. To what temperature should the body be raised to double the amount of radiation?

39. A distant star appears red, with a wavelength 7.000×10^{-7} m. What is the surface temperature of that star?

Additional Problems

40. An aluminum pot contains 10.0 kg of water at 100 °C. The bottom of the pot is 15.0 cm in radius and is 3.00 mm thick. If the bottom of the pot is in contact with a flame at a temperature of 170 °C, how much water will boil per minute?

41. Find how much energy is lost in one day through a concrete slab floor on which the den of a house is built. The den is 5.00 m wide and 6.00 m long, and the slab is 15.0 cm thick. The temperature of the ground is 3.00 °C and the temperature of the room is 22.0 °C.

42. A lead bar 2.00 cm by 2.00 cm and 10.0 cm long is welded end to end to a copper bar 2.00 cm by 2.00 cm by 25.0 cm long. Both bars are insulated from the environment. The end of the copper bar is placed in a steam bath while the end of the lead bar is placed in an ice bath. What is the temperature T at the interface of the copper-lead bar? How much thermal energy flows through the bar per minute?

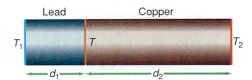

43. Find the amount of thermal energy conducted through a wall, 5.00 m high, 12.0 m long, and 5.00 cm thick, if the wall is made of (a) concrete, (b) brick, (c) wood, and (d) glass. The temperature of the hot wall is 25.0 °C and the cold wall −5.00 °C.

†44. Show that the distribution of solar energy over the surface of the earth is a function of the latitude angle ϕ. Find the energy per unit area per unit time hitting the surface of the earth during the vernal equinox and during the summer solstice at (a) the equator, (b) 30.0° north latitude, (c) 45.0° north latitude, (d) 60.0° north latitude, and (e) 90.0° north latitude. At the vernal equinox the sun is directly overhead at the equator, whereas at the summer solstice the sun is directly overhead at 23.5° north latitude.

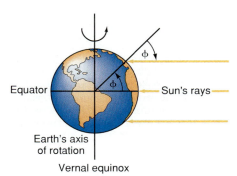

Vernal equinox

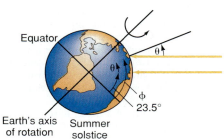

45. An asphalt driveway, 50.0 m² in area and 6.00 cm thick, receives energy from the sun. Using the solar constant of 1.38×10^3 W/m², find the maximum change in temperature of the asphalt if (a) the radiation from the sun hits the driveway normally for a 2.00-hr period and (b) the radiation from the sun hits the driveway at an angle of 35° for the same 2.00-hr period. Take the density of asphalt to be 1219 kg/m³ and the specific heat of asphalt to be 1.02 kcal/kg °C.

†46. Find the amount of radiation from the sun that falls on the planets (a) Mercury, (b) Venus, (c) Mars, (d) Jupiter, and (e) Saturn in units of W/m².

47. If the Kelvin temperature of a blackbody is quadrupled, what happens to the rate of energy radiation?

†48. A house measures 40.0 ft by 30.0 ft by 8.00 ft high. The walls contain 4.00 in. of glass wool. Assume all the heat loss is through the walls of the house. The home thermostat is turned from 70.0 °F down to 60.0 °F for an 8-hr period at night when the outside temperature is 20.0 °F. (a) How much thermal energy can the home owner save by lowering the thermostat? (b) How much energy is used the next morning to bring the temperature of the air in the house back to 70.0 °F? (c) What is the savings in energy now?

†49. An insulated aluminum rod, 1.00 m long and 25.0 cm² in cross-sectional area, has one end in a steam bath at 100 °C and the other end in a cooling container. Water enters the cooling container at an input temperature of 10.0 °C and exits the cooling container at a temperature of 30.0 °C, leaving a mean temperature of 20.0 °C at the end of the aluminum rod. Find the mass of water that must flow through the cooling container per minute to maintain this equilibrium condition.

†50. An aluminum engine, operating at 300 °C is cooled by circulating water over the end of the engine where the water absorbs enough energy to boil. The cooling interface has a surface area of 0.525 m² and a thickness of 1.50 cm. If the water enters the cooling interface of the engine at 100 °C, how much water must boil per minute to cool the engine?

†51. When the surface through which thermal energy flows is not flat, such as in figure 16.6, the equation for heat transfer, equation 16.10, is no longer accurate. With the help of the calculus it can be shown that the amount of thermal energy that flows through the sides of a rectangular annular cylinder is given by

$$\frac{\Delta Q}{\Delta t} = \frac{2\pi k l \Delta T}{\ln(r_2/r_1)}$$

where l is the length of the cylinder, r_1 is the inside radius of the cylinder, and r_2 is the outside radius of the cylinder. Steam at 100 °C flows in a cylindrical copper pipe 5.00 m long, with an inside radius of 10.0 cm and an outside radius of 15 cm. Find the energy lost through the pipe per hour if the outside temperature of the pipe is 30.0 °C.

†52. When the surface through which thermal energy flows is a spherical shell rather than a flat surface, the amount of thermal energy that flows through the spherical surface can be shown to be given by

$$\frac{\Delta Q}{\Delta t} = \frac{4\pi k \Delta T}{(r_2 - r_1)/r_1 r_2}$$

where r_1 is the inside radius of the sphere and r_2 is the outside radius of the sphere.

Consider an igloo as half of a spherical shell. The inside radius is 3.00 m and the outside radius is 3.20 m. If the temperature inside the igloo is 15.0 °C and the outside temperature is −40.0 °C, find the flow of thermal energy through the ice per hour. The thermal conductivity of ice is 4.00×10^{-4} kcal/s m °C.

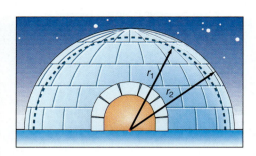

†53. Show that for large values of r_1 and r_2 the solution for thermal energy flow through a spherical shell (problem 52) reduces to the solution for the thermal energy flow through a flat slab.

†54. In problems 51 and 52 assume that you can use the formula for the thermal energy flow through a flat slab. Find $\Delta Q/\Delta t$ and find the percentage error involved by making this approximation.

55. A spherical body of 25.0-cm radius, has an emissivity of 0.45, and is at a temperature of 500.0 °C. How much power is radiated from the sphere?

†56. Newton's law of cooling states that the rate of change of temperature of a cooling body is proportional to the rate at which it gains or loses heat, which is approximately proportional to the difference between its temperature and the temperature of the environment. This is written mathematically as

$$\frac{\Delta T}{\Delta t} = -K(T_{avg} - T_e)$$

where T_{avg} is the average temperature of the body, T_e is the temperature of the environment, and K is a constant. A cup of coffee cools from 98.0 °C to 90.0 °C in 1.5 min. The cup is in a room that has a temperature of 20.0 °C. Find (a) the value of K and (b) how long it will take for the coffee to cool from 90.0 °C to 50.0 °C.

†57. A much more complicated example of heat transfer is one that combines conduction and convection. That is, we want to determine the thermal energy transferred from a hot level plate at 100 °C to air at a temperature of 20.0 °C. The thermal energy transferred is given by the equation

$$Q = hA\Delta T\, t$$

where h is a constant, called the convection coefficient and is a function of the shape, size, and orientation of the surface, the type of fluid in contact with the surface and the type of motion of the fluid itself. Values of h for various configurations can be found in handbooks. If h is equal to 1.78×10^{-3} kcal/s m² °C and A is 2.00 m², find the amount of thermal energy transferred per minute.

†58. Using the same principle of combined conduction and convection used in problem 57, find the amount of thermal energy that will flow through an uninsulated wall 10.0 cm thick of a wood frame house in 1 hr. Assume that both the inside and outside wall have a thickness of 2.00 cm of pine wood and an area of 25.0 m². (*Hint:* First consider the thermal energy loss through the inside wall, then the thermal energy loss through the 10.0-cm air gap, then the thermal energy loss through the outside wall.) The temperatures at the first wall are 18.0 °C and 13.0 °C, and the temperatures at the second wall are 10.0 °C and −6.70 °C. The convection coefficient for a vertical wall is $h = 0.424 \times 10^{-4} (\Delta T)^{1/4}$ cal/s cm² °C.

†59. A thermograph is essentially a device that detects radiation in the infrared range of the electromagnetic spectrum. A thermograph can map the temperature distribution of the human body, showing regions of abnormally high temperatures such as found in tumors. Starting with the Stefan-Boltzmann law show that the ratio of the power emitted from tissue at a slightly higher temperature, $T + \Delta T$, to the power emitted from normal tissue at a temperature T is

$$\frac{P_2}{P_1} = (1 + \Delta T/T)^4$$

Then show that a change of temperature of only 0.9 °C will give an approximate 1.00% increase in the power of the radiation transmitted. Assume that the body temperature is 37.0 °C.

Interactive Tutorials

⌨60. How much thermal energy flows through a glass window per second (Q/s) if the thickness of the window $d = 0.020$ m and its cross-sectional area $A = 2.00$ m². The temperature difference between the window's faces is $\Delta T = 65.0$ °C, and the thermal conductivity of glass is $k = 0.791$ J/(m s °C).

⌨61. Convection. A hot air heating system heats air to a temperature of 125 °C and the return air is at a temperature of 17.5 °C. The fan is capable of moving a volume of 7.50 m³ of air in 1 min, $\Delta V/\Delta t$. The specific heat of air at constant pressure, c, is 1.05×10^3 J/kg °C and take the density of air, ρ to be 1.29 kg/m³. Find the amount of thermal energy transfer per hour from the furnace to the room.

⌨62. Conduction through a compound wall. Find the amount of heat conducted through a compound wall that has a length $L = 8.5$ m and a height $h = 4.33$ m. The wall consists of a thickness of $d_1 = 10.0$ cm of brick, $d_2 = 1.90$ cm of plywood, $d_3 = 10.2$ cm of glass wool, $d_4 = 1.25$ cm of plaster, and $d_5 = 0.635$ cm of oak wood paneling. The inside temperature of the wall is $T_h = 20.0$ °C and the outside temperature of the wall is $T_c = -9.00$ °C. How much thermal energy flows through this wall per day?

⌨63. Radiation. How much energy is radiated in 1 s by an iron sphere 18.5 cm in radius at a temperature of 125 °C? Assume that the sphere radiates as a blackbody of emissivity $e = 1$. What is the wavelength of the maximum intensity of radiation?

17

Thermodynamics

A thermodynamic system in nature.

We can express the fundamental laws of the universe which correspond to the two fundamental laws of the mechanical theory of heat in the following simple form:
1. The energy of the universe is a constant. 2. The entropy of the universe tends toward a maximum.

Rudolf Clausius

17.1 Introduction

Thermodynamics is the study of the relationships between heat, internal energy, and the mechanical work performed by a system. The system considered is usually a heat engine of some kind, although the term can also be applied to living systems such as plants and animals. There are two laws of thermodynamics. The first law of thermodynamics is the law of conservation of energy as applied to a thermodynamic system. We will apply the first law of thermodynamics to a heat engine and study its ramifications. The second law of thermodynamics tells us what processes are, and are not, possible in the operation of a heat engine. The second law is also responsible for telling us in which direction a particular physical process may go. For example a block can slide across a desk and have all of its kinetic energy converted to thermal energy by the work the block does against friction as it is slowed to a stop. However, the reverse process does not happen, that is, the thermal energy in the block does not convert itself into mechanical energy and cause the block to slide across the desk. Using the thermal energy in the block to cause mechanical motion is not a violation of the law of conservation of energy but it is a violation of the second law of thermodynamics.

17.2 The Concept of Work Applied to a Thermodynamic System

Consider what happens to an ideal gas in a cylinder when it is compressed by a constant external force **F,** as shown in figure 17.1(a). The constant force exerted on top of the piston causes it to be displaced a distance Δy, thereby compressing the gas in the cylinder. The work done on the gas by the external force in compressing it is

$$W = F\Delta y \qquad (17.1)$$

This work by the external agent is positive because the external force and the displacement are in the same direction. The external force F and the external pressure p exerted on the gas by the piston are related by

$$F = pA \qquad (17.2)$$

where A is the cross-sectional area of the piston. Substituting equation 17.2 into equation 17.1 gives

$$W = pA\Delta y \qquad (17.3)$$

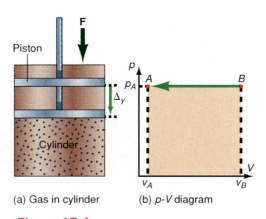

(a) Gas in cylinder (b) p-V diagram

Figure 17.1

Work done in compressing a gas.

which is the work done on the gas by the external agent. If the compression takes place very slowly, the constant external pressure exerted by the piston on the gas is equal to the internal pressure exerted by the gas throughout the process. Thus, equation 17.3 can also be interpreted as the work done by the gas rather than the external agent. This is a departure from the usual way we have analyzed the concept of work. Previously, we have always considered the work as being done by the external agent. From this point on, we will consider all the work to be done on or by the gas itself, not the external agent. The product of the area of the cylinder and the displacement of the gas is equal to the change in volume of the gas. That is,

$$A\Delta y = \Delta V \tag{17.4}$$

the decrease in the volume of the gas. Substituting equation 17.4 into 17.3 gives

$$W = p\Delta V \tag{17.5}$$

Equation 17.5 represents the amount of work done by the gas when a constant external force compresses it by an amount ΔV.

This entire process can be shown on a pressure-volume (p-V) diagram as in figure 17.1(b). The original state of the gas is represented as the point B in the diagram, where it has the volume V_B and the pressure p_A. As the piston moves at constant pressure, the system, the gas in the cylinder, moves from the state at point B to the state at point A [figure 17.1(b)] along a horizontal line indicating that the process is occurring at constant pressure. At point A in the figure, the gas has been compressed to the volume V_A. The change in volume of the gas is seen to be

$$\Delta V = V_A - V_B \tag{17.6}$$

The total work done by the gas in compressing it from the point B to the point A, found from equations 17.5 and 17.6, is

$$W_{BA} = p_A(V_A - V_B)$$

It is important to note here that the product of p_A and $V_A - V_B$ represents the area of the rectangle cross-hatched in figure 17.1(b). Thus, the area under the curve in a p-V diagram always represents a quantity of work. When the area is large, it represents a large quantity of work, and when the area is small the quantity of work likewise is small.

Because V_A is less than V_B, the quantity $V_A - V_B$ is negative. Thus, *when work is done by a gas in compressing it, that work is always negative.* Notice that there are two distinct agents here. The work done *by the external agent* in compressing the gas is positive, but the *work done by the gas* in a compression is negative.

If the gas in the cylinder of figure 17.1(a) is allowed to expand back to the original volume V_B, then the process can be represented on the same p-V diagram of figure 17.1(b) as the same straight line, now going from point A to the point B. The work done by the gas in the expansion from A to B is

$$W = p\Delta V = p_A(V_B - V_A)$$

But now note that since V_B is greater than V_A, the quantity $V_B - V_A$ is now a positive quantity. *Thus, when a gas expands, the work done by the gas is positive.* (The work done on the gas by an external agent during the expansion would be negative. From this point on let us consider only the work done by the gas and forget any external agent.) *Thus, the **work** done by a gas during expansion is positive and the work done by a gas during compression is negative.* In either case, the work done is still the area under the line AB given by the product of the sides of the rectangle p_A and $V_B - V_A$. The areas are the same in both cases, however we consider the area positive when the gas expands and negative when the gas is compressed.

Let us now consider the work done along the different paths of the cyclic process shown in the p-V diagram of figure 17.2. A **cyclic process** is a process that

Thermodynamics

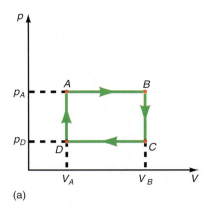

(a)

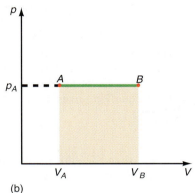

(b)

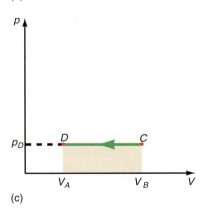

(c)

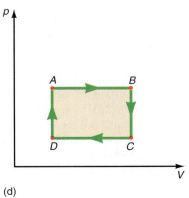

(d)

Figure 17.2

Work done in a cyclic process.

runs in a cycle eventually returning to where it started from. Thus, in figure 17.2(a) the cycle goes from A to B, B to C, C to D, and D back to A. The total work done by the system as it goes through the cycle is simply

$$W_{total} = W_{AB} + W_{BC} + W_{CD} + W_{DA} \qquad (17.7)$$

where

W_{AB} is the work done on the path AB
W_{BC} is the work done on the path BC
W_{CD} is the work done on the path CD
W_{DA} is the work done on the path DA

Let us consider the work done along each path separately. To simplify matters let us first look at the work done along the path BC. The path BC represents a process that is performed at the constant volume V_B. Therefore, $\Delta V = 0$, and no work is performed along BC. Formally,

$$W_{BC} = p(V_B - V_B) = 0 \qquad (17.8)$$

Similarly, along the path DA, the volume is also a constant and therefore ΔV is again zero, and hence the work done must also be zero. Formally,

$$W_{DA} = p(V_A - V_A) = 0 \qquad (17.9)$$

Since the work is given by $p\Delta V$, whenever V is a constant in a process, ΔV is always zero and the work is also zero along that path in the p-V diagram.

The work done along the path AB is

$$W_{AB} = p_A\Delta V = p_A(V_B - V_A) \qquad (17.10)$$

Because the path AB represents an expansion, positive work is done by the gas, as is evidenced by the fact that $V_B - V_A$ is a positive quantity. The work done along the path AB is shown as the area under the line AB in figure 17.2(b).

The work done along the path CD is

$$W_{CD} = p_D\Delta V = p_D(V_A - V_B) \qquad (17.11)$$

Since the path CD represents a compression, work is done on the gas. This work is considered negative, as we can see from the fact that $V_A - V_B$, in equation 17.11, is negative. The work done on the gas is shown as the area under the line CD in figure 17.2(c).

The net work done by the gas in the cyclic process $ABCDA$, found from equation 17.7 with the help of equations 17.8 through 17.11, is

$$W_{total} = W_{AB} + W_{BC} + W_{CD} + W_{DA}$$
$$W_{total} = p_A(V_B - V_A) + 0 + p_D(V_A - V_B) + 0 \qquad (17.7)$$

We can rewrite this to show that the work along CD is negative, that is, $V_A - V_B = -(V_B - V_A)$. Hence,

$$W_{total} = p_A(V_B - V_A) - p_D(V_B - V_A)$$

or

$$W_{total} = (p_A - p_D)(V_B - V_A) \qquad (17.12)$$

Thus, equation 17.12 represents the net work done by the gas in this particular cyclic process. Note that $p_A - p_D$ is one side of the rectangular path of figure 17.2(a) while $V_B - V_A$ is the other side of that rectangle. Hence, their product in equation 17.12 represents the entire area of the rectangle enclosed by the thermodynamic path $ABCDA$ and is shown as the cross-hatched area in figure 17.2(d). Another way to visualize this total area, and hence total work, is to subtract the area in figure 17.2(c), the negative work, from the area in figure 17.2(b), the positive work, and we again get the area bounded by the path *ABCDA*. Although this result was derived for a simple rectangular thermodynamic path, it is true in general. *Thus, in any cyclic process, the net work done by the system is equal to the area enclosed by the cyclic thermodynamic path in a p-V diagram.* Therefore, to get as much work as possible out of a system, the enclosed area must be as large as possible. The net work is positive if the cycle proceeds clockwise, in the *p-V* diagram, and negative if the cycle proceeds counterclockwise. Finally, we should note that the process *AB* takes place at the constant pressure p_A. *A process that takes place at a constant pressure is called an* **isobaric process.** Hence, the process *CD* is also an isobaric process because it takes place at the constant pressure p_D. Process *BC* takes place at the constant volume V_B, and process *DA* takes place at the constant volume V_A. *A process that takes place at constant volume is called an* **isochoric or isometric process.**

There is another type of process that is very important in thermodynamic systems, the isothermal process. *An* **isothermal process** *is a process that occurs at a constant temperature, that is,* $\Delta T = 0$ *for the process.* A picture of an isotherm can be drawn on a *p-V* diagram by using the equation of state for an ideal gas, the working substance in the system. Thus, the ideal gas equation, given by equation 15.23, is

$$pV = nRT$$

Because *n* and *R* are constants, if *T* is also a constant, then the entire right-hand side of equation 15.23 is a constant. We can then write equation 15.23 as

$$pV = \text{constant} \qquad (17.13)$$

If we plot equation 17.13 on a *p-V* diagram, we obtain the hyperbolic curves of figure 17.3. Each curve is called an isotherm and in the figure, T_3 is greater than T_2, which in turn is greater than T_1.

Let us now consider the new cyclic process shown in figure 17.4, in which an ideal gas in a cylinder expands against a piston isothermally. This is shown as the path *AC* in the *p-V* diagram. To physically carry out the isothermal process along the path *AC,* the cylinder is surrounded by a constant temperature heat reservoir. The cylinder either absorbs heat from, or liberates heat to, the reservoir in order to maintain the constant temperature. When the isothermal process is finished the heat reservoir is removed. The gas is then compressed at the constant pressure p_D at point *C* until it reaches the point *D*. The pressure of the gas is then increased from p_D to p_A while the volume of the gas in the cylinder is kept constant. This is shown as the path *DA* in the *p-V* diagram. Now let us assume that the points *A, C,* and *D* are the same points that were considered in figure 17.2(a). Recall that the net work done by the system is equal to the area enclosed by the cyclic path. Thus, the net work done in this process is equal to the cross-hatched area within the path *ACDA* shown in figure 17.4.

It is important to compare figure 17.2(d) with figure 17.4. Remember the points *A, C,* and *D* in figure 17.4 are the same as the points *A, C,* and *D* in figure 17.2(d). But the area under the enclosed curve in figure 17.2(d) is greater than the enclosed area in figure 17.4. Hence, a greater amount of work is done by the system in following the cyclic path *ABCDA* than the cyclic path *ACDA. Thus, the work that the system does depends on the thermodynamic path taken.* Even though both processes started at point *A* and returned to the same point *A,* the work done by the system is different in each case. This result is succinctly stated as: *the work done depends on the path taken, and work is a path dependent quantity.*

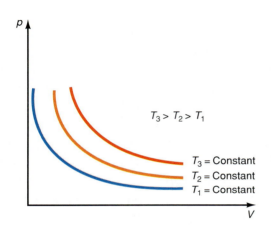

Figure 17.3

Isotherms on a *p-V* diagram.

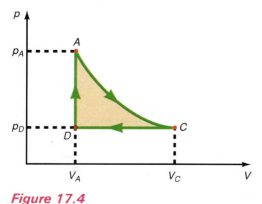

Figure 17.4

Cyclic process with an isothermal expansion.

Example 17.1

Work done in a thermodynamic cycle. One mole of an ideal gas goes through the thermodynamic cycle shown in figure 17.2(a). If $p_A = 2.00 \times 10^4$ Pa, $p_D = 1.00 \times 10^4$ Pa, $V_A = 0.250$ m³, and $V_B = 0.500$ m³, find the work done along the path (a) AB, (b) BC, (c) CD, (d) DA, and (e) $ABCDA$.

Solution

a. The work done along the path AB, found from equation 17.10, is

$$W_{AB} = p_A(V_B - V_A)$$
$$= (2.00 \times 10^4 \text{ Pa})(0.500 \text{ m}^3 - 0.250 \text{ m}^3)$$
$$= 5.00 \times 10^3 \frac{\text{N m}^3}{\text{m}^2} = 5.00 \times 10^3 \text{ N m}$$
$$= 5.00 \times 10^3 \text{ J}$$

b. The work done along the path BC, found from equation 17.8, is

$$W_{BC} = p(V_B - V_B) = 0$$

c. The work done along path CD, given by equation 17.11, is

$$W_{CD} = p_D(V_A - V_B)$$
$$= (1.00 \times 10^4 \text{ Pa})(0.250 \text{ m}^3 - 0.500 \text{ m}^3)$$
$$= -2.50 \times 10^3 \text{ J}$$

Note that the work done in compressing the gas is negative.

d. The work done along path DA, given by equation 17.9, is

$$W_{DA} = p(V_A - V_A) = 0$$

e. The total work done along the entire path $ABCDA$, found from equation 17.7, is

$$W_{\text{total}} = W_{AB} + W_{BC} + W_{CD} + W_{DA}$$
$$= 5.00 \times 10^3 \text{ J} + 0 - 2.50 \times 10^3 \text{ J} + 0$$
$$= 2.50 \times 10^3 \text{ J}$$

17.3 Heat Added to or Removed from a Thermodynamic System

We saw in chapter 14 that the amount of heat added or removed from a body is given by

$$Q = mc\Delta T \tag{14.6}$$

Equation 14.6 can also be applied to the heat added to, or removed from, a gas, if two stipulations are made. First, we saw in chapter 15 that it is more convenient to express the mass m of a gas in terms of the number of moles n of the gas. The total mass m of the gas is the sum of the masses of all the molecules of the gas. That is, m is equal to the mass of one molecule times the total number of molecules in one mole of the substance, times the total number of moles. That is

$$m = m_0 N_A n \tag{17.14}$$

where m_0 is the mass of one molecule; N_A is Avogadro's number, the number of molecules in one mole of a substance; and n is the number of moles of the gas. Notice in equation 15.47, the product of the mass of one molecule times Avogadro's number is called the **molecular mass** of the substance M, that is,

$$M = m_0 N_A \tag{17.15}$$

The molecular mass is thus the mass of one mole of the gas. Substituting equation 17.15 into equation 17.14 gives for the mass of the gas

$$m = nM \tag{17.16}$$

Equation 17.16 says that the mass of the gas is equal to the number of moles of the gas times the molecular mass of the gas. Substituting equation 17.16 for the mass m of the gas into equation 14.6, gives

$$Q = nMc\Delta T \tag{17.17}$$

The product Mc is defined as the **molar specific heat** of the gas, or molar heat capacity, and is represented by the capital letter C. Hence,

$$C = Mc \tag{17.18}$$

The heat absorbed or lost by a gas undergoing a thermodynamic process is found by substituting equation 17.18 into equation 17.17. Thus,

$$Q = nC\Delta T \tag{17.19}$$

The second stipulation for applying equation 14.6 to gases has to do with the specific process to which the gas is subjected. Equation 14.6 was based on the heat absorbed or liberated from a solid or a liquid body that was under constant atmospheric pressure. In applying equation 17.19, which is the modified equation 14.6, we must specify the process whereby the temperature change ΔT occurs. Figure 17.5 shows some possible processes. Let us start at the point A in the p-V diagram of figure 17.5. The temperature at point A is T_0 because point A is on the T_0 isotherm. Heat can be added to the system such that the temperature of the gas rises to T_1. But, as we can see from figure 17.5, there are many different ways to get to the isotherm T_1. The thermodynamic paths AB, AC, AD, AE, or an infinite number of other possible paths can be followed to arrive at T_1. Therefore, there can be an infinite number of specific heats for gases. Let us restrict ourselves to only two paths, and hence only two specific heats. The first path we consider is the path AB, which represents a process taking place at constant volume. The second path is path AE, which represents a process taking place at constant pressure. We designate the molar specific heat for a process occurring at constant volume by C_v, whereas we designate the molar specific heat for a process occurring at a constant pressure by C_p. It is found experimentally that for a monatomic ideal gas such as helium or argon, $C_v = 2.98$ cal/mole K = 12.5 J/mole K, whereas $C_p = 4.97$ cal/mole K = 20.8 J/mole K.

The heat absorbed by the gas as the system moves along the thermodynamic path AB in figure 17.5 is

$$Q_{AB} = nC_v\Delta T = nC_v(T_1 - T_0) \tag{17.20}$$

The heat absorbed by the gas as the system moves along the path AE is given by

$$Q_{AE} = nC_p\Delta T = nC_p(T_1 - T_0) \tag{17.21}$$

Although the system ends up at the same temperature T_1 whether the path AB or AE is traveled, the heat that is absorbed along each path is different because C_p and C_v have different values. *Thus, the **heat** absorbed or liberated **in a thermodynamic process** depends on the path that is followed. That is, heat like work is path dependent.* Although demonstrated for a gas, this statement is true in general.

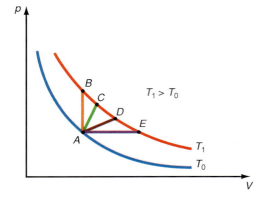

Figure 17.5

The specific heat for a gas depends on the path taken in a p-V diagram.

Example 17.2

The heat absorbed along two different thermodynamic paths. Compute the amount of heat absorbed by 1 mole of He gas along path (a) *AB* and (b) *AE*, of figure 17.5, if $T_1 = 400$ K and $T_0 = 300$ K.

Solution

a. The heat absorbed along path *AB*, given by equation 17.20, is

$$Q_{AB} = nC_v\Delta T = nC_v(T_1 - T_0)$$

$$= (1 \text{ mole})\left(2.98 \frac{\text{cal}}{\text{mole K}}\right)(400 \text{ K} - 300 \text{ K})$$

$$= 298 \text{ cal}$$

b. The heat absorbed along the path *AE*, given by equation 17.21, is

$$Q_{AE} = nC_p\Delta T = nC_p(T_1 - T_0)$$

$$= (1 \text{ mole})\left(4.97 \frac{\text{cal}}{\text{mole K}}\right)(400 \text{ K} - 300 \text{ K})$$

$$= 497 \text{ cal}$$

Thus, a greater quantity of heat is absorbed in the process that occurs at constant pressure. This is because at constant pressure the volume expands and some of the heat energy is used to do work, but at constant volume no work is accomplished.

17.4 The First Law of Thermodynamics

Recall from the kinetic theory of gases studied in chapter 15 that the mean kinetic energy of a molecule, found from equation 15.45, is

$$\text{KE}_{avg} = \tfrac{1}{2}mv_{avg}^2 = \tfrac{3}{2}kT$$

Thus, a change in the absolute temperature of a gas shows up as a change in the average energy of a molecule. If the average kinetic energy of one molecule of the gas is multiplied by N, the total number of molecules of the gas present in the thermodynamic system (i.e., the cylinder filled with gas), then this product represents the total internal energy of this quantity of gas. Recall that the internal energy of a body was defined in chapter 14 as the sum of the kinetic energies and potential energies of all the molecules of the body. Because the molecules of a gas are moving so rapidly and are widely separated on the average, only a few are near to each other at any given time and it is unnecessary to consider any intermolecular forces, and hence potential energies of the molecules. *Thus, the total kinetic energy of all the molecules of a gas constitutes the total **internal energy of the gas.*** We designate this internal energy of the gas by the symbol U. The internal energy of the gas is given by

$$U = (\text{total number of molecules})(\text{mean KE of each molecule})$$

$$= N\text{KE}_{avg}$$

$$U = N(\tfrac{3}{2}kT) \tag{17.22}$$

But recall from equation 15.44 that

$$k = \frac{R}{N_A}$$

Substituting equation 15.44 into equation 17.22 gives for the internal energy of an ideal gas

$$U = N\frac{3}{2}\frac{R}{N_A}T$$

But the total number of molecules N was given by

$$N = nN_A \qquad \text{(15.24)}$$

Thus,

$$U = nN_A\frac{3}{2}\frac{R}{N_A}T$$

and

$$U = \tfrac{3}{2}nRT \qquad \text{(17.23)}$$

From equation 17.23 we see that a change in temperature is thus associated with a change in the internal energy of the gas, that is,

$$\Delta U = \tfrac{3}{2}nR\Delta T \qquad \text{(17.24)}$$

Let us now consider the thermodynamic system shown in the p-V diagram of figure 17.6. The isotherm going through point B is labeled T_B, whereas the one that goes through points A and C is labeled T_{AC}, and finally, the isotherm that goes through point D is called T_D. Before the entire system is considered, let us first consider a process that proceeds isothermally from A to C. Since the path AC is an isotherm, the temperature is constant and thus $\Delta T = 0$. But from equation 17.24, the change in internal energy ΔU must also be zero. That is, *an isothermal expansion occurs at constant internal energy.* But how can this be? As the gas expands along AC it is doing work. If the internal energy is constant, where does the energy come from to perform the work that is being done by the gas? Obviously energy must somehow be supplied in order for the gas to do work. ***Thus, a quantity of heat Q must be supplied to the system in order for the system to do work along an isothermal path.*** Hence, for an isothermal process,

$$Q = W \qquad \text{(17.25)}$$

Let us now consider the portion of the process that is along path BC in figure 17.6. The process BC is performed at constant volume, thus, $\Delta V = 0$ along this path. Because the amount of work done by the gas is given as $W = p\Delta V$, if $\Delta V = 0$, then the work done along the path BC must also be zero. But the temperature T_{AC} at point C is less than the temperature T_B at the point B. There has been a drop in temperature between points B and C and hence a decrease in the

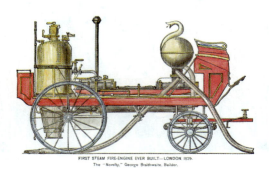

The steam engine is a thermodynamic system.

Figure 17.6

A thermodynamic system on a p-V diagram.

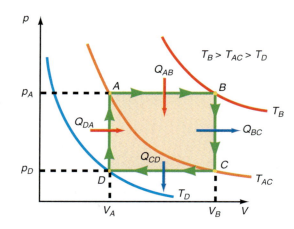

Thermodynamics

internal energy of the system. Since the loss of energy didn't go into work, because $\Delta V = 0$, heat must have been taken away from the system along path *BC*. *The decrease in the internal energy of the system along an isometric path is caused by the heat removed from the system along BC*, that is,

$$\Delta U = Q \qquad (17.26)$$

But the heat removed from the system during a constant volume process was shown in equation 17.20 to be

$$Q = nC_v\Delta T \qquad (17.27)$$

Since the heat removed is equal to the loss in internal energy by equation 17.26, we can write the change in internal energy from equations 17.26 and 17.27 as

$$\Delta U = nC_v\Delta T \qquad (17.28)$$

Equation 17.28 is a general statement governing the change in internal energy during any process, not only the one at constant volume from which equation 17.28 was derived. Recall from equation 17.23, a result from the kinetic theory of gases, that *U*, the internal energy, is only a function of temperature. In fact if ΔU from equation 17.24 is equated to ΔU from equation 17.28, we get

$$\tfrac{3}{2}nR\Delta T = nC_v\Delta T$$

Solving for C_v, the theoretical value of the molar specific heat capacity at constant volume is found to be

$$C_v = \tfrac{3}{2}R \qquad (17.29)$$

Using the value of $R = 1.987$ cal/mole K found in chapter 15, the value of C_v, calculated from equation 17.29, is $C_v = 2.98$ cal/mole K, which agrees with the experimental value.

The two special cases given by equations 17.25 and 17.26 can be combined into one general equation that contains 17.25 and 17.26 as special cases. This general equation is

$$Q = \Delta U + W \qquad (17.30)$$

and is *called **the first law of thermodynamics.*** Thus, we can derive equation 17.25 from 17.30 for an isothermal path because then the change in internal energy $\Delta U = 0$. We can derive equation 17.26 from equation 17.30 for a constant volume thermodynamic path, because then $\Delta V = 0$, and hence $W = 0$. *The first law of thermodynamics, equation 17.30, says that the heat Q, added to a system will show up either as a change in internal energy ΔU of the system and/or as work W performed by the system.* From this analysis we can see that the first law of thermodynamics is just the law of conservation of energy. Equation 17.30 is quite often written in the slightly different form:

$$\Delta U = Q - W \qquad (17.31)$$

which is *also called the first law of thermodynamics. The first law of thermodynamics can be also stated as the change in the internal energy of the system equals the heat added to the system minus the work done by the system on the outside environment.* Perhaps the best way to see the application of the first law to a thermodynamic system is in an example.

Example 17.3

Applying the first law of thermodynamics. Two moles of an ideal gas are carried around the thermodynamic path *ABCDA* in figure 17.6. Here $T_D = 150$ K, $T_{AC} = 300$ K, $T_B = 600$ K, and $p_A = 2.00 \times 10^4$ Pa, while $p_D = 1.00 \times 10^4$ Pa. The volume $V_A = 0.250$ m³, while $V_B = 0.500$ m³. Find the work done, the heat lost or absorbed, and the internal energy of the system for the thermodynamic paths (a) AB, (b) BC, (c) CD, (d) DA, and (e) $ABCDA$.

Solution
---

a. The work done by the expanding gas along the path AB is

$$W = p\Delta V$$

$$W_{AB} = p_A(V_B - V_A)$$

$$= \left(2.00 \times 10^4 \frac{\text{N}}{\text{m}^2}\right)(0.500 \text{ m}^3 - 0.250 \text{ m}^3)$$

$$= 5.00 \times 10^3 \text{ J}$$

The heat absorbed by the gas along path AB is

$$Q = nC_p\Delta T$$

$$Q_{AB} = nC_p(T_B - T_{AC})$$

$$= (2 \text{ moles})\left(4.97 \frac{\text{cal}}{\text{mole K}}\right)(600 \text{ K} - 300 \text{ K})$$

$$= (2.98 \times 10^3 \text{ cal})\left(\frac{4.185 \text{ J}}{1 \text{ cal}}\right)$$

$$= 1.25 \times 10^4 \text{ J}$$

The change in internal energy along path AB, found from the first law equation 17.31, is

$$\Delta U_{AB} = Q_{AB} - W_{AB}$$

$$= 1.25 \times 10^4 \text{ J} - 5.00 \times 10^3 \text{ J}$$

$$= 7.50 \times 10^3 \text{ J}$$

Thus, there is a gain of internal energy along the path AB.

b. The work done along path BC is

$$W = p\Delta V$$

$$W_{BC} = p(V_B - V_B) = 0$$

$$= 0$$

The heat lost along path BC is

$$Q_{BC} = nC_v\Delta T = nC_v(T_{AC} - T_B)$$

$$= (2 \text{ moles})\left(2.98 \frac{\text{cal}}{\text{mole K}}\right)(300 \text{ K} - 600 \text{ K})$$

$$= (-1.79 \times 10^3 \text{ cal})\left(\frac{4.185 \text{ J}}{1 \text{ cal}}\right)$$

$$= -7.50 \times 10^3 \text{ J}$$

The loss of internal energy in dropping from 600 K at B to 300 K at C is found from the first law as

$$\Delta U_{BC} = Q_{BC} - W_{BC}$$

$$= -7.50 \times 10^3 \text{ J} - 0$$

$$= -7.50 \times 10^3 \text{ J}$$

c. The work done during the compression along the path CD is

$$W_{CD} = p\Delta V = p_D(V_A - V_B)$$

$$= \left(1.00 \times 10^4 \frac{N}{m^2}\right)(0.250\ m^3 - 0.500\ m^3)$$

$$= -2.50 \times 10^{-3}\ J$$

The heat lost along the path CD is

$$Q_{CD} = nC_p\Delta T = nC_p(T_D - T_{AC})$$

$$= (2\ moles)\left(4.97\ \frac{cal}{mole\ K}\right)(150\ K - 300\ K)$$

$$= -6.24 \times 10^3\ J$$

The change in internal energy along the path CD, found from the first law, is

$$\Delta U_{CD} = Q_{CD} - W_{CD}$$

$$= -6.24 \times 10^3\ J - (-2.50 \times 10^3\ J)$$

$$= -3.74 \times 10^3\ J$$

Note that the internal energy decreased, as expected, since the temperature decreased from 300 K to 150 K.

d. The work done along the path DA is

$$W_{DA} = p\Delta V = p(V_A - V_A) = 0$$

The heat added along the path DA is

$$Q_{DA} = nC_v\Delta T = nC_v(T_{AC} - T_D)$$

$$= (2\ moles)\left(2.98\ \frac{cal}{mole\ K}\right)(300\ K - 150\ K)$$

$$= (8.94\ cal)\left(4.185\ \frac{J}{cal}\right)$$

$$= 3.74 \times 10^3\ J$$

The change in internal energy along DA is

$$\Delta U_{DA} = Q_{DA} - W_{DA}$$

$$= 3.74 \times 10^3\ J$$

e. The net work done throughout the cycle $ABCDA$ is

$$W_{ABCDA} = W_{AB} + W_{BC} + W_{CD} + W_{DA}$$

$$= 5.00 \times 10^3\ J + 0 - 2.50 \times 10^3\ J + 0$$

$$= 2.50 \times 10^3\ J$$

The net heat added throughout the cycle $ABCDA$ is

$$Q_{ABCDA} = Q_{AB} + Q_{BC} + Q_{CD} + Q_{DA}$$

$$= 1.25 \times 10^4\ J - 7.50 \times 10^3\ J - 6.24 \times 10^3\ J + 3.74 \times 10^3\ J$$

$$= 2.50 \times 10^3\ J$$

Note that Q_{AB} and Q_{DA} are positive quantities, which means that heat is being added to the system along these two paths. Also note that Q_{BC} and Q_{CD} are negative quantities, which means that heat is being taken away from the system along these two paths. *In general, Q is always positive when heat is added to the system and negative when heat is removed from the system.*

This effect is seen in figure 17.6 by drawing lines entering the enclosed thermodynamic path when heat is added to the system, and lines emanating from the enclosed path when heat is taken away from the system. This is a characteristic of all engines operating in a cycle, that is, heat is always added and some heat is always rejected. The net change in internal energy throughout the cycle $ABCDA$ is

$$\Delta U_{ABCDA} = \Delta U_{AB} + \Delta U_{BC} + \Delta U_{CD} + \Delta U_{DA} \qquad (17.32)$$

$$= 7.50 \times 10^3 \text{ J} - 7.50 \times 10^3 \text{ J} - 3.74 \times 10^3 \text{ J} + 3.74 \times 10^3 \text{ J}$$

$$= 0$$

Note that the total change in internal energy around the entire cycle is equal to zero. This is a very reasonable result because the internal energy of a system depends only on the temperature of the system. If we go completely around the cycle, we end up at the same starting point with the same temperature. Since $\Delta T = 0$ around the cycle, $\Delta U = nC_v\Delta T$ must also equal zero around the cycle.

Applying the first law to the entire cycle we have

$$\Delta U_{ABCDA} = Q_{ABCDA} - W_{ABCDA}$$

But as just seen, $\Delta U_{ABCDA} = 0$, therefore,

$$Q_{ABCDA} = W_{ABCDA} \qquad (17.33)$$

That is, the energy for the net work done by the system comes from the net heat applied to the system. Looking at the calculations, we see that this is indeed the case since

$$Q_{ABCDA} = 2.50 \times 10^3 \text{ J}$$

while

$$W_{ABCDA} = 2.50 \times 10^3 \text{ J}$$

Another very interesting thing can be learned from this example. Look at the change in internal energy from the point A to the point C, and note that regardless of the path chosen, the change in internal energy is the same. Thus, from our calculations,

$$\Delta U_{AC} = \Delta U_{AB} + \Delta U_{BC} = 7.50 \times 10^3 \text{ J} - 7.50 \times 10^3 \text{ J} = 0$$

and

$$\Delta U_{AC} = -\Delta U_{AD} - \Delta U_{DC} = -3.74 \times 10^3 \text{ J} + 3.74 \times 10^3 \text{ J} = 0$$

Along the isothermal path AC

$$\Delta U_{AC} = 0$$

because if T is constant, U is constant. *Thus, regardless of the path chosen between two points on a p-V diagram, ΔU is always the same.* (It will not always be zero, as in this case where the points A and C happen to lie along the same isotherm, but whatever its numerical value, ΔU is always the same.)

What is especially interesting about this fact is that *the work done depends on the path taken, the heat absorbed or liberated depends on the path taken, but their difference $Q - W$, which is equal to ΔU is independent of the path taken. That is, ΔU depends only on the initial and final states of the thermodynamic system and not the path between the initial and final states.*

Thus, the internal energy is to a thermodynamic system what the potential energy is to a mechanical system. (Recall from chapter 7, section 7.7 that the work done, and hence the potential energy, was the same whether an object was lifted to a height h, or moved up a frictionless inclined plane to the same height h. That is, the potential energy was independent of the path taken.)

The thermodynamic system considered in figure 17.6 represents an engine of some kind. That is, heat is added to the engine and the engine does work. To compare one engine with another it is desirable to know how efficient each engine is. *The **efficiency** of an engine can be defined in terms of what we get out of the system compared to what we put into the system.* Heat, Q_{in}, is put into the engine, and work, W, is performed by the engine, hence the efficiency of an engine can be defined as

$$\text{Eff} = \frac{\text{Work out}}{\text{Heat in}} = \frac{W}{Q_{in}} \qquad (17.34)$$

Example 17.4

The efficiency of an engine. In example 17.3, 2.50×10^3 J of work was done by the system, whereas the heat added to the system was the heat added along paths *AB* and *DA*, which is equal to 1.25×10^4 J $+ 3.74 \times 10^3$ J, which is equal to 1.62×10^4 J. Find the efficiency of that engine.

Solution

The efficiency of the engine, found from equation 17.34, is

$$\text{Eff} = \frac{W}{Q_{in}} = \frac{2.50 \times 10^3 \text{ J}}{1.62 \times 10^4 \text{ J}} = 0.15$$
$$= 15\%$$

Thus, the efficiency of the engine represented by the thermodynamic cycle of figure 17.6 is only 15%. This is not a very efficient engine. We will discuss the maximum possible efficiency of an engine when we study the Carnot cycle in section 17.8.

Before leaving this section, however, let us take one more look at the change in the internal energy of the system along the path *ABC*. We have already seen that since the initial and final states lie on the same isotherm, the change in internal energy is zero. There is still, however, some more important physics to be obtained by further considerations of this path. The change in internal energy along the path *ABC* is given by

$$\Delta U_{ABC} = \Delta U_{AB} + \Delta U_{BC}$$

But from the first law we can write this as

$$\Delta U_{ABC} = Q_{AB} - W_{AB} + Q_{BC} - W_{BC} \qquad (17.35)$$

But as we have already seen

$$
\left.
\begin{aligned}
Q_{AB} &= nC_p(T_B - T_{AC}) \\
W_{AB} &= p_A(V_B - V_A) \\
Q_{BC} &= nC_v(T_{AC} - T_B) \\
W_{BC} &= p(V_B - V_B) = 0
\end{aligned}
\right\} \qquad (17.36)
$$

Substituting all these terms into equation 17.35, gives

$$\Delta U_{ABC} = nC_p(T_B - T_{AC}) - p_A(V_B - V_A) + nC_v(T_{AC} - T_B)$$
$$\Delta U_{ABC} = nC_p(T_B - T_{AC}) - nC_v(T_B - T_{AC}) - p_A V_B + p_A V_A \qquad (17.37)$$

But from the ideal gas equation,

$$p_A V_A = nRT_{AC} \qquad (17.38)$$

and

$$p_A V_B = p_B V_B = nRT_B \tag{17.39}$$

Substituting equations 17.38 and 17.39 back into equation 17.37, we get

$$\Delta U_{ABC} = nC_p(T_B - T_{AC}) - nC_v(T_B - T_{AC}) - nRT_B + nRT_{AC}$$

$$= nC_p(T_B - T_{AC}) - nC_v(T_B - T_{AC}) - nR(T_B - T_{AC})$$

$$\Delta U_{ABC} = (C_p - C_v - R)n(T_B - T_{AC}) \tag{17.40}$$

However, we have already determined that ΔU_{ABC} is equal to zero. Hence, equation 17.40 implies that

$$C_p - C_v - R = 0$$

or

$$C_p - C_v = R \tag{17.41}$$

Thus we have determined a theoretical relation between the molar specific heat capacities and the universal gas constant R. Since it has already been shown that $C_v = \frac{3}{2}R$ in equation 17.29, C_p can now be solved for in equation 17.41 to obtain

$$C_p = C_v + R$$

$$= \tfrac{3}{2}R + R$$

$$C_p = \tfrac{5}{2}R \tag{17.42}$$

Using the value of $R = 1.987$ cal/mole K found in chapter 15, the value of C_p is 4.97 cal/mole K, which agrees with the experimental value of C_p for a monatomic gas.

17.5 Some Special Cases of the First Law of Thermodynamics

Although we have already discussed the first law of thermodynamics pretty thoroughly, let us summarize some of the results into special cases.

An Isothermal Process

An isothermal process is a process that occurs at constant temperature. Thus, $\Delta T = 0$. But $\Delta U = nC_v\Delta T$. Therefore, if $\Delta T = 0$, then $\Delta U = 0$. The first law then becomes

$$\Delta U = 0 = Q - W$$

$$Q = W \tag{17.43}$$

In an isothermal process, heat added to the system shows up as mechanical work done by the system.

An Adiabatic Process

*An **adiabatic process** is a process that occurs without an exchange of heat between the system and its environment.* That is, heat is neither added to nor taken away from the system during the process. Thus, $Q = 0$ in an adiabatic process. The first law of thermodynamics for an adiabatic process becomes

$$\Delta U = Q - W$$

$$W = -\Delta U \tag{17.44}$$

Thus, *in an adiabatic process, the energy for the work done by the gas comes from a loss in the internal energy of the gas.*

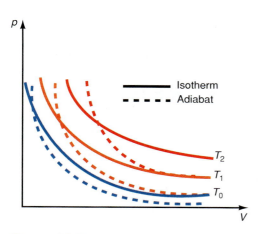

Figure 17.7

Adiabats and isotherms on a *p-V* diagram.

An example of an adiabatic process is the process of cloud formation in the atmosphere, which we will discuss in the section "Have you ever wondered" at the end of this chapter.

Some processes that are not strictly speaking adiabatic can be treated as adiabatic processes because the process occurs so rapidly that there is not enough time for the system to exchange any significant quantities of heat with its environment.

An adiabatic process can be drawn as the dashed line on the *p-V* diagram in figure 17.7. Note that the adiabatic line has a steeper slope than the isotherm. Although the equation for the adiabat cannot be derived without the use of the calculus, we will state the result here for completeness:

$$pV^{\gamma} = \text{constant} \tag{17.45}$$

where γ is equal to the ratio of the molar specific heats. Thus,

$$\gamma = \frac{C_p}{C_v} \tag{17.46}$$

The adiabatic process is essential to the study of the Carnot cycle in section 17.8.

Isochoric Process or Isometric Process

An isochoric process is a process that occurs at constant volume, that is, $\Delta V = 0$. Since the work done, W, is equal to $p\Delta V = 0$, then W must also be zero. The first law of thermodynamics for an isochoric process therefore becomes

$$Q = \Delta U \tag{17.47}$$

Thus, the heat added to a system during an isochoric process shows up as an increase in the internal energy of the system.

An Isobaric Process

An isobaric process is a process that occurs at constant pressure, that is, $\Delta p = 0$. Since the pressure is a constant for an isobaric process, the work done in an isobaric process is given by the product of the constant pressure p and the change in volume ΔV. That is,

$$W = p\Delta V$$

If the process is not an isobaric one then the pressure p has to be an average value of the pressure along the thermodynamic path to give the average amount of work done on that path.

A Cyclic Process

A cyclic process is one that always returns to its initial state. The process studied as *ABCDA* in figure 17.6 is an example of a cyclic process. *Because the system always returns to the original state, ΔU is always equal to zero for a cyclic process.* That is,

$$\Delta U = 0$$

Hence, the first law of thermodynamics for a cyclic process becomes

$$W = Q \tag{17.48}$$

Thus, the work done by the system in the cyclic process is equal to the heat added to the system on a portion of the cycle minus the heat removed on the remainder of the cycle.

17.6 The Gasoline Engine

The thermodynamic system studied so far is somewhat idealistic. In order to be more specific, let us consider the thermodynamic process that occurs in the gasoline engine of an automobile. The engine usually consists of four, six, or eight cylinders.

Figure 17.8

The gasoline engine cycle.

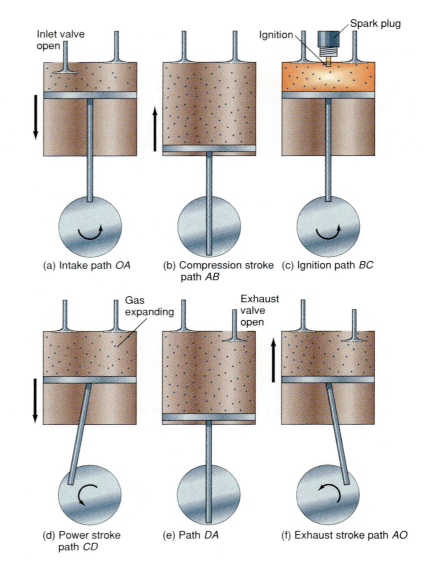

(a) Intake path *OA*

(b) Compression stroke path *AB*

(c) Ignition path *BC*

(d) Power stroke path *CD*

(e) Path *DA*

(f) Exhaust stroke path *AO*

Figure 17.9

The Otto cycle.

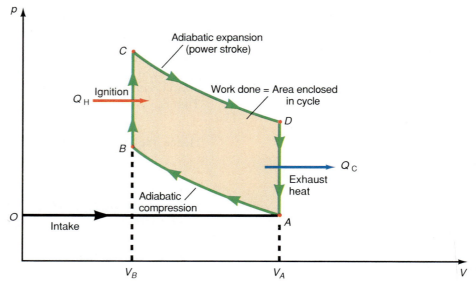

Each cylinder has an inlet valve, an exhaust valve, a spark plug, and a movable piston, which is connected to the crankshaft by a piston rod. The operation of one of these cylinders is shown schematically in figure 17.8. The gasoline engine is approximated by an **Otto cycle** and is shown on the p-V diagram of figure 17.9.

Figure 17.8(a) shows the first stroke of the engine, which is called the intake stroke. The inlet valve opens and a mixture of air and gasoline is drawn into the cylinder as the piston moves downward. Because the inlet valve is open during this first stroke, the air pressure inside the cylinder is the same constant value as atmospheric pressure and is thus shown as the isobaric path OA in figure 17.9. When the cylinder is completely filled to the volume V_A with the air and gasoline mixture, point A, the inlet valve, closes and the compression stroke starts, figure 17.8(b). The piston moves upward very rapidly causing an adiabatic compression of the air-gas mixture. This is shown as the adiabatic path AB in figure 17.9. When the piston is at its highest point (its smallest volume V_B), a spark is applied to the mixture by the spark plug. This spark causes ignition of the air-gas mixture (a small explosion of the mixture), and a great deal of heat is supplied to the mixture by the explosion. This supply of heat is shown as Q_H on the path BC of figure 17.9. The explosion occurs so rapidly that it takes a while to overcome the inertia of the piston to get it into motion. Hence, for this small time period, the pressure and temperature in the cylinder rises very rapidly at approximately constant volume. This is shown as the path BC in figure 17.9. At the point C the force of the air-gas mixture is now able to overcome the inertia of the piston, and the piston moves downward very rapidly during the power stroke, figure 17.8(d). Because the piston moves very rapidly, this portion of the process can be approximated by the adiabatic expansion of the gas shown as CD in figure 17.9. As the piston moves down rapidly this downward motion of the piston is transferred by the piston rod to the crankshaft of the engine causing the crankshaft to rotate. That is, the piston rod is connected off-center to the crankshaft. Thus, when the piston rod moves downward it creates a torque that causes the crankshaft to rotate. The rotating crankshaft is connected by a series of gears to the rear wheels of the car thus causing the wheels to turn and the car to move. At the end of this power stroke the piston has moved down to the greatest volume V_A. At this point D, the exhaust valve of the cylinder opens and the higher pressure at D drops very rapidly to the outside pressure at A, and a good deal of heat Q_C is exhausted out through the exhaust valve. As the piston now moves upward in figure 17.8(f) all the remaining used gas-air mixture is dumped out through the exhaust valve. This is shown as the path AO in figure 17.9. At the position O, the exhaust valve closes and the inlet valve opens allowing a new mixture of air and gasoline to enter the cylinder. The process now starts over again as the same cycle $OABCDAO$ of figure 17.9. The net result of the entire cycle is that heat Q_H is added along path BC, work is done equal to the area enclosed by the cyclic path, and heat Q_C is exhausted out of the system. Thus, heat has been added to the system and the system performed useful work. Four, six, or eight of these cylinders are ganged together with the power stroke of each cylinder occurring at a different time for each cylinder. This has the effect of smoothing out the torque on the crankshaft, causing a more constant rotation of the crankshaft. Unfortunately practical limitations, such as compression ratio, friction, cooling, and so on, cause the efficiency of the gasoline engine, which uses the Otto cycle, to be limited to about 20% to 25%.

17.7 The Ideal Heat Engine

There are many heat engines in addition to the gasoline engine, but they all have one thing in common: *every engine absorbs heat from a source at high temperature, performs some amount of mechanical work, and then rejects some heat at a lower temperature.* This process can be visualized with the schematic diagram for an **ideal heat engine,** and is shown in figure 17.10. The engine is represented by the circle in the diagram. The engine absorbs the quantity of heat Q_H from a hot-temperature reservoir, at a temperature T_H. (In the gasoline engine, the quantity of heat Q_H was

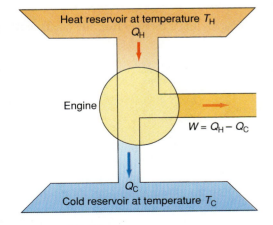

Figure 17.10

An idealized heat engine.

supplied by the combustion of the air-gasoline mixture.) Some of this absorbed heat energy is converted to work, which is shown as the pipe coming out of the engine at the right. This corresponds to the work done during the power stroke of the gasoline engine. The rest of the original absorbed heat energy is dumped as exhaust heat Q_C into the low-temperature reservoir. (In the gasoline engine this is the hot exhaust gas that is rejected to the cooler environment outside the engine.)

Because the engine operates in a cycle, $\Delta U = 0$, and as we have already seen, the net work done is equal to the net heat absorbed by the engine, that is,

$$W = Q$$

But the net heat absorbed is equal to the difference between the total heat absorbed Q_H at the hot reservoir, and the heat rejected Q_C at the cold reservoir, that is,

$$Q = Q_H - Q_C$$

Thus, the work done by the engine is equal to the difference between the heat absorbed from the hot reservoir and the heat rejected to the cold reservoir

$$W = Q_H - Q_C \tag{17.49}$$

The efficiency of a heat engine can also be defined from equation 17.34 as

$$\text{Eff} = \frac{W}{Q_{in}} = \frac{W}{Q_H} = \frac{Q_H - Q_C}{Q_H} \tag{17.50}$$

$$\text{Eff} = 1 - \frac{Q_C}{Q_H} \tag{17.51}$$

Thus, to make any heat engine as efficient as possible it is desirable to make Q_H as large as possible and Q_C as small as possible. It would be most desirable to have $Q_C = 0$, then the engine would be 100% efficient. Note that this would not be a violation of the first law of thermodynamics. However, as we will see in section 17.8, such a process is not possible.

Before leaving this section we should note that a **refrigerator,** or a heat pump, is a heat engine working in reverse. A refrigerator is represented schematically in figure 17.11, where the refrigerator is represented as the circle in the diagram. Work W is done on the refrigerator, thereby extracting a quantity of heat Q_C from the low-temperature reservoir and exhausting the large quantity of heat Q_H to the hot reservoir. The total heat energy exhausted to the high-temperature reservoir Q_H is the sum of the work done on the engine plus the heat Q_C extracted from the low-temperature reservoir. Thus,

$$Q_H = W + Q_C$$

We define the equivalent of an efficiency for a refrigerator, the coefficient of performance, as

$$\text{Coefficient of performance} = \frac{\text{Heat removed}}{\text{Work done}}$$

$$\text{Coefficient of performance} = \frac{Q_C}{W} \tag{17.52}$$

$$\text{Coefficient of performance} = \frac{Q_C}{Q_H - Q_C} \tag{17.53}$$

17.8 The Carnot Cycle

As we saw in section 17.7, it is desirable to get the maximum possible efficiency from a heat engine. Sadi Carnot (1796–1832) showed that the maximum efficiency of any heat engine must follow a cycle consisting of the isothermal and adiabatic paths shown in the p-V diagram in figure 17.12, and now called the **Carnot cycle.** The cycle begins at point A. Let us now consider each path individually.

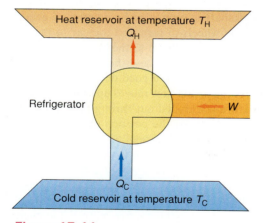

Figure 17.11

An ideal refrigerator.

Thermodynamics

Figure 17.12

A *p-V* diagram for a Carnot cycle.

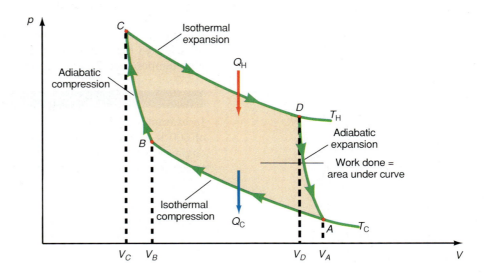

Path AB: An ideal gas is first compressed isothermally along the path *AB*. Since *AB* is an isotherm, $\Delta T = 0$ and hence $\Delta U = 0$. The first law therefore says that $Q = W$ along path *AB*. That is, the work W_{AB} done on the gas is equal to the heat removed from the gas Q_C, at the low temperature, T_C.

Path BC: Path *BC* is an adiabatic compression and hence $Q = 0$ along this path. The first law therefore becomes $\Delta U = W$. That is, the work W_{BC} done on the gas during the compression is equal to the increase in the internal energy of the gas as the temperature increases from T_C to T_H.

Path CD: Path *CD* is an isothermal expansion. Hence, $\Delta T = 0$ and $\Delta U = 0$. Therefore, the first law becomes $W = Q$. That is, the heat added to the gas Q_H at the high temperature T_H is equal to the work W_{CD} done by the expanding gas.

Path DA: Path *DA* is an adiabatic expansion, hence $Q = 0$ along this path. The first law becomes $\Delta U = W$. Thus, the energy necessary for the work W_{DA} done by the expanding gas comes from the decrease in the internal energy of the gas. The gas decreases in temperature from T_H to T_C.

The net effect of the Carnot cycle is that heat Q_H is absorbed at a high temperature T_H, mechanical work W is done by the engine, and waste heat Q_C is exhausted to the low-temperature reservoir at a temperature T_C. The net work done by the Carnot engine is

$$W = Q_H - Q_C$$

The efficiency is given by the same equations 17.50 and 17.51 as we derived before. That is,

$$\text{Eff} = 1 - \frac{Q_C}{Q_H} \tag{17.51}$$

Lord Kelvin proposed that the ratio of the heat rejected to the heat absorbed could serve as a temperature scale. Kelvin then showed that for a Carnot engine

$$\frac{Q_C}{Q_H} = \frac{T_C}{T_H} \tag{17.54}$$

where T_C and T_H are the Kelvin or absolute temperatures of the gas. With the aid of equation 17.54, we can express the efficiency of a Carnot engine as

$$\text{Eff} = 1 - \frac{T_C}{T_H} \tag{17.55}$$

The importance of equation 17.55 lies in the fact that the Carnot engine is the most efficient of all engines. If the efficiency of a Carnot engine can be determined, then the maximum efficiency possible for an engine operating between the high temperature T_H and the low temperature T_C is known.

Example 17.5

In examples 17.3 and 17.4 the engine operated between a maximum temperature of 600 K and a minimum temperature of 150 K. The efficiency of that particular engine was 15%. What would the efficiency of a Carnot engine be, operating between these same temperatures?

Solution

The efficiency of the Carnot engine, found from equation 17.55, is

$$\text{Eff} = 1 - \frac{T_C}{T_H} = 1 - \frac{150 \text{ K}}{600 \text{ K}} = 0.75$$

Therefore, the maximum efficiency for any engine operating between these temperatures cannot be higher than 75%. Obviously the efficiency of 15% for the previous cycle is not very efficient.

17.9 The Second Law of Thermodynamics

There are several processes that occur regularly in nature, but their reverse processes never occur. For example, we can convert the kinetic energy of a moving car to heat in the brakes of the car as the car is braked to a stop. However, we cannot heat up the brakes of a stopped car and expect the car to start moving. That is, we cannot convert the heat in the brakes to kinetic energy of the car. Thus, mechanical energy can be converted into heat energy but heat energy cannot be completely converted into mechanical energy. As another example, a hot cup of coffee left to itself always cools down to room temperature, never the other way around. There is thus a kind of natural direction followed by nature. That is, processes will proceed naturally in one direction, but not in the opposite direction. Yet in any of these types of processes there is no violation of the first law of thermodynamics regardless of which direction the process occurs. This unidirectionality of nature is expressed as **the second law of thermodynamics** and tells which processes will occur in nature. The second law will first be described in terms of the ideal heat engine and refrigerator studied in section 17.7.

The Kelvin-Planck Statement of the Second Law

No process is possible whose sole result is the absorption of heat from a reservoir at a single temperature and the conversion of this heat energy completely into mechanical work. This statement is shown schematically in figure 17.13. That is, the diagram in figure 17.13 cannot occur in nature. Observe from figure 17.13 that heat Q_H is absorbed from the hot reservoir and converted completely into work. In figure 17.10 we saw that there had to be an amount of heat Q_C exhausted into the cold reservoir. Thus the **Kelvin-Planck statement of the second law of thermodynamics** says that there must always be a quantity of heat Q_C exhausted from the engine into a lower temperature reservoir.

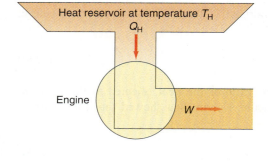

Figure 17.13

Kelvin-Planck violation of the second law.

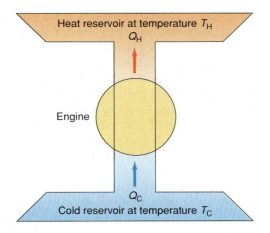

The Clausius Statement of the Second Law of Thermodynamics

No process is possible whose sole result is the transfer of heat from a cooler to a hotter body. The **Clausius statement of the second law of thermodynamics** can best be described by the refrigerator of figure 17.11. Work was done on the refrigerator to draw heat Q_C out of the cold reservoir to then deliver it to the hot reservoir. The Clausius statement says that work must always be done to do this. The violation of this Clausius statement of the second law is shown in figure 17.14. This statement of the second law of thermodynamics is essentially an observation of nature. Thermal energy flows from hot reservoirs (hot bodies) to cold reservoirs (cold bodies). The reverse process where heat flows from a cold body to a hot body without the application of some kind of work does not occur in nature. Thus, the second law of thermodynamics says that such processes are impossible, and the diagram in figure 17.14 cannot occur in nature.

17.10 Entropy

The second law of thermodynamics has been described in terms of statements about which processes are possible and which are not possible. It would certainly be more desirable to put the second law on a more quantitative basis. In 1865, Clausius introduced the concept of entropy to indicate what processes are possible and what ones are not. When a thermodynamic system changes from one equilibrium state to another in a series of small increments such that the system always moves through a series of equilibrium states, the system is said to go through a *reversible process*. A reversible process can be drawn as a continuous line on a *p-V* diagram. All the processes that have been considered are reversible processes. When a thermodynamic system changes from one equilibrium state to another along a reversible path, there is a change in entropy, ΔS of the system given by

$$\Delta S = \frac{\Delta Q}{T} \tag{17.56}$$

where ΔQ is the heat added to the system, and T is the absolute temperature of the system.

Example 17.6

Find the change in entropy when 5.00 kg of ice at 0.00 °C are converted into water at 0.00 °C.

Solution

The heat absorbed by the ice in melting is found from

$$\Delta Q = mL_f = (5.00 \text{ kg})(80.0 \text{ kcal/kg}) = 400 \text{ kcal}$$

The process takes place at 0 °C which is equal to 273 K. The change in entropy, found from equation 17.56, is

$$\Delta S = \frac{\Delta Q}{T}$$
$$= \frac{400 \text{ kcal}}{273 \text{ K}}$$
$$= 1.47 \text{ kcal/K}$$

Whenever heat is added to a system, ΔQ is positive, and hence, ΔS is also positive. If heat is removed from a system, ΔQ is negative, and therefore, ΔS is also negative. When the ice melts there is a positive increase in entropy.

Entropy is a very different concept than the concept of energy. For example, in a gravitational system, a body always falls from a region of high potential energy to low potential energy, thereby losing potential energy. In contrast, *in an isolated thermodynamic system, the system always changes from values of low entropy to values of high entropy, thereby increasing the entropy of the system. Therefore, the concept of entropy can tell us in which direction a process will proceed.* For example, if an isolated thermodynamic system is in a state A, and we wish to determine if it can naturally go to state B by itself, we first measure the initial value of the entropy at A, S_i, and the final value of the entropy at B, S_f. The system will move from A to B only if there is an increase in the entropy in moving from A to B. That is, the process is possible if

$$\Delta S = S_f - S_i > 0 \tag{17.57}$$

If ΔS is negative for the proposed process, the system will not proceed to the point B. *The second law of thermodynamics can also be stated as: the entropy of an isolated system increases in every natural process, and only those processes are possible for which the entropy of the system increases or remains a constant.* The entropy of a nonisolated system may either increase, or decrease, depending on whether heat is added to or taken away from the system. If ΔQ is equal to zero, such as in an adiabatic process, then ΔS also equals zero. Hence, an adiabatic process is also an *isoentropic process.* Just as the change in internal energy of a system from state A to state B is independent of the path taken to get from A to B, the entropy of a system is also independent of the path taken.

Note from the form of equation 17.56 that the temperature T must be a constant. If the temperature is not a constant, as is the case in most processes, the calculus must be used to evaluate the entropy of the system. In some cases an average temperature of the system can be used in equation 17.56 to evaluate the entropy.

Example 17.7

Find the change in entropy when 5.00 kg of ice at $-5.00\ °C$ is warmed to $0.00\ °C$.

Solution

The heat added to the ice is found from

$$\Delta Q = mc\Delta T = mc(T_f - T_i)$$

$$= (5.00\ \text{kg})\left(0.500\ \frac{\text{kcal}}{\text{kg °C}}\right)[0\ °C - (-5.00\ °C)]$$

$$= 12.5\ \text{kcal}$$

We can use equation 17.56 to evaluate the change in entropy of the ice if an average temperature of $-2.50\ °C = 270.5\ K$ is used. Thus,

$$\Delta S = \frac{\Delta Q}{T} \tag{17.56}$$

$$= \frac{12.5\ \text{kcal}}{270.5\ K}$$

$$= 0.0462\ \text{kcal/K}$$

Example 17.8

Find the change in entropy when 5.00 kg of ice at -5.00 °C are converted to water at 0.00 °C.

We can find the change in entropy by dividing the problem into two parts. First, we find the change in entropy in warming the ice to 0.00 °C and then we find the change in entropy in melting the ice. We have already found the change in entropy for these two processes in examples 17.6 and 17.7. The total change in entropy is the sum of the change in entropy for the two processes. Therefore,

$$\Delta S = \Delta S_1 + \Delta S_2$$
$$= 0.0462 \text{ kcal/K} + 1.47 \text{ kcal/K}$$
$$= 1.52 \text{ kcal/K}$$

†17.11 Statistical Interpretation of Entropy

As we have seen in sections 17.9 and 17.10, the second law of thermodynamics is described in terms of statements about which processes in nature are possible and which are not possible. Clausius introduced the concept of entropy to put the second law on a more quantitative basis. *He stated the second law as: the entropy of an isolated system increases in every natural process, and only those processes are possible for which the entropy of the system increases or remains a constant.* But this analysis was done on a macroscopic level, that is, a large-scale level, where concepts of temperature, pressure, and volume were employed. But the gas, the usual working substance discussed, is made up of billions of molecules, as shown in the kinetic theory of gases. Ludwig Boltzmann's approach to the second law of thermodynamics is a further extension of the kinetic theory, and is called *statistical mechanics*. Boltzmann looked at the molecules of the gas and asked what the most probable distribution of these molecules is. There is a certain order to the distribution of the molecules, with some states more probable than others. Thus, statistical mechanics deals with probabilities.

As an example, let us consider the gas molecules in figure 17.15(a). The molecules are contained in the left-hand side of a box by a partition located in the center of the box. When the partition is removed some of the molecules move to the right-hand side of the box until an equilibrium condition is reached whereby there are the same number of molecules in both sides of the box, figure 17.15(b). We now ask, can all the gas molecules in the entire box of figure 17.15(b) move to the left and be found in the original state shown in figure 17.15(a)? We know from experience that this never happens. This would be tantamount to all the gas molecules in the room that you are now sitting in moving completely to the other side of the room, leaving you in a vacuum. This just does not happen in life. However, if it did it would not violate the first law of thermodynamics. But the second law says some processes do not occur. This is certainly one of them. Notice that the first case in which all the molecules are in the left-hand side of the box is more orderly than the second case where the molecules are distributed over the entire box. (If the volume of the box is larger, there are more random paths for the molecules to follow and hence more disorder.)

As another example of order and disorder, let us drop a piece of clay. When the clay is dropped, superimposed over the thermal motion of the molecules of the clay is the velocity of the clay toward the ground. That is, all the molecules have a motion toward the ground, which is an ordered motion. When the clay hits the ground

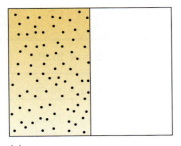
(a)

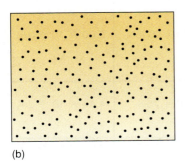
(b)

Figure 17.15

Gas molecules in a partitioned box.

and sticks to it, the kinetic energy of the falling molecules shows up as thermal energy of the clay molecules, which is a random or disordered motion of the molecules. Hence there is a transformation from order to disorder in the natural process of a collision of a falling object. Now as we know, the clay cannot gather together all the random thermal motion of the clay molecules and convert them to ordered translational motion upward, and hence the clay by itself cannot move upward. Thus the concept of which processes can occur in nature can also be measured by the amount of order or disorder between the initial and final states of the system. *Using the concept of order, the second law of thermodynamics can also be stated as: an isolated system in a state of relative order will always pass to a state of relative disorder until it reaches the state of maximum disorder, which is thermal equilibrium.*

Let us return to our example of the gas molecules in a box. Although normally there are billions of molecules in the box, to simplify our discussion let us assume that there are only four molecules present. They are numbered consecutively 1, 2, 3, and 4. Let us ask how many ways we can distribute these four molecules between the left- and right-hand sides of the box. First we could place the four molecules all in the left-hand side of the box as shown in table 17.1. Thus there is only one way we can place the four molecules into the left-hand side of the box. *Let us designate the number of ways that the four molecules can be distributed as N_i* and note that $N_1 = 1$. Next we see how many ways we can place one molecule in the right-hand side of the box and three in the left-hand side. That is, first we place molecule 1 in the right-hand side of the box, and see that that leaves molecules 2, 3, 4, in the left-hand side. Then we place molecule 2 in the right-hand side and see that we then have molecules 1, 3, 4 in the left-hand side. Continuing in this way we see from the table that there are four ways to do this. Thus, we designate the number of ways we can place one molecule in the right-hand side of the box and three in the left-hand side as N_2 and see that this is equal to 4. Next we see how many ways

Table 17.1

Possible Distributions of Four Molecules in a Box

Left	Right	N_i	$P_i = N_i/N$	$S = k \ln P$
1 2 3 4		1	1/16 = 6.25%	2.53×10^{-23} J/K

2 3 4
1   3 4
1 2   4
1 2 3
	1
	  2
	    3
	      4
		4	4/16 = 25%	4.44×10^{-23} J/K

1 2
1   3
  2 3
1   4
  2   4
    3 4
	3 4
	  2 4
	1 4
	  2 3
	1 3
	1 2
		6	6/16 = 37.5%	5.00×10^{-23} J/K

1
  2
    3
      4
	2 3 4
	1 3 4
	1 2 4
	1 2 3
		4	4/16 = 25%	4.44×10^{-23} J/K

	1 2 3 4	1	1/16 = 6.25%	2.53×10^{-23} J/K

$\Sigma N_i = N = 16 \, ; \, \Sigma P_i = 100\%$

we can place two molecules in the right-hand side of the box and two in the left-hand side. From the table we see that there are six ways, and we call this $N_3 = 6$. Next we see how many ways we can place one molecule in the left-hand side of the box and three in the right-hand side. Again from the table we see that there are four ways to do this, and we call this $N_4 = 4$. Finally we ask how many ways can the four molecules be placed in the right-hand side of the box and again we see from the table there is only one way. We call this $N_5 = 1$. Thus there are five possible ways (for a total of 16 possible states) that the four molecules could be distributed between the left- and right-hand sides of the box.

But which of all these possibilities is the most probable? *The probability that the molecules are in the state that they are in compared with all the possible states they could be in is given by*

$$P = \frac{N_i}{N} \times 100\%$$

where N_i is the number of states that the molecules could be in for a particular distribution and N is the total number of possible states. As we see from the table, there are 16 possible states that the four molecules could be in. Hence the probability that the molecules are in the state where all four are on the left-hand side is

$$P = \frac{N_1}{N} \times 100\% = \frac{1}{16} \times 100\% = 6.25\%$$

That is, there is a 6.25% probability that all four molecules will be found in the left-hand side of the box.

The probability that the distribution of the four molecules has three molecules in the left-hand side and one in the right-hand side is found by observing that there are four possible ways that the molecules can be distributed and hence $N_2 = 4$. Therefore,

$$P = \frac{N_2}{N} \times 100\% = \frac{4}{16} \times 100\% = 25\%$$

Thus there is a 25% probability that there are three molecules in the left-hand side and one molecule in the right-hand side. Continuing in this way the probabilities that the molecules will have the particular distribution is shown in table 17.1. Thus there is a 37.5% probability that the distribution has two molecules in each half of the box, a 25% probability that the distribution has one molecule in the left half of the box and three in the right half of the box, and finally a 6.25% probability that the distribution has no molecules in the left half of the box and four in the right half of the box.

Notice that the first and last distributions (all molecules either on the left side or on the right side), are the most ordered and they have the lowest probability, 6.25%, for the distribution of the molecules. Also notice that the third distribution where there are two molecules on each side of the box has the greatest disorder and also the highest probability that this is the way that the molecules will be distributed. Notice that the distribution with the greatest possible number of states gives the highest probability. These ideas led Boltzmann to define the entropy of a state as

$$S = k \ln P \tag{17.58}$$

where k is a constant, that later turned out to be the *Stefan-Boltzmann constant*, which is equal to 1.38×10^{-23} J/K; ln is the natural logarithm; and P is the probability that the system is in the state specified. Thus in our example, the entropy of the first distribution is computed as

$$S = k \ln P = (1.38 \times 10^{-23} \text{ J/K})(\ln 6.25)$$
$$= 2.53 \times 10^{-23} \text{ J/K}$$

The entropy for each possible distribution is computed and shown in table 17.1. Notice that the most disordered state (two molecules on each side of the box) has the highest value of entropy. If we were to start the system with the four molecules in the left-hand box, entropy = 2.53×10^{-23} J/K, the system would move in the direction of maximum entropy, 5.00×10^{-23} J/K, the state with two molecules on each side of the box. As before, natural processes move in the direction of maximum entropy. The actual values of the computed entropy for this example are extremely small, because we are dealing with only four molecules. If we had only one mole of a gas in the box we would have 6.02×10^{23} molecules in the box, an enormous number compared with our four molecules. In such a case the numerical values of the entropy would be much higher. However there would still be the same type of distributions. The state with the greatest disorder, the same number of molecules on each side of the box, would be the state with the greatest value of the entropy. The state with all the molecules on one side of the box would have a finite but vanishingly small value of probability.

Hence the original problem we stated in figure 17.15(a), with the gas in the left partition has the smallest entropy while the gas on both sides of the box in figure 17.15(b), has the greatest entropy. The process flows from the state of lowest entropy to the one of highest entropy. It is interesting to note that it is not impossible for the gas molecules on both sides of the box to all move to the left-hand side of the box, but the probability is so extremely small that it would take a time greater than the age of the universe for it to happen. Hence, effectively it will not happen.

The state of maximum entropy is the state of maximum disorder and is the state where all the molecules are moving in a completely random motion. This state is, of course, the state of thermal equilibrium. We have seen throughout our study of heat that whenever two objects at different temperatures are brought together, the hot body will lose thermal energy to the cold body until the hot and cold body are at the same equilibrium temperature. Thus, as all bodies tend to equilibrium they all approach a state of maximum entropy. Hence, the universe itself tends toward a state of maximum entropy, which is a state of thermal equilibrium of all the molecules of the universe. This is a state of uniform temperature and density of all the atoms and molecules in the universe. No physical, chemical, or biological processes would be capable of occurring, however, because a state of total disorder cannot do any additional work. This ultimate state of the universe is sometimes called the *heat death of the universe*.

One final thought about entropy, and that is the idea of a direction for time. All the laws of physics, except for the second law of thermodynamics, are invariant to a change in the direction of time. That is, for example, Newton's laws of motion would work equally well if time were to run backward. If a picture were taken of a swinging pendulum with a video camera, and then played first forward and then backward, we could not tell from the picture which picture is running forward in time and which is running backward in time. They would both appear the same. On the other hand, if we take a video of a dropped coffee cup that hits the ground and shatters into many pieces we can certainly tell the difference between running the video forward or backward. When the video is run backward we would see a shattered coffee cup on the floor come together and repair itself and then jump upward onto the table. Nature does not work this way, so we know the picture must be running backward. Now before the coffee cup is dropped, we have a situation of order. When the cup is dropped and shattered we have a state of disorder. Since natural processes run from a state of order (low entropy) to one of disorder (high entropy), we can immediately see the time sequence that must be followed in the picture. The correct sequence to view the video picture is to start where the coffee cup has its lowest entropy (on the table initially at rest) and end where the cup has its maximum entropy (on the floor broken into many pieces). Hence the concept of entropy gives us a direction for time. In any natural process, the initial state has the lowest entropy, and the final state has the greatest entropy. *Thus time flows in the direction of the increase in entropy.* Stated another way, the past is the state of lowest entropy and the future is the state of highest entropy. *Thus entropy is sometimes said to be time's arrow, showing its direction.* Because of this, there have been many speculative ideas attributed to time. What happens to time when the universe reaches its state of maximum entropy? Since time flows from low entropy to high entropy, what happens when there is no longer a change in entropy? Would there be an infinite present? An eternity?

"Have you ever wondered . . . ?"

An Essay on the Application of Physics

Meteorology—The Physics of the Atmosphere

Have you ever wondered, while watching the weather forecast on your local TV station, what all those lines and arrows were on those maps? It looked something like figure 1.

If we were to look at the television screen more closely we would see a map of the United States. At every weather station throughout the United States, the atmospheric pressure is measured and recorded on a weather map. On that map, a series of lines, connecting those pressures that are the same, are drawn. These lines are called *isobars* and can be seen in figure 2. An isobar is a line along which the pressure is constant. An isobar is analogous to a contour line that is drawn on a topographical map to indicate a certain height above mean sea level. As an example, consider the mountain and valley shown in figure 3(a). A series of contour lines are drawn around the mountain at constant heights above sea level. The first line is drawn at a height $H = 200$ m above sea level. Everywhere on this line the height is exactly 200 m above sea level. The next contour line is drawn at $H = 400$ m. Everywhere on this line the height is exactly 400 m above sea level. Between the 200 m contour and the 400 m contour line the height varies between 200 m and 400 m. The contour line for 600 m is also drawn in the figure. The very top of the mountain is greater than the 600 m and is the highest point of the mountain. The contour lines showing the valley are drawn at -200 m, -400 m, and -600 m. The -200 m contour line shows that every point on this line is 200 m below sea level. The bottom of the valley is the lowest point in the valley. If we were to look down on the mountain and valley from above, we would see a series of concentric circles representing the contour lines as they are shown in figure 3(b). (On a real mountain and valley the contours would probably not be true circles.)

The isobars are to a weather map as contour lines are to a topographical map. The isobars represent the pressure of the atmosphere. By drawing the isobars, a picture of the pressure field is obtained. Normal atmospheric pressure is 29.92 in. of Hg or 1013.25 mb. But remember that normal is an average of abnormals. At any given time, the pressure in the atmosphere varies slightly from this normal value. If the atmospheric pressure is

GRIN & BEAR IT

"And as you can see in the Midwest, we've got one arrow sticking in the side of another arrow."

Figure 1

Your local TV weatherman. Reprinted with special permission of North America Syndicate, Inc.

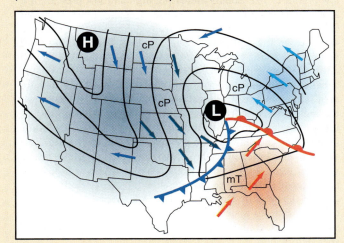

Figure 2

A weather map. Reprinted with special permission of North America Syndicate, Inc.

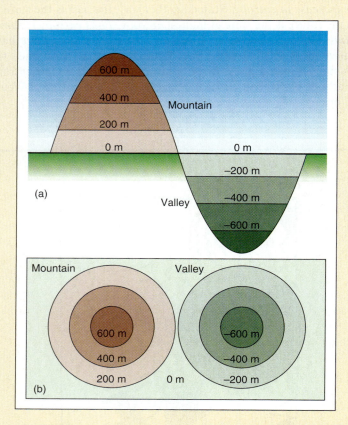

Figure 3

Contour lines on a topographical map.

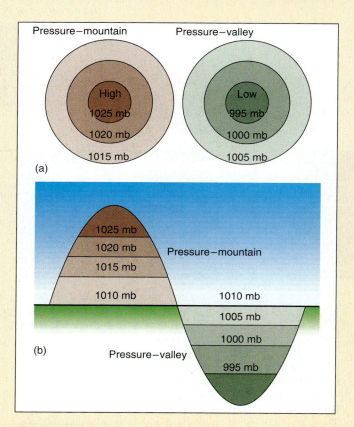

Figure 4

High and low atmospheric pressure.

greater than normal at your location, then you are in a region of high pressure. If, on the other hand, the atmospheric pressure is less than normal at your location, then you are in a region of low pressure. The isobars indicating high and low pressure are shown in figure 4(a). The high-pressure region can be visualized as a mountain and the low-pressure region as a valley in figure 4(b). Air in the high-pressure region flows down the pressure mountain into the low-pressure valley, just as a ball would roll down a real mountain side into the valley below. This flow of air is called *wind*. Hence, air always flows out of a high-pressure area into a low-pressure area. The force on a ball rolling down the mountain is the component, acting down the mountain, of the gravitational force on the ball. The force on a parcel of air is caused by the difference in pressure between the higher pressure and the lower pressure. This force is called the *pressure gradient force* (PGF) per unit mass, and it is directed from the high-pressure area to the low-pressure area. It is effectively the slope of the pressure mountain-valley. A large pressure gradient, corresponding to a steep slope, causes large winds, whereas a small pressure gradient, corresponding to a shallow slope, causes very light winds.

If the earth were not rotating, the air would flow perpendicular to the isobars. However, the earth does rotate, and the rotation of the earth causes air to be deflected to the right of its

original path. The deflection of air to the right of its path in the northern hemisphere is called the *Coriolis effect*. The Coriolis effect arises because the rotating earth is not an inertial coordinate system. For small-scale motion the rotating earth approximates an inertial coordinate system. However, for large-scale motion, such as the winds, the effect of the rotating earth must be taken into account. It is taken into account by assuming that there is a fictitious force, called the *Coriolis force* (CF) that acts to the right of the path of a parcel of air in its motion through the atmosphere. The equation for the Coriolis force is

$$\text{CF} = 2v\Omega \sin \phi \qquad \textbf{(17H.1)}$$

where CF is the Coriolis force per unit mass of air, v is the speed of the wind, Ω is the angular velocity of the earth, and ϕ is the latitude. Thus, the Coriolis force depends on the speed of the air (the greater the speed the greater the force) and the latitude angle ϕ. At the equator, $\phi = 0$ and $\sin \phi = 0$, and hence there is no force of deflection at the equator. For $\phi = 90°$, $\sin \phi = 1$, hence the maximum force and deflection occur at the pole.

Let us describe the motion of the air as it moves toward the low-pressure area. The air starts on its motion at the point A, figure 5(a), along a path that is perpendicular to the isobars. But the air is deflected to the right of its path by the Coriolis force, and ends up at the position B. At B, the pressure gradient force

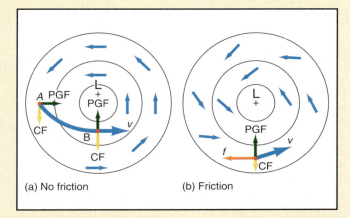

Figure 5

A low-pressure area.

(a) No friction   (b) Friction

is still acting toward the center of the low-pressure area, while the Coriolis force, acting to the right of the path, is opposite to the pressure gradient force. An approximate balance[1] exists between the two forces and the air parcel now moves parallel to the isobars. Notice that the air moves counterclockwise in a low-pressure area.

As the air moves over the ground, there is a frictional force f that acts on the air, is directed opposite to the direction of motion of the air, and is responsible for the slowing down of the air. This is shown in figure 5(b). But, as seen in equation 17H.1, the Coriolis force is a function of the wind speed. If the wind speed decreases because of friction, the Coriolis force also decreases.

---

1. A more detailed analysis by Newton's second law would give

$$a = \frac{F}{M} = PGF + CF$$

Since the air parcel is moving in a circle of radius r, with a velocity v, the acceleration is the centripetal acceleration given by v^2/r. Hence Newton's second law should be written as

$$\frac{v^2}{r} = PGF + CF$$

But in very large scale motion, such as over a continent, $v^2/r \approx 0.1 \times 10^{-3}$ m/s^2, while the PGF $\approx 1.1 \times 10^{-3}$ m/s^2. Thus the centripetal acceleration is about 1/10 of the acceleration caused by the pressure gradient force, and in this simplified analysis is neglected. The second law then becomes

$$0 = PGF + CF$$

or

$$PGF = -CF$$

Hence the force on the air parcel is balanced between the pressure gradient force and the Coriolis force. The wind that results from the balance between the PGF and the CF is called the *geostrophic wind*. For a more accurate analysis and especially in smaller sized pressure systems such as hurricanes and tornadoes this assumption cannot be made and the centripetal acceleration must be taken into account.

Hence, there is no longer the balance between the pressure gradient force and the Coriolis force and the air parcel now moves toward the low-pressure area. The combined result of the pressure gradient force, the Coriolis force, and the frictional force, causes the air to spiral into the low-pressure area, as seen in figure 5(b).

The result of the above analysis shows that air spirals counterclockwise into a low-pressure area at the surface of the earth. But where does all this air go? It must go somewhere. The only place for it to go is upward into the atmosphere. Hence, there is vertical motion upward in a low-pressure area.

Now recall from chapter 13 that the pressure of the air in the atmosphere decreases with altitude. Hence, when the air rises in the low-pressure area it finds itself in a region of still lower pressure aloft. Therefore, the rising air from the surface expands into the lower pressure aloft. But as seen in this chapter, for a gas to expand the gas must do work. Since there is no heat added to, or taken away from this rising air, $\Delta Q = 0$, and the air is expanding adiabatically. But as just shown in equation 17.44, the work done in the expansion causes a decrease in the internal energy of the gas. Hence, the rising air cools as it expands because the energy necessary for the gas to expand comes from the internal energy of the gas itself. Hence the temperature of the air decreases as the air expands and the rising air cools.

The amount of water vapor in the air is called humidity. The maximum amount of water vapor that the air can hold is temperature dependent. That is, at high temperatures the air can hold a large quantity of water vapor, whereas at low temperatures it can only hold a much smaller quantity. If the rising air cools down far enough it reaches the point where the air has all the water vapor it can hold. At this point the air is said to be saturated and the relative humidity of the air is 100%. If the air continues to rise and cool, it cannot hold all this water vapor. Hence, some of the water vapor condenses to tiny drops of water. These drops of water effectively float in the air. (They are buoyed up by the rising air currents.) The aggregate of all these tiny drops of water suspended in the air is called a cloud. Hence, clouds are formed when the rising air is cooled to the condensation point. If the rising and cooling continue, more and more water vapor condenses until the water drops get so large that they fall and the falling drops are called rain. *In summary, associated with a low-pressure area in the atmosphere is rising air. The cooling of this adiabatically expanding air causes the formation of clouds, precipitation, and general bad weather.* Thus, when the weatherman says that low pressure is moving into your area, as a general rule, you can expect bad weather.

Everything we said about the low-pressure area is reversed for a high-pressure area. The pressure gradient force points away from the high-pressure area. As the air starts out of the high-pressure area at the point A, figure 6(a), it is moving along a path that is perpendicular to the isobars. The Coriolis force now acts on the air and deflects it to the right of its path. By the time the air reaches the point B, the pressure gradient force is approximately balanced by the Coriolis force,[2] and the air moves

---

2. The same approximation for the balance between the PGF and the CF used in the analysis of the low-pressure area is also made for the high-pressure area.

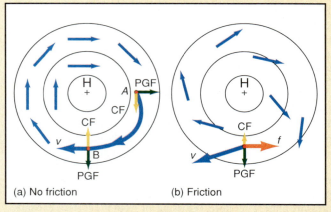

Figure 6

A high-pressure area.

(a) No friction (b) Friction

parallel to the isobars. Thus, the air flows clockwise around the high-pressure area. The frictional force slows down the air and causes the Coriolis force to decrease in size. The pressure gradient force is now greater than the Coriolis force, and the air starts to spiral out of the high-pressure area, figure 6(b).

From what we have just seen, air spirals out of a high-pressure area at the surface of the earth. But if all the air that was in the high-pressure area spirals out, what is left within the high-pressure area? If the air is not replenished, the area would become a vacuum. But this is impossible. Therefore, air must come from somewhere to replenish the air spiraling out of the high. The only place that it can come from is from the air aloft. That is, air aloft moves downward into the high-pressure area at the surface. Thus, there is vertical motion downward in a high-pressure area.

As the air aloft descends, it finds itself in a region of still higher pressure and is compressed adiabatically. Thus, work is done on the gas by the atmosphere and this increase in energy shows up as an increase in the internal energy of the air, and hence an increase in the temperature of the descending air. Thus, the air warms up adiabatically as it descends. Because warmer air can hold more water vapor than colder air, the water droplets that made up the clouds evaporate into the air. As more and more air descends, more and more water droplets evaporate into the air until any clouds that were present have evaporated, leaving clear skies. *Hence, high-pressure areas are associated with clear skies and, in general, good weather.* So when the weatherman tells you that high pressure is moving into your area, you can usually expect good weather.

Now when you look at your TV weather map, look for the low- and high-pressure areas. If the low-pressure area is moving into your region, you can expect clouds and deteriorating weather. If the high-pressure area is moving into your region, you can expect improving weather with clear skies.

Those other lines on the weather map are called *fronts*. A front is a boundary between two different air masses. An air mass is a large mass of air having uniform properties of temperature and moisture throughout the horizontal. Air sitting over the vast

regions of Canada has the characteristic of being cold and dry. This air mass is called a *continental polar air mass* and is designated as a cP air mass. Air sitting over the southern ocean areas and the Gulf of Mexico has the characteristic of being hot and humid. This air mass is called a maritime tropical, mT, air mass. These two air masses interact at what is called the *polar front*. Much of your weather is associated with this polar front. If the continental polar air mass is moving forward, the polar front is called a *cold front*. On a weather map the cold front is shown either as a blue line or, if the presentation is in black and white, a black line with little triangles on its leading edge showing the direction of motion. If the continental polar air mass is retreating northward, the polar front is called a *warm front*. On a weather map the warm front is shown either as a red line or, if the presentation is in black and white, a black line with little semicircles on its edge showing the direction in which the front is retreating. The center of the polar front is embedded in the low-pressure area.

With all this background, let us now analyze the weather map of figure 2. Notice that there is a large low-pressure area over the eastern half of the United States. In general, the poorer weather will be found in this region. A high-pressure area is found across the western half of the United States. In general, good weather will be found in this region. The polar front can also be seen in figure 2. The cold front is the boundary between the cold continental polar air that came out of Canada and the warm moist maritime tropical air that has moved up from the gulf. The arrows on the map indicate the velocity of the air. The cold dry cP air, being heavier than the warm tropical mT air, pushes underneath the mT air, driving it upward. The moisture in the rising tropical air condenses and forms a narrow band of clouds along the length of the cold front. The precipitation usually associated with the cold front is showery.

The warm front is the boundary of the retreating cool air and the advancing warm moist air. The mT air, being lighter than the retreating cP air, rises above the colder air. The sloping front of the retreating air is much shallower than the slope of the advancing cold front. Therefore, the mT air rises over a very large region and gives a very vast region of clouds and precipitation. Thus the weather associated with a warm front is usually more extensive than the weather associated with a cold front.

Your weather depends on where you are with respect to the frontal systems. If you are north of the warm front in figure 2, such as in Illinois, Ohio, or Pennsylvania, the temperatures will be cool, the winds will be from the southeast, the sky will be cloudy, and you will be getting precipitation. If you are south of the warm front and in advance of the cold front, such as in Alabama, Georgia, South Carolina, and Florida, the temperature will be warm, the humidity high, winds will be from the southwest and you will usually have nice weather. If the cold front has already passed you by, such as in Kansas, Oklahoma, Texas, and Arkansas, the skies will be clear or at least clearing, the temperature will be cold, the humidity will be low, the winds will be from the northwest, and in general you will have good weather.

All the highs, lows, and fronts, move across the United States from the west toward the east. So the weather that you get today will change as these weather systems move toward you.

The Language of Physics

Thermodynamics
The study of the relationships between heat, internal energy, and the mechanical work performed by a system. The system is usually a heat engine of some kind (p. 479).

Work
The work done by a gas during expansion is positive and the work done by a gas during compression is negative. The work done is equal to the area under the curve in a *p-V* diagram. The work done depends on the thermodynamic path taken in the *p-V* diagram (p. 480).

Cyclic process
A process that runs in a cycle eventually returning to where it started from. The net work done by the system during a cyclic process is equal to the area enclosed by the cyclic thermodynamic path in a *p-V* diagram. The net work is positive if the cycle proceeds clockwise, and negative if the cycle proceeds counterclockwise on the *p-V* diagram. The total change in internal energy around the entire cycle is equal to zero. The energy for the net work done by the system comes from the net heat applied to the system (p. 480).

Isobaric process
A process that takes place at a constant pressure (p. 482).

Isochoric or isometric processes
A process that takes place at constant volume. The heat added to a system during an isochoric process shows up as an increase in the internal energy of the system (p. 482).

Isothermal process
A process that takes place at constant temperature (p. 482).

Molecular mass
The molecular mass of any substance is equal to the mass of one molecule of that substance times the total number of molecules in one mole of the substance (Avogadro's number). *Thus, the molecular mass of any substance is equal to the mass of one mole of that substance.* Hence, the mass of a gas is equal to the number of moles of a gas times the molecular mass of the gas (p. 484).

Molar specific heat
The product of the specific heat of a substance and its molecular mass (p. 484).

Heat in a thermodynamic process
The heat absorbed or liberated in a thermodynamic process depends on the path that is followed in a *p-V* diagram. Thus, heat, like work, is path dependent. Heat is always positive when it is added to the system and negative when it is removed from the system (p. 484).

Internal energy of a gas
The internal energy of a gas is equal to the sum of the kinetic energy of all the molecules of a gas. A change in temperature is associated with a change in the internal energy of a gas. Hence, an isothermal expansion occurs at constant internal energy. Regardless of the path chosen between two points in a *p-V* diagram, the change in internal energy is always the same. Thus, the internal energy of the system is independent of the path taken in a *p-V* diagram; it depends only on the initial and final states of the thermodynamic system (p. 485).

The first law of thermodynamics
The heat added to a system will show up either as a change in internal energy of the system or as work performed by the system. It is also stated in the form: the change in the internal energy of the system equals the heat added to the system minus the work done by the system on the outside environment. The first law is really a statement of the law of conservation of energy applied to a thermodynamic system (p. 487).

Efficiency
The efficiency of an engine can be defined in terms of what we get out of the system compared to what we put into the system. It is thus equal to the ratio of the work performed by the system to the heat put into the system. It is desirable to make the efficiency of an engine as high as possible (p. 491).

Adiabatic process
A process that occurs without an exchange of heat between the system and its environment. That is, heat is neither added nor taken away from the system during the process (p. 492).

Otto cycle
A thermodynamic cycle that is approximated in the operation of the gasoline engine (p. 495).

Ideal heat engine
An idealized engine that shows the main characteristics of all engines, namely, every engine absorbs heat from a source at high temperature, performs some amount of mechanical work, and then rejects some heat at a lower temperature (p. 495).

Refrigerator
A heat engine working in reverse. That is, work is done on the refrigerator, thereby extracting a quantity of heat from a low-temperature reservoir and exhausting a large quantity of heat to a hot reservoir (p. 496).

Carnot cycle
A thermodynamic cycle of a Carnot engine, consisting of two isothermal and two adiabatic paths in a *p-V* diagram. The Carnot engine is the most efficient of all engines (p. 496).

The second law of thermodynamics
The second law of thermodynamics tells us which processes are possible and which are not. The concept of entropy is introduced to give a quantitative basis for the second law. It is equal to the ratio of the heat added to the system to the absolute temperature of the system, when a thermodynamic system changes from one equilibrium state to another along a reversible path. In an isolated system, the system always changes from values of low entropy to values of high entropy, and only those processes are possible for which the entropy of the system increases or remains a constant (p. 498).

Kelvin-Planck statement of the second law of thermodynamics
No process is possible whose sole result is the absorption of heat from a reservoir at a single temperature and the conversion of this heat energy completely into mechanical work (p. 498).

Clausius statement of the second law of thermodynamics
No process is possible whose sole result is the transfer of heat from a cooler to a hotter body (p. 499).

Summary of Important Equations

Work done by a gas
$$W = p\Delta V \tag{17.5}$$

Mass of the gas
$$m = m_0 N_A n \tag{17.14}$$

Molecular mass
$$M = m_0 N_A \tag{17.15}$$

Mass of the gas
$$m = nM \tag{17.16}$$

Molar specific heat
$$C = Mc \tag{17.18}$$

Heat absorbed or liberated by a gas at constant volume
$$Q = nC_v\Delta T \tag{17.20}$$

Heat absorbed or liberated by a gas at constant pressure
$$Q = nC_p\Delta T \tag{17.21}$$

Internal energy of an ideal gas
$$U = \tfrac{3}{2}nRT \tag{17.23}$$

Change in internal energy of an ideal gas
$$\Delta U = \tfrac{3}{2}nR\Delta T \tag{17.24}$$

Change in internal energy of an ideal gas
$$\Delta U = nC_v\Delta T \tag{17.28}$$

Molar specific heat at constant volume
$$C_v = \tfrac{3}{2}R \tag{17.29}$$

Molar specific heat at constant pressure
$$C_p = \tfrac{5}{2}R \tag{17.42}$$

First law of thermodynamics
$$\Delta U = Q - W \tag{17.31}$$

Adiabatic process
$$Q = 0$$

First law for adiabatic process
$$W = -\Delta U \tag{17.44}$$

Isochoric process
$$\Delta V = 0$$

First law for isochoric process
$$Q = \Delta U \tag{17.47}$$

Isobaric process
$$\Delta p = 0$$

Cyclic process
$$\Delta U = 0$$

First law for cyclic process
$$W = Q \tag{17.48}$$

Efficiency of any engine
$$Eff = \frac{W}{Q_{in}} = \frac{W}{Q_H}$$
$$Eff = \frac{Q_H - Q_C}{Q_H} \tag{17.50}$$
$$Eff = 1 - \frac{Q_C}{Q_H} \tag{17.51}$$

Efficiency of a Carnot engine
$$Eff = 1 - \frac{T_C}{T_H} \tag{17.55}$$

Entropy
$$\Delta S = \frac{\Delta Q}{T} \tag{17.56}$$
$$S = k \ln P \tag{17.58}$$

Questions for Chapter 17

1. Discuss the difference between the work done by the gas and the work done on the gas in any thermodynamic process.
2. Why is the work done in a thermodynamic process a function of the path traversed in the p-V diagram?
3. Define the following processes: isobaric, isothermal, isochoric, adiabatic, cyclic, and isoentropic.
4. How is it possible that a solid and a liquid have one value for the specific heat and a gas can have an infinite number of specific heats?
5. Discuss the first and second laws of thermodynamics.
6. Describe what is meant by the statement, "the internal energy of a thermodynamic system is conservative."
7. Figure 17.7 shows a plot of isotherms and adiabats on a p-V diagram. Explain why the adiabats have a steeper slope.
†8. Discuss the thermodynamic process in a diesel engine, and draw the process on a p-V diagram.
†9. Use the first law of thermodynamics to describe a solar heating system.
10. Can you use a home refrigerator to cool the home in the summer by leaving the door of the refrigerator open?
†11. Why is a heat pump not very efficient in very cold climates?
†12. Show how equation 17.54 could be used as the basis of a temperature scale.
13. Is it possible to connect a heat engine to a refrigerator such that the work done by the engine is used to drive the refrigerator, and the waste heat from the refrigerator is then given to the engine, to drive the engine thus making a perpetual motion machine?
14. Discuss the concept of entropy and how it can be used to determine if a thermodynamic process is possible.
†15. Discuss the statements: (a) entropy is sometimes called time's arrow and (b) the universe will end in a heat death when it reaches its state of maximum entropy.

Problems for Chapter 17

17.2 The Concept of Work Applied to a Thermodynamic System

1. How much work is done by an ideal gas when it expands at constant atmospheric pressure from a volume of 0.027 m³ to a volume of 1.00 m³?
2. What is the area of the cross-hatched area in the p-V diagram? What is the work done in going from A to B?

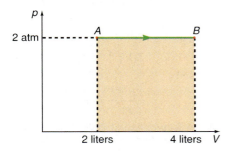

3. What is the net work done in the triangular cycle ABC?

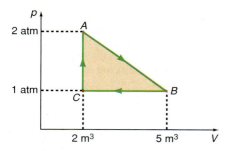

4. How much work is done in the cycle ABCDA?

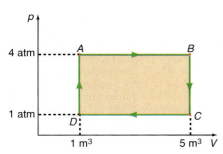

5. One mole of an ideal gas goes through the cycle shown. If p_A = 2.00 × 10⁵ Pa, p_D = 5.00 × 10⁴ Pa, V_B = 2.00 m³, and V_A = 0.500 m³, find the work done along the paths (a) AB, (b) BC, (c) CD, (d) DA, and (e) ABCDA.

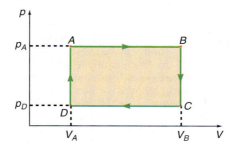

17.3 Heat Added to or Removed from a Thermodynamic System

6. What is the mass of 4.00 moles of He gas?
7. Find the amount of heat required to raise the temperature of 5.00 moles of He, 10.0 °C, at constant volume.
8. Find the amount of heat required to raise the temperature of 5.00 moles of He, 10.0 °C, at constant pressure.
9. Compute the amount of heat absorbed when one mole of a monatomic gas, at a temperature of 200 K, rises to a temperature of 400 K (a) isochoricly and (b) isobaricly.

17.4 The First Law of Thermodynamics

10. What is the total internal energy of 3.00 moles of an ideal gas at (a) 273 K and (b) 300 K?
11. What is the change in internal energy of 3.00 moles of an ideal gas when it is heated from 273 K to 293 K?
12. Find the change in the internal energy of 1 mole of an ideal gas when heated from 300 K to 500 K.
13. In a thermodynamic system, 500 J of work are done and 200 J of heat are added. Find the change in the internal energy of the system.
14. In a certain process, the temperature rises from 50.0 °C to 150.0 °C as 1000 J of heat energy are added to 4 moles of an ideal gas. Find the work done by the gas during this process.
15. In a thermodynamic system, 200 J of work are done and 500 J of heat are added. Find the change in the internal energy of the system.
16. In a certain process with an ideal gas, the temperature drops from 120 °C to 80.0 °C as 2000 J of heat energy are removed from the system and 1000 J of work are done by the gas. Find the number of moles of the gas that are present.

17. Four moles of an ideal gas are carried through the cycle ABCDA of figure 17.6. If T_D = 100 K, T_{AC} = T_A = T_C = 200 K, T_B = 400 K, p_A = 0.500 × 10⁵ Pa, and p_D = 2.50 × 10⁴ Pa, use the ideal gas equation to determine the volumes V_A and V_B.
†18. In problem 17 find the work done, the heat lost or absorbed, and the change in internal energy of the gas for the paths (a) AB, (b) BC, (c) CD, (d) DA, and (e) ABCDA.
19. In a thermodynamic system, 700 J of work are done by the system while the internal energy drops by 450 J. Find the heat transferred to the gas during this process.
20. If 5.00 J of work are done by a refrigerator and 8.00 J of heat are exhausted into the hot reservoir, how much heat was removed from the cold reservoir?
21. A heat engine is operating at 40.0% efficiency. If 3.00 J of heat are added to the system, how much work is the engine capable of doing?

17.5 Some Special Cases of the First Law of Thermodynamics

22. If the temperature of 2.00 moles of an ideal gas increases by 40.0 K during an isochoric process, how much heat was added to the gas?
23. If 800 J of thermal energy are removed from 8 moles of an ideal gas during an isochoric process, find the change in temperature in degrees (a) Kelvin, (b) Celsius, and (c) Fahrenheit.
24. If 3.00 J of heat are added to a gas during an isothermal expansion, how much work is the system capable of doing during this process?
25. During an isothermal contraction, 55.0 J of work are done on an ideal gas. How much thermal energy was extracted from the gas during this process?
†26. A monatomic gas expands adiabatically to double its original volume. What is its final pressure in terms of its initial pressure?
†27. One mole of He gas at atmospheric pressure is compressed adiabatically from an initial temperature of 20.0 °C to a final temperature of 100 °C. Find the new pressure of the gas.
28. If 50.0 J of work are done on one mole of an ideal gas during an adiabatic compression, what is the temperature change of the gas?

17.6 The Gasoline Engine

†29. The crankshaft of a gasoline engine rotates at 1200 revolutions per minute. The area of each piston is 80.0 cm² and the length of the stroke is 13.0 cm. If the average pressure during the power stroke is 7.01×10^5 Pa, find the power developed in each cylinder. (*Hint:* remember that there is only one power stroke for every two revolutions of the crankshaft.)

17.7 The Ideal Heat Engine

30. An engine operates between room temperature of 20.0 °C and a cold reservoir at 5.00 °C. Find the maximum efficiency of such an engine.
31. What is the efficiency of a Carnot engine operating between temperatures of 300 K and 500 K?
32. A Carnot engine is working in reverse as a refrigerator. Find the coefficient of performance if the engine is operating between the temperatures −10.5 °C and 35.0 °C.
33. A Carnot refrigerator operates between −10.0 °C and 25.0 °C. Find how much work must be done per kcal of heat extracted.
34. Calculate the efficiency of an engine that absorbs 500 J of thermal energy while it does 250 J of work.

17.10 Entropy

35. Find the change in entropy if 10.0 kg of ice at 0.00 °C is converted to water at +10.0 °C.
36. A gas expands adiabatically from 300 K to 350 K. Find the change in its entropy.
†37. Find the total change in entropy if 2.00 kg of ice at 0.00 °C is mixed with 25.0 kg of water at 20.0 °C.
38. Find the change in entropy when 2.00 kg of steam at 110 °C is converted to water at 90.0 °C.
39. A gas expands isothermally and does 500 J of work. If the temperature of the gas is 35.0 °C, find its change in entropy.

Additional Problems

40. In the thermodynamic system shown in the diagram, (a) 50.0 J of thermal energy are added to the system, and 20.0 J of work are done by the system along path *abc*. Find the change in internal energy along this path. (b) Along path *adc*, 10.0 J of work are done by the system. Find the heat absorbed or liberated from the system along this path. (c) The system returns from state *c* to its initial state *a* along path *ca*. If 15.0 J of work are done on the system find the amount of heat absorbed or liberated by the system.

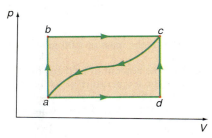

41. Draw the following process on a *p-V* diagram. First 8.00 m³ of air at atmospheric pressure are compressed isothermally to a volume of 4.00 m³. The gas then expands adiabatically to 8.00 m³ and is then compressed isobarically to 4.00 m³.
42. In the diagram shown, one mole of an ideal gas is at atmospheric pressure and a temperature of 250 K at position *a*. (a) Find the volume of the gas at *a*. (b) The pressure of the gas is then doubled while the volume is kept constant. Find the temperature of the gas at position *b*. (c) The gas is then allowed to expand isothermally to position *c*. Find the volume of the gas at *c*.

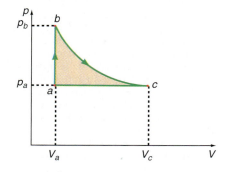

†43. Repeat problem 42, but for part (c) let the gas expand adiabatically to atmospheric pressure. Find the volume of the gas at this point. Show this point on the diagram.
†44. It was stated in equation 17.45 that for an adiabatic process with an ideal gas,

$$pV^\gamma = \text{constant}$$

Show that when an ideal gas in an initial state, with pressure p_1, volume V_1, and temperature T_1, undergoes an adiabatic process to a final state that is described by pressure p_2, volume V_2, and temperature T_2, that

$$p_1 V_1^\gamma = p_2 V_2^\gamma$$

and

$$T_1 V_1^{\gamma-1} = T_2 V_2^{\gamma-1}$$

and

$$\frac{T_1^{\gamma/(\gamma-1)}}{p_1} = \frac{T_2^{\gamma/(\gamma-1)}}{p_2}$$

45. A lecture hall at 20.0 °C contains 100 students whose basic metabolism generates 100 kcal/hr of thermal energy. If the size of the hall is 15.0 m by 30.0 m by 4.00 m, what is the increase in temperature of the air in the hall at the end of 1 hr? It is desired to use an air conditioner to cool the room to 20.0 °C. If the air conditioner is 45.0% efficient, what size air conditioner is necessary?

Interactive Tutorials

▣ 46. A thermodynamic cycle. Three moles of an ideal gas are carried around the thermodynamic cycle *ABCDA* shown in figure 17.6. Find the work done, the heat lost or absorbed, and the internal energy of the system for the thermodynamic paths (a) *AB*, (b) *BC*, (c) *CD*, (d) *DA*, and (e) *ABCDA*. The temperatures are $T_D = 147$ K, $T_{AC} = 250$ K, and $T_B = 425$ K. The pressures are $p_A = 5.53 \times 10^4$ Pa and $p_D = 3.25 \times 10^4$ Pa. The volumes are $V_A = 0.113$ m³ and $V_B = 0.192$ m³. (f) Find the efficiency of this system.

Electricity and Magnetism

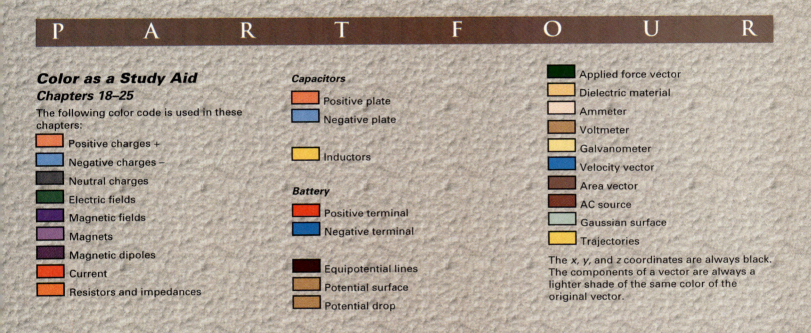

Color as a Study Aid
Chapters 18–25

The following color code is used in these chapters:

- Positive charges +
- Negative charges –
- Neutral charges
- Electric fields
- Magnetic fields
- Magnets
- Magnetic dipoles
- Current
- Resistors and impedances

Capacitors

- Positive plate
- Negative plate

- Inductors

Battery

- Positive terminal
- Negative terminal

- Equipotential lines
- Potential surface
- Potential drop

- Applied force vector
- Dielectric material
- Ammeter
- Voltmeter
- Galvanometer
- Velocity vector
- Area vector
- AC source
- Gaussian surface
- Trajectories

The x, y, and z coordinates are always black. The components of a vector are always a lighter shade of the same color of the original vector.

18

Electrostatics

A steel ball is dropped into a high-energy electric field.

The power which electricity of tension possesses of causing an opposite electrical state in its vicinity has been expressed by the general term Induction.

Michael Farday

In chapter 1 we attempted to explain the world in terms of the fewest number of fundamental quantities. Up to now only length, mass, and time have concerned us. Now we will consider the fourth fundamental quantity, electric charge. Our knowledge of electric charge is not new. The earliest known experiments on electrostatics were performed by Thales of Miletus (ca. 624–547 B.C.) around 600 B.C., when he found that amber, when rubbed with fur, attracted light objects. Today, we say that the amber possesses an electrical charge. (The word *electric* is derived from the Greek word *elektron,* meaning amber.) *The study of electric charges at rest under the action of electric forces is called* **electrostatics.**

18.1 Separation of Electric Charge by Rubbing

To help understand some of the characteristics of electric charge, let us perform the following experiment. We attach a pith ball, a spongy material like cork, by a piece of string to the ceiling, and it hangs freely, as shown in figure 18.1(a). We move an amber rod toward the pith ball, eventually touching it [figure 18.1(b)]. We observe that nothing happens to the pith ball during the touching process. Now we rub the amber rod with a piece of fur and then cause the rod to touch the pith ball, figure 18.1(c). This time the pith ball flies away, as seen in figure 18.1(d).

How can this behavior be explained? When the rod was not rubbed and was touched to the pith ball, nothing happened to the pith ball. But when the rod was rubbed with fur and touched the pith ball, the pith ball was repelled from the rod. We must conclude that something happened to the rod during the rubbing process. Let us assume that rubbing the rod caused an amount of something, which we will call *electric charge,* to be deposited on the rod.

The type of charge deposited on the rod could be arbitrarily called an "A" charge, or a "B" charge, or a "blue" charge, or a "red" charge, but we will, again arbitrarily, call it negative charge. When the negatively charged rod touches the pith ball, some of this negative charge on the rod flows onto the pith ball, leaving the pith ball negatively charged. The pith ball then flies away from the rod, indicating that like charges repel each other. It is because like charges repel each other that the original charges on the rod flowed to the pith ball.

The thought that should now occur to us is, "Does rubbing any material generate electric charge?" Let us repeat the experiment, only this time with a glass rod. First, we touch the unrubbed glass rod to a new pith ball. As before nothing happens to the pith ball. Then we rub the glass rod with silk and touch it to the pith ball, figure 18.1(e). We see that the pith ball flies away, as in figure 18.1(f). Electric

Figure 18.1

Separation of charge by rubbing.

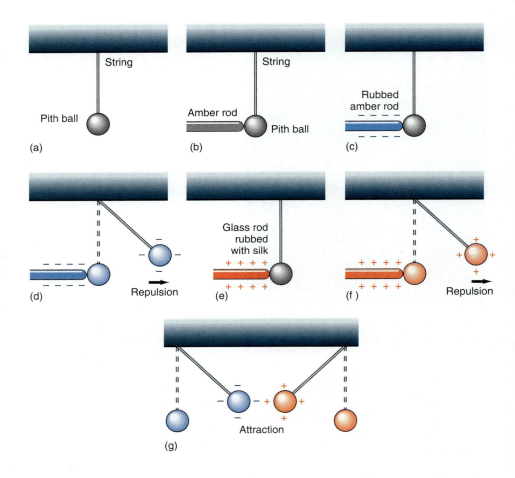

Lightning is a well-known and dramatic example of electricity.

charge has again been generated by rubbing the rod. But is this the same charge that was generated when the amber rod was rubbed? At this point we cannot answer the question. To be completely general, let us assume that it is a different electric charge, one that we will arbitrarily call positive charge. The positive charge appears to flow from the glass rod to the pith ball during contact, leaving the pith ball also positively charged. The pith ball then flies away from the glass rod, again indicating that like charges repel each other. Is this really a different charge, that is, a positive one? The experimental results are the same for both rods. Both pith balls are repelled from the rod.

Let us now hang two separate pith balls from the ceiling, and charge the one on the left negatively by touching it with the rubbed amber rod. The pith ball to the right we charge positively by touching it with the glass rod rubbed with silk. If the charges on both balls are really the same, then the balls should repel each other, just as the balls were repelled from the rods in figures 18.1(d) and 18.1(f). We observe, however, that the two pith balls are attracted to each other, as shown in figure 18.1(g). Therefore rubbing the glass rod with silk does indeed produce a different charge, a positive one, than that produced by rubbing an amber rod with fur, the negative charge. Also note that the unlike charges attract each other. This experiment can be continued by rubbing different rods with various materials, but only these two types of charges, negative and positive, are ever found. We conclude that there are two and only two types of electric charges in nature—negative charge and positive charge. As a result of these experiments, ***the fundamental principle of electrostatics*** can be stated: *Like electric charges repel each other, whereas unlike electric charges attract each other.* To give a more modern description of electrostatics we need to discuss atomic structure.

18.2 Atomic Structure

In an attempt to find simplicity in nature, the Greek philosophers Leucippus and Democritus suggested in the fifth century B.C. that matter is composed of very small particles called atoms. The word *atom* comes from a Greek word that means "that which is indivisible." However, it was not until the early nineteenth century that John Dalton (1766–1844), an English chemist, proposed that to every known chemical element there corresponds an atom of matter. Every material in the world is just some combination of these indivisible atoms. However, in 1897, J. J. Thomson (1856–1940), an English physicist, discovered the *electron,* a negatively charged particle having a mass $m_e = 9.1095 \times 10^{-31}$ kg. Although it had been known that there was such a thing as negative electrical charge, it was not known what the carrier of that negative charge was. This newly discovered electron, however, was the basic or elementary particle carrying the smallest amount of negative charge. All other negative charges that occur in electrostatic experiments are multiples of the electronic charge.

The finding of the electron, however, presented a rather difficult problem. Where did it come from? The only place it could come from is the interior of the indivisible atom, which could not then be indivisible. The indivisible atom must have some structure. Because the atom is generally neutral, there must be some positive charge within the atom to neutralize the negative electron. In the early 1900s Ernest Rutherford (1871–1937), a British physicist, bombarded atoms with alpha particles (positively charged particles) and by observing the effects of the collision, developed the nuclear model of the atom. His model of the atom consisted of a small, dense, positively charged nucleus with negative electrons orbiting about it, somewhat in the manner of the planets orbiting about the sun in the solar system. Rutherford found this positive particle of the nucleus and named it the proton in 1919. The proton has a positive charge equal in magnitude to the charge on the electron. The mass of the *proton, $m_p = 1.6726 \times 10^{-27}$* kg, is about 1836 times greater than the electron mass.

In 1920 Rutherford suggested that there is probably another particle within the nucleus, a neutral one, to which he gave the name the *neutron.* The neutron was discovered some twelve years later in 1932 by the English physicist, James Chadwick. In terms of these particles, or building blocks, the different atoms are formed. The difference between one chemical element and another is in the number of protons and electrons within it. As seen in figure 18.2(a), the chemical element hydrogen contains a nucleus that consists of one proton about which orbits the lighter electron.

Sometimes the electron is to the right and sometimes to the left of the nucleus. Sometimes it is above and sometimes below the nucleus. *By symmetry, the electron's mean position coincides with the position of the positive nucleus. Therefore, the atom as a whole acts as though it were electrically neutral.*

The next chemical element helium is formed by the addition of another proton to the nucleus, and another electron to the orbit, figure 18.2(b). Two neutrons are also found in the nucleus of helium. The next chemical element, lithium, is formed by the addition of another proton, another electron, and another neutron, figure 18.2(c). In this way, all of the chemical elements are formed, although in the higher elements there are usually more neutrons than protons. Because each element contains the same number of electrons and protons, each element is electrically neutral.

Although this model of the atom is quite useful, it is not completely correct. An electron moving in a circle is an accelerated charge, and it has been found that whenever a charge is accelerated, it radiates energy. Therefore the radiating electron should lose energy and spiral into the nucleus, and the atom should cease to exist. The world, which is made up of atoms, should also cease to exist. Since the world continues to exist, the above model of the atom cannot be completely correct.

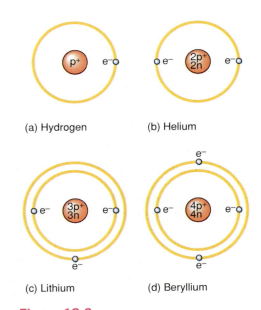

(a) Hydrogen

(b) Helium

(c) Lithium

(d) Beryllium

Figure 18.2

Atomic structure.

Table 18.1

Quarks and Their Charges

Quark	Charge (Fraction of Electron Charge)
u	2/3
d	−1/3
s	−1/3
c	2/3
b	−1/3
t	2/3

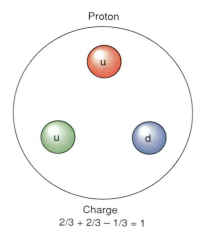

Proton

Charge
2/3 + 2/3 − 1/3 = 1

Figure 18.3

The quark configuration of a proton.

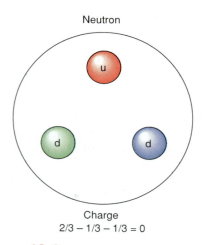

Neutron

Charge
2/3 − 1/3 − 1/3 = 0

Figure 18.4

The quark configuration of a neutron.

Also since like charges repel each other, the protons in the nucleus should also repel each other and the nucleus should blow itself apart. Hence, the whole world should blow itself apart. But it does not. Therefore, there must be some other force within the nucleus holding the protons together. This force is called the *strong nuclear force.*

Because the electron is the smallest unit of charge ever found, the fundamental unit of charge, the coulomb, named after the French physicist Charles A. de Coulomb (1736–1806), is defined in terms of a certain number of these electronic charges. That is, 1 coulomb of charge is equal to 6.242×10^{18} electronic charges, and the charge on one electron is 1.60219×10^{-19} coulomb. We will abbreviate the unit for electrical charge, the coulomb, as a capital C, in keeping with the SI convention that units named after a person are symbolized by capital letters. *Because electric charge only comes in multiples of the electronic charge, it is said that electric charge is quantized. Also, the total net charge of any system is constant, a result known as the law of conservation of electric charge.*

Although no electric charges have ever been found carrying fractional portions of the electronic charge, the latest hypothesis in elementary particle physics, put forth by Murray Gell-Mann in 1964, is that protons and neutrons are made up of more elementary particles called **quarks.** It is presently proposed that there are six quarks, the (1) up (u), (2) down (d), (3) strange (s), (4) charm (c), (5) bottom (b), and (6) top (t) quarks. The charges on these quarks are fractional as shown in table 18.1.

A proton is assumed to be made of two up quarks and one down quark, as shown in figure 18.3. The neutron is assumed to be made of one up quark and two down quarks, as shown in figure 18.4. Of course, individual quarks have not yet been found, and indeed most theories of particle physics predict that an isolated single quark cannot exist. There is, however, strong indirect evidence from scattering experiments that quarks do indeed exist. The quark model has also predicted the existence of other elementary particles that have been found, indicating that the quark hypothesis is on very good experimental ground. Of course, if the existence of quarks is definitely confirmed, the next question that would then have to be asked is, "Of what are quarks made?"

In terms of the atomic concept of electric charge, the generation of charge by rubbing is explained as follows. When an amber rod is rubbed with fur, the rubbing causes electrons to be stripped from the fur atoms and deposited on the amber rod, making the amber rod negative. The fur, which now has a deficiency of electrons, becomes positive. Hence, charge has not been created, it has merely been separated. When the pith ball is touched by the negative amber rod, electrons flow from the rod to the pith ball, leaving the pith ball negatively charged.

When the glass rod is rubbed with silk, electrons are stripped from the atoms in the glass rod and are deposited on the silk cloth making the silk cloth negative. The glass rod, having a deficiency of electrons, becomes positive. When the glass rod touches the neutral pith ball, electrons from the originally neutral pith ball flow onto the glass rod, neutralizing some of its positive charge. The deficiency of electrons on the pith ball causes the pith ball to become positive. This correctly explains the previous experiment where, at that time, we incorrectly assumed that positive charges flowed from the glass rod to the pith ball. In all these cases of charging bodies by rubbing, the charge that moves from one body to another is always the negative electronic charge. Recall that the electron is 1/1836 lighter than the proton and hence easier to move. Also the heavier protons are tightly bound into the nucleus and are not as easy to detach from the atom as the weakly bound electrons.

We use the letter q to represent the electric charge on a body and, in general, the net charge q on a body consisting of both protons and electrons is given by

$$q = (N_p - N_e)e$$

where N_p is equal to the number of protons, N_e is equal to the number of electrons, and e is equal to the basic unit of charge, namely 1.60×10^{-19} C.

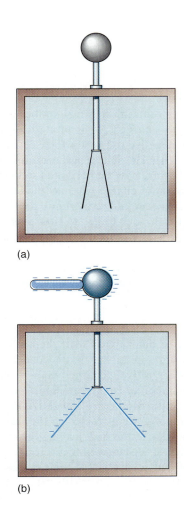

(a)

(b)

Figure 18.5

An electroscope.

Figure 18.6

A conductor and an insulator.

18.3 Measurement of Electric Charge

The Electroscope

One of the traditional means for measuring electric charge is the *electroscope,* as seen in figure 18.5(a). An electroscope is a device made of two strips of thin gold leaf or aluminum foil fastened to the end of a metallic rod. The metallic rod is housed in an insulated enclosure. The top of the rod is connected to a metallic ball at the top of the electroscope. If a negatively charged rod is touched to the metal ball on the top of the electroscope, electrons move from the negative rod to the ball of the electroscope, down through the metallic connecting rod to the aluminum leaves, some moving to the right leaf and some to the left leaf. The negative charge on the aluminum leaves repel each other, thereby separating the leaves, as shown in figure 18.5(b). The amount of separation of the leaves becomes a measure of the quantity of charge. If instead of the negative rod, a positive rod were touched to the electroscope, the leaves would separate in the same way, only this time positive charge would be on each leaf. Electrons on the electroscope would flow to the positive rod leaving a deficiency of negative charges on the leaves of the electroscope. Hence, the leaves would have positive charges on them.

Conductors and Insulators

Two electroscopes are set up, as shown in figure 18.6. When the negatively charged rod is touched to the metal bulb between the two electroscopes, the electroscope to the right shows a charge but the one to the left does not. The reason for this phenomenon can be explained by the connecting material between the metal bulb and the electroscope. The copper wire is called a conductor because it allows the charge deposited on the bulb to be conducted to the electroscope. The rubber band is called an insulator because it does not permit the charge to reach the electroscope. (The rubber band insulates the electroscope from the charged bulb.)

In general, most substances fall into either one of these categories. *Materials that permit the free flow of electric charge through them are called* **conductors.** *Materials that do not permit the free flow of electric charge through them are called* **insulators or dielectrics.** Most metals are good conductors of electric charge, whereas most nonmetals are insulators. (There are a few materials called semiconductors, whose characteristics lie between those of conductors and those of insulators.)

An interesting characteristic of all conductors is that whenever an electric charge is placed on a conducting body, that charge will redistribute itself until all of the charge is on the outside of the body. For example, if electric charges are

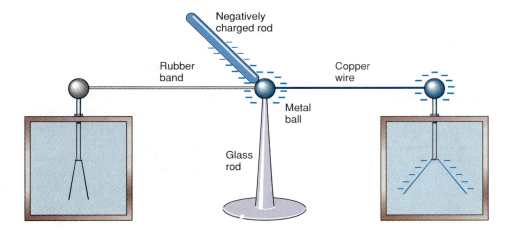

Negatively charged rod

Rubber band

Copper wire

Metal ball

Glass rod

(a) Negative rod far away from metallic sphere

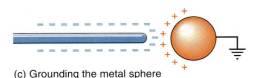

(b) Negative rod brought close to, but not touching, metal sphere

(c) Grounding the metal sphere

(d) The sphere is charged positively

Figure 18.7

Charging by induction.

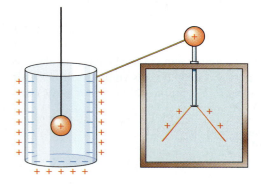

Figure 18.8

The Faraday ice pail.

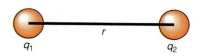

q_1 r q_2

Figure 18.9

Coulomb's law.

placed on a solid metallic sphere, the charges exert forces of repulsion on one another and the charges try to move as far apart as they can. The greatest separation they can achieve is when they are on the outside of the sphere.

Charging by Induction

Most bodies are electrically neutral, that is, they contain the same number of positive charges as negative charges. If a negatively charged rod is brought into the vicinity of an uncharged conducting sphere the negative rod repels the electrons on the sphere, as shown in figures 18.7(a) and 18.7(b). The sphere still contains equal numbers of positive and negative charges but they are now redistributed so that the right side of the sphere has a negative distribution of charge whereas the left side has a positive one. (If the rod is removed, the charge would redistribute itself to its initial neutral configuration.) If you were to touch your finger to the sphere you would provide a path for the electrons to escape to the ground, as shown in figure 18.7(c).

When you remove your finger from the sphere and remove the negative rod, the sphere is charged positively. A positive charge has been induced on the sphere without having touched it with a positive charge.

The Faraday Ice Pail

The amount of electric charge on a body can be determined with the aid of the *Faraday ice pail.* The Faraday ice pail is an empty metal can that is connected to an electroscope, as shown in figure 18.8. When a positively charged body is placed within the pail, but not touching the pail, it induces an equal amount of negative charge inside the pail. Because the pail was originally neutral, negative charges from the outside of the pail, which are free to move, are attracted to the inside of the pail depleting the negative charge on the outside of the pail, leaving the outside of the pail positive. Since the outside wall of the pail is connected to an electroscope, the charge can be measured by the amount of leaf separation of the electroscope. If the charged body is removed from the pail, the electroscope reads zero, because the cause of the induced charge has been removed. If the positively charged body is again placed into the pail, the same results as before are obtained. If the charged body is now touched against the inside wall of the pail, the induced negative charge on the inside of the pail will neutralize the positively charged body. But the positive charge on the outside wall of the pail remains, as can be seen by the reading of the electroscope. If the original body is removed from the ice pail, there is no change in the reading of the electroscope. We can conclude from this demonstration that (1) *a charge cannot exist by itself within a conductor, but rather is found on the outside of the conductor* and (2) *the electroscope leaves have the same net quantity of charge as that placed within the can.* Therefore, the Faraday ice pail becomes a simple means of measuring the electric charge on a body.

18.4 Coulomb's Law

As we have just seen, electric charges exert forces on each other. But what are the magnitudes of these forces? Coulomb invented a torsion balance in 1777 in which a quantity of force could be measured by the amount of twist it produced in a thin wire. In 1785 he used this torsion balance to measure the force between electrical charges. If two point charges, q_1 and q_2, are separated by a distance r between their centers, as in figure 18.9, then Coulomb found that the force between the charges could be stated as: *The force between the point charges q_1 and q_2 is directly proportional to the product of their charges and inversely proportional to the square* *of the distance separating them. The direction of this force lies along the line separating the charges. This result is known as* **Coulomb's law.** We can state Coulomb's law mathematically as

$$F = \frac{kq_1q_2}{r^2} \tag{18.1}$$

where k is a constant depending on the units employed and on the medium in which the charges are located. In a vacuum, $k = 8.9876 \times 10^9$ N m²/C². In air, the value of k is so close to the value of k in a vacuum that we will use the same value for both. To simplify the solution of problems in this text, we will round off the value of k to

$$k = 9.00 \times 10^9 \frac{\text{N m}^2}{\text{C}^2}$$

To simplify more advanced theories of electromagnetism, Coulomb's law is also written in the form

$$F = \frac{1}{4\pi\epsilon_0} \frac{q_1q_2}{r^2}$$

where

$$\frac{1}{4\pi\epsilon_0} = k \tag{18.2}$$

and $\epsilon_0 = 8.854 \times 10^{-12}$ C²/N m² and is called the *permittivity of free space.* If the charges are placed in a medium other than air or a vacuum, then there is a different value for the permittivity, ϵ, for that medium and hence a different value of k. In this text we will use only the simple form, equation 18.1, of Coulomb's law. If the charges are much larger than point charges, Coulomb's law can still be used if the distance separating the charges is quite large compared to the size of the electric charge. Under these circumstances the charges approximate point charges.

Notice the similarity between the form of Coulomb's law of electrostatics and Newton's law of universal gravitation:

$$F = \frac{Gm_1m_2}{r^2} \tag{6.37}$$

Let us compare the electric and gravitational forces between an electron and a proton in a hydrogen atom. The mass of the proton is $m_p = 1.67 \times 10^{-27}$ kg, whereas the mass of the electron is $m_e = 9.11 \times 10^{-31}$ kg. The radius of the lowest energy orbit for the electron is $r = 5.29 \times 10^{-11}$ m. The charge on the electron and proton is 1.60×10^{-19} C. The gravitational force between the electron and the proton is found from Newton's law of universal gravitation, equation 6.37, as

$$F_g = \frac{Gm_pm_e}{r^2}$$

$$= \frac{(6.67 \times 10^{-11} \text{ N m}^2/\text{kg}^2)(1.67 \times 10^{-27} \text{ kg})(9.11 \times 10^{-31} \text{ kg})}{(5.29 \times 10^{-11} \text{ m})^2}$$

$$= 3.63 \times 10^{-47} \text{ N}$$

The electric force between the electron and the proton is found from Coulomb's law of electrostatics, equation 18.1, as

$$F_e = \frac{kq_pq_e}{r^2}$$

$$= \frac{(9.00 \times 10^9 \text{ N m}^2/\text{C}^2)(1.60 \times 10^{-19} \text{ C})^2}{(5.29 \times 10^{-11} \text{ m})^2}$$

$$= 8.23 \times 10^{-8} \text{ N}$$

Although both forces seem quite small, let us compare the relative magnitude of these forces by taking the ratio of the electric force to the gravitational force, that is,

$$\frac{F_e}{F_g} = \frac{8.23 \times 10^{-8} \text{ N}}{3.63 \times 10^{-47} \text{ N}} = 2.27 \times 10^{39}$$

or

$$F_e = (2.27 \times 10^{39})F_g$$

That is, for the electron-proton system discussed here, the electric force is 10^{39} times greater than the gravitational force. Because magnitudes are sometimes hard to visualize in scientific notation for the beginning student, we can also write this number as

$$F_e = 2{,}270{,}000{,}000{,}000{,}000{,}000{,}000{,}000{,}000{,}000{,}000{,}000{,}000 \; F_g$$

Therefore, on the atomic level the gravitational force is an extremely weak force, whereas the electric force is very large. Hence, in the solution of electrostatics problems the gravitational force can be ignored when compared to the electric force.

On an atomic level, the gravitational force is indeed very weak. By contrast, however, when extremely large masses are involved, such as in a large dying star, the gravitational force is so great, that electrons are driven right into the nuclei of atoms, converting the nuclear protons into neutrons, and creating a neutron star. The density of the matter in such a neutron star is so great that one tablespoon of that matter would weigh 10 billion tons if it were located in the gravitational field at the surface of the earth.

Let us now consider the electrostatic force between the two protons in a helium nucleus. The protons are separated by a distance of about 2.40×10^{-15} m. The force between the two protons is

$$F_{pp} = \frac{kq_p q_p}{r^2}$$

$$= \frac{(9.00 \times 10^9 \text{ N m}^2/\text{C}^2)(1.60 \times 10^{-19} \text{ C})^2}{(2.40 \times 10^{-15} \text{ m})^2}$$

$$= 40.0 \text{ N}$$

If we compare this repulsive force between the two protons in the helium nucleus to the attractive electrostatic force between the electron and the proton in the hydrogen atom, we see that their ratio is

$$\frac{F_{pp}}{F_{ep}} = \frac{40.0 \text{ N}}{8.23 \times 10^{-8} \text{ N}} = 4.86 \times 10^8$$

or

$$F_{pp} = 4.86 \times 10^8 \; F_{ep}$$

The force between the two protons in the helium nucleus is an enormous force and hence the helium nucleus should blow itself apart. The fact that it does not is an indication of the existence of another force, the strong nuclear force, that holds the protons together within the nucleus. From the calculation, the nuclear force holding the nucleus together is at least 10^8 times greater than the electric force holding the atom together, that is,

$$F_N = 10^8 \; F_A$$

Let us consider some examples of the use of Coulomb's law.

Figure 18.10

An example of Coulomb's law.

Example 18.1

Coulomb's law for two point charges. A point charge, $q_1 = 2.00\ \mu C$ is placed 0.500 m from another point charge $q_2 = -5.00\ \mu C$, as shown in figure 18.10. Calculate the magnitude and direction of the force on each charge.

Solution

The force acting on charge q_1 is a force of attraction caused by the negative charge of q_2. We will call this force $\mathbf{F}_{12}$ (force on charge 1 caused by charge 2). The magnitude of the force, F_{12}, found by equation 18.1, is

$$F_{12} = \frac{kq_1q_2}{r^2}$$

$$= \frac{(9.00 \times 10^9\ \text{N m}^2/\text{C}^2)(2.00 \times 10^{-6}\ \text{C})(5.00 \times 10^{-6}\ \text{C})}{(0.500\ \text{m})^2}$$

$$= 0.360\ \text{N to the right}$$

The force acting on charge q_2, $\mathbf{F}_{21}$, is a force of attraction caused by charge q_1. The magnitude of the force, F_{21}, found from Coulomb's law, is

$$F_{21} = \frac{kq_2q_1}{r^2}$$

$$= \frac{(9.00 \times 10^9\ \text{N m}^2/\text{C}^2)(5.00 \times 10^{-6}\ \text{C})(2.00 \times 10^{-6}\ \text{C})}{(0.500\ \text{m})^2}$$

$$= 0.360\ \text{N to the left}$$

Note that the magnitudes of the forces on q_1 and q_2 are identical, but their directions are opposite. This could have been deduced immediately from Newton's third law, because if charge 1 exerts a force on charge 2, then charge 2 must exert an equal but opposite force on charge 1.

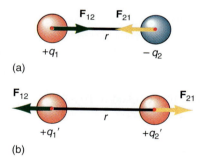

(a)

(b)

Figure 18.11

Charged spheres.

Example 18.2

The effect of touching two charged spheres. Two identical metal spheres are placed 0.200 m apart as shown in figure 18.11. A charge of 9.00 μC is placed on one sphere while a charge of $-3.00\ \mu C$ is placed on the other. (a) What is the force on each of the spheres? (b) If the two spheres are brought together and touched and then returned to their original positions, what is the force on each sphere?

Solution

a. The attractive force on sphere 1, found from Coulomb's law, is

$$F_{12} = \frac{kq_1q_2}{r^2}$$

$$= \frac{(9.00 \times 10^9\ \text{N m}^2/\text{C}^2)(9.00 \times 10^{-6}\ \text{C})(3.00 \times 10^{-6}\ \text{C})}{(0.200\ \text{m})^2}$$

$$= 6.08\ \text{N}$$

The force on sphere 2 is equal and opposite to this.

b. When the spheres are touched together, the $-3.00\ \mu C$ of charge q_2 neutralizes $+3.00\ \mu C$ of charge q_1, leaving a net charge of

$$9.00\ \mu C - 3.00\ \mu C = 6.00\ \mu C$$

This charge of 6.00 μC is then equally distributed between sphere one and sphere two, giving $q_1' = q_2' = 3.00\ \mu C$. Since the charges on both spheres are now positive, the force on each sphere is now repulsive. When the spheres are removed to the original distance the magnitude of the force on each sphere is now

$$F = \frac{kq_2'q_1'}{r^2} = \frac{(9.00 \times 10^9 \text{ N m}^2/\text{C}^2)(3.00 \times 10^{-6} \text{ C})^2}{(0.200 \text{ m})^2}$$

$$= 2.03 \text{ N}$$

18.5 Multiple Charges

If there are three or more charges present, then the force on any one charge is found by the vector addition of the forces associated with the other charges. That is, the resultant force on any one charge is given by

$$\mathbf{F} = \mathbf{F}_1 + \mathbf{F}_2 + \mathbf{F}_3 + \mathbf{F}_4 + \cdots \qquad (18.3)$$

Example 18.3

Multiple charges in a line. Three charges are placed on the line, as shown in figure 18.12. Find the resultant force on each charge if $q_1 = 1.00 \ \mu\text{C}$, $q_2 = -2.00 \ \mu\text{C}$, and $q_3 = 3.00 \ \mu\text{C}$. The separation of the charges are $r_{12} = 0.500$ m and $r_{23} = 0.500$ m.

Solution

The resultant force on each charge is equal to the vector sum of all the forces acting on that charge. The force on charge 1 is

$$\mathbf{F}_1 = \mathbf{F}_{12} + \mathbf{F}_{13}$$

where $\mathbf{F}_{12}$ is the force on charge 1 caused by charge 2 and $\mathbf{F}_{13}$ is the force on charge 1 caused by charge 3, as shown in figure 18.13. Because q_2 is negative, $\mathbf{F}_{12}$ is a force of attraction to the right, and since q_3 is positive, $\mathbf{F}_{13}$ is a force of repulsion to the left. The total force on charge 1 is therefore

$$F_1 = F_{12} - F_{13}$$

where

$$F_{12} = \frac{kq_1q_2}{r_{12}^2}$$

$$= \frac{(9.00 \times 10^9 \text{ N m}^2/\text{C}^2)(1.00 \times 10^{-6} \text{ C})(2.00 \times 10^{-6} \text{ C})}{(0.500 \text{ m})^2}$$

$$= 0.0720 \text{ N to the right}$$

while

$$F_{13} = \frac{kq_1q_3}{r_{13}^2}$$

$$= \frac{(9.00 \times 10^9 \text{ N m}^2/\text{C}^2)(1.00 \times 10^{-6} \text{ C})(3.00 \times 10^{-6} \text{ C})}{(1.00 \text{ m})^2}$$

$$= 0.0270 \text{ N to the left}$$

Therefore,

$$F_1 = F_{12} - F_{13} = 0.0720 \text{ N} - 0.0270 \text{ N}$$

$$= 0.0450 \text{ N to the right}$$

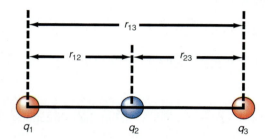

Figure 18.12

Multiple charges all on same line.

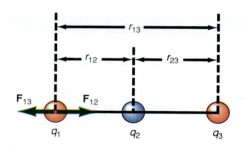

Figure 18.13

Force on charge 1.

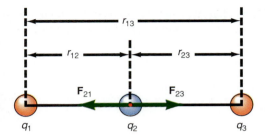

Figure 18.14

Force on charge 2.

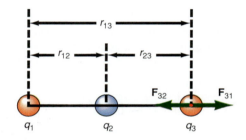

Figure 18.15

Force on charge 3.

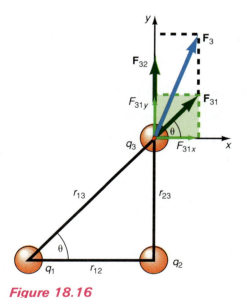

Figure 18.16

Multiple charges not all on the same line.

That is, the resultant force on charge 1 is a force of 0.0450 N to the right.

The force on charge 2 is

$$F_2 = F_{21} + F_{23}$$

or, as can be seen from figure 18.14,

$$F_2 = F_{23} - F_{21}$$

From Newton's third law

$$F_{21} = -F_{12} = -0.0720 \text{ N to the left}$$

and

$$F_{23} = \frac{kq_2q_3}{r_{23}^2}$$

$$= \frac{(9.00 \times 10^9 \text{ N m}^2/\text{C}^2)(2.00 \times 10^{-6} \text{ C})(3.00 \times 10^{-6} \text{ C})}{(0.500 \text{ m})^2}$$

$$= 0.216 \text{ N to the right}$$

Therefore, the net force on charge 2 is

$$F_2 = F_{23} - F_{21} = 0.216 \text{ N} - 0.0720 \text{ N}$$

$$= 0.144 \text{ N to the right}$$

The force on charge 3 is

$$F_3 = F_{31} + F_{32}$$

and, as can be seen in figure 18.15, we can write this as

$$F_3 = F_{31} - F_{32}$$

But by Newton's third law

$$F_{31} = -F_{13} = 0.0270 \text{ N to the right}$$

and

$$F_{32} = -F_{23} = -0.2160 \text{ N to the left}$$

The net force on charge 3 is therefore

$$F_3 = 0.0270 \text{ N} - 0.2160 \text{ N}$$

$$= -0.189 \text{ N to the left}$$

Example 18.4

Multiple charges not on a line. Find the resultant force on charge q_3 in figure 18.16 if $q_1 = 13.0 \ \mu\text{C}$, $q_2 = 5.00 \ \mu\text{C}$, $q_3 = 4.00 \ \mu\text{C}$, $r_{12} = 0.500$ m, and $r_{23} = 0.800$ m.

Solution

The resultant force on charge q_3 is found as

$$F_3 = F_{31} + F_{32}$$

where $\mathbf{F}_{31}$ is the force on charge 3 caused by charge 1, and $\mathbf{F}_{32}$ is the force on charge 3 caused by charge 2. We find the distance r_{13} from figure 18.16 and the Pythagorean theorem as

$$r_{13} = \sqrt{(r_{12})^2 + (r_{23})^2}$$
$$= \sqrt{(0.500 \text{ m})^2 + (0.800 \text{ m})^2}$$
$$= 0.943 \text{ m}$$

The magnitude F_{31}, found from Coulomb's law, is

$$F_{31} = \frac{kq_3q_1}{r_{13}^2}$$
$$= \frac{(9.00 \times 10^9 \text{ N m}^2/\text{C}^2)(4.00 \times 10^{-6} \text{ C})(13.0 \times 10^{-6} \text{ C})}{(0.943 \text{ m})^2}$$
$$= 0.526 \text{ N}$$

while the magnitude of $\mathbf{F}_{32}$ is

$$F_{32} = \frac{kq_3q_2}{r_{23}^2}$$
$$= \frac{(9.00 \times 10^9 \text{ N m}^2/\text{C}^2)(4.00 \times 10^{-6} \text{ C})(5.00 \times 10^{-6} \text{ C})}{(0.800 \text{ m})^2}$$
$$= 0.281 \text{ N}$$

The addition of the two vectors is an example of the addition of vectors discussed in chapter 2. The magnitude of the resultant vector is found, as before, as

$$F_3 = \sqrt{(F_{3x})^2 + (F_{3y})^2}$$

The vector $\mathbf{F}_{31}$ has an x-component given by

$$F_{31x} = F_{31} \cos \theta$$

The angle θ is found from the geometry of the diagram as

$$\theta = \tan^{-1}\frac{0.800 \text{ m}}{0.500 \text{ m}} = 58.0°$$

Therefore,

$$F_{31x} = F_{31} \cos \theta = (0.526 \text{ N}) \cos 58.0° = 0.279 \text{ N}$$

Similarly,

$$F_{31y} = F_{31} \sin \theta = (0.526 \text{ N}) \sin 58.0° = 0.446 \text{ N}$$

The vector $\mathbf{F}_{32}$ has only one component and that is in the y-direction. Therefore,

$$F_{3x} = F_{31x} = 0.279 \text{ N}$$

and

$$F_{3y} = F_{32} + F_{31y} = 0.281 \text{ N} + 0.446 \text{ N} = 0.727 \text{ N}$$

The magnitude of the resultant force on charge 3 is therefore

$$F_3 = \sqrt{(F_{3x})^2 + (F_{3y})^2} = \sqrt{(0.279 \text{ N})^2 + (0.727 \text{ N})^2}$$
$$= 0.779 \text{ N}$$

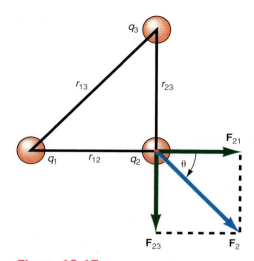

Figure 18.17

Force on charge 2.

The angle ϕ that $\mathbf{F_3}$ makes with the x-axis is found from

$$\phi = \tan^{-1}\frac{F_{3y}}{F_{3x}} = \tan^{-1}\frac{0.727 \text{ N}}{0.279 \text{ N}}$$

$$= 69.0°$$

<hr>

Example 18.5

More multiple charges. Find the resultant force on charge q_2 in figure 18.17 if $q_1 = 3.00 \ \mu C$, $q_2 = 5.00 \ \mu C$, $q_3 = 4.00 \ \mu C$, $r_{12} = 0.500$ m, and $r_{23} = 0.500$ m.

Solution

We find the resultant force on charge 2 from the vector sum

$$\mathbf{F_2} = \mathbf{F_{21}} + \mathbf{F_{23}}$$

Because $\mathbf{F_{21}}$ is perpendicular to $\mathbf{F_{23}}$, the magnitude of the resultant force is

$$F_2 = \sqrt{(F_{21})^2 + (F_{23})^2}$$

where

$$F_{21} = \frac{kq_2q_1}{r_{12}^2}$$

$$= \frac{(9.00 \times 10^9 \text{ N m}^2/\text{C}^2)(5.00 \times 10^{-6} \text{ C})(3.00 \times 10^{-6} \text{ C})}{(0.500 \text{ m})^2}$$

$$= 0.540 \text{ N}$$

and

$$F_{23} = \frac{kq_2q_3}{r_{23}^2}$$

$$= \frac{(9.00 \times 10^9 \text{ N m}^2/\text{C}^2)(5.00 \times 10^{-6} \text{ C})(4.00 \times 10^{-6} \text{ C})}{(0.500 \text{ m})^2}$$

$$= 0.720 \text{ N}$$

The magnitude of the resultant force on charge 2 is therefore

$$F_2 = \sqrt{(F_{21})^2 + (F_{23})^2} = \sqrt{(0.540 \text{ N})^2 + (0.720 \text{ N})^2}$$

$$= 9.00 \text{ N}$$

The direction of the resultant force is determined by

$$\theta = \tan^{-1}\frac{F_{23}}{F_{21}} = \tan^{-1}\frac{0.720 \text{ N}}{0.540 \text{ N}}$$

$$= 53.1°$$

<hr>

Example 18.6

Charged pith balls. Two equally charged pith balls are separated by 0.100 m as shown in figure 18.18(a). Find the charge on each ball and the tension in the string if the mass of each ball is 5.00×10^{-3} kg, and the length l of the string is 0.250 m.

Solution

Let us consider the forces acting on the ball at the right. The forces acting are the tension $\mathbf{T}$ in the string, the weight $\mathbf{w}$ of the ball, and the electric force of repulsion on the ball $\mathbf{F_e}$ as shown in figure 18.18(b).

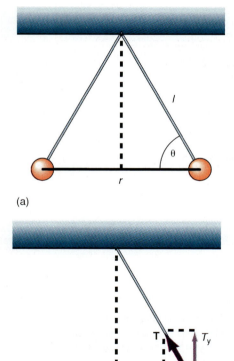

Figure 18.18

Charged pith balls.

Since the ball is in equilibrium the first condition of equilibrium is applied as

$$\Sigma F_y = 0, \qquad \Sigma F_x = 0$$

$$T_y - w = 0, \qquad F_e - T_x = 0$$

$$T \sin \theta = w, \qquad F_e = T \cos \theta$$

$$\frac{F_e}{w} = \frac{T \cos \theta}{T \sin \theta} = \frac{1}{\tan \theta}$$

$$F_e = \frac{w}{\tan \theta}$$

$$\frac{kq^2}{r^2} = \frac{mg}{\tan \theta}$$

Solving for the charge q on each ball we get

$$q = \sqrt{\frac{r^2 mg}{k \tan \theta}}$$

The angle θ, found from the geometry of the figure 18.18(a), is

$$\theta = \cos^{-1}\frac{r/2}{l}$$

$$= \cos^{-1}\frac{5.00 \text{ cm}}{25.0 \text{ cm}} = 78.5°$$

Therefore, the charge on each ball is

$$q = \sqrt{\frac{r^2 mg}{k \tan \theta}}$$

$$= \sqrt{\frac{(0.100 \text{ m})^2(5.00 \times 10^{-3} \text{ kg})(9.80 \text{ m/s}^2)}{(9.00 \times 10^9 \text{ N m}^2/\text{C}^2)\tan 78.5°}}$$

$$= 1.05 \times 10^{-7} \text{ C}$$

The tension in the string is

$$T \sin \theta = w$$

$$T = \frac{w}{\sin \theta} = \frac{mg}{\sin \theta}$$

$$= \frac{(5.00 \times 10^{-3} \text{ kg})(9.80 \text{ m/s}^2)}{\sin 78.5°}$$

$$= 5.00 \times 10^{-2} \text{ N}$$

The Language of Physics

Electrostatics
The study of electric charges at rest under the action of electric forces (p. 517).

The fundamental principle of electrostatics
Like electric charges repel each other, whereas unlike electric charges attract each other (p. 518).

Quarks
Elementary particles of matter. There are six quarks. They are up, down, strange, charm, bottom, and top. The proton and neutron are made of quarks, but the electron is not (p. 520).

Conductors
Materials that permit the free flow of electric charge through them (p. 521).

Insulators or dielectrics
Materials that do not permit the free flow of electric charge through them (p. 521).

Coulomb's law
The force between point charges q_1 and q_2 is directly proportional to the product of their charges and inversely proportional to the square of the distance separating them. The direction of the force lies along the line separating the charges (p. 523).

Summary of Important Equations

Coulomb's law

$$F = \frac{kq_1q_2}{r^2} \qquad \text{(18.1)}$$

Force caused by multiple charges

$$\mathbf{F} = \mathbf{F_1} + \mathbf{F_2} + \mathbf{F_3} + \mathbf{F_4} + \cdots \qquad \text{(18.3)}$$

Questions for Chapter 18

1. Describe the process of charging by induction.
†2. What did Ben Franklin have to do with the classification of electrical charge?
3. The difference between one chemical element and another is the number of protons that each contains. Is it possible for one chemical element to have the same number of neutrons as another chemical element?
4. Discuss the planetary model of the atom and state its good points and its limitations.

†5. Is it possible for a quark to exist if it has not been seen? If it cannot be isolated does it make any sense to describe particles as though they were made up of quarks?
†6. Describe the process of lightning in the atmosphere. How do the earth and the clouds pick up electric charge? If air is an insulator, how can a lightning bolt ever reach the ground?

7. Can you think of a way that you could use electrostatics to measure the humidity of the atmosphere?
8. How could you paint a metallic object with a minimum of paint using the principles of electrostatics?
†9. Describe the phenomenon known as "St. Elmo's Fire" in terms of electrostatics.
10. Why do clothes taken from a clothes dryer sometimes cling to the body?

Problems for Chapter 18

18.4 Coulomb's Law

1. How many electrons are contained in a charge of 3 μC?
2. A point charge of 4.00 μC is placed 25.0 cm from another point charge of -5.00 μC. Calculate the magnitude and direction of the force on each charge.
3. A point charge q_1 of 2.53 μC is placed 1.00 m in front of a second point charge q_2 of 8.64 μC. Find the magnitude and direction of the force on q_1.
4. If the force of repulsion between two protons is equal to the weight of the proton, how far apart are the protons?
5. The force between two point charges is 3.5×10^{-2} N. What is the force if the distance separating the charges is doubled?
6. What equal positive charges would have to be placed on the earth and the moon to neutralize the gravitational force between them?
7. What is the velocity of an electron in the hydrogen atom if the centripetal force is supplied by the coulomb force between the electron and proton? The radius of the electron orbit is 5.29×10^{-11} m.
8. Find the electrical force on an alpha particle when it is 6.00×10^{-11} m from an aluminum nucleus. (*Hint:*

find the number of protons in an alpha particle and the number of protons in an aluminum nucleus.)
9. Two identical metal spheres are placed 15.0 cm apart. A charge of 6.00 μC is placed on one sphere, whereas a charge of -2.00 μC is placed on the other. What is the force on each sphere? If the two spheres are brought together and touched and then separated to their original separation, what will be the force on each sphere?
10. Two identical metal spheres attract each other with a force of 5.00×10^{-6} N when they are 5.00 cm apart. The spheres are then touched together and then removed to the original separation where now a force of repulsion of 1.00×10^{-6} N is observed. What is the charge on each sphere after touching and before touching?
11. Two point charges repel each other with a force of 3.00×10^{-5} N when they are 20.0 cm apart. Find the force if the distance is reduced to 5.00 cm.

18.5 Multiple Charges

12. How far from a charge $q_1 = 3$ μC should you place a charge of $q_2 = 9$ μC, such that the charge q_2 experiences the force 3.98×10^{-2} N?

13. Three charges $q_1 = 2.00$ μC, $q_2 = 5.00$ μC, and $q_3 = 8.00$ μC are placed along the x-axis at 0.00 cm, 45.0 cm, and 72.4 cm, respectively. Find the force on each charge.
14. Three identical charges $q_1 = q_2 = q_3 = 1.00$ μC are placed along the x-axis at $x = 0, 0.750$, and 2.00 m. What is the magnitude and direction of the resultant force on each charge?
15. Three charges of 2.00 μC, -4.00 μC, and 6.00 μC are placed on the same line, each 15.0 cm apart. Find the resultant force on each charge.
16. Two charges are separated as shown. Where should a third charge be placed on the line between them such that the resultant force on it will be zero? Does it matter if the third charge is positive or negative?

$q_1 = 4 \times 10^{-6}$C $\qquad$ $D = 50$ cm $\qquad$ $q_2 = 8 \times 10^{-6}$C

17. Repeat problem 16 but now let charge q_1 be negative, and find any position on the line, either to the left of q_1, between q_1 and q_2, or to the right of q_2, where a third charge can be placed that experiences a zero resultant force.

†18. Three charges of 2.00 μC, -4.00 μC, and 6.00 μC are placed at the vertices of an equilateral triangle of length 10.0 cm on a side. Find the resultant force on each charge.

†19. If $q_1 = 5.00\ \mu$C $= q_2 = q_3 = q_4$ are located on the corners of a square of length 20.0 cm, find the resultant force on q_3.

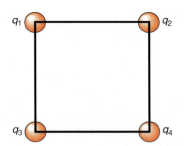

†20. Charges of 2.54 μC, $-7.86\ \mu$C, 5.34 μC, and $-3.78\ \mu$C are placed on the corners of a square of side 23.5 cm. Find the resultant force on the first charge.

†21. Find the force on charge q_3 in the diagram. The distance separating charges q_1 and q_2 is 5.00 cm.

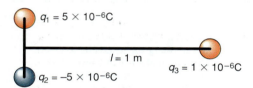

$q_1 = 5 \times 10^{-6}$C

$l = 1$ m

$q_2 = -5 \times 10^{-6}$C

$q_3 = 1 \times 10^{-6}$C

†22. Find the resultant force on charge q_3 in the diagram if $q_1 = 2.00\ \mu$C, $q_2 = -7\ \mu$C, and $q_3 = 5.00\ \mu$C.

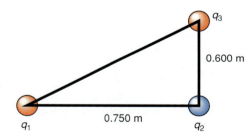

q_3

0.600 m

q_1

0.750 m

q_2

Additional Problems

†23. Find the force on charge $q_5 = 5.00$ μC, located at the center of a square 25.0 cm on a side if $q_1 = q_2 = 3.00$ μC and $q_3 = q_4 = 6.00\ \mu$C.

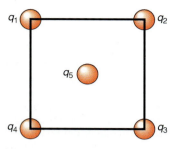

q_1 q_2

q_5

q_4 q_3

24. Four electrons are located at the corners of a rectangle, 4.00 cm by 3.00 cm, one electron at each corner. Find the magnitude and direction of the net force on each electron due to the other three.

†25. Two small, equally charged spheres of mass 0.500 g are suspended from the same point by a silk fiber 50.0 cm long. The repulsion between them keeps them 15.0 cm apart. What is the charge on each sphere?

†26. Two pith balls of 10.0-g mass are hung from ends of a string 25.0 cm long, as shown. When the balls are charged with equal amounts of charge, the threads separate to an angle of 30.0°. What is the charge on each ball?

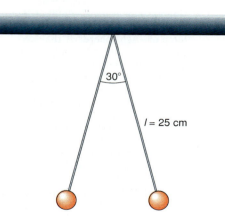

30°

$l = 25$ cm

†27. Two 10.0-g pith balls are hung from the ends of two 25.0-cm long strings as shown. When an equal and opposite charge is placed on each ball, their separation is reduced from 10.0 cm to 8.00 cm. Find the tension in each string and the charge on each ball.

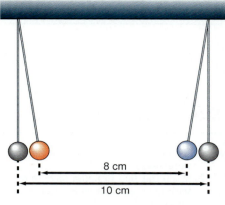

8 cm

10 cm

28. Suppose that equal charges are to be placed on the earth and on a 54.5-kg woman so as to render the woman effectively weightless. How much would the charge on each body have to be?

†29. Find the force on q_3 in the diagram if $q_1 = 8.00\ \mu$C, $q_2 = -8.00\ \mu$C, and $q_3 = 8.00\ \mu$C.

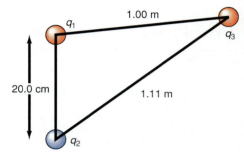

q_1 1.00 m q_3

20.0 cm   1.11 m

q_2

†30. A charge of 15.0 μC is on a metallic sphere 10.0-cm radius. It is then touched to a sphere of 5.00-cm radius, until the surface charge density is the same on both spheres. What is the charge on each sphere after they are separated?

†31. Two small spheres carrying charges $q_1 = 7.00\ \mu$C and $q_2 = 5.00\ \mu$C are separated by 20.0 cm. If q_2 were free to move, what would its initial acceleration be? Sphere 2 has the mass $m_2 = 15.0$ g.

†32. Where should a fourth charge, $q_4 = 3.00 \ \mu C$, be placed to give a net force of zero on charge q_3? Charges $q_1 = 2.00 \ \mu C$, $q_2 = 4.00 \ \mu C$, and $q_3 = 2.00 \ \mu C$.

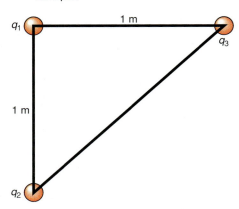

†33. Charge $q_1 = 3.00 \ \mu C$ is located at the coordinates $(0,2)$ and charge $q_2 = 6.00 \ \mu C$ is located at the coordinates $(1,0)$ of a Cartesian coordinate system. Find the coordinates of a third charge that will experience a zero net force.

†34. The configuration of a positive charge q separated by a distance $2a$ from a negative charge $-q$, is called an electric dipole. Show that the force exerted by an electric dipole on a point charge q_0, located as shown in the diagram varies as $1/r^3$ while the force between a point charge q and the point charge q_0 varies as $1/r^2$. Which force is the weaker?

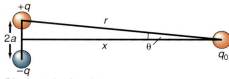

Dipole and point charge

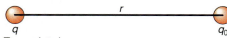

Two point charges

35. A charge of 3.00 μC is located at the origin and a charge of $-3.00 \ \mu C$ is located on the x-axis at the point $x = 10.0$ cm, forming an electric dipole. Find the force exerted by this dipole on a proton located on the x-axis at the point $x = 40.0$ cm.

36. A charge of 3.00 μC is located at the origin and a charge of $-3.00 \ \mu C$ is located on the x-axis at the point $x = 10.0$ cm, forming an electric dipole. Find the force exerted by this dipole on a proton located at the point $x = 25.0$ cm and $y = 35.0$ cm.

†37. A plastic rod 50.0 cm long has a charge $+q_1 = 2.00 \ \mu C$ at each end. The rod is then hung from a string and placed so that each charge is only 5.00 cm from negative charges $q_2 = -10.0 \ \mu C$, as shown in the diagram. Find the torque acting on the string.

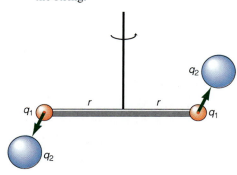

38. A 10.0-g object with a 10.0-μC charge is suspended from a vertical spring of spring constant 50 N/m. A charge of $-5.00 \ \mu C$ is put on the ground directly below the vertical spring. As a result of the presence of the negative charge, the spring stretches by 18.0 cm relative to its equilibrium position before the negatively charged mass was placed below it. Find the distance between the two charged masses in this new equilibrium position. Assume that they are point objects.

†39. A charge of 5.00 μC is uniformly distributed over a copper ring 2.00 cm in radius. What force will this ring exert on a point charge of 8.00 μC that is placed 3.00 m away from the ring? Indicate what assumptions you make to solve this problem.

†40. A charge of 2.50 μC is placed at the center of a hollow sphere of charge of 8.00 μC. What is the resultant force on the charge placed at the center of the sphere? Indicate what assumptions you make to solve this problem.

Interactive Tutorials

⊟ 41. Coulomb's law. Two charges $q_1 = 2.00 \times 10^{-6}$ C and $q_2 = 3.00 \times 10^{-6}$ C are separated by a distance $r = 1.00$ m. Calculate the electrostatic force F of repulsion acting on charge 1 as the distance of separation is increased from $r = 1$ to $r = 10$ m. Show how the force F varies with the distance r.

⊟ 42. Two electrons of charge $q_1 = q_2 = e = 1.60 \times 10^{-19}$ C are positioned at the coordinates $(0,1)$ and $(0,-1)$ (in meters) of a Cartesian coordinate system. Calculate the net force F_3 on an electron q_3 as it is moved from $x = 0$ to $x = 10.0$ m along the x-axis.

⊟ 43. Coulomb's law and multiple charges. Two charges $q_1 = 8.32 \times 10^6$ C and $q_2 = -2.55 \times 10^6$ C lie on the x-axis and are separated by the distance $r_{12} = 0.823$ m. A third charge $q_3 = 3.87 \times 10^6$ C is located a distance $r_{23} = 0.475$ m from charge q_2, and the line between charge 2 and 3 makes an angle $\phi = 60.0°$ with respect to the x-axis. Find the resultant force on (a) charge 3, (b) charge 2, and (c) charge 1.

⊟ 44. Coulomb's law and a continuous charge distribution. A rod of charge of length $L = 0.100$ m lies on the x-axis. One end of the rod lies at the origin and the other end is on the positive x-axis. A charge $q' = 7.36 \times 10^{-6}$ C is uniformly distributed over the rod. Find the force exerted on a point charge $q = 2.95 \times 10^{-6}$ C that lies on the x-axis at distance $x_0 = 0.175$ m from the origin of the coordinate system.

19

The potential of an electric dipole.

I wish to give an account of some investigations which have led to the conclusion that the carriers of negative electricity are bodies, which I have called corpuscles, having a mass very much smaller than that of the atom of any known element and are of the same character from whatever source the negative electricity may be derived.

J. J. Thomson

Electric Fields

19.1 The Electric Field

In chapter 18, we discussed Coulomb's law of electrostatics and saw that, if a charge q_1 is brought into the neighborhood of another charge q_2, a force is exerted on q_1. The magnitude of that force is given by Coulomb's law. However, we should ask what is the mechanism that transmits the force from q_2 to q_1? Coulomb's law states only that there is a force; it says nothing about the mechanism by which the force is transmitted, and it assumes that the force is transmitted instantaneously. Such a force is called an "action at a distance" since it does not explain how the force travels through that distance. Even the ancient Greek philosophers would not have accepted such an idea, it seems too much like magic.

To overcome this shortcoming of Coulomb's law, Michael Faraday (1791–1867) introduced the concept of an electric field. He stated that *it is an intrinsic property of nature that an **electric field** exists in the space around an electric charge. This electric field is considered to be a force field that exerts a force on charges placed in the field.* For example, around the charge q_2 there exists an electric field. When the charge q_1 is brought into the neighborhood of q_2, the electric field of q_2 interacts with q_1, thereby exerting a force on q_1. The electric field becomes the mechanism for transmitting the force from q_2 to q_1, thereby eliminating the "action at a distance" principle.

Because the electric field is considered to be a force field, the existence of an electric field and its strength is determined by the effect it produces on a positive point charge q_0 placed in the region where the existence of the field is suspected. If the point charge, called a test charge, experiences an electrical force acting on it, then it is said that the test charge is in an electric field. *The electric field is measured in terms of a quantity called the **electric field intensity**. The magnitude of the electric field intensity is defined as the ratio of the force F acting on the small test charge q_0 to the small test charge itself. The direction of the electric field is in the direction of the force on the positive test charge.* We can write this as

$$E = \frac{F}{q_0} \tag{19.1}$$

that is, the force acting per unit charge. The SI unit of electric field intensity is a newton per coulomb, abbreviated as N/C. Note also at this point that the small positive test charge q_0 must be small enough so that it does not appreciably distort the electric field that we are trying to measure. (In the measurement of any physical quantity the instruments of measurement should be designed to interfere as little as possible with the quantity being measured.)

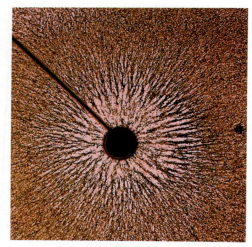

The electric field lines of a single positive point charge. Small bits of thread are suspended in oil and become aligned with the electric field. Note that as we move away from the charge, the lines are farther apart, indicating that the electrical field becomes weaker.

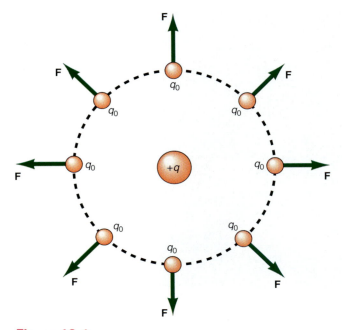

Figure 19.1

The force exerted by a positive point charge.

19.2 The Electric Field of a Point Charge

The electric field of a positive point charge q can be determined by following the definition in equation 19.1. A positive point charge q is shown in figure 19.1. A very small positive test charge q_0 is placed in various positions around the positive charge q. Because like charges repel each other, the positive test charge q_0 experiences a force of repulsion from the positive point charge q. Thus, the force acting on the test charge, and hence the direction of the electric field, is always directed radially away from the point charge q. We find the magnitude of the electric field intensity of a point charge from equation 19.1, with the force found from Coulomb's law,

$$F = \frac{kqq_0}{r^2} \tag{18.1}$$

That is,

$$E = \frac{F}{q_0} = \frac{(kqq_0)/r^2}{q_0}$$

$$E = \frac{kq}{r^2} \tag{19.2}$$

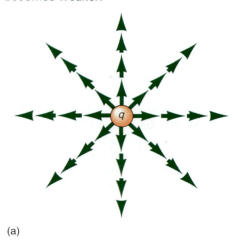

(a)

Equation 19.2 is the equation for the magnitude of the electric field intensity due to a point charge. We have already seen that the direction of the electric field is radially away from the positive point charge. Drawing a picture of the total electric field of a positive point charge is slightly difficult because the electric field is a vector, and as we recall from the study of vectors, a vector is a quantity that has both magnitude and direction. Since the magnitude of the electric field of the point charge varies with the distance r from the charge, a picture of the total electric field would have to be a series of discrete electric vectors, each one pointing radially away from the positive point charge, but the length of each vector would vary depending on how far we are away from the point charge. This is shown in figure 19.2(a). To simplify the picture of the electric field, a series of continuous lines is drawn from the positive point charge to indicate the total electric field, as in figure 19.2(b). Michael Faraday called these continuous lines lines of force because they are in the direction of the force that acts on a positive point charge placed in the field.

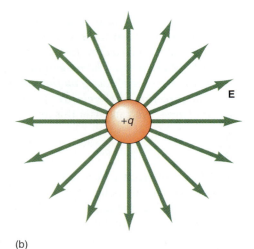

(b)

Figure 19.2

Electric field of a positive point charge.

Electricity and Magnetism

They are everywhere tangent to the direction of the electric field. We must understand, however, that the magnitude of the electric field varies along the lines shown. The greater the distance from the point charge, the smaller the magnitude of the electric field. With these qualifications in mind, we will say that the electric field of a positive point charge is shown in figure 19.2(b). (We will use this same technique to depict the electric fields in the rest of this book.) Note that the electric field always emanates from a positive charge. *The electric field intensity of a point charge is directly proportional to the charge that creates it, and inversely proportional to the square of the distance from the point charge to the position where the field is being evaluated.*

The electric field intensity of a negative point charge is found in the same way. But because unlike charges attract each other, the force between the negatively charged point source and the small positive test charge is one of attraction. Hence, the force is everywhere radially inward toward the negatively charged point source and the electric field is also. Therefore, the electric field of a negative point charge is as shown in figure 19.3. The magnitude of the electric field intensity of a negative point charge is also given by equation 19.2.

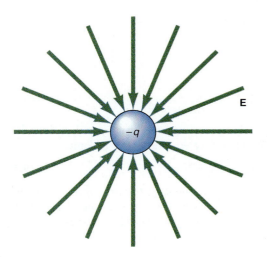

Figure 19.3

Electric field of a negative point charge.

Example 19.1

The electric field of a point charge. Find the magnitude of the electric field intensity at a distance of 0.500 m from a 3.00 μC charge.

Solution

The electric field intensity, found from equation 19.2, is

$$E = \frac{kq}{r^2} = \left(\frac{9.00 \times 10^9 \text{ N m}^2}{\text{C}^2}\right)\frac{(3.00 \times 10^{-6} \text{ C})}{(5.00 \times 10^{-1} \text{ m})^2}$$

$$= 1.08 \times 10^5 \text{ N/C}$$

If the electric field intensity is known in a region, then we determine the force on any charge q placed in that field from equation 19.1 as

$$\mathbf{F} = q\mathbf{E} \tag{19.3}$$

Example 19.2

The force on a charge in an electric field. A point charge, $q = 5.64$ μC, is placed in an electric field of 2.55×10^3 N/C. Find the force acting on the charge.

Solution

The force acting on the point charge, found from equation 19.3, is

$$F = qE = (5.64 \times 10^{-6} \text{ C})(2.55 \times 10^3 \text{ N/C})$$

$$= 1.44 \times 10^{-2} \text{ N}$$

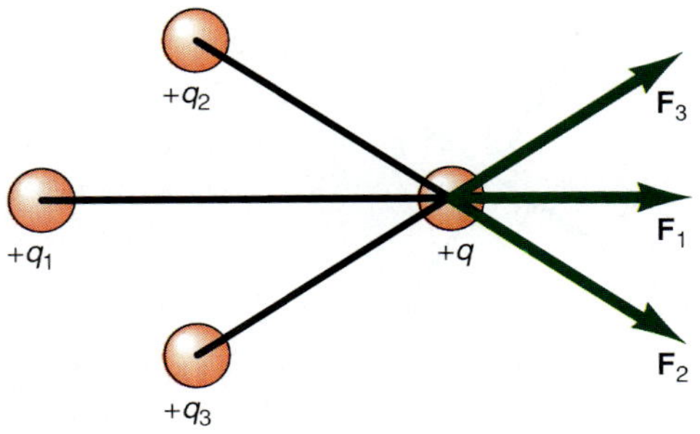

19.3 Superposition of Electric Fields

When more than one charge is present, as in figure 19.4, the force on an arbitrary charge q is the vector sum of the forces produced by each charge, that is,

$$\mathbf{F} = \mathbf{F}_1 + \mathbf{F}_2 + \mathbf{F}_3 + \cdots \tag{19.4}$$

Figure 19.4

Multiple charges.

But if charge q_1 produces a field $\mathbf{E}_1$, then the force on charge q produced by $\mathbf{E}_1$ is found from equation 19.3 as

$$\mathbf{F}_1 = q\mathbf{E}_1 \qquad (19.5)$$

Similarly, the force on charge q produced by $\mathbf{E}_2$ is

$$\mathbf{F}_2 = q\mathbf{E}_2 \qquad (19.6)$$

and finally

$$\mathbf{F}_3 = q\mathbf{E}_3 \qquad (19.7)$$

Substituting equations 19.5, 19.6, and 19.7 into equation 19.4 gives

$$\mathbf{F} = q\mathbf{E}_1 + q\mathbf{E}_2 + q\mathbf{E}_3 + \cdots$$

Dividing each term by q gives

$$\frac{\mathbf{F}}{q} = \mathbf{E}_1 + \mathbf{E}_2 + \mathbf{E}_3 + \cdots \qquad (19.8)$$

But $\mathbf{F}/q$ is the resultant force per unit charge acting on charge q and is thus the total resultant electric field intensity $\mathbf{E}$. Therefore equation 19.8 becomes

$$\mathbf{E} = \mathbf{E}_1 + \mathbf{E}_2 + \mathbf{E}_3 + \cdots \qquad (19.9)$$

Equation 19.9 is the mathematical statement of the principle of the **superposition of electric fields**, which states that when more than one charge contributes to the electric field, the resultant electric field is the vector sum of the electric fields produced by the various charges.

Example 19.3

The electric field of two positive charges. If two equal positive charges, $q_1 = q_2 = 2.00\ \mu C$ are situated as shown in figure 19.5, find the resultant electric field intensity at point A. Note that angle $\theta_1 = 35.0°$ and $\theta_2 = 55°$.

Solution

The magnitude of the electric field intensity produced by q_1, found from equation 19.2, is

$$E_1 = \frac{kq_1}{r_1^2} = \frac{(9.00 \times 10^9\ \text{N m}^2/\text{C}^2)(2.00 \times 10^{-6}\ \text{C})}{(0.819\ \text{m})^2}$$

$$= 2.68 \times 10^4\ \text{N/C}$$

A large static charge causes a person's hair to stand on end. Note the hairs are in line with the electric field.

Figure 19.5

Finding the electric field caused by two point charges.

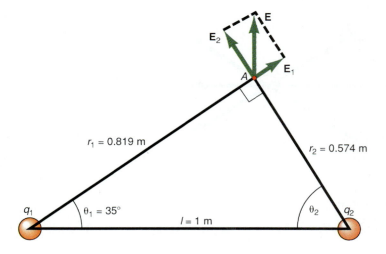

The magnitude of the electric field intensity produced by q_2 is

$$E_2 = \frac{kq_2}{r_2^2} = \frac{(9.00 \times 10^9 \text{ N m}^2/\text{C}^2)(2.00 \times 10^{-6} \text{ C})}{(0.574 \text{ m})^2}$$

$$= 5.46 \times 10^4 \text{ N/C}$$

We find the resultant electric field from equation 19.9 as the vector addition

$$\mathbf{E} = \mathbf{E_1} + \mathbf{E_2}$$

which we find by the component method of vector addition. The magnitude of the resultant vector is

$$E = \sqrt{E_x^2 + E_y^2} \qquad (19.10)$$

where E_x is the x-component of the resultant vector, E_y is the y-component of the resultant vector, and they are as shown in figure 19.6. The x-components of the electric field intensities are

$$E_{1x} = E_1 \cos \theta_1 = (2.68 \times 10^4 \text{ N/C})\cos 35.0°$$

$$= 2.20 \times 10^4 \text{ N/C}$$

$$E_{2x} = E_2 \cos \theta_2 = (5.46 \times 10^4 \text{ N/C})\cos 55.0°$$

$$= 3.13 \times 10^4 \text{ N/C}$$

and the x-component of the resultant field is

$$E_x = E_{1x} - E_{2x} = 2.20 \times 10^4 \text{ N/C} - 3.13 \times 10^4 \text{ N/C}$$

$$= -0.930 \times 10^4 \text{ N/C}$$

The y-components of the electric field intensities are

$$E_{1y} = E_1 \sin \theta_1 = (2.68 \times 10^4 \text{ N/C})\sin 35.0°$$

$$= 1.54 \times 10^4 \text{ N/C}$$

$$E_{2y} = E_2 \sin \theta_2 = (5.46 \times 10^4 \text{ N/C})\sin 55.0°$$

$$= 4.47 \times 10^4 \text{ N/C}$$

and the y-component of the resultant field is

$$E_y = E_{1y} + E_{2y} = 1.54 \times 10^4 \text{ N/C} + 4.47 \times 10^4 \text{ N/C}$$

$$= 6.01 \times 10^4 \text{ N/C}$$

Figure 19.6

The resultant electric field.

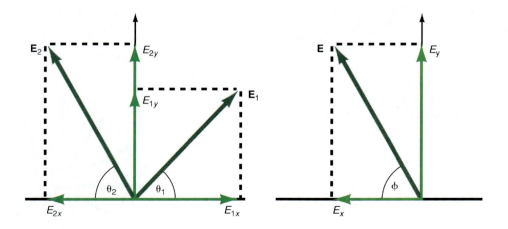

The electric field lines for two equal positive point charges. Small bits of thread are suspended in oil and become aligned with the electric field.

The magnitude of the resultant electric field intensity at point A, found from equation 19.10, is

$$E = \sqrt{E_x^2 + E_y^2} = \sqrt{(-0.930 \times 10^4 \text{ N/C})^2 + (6.01 \times 10^4 \text{ N/C})^2}$$

$$= 6.08 \times 10^4 \text{ N/C}$$

The direction of the electric field vector is found from

$$\tan \phi = \frac{E_y}{E_x} = \frac{6.01 \times 10^4 \text{ N/C}}{-0.930 \times 10^4 \text{ N/C}} = -6.46$$

$$\phi = \tan^{-1} \frac{E_y}{E_x} = -81.2°$$

Because E_x is negative, the angle ϕ lies in the second quadrant. The angle that the vector $\mathbf{E}$ makes with the positive x-axis is $\phi + 180° = 98.8°$.

Thus, by the principle of superposition, the electric field can be determined at any point for any number of charges. However, we have just determined the field at only one point. If we want to see a picture of the entire electric field, as for the point charge, we must evaluate E vectorially at an extremely large number of points. This problem can be readily solved by the use of a computer. Equation 19.9 gives the electric field for any number of discrete charges. To find the electric field of a continuous distribution of charge, we need to use the calculus and the sum in equation 19.9 becomes an integration.

Electric Field of Two Like Charges

The total electric field at any point A, caused by two positive charges, q_1 and q_2, separated by a distance d, is

$$\mathbf{E} = \mathbf{E_1} + \mathbf{E_2}$$

Since it is not obvious what this combined electric field looks like, a good picture of it can be obtained by the graphical method of the vector addition of fields. *The graphical method of vector addition is based on the superposition principle. The resultant electric field of two different charge distributions is simply the vector sum of the individual electric fields due to each charge distribution.* To determine the electric field of two positive point charges, draw the electric field of each point charge and then superimpose one field over the other.

At every point where the electric field lines of charge 1 intersect the electric field lines of charge 2, the two separate fields are added vectorially. This can be done by drawing lines in red through every intersection in the direction of the sum, as shown in figure 19.7. The total electric field of two positively charged particles is shown in figure 19.8.

Figure 19.7

Superposition of the electric fields produced by two positive point charges.

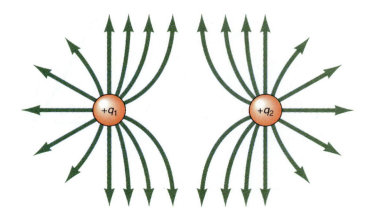

Figure 19.8

Electric field of two positively charged particles.

Electricity and Magnetism

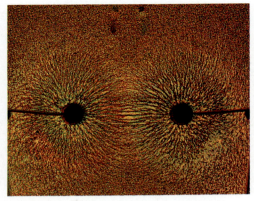

The electric field lines for two equal but opposite point charges. Small bits of thread are suspended in oil and become aligned with the electric field.

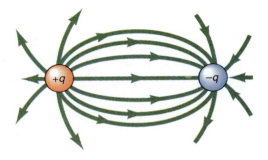

Figure 19.10

Electric field of two unlike point charges.

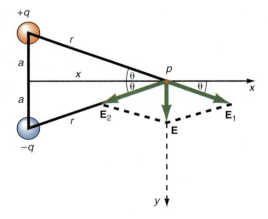

Figure 19.11

The electric dipole.

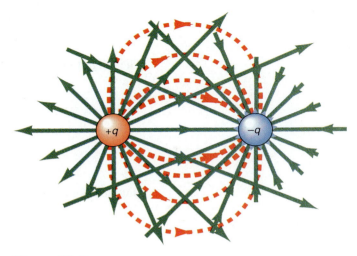

Figure 19.9

Superposition of electric fields of two unlike point charges.

The Electric Field of Two Unlike Charges—The Electric Dipole

The electric field of two equal but unlike charges can also be determined by the graphical addition of electric fields, as shown in figure 19.9. After drawing the electric field of the positive charge, which is a radial field directed away from the center of the charge, draw the electric field of the negative charge, which is also a radial field, but is directed toward the center of the negative charge. At every point where the electric field lines of charge 1 intersect the electric field lines of charge 2, they are added vectorially. This can be done by drawing lines in red through every intersection in the direction of the sum, as shown in figure 19.9. This is the resultant electric field of two unlike charges. The more lines drawn for the original fields, the more detail is apparent in the resultant field. The total resultant field for two unlike point charges is shown in figure 19.10. This configuration of two closely spaced, equal but opposite point charges is a very important one and we give it the special name *electric dipole*.

Example 19.4

Electric field of a dipole. Find the electric field intensity at a point P along the perpendicular bisector of an electric dipole, as shown in figure 19.11. Assume that $x \gg a$.

Solution

The resultant electric field intensity is found by the superposition principle as

$$\mathbf{E} = \mathbf{E_1} + \mathbf{E_2}$$

The magnitudes of E_1 and E_2 are found as

$$E_1 = \frac{kq}{r^2} = E_2 = \frac{kq}{r^2}$$

The x-component of the resultant field at the point P is

$$E_x = E_{1x} - E_{2x} = E_1 \cos\theta - E_2 \cos\theta = 0$$

since $E_1 = E_2$.

The y-component of the resultant field at the point P is

$$E_y = E_{1y} + E_{2y} = E_1 \sin\theta + E_2 \sin\theta$$

Because $E_1 = E_2$, this reduces to

$$E_y = E_1 \sin \theta + E_1 \sin \theta = 2E_1 \sin \theta$$

The magnitude of the resultant field is

$$E = \sqrt{E_x^2 + E_y^2} = \sqrt{0 + E_y^2} = E_y$$

Therefore,

$$E = 2E_1 \sin \theta$$

$$E = \frac{2kq}{r^2} \sin \theta \qquad (19.11)$$

But from figure 19.11 we see that

$$\sin \theta = \frac{a}{r}$$

Therefore,

$$E = \frac{2kq}{r^2} \left(\frac{a}{r} \right)$$

$$E = \frac{k2aq}{r^3} \qquad (19.12)$$

but from the diagram we see that

$$r = \sqrt{a^2 + x^2}$$

Thus,

$$E = \frac{k(2aq)}{(a^2 + x^2)^{3/2}} \qquad (19.13)$$

Because we assumed in the problem that x is very much greater than a, as a first approximation we can let

$$(a^2 + x^2)^{3/2} = (0 + x^2)^{3/2} = x^3$$

Therefore, the magnitude of the electric field intensity along the perpendicular bisector, a distance x from the dipole, is

$$E = \frac{k(2aq)}{x^3} \qquad (19.14)$$

The direction of the electric field, as we can see from figure 19.11, is parallel to the axis of the dipole, pointing from the positive charge toward the negative charge. We call the quantity $2aq$, which is the product of the charge q times the distance separating the charges $2a$, the *electric dipole moment p*. That is,

$$p = 2aq \qquad (19.15)$$

The SI unit for the electric dipole is a coulomb meter, abbreviated C m. Note that the electric field of the dipole varies as $1/x^3$, whereas the electric field of a point charge varies as $1/x^2$. Thus the electric field of a dipole decreases faster with distance than the electric field of a point charge.

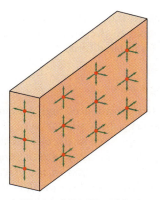

(a) Electric field of individual point charges on the plate

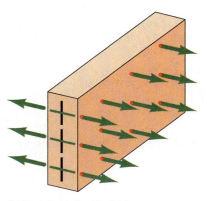

(b) Resultant electric field

Figure 19.12

Charged conducting plate.

Figure 19.13

Electric field of a parallel plate capacitor.

19.4 The Electric Field of a Charged Conducting Plate

Because any charge placed on a conducting body resides on its outer surface, the charged conducting plate has charges only on its outer surface. The electric field inside the conducting plate is zero. Any one point on the surface has the electric field of a point charge, radially outward in every direction. But because of the symmetry of all the charges surrounding the point charge, all the radial electric fields cancel except for those that are perpendicular to the plate. Therefore, the electric field of a charged conducting plate is perpendicular to the plate, as shown in figure 19.12. If the plate is charged negatively, the electric field is again perpendicular to the plate, but this time it points toward the plate. Note that the electric field lines are parallel to each other.

19.5 The Electric Field of Two Parallel Charged Conducting Plates

If two parallel equally charged plates, one positive and one negative, are arranged as in figure 19.13(a), we call the configuration a *parallel plate capacitor*. Figure 19.13(b) shows the superposition of the two electric fields. The electric field of the positive plate is emanating from the plate, whereas the electric field of the negative plate is directed into the plate. Outside the plates, the electric field of one plate is equal and opposite to the electric field of the other plate, and hence they cancel each other, whereas between the plates, the electric fields of the two plates point in the same direction and hence add. Therefore, the electric field of a parallel plate capacitor is a uniform field between the plates, as shown in figure 19.13(c). At the top and bottom edges of the plates the field spreads out to a slight extent. But if the plates are sufficiently large, and we stay away from the edges, this fringing can be neglected and the field between the plates is uniform and parallel.

19.6 Electric Potential Energy and the Potential

Although we define the electric field as a force on a unit test charge, it is extremely difficult to measure fields in this way. Let us return to the energy concepts developed in chapter 7, and apply them to electric charges in an electric field to develop another way to describe electric fields.

Recall that we defined potential energy as the energy that a body possessed by virtue of its position, and that potential energy was equal to the work that had

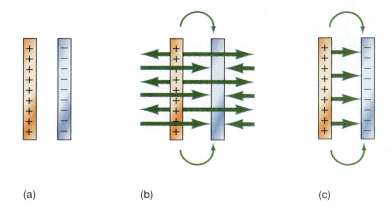

(a)   (b)   (c)

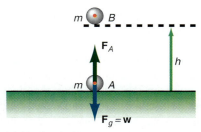

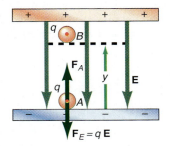

(a) Mechanical

(b) Electrical

Figure 19.14

The potential energy.

The electric field lines between the plates of a parallel-plate capacitor. Small bits of thread are suspended in oil and become aligned with the electric field. Note that the lines are equally spaced, indicating that the electric field there is uniform.

to be done to place the body in that position. For example, the gravitational potential energy of the body at point B, figure 19.14(a), is equal to the work done against gravity, the gravitational field, to lift the body, with an applied force F_A, from point A to point B, through the distance h. That is,

$$\text{PE} = W = F_A y = F_g\, y = wh = mgh$$

When the body was released, it fell from that position of high potential energy to the ground where it has zero potential energy. In a similar way, if an electric charge is placed in a parallel, uniform electric field, such as the electric field between the plates of a parallel plate capacitor, *the potential energy of that charge can be defined as the energy it possesses by virtue of its position in the electric field. The potential energy is equal to the work that must be done to place that charge into that position in the electric field.* Figure 19.14(b) shows a charge in a uniform electric field that emanates from the positive plate at the top, points downward, and terminates on the negative plate at the bottom. The potential energy that a positive charge has at position B is the work that must be done by an external agent to move that charge from the bottom plate A to the position B. That is,

$$\text{PE} = W = F_A y \tag{19.16}$$

But the magnitude of the applied external force F_A is just equal to the magnitude of the electric force on the charge, that is,

$$F_A = F_E = qE \tag{19.17}$$

Substituting equation 19.17 into equation 19.16 yields the potential energy of the charge at B as

$$\text{PE} = qEy \tag{19.18}$$

Example 19.5

The potential energy of a charge in a uniform electric field. A point charge of 8.00 pC is placed 5.00 mm above the negative plate of a parallel plate capacitor that has an electric field intensity of 4.00×10^4 N/C. Find the potential energy of the point charge at this location.

Solution

The potential energy of the point charge, found from equation 19.18, is

$$
\begin{aligned}
\text{PE} &= qEy \\
&= (8.00 \times 10^{-12}\ \text{C})(4.00 \times 10^4\ \text{N/C})(5.00 \times 10^{-3}\ \text{m}) \\
&= 1.60 \times 10^{-9}\ \text{J}
\end{aligned}
$$

Just as a mass dropped from a height h above the surface of the earth has its potential energy converted to kinetic energy of motion, the positive electric charge released from position B falls toward the bottom plate converting its potential energy to kinetic energy.

The Potential

Because the potential energy depends on the charge q, and it is sometimes difficult to work directly with electric charges, it is desirable to define a new quantity that is independent of the charge. Hence, *the electric potential V is defined as the potential energy per unit charge,* that is,

$$V = \frac{\text{PE}}{q} = \frac{W}{q} \tag{19.19}$$

Electricity and Magnetism

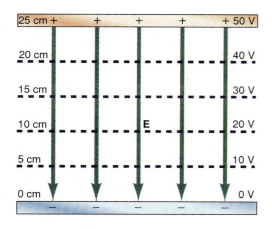

Figure 19.15

The potential field between charged parallel plates.

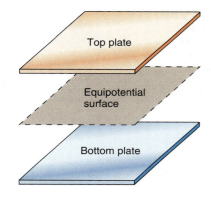

Figure 19.16

An equipotential surface.

Note the analogy between the potential being a potential energy per unit charge and the electric field being a force per unit charge. The SI unit of potential is defined as the volt, where

$$1 \text{ volt} = 1 \frac{\text{joule}}{\text{coulomb}} = 1 \text{ J/C}$$

For the parallel plate configuration of figure 19.14(b), the potential becomes

$$V = \frac{\text{PE}}{q} = \frac{qEy}{q}$$

$$V = Ey \tag{19.20}$$

Just as the electric field exists in a region around an electric charge, we can also talk about a *potential field* existing in a region about an electric charge. Whereas the electric field is a vector field, the potential field is a scalar field. As an example, if the electric field intensity between the plates of figure 19.15 is 200 N/C, let us find the potential field at intervals of 5.00 cm. The distance y is measured from the negative plate, therefore the potential at the negative plate becomes

$$V_0 = Ey = (200 \text{ N/C})(0) = 0$$

The negative plate is therefore the zero of our potential. At a height of 5.00 cm above the negative plate the potential is

$$V_5 = Ey = (200 \text{ N/C})(0.0500 \text{ m})$$
$$= 10.0 \text{ N m/C} = 10.0 \text{ J/C}$$
$$= 10.0 \text{ V}$$

Therefore, on a line 5.00 cm above the bottom plate the potential is 10.0 V everywhere. *This dotted line in the figure is called an **equipotential line**, a line of equal or constant potential.* If we drew this figure in three dimensions, it would be an **equipotential surface,** as shown in figure 19.16. The potentials at 10.0 cm, 15.0 cm, 20.0 cm, and 25.0 cm are found as

$$V_{10} = Ey = (200 \text{ N/C})(0.100 \text{ m}) = 20.0 \text{ V}$$
$$V_{15} = Ey = (200 \text{ N/C})(0.150 \text{ m}) = 30.0 \text{ V}$$
$$V_{20} = Ey = (200 \text{ N/C})(0.200 \text{ m}) = 40.0 \text{ V}$$
$$V_{25} = Ey = (200 \text{ N/C})(0.250 \text{ m}) = 50.0 \text{ V}$$

All these equipotentials are shown in figure 19.15 as dotted lines.

A very interesting result can be observed in figure 19.15. *The equipotential lines are everywhere perpendicular to the electric field lines.* Although we have determined this for the field between the parallel plates of a capacitor, it is true in general for any electric field and its equipotentials.

The numerical value of the potential, like mechanical potential energy, depends on a particular reference position. In this example, the potential was zero at the negative plate because the value of y for equation 19.20 was zero there. In practical problems, such as the electrical wiring in a home or office, the zero of potential is usually taken to be that of the surface of the earth, literally the "ground." When you plug an appliance into a wall outlet in your home, it sees a potential of 120 V. One wire, the "hot" wire, is 120 V above the other wire, the ground or zero wire. (The ground wire, is truly a ground wire. If you look into the electric box servicing your home you will see the ground wire, usually an uninsulated wire. If you follow this ground wire, you will see that it is connected to the cold water pipe of your home. The cold water pipe is eventually buried in the ground. Thus, the zero of potential in your home is in fact the potential of the ground.) When dealing with individual point charges the zero of potential is usually taken at infinity, a step that we will justify later.

The introduction of energy concepts into the description of electric fields now pays its dividend. Solving equation 19.20 for E, the electric field intensity between the plates, gives

$$E = \frac{V}{y} \qquad \text{(19.21)}$$

That is, knowing the potential between the plates, something that is easily measured with a voltmeter, and the distance separating the plates, we obtain a new indirect way of determining the electric field intensity.

When equation 19.21 is used to determine the electric field, an equivalent unit of electric field intensity, the volt per meter, can be used. To see that this is equivalent note that

$$\frac{\text{volt}}{\text{meter}} = \frac{\text{joule/coulomb}}{\text{meter}} = \frac{\text{newton meter}}{\text{coulomb meter}} = \frac{\text{newton}}{\text{coulomb}}$$

The Potential Difference

Instead of knowing the actual potential at a particular point, we sometimes want to know the difference in potential between two points, A and B. If the point A is at the ground potential, then the potential and potential difference is the same. If A is not the ground potential, then the **potential difference** between point A and point B is just the difference between the potential at B and the potential at A, as shown in figure 19.17. From equation 19.20

$$V_B = Ey_B$$

while

$$V_A = Ey_A$$

Therefore, the potential difference between points A and B is

$$\Delta V = V_B - V_A = Ey_B - Ey_A$$
$$= E(y_B - y_A)$$
$$\Delta V = E\Delta y \qquad \text{(19.22)}$$

In general, then, we can find the magnitude of the electric field intensity from equation 19.22 as

$$E = \frac{\Delta V}{\Delta y} \qquad \text{(19.23)}$$

Equation 19.23 gives the average value of the magnitude of the electric field intensity. (Note that if the plate configuration is such that the field is in the x-direction,

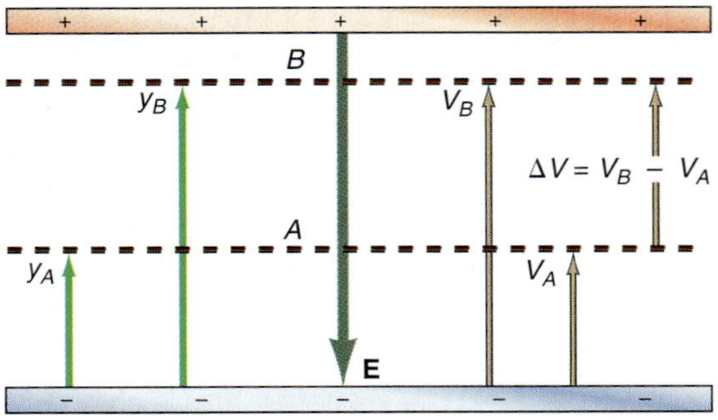

Figure 19.17

The potential difference.

then $E = \Delta V / \Delta x$.) For the parallel plate configuration, the electric field intensity is a constant and hence the average value is the same as the constant value. When dealing with fields that are not constant, equation 19.23 gives us an average value of the field over the interval Δy. To make the average value closer to the actual value of E at a point, we would have to make the interval Δy smaller and smaller until in the limit the actual value of E at any point is given by

$$E = \lim_{\Delta y \to 0} \frac{\Delta V}{\Delta y} \tag{19.24}$$

This is similar to the difference between the average velocity and the instantaneous velocity studied in chapter 3.

Note from figure 19.17 and equation 19.23 that ΔV is positive and Δy is positive, therefore $\Delta V / \Delta y$ is a positive quantity. The vector $\mathbf{E}$ is a negative quantity since it points downward in the figure. If a unit vector $\mathbf{r_0}$ (a unit vector is a vector that has a magnitude of one that is used to specify a particular direction) is defined that points upward from the point A to the point B, then the electric field intensity vector becomes

$$\mathbf{E} = -\frac{\Delta V}{\Delta y} \mathbf{r_0} \tag{19.25}$$

Equation 19.25 says that the magnitude of the electric field intensity is given by $\Delta V / \Delta y$ and its direction by $-\mathbf{r_0}$.

Example 19.6

Determining the electric field from the potential difference. The potential difference between two plates of a parallel plate capacitor is 400 V. If the plate separation is 1.00 mm, what is the magnitude of the electric field intensity between the plates?

Solution

The magnitude of the electric field intensity, determined from equation 19.23, is

$$E = \frac{\Delta V}{\Delta y} = \frac{400 \text{ V}}{1.00 \times 10^{-3} \text{ m}}$$

$$= 4.00 \times 10^5 \text{ V/m}$$

Example 19.7

Force on an electron. What force would act on an electron placed in the field of example 19.6?

Solution

We find the magnitude of the force from

$$F = qE = (1.60 \times 10^{-19} \text{ C})\left(4.00 \times 10^5 \frac{\text{V}}{\text{m}}\right)\left(\frac{\text{J/C}}{1\text{V}}\right)\left(\frac{\text{Nm}}{1\text{J}}\right)$$

$$= 6.40 \times 10^{-14} \text{ N}$$

The direction of the force on the electron would be from the negative plate to the positive plate.

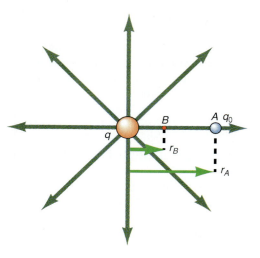

Figure 19.18

Work done in moving a test charge q_0 in the electric field of a point charge.

19.7 Potential of a Point Charge

To determine the potential of a point charge q at a particular location we need to compute the work done in moving a test charge q_0 to that position. As an example, let us move a charge q_0 from position A to position B, as in figure 19.18. Proceeding as in section 19.6, the potential is

$$V = \frac{PE}{q_0} = \frac{\text{Work done}}{q_0} = \frac{Fr}{q_0} \qquad (19.26)$$

The force acting on the charge at position A, given by Coulomb's law, is

$$F_A = \frac{kqq_0}{r_A^2}$$

whereas at position B, the force is

$$F_B = \frac{kqq_0}{r_B^2}$$

Notice that the force varies as $1/r^2$. The equation for the work done, $W = Fr$, is based on a constant force, and the force is no longer a constant. However, if the two points A and B are not too far apart, then as a first approximation we can use a geometric average as

$$r_A^2 = r_B^2 = r_A r_B$$

Therefore the average force between A and B is

$$F_{\text{avg}} = \frac{kqq_0}{r_A r_B} \qquad (19.27)$$

The distance r through which this force is acting is seen from figure 19.18 to be

$$r = r_A - r_B \qquad (19.28)$$

Using equations 19.27 and 19.28, the average work done in moving the charge from A to B is

$$W = F_{\text{avg}} r = \frac{kqq_0(r_A - r_B)}{r_A r_B}$$

$$= kqq_0\left(\frac{r_A}{r_A r_B} - \frac{r_B}{r_A r_B}\right)$$

$$= kqq_0\left(\frac{1}{r_B} - \frac{1}{r_A}\right) \qquad (19.29)$$

The difference in potential between point A and point B is

$$V_{AB} = V_B - V_A = \frac{W}{q_0} = kq\left(\frac{1}{r_B} - \frac{1}{r_A}\right) \qquad (19.30)$$

Although derived on the assumption that point A and point B are relatively close, equation 19.30 turns out to be a general relation regardless of the closeness of A and B.

To set the zero of potential, we let $V_A = 0$ when r_A approaches infinity. Equation 19.30 then becomes

$$V = \frac{kq}{r} \qquad (19.31)$$

Equation 19.31 gives the potential at any point that is at a distance r from a point charge q. The subscript B has been dropped from equation 19.31 to make it more general. However, V_B is the potential V at any position $r_B = r$. Notice from equation

Figure 19.19

Plot of the potential function for a positive point charge.

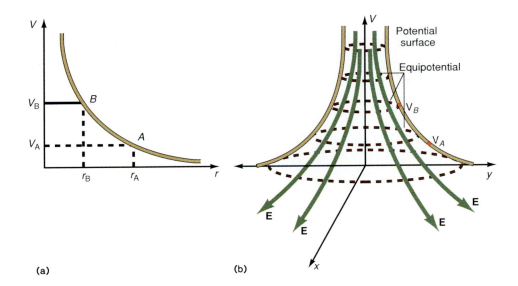

(a) (b)

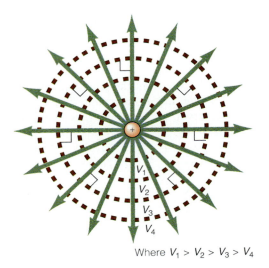

Where $V_1 > V_2 > V_3 > V_4$

Figure 19.20

Two-dimensional picture of the potential field of a positive point charge.

Figure 19.21

The potential well of the negative point charge.

19.31 that the potential varies as $1/r$. For a constant distance r from the point charge, the potential is a constant. A plot of the potential V versus the distance r from the point charge, is shown in figure 19.19(a). The location of the points A and B, the radii r_A and r_B, and the potentials V_A and V_B are shown in the figure. We can see that as r_A approaches infinity, the potential V_A approaches zero. Figure 19.19(b) is a plot of this same potential function, except it is shown with respect to a two-dimensional x,y space. The points A and B, and the potentials V_A and V_B are also shown. This diagram is generated by rotating the diagram of figure 19.19(a) about the potential V as an axis. Notice that the potential looks like a hill or a mountain and work must be done to bring another positive charge up the potential hill.

If figure 19.19(b) is projected onto the x-y plane we obtain a two-dimensional picture of the potential field of a point charge, as shown in figure 19.20. Notice that it consists of a family of concentric circles, around the point charge. Each of these circles is an equipotential line. Note that the electric field is everywhere perpendicular to the equipotential lines. In three dimensions, there would be a family of equipotential spheres surrounding the point charge.

The equipotential surfaces for a negative charge are shown in figure 19.21. A positive charge would have to be held back to prevent it from falling down the potential well.

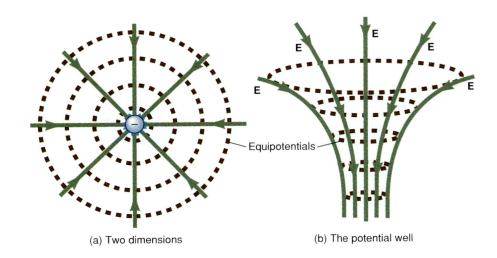

(a) Two dimensions (b) The potential well

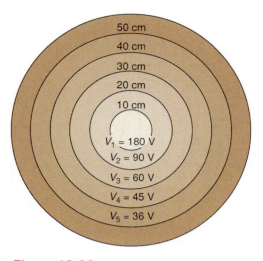

Figure 19.22

Finding the potential field of a point charge.

Example 19.8

Find the potential field for a positive point charge of 2.00 nC at $r = 10.0, 20.0, 30.0, 40.0,$ and 50.0 cm.

Solution

The potential for a positive point charge, found from equation 19.31, is

$$V_1 = \frac{kq}{r_1} = (9.00 \times 10^9 \text{ N m}^2/\text{C}^2)\frac{(2.00 \times 10^{-9} \text{ C})}{0.100 \text{ m}} = 180 \text{ V}$$

$$V_2 = \frac{kq}{r_2} = (9.00 \times 10^9 \text{ N m}^2/\text{C}^2)\frac{(2.00 \times 10^{-9} \text{ C})}{0.200 \text{ m}} = 90.0 \text{ V}$$

$$V_3 = \frac{kq}{r_3} = (9.00 \times 10^9 \text{ N m}^2/\text{C}^2)\frac{(2.00 \times 10^{-9} \text{ C})}{0.300 \text{ m}} = 60.0 \text{ V}$$

$$V_4 = \frac{kq}{r_4} = (9.00 \times 10^9 \text{ N m}^2/\text{C}^2)\frac{(2.00 \times 10^{-9} \text{ C})}{0.400 \text{ m}} = 45.0 \text{ V}$$

$$V_5 = \frac{kq}{r_5} = (9.00 \times 10^9 \text{ N m}^2/\text{C}^2)\frac{(2.00 \times 10^{-9} \text{ C})}{0.500 \text{ m}} = 36.0 \text{ V}$$

These equipotential lines are drawn in figure 19.22.

Determining the electric field of a point charge from the potentials by the use of equations 19.23 and 19.24 is very enlightening. Let us determine E at 30 cm from the point charge. Let us start with the interval $\Delta r = r_5 - r_1$, centered at 30.0 cm, and continually make it smaller. That is,

$$E_{30} = \frac{\Delta V}{\Delta r} = \frac{V_5 - V_1}{r_5 - r_1} = \frac{36.0 \text{ V} - 180 \text{ V}}{0.500 \text{ m} - 0.100 \text{ m}} = -360 \text{ V/m}$$

$$E_{30} = \frac{\Delta V}{\Delta r} = \frac{V_4 - V_2}{r_4 - r_2} = \frac{45.0 \text{ V} - 90.0 \text{ V}}{0.400 \text{ m} - 0.200 \text{ m}} = -225 \text{ V/m}$$

$$E_{30} = \frac{\Delta V}{\Delta r} = \frac{V_{3.5} - V_{2.5}}{r_{3.5} - r_{2.5}} = \frac{51.4 \text{ V} - 72.0 \text{ V}}{0.350 \text{ m} - 0.250 \text{ m}} = -206 \text{ V/m}$$

$$E_{30} = \frac{\Delta V}{\Delta r} = \frac{V_{3.2} - V_{2.8}}{r_{3.2} - r_{2.8}} = \frac{56.3 \text{ V} - 64.3 \text{ V}}{0.320 \text{ m} - 0.280 \text{ m}} = -201 \text{ V/m}$$

$$E_{30} = \frac{\Delta V}{\Delta r} = \frac{V_{3.1} - V_{2.9}}{r_{3.1} - r_{2.9}} = \frac{58.1 \text{ V} - 62.1 \text{ V}}{0.310 \text{ m} - 0.290 \text{ m}} = -200 \text{ V/m}$$

A direct computation of the electric field of a point charge at $r = 30.0$ cm using equation 19.2, gives

$$E_{30} = \frac{kq}{r^2} = (9.00 \times 10^9 \text{ N m}^2/\text{C}^2)\frac{(2.00 \times 10^{-9} \text{ C})}{(0.300 \text{ m})^2} = 200 \text{ V/m}$$

It is obvious that when the limit of $\Delta V/\Delta r$ is taken as Δr gets smaller and smaller, we do indeed approach the actual value of E at that point. When Δr is quite large (0.400 m), then $\Delta V/\Delta r = 360$ V/m, which is certainly not a good approximation for E. But as Δr was made smaller and smaller to 0.200 m, 0.100 m, 0.040 m, and 0.020 m, E became 225, 206, 201, and finally 200 V/m, which is the correct value for the electric field E_{30}. It might seem rather useless and unnecessary to use $\Delta V/\Delta r$ to compute the electric field when it can be computed directly from the equation $E = kq/r^2$. The important thing that concerns us here is the method, not the actual value of the potential, because *if a potential field is given, without any knowledge about either the charge distribution that created the field, or the field itself, the electric field can still be computed from*

$$E = \lim_{\Delta r \to 0} \frac{\Delta V}{\Delta r}$$

19.8 Superposition of Potentials

*The principle of **superposition of potentials** is stated: If there are a number of point charges present, the total potential at any arbitrary point is the sum of the potentials for each point charge.* That is,

$$V = V_1 + V_2 + V_3 + \cdots \tag{19.32}$$

This is, of course, the same superposition principle encountered in section 19.3. However, *because the potentials are scalar quantities they add according to the rules of ordinary arithmetic.* Recall that the superposition of the electric field for a number of point charges consisted in the process of vector addition. Thus, the computation of the total potential of a number of point charges is much simpler than the computation of the vector resultant of the electric field of a number of point charges by the superposition principle.

Example 19.9

Superposition of potentials. Find the potential at point A in figure 19.23 if $q_1 = 2.00 \ \mu C$, $q_2 = -6.00 \ \mu C$, and $q_3 = 8.00 \ \mu C$.

Solution

By the superposition principle the potential at point A is

$$V = V_1 + V_2 + V_3$$
$$= \frac{kq_1}{r_1} + \frac{kq_2}{r_2} + \frac{kq_3}{r_3}$$
$$= (9.00 \times 10^9 \ \text{N m}^2/\text{C}^2)\frac{(2.00 \times 10^{-6} \ \text{C})}{1.00 \ \text{m}}$$
$$- (9.00 \times 10^9 \ \text{N m}^2/\text{C}^2)\frac{(6.00 \times 10^{-6} \ \text{C})}{\sqrt{2.00} \ \text{m}}$$
$$+ (9.00 \times 10^9 \ \text{N m}^2/\text{C}^2)\frac{(8.00 \times 10^{-6} \ \text{C})}{1.00 \ \text{m}}$$
$$= 1.80 \times 10^4 \ \frac{\text{N m}}{\text{C}} - 3.82 \times 10^4 \ \frac{\text{J}}{\text{C}} + 7.20 \times 10^4 \ \text{V}$$
$$= 5.18 \times 10^4 \ \text{V}$$

Notice that the second term in the computation was negative because q_2 was negative.

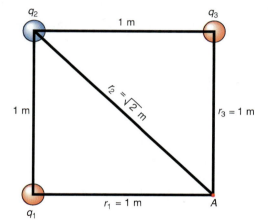

Figure 19.23
Finding the potential of multiple point charges.

Example 19.10

Work done in moving a charge from infinity. How much work is required to bring a charge of $q = 3.50 \ \mu C$ from infinity to point A in example 19.9?

Solution

Recall from the definition of the potential that the potential is equal to the potential energy per unit charge or the work done per unit charge. Thus, the work done is equal to the charge multiplied by the potential. The zero of potential for a point

Figure 19.24

Two like charges.

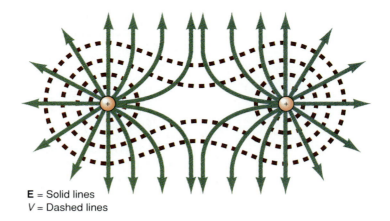

E = Solid lines
V = Dashed lines

charge was taken at infinity, so if a point charge is moved from infinity to a point such as *A,* the work done is simply the charge q multiplied by the potential at *A,* that is,

$$W = qV = (3.50 \times 10^{-6} \text{ C})(5.18 \times 10^4 \text{ V})$$

$$= 0.181 \text{ J}$$

The potential at any point in the field of two like charges, found by the superposition principle, is

$$V = V_1 + V_2$$

$$= \frac{kq_1}{r_1} + \frac{kq_2}{r_2}$$

The equipotential lines are shown in figure 19.24 as dashed lines and the electric field lines are drawn perpendicular to them.

The potential at any point in the field of a dipole (two equal but opposite charges), found from the superposition principle, is

$$V = V_1 + V_2$$

$$= \frac{kq_1}{r_1} - \frac{kq_2}{r_2}$$

The equipotential lines are shown as the dashed lines in figure 19.25, while the electric field lines are drawn everywhere perpendicular to them.

19.9 Dynamics of a Charged Particle in an Electric Field

When an electric charge q is placed in an electric field, it experiences a force given by

$$\mathbf{F} = q\mathbf{E} \tag{19.3}$$

If the electric charge is free to move, then it will experience an acceleration given by Newton's second law as

$$\mathbf{a} = \frac{\mathbf{F}}{m} = \frac{q\mathbf{E}}{m} \tag{19.33}$$

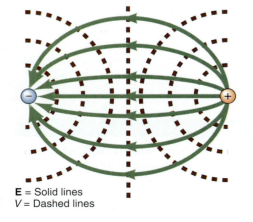

E = Solid lines
V = Dashed lines

Figure 19.25

Two unlike charges.

If the electric field is a constant field, then the acceleration is a constant and the kinematic equations developed in chapter 3 can be used to describe the motion of the charged particle. The position of the particle at any instant of time can be found by

$$\mathbf{r} = \mathbf{v_0}t + \tfrac{1}{2}\mathbf{a}t^2 \tag{3.34}$$

and its velocity by

$$\mathbf{v} = \mathbf{v_0} + \mathbf{a}t \tag{3.35}$$

where the acceleration $\mathbf{a}$ is given by equation 19.33.

Example 19.11

Projectile motion of a charged particle in an electric field. A proton is fired at an initial velocity of 150 m/s at an angle of 60.0° above the horizontal into a uniform electric field of 2.00×10^{-4} N/C between two charged parallel plates, as shown in figure 19.26. Find (a) the total time the particle is in motion, (b) its maximum range, and (c) its maximum height. Neglect any gravitational effects.

Solution

The acceleration of the proton, determined from equation 19.33, is

$$a = \frac{qE}{m} = \frac{(1.60 \times 10^{-19}\text{ C})(2.00 \times 10^{-4}\text{ N/C})}{1.67 \times 10^{-27}\text{ kg}}$$

$$= 1.92 \times 10^4 \text{ m/s}^2$$

The acceleration is in the direction of $\mathbf{E}$, which is in the negative y-direction. There is no acceleration in the x-direction because the component of $\mathbf{E}$ in the x-direction is zero. The kinematic equations in component form are

$$y = v_{0y}t - \tfrac{1}{2}at^2$$
$$x = v_{0x}t$$
$$v_y = v_{0y} - at$$
$$v_x = v_{0x}$$
$$v_y^2 = v_{0y}^2 - 2ay$$

Notice that they have the same form as they did in the discussion of the problem of projectile motion on the surface of the earth.

a. To determine the total time that the particle is in flight, we make use of the fact that when the time t is equal to the total time that the particle is in flight, $t = t_t$, the projectile has returned to the bottom plate and its height is equal to zero, that is, $y = 0$. Thus,

$$0 = v_{0y}t_t - \frac{1}{2}at_t^2$$

$$t_t = \frac{2v_{0y}}{a} = \frac{2v_0 \sin \theta}{a}$$

$$= \frac{2(150\text{ m/s})\sin 60.0°}{1.92 \times 10^4 \text{ m/s}^2}$$

$$= 1.35 \times 10^{-2} \text{ s}$$

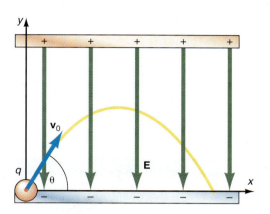

Figure 19.26

Dynamics of a charged particle in an external electric field.

b. The range of the particle is the maximum distance that the projectile moves in the x-direction, hence

$$x_{max} = v_{0x}t_t = (v_0 \cos \theta)t_t$$
$$= [(150 \text{ m/s})\cos 60.0°](1.35 \times 10^{-2} \text{ s})$$
$$= 1.01 \text{ m}$$

c. The maximum height of the projectile is determined by using the fact that when $y = y_{max}$, the projectile is at the top of the trajectory and $v_y = 0$. Hence,

$$v_y^2 = v_{0y}^2 - 2ay$$
$$0 = v_{0y}^2 - 2ay_{max}$$
$$y_{max} = \frac{v_{0y}^2}{2a} = \frac{[(150 \text{ m/s})\sin 60.0°]^2}{2(1.92 \times 10^4 \text{ m/s}^2)}$$
$$= 0.439 \text{ m}$$

To obtain the same projectile motion for an electron the direction of the electric field **E** would have to be reversed because the force, $\mathbf{F} = q\mathbf{E}$, and would be reversed when q is negative.

Example 19.12

Motion of an electron in a cathode ray oscilloscope. A cathode ray oscilloscope works by deflecting a beam of electrons as they pass through a uniform electric field between two parallel plates, as shown in figure 19.27. The deflected beam falls on a fluorescent screen, where it is visible as a dot. If the plates are 2.00 cm long, the initial velocity of the electron is $v_0 = 3.00 \times 10^7$ m/s, and the electric field between the plates is $E = 2.00 \times 10^4$ N/C, (a) what is the y-position of the electron as it leaves the field? (b) If the distance from the end of the plates to the screen is 60.0 cm, find the actual y-position on the screen.

Figure 19.27

The dynamics of an electron in an oscilloscope.

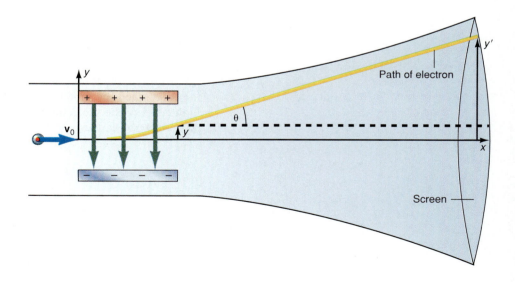

a. Note that the electron is deflected opposite to the direction of the field. This is because the electron is negative. The time it takes for the electron to traverse the field is found from the kinematic equation

$$x = v_{0x}t$$

$$t = \frac{x}{v_{0x}} = \frac{0.0200 \text{ m}}{3.00 \times 10^7 \text{ m/s}}$$

$$= 6.67 \times 10^{-10} \text{ s}$$

The acceleration of the electron in the field, found from equation 19.33, is

$$a = \frac{qE}{m}$$

$$= \frac{(1.60 \times 10^{-19} \text{ C})(2.00 \times 10^4 \text{ N/C})}{9.11 \times 10^{-31} \text{ kg}}$$

$$= 3.51 \times 10^{15} \text{ m/s}^2$$

Therefore, the y-position of the electron as it leaves the field is

$$y = v_{0y}t + \tfrac{1}{2}at^2$$

$$= 0 + \tfrac{1}{2}(3.51 \times 10^{15} \text{ m/s}^2)(6.67 \times 10^{-10} \text{ s})^2$$

$$= 7.81 \times 10^{-4} \text{ m}$$

$$= 0.781 \text{ mm}$$

b. As the electron leaves the field it travels in a straight line making an angle θ with the x-axis, as shown in figure 19:28. At the moment it leaves, its x-component of velocity is $v_x = v_{0x} = v_0$, whereas its y-component of velocity is found from

$$v_y = v_{0y} + at = 0 + (3.51 \times 10^{15} \text{ m/s}^2)(6.67 \times 10^{-10} \text{ s})$$

$$= 2.34 \times 10^6 \text{ m/s}$$

The angle θ at which the electron leaves the field is found from

$$\theta = \tan^{-1}\frac{v_y}{v_x}$$

$$= \tan^{-1}\frac{2.34 \times 10^6 \text{ m/s}}{3.00 \times 10^7 \text{ m/s}}$$

$$= 4.46°$$

If the distance from the end of the plates to the screen is 60.0 cm, then the deflection on the screen, found from figure 19.28, is

$$y_0 = x \tan \theta$$

$$= (60.0 \text{ cm})\tan 4.46°$$

$$= 4.68 \text{ cm}$$

Adding this to y gives the actual deflection y' from the center of the screen as

$$y' = y_0 + y = 4.68 \text{ cm} + 0.078 \text{ cm} = 4.76 \text{ cm}$$

By varying the value of the electric field between the plates any deflection can be obtained in the y-direction on the screen. When the electric field is reversed in sign, the deflection is in the negative y-direction on the screen. If another pair of

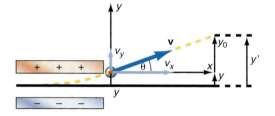

Figure 19.28

Location of electron as it hits the screen.

Figure 19.29

An oscilloscope with two sets of plates.

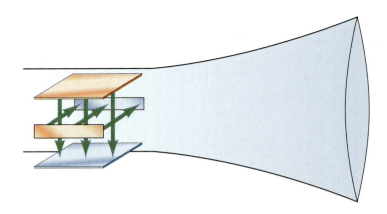

plates is placed vertically around the first set of plates, as shown in figure 19.29, and an electric field is also set up between them, the electron can be deflected anywhere on the *x*-axis. The combination of both plates can move the dot anywhere on the screen.

Example 19.13

The electron in the hydrogen atom. In the Bohr model of the hydrogen atom, an electron orbits the proton in a circular orbit. If the radius of the orbit is $r = 0.529 \times 10^{-10}$ m, what is its velocity?

Solution

If the electron is in a circular orbit, there must be a centripetal force acting on the electron directed toward the center of the orbit. This centripetal force is supplied by the Coulomb electric force. Therefore,

$$F_c = F_e$$

$$\frac{m_e v^2}{r} = \frac{kq_1 q_2}{r^2}$$

$$v = \sqrt{\frac{kq^2}{m_e r}} \tag{19.34}$$

$$= \sqrt{\frac{(9.00 \times 10^9 \ \text{N m}^2/\text{C}^2)(1.60 \times 10^{-19} \ \text{C})^2}{(9.11 \times 10^{-31} \ \text{kg})(0.529 \times 10^{-10} \ \text{m})}}$$

$$= 2.19 \times 10^6 \ \text{m/s}$$

If an electron is placed in a uniform field, as shown in figure 19.30, it possesses the potential energy

$$PE = qV \tag{19.35}$$

If the electron is released it falls toward the positive plate, losing potential energy but acquiring kinetic energy. Its gain in kinetic energy is equal to its loss in potential energy, that is,

$$KE = PE = qV \tag{19.36}$$

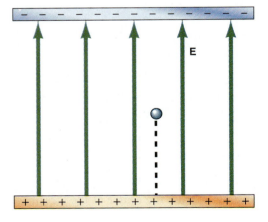

Figure 19.30

Converting potential energy to kinetic energy.

We use this result to establish a new unit of energy. If the electron falls through a potential difference of 1 V, we say that it has acquired an energy of 1 electron volt, abbreviated, eV, where

$$KE = qV$$
$$1 \text{ electron volt} = (1.6 \times 10^{-19} \text{ C})(1 \text{ V})$$
$$1 \text{ eV} = 1.6 \times 10^{-19} \text{ J}$$

The electron volt is a very small unit of energy, but it is used frequently in atomic and nuclear physics.

19.10 The Battery—Source of Potential Differences

A zinc (Zn) rod and a copper (Cu) rod are immersed in a dilute solution of sulfuric acid (H_2SO_4), as shown in figure 19.31. We call this combination an electrolytic cell. When two or more of these cells are connected together, we call the combination a **battery.** The zinc and copper rods are called *electrodes* and the solution of H_2SO_4 in water is called the *electrolyte.* At the zinc electrode, positive zinc ions, Zn^{++}, enter the solution leaving the zinc electrode negative. The sulfuric acid, H_2SO_4, is split up into two positive hydrogen ions, H^+, and a negative SO_4^{--} ion. Each positive hydrogen ion picks up an electron at the originally neutral copper electrode, leaving the copper electrode positive. Whenever two of these hydrogen ions combine with two electrons, hydrogen gas, H_2, is formed at the copper electrode and most of it bubbles up out of the solution. *The net effect of these chemical reactions is to leave the zinc electrode with an excess of negative charge and the copper electrode with a surplus of positive charge. Thus, a potential difference V has been established between the two electrodes.* If the switch S is closed, electrons will flow along the external wire from the negative zinc electrode, through the switch, through the lamp, thereby lighting it, and back to the positive copper electrode. When the electrons return to the copper electrode, they supply more electrons to combine with H^+ ions. As more H^+ ions combine with electrons to form hydrogen gas, H_2, the solution becomes less positive allowing more Zn^{++} ions to go into solution. As each Zn^{++} ion goes into solution it leaves two more electrons at the zinc electrode, which can again flow through the external wire. The cell continues supplying electrons in this way until the electrodes are decomposed or all the hydrogen is gone. The zinc-copper cell just described can supply a potential difference of 1.10 V across its terminals. This particular cell has many drawbacks because it has a by-product of hydrogen gas, which is very flammable. Also, with time, some of the hydrogen gas forms an insulating layer on the copper rod, which lowers the potential difference of the cell because the cell now looks as though the electrodes are zinc and hydrogen.

The more familiar dry cell battery acts on the same principle, but the electrodes are zinc and carbon. The electrolyte is a paste of ammonium chloride, NH_4Cl, and manganese dioxide, MnO_2. The manganese dioxide absorbs the hydrogen gas, thereby prolonging the life of the battery. The potential difference between the terminals of the dry cell is 1.50 V. The battery is therefore a device that converts chemical energy into electrical energy, and in the process supplies a potential difference that is available for many electrical applications.

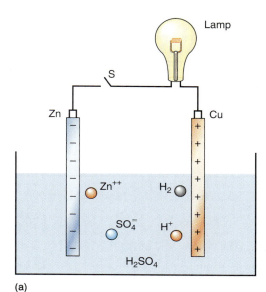

(a)

(b)

Figure 19.31

The electrolytic cell.

The Language of Physics

Electric field
It is an intrinsic property of nature that an electric field exists in the space around an electric charge. The electric field is considered to be a force field that exerts a force on charges placed in the field (p. 535).

Electric field intensity
The electric field is measured in terms of the electric field intensity. The magnitude of the electric field intensity is defined as the ratio of the force acting on a small test charge to the magnitude of the small test charge. The direction of the electric field is in the direction of the force on the positive test charge (p. 535).

Electric field intensity of a point charge
The electric field of a point charge is directly proportional to the charge that creates it, and inversely proportional to the square of the distance from the point charge to the position where the field is being evaluated (p. 537).

Superposition of electric fields
When more than one charge contributes to the electric field, the resultant electric field is the vector sum of the electric fields produced by each of the various charges (p. 538).

Electric potential
The electric potential is defined as the potential energy per unit charge. It is measured in volts (p. 544).

Equipotential line
A line along which the electric potential is the same everywhere (p. 545).

Equipotential surface
A surface along which the electric potential is the same everywhere (p. 545).

Potential difference
The difference in electric potential between two points (p. 546).

Superposition of potentials
When there are several charges present, the total potential at any arbitrary point is the algebraic sum of the potentials for each of the various point charges (p. 551).

Battery
A battery is a combination of electrolytic cells that supplies a potential difference and in the process converts chemical energy to electrical energy (p. 557).

Summary of Important Equations

Definition of the electric field intensity
$$E = \frac{F}{q_0} \tag{19.1}$$

Electric field intensity of a point charge
$$E = \frac{kq}{r^2} \tag{19.2}$$

Force on a charge q in an electric field
$$F = qE \tag{19.3}$$

Superposition principle
$$F = F_1 + F_2 + F_3 + \cdots \tag{19.4}$$
$$E = E_1 + E_2 + E_3 + \cdots \tag{19.9}$$
$$V = V_1 + V_2 + V_3 + \cdots \tag{19.32}$$

Electric dipole moment
$$p = 2aq \tag{19.15}$$

Potential energy of a charge in a uniform electric field
$$PE = qEy \tag{19.18}$$

The potential
$$V = \frac{PE}{q} = \frac{W}{q} \tag{19.19}$$

Potential between parallel plates
$$V = Ey \tag{19.20}$$

The potential difference
$$\Delta V = E\Delta y \tag{19.22}$$

Average value of the electric field intensity
$$E = \frac{\Delta V}{\Delta y} \tag{19.23}$$

The electric field intensity
$$E = \lim_{\Delta y \to 0} \frac{\Delta V}{\Delta y} \tag{19.24}$$

Potential of a point charge
$$V = \frac{kq}{r} \tag{19.31}$$

Acceleration of a charged particle in an electric field
$$a = \frac{F}{m} = \frac{qE}{m} \tag{19.33}$$

Questions for Chapter 19

1. Describe as many different types of fields as you can.
†2. Because you cannot really see an electric field, is anything gained by using the concept of a field rather than an "action at a distance" concept?
†3. Is there any experimental evidence that can substantiate the existence of an electric field rather than the concept of an "action at a distance"?
4. Is the force of gravity also an "action at a distance"? Should a gravitational field be introduced to explain gravity? What is the equivalent gravitational charge?
†5. If there are positive and negative electrical charges, could there be positive and negative masses? If there were, what would their characteristics be?
†6. Michael Faraday introduced the concept of lines of force to explain electrical interactions. What is a line of force and how is it like an electric field line? Is there any difference?
7. If the electric potential is equal to zero at a point, must the electric field also be zero there?
8. Can two different equipotential lines ever cross?
9. If the electric potential is a constant, what does this say about the electric field?
10. If there are electrical charges at rest on a conducting sphere, what can you say about the potential at any part of the sphere?

Problems for Chapter 19

19.2 The Electric Field of a Point Charge

1. Find the electric field 2.00 m from a point charge of 3.00 pC.
2. A point charge, $q = 3.75 \mu C$, is placed in an electric field of 250 N/C. Find the force on the charge.

19.3 Superposition of Electric Fields

3. Find the electric field at point A in the diagram if (a) $q_1 = 2.00 \mu C$ and $q_2 = 3.00 \mu C$ and (b) $q_1 = 2.00 \mu C$ and $q_2 = -3.00 \mu C$.

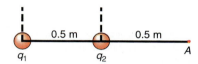

†4. A point charge of $+2.00 \mu C$ is 30.0 cm from a charge of $+3.00 \mu C$. Where is the electric field between the charges equal to zero? What is the value of the potential there?
5. A charge $q_1 = -5.00 \mu C$ is at the origin, while a second charge $q_2 = 3.00 \mu C$ is located on the x-axis at the point $x = 5.00$ cm. At what point on the x-axis is the electric field zero?
†6. Find the electric field at the apex of the triangle shown in the diagram if $q_1 = 2.00 \mu C$ and $q_2 = 3.00 \mu C$. What force would act on a 6.00 μC charge placed at this point?

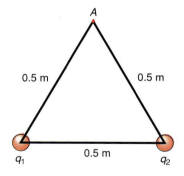

†7. Find the electric field at point A in the diagram if $q_1 = 2.00 \mu C$ and $q_2 = -3.00 \mu C$.

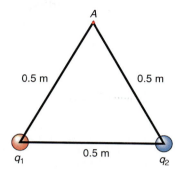

8. Electrons are located at the points (10.0 cm,0), (0,10.0 cm), and (10.0 cm,10.0 cm). Find the magnitude and direction of the electric field at the origin.
9. Find the electric dipole moment of a charge of 4.50 μC separated by 5.00 cm from a charge of $-4.50 \mu C$.
†10. If a charge of 2.00 μC is separated by 4.00 cm from a charge of -2.00 μC, find the electric field at a distance of 5.00 m, perpendicular to the axis of the dipole.
†11. Charges of 2.00 μC, 4.00 μC, -6.00 μC, and 8.00 μC are placed at the corners of a square of 50.0 cm length. Find the electric field at the center of the square.
†12. (a) Find the electric field at point A in the diagram if charges $q_1 = 2.63$ μC and $q_2 = -2.63$ μC, $d = 10.0$ cm, $r_1 = 50.0$ cm, $r_2 = 42.2$ cm, $\theta_1 = 35.0°$, and $\theta_2 = 42.8°$. (b) Find the force on a charge of 1.75 μC if it is placed at point A.

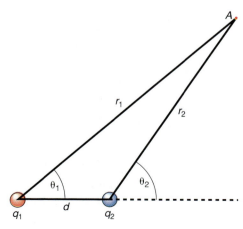

Diagram for problems 12, 13, 27, 28 and 44.

†13. Find the electric field at point A in the diagram if charges $q_1 = 2.63$ μC and $q_2 = -2.63$ μC, $d = 10.0$ cm, $r_1 = 50.0$ cm, $\theta_1 = 25.0°$. (*Hint:* first find r_2 by the law of cosines, then with r_2 known, use the law of cosines again to find the angle θ_2.)

19.6 Electric Potential Energy and the Potential

14. Two charged parallel plates are separated by a distance of 2.00 cm. If the potential difference between the plates is 300 V, what is the value of the electric field between the plates?
†15. A charge of 3.00 pC is placed at point A in the diagram. The electric field is 200 N/C downward. Find the work done in moving the charge along the path ABC and the work done in going from A to C directly.

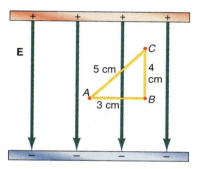

19.7 Potential of a Point Charge

16. A charge of 1.53×10^{-8} C is placed at the origin of a coordinate system. (a) Find the potential at point A located on the x-axis at $x = -5.00$ cm, and at point B located at $x = 20.0$ cm. (b) Find the difference in potential between points A and B.
17. Repeat problem 16 but with point A located at the coordinates $x = 0$ and $y = -5.00$ cm.
18. How much work is done in moving a charge of 3.00 μC from a point where the potential is 50.0 V to another point where the potential is (a) 150 V and (b) -150 V?
19. The potential difference between two terminals of a battery is 12.0 V. How much work is done by the battery in transferring 200 C of charge from one terminal to the other?
20. Find the potential 2.00 m from a point charge of 3.00 μC. How much work is required to bring a charge of 2.00 μC to this point from infinity?

19.8 Superposition of Potentials

21. Find the potential at the apex of the equilateral triangle shown in the diagram if (a) $q_1 = 2.00 \ \mu C$ and $q_2 = 3.00 \ \mu C$ and (b) if $q_1 = 2.00 \ \mu C$ and $q_2 = -3.00 \ \mu C$.

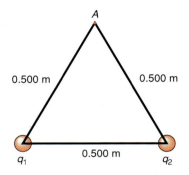

22. Electrons are located at the points (10.0 cm,0), (0,10.0 cm), and (10.0 cm,10.0 cm). Find the value of the electric potential at the origin.
23. A point charge of 2.00 μC is 30.0 cm from a charge of $-3.00 \ \mu C$. Where is the potential between the two charges equal to zero? How much work would be required to bring a charge of 4.00 μC to this point from infinity?
24. If a charge q_1 of $+2.00 \ \mu C$ is separated by 4.00 cm from a charge q_2 of $-2.00 \ \mu C$, find the potential at a distance of 5.00 m on the perpendicular bisector of the dipole axis. What is the electric dipole moment?
†25. Charges of 2.00 μC, 4.00 μC, -6.00 μC, and 8.00 μC are placed at the corners of a square of 50.0-cm sides. Find the potential at the center of the square.
26. Find the potential of a point charge of 5.00×10^{-10} C at distances of 10.0, 20.0, 30.0, 40.0, and 50.0 cm.
27. (a) Find the potential at point A in the diagram (see previous page) if charges $q_1 = 2.63 \ \mu C$ and $q_2 = -2.63 \ \mu C$, $d = 10.0$ cm, $r_1 = 50.0$ cm, $r_2 = 42.2$ cm, $\theta_1 = 35.0°$, and $\theta_2 = 42.8°$. (b) How much work is necessary to bring a charge of 1.75 μC from infinity to the point A?
28. Find the potential at point A in the diagram (see previous page) if charges $q_1 = 2.63 \ \mu C$ and $q_2 = -2.63 \ \mu C$, $d = 10.0$ cm, $r_1 = 50.0$ cm, $\theta_1 = 25.0°$. (*Hint:* first find r_2 by the law of cosines, then with r_2 known, use the law of cosines again to find the angle θ_2.)

19.9 Dynamics of a Charged Particle in an Electric Field

29. The parallel plates of a cathode ray oscilloscope are 1.00 cm apart. A voltage difference of 1000 V is maintained between the plates. (a) What is the electric field between the plates? (b) What force would act on an electron in this field? (c) What would be its acceleration?
30. An electron experiences an acceleration of 5.00 m/s² in an electric field. Find the magnitude of the electric field.
31. An electron is initially at rest at the opening of two parallel plates, as shown in the diagram. The plates are separated by a distance of 5.00 mm, and a potential difference of 150 V is maintained between the plates. (a) What is the initial potential energy of the electron? (b) What is the kinetic energy of the electron when it reaches the opposite side?

†32. An electron with an initial velocity of 1.00×10^6 m/s enters a region of a uniform electric field of 50.0 N/C, as shown in the diagram. How far will the electron move before coming to rest and reversing its motion?

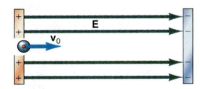

33. Find the speed of an electron and a proton accelerated through a field of 200 N/C for a distance of 2.00 cm.
34. Express the kinetic energy of an electron moving at a speed of 3.00×10^5 m/s, in electron volts.

Additional Problems

35. A point charge of 3.00 pC is located at the origin of a coordinate system. (a) What is the electric field at $x = 50.0$ cm? (b) What is the potential at $x = 50.0$ cm? (c) What force would act on a 2.00-μC charge placed at $x = 50.0$ cm?

36. A point charge of 2.00 μC is 30.0 cm from a charge of 3.00 μC. Find the electric field half way between the charges. Find the potential half way between the charges. How much work would be done in bringing a 4.00-μC charge to this point from infinity?
37. From symmetry considerations what should the electric field be at the center of a ring of charge? What assumptions did you make? Draw a diagram to substantiate your assumptions.
38. An electron is placed between two charged parallel horizontal plates. What must the value of the electric field be in order that the electron be in equilibrium between the electric force and the gravitational force?
39. Two metal plates are oriented horizontally and are 6.50 mm apart. They are connected to a source of electric potential, so that the top plate is positively charged and the bottom plate is negatively charged. What must be the potential difference between the plates if the electric force on an electron between the plates is to balance the weight of the electron, leaving the electron in equilibrium?
40. In the Bohr theory of the hydrogen atom the electron circles the proton in a circular orbit of 5.29×10^{-11} m radius. Find the electric potential, produced by the proton, at this orbital radius. From the definition of the potential, determine the potential energy of the electron in this orbit.
†41. How much work is necessary to assemble three charges from infinity to each apex of the equilateral triangle of 0.500 m on a side shown in the diagram if $q_1 = 3.00 \ \mu C$, $q_2 = 4.50 \ \mu C$, and $q_3 = 6.53 \ \mu C$. Can you now talk about the potential energy of this charge configuration?

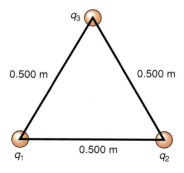

42. How much work must be done to assemble four protons at the corners of a square of edge 10.0 cm? (Assume that the protons start out very far apart.)

43. (a) Find the potential at the points A and B shown in the diagram. (b) Find the potential difference between the points A and B. (c) Find the work required to move a charge of 1.32 μC from point A to point B. (d) Find the work required to move the same charge from point B to point A.

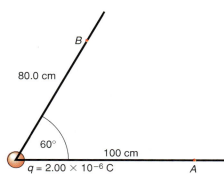

80.0 cm

60°

100 cm

$q = 2.00 \times 10^{-6}$ C A

†**44.** (a) Find the potential for an electric dipole at point A in the diagram (see page 559) if charges $q_1 = 2.00$ μC and $q_2 = -2.00$ μC, $d = 10.0$ cm, $r_1 = 50.0$ cm, $r_2 = 42.2$ cm, $\theta_1 = 35.0°$, and $\theta_2 = 42.8°$. (b) Repeat part a with q_1 now equal to 6.00 μC. Do you get the same result if you superimpose the potential field of the dipole of part a with the potential field of a point charge of 4.00 μC located at the same place as the original charge q_1?

†**45.** Find the potential of a point charge of 3.00 pC at 10.0, 20.0, 25.0, 28.0, 29.0, 30.0, 31.0, 32.0, 35.0, 40.0, and 50.0 cm. Calculate the electric field at 30.0 cm by taking intervals of Δr from 40.0 cm down to 2.00 cm in the formula $E = \Delta V/\Delta r$. What is the value of E at 30.0 cm as computed by $E = kq/r^2$?

†**46.** An electron enters midway through a uniform electric field of 200 N/C at an initial velocity of 400 m/s, as shown in the diagram. If the plates are separated by a distance of 2.00 cm, how far along the x-axis will the electron hit the bottom plate?

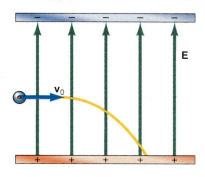

E

v_0

Interactive Tutorials

47. Calculate the value of the electric field E every meter along the line that connects the positive charges $q_1 = 2.40 \times 10^{-6}$ C located at the point (0,0) and $q_2 = 2.00 \times 10^{-6}$ C located at the point (10,0) meters.

48. Multiple charges and the electric field. Two charges $q_1 = 8.32 \times 10^{-6}$ C and $q_2 = -2.55 \times 10^{-6}$ C lie on the x-axis and are separated by the distance $r_{12} = 0.823$ m. (a) Find the resultant electric field at the point A, a distance $r = 0.475$ m from charge q_2, caused by the two charges q_1 and q_2. The line between charge 2 and the point A makes an angle $\phi = 60.0°$ with respect to the positive x-axis. (b) Find the force F acting on a third charge $q = 3.87 \times 10^{-6}$ C, if it is placed at the point A. See figure 19.5 for a picture of a similar problem.

49. The electric potential for multiple charges. Two charges $q_1 = 8.32 \times 10^{-6}$ C and $q_2 = -2.55 \times 10^{-6}$ C lie on the x-axis and are separated by the distance $r_{12} = 0.823$ m. (a) Find the resultant electric potential at the point A, a distance $r = 0.475$ m from charge q_2, caused by the two charges q_1 and q_2. The line between charge 2 and the point A makes an angle $\phi = 60.0°$ with respect to the $+x$-axis. (b) How much work is required to bring a charge $q = 3.87 \times 10^{-6}$ C from infinity to the point A?

50. The electric field of a continuous charge distribution. A rod of charge of length $L = 0.100$ m lies on the x-axis. One end of the rod lies at the origin and the other end is on the positive x-axis. A charge $q' = 7.36 \times 10^{-6}$ C is uniformly distributed over the rod. (a) Find the electric field at the point A that lies on the x-axis at a distance $x_0 = 0.175$ m from the origin of the coordinate system. (b) Find the force F that would act on a charge $q = 2.95 \times 10^{-6}$ C when placed at the point A.

51. The electric potential of a continuous charge distribution. A rod of charge of length $L = 0.100$ m lies on the x-axis. One end of the rod lies at the origin and the other end is on the positive x-axis. A charge $q' = 7.36 \times 10^{-6}$ C is uniformly distributed over the rod. (a) Find the electric potential at the point A that lies on the x-axis at a distance $x_0 = 0.175$ m from the origin of the coordinate system. (b) Find the work done to bring a charge $q = 2.95 \times 10^{-6}$ C from infinity to the point A.

20

Electric Currents and DC Circuits

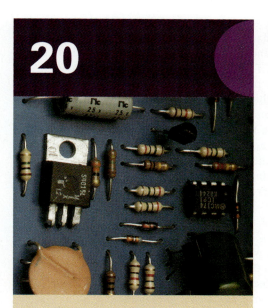

Close-up of electronic circuit board.

What we call physics comprises that group of natural sciences which base their concepts on measurements; and whose concepts and propositions lend themselves to mathematical formulation. Its realm is accordingly defined as that part of the sum total of our knowledge which is capable of being expressed in mathematical terms.

Albert Einstein

Before 1800 the study of electricity was limited to electrostatics, the study of charges at rest. It was impossible to obtain large amounts of electric charge for any continuous period of time. In 1800, Alessandro Volta (1745–1827), an Italian physicist, invented the "Voltaic pile," later to be called an *electric battery*. The battery converted chemical energy into electrical energy thereby supplying a potential difference and relatively large quantities of charge that could flow from the battery for relatively long periods of time. With the invention of the battery, applications of electricity grew by leaps and bounds.

20.1 Electric Current

A battery is used to supply a potential difference between two parallel plates. The space between the plates is either a vacuum or is filled with air. The negative plate has a small hole in it to allow the introduction of an electron into the uniform electric field, as shown in figure 20.1. The electron immediately experiences a force in the electric field, given by

$$\mathbf{F} = q\mathbf{E} = -e\mathbf{E}$$

where e is the charge on the electron. The motion of the electron can be determined by the techniques discussed in section 19.9. The electron will experience the acceleration:

$$\mathbf{a} = \frac{\mathbf{F}}{m} = -\frac{e\mathbf{E}}{m}$$

The position and velocity of the electron are given by the kinematic equations

$$r = v_0 t + \tfrac{1}{2}\mathbf{a}t^2$$

and

$$\mathbf{v} = \mathbf{v_0} + \mathbf{a}t$$

If a battery is connected across the ends of a length of wire, the flow of electrons is much more complicated because of the molecular structure of the wire itself. The atoms of the metal wire are arranged in a lattice structure, as shown in figure 20.2(a). Metals can be thought of as an array of positive ions surrounded by a more or less uniform sea of negatively charged electrons that help hold the ions in place. The outer electron of the atom in the metal wire is loosely bound, and in the lattice structure moves about freely. The free electron of each atom moves about

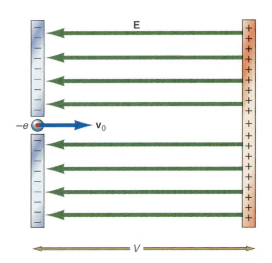

Figure 20.1

Motion of an electron in air or a vacuum.

Figure 20.2

Motion of an electron in a wire.

in a random manner within the lattice structure due to the constant interaction between the moving electron and each fixed positive ionized atom it sees as it moves about. These electrons resemble the random motion of gas molecules and are sometimes referred to as the *electron gas*. They have no net directed motion along the wire. When a potential difference is applied across the ends of the wire, as in figure 20.2(b), an electric field is set up within the wire. The electron in this field starts to move, but the motion is not the simple motion of the electron in figure 20.1 because of the constant interactions with the positive ions of the lattice. The electron slowly drifts with an average velocity v_d, the drift velocity, in the expected direction as seen in figure 20.2(c). Because of the constant interaction with the atoms of the lattice structure, it takes almost 30 s in some metals for the electron to drift about 1 cm. This is extremely slow as compared to the motion that would be experienced by the electron in figure 20.1.

Because of this complexity of the motion of the electron within the solid wire, we will make no further analysis of the motion of the electron by Newton's laws and the kinematic equations; we will adopt a much simpler procedure. When a potential difference is applied across the ends of a wire and electrons start to drift within the wire, we will say that an electric current exists in the wire. *The electric current is defined as the amount of electric charge that flows through a cross section of the wire per unit time.* We write this mathematically as

$$I = \frac{q}{t} \qquad (20.1)$$

The SI unit of current is the ampere, named after the French physicist, André Marie Ampère (1775–1836). *An ampere of current is a flow of one coulomb of charge per second.* That is,

$$1 \text{ ampere} = \frac{1 \text{ coulomb}}{\text{second}}$$

We abbreviate this as

$$1 \text{ A} = 1 \text{ C/s}$$

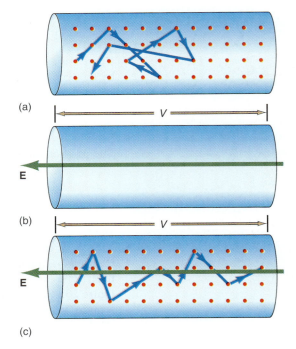

(a)

(b)

(c)

Because one coulomb of charge represents 6.242×10^{18} electronic charges, the ampere is the flow of 6.242×10^{18} electrons per second. Sometimes we use a smaller unit of current, the milliampere, abbreviated mA.

$$1 \text{ mA} = 10^{-3} \text{ A}$$

Because electric current was defined before any knowledge of the existence of electrons, the original current was assumed to be a flow of positive charges. (It is interesting to note that it was Benjamin Franklin [1706–1790] an American statesman and scientist who first called the charges positive and negative. He assumed it was the positive charges that moved.) Although it is now known that this is incorrect for the conduction through a solid, it is usual to continue with this historical convention and *define the current to be a flow of positive charges from a position of high potential to one of low potential. This current is called* **conventional current** to distinguish it from the actual electron flow. The flow of electrons is called the **electron current.** In all the cases considered, a flow of positive charges in one direction (conventional current) is completely equivalent to a flow of negative charges in the opposite direction (electron current). In gases and electrolytic liquids, both positive and negative charges flow in opposite directions, but both contribute to the positive current.

The flow of charges in a wire is represented in what is called a *circuit diagram,* and the simplest one is shown in figure 20.3. This circuit consists of the battery and a very long wire, which is connected to both terminals of the battery. The battery, the supplier of a potential difference, is shown as the two horizontal lines. The longer line represents the positive terminal of the battery, the point of highest potential, and is labeled with a positive sign ($+$). The shorter line represents the negative terminal of the battery and is labeled with a negative sign in the diagram ($-$). This negative terminal is arbitrarily chosen to be the point of zero potential. The line from the positive terminal to the negative terminal is the external circuit and here represents the length of wire connected to the two terminals of the battery. The potential difference maintained by the battery is labeled V in the figure.

The flow of charge in the circuit is analogous to dropping a ball from a height h and is shown in figure 20.4. At the height h, the ball has a positive potential energy with respect to the ground, which is at zero potential energy. The ball falls from a position of high potential energy to the ground, as shown in figure 20.4(a). The electrical circuit of figure 20.3 is pulled apart in figure 20.4(b) to show the analogy more clearly. In the electrical circuit, a positive charge leaves the positive terminal of the battery where it has the potential V, and "falls" through the connecting wire to the "ground" or zero of potential. This analogy between the mechanical case and the electrical case is even clearer if you recall that we defined the potential as the potential energy per unit charge. So when a positive charge flows or "falls" from high potential to low potential it is falling from a position of high potential energy to a position of low potential energy. *Whenever a potential difference exists between any two points in a circuit, positive charge flows from the point of high potential to the point of low potential. The resulting current is called a* **direct current** *(DC) since the charges flow in only one direction.*

20.2 Ohm's Law

If a wire is connected to both terminals of a battery, charges will flow. To determine the current produced by these flowing charges, a device called an ammeter is placed in the circuit and is shown as A in figure 20.5(a). We will discuss the physics of the ammeter in chapter 22, when we discuss the effects of an external magnetic field on wires carrying a current. For now, *the* **ammeter** *is a device that displays the amount of current in a circuit.* Another device, called a voltmeter, the physics of which is quite similar to the ammeter, is connected across the battery and is shown as V in figure 20.5(a). *The* **voltmeter** *reads the potential difference between any*

Figure 20.3

A circuit diagram.

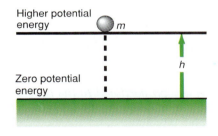

(a) Mechanical

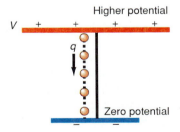

(b) Electrical

Figure 20.4

Analogy of mechanical and electrical motion.

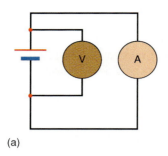 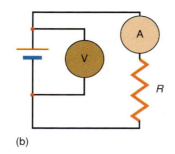 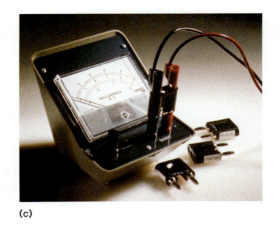

(a) (b) (c)

Figure 20.5

Experimental determination of Ohm's law.

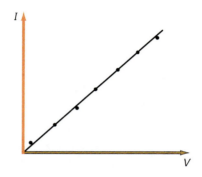

Figure 20.6

Ohm's law.

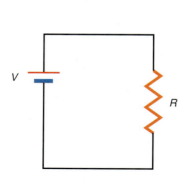

Figure 20.7

Simple application of Ohm's law.

two points in an electrical circuit. In figure 20.5(a) the voltmeter reads the potential difference across the terminals of the battery. A picture of a typical laboratory meter is shown in figure 20.5(c).

For a particular battery maintaining a potential difference V_1, a current I_1 is recorded by the ammeter. When a larger battery of potential difference V_2 replaces battery V_1, a larger current I_2 is observed in the circuit. By successively using different batteries, we observe different currents. If we plot the current recorded by the ammeter in the circuit against the applied battery voltage, we obtain a linear relationship, as shown in figure 20.6. This implies that the current in the circuit is directly proportional to the applied voltage, that is,

$$I \propto V \tag{20.2}$$

To make an equality out of this proportionality, we introduce a constant of proportionality $(1/R)$ and the proportionality 20.2 becomes the equation

$$I = \frac{V}{R} \tag{20.3}$$

where R is called the resistance of the wire. Equation 20.3 is called Ohm's law after Georg Simon Ohm (1787–1854), a German physicist who discovered the relation in 1827. *Ohm's law states that the current in a circuit is directly proportional to the applied potential difference V and inversely proportional to the resistance R of the circuit.* The SI unit of resistance is the ohm, designated by the Greek letter Ω, where

$$1 \text{ ohm} = 1 \frac{\text{volt}}{\text{ampere}} = 1 \text{ V/A}$$

(For completeness, we should note that there are some materials that do not obey Ohm's law. That is, a graph of the current versus the voltage is not a straight line. For such materials, the resistance is not a constant, but varies with voltage. However, the resistance for any voltage can still be defined as the ratio of the voltage to the current. We will not be concerned with such materials in this book.) Every wire has some resistance and the resistance of a circuit is shown in the circuit diagram in figure 20.5(b) as a saw tooth symbol that is labeled as R. The wire, or any other resistive material, is called a *resistor*. *Ohm's law enables us to determine the current in a circuit when the resistance of the circuit and the applied voltage are known.*

Example 20.1

Finding the current by Ohm's law. A 6.00-V battery is applied to a circuit having a resistance of 10.0 Ω. Find the current in the circuit (see figure 20.7).

The current in the circuit, found by Ohm's law, equation 20.3, is

$$I = \frac{V}{R} = \frac{6.00 \text{ V}}{10.0 \text{ }\Omega} = 0.600 \frac{\text{V}}{\text{V/A}}$$

$$= 0.600 \text{ A}$$

Example 20.2

Finding the resistance by Ohm's law. A potential difference of 12.0 V is applied to a circuit and a current of 0.300 A is observed. What is the resistance of the circuit?

Solution

The resistance of the circuit, found from the rearrangement of Ohm's law, is

$$R = \frac{V}{I} = \frac{12.0 \text{ V}}{0.300 \text{ A}} = 40.0 \text{ }\Omega$$

20.3 Resistivity

The concept of the resistance of a wire can be more easily understood by looking at the lattice structure of the wire as in figure 20.2(c). When a potential difference is applied to the ends of the wire the electrons slowly drift along the wire. The greater the length of the wire, the longer it takes for the electrons to drift along the wire. Since the current is the flow of charge per unit time, a longer period of time implies a smaller current in the wire. This reduced current is viewed from Ohm's law as an increase in the resistance of the wire. Therefore, *the resistance of a wire is directly proportional to the length l of the wire, that is,*

$$R \propto l \tag{20.4}$$

The larger the cross-sectional area of the wire the greater are the number of electrons that can pass through it per unit time. Therefore, the greater the area, the larger the current in the wire. Viewed from Ohm's law, equation 20.3, this increased current implies a smaller value of resistance. *Hence, the resistance of a wire is inversely proportional to the cross-sectional area of the wire, that is,*

$$R \propto \frac{1}{A} \tag{20.5}$$

We can combine proportionalities 20.4 and 20.5 into the one proportionality

$$R \propto \frac{l}{A} \tag{20.6}$$

To make an equation out of this, we need a constant of proportionality. The constant of proportionality should depend on the material that the wire is made of, because the electrons in the wire constantly interact with the atoms of the lattice. Different atoms exert different forces on the free electrons, and hence affect the drift velocity of the electrons. The proportionality constant is called the *resistivity* ρ, and proportionality 20.6 becomes the equation

$$R = \rho \frac{l}{A} \tag{20.7}$$

Equation 20.7 says that the resistance of a wire is directly proportional to its resistivity and its length, and inversely proportional to its cross-sectional area. A

Table 20.1
Resistivities (ρ) of Some Different Materials at 20 °C and Mean Temperature Coefficient of Resistivity (α)

Material	ρ, Ω m	α °C^{-1}
Aluminum	2.82×10^{-8}	3.9×10^{-3}
Brass	7.00×10^{-8}	2.00×10^{-3}
Carbon (graphite)	3500×10^{-8}	-0.5×10^{-3}
Copper	1.72×10^{-8}	3.93×10^{-3}
Gold	2.44×10^{-8}	3.40×10^{-3}
Iron	9.71×10^{-8}	5.20×10^{-3}
Lead	20.6×10^{-8}	4.30×10^{-3}
Mercury	98.4×10^{-8}	8.90×10^{-3}
Platinum	10.6×10^{-8}	3.90×10^{-3}
Silver	1.59×10^{-8}	3.80×10^{-3}
Tungsten	5.51×10^{-8}	4.50×10^{-3}
Amber	5.00×10^{14}	
Bakelite	$2 \times 10^5 - 2 \times 10^{14}$	
Glass	$10^{13} - 10^{14}$	
Hard Rubber	$10^{13} - 10^{16}$	
Mica	$10^{11} - 10^{15}$	
Wood	$10^8 - 10^{11}$	

table of resistivities for various materials is shown in table 20.1. The SI unit of resistivity is an ohm meter, abbreviated Ω m. Everything else being equal, materials with relatively small values of resistivity ρ, such as metals, have small resistances and hence make good conductors of electricity. Materials with large values of ρ, the nonmetals, have large resistances and therefore make poor conductors. Poor conductors are, however, good insulators. A good conductor has a resistivity of the order of 10^{-8} Ω m, whereas a good insulator has a resistivity value of the order of 10^{13}–10^{15} Ω m. Good conductors of electricity are also good conductors of heat. This is because the highly mobile electrons are also carriers of thermal energy in conductors. For the same reason, good electrical insulators are also good thermal insulators.

Example 20.3

The resistance of a spool of wire. Find the resistance of a spool of copper wire, 500 m long with a diameter d of 0.644 mm.

Solution

The cross-sectional area of the wire is given by

$$A = \frac{\pi d^2}{4}$$

$$= \frac{\pi (0.644 \times 10^{-3} \text{ m})^2}{4}$$

$$= 3.26 \times 10^{-7} \text{ m}^2$$

Electricity and Magnetism

The resistance of the wire spool, found from equation 20.7, is

$$R = \rho \frac{l}{A}$$

$$= (1.72 \times 10^{-8} \ \Omega \ m) \left(\frac{500 \ m}{3.26 \times 10^{-7} \ m^2} \right)$$

$$= 26.4 \ \Omega$$

Example 20.4

The resistance of an amber rod. Find the resistance of an amber rod 30.0 cm long by 2.00 cm high and 2.50 cm thick, as shown in figure 20.8.

Solution

The resistance from one end of the rod to the other, found from equation 20.7, is

$$R = \rho \frac{l}{A} = (5.00 \times 10^{14}) \frac{0.300 \ m}{(0.0200 \ m)(0.0250 \ m)}$$

$$= 3.00 \times 10^{17} \ \Omega$$

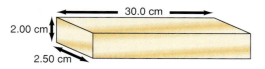

Figure 20.8

The resistance of an amber rod.

20.4 The Variation of Resistance with Temperature

It is found experimentally that the resistivity of a material is not constant but rather varies with temperature. A graph of the variation of the resistivity with temperature for a metal is shown in figure 20.9(a). Notice that the graph is a curve, which means that the variation is not linear. However, a straight line can be drawn that approximates the curve for a range of temperatures, as shown in figure 20.9(b). The slope m of this straight line is given as

$$m = \frac{\Delta \rho}{\Delta t} = \frac{\rho - \rho_0}{t - t_0}$$

where ρ is the resistivity of the material at the temperature t and ρ_0 is the resistivity of the material at the temperature t_0. Rearranging the equation we get

$$\rho - \rho_0 = m(t - t_0)$$

Solving for the resistivity ρ, gives

$$\rho = \rho_0 \left[1 + \frac{m}{\rho_0}(t - t_0) \right] \tag{20.8}$$

Let us now define a new constant α, called the *mean temperature coefficient of resistivity,* as

$$\alpha = \frac{m}{\rho_0}$$

Thus, by measuring the slope of the curve in the area of interest, and dividing by the resistivity ρ_0 at the reference temperature t_0, we obtain the mean temperature coefficient of resistivity α. The value of α for various materials is shown in table 20.1. Notice that since the slope m is a ratio of resistivity to temperature, the tem-

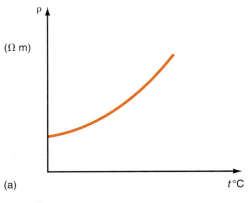

(a)

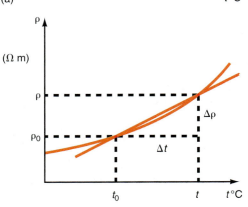

(b)

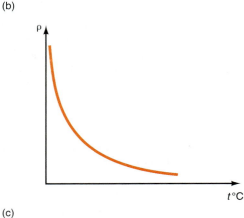

(c)

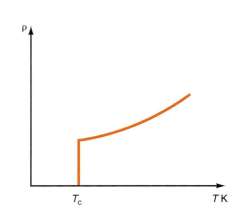

(d)

Figure 20.9

The variation of resistivity with temperature.

perature coefficient of resistivity α, which is equal to that slope divided by the resistivity, has units of $°C^{-1}$. With this new temperature coefficient of resistivity, we can write equation 20.8 as

$$\rho = \rho_0[1 + \alpha(t - t_0)] \tag{20.9}$$

Equation 20.9 gives the resistivity of a material at the temperature t when the resistivity of the material ρ_0 is known at the reference temperature t_0.

Example 20.5

Temperature dependence of resistivity. The resistivity of copper at 20.0 °C is 1.72×10^{-8} Ω m. Find its resistivity at 200 °C.

Solution

The temperature coefficient of resistivity for copper, found from table 20.1, is 3.93×10^{-3} $°C^{-1}$. The resistivity of copper at 200 °C, found from equation 20.9, is

$$\rho = \rho_0[1 + \alpha(t - t_0)]$$
$$= (1.72 \times 10^{-8} \ \Omega \ m)[1 + (3.93 \times 10^{-3} \ °C^{-1})(200 - 20.0) \ °C]$$
$$= 2.94 \times 10^{-8} \ \Omega \ m$$

Because the resistivity of a wire changes with temperature, the resistance of that wire also changes with temperature. If we multiply both sides of equation 20.9 by the length of the wire and divide by the cross-sectional area of the wire, we obtain

$$\rho \frac{l}{A} = \rho_0 \frac{l}{A}[1 + \alpha(t - t_0)] \tag{20.10}$$

However, $\rho l/A$ is equal to the resistance R from equation 20.7. Therefore, we can write equation 20.10 as

$$R = R_0[1 + \alpha(t - t_0)] \tag{20.11}$$

Equation 20.11 gives the resistance of a resistor R, at the temperature t, if the resistance R_0, at the reference temperature t_0, is known.

Example 20.6

Temperature dependence of resistance. If the resistance of a copper resistor is 50.0 Ω at 20.0 °C, find its resistance at 200 °C.

Solution

The temperature coefficient of resistivity for copper, found from table 20.1, is 3.93×10^{-3} $°C^{-1}$. The resistance of the copper resistor at 200 °C, found from equation 20.11, is

$$R = R_0[1 + \alpha(t - t_0)]$$
$$= (50.0 \ \Omega)[1 + (3.93 \times 10^{-3} \ °C^{-1})(200 - 20.0) \ °C]$$
$$= 85.4 \ \Omega$$

We should note that not all materials have the same kind of variation of resistivity with temperature. A group of materials known as *semiconductors* have a temperature variation as shown in figure 20.9(c).

When the resistivity of metals is plotted against the absolute temperature, a very strange phenomenon occurs at very low temperatures, as shown in figure 20.9(d). At a certain temperature, known as the critical temperature T_c, the resistivity, and hence the resistance of the material, drops to zero. The material is then said to be superconducting, and the material is called a superconductor. This phenomenon was first discovered by Kamerlingh Onnes in the Netherlands in 1911. He found that for mercury, the resistivity effectively dropped to zero when the temperature was reduced to 0.05 K. Researchers have since found newer materials that become superconducting at much higher temperatures. By the 1960s the critical temperature for superconduction was found to be as high as 20 K. Steady research into newer materials has raised the critical temperature even further. By the end of 1986 the critical temperature of 40 to 50 K had been reached with an oxide of barium, lanthanum, and copper. By March 1987 superconductivity had been made to occur at temperatures above 90 K.

The reason for the great importance of superconductivity lies in the fact that once the resistance of a material drops to zero, the energy dissipated in the resistance also drops to zero and the current continues to exist forever. (We will see that the energy dissipated in a resistor is given by I^2R. If R is zero, no energy is lost.) If newer superconducting materials can be found that have still higher critical temperatures, a time will come when superconductivity will be a practical new technology that will revolutionize the entire electrical industry. The cost of delivering electricity will be greatly reduced by making essentially lossless transmission lines. Some superconducting thin films will be used in new devices for electronic components in computers. Electric motors will run with a minimum of energy. The change will be as great as it was with the discovery of the transistor and the subsequent use of the microchip in electrical components.

20.5 Conservation of Energy and the Electric Circuit—Power Expended in a Circuit

Consider the simple resistive circuit in figure 20.10. The power supplied by the battery is the work it does per unit time, that is

$$P = \frac{W}{t} \qquad (20.12)$$

But the work done within the battery is done by chemical means, and its final result is to move a positive charge q from the negative terminal inside the battery to the positive terminal within the battery. That is, a charge q had to be "lifted" from the point of zero potential to the point of higher potential V within the battery. Recall from the definition of the potential that the potential is the potential energy per unit charge,

$$V = \frac{PE}{q} = \frac{W}{q} \qquad (19.19)$$

where W is the work that must be done to give the charge its potential energy. From equation 19.19, the work done within the battery is simply

$$W = qV \qquad (20.13)$$

The power supplied by the battery is found from equations 20.12 and 20.13 as

$$P = \frac{W}{t} = \frac{qV}{t} \qquad (20.14)$$

But

$$I = \frac{q}{t} \qquad (20.1)$$

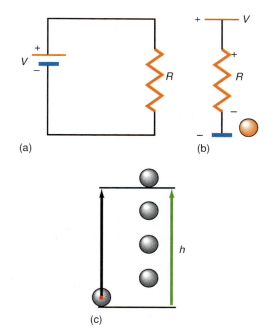

Figure 20.10

Conservation of energy in an electric circuit.

the current coming out of the battery. Combining equation 20.1 with equation 20.14 gives the power supplied by the battery as

$$P = IV \qquad (20.15)$$

The conservation of energy applied to the circuit can be expressed as

$$\text{Energy supplied to the circuit} = \text{Energy consumed in the circuit} \qquad (20.16)$$

Since the power is energy per unit time, if we divide both sides of equation 20.16 by the time t, we get

$$\text{Power supplied to the circuit} = \text{Power consumed in the circuit} \qquad (20.17)$$

Therefore, the power consumed in the circuit is

$$\text{Power consumed} = \text{Power supplied} = IV$$

If the circuit contains only a resistor, then the voltage across the resistor is given by Ohm's law as $V = IR$. Therefore, the power consumed or dissipated in the resistor is

$$P = IV = I(IR)$$

$$P = I^2R \qquad (20.18)$$

*Equation 20.18 gives the rate at which energy is dissipated in the resistor with time. This energy that is lost as the charge "falls" through the resistor shows up as heat in the resistor, and is usually referred to as **Joule heat.*** We can also express equation 20.18 as

$$P = \frac{V^2}{R} \qquad (20.19)$$

by substituting $I = V/R$ into equation 20.18.

Recall from chapter 7, the unit of power is the watt, where

$$1 \text{ watt} = 1 \frac{\text{joule}}{\text{second}} = 1 \text{ J/s}$$

Checking the units for the power supplied to a circuit we get

$$P = IV = \text{ampere volt} = \frac{(\text{coulomb})(\text{joule})}{\text{second coulomb}} = \frac{\text{joule}}{\text{second}} = \text{watt}$$

which we can abbreviate as

$$1 \text{ W} = 1 \text{ A V} = (1 \text{ C/s})(1 \text{ J/C}) = 1 \text{ J/s} = 1 \text{ W}$$

Thus, in electrical circuits, we represent the unit for power as

$$\text{watt} = \text{ampere volt} = \text{A V}$$

because it is equivalent to the previous definition.

When the charge q leaves the battery, it is at the potential V. As it "falls" through the resistor, figure 20.10(b), it loses energy as it falls to a position of lower potential. Therefore, we say that there is a potential drop across the resistor.

The potential drop across the resistor is given by Ohm's law as

$$V = IR \qquad (20.20)$$

It is sometimes convenient to mark the position of the resistor that is at the highest potential with a plus ($+$) sign, and the point of the resistor that is at the lowest potential with a negative sign ($-$) to remind ourselves that the charge will fall from a plus ($+$) to a minus ($-$) potential. This is shown in figure 20.10(b).

The mechanical equivalent of the circuit of figure 20.10(a) is shown in figure 20.10(c). A ball of mass m is lifted to a height h by a person. The person who does the work lifting the mass is equivalent to the battery. The ball has potential energy at the top. We assume that the medium that the ball will fall through is resistive. Therefore, the ball will not fall freely, continually accelerating, but rather will be slowed down by the friction of the medium until the ball moves at a constant velocity, its terminal velocity. This is similar to the charge moving at the constant drift velocity. As the ball falls and loses potential energy, the lost potential energy shows up as heat generated by the frictional forces between the ball and the medium. This is similar to the loss of energy of the charge through Joule heating as it "falls" through the resistor.

Example 20.7

A 60-W light bulb. A 60.0-W light bulb is screwed into a 120-V lamp outlet. (a) What current will flow through the bulb and what is the resistance of the bulb? (b) What is the power dissipated in the Joule heating of the wire in the bulb?

Solution

a. Because the power rating of the bulb is known, we can find the current from equation 20.15 as

$$P = IV$$

$$I = \frac{P}{V} = \frac{60.0 \text{ W}}{120 \text{ V}} = 0.500 \, \frac{\text{A V}}{\text{V}}$$

$$= 0.500 \text{ A}$$

The resistance of the bulb, found from Ohm's law, is

$$R = \frac{V}{I} = \frac{120 \text{ V}}{0.500 \text{ A}} = 240 \, \Omega$$

b. The power dissipated in the Joule heating of the wire in the bulb, found from equation 20.18, is

$$P = I^2R = (0.500 \text{ A})^2(240 \, \Omega)$$

$$= 60.0 \text{ W}$$

Note that the power supplied is equal to the power dissipated as expected.

Example 20.8

A 120-W light bulb. What is the current and resistance of a 120-W light bulb connected to a 120-V source?

Solution

We find the current from

$$I = \frac{P}{V} = \frac{120 \text{ W}}{120 \text{ V}} = 1.00 \text{ A}$$

We determine the resistance of the bulb by Ohm's law as

$$R = \frac{V}{I} = \frac{120 \text{ V}}{1.00 \text{ A}} = 120 \, \Omega$$

Up to now, we studied a circuit that contained only one resistor. Suppose there are several resistors in the circuit. Is there a difference in the circuit if these

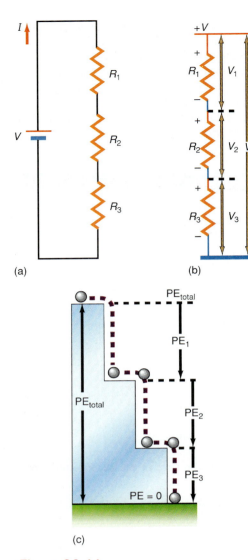

(a)

(b)

(c)

Figure 20.11

Resistors in series.

resistors are connected in different ways? The answer is yes and we will now study different combinations of these resistors—in particular, resistors in series, resistors in parallel, and combinations of resistors in series and parallel.

20.6 Resistors in Series

A typical circuit having resistors in series is shown in figure 20.11(a). *The characteristic of a series circuit is that the same current that flows from the battery flows through each resistor. That is, the current I is the same everywhere in the series circuit.* Figure 20.11(b) displays the circuit unfolded, showing how the positive charge will fall from high potential to low potential. Figure 20.11(c) shows a mechanical analogue of the circuit of resistors in series. A ball is picked up from the ground and placed at the top of a three-step stairway. At the top step the ball has its maximum potential energy. If the ball is given a slight push it rolls off the top step dropping down to the first step, losing an amount of potential energy PE_1. This lost potential energy is first converted to kinetic energy as the ball falls, and the kinetic energy is then converted to thermal energy in the collision of the ball with the step. The ball then rolls off the first step dropping down to the second step, losing an amount of potential energy PE_2. It then rolls off the second step, falling down to the third and last step at the ground level. This time it loses an amount of potential energy PE_3. The law of conservation of energy says that the total energy given to the ball to place it at the top step must be equal to the total energy that it loses as it falls from step to step to the ground. That is,

$$PE_{top} = PE_1 + PE_2 + PE_3 \qquad (20.21)$$

The electrical circuit, figure 20.11(b), is analogous to this mechanical staircase. The charge q is raised to the potential V by the chemical action of the battery. As the charge "falls" through the first resistor, it drops in potential by V_1. As it falls through the second resistor it drops in potential by V_2. The charge experiences another potential drop, V_3, as it falls through R_3. *By the law of conservation of energy the potential supplied to the charge by the battery must be equal to the potential that the charge loses as it falls through the resistors.* Therefore,

$$V = V_1 + V_2 + V_3 \qquad (20.22)$$

The voltage drop across each resistor in figure 20.11, given by Ohm's law, equation 20.20, is

$$V_1 = IR_1$$
$$V_2 = IR_2$$
$$V_3 = IR_3$$

Replacing these equations into equation 20.22 gives

$$V = IR_1 + IR_2 + IR_3 \qquad (20.23)$$

Dividing both sides of equation 20.23 by I gives

$$\frac{V}{I} = R_1 + R_2 + R_3 \qquad (20.24)$$

where V is the total potential applied to the circuit and I is the total current in the circuit, so V/I should, by Ohm's law, equal R, the total resistance of the circuit, that is

$$\frac{V}{I} = R \qquad (20.25)$$

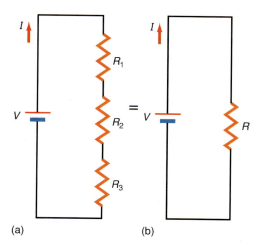

Figure 20.12
Equivalent circuit for resistors in series.

(a)   (b)

From equations 20.24 and 20.25, it is obvious that

$$R = R_1 + R_2 + R_3 \qquad \text{(20.26)}$$

That is, *the sum of the three resistances in the series circuit is equivalent to one resistance R called the **equivalent resistance.*** Figure 20.12(a) is equivalent to the simpler circuit shown in figure 20.12(b), with R, the equivalent resistance, given by equation 20.26. That is, the three resistors R_1, R_2, and R_3 could be replaced by the one equivalent resistor R without detecting any electrical change in the circuit. Although equation 20.26 was derived for three resistors it is obvious that it holds for the sum of any number of resistors in series.

The fact that the equivalent resistance of resistors in series is just the sum of the individual resistances should not be too surprising. Because, if each of the resistors were a wire of the same material, same cross-sectional area, but with different lengths l_1, l_2, and l_3, respectively, then the length of the wire when the three resistors are connected in series is just

$$l = l_1 + l_2 + l_3$$

Multiplying each term by ρ/A gives

$$\rho\frac{l}{A} = \rho\frac{l_1}{A} + \rho\frac{l_2}{A} + \rho\frac{l_3}{A}$$

But using equation 20.7, $R = \rho l/A$ gives

$$R = R_1 + R_2 + R_3$$

which is the same result as before, equation 20.26. This derivation may be easier to see but it does not give the same insight as is obtained by using the law of conservation of energy.

Example 20.9

Resistors in series. Resistors $R_1 = 20.0\ \Omega$, $R_2 = 30.0\ \Omega$, and $R_3 = 40.0\ \Omega$ are connected in series to a 6.00-V battery. Find the equivalent resistance of the circuit and the current in this series circuit.

Solution

The equivalent resistance, given by equation 20.26, is

$$R = R_1 + R_2 + R_3$$
$$= 20.0\ \Omega + 30.0\ \Omega + 40.0\ \Omega$$
$$= 90.0\ \Omega$$

The current is found by Ohm's law, with R the equivalent resistance, that is,

$$I = \frac{V}{R} = \frac{6.00\ \text{V}}{90.0\ \Omega} = 0.0667\ \text{A}$$

Example 20.10

Power dissipated in resistors in series. Find the power supplied and the power dissipated in each resistor in example 20.9.

Solution

The power supplied by the battery is

$$P = VI = (6.00\ \text{V})(0.0667\ \text{A}) = 0.400\ \text{W}$$

The power dissipated in each resistor is

$$P_1 = I^2R_1 = (0.0667 \text{ A})^2(20.0 \text{ }\Omega) = 0.0890 \text{ W}$$

$$P_2 = I^2R_2 = (0.0667 \text{ A})^2(30.0 \text{ }\Omega) = 0.133 \text{ W}$$

$$P_3 = I^2R_3 = (0.0667 \text{ A})^2(40.0 \text{ }\Omega) = 0.178 \text{ W}$$

The total power dissipated in all resistors is

$$P = P_1 + P_2 + P_3$$

$$= 0.0890 \text{ W} + 0.133 \text{ W} + 0.178 \text{ W}$$

$$= 0.400 \text{ W}$$

Note that the power supplied to the circuit is equal to the power dissipated in the resistors.

Example 20.11

Potential drop across resistors in series. Find the potential drop across each resistor of examples 20.9 and 20.10.

Solution

The potential drop across each resistor, found by Ohm's law, is

$$V_1 = IR_1 = (0.0667 \text{ A})(20.0 \text{ }\Omega) = 1.33 \text{ V}$$

$$V_2 = IR_2 = (0.0667 \text{ A})(30.0 \text{ }\Omega) = 2.00 \text{ V}$$

$$V_3 = IR_3 = (0.0667 \text{ A})(40.0 \text{ }\Omega) = 2.67 \text{ V}$$

Notice that the sum of the potential drops, $V_1 + V_2 + V_3$, is equal to 6.00 V, which is equal to the applied voltage of 6.00 V.

20.7 Resistors in Parallel

A typical circuit with resistors connected in parallel is shown in figure 20.13(a). Resistors in a **parallel circuit** are connected such that the top of each resistor is connected to the same point A, whereas the bottom of each resistor is connected to the same point B. Therefore, the potential difference between A and B is the same as the potential difference across each resistor. We assume that the resistance of the connecting wires is negligible compared to the resistors in the circuit, and can be ignored. Therefore, the potential across AB is the same as the potential V supplied by the battery. Consequently, *the characteristic of resistors connected in parallel is that the potential difference is the same across every resistor.* That is,

$$V = V_1 = V_2 = V_3 \qquad (20.27)$$

When the total current I from the battery reaches the junction A, it divides into three parts; I_1 goes through resistor R_1, I_2 goes through resistor R_2, and I_3 goes through R_3. Because none of the charge disappears,

$$I = I_1 + I_2 + I_3 \qquad (20.28)$$

Equation 20.28 is a statement of *the law of conservation of electric charge. Electric charge can neither be created nor destroyed and hence the electric charges entering a junction must be equal to the electric charges leaving a junction. Thus, the electric current entering a junction is equal to the electric current leaving a junction.*

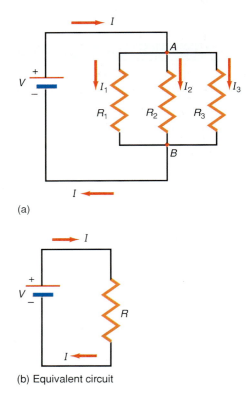

(a)

(b) Equivalent circuit

Figure 20.13

Resistors in parallel.

At the junction B these three currents again combine to form the same total current I that entered the junction A. The current in each resistor can be found by applying Ohm's law to that resistor. That is,

$$I_1 = \frac{V_1}{R_1} \tag{20.29}$$

$$I_2 = \frac{V_2}{R_2} \tag{20.30}$$

$$I_3 = \frac{V_3}{R_3} \tag{20.31}$$

Replacing these values of the currents back into the current equation, 20.28, gives

$$I = \frac{V_1}{R_1} + \frac{V_2}{R_2} + \frac{V_3}{R_3} \tag{20.32}$$

But, the potentials are equal by equation 20.27 (i.e., $V = V_1 = V_2 = V_3$). Hence, equation 20.32 becomes

$$I = \frac{V}{R_1} + \frac{V}{R_2} + \frac{V}{R_3}$$

Dividing both sides of the equation by V, gives

$$\frac{I}{V} = \frac{1}{R_1} + \frac{1}{R_2} + \frac{1}{R_3} \tag{20.33}$$

But the left-hand side of equation 20.33 contains the total current I in the circuit, divided by the total voltage V applied to the circuit, and by Ohm's law is simply

$$\frac{I}{V} = \frac{1}{R} \tag{20.34}$$

where R is the total resistance of the entire circuit. Equating 20.34 to equation 20.33 gives

$$\frac{1}{R} = \frac{1}{R_1} + \frac{1}{R_2} + \frac{1}{R_3} \tag{20.35}$$

Equation 20.35 says that the reciprocal of the total resistance of the circuit is equivalent to the sum of the reciprocals of each parallel resistance. We call R the equivalent resistance of the resistors in parallel. The resistor R could replace the three resistors R_1, R_2, and R_3, and the circuit would still behave the same way electrically.

Although we derived equation 20.35 from a circuit with only three resistors in parallel, the form is completely general. If there were n resistors in parallel, the equivalent resistance would be found from

$$\frac{1}{R} = \frac{1}{R_1} + \frac{1}{R_2} + \frac{1}{R_3} + \frac{1}{R_4} + \cdots + \frac{1}{R_n} \tag{20.36}$$

Example 20.12

Equivalent resistance of resistors in parallel. Three resistors $R_1 = 20.0 \, \Omega$, $R_2 = 30.0 \, \Omega$, and $R_3 = 40.0 \, \Omega$ are connected in parallel to a 6.00-V battery, as shown in figure 20.13. Find the equivalent resistance of the circuit.

Solution

From equation 20.35 the equivalent resistance is

$$\frac{1}{R} = \frac{1}{R_1} + \frac{1}{R_2} + \frac{1}{R_3}$$

$$= \frac{1}{20.0 \, \Omega} + \frac{1}{30.0 \, \Omega} + \frac{1}{40.0 \, \Omega} = \frac{0.108}{\Omega}$$

$$R = 9.26 \, \Omega$$

Notice that in this calculation $1/R = 0.108$, and to get the actual value of R we need to take the reciprocal of 0.108, which yields the correct value of the resistance as 9.26 Ω. Failure to take the final reciprocal is a very common error.

Compare this result with example 20.9 in which the same three resistors were connected in series. There, the total equivalent resistance was 90.0 Ω when the resistors were connected in series, whereas here the same resistors connected in parallel give an equivalent resistance of only 9.26 Ω. It is clear from these two examples that the way the resistors are connected in a circuit makes a great deal of difference in their effect on the circuit. *Note that when the resistors are connected in parallel, the equivalent resistance is always less than the smallest of the original resistances.* In this case the total resistance, 9.26 Ω, is less than the smallest resistance of 20.0 Ω. The equivalent circuit of the parallel circuit in figure 20.13(a) is shown in figure 20.13(b), where R is the equivalent resistance, found in equation 20.35.

Example 20.13

The total current in the parallel circuit. Find the total current coming from the battery in the example 20.12.

Solution

The total current in the circuit, found from Ohm's law with R as the equivalent resistance of the circuit, is

$$I = \frac{V}{R} = \frac{6.00 \, \text{V}}{9.26 \, \Omega} = 0.648 \, \text{A}$$

Notice that the current from the battery is almost ten times greater when the resistors are connected in parallel then when connected in series. This is easily explained, since the resistance in parallel is only about one-tenth of the resistance when the resistors are in series.

Example 20.14

The current in each parallel resistor. Find the current through each resistor in example 20.12.

Solution

Because the voltage across each resistor is 6.00 V, we can find the current from Ohm's law applied to each resistor as given in equations 20.29, 20.30, and 20.31. Namely,

$$I_1 = \frac{V}{R_1} = \frac{6.00 \text{ V}}{20.0 \text{ }\Omega} = 0.300 \text{ A}$$

$$I_2 = \frac{V}{R_2} = \frac{6.00 \text{ V}}{30.0 \text{ }\Omega} = 0.200 \text{ A}$$

$$I_3 = \frac{V}{R_3} = \frac{6.00 \text{ V}}{40.0 \text{ }\Omega} = 0.150 \text{ A}$$

Note that the current through each resistor is different, but the sum of $I_1 + I_2 + I_3 = 0.650$ A is equal, within round-off errors of our calculations, to the total current of 0.648 A flowing from the battery in the circuit. Also note that the resistor with the smallest value of resistance has the largest value of current passing through it. This is sometimes stated as: *The current always takes the path of least resistance.*

Example 20.15

Power supplied to a parallel circuit. What is the power supplied to the circuit in example 20.12?

Solution

The power supplied by the battery is

$$P = IV = (0.648 \text{ A})(6 \text{ V}) = 3.89 \text{ W}$$

Example 20.16

Power dissipated in a parallel circuit. What is the power dissipated in each resistor in example 20.14?

Solution

The power dissipated in each resistor is

$$P_1 = I_1^2 R_1 = (0.300 \text{ A})^2(20.0 \text{ }\Omega) = 1.80 \text{ W}$$

$$P_2 = I_2^2 R_2 = (0.200 \text{ A})^2(30.0 \text{ }\Omega) = 1.20 \text{ W}$$

$$P_3 = I_3^2 R_3 = (0.150 \text{ A})^2(40.0 \text{ }\Omega) = 0.900 \text{ W}$$

Again note that the sum of the powers dissipated in each resistor

$$P_1 + P_2 + P_3 = 1.80 \text{ W} + 1.20 \text{ W} + 0.900 \text{ W}$$

$$= 3.90 \text{ W}$$

is the same as the total power supplied to the circuit by the battery, within round-off error.

20.8 Combinations of Resistors in Series and Parallel

A typical circuit showing a simple combination of resistors connected in series and parallel is shown in figure 20.14. Let us determine the current through each resistor and the voltage drop across it. To do so, we use the techniques of sections 20.6 and 20.7. Resistors R_2 and R_3 are connected in parallel. Their equivalent resistance R_{23}, found from equation 20.36, is

$$\frac{1}{R_{23}} = \frac{1}{R_2} + \frac{1}{R_3}$$

$$= \frac{1}{30.0\ \Omega} + \frac{1}{40.0\ \Omega} = \frac{0.0583}{\Omega}$$

$$R_{23} = 17.2\ \Omega$$

The circuit of figure 20.14(a) can be replaced by its equivalent circuit, figure 20.14(b), where R_{23} is in series with R_1. The equivalent resistance of R_1 and R_{23} in series, found from equation 20.26, is

$$R_{123} = R_1 + R_{23}$$

$$= 20.0\ \Omega + 17.2\ \Omega$$

$$= 37.2\ \Omega$$

The circuit of figure 20.14(b) can now be replaced by the equivalent circuit, figure 20.14(c), containing only one resistor, the equivalent resistor $R_{123} = 37.2\ \Omega$.

The current from the battery is now easily determined by Ohm's law as

$$I = \frac{V}{R_{123}} = \frac{6.00\ V}{37.2\ \Omega} = 0.161\ A$$

Since the resistor R_1 is in series with the battery, this same current flows through it (i.e., $I_1 = I = 0.161$ A). However, this is not the current through R_2 and R_3 because they are in parallel and the current divides as it enters the two paths. In order to determine I_2 and I_3 we must first determine the voltage drop across R_2 and R_3.

The voltage drop across R_1, found from Ohm's law, is

$$V_1 = I_1 R_1 = (0.161\ A)(20.0\ \Omega) = 3.22\ V$$

The total applied potential is equal to the sum of the potential drops, that is,

$$V = V_1 + V_2$$

Figure 20.14

Combinations of resistors in a circuit.

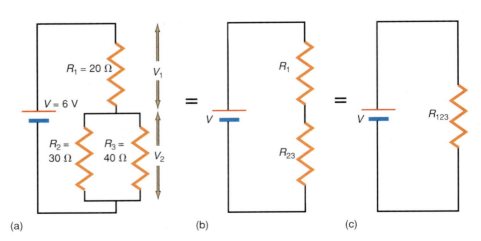

(a)　　　(b)　　　(c)

The potential drop V_2 across the parallel resistors is

$$V_2 = V - V_1$$
$$= 6.00 \text{ V} - 3.22 \text{ V}$$
$$= 2.78 \text{ V}$$

The potential drop V_3 is also equal to 2.78 V since R_2 and R_3 are connected in parallel. With the potential drop across the parallel resistors known, the current in each resistor, found from Ohm's law, is

$$I_2 = \frac{V_2}{R_2} = \frac{2.78 \text{ V}}{30.0 \text{ } \Omega} = 0.0927 \text{ A}$$

$$I_3 = \frac{V_3}{R_3} = \frac{2.78 \text{ V}}{40.0 \text{ } \Omega} = 0.0695 \text{ A}$$

Note that the sum of the currents

$$I_2 + I_3 = 0.0927 \text{ A} + 0.0695 \text{ A} = 0.162 \text{ A}$$

is equal, within round-off errors of our calculations, to the total current in the circuit, 0.161 A.

The power delivered to the circuit is

$$P = IV = (0.161 \text{ A})(6.00 \text{ V}) = 0.966 \text{ W}$$

The power dissipated in each resistor is

$$P_1 = I_1^2 R_1 = (0.161 \text{ A})^2 (20.0 \text{ } \Omega) = 0.518 \text{ W}$$

$$P_2 = I_2^2 R_2 = (0.0927 \text{ A})^2 (30.0 \text{ } \Omega) = 0.258 \text{ W}$$

$$P_3 = I_3^2 R_3 = (0.0695 \text{ A})^2 (40.0 \text{ } \Omega) = 0.193 \text{ W}$$

The sum of the power dissipated in the resistors (0.969 W) is equal, within round-off error, to the supplied power of 0.966 W.

20.9 The Electromotive Force and the Internal Resistance of a Battery

In the early days of electricity, in order to keep an analogy with Newtonian mechanics, it was assumed that if a charge moved through a wire, there must be some force pushing on the charge to cause it to move through the wire. This force acting on the charge was called an **electromotive force.** The battery, the supplier of this force, was called a *seat of electromotive force,* abbreviated emf, and pronounced as the individual letters e-m-f. The emf is usually written as the script letter $\mathscr{E}$. Today of course, we know that this emf is really a misnomer. The battery is supplying a potential difference. However, the name emf still remains. The emf is measured in volts, since the emf is a potential difference. You will hear comments such as, a battery has an emf of 6 V, or the generator supplies an emf of 120 V, and the like.

If there is no current being drawn from the battery, the emf of the battery and the potential difference between its two terminals are the same. When a current exists, however, the potential difference between the terminals is always less than the emf of the battery. The reason for this is that every battery has an internal resistance, which is a characteristic of the battery and cannot be eliminated. The internal resistance of the battery acts as though it were a resistor in series with the battery itself. It is usually represented in a circuit diagram as the lower case r in

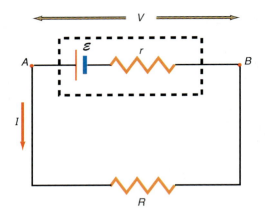

Figure 20.15
The internal resistance of a battery.

figure 20.15. When a current exists in the circuit there is a potential drop, equal to the product Ir, across the internal resistance. The potential difference between the terminals of the battery, becomes

$$V = \mathscr{E} - Ir \qquad (20.37)$$

which is, of course, less than the emf of the battery.[1]

Example 20.17

The potential and emf of a battery. A battery has an emf of 1.50 V and when connected to a circuit supplies a current of 0.0100 A. If the internal resistance of the battery is 2.00 Ω, what is the potential difference across the battery?

Solution

The potential difference across the battery, given by equation 20.37, is

$$V = \mathscr{E} - Ir$$
$$= 1.50 \text{ V} - (0.0100 \text{ A})(2.00 \text{ }\Omega)$$
$$= 1.48 \text{ V}$$

Notice that when the current in the circuit is relatively small, the potential drop across the internal resistance is also quite small, and the terminal voltage is very close to the emf of the battery.
  If the current in the circuit is much larger, as for example $I = 0.500$ A then the terminal voltage is

$$V = \mathscr{E} - Ir$$
$$= 1.50 \text{ V} - (0.500 \text{ A})(2.00 \text{ }\Omega)$$
$$= 0.500 \text{ V}$$

which is very much less than the emf of the battery.

Example 20.18

The emf of a battery connected to a circuit. A battery has an emf of 1.50 V and an internal resistance of 3.00 Ω. It is connected as shown in figure 20.15 to a resistance of 500 Ω. Find the current in the circuit, and the terminal voltage of the battery.

Solution

The internal resistance r is in series with the load resistor R, so the total resistance in the circuit is just the sum of the two of them. Using Ohm's law to determine the current we have

$$I = \frac{\mathscr{E}}{R + r} = \frac{1.50 \text{ V}}{500 \text{ }\Omega + 3.00 \text{ }\Omega} = 2.98 \times 10^{-3} \text{ A}$$

The terminal voltage across the battery is

$$V = \mathscr{E} - Ir$$
$$= 1.50 \text{ V} - (2.98 \times 10^{-3} \text{ A})(3.00 \text{ }\Omega)$$
$$= 1.49 \text{ V}$$

---

1. We should note that if the battery is being charged by an external source, then the terminal voltage would be given by $V = \mathscr{E} + Ir$.

Because the value of r is usually very small, relative to the resistance R of the circuit, we neglect it in many problems. In such cases, we assume that the terminal voltage and the emf are the same. Whether we can make this assumption or not, depends on the values of r and R and the accuracy that we are willing to accept in the solution to the problem. In example 20.18, neglecting r gives a current of 3.00 $\times$ 10^{-3} A, an error of about 0.7%. If that amount of error is acceptable in the problem, the internal resistance can be ignored. If that error is not acceptable then the internal resistance must be taken into account.

Batteries in Series

Since the emf of most batteries is only 1.50 V, we need to place many of them in series to get a higher emf. If three 1.50-V batteries are connected in series, as in figure 20.16, with the positive terminal of one battery connected to the negative terminal of the next battery and so on, the emf of the combination is 3(1.50 V) = 4.50 V. *In general, when batteries are connected in series, the total emf is the sum of the emf's of each battery.* That is,

$$\mathcal{E} = \mathcal{E}_1 + \mathcal{E}_2 + \mathcal{E}_3 + \cdots + \mathcal{E}_n \qquad (20.38)$$

By connecting any number of batteries in series, any larger emf may be obtained. Because each battery is in series, the current through each battery is the same, and the total internal resistance of the combination is just the sum of the internal resistance of each battery.

Batteries in Parallel

Three identical batteries, of negligible internal resistance, are connected in parallel in figure 20.17. The negative terminals of all the batteries are all connected together, and the positive terminals of all the batteries are all connected to one another. *Since the batteries are in parallel, the potential difference across the combination is the same as the emf of each battery,* that is,

$$\mathcal{E} = \mathcal{E}_1 = \mathcal{E}_2 = \mathcal{E}_3 \qquad (20.39)$$

At junction A, the three battery currents combine to give the circuit current

$$I = I_1 + I_2 + I_3 \qquad (20.40)$$

Because we assume the batteries to be identical ($I_1 = I_2 = I_3$), the total circuit current available is three times the current available from any one battery. By a suitable combination of batteries in series and parallel, batteries of any voltage and current can be made.

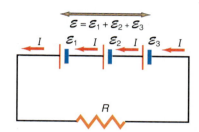

Figure 20.16

Batteries in series.

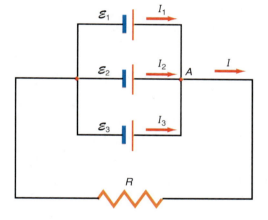

Figure 20.17

Identical batteries in parallel.

Example 20.19

Batteries in parallel. Three identical 6.00-V batteries are connected in parallel to a resistance of 500 Ω, as in figure 20.17. Find the current in the circuit and the current coming from each battery. Assume the internal resistance of the batteries to be zero.

Solution

The total current I in the circuit, found from Ohm's law, is

$$I = \frac{\mathcal{E}}{R} = \frac{6.00 \text{ V}}{500 \ \Omega} = 0.0120 \text{ A}$$

This total current is supplied by three batteries so that, $I = 3I_0$, where I_0 is the actual current from any one battery. Therefore, the battery current is

$$I_0 = \frac{I}{3} = \frac{0.0120 \text{ A}}{3} = 4.00 \times 10^{-3} \text{ A}$$

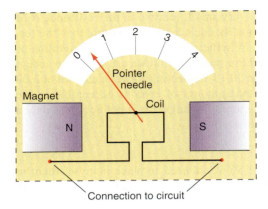

Figure 20.18

The galvanometer.

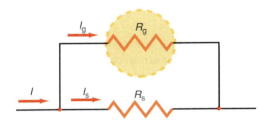

Figure 20.19

Converting a galvanometer to an ammeter.

20.10 Making an Ammeter and Voltmeter from a Galvanometer

The **galvanometer** is the basic device for the measurement of electric current. We will discuss the physical principles underlying the galvanometer more fully in chapter 22. In its simplest form, the galvanometer consists of a coil of wire placed in a magnetic field. When current exists in the galvanometer coil, a torque acts on the coil causing it to rotate. A needle is attached to the coil, as shown in figure 20.18. As the coil rotates, the needle rotates with it, and moves across the scale of the galvanometer. A counter torque is supplied by springs such that the needle comes to rest somewhere on the scale. The deflection of the coil is directly proportional to the current in it. Once the galvanometer is calibrated, the deflection of the needle on the scale is a measure of the current in the circuit. Full-scale deflection of the coil usually occurs with currents of only a few milliamps. The resistance of the coil is usually of the order of 1 to 100 Ω.

The Ammeter

Let us consider a galvanometer whose coil resistance R_g is 15.0 Ω and gives full-scale deflection for a current I_g of 20.0 mA. If the scale is labeled from 0 to 20, the galvanometer is an ammeter that indicates currents in a circuit up to a maximum of 20.0 mA. The ammeter is placed in series in the circuit so that the same charge that flows through the circuit also flows through the ammeter.

However, an ammeter that can read currents only up to a maximum of 20.0 mA is not very useful. It would certainly be desirable if the range of this ammeter could be increased, such that it could read currents up to a maximum of, let us say, 10.0 A. To do this, let us picture the ammeter as a resistance R_g that can carry a maximum current of 20.0 mA. If we place a small resistance R_s, called the shunt resistor, in parallel with R_g, as shown in figure 20.19, then the maximum circuit current of 10.0 A splits at the junction of the two parallel resistors. If we choose R_s properly, 20.0 mA of the total current of 10.00 A goes through the galvanometer, whereas the rest, 10.00 A − 0.0200 A = 9.98 A, is shunted around the galvanometer. However, since the galvanometer still receives 20.0 mA of current, there is full-scale deflection of the galvanometer needle even though there is a total current of 10.0 A in the circuit. We recalibrate the scale of the galvanometer, from 0 to 10.0 A, indicating 10.0 A for full-scale deflection. The galvanometer so modified is now an ammeter capable of reading any current in a circuit up to a maximum of 10.0 A.

Let us formalize the method for obtaining full-scale deflection for any desired current. Because the resistors R_s and R_g are in parallel, the voltage drop across them is equal. That is,

$$V_g = V_s \qquad (20.41)$$

The maximum current that can exist in the galvanometer is I_g. Therefore, the voltage drop across the galvanometer is

$$V_g = I_g R_g \qquad (20.42)$$

The voltage drop across the shunt resistor is

$$V_s = I_s R_s \qquad (20.43)$$

where R_s is the resistance of the shunt and I_s is the current that goes through the shunt. But as we can see in figure 20.19, the total current I in the circuit is the sum of the current in the galvanometer I_g and the current in the shunt I_s. That is,

$$I = I_g + I_s$$

Solving for the shunt current,

$$I_s = I - I_g$$

Substituting this value of I_s into equation 20.43 gives

$$V_s = I_s R_s = (I - I_g) R_s \qquad \text{(20.44)}$$

Equating equations 20.44 and 20.42 gives

$$(I - I_g) R_s = I_g R_g$$

The resistance of the shunt becomes

$$R_s = \frac{I_g}{(I - I_g)} R_g \qquad \text{(20.45)}$$

Equation 20.45 gives the value of the shunt resistor that must be placed in parallel with the galvanometer to convert the galvanometer into an ammeter that will show a full-scale deflection when the current in the circuit is I. The range of the ammeter has thus been increased from a maximum of I_g to a maximum current I. Thus, a galvanometer can be converted to an ammeter by placing a suitable low-resistance shunt in parallel with the galvanometer.

Example 20.20

Converting a galvanometer to an ammeter. A galvanometer that has an internal resistance R_g of 15.0 Ω and a full-scale deflection at 20.0 mA is to be converted into an ammeter that will read up to a maximum current of 10.00 A. Find the value of the shunt resistor that should be placed in parallel with the galvanometer.

Solution

The value of the shunt resistor, found from equation 20.45, is

$$R_s = \frac{I_g}{(I - I_g)} R_g$$

$$= \left(\frac{0.0200 \text{ A}}{10.00 \text{ A} - 0.0200 \text{ A}} \right) (15.0 \ \Omega)$$

$$= 0.0301 \ \Omega$$

Therefore, when a current of 10.00 A flows in the circuit, a current of 9.98 A flows through the 0.0301 Ω shunt resistor, and a current of 0.0200 A flows through the galvanometer, indicating full-scale deflection.

Example 20.21

Total resistance of an ammeter. Find the total resistance of the ammeter in the previous example.

Solution

Since resistors R_g and R_s are in parallel, their combined resistance, which is the resistance of the ammeter R_A, is

$$\frac{1}{R_A} = \frac{1}{R_g} + \frac{1}{R_s}$$

$$= \frac{1}{15.0 \ \Omega} + \frac{1}{0.0301 \ \Omega}$$

$$R_A = 0.0300 \ \Omega$$

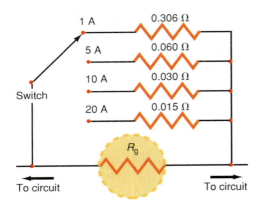

Figure 20.20

Multirange ammeter.

Table 20.2

Shunt Resistors for an Ammeter with $I_g = 0.0200$ A and $R_g = 15.0 \, \Omega$

I (A)	R_s (Ω)
20.0	0.015
10.0	0.030
5.00	0.060
1.00	0.306

Figure 20.21

Converting a galvanometer to a voltmeter.

which is relatively small and allows us to neglect the resistance of the ammeter in a great many circuit problems. Notice that this value is essentially the resistance of the shunt resistor. Because of the small value of resistance, the ammeter usually does not significantly affect the circuit in which it is inserted. This also implies that an ammeter can be easily burned out if it is not protected in a circuit.

Multirange Ammeter An ammeter can be used to measure a large range of currents by connecting it to a many-position switch and various shunt resistors, as shown in figure 20.20. If equation 20.45 is solved for various values of shunt resistors for various maximum circuit currents, we get the values in table 20.2. These values are used for the shunt resistors in figure 20.20.

Example 20.22

Length of a wire for a shunt resistor. What length of #26 B&S gauge copper wire is necessary to make a 0.0300-Ω resistor shunt?

Solution

#26 B&S gauge wire has a diameter of 4.05×10^{-4} m and its cross-sectional area is

$$A = \frac{\pi d^2}{4}$$

$$= \frac{\pi (4.05 \times 10^{-4} \text{ m})^2}{4}$$

$$= 1.29 \times 10^{-7} \text{ m}^2$$

The length of wire needed, found from equation 20.7, is

$$l = \frac{RA}{\rho} = \frac{(0.0300 \ \Omega)(1.29 \times 10^{-7} \text{ m}^2)}{1.72 \times 10^{-8} \ \Omega \text{ m}} = 0.224 \text{ m}$$

That is, a 22.4-cm length of #26 B&S gauge copper wire will provide the shunt resistance of 0.0300 Ω.

The Voltmeter

The galvanometer can also be used as a voltmeter. Using the same galvanometer as before, the resistance of the galvanometer R_g is 15.0 Ω, and full-scale deflection again occurs for a current I_g of 20.0 mA. The voltage drop across the galvanometer is

$$V_g = I_g R_g = (20.0 \times 10^{-3} \text{ A})(15.0 \ \Omega) = 0.300 \text{ V}$$

By changing the scale of the galvanometer so that maximum deflection of the scale reads 0.300 V, we have a voltmeter, as shown in figure 20.21(a), that can read up to a maximum of 0.300 V. However, a voltmeter that can only read to a maximum of 0.300 V is not very useful. *The voltmeter can be converted to one of larger scale by placing a resistor R in series with the galvanometer,* as shown in figure 20.21(b).

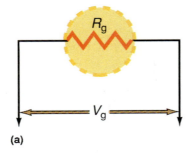

(a)

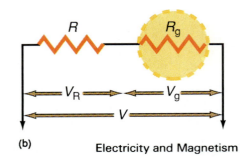

(b)

Electricity and Magnetism

The value of the series resistor R depends on the maximum voltage V that is to be measured. The basic principle involved is to drop the 0.300 V across the galvanometer and drop the remaining voltage across the series resistor R. To determine this value of the series resistor R, we proceed as follows. The voltage drop across the combination is

$$V = V_R + V_g$$

Solving for the voltage drop across the series resistor yields

$$V_R = V - V_g \qquad (20.46)$$

The maximum permissible voltage across the galvanometer for full-scale deflection is

$$V_g = I_g R_g \qquad (20.47)$$

Since R is in series with R_g, the current in the series resistance R must also be I_g. The voltage drop across the series resistor is

$$V_R = I_g R \qquad (20.48)$$

Substituting equations 20.47 and 20.48 into equation 20.46 gives

$$I_g R = V - I_g R_g$$

Solving for resistance of the series resistor R, we get

$$R = \frac{V}{I_g} - R_g \qquad (20.49)$$

Equation 20.49 gives the value of the resistor that should be connected in series with the galvanometer to convert the galvanometer into a voltmeter that gives full-scale deflection for any voltage V.

Example 20.23

Converting a galvanometer to a voltmeter. Convert the above galvanometer to a voltmeter that gives full-scale deflection for $V = 100$ V.

Solution

To convert the galvanometer to a voltmeter, a resistor R, given by equation 20.49, must be placed in series with the galvanometer. Therefore,

$$R = \frac{V}{I_g} - R_g = \frac{100 \text{ V}}{0.0200 \text{ A}} - 15.0 \text{ }\Omega$$

$$= 4985 \text{ }\Omega$$

As a check, note that the voltage across this series resistor is

$$V_R = I_g R = (0.0200 \text{ A})(4985 \text{ }\Omega)$$

$$= 99.7 \text{ V}$$

while the voltage drop across the galvanometer is

$$V_g = I_g R_g = (0.0200 \text{ A})(15.0 \text{ }\Omega) = 0.300 \text{ V}$$

When a potential difference of 100 V is placed across this combination of the galvanometer and its associated series resistor, 99.7 V are dropped across the series resistor while 0.3 V are dropped across the galvanometer. The galvanometer now shows full-scale deflection for 100 V. If a new scale from 0 to 100 is printed on the face of the galvanometer, a voltmeter that will read potential differences from 0 up to 100 V is obtained.

The total resistance of this voltmeter is the sum of the series resistor and the galvanometer resistor

$$R_v = R + R_g \qquad (20.50)$$

For this example,

$$R_v = 4985 \, \Omega + 15.0 \, \Omega = 5000 \, \Omega$$

Note that the resistance of a voltmeter is quite high as compared to the ammeter, which is quite low. *A voltmeter of any range V can be obtained by placing that value of V into equation 20.49 and solving for the value of the resistor that must be placed in series with the galvanometer.* In general, when making a voltmeter, we want to have the resistance of the voltmeter as large as possible. Because voltmeters are always placed in parallel in a circuit, the larger the resistance of the voltmeter, the smaller the current through it, and the less it disturbs the circuit in which the measurement is being made.

Multirange Voltmeter A galvanometer can be converted into a multirange voltmeter by connecting it to a many-position switch and various resistors, as shown in figure 20.22. To make a multirange voltmeter of 25, 50, 100, 500, and 1000 V, these values of V are placed into equation 20.49 and the values of the series resistors, R, are calculated. The results are shown in table 20.3 and figure 20.22. A summary of the characteristics of meters and how they are used in a circuit is shown in table 20.4.

Figure 20.22

A multirange voltmeter.

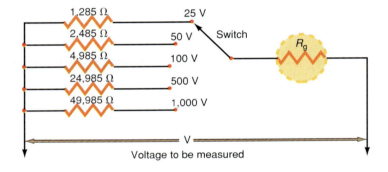

Table 20.3

Series Resistors for a Voltmeter, $I_g = 0.0200$ A and $R_g = 15.0 \, \Omega$

V (V)	R (Ω)
25.0	1,285
50.0	2,485
100.0	4,985
500.0	24,985
1000.0	49,985

Table 20.4

Comparison of Meter Characteristics

	Ammeter	Voltmeter
Used to measure	current	potential
Internal resistance	low	high
Internal resistor connection to galvanometer	parallel	series
How meter is connected to circuit element	series	parallel

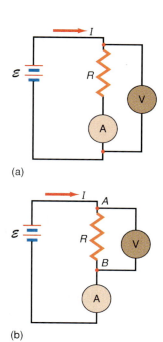

(a)

(b)

Figure 20.23

Techniques for measuring voltage and current in a simple circuit.

20.11 The Wheatstone Bridge

A standard technique to determine the value of a particular resistor is to connect it into a circuit, such as in figure 20.23(a). It is usually assumed that the resistance of the ammeter is negligible, and the resistance of the voltmeter is so large that negligible current flows through it. Therefore, the current I in the resistor is measured, and the voltage V across it is also measured. The value of the resistance R is easily determined from Ohm's law as

$$R = \frac{V}{I} \qquad (20.51)$$

As long as our assumptions hold, equation 20.51 is valid. Now let's look a little closer at our assumptions. If the resistance of the ammeter is anywhere near the order of magnitude of the resistance R, the assumptions break down. Although the ammeter does measure the current I through R, there is now a voltage drop V_A across the ammeter, and the voltmeter reads the voltage drop

$$V = V_R + V_A$$

The calculated resistance is

$$R = \frac{V_R + V_A}{I} \qquad (20.52)$$

which gives a value for R greater than the exact value found by equation 20.51 when the effects of the ammeter can be neglected.

To remedy this, we might say, let's change the measuring technique to the one shown in figure 20.23(b). With this second technique, the voltmeter reads only the voltage V_R across the resistor. At first glance, it might seem that the problem is solved. But if we look carefully we note that the current I in the circuit splits at junction A, where the resistor and voltmeter are in parallel. A current I_R flows through the resistor while a current I_v flows through the voltmeter. The ammeter reads the current

$$I = I_R + I_v$$

The computed resistance R therefore becomes

$$R = \frac{V_R}{I_R + I_v} \qquad (20.53)$$

But this is less than that computed from the exact value of equation 20.51, when the effects of the voltmeter can be neglected in the circuit.

This analysis of the measurements of a simple circuit again points out the importance of the assumptions made in deriving any equation. As long as the initial assumptions hold, the derived equations hold. When there is a breakdown in the assumptions, the derived equations are no longer valid. Note that when the assumptions of a negligible resistance for the ammeter, and a negligible current in the voltmeter (very high resistance voltmeter), are correct, equations 20.52 and 20.53 reduce to 20.51.

When the previous assumptions do not hold, it is still possible to make accurate measurements of resistance by means of a circuit that is called the **Wheatstone bridge,** figure 20.24(a). Its virtue is that it is a null detection method that does not depend on the characteristics of the galvanometer, but uses it merely to be a sensitive detector of the condition of current versus no current. The values of resistors R_1, R_3, and R_4, in the circuit are assumed to be known, whereas R_2 is the unknown resistor. A battery of emf $\mathscr{E}$ is applied to the circuit and a current I emanates from the battery. At the junction A, the current I splits into two currents, I_1 and I_3.

$$I = I_1 + I_3$$

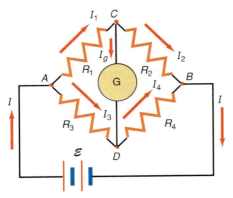

(a) Unbalanced

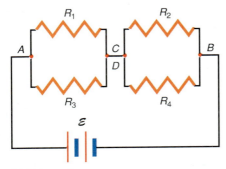

(b) Balanced

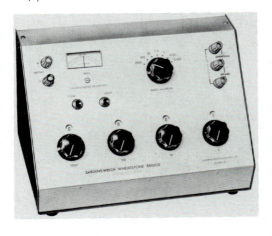

Figure 20.24

The Wheatstone bridge.

When current I_1 reaches the junction C it splits into two currents, I_2 through R_2 and I_g the current through the galvanometer. That is,

$$I_1 = I_2 + I_g \qquad (20.54)$$

The galvanometer needle deflects indicating that there is a current in it. At junction D the galvanometer current I_g combines with current I_3 to form current I_4, which passes through resistor R_4. That is,

$$I_3 + I_g = I_4 \qquad (20.55)$$

At junction B, currents I_2 and I_4 combine to form the original current I.

A current I_g exists in the galvanometer, because there is a difference of potential between points C and D in the circuit. This difference in potential is caused by the different currents in the different resistors. The values of these resistors R_1, R_3, and R_4 are varied until the galvanometer reads zero (i.e., $I_g = 0$). When this occurs the bridge is said to be balanced. When the galvanometer reads zero, it means that there is no longer a difference in potential between points C and D. (Remember a current exists in a conductor when there is a difference in potential. If there is no current, there is no difference in potential.) The circuit in figure 20.24(a) can now be replaced by the one in figure 20.24(b). The points C and D have coalesced electrically into one point, because they are at the same potential. It is now obvious from figure 20.24(b) that R_1 is now in parallel with R_3 and therefore the voltages across them are equal, that is,

$$V_1 = V_3$$

Therefore, from Ohm's law,

$$I_1 R_1 = I_3 R_3 \qquad (20.56)$$

It is also obvious that R_2 is now in parallel with R_4 and hence the voltages across them are also equal, that is,

$$V_2 = V_4$$

$$I_2 R_2 = I_4 R_4 \qquad (20.57)$$

Dividing equation 20.57 by equation 20.56, we obtain

$$\frac{I_2 R_2}{I_1 R_1} = \frac{I_4 R_4}{I_3 R_3} \qquad (20.58)$$

However, since the bridge is balanced, $I_g = 0$, and equation 20.54 reduces to

$$I_1 = I_2$$

and equation 20.55 to

$$I_3 = I_4$$

Therefore, the currents in equation 20.58 cancel, leaving the simple ratio of the resistors

$$\frac{R_2}{R_1} = \frac{R_4}{R_3}$$

Solving for the unknown resistor R_2, we obtain

$$R_2 = \frac{R_4}{R_3} R_1 \qquad (20.59)$$

The Wheatstone bridge is a very accurate means of determining the resistance of an unknown resistor. A picture of a typical Wheatstone bridge is shown in figure 20.24(c).

Example 20.24

The Wheatstone bridge. A Wheatstone bridge is balanced when $R_1 = 5.00\ \Omega$, $R_3 = 10.0\ \Omega$, and $R_4 = 4.00\ \Omega$. Find the unknown resistance R_2.

Solution

The value of the unknown resistor, found from equation 20.59, is

$$R_2 = \frac{R_4}{R_3}R_1 = \frac{(4.00\ \Omega)}{(10.0\ \Omega)}(5.00\ \Omega) = 2.00\ \Omega$$

20.12 Kirchhoff's Rules

Quite often we find circuits in which the resistors are not simply in series or parallel, or their combinations. An example of such a circuit is shown in figure 20.25. The locations of the different batteries in the circuit prevents the resistors from being in parallel. In order to solve this more complicated circuit problem, two rules established by the German physicist Gustav Robert Kirchhoff (1824–1887) are used. *Kirchhoff's first rule is: The sum of the currents entering a junction is equal to the sum of the currents leaving the junction.* That is,

$$\Sigma I_{\text{entering junction}} = \Sigma I_{\text{leaving junction}} \tag{20.60}$$

This first rule is a statement of the law of conservation of electric charge.[2]

In order to apply the first rule, we must make an assumption about the currents in the circuit. We assume that current I_1 emanates from battery $\mathscr{E}_1$, I_2 from $\mathscr{E}_2$, and I_3 from $\mathscr{E}_3$, all in the directions shown in figure 20.25. If the actual directions of the currents are as assumed, the numerical values of the currents found by the equations will be positive. If the direction of the currents is in the opposite direction from those assumed, the numerical values of the currents found by the equations will be negative, indicating that the direction of the currents is really reversed. Using Kirchhoff's first rule at junction A, gives

$$I_2 = I_1 + I_3 \tag{20.61}$$

Kirchhoff's second rule states that the change in potential around a closed loop is equal to zero. This can also be stated in the form, around any closed loop, the sum of the potential rises plus the sum of the potential drops is equal to zero. That is,

$$\Delta V_{\text{closed loop}} = 0 \tag{20.62}$$

Figure 20.25

A circuit problem solved by Kirchhoff's rules.

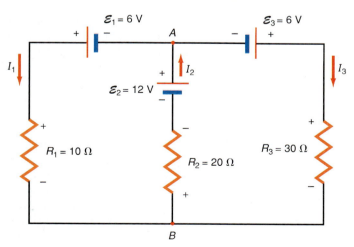

2. We should note that Kirchhoff's first rule is sometimes stated in the equivalent form that the algebraic sum of the currents toward a junction is zero. That is, currents entering the junction are positive and currents leaving the junction are negative. In this notation, equation 20.60 would take the equivalent form $\Sigma I = 0$.

Kirchhoff's second rule is really another statement of the law of conservation of energy. It says that *if a positive charge is carried completely around the loop of the circuit, in either a clockwise or counterclockwise direction, on returning to its original point, it will have the same value of V with which it started. Hence, there is no change in potential when the charge returns to the same place where it started from.* The mechanical analogue is that of a person standing on, let's say, the fifth floor of a ten-story building. The person has a certain potential energy at that point. The person now goes for a walk up and down the stairway. If the person goes up the stairs he increases his potential energy. If he goes down the stairs he decreases his potential energy. But when he returns to the fifth floor, from whence he started, he has the same potential energy that he started with.

Because charge flows from a point of high potential to one at a lower potential, it is convenient to place a plus ($+$) sign on the side of a resistor that the current enters and a negative sign ($-$) on the side of the resistor that the current leaves, as shown in figure 20.25. If the charge that is carried around the circuit traverses the loop of the circuit in the direction of the current, then as it passes the resistor it goes from a position of higher potential to one of lower potential, and thereby loses potential as it crosses the resistor. That is, there is a potential drop across the resistor, which we will designate as $-IR$. If the carried charge passes through a resistor in the direction that is opposite to the direction of the current, then it is going from a position of lower potential to one of higher potential and hence it experiences a potential rise across the resistor, which we will designate as $+IR$.

In traversing a battery, if the carried charge comes out of the positive terminal of the battery, it is at a positive potential, and this we designate as $+\mathscr{E}$. If it comes out of the negative terminal of the battery, when traversing the circuit, it has dropped in potential and this we designate as $-\mathscr{E}$. A good mechanical analogy here would be that the battery is like an elevator in a ten-story building. When the person enters at the ground floor, he is carried up to the fifth floor by the elevator. The elevator has caused his potential energy to increase. The stairs are like the resistors. When the person walks up and down the stairs from the fifth floor, he gains or loses potential energy. If he enters the elevator at any floor along his excursion and rides to another floor, he can gain or lose additional potential energy.

With the help of these conventions, we can now apply Kirchhoff's second rule to figure 20.25. Let us carry a charge around the left-hand loop in a counterclockwise direction, starting at the positive terminal of $\mathscr{E}_1$. *The sum of the potential rises and drops as the loop is traversed is*

$$\mathscr{E}_1 - I_1R_1 - I_2R_2 + \mathscr{E}_2 = 0 \qquad \textbf{(20.63)}$$

Note that $\mathscr{E}_1$ and $\mathscr{E}_2$ are positive since the charge emanates from the positive terminals in traversing the loop counterclockwise, and that the products, I_1R_1 and I_2R_2, are both negative since the charge dropped in potential as it traversed these resistors in the counterclockwise direction. The decision to traverse the loop counterclockwise was completely arbitrary. If we wanted to traverse the loop in a clockwise direction, the sum of the potential drops and rises would yield

$$-\mathscr{E}_1 - \mathscr{E}_2 + I_2R_2 + I_1R_1 = 0 \qquad \textbf{(20.64)}$$

In this case the emf's are negative because in traversing the loop in a clockwise direction, the carried charge emerges from the negative terminals of the batteries. The products of the IRs across the resistors are now positive, and are called *potential rises,* because in traversing the loop in a clockwise direction, the carried charge entered the resistors at their lowest potential ($-$) and emerged at their highest potential ($+$). No new information is obtained in this way, however, because if we multiply each term in equation 20.64 by a negative one, we get back equation 20.63. Therefore, it does not matter whether the loop is traversed in a clockwise or counterclockwise direction.

Recall one of the fundamental principles of algebra that in the solution of equations there must be as many equations as there are unknowns in order to solve for all the unknowns. At this point, there are three unknowns, the currents I_1, I_2, and I_3 and only two equations 20.61 and 20.63. We need one additional equation to solve the problem. We can obtain the third equation by applying Kirchhoff's second rule to the right-hand loop in figure 20.25. This loop will be traversed in a counterclockwise direction starting at the negative terminal of the battery $\mathscr{E}_3$. The sum of the potential rises and drops around the right-hand loop is

$$-\mathscr{E}_3 - \mathscr{E}_2 + I_2R_2 + I_3R_3 = 0 \qquad (20.65)$$

Again note that by our convention, in traversing the loop in a counterclockwise direction, the carried charge emerged from the negative terminals of the batteries, and hence the emf's are negative. The products of the IRs are positive and are potential rises because the resistors were entered at the position of their lowest potential and exited at the position of their highest potential.

There are now three equations in the three unknown currents, I_1, I_2, and I_3, and we must solve them simultaneously. It is more convenient to place the numerical values of the emf's and the resistors into the equations and then solve them simultaneously. The equations are

Algebraically   *Numerically*

$I_2 = I_1 + I_3$ $I_2 = I_1 + I_3$ $\qquad (20.61)$

$\mathscr{E}_1 - I_1R_1 - I_2R_2 + \mathscr{E}_2 = 0$ $6.00 - 10.0I_1 - 20.0I_2 + 12.0 = 0$ $\qquad (20.66)$

$-\mathscr{E}_3 - \mathscr{E}_2 + I_2R_2 + I_3R_3 = 0$ $-6.00 - 12.0 + 20.0I_2 + 30.0I_3 = 0$ $\qquad (20.67)$

We will find the solution to the three equations by the method of substitution. We eliminate the current I_2 from the equations by substituting equation 20.61 into both equations 20.66 and 20.67, that is,

$$6.00 - 10.0I_1 - 20.0(I_1 + I_3) + 12.0 = 0$$

$$-6.00 - 12.0 + 20.0(I_1 + I_3) + 30.0I_3 = 0$$

Combining terms, the loop equations become

$$18.0 - 30.0I_1 - 20.0I_3 = 0 \qquad (20.68)$$

and

$$-18.0 + 20.0I_1 + 50.0I_3 = 0 \qquad (20.69)$$

The problem is now reduced to solving two equations for the two unknowns I_1 and I_3. The easiest way to solve these equations is to multiply one of the equations by a factor that will make one current term in equation 20.69 equal to a current term in equation 20.68. Then the two equations are either added or subtracted to eliminate the common term, leaving one equation, in one unknown current, which is then easily solved. For example, if we multiply both sides of equation 20.69 by 1.50, one term becomes $+30.0I_1$, and when this equation is added to equation 20.68, the I_1 term is eliminated. That is,

$$(1.50)(-18.0 + 20.0I_1 + 50.0I_3) = (1.50)(0)$$

gives

$$-27.0 + 30.0I_1 + 75.0I_3 = 0 \qquad (20.70)$$

Adding this to equation 20.68 gives

$$\begin{array}{r} 18.0 - 30.0I_1 - 20.0I_3 = 0 \qquad (20.68) \\ \text{ADD} \quad -27.0 + 30.0I_1 + 75.0I_3 = 0 \qquad (20.70) \\ \hline -9.00 + \quad 0 \quad + 55.0I_3 = 0 \end{array}$$

Solving for the current I_3, we obtain

$$55.0I_3 = 9.00$$
$$I_3 = 0.164 \text{ A}$$

Substituting this value for the current I_3 back into equation 20.68 allows us to solve for the current I_1. That is,

$$18.0 - 30.0I_1 - 20.0(0.164) = 0$$
$$I_1 = 0.491 \text{ A}$$

At this point the currents I_1 and I_3 could be substituted back into equation 20.61 to find the current I_2. However, we will not do this. Instead we will evaluate the current I_2 from one of the loop equations and use equation 20.61 as a check on our arithmetic. The reason for doing this check is that it is very easy to make a mistake in the algebra and/or the arithmetic in the problem.

Substituting the value of I_1 just found, back into equation 20.66, yields

$$6.00 - 10.0(0.491) - 20.0I_2 + 12.0 = 0$$

Solving for I_2,

$$I_2 = 0.655 \text{ A}$$

As a check, place the values of the currents just found back into equation 20.61:

$$I_2 = I_1 + I_3$$
$$0.655 \text{ A} = 0.491 \text{ A} + 0.164 \text{ A} = 0.655 \text{ A}$$

Thus,

$$0.655 \text{ A} = 0.655 \text{ A} \quad \text{CHECK}$$

The current equation thus acts as a good check on our calculations. The problem is now solved, since we know the three currents.

It is interesting and informative to plot the change in potential that a positive charge would encounter as it traversed each battery and resistor as it moves around the loop. Considering the first loop

$$\mathscr{E}_1 - I_1R_1 - I_2R_2 + \mathscr{E}_2 = 0$$
$$6.00 - 10.0I_1 - 20.0I_2 + 12.0 = 0$$
$$6.00 - 10.0(0.491) - 20.0(0.655) + 12.0 = 0$$
$$6.00 \text{ V} - 4.91 \text{ V} - 13.09 \text{ V} + 12.0 \text{ V} = 0 \qquad \textbf{(20.63)}$$

These terms are plotted in figure 20.26. Emerging from the positive terminal of the first battery the positive charge is at the potential $\mathscr{E}_1$, which is $+6.00$ V above ground, the zero of potential. As the first resistor is traversed, there is a potential drop V_{R1} of 4.91 V, leaving the charge at a potential of $6.00 \text{ V} - 4.91 \text{ V} = 1.09 \text{ V}$ above ground. This is the value of the potential as the charge enters resistor R_2. Across R_2, the charge experiences a potential drop of 13.09 V. Thus, on leaving R_2 the potential is at a negative value of $1.09 \text{ V} - 13.09 \text{ V} = -12.0 \text{ V}$. At that point, the charge enters the negative terminal of battery 2 at a potential of -12 V. As battery 2 is traversed, chemical energy is converted into electrical energy until, when the charge emerges from the positive terminal of the battery, it is 12.0 V higher in potential then when it entered and it is now at a zero potential. The charge now enters the negative terminal of battery 1, $-\mathscr{E}_1$, at a 0.00 V potential and is raised 6.00 V by chemical means as it emerges from the positive terminal of $\mathscr{E}_1$. The charge is now back to where it started. As we can see from figure 20.26, *the net change in potential as the charge traversed the closed loop is zero.*

Figure 20.26

Plot of the potential drops and rises as the left loop of figure 20.25 is traversed.

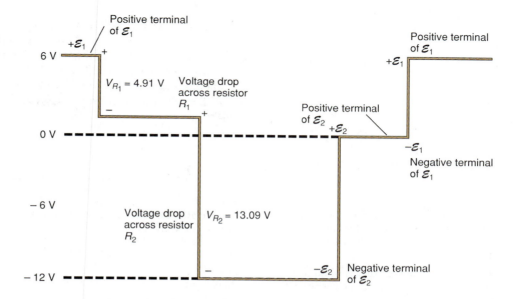

Figure 20.27

Plot of potential drops and rises around the right-hand loop of figure 20.25.

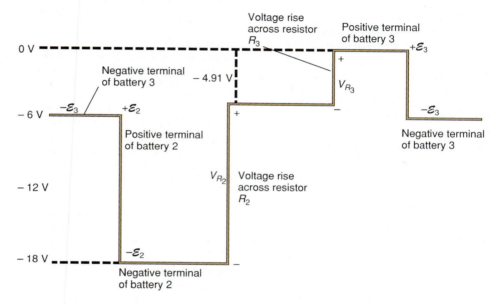

A plot of the potential drops and rises experienced by a charge as it traverses the right-hand loop is shown in figure 20.27. The loop equation is

$$-\mathscr{E}_3 - \mathscr{E}_2 + I_2 R_2 + I_3 R_3 = 0$$

$$-6.00 - 12.00 + 20.0 I_2 + 30.0 I_3 = 0$$

$$-6.00 \text{ V} - 12.00 \text{ V} + 13.09 \text{ V} + 4.91 \text{ V} = 0 \qquad (20.65)$$

Starting at the negative terminal of battery 3, the charge is 6.00 V below ground. The charge now enters the positive terminal of battery 2. When it emerges from the negative terminal of battery 2 it is 12.0 V more negative, or 18.0 V below ground. As the charge enters resistor R_2 it is at the end of the resistor's lowest potential, and when it emerges at the other end of the resistor it is at the resistor's highest potential. The charge, therefore, experienced a potential rise of 13.09 V, leaving it at a potential of -18.0 V $+ 13.09$ V $= -4.91$ V. The charge now enters resistor R_3, where it experiences a potential rise of $+4.91$ V, leaving it at the zero of potential and the positive terminal of battery $\mathscr{E}_3$. As the battery is traversed a potential drop of 6.00 V is experienced as the charge emerges from the negative terminal of $\mathscr{E}_3$. The charge is back where it started and, as we can see from figure 20.27, the total change in potential in traversing the closed loop is zero. It is not necessary to draw these potential diagrams for every Kirchhoff's rules problem. The student should, however, draw at least one potential diagram to help in the understanding of Kirchhoff's second rule.

The Language of Physics

Electric current
Electric current is defined as the amount of electric charge that flows through a cross section of a wire per unit time (p. 564).

Ampere
The SI unit of current that is equal to a flow of one coulomb of charge per second (p. 564).

Conventional current
A flow of positive charges in a circuit from a position of high potential to one of low potential (p. 565).

Electron current
The actual current in a circuit; it is a flow of electrons from a position of low potential to one of high potential (p. 565).

Direct current
An electric current in which the electric charges flow in only one direction in a circuit (p. 565).

Ammeter
A device that measures the amount of current in a circuit. It is constructed by placing a low-resistance shunt in parallel with a galvanometer (p. 565).

Voltmeter
A device that measures the potential difference between any two points in an electric circuit. It is constructed by placing a high resistance in series with a galvanometer (p. 565).

Ohm's law
For metallic conductors, the current in a circuit is directly proportional to the applied potential difference and inversely proportional to the resistance in the circuit (p. 566).

Joule heat
The energy that is lost as a charge "falls" through a resistor shows up as heat in the resistor (p. 572).

Series circuit
A circuit in which each element of the circuit is connected to an adjacent element of the circuit such that the same amount of charge flows through each and every circuit element. For a resistive circuit, the current is the same through each resistor (p. 574).

Equivalent resistance
A single resistor, whose value is equal to the combined resistance of the individual resistors in the circuit. For resistors in series, the equivalent resistance is equal to the sum of the individual resistances. For resistors in parallel, the reciprocal of the equivalent resistance is equal to the sum of the reciprocals of the individual resistances (p. 575).

Parallel circuit
A circuit in which the circuit elements are connected in such a way that the potential difference across all the elements of the circuit is the same. For a resistive circuit, the potential difference across the resistors is equal (p. 576).

Law of conservation of electric charge
Electric charge can neither be created nor destroyed (p. 576).

Electromotive force (emf)
A potential difference that is supplied by a battery. If the internal resistance of the battery is relatively small, then the emf is equal to the terminal voltage of the battery (p. 581).

Galvanometer
A device that indicates that there is a current in a circuit. By appropriate construction, a galvanometer can be made into an ammeter or a voltmeter (p. 584).

Wheatstone bridge
An electric circuit that is used to measure the value of an unknown resistor very accurately (p. 589).

Kirchhoff's first rule
The sum of the currents in a circuit entering a junction is equal to the sum of the currents leaving the junction (p. 591).

Kirchhoff's second rule
The change in potential around a closed loop of a circuit is equal to zero. This can also be stated in the form, the sum of the potential rises and potential drops around any closed loop of a circuit is equal to zero (p. 591).

Summary of Important Equations

Current

$$I = \frac{q}{t} \tag{20.1}$$

Ohm's law

$$I = \frac{V}{R} \tag{20.3}$$

Resistance of a wire

$$R = \rho \frac{l}{A} \tag{20.7}$$

Work done within a battery

$$W = qV \tag{20.13}$$

Power

$$P = IV \tag{20.15}$$

Power dissipated in a resistor

$$P = I^2R \tag{20.18}$$

Resistors in series

$$V = V_1 + V_2 + V_3 \tag{20.22}$$
$$I = I_1 = I_2 = I_3$$
$$R = R_1 + R_2 + R_3 \tag{20.26}$$
$$P = P_1 + P_2 + P_3$$

Resistors in parallel

$$V = V_1 = V_2 = V_3 \tag{20.27}$$
$$I = I_1 + I_2 + I_3 \tag{20.28}$$
$$\frac{1}{R} = \frac{1}{R_1} + \frac{1}{R_2} + \frac{1}{R_3} \tag{20.35}$$
$$P = P_1 + P_2 + P_3$$

Terminal voltage of a battery

$$V = \mathcal{E} - Ir \tag{20.37}$$

Batteries in series

$$\mathcal{E} = \mathcal{E}_1 + \mathcal{E}_2 + \mathcal{E}_3$$
$$+ \cdots + \mathcal{E}_n \tag{20.38}$$

Identical batteries in parallel

$$\mathcal{E} = \mathcal{E}_1 = \mathcal{E}_2 = \mathcal{E}_3 \tag{20.39}$$
$$I = I_1 + I_2 + I_3 \tag{20.40}$$

Resistance for an ammeter shunt

$$R_s = \left(\frac{I_g}{I - I_g} \right) R_g \tag{20.45}$$

Series resistor for a voltmeter

$$R = \frac{V}{I_g} - R_g \tag{20.49}$$

Wheatstone bridge

$$R_2 = \frac{R_4}{R_3} R_1 \tag{20.59}$$

Kirchhoff's first rule

$$\Sigma\, I_{\text{entering junction}}$$
$$= \Sigma\, I_{\text{leaving junction}} \tag{20.60}$$

Kirchhoff's second rule

$$\Delta V_{\text{closed loop}} = 0 \tag{20.62}$$

Questions for Chapter 20

1. Can the flow of electric charge in a circuit be compared to the flow of a fluid in a pipe? Point out where it is a good analogy and where it is a bad analogy.
2. Can a resistor be used as a thermometer?
†3. If a metal consists of positive ions surrounded by a sea of electrons, how does the metal stay together?
4. If the drift velocity of an electron in a wire is so small that it takes almost 30 s for it to move about 1 cm, why is no delay observed on an ammeter when a battery is connected to a circuit?
5. Describe the flow of the positive charges of a conventional current with the flow of electrons in an electron current in a circuit. What are the advantages and disadvantages of using conventional current?

6. Why would you not want to hook up the simple circuit shown in figure 20.3? Estimate the value of the resistance of the wire. Using the concept of Ohm's law, would a very large current exist? Would this be called "shorting" the battery?
7. Compare a positive charge falling through a potential difference with a mass falling in a gravitational field. Why is this a pretty good analogy? Try to make a similar analogy with the motion of an electron in a circuit.
†8. A black box is a device that is contained in a closed box, so that you cannot see inside the box to see what the device is. There is an input wire to the box and an output wire from the box. All you can do is to infer characteristics about the device from the information obtained from the input and output connections. A range of different potentials is applied to this black box, and the current coming out of the box is measured. A plot of the potential versus the current is a curve rather than a straight line. What are some of the characteristics of this black box?

9. A copper block is 35 cm long, 2 cm thick, and 2 cm wide. Does the block have the same resistance from end to end as it does from side to side? Does it depend in any way on the size of the connection to the block?
†10. In the 1970s when copper became very expensive, some new homes were built with aluminum wires rather than copper wires. If the aluminum wire is connected to a copper outlet, and the coefficient of linear expansion for aluminum and copper are different, what hazard could this produce?
11. At very low temperatures some materials become superconductors, that is, their electrical resistance becomes essentially zero. What are the advantages and disadvantages of superconductors?
12. How would you hook up three resistors in a circuit to get (a) the maximum current and (b) the minimum current?

Problems for Chapter 20

20.1 Electric Current

1. How many electrons are associated with a current of 5.00 A?
2. A wire carries a current of 7.50 A for a period of 30.0 min. How much charge flows through a cross section of the wire? How many electrons does this represent?
3. Suppose that the region between two oppositely charged parallel plates is occupied by both singly charged positive ions and by electrons. If 5×10^6 electrons flow from the negative plate to the positive plate in 4.00 s, while 6×10^6 ions flow from the positive plate to the negative plate in the same time, find the value of the net current flowing from the positive plate to the negative plate.

20.2 Ohm's Law

4. A 500-Ω resistor is connected to a 12.0-V battery. Find the current through the resistor.
5. What value of resistance is necessary to get a current of 2.50 A when it is connected to a 120-V source?
6. If a current of 8.75 mA passes through a resistor of 550 Ω, find the voltage drop across the resistor.

20.3 Resistivity

7. Find the resistance of a 100-m spool of #22 B&S gauge copper wire (diameter 6.44×10^{-4} m).
8. The connecting wires in an electrical experiment are 1.00 m long and are made of copper. Their diameter is 6.44×10^{-4} m. What is the resistance of the connecting wires? Is it reasonable to neglect their effect in analyzing a circuit?
9. Does doubling both the length and the diameter of a wire have any effect on its resistance?
10. What length of #22 B&S gauge copper wire is needed to make a resistance of 500 Ω?
11. Find the resistance of an aluminum bar 1.00 m long, 2.00 cm wide, and 1.00 cm high from (a) end to end and (b) from the 1.00-cm side to the other 1.00-cm side.
12. A wire 2.00 m long and 1.00 mm in diameter has resistance of 0.500 Ω. A second wire 4.00 m long and 2.00 mm in diameter has twice the resistivity of the first wire. Find the resistance of the second wire.

20.4 The Variation of Resistance with Temperature

13. If the resistance of a resistor is 20.00 Ω at 20.00 °C, find its resistance at 200.00 °C. The temperature coefficient of resistance α for copper is 3.93×10^{-3}/°C.
14. If the resistance of a copper wire is 500 Ω at 20 °C, find its resistance at 100 °C.
15. A copper wire has a resistance of 20.00 Ω at room temperature. The wire is placed in an oven and the resistance is then measured as 30.0 Ω. Find the temperature of the oven.
16. The resistance of a copper wire changes from 200 Ω to 210 Ω. What was the change in temperature of the wire?
17. A copper resistor has a resistance of 500 Ω at a temperature of 20.0 °C. What is the resistance if the temperature doubles?

20.5 Conservation of Energy and the Electric Circuit—Power Expended in a Circuit

18. An electric toaster is rated at 1200 W. If it is connected to a 120-V line, what current does it draw? What is the resistance of the toaster? How much energy is used if the toaster is "on" for 1.00 min?
19. A stereo amplifier is rated at 60.0 W. If it is connected to a 120-V line, how much current does it draw? If the stereo is "on" for 5 hr, how much energy is used?
20. A 60.0-W light bulb is connected to a 120-V outlet. What is the current through the bulb? How many electrons flow through the bulb per second?
21. A power line carries 1000 A and has a resistance of 20.0 Ω. How much energy is lost per second in terms of Joule heat?
22. How much power is consumed in starting a car if 200 A is drawn from a 12.0-V battery?

20.6 Resistors in Series

23. Find the equivalent resistance of three resistors of 100, 200, and 300 Ω when connected in series.
24. Find the equivalent resistance of four resistors of 250, 400, 186, and 375 Ω when connected in series.

20.7 Resistors in Parallel

25. Find the equivalent resistance of three resistors of 100, 200, and 300 Ω when connected in parallel.
26. Find the equivalent resistance of four resistors of 250, 400, 186, and 375 Ω when connected in parallel.

20.8 Combinations of Resistors in Series and Parallel

27. Find the equivalent resistance of the circuit in the diagram.

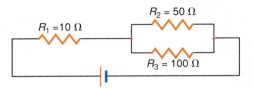

28. Find the equivalent resistance of the circuit in the diagram if $R_1 = 30.0$ Ω, $R_2 = 40.0$ Ω, $R_3 = 50.0$ Ω, $R_4 = 70.0$ Ω, $R_5 = 300$ Ω, $R_6 = 200$ Ω, and $R_7 = 100$ Ω.

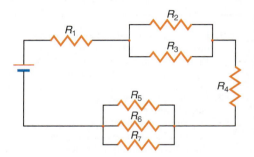

†29. In the diagram find (a) the equivalent resistance of the circuit, (b) the current flowing from the battery, (c) the voltage drop across R_1, (d) the voltage drop across R_2 and R_3, (e) the current through each resistor, (f) the power supplied to the circuit, and (g) the power dissipated in each resistor. The emf of the battery is 12.0 V.

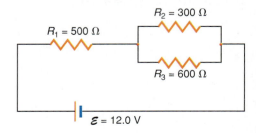

30. Find the current through each resistor in the diagram.

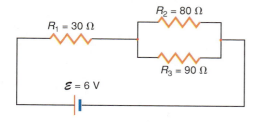

$R_1 = 30\ \Omega$
$R_2 = 80\ \Omega$
$R_3 = 90\ \Omega$
$\mathcal{E} = 6$ V

†31. In the diagram, find (a) the equivalent resistance of the resistors in parallel, (b) the equivalent resistance of the circuit, (c) the current from the battery, (d) the voltage drop across R_1, (e) the voltage drop across R_2, R_3, and R_4, (f) the current through each resistor, and (g) the power dissipated in each resistor.

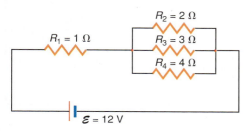

$R_1 = 1\ \Omega$
$R_2 = 2\ \Omega$
$R_3 = 3\ \Omega$
$R_4 = 4\ \Omega$
$\mathcal{E} = 12$ V

32. Find the voltage drop across R_3 in the diagram.

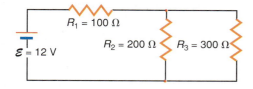

$R_1 = 100\ \Omega$
$R_2 = 200\ \Omega$
$R_3 = 300\ \Omega$
$\mathcal{E} = 12$ V

33. Find (a) the equivalent resistance of the circuit shown in the diagram if $R_1 = 50.0\ \Omega$, $R_2 = 80.0\ \Omega$, $R_3 = 150\ \Omega$, $R_4 = 30.0\ \Omega$, $R_5 = 200\ \Omega$, and $R_6 = 300\ \Omega$ and (b) the current through each resistor.

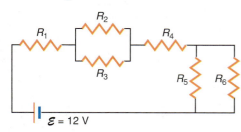

R_1
R_2
R_3
R_4
R_5
R_6
$\mathcal{E} = 12$ V

†34. In the diagram find (a) the equivalent resistance, (b) the current from the battery, (c) the current through each resistor, (d) the voltage drop across each resistor, (e) the power supplied to the circuit, and (f) the power dissipated in each resistor.

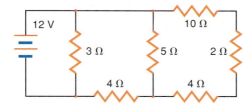

12 V
10 Ω
3 Ω
5 Ω
2 Ω
4 Ω
4 Ω

20.9 The Electromotive Force and the Internal Resistance of a Battery

35. A 6.00-V battery is connected to a 100-Ω resistor. A voltmeter placed across the resistor measures 5.60 V. Find the internal resistance of the battery.

36. A 12-V battery has an internal resistance of 5.5 Ω. Find the voltage applied to a circuit of 50 Ω.

37. Six 1.50-V batteries are connected in series, with the positive terminal of one battery connected to the negative terminal of the next. What is the emf of the combination? If the internal resistance of each battery is 2.00 Ω, find the terminal voltage when there is a current of 200 mA in the circuit.

38. Six 1.50-V batteries are connected in series to a resistance of 50.0 Ω. Find the current in the circuit (a) neglecting the internal resistance of the battery and (b) taking the 2.00 Ω resistance of each battery into account.

39. Three 12-V batteries are connected in series to a 35-Ω resistor. If the internal resistance of each battery is 10 Ω, 20 Ω, and 30 Ω, respectively, find the voltage drop across the 35-Ω resistor.

40. When a battery is connected to an external resistance of 10.0 Ω, the current through the external resistance is 0.100 A. When the same battery is connected to an external resistance of 5 Ω, the current is 0.150 A. Find the emf of the battery and the internal resistance of the battery.

20.10 Making an Ammeter and Voltmeter from a Galvanometer

41. A galvanometer has an internal resistance of 3.00 Ω and gives full-scale deflection for a current of 10.0 mA. What shunt resistance is necessary to convert this to an ammeter that can read a maximum current of 5.00 A?

42. A galvanometer has an internal resistance of 3.00 Ω and gives full-scale deflection for a current of 10.0 mA. What series resistor is necessary to convert this to a voltmeter that can read a maximum voltage of 150 V?

20.11 The Wheatstone Bridge

43. The voltmeter reads 12.0 V and the ammeter reads a current of 5.45×10^{-2} A in the circuit shown. The ammeter has a resistance of 20.0 Ω and the voltmeter has a resistance of 5000 Ω. Taking the resistance of the ammeter and voltmeter into account, find the value of the resistor R.

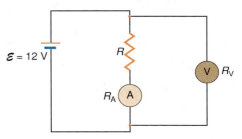

$\mathcal{E} = 12$ V
R
V R_V
R_A A

44. Find the value of R_2 in the diagram if the bridge is balanced.

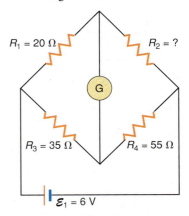

$R_1 = 20\ \Omega$
$R_2 = ?$
G
$R_3 = 35\ \Omega$
$R_4 = 55\ \Omega$
$\mathcal{E}_1 = 6$ V

45. Find the equivalent resistance of the Wheatstone bridge in problem 44 when the bridge is balanced. Find the current through each resistor.

20.12 Kirchhoff's Rules

†46. In the accompanying diagram (a) find the current through each resistor and (b) plot the potential as you traverse each loop.

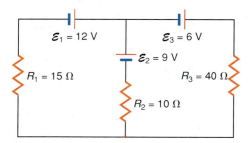

†47. Two batteries, with emf's and internal resistances shown, are connected in parallel. Using Kirchhoff's rules, find the current through R and the voltage drop across R.

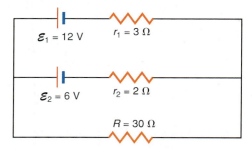

†48. Find the current through each resistor in the diagram.

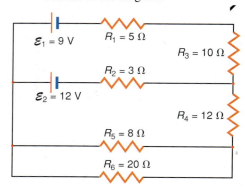

†49. Using Kirchhoff's rules, find the current through each resistor in the diagram.

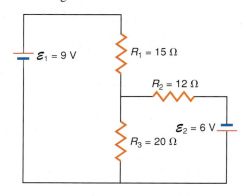

50. Three 12.0-V batteries with different internal resistances are connected in parallel as shown in the diagram. If $r_1 = 10.0\ \Omega$, $r_2 = 20.0\ \Omega$, and $r_3 = 30.0\ \Omega$, find the voltage drop across AB.

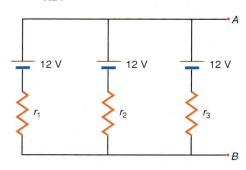

Additional Problems

†51. Find the potential difference across AB in the accompanying diagram if $R_1 = 10.0\ \Omega$, $R_2 = 20.0\ \Omega$, $R_3 = 30.0\ \Omega$, and the current through R_3 is 0.300 A.

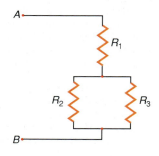

52. Find the voltage drop across R_2 if $R_1 = 20.0\ \Omega$, $R_2 = 30.0\ \Omega$, $R_3 = 50.0\ \Omega$, and the emf is 6.00 V in the diagram.

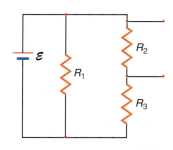

53. Find the current through each of the seven resistors, and the potential drop across each resistor in the circuit in the diagram.

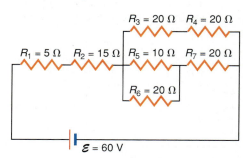

†54. Find the equivalent resistance of the circuit shown in the diagram if $R_1 = 10.0\ \Omega$, $R_2 = 20.0\ \Omega$, $R_3 = 30.0\ \Omega$, $R_4 = 40.0\ \Omega$, $R_5 = 50.0\ \Omega$, $R_6 = 60.0\ \Omega$, and $R_7 = 70.0\ \Omega$.

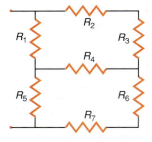

†55. If $R_1 = 10.0\ \Omega$, $R_2 = 20.0\ \Omega$, $R_3 = 30.0\ \Omega$, and $R_4 = 40.0\ \Omega$ in the circuit shown, find (a) the equivalent resistance of the circuit and, if $\mathscr{E}_1 = 12.0$ V, (b) the current through each resistor.

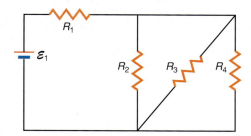

56. (a) What is the resistance of 100 resistors of $10.0\ \Omega$ each if they are all in series? (b) What is the resistance of 100 resistors of $10.0\ \Omega$ each if they are all in parallel?

†57. A 200-W immersion heater is placed in a beaker of water containing 0.500 liters at room temperature. How long will it take for the water to boil?

†58. An electrically insulated coil of wire is immersed in 100.0 g of water in a calorimeter cup at room temperature. The wire is connected to a 120-V source and a current of 2.00 A is observed in the wire. Find the temperature of the water after 60.0 s.

†59. A voltage divider is shown in the diagram. The resistor R is a variable slide wire resistor. Show that by sliding the arrow contact in the figure along the slide wire any fraction of the original applied voltage can be obtained. In particular show that the voltage out of the divider is given by

$$V_{out} = \frac{l_{out}}{l_{in}} V_{in}$$

where l_{in} is the original length of the wire resistor and l_{out} is the shorter length of the wire resistor that the output connector is attached to.

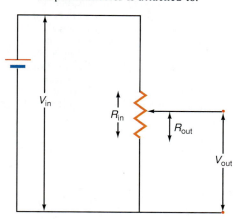

†60. The Wheatstone bridge shown in the diagram is *not* balanced. Using Kirchhoff's rules find the current through each resistor if $R_1 = 10.0\ \Omega$, $R_2 = 20.0\ \Omega$, $R_3 = 30.0\ \Omega$, $R_4 = 40.0\ \Omega$, $R_g = 50.0\ \Omega$, and $\mathscr{E}_1 = 6.00$ V. Find the equivalent resistance of the circuit.

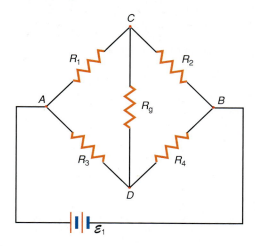

†61. In problem 60, find the potential differences across points (a) *AC*, (b) *AD*, (c) *CD*, (d) *CB*, and (e) *DB*.

62. Resistors in series. Three resistors, R_1 = 25.0 Ω, R_2 = 45.5 Ω, and R_3 = 83.5 Ω, are connected in series to a 50.0-V battery. Find (a) the equivalent resistance R, (b) the current coming from the battery, (c) the current through each resistor, (d) the voltage drop across each resistor, (e) the power lost across each resistor, and (f) the power supplied to the circuit.

63. Resistors in parallel. Three resistors, R_1 = 25.0 Ω, R_2 = 45.5 Ω, and R_3 = 83.5 Ω, are connected in parallel to a 50.0-V battery. Find (a) the equivalent resistance R, (b) the current coming from the battery, (c) the voltage drop across each resistor, (d) the current through each resistor, (e) the power lost across each resistor, and (f) the power supplied to the circuit.

64. Combination of resistors in series and parallel. Resistor R_1 = 25.0 Ω, is in series with the parallel resistors R_2 = 45.5 Ω and R_3 = 83.5 Ω, as in figure 20.14(a). The resistors are connected to a 50.0-V battery. Find (a) the equivalent resistance R, (b) the current coming from the battery, (c) the voltage drop across each resistor, (d) the current through each resistor, (e) the power lost across each resistor, and (f) the power supplied to the circuit.

65. A galvanometer has an internal resistance of R_g = 3.00 Ω and gives full-scale deflection for a current of I_g = 10.0 mA. Calculate (a) the value of the shunt resistor R_s, connected in parallel, that converts the galvanometer into a I = 5.00-A ammeter and (b) the value of the resistor R, connected in series, that converts the galvanometer into a V = 100-V voltmeter.

Capacitance

Industrial capacitors.

Most of the fundamental ideas of science are essentially simple, and may, as a rule, be expressed in a language comprehensible to everyone.

Albert Einstein

21.1 Introduction

*Whenever two nearby conductors of any size or shape carry equal and opposite charges, the combination of these conducting bodies is called a **capacitor.*** Because the isolated conducting bodies have equal but opposite charges on them, an electric field exists in the space between them. *The importance of the capacitor lies in the fact that energy can be stored in the electric field between the two conducting bodies.* Some simple examples of capacitors are shown in figure 21.1.

Figure 21.1(a) is a parallel plate capacitor, which consists of two metal plates separated by a distance d. A positive charge $+q$ is placed on one of the plates, let us say the left one, and a charge $-q$ is placed on the right conducting plate. Neglecting any edge corrections, there is a uniform electric field **E** between the two charged plates.

Figure 21.1(b) is a coaxial cylindrical capacitor. As the name implies, it consists of two coaxial cylinders. The inner cylinder has a positive charge $+q$ placed on it, whereas a charge $-q$ is placed on the outer cylinder. The electric field fills the space between the cylinders, as shown in figure 21.1(b).

A concentric spherical capacitor is shown in figure 21.1(c). The inner sphere has a positive charge $+q$ placed on it, and a charge $-q$ is placed on the outer sphere. A spherical electric field exists between the two conducting spheres. In all of these cases, energy is stored in the electric field between the plates. Let us now go into more detail on the parallel plate capacitor.

21.2 The Parallel Plate Capacitor

A parallel plate capacitor is connected to a battery, as shown in figure 21.2. Although the actual charge carriers in a metal are electrons, we will continue with the convention introduced in section 20.1 that it is the positive charges that are moving in the circuit. Recall that a flow of negative charges in one direction is equivalent to a flow of positive charges in the opposite direction. Hence, using the concepts of conventional current described in chapter 20, positive electric charge flows from the battery of potential V to the left-hand plate of the capacitor when the switch S is closed. The charge distributes itself over the left-hand plate. This positive charge on the left-hand plate induces an equal but negative charge on the right-hand plate. The positive charge on the originally neutral right-hand plate is pushed into the battery. The net result of applying the battery to the capacitor is that a charge of $+q$ is deposited on the left-hand plate and $-q$ on the right-hand plate and a uniform electric field has been set up between the plates. In effect, the battery has supplied

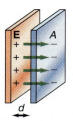

(a) Parallel plate capacitor

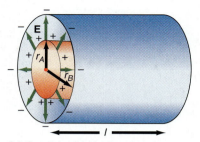

(b) Coaxial cylindrical capacitor

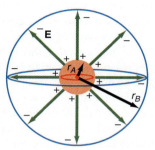

(c) Concentric spherical capacitor

Figure 21.1

Some simple capacitors.

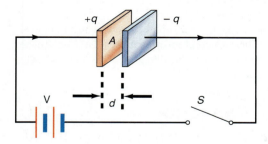

Figure 21.2

The parallel plate capacitor.

energy to move positive charges from the right-hand plate and transferred them, through the battery, to the left-hand plate. The relationship between the potential V across the plates, the electric field E between the plates, and the plate separation d was found in chapter 19 to be

$$V = Ed \qquad (19.20)$$

The electric field between the plates is the superposition of the electric field of the positive plate and the electric field of the negative plate. The actual value of the electric field between the plates is found from a relation known as Gauss's law, which we will derive in section 25.2. The electric field is found to be

$$E = \frac{q}{\epsilon_0 A} \qquad (21.1)$$

where q is the charge on the plates of cross-sectional area A and ϵ_0 is the permittivity of free space introduced in chapter 18. Substituting equation 21.1 into equation 19.20, gives

$$V = \frac{qd}{\epsilon_0 A}$$

Solving for q,

$$q = \left(\frac{\epsilon_0 A}{d}\right) V \qquad (21.2)$$

Notice from equation 21.2 that the charge q on the plate is directly proportional to the potential between the plates, which is of course supplied by the battery. The greater the battery voltage, the greater will be the charge on the plate; the smaller the voltage, the smaller the charge.

Let us now look carefully at the term in parentheses in equation 21.2. Notice that it is a constant and is a function of the geometry of the capacitor. Recall from chapter 18 that ϵ_0 is called the permittivity of free space and has approximately the same value for air as for a vacuum. This term is a function of the medium between the plates, which in this case is air. In equation 21.2, A is the cross-sectional area of a plate of the capacitor and d is the separation between the two plates. Because all these terms are constant for a particular capacitor, they are set equal to a new constant C, called the *capacitance* of the capacitor. Therefore, for a parallel plate capacitor

$$C = \frac{\epsilon_0 A}{d} \qquad (21.3)$$

The capacitance is thus a function of the geometry of the capacitor itself. The larger the area of the plates A, the greater will be the value of the capacitance. The greater the plate separation d, the smaller will be the capacitance. A parallel plate capacitor of any value can be obtained by proper selection of area, plate separation, and the medium between the plates. So far, our discussion has been limited to a capacitor with air between the plates. In section 21.7 an insulating material will be placed between the plates.

A slightly more sophisticated analysis would yield for the capacitance of the coaxial cylinder of figure 21.1

$$C = \frac{2\pi\epsilon_0 l}{\ln(r_B/r_A)} \qquad (21.4)$$

and

$$C = \frac{4\pi\epsilon_0 r_A r_B}{r_B - r_A} \qquad (21.5)$$

for the capacitance of a spherical capacitor, figure 21.1(c). (The ln in equation 21.4 is the natural logarithm function.) Notice from equations 21.3, 21.4, and 21.5 that each is a function of the geometry of the particular capacitor.

The introduction of the concept of the capacitance allows us to write equation 21.2 in the more general form

$$q = CV \qquad (21.6)$$

The charge q on a capacitor is directly proportional to the potential between the plates, and the constant of proportionality is the capacitance of the particular capacitor. The capacitance can then be defined in general, from equation 21.6, as

$$C = \frac{q}{V} \qquad (21.7)$$

The SI unit for capacitance is defined from equation 21.7 to be a farad, F, where

$$1 \text{ farad} = \frac{1 \text{ coulomb}}{\text{volt}}$$

This is abbreviated as

$$1 \text{ F} = 1 \text{ C/V}$$

This unit is named in honor of Michael Faraday (1791–1867), an English physicist, and as in all SI units named for a person, the abbreviation is capitalized, but the unit itself is not. If a charge of 1 C is placed on the plates and the potential difference between the plates is 1 V, the capacitance is then defined to be one farad. A capacitance of 1 F is extremely large, and the smaller units of microfarads, μF, or picofarads, pF, are usually used.

$$1 \text{ } \mu F = 10^{-6} \text{ F}$$
$$1 \text{ pF} = 10^{-12} \text{ F}$$

Example 21.1

Capacitance of a parallel capacitor. A parallel plate capacitor consists of two metal disks, 5.00 cm in radius. The disks are separated by air and are a distance of 4.00 mm apart. A potential of 50.0 V is applied across the plates by a battery. Find (a) the capacitance of the capacitor and (b) the charge on the plate.

Solution

a. The area of the plate is

$$A = \pi r^2 = \pi(0.0500 \text{ m})^2 = 7.85 \times 10^{-3} \text{ m}^2$$

The capacitance, found from equation 21.3, is

$$C = \epsilon_0 \frac{A}{d} = \left(8.85 \times 10^{-12} \frac{C^2}{N \text{ m}^2}\right)\left(\frac{7.85 \times 10^{-3} \text{ m}^2}{4.00 \times 10^{-3} \text{ m}}\right)$$

$$= 17.4 \times 10^{-12} \frac{C^2}{N \text{ m}} \left(\frac{N \text{ m}}{J}\right)$$

$$= 17.4 \times 10^{-12} \frac{C}{J/C} \left(\frac{J/C}{V}\right)\left(\frac{F}{C/V}\right)$$

$$= 17.4 \times 10^{-12} \text{ F}$$

$$= 17.4 \text{ pF}$$

Note how the conversion factors have been carried through in the example to show that the capacitance does indeed come out to have the unit of farads.

b. The charge on the plate is determined from equation 21.6 as

$$q = CV$$
$$= (17.4 \times 10^{-12} \text{ F})(50 \text{ V})$$
$$= 8.7 \times 10^{-10} \text{ C}$$

21.3 Energy Stored in a Capacitor

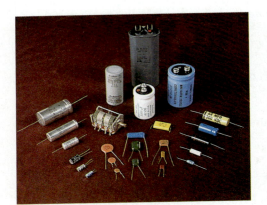

A collection of capacitors used in various applications.

As shown in figure 21.2, *using the concept of conventional current, the net effect of charging a capacitor by a battery is to take a positive charge q from the right-hand plate, move it through the battery, and deposit it on the left-hand plate. The work done by the battery in moving the charge from the negative plate, or ground plate, to the positive plate is converted to electrical potential energy of the charge.* The mechanical analogue is that of a person who does work by lifting a bowling ball from the floor to a table, where the ball now has potential energy with respect to the floor. *Since the charge creates an electric field between the plates, this energy associated with the charge may be viewed as residing in the electric field between the plates. Thus, the energy stored in the capacitor is equal to the work done to charge the capacitor.* The work done by the battery in moving a charge q through the battery was shown in chapter 20 to be

$$W = qV \tag{20.13}$$

But in charging the capacitor the rate at which work is done is not a constant. At the instant that the switch in figure 21.2 is closed, the initial potential V_i across the capacitor is zero. A small charge $+q_i$ is then placed on the left-hand plate, which then induces the charge $-q_i$ on the right-hand plate. An electric field E_1 is established between the plates and a potential V_1 appears across the plates. When the next charge q_1 is brought to the left-hand plate, an amount of work, $W_1 = q_1V_1$, must be done by the battery. With the new charge on the left-hand plate, $q_i + q_1$, a new electric field E_2 is established between the plates and a new potential V_2, which is greater than V_1, appears across the plates. When the next charge q_2 is brought to the left-hand plate, the work done is $W_2 = q_2V_2$. Since V_2 is greater than V_1, the work W_2 must be greater than W_1. With the new charge on the left-hand plate, $q_i + q_1 + q_2$, a new electric field E_3 is established between the plates, and a new potential V_3, which is greater than V_2, appears across the plates. When the next charge q_3 is brought to the left-hand plate, the work done is $W_3 = q_3V_3$. However, because V_3 is greater than V_2, the work done W_3 is greater than the work W_2. It is obvious that a different amount of work must be done to move each charge to the plate, because as each charge is placed on the plate, a new potential appears across the plates and the product of qV is different for each charge. Hence, the total work necessary to charge the capacitor is equal to the sum of all the products of qV for an extremely large number of charges. We can greatly simplify the problem, however, by computing the average amount of work done in charging the capacitor by noting that the average potential V_{avg} that a charge sees is

$$V_{avg} = \frac{V_f + V_i}{2}$$

where V_f is the final value of the potential across the capacitor and is equal to the terminal voltage V of the battery and V_i is the initial potential that the charge sees and is zero. The work done in charging the capacitor becomes

$$W = qV_{avg} = q\left(\frac{V_f + V_i}{2}\right) = q\left(\frac{V + 0}{2}\right)$$

$$W = \frac{1}{2}qV \tag{21.8}$$

Equation 21.8 gives the energy that is stored in the electric field of the capacitor.
Because the energy stored in the capacitor is equal to the work done to charge the
capacitor, we will now use the letter W to designate the energy stored in a capacitor,
since the letter E, usually associated with energy, is now being used for the electric
field intensity.

Using equation 21.6, $q = CV$, we can also write the energy stored in a
capacitor as

$$W = \tfrac{1}{2}qV = \tfrac{1}{2}(CV)V$$

$$W = \tfrac{1}{2}CV^2 \qquad (21.9)$$

Rearranging equation 21.6 into the form, $V = q/C$, we can also write the energy
stored in the capacitor as

$$W = \frac{1}{2}qV = \frac{1}{2}q\frac{q}{C}$$

$$W = \frac{1}{2}\frac{q^2}{C} \qquad (21.10)$$

Equations 21.8, 21.9, and 21.10 are different ways of expressing the energy that is
stored in the capacitor. This stored energy can be related to the electric field intensity between the plates of a parallel plate capacitor by using equations 21.9, 21.3,
and 19.20 as

$$W = \frac{1}{2}CV^2 = \frac{1}{2}\frac{\epsilon_0 A}{d}(Ed)^2$$

$$= \frac{1}{2}\epsilon_0 A d E^2$$

The product of A, the cross-sectional area of the plates, and d, the separation of the plates, is the volume between the plates that the electric field occupies,
and as such is the volume of the electric field. This allows us to define a quantity
called the **energy density** u_E as

$$u_E = \frac{\text{total energy}}{\text{volume}}$$

$$= \frac{W}{Ad} = \frac{1}{2}\frac{\epsilon_0 A d E^2}{Ad} \qquad (21.11)$$

$$u_E = \tfrac{1}{2}\epsilon_0 E^2 \qquad (21.12)$$

The energy density of the electric field, equation 21.12, was derived from the parallel plate capacitor, but it holds true in general for any electric field. Notice that
the energy density depends upon the electric field intensity. It is therefore certainly
appropriate to think of the energy as residing in the electric field.

Example 21.2

The energy stored in a capacitor. Find the energy stored in the capacitor of example
21.1.

The energy stored in the capacitor, given by equation 21.9, is

$$W = \frac{1}{2}CV^2 = \frac{1}{2}(17.4 \times 10^{-12} \text{ F})(50.0 \text{ V})^2$$

$$= 2.18 \times 10^{-8} \frac{\text{C}}{\text{V}}\text{V}^2\left(\frac{\text{J}}{\text{CV}}\right)$$

$$= 2.18 \times 10^{-8} \text{ J}$$

Example 21.3

The energy density in the electric field. Find the energy density in the electric field between the plates of the parallel plate capacitor in example 21.2.

Since we already know the energy stored in the capacitor, W, the energy density, found from equation 20.18, is

$$u_E = \frac{W}{\text{volume}} = \frac{W}{Ad} = \frac{2.18 \times 10^{-8} \text{ J}}{(7.85 \times 10^{-3} \text{ m}^2)(4.00 \times 10^{-3} \text{ m})}$$

$$= 6.94 \times 10^{-4} \text{ J/m}^3$$

As a check, let us find the electric field between the plates of the capacitor and then use equation 21.12 to find the energy density. The electric field between the plates, found from equation 19.20, is

$$E = \frac{V}{d} = \frac{50.0 \text{ V}}{4.00 \times 10^{-3} \text{ m}} = 1.25 \times 10^4 \text{ V/m}$$

and the energy density from equation 21.12 is

$$u_E = \frac{1}{2}\epsilon_0 E^2$$

$$= \frac{1}{2}\left(8.85 \times 10^{-12} \frac{\text{C}}{\text{N m}^2}\right)(1.25 \times 10^4 \text{ V/m})^2$$

$$= 6.94 \times 10^{-4} \text{ J/m}^3$$

which is the same energy density determined in the previous calculation.

21.4 Capacitors in Series

If there are several capacitors in a circuit, another capacitor can be found that is equivalent to the entire combination present. As an example consider figure 21.3(a), which is a schematic diagram of a circuit containing three capacitors in series. (Note that the symbol used for a capacitor in a circuit diagram is two parallel lines, symbolic of the parallel plate capacitor.) Again using the concept of conventional current, that is, a flow of positive charges, the battery causes charge $+q$ to flow to the left-hand plate of capacitor C_1. This charge induces a charge $-q$ in the right-hand plate of C_1. In this induction process, positive charges that were originally on the right-hand plate of C_1 are forced to move away from the right-hand plate. The only place for them to go is the left-hand plate of C_2, where they now distribute themselves. They in turn induce a charge $-q$ on the right-hand plate of C_2. This causes a charge of $+q$ to appear on the left-hand plate of C_3, which in turn induces a charge $-q$ on the right-hand plate of C_3. The positive charges that were on the right-hand

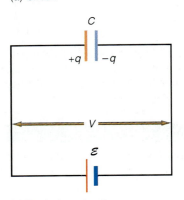

(a) Circuit

(b) Equivalent circuit

Figure 21.3

Capacitors in series.

charges that were on the right-hand plate of C_3 are now returned to the negative terminal of the battery. The net effect, therefore, is that *when capacitors are connected in series, each capacitor has the same charge q on its plates.*

The potential drop across each capacitor, given by equation 21.6, is

$$V_1 = \frac{q}{C_1} \tag{21.13}$$

$$V_2 = \frac{q}{C_2} \tag{21.14}$$

$$V_3 = \frac{q}{C_3} \tag{21.15}$$

The total potential V across the three capacitors is equal to the sum of the potential drops across each capacitor, that is,

$$V = V_1 + V_2 + V_3$$

Substituting equations 21.13, 21.14, and 21.15 into this equation gives

$$V = \frac{q}{C_1} + \frac{q}{C_2} + \frac{q}{C_3}$$

or

$$V = q\left(\frac{1}{C_1} + \frac{1}{C_2} + \frac{1}{C_3}\right) \tag{21.16}$$

Dividing both sides of equation 21.16 by q gives

$$\frac{V}{q} = \left(\frac{1}{C_1} + \frac{1}{C_2} + \frac{1}{C_3}\right) \tag{21.17}$$

A rearrangement of equation 21.6 for a single capacitor gives

$$\frac{V}{q} = \frac{1}{C} \tag{21.18}$$

Just as we introduced the concept of an equivalent resistor in resistive circuits, we now introduce an *equivalent capacitor* for a circuit containing capacitors, as seen in figure 21.3(b). From equations 21.17 and 21.18, *the equivalent capacitance for capacitors in series* is given by

$$\frac{1}{C} = \frac{1}{C_1} + \frac{1}{C_2} + \frac{1}{C_3} \tag{21.19}$$

That is, capacitors C_1, C_2, and C_3 can be replaced in the series circuit by the one equivalent capacitor C. Hence, the reciprocal of the equivalent capacitance of capacitors in series is equal to the sum of the reciprocals of each capacitor.

Example 21.4

Capacitors in series. Three capacitors of 3.00 μF, 6.00 μF, and 9.00 μF are connected in series to a 50.0-V battery. Find (a) the equivalent capacitance, (b) the charge on each capacitor, (c) the voltage drop across each capacitor, (d) the energy stored in each capacitor, and (e) the total energy stored in the circuit.

a. The equivalent capacitance, found from equation 21.19, is

$$\frac{1}{C} = \frac{1}{C_1} + \frac{1}{C_2} + \frac{1}{C_3}$$

$$= \frac{1}{3.00 \times 10^{-6} \text{ F}} + \frac{1}{6.00 \times 10^{-6} \text{ F}} + \frac{1}{9.00 \times 10^{-6} \text{ F}} = 6.11 \times 10^5 \text{ F}^{-1}$$

$$C = 1.64 \times 10^{-6} \text{ F} = 1.64 \; \mu\text{F}$$

b. Because the capacitors are in series the same charge is on each capacitor and is found by

$$q = CV = (1.64 \times 10^{-6} \text{ F})(50.0 \text{ V})$$

$$= 8.18 \times 10^{-5} \text{ C}$$

c. The voltage drop across each capacitor, found from equations 21.13, 21.14, and 21.15, is

$$V_1 = \frac{q}{C_1} = \frac{8.18 \times 10^{-5} \text{ C}}{3.00 \times 10^{-6} \text{ F}} = 27.3 \text{ V}$$

$$V_2 = \frac{q}{C_2} = \frac{8.18 \times 10^{-5} \text{ C}}{6.00 \times 10^{-6} \text{ F}} = 13.6 \text{ V}$$

$$V_3 = \frac{q}{C_3} = \frac{8.18 \times 10^{-5} \text{ C}}{9.00 \times 10^{-6} \text{ F}} = 9.09 \text{ V}$$

Note that $V_1 + V_2 + V_3 = 50.0$ V, which is equal to the applied voltage of 50.0 V.

d. The energy stored in each capacitor, found from equation 21.10, is

$$W_1 = \frac{1}{2}\frac{q^2}{C_1} = \frac{1}{2}\frac{(8.18 \times 10^{-5} \text{ C})^2}{(3.00 \times 10^{-6} \text{ F})} = 1.12 \times 10^{-3} \text{ J}$$

$$W_2 = \frac{1}{2}\frac{q^2}{C_2} = \frac{1}{2}\frac{(8.18 \times 10^{-5} \text{ C})^2}{(6.00 \times 10^{-6} \text{ F})} = 0.558 \times 10^{-3} \text{ J}$$

$$W_3 = \frac{1}{2}\frac{q^2}{C_3} = \frac{1}{2}\frac{(8.18 \times 10^{-5} \text{ C})^2}{(9.00 \times 10^{-6} \text{ F})} = 0.372 \times 10^{-3} \text{ J}$$

Note that $W_1 + W_2 + W_3 = 2.05 \times 10^{-3}$ J.

e. We can find the total energy stored in the circuit by using the equivalent circuit of figure 21.3(b) as

$$W = \tfrac{1}{2}CV^2 = \tfrac{1}{2}(1.64 \times 10^{-6} \text{ F})(50.0 \text{ V})^2$$

$$= 2.05 \times 10^{-3} \text{ J}$$

Note that it checks with the result from part d.

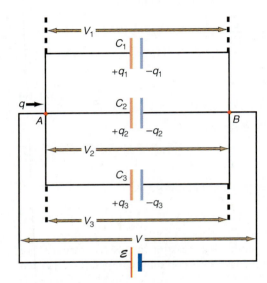

(a) Circuit

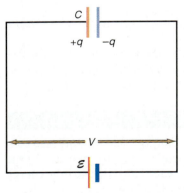

(b) Equivalent circuit

Figure 21.4

Capacitors in parallel.

21.5 Capacitors in Parallel

Consider the three capacitors connected in parallel in figure 21.4(a). Using the concept of conventional current, the battery causes a positive charge q to flow toward the capacitors. When it arrives at the junction A, it divides up into three different charges, q_1, q_2, and q_3, which are then deposited on the left-hand plates of capacitors C_1, C_2, and C_3, respectively. *Thus, when capacitors are connected in parallel there is a different amount of charge deposited on the plates of each capacitor.* These positive charges induce their negative counterparts on the right-hand plate. The positive charges that were originally on the neutral right-hand plate are returned to the battery. Because charge is conserved

$$q = q_1 + q_2 + q_3 \tag{21.20}$$

But the relation between the charge and potential difference across each capacitor is given by equation 21.6 as

$$q_1 = C_1 V_1$$

$$q_2 = C_2 V_2$$

$$q_3 = C_3 V_3$$

Substituting these values of charge in equation 21.20, gives

$$q = C_1 V_1 + C_2 V_2 + C_3 V_3 \tag{21.21}$$

However, *since the battery is in parallel with each capacitor the potential drop across each capacitor is the same,* that is,

$$V = V_1 = V_2 = V_3$$

Equation 21.21 therefore becomes

$$q = (C_1 + C_2 + C_3)V \tag{21.22}$$

An equivalent circuit is now introduced, as shown in figure 21.4(b). For this equivalent circuit

$$q = CV$$

If this equivalent capacitance C is equal to the capacitance of equation 21.22, then the two circuits are in fact equivalent. *Therefore,*

$$C = C_1 + C_2 + C_3 \tag{21.23}$$

is the equivalent capacitance of capacitors in parallel. Hence, the equivalent capacitance of **capacitors in parallel** *is equal to the sum of the individual capacitances.*

Example 21.5

Capacitors in parallel. Three capacitors of 3.00 μF, 6.00 μF, and 9.00 μF are connected in parallel across a battery of 50.0 V, as shown in figure 21.4(a). Find (a) the equivalent capacitance, (b) the total charge on the equivalent capacitor, (c) the voltage drop across each capacitor, (d) the charge on each capacitor, (e) the energy stored in each capacitor, and (f) the total energy stored in the circuit.

Solution

a. The equivalent capacitance, found from equation 21.23, is

$$C = C_1 + C_2 + C_3$$

$$= 3.00 \ \mu\text{F} + 6.00 \ \mu\text{F} + 9.00 \ \mu\text{F}$$

$$= 18.00 \ \mu\text{F}$$

Again note the difference between capacitors connected in series and parallel. When the same three capacitors were connected in series in example 21.4, the equivalent capacitance was then 1.64 μF, very much different from the 18.00 μF when they are in parallel.

b. The total charge on the equivalent capacitor is found from

$$q = CV$$
$$= (18.00 \times 10^{-6} \text{ F})(50.0 \text{ V})$$
$$= 9.00 \times 10^{-4} \text{ C}$$

This is the total charge drawn from the battery.

c. The voltage drop across each capacitor is found by inspection. Since the capacitors are all in parallel with the battery, the voltage drop across them is just equal to the battery voltage. Therefore,

$$V_1 = V_2 = V_3 = V = 50.0 \text{ V}$$

d. The charge on each capacitor is found from

$$q_1 = C_1V_1 = C_1V = (3.00 \times 10^{-6} \text{ F})(50.0 \text{ V}) = 1.50 \times 10^{-4} \text{ C}$$
$$q_2 = C_2V_2 = C_2V = (6.00 \times 10^{-6} \text{ F})(50.0 \text{ V}) = 3.00 \times 10^{-4} \text{ C}$$
$$q_3 = C_3V_3 = C_3V = (9.00 \times 10^{-6} \text{ F})(50.0 \text{ V}) = 4.50 \times 10^{-4} \text{ C}$$

Note that $q_1 + q_2 + q_3 = 9.00 \times 10^{-4}$ C, which is equal to the total charge q drawn from the battery, which we found in part b.

e. The energy stored in each capacitor is

$$W_1 = \tfrac{1}{2}C_1V^2 = \tfrac{1}{2}(3.00 \times 10^{-6} \text{ F})(50.0 \text{ V})^2$$
$$= 3.75 \times 10^{-3} \text{ J}$$
$$W_2 = \tfrac{1}{2}C_2V^2 = \tfrac{1}{2}(6.00 \times 10^{-6} \text{ F})(50.0 \text{ V})^2$$
$$= 7.50 \times 10^{-3} \text{ J}$$
$$W_3 = \tfrac{1}{2}C_3V^2 = \tfrac{1}{2}(9.00 \times 10^{-6} \text{ F})(50.0 \text{ V})^2$$
$$= 11.25 \times 10^{-3} \text{ J}$$

Note that the total energy stored in the capacitors when they are in parallel is

$$W_1 + W_2 + W_3 = 22.5 \times 10^{-3} \text{ J}$$

which is much greater than when the same capacitors were connected in series.

f. The total energy stored in the circuit can be found from the equivalent circuit as

$$W = \tfrac{1}{2}CV^2 = \tfrac{1}{2}(18.00 \times 10^{-6} \text{ F})(50.0 \text{ V})^2$$
$$= 22.5 \times 10^{-3} \text{ J}$$

Again, note that $W_1 + W_2 + W_3 = W$.

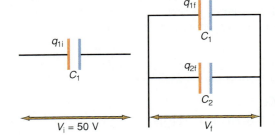

Figure 21.5

Connecting a charged capacitor to an uncharged one.

Example 21.6

Redistributing the charge. A 6.00-μF capacitor is momentarily connected to a 50.0-V battery. The battery is then disconnected and a 9.00-μF capacitor is connected in parallel to the 6.00-μF capacitor (figure 21.5). Find (a) the original charge on the first capacitor, (b) the charge on each capacitor after they are connected in parallel, and (c) the voltage across the combination.

a. The initial charge q_{1i} on the first capacitor is found from

$$q_{1i} = C_1 V_{1i}$$

$$= (6.00 \times 10^{-6}\ \text{F})(50.0\ \text{V})\left(\frac{\text{C/V}}{\text{F}}\right)$$

$$= 3.00 \times 10^{-4}\ \text{C}$$

b. When the second capacitor is connected in parallel to the first, the initial charge on the first capacitor q_{1i} redistributes itself between C_1 and C_2, such that there is now a final charge q_{1f} on C_1 and a final charge q_{2f} on C_2. By the law of conservation of electric charge

$$q_{1i} = q_{1f} + q_{2f}$$

Because the two capacitors are now in parallel, the voltage across them is equal and is expressed as V_f. This final voltage is therefore,

$$V_f = \frac{q_{1f}}{C_1} = \frac{q_{2f}}{C_2}$$

the ratio of the charge distribution can be found as

$$\frac{q_{1f}}{q_{2f}} = \frac{C_1}{C_2} = \frac{6.00\ \mu\text{F}}{9.00\ \mu\text{F}} = \frac{2}{3}$$

or

$$q_{1f} = \tfrac{2}{3} q_{2f}$$

That is, the final charge on C_1 is two-thirds of the final charge on C_2. To find the exact amount of charge, we substitute this relation back into the law of conservation of charge to obtain

$$q_{1i} = q_{1f} + q_{2f} = \tfrac{2}{3} q_{2f} + q_{2f} = \tfrac{5}{3} q_{2f}$$

Therefore, the final charge on C_2 is

$$q_{2f} = \tfrac{3}{5} q_{1i} = \tfrac{3}{5}(3.00 \times 10^{-4}\ \text{C}) = 1.80 \times 10^{-4}\ \text{C}$$

and the final charge on C_1 is

$$q_{1f} = \tfrac{2}{3} q_{2f} = \tfrac{2}{3}(1.80 \times 10^{-4}\ \text{C}) = 1.20 \times 10^{-4}\ \text{C}$$

The initial charge of 3.00×10^{-4} C redistributes itself such that there is now 1.20×10^{-4} C on C_1 and 1.80×10^{-4} C on C_2.

c. The voltage across the capacitors is found from

$$V_{1f} = \frac{q_{1f}}{C_1} = \frac{1.20 \times 10^{-4}\ \text{C}}{6.00 \times 10^{-6}\ \text{F}} = 20.0\ \text{V}$$

21.6 Combinations of Capacitors in Series and Parallel

If capacitors are connected in combinations of series and parallel, as shown in figure 21.6(a), the charge on each capacitor can be found by the techniques of sections 21.4 and 21.5. Since capacitor C_2 is parallel to capacitor C_3, its equivalent capacitance, found from equation 21.23, is

$$C_{23} = C_2 + C_3$$

$$= 6.00\ \mu\text{F} + 9.00\ \mu\text{F}$$

$$= 15.00\ \mu\text{F}$$

Figure 21.6

Combination of capacitors in series and
parallel and the equivalent circuit.

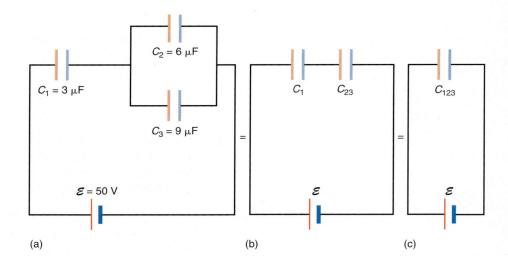

(a)   (b)   (c)

Figure 21.6(a) is now equivalent to figure 21.6(b), here $C_{23} = 15.00$ μF and is in series with C_1. The equivalent capacitance of capacitors C_1 and C_{23} in series, found from equation 21.19, is

$$\frac{1}{C_{123}} = \frac{1}{C_1} + \frac{1}{C_{23}}$$

$$\frac{1}{C_{123}} = \frac{1}{3.00 \ \mu F} + \frac{1}{15.00 \ \mu F} = \frac{0.400}{\mu F}$$

and

$$C_{123} = 2.50 \ \mu F$$

Figure 21.6(b) is now equivalent to figure 21.6(c), with $C_{123} = 2.50$ μF. The total charge released from the battery becomes

$$q = C_{123}V$$

$$= (2.50 \times 10^{-6} \text{ F})(50.0 \text{ V})\left(\frac{\text{C/V}}{\text{F}}\right)$$

$$= 1.25 \times 10^{-4} \text{ C}$$

Since C_1 is in series with the battery this total charge is deposited on the plates of C_1. Therefore,

$$q_1 = q = 1.25 \times 10^{-4} \text{ C}$$

The voltage drop across C_1 is found as

$$V_1 = \frac{q_1}{C_1}$$

$$= \frac{1.25 \times 10^{-4} \text{ C}}{3.00 \times 10^{-6} \text{ F}}\left(\frac{\text{F}}{\text{C/V}}\right)$$

$$= 41.67 \text{ V}$$

The voltage drop across C_{23} is

$$V_{23} = V - V_1$$

$$= 50.0 \text{ V} - 41.67 \text{ V} = 8.33 \text{ V}$$

Since C_2 and C_3 are in parallel, the voltages across each capacitor are equal to each other and to the voltage V_{23}. That is,

$$V_2 = V_3 = V_{23}$$

The charge on capacitor C_2 is found from

$$q_2 = C_2 V_2$$
$$= (6.00 \times 10^{-6} \text{ F})(8.33 \text{ V})\left(\frac{\text{C/V}}{\text{F}}\right)$$
$$= 5.00 \times 10^{-5} \text{ C}$$

And the charge on capacitor C_3 is

$$q_3 = C_3 V_3$$
$$= (9.00 \times 10^{-6} \text{ F})(8.33 \text{ V})$$
$$= 7.50 \times 10^{-5} \text{ C}$$

Note that the charge displaced from capacitor C_1 has been distributed to capacitors C_2 and C_3 and that

$$q_1 = 1.25 \times 10^{-4} \text{ C} = q_2 + q_3$$
$$= 5.00 \times 10^{-5} \text{ C} + 7.50 \times 10^{-5} \text{ C}$$
$$= 1.25 \times 10^{-4} \text{ C}$$

as it must by the law of conservation of charge.

The charge on other simple combinations of capacitors can be found using the same techniques.

21.7 Capacitors with Dielectrics Placed between the Plates

Consider the parallel plate capacitor C_0 with air between the plates, shown in figure 21.7(a). A battery is momentarily connected to the plates of the capacitor, thereby leaving a charge q on the plates and establishing a potential difference V_0 between them. The battery is then removed. An electrometer, which is essentially a very sensitive voltmeter, is connected across the plates and reads the potential difference V_0. The relationship between the charge, the potential, and the capacitance is, of course,

$$C_0 = \frac{q}{V_0} \tag{21.24}$$

A very interesting phenomenon occurs when an insulating material, a *dielectric,* is placed between the plates of the capacitor, as shown in figure 21.7(b). The voltage, as recorded by the electrometer, drops to a lower value V_d. Since the charge on the plates remains the same (there is no place for it to go because the battery was disconnected), the only way the potential between the plates can change is for the capacitance to change. The capacitance with the dielectric between the plates can be written as

$$C_d = \frac{q}{V_d} \tag{21.25}$$

where C_d is the capacitance of the capacitor and V_d is the potential between the plates when there is a dielectric between them. Dividing equation 21.25 by equation 21.24 gives

$$\frac{C_d}{C_0} = \frac{q/V_d}{q/V_0} = \frac{V_0}{V_d} \tag{21.26}$$

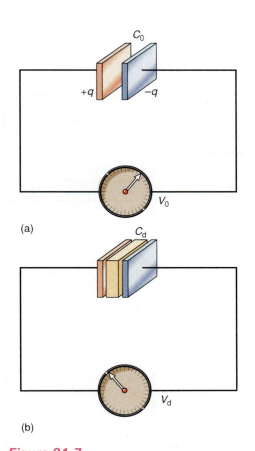

(a)

(b)

Figure 21.7

Placing a dielectric between the plates of a capacitor.

and

$$C_d = \frac{V_0}{V_d} C_0 \qquad (21.27)$$

Because it is observed experimentally that V_d is less than V_0, the ratio V_0/V_d is greater than 1 and hence C_d must be greater than C_0. Therefore, *placing a dielectric between the plates of a capacitor increases the capacitance of that capacitor.* This should not come as too great a surprise since we saw earlier that the capacitance is a function of the geometrical configuration of the capacitor. Placing an insulator between its plates is a significant change in the capacitor configuration.

The effect of the dielectric between the plates is accounted for by the introduction of a new constant, called *the **dielectric constant** κ and it is defined as the ratio of the capacitance with a dielectric between the plates to the capacitance with a vacuum between the plates,* that is,

$$\kappa = \frac{C_d}{C_0} \qquad (21.28)$$

The dielectric constant depends on the material that the dielectric is made of, and some representative values are shown in table 21.1. Note that κ for air is very close to the value of κ for a vacuum, which allows us to treat a great many of electric problems in air as if they were in vacuum.

The capacitance of any **capacitor with a dielectric** is easily found from the juxtaposition of equation 21.28 to

$$C_d = \kappa C_0 \qquad (21.29)$$

For example, the capacitance of a parallel plate capacitor is found from equations 21.3 and 21.29 to be

$$C_d = \kappa \epsilon_0 \frac{A}{d} \qquad (21.30)$$

Table 21.1
Dielectric Constant and Dielectric Strength

Material	Dielectric Constant (κ)	Dielectric Strength ($\times 10^6$ V/m)
Vacuum	1.00000	∞
Air	1.00059	3
Carbon tetrachloride	2.238	
Water	80.37	3
Plexiglass	3.12	40
Paper	3.5	40
Mica	5.40	200
Amber	2.7	
Pyrex glass	4.5	30
Bakelite	4.8	24
Ethyl alcohol	26.0	
Wax paper	2.2	
Benzene	2.3	
Rubber	2.94	21

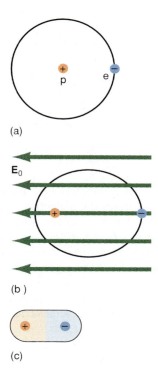

(a)

E_0

(b)

(c)

Figure 21.8

Placing an atom in a uniform electric field.

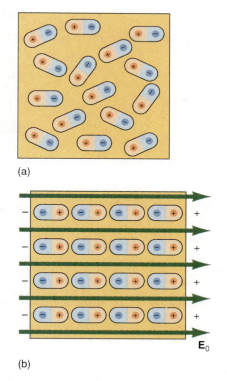

(a)

(b)

E_0

Figure 21.9

A polor dielectric.

Quite often a new constant is used, called the *permittivity of the medium* and is defined as

$$\epsilon = \kappa\epsilon_0$$
(21.31)

Atomic Description of a Dielectric

As was mentioned in chapter 18, all atoms are electrically neutral. We can see the reason for this from figure 21.8(a). The positive charge is located at the center of the atom. The electron revolves around the nucleus. Sometimes it is to the right of the nucleus, sometimes to the left, sometimes it is above the nucleus, sometimes below. Because of this symmetry, its *mean* position appears as if it were at the center of the nucleus. Therefore, it appears as though there is a positive charge and a negative charge at the center of the atom. These equal but opposite charges give the effect of neutralizing each other, and hence the atom does not appear to have any charges and appears neutral. The electric field of the positive charge is radially outward, whereas the average electric field of the negative charge is radially inward. These equal but opposite average electric fields have the effect of neutralizing each other, and hence the atom also does not appear to have any electric field associated with it. Because of this basic symmetry of the atom, all atoms are electrically neutral.

If the atom is now placed in a uniform external electric field E_0, we observe a change in this symmetry, as shown in figure 21.8(b). The electron of the atom finds itself in an external field E_0 and experiences the force

$$F = qE_0 = -eE_0$$

Because the electronic charge is negative, the force on the electron is opposite to the direction of the external field E_0. When the electron is to the right of the nucleus, the external electric field pulls it even further to the right, away from the nucleus. When the electron is to the left of the nucleus, the external field again pulls it toward the right, this time toward the nucleus. The overall effect of this external field is to change the orbit of the electron from a circle to an ellipse, as shown in figure 21.8(b). The mean position of the negative electron no longer coincides with the positive nucleus, but because of the new symmetry is displaced to the right of the nucleus, as shown in figure 21.8(c) The apparently neutral atom has been converted to an electric dipole by the uniform external electric field $\mathbf{E_0}$.

When two or more atoms combine they form a molecule. Many molecules have symmetric charge distributions and therefore also appear electrically neutral. Some molecules, however, do not have symmetric charge distributions and therefore exhibit an electric dipole moment. Such molecules are called *polar molecules*. A slab of material made from polar molecules is shown in figure 21.9(a). In general, the electric dipoles are arranged randomly. If the dielectric is placed in an external electric field, electric forces act on the dipoles until the dipoles become aligned with the field, as shown in figure 21.9(b).

The net effect is that a dielectric material made of dipole molecules has those dipoles aligned in an external electric field. A dielectric material not composed of polar molecules has electric dipoles induced in them by the external electric field, as in figure 21.8. These dipoles are in turn aligned by the external electric field. Therefore, all insulating materials become polarized in an external electric field. The overall effect is to have positive charges on the right of the slab in figure 21.9(b) and negative charges to the left. These charges are not free charges, but are bound to the atoms of the dielectric. They are merely the displaced charges and not a flow of charges.

With this digression into the atomic nature of dielectrics, we can now explain the effect of a dielectric placed between the plates of a capacitor. Figure 21.10(a) shows a charged capacitor with air between the plates. A uniform electric field E_0 is found between the plates. A slab of dielectric material is now placed

Figure 21.10

Electric field between the plates of the capacitor.

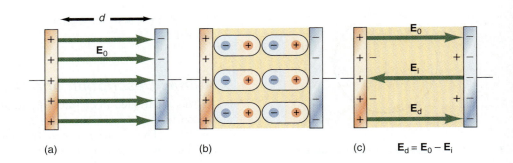

(a)　　　　　　　　(b)　　　　　　　　(c)　　$E_d = E_0 - E_i$

between the plates, as shown in figure 21.10(b), and completely fills the space between the plates. The uniform electric field E_0 polarizes the dielectric by aligning the electric dipoles, figure 21.10(b). The net effect is to displace positive bound charges to the right of the slab and negative bound charges to the left of the slab. These induced bound charges have an electric field E_i associated with them. The induced electric field E_i emanates from the positive bound charges on the right of the dielectric and terminates on the negative bound charges on the left of the dielectric, and, as we can see from figure 21.10(c), E_i is opposite to the original electric field of the capacitor. Hence, the total electric field within the dielectric is

$$E_d = E_0 - E_i \qquad (21.32)$$

Therefore, placing a dielectric between the plates of a capacitor reduces the electric field between the plates of the capacitor.

The potential difference V_0 across the plates of the capacitor, when there is no dielectric between the plates, is given by equation 19.20 as

$$V_0 = E_0 d \qquad (21.33)$$

While the potential difference V_d across the capacitor when there is a dielectric between the plates is

$$V_d = E_d d \qquad (21.34)$$

Dividing equation 21.33 by equation 21.34 gives

$$\frac{V_0}{V_d} = \frac{E_0}{E_d} \qquad (21.35)$$

But from equation 21.26 the ratio of the capacitance with a dielectric to the capacitance without a dielectric was

$$\frac{C_d}{C_0} = \frac{V_0}{V_d} \qquad (21.26)$$

Combining this result with equation 21.35 gives

$$\frac{C_d}{C_0} = \frac{V_0}{V_d} = \frac{E_0}{E_d}$$

Recall that the ratio of C_d/C_0 was defined in equation 21.28 as the dielectric constant, κ. Therefore,

$$\frac{C_d}{C_0} = \frac{V_0}{V_d} = \frac{E_0}{E_d} = \kappa \qquad (21.36)$$

The effect of the dielectric between the plates is summarized from equation 21.36 as follows:

1. The capacitance with a dielectric increases to

$$C_d = \kappa C_0 \qquad (21.29)$$

2. The potential difference between the plates decreases to

$$V_d = \frac{V_0}{\kappa} \qquad (21.37)$$

and

3. the electric field between the plates decreases to

$$E_d = \frac{E_0}{\kappa} \qquad (21.38)$$

The value of the induced electric field E_i within the dielectric can be found from equations 21.38 and 21.32 as

$$E_d = E_0 - E_i = \frac{E_0}{\kappa}$$

Solving for E_i gives

$$E_i = E_0 - \frac{E_0}{\kappa} = E_0\left(1 - \frac{1}{\kappa}\right)$$

$$E_i = E_0\left(\frac{\kappa - 1}{\kappa}\right) \qquad (21.39)$$

The presence of a dielectric between the plates of a capacitor not only increases the capacitance of the capacitor, but it also allows a much larger voltage to be applied across the plates. When the voltage across the plates of an air capacitor exceeds 3.00×10^6 V/m, the air between the plates is no longer an insulator, but conducts electric charge from the positive plate to the negative plate as a *spark* or *arc discharge*. The dielectric is said to break down. The capacitor is now useless as a capacitor since it now conducts electricity through it and is said to be *short circuited*. *The value of the potential difference per unit plate separation when the dielectric breaks down is called the* **dielectric strength** *of the medium,* and is given in the second column of table 21.1. If mica is placed between the plates, the breakdown voltage rises to 200×10^6 V/m, a value that is 66 times greater than the value for air.

Example 21.7

A capacitor with a dielectric. A parallel plate capacitor, having a plate area of 25.0 cm² and a plate separation of 2.00 mm, is charged by a 100-V battery. The battery is then removed. Find (a) the capacitance of the capacitor and (b) the charge on the plates of the capacitor. A slab of mica is then placed between the plates of the capacitor. Find (c) the new value of the capacitance, (d) the potential difference across the capacitor with the dielectric, (e) the initial electric field between the plates, (f) the final electric field between the plates, (g) the initial energy stored in the capacitor, and (h) the final energy stored in the capacitor.

Solution

a. The original capacitance of the capacitor, found from equation 21.3, is

$$C_0 = \epsilon_0\frac{A}{d} = \left(8.85 \times 10^{-12}\frac{C^2}{N\,m^2}\right)\left(\frac{25.0\ cm^2}{0.002\ m}\right)\left(\frac{1\ m}{10^2\ cm}\right)^2$$

$$= \left(1.11 \times 10^{-11}\frac{C^2}{N\,m}\right)\left(\frac{N\,m}{J}\right)$$

$$= \left(1.11 \times 10^{-11}\frac{C}{J/C}\right)\left(\frac{J/C}{V}\right)\left(\frac{F}{C/V}\right)$$

$$= 1.11 \times 10^{-11}\ F = 11.1 \times 10^{-12}\ F$$

$$= 11.1\ pF$$

b. The charge on the plates of the capacitor, found from equation 21.6, is

$$q = C_0 V_0 = (11.1 \times 10^{-12} \text{ F})(100 \text{ V})\left(\frac{C/V}{F}\right)$$

$$= 11.1 \times 10^{-10} \text{ C}$$

This charge remains the same throughout the rest of the problem because the battery was disconnected.

c. The value of the capacitance when a mica dielectric is placed between the plates, found from equation 21.29 and table 21.1, is

$$C_d = \kappa C_0$$

$$= (5.40)(11.1 \text{ pF})$$

$$= 59.9 \text{ pF}$$

d. The new potential difference between the plates, found from equation 21.37, is

$$V_d = \frac{V_0}{\kappa} = \frac{100 \text{ V}}{5.40} = 18.5 \text{ V}$$

This could also have been found by equation 21.25 as

$$V_d = \frac{q}{C_d} = \frac{11.1 \times 10^{-10} \text{ C}}{59.9 \times 10^{-12} \text{ F}} = 18.5 \text{ V}$$

e. The initial electric field between the plates, found from equation 21.33, is

$$E_0 = \frac{V_0}{d} = \frac{100 \text{ V}}{0.002 \text{ m}} = 5.00 \times 10^4 \text{ V/m}$$

f. The final electric field between the plates, when the dielectric is present, found from equation 21.38, is

$$E_d = \frac{E_0}{\kappa} = \frac{5.00 \times 10^4 \text{ V/m}}{5.40} = 9.25 \times 10^3 \text{ V/m}$$

This could also have been found from equation 21.34 as

$$E_d = \frac{V_d}{d} = \frac{18.5 \text{ V}}{0.002 \text{ m}} = 9.25 \times 10^3 \text{ V/m}$$

g. The initial energy stored in the capacitor, found from equation 21.9, is

$$W_0 = \frac{1}{2} C_0 V_0^2 = \frac{1}{2} (11.1 \times 10^{-12} \text{ F})(100 \text{ V})^2$$

$$= 5.55 \times 10^{-8} \frac{C}{V} V^2 \left(\frac{J}{CV}\right)$$

$$= 5.55 \times 10^{-8} \text{ J}$$

h. The final energy stored in the capacitor when there is a dielectric between the plates is

$$W_d = \tfrac{1}{2} C_d V_d^2 \qquad\qquad \textbf{(21.40)}$$

The energy can be computed directly from equation 21.40, but it is instructive to do the following. The capacitance C_d can be substituted from equation 21.29 and the potential difference can be substituted from equation 21.37 into equation 21.40 to yield

$$W_d = \frac{1}{2}C_d V_d^2 = \frac{1}{2}(\kappa C_0)\left(\frac{V_0}{\kappa}\right)^2$$

$$= \frac{1}{\kappa}\left(\frac{1}{2}C_0 V_0^2\right)$$

$$W_d = \frac{W_0}{\kappa} \tag{21.41}$$

This is the energy stored in the capacitance with a dielectric between the plates. Substituting the numbers into equation 21.41 gives

$$W_d = \frac{W_0}{\kappa} = \frac{5.55 \times 10^{-8} \text{ J}}{5.40}$$

$$= 1.03 \times 10^{-8} \text{ J}$$

The energy stored in the capacitor with a dielectric is less than without the dielectric. This, at first, may seem to be a violation of the law of conservation of energy, since there is a decrease in energy. But this decrease in energy can be accounted for by the work done in moving the dielectric slab between the plates of the capacitor.

Example 21.8

A capacitor connected to a battery. A 6.00-μF air capacitor is connected to a 100-V battery (figure 21.11). A piece of mica is now placed between the plates of the capacitor. Find (a) the original charge on the plates of the capacitor, (b) the new value of the capacitance with the dielectric, and (c) the new charge on the plates.

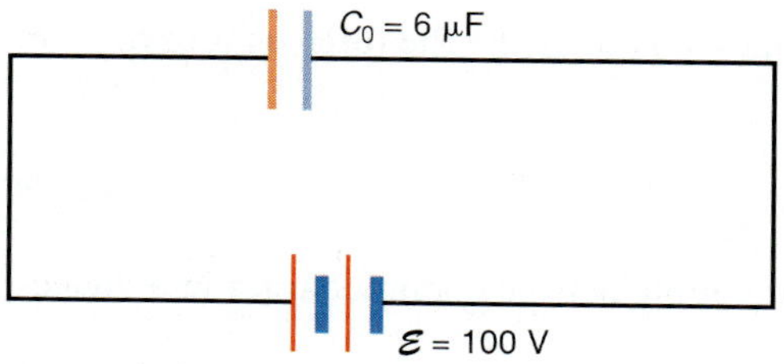

Solution

a. The original charge on the plates is found from

$$q_0 = C_0 V_0 = (6.00 \times 10^{-12} \text{ F})(100 \text{ V})\left(\frac{C/V}{F}\right)$$

$$= 6.00 \times 10^{-10} \text{ C}$$

b. The new value of the capacitance is found from

$$C_d = \kappa C_0 = (5.40)(6.00 \times 10^{-12} \text{ F})$$

$$= 32.4 \times 10^{-12} \text{ F}$$

$$= 32.4 \text{ pF}$$

c. In this example, the battery is *not* disconnected from the capacitor. Therefore the potential across the plates remains the same whether the capacitor has a dielectric or not. Because the capacitance of the capacitor has changed, the charge on the plates must now change and is found from

$$q_d = C_d V_d$$

$$= (32.4 \times 10^{-12} \text{ F})(100 \text{ V})$$

$$= 32.4 \times 10^{-10} \text{ C}$$

By keeping the potential across the plates constant, more charge was drawn from the battery. A clear distinction must be made between this example and

$C_0 = 6 \ \mu\text{F}$

$\mathcal{E} = 100 \text{ V}$

Figure 21.11

Placing a dielectric between the plates while maintaining a constant potential between the plates.

the previous one. In example 21.7 the charge remained constant because the battery was disconnected, and the potential varied by the introduction of the dielectric. In this example, the potential remained a constant, because the battery was *not disconnected,* and the charge varied with the introduction of the dielectric.

Dielectrics and Coulomb's Law

The electric field in a vacuum was defined as $E = F/q_0$ in equation 19.1. Calling the electric field when no dielectric is present E_0, equation 19.1 becomes

$$E_0 = \frac{F}{q} \tag{21.42}$$

But if a dielectric medium is present the electric field within the dielectric was given by equation 21.38 as

$$E_d = \frac{E_0}{\kappa}$$

Substituting equation 21.42 into equation 21.38 gives

$$E_d = \frac{F}{\kappa q} \tag{21.43}$$

Comparing equations 21.43 with equation 21.42 prompts us to define the electric field in a dielectric as the ratio of the force acting on a particle within the dielectric to the charge, that is,

$$E_d = \frac{F_d}{q} \tag{21.44}$$

Comparing equations 21.43 and 21.44 allows us to define the force on a particle in a dielectric as

$$F_d = \frac{F}{\kappa} \tag{21.45}$$

Using Coulomb's law, we can now find the force between point charges in a dielectric medium as

$$F_d = \frac{F}{\kappa} = \frac{1}{4\pi\epsilon_0\kappa}\frac{q_1 q_2}{r^2}$$

Recall from equation 21.31 that the permittivity of the medium is defined as

$$\epsilon = \kappa\epsilon_0$$

In general then, Coulomb's law should be expressed as

$$F = \frac{1}{4\pi\epsilon}\frac{q_1 q_2}{r^2} \tag{21.46}$$

As the special case for a vacuum $\kappa = 1$, and equation 21.46 reduces to

$$F = \frac{1}{4\pi\epsilon_0}\frac{q_1 q_2}{r^2}$$

the form of Coulomb's law used in chapter 18.

As an example of the effect of the dielectric on the force between charges consider the salt sodium chloride, NaCl, which is common table salt. Although NaCl consists of a lattice structure of Na$^+$ and Cl$^-$ ions, let us just consider the electric force between one Na$^+$ ion, and one Cl$^-$ ion separated by a distance of 1.00×10^{-7} cm. The force between the ions is given by Coulomb's law as

$$F_0 = \frac{1}{4\pi\epsilon_0}\frac{q_1 q_2}{r^2}$$

$$= \left(9.00 \times 10^9\ \frac{\text{N m}^2}{\text{C}^2}\right)\frac{(1.60 \times 10^{-19}\ \text{C})^2}{(1.00 \times 10^{-9}\ \text{m})^2}$$

$$= 2.30 \times 10^{-10}\ \text{N}$$

If this salt is now placed in water the force between the ions becomes

$$F_d = \frac{1}{4\pi\epsilon_0 \kappa}\frac{q_1 q_2}{r^2}$$

$$= \left[\frac{9.00 \times 10^{-9}\ \text{N m}^2}{(80.0)\quad \text{C}^2}\right]\frac{(1.60 \times 10^{-19}\ \text{C})^2}{(1.00 \times 10^{-9}\ \text{m})^2}$$

$$= 0.0288 \times 10^{-10}\ \text{N}$$

This could also be found as

$$F_d = \frac{F}{\kappa} = \frac{2.30 \times 10^{-10}\ \text{N}}{80.0} = 0.0288 \times 10^{-10}\ \text{N}$$

The force between the ions in the water solution has decreased by a factor of $1/80$. This is why NaCl readily dissolves in water. The force between the ions becomes small enough for simple thermal energy to pull the ions apart. In fact, most chemicals are soluble in water because of this high value of 80 for the dielectric constant. If the NaCl salt is placed in benzene, $\kappa = 2.3$, the force between the Na$^+$ and Cl$^-$ ions is not reduced enough to allow the ions to disassociate, and hence NaCl does not dissolve in benzene.

The water molecule, H$_2$O, is composed of two atoms of hydrogen and one atom of oxygen. The molecular configuration is not spherically symmetric, but is rather asymmetric, as shown in figure 21.12(a). The oxygen atom has picked up two electrons, one from each hydrogen atom, and is therefore charged doubly negative. Each hydrogen atom, having lost an electron to the oxygen, is positive. The mean position of the positive charge does not coincide with the location of the negative charge but lies below it in the diagram. Hence, the water molecule behaves as an electric dipole, as seen in figure 21.12(b), and it is a polar molecule. When a material, such as the sodium chloride, is placed in water, these water dipoles surround the Na$^+$ and Cl$^-$ ions, as shown in figure 21.13. The negative ends of the water molecule are attracted to the positive ions, whereas the positive end of the water molecules are attracted to the negative ion. These dipoles that surround the ions, shield the ions and have the effect of decreasing the effective charge of the ions and hence decrease the coulomb force between them. Liquids with large values of κ are the most effective in decreasing the force between the separated ions and are therefore the best solvents for ionic crystals.

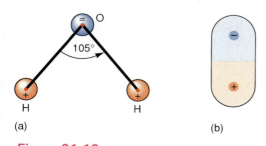

(a)   (b)

Figure 21.12

The water molecule.

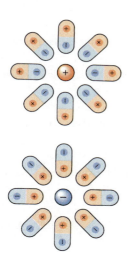

Figure 21.13

Water dipoles surround ions placed in water.

The Language of Physics

Capacitor

Two conductors of any size or shape carrying equal and opposite charges are called a capacitor. The charge on the capacitor is directly proportional to the potential difference between the plates. The importance of the capacitor lies in the fact that energy can be stored in the electric field between the two bodies (p. 603).

Energy density

The energy per unit volume that is stored in the electric field (p. 607).

Capacitors in series

When capacitors are connected in series, each capacitor has the same charge on its plates. The reciprocal of the equivalent capacitance is equal to the sum of the reciprocals of each capacitor (p. 609).

Capacitors in parallel

When capacitors are connected in parallel there is a different amount of charge deposited on the plates of each capacitor, but the potential difference is the same across each of the parallel capacitors. The equivalent capacitance is equal to the sum of the individual capacitances (p. 611).

Dielectric constant

The dielectric constant is defined as the ratio of the capacitance with a dielectric between the plates to the capacitance with air or vacuum between the plates (p. 616).

Capacitors with a dielectric

Placing a dielectric, an insulator, between the plates of a capacitor increases the capacitance of that capacitor; decreases the electric field and the potential difference between the plates; decreases the amount of energy that can be stored in the capacitor; and increases the dielectric strength of the capacitor (p. 616).

Dielectric strength

The value of the potential difference per unit plate separation when the dielectric breaks down (p. 619).

Summary of Important Equations

Capacitance of parallel plate capacitor

$$C = \frac{\epsilon_0 A}{d} \tag{21.3}$$

Capacitance for coaxial cylindrical capacitor

$$C = \frac{2\pi\epsilon_0 l}{\ln(r_B/r_A)} \tag{21.4}$$

Capacitance for concentric spherical capacitor

$$C = \frac{4\pi\epsilon_0 r_A r_B}{r_B - r_A} \tag{21.5}$$

Charge on a capacitor

$$q = CV \tag{21.6}$$

Definition of capacitance

$$C = \frac{q}{V} \tag{21.7}$$

Energy stored in a capacitor

$$W = \tfrac{1}{2}qV \tag{21.8}$$

$$W = \tfrac{1}{2}CV^2 \tag{21.9}$$

$$W = \tfrac{1}{2}\frac{q^2}{C} \tag{21.10}$$

Energy density in electric field

$$u_E = \tfrac{1}{2}\epsilon_0 E^2 \tag{21.12}$$

Equivalent capacitance of capacitors in series

$$\frac{1}{C} = \frac{1}{C_1} + \frac{1}{C_2}$$
$$+ \frac{1}{C_3} + \cdots \tag{21.19}$$

Equivalent capacitance of capacitors in parallel

$$C = C_1 + C_2$$
$$+ C_3 + \cdots \tag{21.23}$$

Capacitance with dielectric

$$C_d = \frac{q}{V_d} \tag{21.25}$$

Definition of dielectric constant

$$\kappa = \frac{C_d}{C_0} \tag{21.28}$$

Capacitance of capacitor with a dielectric

$$C_d = \kappa C_0 \tag{21.29}$$

Permittivity of the dielectric medium

$$\epsilon = \kappa\epsilon_0 \tag{21.31}$$

The effect of a dielectric between the plates

$$\frac{C_d}{C_0} = \frac{V_0}{V_d} = \frac{E_0}{E_d} = \kappa \tag{21.36}$$

Potential in the dielectric

$$V_d = \frac{V_0}{\kappa} \tag{21.37}$$

Electric field within the dielectric

$$E_d = \frac{E_0}{\kappa} \tag{21.38}$$

Energy stored in dielectric capacitor

$$W_d = \frac{W_0}{\kappa} \tag{21.41}$$

Force on a particle in a dielectric medium

$$F_d = \frac{F}{\kappa} \tag{21.45}$$

Coulomb's law within a dielectric medium

$$F = \frac{1}{4\pi\epsilon}\frac{q_1 q_2}{r^2} \tag{21.46}$$

Questions for Chapter 21

1. Can you speak of the capacitance of an isolated conducting sphere?
†2. How does the direction signal device on your car make use of a capacitor?
3. If an ammeter is connected in series to the circuit of figure 21.3, what will it indicate? Why?
4. Discuss the statement, "If a capacitor contains equal amounts of positive and negative charge it should be neutral and have no effect whatsoever."

†5. In tuning in your favorite radio station, you adjust the tuning knob, which is connected to a variable capacitor. How does this variable capacitor work?
6. If the potential difference across a capacitor is doubled what does this do to the energy that the capacitor can store?
7. Is more energy stored in capacitors when they are connected in parallel or in series? Why?

8. If the applied potential difference across a capacitor exceeds the dielectric strength of the medium between the plates, what happens to the capacitor?
†9. Sketch what you think a plot of the charge on a capacitor versus time would look like, when it is charging and when it is discharging.
10. Can a coaxial cable have a capacitance associated with it?

Problems for Chapter 21

21.2 The Parallel Plate Capacitor

1. Find the charge on a 4.00-μF capacitor if it is connected to a 12.0-V battery.
2. If the charge on a 9.00-μF capacitor is 5.00×10^{-4} C, what is the potential across it?
3. How much charge must be removed from a capacitor such that the new potential across the plates is 1/2 of the original potential?
4. A charge of 6.00×10^{-4} C is found on a capacitor when a potential of 120 V is placed across it. What is the value of the capacitance of this capacitor?
5. A parallel plate capacitor has plates that are 6.00 cm by 4.00 cm and are separated by 8.00 mm. Find the capacitance of the capacitor.
6. If a 5.00-μF parallel plate capacitor has its plate separation doubled while its cross-sectional area is tripled, determine the new value of its capacitance.
7. You are asked to design your own parallel plate capacitor that is capable of holding a charge of 3.60×10^{-12} C when placed across a potential difference of 12.0 V. If the area of the plates is equal to 100 cm² what must the plate separation be? What is the value of the capacitance of this capacitor?
8. You are asked to design a parallel plate capacitor that can hold a charge of 9.87 pC when a potential difference of 24.0 V is applied across the plates. Find the ratio of the area of the plates to the plate separation, and then pick a reasonable set of values for them.
9. A cylindrical capacitance 10.0 cm long has an inner radius of 0.500 mm and an outside radius of 5.00 mm. Find the capacitance of the capacitor.

10. A cylindrical capacitor 1.00 m long has radii 20.0 cm and 50.0 cm. Find its capacitance.
11. A spherical capacitor has radii 20.0 cm and 50.0 cm. Find its capacitance.

21.3 Energy Stored in a Capacitor

12. A 7.00-μF capacitor is connected to a 400-V source. Find (a) the charge on the capacitor and (b) the energy stored in the capacitor.
13. How much energy can be stored in a capacitor of 9.45 μF when it is placed across a potential difference of 120 V? How much charge will be on this capacitor?
14. What value of capacitance is necessary to store 1.73×10^{-3} J of energy when it is placed across a potential difference of 24.0 V?

21.4 Capacitors in Series

15. Find the equivalent capacitance of 2.00-μF, 4.00-μF, and 8.00-μF capacitors connected in series.
16. Find (a) the equivalent capacitance, (b) the charge on each capacitor, (c) the voltage across each capacitor, and (d) the energy stored in each capacitor in the diagram.

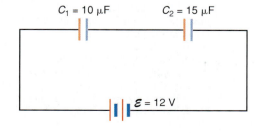

21.5 Capacitors in Parallel

17. Find the equivalent capacitance of 2.00-μF, 4.00-μF, and 8.00-μF capacitors connected in parallel.
18. Find the charge on each capacitor and the energy stored in each capacitor in the diagram.

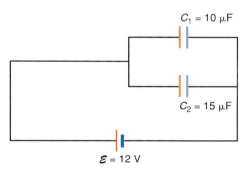

19. A 3.00-μF capacitor is initially connected to a battery and charged to 100 V. It is then removed and connected in parallel to a 15.0-μF capacitor that was uncharged. Find (a) the initial charge on the first capacitor, (b) the charge on each capacitor after being connected in parallel, (c) the initial energy stored in the first capacitor, and (d) the energy stored in each capacitor after they are connected in parallel.

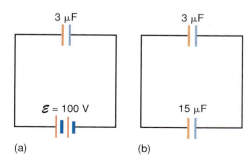

(a)　　　　(b)

20. A 40.0-μF capacitor and an 80.0-μF capacitor are initially connected in parallel across a 10.0-V potential difference. The two capacitors are then disconnected from each other and from the potential source, and reconnected with plates of unlike sign connected together. Find the charge on each capacitor and the potential difference across each capacitor after the two capacitors are so connected.

21.6 Combinations of Capacitors in Series and Parallel

†21. Find (a) the equivalent capacitance of the circuit, (b) the total charge drawn from the battery, (c) the voltage across each capacitor, (d) the charge on each capacitor, and (e) the energy stored in each capacitor in the accompanying diagram.

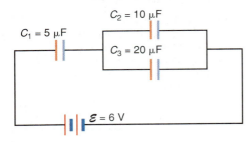

22. Find the equivalent capacitance of the circuit diagram if $C_1 = 1.00$ μF, $C_2 = 2.00$ μF, $C_3 = 3.00$ μF, $C_4 = 4.00$ μF, and $C_5 = 5.00$ μF.

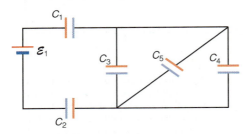

†23. In the accompanying diagram find (a) the equivalent capacitance, (b) the charge on capacitor C_2, and (c) the voltage drop across C_4, if $C_1 = 10.0$ μF, $C_2 = 20.0$ μF, $C_3 = 30.0$ μF, $C_4 = 40.0$ μF, and $\mathcal{E} = 12.0$ V.

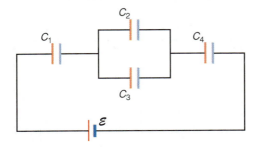

24. The network of capacitors in the diagram is connected across a potential difference $V = 3.00$ V. If $C_1 = C_2 = 30.0$ μF, $C_3 = 60.0$ μF, and $C_4 = C_5 = C_6 = 20.0$ μF, find (a) the equivalent capacitance of the network, (b) the charge on each capacitor, (c) the potential across each capacitor, and (d) the energy stored in each capacitor.

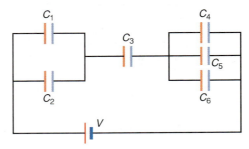

†25. Find the equivalent capacitance and the charge on each capacitor in the diagram if $C_1 = 5.00$ μF, $C_2 = 15.00$ μF, $C_3 = 8.00$ μF, $C_4 = 9.00$ μF, and $\mathcal{E} = 12.0$ V.

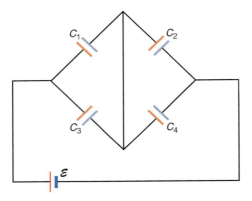

21.7 Capacitors with Dielectrics Placed between the Plates

26. A 5.00-μF parallel plate capacitor has air between the plates. When an insulating material is placed between the plates, the capacitance increases to 13.5 μF. Find the dielectric constant of the insulator.

27. If the dielectric constant of rubber is 2.94, what is its permittivity?

28. Find the capacitance of a parallel plate capacitor having an area of 75.0 cm², a plate separation of 3.00 mm, and a strip of bakelite placed between the plates.

29. A 350-μF capacitor is constructed with bakelite between its plates. If the bakelite is removed and replaced by amber, find the new capacitance.

30. A 0.100-mm piece of wax paper is to be placed between two pieces of aluminum foil to make a parallel plate capacitor of 5 μF. What must the area of the foil be?

31. A parallel plate capacitor has plates that are 5.00 cm by 6.00 cm and are separated by an air gap of 1.50 mm. Calculate the maximum voltage that can be applied to the capacitor before dielectric breakdown.

32. A 1.00-mm piece of wax paper is placed between two pieces of aluminum foil that are 10.0 cm by 10.0 cm. How much energy can be stored in this capacitor if it is connected to a 24.0 V battery?

33. If 1.73×10^{-3} J of energy can be stored in a capacitor with air between the plates, how much energy can be stored in the capacitor if a slab of mica is placed between the plates?

34. What is the maximum charge that can be placed on the plates of an air capacitor of 9.00 μF before breakdown if the plate separation is 1.00 mm? If a strip of mica is placed between the plates, what is the maximum charge?

35. A parallel plate capacitor has a value of 3.00 pF. A 1.00-mm piece of plexiglass is placed between the plates of the capacitor, which is then connected to a 12.0-V battery. Find (a) the electric field E_0 without the dielectric between the plates and (b) the induced electric field in the dielectric E_i.

36. A sample of NaCl is placed into a solution of ethyl alcohol. If the distance between the Na$^+$ and the Cl$^-$ ions is 1.00×10^{-7} cm, find the force on the ions.

Additional Problems

†37. Show that if the radii of a spherical capacitor are very large, the equation for the capacitance of a spherical capacitor reduces to the equation for a parallel plate capacitor.

†38. Find the formula for the capacitance of an isolated sphere. (*Hint:* use the relation for a concentric spherical capacitor and let the radius of the outer sphere go to infinity.)

†39. Consider the earth to be an isolated sphere. Find the capacitance of the earth. If the electric field of the earth is measured to be 100 V/m, what charge must be on the surface of the earth? (Note that the fair weather electric field points downward, indicating an effective negative charge on the solid sphere.)

40. Determine all the values of capacitance that you can obtain from the three capacitors of 3.00 μF, 6.00 μF, and 9.00 μF.

41. (a) Find the equivalent capacitance of 50 identical 2.50-μF capacitors in series. (b) Find the equivalent capacitance of 50 identical 2.50-μF capacitors in parallel.

†42. A potential of 24.0 V is applied across capacitor $C_1 = 2.50$ μF. Find the charge on capacitors C_2 and C_3 if $C_2 = 6.00$ μF and $C_3 = 4.00$ μF.

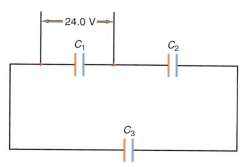

†43. In the accompanying diagram find (a) the equivalent capacitance, (b) the voltage drop across each capacitor, and (c) the charge on each capacitor if $C_1 = 3.00$ μF, $C_2 = 6.00$ μF, $C_3 = 9.00$ μF, $C_4 = 12.00$ μF, $C_5 = 15.00$ μF, $C_6 = 18.00$ μF, and $\mathscr{E}_1 = 12.0$ V.

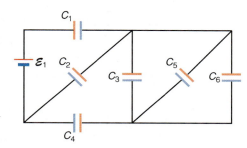

†44. Find the potential drop across AB in the diagram if $C_1 = 3.00$ μF, $C_2 = 6.00$ μF, $C_3 = 9.00$ μF, $C_4 = 12.00$ μF, and $\mathscr{E} = 24.0$ V.

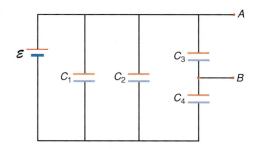

†45. In the accompanying circuit, find (a) the final value of current recorded by an ammeter that is first placed in series with the resistor R, then the capacitor C_1, and finally the capacitor C_2; (b) find the voltage drop recorded by a voltmeter when placed across the resistor R, the capacitor C_1, and the capacitor C_2; and (c) find the charge on C_1 and C_2.

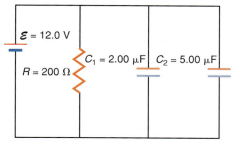

46. A cylindrical capacitor is made by wrapping a sheet of aluminum foil around a hollow cardboard core 2.50 cm in diameter. A 1.00 mm thick piece of wax paper is then wrapped around the foil and then another sheet of aluminum foil is wrapped around the wax paper. The length of the core is 25.0 cm. Find the capacitance of this cylindrical capacitor.

†47. Find (a) the capacitance of a parallel plate capacitor having an area of 40.0 cm² and separated by air 2.00 mm thick, (b) the capacitance if 2.00 mm of mica is placed between the plates, (c) if the capacitor is connected to a 6.00-V battery, find the charge on the capacitor with air and with mica between the plates, (d) the electric field between the plates, and (e) the energy stored in each capacitor.

†48. The electric field between the plates of a parallel plate capacitor filled with air is $E_0 = q/(\epsilon_0 A)$. When a dielectric is placed between the plates, the induced electric field within the dielectric is given by $E_i = q'/(\epsilon_0 A)$, where q' is the bound charge. (a) Show that the bound charge on the dielectric is given by

$$q' = q\left(1 - \frac{1}{\kappa}\right)$$

(b) If a piece of mica is placed between the plates of a capacitor of 6.00 μF and is then connected to the plates of a 24.0-V battery, find the charge q and q'.

49. Two identical parallel plate capacitors, one with air between the plates, and the other with mica, are connected to a 100-V battery, as shown in the diagram. If $C_1 = 3.00$ μF and the plate separation is 0.500 mm, find (a) the charge on each plate, (b) the electric field between the plates, and (c) the energy stored in each capacitor.

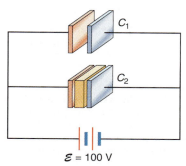

50. A cumulonimbus cloud is 5.00 km long by 5.00 km wide and has its base 1.00 km above the surface of the earth, as shown in the diagram. Consider the cloud and earth to be a parallel plate capacitor with air as the dielectric. (a) Find the capacitance of the cloud-earth combination. (b) Find the potential difference between the cloud and the earth when lightning occurs. (c) Calculate the charge that must be on the base of the cloud when lightning occurs.

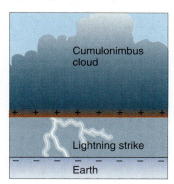

†51. A parallel plate capacitor with air between the plates has a plate separation d_0 and area A. A sheet of aluminum of negligible thickness is now placed midway between the parallel plates. Find the capacitance (a) before and (b) after the aluminum sheet is placed between the plates.

†52. A parallel plate capacitor with air between the plates has a plate separation d_0 and area A. A half of a sheet of aluminum of negligible thickness is now placed midway between the parallel plates as shown. Show what this combination is equivalent to and find the equivalent capacitance.

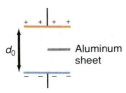

d_0 ▨▨▨ Aluminum
              sheet

†53. A parallel plate capacitor with dielectric constant κ_1 is connected in series with a similar parallel plate capacitor except that it has a dielectric constant κ_2 between the plates, as shown in the diagram. (a) Find a single equation for the equivalent capacitor. (b) What is the value of an equivalent dielectric that could be placed in one of the original capacitors to give an equivalent capacitor?

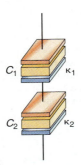

C_1 κ_1

C_2 κ_2

†54. A parallel plate capacitor has a moveable top plate that executes simple harmonic motion at a frequency f. The displacement of the top plate varies such that the plate separation varies from d to $2d$. Show that the voltage across MN varies as

$$V = V_0(1 + \tfrac{1}{2}\cos 2\pi ft)$$

where V_0 is the voltage across the capacitor when the plates are at a constant separation d.

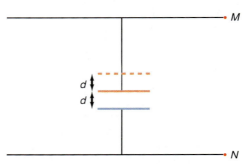

• M

d

d

• N

†55. One model of a red blood cell considers the cell to be a spherical capacitor. The radius of a blood cell is about 6.00×10^{-4} cm and the membrane wall of 0.100×10^{-4} cm is considered to be a dielectric material with a dielectric constant of 5.00. Find (a) the capacitance of the cell. (b) If a potential difference of 100 mV is measured across the blood cell, find the charge on a red blood cell.

†56. When you tune in your favorite radio station, you use a variable capacitor in the radio. A variable capacitor consists of a series of ganged metal plates, as shown in the diagram. By rotating the plates, the effective area of the capacitor can be changed, thus changing the capacitance of the capacitor, which in turn changes the frequency of the tuned circuit. Show that the equivalent capacitance of the capacitor shown is

$$C = \frac{5\epsilon_0 A}{d}$$

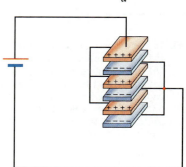

Interactive Tutorials

57. A parallel plate capacitor has plates that have an area $A = 83.5$ cm² and a plate separation $d = 6.00$ mm. The space between the plates is filled with air. The capacitor is connected to a 12.0-V battery. Find (a) the capacitance C of the capacitor, (b) the charge q deposited on its plates, and (c) the energy stored in the capacitor.

58. Capacitors in series. Three capacitors, $C_1 = 2.55 \times 10^{-6}$ F, $C_2 = 5.35 \times 10^{-6}$ F, and $C_3 = 8.55 \times 10^{-6}$ F are connected in series to a 100-V battery. Find (a) the equivalent capacitance, (b) the charge on each capacitor, (c) the voltage drop across each capacitor, (d) the energy stored in each capacitor, and (e) the total energy stored in the circuit.

59. Capacitors in parallel. Three capacitors, $C_1 = 2.55 \times 10^{-6}$ F, $C_2 = 5.35 \times 10^{-6}$ F, and $C_3 = 8.55 \times 10^{-6}$ F are connected in parallel to a 100-V battery. Find (a) the equivalent capacitance, (b) the total charge on the equivalent capacitor, (c) the voltage drop across each capacitor, (d) the charge on each capacitor, (e) the energy stored in each capacitor, and (f) the total energy stored in the circuit.

60. Combination of capacitors in series and parallel. Capacitor $C_1 = 2.55 \times 10^{-6}$ F is in series with capacitors $C_2 = 5.35 \times 10^{-6}$ F and $C_3 = 8.55 \times 10^{-6}$ C, which are in parallel with each other. The entire combination is connected to a 100-V battery. A similar schematic is shown in figure 21.6. Find (a) the equivalent capacitance of the combination, (b) the total charge q on the equivalent capacitor, (c) the charge q_1 on capacitor C_1, (d) the voltage drop across each capacitor, (e) the charge on capacitors C_2 and C_3, (f) the energy stored in each capacitor, and (g) the total energy stored in the circuit.

61. A capacitor with a dielectric. A parallel plate capacitor, having a plate area $A = 3.8 \times 10^{-3}$ m² and a plate separation $d = 1.75$ mm, is charged by a battery to $V_0 = 85.0$ V. The battery is then removed. Find (a) the capacitance of the capacitor and (b) the charge on the plates of the capacitor. A slab of mica is then placed between the plates of the capacitor. Find (c) the new value of the capacitance, (d) the potential difference across the capacitor with the dielectric, (e) the initial electric field between the plates, (f) the final electric field between the plates, (g) the initial energy stored in the capacitor, and (h) the final energy stored in the capacitor.

Magnetism

The magnetic field of a bar magnet.

What is the fundamental hypothesis of science, the fundamental philosophy? [It is the following:] *the sole test of the validity of any idea is experiment.*

Richard P. Feynman

22.1 The Force on a Charge in a Magnetic Field—The Definition of the Magnetic Field B

In addition to the existence of electric fields in nature, there are also magnetic fields. Most students have seen and played with a simple bar magnet, observing that the magnet attracts nails, paper clips, and the like. Some may have even placed the bar magnet under a piece of paper and sprinkled iron filings on the paper, observing the characteristic magnetic field of a bar magnet, figure 22.1. One end of the magnet is called a *north pole,* while the other end is called a *south pole.* The magnetic field is defined to emerge from the north pole of the magnet and enter at the south pole. A compass needle, a tiny bar magnet, placed in a magnetic field, aligns itself with the field. The designation of poles as north and south is arbitrary, just as electric charges are arbitrarily called positive and negative. Just as the combination of a positive and a negative electric charge is called an electric dipole, a bar magnet, consisting as it does of a north and a south magnetic pole, is sometimes called a *magnetic dipole.* The force between magnets is similar to the force between electric charges and can be stated as the ***fundamental principle of magnetism:*** *like magnetic poles repel, while unlike magnetic poles attract.* The earth has a magnetic field and when the north pole of a compass needle points in a northerly direction on the surface of the earth, it is really being attracted toward a south magnetic pole, because unlike poles attract. Hence, what is usually called the north magnetic pole of the earth is really a south pole. (Because compass needles always point toward that pole, it is sometimes erroneously called the north magnetic pole. Thus a pilot or navigator would refer to it as magnetic north.) This south magnetic pole is displaced about 1300 miles from the north geographic pole of the earth. Similarly, when the south pole of a compass needle points in a southerly direction on the surface of the earth, it is really being attracted toward a north magnetic pole. The north magnetic pole is displaced about 1200 miles from the south geographic pole of the earth. We will use the symbol **B** to designate the magnetic field. Later in this chapter we will see how magnetic fields are generated, but for now let us accept the fact that magnetic fields do indeed exist.

Recall from chapter 19 that we determined the existence of an electric field and its strength by the effect it produced on a small positive test charge q_0 placed

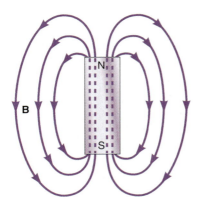

Figure 22.1

The magnetic field of a bar magnet.

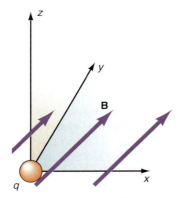

(a) Charge at rest—no force

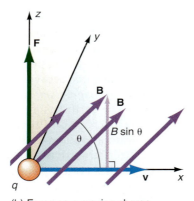

(b) Force on a moving charge

Figure 22.2

A charge in a magnetic field.

in the region where we assumed the field to exist. If the test charge experienced a force, we said that the charge was in an electric field, and defined the electric field as

$$E = \frac{F}{q_0} \qquad (19.1)$$

It is desirable to define the magnetic field in a similar way. A positive charge q is placed at rest in a uniform magnetic field **B,** as shown in figure 22.2(a). But to our surprise, nothing happens to the charge. We observe no force acting on the charge. The experiment is repeated, but now the charge is fired into the magnetic field with a velocity **v.** Now we observe that a force does indeed act on the charge figure 22.2(b). *The force, however, acts at an angle of 90° to the plane determined by the velocity vector v and the magnetic field B.* In fact, it is found experimentally that *the magnitude of the force **F** acting on the moving charge is given by the product of qv, and the component of **B** that is perpendicular to v.*

$$F = qvB_\perp$$

The perpendicular component of **B,** found from figure 22.2, is

$$B_\perp = B \sin \theta$$

The magnitude of the force is therefore given by

$$F = qvB \sin \theta \qquad (22.1)$$

The angle θ is the angle between the velocity vector **v** and the magnetic field vector **B.** An easy way to remember the direction of the force on a charge in a magnetic field is to use *the right-hand rule. Point the fingers of your right hand in the direction of the velocity vector v with your palm facing in the direction of the magnetic field vector B. Rotate your right hand from v to B and your thumb will point in the direction of the force vector F,* in this case upward. If the angle θ between *v* and **B** is 90° then the magnitude of the force simplifies to

$$F = qvB \qquad (22.2)$$

We now use this result to define the magnitude of the magnetic field B as

$$B = \frac{F}{qv} \qquad (22.3)$$

This definition is now similar to the definition of the magnitude of the electric field. *The magnitude of the **magnetic field B,** called the **magnetic induction** or the **magnetic flux density,** is defined as the force per unit charge, per unit velocity, provided that v is perpendicular to B.* The SI unit for the magnetic induction, defined from equation 22.3, is a **tesla,** named after Nikola Tesla (1856–1943), where

$$1 \text{ tesla} = 1 \frac{\text{newton}}{\text{coulomb m/s}}$$

abbreviated as

$$1 \text{ T} = 1 \frac{\text{N}}{\text{C m/s}}$$

To get an idea of the size of the tesla, note that the earth's magnetic field is about 1/20,000 tesla. The relation of the tesla with other equivalent units for the magnetic field is

$$\text{tesla} = \frac{\text{N}}{\text{A m}} = \frac{\text{weber}}{\text{m}^2} = 10^4 \text{ gauss}$$

The gauss is an older cgs unit that is still occasionally used because of its convenient size. The weber is a unit of magnetic flux that will be discussed in chapter 23.

Magnetic field pattern of a bar magnet as shown by iron fillings on a piece of paper.

From the definition of the magnetic force it is obvious that if the charge has a velocity that is parallel to the magnetic field, then the angle θ is zero, and hence

$$F = qvB \sin 0° = 0 \quad \text{(for } \mathbf{v} \parallel \mathbf{B})$$

Of course, if the velocity of the charge is zero then the force is also zero. The magnetic field manifests itself to a charge only when the charge is in motion with respect to the field. The maximum force occurs when $\mathbf{v}$ is at an angle of 90° to $\mathbf{B}$, as shown in equation 22.2. We should also note that equation 22.1 was defined with q being a positive charge. If the charge in motion is a negative particle, such as an electron, q is negative and the force on the negative particle is in the opposite direction to the force on the positive particle.

Example 22.1

The force on a positive charge in a magnetic field. A proton is fired into a uniform magnetic field $\mathbf{B}$, of magnitude 0.500 T, at a speed of 300 m/s at an angle of 30.0° to $\mathbf{B}$. Find the force and the acceleration of the proton.

Solution

The magnitude of the force acting on the proton, found from equation 22.1, is

$$F = qvB \sin \theta$$
$$= (1.60 \times 10^{-19} \text{ C})(300 \text{ m/s})(0.500 \text{ T}) \sin 30.0°$$
$$= 1.2 \times 10^{-17} \text{ C (m/s)T}\left[\frac{\text{N/(C m/s)}}{\text{T}}\right]$$
$$= 1.20 \times 10^{-17} \text{ N}$$

Note that we have used the conversion factor for a tesla to make the force come out in the unit of newtons.

The direction of the force is perpendicular to the plane of $\mathbf{v}$ and $\mathbf{B}$, as shown in figure 22.2(b). The magnitude of the acceleration of the proton, found from Newton's second law, is

$$a = \frac{F}{m_p} = \frac{1.20 \times 10^{-17} \text{ N}}{1.67 \times 10^{-27} \text{ kg}} = 7.19 \times 10^9 \text{ m/s}^2$$

As long as the particle stays within the magnetic field, at the same angle θ, the magnitude of the force and the magnitude of the acceleration is a constant.

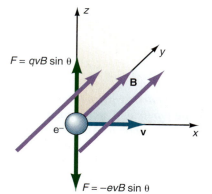

Figure 22.3

Force on a negative charge in a magnetic field.

Example 22.2

A force on an electron in a magnetic field. An electron is fired into a uniform magnetic field with a velocity $\mathbf{v}$, as shown in figure 22.3. Find the direction of the magnetic force.

Solution

The direction of the force is perpendicular to the plane of $\mathbf{v}$ and $\mathbf{B}$. However, since the magnitude of the force is given by

$$F = qvB \sin \theta$$

and the electron has a negative charge, $q = -e$, the force on the electron is

$$F = -evB \sin \theta$$

As we can see from figure 22.3, the force **F** points in the negative direction and points downward. (Recall that the direction of the force on a charge in a magnetic field is found by the right-hand rule. That is, point the fingers of your right hand in the direction of the velocity vector **v** with your palm facing the magnetic field vector **B**. Rotate your right hand from **v** to **B** and your thumb will point in the direction of the force vector, in this case upward.) But since the force is minus e times $vB \sin \theta$, for the negative charge, the force **F** points downward.

Because a charge can experience a force in either an electric or a magnetic field, the total force on a particle in an electromagnetic field is the superposition of both forces, or

$$\mathbf{F} = \mathbf{F_e} + \mathbf{F_m} \tag{22.4}$$

where

$$F_e = qE$$

and

$$F_m = qvB \sin \theta$$

This relation is known as the **Lorentz force** law.

An interesting result occurs if we look at the work done on the particle in such a field. The work is defined as

$$W = Fx \cos \theta$$

But the total force acting on the charge was given by equation 22.4. Hence the work done by the Lorentz force is

$$W = F_e x \cos \theta_1 + F_m x \cos \theta_2 \tag{22.5}$$

The angle θ_1 is the angle between the direction of the electric force F_e and the displacement x, whereas the angle θ_2 is the angle between the direction of the magnetic force F_m and the displacement x. Let us consider the second term. Since x is the displacement of the moving charge from its initial position to its next position, it is always in the direction of **v** at any point. Since the magnetic force is perpendicular to **v** it is also perpendicular to the displacement x. The angle θ_2 between the magnetic force F_m and the displacement x is therefore 90°. Hence, the work done by the magnetic force is zero. That is,

$$W_m = F_m x \cos 90° = 0$$

Therefore, equation 22.5 reduces to

$$W = W_e = F_e x \cos \theta_1$$

or

$$W = qEx \cos \theta_1 \tag{22.6}$$

Equation 22.6 says that only the work done on the particle by the electric field can change the energy, and hence the speed, of the particle. The magnetic field can only change the direction of the velocity vector, but not its speed. Therefore, a particle moving in a magnetic field only, always moves at a constant speed.

If a charged particle q enters a uniform magnetic field **B** with a velocity **v**, as shown in figure 22.4, the magnetic force **F** acts on the particle, deflecting it from its straight line motion. The particle follows the curved path until it emerges from the magnetic field. It then moves in a straight line at a constant velocity. Uniform magnetic fields are sometimes used in this way to deflect charged particles in a cathode ray oscilloscope. (Note that in figure 22.4 the symbol X represents the tail

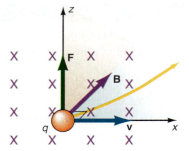

Figure 22.4

Deflecting a charged particle in a magnetic field.

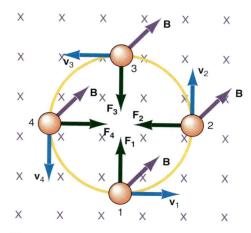

Figure 22.5

Deflecting a charged particle into a circular path.

of the arrow of the vector **B**, indicating that **B** is *into the page*. If **B** were *out of the page*, it would be represented by dots (·), indicating that the tip of the arrow of the vector **B** is coming out of the page.)

If the uniform magnetic field covers a large enough area, and the velocity vector makes an angle of 90° with the magnetic field, the particle stays in the magnetic field and moves in a circle. Consider the particle, of charge q, entering the uniform magnetic field **B** in figure 22.5. The magnetic field is everywhere into the paper and is perpendicular to **v**. When the charge is at position 1, moving at a velocity v_1, it experiences a force F_1, which acts upward. (Recall that the direction of the force on a charge in a magnetic field is found by the right-hand rule. Point the fingers of your right hand in the direction of the velocity vector **v** with your palm facing the magnetic field vector, **B**. Then rotate your right hand from **v** to **B** and your thumb will point in the direction of the force vector.) This force deflects the charge from its straight line motion and it follows the yellow path. All along that path a force is always acting perpendicular to the path. When the charge is at position 2, the force on it is now toward the left, again deflecting the direction of motion of the charge as shown. At position 3, the force acts downward, whereas at position 4 it acts toward the right. The particle is again deflected until it returns to its initial position 1. At every point on the trajectory of the particle, the force always acts perpendicular to the velocity vector **v**. Hence in this case, the magnetic force $F_m = qvB \sin \theta$ is a centripetal force. The particle moves in a circular orbit in the uniform magnetic field at constant speed v. (The speed v remains constant because only an electric field can change the particle's speed.)

The magnetic force supplies the necessary centripetal force for the particle to move in the circular path, that is,

$$F_c = F_m$$

In this case **v** and **B** are perpendicular and hence $\theta = 90°$, the $\sin 90° = 1$, and

$$F_m = qvB \qquad (22.7)$$

Equating the centripetal force, equation 6.14, to the magnetic force, equation 22.7, gives

$$\frac{mv^2}{r} = qvB \qquad (22.8)$$

Solving equation 22.8 for r, the radius of the orbit, gives

$$r = \frac{mv}{qB} \qquad (22.9)$$

Since mv is the momentum of the charged particle, the radius of the orbit is directly proportional to the momentum of the particle. The larger the momentum, the larger the circular orbit. The orbital radius is inversely proportional to the charge of the particle q and the magnetic field B. A large magnetic field produces a small circular orbit. If the angle θ between v and B is not 90°, a component of the velocity will be in the direction of the magnetic field. In that case, the motion does not remain in a plane and the trajectory is a helix in three dimensions.

Example 22.3

The radius of the orbit of an electron in a magnetic field. Find the radius of the orbit of an electron moving at a speed of 2.00×10^7 m/s in a uniform magnetic field of 1.20×10^{-3} T.

The radius of the orbit, found from equation 22.9, is

$$r = \frac{mv}{qB} = \frac{(9.11 \times 10^{-31} \text{ kg})(2.00 \times 10^7 \text{ m/s})}{(1.61 \times 10^{-19} \text{ C})(1.20 \times 10^{-3} \text{ T})}$$

$$= \left(9.43 \times 10^{-2} \frac{\text{kg m/s}}{\text{C T}}\right)\left(\frac{\text{T}}{\frac{\text{N}}{\text{C m/s}}}\right)\left(\frac{\text{N}}{\text{kg m/s}^2}\right)$$

$$= 9.43 \times 10^{-2} \text{ m}$$

$$= 9.43 \text{ cm}$$

Note how the conversion factors were used to give the radius in the correct units.

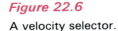

If an electric and a magnetic field are superimposed at right angles to each other, as shown in figure 22.6, the arrangement acts as a **velocity selector.** The force on charge q, given by the Lorentz force, is

$$\mathbf{F} = \mathbf{F_e} + \mathbf{F_m} \qquad (22.4)$$

The magnetic force, $F_m = qvB \sin \theta$, acts upward while the electric force $\mathbf{F_e} = q\mathbf{E}$ acts downward. If the electric force is exactly equal in magnitude to the magnetic force, the forces cancel each other out and the net force $\mathbf{F}$ on the charged particle is zero. The particle is not deviated from its straight line motion. The requirement for the velocity to be undeviated as it moves through the combined fields, obtained from equation 22.4, is

$$0 = -qE + qvB \sin 90°$$

$$qE = qvB$$

and

$$v = \frac{E}{B} \qquad (22.10)$$

Therefore, particles whose speed v is given by equation 22.10 pass through the combined fields undeflected and pass through the slit in the screen in figure 22.6. All particles with a different velocity have a net force acting on them and are deflected from their straight line motion and do not pass through the slit in the screen.

Figure 22.6

A velocity selector.

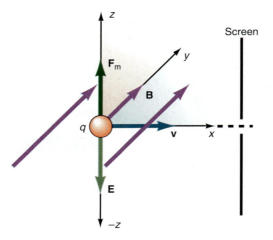

Example 22.4

A velocity selector. Alpha particles ranging in speeds from 1000 m/s to 2000 m/s enter an electromagnetic field, as in figure 22.6, where the electric intensity is 300 V/m and the magnetic induction is 0.200 T. Which particles will move undeviated through the field?

Solution

The alpha particles moving at a speed given by equation 22.10 are selected to pass straight through the field, that is,

$$v = \frac{E}{B}$$

$$= \frac{300 \text{ V/m}}{0.200 \text{ T}}$$

$$= 1500 \text{ m/s} \tag{22.10}$$

Only particles moving at this speed move straight through the electromagnetic field; all others are deflected.

22.2 Force on a Current-Carrying Conductor in an External Magnetic Field

If a wire carrying a current I is placed in an external magnetic field **B,** as shown in figure 22.7, a force acts on the wire. This force results from the magnetic force acting on a charged particle in a magnetic field. If the wire is carrying a current, then there are charges in motion within the wire. These charges are moving with a drift velocity v_d in the direction of the current flow. Any one of these charges q experiences the force

$$F_q = qv_dB \sin \theta \tag{22.11}$$

where θ is the angle between the drift velocity v_d and the direction of the magnetic field **B.** This force on an individual charge causes the charge to interact with the lattice structure of the wire, exerting a force on the lattice and hence on the wire itself. We can write the drift velocity of the moving charge as

$$v_d = \frac{\ell}{t}$$

where ℓ is a small length of the wire in the direction of the current, as shown in figure 22.7, and t is the time. Substituting this drift velocity into equation 22.11, gives

$$F_q = q\left(\frac{\ell}{t}\right)B \sin \theta = \frac{q}{t}\ell B \sin \theta$$

where θ is now the angle between ℓ and **B,** which is the same angle between v_d and **B.** The net force on the wire is the sum of the individual forces associated with each charge carrier, that is,

$$F = \Sigma_q F_q = \Sigma_q \frac{q}{t}\ell B \sin \theta$$

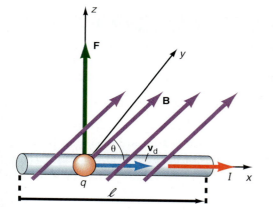

Figure 22.7

Force on a current-carrying wire in an external magnetic field.

But $\Sigma_q\, q/t$ is equal to all the charges passing through a plane of the wire per unit time and is defined as the current in the circuit I. Hence *a wire carrying a current I in any external magnetic field B, experiences a force given by*

$$F = I\ell B \sin\theta \qquad\qquad (22.12)$$

The direction of the force is again found by the right-hand rule. Hence, point the fingers of your right hand in the direction of the length vector $\boldsymbol{\ell}$, which is in the direction of the current, with your palm facing the magnetic field vector **B**, which points into the page in figure 22.7. Rotate your right hand from $\boldsymbol{\ell}$ to **B** and your thumb points in the direction of the force vector, in this case upward. If the direction of the current flow is reversed, $\boldsymbol{\ell}$ is reversed and rotating the first vector $\boldsymbol{\ell}$ toward the second vector **B**, your thumb now points downward.

If equation 22.12 is solved for B, another set of units for the magnetic induction B is determined, namely

$$B = \frac{F}{I\ell}$$

Thus,

$$1\ \text{tesla} = \frac{\text{newton}}{\text{ampere meter}}$$

abbreviated as

$$1\ \text{T} = 1\ \frac{\text{N}}{\text{A m}}$$

Example 22.5

The force on current-carrying wire in a magnetic field. A 20.0-cm wire carrying a current of 10.0 A is placed in a uniform magnetic field of 0.300 T, as shown in figure 22.7. If the wire makes an angle of 40.0° with the vector **B**, find the direction and magnitude of the force on the wire.

Solution

The direction of the force is found by rotating the vector $\boldsymbol{\ell}$ toward the vector **B**. The thumb points upward, indicating that the direction of the force is also upward. The magnitude of the force, found from equation 22.12, is

$$F = I\ell B \sin\theta$$
$$= (10.0\ \text{A})(0.200\ \text{m})(0.300\ \text{T}) \sin 40.0°$$
$$= 0.386\ \text{A m T}\left(\frac{\text{N/A m}}{\text{T}}\right)$$
$$= 0.386\ \text{N}$$

22.3 Generation of a Magnetic Field

Although magnets have been known for hundreds of years, it was not until 1820 that Hans Christian Oersted (1777–1851) discovered a relation between electric currents and magnetic fields. If a series of compasses are placed around a wire that is not carrying a current, all the compass needles point toward the north, the direction of the earth's magnetic field, as shown in figure 22.8(a). If, however, a current I is sent through the wire, the compass needles no longer point to the north. Instead they point in a direction that is everywhere tangential to a circle drawn around the wire, passing through each compass, as shown in figure 22.8(b). Because a compass needle always aligns itself in the direction of a magnetic field, the current

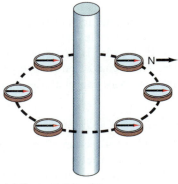

(a) No current in wire

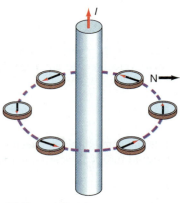

(b) Current I in wire

Figure 22.8

The creation of a magnetic field by an electric current.

in the wire has created a circular magnetic field directed counterclockwise around the wire. If the direction of the current is reversed, the direction of the magnetic field is also reversed and all the compass needles point in a clockwise direction. *The direction of the magnetic field around a long straight wire carrying a current I is easily determined by the so-called **right-hand wire rule.**[1] Grasp the wire with the right hand, with the thumb in the direction of the current flow, the fingers will curl around the wire in the direction of the magnetic field.*

The observation that electric currents can create a magnetic field was responsible for linking the then two independent sciences of electricity and magnetism into the one unified science of electromagnetism. In fact, we will see later that all magnetic fields are caused by the flow of electric charge.

22.4 The Biot-Savart Law

*The **Biot-Savart law** relates the amount of magnetic field ΔB produced by a small element $\Delta\ell$ of a wire carrying a current I and is given by[2]*

$$\Delta B = \frac{\mu I \Delta\ell}{4\pi r^2}\sin\theta \qquad (22.13)$$

and is shown in figure 22.9(a). Here θ is the angle between the small element of length $\Delta\ell$ and the position vector **r**. The Greek letter μ is a constant called the *permeability of the medium.* In a vacuum or air, it is called the permeability of free space, and is denoted by μ_0, where

$$\mu_0 = 4\pi \times 10^{-7}\,\frac{\text{T m}}{\text{A}}$$

The direction of ΔB is found by the right-hand rule by rotating the vector $\Delta\ell$ toward the vector **r**, and the thumb will point in the direction of ΔB, and is shown in figure 22.9(a). The Biot-Savart law says that the small current-carrying element $\Delta\ell$ produces a small amount of magnetic field ΔB at the point P. But the entire length of wire can be cut up into many $\Delta\ell$'s, and each contributes to the total magnetic field at **P**, as is shown in figure 22.9(b). Therefore, *the total magnetic field at point P is the vector sum of all the ΔB's associated with each current element,* that is,

$$\mathbf{B} = \Sigma\,\Delta\mathbf{B} \qquad (22.14)$$

The computation of the magnetic field from equation 22.14 can be quite complicated for most problems because of the vector sum. In general, the summation sign in equation 22.14 should be replaced by an integral sign and the determination of the magnetic field for most cases entails the use of the calculus. However, some problems can be easily solved by the Biot-Savart law, and we will solve one in section 22.5.

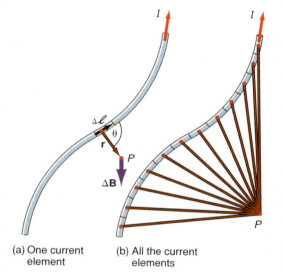

(a) One current element (b) All the current elements

Figure 22.9

The magnetic field produced by a current element.

22.5 The Magnetic Field at the Center of a Circular Current Loop

To determine the magnetic field at the center of a circular current loop, figure 22.10, we use the Biot-Savart law. A small element of the wire $\Delta\ell$ produces an element of magnetic field ΔB at the center of the wire. But ΔB points upward at the center of the circle for every current element, as seen in figure 22.10. The total magnetic field **B**, which is the sum of all the ΔB's, must also point upward at the center of the

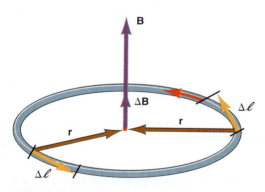

Figure 22.10

The magnetic field at the center of a circular current loop.

1. Note that this is not the same right-hand rule we use to show the direction of force on a charge or a current-carrying wire in a magnetic field.
2. It should be noted that in some books the constant $\mu/4\pi$ in equation 22.13 is written as the constant k', so that the Biot-Savart law can also be written in the form

$$\Delta B = \frac{k' I \Delta\ell}{r^2}\sin\theta$$

loop. Therefore, the magnetic field at the center of the current loop is perpendicular to the plane formed by the loop, and points upward. Since we now know the direction of the vector **B** we can reduce equation 22.14 to the scalar form

$$B = \Sigma \, \Delta B \tag{22.15}$$

The magnitude of ΔB, found from equation 22.13, is

$$\Delta B = \frac{\mu_0 I \Delta \ell \sin \theta}{4\pi r^2}$$

Since $\Delta \ell$ is perpendicular to **r**, the angle θ is equal to 90°, and the sin 90° = 1. Therefore, ΔB becomes

$$\Delta B = \frac{\mu_0 I \Delta \ell}{4\pi r^2} \tag{22.16}$$

Substituting equation 22.16 into equation 22.15 gives the magnitude of the magnetic induction **B** as

$$B = \Sigma \, \Delta B = \Sigma \, \frac{\mu_0 I \Delta \ell}{4\pi r^2}$$

Because the loop is a circle of constant radius r and $\mu_0 I/4\pi$ is a constant, these terms are in every term of the summation and can be factored out of the summation to yield

$$B = \frac{\mu_0 I}{4\pi r^2} \, \Sigma \, \Delta \ell \tag{22.17}$$

But the summation of all the $\Delta \ell$'s is simply the circumference of the wire, that is,

$$\Sigma \, \Delta \ell = 2\pi r$$

Therefore, equation 22.17 becomes

$$B = \frac{\mu_0 I}{4\pi r^2}(2\pi r)$$

Cancelling like terms, this becomes

$$B = \frac{\mu_0 I}{2r} \tag{22.18}$$

Equation 22.18 gives the magnitude of the magnetic field at the center of a circular current loop and, as seen in figure 22.10, the direction of the magnetic field is upward. Notice that the magnetic field at the center of the circular current loop is directly proportional to the current I—the larger the current, the larger the magnetic field—and inversely proportional to the radius of the loop—the larger the radius, the smaller the magnetic field. If the loop has N turns of wire, the magnetic field at the center is

$$B = \frac{\mu_0 N I}{2r} \tag{22.19}$$

The magnetic field just found is the magnetic field at the *center* of the current loop; the magnetic field *around* the loop is shown in figure 22.11. Note that it looks something like the magnetic field of a bar magnet, where the top of the loop would be the north pole.

Example 22.6

The magnetic field at the center of a circular current loop. Find the magnetic field at the center of a circular current loop of 0.500 m radius, carrying a current of 7.00 A.

Figure 22.11

The magnetic field of a current loop.

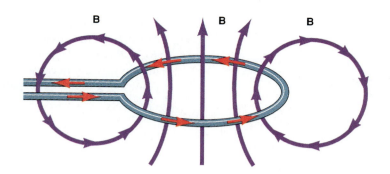

Solution

The magnetic field at the center of the loop, found from equation 22.18, is

$$B = \frac{\mu_0 I}{2r} = \left(\frac{4\pi \times 10^{-7} \text{ T m}}{\text{A}}\right)\frac{(7.00 \text{ A})}{2(0.500 \text{ m})}$$

$$= 8.80 \times 10^{-6} \text{ T}$$

Example 22.7

A circular loop of N turns. Find the magnetic field at the center of a circular current loop of 10 turns, with a radius of 5.00 cm, carrying a current of 10.0 A.

Solution

The magnetic field, found from equation 22.19, is

$$B = \frac{\mu_0 NI}{2r} = \left(\frac{4\pi \times 10^{-7} \text{ T m}}{\text{A}}\right)\frac{(10)(10.0 \text{ A})}{2(0.0500 \text{ m})}$$

$$= 1.26 \times 10^{-3} \text{ T}$$

22.6 Ampère's Circuital Law

Although the Biot-Savart law can be used to determine the magnetic field for different current distributions, another technique for the computation of magnetic fields is Ampère's circuital law. *Ampère's law* states: *along any arbitrary path encircling a total current I, the sum of the product of the component of the magnetic field that is parallel to the path $B_\parallel$ with the element of length $\Delta\ell$ of the path, is equal to the permeability μ_0 times the total current I enclosed by the path.* That is,

$$\Sigma B_\parallel \Delta\ell = \mu_0 I \tag{22.20}$$

Note that Ampère's law is a fundamental law based on experiments and cannot be derived. Ampère's law is especially helpful in problems with symmetry and will be used in the following subsections.

The Magnetic Field around a Long Straight Wire

To determine the magnetic field around a long straight wire by the Biot-Savart law is a little complicated because it entails the use of the calculus. However, because of the symmetry of the magnetic field around a long straight wire, we can find a simple solution to the magnetic field using Ampère's law. We noted in section 22.3 that the magnetic field around a long straight wire was found by experiment to be circular, as shown in figure 22.12(a). To apply Ampère's law, we must draw an arbitrary path around the wire containing the current I. The most symmetrical path

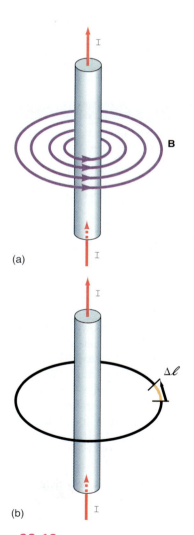

(a)

(b)

Figure 22.12

The magnetic field around a long straight wire.

that can be drawn about the wire is a concentric circle, as shown in figure 22.12(b). We divide this circular path into elements $\Delta \ell$ and then compute the component $B_\parallel = B \cos \theta$, where θ is the angle between $\mathbf{B}$ and $\Delta \ell$, for each element $\Delta \ell$, and then multiply by $\Delta \ell$. Because the magnetic field is circular, $\mathbf{B}$ and $\Delta \ell$ are parallel at every point along the circular path and θ is equal to 0. Ampère's law becomes

$$\Sigma B_\parallel \Delta \ell = \Sigma B \Delta \ell \cos \theta = \mu_0 I$$

$$\Sigma B \Delta \ell \cos 0 = \Sigma B \Delta \ell = \mu_0 I$$

$$B \Sigma \Delta \ell = \mu_0 I$$

But the sum of all the $\Delta \ell$'s (i.e., $\Sigma \Delta \ell$) is the circumference of the circular path. Hence,

$$\Sigma \Delta \ell = 2\pi r$$

Therefore,

$$B(2\pi r) = \mu_0 I$$

Thus, *the magnitude of the magnetic field around a long straight wire is*

$$B = \frac{\mu_0 I}{2\pi r} \tag{22.21}$$

Note that if there is more than one long straight wire present, the total magnetic field at a point will be equal to the vector sum of the magnetic fields of each of the wires. (As an example, see problem 25.)

Example 22.8

The magnetic field of a long, straight wire. A long straight wire is carrying a current of 15.0 A. Find the magnetic field 30.0 cm from the wire.

Solution

The magnetic field around the wire, found from equation 22.21, is

$$B = \frac{\mu_0 I}{2\pi r} = \frac{(4\pi \times 10^{-7} \text{ T m/A})(15.0 \text{ A})}{2\pi(0.300 \text{ m})}$$

$$= 1.00 \times 10^{-5} \text{ T}$$

The Magnetic Field inside a Solenoid

A *solenoid* is a long coil of wire with many turns and is shown schematically in figure 22.13(a). Note that the magnetic field of a solenoid looks like the magnetic field of a bar magnet. The magnetic field is uniform and intense within the coils, but is so small outside the coil, compared with the intensity inside the coil, that as an approximation the magnetic field is taken to be zero outside the coil. The magnetic field inside the solenoid lies along the axis of the coil, as seen in figure 22.13(a). The value of B inside the solenoid is found from Ampère's law, by adding the values of $B\Delta \ell \cos \theta$ along the rectangular path $ABCD$ in figure 22.13(b). That is,

$$\Sigma B \Delta \ell \cos \theta = \mu_0 I$$

$$B\ell_{AB} \cos \theta_1 + B\ell_{BC} \cos \theta_2 + B\ell_{CD} \cos \theta_3 + B\ell_{DA} \cos \theta_4 = \mu_0 I_{\text{total}} \tag{22.22}$$

But since $\mathbf{B} \approx 0$ outside the solenoid,

$$B\ell_{CD} \cos \theta_3 = 0 \tag{22.23}$$

Because $\mathbf{B}$ is perpendicular to the paths $\boldsymbol{\ell}_{BC}$ and $\boldsymbol{\ell}_{DA}$, the angles θ_2 and θ_4 are equal to 90°. Therefore,

$$B\ell_{BC} \cos 90° = 0 \tag{22.24}$$

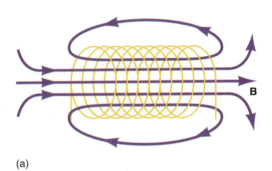

(a)

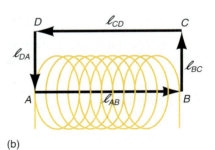

(b)

Figure 22.13

The magnetic field of a solenoid.

Electricity and Magnetism

and

$$B\ell_{DA} \cos 90° = 0 \qquad (22.25)$$

Now **B** is parallel to **ℓ**$_{AB}$, therefore their product becomes

$$B\ell_{AB} \cos 0° = B\ell_{AB} \qquad (22.26)$$

Substituting equations 22.23 through 22.26 into equation 22.22, gives

$$B\ell_{AB} = \mu_0 I_{\text{total}} \qquad (22.27)$$

The term I_{total} represents the total amount of current contained within the path *ABCD*. Each turn of wire carries a current I, but there are N turns of wire in the solenoid. Therefore, the total current contained within the path is

$$I_{\text{total}} = NI \qquad (22.28)$$

A more convenient unit for the number of turns of wire in the coil is the number of turns per unit length, n. Since the coil has a length ℓ_{AB}, the total number of turns N can be expressed as

$$N = n\ell_{AB} \qquad (22.29)$$

Substituting equations 22.29 and 22.28 into equation 22.27, yields

$$B\ell_{AB} = \mu_0 I_{\text{total}} = \mu_0 NI = \mu_0 n\ell_{AB}I$$

Simplifying, the magnetic field within the solenoid becomes

$$B = \mu_0 nI \qquad (22.30)$$

Equation 22.30 gives the magnitude of the magnetic field inside a solenoid and, as seen in figure 22.13, the direction of the magnetic field is along the axis of the solenoid. Notice that the magnetic field within the solenoid can be increased by increasing the current I in the wires, and/or by increasing the number of turns of wire per unit length.

Example 22.9

Magnetic field inside a solenoid. A solenoid 15.0 cm long is composed of 300 turns of wire. If there is a current of 5.00 A in the wire, what is the magnetic field inside the solenoid?

Solution

The number of turns of wire per unit length is

$$n = \frac{N}{\ell} = \frac{300 \text{ turns}}{0.150 \text{ m}} = 2000 \text{ turns/m}$$

The magnetic field inside the solenoid, found from equation 22.30, is

$$B = \mu_0 nI = \left(4\pi \times 10^{-7} \frac{\text{T m}}{\text{A}}\right)\left(\frac{2000 \text{ turns}}{\text{m}}\right)(5.00 \text{ A})$$

$$= 1.26 \times 10^{-2} \text{ T}$$

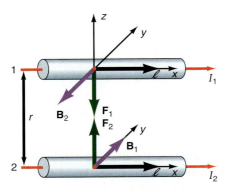

(a)

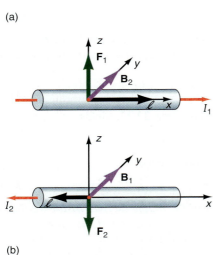

(b)

Figure 22.14

Force on parallel conductors carrying currents.

22.7 Force between Parallel, Current-Carrying Conductors— The Definition of the Ampere

Two conductors, carrying currents I_1 and I_2, respectively, are separated by a distance r, as shown in figure 22.14. The current I_2 in wire 2 produces a magnetic field $\mathbf{B_2}$ at the location of wire 1. Its magnitude, given by equation 22.21, is

$$B_2 = \frac{\mu_0 I_2}{2\pi r} \tag{22.31}$$

The direction of this magnetic field is found using the right-hand wire rule. If you grasp the wire with your thumb in the direction of the current, then your fingers curl in the direction of the magnetic field. In this case, $\mathbf{B_2}$ at the location of wire 1 has a direction coming out of the paper, as shown in figure 22.14(a). Since wire 1 is a current-carrying wire in the external magnetic field $\mathbf{B_2}$, it experiences a force given by equation 22.12 as

$$F_1 = I_1 \ell B_2 \sin \theta$$

Using the same technique as in section 22.2, the direction of the force is again found by the right-hand rule. The fingers of your right hand are pointed in the direction of the length vector $\boldsymbol{\ell}$, which is in the direction of the current, with your palm facing the magnetic field vector $\mathbf{B_2}$, which points out of the page in figure 22.14. Rotate your right hand from $\boldsymbol{\ell}$ to $\mathbf{B_2}$ and your thumb points in the direction of the force vector $\mathbf{F_1}$, which is downward in figure 22.14(a). Hence wire 1 experiences a force of attraction toward wire 2. The angle θ between $\boldsymbol{\ell}$ and $\mathbf{B_2}$ is 90° and since sin 90° = 1, the magnitude of $\mathbf{F_1}$ is

$$F_1 = I_1 \ell B_2$$

Substituting for the magnetic field B_2 from equation 22.31, gives

$$F_1 = I_1 \ell \left(\frac{\mu_0 I_2}{2\pi r} \right)$$

Rearranging terms, the force on wire 1 caused by wire 2 is

$$F_1 = \frac{\mu_0 \ell I_1 I_2}{2\pi r} \tag{22.32}$$

Just as there is a force on wire 1, there is also a force on wire 2, because the current flowing in wire 1 produces a magnetic field $\mathbf{B_1}$. Wire 2 finds itself in this external magnetic field and hence experiences the force

$$F_2 = I_2 \ell B_1 \sin \theta$$

The direction of the magnetic field $\mathbf{B_1}$ produced by wire 1 is found from the right-hand wire rule, and is shown going into the paper at the location of wire 2 in figure 22.14(a). The direction of the force $\mathbf{F_2}$ is found using the right-hand rule. The fingers of your right hand are pointed in the direction of the length vector $\boldsymbol{\ell}$, which is in the direction of the current. Your palm should be facing the magnetic field vector $\mathbf{B_1}$, which points into the page in figure 22.14. Rotate your right hand from $\boldsymbol{\ell}$ to $\mathbf{B_1}$ and your thumb points in the direction of the force vector $\mathbf{F_2}$, which is upward in figure 22.14(a). Hence the force $\mathbf{F_2}$ acting on wire 2 is one of attraction toward wire 1. Since $\boldsymbol{\ell}$ and $\mathbf{B_1}$ are perpendicular, the magnitude of the force $\mathbf{F_2}$ is

$$F_2 = I_2 \ell B_1 \tag{22.33}$$

The magnetic field B_1, found from equation 22.21, is

$$B_1 = \frac{\mu_0 I_1}{2\pi r} \tag{22.34}$$

Substituting equation 22.34 into 22.33, we obtain

$$F_2 = I_2 \ell \left(\frac{\mu_0 I_1}{2\pi r} \right)$$

After rearranging terms, the magnitude of the force on wire 2, caused by the current flowing in wire 1, is

$$F_2 = \frac{\mu_0 \ell I_1 I_2}{2\pi r} \qquad \qquad (22.35)$$

Observe from equations 22.32 and 22.35 that

$$F_1 = F_2$$

And, from figure 22.14,

$$\mathbf{F_1} = -\mathbf{F_2}$$

Thus the force on wire 1 is equal and opposite to the force on wire 2, as it should be according to Newton's third law.

If one of the currents is reversed, such as I_2 shown in figure 22.14(b), the magnitudes of the forces do not change but their directions do. Changing the direction of I_2, causes $\mathbf{B_2}$ to point into the page in figure 22.14(b). The direction of the force $\mathbf{F_1}$ is again found using the right-hand rule. The fingers of your right hand are pointed in the direction of the length vector $\boldsymbol{\ell}$. Rotate your right hand from $\boldsymbol{\ell}$ to $\mathbf{B_2}$ and your thumb points in the direction of the force vector $\mathbf{F_1}$, which is upward in figure 22.14(b), indicating that there is a repulsive force acting on wire 1. Because of the new direction of the current in wire 2, $\boldsymbol{\ell}$ points toward the left, and the force now acts downward in figure 22.14(b). Hence, the force $\mathbf{F_2}$ acts downward and is now a repulsive force.

In summary, *if the currents in the parallel wires act in the same direction, the force on the wires is attractive; if the currents in the wires act in opposite directions, the force on the wires is repulsive.*

In chapter 1, we noted that electric charge is a fundamental characteristic of matter and should be listed as the fourth fundamental quantity along with length, mass, and time. However, because electric charge is relatively difficult to measure, electric current, which is relatively easy to measure, is used as the fourth fundamental quantity. (Obviously charge is more fundamental than current in the description of nature because charges can exist at rest, which is equivalent to no current at all. That is, a charge can exist even when a current does not.) *The fundamental unit of electricity is defined as the* **ampere,** *where the ampere is that constant current that, if maintained in two straight parallel conductors of infinite length, of negligible circular cross section, and placed 1 m apart in a vacuum, would produce between these conductors a force equal to* 2×10^{-7} *N/m.*

This force per unit length can be obtained from equation 22.35 with $I_1 = I_2 = I$, and $r = 1$ m, as

$$\frac{F}{\ell} = \frac{\mu_0 I^2}{2\pi(1 \text{ m})}$$

For $I = 1$ A, the force is

$$\frac{F}{\ell} = \frac{(4\pi \times 10^{-7} \text{ T m/A})(1 \text{ A})^2}{2\pi(1 \text{ m})} = 2 \times 10^{-7} \text{ T A}$$

$$= (2 \times 10^{-7} \text{ T A}) \left[\frac{\text{N}/(\text{A m})}{\text{T}} \right]$$

$$= 2 \times 10^{-7} \frac{\text{N}}{\text{m}}$$

That is, the amount of current that causes this force is defined as the ampere.

Example 22.10

Force on parallel conducting wires. Two straight parallel conductors 30.0 cm long are separated by a distance of 10.0 cm. If one carries a current of 5.00 A to the right while the second carries a current of 7.00 A to the left, find the force on the first wire.

Solution

The magnitude of the force on the first wire, found from equation 22.32, is

$$F_1 = \frac{\mu_0 \ell I_1 I_2}{2\pi r}$$

$$= \frac{(4\pi \times 10^{-7} \text{ T m/A})(0.300 \text{ m})(5.00 \text{ A})(7.00 \text{ A})}{2\pi(0.100 \text{ m})}$$

$$= 2.10 \times 10^{-5} \text{ N}$$

and is repulsive.

22.8 Torque on a Current Loop in an External Magnetic Field— The Magnetic Dipole Moment

Let us place a rectangular coil of wire in a uniform magnetic field **B,** as shown in figure 22.15(a). Notice that the magnetic field emanates from the north pole of the magnet and enters the south pole of the magnet. A current, I, is set up in the coil by an external battery in the direction indicated. Any segment of the coil now represents a current-carrying wire in an external magnetic field and thus experiences a force on it given by equation 22.12,

$$F = I\ell B \sin \theta$$

Let us divide the coil into sections of length ℓ_{ab}, ℓ_{bc}, ℓ_{cd}, and ℓ_{da} and compute the force on each segment.

Segment ab: The force F_{ab} acting on segment $\boldsymbol{\ell}_{ab}$ is

$$F_{ab} = I\ell_{ab}B \sin \theta$$

Since $\boldsymbol{\ell}_{ab}$ is in the direction of the current in segment *ab*, it points downward in figure 22.15(a), while the magnetic field **B** points toward the right. The direction of the force is found by using the right-hand rule. Point the fingers of your right hand in the direction of the length vector **ℓ,** which in this case is downward, with the palm of your hand facing in the direction of the magnetic field vector **B,** which points to the right in figure 22.15. Rotate your right hand from **ℓ** to **B** and your thumb points in the direction of the force vector **F**$_{ab}$, which is outward as shown in the side view of the coil in figure 22.15(a) and from a top view in figure 22.15(b). The angle θ between **ℓ** and **B** is 90°. Therefore the magnitude of the force is

$$F_{ab} = I\ell_{ab}B \sin 90° = I\ell_{ab}B \tag{22.36}$$

Segment bc: The force F_{bc} acting on segment $\boldsymbol{\ell}_{bc}$ is

$$F_{bc} = I\ell_{bc}B \sin \theta$$

Figure 22.15

A coil in a magnetic field.

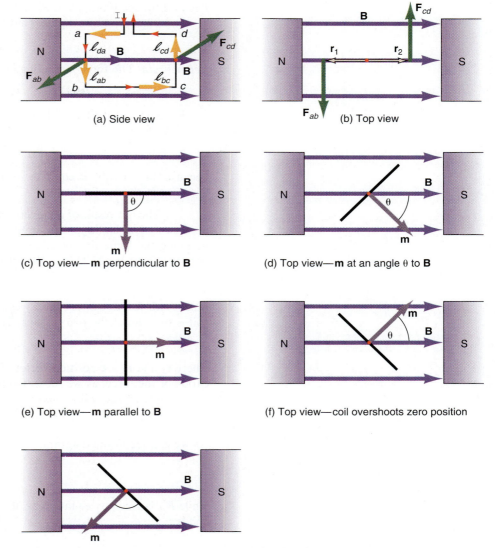

(a) Side view

(b) Top view

(c) Top view—**m** perpendicular to **B**

(d) Top view—**m** at an angle θ to **B**

(e) Top view—**m** parallel to **B**

(f) Top view—coil overshoots zero position

(g) Top view—reversing the current reverses the direction of **m**

Since $\boldsymbol{\ell}_{bc}$ is in the direction of the current flow, it points to the right in figure 22.15(a), parallel to the magnetic field **B**. Hence the angle between $\boldsymbol{\ell}_{bc}$ and B is zero. Therefore the force on wire bc is

$$F_{bc} = I\ell_{bc}B \sin 0° = 0$$

Thus, there is no force acting on the bottom of the wire coil in the configuration shown in figure 22.15(a).

Segment cd: The force F_{cd} acting on segment $\boldsymbol{\ell}_{cd}$ is

$$F_{cd} = I\ell_{cd}B \sin \theta$$

Since $\boldsymbol{\ell}_{cd}$ points upward and **B** points to the right in figure 22.15(a), upon rotating your right hand from $\boldsymbol{\ell}_{cd}$ to **B,** your thumb, and the force F_{cd}, points into the page in the side view of figure 22.15(a) and the top view of figure 22.15(b). The magnitude of F_{cd} is

$$F_{cd} = I\ell_{cd}B \sin 90° = I\ell_{cd}B \qquad (22.37)$$

Because the lengths ℓ_{ab} and ℓ_{cd} are equal, the forces, found from equations 22.36 and 22.37, are also equal. That is,

$$F_{ab} = F_{cd}$$

Segment da: The force F_{da} acting on the upper segment ℓ_{da}, is found from

$$F_{da} = I\ell_{da}B \sin \theta$$

But ℓ_{da} points to the left and makes an angle of 180° with the magnetic field **B.** Therefore,

$$F_{da} = I\ell_{da}B \sin 180° = 0$$

because the sin 180° is equal to zero. Hence, there is no force acting on the top wire in this configuration.[3]

The net result of a current in a coil in an external magnetic field is to produce two equal and opposite forces acting on the coil. But since the forces do not lie along the same line of action, the forces cause a torque to act on the coil as observed in figure 22.15(b). The torque acting on the coil is given by

$$\tau = rF \sin \theta$$

Since both forces can produce a counterclockwise torque, the total torque on the coil is the sum of the two torques. That is,

$$\tau = r_1 F_{ab} \sin \theta + r_2 F_{cd} \sin \theta$$

where r_1 and r_2 are position vectors from the axis of the coil to the point of application of the force, as we see in figure 22.15(b). Since $r_1 = r_2$ and $F_{ab} = F_{cd}$, the total torque on the coil is therefore the counterclockwise torque

$$\tau = 2r_1 F_{ab} \sin \theta$$

As we can see in the diagram,

$$r_1 = \frac{\ell_{bc}}{2}$$

while F_{ab} is given by equation 22.36. Therefore the magnitude of the torque on the coil is

$$\tau = 2r_1 F_{ab} \sin \theta = 2\frac{\ell_{bc}}{2}I\ell_{ab}B \sin \theta \qquad \textbf{(22.38)}$$

Note that in equation 22.38 the angle θ is the angle between the radius vector **r** and the force vector **F,** and the product of ℓ_{bc} and ℓ_{ab} is the area of the loop, that is

$$\ell_{bc}\ell_{ab} = A$$

Equation 22.38 simplifies to

$$\tau = IAB \sin \theta \qquad \textbf{(22.39)}$$

Although we have derived equation 22.39 for a rectangular current loop, it can be proved, using the calculus, that it is a perfectly general result for any planar loop having an area A, regardless of the shape of the loop. *The torque on the coil in a magnetic field depends on the current I in the coil, the area A of the coil, the intensity of the magnetic field B, and the angle θ between r and F_{ab}.*

3. Note that if the coil is oriented at an angle different than shown in figure 22.15(a), there is a force on the bottom and top wires of the coil. However the force on the bottom wire is downward and the force on the top wire is upward. The forces are equal and opposite and lie on the axis of rotation of the coil and do not create a torque to rotate the coil.

Figure 22.16

The magnetic dipole moment of a current loop.

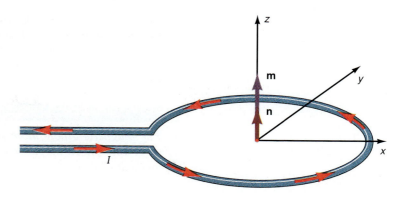

Since we saw in section 22.5 that there is a magnetic field at the center of a current loop, and this magnetic field looks like the magnetic field of a bar magnet, *a magnetic dipole moment* **m** *of a current loop is defined as*

$$\mathbf{m} = IA\mathbf{n} \tag{22.40}$$

The magnetic dipole moment is shown in figure 22.16. Here **n** is a unit vector that determines the direction of **m,** and is itself determined by the direction of the current. If the right hand curls around the loop in the direction of the current, the thumb points in the direction of the unit vector **n,** and hence, in the direction of the magnetic dipole moment **m.** Reversing the direction of the current reverses the direction of the magnetic dipole moment. If the coil consists of N loops of wire, the magnetic dipole moment is

$$\mathbf{m} = NIA\mathbf{n} \tag{22.41}$$

Figure 22.15(c) is a top view of the coil, showing the magnetic dipole moment **m** perpendicular to the coil. We can now write the torque acting on the coil, equation 22.39, in terms of the magnetic dipole moment of the coil:

$$\tau = mB \sin \theta \tag{22.42}$$

Note that the angle θ in equation 22.42, the angle between the magnetic dipole moment vector **m** and the magnetic field vector **B,** is the same angle θ that was between the radius vector **r** and the force vector **F.** When **m** is perpendicular to the magnetic field **B,** θ is equal to 90°, and sin 90° = 1. Therefore the torque acting on the coil is at its maximum value and acts to rotate the coil counterclockwise in figure 22.15(c). As the coil rotates, the angle θ decreases, figure 22.15(d), until the magnetic dipole moment becomes parallel to the magnetic field ($\theta = 0$), as shown in figure 22.15(e). At this point the torque acting on the coil becomes zero, because the sin θ term in equation 22.42 is zero. Because of the inertia of the coil, however, the coil does not quite stop at this position but overshoots it, as shown in figure 22.15(f). But now the torque is reversed and the coil rotates clockwise until **m** is again parallel to **B** and $\tau = 0$ again. The coil may oscillate one or two times but because of friction it eventually stops with its magnetic dipole moment parallel to the magnetic field. *In summary, the magnetic field causes a torque to act on the coil until the magnetic dipole moment of the coil is aligned with the magnetic field.*

This result should not come as too much of a surprise, because this is exactly what happens with a compass needle. The compass needle is a tiny bar magnet with a magnetic dipole moment. The earth's magnetic field acts on this dipole to align it with the earth's magnetic field. On the earth's surface, the earth's magnetic field points toward the north and the compass needle also points toward the north.

The torque vector τ is perpendicular to the plane of the rotation and lies along the axis of rotation. Its direction can be determined by the right-hand rule. Point the fingers of your right hand in the direction of the magnetic dipole moment

m, with your palm facing the magnetic field **B.** Rotate your right hand from **m** to **B.** Your thumb will point in the direction of the torque vector τ. When the torque vector τ points out of the plane of the rotation, the rotation will be counterclockwise. When the torque vector τ points into the plane of rotation, the rotation will be clockwise.

Example 22.11

The magnetic dipole moment of a coil. A circular coil, consisting of 10 turns of wire, 10.0 cm in diameter carries a current of 2.00 A. Find the magnetic dipole moment of the coil.

Solution

The area of the coil is

$$A = \frac{\pi d^2}{4} = \frac{\pi(0.100 \text{ m})^2}{4} = 7.85 \times 10^{-3} \text{ m}^2$$

The magnitude of the magnetic dipole moment of the coil, found from equation 22.41, is

$$m = NIA$$
$$= (10)(2.00 \text{ A})(7.95 \times 10^{-3} \text{ m}^2)$$
$$= 0.157 \text{ A m}^2$$

Notice that the unit for magnetic dipole moment is an ampere meter squared.

Example 22.12

The torque on a magnetic dipole in an external magnetic field. The coil of example 22.11 is placed in a uniform magnetic field of 0.500 T. Find the maximum torque on the coil.

Solution

The torque acting on the coil, found from equation 22.42, is

$$\tau = mB \sin \theta$$

and the maximum torque occurs when $\theta = 90°$. Therefore, the maximum torque acting on the coil is

$$\tau_{\text{max}} = mB = (0.157 \text{ A m}^2)(0.500 \text{ T})$$
$$= 7.85 \times 10^{-2} \text{ A m}^2 \text{ T} \left[\frac{\text{N/(A m)}}{\text{T}} \right]$$
$$= 7.85 \times 10^{-2} \text{ m N}$$

22.9 Applications of the Torque on a Current Loop in an External Magnetic Field

The fact that a torque acts on a current loop in a magnetic field gives rise to a number of important applications. Among them are (1) the DC motor, (2) the galvanometer, (3) the ammeter, and (4) the voltmeter.

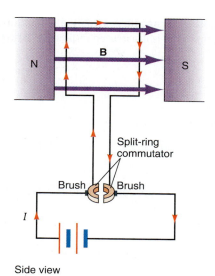

Figure 22.17

The DC motor.

The DC Motor

We saw in section 22.8 that when a current flows in a coil in an external magnetic field **B,** a torque arises, given by equation 22.42, that causes the magnetic dipole moment of the coil to be aligned with the magnetic field, as in figures 22.15(c), 22.15(d), and 22.15(e). Because of the inertia of the coil, the coil passes the aligned position, as shown in figure 22.15(f). Then the torque is reversed and the coil rotates back to the aligned position. When the coil overshoots the aligned position, figure 22.15(f), if the direction of the current were to simultaneously change, the direction of the magnetic dipole moment of the coil would be reversed, as shown in figure 22.15(g). The torque τ would again point upward, indicating that the coil will continue to rotate in a counterclockwise direction away from the aligned position. If we continue to change the direction of the current, and hence the direction of the magnetic dipole moment, every time the magnetic dipole moment passes the aligned position, the coil will rotate indefinitely. To accomplish this change in current every 180° of rotation we use a split-ring commutator, as shown schematically in figure 22.17. In the position shown, current from the battery flows through the coil as indicated. When the first split ring rotates past the brush, the other side of the coil gets connected to the positive terminal of the battery and the current flows in the opposite direction. The split ring is thus responsible for changing the direction of the current, and hence the direction of the magnetic dipole moment. The coil continues to rotate as long as the current flows and this is the essence of a DC motor.

The Galvanometer

A torque acting on a current loop in an external magnetic field is the basis for the d'Arsonval galvanometer named after the French physicist Jacques Arsène d'Arsonval (1851–1940). The galvanometer is used to measure current in a circuit. A simplified version of a galvanometer is shown schematically in figure 22.18. With a current I in the coil, a torque τ is produced in the coil, whose magnitude is given by

$$\tau = mB \sin \theta \qquad (22.42)$$

This torque causes the coil to rotate. A spring coil is attached to the bottom of the coil and causes an opposite reaction torque to be developed given by

$$\tau_{\text{spring}} = K\phi$$

where K is the torsional constant of the spring and ϕ is the rotational angle turned through by the coil. The coil rotates until there is an equilibrium between the torque

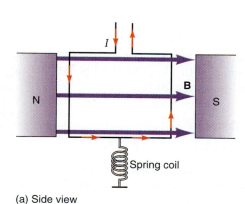

(a) Side view

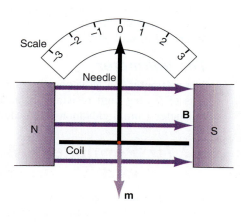

(b) Top view

(c)

Figure 22.18

The galvanometer.

Chapter 22   Magnetism

651

on the coil and the restoring torque of the spring. A needle is attached to the top of the coil and moves over a scale as the coil turns. This is illustrated in figure 22.18(b). At equilibrium,

$$\tau_{spring} = \tau_{coil}$$

Therefore,

$$K\phi = mB \sin \theta$$

In the position shown in figure 22.18, $\theta = 90°$, and $\sin 90° = 1$. Therefore,

$$K\phi = mB$$

But the magnitude of the magnetic dipole moment of the coil is found from equation 22.40 as

$$m = IA$$

Therefore, the angle that the coil turns through before coming to rest is

$$\phi = \frac{IAB}{K} = \left(\frac{AB}{K}\right)I \tag{22.43}$$

The angle ϕ thus depends on the area A of the coil, the magnetic field B, the torsional constant K of the spring, and the current flowing in the loop. Since A, B, and K are constants for a particular loop, *the angle that the coil turns through, equation 22.43, is directly proportional to the current flowing in the coil.* This gives us a means for measuring the current in a circuit. Because of the uniform magnetic field B shown in figure 22.18, equation 22.43 is good for only very small deflections. In actual galvanometers, a cylindrical core is placed within the coil, and the pole faces of the magnet are usually curved so as to supply a radial magnetic field. Thus the coil always has **m** perpendicular to **B,** the swing of the coil is independent of $\sin \theta$, and the swing is proportional to the current I throughout the range of the meter.

The galvanometer, however, reads only relative currents (that is, where I_1 is twice as great as I_0, I_2 is five times as large as I_0, etc.). The galvanometer must be calibrated to read a current in amperes. Most galvanometers are rated in terms of their electrical resistance R_g and the amount of current I_g that is necessary for a full-scale deflection of the galvanometer. A typical student galvanometer is shown in figure 22.18(c).

The Ammeter

The ammeter is a galvanometer that is calibrated to read currents directly. A low-resistance shunt is connected in parallel with the galvanometer to divert most of the current from flowing through the coil. Making an ammeter from a galvanometer is described in section 20.10.

The Voltmeter

The voltmeter is a galvanometer that is calibrated to read voltages directly. It consists of a very high resistance placed in series with the galvanometer. Making a voltmeter from a galvanometer is described in section 20.10.

22.10 Permanent Magnets and Atomic Magnets

The magnetic field of a bar magnet is shown in figure 22.19(a). The magnetic field emanates from the north pole of the magnet and enters at the south pole. The field is similar to the field of an electric dipole, shown in figure 22.19(b). The electric field emanates from the positive electric charge and terminates at the negative electric charge. Since both negative and positive electric charges can exist separately, it is reasonable to ask if magnetic poles can exist separately. The simplest test would

Figure 22.19

Some magnetic fields.

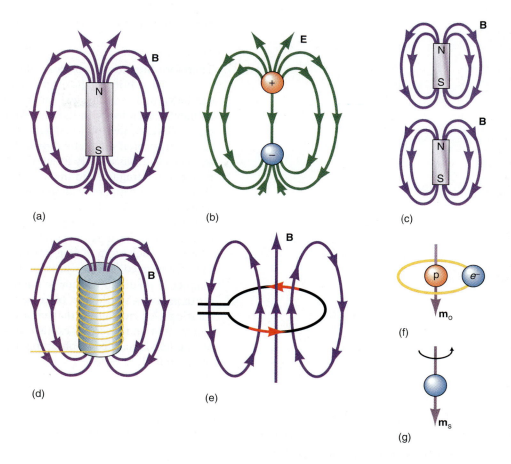

be to cut the bar magnet in figure 22.19(a) in half, expecting to obtain one isolated north pole and one isolated south pole. When we perform the experiment, however, two smaller bar magnets result, each with a north pole and a south pole, as seen in figure 22.19(c). No matter how many times we subdivide the bar magnet, we never obtain an isolated magnetic pole; we always get a dipole.

The magnetic field of a solenoid, figure 22.19(d), appears to be the same as the magnetic field of a bar magnet. So, perhaps all magnetic fields are caused by electric currents. The magnetic field of a single current loop is shown in figure 22.19(e). Looking at a picture of the atom, we find the negative electron orbiting around the positive nucleus. The orbiting electron, figure 22.19(f), looks exactly like a current loop, and there is, therefore, a *magnetic dipole moment of the atom,* caused by the orbital electron. This orbital magnetic dipole moment is

$$m_o = IA$$

The atomic current is equal to the charge on the electron divided by the time T for the electron to make one complete orbit, that is,

$$I = \frac{-e}{T} = -ef$$

where f, the reciprocal of the period, is the frequency, or the number of times the electron orbits the nucleus per second. Assuming the orbit to be a circle, it has an area

$$A = \pi r^2$$

Hence, the orbital magnetic moment of an atom is

$$m_o = IA = -ef\pi r^2$$

Therefore the atom itself looks like a tiny bar magnet.

In addition to the orbital magnetic moment of the atom, the electron, which can be viewed as a charged sphere spinning on its axis, also has a magnetic dipole moment associated with its spin. We represent the spin magnetic dipole moment by m_s. In general, the total magnetic dipole moment of an atom is equal to the vector sum of its orbital magnetic dipole moment m_o and its spin magnetic dipole moment m_s. The electrons usually fill up the shells of an atom with one spin magnetic dipole moment pointing up, and the next down. Thus, when an atom has completely closed shells, it has no magnetic dipole moment. Therefore, most chemical elements do not display magnetic behavior. In iron, cobalt, and nickel, the electrons do not pair off to permit their spin magnetic dipole moments to cancel. In iron, for example, five of its electrons have parallel spins, giving it a large resultant magnetic dipole moment. Therefore each atom of iron is a tiny atomic bar magnet. When these atomic magnets are aligned in a bar of iron, we have the common bar magnet. Thus, the bar magnet is made up of atomic currents. This is the reason we can never isolate a north pole or a south pole, because they do not exist. What does exist is an orbiting, rotating electric charge that creates a magnetic dipole moment. Some physicists believe that under certain exotic conditions magnetic monopoles might exist. To date no one has succeeded in observing one, but elementary particle physicists continue the search.

The Language of Physics

Fundamental principle of magnetism
Like magnetic poles repel each other; unlike magnetic poles attract each other (p. 631).

Magnetic field
The field of force in the neighborhood of a magnetized body, or a current-carrying wire. It is measured by the magnetic induction or magnetic flux density (p. 632).

Magnetic induction or magnetic flux density
Is equal to the force per unit charge per unit velocity that acts on a charge that is moving perpendicular to the magnetic field. It is also the force acting on a wire of unit length carrying a unit current, when placed in the magnetic field (p. 632).

Right-hand rule
An easy way to remember the direction of the force on a charge in a magnetic field. Point the fingers of your right hand in the direction of the velocity vector v with your palm facing the magnetic field vector B. Rotate your right hand from v to B and your thumb will point in the direction of the force vector F (p. 632).

Tesla
The SI unit for the magnetic field (p. 632).

Lorentz force
The total force acting on a moving charged particle is composed of the electric force and the magnetic force. The magnetic field can only change the direction of the velocity vector of the moving charge but not its speed. A charged particle moving in a magnetic field only, always moves at a constant speed (p. 634).

Velocity selector
A device formed from the superposition of an electric and a magnetic field that are at right angles to each other. By a suitable choice of the directions of the fields, the magnetic force acts upward while the electric force acts downward. If the electric force is exactly equal in magnitude to the magnetic force, the forces cancel each other out and the net force on a charged particle moving through this combined field is zero. The particle is not deviated from its straight line motion. The condition for a particle to pass through the combined fields undeflected is that the speed v of the particle must equal the ratio of E/B. Only particles moving at this speed move straight through the electromagnetic field, all others are deflected from their straight line motion (p. 636).

Right-hand wire rule
To determine the direction of the magnetic field around a wire carrying a current, grasp the wire with the right hand, with the thumb in the direction of the current, the fingers will curl around the wire in the direction of the magnetic field (p. 639).

Biot-Savart law
A law that relates the amount of magnetic field generated by a small element of wire carrying a current (p. 639).

Ampère's law
Along any arbitrary path encircling a total current, the sum of the product of the parallel component of the magnetic induction with the element of length of the path is equal to the permeability times the total current enclosed by the path (p. 641).

Ampere
The fundamental unit of electric current. It is that constant current that, if maintained in two straight parallel conductors of infinite length, of negligible circular cross section, and placed 1 m apart in a vacuum, would produce between these conductors a force equal to 2×10^{-7} N/m (p. 645).

Summary of Important Equations

Force on a charged particle in an external magnetic field

$$F = qvB \sin \theta \tag{22.1}$$

Definition of the magnetic induction

$$B = \frac{F}{qv} \tag{22.3}$$

Lorentz force where

$$\mathbf{F} = \mathbf{F_e} + \mathbf{F_m} \tag{22.4}$$

and

$$F_e = qE$$
$$F_m = qvB \sin \theta$$

Force on a current-carrying conductor in an external magnetic field

$$F = I\ell B \sin \theta \tag{22.12}$$

Biot-Savart law

$$\Delta B = \frac{\mu_0 I \Delta \ell}{4\pi r^2} \sin \theta \tag{22.13}$$

Total magnetic field

$$\mathbf{B} = \Sigma \Delta \mathbf{B} \tag{22.14}$$

Magnetic field at the center of a circular current loop

$$B = \frac{\mu_0 I}{2r} \tag{22.18}$$

Ampère's circuital law

$$\Sigma B_\parallel \Delta \ell = \mu_0 I \tag{22.20}$$

Magnetic field around a long straight wire

$$B = \frac{\mu_0 I}{2\pi r} \tag{22.21}$$

Magnetic field inside a solenoid

$$B = \mu_0 n I \tag{22.30}$$

Force on wire 1 caused by the magnetic field of wire 2

$$F_1 = \frac{\mu_0 \ell I_1 I_2}{2\pi r} \tag{22.32}$$

Torque on a current loop in an external magnetic field

$$\tau = IAB \sin \theta \tag{22.39}$$

Magnetic dipole moment of a current loop

$$\mathbf{m} = IA\mathbf{n} \tag{22.40}$$
$$\mathbf{m} = NIA\mathbf{n} \tag{22.41}$$

Torque on a current loop in an external magnetic field

$$\tau = mB \sin \theta \tag{22.42}$$

Questions for Chapter 22

†1. Should there be a law similar to Coulomb's law of electrostatics that shows the force between magnetic poles? What would be the advantages and disadvantages of such a law?

†2. In the very early days of nuclear physics, nuclear radiation was described in terms of alpha, beta, and gamma particles. How did Rutherford use a magnetic field to distinguish among these particles?

3. A charge in motion in a long straight wire with a drift velocity v_d generates a magnetic field around the wire. If you were to move parallel to the wire

at the same velocity, would you still observe a magnetic field? If not, where did the field go?

4. Since a moving electric charge creates a magnetic field, does moving a magnet create an electric field?

5. Electric fields begin on positive electric charges and end on negative electric charges. If there are no isolated north and south poles, do magnetic fields begin and end anywhere or are they always continuous? Describe the magnetic field of a bar magnet from this point of view.

6. What is meant by an electromagnet?

7. How can a solenoid be used as a switch? Give an example of such a use in your car.

†8. How can you use a magnetic field to separate isotopes of a chemical element?

9. How can you make a bar magnet?

†10. What causes the earth's magnetic field? Is the field constant or does it change with time? Is it possible for the earth's poles to flip, that is, for the north pole to become the south and vice versa?

11. If you heat a bar magnet it loses its magnetism. Why?

†12. What does magnetism have to do with the Van Allen belts?

†13. What is magnetohydrodynamics?

Problems for Chapter 22

22.1 The Force on a Charge in a Magnetic Field—The Definition of the Magnetic Field B

1. A proton, moving at a speed of 1.62×10^3 m/s, enters a magnetic field of 0.250 T at an angle of 43.5°. Find the force acting on the proton.

2. An electron, moving at a speed of 3.00×10^6 m/s, enters a magnetic field of 0.200 T at an angle of 35.0°. Find the force acting on the electron.

3. An electron, moving at a speed of 3.00×10^5 m/s, enters a magnetic field of 0.250 T, at an angle of 30.0°. Find (a) the force on the electron and (b) the acceleration of the electron.

4. What is the force on a proton moving north to south at a speed of 3.00×10^5 m/s in the earth's magnetic field

if the vertical component of the earth's magnetic field at that location is 25.0×10^{-6} T?

5. How fast must a proton move in a magnetic field of 2.50×10^{-3} T such that the magnetic force is equal to its weight?

6. Find the value of a magnetic field such that an electron moving at a speed of 2.50×10^4 m/s experiences a maximum force of 2.00×10^{-17} N.

7. An electron is accelerated through a potential difference of 1000 V. It then enters perpendicularly to a uniform magnetic field of 0.200 T. Find the radius of the circular orbit of the electron.

8. Find the value of the magnetic field necessary to cause a proton moving at a speed of 2.50×10^3 m/s to go into a circular orbit of 15.5-cm radius.

9. An electron has an energy of 100 eV as it enters a magnetic field of 3.50×10^{-2} T. Find the radius of the orbit.

10. A velocity selector has a magnetic field of 0.300 T. If a perpendicular electric field of 10,000 V/m is applied, what will be the speed of the particles that will pass through the selector?

11. Find the value of the magnetic field that is necessary for a particle, moving at a speed of 2.5×10^3 m/s, to move straight through an electric field of 500 N/C in a velocity selector.

12. Find the necessary value of the magnetic field in a velocity selector that has an electric field of 500 V/m such that an electron will have an orbital radius of 25.0 cm when it enters a secondary region where the magnetic field is 1.20×10^{-3} T.

22.2 Force on a Current-Carrying Conductor in an External Magnetic Field

13. A wire 35.0 cm long, carrying a current of 3.50 A, is placed at an angle of 40.0° in a uniform magnetic field of 0.002 T. Find the force on the wire.
14. What is the maximum force due to the horizontal component of the earth's magnetic field (20.0×10^{-6} T) acting on a 20.0-cm wire, carrying a current of 5.00 A?
15. Find the value of the magnetic field that will cause a maximum force of 7.00×10^{-3} N on a 20.0-cm straight wire carrying a current of 10.0 A.

22.5 The Magnetic Field at the Center of a Circular Current Loop

16. A circular loop of wire of radius 5.00 cm carries a current of 3.00 A. Find the magnetic induction at the center of the current loop.
17. A circular current loop of 10 turns carries a current of 5.00 A. If the radius of the loop is 5.00 cm, find the magnetic field at the center of the loop.
18. It is desired to neutralize the vertical component of the earth's magnetic field (20.0×10^{-6} T) at a particular point. A flat circular coil is mounted horizontally over this point. If the coil has 10 turns and has a radius of 10.0 cm, what current is necessary and in what direction should it flow through the coil?
19. How many loops of wire are necessary to give a magnetic field of 1.50×10^{-3} T at the center of a circular current loop carrying a current of 10.0 A, if the radius of the loop is 5.00 cm?

22.6 Ampère's Circuital Law

20. A long straight wire carries a current of 10.0 A. Find the magnetic field 5.00 cm from the wire.
21. A power line 10.0 m high carries a current of 200 A. Find the magnetic field of the wire at the ground.
22. A long straight wire carries a current of 10.0 A. How far from the wire will the magnetic field be (a) 1.00 T, (b) 0.100 T, (c) 1.00×10^{-2} T, and (d) 1.00×10^{-3} T?
23. What current is necessary to generate a magnetic field of 0.100 T at a distance of 10.0 cm from a long straight wire?
24. Find the magnetic field at the position A of a long straight wire carrying a current of 10.0 A.

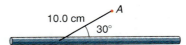

25. Two long parallel wires each carry a current of 5.00 A. If the wires are 15.00 cm apart, find the magnetic field midway between them if (a) the currents are in the same direction and (b) the currents are in the opposite direction.
26. Two parallel wires 10.0 cm apart carry currents of 10.0 A each. Find the magnetic field 5.00 cm to the left of wire 1, 5.00 cm to the right of wire 1, and 15.00 cm to the right of wire 1, if (a) the currents are in the same direction and (b) the currents are in the opposite directions.

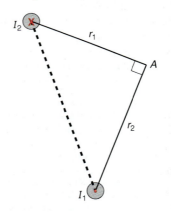

27. Find the magnetic field at point A in the diagram if I_1 (15.0 A) is a current in a wire coming out of the page and I_2 (10.0 A) is a current in a wire going into the page. The distances to point A are $r_1 = 5.00$ cm and $r_2 = 10.00$ cm.

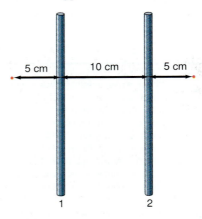

28. A solenoid is 20.0 cm long and carries 500 turns of wire. If the current in the solenoid is 2.00 A, find the magnetic field inside the solenoid.
29. You are asked to design a solenoid that will give a magnetic field of 0.100 T, yet the current must not

exceed 10.0 A. Find the number of turns per unit length that the solenoid should have.

22.7 Force between Parallel, Current-Carrying Conductors—The Definition of the Ampere

30. Two parallel wires 15.0 cm apart carry currents of 10.0 A in the same direction. If the wires are 25.0 cm long, (a) find the magnitude and direction of the force on each wire. (b) If the direction of one current is reversed, find the force on each wire.
31. Wire 1 carries a current of 5.00 A; wire 2, 3.00 A; and wire 3, 7.00 A. Find the total force per unit length on wire 2.

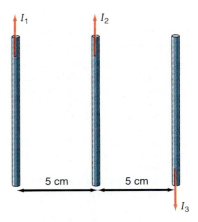

32. Two horizontal parallel wires are placed one above the other. Find the equal currents in the two parallel wires, whose repulsive force will just balance the gravitational force on the top wire if the top wire has a mass of 5.00 g, is 20.0 cm long, and is 4.00 cm above the bottom wire.

22.8 Torque on a Current Loop in an External Magnetic Field—The Magnetic Dipole Moment

33. A coil of wire of 10.0-cm radius carries a current of 5.00 A. Find its magnetic dipole moment.
34. A coil of wire has a magnetic dipole moment of 25.0 A m². It is placed perpendicular to the horizontal magnetic field of the earth of 20.0×10^{-6} T. What torque will act on the coil?
35. A coil of wire 30.0 cm in diameter is placed at an angle of 60.0° in a magnetic field of 2.50×10^{-2} T and experiences a torque of 3.25×10^{-3} m N. Find (a) the magnetic dipole moment of the coil and (b) the current in the coil.

36. A rectangular galvanometer coil 2.00 cm by 1.50 cm has 10 turns of wire. If the current through the coil is 3.00 mA, find its magnetic dipole moment. The coil is placed in a magnetic field of 0.300 T. Find the torque on the coil when the magnetic dipole moment makes an angle of 30.0° with the magnetic field.

37. A coil of 3.00-cm radius, carrying a current of 2.00 A, is placed within a solenoid that carries a current of 3.00 A. If the solenoid has 5000 turns per meter, find the torque on the coil.

†38. A galvanometer coil, 5.00 cm² in area, is placed in a magnetic field of 1.00×10^{-3} T. If the coil deflects 40.0° when it carries its maximum current of 200 mA, find the torsion constant of the spring in the galvanometer.

Additional Problems

†39. An electron moving at an initial velocity of 6.00×10^5 m/s enters a uniform magnetic field of 2.50×10^{-6} T directed into the paper. The length of the magnetic field is 10.0 cm, as shown in the diagram. Find (a) the force on the electron, (b) the magnitude and direction of the acceleration of the electron, (c) the time that the electron remains in the field, and (d) the amount of deflection of the electron as it leaves the magnetic field.

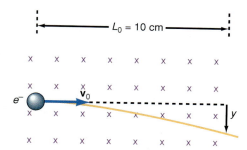

40. An electron moves at a speed of 4.00×10^5 m/s parallel to a long straight wire carrying a 12.0-A current. The velocity of the electron is in the same direction as the current in the wire. If the electron is 5.00 cm from the current, find the magnitude and direction of the force on the electron due to the magnetic field of the current.

41. The diagram shows the end view of four wires each carrying a current $I = 10.0$ A. Wires 1, 2, and 3 have currents coming out of the page, whereas wire 4 has a current going into the page. Find the magnetic field at the center of the square A.

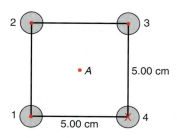

42. A 5000-turn solenoid carries a 2.45-A current. A proton is located at the center of the coil, with a velocity of 7.50×10^4 m/s making an angle of 30.0° with the axis of the solenoid. Determine the magnitude of the force on the proton due to the magnetic field of the solenoid.

43. A small element of current is located at the center of a flat coil, as shown in the diagram. The coil has 50 turns, it is 50.0 cm in radius, and it carries a current of $I = 350$ mA in the clockwise direction. The element of current is 2.00 mm long and it carries a current of $I = 10.0$ mA to the right. Compute the magnitude and direction of the force on the element of current due to the magnetic field of the coil.

44. You are asked to design a circular coil that will have a value of 3.00×10^{-2} T at its center. Find the ratio of the current in the coil to the radius of the coil that will give this value of B. Pick a reasonable value for the combination of I and r such that the current is not too large nor the radius too small. Would it be desirable to introduce more than one loop of wire? What would be a better combination of I, r, and N?

45. An electron, located halfway between two long straight wires, moves at 2.00×10^6 m/s to the left. The top wire carries a 5.00-A current out of the paper, the bottom wire carries a current of 2.00 A into the paper, and

the wires are 6.00 cm apart. Find the magnitude and direction of the force on the electron due to the magnetic fields of the current in the wires.

†46. In the Bohr model of the hydrogen atom, a negative electron orbits about a positive proton at a radius of 5.29×10^{-11} m and at a speed of 2.19×10^6 m/s. (a) How long does it take for the electron to revolve around the proton? (b) From the definition of current show that the orbiting electron constitutes a current and determine its magnitude. (c) Since the orbiting electron looks like a current loop, determine the magnetic field at the location of the proton caused by the orbiting electron.

†47. Everyone knows that the earth revolves around the sun. Yet, in our everyday experience the sun is seen to rise in the eastern sky and set in the western sky, as though the sun revolved around the earth. The observed motion seems to depend on the frame of reference of the observer. When the frame of reference is placed on the earth, the sun appears to revolve around the earth. In a similar way, if the frame of reference is placed on the orbiting electron in the Bohr theory of the atom, it appears as though the proton is moving in a circular orbit about the electron. Find the value of the magnetic field at the position of the electron, caused by the proton.

48. Find the magnetic fields at the center of the following two coils: (a) a 100-turn coil 20.0 cm long with 0.200-cm diameter, carrying a current of 2.00 A, and (b) a 100-turn coil 20.0 cm in diameter and 0.200 cm thick, carrying a current of 2.00 A.

†49. You are asked to design a solenoid by wrapping insulated copper wire around the hollow cardboard core of an empty roll of paper towels. The completed solenoid will then be connected to a 12.0-V battery and the maximum current that can flow in the circuit is 10.0 A. (a) What is the minimum value of the resistance of this solenoid? (b) If the wire used is #22 S&W gauge copper wire, which has a diameter of 6.44×10^{-4} m, what is the minimum length of this wire? (c) With the above restrictions on I and ℓ, how many turns of wire can you have if the diameter of the solenoid is 4.00 cm? (d) If the length of the cardboard core is 28.0 cm, and with all of the above restrictions, find the value of B inside the solenoid. (e) In this design, what factors might be changed to increase the value of B?

†50. A solenoid is filled with iron, which has a permeability of $\mu = 4.80 \times 10^{-3}$ T m/A. The solenoid has 200 turns of wire, is 30.0 cm long, and carries a current of 5.00 A. Find the value of B inside the solenoid. (*Hint:* just as the electric field in a vacuum was characterized by the permittivity of free space ϵ_0, and an electric field in a different medium was characterized by ϵ, the permittivity of the medium the magnetic field in a vacuum is characterized by the permeability μ_0, while the magnetic field in a medium is characterized by the permeability of that medium μ.)

†51. You are to design a galvanometer that has an internal resistance $R_g = 20.0 \ \Omega$ and gives full-scale deflection for a current $I_g = 20.0$ mA. (a) Find the length of #22 B&S gauge copper wire necessary to give this resistance. (b) If this length of wire is to be made into a 500 turn loop, find the radius of the loop. (c) Find the area A of this loop. (d) Find the magnetic dipole moment of this loop. (e) If this loop is placed in a magnetic field of 2.50×10^{-2} T, find the maximum torque on the coil. (f) If the galvanometer needle is to rotate through an angle of $\pi/2$ rad, find the torsion constant of the spring that will supply the countertorque.

52. A long straight wire carries a current of 10.0 A as shown in the diagram. The loop carries a current of 2.00 A. Find (a) the force per unit length on wire *ab*, (b) the force per unit length on wire *cd*, and (c) the torque on the loop.

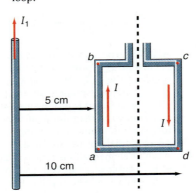

†53. The coil in a DC motor is called an *armature*. The armature, 15.0 cm by 10.0 cm, with 100 turns of wire, carries a current of 3.50 A in a magnetic field of 0.250 T. Find the maximum torque exerted by the motor. If the armature rotates at 1000 rpm, what is the power rating of the motor?

†54. A *toroid* is essentially a solenoid that is bent into a circle. As seen in the diagram, it looks like a doughnut. Use Ampère's law to show that the magnetic field within the toroid is given by

$$B = \frac{\mu_0 NI}{2\pi r}$$

where N is the number of turns of wire and r is the distance from the center of the toroid to the position where the magnetic field is desired. Note that the magnitude of the magnetic field varies with r.

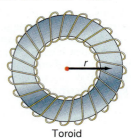

Toroid

55. Every chemical element has isotopes. An *isotope* has the same number of protons as the element but a different number of neutrons. Hence an isotope has a different mass than the parent element. The different masses of the isotopes can be found with a mass spectrometer, as shown in the diagram. Ions of the isotope are fired into the velocity selector, and those of velocity v enter into the mass spectrometer where there is a magnetic field B into the paper. The path of the ions is bent into a circular path of radius R. (a) Find the equation for the mass of any isotope. (b) The chemical element oxygen has eight protons and eight neutrons. It has two stable isotopes, one with nine neutrons and the other with ten neutrons. Find the radii of the path of the isotopes in the spectrometer in terms of the radius of the oxygen element.

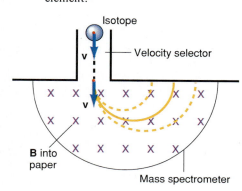

Interactive Tutorials

56. Force on a charge in a magnetic field. An electron is fired into a uniform magnetic field, $B = 0.255$ T, at a speed $v = 185$ m/s and at an angle $\theta = 43.5°$ to the direction of the magnetic field **B**. Find (a) the force F acting on the charge q and (b) the acceleration a of the charge.

57. Force on a current-carrying wire in a magnetic field. A length of wire $L = 42.0$ cm carries a current $I = 15.0$ A. The wire is placed in a uniform magnetic field, $B = 0.365$ T. If the wire makes an angle $\theta = 55.5°$ with the direction of the magnetic field **B**, find the magnitude of the force **F** on the wire.

58. The magnetic field at the center of a circular current loop. Find the magnetic field at the center of a circular current loop of $N = 10$ turns with a radius $r = 5.00$ cm that carries a current $I = 10.0$ A.

59. The magnetic field of a long straight wire. Find the magnetic field at a distance $r = 10.0$ cm from a long straight wire that carries a current $I = 12.5$ A.

60. The magnetic field inside a solenoid. Find the magnetic field inside a solenoid that has a length $L = 25.0$ cm and is composed of $N = 1000$ turns of wire. The current in the solenoid is $I = 7.50$ A.

61. The torque on a current loop in an external magnetic field. A circular coil, consisting of $N = 15$ turns of wire, having a diameter $d = 12.5$ cm, carries a current $I = 8.00$ A. The coil is placed in a uniform magnetic field $B = 0.385$ T. Find (a) the magnetic dipole moment m of the coil and (b) the maximum torque τ_{max}. (c) Plot the torque acting on the coil as a function of the angle θ that the magnetic dipole moment of the coil makes with the magnetic field.

23 Electromagnetic Induction

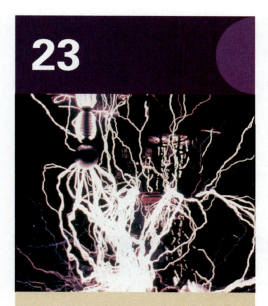

Tesla coils make lightning.

For us, who took in Faraday's ideas so to speak with out mother's milk, it is hard to appreciate their greatness and audacity.

Albert Einstein

23.1 Introduction

Since a current in a wire produces a magnetic field, it is logical to ask if the reverse process is possible. That is, is there any way in which a magnetic field can produce a current? The answer to the question is yes, and the two men who were responsible for the discovery are Michael Faraday (1791–1867), an English physicist, and Joseph Henry (1797–1878), an American physicist. Before investigating this phenomenon we need first to define the concept of magnetic flux.

23.2 Magnetic Flux

Flux is a quantitative measure of the number of lines of a vector field that passes perpendicularly through a surface. Figure 23.1(a), shows a magnetic field **B** passing through a portion of a surface of area **A.** The area of the surface is represented by a vector **A,** whose magnitude is the area A of the surface, and whose direction is perpendicular to the surface. Here θ is the angle between **B** and **A.** *The magnetic flux is defined as*

$$\Phi_M = BA \cos \theta \qquad (23.1)$$

*and is a quantitative measure of the number of lines of **B** that pass normally through the surface area A*. The number of lines represents the strength of the field. The vector **B**, at the point P of figure 23.1(a), can be resolved into the components $B_\perp$, the component perpendicular to the area, and $B_\parallel$ the parallel component. The perpendicular component is

$$B_\perp = B \cos \theta$$

whereas the parallel component is

$$B_\parallel = B \sin \theta$$

The parallel component $B_\parallel$ lies in the surface itself and therefore does not pass through the surface, whereas the perpendicular component $B_\perp$ completely passes through the surface at the point P. The product of the perpendicular component and the area

$$B_\perp A = (B \cos \theta)A = BA \cos \theta = \Phi_M$$

is therefore a quantitative measure of the number of lines of B passing normally through the entire area A. If the angle θ in equation 23.1 is zero, then **B** is parallel to the vector **A** and all the lines of **B** pass normally through the area A, as seen in

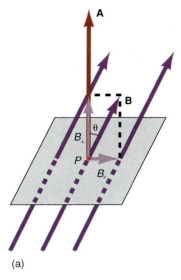

(a)

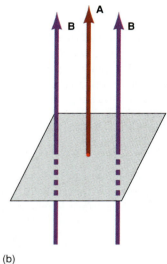

(b)

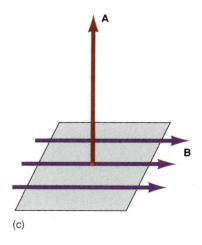

(c)

Figure 23.1

Magnetic flux.

figure 23.1(b). If the angle θ in equation 23.1 is 90° then **B** is perpendicular to the area vector **A,** and none of the lines of **B** pass through the surface A, as seen in figure 23.1(c). The concept of flux is a very important one, and one that is used frequently.

One of the units for the magnetic field was shown, in chapter 22, to be 1 tesla = weber/m². This unit was listed in anticipation of the introduction of the magnetic flux. We will now define the unit for magnetic flux from the formula as

$$\Phi_M = BA$$

$$\Phi_M = \frac{weber}{m^2} \, m^2 = weber$$

Hence, the unit for magnetic flux is the weber. The abbreviation that we use for the weber is Wb.

Example 23.1

Magnetic flux. A magnetic field of 5.00×10^{-2} T passes through a plane 25.0 cm by 35.0 cm at an angle of 40.0° to the normal. Find the magnetic flux Φ_M passing through the plane.

Solution

The area of the plane is

$$A = (0.250 \text{ m})(0.350 \text{ m}) = 8.75 \times 10^{-2} \text{ m}^2$$

The magnetic flux, found from equation 23.1, is

$$\Phi_M = BA \cos \theta$$

$$= (5.00 \times 10^{-2} \text{ T})(8.75 \times 10^{-2} \text{ m}^2)(\cos 40.0°)\left(\frac{\text{Wb/m}^2}{1 \text{ T}}\right)$$

$$= 3.35 \times 10^{-3} \text{ Wb}$$

23.3 Motional emf and Faraday's Law of Electromagnetic Induction

Let us consider the following experiment, as outlined in figure 23.2(a). Two parallel metal rails are separated by a distance ℓ. A metal wire rests on the two rails. A uniform magnetic field **B** is applied such that its direction is into the paper as shown.

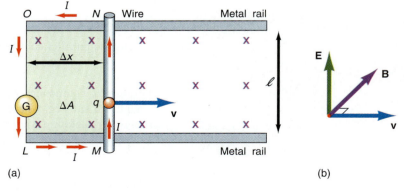

(a)

(b)

Figure 23.2

Motional emf.

Electricity and Magnetism

(Recall that the X's in the figure represent the tail of the arrow representing the vector as it goes into the plane.) A galvanometer G is connected across the two rails. The galvanometer reads zero, indicating that there is no current in the circuit, which consists of the rails and wires; that is, the circuit is the electrical path designated *LMNOL* in figure 23.2(a). *The metal wire MN is now pulled along the rails at a velocity v to the right. The galvanometer now indicates that a current is flowing in the circuit. Somehow, the motion of the wire through the magnetic field generated an electric current.* Let us analyze the cause of this current.

As the wire *MN* is moved to the right, any charge q within the wire experiences the force

$$F = qvB \sin \theta \tag{22.1}$$

as was shown in section 22.1.[1] If both sides of equation 22.1 are divided by q, we have

$$\frac{F}{q} = vB \sin \theta \tag{23.2}$$

But an electrostatic field was originally defined as

$$E = \frac{F}{q_0} \tag{19.1}$$

where F was the force acting on a test charge q_0 placed at rest in the electrostatic field. It is therefore reasonable now to *define an induced electric field E by equation 23.2 as*

$$E = \frac{F}{q} = vB \sin \theta \tag{23.3}$$

This is a different type of field than the electrostatic field. *The induced electric field exists only when the charge is in motion at a velocity v. When $v = 0$, the induced electric field is also zero as can be seen by equation 23.3. The induced electric field is the cause of the current in the wire. The direction of the induced electric field in figure 23.2(b), and hence the direction that a positive charge q within the wire will move, is found by the right-hand rule. Point the fingers of your right hand in the direction of the velocity vector v, with your palm facing the magnetic field vector **B**. Rotate your right hand from v to **B** and your thumb points in the direction of the force vector, and hence the electric field vector, in this case upward.* Hence, the direction of the current is in the direction $M \rightarrow N \rightarrow O \rightarrow L \rightarrow M$, as shown in figure 23.2(a). The magnitude of the induced electric field, found from equation 23.3, is

$$E = vB \sin \theta \tag{23.4}$$

For the particular problem considered here, the angle between **v** and **B** is 90°, and the $\sin \theta = \sin 90° = 1$. Therefore, the induced electric field is

$$E = vB \tag{23.5}$$

We saw in chapter 19, equation 19.20, that for a uniform electric field,

$$E = \frac{V}{y}$$

1. Note that the angle θ in equation 22.1 is the angle between the velocity vector **v** and the magnetic field vector **B**, and should not be confused with the angle θ between the area vector **A** and the magnetic field vector **B** in the determination of the magnetic flux, equation 23.1.

where V is the potential difference between two points and y is the distance between them. For the connecting wire MN, we can assume the induced electric field within the wire is uniform, and designate the induced potential V between M and N by $\mathscr{E}$, which we call an induced emf. The distance y between M and N is the length ℓ of the wire. Therefore, we can write the induced electric field as

$$E = \frac{\mathscr{E}}{\ell} \qquad \text{(23.6)}$$

Equating 23.6 to 23.5, gives

$$\frac{\mathscr{E}}{\ell} = vB$$

The induced emf $\mathscr{E}$ in the wire is therefore

$$\mathscr{E} = vB\ell \qquad \text{(23.7)}$$

If the circuit has a resistance R, then there is an induced current in the circuit given by Ohm's law as

$$I = \frac{\mathscr{E}}{R}$$

This is the current that is recorded by the galvanometer.

Example 23.2

An induced emf. The wire MN in figure 23.2(a) moves with a velocity of 50.0 cm/s to the right. If $\ell = 25.0$ cm, $B = 0.250$ T, and the total electric resistance of the circuit is 35.0 Ω, find (a) the induced emf in the circuit, (b) the current in the circuit, and (c) the direction of the current.

Solution

a. The induced emf in the circuit, found from equation 23.7, is

$$\mathscr{E} = vB\ell = \left(\frac{0.500 \text{ m}}{\text{s}}\right)(0.250 \text{ T})(0.250 \text{ m}) \left[\frac{\text{N}/(\text{A m})}{\text{T}}\right]$$

$$= 3.13 \times 10^{-2} \frac{\text{m N}}{\text{s A}} \left(\frac{\text{J}}{\text{N m}}\right)$$

$$= 3.13 \times 10^{-2} \frac{\text{J}/\text{s}}{\text{A}} \left(\frac{\text{A V}}{\text{J}/\text{s}}\right)$$

$$= 3.13 \times 10^{-2} \text{ V}$$

Notice how the units are manipulated to give the correct units for the final answer.

b. The current in the circuit, found from Ohm's law, is

$$I = \frac{\mathscr{E}}{R} = \frac{3.13 \times 10^{-2} \text{ V}}{35.0 \text{ }\Omega} = 8.94 \times 10^{-4} \text{ A}$$

c. The current in the circuit is counterclockwise, as in figure 23.2(a).

Further insight into this induced emf can be obtained by noting that the speed of the wire is $v = \Delta x / \Delta t$, where Δx is the distance the wire moves to the right in the time Δt. We can write the product of v and ℓ in equation 23.7 as

$$v\ell = \frac{\Delta x}{\Delta t} \ell \qquad \text{(23.8)}$$

But

$$(\Delta x)\ell = \Delta A \qquad (23.9)$$

is the area of the loop swept out as the wire is moved to the right and is shown in figure 23.2(a). Substituting equation 23.9 into equation 23.8 gives

$$v\ell = \frac{\Delta A}{\Delta t}$$

And the induced emf, equation 23.7, becomes

$$\mathscr{E} = B\frac{\Delta A}{\Delta t} \qquad (23.10)$$

But recall that we defined the magnetic flux Φ_M as

$$\Phi_M = BA \cos \theta \qquad (23.1)$$

For a constant magnetic field B and constant angle θ between the magnetic field vector $\mathbf{B}$ and the area vector $\mathbf{A}$, the only way for the magnetic flux to change is for the area A to change. That is, the change in the magnetic flux is

$$\Delta \Phi_M = B\Delta A \cos \theta$$

The rate at which the magnetic flux changes with time is

$$\frac{\Delta \Phi_M}{\Delta t} = B \frac{\Delta A}{\Delta t} \cos \theta \qquad (23.11)$$

The original area vector $\mathbf{A_1}$ points upward and out of the plane of the paper. As the area of the loop increases, the area vector increases from $\mathbf{A_1}$ to $\mathbf{A_2}$, the change in area, $\Delta \mathbf{A} = \mathbf{A_2} - \mathbf{A_1}$, is positive and also points upward and out of the plane of the paper. Since $\mathbf{B}$ points downward into the paper, the angle between the vectors $\mathbf{B}$ and $\Delta \mathbf{A}$ is 180°. Using this fact in equation 23.11 we get

$$\frac{\Delta \Phi_M}{\Delta t} = B\frac{\Delta A}{\Delta t} \cos 180° = -B\frac{\Delta A}{\Delta t}$$

Therefore,

$$B\frac{\Delta A}{\Delta t} = -\frac{\Delta \Phi_M}{\Delta t} \qquad (23.12)$$

Combining equations 23.10 and 23.12 gives an extremely important relationship known as Faraday's law, namely

$$\mathscr{E} = -\frac{\Delta \Phi_M}{\Delta t} \qquad (23.13)$$

Faraday's law of electromagnetic induction states that *whenever the magnetic flux changes with time, there is an induced emf.*

In the case considered, B was a constant and the area changed with time. It is also possible to keep the area a constant but change the magnetic field B with time. That is, if wire MN remains fixed in figure 23.2(a), but the magnetic field is changed with time, a current is indicated on the galvanometer. The reason for this is that the magnetic flux Φ_M, given by equation 23.1, can also change if B changes. That is,

$$\Delta \Phi_M = \Delta B \, A \cos \theta$$

In general the change in the magnetic flux caused by a change in area ΔA and a change in the magnetic field ΔB, is

$$\Delta \Phi_M = B\Delta A \cos \theta + A\Delta B \cos \theta$$

and we can also write Faraday's law, equation 23.13, as

$$\mathcal{E} = -\frac{\Delta \Phi_M}{\Delta t} = -B\frac{\Delta A}{\Delta t} \cos \theta - A\frac{\Delta B}{\Delta t} \cos \theta \qquad (23.14)$$

Faraday's law, in this form, says that an emf can be induced either by changing the area of the loop of wire with time, while the magnetic field and the angle θ remains constant, or by keeping the area of the loop of wire and the angle θ constant and changing the magnetic field with time. It is also possible to keep the magnitudes *B* and *A* constant, but vary the angle θ between them. This also causes a change in the magnetic flux and hence an induced emf. We will consider this case later in the chapter.

Note here that although Faraday's law was derived on the basis of motional emf, it is true in general. For the general case where there are *N* loops of wire, Faraday's law of induction becomes

$$\mathcal{E} = -N\frac{\Delta \Phi_M}{\Delta t} \qquad (23.15)$$

Example 23.3

Inducing an emf by changing the magnetic field with time. The wire *MN* in figure 23.2(a) is fixed 10.0 cm away from the galvanometer wire *OL*. The magnetic field varies from 0 to 0.500 T in a time of 2.00×10^{-3} s. If the resistance of the circuit is 35.0 Ω and the length of the wire $\ell = 25.0$ cm, find (a) the induced emf, (b) the current in the circuit while the magnetic field is changing with time and, (c) the induced emf if the magnetic field remains at a constant 0.500 T.

Solution

a. The induced emf, found from Faraday's law, equation 23.14, is

$$\mathcal{E} = -\frac{\Delta \Phi_M}{\Delta t} = -B\frac{\Delta A}{\Delta t} \cos \theta - A\frac{\Delta B}{\Delta t} \cos \theta$$

Because the wire *MN* is fixed, the area of the loop does not change with time (i.e., $\Delta A = 0$). However, *B* is changing with time, and the induced emf is therefore

$$\mathcal{E} = -A\frac{\Delta B}{\Delta t} \cos \theta$$

The changing magnetic field is

$$\Delta \mathbf{B} = \mathbf{B_2} - \mathbf{B_1}$$

where $\mathbf{B_1}$ is the initial magnetic field and $\mathbf{B_2}$ is the final magnetic field as shown in figure 23.3(a). Since the initial magnetic field $\mathbf{B_1}$ is zero, $\Delta \mathbf{B}$ has the direction of $\mathbf{B_2}$, which is perpendicular to the paper and into the paper. The angle between $\mathbf{A}$ and $\Delta \mathbf{B}$ is thus 180°. Therefore

$$\mathcal{E} = -A\frac{\Delta B}{\Delta t} \cos 180° = A\frac{\Delta B}{\Delta t} \qquad (23.16)$$

Figure 23.3

Changing the magnetic field with time.

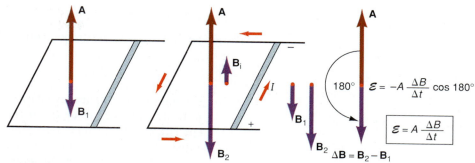

(a) **B** increasing
A constant

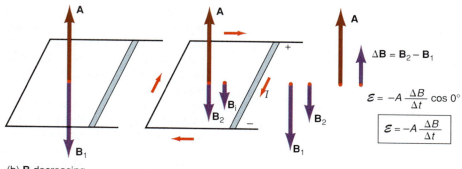

(b) **B** decreasing
A constant

The induced emf, found from equation 23.16, is

$$\mathscr{E} = A\frac{\Delta B}{\Delta t} = (0.100 \text{ m})(0.250 \text{ m})\left(\frac{0.500 \text{ T} - 0 \text{ T}}{2.00 \times 10^{-3} \text{ s}}\right)$$

$$= \left(6.25 \frac{\text{T m}^2}{\text{s}}\right)\left[\frac{\text{N}/(\text{A m})}{\text{T}}\right]$$

$$= 6.25 \frac{\text{N m}}{\text{s A}}\left(\frac{\text{J}}{\text{N m}}\right)\left(\frac{\text{A}}{\text{C}/\text{s}}\right)$$

$$= 6.25 \frac{\text{J}}{\text{C}}\left(\frac{\text{V}}{\text{J}/\text{C}}\right)$$

$$= 6.25 \text{ V}$$

Notice how the units have been converted.

b. The induced current, found from Ohm's law, is

$$I = \frac{\mathscr{E}}{R} = \frac{6.25 \text{ V}}{35.0 \text{ }\Omega} = 0.179 \text{ A}$$

c. When the magnetic field remains constant at 0.500 T, there is no changing magnetic field, $\Delta B = 0$, and there is no induced emf. That is,

$$\mathscr{E} = A\frac{\Delta B}{\Delta t} = 0$$

Thus, for a loop of wire of constant area, the only induced emf occurs when there is a changing magnetic field with time.

23.4 Lenz's Law

Faraday's law of electromagnetic induction was derived in section 23.3 as

$$\mathscr{E} = -\frac{\Delta \Phi_M}{\Delta t} \qquad (23.13)$$

The minus sign in Faraday's law is very important. It was obtained automatically in the derivation by using the angle between the direction of the vectors **B** and Δ**A.** The effect of that minus sign gives rise to a relation known as **Lenz's law** and is stated as: *The direction of an induced emf is such that any current it produces, always opposes, through the magnetic field of the induced current, the change inducing the emf.* As an example, consider figure 23.3(a), which is an example of the magnetic field **B** increasing with time. The induced emf and current flow is as shown. Look at the induced current and recall that whenever a current flows in a wire a magnetic field is found around that wire. Therefore, there is an induced magnetic field B_i inside the loop that points upward, opposing the increasing magnetic field B_2 that points downward. The induced magnetic field upward opposes the increasing magnetic field downward, which was the cause of the induced emf, the induced current, and the induced magnetic field. Hence, the effect—the induced magnetic field upward—opposes the cause—the increasing magnetic field pointing downward.

Lenz's law is also a statement of the law of conservation of energy. If the induced magnetic field is in the same direction as the changing magnetic field, the total magnetic field increases even more, which would in turn induce a larger magnetic field, which would again increase the total magnetic field. The process would continue forever, gaining energy all the time without any work being done by a source, which is of course impossible by the law of conservation of energy. Figure 23.3(b) is an example of the magnetic field **B** decreasing with time. Note how the results changed from figure 23.3(a).

Lenz's law is a convenient tool because it allows us to determine the direction of induced current flow rather simply when the configuration of the changing magnetic field is not simple. As an example, consider the loop of wire in figure 23.4(a). It is connected to the galvanometer and when the bar magnet near the coil is stationary, the galvanometer reads zero. When the north pole of the bar magnet is pushed through the loop, the magnetic field within the loop changes with time and by Faraday's law an emf is induced within the loop, which causes a current. The direction of this current flow is easily determined by Lenz's law. Because the cause of the induced emf is the motion of a north magnetic pole toward the loop, the induced current within the loop causes a magnetic field that tends to oppose that north magnetic pole. Hence, the induced magnetic field looks like the field of a north pole to oppose the north pole of the bar magnet. (Recall that opposite magnetic poles attract while like magnetic poles repel.) Since a magnetic field emanates from a north pole and enters a south pole, the induced magnetic field B_i must come out of the loop toward the bar magnet. If the right hand is wrapped around any portion of the loop of wire, such that the fingers point in the direction of the induced magnetic field, the thumb points in the direction of the current. Thus the direction of the induced current in the loop is as shown in the diagram of figure 23.4(a).

If the bar magnet is pulled away from the loop, as in figure 23.4(b), then the induced magnetic field should tend to oppose the pulling away of the magnet. Therefore, the face of the loop behaves like a south magnetic pole in order to attract the bar magnet. The magnetic field of the induced south pole B_i points into the loop. Using the right-hand rule, and wrapping the hand around the wire with the fingers pointing in the direction of the induced magnetic field B_i, the thumb points in the direction of the current, which is shown in figure 23.4(b). Notice that when the magnet is pushed toward the loop, the direction of the induced current is counterclockwise in the loop; when the magnet is pulled away from the loop, the direction of the current is clockwise.

Figure 23.4

The determination of the direction of an
induced current by Lenz's law.

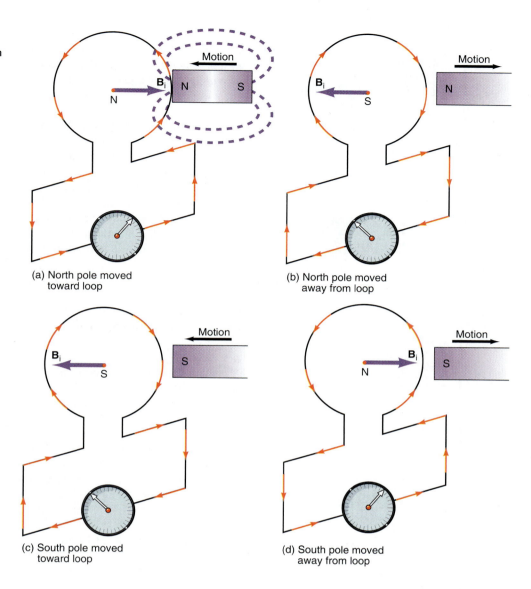

(a) North pole moved
toward loop

(b) North pole moved
away from loop

(c) South pole moved
toward loop

(d) South pole moved
away from loop

The case of moving a south magnetic pole toward the loop is shown in figure
23.4(c), and pulling it away from the loop is shown in figure 23.4(d). The direction
of the induced current is easily found by this analysis of Lenz's law.

23.5 The Induced emf in a Rotating Loop of Wire in a Magnetic Field— Alternating emf's and the AC Generator

*In chapter 22, we saw that when a current-carrying loop of wire is placed in an
external magnetic field, a torque is produced on the loop, causing it to rotate in
the magnetic field.* Recall that this was the essence of the electric motor. Let us
now look at the inverse problem and ask *what happens if a loop of wire is mechanically rotated in an external magnetic field? Can a current be induced in
the loop?* The answer is, of course, yes, and the device so described is an electric
generator.

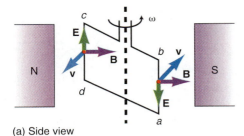

(a) Side view

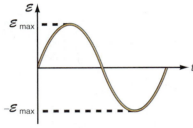

(b) Top view

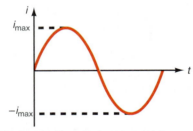

(c) Induced emf varies sinusoidally

(d) Induced current varies sinusoidally

Figure 23.5

Rotating loop of wire in an external magnetic field.

Consider the loop of wire shown in figures 23.5(a) and 23.5(b). The loop is made to rotate with an angular speed ω. The portion of the wire loop ab is moving with a velocity v through the magnetic field $\mathbf{B}$. An electric field is induced in the wire ab, given by equation 23.4 as

$$E = vB \sin \theta$$

where θ is the angle between the velocity vector $\mathbf{v}$ and the magnetic field $\mathbf{B}$ and is seen in figure 23.5(b). Notice that θ is also the angle that the loop turns through. The direction of the induced electric field $\mathbf{E}$ in figure 23.5(a), and hence the direction that a positive charge q within the wire moves is found by the right-hand rule. Point the fingers of your right hand in the direction of the velocity vector $\mathbf{v}$, with your palm facing the magnetic field vector $\mathbf{B}$. Rotate your right hand from $\mathbf{v}$ to $\mathbf{B}$ and your thumb points in the direction of the force vector, and hence the electric field vector, in this case downward. The induced current flows in this direction. The induced electric field in wire ab is also given by equation 23.6 as

$$E = \frac{\mathscr{E}}{\ell}$$

where ℓ is the length of wire from a to b. Combining equations 23.6 and 23.4 gives, for the induced emf in wire ab,

$$\mathscr{E}_{ab} = vB\ell \sin \theta \qquad (23.17)$$

Wire cd has an induced electric field that points upward and is seen in figure 23.5(a). Therefore, an induced current in wire cd flows upward in the direction of the electric field. The induced emf in wire cd is similar to wire ab and is

$$\mathscr{E}_{cd} = vB\ell \sin \theta \qquad (23.18)$$

Along wires bc and da the induced electric field is perpendicular to the wire itself and cannot cause any current to flow along the length of the wires bc or da. Thus, the emf $\mathscr{E}_{bc} = \mathscr{E}_{da} = 0$. The total emf around the loop is, therefore, the sum of the induced emf along ab, $\mathscr{E}_{ab}$, and the induced emf along cd, $\mathscr{E}_{cd}$. That is,

$$\mathscr{E} = \mathscr{E}_{ab} + \mathscr{E}_{cd} \qquad (23.19)$$

Substituting equations 23.17 and 23.18 into equation 23.19, gives

$$\mathscr{E} = vB\ell \sin \theta + vB\ell \sin \theta$$

or

$$\mathscr{E} = 2vB\ell \sin \theta \qquad (23.20)$$

The coil is rotating at an angular speed ω and the linear speed v of wire ab and cd is

$$v = \omega r \qquad (9.2)$$

Substituting v from equation 9.2 back into equation 23.20 gives, for the induced emf in the coil,

$$\mathscr{E} = 2(\omega r)B\ell \sin \theta$$

But $2r$ is the length of wire bc, and since ℓ is the length of wire ab, the product $2r\ell$ is the area of the loop, that is,

$$A = 2r\ell$$

The total induced emf in the coil is thus

$$\mathscr{E} = \omega AB \sin \theta \qquad (23.21)$$

Because the coil is rotating at an angular speed ω, the angle θ turned through at any time is

$$\theta = \omega t$$

The total emf induced in the coil at any instant of time is

$$\mathscr{E} = \omega AB \sin \omega t \tag{23.22}$$

Equation 23.22 says that to get a large induced emf in a coil, the coil should move at a high angular speed ω, have a relatively large area A, and be placed in a large magnetic field B.

Equation 23.22 also points out that the induced emf $\mathscr{E}$ varies sinusoidally with time, as shown in figure 23.5(c). The maximum emf occurs when $\sin \omega t$ is equal to 1 and is denoted by $\mathscr{E}_{max}$, where

$$\mathscr{E}_{max} = \omega AB \tag{23.23}$$

We can then write the induced emf as

$$\mathscr{E} = \mathscr{E}_{max} \sin \omega t \tag{23.24}$$

If the coil is connected to a circuit that has a resistance R, the current that flows from the coil to the circuit, given by Ohm's law, is

$$i = \frac{\mathscr{E}}{R} = \frac{\mathscr{E}_{max}}{R} \sin \omega t \tag{23.25}$$

Equation 23.25 indicates that the current in the circuit also varies sinusoidally with time, and this is shown in figure 23.5(d). The current starts at zero and increases to a maximum value given by

$$i_{max} = \frac{\mathscr{E}_{max}}{R}$$

The variation of current with time is thus

$$i = i_{max} \sin \omega t \tag{23.26}$$

The variation of this current can be more easily understood by studying figure 23.6. Figure 23.6(a) shows the induced current as a function of θ, the angle through which the loop turns. Figure 23.6(b) shows the position of the rotating coil at various positions. At position 0, of figure 23.6(b), the velocity $\mathbf{v}$ is parallel to the magnetic field $\mathbf{B}$, $\theta = 0$, and hence the induced electric field ($E = vB \sin \theta$) is zero. Thus, the induced emf and current in the coil at this position is zero. At position 1, the coil has rotated through 45°. The electric field points downward in wire *ab,* and upward in wire *cd.* Hence, current flows in the coil in the direction from *b* to *a* to *d* to *c.* The magnitude of the current at position 1 is $0.707\, i_{max}$. At position 2, the coil has rotated through 90°. The directions of the electric fields in the wires are the same as at position 1, and hence the current flows in the same direction. The current flowing in the coil is now at its maximum value. At position 3, the coil has rotated through an angle of 135°. The direction of the electric fields and hence the currents are still the same. However, the current now decreases to $i_3 = i_{max} \sin 135°$ $= 0.707\, i_{max}$. At position 4, the coil has rotated through an angle of 180°. Now $\mathbf{v}$ is antiparallel to $\mathbf{B}$, $\theta = 180°$, and hence ($vB \sin 180°$) is zero. Therefore, at this position, the electric field within the wire is zero and so is the current. At position 5, the coil has rotated through an angle of 225°. Now $\mathbf{E}$ in wire *ab* points upward, changing the direction of the electric field and hence the direction of the current. This is shown in figure 23.6(a) as the negative current $-0.707\, i_{max}$. The negative current means that the direction of the current has been changed from its original

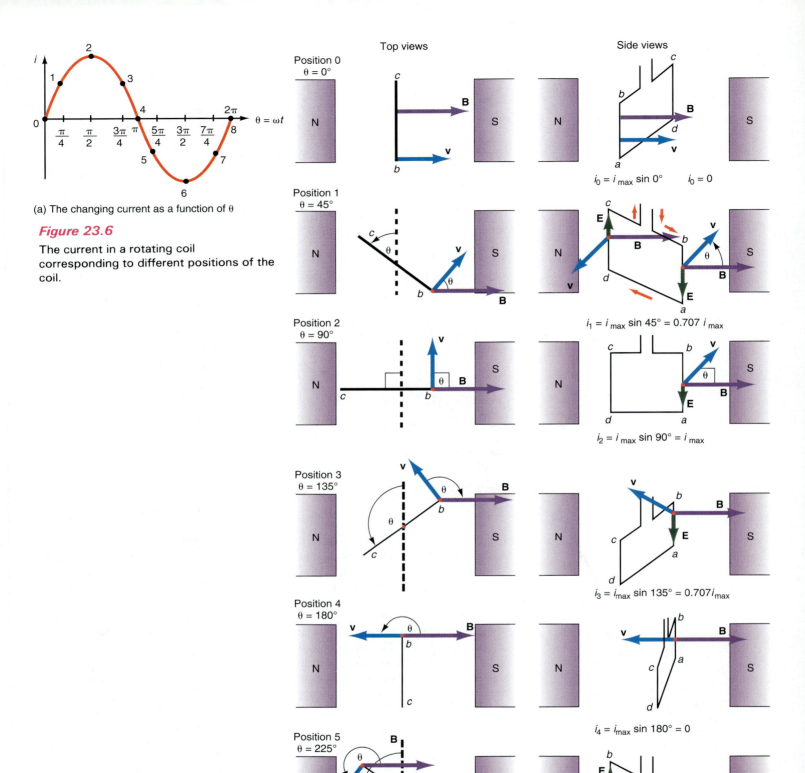

(a) The changing current as a function of θ

Figure 23.6

The current in a rotating coil corresponding to different positions of the coil.

Top views

Side views

Position 0
$\theta = 0°$

$i_0 = i_{max} \sin 0°$ $i_0 = 0$

Position 1
$\theta = 45°$

$i_1 = i_{max} \sin 45° = 0.707\, i_{max}$

Position 2
$\theta = 90°$

$i_2 = i_{max} \sin 90° = i_{max}$

Position 3
$\theta = 135°$

$i_3 = i_{max} \sin 135° = 0.707 i_{max}$

Position 4
$\theta = 180°$

$i_4 = i_{max} \sin 180° = 0$

Position 5
$\theta = 225°$

$i_5 = i_{max} \sin 225° = -0.707\, i_{max}$

(b) Positions of the rotating coil

(Figure continued)

Figure 23.6
Continued

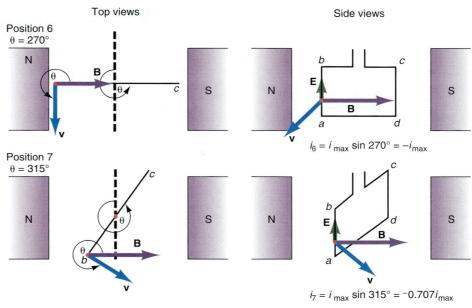

Top views   Side views

Position 6
$\theta = 270°$

$i_6 = i_{max} \sin 270° = -i_{max}$

Position 7
$\theta = 315°$

$i_7 = i_{max} \sin 315° = -0.707 i_{max}$

(b) Positions of the rotating coil

direction to the opposite direction. At position 6, the coil has rotated through an angle of 270°. Here **v** is perpendicular to **B,** giving the maximum current in the negative direction. At position 7, the coil has rotated through an angle of 315° and the current has decreased to $-0.707\, i_{max}$. When the coil rotates the last 45°, it finds itself back at position 0; the electric field and the current is again zero. The cycle repeats itself as the coil is rotated and a current that alternates in direction and varies in magnitude is obtained.

This rotating loop of wire in an external magnetic field is called an AC generator. The AC stands for alternating current, because the current alternates sinusoidally from one direction to another. The current alternates sinusoidally because the induced emf in the coil alternates sinusoidally. In addition to changing in direction, the magnitude of the AC current continually changes. This is very different from a DC current, which remains at a constant magnitude. Notice that in the AC generator, the magnitudes of the vectors **B** and **A** are constant, but the direction between the two vectors is constantly changing with time. Because the magnetic flux is equal to $BA \cos \theta$, if the angle θ changes with time, there is also a changing magnetic flux, and hence, an induced emf. Thus the AC generator is based on the fact that the angle θ between the velocity vector **v** and the magnetic field vector **B** changes with time to cause the changing flux that creates the induced electric field, potential, and current.

From a practical standpoint it is usually desirable to obtain an even larger emf than that obtained from equation 23.22. This can be accomplished by increasing the number of turns constituting the coil to N turns. For example, if two turns are used, the voltage and current are doubled, if ten turns are used, the voltage and current increase tenfold. For a coil of N turns, equations 23.22 and 23.23 are multiplied by the factor N, and the voltage and current increase N times. Thus the induced emf is

$$\mathscr{E} = N\omega AB \sin \omega t \qquad (23.27)$$

Example 23.4

The AC generator. A coil of 100 turns of wire, with a cross-sectional area of 100 cm², is rotated at an angular speed of 377 rad/s in an external magnetic field of 0.450 T. Find (a) the maximum induced emf, (b) the frequency of the alternating emf, and (c) if the resistance of the circuit is 100 Ω, find the current flow and the maximum current.

Solution

a. The induced emf, found from equation 23.27, is

$$\mathscr{E} = N\omega AB \sin \omega t$$

and the maximum emf is

$$\mathscr{E}_{max} = N\omega AB$$

$$= (100)\left(377 \, \frac{rad}{s}\right)(100 \, cm^2)\left(\frac{1 \, m^2}{10^4 \, cm^2}\right)(0.450 \, T)$$

$$= 170 \, \frac{T \, m^2}{s}\left[\frac{N/(A \, m)}{T}\right]\left(\frac{A}{C/s}\right)\left(\frac{J}{N \, m}\right)\left(\frac{V}{J/C}\right)$$

$$= 170 \, V$$

Notice how the conversion factors of units cancel, leaving the correct unit for the answer.

b. The frequency of the alternating emf is related to the angular speed of the loop by equation 11.9 as

$$\omega = 2\pi f$$

$$f = \frac{\omega}{2\pi}$$

or

$$f = \frac{377 \, rad/s}{2\pi} = 60.0 \, cycles/s$$

Thus, a 60.0 cycle/s (60.0 Hz) frequency is obtained from a generator by rotating the coil at an angular speed of 377 rad/s.

c. The current, found from equation 23.25, is

$$i = \frac{\mathscr{E}_{max}}{R} \sin \omega t$$

$$= \frac{170 \, V}{100 \, \Omega} \sin \omega t$$

$$= (1.70 \, A) \sin \omega t$$

and the maximum current is 1.70 A.

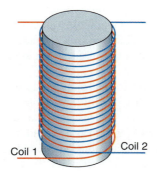

Figure 23.7

Two coaxial solenoids.

23.6 Mutual Induction

Consider the two coaxial solenoids of the same size shown in figure 23.7. (One blue wire and one red wire are wrapped around a hollow roll of cardboard.) If the current i_1 in coil 1 is changed with time, this changes the magnetic field B_1 of solenoid 1 with time. But solenoid 2 is in the magnetic field of coil 1 and any change in B_1 causes an induced emf $\mathscr{E}_2$ in coil 2. From Faraday's law of induction, we can state this as

$$\mathscr{E}_2 = -N_2\frac{\Delta\Phi_M}{\Delta t}$$

$$= -N_2\frac{\Delta(B_1A)}{\Delta t}$$

$$= -N_2A\frac{\Delta B_1}{\Delta t} \tag{23.28}$$

The magnetic field inside solenoid 1, found from equation 22.30, is

$$B_1 = \mu_0\frac{N_1}{\ell}i_1$$

where, as you recall, μ_0 is the permeability of empty space, N is the number of turns, and ℓ is the length of the solenoid. When i_1 changes with time, B_1 changes as

$$\Delta B_1 = \mu_0\frac{N_1}{\ell}\Delta i_1 \tag{23.29}$$

Substituting equation 23.29 into equation 23.28, gives

$$\mathscr{E}_2 = \frac{-N_2A[\mu_0(N_1/\ell)\Delta i_1]}{\Delta t}$$

or

$$\mathscr{E}_2 = -\mu_0A\frac{N_1N_2}{\ell}\frac{\Delta i_1}{\Delta t}$$

Let us define the quantity

$$\frac{\mu_0AN_1N_2}{\ell} = M \tag{23.30}$$

as the *coefficient of mutual induction* for the coaxial solenoids. Faraday's law now becomes

$$\mathscr{E}_2 = -M\frac{\Delta i_1}{\Delta t} \tag{23.31}$$

Equation 23.31 says that changing the current i_1 in coil 1 with time induces an emf in coil 2. This process is called **mutual induction.** Note that M in equation 23.30 is only a function of the geometry of solenoids 1 and 2, and is a constant for the particular set of solenoids chosen. For configurations other than the coaxial solenoids, it is very difficult, sometimes almost impossible, to determine M in this way. However, even in these difficult cases, M can always be found by measuring the quantities $\Delta i_1/\Delta t$ and $\mathscr{E}_2$. Then we find M from equation 23.31 as

$$M = \frac{-\mathscr{E}_2}{\Delta i_1/\Delta t} \tag{23.32}$$

Hence, equation 23.31 can be used to determine the induced emf in coil 2, produced by the changing current in coil 1, regardless of the actual geometrical configuration.

If the emf in equation 23.32 is expressed in volts, the current in amperes, and the time in seconds, then the unit for mutual inductance M is called a henry and is given by

$$1 \text{ henry} = \frac{\text{volt}}{\text{ampere/second}}$$

and abbreviated as

$$1 \text{ H} = \frac{\text{V}}{\text{A/s}}$$

By using a similar analogy, the current in coil 2 can be varied and an emf induced in coil 1 given by

$$\mathcal{E}_1 = -N_1 \frac{\Delta \Phi_M}{\Delta t} = -N_1 A \frac{\Delta B_2}{\Delta t}$$

where

$$\Delta B_2 = \mu_0 \frac{N_2}{\ell} \Delta i_2$$

and

$$\mathcal{E}_1 = -N_1 A \mu_0 \frac{N_2}{\ell} \frac{\Delta i_2}{\Delta t}$$

$$= -\frac{\mu_0 A N_1 N_2}{\ell} \frac{\Delta i_2}{\Delta t} \qquad (23.33)$$

Notice that the coefficient of $\Delta i_2 / \Delta t$ in equation 23.33 is the coefficient of mutual induction found in equation 23.30. Therefore, *the induced emf in coil 1 caused by a changing current in coil 2 is given by*

$$\mathcal{E}_1 = -M \frac{\Delta i_2}{\Delta t} \qquad (23.34)$$

Example 23.5

Coefficient of mutual inductance of coaxial solenoids. Find the coefficient of mutual inductance of two coaxial solenoids of 5.00-cm radius and 30.0 cm long if one coil has 10 turns and the second has 1000 turns.

Solution

The area of the solenoid is found as

$$A = \pi r^2 = \pi (0.0500 \text{ m})^2 = 7.85 \times 10^{-3} \text{ m}^2$$

The coefficient of mutual inductance, found from equation 23.30 is

$$M = \frac{\mu_0 A N_1 N_2}{\ell}$$

$$= \frac{(4\pi \times 10^{-7} \text{ T m/A})(7.85 \times 10^{-3} \text{ m}^2)(10)(1000)}{0.300 \text{ m}}$$

and

$$M = 3.29 \times 10^{-4} \frac{\text{T m}^2}{\text{A}} \left[\frac{\text{N/(C m/s)}}{\text{T}} \right]$$

$$= 3.29 \times 10^{-4} \frac{\text{N m}}{\text{A C/s}} \left(\frac{\text{J}}{\text{N m}} \right) \left(\frac{\text{V}}{\text{J/C}} \right)$$

$$= 3.29 \times 10^{-4} \frac{\text{V}}{\text{A/s}} \left[\frac{\text{H}}{\text{V/(A/s)}} \right]$$

$$= 3.29 \times 10^{-4} \text{ H}$$

Notice how the unit conversion factors are used to show the correct unit for the answer.

Example 23.6

The induced emf in the second coil. If the current in the first coil of example 23.5 changes by 2.00 A in 0.001 s, what is the induced emf in the second coil?

Solution

The induced emf in the second coil, found from equation 23.31, is

$$\mathscr{E}_2 = -M\frac{\Delta i_1}{\Delta t} = -(3.29 \times 10^{-4}\ \mathrm{H})\left(\frac{2.00\ \mathrm{A}}{0.001\ \mathrm{s}}\right)$$

$$= -6.58 \times 10^{-1}\ \mathrm{H}\left[\frac{\mathrm{V/(A/s)}}{\mathrm{H}}\right]\left(\frac{\mathrm{A}}{\mathrm{s}}\right)$$

$$= -0.658\ \mathrm{V}$$

23.7 Self-Induction

Not only does a changing magnetic flux in one coil induce an emf in an adjacent coil, it even induces an emf in itself. That is, if the current is varied in a single solenoid, the changing current produces a changing magnetic field and the changing magnetic field induces an emf in the single solenoid itself. This process is called **self-induction** *and can be explained by Faraday's law of electromagnetic induction.*

A single solenoid is shown in figure 23.8. If the two ends of the solenoid coil are attached to the AC generator of section 23.5 (shown as a circle with a sine wave in it in figure 23.8), the varying current in the coil causes a varying magnetic field to exist in the coil. Because the magnetic field of a solenoid is given by

$$B = \mu_0 n i \tag{23.35}$$

changing the current i causes the magnetic field to change by

$$\Delta B = \mu_0 n \Delta i \tag{23.36}$$

The induced emf, given by Faraday's law, is

$$\mathscr{E} = -N\frac{\Delta \Phi_\mathrm{M}}{\Delta t} = -N\frac{\Delta(BA)}{\Delta t}$$

Because the area A of the coil does not change, this becomes

$$\mathscr{E} = -NA\frac{\Delta B}{\Delta t} \tag{23.37}$$

The changing magnetic field within the solenoid is given in equation 23.36, and substituting it into equation 23.37, gives

$$\mathscr{E} = -NA\mu_0 n\frac{\Delta i}{\Delta t} \tag{23.38}$$

The total number of turns N is related to n the number of turns per unit length by

$$N = n\ell$$

where ℓ is the length of the solenoid. Substituting this into equation 23.38 gives

$$\mathscr{E} = -(\mu_0 A \ell n^2)\frac{\Delta i}{\Delta t}$$

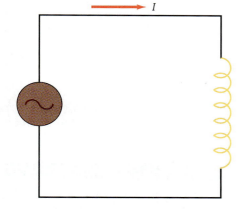

Figure 23.8
Self-induction.

Note that the coefficient of $\Delta i/\Delta t$ is a constant that depends only on the geometry of the solenoid coil. This constant is called the *self-inductance* of the solenoid coil and is designated by L, where

$$L = \mu_0 A \ell n^2 \tag{23.39}$$

The self-induced emf is now

$$\mathcal{E} = -L\frac{\Delta i}{\Delta t} \tag{23.40}$$

Equation 23.40 says that changing the current in a coil induces an emf in the coil, and the minus sign says that the induced emf acts to oppose the cause of the induced emf. That is, if the applied emf is positive, the self-induced emf is negative; if the applied emf is negative, the self-induced emf is positive. Or stated another way, if the current is increasing in the coil, the direction of the induced emf is opposite to that of the current. If the current is decreasing, the induced emf is in the same direction as the original current.

In many cases it is very difficult, if not impossible, to derive L as we did for the special case of the solenoid in equation 23.39. However, in all cases, equation 23.40 holds and can be used to define the self-inductance as

$$L = -\frac{\mathcal{E}}{\Delta i/\Delta t} \tag{23.41}$$

Note that $\mathcal{E}$ and $\Delta i/\Delta t$ can be measured experimentally for any coil configuration and L can be determined from equation 23.41. The self-inductance L is usually called the *inductance of the coil,* and is measured in henries, just as the mutual inductance was, that is,

$$1 \text{ henry} = 1 \frac{V}{A/s}$$

A circuit element in which a self-induced emf accompanies a changing current is called an **inductor.** The inductance L of the inductor is represented in a circuit diagram by the symbol for a coil.

Example 23.7

The inductance of a coil. A battery is connected through a switch to a solenoid coil. The coil has 50 turns/cm, a diameter d of 10.0 cm, and is 50.0 cm long. When the switch is closed, the current goes from 0 to its maximum value of 3.00 A in 0.002 s. Find the inductance of the coil and the induced emf in the coil during this period.

Solution

The cross-sectional area of the coil is

$$A = \frac{\pi d^2}{4} = \frac{\pi(0.100 \text{ m})^2}{4} = 7.85 \times 10^{-3} \text{ m}^2$$

and 50 turns per centimeter of the coil is equal to 5000 turns per meter. The inductance of the solenoid coil is now found from equation 23.39 as

$$L = \mu_0 A \ell n^2$$

$$= (4\pi \times 10^{-7} \text{ T m/A})(7.85 \times 10^{-3} \text{ m}^2)(0.500 \text{ m})\left(\frac{5000}{1 \text{ m}}\right)^2$$

$$= 0.123 \text{ H}$$

The induced emf in the coil, found from equation 23.40, is

$$\mathcal{E} = -L\frac{\Delta i}{\Delta t}$$

$$= -(0.123 \text{ H})\left(\frac{3.00 \text{ A} - 0 \text{ A}}{0.002 \text{ s}}\right)$$

$$= -185 \text{ V}$$

The minus sign on $\mathcal{E}$ indicates that it is opposing the battery voltage V.

Example 23.8

The induced emf when the current is constant. What is the induced emf in example 23.7 after 0.002 s?

Solution

Since the maximum current in the circuit, 3.00 A, is attained after 0.002 s, there is no longer a change in current with time (i.e., $\Delta i/\Delta t = 0$). Therefore, the induced emf is given by

$$\mathcal{E} = -L\frac{\Delta i}{\Delta t} = -(0.123 \text{ H})(0) = 0$$

When the current is steady there is no longer a self-induced emf in the circuit.

Example 23.9

The induced emf when the current decays. The switch in examples 23.7 and 23.8 is now opened. If the current decays at the same rate at which it rose in the circuit, find the induced emf in the circuit.

Solution

The induced emf, given by equation 23.40, is

$$\mathcal{E} = -L\frac{\Delta i}{\Delta t}$$

$$= -(0.123 \text{ H})\left(\frac{0 - 3.00 \text{ A}}{0.002 \text{ s}}\right)$$

$$= +185 \text{ V}$$

Note that $\mathcal{E}$ is now positive because the current is going from 3.00 A to 0 A and hence Δi is negative. This positive induced emf is opposing the decay of the current in the circuit.

23.8 The Energy Stored in the Magnetic Field of an Inductor

In section 21.3, we saw that energy can be stored in the electric field between the plates of a capacitor. In a similar manner energy can be stored in the magnetic field of the coils of an inductor. When the switch in figure 23.9(a) is closed, the total applied voltage V is impressed across the ends of the solenoid coil. The initial current i is zero. In order for the current to increase, an amount of charge Δq must be taken

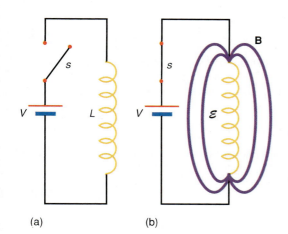

(a)   (b)

Figure 23.9

Energy is stored in the magnetic field of an inductor.

out of the positive side of the battery and moved against the induced emf $\mathscr{E}$ in the coil. The amount of work done by the battery in moving this small amount of charge is

$$\Delta W = \Delta q\, \mathscr{E} \tag{23.42}$$

But the magnitude of the induced emf, given by equation 23.40, is

$$\mathscr{E} = L\frac{\Delta i}{\Delta t}$$

Therefore, the small amount of work done is

$$\Delta W = \Delta q\, L\frac{\Delta i}{\Delta t}$$

Rearranging terms,

$$\Delta W = \frac{\Delta q}{\Delta t}\, L\Delta i$$

$$= iL\Delta i$$

where we have used the fact that $\Delta q/\Delta t = i$, the current. This current i is not, however, a constant, but varies with time. Therefore, the small amount of work done is not a constant, but varies with the current i in the circuit. The total work done is equal to the average current times the term $L\Delta i$, that is,

$$W = i_{\text{avg}}L\Delta i \tag{23.43}$$

But the average current in the circuit is

$$i_{\text{avg}} = \frac{0 + I}{2} = \frac{I}{2} \tag{23.44}$$

where 0 is the initial current and I is the final current. The change in current becomes

$$\Delta i = I - 0 = I \tag{23.45}$$

Substituting equations 23.45 and 23.44 into equation 23.43 yields

$$W = \left(\frac{I}{2}\right)L(I)$$

and

$$W = \tfrac{1}{2}LI^2 \tag{23.46}$$

This work W, done by the battery on the charge, shows up as potential energy of the charge. This energy is said to reside in the magnetic field of the coil, figure 23.9(b), and is designated as U_{M}. Thus, *the energy stored in the inductor is given by*

$$U_{\text{M}} = \tfrac{1}{2}LI^2 \tag{23.47}$$

This stored energy can also be expressed in terms of the magnetic field B by recalling that for a solenoid

$$L = \mu_0 A \ell n^2 \tag{23.39}$$

and

$$B = \mu_0 nI \tag{23.35}$$

Solving equation 23.35 for the current I, we get

$$I = \frac{B}{\mu_0 n} \tag{23.48}$$

Substituting equations 23.39 and 23.48 into equation 23.47, gives

$$U_M = \frac{1}{2}LI^2 = \frac{1}{2}(\mu_0 A \ell n^2)\left(\frac{B}{\mu_0 n}\right)^2$$

$$U_M = \frac{1}{2}\frac{B^2 A \ell}{\mu_0} \qquad (23.49)$$

Equation 23.49 gives the energy stored in the magnetic field of a solenoid.

The energy density is defined as the energy per unit volume and can be represented as

$$u_M = \frac{U_M}{V} = \frac{1}{2}\frac{(B^2 A \ell)/\mu_0}{V}$$

But the volume of the solenoid is $V = A\ell$. Therefore,

$$u_M = \frac{1}{2}\frac{B^2}{\mu_0} \qquad (23.50)$$

Equation 23.50 gives the magnetic energy density, or the energy per unit volume that is stored in the magnetic field. Although equations 23.46 and 23.50 were derived for a solenoid, they are perfectly general and apply to any inductor. Note the similarity with the energy density for an electric field.

$$u_E = \frac{1}{2}\epsilon_0 E^2 \qquad (21.12)$$

Example 23.10

The energy stored in the magnetic field of an inductor. What is the energy stored in the magnetic field in example 23.7?

Solution

In that example we found the inductance to be $L = 0.123$ H and the final current was 3.00 A. Therefore, the energy stored in the magnetic field, found from equation 23.46, is

$$U_M = \frac{1}{2}LI^2$$

$$= \frac{1}{2}(0.123 \text{ H})(3.00 \text{ A})^2$$

$$= 0.554 \text{ H A}^2\left[\frac{V/(A/s)}{H}\right]$$

$$= 0.554 \frac{C}{s} \text{ V s}\left(\frac{J/C}{V}\right)$$

$$= 0.554 \text{ J}$$

The Language of Physics

Magnetic flux
A quantitative measure of the number of lines of the magnetic field that passes normally through a surface. It is measured in webers (p. 661).

Faraday's law of electromagnetic induction
Whenever the magnetic flux through a coil changes with time, an emf is induced in the coil. The magnetic flux can be changed by changing the magnetic field, the area of the loop, or the direction between the magnetic field and the area vector (p. 665).

Lenz's law
The direction of an induced emf is such that any current it produces, always opposes, through the magnetic field of the induced current, the change inducing the emf (p. 668).

AC generator
A device in which a coil of wire is manually rotated in an external magnetic field. Because the magnetic flux through the coil changes with time, the rotating coil has an alternating, sinusoidally varying, emf and current induced in the coil. The AC generator is a source of alternating current (p. 673).

Mutual induction
Changing the magnetic flux in one coil induces an emf in an adjacent coil (p. 675).

Self-induction
Changing the magnetic flux in a coil induces an emf in that coil. The induced emf opposes the changing magnetic flux (p. 677).

Inductor
A circuit element in which a self-induced emf accompanies a changing current (p. 678).

Summary of Important Equations

Magnetic flux
$$\Phi_M = BA \cos\theta \tag{23.1}$$

Induced electric field
$$E = \frac{F}{q} = vB \sin\theta \tag{23.3}$$

Magnitude of induced electric field
$$E = vB \sin\theta \tag{23.4}$$

Faraday's law of induction
$$\mathcal{E} = -\frac{\Delta\Phi_M}{\Delta t} \tag{23.13}$$

$$\mathcal{E} = -B\frac{\Delta A}{\Delta t}\cos\theta$$
$$\quad - A\frac{\Delta B}{\Delta t}\cos\theta \tag{23.14}$$

Faraday's law for N loops
$$\mathcal{E} = -N\frac{\Delta\Phi_M}{\Delta t} \tag{23.15}$$

Induced emf in a rotating coil
$$\mathcal{E} = \omega AB \sin\omega t \tag{23.22}$$

An alternating emf
$$\mathcal{E} = \mathcal{E}_{max} \sin\omega t \tag{23.24}$$

An alternating current
$$i = i_{max} \sin\omega t \tag{23.26}$$

emf induced in coil 2 by changing current in coil 1
$$\mathcal{E}_2 = -M\frac{\Delta i_1}{\Delta t} \tag{23.31}$$

Mutual inductance
$$M = \frac{-\mathcal{E}_2}{\Delta i_1/\Delta t} \tag{23.32}$$

Inductance of a solenoid
$$L = \mu_0 A\ell n^2 \tag{23.39}$$

Self-induced emf of a coil
$$\mathcal{E} = -L\frac{\Delta i}{\Delta t} \tag{23.40}$$

Inductance
$$L = \frac{-\mathcal{E}}{\Delta i/\Delta t} \tag{23.41}$$

Energy stored in magnetic field of an inductor
$$U_M = \tfrac{1}{2}LI^2 \tag{23.47}$$

Energy stored in magnetic field of a solenoid
$$U_M = \frac{1}{2}\frac{B^2 A\ell}{\mu_0} \tag{23.49}$$

Magnetic energy density
$$u_M = \frac{1}{2}\frac{B^2}{\mu_0} \tag{23.50}$$

Questions for Chapter 23

1. Describe the concept of magnetic flux. Could you also define an electric flux for an electric field?
†2. If changing the magnetic flux with time induces an electric field, does changing the electric flux with time induce a magnetic field?
3. If the metal wire MN in figure 23.2 were replaced with a wooden stick, how would this affect the experiment?

4. Show that if the area vector in figure 23.3 were defined in the opposite direction, the analysis would not be consistent with Lenz's law. Therefore, show that the choice of direction for the area vector **A** cannot be arbitrary.
5. Is it possible to change both the area of a loop and the magnetic field passing through the loop and still not have an induced emf in the loop?
6. Discuss Lenz's law.

7. Can an electric motor be used to drive an AC generator, with the output from the generator being used to operate the motor?
8. Discuss how energy is stored in the magnetic field of a coil. Compare this to the way energy is stored in the electric field of a capacitor.
†9. If changing an electric field with time produces a magnetic field, and changing a magnetic field with time produces an electric field, is it possible for these changing fields to couple together to propagate through space?

Electricity and Magnetism

Problems for Chapter 23

23.2 Magnetic Flux

1. A rectangular coil 6.00 cm by 8.00 cm is located in a uniform magnetic field of 0.250 T. Find the flux through the coil when the plane of the coil is (a) perpendicular to **B**, (b) parallel to **B**, and (c) makes an angle of 60.0° with **B**.

2. A circular coil, 6.00 cm in diameter, is placed in a uniform magnetic field of 0.300 T. Find the flux through the coil when the coil makes an angle of 53.0° with **B**. What is the flux if the angle is increased to 90.0°?

23.3 Motional emf and Faraday's Law of Electromagnetic Induction

3. The magnetic flux through a coil of 10 turns, changes from 5.00×10^{-4} Wb to 5.00×10^{-3} Wb in 1.00×10^{-2} s. Find the induced emf in the coil.

4. A circular coil 6.00 cm in diameter is placed in a uniform magnetic field of 0.500 T. If B drops to 0 in 0.002 s find the maximum induced emf in the coil.

5. A 25-turn circular coil 6.00 cm in diameter is placed in, and perpendicular to, a magnetic field that is changing at 2.50×10^{-2} T/s. (a) Find the induced emf in the coil. (b) If the resistance of the coil is 25.0 Ω, find the induced current in the coil.

6. A circular coil 5.00 cm in diameter has a resistance of 2.00 Ω. If a current of 4.00 A is to flow in the coil, at what rate should the magnetic field change with time if (a) the coil is perpendicular to the magnetic field and (b) if the coil makes an angle of 30° with the field?

7. In the diagram, the magnetic field **B** is a constant and points downward. The wire moves to the right at the velocity **v**. Hence the area vector **A₁**, which points upward, increases to **A₂**, which also points upward. Find (a) the direction of Δ**A**, (b) the angle between Δ**A** and **B**, (c) the induced emf, and (d) the direction of the induced current.

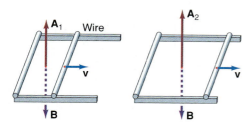

8. In the diagram, the magnetic field **B** is a constant and points downward. The wire moves to the left at the velocity −**v**. Hence the area vector **A₁**, which points upward, decreases to **A₂**. Find (a) the direction of Δ**A**, (b) the angle between Δ**A** and **B**, (c) the induced emf, and (d) the direction of the induced current.

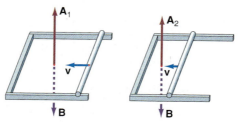

9. The wire AB in the diagram moves to the right with a velocity of 25.0 cm/s. If $\ell = 20.0$ cm, $B = 0.300$ T, and the total resistance of the circuit is 50.0 Ω, find (a) the induced emf in wire AB, (b) the current flowing in the circuit, and (c) the direction of the current.

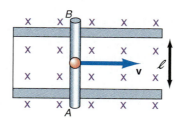

10. Repeat problem 9 if the wire AB is moving to the left with a velocity of 25.0 cm/s.

11. If wire AB of problem 9 moves with a velocity of 50.0 m/s to the right, at what rate should **B** change when the wire is at a distance of 5.00 cm from the left end of the loop, such that there will be no induced emf in the circuit?

12. Wire MN ($l = 25.0$ cm) of figure 23.2(a) is fixed 20.0 cm away from the galvanometer wire OL. The resistance of the circuit is 50.0 Ω. (a) If the magnetic field varies from 0 to 0.350 T in a time of 0.030 s, find the induced emf and current in the circuit during this time. (b) If the magnetic field remains constant at 0.350 T, find the induced emf and current in the coil. (c) If the magnetic field decays from 0.350 T to 0 T in 0.0200 s, find the induced emf, the current, and its direction in the wire.

13. An airplane is flying at 200 knots through an area where the vertical component of the earth's magnetic field is 3.50×10^{-5} T. If the wing span is 12.0 m, what is the difference in potential from wing tip to wing tip? Could this potential difference be used as a source of current to operate the aircraft's equipment?

14. A train is traveling at 40.0 km/hr in a location where the vertical component of the earth's magnetic field is 3.00×10^{-5} T. If the axle of the wheels is 1.50 m apart, what is the induced potential difference across the axle?

15. A coil of 10 turns and 35.0-cm² area is in a perpendicular magnetic field of 0.0500 T. The coil is then pulled completely out of the field in 0.100 s. Find the induced emf in the coil as it is pulled out of the field.

16. It is desired to determine the magnitude of the magnetic field between the poles of a large magnet. A rectangular coil of 10 turns, with dimensions of 5.00 cm by 8.00 cm, is placed perpendicular to the magnetic field. The coil is then pulled out of the field in 0.0500 s and an induced emf of 0.0250 V is observed in the coil. What is the value of the magnetic field between the poles?

17. It is desired to determine the magnetic field of a bar magnet. The bar magnet is pushed through a 10.0-cm diameter circular coil in 2.50×10^{-3} s and an emf of 0.750 V is obtained. Find the magnetic field of the bar magnet.

18. The flux through a 20-turn coil of 20.0 Ω resistance changes by 5.00 Wb/m² in a time of 0.0200 s. Find the induced current in the coil.

†19. Show that whenever the flux through a coil of resistance R changes by $\Delta\Phi_M$, there will always be an induced charge in the loop given by

$$\Delta Q = \frac{-N\Delta\Phi_M}{R}$$

This induced charge becomes the induced current in the coil.

23.4 Lenz's Law

20. What is the direction of the induced current in the solenoid if a north magnetic pole is moved toward the solenoid in the diagram?

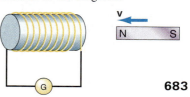

21. What is the direction of the induced current in the solenoid if a south magnetic pole is moved toward the solenoid in the diagram?

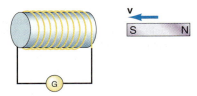

22. Find the direction of the current in the second solenoid when the switch $S1$ in the circuit of solenoid 1 is (a) closed and (b) opened.

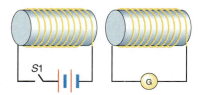

23.5 The Induced emf in a Rotating Loop of Wire in a Magnetic Field— Alternating emf's and the AC Generator

23. A circular coil in a generator has 50 turns, a diameter of 10.0 cm, and rotates in a field of 0.500 T. What must be the angular velocity of the coil if it is to generate a maximum voltage of 50.0 V?

†24. A 50.0-Ω circular coil of wire of 20 turns, 5.00 cm in diameter, is rotated at an angular velocity of 377 rad/s in an external magnetic field of 2.00 T. Find (a) the maximum induced emf, (b) the frequency of the alternating emf, (c) the maximum current in the coil, and (d) the instantaneous current at 5.00 s.

25. An AC generator consists of a coil of 200 turns, 10.0 cm in diameter. If the coil rotates at 500 rpm in a magnetic field of 0.250 T, find the maximum induced emf.

†26. You are asked to design an AC generator that will give an output voltage of 156 V maximum at a frequency of 60.0 Hz. (a) Find the product of N, the number of turns in the coil, B, the magnetic field, and A, the area of the coil, that will give this value of voltage. (b) Pick a reasonable set of values for N, B, and A to satisfy the requirement.

27. At what angular speed should a coil of 1.00-m² area be rotated in the earth's magnetic field in a region where B is 3.50×10^{-5} T in order to generate a maximum emf of 1.50 V?

28. A circular coil of wire of 25.0-cm² area is placed in a magnetic field of 0.0500 T. What will the induced emf in the coil be if the coil is rotated at 20.0 rad/s (a) about an axis that is aligned with the magnetic field and (b) about an axis that is perpendicular to the magnetic field?

29. An AC generator is designed with a circular cross section of radius 1.25 cm to produce 120 V. What would the radius of the generator coil have to be if it is to produce 240 V, assuming that the magnetic field, the frequency, and the number of turns remains constant?

30. An AC generator produces 10.0 V when the coil rotates at 500 rpm. Find the emf when the coil is made to rotate at 1500 rpm.

23.6 Mutual Induction

31. An emf of 0.800 V is observed in a circuit when a nearby circuit has a current change at the rate of 200 A/s. What is the mutual inductance of the circuits?

32. Find the mutual inductance of two coaxial solenoids of 10.0-cm radius, 20.0 cm in length, with 120 turns in coil 1 and 200 turns in coil 2.

33. Two coils have a mutual inductance of 5.00 mH. If the current in the first coil changes by 3.00 A in 0.0200 s, what is the induced emf in the second coil?

23.7 Self-Induction

34. What is the self-induced emf in a coil of 5.00 H if the current through it is changing at the rate of 150 A/s?

35. A coil has an inductance of 5.00 mH. At what rate should current change in the coil to give an induced emf of 100 V?

36. The current through a 10.0-mH inductor changes at the rate of 250 A/s. Find the change of flux in the coil.

23.8 The Energy Stored in the Magnetic Field of an Inductor

37. How much energy is stored in the magnetic field of a coil of 5.00-mH inductance, carrying a current of 5.00 A?

38. A solenoid has 2500 turns/m, a diameter of 10.0 cm, and a length of 20.0 cm. If the current varies from 0 to 10.0 A, how much energy is stored in the magnetic field of the solenoid?

39. If 80.0 J of energy are stored in an inductor when the current changes from 5.00 A to 15.0 A, find the inductance of the inductor.

Additional Problems

†40. A 10.0-cm bar magnet has a value of $B = 2.50 \times 10^{-3}$ T. It is 30.0 cm above a circular coil of radius 5.00 cm and 20 turns. The magnet is dropped from rest. Find (a) the velocity of the north pole of the bar magnet as it arrives at the coil, (b) the time it takes for the leading edge of the bar magnet to go through the 1.00-mm thickness of the coil, (c) the induced emf in the coil as the leading edge of the bar magnet enters the coil, (d) the emf in the coil while the entire bar magnet goes through the coil, and (e) the emf in the coil while the trailing edge of the bar magnet goes through the coil. State the assumptions you make in the solution of the problem.

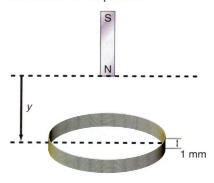

41. In the diagram, the two coils are wound in opposite senses about a common core. The resistor r is variable. Find the direction of the current in the resistor R when (a) r is increased and (b) r is decreased.

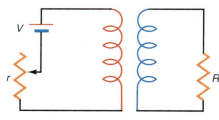

†42. A rod 25.0 cm long is rotated at an angular velocity of 20.0 rad/s in a uniform magnetic field of 4.5×10^{-3} T that points into the page. Find the induced emf from end to end of the rod.

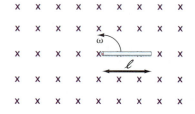

43. A circular loop of radius 5.00 cm is placed in a uniform external magnetic field of 0.0400 T, directed into the page, as shown in the diagram. The circular loop is connected to a resistor, $R = 10.0 \, \Omega$, by wires. The loop is now allowed to collapse until its area becomes effectively zero in 0.010 s. Find the magnitude and direction of the induced current in the resistor R.

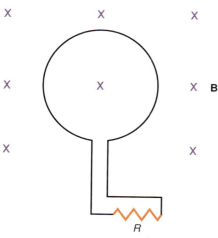

44. Find the induced emf in the inside loop when the current changes in the outside loop at the rate $\Delta i / \Delta t$.

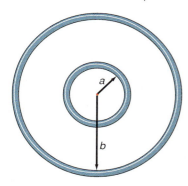

45. A square loop of wire, 2.00 mm on each side, is placed at the center of a flat coil of 10 turns and 10.0 cm radius. The current in the coil increases from 0 to 1.00 A in 2.50 s. If the resistance of the loop is 0.15 Ω, find the emf induced in the square loop.

†46. A rotating coil of wire in a magnetic field is used to create a sparking device. The coil is connected by brushes to two wires C and D, the ends of which are separated by 1.00 mm. (a) Find the necessary emf

between points C and D that will permit dielectric breakdown of the air gap between these points. (b) If $N = 200$ turns, the length of the coil is 10.0 cm, its width is 8.00 cm, and the magnetic field is 8.50×10^{-1} T, find the angular velocity ω of the coil to give the induced emf necessary for sparking.

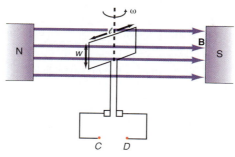

47. A 7.50-cm diameter circular loop of wire is initially oriented so that an external magnetic field of 4.00 T is normal to the plane of the loop. In 2.00 s, the loop rotates about an axis in its own plane so that it ultimately makes an angle of 45° with the field. During the *same* 2.00 s, the magnitude of the external field rises to 5.66 T. Determine the emf induced in the loop.

48. The current through a coil changes from 300 mA to 150 mA in 5.00×10^{-3} s. An induced emf of 2.00×10^{-2} V is obtained. Find (a) the inductance of the coil, (b) the initial energy in the field, and (c) the final energy in the field.

49. An air solenoid has an inductance of 5.00 mH. Find its inductance if a piece of iron ($\mu = 800 \, \mu_0$) is placed within the solenoid.

50. A 5.00-mH inductor and a 50.0-Ω resistor are connected in series to a 24.0-V battery. Find (a) the final current in the circuit and (b) the energy stored in the magnetic field of the inductor.

†51. A coil 5.00 cm in diameter is placed inside a solenoid of 1000 turns/m. (a) If the current in the solenoid rises to 10.0 A in 2.00×10^{-2} s, find the induced emf in the inner coil. (b) When the current in the solenoid is constant at 10.0 A, what is the magnetic flux through the inner coil? (c) If the inner coil is now rotated at 20.0 rad/s, what is the induced emf in the coil? (d) If the inner coil is connected to a circuit with a resistance of 30.0 Ω, find the maximum current that will flow in the circuit.

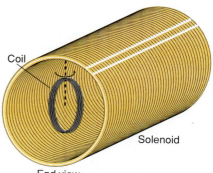

Solenoid

End view

†52. A 5.00-μF capacitor is charged to a potential of 100 V and is then connected to a coil of 7.00-mH inductance at points A and B in the diagram. (a) Find the original energy in the capacitor. (b) As the capacitor discharges, what is the maximum current in the coil? (c) As the capacitor discharges, where does the energy in the capacitor go? (d) What will happen after the capacitor completely discharges?

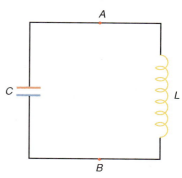

†53. In the derivation of equation 23.18, the induced emf in wire cd was given by $\mathscr{E}_{cd} = vB\ell \sin \theta$. Using figure 23.5, find the angle between $\mathbf{v}$ and $\mathbf{B}$ for wire cd. Show that the sine of that angle reduces to the sine of the angle θ.

†54. It was shown in figure 23.2 that when the wire MN is moved to the right a current is induced in the wire from M to N. Show that this wire is a current-carrying wire in an external magnetic field, and as such experiences a force. Show that this force opposes the original motion and tends to slow down the motion of the wire and is a manifestation of Lenz's law.

†55. As a variation of the motional emf studied in figure 23.2, let the rails be placed on an inclined plane as shown. Find how the induced emf varies with time.

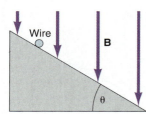

Side view

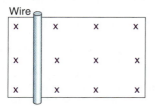

Top view

Interactive Tutorials

⌂56. Magnetic flux. A uniform magnetic field, $B = 3.55 \times 10^{-2}$ T, passes through a plane surface of area $A = 9.35 \times 10^{-2}$ m² at an angle $\theta = 53.5°$ to the normal of the surface. Find the magnetic flux Φ_M passing through the plane surface.

⌂57. Faraday's law. A coil of area $A = 0.035$ m² is connected to a galvanometer. A magnetic field varies from an initial value $B_i = 0.200$ T to a final value $B_f = 0.500$ T in a time of 1.50×10^{-3} s. The initial and final values of the magnetic field vector points out of the area in the direction of the area vector **A**. If the resistance of the circuit is 20.5 Ω, find (a) the induced emf in the circuit and (b) the current in the circuit while the magnetic field is increasing with time.

⌂58. An AC generator. A coil with $N = 200$ turns of wire, with a cross-sectional area $A = 0.015$ m², is rotated at an angular speed of $\omega = 377$ rad/s in an external magnetic field $B = 0.225$ T. Find (a) the maximum induced emf $\mathcal{E}_{max}$, (b) the frequency f of the alternating emf, and (c) if the resistance of the circuit is $R = 100$ Ω, find the maximum current i_{max}. (d) Write the equation for the induced alternating emf $\mathcal{E}$ and the induced alternating current i and make a plot of both.

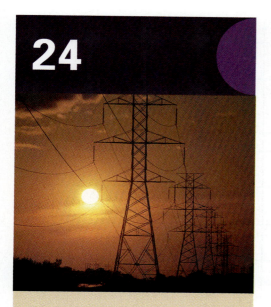

24

Alternating Current Circuits

The distribution of electron power.

History teaches us that the searching spirit of man required thousands of years for the discovery of the fundamental principles of the sciences, on which the superstructure was then raised in a comparatively short time. But these very fundamental propositions are nevertheless so clear and simple, that the discover of them reminds us, in more than one respect, of Columbus's egg.

Julius Robert Mayer

24.1 Introduction

Up to now the only electric circuits considered were DC circuits. It is only natural that this historical sequence should be followed because the original power source available for a circuit was the DC battery. However, with the knowledge gained through electromagnetic induction, it soon became apparent that by mechanically rotating a loop of wire in an external magnetic field, a sinusoidally varying electromagnetic force could be obtained in the rotating loop of wire. This varying emf gives rise to a varying alternating current. This new AC generator was described in section 23.5. Recall that the alternating emf was given by

$$\mathscr{E} = \omega AB \sin \omega t \tag{23.22}$$

or

$$\mathscr{E} = \mathscr{E}_{\text{max}} \sin \omega t \tag{23.24}$$

where

$$\mathscr{E}_{\text{max}} = \omega AB \tag{23.23}$$

Here ω is the angular speed of the rotating coil, A is the cross-sectional area of the coil, and B is the magnetic field that the coil is rotated through.

The resulting alternating current in the coil was found to be

$$i = \frac{\mathscr{E}_{\text{max}}}{R} \sin \omega t = i_{\text{max}} \sin \omega t \tag{23.26}$$

The angular speed ω of the current loop is related to the frequency f of the alternating current by

$$\omega = 2\pi f \tag{24.1}$$

24.2 The Effective Current and Voltage in an AC Circuit

In contrast to a direct current, in which charge flows in one direction, in an alternating current the charge flows first in one direction and then in the opposite direction. The alternating current i varies sinusoidally with time, as shown in figure 24.1(a).

The current i starts from zero and increases in one direction until it reaches its maximum value i_{max}. It then decreases to zero. It starts to increase again, but

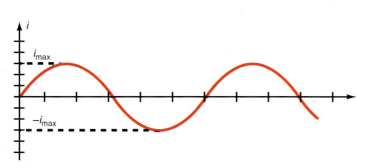

(a) The alternating current

Figure 24.1

An alternating current.

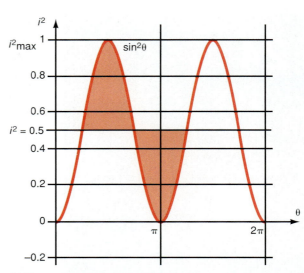

(b) The square of an alternating current

now the charge is flowing in the opposite direction. This is shown as the negative portion of the sine wave. The current increases negatively to $-i_{max}$ and then decreases to zero, where the cycle begins again. We can express the alternating current i as

$$i = i_{max} \sin(2\pi ft) \qquad (24.2)$$

where f is the frequency of the alternating current and t is the time.

With a DC current, the current is specified as having a certain value, let us say 5 A. It always has that value because it is a constant current. We would also like to describe an AC current by a single number, such as a current of 5 A. However, with an AC current, we cannot say that the current has only one value; it has many values because the current is constantly changing with time. It would certainly be desirable to be able to refer to a single value for the alternating current. We therefore need to find a different way to describe the alternating current in order to consider it as having a single value, like a DC current. We cannot simply take the average of an AC current, because the average current is zero; that is, half the time, the current is positive, half the time, the current is negative. We need to look at some other way to describe the effects of the alternating current.

Consider the heating effect of the alternating current. The power P dissipated in a resistor R in a DC circuit carrying a current I was found in chapter 20 to be

$$P = I^2 R \qquad (20.18)$$

Using the same procedure, the power dissipated in an AC circuit is

$$P = i^2 R = [i_{max} \sin(2\pi ft)]^2 R$$
$$= i^2_{max} \sin^2(2\pi ft)\ R \qquad (24.3)$$

Equation 24.3 shows that the power dissipated across the resistor is a function of the time t. That is, the power dissipated is not a constant as in the DC circuit but varies with time. Let us, instead, consider the average value of the power generated, that is,

$$P_{avg} = i^2_{max}\ [\sin^2(2\pi ft)]_{avg} R$$

Although the average value of $\sin(2\pi ft)$ is zero, the average value of its square is not, as we can see from figure 24.1(b). We find the average value of $\sin^2(2\pi ft)$ by using the trigonometric identity

$$\sin^2\theta = \tfrac{1}{2}(1 - \cos 2\theta)$$

Applying that identity to the problem

$$\sin^2(2\pi ft) = \tfrac{1}{2}[1 - \cos(4\pi ft)]$$

Its average value is

$$[\sin^2(2\pi ft)]_{\text{avg}} = \tfrac{1}{2}\{1 - [\cos(4\pi ft)]_{\text{avg}}\}$$

However, the average value of the cosine is zero. Hence,

$$[\sin^2(2\pi ft)]_{\text{avg}} = \tfrac{1}{2}$$

We can also see this in figure 24.1(b) by noting that the straight line labeled $i^2 = 0.5$ cuts the $\sin^2\theta$ curve in half. Thus the shaded area above 0.5 is seen to be equal to the shaded area below 0.5. Using this value of $\tfrac{1}{2}$ for the average value of the $\sin^2(2\pi ft)$, the average power expended in an AC circuit is found from equation 24.3 as

$$P_{\text{avg}} = \tfrac{1}{2}i^2_{\text{max}}R \qquad (24.4)$$

This average power expended in an AC circuit, with a current varying from $-i_{\text{max}}$ to $+i_{\text{max}}$, is not equivalent to the power expended in a DC circuit carrying the constant current i_{max}. To find the equivalence between the power expended in an AC circuit and the power expended in a DC circuit, we introduce an effective current. *An **effective current** i_{eff} is defined as a constant current that generates heat in a resistor R at the same rate as an alternating current.* Therefore, the average power generated in an AC circuit is equal to the power generated in a DC circuit carrying an effective current i_{eff}, that is,

$$(P)_{\text{DC}} = (P_{\text{avg}})_{\text{AC}}$$
$$i^2_{\text{eff}}R = \tfrac{1}{2}i^2_{\text{max}}R$$

Hence,

$$i_{\text{eff}} = \sqrt{\tfrac{1}{2}}\, i_{\text{max}} = 0.707\, i_{\text{max}} \qquad (24.5)$$

Equation 24.5 gives the effective value of an alternating current. The effective value is equivalent to a constant current of this value. It is 70.7% of the maximum or peak value of the AC current. Even though the current is constantly changing with time, the current can be referred to in terms of a single value, its effective or equivalent value. The effective value of an AC current is equivalent to a constant DC current of this value.

If an alternating current is applied to a resistor, the voltage drop across the resistance R is

$$V = iR$$

Using the definition of the alternating current in equation 24.2, $i = i_{\text{max}} \sin(2\pi ft)$, we can write this as

$$V = i_{\text{max}}R\, \sin(2\pi ft)$$

But $i_{\text{max}}R = V_{\text{max}}$. Hence, the alternating voltage across the resistor is given by

$$V = V_{\text{max}} \sin(2\pi ft) \qquad (24.6)$$

Using the same analogy as for current, we obtain

$$V_{\text{eff}} = \sqrt{\tfrac{1}{2}}\, V_{\text{max}} = 0.707\, V_{\text{max}} \qquad (24.7)$$

*Equation 24.7 gives the effective value of the voltage in an AC circuit. The **effective voltage** is a constant value of the voltage that produces the same effect as the alternating voltage. Thus, the constantly changing alternating voltage is described in terms of the single valued, constant effective voltage.* Whenever we discuss current or voltage in an AC circuit, we mean its effective value. AC ammeters and voltmeters measure the effective value of current and voltage, respectively. *Since* *this effective voltage or current is the square root of the mean or average value of the voltage or current squared, it is often called the root-mean-squared voltage or current, or simply the rms voltage or current.* Hence, an alternative notation is occasionally used with I_{eff} replaced by I_{rms}, and V_{eff} replaced by V_{rms}.

Example 24.1

The effective value of an AC current. An AC current in a circuit varies from -3.50 A to $+3.50$ A. Find the effective value of this current.

Solution

The effective value of the current, found from equation 24.5, is

$$i_{eff} = 0.707 i_{max}$$
$$= (0.707)(3.50 \text{ A})$$
$$= 2.47 \text{ A}$$

Thus, even though the current is varying with time, if an AC ammeter were placed in the circuit it would read the single value of 2.47 A.

Example 24.2

Finding the maximum value when the effective value is known. What is the maximum voltage in a 60.0-Hz, 120-V line?

Solution

The effective voltage is 120 V and the maximum voltage, found from equation 24.7, is

$$V_{eff} = 0.707 \ V_{max}$$
$$V_{max} = \frac{V_{eff}}{0.707} = \frac{120 \text{ V}}{0.707}$$
$$= 170 \text{ V}$$

Hence, even though an AC voltmeter would read the single value of 120 V, the actual voltage would be varying between -170 V and $+170$ V.

In the analysis of an AC circuit, we will depart from our usual custom of studying the simplest cases first and then building up to the most general case. Instead, we will start with the general case, an *RLC* series circuit (i.e., a circuit with a resistance *R*, an inductor *L*, and a capacitor *C*, connected in series) and take as special cases the simpler examples of an *RC* circuit and an *RL* circuit. By now, the student should be advanced enough to treat the general case first, and to be able to see the beauty in the general case, in that it contains all the other examples as special cases. The *LC* circuit will be left as a problem for the student to do.

24.3 An *RLC* Series Circuit

Consider a circuit containing a resistor *R*, an inductor *L*, and a capacitor *C*, connected in series. This combination, shown in figure 24.2, is called an ***RLC*** **series**

Figure 24.2

An *RLC* series circuit.

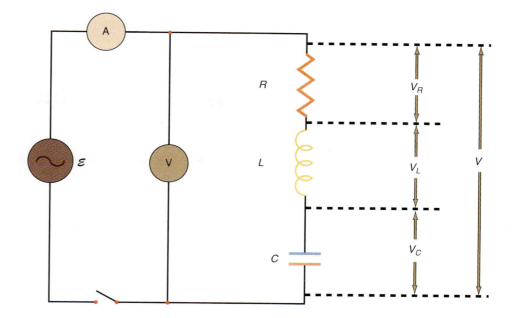

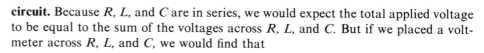

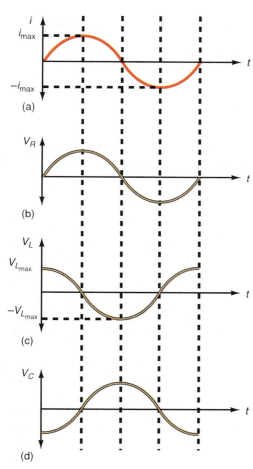

Figure 24.3

Phase relations in a series *RLC* circuit.

circuit. Because R, L, and C are in series, we would expect the total applied voltage to be equal to the sum of the voltages across R, L, and C. But if we placed a voltmeter across R, L, and C, we would find that

$$V \neq V_R + V_L + V_C \tag{24.8}$$

The effective voltages do not add up algebraically because the sinusoidal voltages are not in phase with one another.

Phase Relations

The different phase relations can be seen with the help of figure 24.3. To understand these phase relations it helps to see why the curves take on their maximum, minimum, and zero values at the location where they do.

Phase Relationship for a Resistor Figure 24.3(a) shows the AC current in the circuit. Part b of the diagram shows the voltage drop across the resistor. Since $V_R = iR$, the current is multiplied by the constant R. This cannot change the location of the maximum or minimum values of the current. Hence, the voltage is everywhere in phase with the current i.

Phase Relationship for an Inductor Recall that the voltage drop across an inductor is given by

$$V_L = L\frac{\Delta i}{\Delta t} \tag{24.9}$$

Therefore, the maximum, zero, and minimum values of V_L depend on the rate at which the current changes with time in the circuit, that is, $\Delta i/\Delta t$. Note that $\Delta i/\Delta t$ is the slope of the curve obtained when the current is plotted as a function of the time and that $\Delta i/\Delta t$ is equal to zero at either a maximum or minimum of the graph of current versus time, that is, when the slope is horizontal. Hence,

$$V_L = 0 \quad \text{when } i = \pm i_{\text{max}}$$

Remember that the current is alternating between positive and negative values, and thus there are two places where $V_L = 0$, one at $i = +i_{\text{max}}$ and one at $i = -i_{\text{max}}$. Also, V_L takes on its greatest value when $\Delta i/\Delta t$ is at its maximum. This occurs when the current is changing from positive to negative or vice versa; that is, when $i = 0$. When $t = 0$, the slope $\Delta i/\Delta t$ is positive, and hence V_L is positive, as seen from equation 24.9. Thus,

$$V = +V_{Lmax} \quad \text{when } t = 0 \text{ and } i = 0$$

When $t = T/2$ (where T is the time period), $i = 0$, but the slope $\Delta i/\Delta t$ is negative; hence, from equation 24.9, V_L is negative. That is,

$$V = -V_{Lmax} \quad \text{when } t = T/2 \text{ and } i = 0$$

Knowing the locations of the maximum, minimum, and zero values of the voltage across the inductor allows us to draw the entire sinusoidal voltage as a function of time. The result of this analysis is shown in figure 24.3(c). *Note that the voltage across the inductance leads the voltage across the resistance and the current by 90°.* In other words, V_L leads V_R or i, because the peak of V_L occurs before the peak of V_R.

Phase Relationship for a Capacitor The potential drop across the capacitor is

$$V_C = \frac{q}{C} \tag{24.10}$$

The value of V_C is zero whenever the charge q on the capacitor is equal to zero. The charge q is zero when the current is at its maximum. That is, the current is "on," but the charge has not yet had time to reach the capacitor and build up its charge. Thus, all the charge is in the circuit and none is on the capacitor. Hence,

$$V_C = 0 \quad \text{when } i = \pm i_{max}$$

Remember that the current is alternating between positive and negative values, and thus there are two places where $V_C = 0$, one at $i = +i_{max}$ and one at $i = -i_{max}$. As time continues, charge starts to build up on the capacitor, and a voltage V_C starts to appear across the capacitor. When the current reaches zero, the entire charge has piled up on the capacitor, and the capacitor has its maximum charge, $\pm q_{max}$. That is, all the charge is now on the capacitor and none is in the circuit. Hence,

$$V_C = \pm V_{Cmax} \quad \text{when } q = \pm q_{max} \quad \text{which occurs when } i = 0$$

The variation of the potential drop across the capacitor is shown in figure 24.3(d).

To understand why V_C is negative rather than positive when $t = 0$, refer to figure 24.4, which shows the variation of the charge on the capacitor with the current in the circuit. Let us start at the point b when the instantaneous current is at a maximum. At this moment in time, the charge q on the capacitor is zero, and hence $V_C = 0$. As the current decreases in value from b to c, the charge piles up on the capacitor. This appears as q increasing from b to c. At c the entire charge q is on the plates of the capacitor, and there is no charge moving about in the circuit, that is, the current $i = 0$. This charge on the plates now starts to flow off the plates in the opposite direction, and as such appears as a negative current between c and d. Of course, q is seen to be decreasing from c to d. At d there is no more charge q on the plates of the capacitor (it is now all in the circuit). Thus, the current is maximum at d and in the opposite direction from which it was at b. This negative current starts to add negative charges to the original plate, and this is seen as the charge q going negative between d and e. At e the current has decreased to zero, and the capacitor is fully charged. The charge now starts to flow in the positive direction between e and f, and the charge on the plates decreases to zero at f.

The cycle now starts over again as $b \rightarrow c \rightarrow d \rightarrow e \rightarrow f \rightarrow \cdots$. Because we started the cycle at b, the curve for the charge q can be dashed back to a, as shown. Since $V_C = q/C$, the V_C curve follows the charge curve exactly. Hence, we

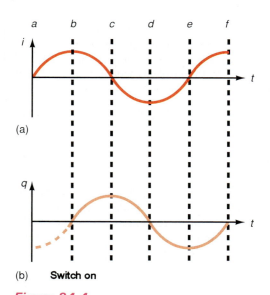

(a)

(b)   **Switch on**

Figure 24.4

Variation of charge on a capacitor in an AC circuit.

see that the V_C *curve is one-quarter of a cycle behind the current curve. We can therefore say that the voltage drop across a capacitor lags behind the voltage across the resistor or the current in the AC circuit by 90°.*

An old mnemonic device to remember whether the emf across an inductor or capacitor leads or lags the current in an AC circuit, is the phrase "ELI the ICE man." Meaning, in an inductor L, the emf $\mathscr{E}$ leads the current I in the circuit, whereas in a capacitor C, the emf $\mathscr{E}$ lags behind the current I in the circuit.

Ohm's Law for an RLC Series Circuit

Equation 24.8 stated that the applied effective voltage in the circuit was not equal to the algebraic sum of the effective voltages, that is,

$$V \neq V_R + V_L + V_C \tag{24.8}$$

If we were to place a voltmeter across the circuit, we would not read the algebraic sum of these values, because, as just seen, the voltage across the inductor V_L and the voltage across the capacitor V_C are out of phase with the voltage across the resistor V_R. The instantaneous values of V_R, V_L, and V_C do add up by Kirchhoff's rule. It is the phase relations that causes the effective values not to add up algebraically. Since V_L and V_C are 180° out of phase with each other, as seen in figure 24.3, and V_L is 90° out of phase with V_R, we can treat each of these voltages as though they were vectors, as shown in figure 24.5(a). Taking the x-axis as the reference direction, we place V_R along the x-axis. V_L then lies along the positive y-axis, which, if we think of a counterclockwise rotation of the vector, is 90° before the x-axis, and V_C lies along the negative y-axis, which is 90° after the x-axis. The sum of these voltages becomes the vector sum shown in figure 24.5(b) and its magnitude is given by

$$V = \sqrt{V_R^2 + (V_L - V_C)^2} \tag{24.11}$$

The voltage across RLC is, of course, also equal to the impressed voltage from the AC source. The angle ϕ between the applied voltage V and the voltage across the resistor V_R is called the phase angle, and is found from the figure to be

$$\phi = \tan^{-1}\left(\frac{V_L - V_C}{V_R}\right) \tag{24.12}$$

Recall that the voltage across the resistor is in phase with the current in the circuit. Hence, *the phase angle is a measure of how much the applied voltage V leads or lags the AC current in the circuit.* If ϕ is above the x-axis, ϕ is positive, V_L is greater than V_C, and the voltage leads the current by ϕ degrees. A circuit that has a positive value of ϕ is called an *inductive circuit*. If ϕ lies below the x-axis, ϕ is negative, V_C is greater than V_L, and the voltage lags the current by ϕ degrees. A circuit that has a negative value of ϕ is called a *capacitive circuit*.

Just as the potential drop across the resistor is given by Ohm's law as

$$V_R = iR \tag{24.13}$$

the effective voltage drop across the inductor can be given by

$$V_L = iX_L \tag{24.14}$$

where X_L is defined as the **inductive reactance**. The inductive reactance X_L is the inductive analogue to resistance, that is, it also has the effect of impeding current in a circuit. The inductive reactance, which is measured in ohms, is

$$X_L = 2\pi f L \tag{24.15}$$

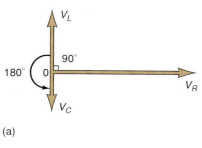

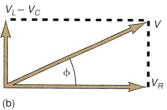

Figure 24.5

Vector diagram of the voltages in an AC circuit.

where f is the frequency of the alternating current and L is the inductance of the circuit, measured in henries (H). A derivation of the inductive reactance, equation 24.15, requires the use of the calculus and is, therefore, omitted here. Note that in practice, most inductors also have some resistance. We will assume here that the resistance is small enough to be ignored. In the laboratory, it is a good idea to measure the resistance of the coil you are using to see if the assumption is valid.

The potential drop across the capacitor, using the same analogy, is given by

$$V_C = iX_C \tag{24.16}$$

where X_C is the **capacitive reactance.** The capacitive reactance also measured in ohms, is

$$X_C = \frac{1}{2\pi f C} \tag{24.17}$$

where f is again the frequency of the alternating current and C is the capacitance of the circuit element, measured in farads (F). The capacitive reactance, like the inductive reactance, cannot be derived without the use of the calculus. So let us accept them as though they were experimental results.

Substituting equations 24.13, 24.14, and 24.16 back into equation 24.11, yields

$$V = \sqrt{(iR)^2 + (iX_L - iX_C)^2}$$

or

$$\frac{V}{i} = \sqrt{R^2 + (X_L - X_C)^2}$$

Just as V/i was defined as the resistance R of a DC circuit, we will now define the ratio as the **impedance** Z of an AC circuit, that is,

$$Z = \sqrt{R^2 + (X_L - X_C)^2} \tag{24.18}$$

With this definition, Ohm's law for an AC circuit takes the simple form

$$i = \frac{V}{Z} \tag{24.19}$$

Therefore, in order to solve for the current in an AC circuit, we reduce it to the same simple form as a DC circuit. Instead of simply knowing the resistance, as in a DC circuit, we must now know the impedance of an AC circuit for its solution. The impedance is a function of the frequency of the AC circuit, through its inductive and capacitive reactances. It is often helpful to draw an impedance diagram, which can be deduced from the voltage diagram, and is shown in figure 24.6.

Example 24.3

An RLC series circuit. The *RLC* series circuit shown in figure 24.7 has a resistance $R = 400\ \Omega$, an inductor $L = 5.00$ H, a capacitor $C = 3.00\ \mu$F, and they are connected to a 110-V, 60.0-Hz line. Find (a) the inductive reactance X_L, (b) the capacitive reactance X_C, (c) the impedance Z of the circuit, (d) the current i in the circuit, (e) the voltage drop V_R across R, (f) the voltage drop V_L across L, (g) the voltage drop V_C across C, (h) the total voltage V across RLC, and (i) the phase angle ϕ.

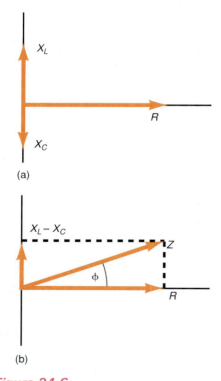

Figure 24.6

Impedance diagram for an AC circuit.

Figure 24.7

An example of an *RLC* circuit.

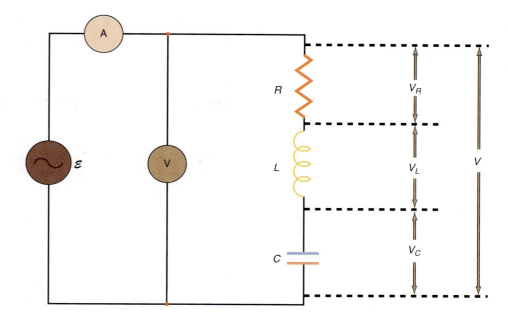

Solution

a. The inductive reactance, found from equation 24.15, is

$$X_L = 2\pi f L$$

$$= 2\pi \left(60.0 \, \frac{1}{s}\right)(5.00 \text{ H})\left[\frac{\text{V}/(\text{A}/\text{s})}{\text{H}}\right]$$

$$= 1890 \, \frac{\text{V}}{\text{A}} \left(\frac{\Omega}{\text{V}/\text{A}}\right)$$

$$= 1890 \, \Omega$$

b. The capacitive reactance, found from equation 24.17, is

$$X_C = \frac{1}{2\pi f C} = \frac{1}{2\pi(60.0 \, 1/\text{s})(3.00 \times 10^{-6} \text{ F})} \left(\frac{\text{F}}{\text{C}/\text{V}}\right)$$

$$= 884 \, \frac{\text{V}}{\text{C}/\text{s}} \left(\frac{\text{C}/\text{s}}{\text{A}}\right)\left(\frac{\Omega}{\text{V}/\text{A}}\right)$$

$$= 884 \, \Omega$$

c. The impedance of the circuit, found from equation 24.18, is

$$Z = \sqrt{R^2 + (X_L - X_C)^2}$$

$$= \sqrt{(400 \, \Omega)^2 + (1890 \, \Omega - 884 \, \Omega)^2}$$

$$= 1080 \, \Omega$$

d. The effective current i in the circuit, found from equation 24.19, is

$$i = \frac{V}{Z}$$

$$= \frac{110 \text{ V}}{1080 \, \Omega}$$

$$= 0.102 \text{ A}$$

e. The voltage drop across R, found from equation 24.13, is

$$V_R = iR$$
$$= (0.102 \text{ A})(400 \text{ } \Omega)$$
$$= 40.8 \text{ V}$$

f. The voltage drop across L, found from equation 24.14, is

$$V_L = iX_L$$
$$= (0.102 \text{ A})(1890 \text{ } \Omega)$$
$$= 193 \text{ V}$$

g. The voltage drop across C, found from equation 24.16, is

$$V_C = iX_C$$
$$= (0.102 \text{ A})(884 \text{ } \Omega)$$
$$= 90.2 \text{ V}$$

h. The total voltage across R, L, and C in series, found from equation 24.11, is

$$V = \sqrt{V_R^2 + (V_L - V_C)^2}$$
$$= \sqrt{(40.8 \text{ V})^2 + (193 \text{ V} - 90.2 \text{ V})^2}$$
$$= 110 \text{ V}$$

which is, of course, equal to the applied voltage. Notice that the voltages are added vectorially and not algebraically.

i. The phase angle, found from equation 24.12, is

$$\phi = \tan^{-1}\left(\frac{V_L - V_C}{V_R}\right)$$
$$= \tan^{-1}\left(\frac{193 \text{ V} - 90.2 \text{ V}}{40.8 \text{ V}}\right)$$
$$= 68.4°$$

This means that the applied voltage leads the current in the circuit by 68.4°, and the phase relation is shown in figure 24.8. Since ϕ is a positive angle the circuit is called an inductive circuit.

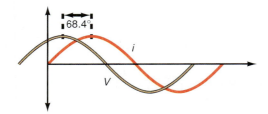

Figure 24.8

The phase angle between the applied voltage and the current in the AC circuit.

Figure 24.9

An *RC* series circuit.

Example 24.4

An RC series circuit. A 110-V, 60.0-Hz, AC line is connected across a resistance of 1000 Ω and a capacitor of 1.00 μF, as shown in figure 24.9. Find (a) the capacitive reactance, (b) the impedance, (c) the current in the circuit, (d) the voltage drop V_R across the resistor, (e) the voltage drop V_C across the capacitor, (f) the total voltage across R and C, and (g) the phase angle ϕ between the voltage and current.

Solution

a. The capacitive reactance X_C, found from equation 24.17, is

$$X_C = \frac{1}{2\pi fC} = \frac{1}{2\pi(60.0 \text{ 1/s})(1.00 \times 10^{-6} \text{ F})}$$
$$= 2650 \text{ } \Omega$$

b. The impedance Z is found from equation 24.18. Because there is no induction in this RC circuit, $X_L = 0$. Therefore, the impedance becomes

$$Z = \sqrt{R^2 + X_C^2}$$

$$= \sqrt{(1000 \ \Omega)^2 + (2650 \ \Omega)^2}$$

$$= 2830 \ \Omega$$

c. The effective current i in the circuit comes from Ohm's law, equation 24.19, and is

$$i = \frac{V}{Z} = \frac{110 \ \text{V}}{2830 \ \Omega}$$

$$= 3.89 \times 10^{-2} \ \text{A}$$

d. The voltage drop V_R across the resistor, found from equation 24.13, is

$$V_R = iR = (3.89 \times 10^{-2} \ \text{A})(1000 \ \Omega)$$

$$= 38.9 \ \text{V}$$

e. The voltage drop V_C across the capacitor, found from equation 24.16, is

$$V_C = iX_C$$

$$= (3.89 \times 10^{-2} \ \text{A})(2650 \ \Omega)$$

$$= 103 \ \text{V}$$

f. The total voltage drop across R and C in series is found from equation 24.11. Since there is no inductance in this circuit, $V_L = 0$. Therefore,

$$V = \sqrt{V_R^2 + V_C^2}$$

$$= \sqrt{(38.9 \ \text{V})^2 + (103 \ \text{V})^2}$$

$$= 110 \ \text{V}$$

Note that the voltage across R and C in series is the same as the applied voltage, which it should be. Because of the phase difference of the voltages, they add as vectors rather than as algebraic quantities.

g. The phase angle ϕ between the voltage and the current in the circuit is found from equation 24.12 with $V_L = 0$. Therefore,

$$\phi = \tan^{-1}\left(\frac{-V_C}{V_R}\right)$$

$$= \tan^{-1}\left(\frac{-103 \ \text{V}}{38.9 \ \text{V}}\right)$$

$$= -69.3°$$

This phase angle is represented in figures 24.10(a) and 24.10(b). The voltage in the circuit lags the current in the circuit by 69.3°. Since ϕ is a negative quantity, the circuit is called a *capacitive circuit*.

Example 24.5

An RL series circuit. A 110-V, 60-Hz, AC line is connected across a resistance of 1000 Ω and an inductor of 5.00 H, as shown in figure 24.11. Find (a) the inductive reactance, (b) the impedance, (c) the current in the circuit, (d) the voltage drop V_R across the resistor, (e) the voltage drop V_L across the inductor, (f) the total voltage drop V across R and L, and (g) the phase angle between the voltage V and the current i.

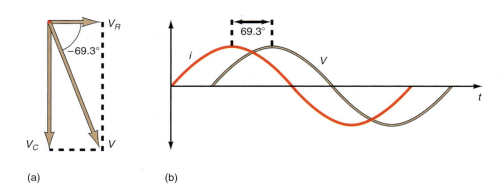

Figure 24.10

Phase relation for an *RC* circuit.

V_R

$-69.3°$

V_C

V

(a)

69.3°

i

V

t

(b)

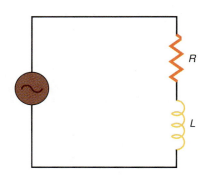

R

L

Figure 24.11

An *RL* series circuit.

a.   The inductive reactance X_L, found from equation 24.15, is

$$X_L = 2\pi f L = 2\pi(60 \ 1/s)(5.00 \ H)$$

$$= 1890 \ \Omega$$

b.   The impedance Z is found from equation 24.18, with $X_C = 0$. Therefore,

$$Z = \sqrt{R^2 + X_L^2}$$

$$= \sqrt{(1000 \ \Omega)^2 + (1890 \ \Omega)^2}$$

$$= 2140 \ \Omega$$

c.   The current i in the circuit, found from Ohm's law, equation 24.19, is

$$i = \frac{V}{Z} = \frac{110 \ V}{2140 \ \Omega}$$

$$= 5.14 \times 10^{-2} \ A$$

d.   The voltage drop V_R across the resistor, found from equation 24.13, is

$$V_R = iR = (5.14 \times 10^{-2} \ A)(1000 \ \Omega)$$

$$= 51.4 \ V$$

e.   The voltage drop V_L across the inductor, found from equation 24.14, is

$$V_L = iX_L = (5.14 \times 10^{-2} \ A)(1890 \ \Omega)$$

$$= 97.1 \ V$$

f.   The total voltage drop V across R and L is found from equation 24.11 with $V_C = 0$. Therefore,

$$V = \sqrt{V_R^2 + V_L^2}$$

$$= \sqrt{(51.4 \ V)^2 + (97.1 \ V)^2}$$

$$= 110 \ V$$

Note that the voltage drop across R and L is the same as the applied voltage V, as is expected.

g.   The phase angle ϕ between the current and voltage in the circuit is found from equation 24.12 with $V_C = 0$. Therefore,

$$\phi = \tan^{-1}\frac{V_L}{V_R} = \tan^{-1}\frac{97.1 \ V}{51.4 \ V}$$

$$= 62.1°$$

Figure 24.12

Phase relations in an *RL* circuit.

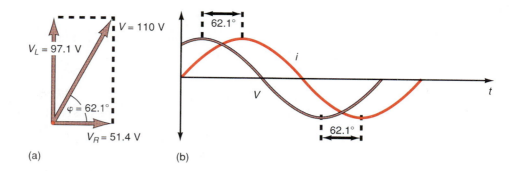

(a)   (b)

Because this is a positive angle it means that the voltage leads the current in the circuit by 62.1° and the circuit is an *inductive circuit*. This is shown in figures 24.12(a) and 24.12(b).

24.4 Resonance in an *RLC* Series Circuit

Not only does the instantaneous value of the AC current change with time, but the effective value of the AC current even changes with the frequency of the AC source. When is this current a maximum in the *RLC* series circuit? The current is given by Ohm's law as

$$i = \frac{V}{Z} = \frac{V}{\sqrt{R^2 + (X_L - X_C)^2}} \qquad (24.20)$$

Notice from the form of the equation that the current i is a maximum whenever the impedance Z is a minimum. This condition is called **resonance.** As we can see from equation 24.20, Z is a minimum whenever $X_L = X_C$. In that case the impedance Z is equal to the resistance R. To determine the frequency at which resonance occurs, we set the inductive reactance equal to the capacitive reactance and get

$$X_L = X_C$$

$$2\pi fL = \frac{1}{2\pi fC}$$

Solving for the **resonant frequency** we get

$$f_0 = \frac{1}{2\pi\sqrt{LC}} \qquad (24.21)$$

where f_0 is the frequency at which resonance occurs. Notice that it depends on the value of the inductance L and capacitance C. At this frequency, the maximum current that can flow in the circuit is obtained. If we plot the current in the circuit as a function of the frequency of the impressed voltage source, we obtain figure 24.13. At the frequency f_0, the maximum current is obtained. At resonance, V_L and V_C are equal, but are 180° out of phase. Resonance is used in the tuning circuit of a radio and in variable oscillators such as are found in physics laboratories.

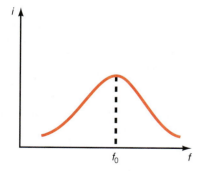

Figure 24.13

Resonant frequency of an *RLC* series circuit.

Example 24.6

Resonant frequency. Find the resonant frequency of the *RLC* series circuit in example 24.3.

The resonant frequency, found from equation 24.21, is

$$f_0 = \frac{1}{2\pi\sqrt{LC}}$$

$$= \frac{1}{2\pi\sqrt{(5.00\ \text{H})(3.00\times10^{-6}\ \text{F})}}\ \sqrt{\frac{\text{H}}{\text{V}/(\text{A/s})}}\ \sqrt{\frac{\text{F}}{\text{C/V}}}$$

$$= 41.1\ \sqrt{\frac{\text{A}}{\text{s C}}\left(\frac{\text{C/s}}{\text{A}}\right)} = 41.1\ \frac{1}{\sqrt{\text{s}^2}}$$

$$= 41.1\ \text{cycles/s} = 41.1\ \text{Hz}$$

24.5 Power in an AC Circuit

A current in an inductor causes a magnetic field to exist within that coil as discussed in chapter 22. In chapter 23 we saw that the work done in moving a charge through a coil shows up as stored energy in the magnetic field of the coil. If the coil is a perfect inductor, that is, its resistance is zero or at least negligible, all the energy applied to the coil is stored in the magnetic field of the coil and no energy is lost in the coil. When the applied emf goes to zero, the energy in the magnetic field of the coil is returned to the circuit by a current. Hence, there is no energy lost in a perfect inductor. Whatever energy the inductor absorbs in one half of a cycle, it returns to the circuit in the next half-cycle.

The capacitor acts similarly. As charge is stored on the plates of the capacitor, the energy used to place that charge on the capacitor is stored in the electric field between the plates of the capacitor. The energy absorbed by the electric field in one half-cycle is returned to the circuit in the next half-cycle. Therefore, there is no power consumed by the inductor or the capacitor in an AC circuit. Only across the resistance R of the circuit, is power dissipated by Joule heating.

The power lost across the resistor is given by

$$P = iV_R$$

But as we can see in figure 24.5,

$$V_R = V\cos\phi$$

where ϕ is the phase angle between the voltage and the current i in the circuit. Therefore, the power dissipated in an AC circuit is

$$P = iV\cos\phi \qquad (24.22)$$

The quantity $\cos\phi$ is called the power factor of the circuit, and is given by

$$\text{PF} = \cos\phi \qquad (24.23)$$

We can obtain further insight into the meaning of the power factor from the geometry in figure 24.6 as

$$\text{PF} = \cos\phi = \frac{R}{Z} \qquad (24.24)$$

If we multiply both numerator and denominator of equation 24.24 by i^2, we get

$$\text{PF} = \cos\phi = \frac{i^2R}{i^2Z} = \frac{i^2R}{iV_\text{applied}}$$

The quantity i^2R is the power dissipated by the circuit, whereas $iV_{applied}$ is the power supplied to the circuit by the AC generator. Therefore,

$$\text{PF} = \cos\phi = \frac{\text{Power consumed}}{\text{Power supplied}} \qquad (24.25)$$

*Hence, the **power factor** is the ratio of how much power is actually used in the circuit to the total power that must be supplied to the circuit.* The power factor is quite often expressed as a percentage. For example, if the phase angle $\phi = 30.0°$, the power factor is

$$\text{PF} = \cos 30.0° = 0.866 = 86.6\%$$

This means that 86.6% of the power supplied to the circuit is actually consumed in the circuit. The rest is returned to the generator. A power factor close to 1 (or 100%) is usually desired for AC circuits. If a circuit has a power factor of 50%, and 500 W of power are actually used in the circuit, the AC generator would have to be able to deliver 1000 W. Thus, a larger and more expensive generator would be necessary to supply energy to a circuit with a low power factor. If the power factor was 1 (or 100%) a generator rated at 500 W would be adequate for the circuit.

Example 24.7

Power factor in an AC circuit. In the *RLC* series circuit of example 24.3, find (a) the power factor, (b) the power consumed, and (c) the total power that must be supplied to the circuit.

Solution

a. In example 24.3 the phase angle ϕ was found to be 68.4°. The power factor of the circuit, found from equation 24.23, is

$$\text{PF} = \cos\phi = \cos 68.4°$$
$$= 0.369 = 36.9\%$$

b. The power dissipated in the circuit is

$$P = i^2R = (0.102 \text{ A})^2(400 \text{ }\Omega)$$
$$= 4.16 \text{ W}$$

c. The power applied to the circuit is

$$P_{app} = iV_{app} = (0.102 \text{ A})(110 \text{ V})$$
$$= 11.2 \text{ W}$$

As a check, note that

$$\frac{\text{Power consumed}}{\text{Power supplied}} = \frac{4.16 \text{ W}}{11.2 \text{ W}} = 0.371 = 37.1\%$$

which agrees with the original value, within round-off errors of the calculations.

24.6 An *RLC* Parallel Circuit

An AC circuit with a resistor, inductor, and capacitor in parallel, called an ***RLC* parallel circuit,** is shown in figure 24.14. Since *R, L,* and *C* are in parallel the voltage across each of them is the same. This means that the voltages are not only equal, but they are also in phase with one another. Thus,

$$V = V_R = V_L = V_C$$

Figure 24.14

A parallel *RLC* circuit.

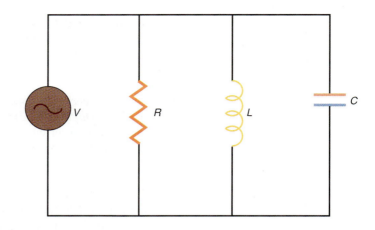

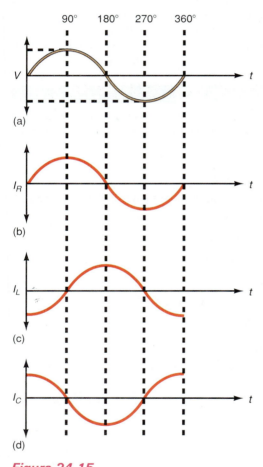

Figure 24.15

Phase relations in a parallel *RLC* circuit.

Because all these voltages are in phase, it is reasonable to expect that some of the currents will not be in phase. The impressed voltage V, or the voltage across the resistor V_R, will be used as the reference to measure all phase differences. The current I_R through R is in phase with V as shown in figures 24.15(a) and 24.15(b). The current through the inductor is out of phase with the current through the resistor. We can see this phase difference by observing the phase relations in the series circuit of figure 24.3, where the voltage V_L leads V_R and the current i by 90°. Therefore, if V_L leads i by 90°, then we can equally well say that i lags V_L by 90°. Hence, the current in the inductor part of the series circuit lags the voltage across the inductor by 90°. For the parallel circuit, the current I_L through the inductor must also lag the voltage across the inductor by 90°. But since the voltage across the inductor is in phase with the voltage across the resistance, which, in turn, is in phase with the current I_R in the resistor, it follows that I_L, the current through the inductor, must lag the current I_R in the resistor by 90°. This is shown in figure 24.15(c).

The current through the capacitor is also out of phase with the current through the resistor. For the series circuit, the voltage V_C lags V_R and i by 90°, as observed in figure 24.3. Therefore, the current in the series circuit leads the voltage across the capacitor by 90°. Thus, in the parallel circuit of figure 24.14, the current through the capacitor must also lead the voltage across the capacitor. Since the voltage across the capacitor is in phase with the voltage across the resistor, which is in phase with the current I_R through the resistor, then the current through the capacitor I_C must lead the current through the resistor I_R by 90°. This is shown in figure 24.15(d). Figure 24.15 can be summarized by saying that

$$I_L \text{ lags } I_R \text{ by } 90°$$

$$I_C \text{ leads } I_R \text{ by } 90°$$

$$I_C \text{ is } 180° \text{ out of phase with } I_L$$

Because the currents in the parallel *RLC* circuit are out of phase with each other, they can be drawn as vectors in figure 24.16(a). The total resultant current I_T, found from the geometry of figure 24.16(b), is

$$I_T = \sqrt{I_R^2 + (I_C - I_L)^2} \qquad (24.26)$$

The individual currents, found by Ohm's law, are

$$I_R = \frac{V}{R} \qquad (24.27)$$

$$I_L = \frac{V}{X_L} \qquad (24.28)$$

$$I_C = \frac{V}{X_C} \qquad (24.29)$$

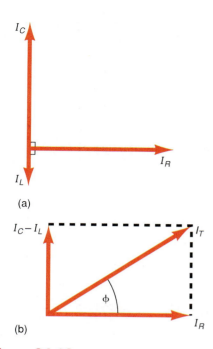

Figure 24.16

Currents in a parallel circuit are added vectorially.

The phase angle ϕ that the total current leads or lags the current in the resistor, and hence the applied voltage, found from figure 24.16(b), is

$$\tan \phi = \frac{I_C - I_L}{I_R}$$

$$\phi = \tan^{-1}\frac{I_C - I_L}{I_R} \qquad (24.30)$$

If we substitute equations 24.27, 24.28, and 24.29 into equation 24.30, we also obtain

$$\phi = \tan^{-1}\frac{\dfrac{V}{X_C} - \dfrac{V}{X_L}}{V/R}$$

$$\phi = \tan^{-1}\left(\frac{R}{X_C} - \frac{R}{X_L}\right) \qquad (24.31)$$

The total impedance of the circuit, found from Ohm's law, is

$$Z = \frac{V}{I_T} \qquad (24.32)$$

Example 24.8

An RLC parallel circuit. A resistor of 400 Ω, an inductor of 5.00 H, and a capacitor of 3.00 μF are connected in parallel to a 110-V, 60.0-Hz line. Find (a) the inductive reactance, (b) the capacitive reactance, (c) the current through the resistor I_R, (d) the current through the inductor I_L, (e) the current through the capacitor I_C, (f) the total current in the circuit I_T, (g) the phase angle ϕ, and (h) the total impedance of the circuit.

Solution

a. The inductive reactance is

$$X_L = 2\pi f L$$
$$= 2\pi(60.0\ 1/s)(5.00\ \text{H})$$
$$= 1890\ \Omega$$

b. The capacitive reactance is

$$X_C = \frac{1}{2\pi f C}$$
$$= \frac{1}{2\pi(60.0\ 1/s)(3.00 \times 10^{-6}\ \text{F})}$$
$$= 884\ \Omega$$

c. The current through the resistor I_R, found from equation 24.27, is

$$I_R = \frac{V}{R}$$
$$= \frac{110\ \text{V}}{400\ \Omega}$$
$$= 0.275\ \text{A}$$

d. The current I_L through the inductor, found from equation 24.28, is

$$I_L = \frac{V}{X_L}$$

$$= \frac{110 \text{ V}}{1890 \ \Omega}$$

$$= 5.82 \times 10^{-2} \text{ A}$$

e. The current I_C through the capacitor, found from equation 24.29, is

$$I_C = \frac{V}{X_C}$$

$$= \frac{110 \text{ V}}{884 \ \Omega}$$

$$= 0.124 \text{ A}$$

f. The total current I_T in the circuit, found from equation 24.26, is

$$I_T = \sqrt{(I_R)^2 + (I_C - I_L)^2}$$

$$= \sqrt{(0.275 \text{ A})^2 + (0.124 \text{ A} - 0.0582 \text{ A})^2}$$

$$= 0.283 \text{ A}$$

g. The phase angle ϕ, found from equation 24.30, is

$$\phi = \tan^{-1}\frac{I_C - I_L}{I_R}$$

$$= \tan^{-1}\left(\frac{0.124 \text{ A} - 0.0582 \text{ A}}{0.275 \text{ A}}\right)$$

$$= 13.5°$$

The total current in the circuit leads the applied voltage by 13.5°.

h. The total impedance of the circuit, found from equation 24.32, is

$$Z = \frac{V}{I_T}$$

$$= \frac{110 \text{ V}}{0.283 \text{ A}}$$

$$= 389 \ \Omega$$

Example 24.9

An RC parallel circuit. A 110-V, 60.0–Hz, AC line is connected in parallel to a resistor of 1000 Ω and a capacitor of 1.00 μF, as shown in figure 24.17. Find (a) the capacitive reactance, (b) the current I_R through the resistor, (c) the current I_C through the capacitor, (d) the total current in the circuit, (e) the phase angle ϕ, and (f) the total impedance of the circuit.

Solution

---

a. The capacitive reactance is found to be

$$X_C = \frac{1}{2\pi f C}$$

$$= \frac{1}{2\pi(60.0 \ 1/\text{s})(1.00 \times 10^{-6} \text{ F})}$$

$$= 2650 \ \Omega$$

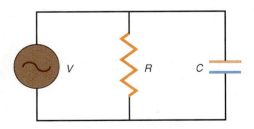

Figure 24.17

A parallel *RC* circuit.

b. The current in the resistor, found from equation 24.27, is

$$I_R = \frac{V}{R}$$

$$= \frac{110 \text{ V}}{1000 \text{ }\Omega}$$

$$= 0.110 \text{ A}$$

c. The current I_C through the capacitor, found from equation 24.29, is

$$I_C = \frac{V}{X_C}$$

$$= \frac{110 \text{ V}}{2650 \text{ }\Omega}$$

$$= 4.15 \times 10^{-2} \text{ A}$$

d. The total current in the circuit is found from equation 24.26 with $I_L = 0$, that is,

$$I_T = \sqrt{I_R^2 + I_C^2}$$

$$= \sqrt{(0.110 \text{ A})^2 + (0.0415 \text{ A})^2}$$

$$= 0.118 \text{ A}$$

e. The phase angle ϕ is found from equation 24.30 with $I_L = 0$, that is,

$$\phi = \tan^{-1}\frac{I_C}{I_R}$$

$$= \tan^{-1}\frac{0.0415}{0.110}$$

$$= 20.7°$$

f. The total impedance of the circuit, found from equation 24.32, is

$$Z = \frac{V}{I_T}$$

$$= \frac{110 \text{ V}}{0.118 \text{ A}}$$

$$= 932 \text{ }\Omega$$

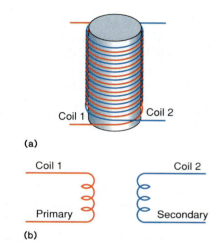

Coil 1   Coil 2

(a)

Coil 1   Coil 2

Primary   Secondary

(b)

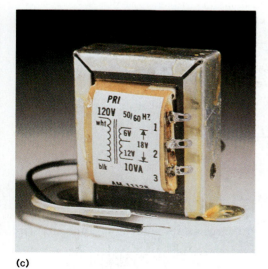

(c)

Figure 24.18

A transformer.

24.7 The Transformer

The **transformer** is a device, based on the principle of mutual induction studied in chapter 23, that changes electrical energy at a particular voltage and current in one coil into a different voltage and current in a second coil. The simplest transformer is the coaxial solenoids shown in figure 23.7 and reproduced here as figure 24.18(a). The schematic diagram for the transformer is shown in figure 24.18(b). The first solenoid coil is called the *primary coil,* whereas coil 2 is called the *secondary.*

If a varying emf is applied across coil 1 causing a changing current in coil 1 and hence a changing magnetic flux, an emf is induced in coil 2 by this changing magnetic flux $\Delta\Phi_m$, and is given by

$$\mathscr{E}_2 = -N_2\frac{\Delta\Phi_m}{\Delta t}$$

The same changing magnetic flux in coil 2 could induce an emf in coil 1 given by

$$\mathscr{E}_1 = -N_1 \frac{\Delta \Phi_m}{\Delta t}$$

Since the changing magnetic flux is the same in each coil

$$\frac{\Delta \Phi_m}{\Delta t} = \frac{-\mathscr{E}_2}{N_2} = \frac{-\mathscr{E}_1}{N_1}$$

Hence,

$$\frac{\mathscr{E}_2}{N_2} = \frac{\mathscr{E}_1}{N_1} \qquad (24.33)$$

Therefore, the emf induced in the secondary of the transformer is given by

$$\mathscr{E}_2 = \frac{N_2}{N_1} \mathscr{E}_1 \qquad (24.34)$$

If N_2, the number of turns in the secondary, is greater than N_1, the number of turns in the primary, then the ratio of N_2/N_1 is greater than 1. Therefore, the induced emf in the secondary $\mathscr{E}_2$ is greater than the impressed emf in the primary $\mathscr{E}_1$. When N_2 is greater than N_1, the transformer is called a *step-up transformer* because it "steps-up" or increases the voltage from the primary to the secondary. When N_2 is less than N_1, the ratio N_2/N_1 is less than 1, and the induced emf in the secondary is less than the applied emf across the primary. In this case, the transformer is called *a step-down transformer* because it has reduced the voltage from the primary to the secondary.

The law of conservation of energy states that the power put into coil 1 is equal to the power taken out of coil 2. Because the product of an emf $\mathscr{E}$ with the current i represents power, we must have

$$\mathscr{E}_2 i_2 = \mathscr{E}_1 i_1 \qquad (24.35)$$

If an alternating emf and current are impressed across coil 1 by an AC generator, then an alternating emf and current are found in coil 2, where the relation between the emfs and currents is given by equation 24.35. (Of course, the emfs and currents are given by their effective values.) Solving for the ratio of the emf of coil 2 to the emf of coil 1 gives

$$\frac{\mathscr{E}_2}{\mathscr{E}_1} = \frac{i_1}{i_2} \qquad (24.36)$$

Equation 24.36 shows that the ratio of the emfs in a transformer is reciprocal to the ratio of the currents.

Note, that if equation 24.33 is written in the form

$$\frac{\mathscr{E}_2}{\mathscr{E}_1} = \frac{N_2}{N_1}$$

then we can combine it with equation 24.36 as

$$\frac{\mathscr{E}_2}{\mathscr{E}_1} = \frac{i_1}{i_2} = \frac{N_2}{N_1} \qquad (24.37)$$

Equation 24.37 shows that the current in the secondary winding of a transformer is given by

$$i_2 = \frac{N_1}{N_2} i_1 \qquad (24.38)$$

Notice, *that if N_2 is greater than N_1 (a step-up transformer), the current in the secondary is less than the current in the primary.* Comparing equations 24.34 and

24.38 shows that the currents are reciprocal to the emfs. This means that in a step-up transformer, $\mathscr{E}_2$ is greater than $\mathscr{E}_1$, but the current i_2 is less than i_1. Hence, we do not get something for nothing from the transformer. The increase in voltage comes with a decrease in current, such that the power into the circuit is equal to the power out. A picture of a typical transformer is shown in figure 24.18(c).

Example 24.10

The voltage in a step-up transformer. A 120-V line from an AC generator is connected to the primary of a transformer of 50 turns. If the secondary of the transformer has 100 turns, what voltage will be found across the secondary of the transformer?

Solution

The voltage across the secondary, found from equation 24.34, is

$$\mathscr{E}_2 = \frac{N_2}{N_1}\mathscr{E}_1 = \left(\frac{100}{50}\right)(120 \text{ V}) = 240 \text{ V}$$

Example 24.11

The current in a step-up transformer. If the current in the primary of example 24.10 is 3.00 A, what is the current in the secondary?

Solution

The current in the secondary, found from equation 24.38, is

$$i_2 = \frac{N_1}{N_2}i_1 = \left(\frac{50}{100}\right)(3.00 \text{ A}) = 1.50 \text{ A}$$

Example 24.12

A step-down transformer. You wish to make a transformer that can take the 120 V from your wall outlet and drop it down to 24.0 V to operate a toy electric train. If there are 100 turns of wire in the primary how many turns do you need for the secondary?

Solution

The number of turns in the secondary can be found from equation 24.37 as

$$\frac{N_2}{N_1} = \frac{\mathscr{E}_2}{\mathscr{E}_1}$$

$$N_2 = \frac{\mathscr{E}_2}{\mathscr{E}_1}N_1$$

$$= \left(\frac{24.0 \text{ V}}{120 \text{ V}}\right)(100 \text{ turns})$$

$$= 20.0 \text{ turns}$$

Having analyzed an AC and a DC circuit, it is obvious that the AC circuit is more difficult to deal with than a DC circuit. What then is the great advantage of AC power over DC power that it is used so extensively? Two of the main reasons for the desirability of AC power are the ease in generating AC power with the AC generator described in chapter 23 and, second, the ease of transmitting this AC power from the power station to the home by means of transformers. Power is lost in transmission lines by Joule heating in the form $P = I^2R$. The smaller the value of the current in the line the smaller is the energy loss. As an example, suppose a particular power station generated 100,000 W by having a current of 20.0 A at 5000 V. If the transmission line has a resistance of 20.0 Ω, the power lost to heat during transmission is

$$P = I^2R = (20.0 \text{ A})^2(20.0 \text{ } \Omega) = 8000 \text{ W}$$

If the power output from the power station is connected to a step-up transformer that has a turns ratio of 1:10 from the primary to the secondary, then the voltage at the secondary becomes

$$\mathscr{E}_2 = \frac{N_2}{N_1}\mathscr{E}_1$$

$$= \left(\frac{10}{1}\right)(5000 \text{ V})$$

$$= 50,000 \text{ V}$$

The current in the secondary is

$$i_2 = \frac{N_1}{N_2} i_1$$

$$= \left(\frac{1}{10}\right)(20.0 \text{ A})$$

$$= 2.00 \text{ A}$$

The power lost in the transmission line due to Joule heating is now

$$P = i^2R$$

$$= (2.00 \text{ A})^2(20.0 \text{ } \Omega)$$

$$= 80.0 \text{ W}$$

Therefore, instead of transmitting the 100,000 W of power at 20.0 A and 5000 V, with a loss of 8000 W, the same 100,000 W can be transmitted at 2.00 A and 50,000 V with a power loss of only 80.0 W, a saving of 7920 W. When the transmission line reaches the local substation another transformer steps-down this voltage to lower values, which are then transmitted to your local electric pole. Another transformer steps-down this voltage further to give you the desired current and voltage for your home.

Before we leave this chapter it is perhaps appropriate to compare the effects of electromagnetic induction when used in a DC circuit and in an AC circuit. In chapter 23 we saw that a changing current in one coil induced an emf in the second coil. In most cases in chapter 23, we considered a DC current rising from zero to some maximum value, remaining constant for a time, and then decreasing to zero. The induced emf existed only for the short time that the current was rising or falling. The transformer reaches its maximum utilization in an AC circuit because the instantaneous value of an AC current is continually changing with time. Hence, there is a continuously alternating emf and current induced in the second coil. The transformer is thus an excellent practical example of electromagnetic induction when used in an AC circuit.

An Essay on the Application of Physics

Metal Detectors at Airports

Have you ever wondered how airport security can determine if a terrorist passenger is trying to sneak a gun or a bomb aboard an aircraft? To determine if a gun, or any piece of iron, is being carried by a passenger, each passenger is required to walk through a metal detector, figure 1(a).

The metal detector is essentially a coil of wire that the passenger walks through. The coil is connected to an AC source, as shown in figure 1(b). An AC voltmeter is connected across the coil to record the voltage V_{L_A} across the coil. The subscript A has been placed on V_L to indicate that this is the voltage across the coil when air is in the space within the coil. The voltage drop across the coil, found by equation 24.14, is

$$V_{L_A} = iX_{L_A} \qquad \textbf{(24H.1)}$$

where X_{L_A} is the inductive reactance given by equation 24.15 as

$$X_{L_A} = 2\pi f L_A \qquad \textbf{(24H.2)}$$

Here L_A is the inductance of the coil with air in the space within the coil. For simplicity in our discussion, let us assume that the coil of the metal detector can be considered to be a solenoid with air in the space inside the coil. The inductance for such a coil is found from equation 23.39 as

$$L_A = \mu_0 A \ell n^2 \qquad \textbf{(24H.3)}$$

where

μ_0 is the permeability of free space or air,
A is the area of the coil,
ℓ is the length of the coil, and
n is the number of turns of wire per unit length.

If an iron rod is now placed within the coil, the permeability μ_0 in equation 24H.3 must be replaced with the permeability μ of the iron rod. Recall that when dealing with electric fields ϵ_0 was the permittivity of free space, and when dealing with different materials the permittivity of the medium was given by $\epsilon = \kappa\epsilon_0$. In the same way, the permeability of a medium is given by

$$\mu = K_m \mu_0 \qquad \textbf{(24H.4)}$$

where K_m is called the *relative permeability of the medium*. For soft iron $K_m \approx 5000$, and the inductance of the coil with iron in it is now given by

$$L_I = \mu A \ell n^2 = (5000\mu_0)A\ell n^2 = 5000 L_A \qquad \textbf{(24H.5)}$$

The inductive reactance of the coil with iron in the space within the coil now becomes X_{L_I},

$$X_{L_I} = 2\pi f L_I = 2\pi f(5000 L_A) = (5000)(2\pi f L_A) = (5000)X_{L_A}$$

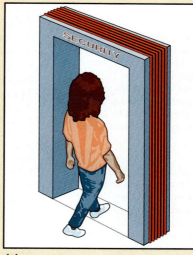

(a)

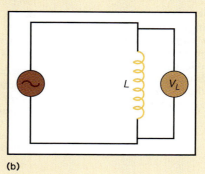

(b)

Figure 1

Every passenger must walk through the metal detector.

The voltage drop across the iron core solenoid is now

$$V_{L_I} = iX_{L_I} = i(5000 X_{L_A}) \qquad \textbf{(24H.6)}$$
$$= 5000(iX_{L_A}) = 5000 V_{L_A}$$

Thus by placing an iron rod into the solenoid the voltage drop across the solenoid has increased by a factor of 5000 over the voltage across the solenoid when air was in the space. Hence moving iron into the coil is relatively easy to detect. This is the principle of the metal detector. In practice, the iron to be detected does not fill up the entire space within the coil and hence the voltage drop across the coil is not as great as in this simplified model. However, the principle is still the same and the introduction of iron into the coil produces an increase in voltage across the coil. In our simplified description we have used a voltmeter to detect the presence of metal. This increased voltage could also be used to trigger an alarm or a bell to indicate the presence of the metal. Hence, if a terrorist passenger tries to carry a concealed gun through the detector coil, the increase in the permeability of the coil due to the presence of the gun causes an alarm to sound, indicating the presence of concealed metal on the passenger.

The Language of Physics

Effective current
That constant current that generates heat in a resistor at the same rate as an alternating current. The effective value of an alternating current is equal to 70.7% of the maximum or peak value of the AC current. An AC ammeter measures the effective current in the circuit. The effective current is sometimes called the rms current (p. 689).

Effective voltage
The constant value of the voltage that produces the same effect as an alternating voltage. The effective value of an alternating voltage is equal to 70.7% of the maximum or peak value of the AC voltage. An AC voltmeter measures the effective voltage. The effective voltage is sometimes called the rms voltage (p. 689).

RLC series circuit
A circuit containing a resistor, an inductor, and a capacitor in series. The voltage across the resistor is in phase with the current in the circuit. The voltage across the inductor in the circuit leads the current in the circuit by 90°, whereas the voltage across the capacitor lags the current in the circuit by 90° (p. 690).

Phase angle
A measure of how much the applied voltage leads or lags the current in an AC circuit (p. 693).

Inductive reactance
The inductive analogue to resistance in an AC circuit; that is, an inductive reactance tends to impede the flow of charge in an AC circuit. The inductive reactance is proportional to the frequency of the AC source; a high-frequency AC source causes a high reactance, whereas a low-frequency source causes a low reactance (p. 693).

Capacitive reactance
The capacitive analogue to resistance in an AC circuit; that is, a capacitive reactance tends to impede the flow of charge in an AC circuit. The capacitive reactance is inversely proportional to the frequency of the AC source; a high-frequency AC source causes a low reactance, whereas a low-frequency source causes a high reactance (p. 694).

Impedance
A measure of the opposition to the flow of charges in an AC circuit. It is composed of the resistance, inductive reactance, and capacitive reactance of the circuit. It is the AC analogue to resistance in a DC circuit (p. 694).

Resonance
That condition in an AC circuit where the maximum current is obtained. It occurs when the inductive reactance is equal to the capacitive reactance (p. 699).

Resonant frequency
The frequency of an AC circuit that causes resonance in the circuit. It depends on the inductance and capacitance of the circuit (p. 699).

Power factor
The ratio of the power consumed in an AC circuit to the power applied to the circuit. It is also equal to the cosine of the phase angle (p. 680).

RLC parallel circuit
An AC circuit in which the resistor, inductor, and capacitor are placed in parallel with each other. In such a circuit, the voltages are all in phase but the currents are out of phase with each other. The current in the capacitor leads the current in the resistor by 90°, whereas the current in the inductor lags the current in the resistor by 90° (p. 701).

Transformer
A device, based on the principle of mutual induction, that changes electrical energy at a particular voltage and current in the primary coil into a different voltage and current in the secondary coil (p. 705).

Summary of Important Equations

AC voltage
$$\mathscr{E} = \mathscr{E}_{max} \sin \omega t \tag{23.24}$$
$$V = V_{max} \sin(2\pi f t) \tag{24.6}$$

AC current
$$i = i_{max} \sin \omega t \tag{23.26}$$
$$i = i_{max} \sin(2\pi f t) \tag{24.2}$$

Relation of angular speed and frequency of sine wave
$$\omega = 2\pi f \tag{24.1}$$

Effective current in an AC circuit
$$i_{eff} = 0.707 i_{max} \tag{24.5}$$

Effective voltage in an AC circuit
$$V_{eff} = 0.707 V_{max} \tag{24.7}$$

Inductive reactance
$$X_L = 2\pi f L \tag{24.15}$$

Capacitive reactance
$$X_C = \frac{1}{2\pi f C} \tag{24.17}$$

Impedance of an AC series circuit
$$Z = \sqrt{R^2 + (X_L - X_C)^2} \tag{24.18}$$

Ohm's law for an AC circuit
$$i = \frac{V}{Z} \tag{24.19}$$

Voltage drop across a resistor
$$V_R = iR \tag{24.13}$$

Voltage drop across an inductor
$$V_L = iX_L \tag{24.14}$$

Voltage drop across a capacitor
$$V_C = iX_C \tag{24.16}$$

Total voltage in an AC series circuit
$$V = \sqrt{V_R^2 + (V_L - V_C)^2} \tag{24.11}$$

Phase angle between voltage and current in an AC series circuit
$$\phi = \tan^{-1}\left(\frac{V_L - V_C}{V_R}\right) \tag{24.12}$$

Resonant frequency of an AC series circuit
$$f_0 = \frac{1}{2\pi\sqrt{LC}} \tag{24.21}$$

Power dissipated in an AC circuit
$$P = iV \cos \phi \tag{24.22}$$

Power factor
$$PF = \cos \phi = \frac{R}{Z} \tag{24.24}$$

Power factor
$$PF = \frac{\text{Power consumed}}{\text{Power supplied}} \tag{24.25}$$

Total current in a parallel RLC circuit
$$I_T = \sqrt{(I_R^2 + (I_C - I_L)^2} \tag{24.26}$$

Current in resistor for parallel RLC circuit
$$I_R = \frac{V}{R} \tag{24.27}$$

Current in inductor for parallel RLC circuit
$$I_L = \frac{V}{X_L} \tag{24.28}$$

Current in capacitor for parallel RLC circuit
$$I_C = \frac{V}{X_C} \tag{24.29}$$

Phase angle for parallel RLC circuit
$$\phi = \tan^{-1}\frac{I_C - I_L}{I_R} \tag{24.30}$$

Total impedance for parallel RLC circuit
$$Z = \frac{V}{I_T} \tag{24.32}$$

Transformer
$$\frac{\mathscr{E}_2}{\mathscr{E}_1} = \frac{i_1}{i_2} = \frac{N_2}{N_1} \tag{24.37}$$

Questions for Chapter 24

†1. When Americans visit Europe, they are told to get an adapter for their personal appliances such as electric razors, hair driers, and the like. Why is this necessary? What is the difference between European electricity and American electricity?

2. What does doubling the frequency do to the inductive reactance? The capacitive reactance?

3. Can you use the d'Arsonval galvanometer to measure AC currents or voltages?

†4. How can you measure an AC current or voltage?

†5. What does increasing the area of the plates of a capacitor do to the resonant frequency in an *RLC* series circuit? How is this used in a variable capacitor of a tuned circuit?

6. What is the phase angle of an *RLC* series circuit when the inductive reactance equals the capacitive reactance? What is the power factor in this case?

†7. How is an *LC* circuit like a mass attached to a vibrating spring? Compare the equation for resonance for a vibrating spring and the *LC* circuit. Can you make an analogue between them?

†8. How can an *RLC* circuit be used to filter signals of a special frequency?

†9. In many transformers, the wires are wrapped around an iron core rather than a hollow air core. Why do you think this is done?

Problems for Chapter 24

24.2 The Effective Current and Voltage in an AC Circuit

1. The effective voltage in an AC circuit is measured as 50.0 V. What is the maximum value of the voltage?

2. The peak-to-peak voltage of an AC voltage is measured on an oscilloscope to be 100 V. What is the effective voltage?

3. The effective current in an AC circuit is measured to be 5.00 A. What is the maximum value of the current?

4. An AC current varies from -5.65 A to $+5.65$ A. Find the effective value of the current. What DC current would give the same effect?

24.3 An *RLC* Series Circuit

5. Find the inductive reactance of a 5.00-H coil when a 400-Hz AC voltage is impressed across it.

6. At what frequency will a 50.0-mH inductor have a reactance of 800 Ω?

7. A 110-V, 60.0-Hz, AC line is connected to a 6.55-mH coil. Find the current through the coil.

8. A coil has an impedance of 800 Ω and an inductive reactance of 600 Ω. Find the resistance of the coil.

9. Find the capacitive reactance of a 10.0-μF capacitor at 60.0 Hz.

10. A 55.5-V, 400-Hz, AC line is connected to a 6-pF capacitor. Find the current in the circuit.

11. At what frequency will the inductive reactance of a 2.00-H inductor be (a) 20 Ω, (b) 200 Ω, and (c) 2000 Ω? Is it easier for an inductor to pass low-frequency or high-frequency signals?

12. At what frequency will the capacitive reactance of a 5.00-μF capacitor be (a) 10 Ω, (b) 100 Ω, and (c) 1000 Ω? Is it easier for a capacitor to pass low-frequency or high-frequency signals?

13. A 2.00-mH inductor is connected to a 110-V, 60.0-Hz line. Find (a) the inductive reactance and (b) the current through the inductor.

14. A 2.00-μF capacitor is connected to a 50.0-V, 40.0-Hz, AC line. Find the current flowing in the capacitor circuit.

15. Find the impedance of a series circuit of $R = 1000$ Ω, $L = 5.00$ mH, and $C = 10.0$ μF, if the AC source is at a frequency of 60.0 Hz.

16. A resistor $R = 500$ Ω, an inductor $L = 20.0$ mH, and a capacitor $C = 6.00$ μF are connected in series. Find the impedance if the source is 110 V at 400 Hz.

†17. An *RLC* series circuit has $R = 1800$ Ω, $L = 4.89$ mH, $C = 4.78$ μF, and $f = 60.0$ Hz. Find (a) the phase angle between the applied voltage and the current in the circuit, (b) the phase angle between the voltage across the inductor and the applied voltage, and (c) the phase angle between the voltage across the capacitance and the applied voltage.

18. If the impedance of an *RLC* series circuit of $R = 800$ Ω, $L = 50.0$ mH, and $C = 3.00$ μF is 1178 Ω, find the phase angle between the current in the circuit and the applied voltage.

19. In an *RLC* series circuit, $R = 200$ Ω, $C = 10.0$ μF, and the frequency $f = 70.0$ Hz. What inductance would result in the potential across the *RLC* combination leading the current by 30.0°?

20. A 240-V, 50.0-Hz, AC line is connected in series with a resistor of 958 Ω and a capacitor of 9.50 μF. Find (a) the impedance of the circuit and (b) the phase angle of the circuit.

21. A 120-V, 50.0-Hz, AC line is connected to an *RL* series circuit of $R = 280$ Ω and $L = 5.75$ mH. Find the impedance of the circuit and the phase angle between the applied voltage and the current in the circuit.

22. In an *RLC* series circuit, $V_L = 12.0$ V, $V_C = 24.0$ V, and $R = 160$ Ω. If the potential across the *RLC* combination lags the current by 42.0°, find the current in the circuit.

23. A 400-Ω resistor is connected in series with an unknown inductor to a 110-V, 60.0-Hz, AC line. If the current in the circuit is 0.194 A, what is the value of the inductance of the inductor?

24.4 Resonance in an *RLC* Series Circuit

24. A 2.00-mH inductor is connected in series with a 10.0-μF capacitor to an AC line of variable frequency. At what frequency will resonance occur?

25. What value of inductance is necessary to tune an *RLC* circuit to a frequency of 106.2 MHz if the capacitor has a value of 5.00 pF?

26. At what frequency is the inductive reactance equal to the capacitive reactance if $X_L = 850$ Ω and $C = 6.00$ μF?

27. In an *RLC* series circuit, $\mathscr{E} = 120$ V, $C = 15.0$ pF, and the resonant frequency is 50.0 Hz. (a) Find the inductance in the circuit. (b) If the current is 10.0 mA at resonance, find the resistance in the circuit. (c) If the resonant frequency is changed to 60.0 Hz by varying the capacitance, find the new value of the capacitance.

28. If $R = 800\ \Omega$, $L = 50.0$ mH, and $C = 3.00\ \mu$F in an *RLC* series circuit of $\mathscr{E} = 120$ V, plot the current I as a function of the frequency, for $f = 100$ to 800 Hz in intervals of 100 Hz. Find the resonant frequency from the graph and compare with the resonant frequency obtained from the equation.

29. A local FM radio station transmits at a frequency of 94.3 MHz. What value of capacitance is needed in a *LC* series circuit of $L = 1.00$ H in order to have resonance?

24.5 Power in an AC Circuit

30. A 5.00-A current flows in a series *RLC* circuit and leads the 120 V applied voltage by 40.0°. Find the power factor of the circuit and the power.

31. The power factor in an *RLC* series circuit is 0.80. If the resistance of the circuit is 60.0 Ω, find the difference between the reactances, $X_L - X_C$.

24.6 An *RLC* Parallel Circuit

†32. A 220-V, 40.0-Hz, AC line is connected across a parallel *RLC* circuit of $R = 900\ \Omega$, $L = 7.00$ H, and $C = 8.00\ \mu$F. Find (a) the inductive reactance, (b) the capacitive reactance, (c) the current through the resistor, (d) the current through the inductor, (e) the current through the capacitor, (f) the total current in the circuit, (g) the phase angle ϕ, and (h) the total impedance.

†33. A 110-V, 60.0-Hz, AC line is connected across a resistance of 1000 Ω, which is in parallel with an inductor of 5.00 H. Find (a) the inductive reactance, (b) the current I_R through the resistor, (c) the current I_L through the inductor, (d) the total current in the circuit, (e) the phase angle ϕ, and (f) the total impedance of the circuit.

†34. A 110-V, 60.0-Hz, AC line is connected across a 1.00-μF capacitor in parallel with a 5.00-H inductor. Find (a) the capacitive reactance, (b) the inductive reactance, (c) the current I_C through the capacitor, (d) the current I_L through the inductor, (e) the total current I_T in the circuit, (f) the phase angle ϕ, and (g) the total impedance of the circuit.

†35. A 110-V, 60.0-Hz, AC line is connected across a resistance of 500 Ω in parallel with a capacitor of 4.00 μF. Find (a) the capacitive reactance, (b) the current through R, (c) the current through C, (d) the total current in the circuit, (e) the phase angle ϕ, and (f) the total impedance of the circuit.

24.7 The Transformer

36. A transformer has 100 turns in the primary coil and 250 turns in secondary. If the primary of the transformer is connected to an alternating voltage generator of 400 V, what voltage will appear across the secondary?

37. A transformer has 100 turns in the primary coil and 250 turns in the secondary. If there is an alternating current in the primary of the transformer of 3.00 A, what current will appear in the secondary?

38. If there are 200 turns of wire in the primary of a transformer connected to a 120-V outlet, how many turns do you need in the secondary in order to produce 6.00 V there?

39. A transformer has a current in its primary winding of 2.50 A and a current in its secondary winding of 0.750 A. (a) If there are 20 turns in the primary coil, find the number of turns in the secondary coil. (b) If the resistance of the secondary coil is 100 Ω, find the applied emf in the primary coil and the induced emf in the secondary coil. (Assume 100% efficiency.)

40. Find the turns ratio for the transformer in the diagram such that a spark is given off at the points AB. The distance from point A to point B is 5.00 mm.

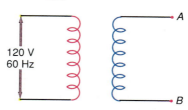

41. A power station generates 200 kW at 100 A and 2000 V. If the transmission line has a resistance of 10.0 Ω, how much power will be lost in transmission? If a step-up transformer of 1:10 is used, what will the power loss be then?

Additional Problems

42. *RC* and *RL* circuits are sometimes used as timing circuits; show that the units of *RC* and L/R have the unit of time.

†43. A 220-V, 40.0-Hz, AC line is connected across a series *RLC* circuit of $R = 900\ \Omega$, $L = 7.00$ H, and $C = 8.00\ \mu$F. Find (a) the inductive reactance; (b) the capacitive reactance; (c) the impedance of the circuit; (d) the current in the circuit; (e) the voltage drop across R, L, and C; (f) the total voltage across the circuit; (g) the phase angle ϕ; and (h) the resonant frequency.

†44. A 60.0-V, 100-Hz, AC line is connected across a series *RLC* circuit of $R = 1000\ \Omega$, $L = 10.0$ H, and $C = 2.00\ \mu$F. Find (a) the inductive reactance, (b) the capacitive reactance, (c) the impedance of the circuit, (d) the current in the circuit, (e) the voltage drop across R, (f) the voltage drop across L, (g) the voltage drop across C, (h) the total voltage drop across the circuit, (i) the phase angle, (j) the power factor, and (k) the resonant frequency.

†45. A 110-V, 60.0-Hz, AC line is connected across a resistor of 1800 Ω, in series with a capacitor of 3.00 μF. Find (a) the capacitive reactance, (b) the impedance, (c) the current in the circuit, (d) the voltage drop across R, (e) the voltage drop across C, (f) the total voltage measured across R and C in series, (g) the phase angle ϕ, and (h) the power factor.

†46. A 120-V, 60.0-Hz, AC line is connected across a series circuit of a resistance of 500 Ω and an inductor of 10.0 H. Find (a) the inductive reactance, (b) the impedance, (c) the current in the circuit, (d) the voltage drop across R, (e) the voltage drop across L, (f) the total voltage drop across the circuit, (g) the phase angle ϕ, and (h) the power factor.

†47. Three inductors are connected in series to an AC source as shown in the diagram. The inductors are shielded so that a changing flux in one inductor does not cause a mutual inductance in another inductor. Show that the equivalent inductance of the inductors in series is given by

$$L = L_1 + L_2 + L_3$$

†48. Three shielded inductors are connected in parallel as shown in the diagram. The inductors are shielded so that a changing flux in one inductor does not cause a mutual inductance in another inductor. Show that the equivalent inductance of the inductors in parallel is given by

$$\frac{1}{L} = \frac{1}{L_1} + \frac{1}{L_2} + \frac{1}{L_3}$$

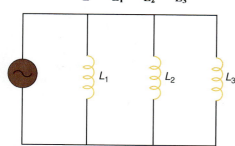

†49. In the circuit shown $R = 1300\ \Omega$, $C_1 = 5.00\ \mu F$, $C_2 = 8.00\ \mu F$, $L = 6.78$ mH, and the applied voltage is 110 V at 60.0 Hz. Find the current through each circuit element.

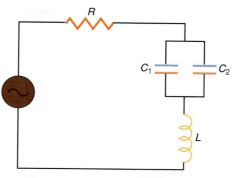

†50. In the circuit shown $R_1 = 978\ \Omega$, $R_2 = 560\ \Omega$, $L_1 = 6.78$ mH, $L_2 = 3.25$ mH, $C = 5.00\ \mu F$, and the applied voltage is 110 V at 60.0 Hz. Find the current through each circuit element.

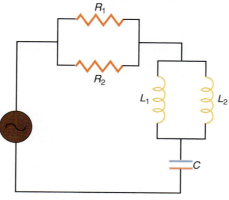

†51. In the circuit shown $R_1 = 535\ \Omega$, $R_2 = 350\ \Omega$, $L_1 = 3.25$ mH, $L_2 = 2.56$ mH, $C_1 = 5.00\ \mu F$, $C_2 = 7.50\ \mu F$, and the applied voltage is 110 V at 60.0 Hz. Find the current through each circuit element.

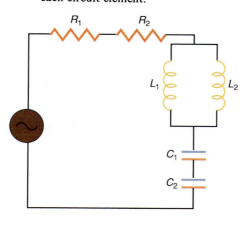

52. The diagram is an example of a low-pass filter. Show that very low-input frequencies pass through the circuit while high frequencies do not. That is, show that high-output voltages are obtained across AB for low-input frequencies, while very low-output voltages are obtained for high-input frequencies.

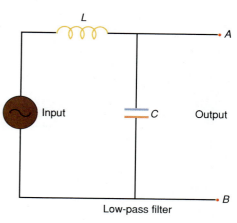

Low-pass filter

53. The diagram is an example of a high-pass filter. Show that very high-input frequencies pass through the circuit while low frequencies do not. That is, show that high-output voltages are obtained across AB for high-input frequencies, while very low-output voltages are obtained for low-input frequencies.

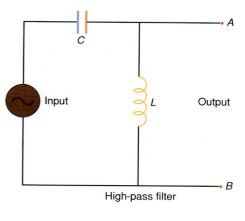

High-pass filter

†54. Find the turns ratio between the primary and secondary coils in the diagram, to produce a spark across AB, a distance of 5.00 mm, and find the value of C to give a spark rate of 5.00 Hz, 10.0 Hz, and 30.0 Hz. The value of the inductance is 5.00 H, $R = 500 \, \Omega$, and the peak voltage is 1200 V.

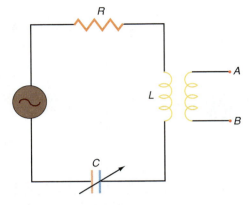

Interactive Tutorials

55. Effective current in an AC circuit. An alternating current of frequency $f = 60$ Hz varies from -2.54 A to 2.54 A. Find the effective value of this current. Plot the current and show the effective current on the plot.

56. *RLC* series circuit. An *RLC* series circuit has a resistance $R = 800 \, \Omega$, inductance $L = 2.00$ H, capacitance $C = 5.55 \times 10^{-6}$ F, and they are connected to an applied voltage $V = 110$ V, operating at a frequency $f = 60$ Hz. Find (a) the inductive reactance X_L, (b) the capacitive reactance X_C, (c) the impedance Z of the circuit, (d) the current i in the circuit, (e) the voltage drop V_R across the resistor, (f) the voltage drop V_L across the inductor, (g) the voltage drop V_C across the capacitor, (h) the total voltage drop across *RLC*, (i) the phase angle ϕ, and (j) the resonant frequency f_0. (k) Plot the curves of the voltage across R, L, and C individually and the voltage across the combination. Show the phase relations for all these voltages.

57. Resonance. An *RLC* series circuit has a resistance $R = 800 \, \Omega$, inductance $L = 2.00$ H, capacitance $C = 5.55 \times 10^{-6}$ F, and they are connected to an applied voltage $V = 110$ V, operating at a frequency $f = 60$ Hz. Plot the current flow i as a function of the frequency f of the AC source. From this graph pick out the resonant frequency of the circuit and compare it to the calculated value for the resonant frequency.

58. *RLC* parallel circuit. An *RLC* parallel circuit has a resistance $R = 800 \, \Omega$, inductance $L = 2.00$ H, capacitance $C = 5.55 \times 10^{-6}$ F, and they are connected to an applied voltage $V = 110$ V, operating at a frequency $f = 60$ Hz. Find (a) the inductive reactance X_L, (b) the capacitive reactance X_C, (c) the current I_R through the resistor, (d) the current I_L through the inductor, (e) the current I_C through the capacitor, (f) the total current I_T in the circuit, (g) the phase angle ϕ, and (h) the impedance Z. (i) Plot the curves of the current I_R, I_L, and I_C individually and the total current I_T. Show the phase relations for all these currents.

59. The transformer. A 120-V line from an AC generator is connected to the primary of a transformer of $N_1 = 150$ turns. (a) If the secondary of the transformer has $N_2 = 25$ turns, find the voltage V_2 that will be found across the secondary of the transformer. (b) If the current in the primary $i_1 = 0.05$ A, find the current i_2 in the secondary.

25 Maxwell's Equations and Electromagnetic Waves

Electromagnetic waves are received by satellite dishes.

Science is the attempt to make the chaotic diversity of our sense-experience correspond to a logically uniform system of thought. In this system single experiences must be correlated with the theoretic structure in such a way that the resulting coordination is unique and convincing.

Albert Einstein

25.1 Introduction

In 1864, James Clerk Maxwell (1831–1879) took all of the then known equations of electricity and magnetism, and with the addition of a new term to one of the equations, combined them into only four equations that could be used to derive all the results of electromagnetic theory. These four equations came to be known as **Maxwell's equations.** The four Maxwell's equations are (1) Gauss's law for electricity, (2) Gauss's law for magnetism, (3) Ampère's law with the addition of a new term called the *displacement current,* and (4) Faraday's law of electromagnetic induction. With these four equations, Maxwell predicted that waves should exist in the electromagnetic field. Thirteen years later, in 1887, Heinrich Hertz (1857–1894) produced and detected these electromagnetic waves. Maxwell also predicted that the speed of these electromagnetic waves should be 3×10^8 m/s. Observing that this is also the speed of light, Maxwell declared that light itself is an electromagnetic wave. In fact it eventually became known that there was an entire spectrum of these electromagnetic waves. They differed only in frequency and wavelength. Finally, it was found that these electromagnetic waves are capable of transmitting energy from one place to another, even through the vacuum of space.

25.2 Gauss's Law for Electricity

In section 23.2, we saw that the magnetic flux is a quantitative measure of the number of lines of the magnetic field **B** that pass normally through a surface area **A.** This was illustrated in figure 23.1 and is written mathematically as

$$\Phi_M = BA \cos \theta \tag{23.1}$$

In a similar way, the electric flux is defined as a quantitative measure of the number of lines of the electric field **E,** that pass normally through the surface area **A,** as shown in figure 25.1. This is written mathematically as

$$\Phi_E = EA \cos \theta \tag{25.1}$$

It might be beneficial to refer back to section 23.2 and figure 23.1 at this time to recall the characteristics of flux.

Let us now consider the amount of electric flux that emanates from a positive point charge. Figure 25.2 shows a positive point charge surrounded by an imaginary spherical surface called a **Gaussian surface.** Let us measure the amount of electric flux through the sphere. The direction of the **E** field is different at every

Figure 25.1
Electric flux.

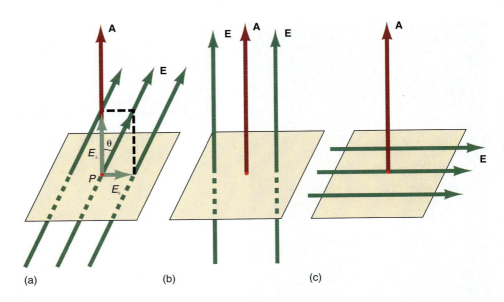

(a)   (b)   (c)

Figure 25.2
Gauss's law for electricity.

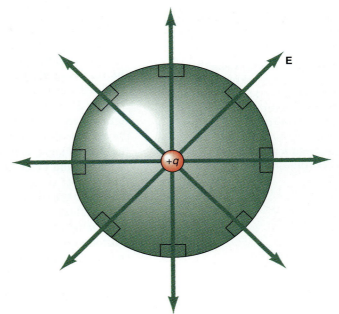

point, however, so equation 25.1 cannot be used in its present form. Instead, we break up the spherical surface into a large number of little surface areas ΔA, and compute the small amount of flux $\Delta\Phi_E$ through each of these little areas:

$$\Delta\Phi_E = E\Delta A \cos\theta$$

The total flux out of the Gaussian surface becomes the sum of all these $\Delta\Phi_E$'s. That is,

$$\Phi_E = \Sigma\,\Delta\Phi_E = \Sigma\,E\Delta A \cos\theta \tag{25.2}$$

The electric vector **E** is everywhere radial from the point charge q, and ΔA is also everywhere radial, hence the angle θ between **E** and ΔA is 0°, and the flux through the Gaussian surface is

$$\Phi_E = \Sigma\,E\Delta A \cos\theta = \Sigma\,E\Delta A \cos 0°$$

Electricity and Magnetism

Thus,

$$\Phi_E = \Sigma \, E\Delta A \qquad (25.3)$$

But we found in chapter 19 that the electric field of a point charge is given by equation 19.2 as

$$E = \frac{kq}{r^2}$$

Substituting the electric field of a point charge from equation 19.2 into equation 25.3, we get, for the electric flux,

$$\Phi_E = \Sigma \, \frac{kq}{r^2}\Delta A \qquad (25.4)$$

Now r is the radius of the spherical Gaussian surface, and for that sphere it is a constant. The terms k and q are also constants and are in every term of the sum in equation 25.4, and can, therefore, be factored out of the sum. Thus,

$$\Phi_E = \frac{kq}{r^2} \, \Sigma \, \Delta A$$

But the sum of all the small elements of area ΔA is equal to the entire surface area of the sphere. Hence, since the surface area of a sphere is $4\pi r^2$, we have

$$\Sigma \, \Delta A = 4\pi r^2$$

Thus, the electric flux from a point charge becomes

$$\Phi_E = \frac{kq}{r^2} \, 4\pi r^2$$

$$= \frac{1}{4\pi\epsilon_0} \frac{q4\pi r^2}{r^2} \qquad (25.5)$$

Where use has been made of the fact that k in equation 25.5 is equal to $1/4\pi\epsilon_0$. Hence, *the electric flux associated with a point charge is*

$$\Phi_E = \frac{q}{\epsilon_0} \qquad (25.6)$$

*Equation 25.6 is **Gauss's law for electricity,** and it says that the electric flux Φ_E through a surface surrounding the point charge q is a measure of the amount of charge q contained within the Gaussian surface.* Although equation 25.6 was derived for the flux emanating from a point charge, it is true, in general, for the flux emanating from any kind of charge distribution. Since Φ_E was initially defined in equation 25.2, it can be combined with equation 25.6 into *the generalization of Gauss's law as*

$$\Phi_E = \Sigma \, E\Delta A \cos \theta = \frac{q}{\epsilon_0} \qquad (25.7)$$

where q is now the net charge contained within the Gaussian surface.[1] *When Φ_E is a positive quantity, the Gaussian surface surrounds a source of positive charge, and electric flux diverges out of the surface.*

---

1. Although we have picked a spherical surface for the Gaussian surface for this problem, the surface does not have to be spherical. In general, the Gaussian surface is picked depending on the symmetry of the problem. Thus, spheres, cylinders, boxes, and the like are used for the Gaussian surface as appropriate.

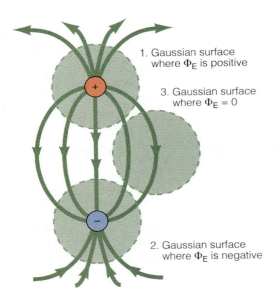

Figure 25.3

Gaussian surfaces and an electric dipole.

1. Gaussian surface where Φ_E is positive

3. Gaussian surface where $\Phi_E = 0$

2. Gaussian surface where Φ_E is negative

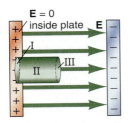

(a) Side view of capacitor

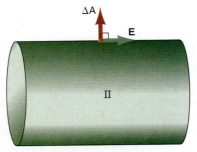

(b) The cylindrical portion of the Gaussian surface (surface II)

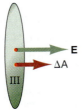

(c) End cap of Gaussian surface (surface III)

Figure 25.4

The electric field between the plates of a parallel plate capacitor.

If the point charge is negative then the electric field goes inward, through the Gaussian surface, to the point charge. The vector **E**, therefore, makes an angle of 180° with the area vectors **ΔA**, and the electric flux is

$$E\Delta A \cos \theta = E\Delta A \cos 180° = -E\Delta A$$

Thus, the flux passing through the Gaussian surface is negative. *Hence, whenever Φ_E is negative, the Gaussian surface surrounds a negative charge distribution and the electric flux converges into the Gaussian surface.* If there is no enclosed charge, $q = 0$, and hence, $\Phi_E = 0$. In this case, whatever electric flux enters one part of a Gaussian surface, the same amount must leave somewhere else. These different possibilities are shown in figure 25.3. Gaussian surface 1 shows the electric flux diverging from the positive point charge; Gaussian surface 2 shows the electric flux converging into the negative point charge; and Gaussian surface 3 shows no enclosed charge, and the amount of electric field entering normally through the surface is equal to the amount leaving the surface normally.

As an example of the application of Gauss's law, let us consider the problem of determining the electric field between the plates of the parallel plate capacitor, shown in figure 25.4(a). We start by drawing a Gaussian surface. For the symmetry of this problem we pick a cylinder for the Gaussian surface, as shown in the diagram. Gauss's law is given by equation 25.7 as

$$\Phi_E = \Sigma\, E\Delta A \cos \theta = \frac{q}{\epsilon_0}$$

The sum in equation 25.7 is over the entire Gaussian surface. We can break the entire surface of the cylinder into three surfaces. Surface I is the end cap on the left-hand side of the cylinder, surface II is the main cylindrical surface, and surface III is the end cap on the right-hand side of the cylinder as shown in figure 25.4(a). The total flux Φ through the Gaussian surface is the sum of the flux through each individual surface. That is,

$$\Phi = \Phi_I + \Phi_{II} + \Phi_{III}$$

Gauss's law becomes

$$\Phi_E = \Sigma_I\, E\Delta A \cos \theta + \Sigma_{II}\, E\Delta A \cos \theta + \Sigma_{III}\, E\Delta A \cos \theta = \frac{q}{\epsilon_0} \qquad (25.8)$$

Because the plate is a conducting body, all charge must reside on its outer surface, hence $E = 0$, inside the conducting body. Since Gaussian surface I lies within the conducting body, the electric field on surface I is zero. Hence the flux through surface I is

$$\Phi_I = \Sigma_I E\Delta A \cos \theta = \Sigma_I(0)\Delta A \cos \theta = 0$$

Along cylindrical surface II, **ΔA** is everywhere perpendicular to the surface, as shown in figure 25.4(b). **E** points toward the right and is everywhere perpendicular to the surface vector **ΔA** on surface II and therefore $\theta = 90°$. Hence the electric flux through surface II is

$$\Phi_{II} = \Sigma_{II} E\Delta A \cos \theta = \Sigma_{II} E\Delta A \cos 90° = 0$$

Surface III is the end cap on the right-hand side of the cylinder, and as can be seen from figure 25.4(c) **E** points toward the right and since the area vector **ΔA** is perpendicular to the surface pointing outward, it also points to the right. Hence **E** and **ΔA** are parallel to each other and the angle θ between **E** and **ΔA** is zero. Hence, the flux through surface III is

$$\Phi_{III} = \Sigma_{III} E\Delta A \cos \theta = \Sigma_{III} E\Delta A \cos 0 = \Sigma_{III} E\Delta A$$

Electricity and Magnetism

The total flux through the Gaussian surface is equal to the sum of the fluxes through the individual surfaces. Hence, Gauss's law becomes

$$\Phi = \Phi_{\text{I}} + \Phi_{\text{II}} + \Phi_{\text{III}} = 0 + 0 + \Sigma_{\text{III}} E \Delta A = \frac{q}{\epsilon_0}$$

But E is constant in every term of the sum and can be factored out of the sum giving

$$E \Sigma_{\text{III}} \Delta A = \frac{q}{\epsilon_0}$$

But $\Sigma \Delta A = A$ the area of the end cap, hence

$$EA = \frac{q}{\epsilon_0}$$

where A is the magnitude of the area of the Gaussian surface that is parallel to the conducting plate and q is the charge enclosed within that Gaussian surface. If we were to make the Gaussian surface larger such that the area of the end cap of the Gaussian surface was equal to the area A of the capacitor plate, then q would be the total charge enclosed on the capacitor plate. Solving for *the electric field between the plates of a parallel plate capacitor gives*

$$E = \frac{q}{\epsilon_0 A} \tag{25.9}$$

where q is the charge on the plates and A is the area of the plates. *Thus, the electric field between the plates of a parallel plate capacitor can be found by Gauss's law and is given by equation 25.9.* Notice that this is the same equation that was stated without proof as equation 21.1, in chapter 21 on capacitors.

It is sometimes convenient to express this result in terms of the surface charge density. The *surface charge density* σ is defined as the charge per unit area, that is,

$$\sigma = \frac{q}{A} \tag{25.10}$$

Combining equation 25.10 with 25.9, gives, for the electric field,

$$E = \frac{\sigma}{\epsilon_0} \tag{25.11}$$

Thus, the electric field between the plates of a parallel plate capacitor is also given by equation 25.11 in terms of the surface charge density and the permittivity of free space.

Example 25.1

The electric field between the plates of a parallel plate capacitor. Find the electric field between the circular plates of a parallel plate capacitor of 5.00-cm radius, if a charge of 8.70×10^{-10} C is placed on the plates.

Solution

The area of the plate is

$$A = \pi r^2 = \pi (0.0500 \text{ m})^2 = 7.85 \times 10^{-3} \text{ m}^2$$

The electric field between the plates of the capacitor, found from equation 25.9, is

$$E = \frac{q}{\epsilon_0 A}$$

$$= \frac{8.70 \times 10^{-10}\ \text{C}}{[8.85 \times 10^{-12}\ \text{C}^2/(\text{N m}^2)](7.85 \times 10^{-3}\ \text{m}^2)}$$

$$= 1.25 \times 10^4\ \text{N/C}$$

25.3 Gauss's Law for Magnetism

Just as there is an electric flux Φ_E associated with an electric field $\mathbf{E}$ passing through a surface area $\mathbf{A}$, there is also a magnetic flux Φ_M associated with a magnetic field $\mathbf{B}$ passing through a surface area $\mathbf{A}$. The magnetic flux was defined in section 23.2, and was given by equation 23.1 as

$$\Phi_M = BA \cos \theta$$

(It is recommended that the student reread section 23.2 at this time.) Because of the similarity of these fluxes it is reasonable to assume that Gauss's law should also apply to magnetism. Gauss's law for electricity was found in equation 25.6 as

$$\Phi_E = \Sigma E \Delta A \cos \theta = \frac{q}{\epsilon_0}$$

Thus, the electric flux is a measure of the electric charge q enclosed within the Gaussian surface. It is therefore reasonable to assume that Gauss's law for magnetism should take the same form as equation 25.6. If the assumption is valid, Gauss's law for magnetism should be written as

$$\Phi_M = \Sigma\, B \Delta A \cos \theta = \text{enclosed magnetic pole} \qquad \textbf{(25.12)}$$

But here we run into a slight difficulty. The magnetic flux should be a measure of the amount of magnetic pole enclosed within the Gaussian surface. But, as has been seen in section 22.10, isolated magnetic poles do not exist. Thus the term for the enclosed magnetic pole on the right-hand side of equation 25.12 must be equal to zero. *Thus Gauss's law for magnetism becomes*

$$\Phi_M = \Sigma\, B \Delta A \cos \theta = 0 \qquad \textbf{(25.13)}$$

Just as we considered Gaussian surfaces at various positions in the field of an electric dipole in figure 25.3, we will consider Gaussian surfaces at various positions in the field of a magnetic dipole, represented by the simple bar magnet in figure 25.5.

In a bar magnet, the magnetic field lines also go through the bar, as shown in figure 25.5. *Hence, the amount of magnetic flux entering the Gaussian surface around the north magnetic pole inside the bar magnet is equal to the amount of magnetic flux coming out of the same Gaussian surface outside of the bar magnet. Thus, the flux into the Gaussian surface is equal to the flux out of the Gaussian surface and the net flux through the Gaussian surface surrounding the north magnetic pole is zero.*

In a similar way, the magnetic field lines entering the south magnetic pole continue through the bar magnet until they emerge at the north magnetic pole. *Hence, a Gaussian surface surrounding the south magnetic pole also has a net flux of zero passing through it.* That is, the flux into the Gaussian surface is equal to the flux out because all magnetic field lines form closed loops. The net flux through any Gaussian surface anywhere in a magnetic field is zero because there can never be any isolated magnetic poles.

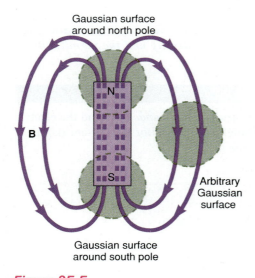

Gaussian surface around north pole

B

Arbitrary Gaussian surface

Gaussian surface around south pole

Figure 25.5

Gaussian surfaces and a magnetic dipole.

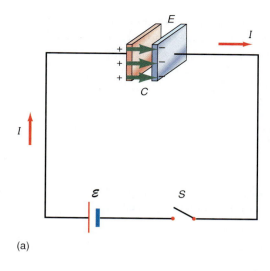

(a)

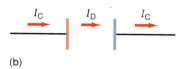

(b)

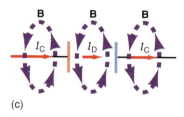

(c)

Figure 25.6

The displacement current.

The electric field vectors in an electrostatic field always begin and end on electric charges. Because there are no isolated magnetic poles, magnetic field vectors do not begin or end on magnetic poles, but rather, are always continuous.

25.4 The Displacement Current and Ampère's Law

In the study of a capacitor in chapter 21 (where we assumed that the current was conventional current, that is a flow of positive charges) we saw that when the switch in the circuit is closed, charge flows from the positive terminal of the battery to one plate of the capacitor, called the positive plate, and charge also flows from the negative plate of the capacitor back to the negative terminal of the battery. This is shown in figure 25.6(a). Until the plates are completely charged, there is a current into the positive plate, and a current out of the negative plate, yet there seems to be no current between the plates. There is thus a discontinuity in the current in the circuit because of the capacitor. As charge is placed on the plates of the capacitor an electric field is set up between the plates. The electric field between the plates of a capacitor, just found by Gauss's law, is

$$E = \frac{q}{\epsilon_0 A} \tag{25.9}$$

As additional charge Δq is added to the positive plate, it causes an additional electric field between the plates given by

$$\Delta E = \frac{\Delta q}{\epsilon_0 A} \tag{25.14}$$

The additional charge Δq just added to the plate came from the current from the battery, and since the current is defined as $I = \Delta q / \Delta t$, we can write the additional charge as

$$\Delta q = I \Delta t \tag{25.15}$$

Substituting equation 25.15 back into equation 25.14, we get

$$\Delta E = \frac{I \Delta t}{\epsilon_0 A} \tag{25.16}$$

Solving equation 25.16 for the current I, we get

$$I_D = \epsilon_0 A \frac{\Delta E}{\Delta t} \tag{25.17}$$

where $\Delta E / \Delta t$ is the rate at which the electric field between the plates changes with time, and we see, from equation 25.17, that it is related to the current entering or leaving the capacitor. *Maxwell said that this changing electric field within the capacitor is equivalent to a current through the capacitor and he called this current the* **displacement current** I_D. *With the concept of the displacement current, there is no discontinuity in the current in the circuit. The usual current in the conducting wires is now called the* **conduction current** I_C. The continuity of current is shown in figure 25.6(b) as the conduction current I_C entering the capacitor, the displacement current I_D through the capacitor, and the conduction current I_C leaving the capacitor.

Example 25.2

The displacement current. At a certain instant, a parallel plate capacitor, rated at 17.4 μF, has a potential of 50.0 V across its plates. The area of the plate is 5.00×10^{-2} m². If it takes a time of 0.500 s to reach this 50.0-V potential, find (a) the charge deposited on the plates of the capacitor, (b) the average conduction current at that time, (c) the average displacement current at that time, and (d) the rate at which the electric field between the plates is changing at that time.

Solution

a. The charge deposited on the plates, found from chapter 21, equation 21.6, is

$$q = CV$$
$$= (17.4 \times 10^{-6} \text{ F})(50.0 \text{ V})$$
$$= 8.70 \times 10^{-4} \text{ C}$$

b. The current in the circuit, corresponding to that amount of charge flowing in 0.500 s, found from the definition of the conduction current, is

$$I_C = \frac{\Delta q}{\Delta t} = \frac{8.70 \times 10^{-4} \text{ C}}{0.500 \text{ s}} = 1.74 \times 10^{-3} \text{ A}$$

c. The displacement current across the capacitor is equal to the conduction current entering the capacitor, therefore

$$I_D = I_C = 1.74 \times 10^{-3} \text{ A}$$

d. The rate at which the electric field between the plates is changing with time is given by rearranging equation 25.17 to

$$\frac{\Delta E}{\Delta t} = \frac{I_D}{\epsilon_0 A}$$
$$= \frac{1.74 \times 10^{-3} \text{ A}}{(8.85 \times 10^{-12} \text{ C}^2/\text{N m}^2)(5.00 \times 10^{-2} \text{ m}^2)} \left(\frac{\text{C/s}}{\text{A}}\right)$$
$$= 3.93 \times 10^9 \frac{\text{N/C}}{\text{s}}$$

Just as there is a magnetic field around a long straight wire carrying a conduction current, there is a magnetic field around the capacitor associated with a displacement current. This is shown in figure 25.6(c).

Ampère's law was stated in chapter 22 as: Along any arbitrary path encircling a total current I, the sum of the product of the parallel component of the magnetic field $B_\parallel$ and an element of length $\Delta \ell$ of the path, is equal to the permeability μ_0 times the total current enclosed by the path. *Maxwell reinterpreted Ampère's law to mean that the total current must be the sum of the conduction current and the displacement current. Thus, Maxwell rewrote Ampère's law as*

$$\Sigma\, B_\parallel \Delta \ell = \mu_0 (I_C + I_D) \tag{25.18}$$

With even deeper insight, Maxwell felt that the magnetic field that he associated with the displacement current is more likely associated with the changing electric field with time. *In Faraday's law, it was shown that a changing magnetic field induces an electric field, it is therefore reasonable to assume that the inverse situation also occurs in nature; that is, that a changing electric field can produce a magnetic field.* Thus, Maxwell rewrote Ampère's law in the form of equation 25.18 but then he added his result for the displacement current found in equation 25.17. With these modifications, *Ampère's law becomes*

$$\Sigma\, B_\parallel \Delta \ell = \mu_0 I_C + \mu_0 \epsilon_0 A \frac{\Delta E}{\Delta t} \tag{25.19}$$

Ampère's law, equation 25.19, says that a magnetic field can be produced by a conduction current or a changing electric field with time.

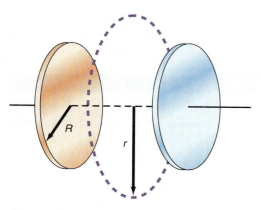

Figure 25.7

The magnetic field caused by the changing electric field.

As a still further generalization of Ampère's law, notice that the term $A\Delta E/\Delta t$ in equation 25.19 is equal to the change in the electric flux with time. That is,

$$A\frac{\Delta E}{\Delta t} = \frac{\Delta \Phi_E}{\Delta t}$$

Hence, Ampère's law can be written in the general form

$$\Sigma B_{\parallel}\Delta \ell = \mu_0 I_C + \mu_0 \epsilon_0 \frac{\Delta \Phi_E}{\Delta t} \qquad (25.20)$$

As an example of the use of Ampère's law, let us determine the magnetic field that exists around a parallel plate capacitor that is caused by the changing electric field within the space between the parallel plates. This is shown in figure 25.7. The parallel plates are circular and have a radius R. Let us determine the magnetic field **B** at a distance r from the center of the capacitor. Within the capacitor there is no conduction current (i.e., $I_C = 0$). Therefore, Ampère's law, equation 25.19, becomes

$$\Sigma B_{\parallel}\Delta \ell = \mu_0 \epsilon_0 A\frac{\Delta E}{\Delta t} \qquad (25.21)$$

Because the magnetic field around a long straight wire carrying a current I is circular, from the point of view of symmetry, it is reasonable to expect that the magnetic field around a displacement current should also be circular, a result that can be proven by experiment. Thus, the magnetic field **B** is parallel to $\Delta \ell$ along the entire circular path shown in figure 25.7. Hence,

$$\Sigma B_{\parallel}\Delta \ell = \Sigma B\Delta \ell \cos 0°$$

$$= \Sigma B\Delta \ell = B\Sigma\Delta \ell$$

But the sum of the path elements ($\Sigma\Delta \ell$) is the circumference of the circle, namely $2\pi r$. Therefore,

$$\Sigma B_{\parallel}\Delta \ell = B2\pi r$$

Substituting this into Ampère's law, equation 25.21, we get

$$B2\pi r = \mu_0 \epsilon_0 A\frac{\Delta E}{\Delta t}$$

But A is the area of the parallel plates and is πR^2. Thus,

$$B2\pi r = \mu_0 \epsilon_0 \pi R^2\frac{\Delta E}{\Delta t}$$

The magnetic field around, and at a distance r from, the capacitor is thus

$$B = \frac{\mu_0 \epsilon_0 R^2}{2r}\frac{\Delta E}{\Delta t} \qquad (25.22)$$

Notice that just as the magnetic field around a long straight wire varied as $1/r$, so also the magnetic field around the capacitor varies as $1/r$. Since μ_0 and ϵ_0 are constants of free space, R is the constant radius of the capacitor plate, and for a fixed value of r from the center of the capacitor to the point where the magnetic field is to be determined, we can write the magnetic field at that point as

$$B = (\text{constant})\frac{\Delta E}{\Delta t} \qquad (25.23)$$

That is, the changing electric field $\Delta E/\Delta t$ is capable of producing a magnetic field B. If the changing electric field within the plates of a capacitor can produce a magnetic field, should not every changing electric field produce a magnetic field? The answer is yes. Hence, there is a symmetry in nature. *Just as a changing magnetic field can produce an electric field (Faraday's law), a changing electric field can produce a magnetic field (Ampère's law as modified by Maxwell).*

Example 25.3

A changing electric field with time creates a magnetic field. Find the magnetic field a distance of 20.0 cm from the center of the parallel plate capacitor in example 25.2.

Solution

The area of the plates of the capacitor ($A = \pi R^2$) was given as 5.00×10^{-2} m², hence the radius of the plate is

$$R = \sqrt{\frac{A}{\pi}} = \sqrt{\frac{5.00 \times 10^{-2} \text{ m}^2}{\pi}} = 0.126 \text{ m}$$

The changing electric field, found in example 25.2, is $\Delta E/\Delta t = 3.93 \times 10^9$ (N/C)/s. Hence, the magnetic field at a distance of 20.0 cm from the center of the capacitor, found from equation 25.22, is

$$B = \frac{\mu_0 \epsilon_0 R^2}{2r} \frac{\Delta E}{\Delta t}$$

$$= \left[\frac{(4\pi \times 10^{-7} \text{ T m/A})(8.85 \times 10^{-12} \text{ C}^2/\text{N m}^2)(0.126 \text{ m})^2}{2(0.0200 \text{ m})} \right][3.93 \times 10^9 \text{ (N/C)/s}]$$

$$= 1.74 \times 10^{-9} \text{ T}$$

Example 25.4

The magnetic field outside the long straight line. Find the magnetic field a distance of 20.0 cm from the long straight wire that is carrying the conduction current, $I_C = 1.74 \times 10^{-3}$ A, into the plate of the capacitor.

Solution

The magnetic field around a long straight wire was given by equation 22.21 as

$$B = \frac{\mu_0 I_C}{2\pi r}$$

$$= \frac{(4\pi \times 10^{-7} \text{ T m/A})(1.74 \times 10^{-3} \text{ A})}{2\pi(0.200 \text{ m})}$$

$$= 1.74 \times 10^{-9} \text{ T}$$

Notice that the magnetic field outside the current-carrying wire is the same as the magnetic field caused by the changing electric field. Of course, this should come as no great surprise because the changing electric field is equivalent to a displacement current and the displacement current is the same as the conduction current. The importance of looking at the problem from the point of view of a changing electric field rather than a displacement current lies in the production and propagation of electromagnetic waves, which we will study shortly.

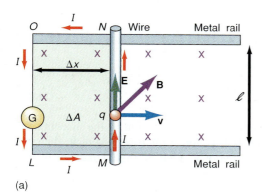

(a)

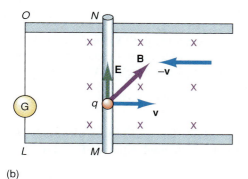

(b)

Figure 25.8

(a) Motion of a wire through a uniform magnetic field and (b) motion of a uniform magnetic field past a wire at rest.

25.5 Faraday's Law

In chapter 23 we discussed **Faraday's law** and saw that an electric field can be produced by changing a magnetic field with time. However, it is appropriate here to discuss some of the ramifications of Faraday's law as they apply to electromagnetic waves. First, in figure 23.2, repeated here as figure 25.8(a), a wire that was in a uniform magnetic field that pointed into the paper was pulled to the right with a velocity **v**. We then found that an induced electric field existed in the wire whose magnitude was given by equation 23.4, namely

$$E = vB \sin \theta$$

The induced electric field was the cause of the induced emf $\mathscr{E}$ given by equation 23.6 as

$$E = \frac{\mathscr{E}}{\ell}$$

where ℓ was the length of the wire in motion. *Now the important thing about the cause of the induced electric field and the induced emf was the motion of the wire through the magnetic field. However, what is the difference between the wire in motion toward the right through a uniform magnetic field that is stationary and a wire that is at rest while a uniform magnetic field moves toward the left at a velocity* $- v$. With a little bit of thought, we can see that they must both give the same result. The important thing is the relative motion between the wire and the magnetic field. Hence, if a uniform magnetic field is propagated toward the left, past the stationary wire, then an induced electric field is found in the wire, as shown in figure 25.8(b). The induced electric field causes the induced emf $\mathscr{E}$ between the points M and N and a current I is observed in the galvanometer, the same as before.

Now if the resistance of the wire MN is increased, the current through it, and hence the current recorded by the galvanometer, decreases. In fact, if the resistance R of wire MN were increased to infinity, no current at all would flow through the wire or the galvanometer. However, the induced electric field would still be present and so would its associated induced emf. If the resistance of MN is increased to infinity, the wire is no longer a conductor, but instead, becomes an insulator. In fact, the wire MN could be replaced by a wooden stick, and the relative motion of the wooden stick with respect to the uniform magnetic field would induce an electric field within the stick. Of course no current would flow through the stick, but the induced electric field would still be there. But what is so special about a wooden stick as an insulator? Suppose the stick were removed entirely and only an air gap for MN is left. The air gap would also act as an insulator. *If, again, the uniform magnetic field was to move past the air gap, MN, at a speed v toward the left, then there must be an induced electric field within the air gap itself, in the same direction as the induced electric field within the conducting wire.* As the magnetic field passes the line MN, the magnetic field on the line changes with time, *thus a changing magnetic field induces an electric field anywhere, that is, in a conductor, in an insulator, or in an air gap.*

To show this application of Faraday's law let us recall Faraday's law from equation 23.13, which is

$$\mathscr{E} = -\frac{\Delta \Phi_{\mathbf{M}}}{\Delta t}$$

We related the induced emf $\mathscr{E}$ to the induced electric field E by equation 23.6 as

$$E = \frac{\mathscr{E}}{\ell}$$

We can rewrite equation 23.6 in the form

$$\mathscr{E} = E\ell$$

Then we can write Faraday's law, equation 23.13, in the form

$$E\ell = -\frac{\Delta\Phi_M}{\Delta t} \qquad (25.24)$$

Equation 23.6 was based on the fact that E was a constant and in the direction of ℓ, which is in the direction of the current. If **E** is not in the direction of **ℓ**, then we must take the component of **E** in the direction of **ℓ**. That is, the component of the electric field parallel to the path, $E_{\parallel}$, is $E \cos \theta$. The left-hand side of equation 25.24 should be generalized to

$$E_{\parallel}\ell = E\ell \cos \theta = -\frac{\Delta\Phi_M}{\Delta t} \qquad (25.25)$$

As a further generalization, if **E** is not constant along the path **ℓ**, then we break the path **ℓ** up into a number of smaller paths $\Delta\ell$ along which **E** is a constant. Then the left-hand side of equation 25.25 becomes the sum of all these $E_{\parallel}\Delta\ell$'s. That is,

$$\Sigma E_{\parallel}\Delta\ell = \Sigma E\Delta\ell \cos \theta = -\frac{\Delta\Phi_M}{\Delta t} \qquad (25.26)$$

Equation 25.26 is the generalization of Faraday's law and in this form it is the fourth of Maxwell's equations. As an example of the application of Faraday's law in the form of equation 25.26, consider the circular loop of wire in the magnetic field of figure 25.9(a). The magnetic field is pointing into the paper and is increasing with time, so that $\Delta\mathbf{B}$ also points into the paper. The changing magnetic field induces a current in the coil such as to oppose the changing magnetic field (Lenz's law). Hence, the induced magnetic field $\mathbf{B}_i$ must point outward from the paper, as shown in figure 25.9(b). Therefore, the current in the loop of wire must be counterclockwise. Since charge flows in the direction of the electric field, there must be an induced electric field in the wire, tangential to the wire, as shown in figure 25.9(a). Hence, **E** is always in the direction of $\Delta\mathbf{\ell}$, and θ is equal to zero. Faraday's law then becomes

$$\Sigma E_{\parallel}\Delta\ell = \Sigma E\Delta\ell \cos \theta = \Sigma E\Delta\ell \cos 0° = \Sigma E\Delta\ell \qquad (25.27)$$

But from the symmetry of the problem, the value of E must be the same at every small path $\Delta\ell$, and can thus be factored out of the sum in equation 25.27. Therefore,

$$\Sigma E_{\parallel}\Delta\ell = E \Sigma\Delta\ell \qquad (25.28)$$

But the sum of $\Delta\ell$ around the circular loop is just the circumference of the loop itself, that is,

$$\Sigma\Delta\ell = 2\pi r \qquad (25.29)$$

Substituting equations 25.28 and 25.29 into equation 25.26, gives

$$E(2\pi r) = -\frac{\Delta\Phi_M}{\Delta t} \qquad (25.30)$$

But

$$\Delta\Phi_M = A\Delta B \cos \theta$$

since the area of the loop is a constant. But **A** points out of the paper while $\Delta\mathbf{B}$ points inward, and therefore $\theta = 180°$. Thus

$$\Delta\Phi_M = A\Delta B \cos 180° = -A\Delta B \qquad (25.31)$$

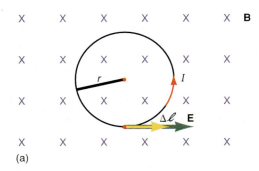

(a)

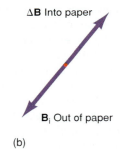

Δ**B** Into paper

B$_i$ Out of paper

(b)

Figure 25.9

The induced electric field in a circular loop in an increasing magnetic field.

Substituting equation 25.31 into equation 25.30, gives

$$E(2\pi r) = + A\frac{\Delta B}{\Delta t}$$

Solving for the induced electric field E,

$$E = \frac{A}{2\pi r}\frac{\Delta B}{\Delta t}$$

The area of the loop is $A = \pi r^2$, thus,

$$E = \frac{\pi r^2}{2\pi r}\frac{\Delta B}{\Delta t}$$

and

$$E = \frac{r}{2}\frac{\Delta B}{\Delta t} \qquad (25.32)$$

Equation 25.32 says that the changing magnetic field with time $\Delta B/\Delta t$ *induces an electric field E around the loop.*

Because r is the radius of the loop and is a constant, we can also write equation 25.32 as

$$E = (\text{constant})\frac{\Delta B}{\Delta t} \qquad (25.33)$$

Let us now compare equation 25.33 with equation 25.23, namely

$$B = (\text{constant})\frac{\Delta E}{\Delta t}$$

Equations 25.33 and 25.23 show that a changing magnetic field with time induces an electric field, whereas a changing electric field with time induces a magnetic field.

Example 25.5

A changing magnetic field with time creates an electric field. If the above loop of wire has a radius of 5.00 cm and the magnetic field changes at the rate of 2.50×10^2 T/s, what is the induced electric field in the loop?

Solution

The induced electric field, given by equation 25.32, is

$$E = \frac{r}{2}\frac{\Delta B}{\Delta t} = \left(\frac{5.00 \times 10^{-2}\text{ m}}{2}\right)(2.50 \times 10^2\text{ T/s})$$

$$= 6.25\,\frac{\text{m}}{\text{s}}\text{T}\left[\frac{\text{N/(A m)}}{\text{T}}\right]\left(\frac{\text{A}}{\text{C/s}}\right)$$

$$= 6.25\text{ N/C}$$

25.6 Maxwell's Equations

The four Maxwell's equations that completely describe all electromagnetic phenomena, have now been developed. They are summarized below:

  I. Gauss's Law for Electricity

$$\Phi_E = \Sigma \, E\Delta A \cos\theta = \frac{q}{\epsilon_0} \tag{25.7}$$

  II. Gauss's Law for Magnetism

$$\Phi_M = \Sigma \, B\Delta A \cos\theta = 0 \tag{25.13}$$

  III. Ampère's Law

$$\Sigma \, B_\| \Delta \ell = \mu_0 I_C + \mu_0\epsilon_0 \frac{\Delta \Phi_E}{\Delta t} \tag{25.20}$$

  IV. Faraday's Law

$$\Sigma \, E_\| \Delta \ell = -\frac{\Delta \Phi_M}{\Delta t} \tag{25.26}$$

The rest of this chapter concerns the propagation of electromagnetic waves in space. Therefore, the charge q in equation 25.7 will be zero and the conduction current I_C in equation 25.20 will also be zero. The change of electric flux in equation 25.20 will be written as

$$\Delta \Phi_E = A\frac{\Delta E}{\Delta t}$$

and the change in magnetic flux in equation 25.26 will be written as

$$\Delta \Phi_M = \frac{A\Delta B}{\Delta t}$$

Hence, *Maxwell's equations for charge-free space can now be written as*

  I. Gauss's Law for Electricity

$$\Sigma \, E\Delta A \cos\theta = 0 \tag{25.34}$$

  II. Gauss's Law for Magnetism

$$\Sigma \, B\Delta A \cos\theta = 0 \tag{25.35}$$

  III. Ampère's Law

$$\Sigma \, B_\| \Delta \ell = \mu_0\epsilon_0 A\frac{\Delta E}{\Delta t} \tag{25.36}$$

  IV. Faraday's Law

$$\Sigma \, E_\| \Delta \ell = -A\frac{\Delta B}{\Delta t} \tag{25.37}$$

The implication of equations 25.34 and 25.35 is that all electric and magnetic fields in charge-free space are continuous, that is, the electric fields do not begin or end on any charges. They also imply that the electric and magnetic flux neither converges nor diverges. Ampère's law, equation 25.36, tells us that a changing electric field produces a magnetic field, and Faraday's law, equation 25.37, says that a changing magnetic field produces an electric field. *The fact that a changing electric field produces a magnetic field, whereas a changing magnetic field produces an electric field, suggests that it should be possible to propagate an electromagnetic wave through empty space.*

25.7 The Production of an Electromagnetic Wave—An Oscillating Dipole

The electrostatic field of a dipole was shown in chapter 19, figure 19.10. It is reproduced here in figure 25.10(a). Recall that the electric field E emanates from the positive point charge and terminates on the negative point charge. This field is an electrostatic field, that is, one that is constant in time. What happens to this field if the charges comprising the dipole are allowed to vary in position with time?

Figure 25.10(b) shows two pieces of wire connected to an AC source. At the instant shown, one positive charge has moved from the AC source and is found at the bottom of the upper wire. As this charge flows from the source, another charge that was initially at the top of the bottom wire has moved into the AC source leaving a negative charge at the top of the bottom wire. At this instant shown, an electric dipole has been created. Only two of the many electric field lines that are present in the space around the dipole are shown in the diagram. A short time later, the positive charge has moved to the midpoint of the upper wire, while the equivalent negative charge is found at the midpoint of the lower wire, figure 25.10(c). The configuration is still that of an electric dipole, but the separation between the charges is increasing with time and the two electric field lines shown in figure 25.10(b) have increased with the increasing separation of the charges, figure 25.10(c). In figure 25.10(d) the charges have momentarily come to a stop at the ends of the wires. We now show only one electric field line and the electric dipole field line has gotten larger. In figure 25.10(e) the charges have reversed direction and are again midway between the ends of the wire. The beginning and ending points of the electric field are starting to come together. In figure 25.10(f) the charges have almost come together. As the alternating emf of the source goes through zero, the electric field line closes on itself, as shown in figure 25.10(g), since there are no longer any charges for the electric field to begin or end on. In figure 25.10(h) the emf has reversed itself and the positive charge is now at the top of the bottom wire, while the negative charge is at the bottom of the top wire. The direction of the electric field line has become reversed. The same process as in figures 25.10(b) through 25.10(g) continues in figures 25.10(i) through 25.10(m), but with the direction of the electric field line reversed.

Our examination of figure 25.10 has not yet given the entire picture of the radiation. As the positive charge moves upward in the top wire of figure 25.10(b) it represents a current in a wire. Associated with such a current in a straight wire is a magnetic field that encircles the wire, as was shown in chapter 22, figure 22.12, and shown here in purple in the top half of figure 25.10b. The magnetic field is also shown as a purple x in the diagram, standing for the tail of the arrow associated with the magnetic field vector, which is going into the paper. The negative charge moving downward in the bottom wire of figure 25.10(b) is equivalent to a positive charge moving upward. Hence, there is also a magnetic field around the bottom wire of figure 25.10(b), as shown. In figure 25.10(c), the magnetic field is in the same direction as in figure 25.10(b). In figure 25.10(d), however, the charge has momentarily come to rest at the end of the wire. Since the charge is at rest here, there is no current and hence there is no magnetic field around the wire. As the charge on the top wire moves midway down the wire, the current is downward and hence the magnetic field associated with that current now comes out of the paper as shown by the purple circled dot in figure 25.10(e), which stands for the tip of the magnetic field vector. The magnetic field continues to come out of the page in figures 25.10(f) and 25.10(g). In figure 25.10(h), a negative charge starts to move up the bottom of the top wire while a positive charge starts to move down the top of the bottom wire. A negative charge moving upward looks exactly like a positive charge moving downward. Therefore, the magnetic field associated with this current comes out of the paper on the right side of the dipole, as shown in purple in figure 25.10(h). The magnetic field continues to come out of the page through figure 25.10(i). In

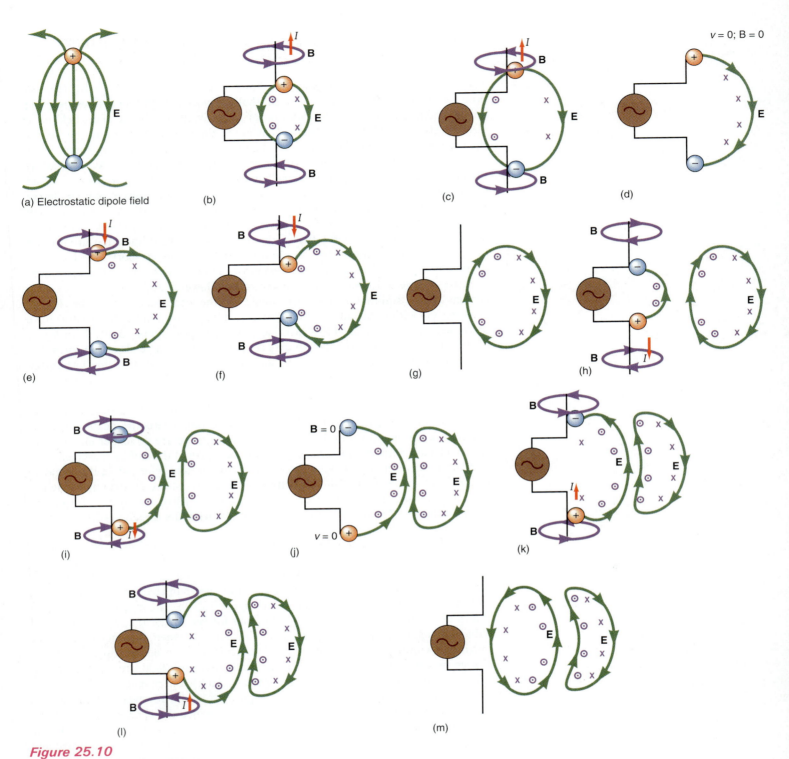

(a) Electrostatic dipole field

(b)

(c)

(d)

(e)

(f)

(g)

(h)

(i)

(j)

(k)

(l)

(m)

Figure 25.10

The generation of an electromagnetic
wave from a dipole.

figure 25.10(j), the charge has momentarily come to rest at the ends of the wire, and the magnetic field then becomes equal to zero. In figure 25.10(k), the negative charge on the top wire is moving downward while the positive charge on the bottom wire is moving upward. This looks like a total current I moving upward in both wires and hence the magnetic field associated with this current goes into the paper at the right of the dipole as shown in purple in figure 25.10(k), and continues in figure 25.10(l). In figure 25.10(m), the current has momentarily become zero, and hence the magnetic field becomes zero. Hence an oscillating dipole has created a varying electric and magnetic field about it. These fields are shown only to the right of the dipole, but they also exist to the left of the dipole, and out of the plane of the drawing, spreading symmetrically outward. But how can these fields propagate through empty space?

25.8 The Propagation of an Electromagnetic Wave

To understand the propagation of an electromagnetic disturbance let us consider the five loops of wire in figure 25.11. A small magnetic field disturbance B_1, pointing into the paper, just passes wire ab of loop 1 at a velocity v_1 toward the right. But B_1 moving to the right with a velocity v_1 is equivalent to moving straight wire ab through the magnetic field with a velocity v toward the left.

An electric field E_1, is induced in wire ab given by

$$E_1 = vB_1 \sin \theta$$

The direction of E_1 is seen pointing downward in wire ab. This electric field E_1 causes a current I_1 in loop 1, as shown in the figure. But the current I_1 in wire dc produces a magnetic field B_2, pointing out of the page, to the left of cd. Because the magnetic field was initially zero around wire cd, when it increases to the value B_2 this means that there was a changing magnetic field with time $(\Delta B / \Delta t)$ around wire cd, that is,

$$\frac{\Delta B}{\Delta t} = \frac{B_2 - 0}{\Delta t}$$

But wire ef of loop 2 finds itself in that changing magnetic field, and by Faraday's law there is now an induced electric E_2 in wire ef of loop 2. This induced electric field is equivalent to wire ef moving to the left in the magnetic field B_2 and hence E_2 is given by

$$E_2 = vB_2 \sin \theta$$

The direction of E_2 points upward, as seen in figure 25.11. This electric field E_2 causes a current I_2 to flow in loop 2, as shown. Current I_2 then causes a magnetic

Figure 25.11

The propagation of an electromagnetic disturbance through several loops of wire.

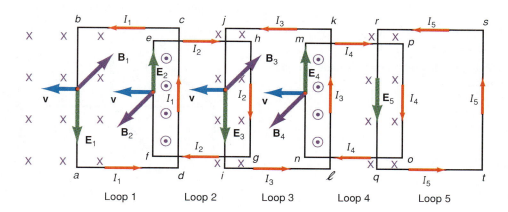

field $\mathbf{B_3}$ to form around wire gh of loop 2, which points into the paper on the left side of gh. Wire ij of loop 3 now finds itself in the magnetic field $\mathbf{B_3}$, but since the original magnetic field around wire ij was zero, the introduction of the new magnetic field $\mathbf{B_3}$ constitutes a changing magnetic field with time, $\Delta\mathbf{B}/\Delta t$. That is,

$$\frac{\Delta\mathbf{B}}{\Delta t} = \frac{\mathbf{B_3} - 0}{\Delta t}$$

But by Faraday's law, this changing magnetic field induces an electric field $\mathbf{E_3}$ in wire ij. This changing magnetic field is equivalent to wire ij moving to the left in field $\mathbf{B_3}$ and the induced electric field $\mathbf{E_3}$ is

$$E_3 = vB_3 \sin\theta$$

and points downward in wire ij, as shown. This induced electric field causes the current I_3 to flow in loop 3 in a counterclockwise direction, as shown. The current I_3 now causes a magnetic field $\mathbf{B_4}$ to form around wire kl. But wire mn of loop 4 now finds itself in the magnetic field $\mathbf{B_4}$. Since the initial magnetic field around wire mn was zero, and it is now $\mathbf{B_4}$, there is a changing magnetic field with time given by

$$\frac{\Delta\mathbf{B}}{\Delta t} = \frac{\mathbf{B_4} - 0}{\Delta t}$$

This changing magnetic field induces an electric field $\mathbf{E_4}$ in wire mn of loop 4. The magnetic field of the current I_4 produces an electric field in loop 5, and so on and on.

The net result of changing the magnetic field at loop 1 is to propagate both an electric and a magnetic field from loop to loop. That is, a changing magnetic field induced an electric field and the changing electric field produced a magnetic field and so on. The directions of $\mathbf{E}$ and $\mathbf{B}$ for each loop are shown in figure 25.12. Note that the electric field vector $\mathbf{E}$ is always perpendicular to the magnetic field vector $\mathbf{B}$. Also note that the current in the first loop is counterclockwise, the second loop clockwise, the third loop counterclockwise, and so on.

Instead of through actual wire loops, an initial disturbance of a magnetic field in the air can be propagated through the air in the same manner. As we saw earlier, an electric field exists when there is a magnetic field changing with time even if there is no conducting path. An actual conducting path for current is not needed because there is a displacement current ($I_\mathrm{D} = \mu_0\epsilon_0 A\Delta E/\Delta t$) to generate the magnetic field. *Thus, the electric and magnetic fields in figure 25.11 would propagate to the right even if the conducting loops were eliminated.*

If the initial disturbance were a sinusoidally varying electric or magnetic field in air, that field would be propagated through the air as a sinusoidal **electromagnetic wave**, as shown in figure 25.13. The electric field vector is everywhere perpendicular to the magnetic field vector and hence *the electric wave is everywhere perpendicular to the magnetic wave.*

Figure 25.12

The electric and magnetic fields for each loop.

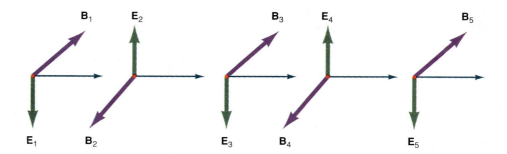

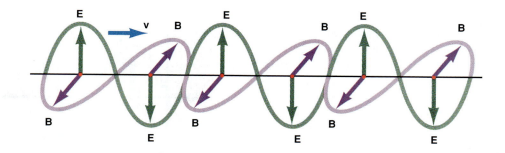

Figure 25.13

An electromagnetic wave.

The electric wave can be represented mathematically, the same as any wave motion, in the form of equation 12.13 as

$$E = E_0 \sin(kx - \omega t) \tag{25.38}$$

where E_0 is the maximum value of the electric field disturbance and is the amplitude of the electric wave and E is the magnitude of the electric field at any position x and time t. The magnetic wave can be written similarly as

$$B = B_0 \sin(kx - \omega t) \tag{25.39}$$

where B_0 is the maximum value of the magnetic field and B is the magnitude of the magnetic field at any position x and time t. Recall from chapter 12 that k is the wave number, given by

$$k = \frac{2\pi}{\lambda} \tag{12.9}$$

while ω is the angular frequency of the wave, given by

$$\omega = 2\pi f \tag{12.12}$$

Here λ is the wavelength of the wave and f is its frequency. Hence, it is logical to assume that there should be a large number of electromagnetic waves, differing only in frequency and wavelength. We will say more about this shortly.

Almost everything said about waves in chapter 12 applies to the electromagnetic waves. That is, electromagnetic waves can be reflected and transmitted. The principle of superposition applies so that any number of electromagnetic waves can be added together. Standing electromagnetic waves are produced and a Doppler effect, slightly different than that for sound waves, is also observed. The main difference between the mechanical waves of chapter 12 and electromagnetic waves is in the medium of the propagation. The medium for the propagation of electromagnetic waves is discussed in detail in chapter 29 on special relativity.

25.9 The Speed of an Electromagnetic Wave

Now that we see that it is possible to generate and propagate electromagnetic waves through space, we want to determine the speed of these electromagnetic waves. To determine this speed let us consider the electromagnetic wave moving toward the right in figure 25.14. At the instant shown, the wave is just starting to move through the wire loop *abcd*, which is lying in the *x–y* plane. Although the electromagnetic wave is moving toward the right, we can just as easily consider it as though the wave were stationary and the loop is moving toward the left at the same speed v. In the short period of time Δt the loop will have moved a distance Δx to the left to the positions a', b', c', and d', as shown. As the loop moves, the flux through the loop changes with time. Let us now apply Ampère's law, equation 25.20, to the loop. We

Figure 25.14

The speed of propagation of an electromagnetic wave.

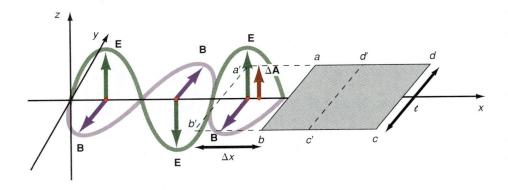

assume that the wire loop has infinite resistance so that the conduction current I_C is zero. Hence, the only current in the loop is the displacement current. Thus,

$$\Sigma\,B_\parallel \Delta \ell = \mu_0 \epsilon_0 \frac{\Delta \Phi_E}{\Delta t} \tag{25.20}$$

The change in the electric flux through the moving loop is

$$\Delta \Phi_E = E \Delta A \cos \theta \tag{25.40}$$

We assume the change in area of the loop ΔA to be small enough that $\mathbf{E}$ is constant throughout that small area. Since the $\mathbf{E}$ vector and $\Delta \mathbf{A}$ vector are parallel

$$E \Delta A \cos \theta = E \Delta A \cos 0° = E \Delta A \tag{25.41}$$

The change in area ΔA is the area of the rectangle $a\,a'\,b\,b'$, which we can determine from figure 25.14 as

$$\Delta A = \ell \Delta x \tag{25.42}$$

Placing the results of equations 25.40, 25.41, and 25.42 back into equation 25.20 for Ampère's law, we get

$$\Sigma\,B_\parallel \Delta \ell = \mu_0 \epsilon_0 E \frac{\Delta A}{\Delta t} = \mu_0 \epsilon_0 E \ell \frac{\Delta x}{\Delta t} \tag{25.43}$$

Because the loop is moving to the left with a speed v, the distance Δx is

$$\Delta x = v \Delta t \tag{25.44}$$

Substituting equation 25.44 into 25.43 gives

$$\Sigma\,B_\parallel \Delta \ell = \mu_0 \epsilon_0 E \ell \frac{v \Delta t}{\Delta t}$$

and

$$\Sigma\,B_\parallel \Delta \ell = \mu_0 \epsilon_0 E \ell v \tag{25.45}$$

Let us now consider the left-hand side of Ampère's law, equation 25.45, and compute $\Sigma\,B_\parallel \Delta \ell$ around the loop $abcd$. The sum of $\Sigma\,B_\parallel \Delta \ell$ around the entire loop is equal to the sum of $B_\parallel \Delta \ell$ along paths ab, bc, cd, and da. That is,

$$\Sigma\,B_\parallel \Delta \ell = \Sigma\,(B_\parallel \Delta \ell)_{ab} + \Sigma\,(B_\parallel \Delta \ell)_{bc} +$$
$$\Sigma\,(B_\parallel \Delta \ell)_{cd} + \Sigma\,(B_\parallel \Delta \ell)_{da} \tag{25.46}$$

Let us consider each term separately.

1. Along path *ab:* **B** is a constant along path *ab* and points in the same direction as $\Delta \ell$, therefore,

$$\Sigma \, (B_{\parallel}\Delta\ell)_{ab} = \Sigma \, B\Delta\ell \cos 0°$$

$$= \Sigma \, B\Delta\ell = B\Sigma\Delta\ell = B\ell \tag{25.47}$$

2. Along path *bc:* **B** is perpendicular to $\Delta\ell$ along that portion of the path *bc* that the wave has reached. Therefore,

$$\Sigma \, (B_{\parallel}\Delta\ell)_{bc} = \Sigma \, B\Delta\ell \cos 90° = 0 \tag{25.48}$$

Along that portion of the path *bc* that the wave has not reached yet, **B** will be zero. Hence, $\Sigma \, (B_{\parallel}\Delta\ell)_{bc} = 0$ along the entire path *bc*.

3. Along path *cd:* The wave has not yet reached path *cd* and hence **B** $= 0$ along this path. Therefore,

$$\Sigma \, (B_{\parallel}\Delta\ell)_{cd} = 0 \tag{25.49}$$

4. Along path *da:* **B** is perpendicular to $\Delta\ell$ along that portion of the path *da* that the wave has reached. Therefore,

$$\Sigma \, (B_{\parallel}\Delta\ell)_{da} = \Sigma B\Delta\ell \cos 90° = 0 \tag{25.50}$$

Along that portion of the path *da* that the wave has not reached yet, **B** will be zero. Hence, $\Sigma \, (B_{\parallel}\Delta\ell)_{da} = 0$ along the entire path *da*.

Substituting equations 25.47 through 25.50 into the left-hand side of Ampère's law, equation 25.46 gives

$$\Sigma B_{\parallel}\Delta\ell = B\ell + 0 + 0 + 0 = B\ell \tag{25.51}$$

Substituting equation 25.51 back into Ampère's law, equation 25.45, we obtain

$$B\ell = \mu_0\epsilon_0 E\ell v$$

Cancelling the ℓ's from both sides of the equation, we get

$$B = \mu_0\epsilon_0 E v \tag{25.52}$$

In dealing with Faraday's law, we already saw that the relation between B and E is

$$E = vB \tag{23.5}$$

Substituting equation 23.5 into equation 25.52 gives

$$B = \mu_0\epsilon_0(vB)v \tag{25.53}$$

$$B = \mu_0\epsilon_0 v^2 B \tag{25.54}$$

Cancelling the B's from both sides of the equation allows us to obtain the speed v of the wave as

$$1 = \mu_0\epsilon_0 v^2$$

$$v = \frac{1}{\sqrt{\mu_0\epsilon_0}} \tag{25.55}$$

Equation 25.55 represents the speed of the electromagnetic wave. Substituting the values of μ_0 and ϵ_0 into equation 25.55 allows us to obtain for the speed of the wave

$$v = \frac{1}{\sqrt{\mu_0\epsilon_0}}$$

$$= \frac{1}{\sqrt{(4\pi \times 10^{-7} \text{ T m/A})(8.85 \times 10^{-12} \text{ C}^2/\text{N m}^2)}}$$

$$= 3.00 \times 10^8 \text{ m/s}$$

But 3.00×10^8 m/s *is the speed of light, usually designated by the letter c. Hence, an electromagnetic wave moves at the speed of light. This result led Maxwell to declare that light itself must be an electromagnetic wave of some appropriate wavelength and frequency,* a prediction since confirmed many times over. With the designation of c as the speed of an electromagnetic wave, equation 23.5 should now be written as

$$E = cB \qquad (25.56)$$

Also note that equation 25.55 should now be written as

$$c = \frac{1}{\sqrt{\mu_0 \epsilon_0}} \qquad (25.57)$$

for the speed of an electromagnetic wave.

Example 25.6

Compare the size of the magnetic field with the electric field. What is the relative strength of the magnetic wave as compared to the electric wave in an electromagnetic wave?

Solution

The strength of the magnetic field at any instant, found from equation 25.56, is

$$B = \frac{E}{c} = \frac{E}{3.00 \times 10^8 \text{ m/s}}$$
$$= (3.33 \times 10^{-9} \text{ s/m})E$$

Thus, the magnetic field is much smaller numerically than the electric field, although, as we will see later, they both carry the same energy.

Satellite television antennas are used to receive both television waves (VHF) and ultra-high frequency television waves (UHF).

25.10 The Electromagnetic Spectrum

We have seen that electromagnetic waves exist and propagate through space at the speed of light. We represent the electric wave by equation 25.38, and the electric field vector **E** depends on the wavelength and frequency of the wave. The wavelength and frequency are not independent but are related by the fundamental equation of wave propagation, equation 12.3 with the speed v replaced by c, and the frequency f replaced by the Greek lower case letter ν (nu). Thus,

$$c = \lambda \nu \qquad (25.58)$$

The use of the letter ν for the frequency of the electromagnetic wave rather than the letter f, that we used previously when waves were discussed in chapter 12, is customary in physics when dealing with electromagnetic radiation in modern physics.

It is evident that an entire series of electromagnetic waves should exist, differing only in frequency and wavelength. Such a group of electromagnetic waves has been found and they are divided into six main categories: radio waves, infrared waves, visible light waves, ultraviolet light, X rays, and gamma rays. *The entire group of electromagnetic waves is called* **the electromagnetic spectrum.** Let us look at some of the characteristics of these waves.

1. *Radio Waves.* Radio waves are usually described in terms of their frequency. AM (amplitude modulated) radio waves are emitted at frequencies from 550 kHz to 1600 kHz. (Recall that the unit kHz is a kilohertz, which is a thousand cycles per second, hence 550 kHz is equal to 550×10^3 cycles per second or 5.50×10^5 cycles/s.) FM (frequency modulated) radio waves, on the other hand, are transmitted in the range of 88 MHz to 108 MHz. (Recall that MHz is a megahertz which is equal to 10^6 Hz.) Television waves are

transmitted in the range of 44 MHz to 216 MHz. Ultra-High frequency (UHF) TV waves are broadcast in the range of 470 MHz to 890 MHz. Microwaves, which are used in radar sets and microwave ovens fall in the range of 1 GHz to 30 GHz. A gigahertz (GHz) is equal to 10^9 cycles/s.

2. *Infrared Waves.* Infrared waves are usually described in terms of their wavelength rather than their frequency. The infrared spectrum extends from approximately 720 nm to 50,000 nm. Recall that the unit nm is a nanometer and is equal to 10^{-9} m. Infrared frequencies can be determined from equation 25.58.

3. *Visible Light.* Visible light occupies a very small portion of the electromagnetic spectrum, from 380 nm to 720 nm. The wavelength of 380 nm corresponds to a violet color, while 720 nm corresponds to a red color.

4. *Ultraviolet Light.* The ultraviolet portion of the spectrum extends from around 10 nm up to about 380 nm. It is this ultraviolet radiation from the sun that causes sunburn and skin cancer.

5. *X Rays.* X rays are very energetic electromagnetic waves. They are usually formed when high-speed charged particles are brought to rest on impact with matter. The x-ray portion of the electromagnetic spectrum lies in the range 0.01 nm up to about 150 nm.

6. *Gamma Rays.* Gamma rays are the most energetic of all the electromagnetic waves and fall in the range of almost 0 to 0.1 nm overlapping the x-ray region. They differ from X rays principally in origin. They are emitted from the nucleus of an atom, whereas X rays are usually associated with processes occurring in the electron shell structure of the atom.

25.11 Energy Transmitted by an Electromagnetic Wave

In chapter 21, we saw that the energy density in the electric field is

$$u_E = \tfrac{1}{2} \epsilon_0 E^2 \tag{21.12}$$

whereas the energy density in a magnetic field was found in chapter 23 to be

$$u_M = \frac{1}{2} \frac{B^2}{\mu_0} \tag{23.50}$$

Hence, the total energy density residing in the electromagnetic field is the sum of the electric energy density and the magnetic energy density, which is simply the sum of equations 21.12 and 23.50. That is,

$$u = u_E + u_M$$

$$u = \frac{1}{2} \epsilon_0 E^2 + \frac{1}{2} \frac{B^2}{\mu_0} \tag{25.59}$$

To show how this energy is distributed, we use equation 25.56 for *B,* and substitute it into equation 25.59 to get

$$u = \frac{1}{2} \epsilon_0 E^2 + \frac{1}{2} \frac{(E/c)^2}{\mu_0}$$

But, substituting for c^2 from equation 25.57, we get

$$u = \frac{1}{2} \epsilon_0 E^2 + \frac{1}{2} \frac{(\mu_0 \epsilon_0) E^2}{\mu_0}$$

Hence,

$$u = \frac{1}{2} \epsilon_0 E^2 + \frac{1}{2} \epsilon_0 E^2 \tag{25.60}$$

The second term on the right-hand side of equation 25.60 represents the magnetic energy density, and since it is equal to the first term, which represents the electric energy density, it is clear that *the total energy of the electromagnetic wave is divided evenly between the electric wave and the magnetic wave.* That is, one half of the total energy of the electromagnetic wave is contained in the electric wave while the other half of the total energy is contained in the magnetic wave. *The total energy density of the electromagnetic field can be written from equation 25.60 as*

$$u = \epsilon_0 E^2 \qquad (25.61)$$

Another quantity that is of great interest is the *intensity of the electromagnetic radiation.* **The intensity of radiation** *is defined as the total energy per unit area per unit time.* Because the total energy per unit time is power, *the intensity of the radiation can also be defined as the power of the electromagnetic wave falling on a unit area.* Thus,

$$\text{Intensity} = \frac{\text{Total energy}}{(\text{area})(\text{time})} \qquad (25.62)$$

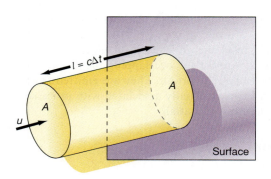

Figure 25.15

Energy intensity.

Equation 25.61 represents the energy density, that is, the energy per unit volume. To obtain the total energy, the energy density u must be multiplied by a volume V of the field. This can be seen more clearly by referring to figure 25.15. The total energy that falls on a unit area A of the surface in a time Δt is all the energy contained in the imaginary cylindrical surface shown in the figure. The volume of the cylinder is

$$V = A\ell = Ac\Delta t$$

Hence, the intensity becomes

$$\text{Intensity} = \frac{uV}{A\Delta t}$$

$$= \frac{uAc\Delta t}{A\Delta t}$$

$$= uc$$

$$\text{Intensity} = \epsilon_0 c E^2 \qquad (25.63)$$

Again using the fact that $E = cB$, we get

$$\text{Intensity} = \epsilon_0 c E (cB)$$

$$= \epsilon_0 c^2 EB$$

We now also use the result that $c^2 = 1/(\mu_0 \epsilon_0)$ to get

$$\text{Intensity} = \epsilon_0 \frac{1}{\mu_0 \epsilon_0} EB$$

$$\text{Intensity} = \frac{1}{\mu_0} EB \qquad (25.64)$$

Example 25.7

Determining the values of E and B for the sun's radiation received at the earth. The solar constant, 1.38×10^3 J/(m² s), is the average intensity of radiation from the sun falling on the top of the earth's atmosphere. What are the average values of the E and B fields associated with this intensity?

The average value of the electric field is found from equation 25.63 as

$$\text{Intensity} = \epsilon_0 c E^2$$

$$E = \sqrt{\frac{\text{Intensity}}{\epsilon_0 c}}$$

$$= \sqrt{\frac{1.38 \times 10^3 \text{ J/(m}^2\text{ s)}}{[8.85 \times 10^{-12} \text{ C}^2/(\text{N m}^2)](3.00 \times 10^8 \text{ m/s})}\left(\frac{\text{N m}}{\text{J}}\right)}$$

$$= 721 \text{ N/C} = 721 \text{ V/m}$$

The average value of B is found from

$$B = \frac{E}{c} = \frac{721 \text{ N/C}}{3.00 \times 10^8 \text{ m/s}} = 2.40 \times 10^{-6} \frac{\text{N}}{\text{A s}}$$

$$= 2.40 \times 10^{-6} \text{ T}$$

Note that the value of B is very much smaller than E, yet each wave contains one half of the total energy of the electromagnetic wave.

The Language of Physics

Maxwell's equations
A set of four equations that completely describe all electromagnetic phenomena (p. 715).

Gaussian surface
An imaginary surface that is placed in space to measure either electric or magnetic flux (p. 715).

Gauss's law for electricity
The electric flux through a Gaussian surface surrounding electric charge is a measure of the amount of electric charge contained within the Gaussian surface. If the charge cannot be measured directly, but the flux can, the charge distribution can be determined by Gauss's law. When the flux is positive, the Gaussian surface surrounds positive charge, and electric flux diverges out of the surface. When the flux is negative, the enclosed charge is negative, and flux converges into the Gaussian surface. For very symmetric problems, Gauss's law can be used to determine the electric field of a particular charge distribution (p. 717).

Gauss's law for magnetism
The magnetic flux passing through a Gaussian surface is always equal to zero because there are no isolated magnetic poles. Thus, magnetic field lines are always continuous, that is, they do not begin or end (p. 720).

Displacement current
A changing electric field in a capacitor is equivalent to a current through the capacitor. This current is called the displacement current (p. 721).

Conduction current
Ordinary current in conducting wires (p. 721).

Ampère's law
A magnetic field can be produced by a conduction current or a changing electric field with time (p. 722).

Faraday's law
An electric field can be produced by changing a magnetic field with time (p. 725).

Electromagnetic waves
Waves that are characterized by a changing electric field and a changing magnetic field. They propagate through space at the speed of light. The electric wave and the magnetic wave are always perpendicular to each other (p. 732).

The electromagnetic spectrum
The complete range of electromagnetic waves, from the longest radio waves down to infrared rays, visible light, ultraviolet light, X rays, and the shortest waves, the gamma rays (p. 736).

Intensity of radiation
The total energy of an electromagnetic wave impinging on a unit area in a unit period of time. It is also represented as the power per unit area (p. 738).

Summary of Important Equations

Electric flux
$$\Phi_E = EA \cos \theta \qquad (25.1)$$

Gauss's law for electricity
$$\Phi_E = \Sigma E \Delta A \cos \theta = \frac{q}{\epsilon_0} \qquad (25.7)$$

Electric field between the plates of a parallel plate capacitor
$$E = \frac{\sigma}{\epsilon_0} \qquad (25.11)$$

$$E = \frac{q}{\epsilon_0 A} \qquad (25.9)$$

Magnetic flux
$$\Phi_M = BA \cos \theta \qquad (23.1)$$

Gauss's law for magnetism
$$\Phi_M = \Sigma B \Delta A \cos \theta = 0 \qquad (25.13)$$

The displacement current
$$I_D = \epsilon_0 A \frac{\Delta E}{\Delta t} \qquad (25.17)$$

Ampère's law
$$\Sigma B_\parallel \Delta \ell = \mu_0 (I_C + I_D) \qquad (25.18)$$

$$\Sigma B_\parallel \Delta \ell = \mu_0 I_C + \mu_0 \epsilon_0 A \frac{\Delta E}{\Delta t} \qquad (25.19)$$

$$\Sigma B_\parallel \Delta \ell = \mu_0 I_C + \mu_0 \epsilon_0 \frac{\Delta \Phi_E}{\Delta t} \qquad (25.20)$$

Faraday's law
$$\Sigma E_\parallel \Delta \ell = -\frac{\Delta \Phi_M}{\Delta t} \qquad (25.26)$$

A changing magnetic field produces an electric field
$$E = (\text{constant}) \frac{\Delta B}{\Delta t} \qquad (25.33)$$

A changing electric field produces a magnetic field
$$B = (\text{constant}) \frac{\Delta E}{\Delta t} \qquad (25.23)$$

Maxwell's equations
$$\Phi_E = \Sigma E \Delta A \cos \theta = \frac{q}{\epsilon_0} \qquad (25.7)$$

$$\Phi_M = \Sigma B \Delta A \cos \theta = 0 \qquad (25.13)$$

$$\Sigma B_\parallel \Delta \ell = \mu_0 I_C + \mu_0 \epsilon_0 \frac{\Delta \Phi_E}{\Delta t} \qquad (25.20)$$

$$\Sigma E_\parallel \Delta \ell = -\frac{\Delta \Phi_M}{\Delta t} \qquad (25.26)$$

Maxwell's equations for charge-free space
$$\Sigma E \Delta A \cos \theta = 0 \qquad (25.34)$$
$$\Sigma B \Delta A \cos \theta = 0 \qquad (25.35)$$

$$\Sigma B_\parallel \Delta \ell = \mu_0 \epsilon_0 A \frac{\Delta E}{\Delta t} \qquad (25.36)$$

$$\Sigma E_\parallel \Delta \ell = -A \frac{\Delta B}{\Delta t} \qquad (25.37)$$

Electric plane wave
$$E = E_0 \sin(kx - \omega t) \qquad (25.38)$$

Magnetic plane wave
$$B = B_0 \sin(kx - \omega t) \qquad (25.39)$$

Wave number
$$k = \frac{2\pi}{\lambda} \qquad (12.9)$$

Angular frequency
$$\omega = 2\pi f \qquad (12.12)$$

Speed of light
$$c = \frac{1}{\sqrt{\mu_0 \epsilon_0}} \qquad (25.57)$$

Relation of electric field to magnetic field
$$E = cB \qquad (25.56)$$

Speed of light
$$c = \lambda \nu \qquad (25.58)$$

Electric energy density
$$u_E = \tfrac{1}{2} \epsilon_0 E^2 \qquad (21.12)$$

Magnetic energy density
$$u_M = \frac{1}{2} \frac{B^2}{\mu_0} \qquad (23.50)$$

Energy density of electromagnetic field
$$u = \frac{1}{2} \epsilon_0 E^2 + \frac{1}{2} \frac{B^2}{\mu_0} \qquad (25.59)$$

$$u = \epsilon_0 E^2 \qquad (25.61)$$

Intensity of radiation
$$\text{Intensity} = \frac{\text{Total energy}}{(\text{area})(\text{time})} \qquad (25.62)$$

$$\text{Intensity} = \epsilon_0 c E^2 \qquad (25.63)$$

$$\text{Intensity} = \frac{1}{\mu_0} EB \qquad (25.64)$$

Questions for Chapter 25

†1. If an electromagnetic wave has energy, should it also have momentum?

2. According to a news report, a woman once washed her cat and then placed it in the microwave oven to dry. Why was this catastrophic for the cat?

3. How does an antenna receive electromagnetic waves?

4. If a radio wave is 1 km long does the radio antenna have to be this long?

5. A student's automobile antenna was stolen from her car. She then took a metal coat hanger and placed it into the empty antenna mount. Would this work to operate her car radio?

†6. How can you take a picture of people at night with an infrared camera?

7. Most people are concerned about receiving too many X rays, but are not concerned about receiving too much visible radiation. Since both radiations are electromagnetic waves, what is the difference?

8. You have no antenna for your FM radio. Will connecting a 1-m length of TV wire to the FM set act as an antenna?

†9. There is growing concern that the earth may be losing its ozone layer. Why should this concern us?

†10. If you could move at the speed of light, what would an electromagnetic wave look like?

†11. What is the difference between a whip antenna and a loop antenna?

Problems for Chapter 25

25.2 Gauss's Law for Electricity

1. What is the magnitude of the electric flux from a point charge of 2.00 μC?
2. Find the total flux passing through the sides of a cube 1.00 m on a side if a point charge of 5.00×10^{-6} C is located at its center.
3. Find the electric field between the plates of a parallel plate capacitor of 2.00×10^{-3} m² if a charge of 6.00 μC is placed on them.
4. Using Gauss's law find the electric field outside the two concentric cylinders in the diagram, if they are carrying equal but opposite charge.

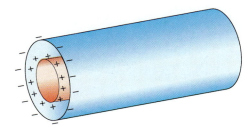

25.4 The Displacement Current and Ampère's Law

5. A displacement current of 5.00 A exists in a parallel plate capacitor that has an area of 7.50 cm². Find the rate at which the electric field changes within the capacitor.
†6. A potential of 100 V is placed across the plates of a parallel plate capacitor rated at 8.50 μF. If it took 0.800 s for this potential to be reached, and if the plates have an area of 25.0×10^{-3} m², find (a) the charge deposited on the plates, (b) the conduction current, (c) the displacement current, (d) the rate at which the electric field changed with time.
†7. Show that the displacement current given by equation 25.17 can also be written as

$$I_D = C \frac{\Delta V}{\Delta t}$$

where C is the capacitance of the capacitor and $\Delta V/\Delta t$ is the rate of change of the voltage across the capacitor.

8. For a parallel plate capacitor of 6.00 μF, what should the value of $\Delta V/\Delta t$ be in order that the displacement current be 3.00 mA?
9. A parallel plate capacitor of 6.00 μF has its applied voltage across the plates changing at the rate of 10,000 V/s. What is its displacement current?

10. If the electric field between the plates of a circular parallel plate capacitor changes at the rate of 4.00 $\times$ 10⁸ (V/m)/s, and if the radius of the capacitor is 10.0 cm, find the magnetic field at (a) $r = 10.0$ cm, (b) $r = 50.0$ cm, and (c) $r = 100$ cm.
†11. Show that the magnetic field at the distance r from the center of a parallel plate capacitor, equation 25.22, can also be written as

$$B = \frac{\mu_0 C \Delta V/\Delta t}{2\pi r}$$

where C is the capacitance of the capacitor and $\Delta V/\Delta t$ is the rate at which the voltage changes across the capacitor.

12. If the voltage that is applied to the parallel plates of a capacitor varies at the rate of 0.500 V/s, find the magnetic field at a distance of 20.0 cm from the center of a 5.00-μF capacitor.

25.8 The Propagation of an Electromagnetic Wave

13. An electric plane wave has a frequency of 90.0 MHz and an amplitude of 0.85 V/m. Write the equation for the electric wave and the magnetic wave.

25.9 The Speed of an Electromagnetic Wave

14. A radar pulse is sent to the moon when the moon is at its mean distance from the earth. How long does it take the pulse to get to the moon and be reflected back to earth?
15. How long does it take to transmit and receive a reflected signal from a satellite that is orbiting Mars when earth and Mars are aligned?
16. A radar set picks up an aircraft in a time of 3.33×10^{-3} s. How far away is the aircraft?

25.10 The Electromagnetic Spectrum

17. What is the range of frequencies for visible light of wavelengths 380 nm to 720 nm?
18. What is the frequency of a 0.100-nm gamma ray?
19. What is the range of frequencies for infrared radiation lying between 720 nm and 50,000 nm?
20. A diathermy machine generates an electromagnetic wave of 6.00-m wavelength. What frequency does this correspond to?

21. An FM radio station broadcasts at 93.4 MHz. What wavelength is associated with this wave?
22. Channel 2 TV operates in a frequency range of 54 to 60 MHz. What range of wavelengths does this represent?

25.11 Energy Transmitted by an Electromagnetic Wave

23. Approximately 60.0% of the solar radiation that impinges on the top of the atmosphere makes it to the surface of the earth. How much energy per square meter hits the surface in 8.00 hr?
†24. If the earth receives 1.38×10^3 J/(m² s) of radiation from the sun, how much energy is radiated from the sun per second? What is the percentage of the sun's energy received on the earth to that radiated by the sun?

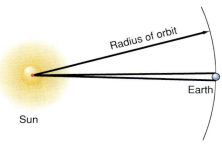

25. Find the intensity of a 100-W incandescent light bulb at a distance of (a) 20.0 cm, (b) 40.0 cm, (c) 60.0 cm, (d) 80.0 cm, and (e) 100.0 cm from the source.
26. Using the results of problem 25, find the energy density at (a) 20.0 cm, (b) 40.0 cm, (c) 60.0 cm, (d) 80.0 cm, and (e) 100.0 cm.
27. Using the results of problems 25 and 26 find the average value of the electric field and the magnetic field at (a) 20.0 cm, (b) 40.0 cm, (c) 60.0 cm, (d) 80.0 cm, and (e) 100.0 cm.
28. Show that if the distance from the source doubles, the intensity of the radiation decreases by a fourth.
29. Find the average value of the electric and magnetic field a distance of 20.0 m from a 100-W incandescent lamp bulb.
30. What is the maximum intensity of an electromagnetic wave whose maximum electric field is 200 N/C?
31. A radio station transmits at 1000 W. Find the value of the electric field at a distance of 10.0 km.
32. Find the intensity associated with an electric wave that has a value of 63.0 V/m.

33. What is the intensity on the surface of a wire 5.00 mm in diameter, 5.00 m long, and having a resistance of 100 Ω when it carries a current of 15.0 A?

Additional Problems

34. A hemispherical surface of diameter $d = 10.0$ cm is placed with its flat side down, as shown in the diagram. A uniform electric field of 625 V/m is directed down throughout this region. Find the electric flux passing downward through the surface.

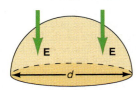

35. The surface charge density of the charge on the plates of a parallel plate capacitor has magnitude $\sigma = 250$ μC/m². Find the electric field between the plates.

36. The displacement current in a parallel plate capacitor is 50.0 mA when the electric field changes at the rate of 16.0 V/(m s). Find the area of each plate.

37. If the intensity of a source of electromagnetic waves is 6.38 W/m² at a distance of 20.0 cm, find the power output of the source.

38. In a velocity selector designed to pass particles of speed 2×10^6 m/s undeflected, the electric field has a magnitude of 740 N/C. Find the total energy density (due to both electric and magnetic fields).

†39. A radio station emits a power of 50,000 W. Assuming that this power is emitted uniformly in all directions, (a) what would be the power received at a radio antenna of 0.0900 m² area, 10.0 miles away? (b) What is the maximum value of the E field picked up by the radio?

†40. You are asked to design a small radio station that will transmit a carrier wave at a frequency of 100 MHz at a power level of 1000 W. (a) If the tuned LC circuit has an inductance of 5.00 mH, what must the value of the capacitance be to generate the 100-MHz signal? (b) What will be the intensity of the radiation at a distance of 20.0 km? (c) What will be the values of E and B at 20.0 km?

†41. If a tuned LC circuit has a capacitance of 7.00×10^{-2} pF and an inductor of 100 μH, (a) what is its natural frequency? (b) If this turned circuit is attached to a dipole antenna, what will the frequency of the electromagnetic wave be? (c) What will its wavelength be?

†42. A ray of light of 400-nm wavelength is traveling in air. It then enters a pool of water where its speed is reduced to 2.26×10^8 m/s. What is the wavelength of the light in the water?

†43. The speed of light in a vacuum was given by equation 25.57. The speed of light in a medium of permittivity ϵ is also given by equation 25.57, but with ϵ_0 replaced by ϵ. Show that the index of refraction, which is defined as the ratio of the speed of light in vacuum to the speed of light in the medium is given by

$$n = \frac{c}{v} = \sqrt{\kappa}$$

where κ is the dielectric constant of the medium.

†44. Using Gauss's law, show that the electric field at a distance r from an infinite line of charge carrying a charge per unit length, $\lambda = q/l$, is given by

$$E = \frac{\lambda}{2\pi\epsilon_0 r}$$

(*Hint:* Draw a cylindrical Gaussian surface around a portion of the infinite line of charge, as shown in the diagram. Then compute $\Sigma E \Delta A \cos \theta$ over the entire cylindrical surface. The entire surface can be broken up into two end caps and the cylindrical surface.)

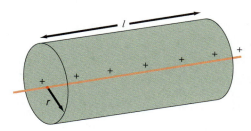

Interactive Tutorials

45. Wavelength-frequency calculator. (a) Calculate the frequency ν of an electromagnetic wave when the wavelength λ is given and (b) calculate the wavelength λ of an electromagnetic wave when the frequency ν is given.

46. Intensity of an electromagnetic wave. A source is radiating electromagnetic waves at a power output $P = 1000$ W. At a distance $r = 2.00$ m from the source, find (a) the intensity I of the radiation, (b) the energy density u of the radiation, (c) the average value of the electric field E, and (d) the average value of the magnetic field B.

Light and Optics

Color as a Study Aid
Chapters 26–28

The following color code is used in these chapters:

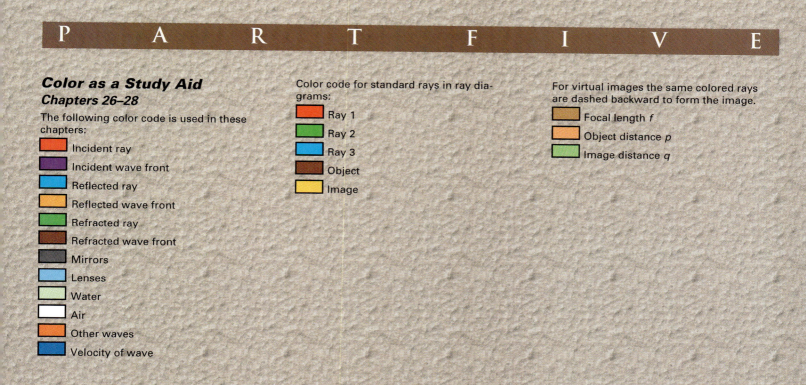

- Incident ray
- Incident wave front
- Reflected ray
- Reflected wave front
- Refracted ray
- Refracted wave front
- Mirrors
- Lenses
- Water
- Air
- Other waves
- Velocity of wave

Color code for standard rays in ray diagrams:

- Ray 1
- Ray 2
- Ray 3
- Object
- Image

For virtual images the same colored rays are dashed backward to form the image.

- Focal length f
- Object distance p
- Image distance q

The Law of Reflection

Using reflections in architecture.

It follows that each little region of a luminous body, such as the Sun, a candle, or a burning coal, generates its own waves of which that region is the center. Thus in the flame of a candle, having distinguished the points A, B, C, concentric circles described about each of these points represent the waves which come from them. And one must imagine the same about every point of the surface and of the part within the flame.

Christian Huygens

26.1 Light as an Electromagnetic Wave

In chapter 25 we saw that light is an electromagnetic wave; a transverse wave that travels through empty space at the speed of 3.00×10^8 m/s or 186,000 miles/s. The relation between the wavelength λ, frequency ν, and speed c of the light is given by the fundamental equation of wave propagation as

$$\lambda\nu = c \tag{26.1}$$

Notice that we are now using the Greek lower case letter ν (nu) to represent frequency rather than the letter f, we used previously. This usage is customary in physics, especially when dealing with electromagnetic radiation in modern physics.

The wavelength of visible light varies from about 3.8×10^{-7} m for violet light to the longer wavelength of red light at 7.2×10^{-7} m. Since the wavelength of light is so small, we commonly use the nanometer (nm), where

$$1 \text{ nm} = 10^{-9} \text{ m}$$

In these units visible light varies from about 380.0 nm to about 720.0 nm. A nonstandard unit for the wavelength of light that is still in common usage is the angstrom abbreviated Å. One angstrom is equal to 10^{-10} m. In this unit the wavelength of light varies from about 3800 to 7200 Å.

A great deal of research went into **optics,** *the study of light,* before its electromagnetic character was known. Hence, an elementary treatment of optics can be followed without recourse to electromagnetic theory. A light wave is represented in figure 26.1. The electric and magnetic vectors are not shown and the magnetic portion of the wave is completely missing.

If a monochromatic point source of light (one of a single wavelength) is turned on at a particular instant, then a spherical wave emanates from the source. A two-dimensional view of the wave is shown in figure 26.2. *A wave front is a line connecting points all having the same phase of vibration and some are shown in figure 26.2. Far away from the source of light the circular fronts look more like plane fronts. The waves are then called plane waves.* These plane waves can then be shown as in figure 26.3. *A line drawn perpendicular to the wave front is called a ray of light and represents the direction of propagation of the light wave.* Note that the ray of light travels in a straight line. The wavelength of light is so small that in a relatively short distance away from the point source of light, the waves appear plane. Even if the source of light is not a point, when we are sufficiently far away from the source, the waves are effectively plane. In all the subsequent discussions we will assume that all the light waves are plane waves.

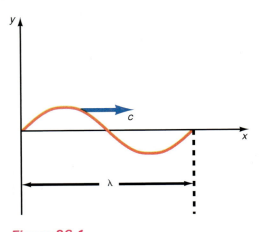

Figure 26.1

A light wave.

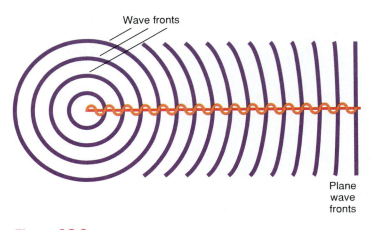

Figure 26.2

The wave front.

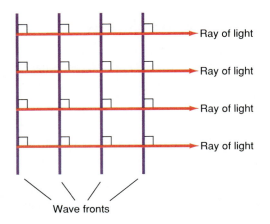

Figure 26.3

Plane waves.

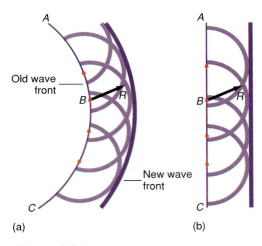

(a)   (b)

Figure 26.4

Huygens' construction.

This analysis of light into waves and rays allows us two different descriptions of light. *When only the light rays are dealt with in the analysis of an optical system, the description is called* **geometrical optics.** *When the analysis of an optical system is done in terms of waves, the description is called* **wave optics** *or* **physical optics.** *Still another description of light is possible by treating light as little bundles of electromagnetic energy, called photons. Such a description is called* **quantum optics** *and will be discussed later in chapter 31.* Except for the analysis of the laws of reflection and refraction by wave motion in sections 26.2, 27.1, and 27.2, chapters 26 and 27 discuss geometrical optics. Chapter 28 deals with wave optics.

At a large distance from a point source of light, the light waves are plane waves propagating as in figure 26.3. If these plane waves were to strike an obstacle of some kind, the wave front becomes distorted. In order to find the new location and shape of the distorted wave front after a passage of time, use is made of a principle, called Huygens' principle, named after Christian Huygens (1629–1695). **Huygens' principle** *states that each point on a wave front may be considered as a source of secondary spherical wavelets. These secondary wavelets propagate in the forward direction at the same speed as the initial wave. The new position of the wave front at a later time is found by drawing the tangent to all of these secondary wavelets at the later time.*

As an example, consider the wave front in figure 26.4(a), labeled *AC*. The points marked on the wave front, *AC*, act as a source of secondary waves. The secondary wavelets travel a distance *R*, given by

$$R = v\Delta t$$

where v is the speed of both the initial wave *AC* and the wavelet and Δt is the time interval considered. The point *B* in figure 26.4 represents the source of one such wavelet. The tangent to all these secondary wavelets is shown as the new wave front in the figure. A similar case is shown in figure 26.4(b) for plane waves. An important application of Huygens' principle is found in the law of reflection.

26.2 The Law of Reflection

Consider a plane wave advancing toward a smooth surface such as a glass or a mirror, as shown in figure 26.5(a). A light ray *AO*, which is perpendicular to the wave fronts, is shown making an angle of incidence *i* with the normal *N*. The side B_1 of the advancing wave front B_1B_2 has just made contact with the surface. While B_2 is

Figure 26.5

The law of reflection.

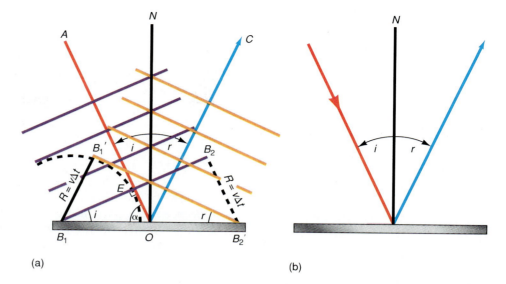

(a)   (b)

moving the distance $R = v\Delta t$, a secondary wavelet emanates from B_1 and also travels the distance $R = v\Delta t$ and is shown in the figure. At the end of the time interval Δt, B_2 will have just made contact with the surface, at the point B_2'. The new reflected wave front is found from Huygens' principle by drawing the new wave front such that it touches the point B_2' and is tangent to the secondary wavelet from B_1. The point of tangency is called B_1'. The new wave front is shown in figure 26.5(a) as the orange line $B_1'B_2'$. The wave is said to have been *reflected*. The new reflected ray OC is drawn perpendicular to the wave front $B_1'B_2'$. The rest of the reflected wave fronts are drawn parallel to $B_1'B_2'$ and perpendicular to ray OC and are shown in the figure. *The reflected ray OC makes an angle r called the angle of reflection, with the normal N.* Because the incident ray AO is perpendicular to the wave front B_1B_2, the angle that the wave front B_1B_2 makes with the surface is the same as the angle of incidence i that incident ray AO makes with the normal N. This can be seen by looking at triangle B_1EO. For the moment, call angle EB_1O, the angle θ. The sum of all the angles in triangle B_1EO must equal $180°$. That is,

$$\theta + \alpha + 90° = 180°$$

but

$$\alpha + i = 90°$$

or

$$\alpha = 90° - i$$

Therefore,

$$\theta + (90° - i) + 90° = 180°$$

and

$$\theta = i$$

as stated. Similarly, since reflected ray OC is perpendicular to reflected wave front $B_1'B_2'$, the reflected wave front $B_1'B_2'$ makes an angle r with the reflecting surface. We see in the figure that in triangle $B_1B_2B_2'$,

$$\sin i = \frac{v\Delta t}{B_1B_2'} \tag{26.2}$$

While from triangle $B_1B_1'B_2'$, we see that

$$\sin r = \frac{v\Delta t}{B_1B_2'} \tag{26.3}$$

Since the right-hand side of equation 26.2 is equal to the right-hand side of equation 26.3, we have

$$\sin i = \sin r$$

or

$$\boxed{i = r} \tag{26.4}$$

*Equation 26.4 is a statement of the first **law of reflection** and it says that the angle of incidence i is equal to the angle of reflection r. The second law of reflection was implied in the derivation and it says that the incident ray, the normal, and the reflected ray all lie in the same plane.*

We have derived the law of reflection by Huygens' principle and the wave nature of light. Having thus established the law, we can simplify the procedure for its use by drawing the ray diagram as in figure 26.5(b). *An incident ray, making an angle i with the normal, is reflected such that the reflected ray makes an angle r, which is equal to the angle i.* In the rest of this chapter, we will apply the law of reflection by using ray diagrams only.

26.3 The Plane Mirror

An object O is placed a distance p (called the *object distance*) in front of a plane mirror RT, as shown in figure 26.6. As you know, if you are the object O, you will see your image in the mirror. How is this image formed and how far behind the mirror is the image located?

To determine how an image is formed, we need consider only two of the many light rays that emanate from the point O. Consider first the ray OA. It makes an angle of incidence i_1 with the normal to the mirror. By the law of reflection, this ray is reflected as AC such that the angle r_1 is equal to the angle i_1. Anyone standing in front of the mirror at C sees this reflected ray, but to this observer it appears to have come from behind the mirror as indicated by the dashed line.

The second incident ray OB makes an angle of incidence i_2 with the mirror and is reflected as ray BD at an angle r_2 such that angle r_2 is equal to angle i_2. To an observer in front of the mirror at D this reflected ray also appears to come from behind the mirror, as indicated by the dashed line. The two reflected rays appear to have come from the point I behind the mirror where the two dashed lines intersect. The point I is called the image of the object O. The distance from the image to the mirror is called the *image distance* and is designated by the letter q. The image is called a virtual image because the light rays do not actually come from that point, they only appear to come from that point. If you were to walk behind the mirror, you would not find any real image there. A real image can be projected onto a screen while a virtual image cannot. In general, a **real image** is one on which all the rays are converging, while a **virtual image** is one from which all the rays are diverging. Thus, rays AC and BD are diverging away from the virtual image I and will never cross on the left-hand side of the mirror.

We should also note that there is a third ray that can be used. This is the ray that is perpendicular to the mirror. Since its angle of incidence is zero, its reflected angle is also zero, and the ray is reflected back along the normal. This ray appears to come from directly behind the mirror and is also indicated by a dashed line that comes from the image point I.

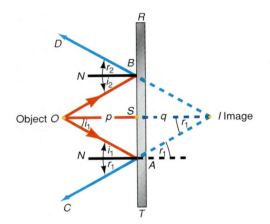

Figure 26.6

A plane mirror.

In the triangle OSA the angle AOS is also equal to i_1 because alternate-interior angles of two parallel lines are equal. Using similar reasoning, angle AIS of triangle SIA is equal to angle r_1. From triangle OSA we have

$$\tan i_1 = \frac{\text{opposite}}{\text{adjacent}} = \frac{SA}{p} \qquad (26.5)$$

while from triangle SIA

$$\tan r_1 = \frac{SA}{q} \qquad (26.6)$$

But the law of reflection states that the angle r_1 is equal to the angle i_1. Equating 26.5 to 26.6 leads to

$$q = p \qquad (26.7)$$

Equation 26.7 says that the image is as far behind the mirror as the object is in front of it.

In general, to describe an optical image three words are necessary: its nature (real or virtual), its orientation (erect, inverted, perverted), and its size (enlarged, true, or reduced). Recall that when you look into a mirror your image is reversed. That is, if you hold up your right hand in front of the mirror, your image appears as though the left hand was held up. This inversion of left-right symmetry is called perversion. Thus, a plane mirror produces a virtual, perverted, true image.

Example 26.1

An image in a plane mirror. If an object 15.0 cm high is placed 20.0 cm in front of a plane mirror, where is the image located and how high is the image?

Solution

The example is illustrated in figure 26.7. A ray from the top and bottom of the object is drawn normal to the mirror. These rays are reflected upon themselves, as shown. The top and bottom of the object appear to come from the dashed lines behind the mirror. Another ray OB is reflected from the mirror as ray BC. If this ray is produced backward it intersects the dashed line from the top of the object at I, producing the image.

From the triangles we have

$$\tan i = \frac{h_o}{p} \qquad (26.8)$$

where h_o is the height of the object and p is the object distance. Also from the figure,

$$\tan r = \frac{h_i}{q} \qquad (26.9)$$

where h_i is the height of the image and q is the image distance. The image distance for the plane mirror, found from equation 26.7, is

$$q = p = 20.0 \text{ cm}$$

And since $i = r$, by the law of reflection, equation 26.9 is equal to equation 26.8, and hence

$$h_i = h_o \qquad (26.10)$$

Equation 26.10 says that for a plane mirror, the height of the image h_i is equal to the height of the object h_o. For this example, $h_o = 15.0$ cm, therefore, h_i is also equal to 15.0 cm. When the height of the image is the same height as the object, the image is said to be true.

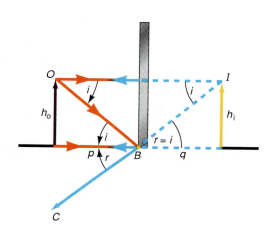

Figure 26.7

Finding the height of an image for a plane mirror.

(a)

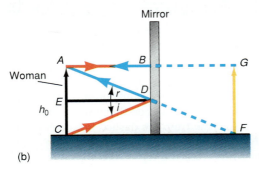

(b)

Figure 26.8

Minimum height of a plane mirror.

Example 26.2

Minimum size of a mirror. What is the minimum height of a plane mirror such that a woman of height h_o can see herself full length?

Solution

The minimum height is most easily found with the help of figure 26.8. The woman is represented by the arrow. Let us assume that the distance from her eyes to the top of her head is small enough to be neglected. The normal ray AB from the top of her head, when produced backward, locates the top of the head of the image. The ray CD, emanating from her feet, strikes the mirror at point D and is reflected back to her eyes at point A. Ray CDA appears to have come from point F, along the ray FDA. A vertical line from F that intersects the produced ray AB at G is the entire image of the woman. The minimum length of the mirror for the woman to see her entire height is just the length BD, which we must now determine.

The tangent of the angle of incidence i is

$$\tan i = \frac{EC}{ED}$$

whereas the tangent of the angle of reflection r is

$$\tan r = \frac{AE}{ED}$$

Because the angle i is equal to the angle r by the law of reflection, these two tangents are equal, and hence,

$$\frac{EC}{ED} = \frac{AE}{ED}$$

or

$$EC = AE \qquad (26.11)$$

But, as we can see from the figure

$$AE + EC = h_o$$

the height of the object. Using equation 26.11, this becomes

$$AE + AE = h_o$$
$$2AE = h_o$$
$$AE = \frac{h_o}{2}$$

But, as we can see from the figure, $AE = BD$, the length of the mirror. Therefore, the minimum height of the mirror is

$$BD = \frac{h_o}{2}$$

That is, the woman needs a mirror that is at least half her size, in order to see the entire length of her body in the mirror.

Determining the image of an object with a plane mirror is, as we have seen, rather simple. But what if the reflecting surface is not a plane? How do we find the image then? We will now consider a spherical reflecting surface and find the image of an object for that spherical surface. But remember there are many other reflecting surfaces that could be considered such as parabolas, cylinders, and so on.

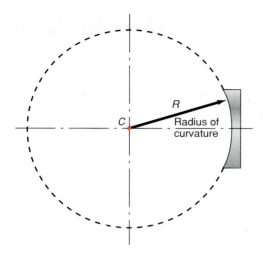

Figure 26.9

A concave spherical mirror.

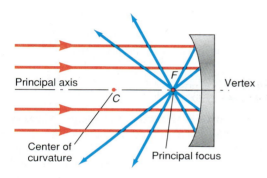

Figure 26.10

Parallel light converges to the principal focus of a concave spherical mirror.

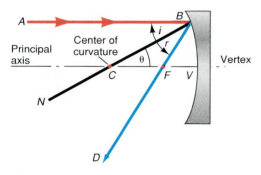

Figure 26.11

Center of curvature and principal focus of a concave spherical mirror.

26.4 The Concave Spherical Mirror

A spherical mirror is a reflecting surface, whose radius of curvature is the radius of the sphere from which the mirror is formed, figure 26.9. Here C is the center of curvature of the mirror and R is its radius of curvature. The line going through the center of the mirror, the *vertex* in figure 26.10, is called the *principal axis,* or *optical axis,* of the mirror. *Light rays that are parallel and close to the principal axis of the concave mirror converge to a point called the principal focus F of the mirror.* We should note here that if a light source were placed at the principal focus, light rays from the principal focus would retrace the same path, coming out parallel to the principal axis after reflection. *This is a principle of optics sometimes referred to as the **principle of reversibility**. It can be stated formally as: if a ray traces a certain path through an optical system in one direction, then a ray sent backward through the system along the same path, traverses the original path and comes out along the line that the original ray entered.* Let us now determine the location of the principal focus of a concave spherical mirror.

Focal Length of a Concave Spherical Mirror

Consider a single ray AB parallel to the principal axis, as shown in figure 26.11. The normal N to the reflecting surface at point B is actually an extension of the radius of curvature since the radius is always perpendicular to its arc. The incoming ray AB, therefore, makes an angle of incidence i with the normal. By the law of reflection, the angle of reflection r is equal to this angle of incidence i and the reflected ray is shown as BD. The point where this reflected ray crosses the principal axis is called the *principal focus* and is designated by the letter F. Because AB is parallel to the principal axis, the angle θ is equal to the angle i. However, the angle i is equal to the angle r by the law of reflection. Therefore, the angle θ is equal to the angle r, and triangle CBF is an isosceles triangle, with side CF equal to side BF. If the incident ray is fairly close to the principal axis, then the length BF is approximately equal to the length VF. That is,

$$CF = BF = VF$$

and, therefore,

$$CF = VF \tag{26.12}$$

From the figure we see that

$$CF + VF = R \tag{26.13}$$

where R is the radius of curvature of the mirror. Combining equations 26.13 and 26.12, gives

$$VF = \frac{R}{2}$$

The distance VF from the vertex of the mirror to the principal focus F is called the focal length f of the mirror. Using this notation,

$$f = \frac{R}{2} \tag{26.14}$$

That is, *the **focal length** of a concave spherical mirror is equal to one-half the radius of curvature of the mirror.* (Remember that, in this derivation we assumed that the incident ray was relatively close to the principal axis. If the width of the mirror is comparable to its radius of curvature, all incident parallel rays are not close to the principal axis, and all these rays, after reflection, do not pass through the principal focus. This defect is called *spherical aberration*. See problem 36 at the end of this chapter.)

Example 26.3

The focal length of a spherical mirror. To what radius should you grind a spherical concave mirror in order to get a focal length of 10.0 cm?

The radius of the mirror, found from equation 26.14, is

$$R = 2f = 2(10.0 \text{ cm}) = 20.0 \text{ cm}$$

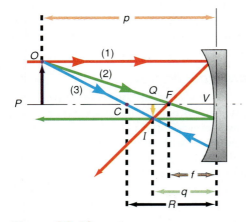

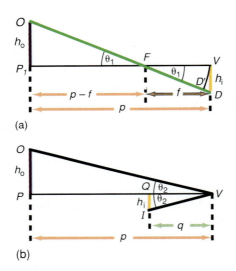

Figure 26.12

Ray diagram for a concave spherical mirror.

Figure 26.13

The geometry for the mirror equation.

Location of an Image by Ray Diagram

Let us find the image produced by an object placed in front of a concave spherical mirror, as shown in figure 26.12. The object *OP*, in brown, is placed a distance *p* in front of the mirror, called the object distance. The center of curvature is located at *C*, and the principal focus *F* is located at a distance $R/2$ from the mirror. *The image is found by drawing the following three standard rays.*

1. *Ray (1), in red, is drawn parallel to the principal axis.* On striking the mirror, it is reflected through the principal focus *F*.
2. *Ray (2), in green, is drawn through the principal focus F.* On striking the mirror, it is reflected parallel to the principal axis because all rays that emanate from the principal focus come out parallel to the principal axis after reflection.
3. *Ray (3), in blue, is drawn through the center of curvature C.* It is reflected upon itself, since it lies along the normal.

The intersection of these three rays defines the top of the image, whereas the principal axis defines the bottom of the image. The image *QI* is shown in yellow in the figure. The distance of this image from the vertex of the mirror is called the image distance and is designated by the letter *q*. The image is real, since the rays of light actually converge at this point. If we were to place a screen at this point, we would see a sharp image of the original object located at this point. Note that the image is inverted.

The Mirror Equation

We can determine the location of the image analytically by making use of the geometry of the ray diagram. Figure 26.13(a) shows a portion of ray (2) of figure 26.12 as it goes through the principal focus and hits the mirror at location *D'*. Since only a small portion of the mirror, which is close to the principal axis, is ever used (in order to avoid spherical aberration), the arc *VD'* of the mirror can be represented by the straight line *VD*. The length of the line *VD* is the same as the height of the image h_i. There are two triangles, *OPF* and *FVD*. The tangents of the angle θ_1 of these triangles are given by

$$\tan \theta_1 = \frac{h_o}{p - f} \tag{26.15}$$

and

$$\tan \theta_1 = \frac{h_i}{f} \tag{26.16}$$

Since the left-hand sides of equations 26.15 and 26.16 are equal, they can be equated, to give

$$\frac{h_o}{p - f} = \frac{h_i}{f}$$

Rearranging terms, we can write

$$\frac{h_o}{h_i} = \frac{p - f}{f} \tag{26.17}$$

We will return to this equation shortly.

Now consider the ray from the object that goes through the vertex V of the mirror, figure 26.13(b). This ray is reflected and intersects the image at the point I. Because V lies along the principal axis, which is normal to the mirror at V, the angle of incidence OVP is equal to the angle of reflection IVQ by the law of the reflection. Let us call these angles θ_2. From figure 26.13(b) we can see that

$$\tan \theta_2 = \frac{h_o}{p}$$

and

$$\tan \theta_2 = \frac{h_i}{q}$$

Since the left-hand side of these two equations are equal they can be equated, to give

$$\frac{h_o}{p} = \frac{h_i}{q}$$

Rearranging terms,

$$\frac{h_o}{h_i} = \frac{p}{q} \tag{26.18}$$

Since the left-hand side of equation 26.17 is equal to the left-hand side of equation 26.18, they can be equated to give

$$\frac{p}{q} = \frac{p - f}{f}$$

$$= \frac{p}{f} - 1$$

Dividing each term by p, gives

$$\frac{1}{q} = \frac{1}{f} - \frac{1}{p} \tag{26.19}$$

Rearranging terms, the equation becomes

$$\frac{1}{f} = \frac{1}{p} + \frac{1}{q} \tag{26.20}$$

Equation 26.20 is called the mirror equation, and it shows the relation between the focal length f of the mirror, the object distance p, and the image distance q.

Example 26.4

Finding the image for a concave spherical mirror. An object 5.00 cm high is placed 30.0 cm in front of a concave spherical mirror of 10.0-cm focal length. Where is the image located?

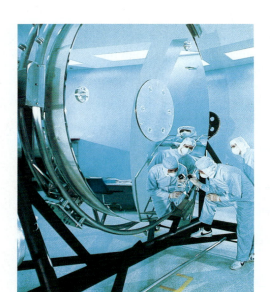

The 2.4-m main mirror from the Hubble Space Telescope.

The location of the image, found from equation 26.19, is

$$\frac{1}{q} = \frac{1}{f} - \frac{1}{p}$$

$$= \frac{1}{10.0 \text{ cm}} - \frac{1}{30.0 \text{ cm}}$$

$$= 0.100 \text{ cm}^{-1} - 0.0333 \text{ cm}^{-1} = 0.0667 \text{ cm}^{-1}$$

and

$$q = 15.0 \text{ cm}$$

Magnification

Another important relationship in an optical system can be obtained by *defining the linear magnification M of the mirror as the ratio of the size of the image h_i to the size of the object h_o.* That is,

$$M = \frac{h_i}{h_o}$$

Thus, the magnification tells how much larger the image is than the object. Using equation 26.18, we can rewrite this as

$$M = \frac{h_i}{h_o} = -\frac{q}{p} \tag{26.21}$$

When we know the magnification M, we find the height of the image h_i from equation 26.21 as

$$h_i = Mh_o \tag{26.22}$$

The minus sign in equation 26.21 is one of convention. Notice from example 26.4 and the ray diagram of figure 26.12, that a positive value of q gives an inverted image. The minus sign in equation 26.21 thus signifies that when M is negative, the image is inverted. We will see a little later that it is possible to have a negative value of q. That case gives us a positive value for M, and thus signifies that the image is erect or right side up.

Example 26.5

The magnification of a spherical mirror. Determine the magnification and size of the image in example 26.4.

The magnification of the system, found from equation 26.21, is

$$M = -\frac{q}{p} = -\frac{15.0 \text{ cm}}{30.0 \text{ cm}} = -0.500$$

The height of the image, found from equation 26.22, is

$$h_i = Mh_o$$

$$= (-0.500)(5.00 \text{ cm})$$

$$= -2.50 \text{ cm}$$

The minus sign indicates that the image is inverted.

To summarize the possible cases of magnification:

if $|M| > 1$ the image is enlarged
if $|M| = 1$ the image is true
if $|M| < 1$ the image is reduced
if $M > 0$ the image is erect
if $M < 0$ the image is inverted.

Some Special Cases of the Concave Spherical Mirror

A great deal of insight into the concave mirror can be obtained by first placing the object at infinity (or at least a very far distance away from the mirror). Now we move the object in toward the mirror and observe what happens to the image.

Case 1: *The Object Is Located at Infinity*

When the object is at a very large distance from the mirror, the object distance p is effectively infinite. The location of the image, found from equation 26.19, is

$$\frac{1}{q} = \frac{1}{f} - \frac{1}{p} = \frac{1}{f} - \frac{1}{\infty}$$

But $1/\infty$ is equal to zero. Therefore,

$$\frac{1}{q} = \frac{1}{f}$$

and

$$q = f \qquad\qquad (26.23)$$

Equation 26.23 says that when the object is located at infinity the image is located at the principal focus. The ray diagram for this case was shown in figure 26.10. The magnification is $M = q/p = q/\infty = 0$ for this case, showing that all the rays converge to a point and the height of the image is $h_i = 0$.

Case 2: *The Object Is between Infinity and the Center of Curvature*

This is a general case, the ray diagram of which was shown in figure 26.12. The image location is found from equation 26.19 when the appropriate values of f and p are introduced. Note from figure 26.12 that the image, which was at F when the object was at infinity, has now moved to the left of F away from the mirror. The image is inverted and real and the magnification ($M = q/p$) is less than 1 because q is less than p. The image is therefore smaller than the object, but still greater than its value of zero, when the image was at the focal point F.

Case 3: *The Object Is Located at the Center of Curvature of the Mirror*

When the object is placed at the center of curvature of the mirror, the object distance p is equal to the radius of curvature R, which is equal to $2f$. The image is found from

$$\frac{1}{q} = \frac{1}{f} - \frac{1}{p}$$

$$= \frac{1}{f} - \frac{1}{R} = \frac{1}{f} - \frac{1}{2f}$$

$$= \frac{1}{2f}$$

and

$$q = 2f = R \qquad\qquad (26.24)$$

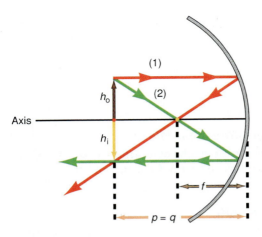

Figure 26.14

Ray diagram for an object placed at the center of curvature of the mirror.

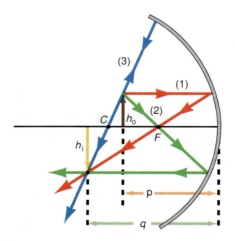

Figure 26.15

An object is placed between the center of curvature *C* and the focal point *F*.

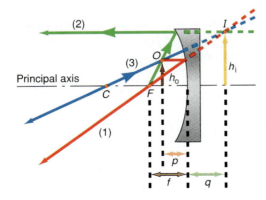

Figure 26.16

Ray diagram for an object placed within the principal focus.

Equation 26.24 says that when the object is located at the center of curvature of the mirror, the image is also located there. Figure 26.14 shows the ray diagram for this case. Because $q = p$ for this case, the magnification becomes

$$M = -\frac{q}{p} = -\frac{p}{p} = -1$$

The minus sign for the value of *M* indicates that the image is inverted, as seen in figure 26.14. The absolute value of the magnification is equal to 1, hence, the height of the image h_i is the same size as the height of the object h_o. Again note that as the object moves closer to the mirror, the image moves farther away from the mirror. The image has gotten bigger until it is now the same size as the object.

Case 4: *The Object Is Located between the Center of Curvature and the Principal Focus of the Mirror*

The image for this case is found by the ray diagram in figure 26.15. Note from the diagram that *the image distance q is now greater than the object distance p. Therefore, the magnification, $M = -q/p$, is negative, which says that the image is inverted, and since the absolute value of the magnification is now greater than 1, the image is now enlarged. Also note that as the object moves closer to the mirror, the image keeps moving farther away from the mirror.* We can find numerical values for *q* from equation 26.19 when the values of *f* and *p* are specified. The magnification is found from equation 26.21 and the height of the image from equation 26.22.

Case 5: *The Object Is Located at the Principal Focus*

When the object is located at the principal focus, $p = f$, and the image, found from equation 26.19, is

$$\frac{1}{q} = \frac{1}{f} - \frac{1}{p} = \frac{1}{f} - \frac{1}{f} = 0$$

The only way for $1/q$ to equal zero is for *q* to become infinite. Therefore,

$$q = \infty$$

That is, *when the object moves into the principal focus, the image moves out to infinity.*

Case 6: *The Object Lies within the Principal Focus*

To see what happens when the object is moved within the principal focus, let us draw a new ray diagram, as shown in figure 26.16.

1. Ray (1) is drawn parallel to the principal axis as before. It is reflected from the mirror and goes through the principal focus *F*.
2. There are an infinite number of rays emanating from the tip of the arrow. Ray (2) is that ray that is emitted from the tip of the arrow but is aligned with an imaginary ray that comes from the principal focus to the tip of the arrow. Ray (2) thus acts as though it is a ray that comes from the principal focus and hence, it is reflected from the mirror parallel to the principal axis.
3. Ray (3) is drawn as though it came from the center of curvature *C*. It is reflected upon itself, as shown.

   Notice that these three rays do not intersect anywhere on the left-hand side of the mirror, and hence they cannot form a real image anywhere on that side of the mirror. However, if we dash these rays backward, through the mirror, we will see that they do intersect on the right-hand side of the mirror. When we look into the mirror, it will appear as though these three rays are emanating from that point on the other side of the mirror. This point, therefore, locates the image; but we now call it a virtual image, since light rays only appear to come from that point. The image is not real; it cannot be projected on any screen. Note that the image is now erect and enlarged. *This example points out very clearly the difference between real and virtual images. For a real image, rays converge to that point; for a virtual image, rays diverge away from that point.* If the mirror equation were solved for the image distance q, it would come out as a negative quantity.

Example 26.6

An object is placed within the principal focus. An object, 5.00 cm high, is placed 7.00 cm in front of a concave spherical mirror of 10.00-cm focal length. Find the location of its image and its final size.

Solution

The image, found from equation 26.19, is

$$\frac{1}{q} = \frac{1}{f} - \frac{1}{p} = \frac{1}{10.00 \text{ cm}} - \frac{1}{7.00 \text{ cm}}$$

$$= 0.1000 \text{ cm}^{-1} - 0.1428 \text{ cm}^{-1} = -0.0428 \text{ cm}^{-1}$$

and

$$q = -23.3 \text{ cm}$$

Notice that for the object placed within the principal focus, the image distance q is negative, as stated previously. A negative value of q indicates that the image is virtual. The magnification is

$$M = -\frac{q}{p} = -\left(\frac{-23.3 \text{ cm}}{7 \text{ cm}}\right) = +3.33$$

Note in this case that since q is negative, the magnification M becomes positive and hence, the image is erect. The final height of the image is

$$h_i = M h_o = (3.33)(5.00 \text{ cm})$$

$$= 16.7 \text{ cm}$$

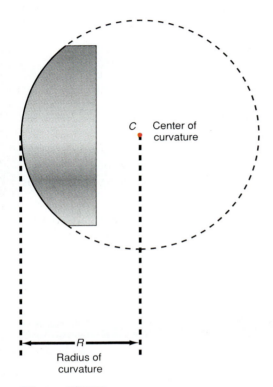

Figure 26.17

A convex spherical mirror.

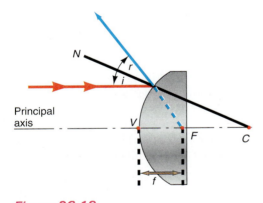

Figure 26.18

A parallel ray of light diverges away from the principal axis for a convex spherical mirror.

A practical example of a concave spherical mirror used in this way is found in a man's shaving mirror or a woman's make-up mirror. The focal length is designed such that when you place your head in front of the mirror, it is located within the principal focus of the mirror. You therefore see your face erect and enlarged.

26.5 The Convex Spherical Mirror

A convex spherical mirror is a reflecting surface formed from a portion of the outside of a sphere, as shown in figure 26.17. The Christmas tree ball decoration is a good example of such a mirror. An incident ray, parallel to the principal axis of a convex spherical mirror, is shown in figure 26.18. The incident ray makes an angle i with the normal to the surface. The normal is the extension of the radius of curvature from the center of curvature C. By the law of reflection, the reflected ray makes an angle r with the normal and the angle r is equal to angle i. The reflected ray is thus seen to diverge away from the principal axis. Hence, for a convex spherical mirror, all rays parallel to the principal axis are reflected such that they diverge away from the principal axis. (Recall that for a concave spherical mirror, parallel rays converged to the principal axis.) If the reflected ray is dashed backward, it intersects the principal axis at the point F. All rays parallel to the principal axis intersect at this same point, which is called the *principal focus* of the convex spherical mirror. Thus, rays that are parallel to the principal axis appear to diverge from the principal

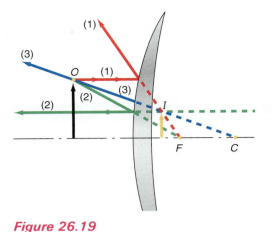

Figure 26.19

Ray diagram for a convex spherical mirror.

focus F. The distance from V, the vertex of the mirror, to F is called the *focal length* f of the mirror. The radius of curvature of the convex spherical mirror is considered to be negative because the center of curvature is on the other side of the mirror. (Recall that for the concave mirror, the center of curvature was on the same side as the reflecting surface and the radius of curvature was positive.) Since the radius of curvature is negative for the convex spherical mirror, the focal length f, which is equal to one-half the radius of curvature, is also negative. With this convention, the mirror equation, equation 26.20, is applicable to either type of mirror.

Location of an Image by a Ray Diagram

The location of the image for a convex spherical mirror is found by drawing the three standard rays in the ray diagram shown in figure 26.19.

1. The first ray (1), in red, is drawn parallel to the principal axis and is reflected from the mirror as if it came from the principal focus F.
2. The second ray (2), in green, is drawn straight toward the principal focus F. It is reflected parallel to the principal axis before it can get to the principal focus. This is another example of the principle of reversibility; all parallel rays are diverged away from the principal focus, while all rays converging to F are reflected parallel to the principal axis.
3. The third ray (3), in blue, is drawn through the center of curvature C of the mirror. Any line through C lies along the radius of curvature of the mirror and is therefore normal to the reflecting surface. Any ray along the normal is reflected upon itself.

As we can see from figure 26.19, the three standard rays diverge away from the principal axis and hence never intersect to the left of the mirror to form a real image. However, if each of these lines are dashed backward behind the mirror, they intersect there to form a virtual image. Also note that the image is erect and reduced in size. If the location of the image were determined by the mirror equation, the image distance q would be negative.

Example 26.7

Finding the image for a convex spherical mirror. An object 5.00 cm high is placed 30.0 cm in front of a convex spherical mirror of −10.0-cm focal length. Where is the image located?

Solution

The location of the image, found from equation 26.19, is

$$\frac{1}{q} = \frac{1}{f} - \frac{1}{p}$$

$$= \frac{1}{-10.0 \text{ cm}} - \frac{1}{30.0 \text{ cm}}$$

$$= -0.100 \text{ cm}^{-1} - 0.0333 \text{ cm}^{-1} = -0.133 \text{ cm}^{-1}$$

$$q = -7.52 \text{ cm}$$

The negative value of q indicates that the image is virtual. Thus, we can use the mirror equation for a convex spherical mirror but the focal length and the image are negative.

Example 26.8

The magnification of a convex spherical mirror. Find the magnification and the size of the image in example 26.7.

The magnification is

$$M = -\frac{q}{p} = -\left(\frac{-7.52 \text{ cm}}{30.0 \text{ cm}}\right) = +0.251$$

Note that M is positive indicating an erect image. The size of the image is

$$h_i = Mh_o = (0.251)(5.00 \text{ cm})$$

$$= 1.26 \text{ cm}$$

If we consider the different possible cases of moving the object from infinity toward the mirror, as in the case of the concave mirror, *we find for a convex spherical mirror that for a real object, the image is always virtual, erect, and reduced in size.*

A convex spherical mirror.

Table 26.1 is a summary of the sign conventions for spherical mirrors. In general, images are real if rays converge to them, whereas images are virtual if rays diverge from them. Objects, on the other hand, are real if rays diverge from them, whereas objects are virtual if rays are converging toward them. It is possible that a real image can be a virtual object for a second optical surface. The focal length of all spherical mirrors is equal to one-half the radius of curvature of the mirror.

Table 26.1
Sign Conventions for Spherical Mirrors

Quantity	Positive	Negative
Convex mirror		R and f
Concave mirror	R and f	
Real image	q	
Virtual image		q
Erect image	M	
Inverted image		M
Real object	p	
Virtual object		p

The Language of Physics

Optics
The study of light (p. 747).

Wave front
A line connecting points all having the same phase of vibration. Wave fronts from a point source of light are spherical. Far away from the source, the waves appear plane (p. 747).

Geometrical optics
The analysis of an optical system in terms of light rays that travel in straight lines (p. 748).

Wave optics or physical optics
The analysis of an optical system in terms of the wave nature of light (p. 748).

Quantum optics
The study of light in terms of little bundles of electromagnetic energy, called photons (p. 748).

Huygens' principle
Each point on a wave front may be considered as a source of secondary spherical wavelets. These secondary wavelets propagate at the same speed as the initial wave. The new position of the wave front at a later time is found by drawing the tangent to all of these secondary wavelets at the later time (p. 748).

The law of reflection
The angle of incidence is equal to the angle of reflection. The incident ray, the normal, and the reflected ray all lie in the same plane (p. 750).

Real image
An image formed by rays converging to a point. A real image can be projected onto a screen (p. 750).

Virtual image
An image formed by rays diverging from a point. A virtual image cannot be projected onto a screen (p. 750).

Optical image
A reproduction of an object by an optical system. To describe an image three words are necessary: its nature (real or virtual), its orientation (erect, inverted, or perverted), and its size (enlarged, true, or reduced) (p. 751).

Spherical mirror
A reflecting surface whose radius of curvature is the radius of the sphere from which the mirror is formed. A plane mirror is a special case of a spherical mirror with a radius of curvature that is infinite (p. 753).

Principle of reversibility
If a ray traces a certain path through an optical system in one direction, then a ray sent backward through the system along the same path, traverses the original path and exits along the line that the original ray entered (p. 753).

Focal length
The point where all rays parallel and close to the principal axis converge. The focal length of a concave spherical mirror is equal to one-half of its radius of curvature (p. 753).

Magnification
The ratio of the size of an image to the size of its object (p. 756).

Summary of Important Equations

The wavelength, frequency, and speed of light
$$\lambda v = c \tag{26.1}$$

Law of reflection
$$i = r \tag{26.4}$$

Focal length of a spherical mirror
$$f = \frac{R}{2} \tag{26.14}$$

Mirror equation
$$\frac{1}{f} = \frac{1}{p} + \frac{1}{q} \tag{26.20}$$

Magnification
$$M = \frac{h_i}{h_o} = -\frac{q}{p} \tag{26.21}$$

Height of image
$$h_i = M h_o \tag{26.22}$$

Questions for Chapter 26

1. If you hold this textbook in front of a plane mirror the letters will all be reversed from left to right. Why are they not also reversed from top to bottom?
2. Explain the difference between a real image and a virtual image.
†3. How can Huygens' principle be justified?
4. What effect does changing the object distance have on the size of an image in a plane mirror?
5. It is easy to see an image formed by the reflection of light from a smooth surface such as a mirror. What type of reflection would occur if the surface were rough?
6. In the barbershop, mirrors are placed on the wall in front and behind the customer. How are the mirrors arranged so that the customer can see the back of his head?
7. Explain the process of formation of an image for a concave spherical mirror as the object starts at infinity and moves toward the mirror. What happens to the magnification?
8. You would like to place a mirror in the corner of your store so that you can see every thing going on in the store. What type of mirror should you use?
†9. Using Huygens' principle, describe what happens to a wave front of light when it hits the edge of a surface.
†10. Two plane mirrors are placed at 90° to each other. Explain how, when you look directly toward the vertex of the two mirrors, you see yourself as others see you. That is, left is left and right is right. Draw a diagram to help in the analysis.
11. How can you make a toy periscope with two plane mirrors?
12. The Hubble Space Telescope, launched by NASA in May 1990 to take pictures of the universe never seen before, developed a serious flaw. The telescope was not able to focus properly. The cause of the problem was spherical aberration of either the primary or secondary mirror. What is spherical aberration and how does it affect the telescope?

Problems for Chapter 26

26.1 Light as an Electromagnetic Wave

1. What is the frequency of (a) violet light of 380.0-nm wavelength and (b) red light of 720.0-nm wavelength?
2. What is the wavelength of the electromagnetic radiation of the following frequencies: (a) 100 kHz, (b) 10.0 MHz, (c) 4.00×10^{14} Hz, and (d) 6.00×10^{15} Hz.
3. How many light waves of 450-nm wavelength can you fit into a distance of 1.00 cm?

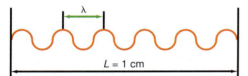

L = 1 cm

4. How long does it take light to reach earth from (a) the sun and (b) the moon?

26.3 The Plane Mirror

5. An object 10.0 cm high is placed 20.0 cm in front of a plane mirror. Where is the image located and how big is it?
6. What is the minimum height of a mirror such that a student, 5'8" tall, can see his entire body?
7. Repeat problem 6, but take into account that the eyes of the student are 4.00 in. below the top of his head.
8. A student stands in front of a plane mirror that is equal to half of her height, but when looking straight ahead, her eyes look directly into the center of the mirror. How far back from the mirror must she move in order to see her entire body in the mirror?

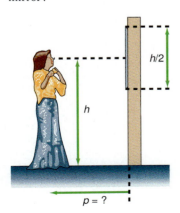

26.4 The Concave Spherical Mirror

9. What is the focal length of a spherical mirror if the radius of curvature is 25.0 cm?
10. An object 10.0 cm high is placed 50.0 cm in front of a concave spherical mirror of 15.0-cm focal length. Find the image by (a) a ray diagram and (b) the mirror equation. Is the image real or virtual? Is the image erect or inverted? What is the size of the image?
†11. Find the image with a concave spherical mirror of 10.0-cm focal length if the object is located at (a) 60.0 cm, (b) 40.0 cm, (c) 20.0 cm, (d) 10.0 cm, and (e) 5.00 cm. Draw a ray diagram for each case.
12. Find the magnification for each case in problem 11.
13. If the object is 5.00 cm high, find the height of the images in each case in problem 12.
†14. A concave spherical mirror has a focal length of 20.0 cm. Find the image distance, magnification, and height of the image when the object is located at (a) 100 cm, (b) 80.0 cm, (c) 40.0 cm, (d) 20.0 cm, and (e) 10.0 cm. Draw a ray diagram for each case.
15. An object is placed 15.0 cm in front of a concave spherical mirror mounted on an optical bench in the laboratory. A screen is moved along the optical bench until the object and image are located at the same point. Find the focal length of the mirror.
16. Find the focal length of a concave spherical mirror that has a magnification of 2.00 when an object is placed 20.0 cm in front of it.
17. An object 10.0 cm high is placed 10.0 cm in front of a concave spherical mirror of 15.0-cm focal length. Find the image by a ray diagram and the mirror equation. How high is the image?
18. Where should an object 5.00 cm high be placed in front of a 25.0-cm concave spherical mirror in order for its image to be erect and 10.0 cm high?
19. For a concave spherical mirror of 20.0-cm focal length, find two locations of an object such that the height of the image is four times the height of the object.
20. An object is placed 40.0 cm in front of a concave spherical mirror and its image is found 25.0 cm in front of the mirror. What is the focal length of the mirror?

21. A concave spherical mirror has a focal length of 15.0 cm. Where should an object be placed such that the height of the image is a quarter of the height of the object?
22. An object is placed 10.0 cm in front of a concave spherical mirror of 15.0-cm focal length. Find the location of the image and its magnification.
23. Find the radius of curvature of a shaving mirror such that when the object is placed 15.0 cm in front of the mirror, the image has a magnification of 2.
24. A concave spherical mirror has a focal length of 15.0 cm. Where should an object be placed to give a magnification of (a) 2.00 and (b) −2.00?

26.5 The Convex Spherical Mirror

25. A reflecting Christmas tree ball has a diameter of 8.00 cm. What is the focal length of such an ornament?
26. An object 10.0 cm high is placed 30.0 cm in front of a convex spherical mirror of −10.0-cm focal length. Find the image by a ray diagram and the mirror equation. Find the height of the image.
†27. Find the image with a convex spherical mirror of 10.0-cm focal length if the object is located at (a) 60.0 cm, (b) 40.0 cm, (c) 20.0 cm, (d) 10.0 cm, and (e) 5.00 cm. Draw a ray diagram for each case.
28. An object is 12.0 cm in front of a convex spherical mirror, and the image is formed 24.0 cm behind the mirror. Find the focal length of the mirror.
29. Where should an object be placed in front of a convex spherical mirror of 15.5-cm focal length in order to get a virtual image with a magnification of one-half?
30. The distance between a real object and a virtual image formed by a convex spherical mirror is 50.0 cm. If the focal length of the mirror is $f = -25.0$ cm, find the two possible positions for the mirror.

Additional Problems

†31. A plane mirror is rotated through an angle θ. Show that the reflected ray will always be rotated through an angle of 2θ.
32. Show that a plane mirror is a special case of a concave spherical mirror whose radius of curvature is infinite. What does the mirror equation reduce to?

†33. Two mirrors make an angle of 90° with each other. Show that if a ray of light is incident on the first mirror at an angle of incidence i, the reflected ray from the second mirror makes an angle of reflection of $90° - i$.

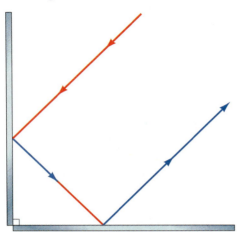

†34. Two mirrors make an angle of θ with each other. Show that if a ray of light is incident on the first mirror at an angle of incidence i, the reflected ray from the second mirror makes an angle of reflection of $\theta - i$.

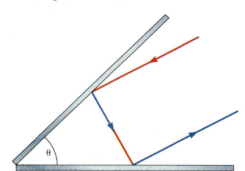

†35. Show that when an object is placed in front of a concave spherical mirror of focal length f and experiences a magnification M, the image is located at the image distance given by

$$q = f(1 - M)$$

36. An object is placed 25.0 cm in front of a concave spherical mirror. The image is found to be a quarter of the size of the object. Find the focal length of the mirror.

37. An object is magnified by a factor of 2 when it is placed 15.0 cm in front of a concave spherical mirror. Find the radius of curvature of the mirror.

38. A dentist uses a small concave spherical mirror to see a cavity in a tooth. If the image is to be magnified by a factor of 3 when the tooth is 2.50 cm in front of the mirror, what should be the focal length of such a mirror?

39. An optical system is designed so that an object for a convex spherical mirror of focal length $f = -15.0$ cm is 20.0 cm *behind* the mirror (a virtual object). Find the image distance and the magnification, and determine whether the image is real or virtual, erect or inverted.

40. Repeat problem 39 with an object located only 5.00 cm *behind* the mirror.

†41. Use a compass to draw a concave spherical mirror 10.0 cm in radius. Draw light rays parallel to the principal axis at every 1.00 cm above and below the principal axis. Using a protractor, carefully measure the angles of incidence and reflection for each of these rays, and see where they cross the principal axis. What does this tell you about the underlying assumption in the mirror equation? How does this relate to spherical aberration of the mirror?

†42. Draw a graph of the image distance q as a function of the object distance p for a concave spherical mirror of focal length 10.0 cm. Show the regions that represent the concave mirror and the convex mirror. Show the regions where the images are real and where they are virtual.

†43. Draw a graph of the magnification M of a concave spherical mirror as a function of the object distance p. Repeat for a convex spherical mirror.

Interactive Tutorials

🖳 44. A tower on earth transmits a laser beam of frequency $f = 5.00 \times 10^{14}$ Hz to a spaceship at a distance $x = 7.40 \times 10^{11}$ m. Calculate (a) the wavelength λ of the laser beam and (b) the time t for the beam to reach the spaceship.

🖳 45. Spherical mirror. An object of height $h_o = 3.00$ cm is placed at the object distance $p = 10.0$ cm of a spherical mirror of radius of curvature $R = 8.00$ cm. Find (a) the focal length f of the mirror, (b) the image distance q, (c) the magnification M, and (d) the height h_i of the resulting image.

🖳 46. Spherical mirror. An object of height $h_o = 8.50$ cm is placed at the object distance $p = 35.0$ cm of a spherical mirror of focal length $f = 15.0$ cm. Find (a) the radius of curvature R of the mirror, (b) the image distance q, (c) the magnification M, and (d) the height h_i of the resulting image.

The Law of Refraction

Internal reflection and refraction of light.

If in other sciences we should arrive at certainty without doubt and truth without error, it behooves us to place the foundations of knowledge in mathematics.

Roger Bacon

27.1 Refraction

In chapter 26, we saw that an incident ray of light is reflected from a piece of reflecting material such that the reflected ray makes the same angle with the normal as the incident ray. If the reflecting surface is a boundary between two different transparent mediums, such as air and glass, some of the incident light is also transmitted into the glass, as shown in figure 27.1. However, it is observed experimentally that this transmitted ray of light is bent as it enters the second medium. *The bending of light as it passes from one medium into another is called* **refraction.** *Refraction of light occurs because light travels at different speeds in different mediums.* Light traveling through a vacuum travels at the speed $c = 3.00 \times 10^8$ m/s. But when light enters a medium there is a complex interaction between the electromagnetic wave (light) and the atomic configuration of the medium. This interaction causes the electromagnetic wave to slow down in the medium. This slowing down of the wave as it goes from a vacuum into the medium causes it to bend. We will use Huygens' principle to show how this is accomplished in section 27.2.

27.2 The Law of Refraction

Let us consider a wave front $B_1 B_2$ of a plane parallel monochromatic wave impinging on the boundary of two different mediums, as shown in figure 27.2. The incident ray makes an angle of incidence i with the normal N. The incident light moves at a speed v_1 in medium 1 and v_2 in medium 2, and we assume that v_1 is greater than v_2. The incident wave has just touched the boundary at B_1. In a time Δt, B_2, the upper portion of the initial wave front, travels a distance $v_1 \Delta t$, and impinges at the boundary of the interface at B_2'. In this same time interval Δt, the wave front at B_1 enters the second medium. By Huygens' principle, a secondary wavelet can be drawn emanating from the point B_1. This wave moves a radial distance $v_2 \Delta t$ in the second medium in the time interval Δt, and is shown as the circle of radius $v_2 \Delta t$ in the figure. The radial distance $v_2 \Delta t$ is less than the distance $v_1 \Delta t$ because v_2 is less than v_1. By Huygens' principle, the line drawn from B_2' that is tangent to the secondary wavelet is the new wave front. The point of tangency is denoted by B_1' and the new wave front in medium 2 is $B_1' B_2'$. The radius from B_1 to B_1', when extended, becomes the refracted ray $B_1 C$. The other refracted rays are drawn parallel to $B_1 C$, as shown in figure 27.2. The angle that the refracted ray makes with the normal is called the *angle of refraction r.*

Figure 27.1

Reflection and refraction of light.

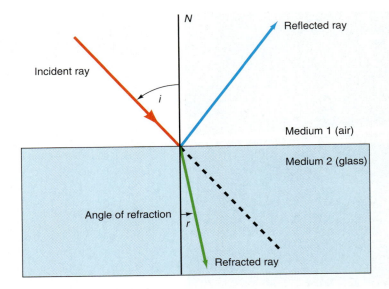

Figure 27.2

The law of refraction by Huygens' principle.

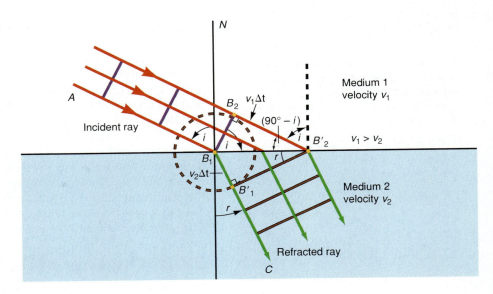

We can obtain the relation between the angles *i* and *r* from the geometry of figure 27.2. Since line B_2B_2' makes an angle *i* with the dashed normal, angle $B_2B_2'B_1$ is equal to $(90° - i)$, and since the sum of the angles in triangle $B_1B_2B_2'$ must equal 180°, it follows that angle $B_2B_1B_2'$ is also equal to the angle *i*. Using similar reasoning, angle $B_1B_2'B_1'$ is equal to *r*. Hence, from the trigonometry of figure 27.2,

$$\sin i = \frac{v_1\Delta t}{B_1B_2'} \tag{27.1}$$

and

$$\sin r = \frac{v_2\Delta t}{B_1B_2'} \tag{27.2}$$

Light and Optics

Table 27.1

Index of Refraction for Various Materials ($\lambda = 589.2$ nm, the D line of sodium)

Substance	n
Air	1.00029
Benzene	1.50
Diamond	2.42
Glass, crown	1.52
Glass, flint	1.57–1.72
Glass, fused quartz	1.46
Glycerine	1.47
Ice	1.31
Plexiglass	1.49
Quartz crystal	1.54
Water	1.33

Let us divide equation 27.1 by equation 27.2 and obtain

$$\frac{\sin i}{\sin r} = \frac{\dfrac{v_1 \Delta t}{B_1 B_2'}}{\dfrac{v_2 \Delta t}{B_1 B_2'}}$$

and

$$\frac{\sin i}{\sin r} = \frac{v_1}{v_2} = \text{constant} = n_{21} \tag{27.3}$$

*Equation 27.3 is the **law of refraction**. It says that the ratio of the sine of the angle of incidence to the sine of the angle of refraction is equal to the ratio of the speed of light in medium 1 to the speed of light in medium 2.* Because the speed of light in medium 1 v_1 is a constant and the speed of light in medium 2 v_2 is a constant, then their ratio v_1/v_2, must also be a constant. This constant is called the *index of refraction* of medium 2 with respect to medium 1 and is denoted by n_{21}.

If medium 1 is a vacuum, then $v_1 = c$, and the index of refraction of the medium with respect to a vacuum is

$$n = \frac{c}{v} \tag{27.4}$$

Since the speed v in any medium is always less than c, the index of refraction, $n = c/v$, is always greater than 1, except for in a vacuum where it is equal to 1. Indices of refraction for various substances are given in table 27.1. Notice that the index of refraction of air is so close to the value 1, the index of refraction of a vacuum, that in many practical situations, air is used in place of a vacuum.

The fact that the speed of light varies from medium to medium has an important effect on the wavelength of light. When an initial wave enters a second medium, its wavelength changes. We can see this from equation 27.3, where

$$n_{21} = \frac{v_1}{v_2}$$

However, the speed of any wave is given by

$$v = \lambda \nu \tag{27.5}$$

where λ is the wavelength and ν is the frequency of the wave. Hence,

$$n_{21} = \frac{v_1}{v_2} = \frac{\lambda_1 \nu}{\lambda_2 \nu} = \frac{\lambda_1}{\lambda_2} \tag{27.6}$$

The frequency of the wave does not change as it goes across the boundary because there are the same number of wave fronts passing from medium 1 into medium 2, per unit time. Therefore,

$$\lambda_2 = \frac{\lambda_1}{n_{21}} \tag{27.7}$$

That is, the wavelength of the light in the second medium λ_2 is less than the initial wavelength λ_1 by the factor $1/n_{21}$. To summarize, the speed of light varies from one medium to another but the frequency of the light remains the same in both mediums. Because of the changing speed of light, the wavelength of the light changes as it goes into the second medium.

The law of refraction can be put in a more convenient form by using equation 27.4. We can write the index of refraction of medium 1 with respect to a vacuum as

$$n_1 = \frac{c}{v_1} \tag{27.8}$$

whereas we can write the index of refraction of medium 2 with respect to a vacuum as

$$n_2 = \frac{c}{v_2} \tag{27.9}$$

Solving for the speeds v_1 and v_2 from equations 27.8 and 27.9, respectively, and substituting them into equation 27.3, gives

$$n_{21} = \frac{v_1}{v_2} = \frac{c/n_1}{c/n_2} = \frac{n_2}{n_1} \tag{27.10}$$

Using equation 27.10, we can write the law of refraction, equation 27.3, as

$$\frac{\sin i}{\sin r} = n_{21} = \frac{n_2}{n_1}$$

or

$$n_1 \sin i = n_2 \sin r \tag{27.11}$$

Equation 27.11 is the form of the law of refraction that we will use in what follows. It is also called Snell's law after its discoverer, Willebrord Snell (1591–1626) a Dutch mathematician who discovered it in 1620, the same year the Pilgrims landed at Plymouth Rock. Note that if a ray lies along the normal, then the angle of incidence i is equal to zero, and hence the angle of refraction r must also be zero, and there is no refraction of this ray.

The stick appears bent because of refraction at the water's surface.

Example 27.1

The refraction of light. A ray of light of 500.0 nm wavelength in air, impinges on a piece of crown glass at an angle of incidence of 35.0°. Find (a) the angle of refraction, (b) the speed of light in the glass, and (c) the wavelength of light in the glass.

Solution

a. We find the angle of refraction from Snell's law, equation 27.11, with the indices of refraction $n_1 = n_{air} = 1.00$ and $n_2 = n_{glass} = 1.52$, found in table 27.1. Therefore,

$$n_1 \sin i = n_2 \sin r$$

$$\sin r = \frac{n_1}{n_2} \sin i$$

$$= \frac{1.00}{1.52} \sin 35.0° = 0.377$$

$$r = \sin^{-1} 0.377$$

and the angle of refraction is

$$r = 22.2°$$

b. The speed of light in the glass, found from equation 27.9, is

$$v_2 = \frac{c}{n_2} = \frac{3.00 \times 10^8 \text{ m/s}}{1.52}$$

$$= 1.97 \times 10^8 \text{ m/s}$$

Light and Optics

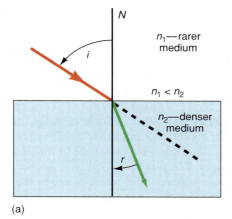

(a)

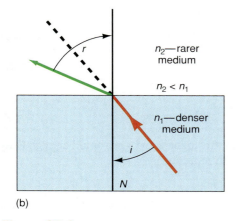

(b)

Figure 27.3

The law of refraction by a ray diagram.

Figure 27.4

Apparent depth of an object immersed in water.

$$\lambda_2 = \frac{\lambda_1}{n_{21}} = \frac{\lambda_1}{n_2/n_1} = \frac{n_1\lambda_1}{n_2}$$

$$= \frac{(1.00)(500.0 \text{ nm})}{1.52}$$

$$= 329 \text{ nm}$$

The law of refraction has been derived by Huygens' principle, treating light as a wave phenomenon. We will now simplify the analysis of refraction by using only the ray model of light. Thus, the refraction of *waves* in figure 27.2 will now be shown as the equivalent refraction of *rays* in figure 27.3. The dashed line in figure 27.3(a) is the direction that the incident ray would have followed if it had not entered the second medium. Note that the incident ray was bent away from its original direction and bent toward the normal. Medium 1 has an index of refraction that is less than medium 2. The medium with the smaller value of the index of refraction is called the *rarer medium,* whereas the medium with the larger value of *n* is called the *denser medium.* Hence, *whenever a ray of light goes from a rarer medium to a denser medium the refracted ray is always bent toward the normal.* The larger the value of the index of refraction n_2, the greater the amount of bending.

By the principle of reversibility, a light ray that reverses the path in figure 27.3(a), goes from a denser medium to a rarer medium and is bent away from the normal, as seen in figure 27.3(b). Hence, *whenever a ray of light goes from a denser medium to a rarer medium, the refracted ray is bent away from the normal.*

27.3 Apparent Depth of an Object Immersed in Water

An interesting example of the refraction of light is the observation of an object when it is under water. A spoon in a glass of water appears to be bent and a fish in water is not where it seems to be. Sometimes in a carnival or a bazaar, a game is played where a glass is placed in water, as in figure 27.4, and the patrons try to throw a quarter into the glass, receiving a prize if they are successful. Of course it is much more difficult than it appears because the glass is not where it seems to be.

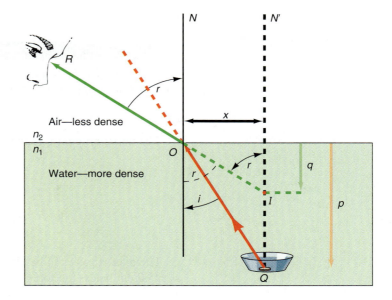

As an example, let the object, the quarter in the glass, be at the bottom of the water at Q, a distance p from the top of the water. A ray from Q makes an angle of incidence i with the normal. Because the ray is going from the more dense medium, water, to the less dense medium, air, the refracted ray is bent away from the normal, as shown. The observer's eye, located at R, sees the ray ROQ, but the person believes that light rays travel in straight lines, and thinks he sees the ray ROI. The person, therefore, assumes that the quarter in the glass is located at I, a distance q below the surface of the water. In a sense, you might say that he is looking at the image of the glass. The vertical ray QN' lies along the normal and hence is not refracted, as shown. These two rays QOR and QN' do not intersect anywhere in medium 2, the air, but when produced backward, the ray ROI intersects QN' at I, and in this sense, I appears as the image of Q. The glass appears to be much closer to the top of the water than it really is. The distance q from the top of the water to I is called the *apparent depth* of the object. Let us now solve for the apparent depth.

From Snell's law of refraction, equation 27.11, we have

$$n_1 \sin i = n_2 \sin r$$

We determine the angles i and r from the geometry of figure 27.4 by observing that

$$\tan i = \frac{x}{p} \qquad \textbf{(27.12)}$$

and

$$\tan r = \frac{x}{q} \qquad \textbf{(27.13)}$$

To simplify the solution, let us assume that the angles i and r are small enough (less than $10°$) to use the small-angle approximation, that is,

$$\sin \theta \approx \tan \theta$$

With this assumption, equations 27.12 and 27.13 can be substituted back into equation 27.11 as

$$n_1 \tan i = n_2 \tan r$$

$$n_1 \frac{x}{p} = n_2 \frac{x}{q}$$

Solving for the apparent depth,

$$q = \frac{n_2}{n_1} p \qquad \textbf{(27.14)}$$

Example 27.2

Apparent depth. A glass is placed in the bottom of a tub of water 25.0 cm deep. What is its apparent depth?

Solution

The apparent depth, found from equation 27.14, is

$$q = \frac{n_2}{n_1} p$$

$$= \left(\frac{1}{1.33}\right)(25.0 \text{ cm})$$

$$= 18.8 \text{ cm}$$

The glass appears to be 18.8 cm below the top of the water when in fact it is really 25.0 cm below the top.

In the solution of the problem of the apparent depth, we made an assumption, just as is often done in physics. The solution is, of course, only as good as the assumption, that is, that the angles i and r are less than 10°. Suppose they are not. What effect does this have on the solution? For example, if x in figure 27.4 is 12.5 cm then the angle i, found from equation 27.12, is

$$i = \tan^{-1}\frac{x}{p} = \tan^{-1}\frac{12.5}{25.0}$$

$$= 26.6°$$

The angle r, found from Snell's law, is

$$\sin r = \frac{n_1}{n_2}\sin i$$

$$= \frac{1.33}{1.00}\sin 26.6°$$

and the angle of refraction is

$$r = 36.5°$$

This is obviously much larger than the assumed limit of 10°. The apparent depth, now found from equation 27.13, is

$$q = \frac{x}{\tan r} = \frac{12.5 \text{ cm}}{\tan 36.5°}$$

$$= 16.9 \text{ cm}$$

If the calculation is repeated with $x = 25.0$ cm, the angle of incidence i is now equal to 45.0°, and the angle of refraction r is now equal to 70.1° and the apparent depth q is 9.05 cm. These values are significantly different from the values obtained with the initial assumptions. This is again a reminder that in trying to represent the physical world in terms of mathematical equations, certain assumptions are made. When these assumptions are valid, the equations are good. When they are not valid the equations no longer represent the physical world and are essentially useless.

27.4 Refraction through Parallel Faces

A ray of light passing through two parallel boundaries is refracted at each of these boundaries, but the final ray is in the same direction as the initial ray, as shown in figure 27.5. An incident ray makes an angle i_1 with the normal N. Because medium 2 is more dense than medium 1, the refracted ray is bent toward the normal in medium 2, making an angle of refraction of r_2 with the normal. Snell's law for the first interface becomes

$$n_1 \sin i_1 = n_2 \sin r_2 \tag{27.15}$$

This refracted ray now becomes the incident ray for the interface between medium 2 and medium 1, and makes an angle i_2 with the normal N'. Since normals N and N' are parallel, angle r_2 is equal to angle i_2. As the ray goes from the more dense medium 2 to the less dense medium 1, the ray is refracted away from the normal through the angle r_1, as shown. Snell's law for the refraction at this interface is

$$n_2 \sin i_2 = n_1 \sin r_1 \tag{27.16}$$

Because angle r_2 is equal to angle i_2, the right-hand side of equation 27.15 is equal to the left-hand side of equation 27.16. Equating them, we obtain

$$n_1 \sin i_1 = n_1 \sin r_1$$

Thus, the angle r_1 is equal to the initial angle i_1, and the final refracted ray comes out parallel to the direction of the original ray, but slightly displaced.

Figure 27.5

Refraction through parallel faces.

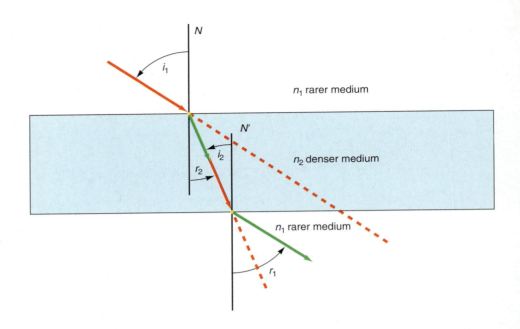

Example 27.3

Refraction through parallel faces. A ray of light in air makes an angle of incidence of 40.0° with the normal to a plate of glass of $n = 1.50$, as in figure 27.5. Find the angle of refraction of the ray in the glass and the angle of refraction of the final ray as it passes from the glass back into the air.

Solution

The angle of refraction for the first interface, found from Snell's law, is

$$n_1 \sin i_1 = n_2 \sin r_2$$

$$\sin r_2 = \frac{n_1}{n_2} \sin i_1$$

$$= \frac{1.00}{1.50} \sin 40.0° = 0.4285$$

and the angle of refraction for the first interface becomes

$$r_2 = 25.4°$$

For the second interface, the law of refraction is

$$n_2 \sin i_2 = n_1 \sin r_1$$

$$\sin r_1 = \frac{n_2}{n_1} \sin i_2 = \frac{1.50}{1.00} \sin 25.4° = 0.6434$$

and the angle of refraction for the second interface is

$$r_1 = 40.0°$$

Note that the final ray makes the same angle as the initial ray, and they are, therefore, parallel.

Light and Optics

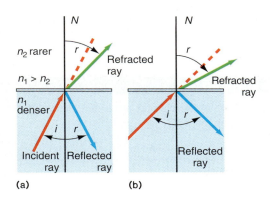

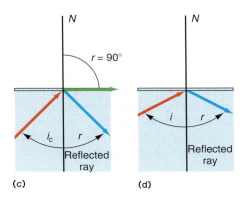

Figure 27.6

Total internal reflection.

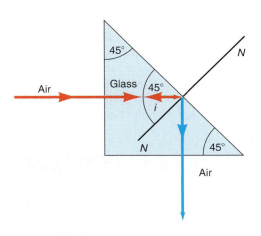

Figure 27.7

Total internal reflection and the prism.

27.5 Total Internal Reflection

As shown in figure 27.1, when an incident ray falls on a boundary of a transparent material, part of the ray is reflected back into the first medium, while part of the ray is refracted through the second medium. If the incident ray is in the dense medium, then the refracted ray is bent away from the normal, as shown in figure 27.6(a). If the angle of incidence is increased, the angle of refraction is also increased, as shown in figure 27.6(b). If the angle of incidence is increased still further, the point is reached where the angle of refraction becomes 90°, as shown in figure 27.6(c). *The angle of incidence that causes the refracted ray to bend through 90° is called **the critical angle of incidence**. When the incident angle becomes greater than the critical angle, no refraction occurs. That is, no light enters the second medium at all; all the light is reflected. This condition is called total internal reflection, because refraction has been eliminated entirely.* We obtain the condition for total internal reflection by finding the critical angle of incidence that allows the angle of refraction to be 90°. This is done by applying Snell's law of refraction:

$$n_1 \sin i_c = n_2 \sin 90°$$

or

$$\sin i_c = \frac{n_2}{n_1} \sin 90° = \frac{n_2}{n_1} \qquad (27.17)$$

Equation 27.17 gives the condition for total internal reflection.

Total internal reflection can only occur when light travels from a denser medium to a rarer medium, because it is only then that the refracted ray is bent away from the normal. For a light ray going in the opposite direction, the refracted ray bends toward the normal and the angle of refraction could never become 90°.

Example 27.4

Critical angle for a glass-air interface. What is the critical angle for a ray of light that goes from glass to air? Take $n_{glass} = 1.50$.

Solution

The critical angle, found from equation 27.17, is

$$\sin i_c = \frac{n_2}{n_1} = \frac{n_{air}}{n_{glass}} = \frac{1.00}{1.50} = 0.667$$

$$i_c = 41.8°$$

Let us now take an ordinary piece of glass, whose index of refraction is 1.50, and cut it into a triangle with angles of 45.0°, 90.0°, and 45.0°, as shown in figure 27.7 An incident ray perpendicular to the first face goes directly into the glass without any refraction, because it is along the normal of the first face. This ray makes an angle of incidence of 45.0° with the normal as it hits the second face. But 45.0° is greater than the critical angle of 41.8° just found in example 27.4. Hence, none of this light crosses the interface into the air, but instead it is totally reflected at this second face. This reflected ray, being normal to the third face, completely passes into the air at the third face without any refraction there. *This ordinary piece of glass, cut into the shape of a triangle, with an angle greater than the critical angle, is called a **prism**.* One of its functions is to completely reflect light. It is even better than a plane mirror for reflection because in a plane mirror some light energy is absorbed by the silver of the mirror, whereas the prism, acting as a device for total internal reflection, reflects everything. No energy is ever transmitted into the other medium at the interface. Prisms are used in binoculars, spectrometers, and reflecting telescopes.

A further example of internal reflection can be found in the area of **fiber optics.** An optical fiber consists of a single flexible glass rod, or an array of them, of high refractive index. Light entering the glass undergoes total internal reflection from the walls of the glass fiber and the light travels down the length of the fiber with little or no absorption. Due to their flexibility, the fibers can be curved into various shapes. But as long as the reflection angle remains greater than the critical angle, the light travels along the length of the fiber.

27.6 Dispersion

In all the previous discussions of refraction we assumed that the incident light was monochromatic, that is, it contained light of only one wavelength. If white light, a mixture of all colors and hence numerous wavelengths, impinges on a transparent surface, we obtain the refraction pattern shown in figure 27.8. The white light is refracted such that each color, or wavelength, is refracted by a different angle. Hence, the white light is dispersed into its constituent wavelengths. *The separation of white light into its component colors is called* **dispersion.** *The band of colors is known as a spectrum.* The spectrum of visible light contains the colors red, orange, yellow, green, blue, and violet.

Violet light, which has the shortest wavelength, is bent the most, whereas red light with the longest wavelength, is bent the least. *Dispersion occurs because the index of refraction is not strictly a constant for a particular material, but rather varies slightly with wavelength.* For example, the index of refraction of crown glass for red light (700.0 nm) is about 1.51, whereas for violet light (400.0 nm) it is about 1.53, which is only a slight difference, yet significant enough to cause the phenomenon of dispersion.

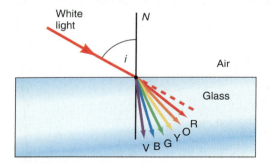

Figure 27.8

Dispersion of light.

a. The angle of refraction for red light, found from Snell's law, is

$$n_1 \sin i = n_2 \sin r_r$$

$$\sin r_r = \frac{n_1}{n_2} \sin i = \frac{1.00}{1.51} \sin 30.0° = 0.331$$

$$r_r = 19.3°$$

b. For violet light, the angle of refraction is

$$n_1 \sin i = n_2 \sin r_v$$

$$\sin r_v = \frac{n_1}{n_2} \sin i = \frac{1.00}{1.53} \sin 30.0° = 0.3267$$

$$r_v = 19.1°$$

Notice that the total angular separation between the red and violet light is only 0.2 degrees for this glass.

Dispersion is most pronounced if a prism is used, as shown in figure 27.9. Because of the angle between the two faces of the prism, the deviation D of the final refracted ray is greater.

Because the index of refraction is a function of the incident wavelength, when defining the index of refraction of a medium, we need to specify the particular wavelength used. When the wavelength is not specified in this book it is assumed that the wavelength is $\lambda = 589.2$ nm, corresponding to the yellow D line of the sodium spectrum.

27.7 Thin Lenses

*An optical **lens** is a piece of transparent material, such as glass or plastic.* Because of its shape, however, light passing through it either converges or diverges to its principal axes. The entire effect of the lens is due to its shape and the index of refraction of the lens.

One such shape of a lens is formed from a piece of plane glass by shaping it into the form of two spherical surfaces with radii of curvature R_1 and R_2, as shown in figure 27.10, where C_1 and C_2 are the centers of the two spherical surfaces. This shaped piece of glass is called a *double convex lens,* because the shape of its surfaces are convex.

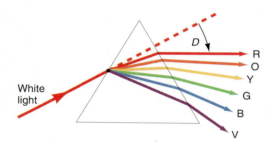

Figure 27.9

Dispersion by a prism.

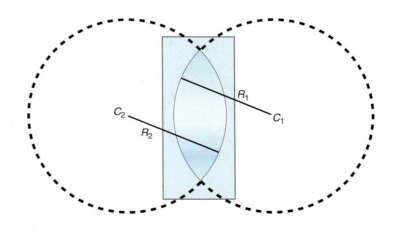

Figure 27.10

Formation of a lens.

Figure 27.11

Refraction by a lens.

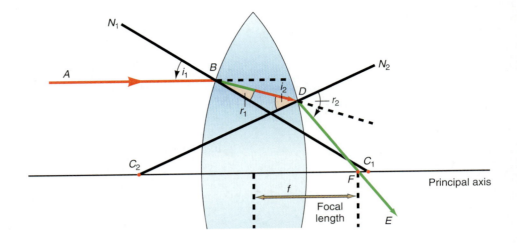

The line going through the center of the lens is called the *principal axis* and is shown in figure 27.11. Consider a ray of light *AB,* parallel and close to the principal axis and impinging on the first surface at the point *B.* The normal N_1 to the surface at *B* is a continuation of the radius of curvature of the first surface, which emanates from the center of curvature C_1 of the first surface. The incident ray *AB* makes an angle of incidence i_1 with the normal N_1. Because the ray is going from the less dense medium, air, to the more dense medium, glass, the refracted ray is bent through an angle r_1 toward the normal, as shown. This refracted ray impinges on the second surface at the point *D.* The normal at *D* is the continuation of the radius of curvature of the second surface and is designated N_2. Ray *BD* makes an angle of incidence i_2 with the normal N_2. Since the ray now goes from glass to air, it is going from a more dense medium to a less dense medium. The refracted ray is, therefore, bent away from the normal N_2 through an angle r_2 as ray *DE. The net effect of the two refractions is to take a ray of light, which is parallel to the principal axis, and bend it such that it crosses the principal axis. The point where this ray crosses the principal axis is called the principal focus, and is designated by the letter F.* The distance from the center of the lens to the principal focus is called the *focal length f* of the lens. It is important to note that for a particular material, it is the shape of the lens that determines the normals and hence the refracted rays and thus the focal length of the particular lens.

The equation that relates the focal length, index of refraction, and radii of curvature of a lens is called the **lensmaker's formula.** It is a rather long derivation, so we will state it without proof as

$$\frac{1}{f} = (n - 1)\left(\frac{1}{R_1} - \frac{1}{R_2}\right) \tag{27.18}$$

where n is the index of refraction of the glass, R_1 is the radius of curvature of the first surface of the lens, R_2 is the radius of curvature of the second surface, and f is the focal length of the lens. By convention, if the surface is convex to the incident ray, R is considered positive. If the surface is concave to the incident ray, R is considered negative. It is assumed that the ray of light is incident from the left. Thus, for a double convex lens, R_1 would be positive and R_2 would be negative. Equation 27.18 says that in order to make a lens of a particular focal length f for a medium of index of refraction n, the medium must be ground to the radii of curvature R_1 and R_2 that satisfies equation 27.18. Note that in the derivation of equation 27.18 (which has been omitted), *we assume that the thickness of the glass lens is negligible compared to the distance to the principal focus and to any object or image distance concerned. Such a lens of negligible thickness is called a* **thin lens.** We will consider only thin lenses in this book.

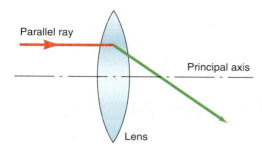

Figure 27.12

Bending of a ray of light by a convex lens.

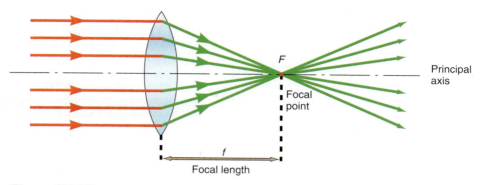

Figure 27.13

A converging lens.

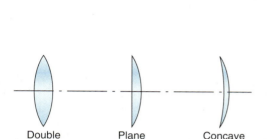

Figure 27.14

Examples of converging lenses.

Since we have already seen that the converging of a parallel ray to the principal axis is caused by the refraction at the two surfaces of the lens, it will not be necessary to go into all this detail every time we consider a lens. Instead, we will say that a ray of light, parallel to the principal axis, is bent in the middle of the lens. It then converges to the principal focus, as shown in figure 27.12.

All rays parallel and close to the principal axis converge to the principal focus, as seen in figure 27.13. Such a lens is called a **converging lens.** If the incident rays are not close to the principal axis, the converging rays do not all intersect at the same point, a result known as *spherical aberration.* We will assume in this book that all rays are sufficiently close to the principal axis so that spherical aberration can be ignored. If parallel light comes in from the right of the lens it converges at the same distance to the left of the lens. Note also that, by the principle of reversibility, light emanating from the principal focus comes out parallel to the principal axis. Figure 27.14 shows some examples of converging lenses. The actual shapes are determined by the radii of curvature, but *an easy way to identify such a lens is to note that a converging lens is always thicker at the center of the lens than it is at the rim.* If any of the thin lenses in figure 27.14 are reversed, they still produce the same effect.

*A **diverging lens** is a lens that takes a bundle of parallel light rays and diverges them away from the principal axes,* as shown in figure 27.15. If the diverging rays are produced backward, they all intersect at the same point *F,* called the principal focus, or focal point, of the diverging lens. *The diverging rays appear to come from the principal focus. The distance from the center of the lens to the focal point is called the focal length f of the diverging lens.* Some examples of diverging lenses are shown in figure 27.16. *One characteristic of all diverging lenses is that they are thinner at the center of the lens than they are at the rim.*

Figure 27.15

A diverging lens.

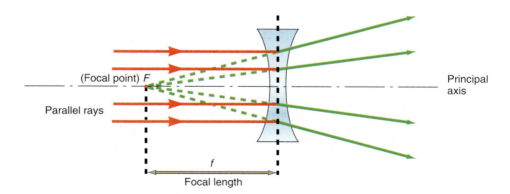

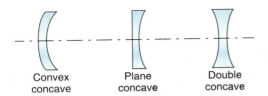

Convex concave   Plane concave   Double concave

Figure 27.16

Examples of diverging lenses.

The focal length of any converging or diverging lens is found from the lens-maker's formula, equation 27.18. However, it is important to remember the convention that R is positive for an incident ray that impinges on a convex surface and negative for one that impinges on a concave surface. A plane surface is a surface whose radius of curvature is infinite, and hence, $1/R = 0$.

Example 27.7

Finding the focal length of a converging lens. A double convex lens has radii of curvature of 25.0 cm. The index of refraction is 1.52. What is the focal length of the lens?

Solution

By our convention, the radius of curvature of the first convex surface is $+25.0$ cm. The inside of the second convex surface is concave to the incident light and its radius of curvature is therefore -25.0 cm. The focal length of the lens, found by the lens-maker's formula, equation 27.18, is

$$\frac{1}{f} = (n-1)\left(\frac{1}{R_1} - \frac{1}{R_2}\right)$$

$$= (1.52 - 1)\left(\frac{1}{25.0\text{ cm}} - \frac{1}{-25.0\text{ cm}}\right) = 0.0416\text{ cm}^{-1}$$

$$f = 24.0\text{ cm}$$

Example 27.8

Finding the focal length of a diverging lens. A double concave lens has radii of curvature of 25.0 cm. The index of refraction is 1.52. What is the focal length of the lens?

Solution

By our convention, the radius of curvature of the first concave surface is -25.0 cm. The inside of the second concave surface is convex to the incident ray, and hence, the radius of curvature is $+25.0$ cm. The focal length of the diverging lens, found from equation 27.18, is

$$\frac{1}{f} = (n-1)\left(\frac{1}{R_1} - \frac{1}{R_2}\right)$$

$$= (1.52 - 1)\left(\frac{1}{-25.0\text{ cm}} - \frac{1}{+25.0\text{ cm}}\right) = -0.0416\text{ cm}^{-1}$$

$$f = -24.0\text{ cm}$$

Note that converging lenses have positive focal lengths while diverging lenses have negative focal lengths.

27.8 Ray Tracing and the Standard Rays

To determine the location of an image formed by a lens, we use the technique of *ray tracing.* An object *OP,* represented by a brown arrow, is placed a distance p in front of a convex lens in figure 27.17. A ray, from the bottom of the object at *P,* travels along the principal axis. This ray goes through the lens undeviated. Hence the bottom of the image must lie somewhere on the principal axis of the lens. To

Figure 27.17

Ray diagram for a convex lens.

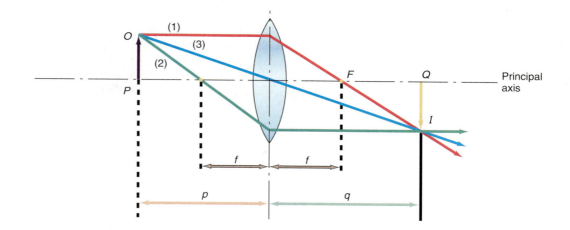

determine the location of the tip of the image, three rays are drawn from the tip of the object. These three rays are called the *standard rays* and are shown in figure 27.17.

The first ray (1), in red, is drawn parallel to the principal axis and after refraction it passes through the principal focus. A second ray (2), in green, is drawn from the top of the object and passes through the first focal point. This ray is refracted through the lens so that it comes out parallel to the principal axis. (Recall that by the principle of reversibility, light emanating from the principal focus comes out parallel to the principal axis.) The third of the standard rays (3), in blue, is drawn from the tip of the arrow and goes through the center of the lens. The refraction experienced at the first surface is exactly compensated by the refraction at the second surface, and the ray passes through the lens undeviated. These three rays intersect at the point *I* and form the real, inverted image *IQ,* in yellow. If a screen were placed at this point a sharp image would be found there. The distance from the object to the lens is called the *object distance p,* whereas the distance from the lens to the image is called the *image distance q.* Thus, we can find any image by drawing a ray diagram.

A ray diagram for a diverging lens is shown in figure 27.18. An object *OP* is placed a distance *p* in front of a diverging lens. A ray (1), in red, is drawn from the object parallel to the principal axis. The diverging lens causes the ray to bend

Figure 27.18

Ray diagram for a concave lens.

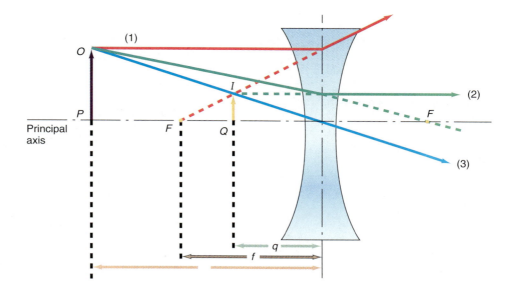

away from the principal axis, as if the ray had come from the principal focus *F*. A second ray (2), in green, is drawn, which would go to the second focal point if there were no lens present. This ray is caused to bend by the diverging lens so that it comes out parallel to the principal axis. A third ray (3), in blue, passes through the center of the lens and is not bent. These three rays do not intersect anywhere to the right of the lens and hence cannot form a real image there. However, if we follow each of the three rays backward, they appear to intersect at the point *I*, and the image is located at that point. The distance from the lens to the image, the image distance, is denoted by the letter *q* in the diagram. We can find an image of any object for any lens by drawing these standard rays.

27.9 The Lens Equation

A mathematical relation between the object distance *p*, the image distance *q*, and the focal length *f*, can be obtained from the geometry of figure 27.19.

In triangle *OPC*,

$$\tan \theta_1 = \frac{OP}{PC} = \frac{h_o}{p}$$

whereas in triangle *QIC*,

$$\tan \theta_1 = \frac{QI}{QC} = \frac{h_i}{q}$$

Equating the tangent of θ_1 from each equation gives

$$\frac{h_o}{p} = \frac{h_i}{q}$$

Rearranging terms, this becomes

$$\frac{h_i}{h_o} = \frac{q}{p} \tag{27.19}$$

We will return to this equation shortly.

In triangle *ACF*,

$$\tan \theta_2 = \frac{AC}{CF} = \frac{h_o}{f}$$

whereas in triangle *IQF*,

$$\tan \theta_2 = \frac{QI}{QF} = \frac{h_i}{q - f}$$

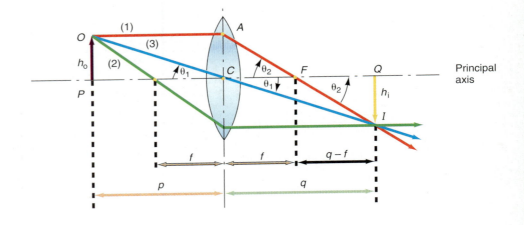

Figure 27.19

Determining the lens equation.

Light and Optics

Equating the tangent of θ_2 from each equation, gives

$$\frac{h_i}{q - f} = \frac{h_o}{f}$$

Rearranging terms, this becomes

$$\frac{h_i}{h_o} = \frac{q - f}{f} \qquad (27.20)$$

Setting equation 27.20 equal to equation 27.19, gives

$$\frac{q - f}{f} = \frac{q}{p}$$

or

$$\frac{q}{f} - 1 = \frac{q}{p}$$

Dividing each term by q, gives

$$\frac{1}{f} - \frac{1}{q} = \frac{1}{p}$$

Solving for $1/f$, we obtain

$$\frac{1}{f} = \frac{1}{p} + \frac{1}{q} \qquad (27.21)$$

Equation 27.21 is **the lens equation** and gives the relation between the object distance p, the image distance q, and the focal length f of the lens. Notice that it is the same formula as the mirror equation found in chapter 26.

The linear magnification M of a lens system is defined as the ratio of the height of the image to the height of the object, that is,

$$M = \frac{\text{height of image}}{\text{height of object}} = \frac{h_i}{h_o} \qquad (27.22)$$

Using equation 27.19, we can also write this as

$$M = \frac{h_i}{h_o} = -\frac{q}{p} \qquad (27.23)$$

The magnification tells how much larger the image is than the object. When we know M, we find the height of the image h_i from equation 27.23, as

$$h_i = Mh_o \qquad (27.24)$$

The minus sign in equation 27.23 is placed there by convention. For a single lens, when the magnification is negative the image is real and inverted, when it is positive the image is virtual and erect. Notice that the lens equation, the magnification equation, and the height of the image equation, are the same equations that were obtained for the spherical mirror in chapter 26.

Example 27.9

A converging lens. An object 5.00 cm high is placed 35.0 cm in front of a converging lens of 10.0 cm focal length. (a) Where is the image located? (b) What is the magnification? (c) What is the size of the image?

a. The image distance, found from the lens equation, equation 27.21, is

$$\frac{1}{f} = \frac{1}{p} + \frac{1}{q}$$

$$\frac{1}{q} = \frac{1}{f} - \frac{1}{p} = \frac{1}{10.0 \text{ cm}} - \frac{1}{35.0 \text{ cm}} = 0.0714 \text{ cm}^{-1}$$

$$q = 14.0 \text{ cm}$$

b. The magnification, found from equation 27.23, is

$$M = -\frac{q}{p} = -\frac{14.0 \text{ cm}}{35.0 \text{ cm}} = -0.400$$

Since M is negative, the image is real and inverted.

c. The size of the image, found from equation 27.24, is

$$h_i = Mh_o = -(0.400)(5.00 \text{ cm}) = -2.00 \text{ cm}$$

The image is thus smaller than the object. The minus sign on h_i simply means that the distance h_i is in the negative y direction, and hence, the image is inverted.

Example 27.10

A diverging lens. An object 5.00 cm high is placed 35.0 cm in front of a diverging lens of -10.0 cm focal length. (a) Where is the image located? (b) What is the magnification? (c) What is the size of the image?

a. The image distance, found from the lens equation, equation 27.21, is

$$\frac{1}{f} = \frac{1}{p} + \frac{1}{q}$$

$$\frac{1}{q} = \frac{1}{f} - \frac{1}{p} = \frac{1}{-10.0 \text{ cm}} - \frac{1}{35.0 \text{ cm}}$$

$$= -0.100 \text{ cm}^{-1} - 0.0285 \text{ cm}^{-1} = -0.1286 \text{ cm}^{-1}$$

$$q = -7.78 \text{ cm}$$

Since the image distance q is negative, the image is virtual and is located 7.78 cm in front of the lens.

b. The magnification, found from equation 27.23, is

$$M = -\frac{q}{p} = -\left(\frac{-7.78 \text{ cm}}{35.0 \text{ cm}}\right) = +0.222$$

Because M is positive the image is virtual and erect.

c. The size of the image, found from equation 27.24, is

$$h_i = Mh_o = (0.222)(5.00 \text{ cm}) = 1.11 \text{ cm}$$

27.10 Some Special Cases for the Convex Lens

Case 1: *The Object Is Located at Infinity*

When the object is very far away from the lens, we can assume that it is at infinity. Setting $p = \infty$ in the lens equation, gives, for the location of the image,

$$\frac{1}{q} = \frac{1}{f} - \frac{1}{p}$$

$$= \frac{1}{f} - \frac{1}{\infty}$$

But $1/\infty = 0$, and the equation reduces to

$$q = f$$

That is, when the object is located at infinity, the image is located at the principal focus. The magnification becomes

$$M = -\frac{q}{p} = -\frac{q}{\infty} = 0$$

which shows that the image is reduced to a point. This case, which was shown in figure 27.13, is used as a simple technique to determine the focal length of a converging lens. A converging lens is held in front of a card and the distance between the card and the lens is varied until a sharp image is obtained on the card of a very distant object (p equal to infinity). The distance between the lens and the card is the image distance q. But as just seen, if the object is at infinity, then $q = f$. Hence, the distance from the lens to the card is the focal length of the converging lens.

Case 2: *The Object Lies between Infinity and 2f, That Is, $2f < p < \infty$*

This is a rather general case and the ray diagram is shown as the continuous lines in figure 27.20. The actual location of the image is found from the lens equation when p and f are specified.

A very interesting result occurs when the object is moved in toward the lens. The ray diagram for this case is shown in dashed lines in figure 27.20. *Notice that as the object moves toward the lens the image moves away from the lens and gets bigger.* However, the height of the image h_i never gets greater than the height of the object h_o, and thus the magnification is always less than 1. (This will be proved in case 3.)

Figure 27.20

The object moves closer to the lens.

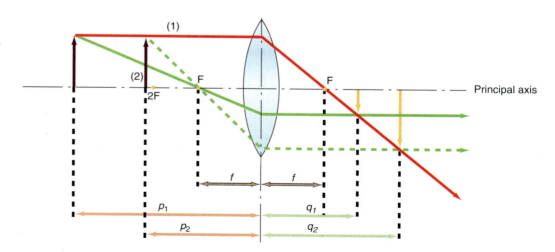

Case 3: The Object Is Located at a Distance of Twice the Focal Length from the Lens, That Is, p = 2f

For this case the image, found from the lens equation, is

$$\frac{1}{f} = \frac{1}{p} + \frac{1}{q} = \frac{1}{2f} + \frac{1}{q}$$

or

$$\frac{1}{q} = \frac{1}{f} - \frac{1}{2f} = \frac{1}{2f}$$

and hence,

$$q = 2f$$

That is, *when the object is placed a distance of 2f to the left of the lens, the image is located at the same distance of 2f to the right side of the lens.* The magnification for this case becomes

$$M = \frac{h_i}{h_o} = -\frac{q}{p} = -\frac{2f}{2f} = -1$$

and

$$h_i = -h_o$$

Hence, *when p = q = 2f, the height of the image is the same size as the height of the object. For any value of p greater than 2f the magnification is always less than 1, and hence the height of the image is less than the height of the object. For any value of p less than 2f, the magnification is greater than 1, and hence the height of the image is always greater than the height of the object.*

Case 4: The Object Is Located between the Focal Length and Twice the Focal Length, That Is, f < p < 2f

It is within this region that the lens acts as a magnifier, because if p is less than 2f, the magnification is greater than 1. The ray diagram is shown in figure 27.21. It is obvious from the diagram that q is greater than p, and hence, the magnification, $|M| = q/p$, is greater than 1, and the image is enlarged. This is also seen in the diagram, that is, the height of the image is greater than the height of the object.

Figure 27.21

Ray diagram for object located between the focal length and twice the focal length.

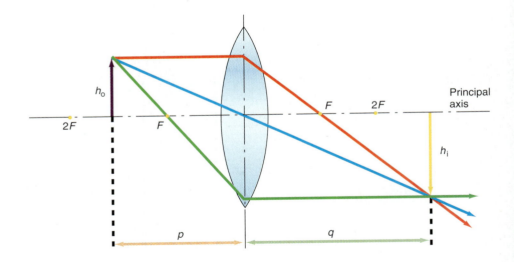

Light and Optics

Figure 27.22

Ray diagram for an object at the principal focus.

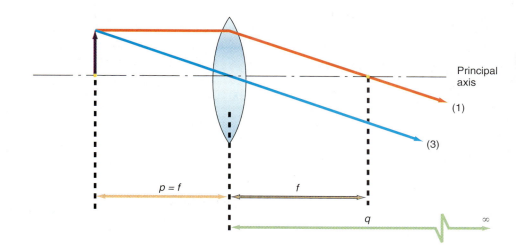

Case 5: *The Object Is Located at the Principal Focus of the Lens, That Is, p = f*

When the object is placed at the principal focus, $p = f$, and the lens equation gives for the location of the image,

$$\frac{1}{q} = \frac{1}{f} - \frac{1}{p} = \frac{1}{f} - \frac{1}{f} = 0$$

The only way for $1/q$ to be equal to zero, is for q to equal infinity, that is,

$$q = \infty$$

The ray diagram for such a case is shown in figure 27.22. The standard ray (2) cannot be drawn, but rays (1) and (3) can, and as we see from the diagram, they become parallel on the right-hand side of the lens and hence never intersect to form a real image there at any finite distance from the lens. The parallel rays only intersect at infinity and it is there where the image may be found. Note that if rays (1) and (3) were dashed backward, they would still be parallel, and hence, could not form a virtual image either.

Case 6: *The Object Is Placed within the Principal Focus, That Is, p < f*

A ray diagram for this case is shown in figure 27.23. Standard ray (1) is drawn parallel to the principal axis and is refracted in the lens such that it passes through the principal focus *F*, as shown. Of all the infinite rays that emanate from the tip of the arrow, standard ray (2) is a ray that is aligned with the dashed straight line from the principal focus to the tip of the arrow of the object. Therefore, ray (2) appears to come from the first principal focus, and hence, is refracted in the lens such that it comes out parallel to the principal axis. The third standard ray (3) is drawn from the tip of the object and goes through the center of the lens undeviated, as shown. The three standard rays do not converge anywhere on the right-hand side of the lens to form a real image there. However, if these standard rays are produced backward they intersect on the left-hand side of the lens and form a virtual image there. Thus a person looking through the lens from the right would see the enlarged, erect, virtual image of the object on the left-hand side of the lens. *When a convex lens is used with the object located within the principal focus, it is called a simple magnifying glass.* In this case, the image distance *q* is always negative.

Figure 27.23

Ray diagram for an object placed within the principal focus.

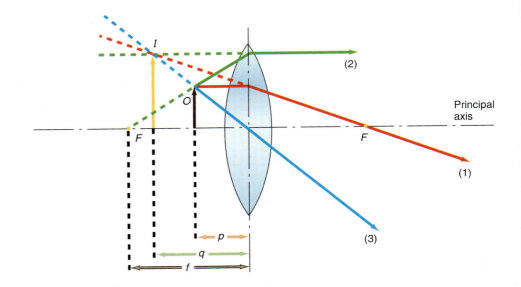

Principal axis

Example 27.11

A simple magnifying glass. An object 5.00 cm high is placed 3.00 cm in front of a convex lens of 5.00-cm focal length. (a) Where is the image located? (b) What is the magnification? (c) How high is the image?

Solution

a. The image distance, found from the lens equation, is

$$\frac{1}{q} = \frac{1}{f} - \frac{1}{p} = \frac{1}{5.00 \text{ cm}} - \frac{1}{3.00 \text{ cm}} = -0.1333 \text{ cm}^{-1}$$

and

$$q = -7.50 \text{ cm}$$

Note that q is a negative quantity.

b. The magnification, found from equation 27.23, is

$$M = -\frac{q}{p} = -\left(\frac{-7.50 \text{ cm}}{3.00 \text{ cm}}\right) = +2.50$$

The magnification is thus positive and indicates that the image is virtual and erect.

c. The image height, found from equation 27.24, is

$$h_i = Mh_o$$
$$= (2.50)(5.00 \text{ cm})$$
$$= 12.5 \text{ cm}$$

Thus the lens used in this way gives an enlarged (magnified 2.5 times), erect, virtual image of the object. The fact that q is negative is an indication that the image is virtual.

In summary, for a converging lens, with the object distance p positive, if the image distance q is positive the image is real and located on the other side of the lens. If the image distance q is negative, the image is on the same side of the lens as the object and is a virtual image.

Figure 27.24

Combination of two convex lenses.

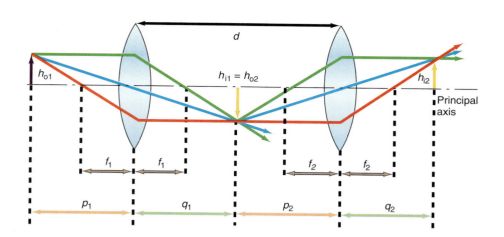

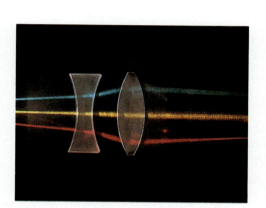

Combination of lenses. Parallel light rays incident on a diverging lens, followed by a converging lens.

27.11 Combinations of Lenses

Most optical systems consist of a combination of two or more lenses, prisms, or mirrors. Figure 27.24 shows a combination of two convex lenses separated by a distance d. The first lens has a focal length f_1 and the second lens has the focal length f_2, as illustrated. An object of height h_{o1} is placed a distance p_1 in front of the first lens. Let us find the location and size of the resulting image of the lens combination. We begin the analysis by considering the first lens only, that is, we temporarily ignore the existence of the second lens. We find the image distance q_1 of the first lens by drawing the standard rays of a ray diagram, as in figure 27.24. This image of the first lens now acts as the object of the second lens. We now completely ignore the existence of the first lens and consider the second lens as though an object has been placed a distance p_2 in front of it. We again find the image from this second lens by drawing the standard rays of a ray diagram, as shown in figure 27.24. Thus, the image distance q_2 of the second lens and the image height h_{i2} are easily found.

In addition to using the ray diagram, we can also find the final image from the lens equation. The image distance of the first lens is found from

$$\frac{1}{q_1} = \frac{1}{f_1} - \frac{1}{p_1} \qquad (27.25)$$

whereas the image distance of the second lens is found from

$$\frac{1}{q_2} = \frac{1}{f_2} - \frac{1}{p_2} \qquad (27.26)$$

However, as we can see from the diagram, the object distance p_2 is

$$p_2 = d - q_1 \qquad (27.27)$$

In most of the cases that we consider in this book, the image distance q_1 turns out to be less than the separation distance d, and the term $(d - q_1)$ is positive, giving a positive object distance p_2. There are some cases where the image distance q_1 is greater than the separation distance d. In these cases, the term $(d - q_1)$ is a negative quantity and hence, the object distance p_2 is also a negative quantity. An object having a negative object distance is called a virtual object. For a virtual object, the rays strike the lens before the object is formed. We will see an example of this later.

The magnification of the lens combination is defined as the ratio of the height of the final image to the height of the initial object. That is,

$$M = \frac{\text{height of image 2}}{\text{height of object 1}} = \frac{h_{i2}}{h_{o1}} \qquad (27.28)$$

But the magnification of the first lens is

$$M_1 = \frac{\text{height of image 1}}{\text{height of object 1}} = \frac{h_{i1}}{h_{o1}} \tag{27.29}$$

whereas the magnification of the second lens is

$$M_2 = \frac{\text{height of image 2}}{\text{height of object 2}} = \frac{h_{i2}}{h_{o2}} \tag{27.30}$$

Thus, from equation 27.30,

$$h_{i2} = h_{o2}M_2 \tag{27.31}$$

and from equation 27.29,

$$h_{o1} = \frac{h_{i1}}{M_1} \tag{27.32}$$

Substituting equations 27.31 and 27.32 back into equation 27.28, and noting that the height of image 1 h_{i1} is equal to the height of object 2 h_{o2}, we get

$$M = \frac{h_{i2}}{h_{o1}} = \frac{h_{o2}M_2}{h_{i1}/M_1} = \frac{h_{o2}M_2}{h_{o2}/M_1}$$

or

$$M = M_1 M_2 \tag{27.33}$$

Equation 27.33 says that the magnification of a combined optical system is equal to the product of the magnification of each individual lens. The height of the final image, obtained from equation 27.28, is

$$h_{i2} = M h_{o1} \tag{27.34}$$

The location and nature of the image of a particular optical system depends on the focal lengths f_1 and f_2, the location of the first object p_1, and the separation d between the two lenses. We will consider different optical systems in section 27.13.

Example 27.12

A combination of lenses. An object 5.00 cm high is placed 15.0 cm in front of a convex lens of 10.0-cm focal length. A second convex lens, also of 10.0-cm focal length, is placed 48.0 cm behind the first lens. Find (a) the image of the lens combination, (b) the magnification of the system, and (c) the height of the final image.

Solution

a. The image of the first lens, found from equation 27.25, is

$$\frac{1}{q_1} = \frac{1}{f_1} - \frac{1}{p_1} = \frac{1}{10.0 \text{ cm}} - \frac{1}{15.0 \text{ cm}} = 0.0333 \text{ cm}^{-1}$$

$$q_1 = 30.0 \text{ cm}$$

The object distance for the second lens, found from equation 27.27, is

$$p_2 = d - q_1 = 48.0 \text{ cm} - 30.0 \text{ cm} = 18.0 \text{ cm}$$

Hence, the final image, found from equation 27.26, is

$$\frac{1}{q_2} = \frac{1}{f_2} - \frac{1}{p_2} = \frac{1}{10.0 \text{ cm}} - \frac{1}{18.0 \text{ cm}} = 0.0444 \text{ cm}^{-1}$$

$$q_2 = 22.5 \text{ cm}$$

Thus, the final image of the lens combination is found 22.5 cm behind the second lens, or 70.5 cm behind the first lens.

b.   The magnification of each lens is found as

$$M_1 = -\frac{q_1}{p_1} = -\frac{30.0 \text{ cm}}{15.0 \text{ cm}} = -2.00$$

$$M_2 = -\frac{q_2}{p_2} = -\frac{22.5 \text{ cm}}{18 \text{ cm}} = -1.25$$

The final magnification of the system is

$$M = M_1 M_2 = (-2.00)(-1.25) = 2.50$$

Note that M_1 is negative indicating that the first lens inverts the first image. The fact that M_2 is also negative means that lens 2 inverted the second image from upside down to erect. The total magnification M is positive indicating that the final image is right side up. In dealing with combinations of lenses we must be careful of the convention adopted for the magnification of a single lens. The convention adopted is that if M is negative, the image is real and inverted, whereas if M is positive, the image is erect and virtual. In this case of the combined lens, M_1 is negative and image 1 is real and inverted, as expected. Also M_2 is negative and image 2 is real and inverted from its original orientation when in front of lens 2, also as expected. The inversion of the inverted image, gives a final image that is erect. Notice however, that the final magnification is positive because it is the product of two negative numbers. But the final image is real not virtual. Hence, the convention that a positive magnification indicates that the image must be virtual cannot be used for a system of multiple lenses.

c.   The final height of the image of the system, found from equation 27.34, is

$$h_{i2} = M h_{o1} = (2.50)(5.00 \text{ cm})$$

$$= 12.5 \text{ cm}$$

27.12 Thin Lenses in Contact

An interesting case results from the lens combination of section 27.11 when the distance d between the two lenses is made equal to zero, thereby placing the two lenses in contact with each other. The lens equation applied to the second lens is

$$\frac{1}{f_2} = \frac{1}{p_2} + \frac{1}{q_2} \tag{27.35}$$

Setting $p_2 = d - q_1$ from equation 27.27 into equation 27.35, yields

$$\frac{1}{f_2} = \frac{1}{d - q_1} + \frac{1}{q_2} \tag{27.36}$$

Setting the separation distance d equal to zero in equation 27.36 yields

$$\frac{1}{f_2} = \frac{1}{-q_1} + \frac{1}{q_2} \tag{27.37}$$

But the lens equation applied to the first lens is given by

$$\frac{1}{q_1} = \frac{1}{f_1} - \frac{1}{p_1}$$

and

$$\frac{1}{-q_1} = -\frac{1}{f_1} + \frac{1}{p_1} \tag{27.38}$$

Substituting equation 27.38 into equation 27.37, yields

$$\frac{1}{f_2} = -\frac{1}{f_1} + \frac{1}{p_1} + \frac{1}{q_2}$$

and

$$\frac{1}{f_1} + \frac{1}{f_2} = \frac{1}{p_1} + \frac{1}{q_2} \qquad (27.39)$$

Let us examine equation 27.39 and see what it is telling us. Here p_1 is the object distance of the first lens, but it is also the object distance of the lens combination, and we will designate it as p_c. Although q_2 is the image distance of the second lens, it is also the image distance of the lens combination, and we can call it q_c. Hence, we can rewrite equation 27.39 as

$$\frac{1}{f_1} + \frac{1}{f_2} = \frac{1}{p_c} + \frac{1}{q_c} \qquad (27.40)$$

The right-hand side of equation 27.40 looks like part of the lens equation, so it is only natural to define the focal length for the two lenses in contact as f_c, and the lens equation for the combination should then be

$$\frac{1}{f_c} = \frac{1}{p_c} + \frac{1}{q_c} \qquad (27.41)$$

Combining equations 27.40 and 27.41, gives

$$\frac{1}{f_c} = \frac{1}{f_1} + \frac{1}{f_2} \qquad (27.42)$$

The focal length of the combination is thus obtained from equation 27.42. Hence, *whenever two lenses are in contact, the reciprocal of the focal length of the combination is equal to the sum of the reciprocals of the focal lengths of the individual lenses.*

Example 27.13

The focal length of two convex lenses in contact. A convex lens of 10.0-cm focal length is placed in contact with a convex lens of 15.0-cm focal length. What is the focal length of the combination?

Solution

The focal length of the combination, found from equation 27.42, is

$$\frac{1}{f_c} = \frac{1}{f_1} + \frac{1}{f_2} = \frac{1}{10.0 \text{ cm}} + \frac{1}{15.0 \text{ cm}} = 0.1666 \text{ cm}^{-1}$$

and

$$f_c = 6.00 \text{ cm}$$

Example 27.14

The focal length of a convex and a concave lens in contact. A 10.0-cm convex lens is placed in contact with a 15.0-cm concave lens. What is the focal length of the combination?

The focal length of the combination is found from equation 27.42, but since the second lens is concave its focal length is negative, that is, $f_2 = -15.0$ cm. Therefore

$$\frac{1}{f_c} = \frac{1}{f_1} + \frac{1}{f_2} = \frac{1}{10.0 \text{ cm}} - \frac{1}{15.0 \text{ cm}} = 0.0333 \text{ cm}^{-1}$$

and

$$f_c = 30.0 \text{ cm}$$

Equation 27.42, although derived for only two lenses, is completely general and *thus, for any number of lens placed in contact the focal length of the combination is*

$$\frac{1}{f_c} = \frac{1}{f_1} + \frac{1}{f_2} + \frac{1}{f_3} + \frac{1}{f_4} + \cdots \qquad (27.43)$$

When dealing with many lenses in contact it is convenient to *define the dioptric power of a lens as the reciprocal of the focal length in meters.* Thus

$$P = \frac{1}{f} \qquad (27.44)$$

The reciprocal of the focal length in meters is called a *diopter.*

Example 27.15

The power of a lens. What is the power of a 15.0-cm focal length lens?

The focal length of the lens is $f = 15.0$ cm $= 0.150$ m. The power of the lens, found from equation 27.44, is

$$P = \frac{1}{f} = \frac{1}{0.150 \text{ m}} = 6.67 \text{ diopters}$$

The advantage of expressing a lens in terms of its dioptric power is evident when the lenses are in contact. For example, if there are four lenses of focal lengths $f_1, f_2, f_3,$ and f_4, then the power of each lens is

$$P_1 = \frac{1}{f_1}$$

$$P_2 = \frac{1}{f_2}$$

$$P_3 = \frac{1}{f_3}$$

$$P_4 = \frac{1}{f_4}$$

The power of these lenses in combination, determined from equation 27.43, is

$$\frac{1}{f_c} = \frac{1}{f_1} + \frac{1}{f_2} + \frac{1}{f_3} + \frac{1}{f_4}$$

and substituting the values of $1/f$, we get

$$\frac{1}{f_c} = P_1 + P_2 + P_3 + P_4$$

The power of the combination is thus

$$P_c = P_1 + P_2 + P_3 + P_4 \qquad (27.45)$$

Expressing a lens in terms of dioptric power is convenient for an optician, who places many different lenses in contact. When a patient is given an eye exam, the optician places different lenses in front of the patient's eyes and asks him to read a chart. When the patient finds the sharpest image, the optician immediately gets the power of the combination of lenses used, P_c, from equation 27.45. The reciprocal of P_c is the focal length of the eyeglasses needed for the patient and a lens of this focal length can be ground using the lensmaker's formula.

27.13 Optical Instruments

Most optical devices consist of an arrangement or combination of lenses. The focal lengths of the lenses and their relative positions determines the function of the optical system. A very brief description of these optical instruments is given here as an example of lens applications. Most optical instruments consist of many lens combinations to eliminate various lens defects. We will look at only combinations of single lenses in order to demonstrate the principle of the device. More detailed discussions can be found in books on optics. The first optical instrument considered is the camera.

The Camera

The basic elements of a camera, figure 27.25(a), are a converging lens, contained in a bellows or tube type mechanism; a roll of photographic film that can be rolled across a focal plane; an aperture that determines the amount of light that enters the camera; and a shutter mechanism that determines the time that the lens is open to light. A schematic diagram of a camera is shown in figure 27.25(b). A ray diagram shows the location of the image. Because the focal length of the lens is a constant, increasing or decreasing the object distance p changes the image distance q, as can easily be determined by the lens equation. Therefore, in order to get a sharp image of any object, we need to be able to change the distance from the lens to the film plane. On studio type cameras, this is accomplished by moving the bellows inward or outward. On small hand cameras, such as a 35 mm camera, there are two con-

(a)

Figure 27.25

A simple camera.

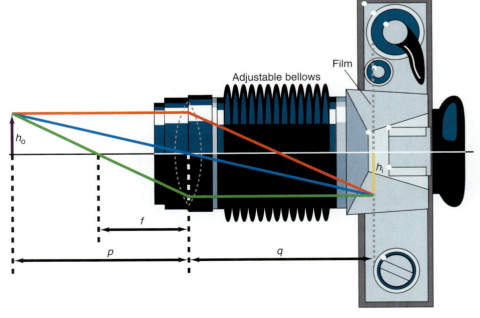

(b)

Light and Optics

centric tubes containing the lenses. Rotating the tubes clockwise or counterclockwise causes the lens to move in or out. The total variation of q is relatively small for a 35 mm camera.

Example 27.16

The change in the image distance for a 35 mm camera. A 35 mm camera has a focal length of 50.0 mm. By how much will the image distance change when photographing one object at infinity and another 1.00 m away?

Solution

When the object is at infinity, the image distance q is equal to the focal length f, as seen in section 27.10, case 1. Therefore,

$$q_1 = f = 50.0 \text{ mm} = 5.00 \text{ cm}$$

When $p_2 = 1.00 \text{ m} = 100 \text{ cm}$, the image distance q_2 found from the lens equation, is

$$\frac{1}{q_2} = \frac{1}{f} - \frac{1}{p_2} = \frac{1}{5.00 \text{ cm}} - \frac{1}{100 \text{ cm}}$$

$$q_2 = 5.26 \text{ cm}$$

The total variation in the image distance q becomes

$$\Delta q = q_2 - q_1 = 5.26 \text{ cm} - 5.00 \text{ cm} = 0.26 \text{ cm}$$

Thus, the lens does not have to move very far at all to get the two different pictures. One of the reasons for this is, of course, the relatively small focal length of the lens. If the focal length of the camera had been 20.0 cm, then the same calculation as above would have given a variation of the image distance Δq of 5.00 cm. Therefore, to obtain a small variation of q a small focal length is used.

Example 27.17

Getting the entire picture into the camera. At what distance from the 35 mm camera of example 27.16 should a man, 2.00 m tall, stand in order for his entire length to be in the picture?

Solution

For 35 mm film, the maximum height that the image h_i can be is 35.0 mm. Therefore, the magnification of the camera is

$$M = \frac{h_i}{h_o} = \frac{3.50 \text{ cm}}{200 \text{ cm}} = 0.0175$$

Since $M = q/p$, we can solve for the object distance, p, as

$$p = \frac{q}{M}$$

Now, as seen above, the image distance q is quite close to the focal length f, so we will assume that they are equal, that is,

$$q = f$$

Hence,

$$p = \frac{q}{M} = \frac{f}{M} = \frac{5.00 \text{ cm}}{0.0175} = 286 \text{ cm} = 2.86 \text{ m}$$

That is, for a 2.00-m tall man to have his entire body in the picture he must stand 2.86 m from the camera. Note that the magnification of a camera is quite small; this is necessary because we usually want to put a large scene onto a small piece of film. The images must be greatly reduced if they are to fit on the film.

The amount of light impinging on the film depends on the speed of the shutter, that is, 1/50, 1/100, 1/200, 1/500 of a second and so forth, and the size of the opening at the lens. The size of the opening is determined by an *iris diaphragm* and is calibrated in terms of the *f*-number. The *f*-number is defined as

$$f\# = \frac{\text{focal length}}{\text{diameter of aperture}} = \frac{f}{d} \qquad (27.46)$$

The standard *f*-numbers of a camera are: $f/2.8$, $f/4$, $f/5.6$, $f/8$, $f/11$, and $f/16$. Thus, an *f*-number of $f/4$ means that the diameter d of the aperture is 1/4 of the value of the focal length. If a 50.0-mm focal length is being used, then an *f*-number of $f/4$ implies that the lens opening is

$$d = \frac{f}{f\#} = \frac{f}{4} = \frac{50.0 \text{ mm}}{4} = 12.5 \text{ mm}$$

The opening for $f/16$ is

$$d = \frac{f}{f\#} = \frac{f}{16} = \frac{50.0 \text{ mm}}{16} = 3.13 \text{ mm}$$

Hence, the larger the *f*-number of a camera, the smaller the lens opening. Thus, a picture taken in bright sunshine would use a larger *f*-number than one taken in subdued lighting, for the same shutter speed.

The Simple Microscope

The simple microscope is the same as the magnifying glass that was treated in section 27.10, case 6. The object is placed within the principal focus of a converging lens and an enlarged, erect, and virtual image is formed.

The Compound Microscope

The compound microscope was invented by Galileo in 1610. It consists of two converging lens. The first lens is called the *objective lens* and is of short focal length. The object to be magnified is placed just outside the focal length f_1 of the objective, so that a real, enlarged image is formed at q_1, figure 27.26(a). The second lens is called the *eyepiece,* and the image of the objective lens becomes the object for the eyepiece lens. The eyepiece is situated such that the object p_2 always lies just within the principal focus of the eyepiece f_2. Thus the eyepiece acts as a simple magnifying glass, giving an enlarged virtual image at q_2. The lens equation is used to find the location of the final image. Thus, q_1 is found from

$$\frac{1}{q_1} = \frac{1}{f_1} - \frac{1}{p_1}$$

Once q_1 is determined, the object distance p_2 is found from

$$p_2 = d - q_1$$

The final image is found from

$$\frac{1}{q_2} = \frac{1}{f_2} - \frac{1}{p_2}$$

The magnification of the entire system is

$$M = M_1 M_2$$

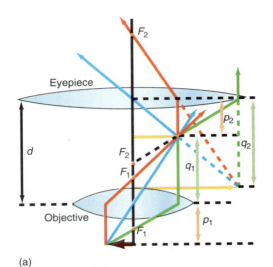

(a)

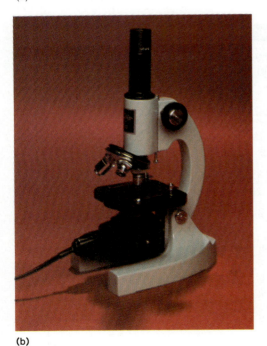

(b)

Figure 27.26

The compound microscope.

Light and Optics

where

$$M_1 = -\frac{q_1}{p_1} \quad \text{and} \quad M_2 = -\frac{q_2}{p_2}$$

A typical student microscope is shown in figure 27.26(b).

Example 27.18

The microscope. A bug 2.00 mm in diameter is placed 1.50 cm in front of a 1.00-cm focal length objective lens of a compound microscope. The eyepiece lens has a focal length of 23.0 cm and the lenses are separated by a distance of 25.0 cm. Find the size of the resulting image.

Solution

We treat the microscope as a system of two lenses in combination. The image of the first lens is the object for the second lens. The image of the first lens, found from the lens equation, equation 27.25, is

$$\frac{1}{q_1} = \frac{1}{f_1} - \frac{1}{p_1}$$

$$= \frac{1}{1.00 \text{ cm}} - \frac{1}{1.50 \text{ cm}}$$

$$q_1 = 3.00 \text{ cm}$$

This image serves as the object of the second lens. The object distance for the second lens, found from equation 27.27, is

$$p_2 = d - q_1$$

$$= 25.0 \text{ cm} - 3.00 \text{ cm}$$

$$= 22.0 \text{ cm}$$

The image of the second lens, found from the lens equation, equation 27.26, is

$$\frac{1}{q_2} = \frac{1}{f_2} - \frac{1}{p_2}$$

$$= \frac{1}{23.0 \text{ cm}} - \frac{1}{22.0 \text{ cm}}$$

$$q_2 = -506 \text{ cm}$$

The magnification of the first lens is found as

$$M_1 = -\frac{q_1}{p_1} = -\frac{3.00 \text{ cm}}{1.50 \text{ cm}} = -2.00$$

whereas the magnification of the second lens is found as

$$M_2 = -\frac{q_2}{p_2} = -\left(\frac{-506 \text{ cm}}{22.0 \text{ cm}}\right) = +23.0$$

The total magnification of the system, found from equation 27.33, is

$$M = M_1 M_2 = (2.00)(23.0) = 46.0$$

The size of the final image, found from equation 27.34, is

$$h_{i2} = M h_{o1} = (46.0)(2.00 \text{ mm})$$

$$= 92.0 \text{ mm} = 9.20 \text{ cm}$$

Notice that because M_1 is negative, the first image is real, whereas M_2 is positive, indicating that the second image is virtual. This is as expected, since the object p_2 lies within the principal focus f_2 of the second lens. The final result of the optical system is to take the initial object, 2.00 mm in size, and make it observable as an image 92.0 mm in size.

Figure 27.27

The astronomical telescope.

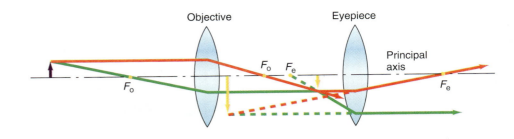

The Astronomical Telescope

The astronomical telescope is another example of a combination of two converging lenses. However, this time the objective is a large converging lens of a long focal length. The astronomical telescope is shown in figure 27.27. Because the object is very far away, the image of the objective is, for all intents and purposes, located at the principal focus of the objective lens f_o. (Recall that when $p = \infty$, $q = f$.) It is shown slightly away from there in figure 27.27 so that the ray diagram is easier to see. The eyepiece is a converging lens of small focal length, f_e, and it is placed such that the image of the first lens falls just within the principal focus of the second lens, thereby forming an enlarged, virtual image. That is, the eyepiece acts as a simple magnifying glass. Since the first image falls approximately at the focal distance of the first lens, and the second lens is placed so that the image of the first lens falls just within the focal length of the second lens, the distance separating the lenses is approximately the sum of the two focal lengths. Hence, the overall length of the telescope is

$$L = f_o + f_e \qquad (27.47)$$

The final image of the astronomical telescope is, of course, inverted but this is not serious for analyzing the heavenly bodies. However, the astronomical telescope is not very good for viewing distant objects on the surface of the earth, because they are all inverted. (The astronomical telescope can be converted to a terrestrial telescope by placing a third converging lens between the objective and the eyepiece.) The magnification of a telescope is usually expressed in terms of the *angular magnification*. The angular magnification M_A of the lens is defined as the ratio of the angle subtended at the eye by the object when the lens is used, to the angle subtended by the unaided eye. We state without proof that the *total angular magnification* of an astronomical telescope when the object is at infinity is given by

$$M_A = \frac{\text{focal length of objective}}{\text{focal length of eyepiece}} = \frac{f_o}{f_e} \qquad (27.48)$$

Example 27.19

An astronomical telescope. Find the angular magnification of an astronomical telescope whose objective has a focal length of 100 cm, and its eyepiece has a focal length of 5.00 cm.

Solution

The magnification, found from equation 27.48 is,

$$M = \frac{f_o}{f_e} = \frac{100 \text{ cm}}{5.00 \text{ cm}} = 20.0$$

Figure 27.28

Ray diagram for a Galilean telescope.

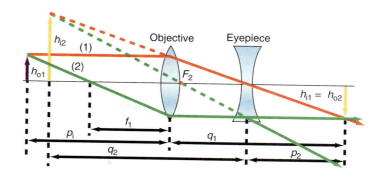

The Galilean Telescope

The Galilean telescope consists of a converging lens of large focal length for the objective and a diverging lens of short focal length for the eyepiece. The advantage of the Galilean telescope is that it gives an erect image. A ray diagram for a Galilean telescope is shown in figure 27.28. Only two of the standard rays are shown to simplify the diagram. The image of the first lens is located at q_1. But note that the second lens is placed before the location of the first image. This is an example of what is called a virtual object, and the object distance p must be written as a negative quantity in the lens equation. The effect of the diverging lens for the eyepiece is to diverge the rays away from the position where they should focus by the objective lens. For example, when ray (2) hits the diverging lens it is parallel to the principal axis. However, a ray parallel to the principal axis is diverged away from the principal axis as though it came from the principal focus of the diverging lens. This ray is now shown as diverging away from the principal axis. In this particular example, the second lens is located at the principal focus of the first lens, hence ray (1) passes through the geometrical center of lens two and is not deviated. We can see from the figure that these two rays diverge on the right-hand side of the second lens and therefore can never form a real image there. However if these rays are produced backward they intersect on the left-hand side of the lens to form an enlarged, erect, virtual image. Thus the Galilean telescope, also called a terrestrial telescope, produces enlarged images that are right side up.

Example 27.20

A Galilean telescope. A Galilean telescope is made of a + 20.0-cm objective lens and a −4.50-cm eyepiece lens. The two lenses are separated by a distance d of 15.0 cm. For an object situated at 1000 cm, find (a) the image distance of the first lens, (b) the object distance for the second lens, (c) the image distance for the second lens.

Solution

a. The image distance of the first lens, found from the lens equation, equation 27.25, is

$$\frac{1}{q_1} = \frac{1}{f_1} - \frac{1}{p_1} = \frac{1}{20.0 \text{ cm}} - \frac{1}{1000 \text{ cm}}$$

$$q_1 = 20.4 \text{ cm}$$

b. The object distance for the second lens, found from equation 27.27, is

$$p_2 = d - q_1 = 15.0 \text{ cm} - 20.4 \text{ cm} = -5.4 \text{ cm}$$

Note that this is a virtual object since p_2 is negative.

c. The image distance for the second lens, found from equation 27.26, is

$$\frac{1}{q_2} = \frac{1}{f_2} - \frac{1}{p_2} = \frac{1}{(-4.50 \text{ cm})} - \frac{1}{(-5.4 \text{ cm})}$$

$$q_2 = -27 \text{ cm}$$

Because q_2 is negative the final image is virtual.

"Have you ever wondered . . . ?"

An Essay on the Application of Physics

Nature's Camera—the Human Eye

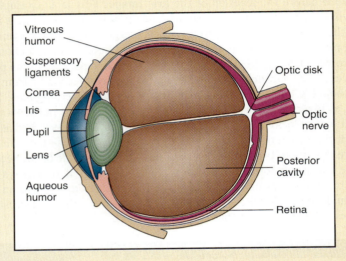

Figure 1

The human eye.

Have you ever wondered how the human eye works? It is one of the greatest marvels of nature, but one that is usually taken for granted. The human eye is very much like a camera. It contains a lens, an iris diaphragm to let the light in, and its film is the retina at the back of the eye, figure 1. The lens focuses incident light onto the retina so a sharp image is observed. The iris diaphragm opens wide in subdued lighting and closes down in very bright light.

In the eye, the distance from the lens to the retina q, does not change as in a camera, instead the shape of the lens is changed by the muscles and ligaments of the eye. *Hence, the eye changes its focal length in order to bring objects at different object distances to a focus at the same image distance q, the distance from the lens to the retina. This changing of the focal length of the eye is called* **accommodation.** A schematic of the normal eye is shown in figure 2(a). The greatest amount of bending of light in the eye occurs at the cornea, because it is here where the greatest difference in the index of refraction occurs. The index of refraction of air is of course equal to one, whereas the index of refraction of the cornea is about 1.35. The index of refraction of the lens is about 1.44; that of the aqueous and vitreous humor is about 1.34.

Some eyes do not have the ability to change the shape of the lens as necessary and this is referred to as a defect in vision. One such defect is called **hyperopia, or farsightedness,** and is shown in figure 2(b). *A farsighted person can see distant objects*

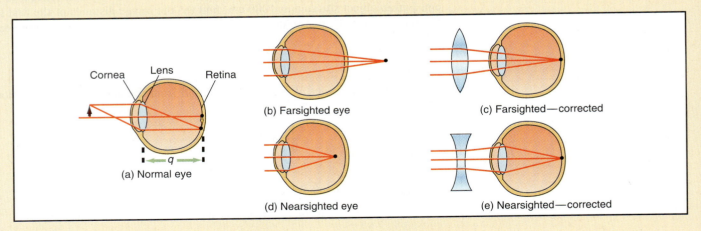

Figure 2

Defects of the human eye.

Light and Optics

clearly, but objects a short distance away are blurred. The lens is not capable of bending the incoming light of near objects to a focus on the retina. Instead the light is focused behind the retina, leaving blurred vision. The defect is easily remedied by placing a converging lens in front of the eye, as shown in figure 2(c). The converging lens brings these incoming rays to a focus at the retina. Farsightedness is a defect that occurs to most people as they age. The lens hardens with age and the muscles weaken and are no longer capable of shaping the lens as needed.

Another defect is **myopia, or nearsightedness,** as illustrated in figure 2(d). *A nearsighted person can see nearby objects clearly, but distant objects are blurred.* The distant objects come to a focus in front of the retina leaving a blurred image on the retina. This defect can be eliminated by placing a diverging lens in front of the eye. The diverging lens causes the rays to converge more slowly and hence the image is formed farther back at the retina, as shown in figure 2(e).

*The minimum distance from the eye at which an object can be seen distinctly is called **the near point of the eye.** For the average person, the near point is about 25.0 cm.* That is, an object at a distance p of 25.0 cm is the minimum distance that an object can be placed to give a distinct image on the retina. If the distance is decreased below this value, the image will be blurred for most people. Remember, this is only an average value, as we get older, the minimum distance of distinct vision increases. For example, the near point can be as close as 6.00 cm for a child and can increase to over a 100 cm for an elderly person.

Another common defect of the eye is called *astigmatism.* Astigmatism occurs in some people, because their eye is not completely spherical. That is, the radius of curvature of the eye in the vertical direction is not the same as the radius of curvature in the horizontal direction. Hence, vertical rays do not converge to the same position as horizontal rays. This defect is usually corrected with a lens of cylindrical curvature.

Although the magnification of a lens was defined in equation 27.22 as $M = h_i/h_o$, when viewing an object with the eye it is sometimes more desirable to talk about the angular magnification of the lens. Figure 3(a) shows an object at three distances from the unaided eye. The size of the object, as determined by the unaided eye, depends on the angle subtended by the object at the eye. Thus, when the object is at position 1, it subtends an angle θ_1; when it is at position 2, it subtends the angle θ_2; and when it is at position 3, it subtends the angle θ_3. As seen from the diagram, $\theta_1 < \theta_2 < \theta_3$. Thus, the closer the object is to the eye, the greater the angle subtended and the easier it is for the eye to see the object. However, the object cannot be moved closer to the eye than the near point, because that is the closest point that the eye can still see the object distinctly. In figure 3(b), the object is placed at the object distance $p_1 = 25.0$ cm in front of a normal eye. The height of the object h_o subtends an angle α at the lens of the eye. The image of the object is focused on the retina of the eye. This is the largest image that the unaided eye can make of the object. If the object is placed within the focal point of a converging lens, the object can be brought even closer

Example 27H.1

Range of focal length for the eye. The distance from the cornea to the retina in the human eye is about 20.0 mm. What should the focal length of the human eye be in order to see clearly an object (a) at infinity and (b) at the near point of most distinct vision, 25 cm?

Solution

a. For the object at infinity, the image is located at the principal focus. Hence

$$f = q = 20.0 \text{ mm} = 2.00 \text{ cm}$$

is the focal length of the eye when it views an object at infinity.

b. For the object at $p = 25.0$ cm, the focal length would have to be

$$\frac{1}{f} = \frac{1}{p} + \frac{1}{q} = \frac{1}{25.0 \text{ cm}} + \frac{1}{2.00 \text{ cm}} = 0.540 \text{ cm}^{-1}$$

$$f = 1.85 \text{ cm}$$

Thus, the eye only has to change its focal length by (2.00 cm − 1.85 cm = 0.15 cm) 15 mm to see its entire range of vision.

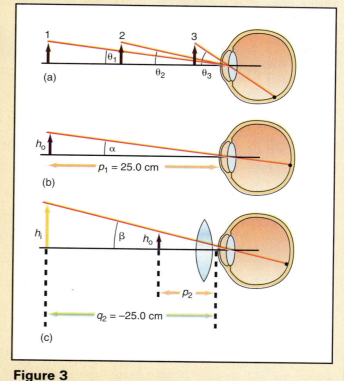

(a)

(b)

(c)

Figure 3

Angular magnification and the simple microscope. A doctor examines a patient's eye.

to the eye and hence will subtend a larger angle β, figure 3(c), than with the unaided eye. Therefore, the object is placed within the focal point, at the position p_2, such that the image distance q_2 is equal to the near point of the eye, the closest position of distinct vision. We assume that the value of the near point $q_2 = 25.0$ cm. The image is virtual, erect, and magnified as seen earlier. *The angular magnification M_A of the lens is defined as the ratio of the angle subtended at the eye by the object when the lens is used, to the angle subtended by the unaided eye.* That is,

$$M_A = \frac{\beta}{\alpha} \qquad \text{(27H.1)}$$

From figure 3(c), we see that

$$\tan \beta = \frac{h_o}{p_2}$$

whereas from figure 3(b), we see that

$$\tan \alpha = \frac{h_o}{p_1} = \frac{h_o}{25.0 \text{ cm}}$$

Let us take the ratio

$$\frac{\tan \beta}{\tan \alpha} = \frac{h_o/p_2}{h_o/25.0 \text{ cm}}$$

$$= \frac{25.0 \text{ cm}}{p_2}$$

Because the angles considered are small, we can use the small-angle approximation, setting the tangent of an angle equal to the angle itself. Hence,

$$\frac{\beta}{\alpha} = \frac{25.0 \text{ cm}}{p_2}$$

But this ratio is the definition of the angular magnification, equation 27H.1, hence

$$M_A = \frac{25.0 \text{ cm}}{p_2} \qquad \text{(27H.2)}$$

The value of p_2, found from the lens equation, is

$$\frac{1}{p_2} = \frac{1}{f} - \frac{1}{q_2}$$

$$\frac{1}{p_2} = \frac{1}{f} - \frac{1}{-25.0 \text{ cm}} = \frac{25.0 \text{ cm} + f}{(25.0 \text{ cm})f} \qquad \text{(27H.3)}$$

And the object distance for the image to be at the near point is

$$p_2 = \frac{(25.0 \text{ cm})f}{25.0 \text{ cm} + f} \qquad \text{(27H.4)}$$

Substituting equation 27H.4 into equation 27H.2, gives

$$M_A = \frac{\beta}{\alpha} = \frac{25.0 \text{ cm}}{p_2} = \frac{25.0 \text{ cm}}{(25.0 \text{ cm})f/(25.0 \text{ cm} + f)}$$

$$= \frac{25.0 \text{ cm}}{f} + 1 \qquad \text{(27H.5)}$$

Equation 27H.5 gives the angular magnification of a converging lens when the image is located at the near point of the eye, which is assumed to be 25.0 cm for the normal eye.

Example 27H.2

The angular magnification of a lens when the image is at the near point of the eye. A converging lens of 10.0-cm focal length is used as a magnifying glass. If the image is to be at the near point of the eye, find the angular magnification of the lens.

Solution

The angular magnification, found from equation 27H.5, is

$$M_A = \frac{25.0 \text{ cm}}{f} + 1$$

$$= \frac{25.0 \text{ cm}}{10.0 \text{ cm}} + 1$$

$$= 3.5$$

The eye is more relaxed when viewing an object at infinity than at the near point. It is, therefore, sometimes more desirable to view an image at infinity than at the near point. For the image to be at infinity, the object must be placed at the principal focus. The lens equation, equation 27H.3, shows that if p is at infinity, then the object distance p_2 is equal to the focal length f. Hence substituting f for p_2 in equation 27H.2, gives

$$M_A = \frac{25.0 \text{ cm}}{f} \qquad \text{(27H.6)}$$

Equation 27H.6 gives the angular magnification of a simple microscope when the image is set for viewing at infinity.

Example 27H.3

The angular magnification of a lens when the image is at infinity. A magnifying glass of 10.0-cm focal length is used to view an object, such that the image is at infinity. Find the angular magnification of the lens.

Solution

The angular magnification, found from equation 27H.6, is

$$M_A = \frac{25.0 \text{ cm}}{f}$$

$$= \frac{25.0 \text{ cm}}{10.0 \text{ cm}} = 2.50$$

Notice that when the image is viewed at the near point the angular magnification is 3.5, whereas when viewed at infinity, it is 2.5. If viewing is for a long time, it might be desirable to view the image at infinity because it is easier on the eyes. If the magnification is the most important consideration, then the image should be viewed at the near point of the eye.

The Language of Physics

Refraction
The bending of light as it travels from one medium into another. It occurs because of the difference in the speed of light in the different mediums. Whenever a ray of light goes from a rarer medium to a denser medium the refracted ray is always bent toward the normal. Whenever a ray of light goes from a denser medium to a rarer medium, the refracted ray is bent away from the normal (p. 765).

Law of refraction
The ratio of the sine of the angle of incidence to the sine of the angle of refraction is a constant. The constant is called the relative index of refraction and it is equal to the ratio of the speed of light in the first medium to the speed of light in the second medium. Because of the changing speed of light, the wavelength of light changes as the light passes into the second medium (p. 767).

The critical angle of incidence
The angle of incidence that causes the refracted ray to bend through 90°. When the incident angle exceeds the critical angle no refraction occurs. In that case, it is called total internal reflection because all the light that strikes the interface is reflected (p. 773).

Prism
A triangular piece of transparent material whose angle exceeds the critical angle. A ray of light falling on one of the smaller sides of the prism enters the prism and is totally reflected from the longer side of the prism. Prisms are also used for analyzing the dispersion of white light into its component colors (p. 773).

Fiber optics
A flexible glass rod of high refractive index. Light entering the glass undergoes total internal reflection from the walls of the glass fiber and the light travels down the length of the fiber with little or no absorption of light (p. 774).

Dispersion
The separation of white light into its component colors. It occurs because the index of refraction of a medium varies slightly with the wavelength of light (p. 774).

Lens
A piece of transparent material, such as glass or plastic, that causes light passing through it to either converge or diverge depending on the shape of the material (p. 775).

Lensmaker's formula
An equation that relates the focal length, index of refraction, and the radii of curvature of the lens (p. 776).

Thin lens
A lens whose thickness is negligible compared to the distance to the principal focus and to any object or image distance (p. 776).

Converging lens
A lens that causes light, parallel to its principal axis, to converge to the principal axis. Converging lenses have positive focal lengths (p. 777).

Diverging lens
A lens that causes light, parallel to its principal axis to diverge away from the principal axis. Diverging lenses have negative focal lengths (p. 777).

The lens equation
An equation that relates the image distance, the object distance, and the focal length of a lens. It has the same form as the mirror equation (p. 781).

Dioptric power of a lens
The reciprocal of the focal length of a lens. The focal length must be expressed in meters. For a combination of any number of lenses in contact, the power of the combination is equal to the sum of the powers of each individual lens (p. 791).

Accommodation
The changing of the focal length of the eye in order to focus an image on the retina of the eye (p. 798).

Hyperopia, or farsightedness
A defect of the eye that causes objects far away to be seen clearly while close objects are blurred. The condition is remedied by placing a converging lens in front of the eye (p. 798).

Myopia, or nearsightedness
A defect of the eye that causes objects close to the eye to be seen clearly, while objects far away are blurred. The condition is remedied by placing a diverging lens in front of the eye (p. 799).

Near point of the eye
The minimum distance from the eye at which an object can be seen distinctly. For the average person the near point is about 25 cm (p. 799).

Angular magnification
The ratio of the angle subtended at the eye by an object, when a lens is used, to the angle subtended by the unaided eye (p. 800).

Summary of Important Equations

The law of refraction
$$\frac{\sin i}{\sin r} = \frac{v_1}{v_2}$$
$$= \text{constant} = n_{21} \quad (27.3)$$
$$n_1 \sin i = n_2 \sin r \quad (27.11)$$

The index of refraction
$$n = \frac{c}{v} \quad (27.4)$$

Speed of a wave in terms of wavelength and frequency
$$v = \lambda \nu \quad (27.5)$$

Index of refraction in terms of wavelengths in two mediums
$$n_{21} = \frac{\lambda_1}{\lambda_2} \quad (27.6)$$

Apparent depth for small angles
$$q = \frac{n_2}{n_1} p \quad (27.14)$$

Critical angle
$$\sin i_c = \frac{n_2}{n_1} \quad (27.17)$$

Lensmaker's formula
$$\frac{1}{f} = (n - 1)$$
$$\left(\frac{1}{R_1} - \frac{1}{R_2} \right) \quad (27.18)$$

Lens equation
$$\frac{1}{f} = \frac{1}{p} + \frac{1}{q} \quad (27.21)$$

Magnification
$$M = \frac{h_i}{h_o} = -\frac{q}{p} \quad (27.23)$$

Height of image
$$h_i = M h_o \quad (27.24)$$

Combination of lenses
Final image
$$\frac{1}{q_2} = \frac{1}{f_2} - \frac{1}{p_2} \quad (27.26)$$

where
$$p_2 = d - q_1 \quad (27.27)$$

and
$$\frac{1}{q_1} = \frac{1}{f_1} - \frac{1}{p_1} \quad (27.25)$$

Magnification
$$M = M_1 M_2 \qquad (27.33)$$

Thin lenses in contact
$$\frac{1}{f_c} = \frac{1}{f_1} + \frac{1}{f_2}$$
$$+ \frac{1}{f_3} + \cdots \qquad (27.43)$$

Power of a lens
$$P = \frac{1}{f \text{ (in meters)}} \qquad (27.44)$$

Power of combinations of lenses
$$P_c = P_1 + P_2 + P_3$$
$$+ P_4 + \cdots \qquad (27.45)$$

f-number of a lens
$$f\# = \frac{f}{d} \qquad (27.46)$$

Angular magnification, astronomical telescope
$$M_A = \frac{f_o}{f_e} \qquad (27.48)$$

Angular magnification, viewing at near point
$$M_A = \frac{25.0 \text{ cm}}{f} + 1 \qquad (27H.5)$$

Angular magnification, viewing at infinity
$$M_A = \frac{25.0 \text{ cm}}{f} \qquad (27H.6)$$

Questions for Chapter 27

†1. Why does a diamond sparkle?
2. When the angle of incidence is equal to zero, the angle of refraction is also zero. Does the wavelength of the light change when going from one medium to another under these circumstances?
3. If you are at the bottom of a pool of water and you look upward into the air at an angle greater than 50° can you see anything in the air?
4. Is it possible to have a ray of light refracted into a medium such that the new wavelength decreases to the point where it is no longer in the visible spectrum? What would you see? Does the eye interpret a wavelength or a frequency?

†5. What does a wide-angle lens and a telephoto lens do when each is attached to a camera?
†6. Describe the optical system used for (a) a slide projector, (b) a movie camera, and (c) an overhead projector.
7. A swimmer forgets to take off her glasses as she enters the pool. When she is under water, wearing the glasses, does the water have any effect on what she can see with the glasses?
8. When you see yourself in a mirror, your right and left hand are interchanged. Yet, if you see a picture of yourself made with a camera, containing lenses, your right and left are not changed. Why?

†9. What are bifocals and why are they used?
†10. What is a mirage and how is it explained by the index of refraction?
†11. How is a rainbow formed?
12. How is a convex lens used, with the help of the sun, to start a fire?

Problems for Chapter 27

27.2 The Law of Refraction

1. A ray of light impinges on a piece of glass ($n_g = 1.52$) at an angle of incidence of 50.0°. Find the angle of refraction.
2. A ray of light passes from water to glass at an angle of incidence of 50.0°. Find the angle of refraction.
3. A ray of light is refracted by an angle of 34.5° as it enters water from glass. Find the angle of incidence.

4. A ray of light in air makes an angle of incidence of 35.0° as it enters an unknown liquid. The refracted ray in the fluid is measured to be 22.5°. Find the index of refraction of the unknown liquid. What substance might it be?

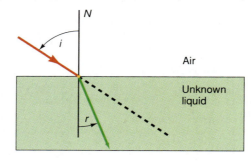

5. Determine the speed of light in (a) water, (b) glycerine, and (c) diamond.
6. If a ray of light of 480.0-nm wavelength in air enters into water, what is the wavelength of the light in the water? Is this light visible?
7. A ray of light of 580.0-nm wavelength in water, enters into the air. What is the wavelength of the light in the air?
8. A ray of light of 700.0-nm wavelength in water enters into the air. What is the wavelength of the light in the air? Will the light be seen in the air?

9. A ray of light of 590.0-nm wavelength in air impinges on a piece of flint glass at an angle of 30.0° with the vertical. The index of refraction of this flint glass is 1.57. Find (a) the angle of refraction, (b) the speed of light in the glass, and (c) the wavelength of light in the glass.

27.3 Apparent Depth of an Object Immersed in Water

10. A rock sits at the bottom of a 3.50-m-deep pool. What is its apparent depth?

11. A fish appears to be at a depth of 120 cm in water. What is its actual depth?

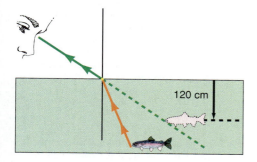

27.4 Refraction through Parallel Faces

12. A ray of light in air makes an angle of incidence of 30.0° with a sheet of glass 0.500 cm thick. Find the distance d that the final ray is displaced from its original direction.

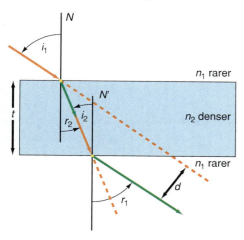

27.5 Total Internal Reflection

13. What is the critical angle of refraction for a light ray going from water to glass?

14. Find the critical angle of refraction for a ray of light passing from glycerine ($n = 1.47$) into air.

15. The critical angle of a ray of light is measured as 41.8° as it goes from an unknown liquid into air. Find the index of refraction of the liquid.

27.7 Thin Lenses

16. A double convex glass lens of index of refraction 1.52, has radii of curvature $R_1 = +50.0$ cm and $R_2 = -25.0$ cm. Find its focal length.

17. A glass lens of index of refraction 1.52, has radii of curvature of $R_1 = +40.0$ cm and $R_2 = -15.0$ cm. What is its focal length and is it a converging or a diverging lens?

18. A glass lens of index of refraction 1.52, has radii of curvature of $R_1 = -40.0$ cm and $R_2 = +15.0$ cm. What is its focal length and is it a converging or a diverging lens?

19. A glass lens of index of refraction 1.52, has radii of curvature of $R_1 = -40.0$ cm and $R_2 = -15.0$ cm. What is its focal length and is it a converging or a diverging lens?

20. A lens is made of transparent material, with radii of curvature $R_1 = 15.0$ cm and $R_2 = 90.0$ cm. If the focal length of the lens is 5.00 cm, find the index of refraction of the material.

27.9 The Lens Equation, and Section 27.10

Some Special Cases for the Convex Lens

†21. An object 3.00 cm high is placed 20.0 cm in front of a converging lens of 15.0 cm focal length. Draw a ray diagram. Find (a) the image distance, (b) the magnification, and (c) the final size of the image. (d) Is the image real or virtual? (e) Is the image erect or inverted?

†22. An object 3.00 cm high is placed 20.0 cm in front of a diverging lens of −15.0 cm focal length. Draw a ray diagram. Find (a) the image distance q, (b) the magnification M, and (c) the height of the image h_i. (d) Is the image real or virtual? (e) Is the image erect or inverted?

23. Where should an object be placed in front of a 20.0-cm lens in order for the image to be the same size as the object?

24. An object 7.00 cm high is placed 5.00 cm in front of a convex lens of 10.0 cm focal length. (a) Draw a ray diagram. Find (b) the image distance, (c) the magnification, and (d) the height of the image.

25. How far in front of a 20.0-cm converging lens should an object be placed in order to produce an image 25.0 cm from the lens on the same side of the lens as the object?

26. An object is placed 15.0 cm in front of a diverging lens of 5.00 cm focal length. Where is the image located and what is its magnification?

27. Where should an object be placed in front of a diverging lens of 10.0 cm focal length in order to give an image a magnification of 0.500?

28. An object 5.00 cm high is placed 20.0 cm in front of a converging lens. The image is measured to be 7.00 cm high. Where is the image located?

29. An object 5.00 cm high is placed 30.0 cm in front of a converging lens of 7.50 cm focal length. What is the size of the image?

27.11 Combination of Lenses

†30. An object 5.00 cm high is placed 20.0 cm in front of a converging lens of 10.0 cm focal length. A second converging lens of 20.0 cm focal length is placed 30.0 cm behind the first lens. Find (a) the image distance for the first lens, (b) the object distance for the second lens, (c) the image distance for the second lens, and (d) the total magnification of the combination. Draw a ray diagram.

†31. An object 5.00 cm high is placed 10.0 cm in front of a 20.0-cm converging lens. A second converging lens, also of 20.0 cm focal length, is placed 20.0 cm behind the first lens. Find (a) the location of the final image and (b) its size. Draw a ray diagram.

†32. A converging lens of +20.0 cm is separated by a distance of 20.0 cm from a diverging lens of −5.00 cm. An object is located 30.0 cm in front of the first lens. Find (a) the image distance of the first lens, (b) the object distance for the second lens, (c) the image distance for the second lens, and (d) the magnification of the system.

27.12 Thin Lenses in Contact

33. A 20.0-cm convex lens is placed in contact with a diverging lens of unknown focal length. The lens combination has a focal length of 30.0 cm. Find the focal length of the diverging lens.

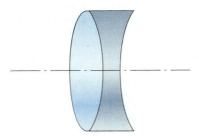

34. What is the power of (a) a 50.0-cm converging lens, (b) a 20.0-cm diverging lens, (c) the converging and diverging lenses in contact, and (d) the focal length of the combination when they are in contact?

35. Ten identical converging thin lenses, each of focal length 6.50 cm, are in contact. Find the focal length of the composite lens.

Additional Problems

36. A ray of light of 590.0-nm wavelength in glycerine impinges on a piece of flint glass at an angle of 30.0° with the vertical. The index of refraction of glycerine is 1.47 and for the flint glass used it is 1.70. Find (a) the angle of refraction, (b) the speed of light in the glycerine, (c) the speed of light in the glass, and (d) the wavelength of light in the glass.

37. Light travels in medium 1 at 1.50×10^8 m/s. The light is incident at an angle of incidence of 40.0° when entering medium 2, where the angle of refraction is 30°. Find the speed of light in medium 2.

38. A ray of light of 590.0-nm wavelength makes an angle of incidence of 40.0° on to the cornea of the eye. The index of refraction of the cornea is 1.35. Find (a) the angle of refraction, (b) the speed of light in the cornea, and (c) the wavelength of light in the cornea.

39. Light travels from glass of index of refraction 1.50 into water of index 1.33. If there exists a refracted ray, find the angle of refraction when (a) the angle of incidence is 55.0° and (b) the angle of incidence is 70.0°.

†40. If an object is placed a distance d_1 in front of the principal focus of a lens and the image is located a distance d_2 beyond the other principal focus, show that the focal length of the lens is given by

$$f = \sqrt{d_1 d_2}$$

41. An object is mounted 60.0 cm in front of a screen. At what two positions will a 12.0 cm focal length lens yield a distinct image on the screen?

42. An optical system creates a virtual object (an object *behind* the lens) for a diverging lens of focal length $f = -8.00$ cm. Find the position of the image and describe it for the following two object distances: (a) $p = -4.00$ cm and (b) $p = -10.0$ cm.

43. A farsighted person has a near point of 60.0 cm. What power lens should be used for eyeglasses such that the person can read this book at a distance of 25.0 cm?

44. A nearsighted person has a far point of 15 cm. (The far point of the eye is the farthest distance that an object can be seen clearly.) What power lens should be used to allow this person to view far distant objects?

45. What is the smallest wavelength in air of visible light that can still be seen on refraction into water?

†46. The lensmaker's formula for a lens of index of refraction n, immersed in a material with an index of refraction of n_m, is given by

$$\frac{1}{f_m} = \frac{n - n_m}{n_m}\left(\frac{1}{R_1} - \frac{1}{R_2}\right)$$

A glass lens, $n_g = 1.70$, has a focal length of 25.0 cm in air. What is its focal length when it is submerged in water?

Interactive Tutorials

47. The law of refraction. A ray of light of wavelength $\lambda_1 = 500.0$ nm in medium 1 (index of refraction of medium one is $n_1 = 1.00$) impinges on a second medium of index of refraction $n_2 = 1.52$, at an angle of incidence $i = 35.0°$. Find (a) the angle of refraction r, (b) the speed of light v_2 in the second medium, and (c) the wavelength λ_2 in the second medium.

48. The critical angle of incidence. A ray of light in the denser medium (index of refraction $n_1 = 1.50$) impinges on a less dense medium of index of refraction $n_2 = 1.00$. Find the critical angle of incidence i_c for the ray of light going from the more dense medium to the less dense medium.

49. Lensmaker's formula. An optical lens has an index of refraction $n = 1.52$ and radii of curvature of $R_1 = +35.0$ cm and $R_2 = -15.0$ cm. Find the focal length of this lens.

50. Thin lens. An object of height $h_o = 8.50$ cm is placed at the object distance $p = 35.0$ cm of a thin lens of focal length $f = 15.0$ cm. Find (a) the image distance q, (b) the magnification M, and (c) the height h_i of the resulting image.

51. Thin lens. An object of height $h_o = 0.500$ cm is placed at the object distance $p = 2.00$ cm of a thin lens of focal length $f = 0.700$ cm. Plot (a) the image distance q and (b) the height h_i of the resulting image as the object is moved toward the lens.

52. A combination of lenses. An object of height $h_{o1} = 5.00$ cm is placed at the object distance $p_1 = 15.0$ cm of a thin lens of focal length $f_1 = 10.0$ cm. A second lens of focal length $f_2 = 10.0$ cm is placed $d = 48.0$ cm behind the first lens. Find (a) the image distance q_1 of the first lens, (b) the magnification M_1 of the first lens, (c) the object distance p_2 of the second lens, (d) the image distance q_2 of the second lens, (e) the magnification M_2 of the second lens, (f) the magnification M of the system, and (g) the height of the final image h_{i2}.

28

Physical Optics

Interference pattern of a pinhole light source.

In making some experiments on the fringes of colors accompanying shadows, I have found so simple and so demonstrative a proof of the general law of the interference of two portions of light, which I have already endeavored to establish, that I think it right to lay before the Royal Society a short statement of the facts, which appear to me to be thus decisive.

Thomas Young

28.1 Introduction

In chapter 25, we saw that light is an electromagnetic wave, and the equation for the electric portion of the plane wave was given by equation 25.38 as

$$E = E_0 \sin(kx - \omega t) \qquad (25.38)$$

In chapter 26, using the fact that light is a wave, we used Huygens' principle to derive the law of reflection. In chapter 27, we again used the wave nature of light to derive the law of refraction by another application of Huygens' principle. However, once these two laws were developed, we treated light as a ray and explained a large number of optical phenomena by geometrical optics. However, we need to ask ourselves, is there any limit to the use of geometrical optics? Is it ever necessary to treat light strictly from its wave nature? The answer to each of these questions is yes. To discuss the experimentally observed phenomena of interference and diffraction, we need to treat light as a wave. The study of light from its wave nature is called *physical optics,* or sometimes *wave optics.*

To see at what point light must be treated as a wave, let us consider a series of plane waves and their associated light rays, as they impinge on the openings in figure 28.1. The light rays are shown perpendicular to the wave fronts, as discussed in chapter 26. *When the opening is very large compared to the wavelength of light, that is, when a $\gg \lambda$, figure 28.1(a), the light passes through the opening as a series of geometrical rays and a sharp image of the opening is found on the screen. This is the region in which light can be treated from the point of view of geometrical optics.* Because the wavelength of red light, the longest of the visible spectrum, is approximately 720.0 nm or 7.2×10^{-5} cm, the size of the opening a does not have to be very large, from a practical viewpoint, to be very much larger than the wavelength λ. This is why geometrical optics gives such a good description of so many optical problems.

When the size of the opening is made smaller and smaller, until the slit opening a becomes the same order of magnitude as the wavelength of light λ, a breakdown in geometrical optics occurs. Instead of the image on the screen becoming smaller and sharper as a approaches the size of the wavelength λ, the image on the screen starts to become larger, and the edges fuzzier, as shown in figure 28.1(b). Instead of light propagating in the straight line motion of geometrical optics, the rays of light bend into the region that would normally be considered as a region of shadow. *This bending of light around an obstacle, into the region that should be a shadow area, is called **diffraction.*** The obstacle in this case is the edge of the opening. We should note here that diffraction is very much different from refraction,

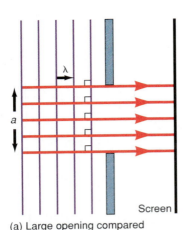

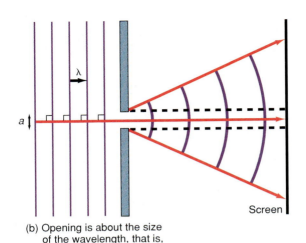

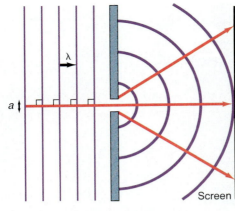

(a) Large opening compared to the wavelength of light $a \gg \lambda$

(b) Opening is about the size of the wavelength, that is, $a \sim \lambda$

(c) Very small opening compared to the wavelength of light $a < \lambda$

Figure 28.1

Transition from geometrical optics to physical optics.

Direction of propagation of wave

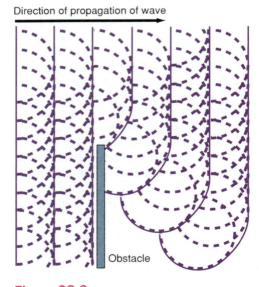

Figure 28.2

The bending of light around an obstacle.

which is the bending of light as it goes from one medium into another. In the case of diffraction there is no different medium. The bending occurs in the same medium.

If the slit opening is made smaller still, until the opening a is less than the wavelength λ, then the diffraction of the light wave becomes even greater, as seen in figure 28.1(c). In fact, in figure 28.1(c), the opening is so small that it now appears as a point source of light, with light emanating radially from the slit.

Diffraction is not limited to light waves but occurs in all wave phenomena such as water waves and sound waves. The greater the wavelength of the wave the greater the amount of diffraction. Since the wavelength of light is very small the bending of the light ray is very small. The wavelength for sound waves, varies from 16.5 m, for a sound of frequency 20.0 Hz, to 1.65 cm, for a frequency of 20,000 Hz. Hence, the wavelength of sound is very much greater than the wavelength of light, which is no larger than 7.2×10^{-5} cm. For large wavelengths such as sound, diffraction is very pronounced. Thus, when someone speaks in another room, the sound waves bend around the doorway so the person can be heard, even though the light from that person cannot bend around the doorway and the person cannot be seen. In fact, the diffraction of sound waves is such a common phenomenon, most people are unaware of it, unless someone points it out to them.

The wave theory of light was first proposed by Christian Huygens and diffraction was first observed by Francesco Maria Grimaldi (1618–1683), a Professor of Mathematics at the University of Bologna. In essence, Grimaldi measured the bright image, such as in figure 28.1(b), and found it to be greater than the size of the opening of the slit. He also placed an obstacle in the path of the beam of light and found that its projected image on a wall was smaller than the obstacle itself. Both of these examples of diffraction can be explained by the application of Huygens' principle to the wave fronts and is shown in figure 28.2. A plane wave front is seen approaching an obstacle. The new position of the wave front is found by drawing new secondary wavelets as described in chapter 26. When the wave fronts hit the obstacle, these waves cannot send out new wavelets in the forward direction. However, *the waves above the top of the obstacle continue to send out secondary wavelets and, as seen in the figure, these secondary wavelets now extend into the normally dark region. Because the new wave front is found by drawing a line tangent to all these wavelets, we see that the wave front bends around the obstacle, into the shadow region.* Thus, Grimaldi's observation that the measured bright image was greater than the size of the opening of the slit can be explained by the light bending at both sides of the slit into both shadow areas. The fact that an obstacle

Figure 28.3

Young's double-slit experiment.

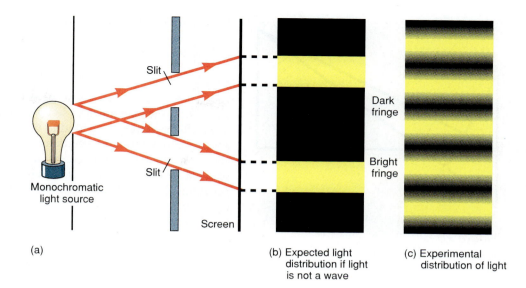

Monochromatic
light source

Slit

Slit

Screen

Dark
fringe

Bright
fringe

(a)

(b) Expected light
distribution if light
is not a wave

(c) Experimental
distribution of light

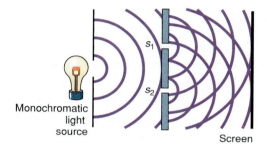

Monochromatic
light
source

Screen

s_1

s_2

Figure 28.4

The waves in Young's double-slit
experiment.

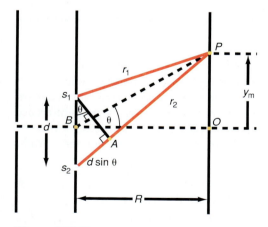

Figure 28.5

Geometry of Young's double-slit
experiment.

in the path of the beam of light formed an image on the wall that was smaller than the obstacle itself, can be explained by the light bending around both sides of the obstacle into the normal shadow area of the obstacle.

The interference of waves is caused by a superposition of waves and was described in chapter 12 on wave motion. The interference of light waves is a direct extension of the superposition of waves and we will treat it in detail in sections 28.2 and 28.3.

28.2 The Interference of Light— Young's Double-Slit Experiment

The first proof that light had wave characteristics was furnished in 1800 by the crucial experiment on the interference of light from a double slit by Thomas Young (1773–1829), an English physicist. The experiment is now referred to as **Young's double-slit experiment.** The experimental set-up is shown in figure 28.3(a). A **monochromatic source of light** (light of a single wavelength) is passed through a small hole in the first opaque screen and falls on the second opaque screen that contains two small slits. If light is not a wave then the light should follow the geometrical paths shown in figure 28.3(a), and thus give the intensity distribution shown in figure 28.3(b). That is, only two light images should appear on the third screen with a large dark shadow between the bright fringes. Instead, the experiment showed that there was a series of bright and dark fringes on the screen, as shown in figure 28.3(c). There is no way that this strange result can be explained by geometrical optics. However, the result can be easily explained if light is a wave.

Young's double-slit experiment is shown again in figure 28.4. Light from the source passes through the narrow slit in the first screen and is diffracted in the same way as shown in figure 28.1(c). When light from the first slit reaches the second set of slits, it is again diffracted as if there was a point source at each of the second slits. The waves from each slit now propagate toward the screen where they interfere, or superimpose, with each other.

The double-slit experiment is redrawn in figure 28.5. The double slits are labeled s_1 and s_2 and are located a distance d apart. If we consider the arbitrary point P, then the light intensity at that point is the result of a superposition of a wave of light from slit 1 and a wave from slit 2. Notice that the wave from slit 2 has to travel a greater distance r_2 than the wave from slit 1, the distance r_1. Hence,

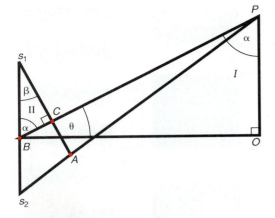

Figure 28.6

A closer look at the angles.

there is a difference in the optical path between path 1 and path 2. It is this path difference that is responsible for the bright and dark fringes. We can write the path difference as

$$\text{Path difference} = PD = r_2 - r_1 \qquad (28.1)$$

A perpendicular is dropped from the center of slit 1 to the path r_2, at the point A. In the experimental setup the distance R from the plane of the slits to the screen is very large, compared to the slit separation d. Therefore, the length AP is approximately the same length as r_1. This is equivalent to rotating the distance r_1 about the point P until r_1 comes into coincidence with the length AP. The arc associated with that rotation is approximately equal to the chord s_1A. Hence, s_1A is perpendicular to BP and s_2P. Two of the triangles of figure 28.5 are redrawn in figure 28.6 in order to show the relationship between the angles in these triangles more clearly. The angle θ is the angle that defines the location of the fringe at the point P and is angle PBO of triangle I. We will call angle BPO angle α, whereas angle POB is a right angle. Thus, in triangle I,

$$\theta + \alpha + 90° = 180°$$

In triangle II, angle s_1BC is equal to the same angle, α, of triangle I because alternate, interior angles of two parallel lines are equal. Angle s_1CB is a right angle. Let us determine the angle β. In triangle II we have

$$\beta + \alpha + 90° = 180°$$

Comparing these two equations, we see that

$$\beta = \theta$$

Thus, angle s_2s_1A is equal to the angle θ, and hence, the side opposite angle θ s_2A is equal to $d \sin \theta$, as shown in figure 28.5. Therefore, we can write the path length r_2 as

$$r_2 = r_1 + d \sin \theta \qquad (28.2)$$

The path difference between waves 1 and 2, equation 28.1, becomes

$$PD = r_2 - r_1 = r_1 + d \sin \theta - r_1$$

$$PD = d \sin \theta \qquad (28.3)$$

We will now analyze the experiment in two ways. First, we will examine a simple intuitive geometrical solution showing where the bright and dark fringes are located, and then we will explore a more detailed analytical solution showing the actual intensity distribution of the light.

Geometrical Solution

As seen in equation 28.3, there is a path difference between wave 1 and wave 2. *Wherever the waves are in phase when they superimpose, there is constructive interference and a bright image or bright fringe on the screen.* If the path difference between the two waves is a whole number of wavelengths, then the two waves are in phase at the screen, and a bright fringe appears. Thus the condition for a bright fringe to appear on the screen is

$$\text{Bright Fringe} \qquad PD = d \sin \theta = m\lambda \qquad (28.4)$$

where m is an integer.

Wherever the waves are out of phase by 180°, or a half of a wavelength ($\lambda/2$), there is destructive interference and a dark region or dark fringe appears on the screen. If the path difference between the two waves is an odd multiple of half

Light and Optics

wavelengths, then the two waves are out of phase at the screen, and a dark fringe appears. Thus the condition for a dark fringe to appear on the screen is

$$\text{Dark Fringe} \qquad \text{PD} = d \sin \theta = (2m - 1)\frac{\lambda}{2} \qquad \textbf{(28.5)}$$

Notice that the term $(2m - 1)$ always gives an odd number for any integer value of m.

Hence, the observed distribution of light, that is, bright and dark fringes, can be explained by the processes of diffraction and interference, both wave manifestations of light.

Analytical Solution

Let us now determine the light intensity distribution analytically. We can represent the wave from slit 1 as the plane wave

$$E_1 = E_0 \sin(kr_1 - \omega t) \qquad \textbf{(28.6)}$$

where E_0 is the amplitude of the wave, k is the wavenumber, and ω is the angular frequency of the wave. These quantities have the same relations as previously defined in chapter 12 on wave motion. We can represent the wave from slit 2 as

$$E_2 = E_0 \sin(kr_2 - \omega t) \qquad \textbf{(28.7)}$$

Since the light comes from the same source, E_0, k, and ω are the same for each wave. If the value of r_2 from equation 28.2 is placed into equation 28.7, we get

$$E_2 = E_0 \sin[k(r_1 + d \sin \theta) - \omega t]$$

or

$$E_2 = E_0 \sin(kr_1 - \omega t + kd \sin \theta) \qquad \textbf{(28.8)}$$

If we compare E_2, equation 28.8, with E_1, equation 28.6, we see that they look alike except for the term, $kd \sin \theta$, in E_2. Recall, from equation 12.38 of chapter 12, that a wave

$$y_2 = A \sin(kx - \omega t + \phi) \qquad \textbf{(12.38)}$$

represents a wave, just like

$$y_1 = A \sin(kx - \omega t) \qquad \textbf{(12.37)}$$

except for the term ϕ. Recall that ϕ is a phase angle and measures how far wave 2 is displaced from wave 1, or how much wave 2 is out of phase with wave 1. Therefore, let the term

$$kd \sin \theta = \phi$$

and since the wavenumber $k = 2\pi/\lambda$, ϕ becomes

$$\phi = \frac{2\pi d}{\lambda} \sin \theta \qquad \textbf{(28.9)}$$

Hence, ϕ is the phase difference between wave 1 and wave 2. We can now write wave 2 as

$$E_2 = E_0 \sin(kr_1 - \omega t + \phi) \qquad \textbf{(28.10)}$$

The resultant wave at the point P can now be determined by the interference of the wave from slit 1 with the wave from slit 2. Thus, the resultant wave at P is given by

$$E = E_1 + E_2 \qquad \textbf{(28.11)}$$

Substituting the waves from equations 28.6 and 28.10 into equation 28.11, gives

$$E = E_0 \sin(kr_1 - \omega t) + E_0 \sin(kr_1 - \omega t + \phi)$$

Factoring out the amplitude E_0 from each term, gives

$$E = E_0 [\sin(kr_1 - \omega t) + \sin(kr_1 - \omega t + \phi)]$$

But this sum looks like the trigonometric identity for the sum of two sine waves given in appendix B and used in chapter 12,

$$\sin B + \sin C = 2 \sin\left(\frac{B + C}{2}\right)\cos\left(\frac{B - C}{2}\right)$$

with

$$B = kr_1 - \omega t$$

and

$$C = kr_1 - \omega t + \phi$$

Hence, the resultant wave at P becomes

$$E = E_0 \left[2 \sin\left(\frac{kr_1 - \omega t + kr_1 - \omega t + \phi}{2}\right)\cos\left(\frac{kr_1 - \omega t - kr_1 + \omega t - \phi}{2}\right)\right]$$

Simplifying,

$$E = 2E_0 \sin\left(kr_1 - \omega t + \frac{\phi}{2}\right)\cos\left(\frac{-\phi}{2}\right)$$

Because the cosine is an even function, we have

$$\cos\left(\frac{-\phi}{2}\right) = \cos\left(\frac{\phi}{2}\right)$$

Thus,

$$E = 2E_0 \cos\left(\frac{\phi}{2}\right)\sin\left(kr_1 - \omega t + \frac{\phi}{2}\right) \qquad (28.12)$$

Equation 28.12 is the resultant wave at the arbitrary point P.

Since bright and dark fringes are observed on the screen, the intensity distribution of this light as it hits the screen must be determined. Recall from chapter 25 on electromagnetic waves that the intensity I is given by

$$I = \epsilon_0 c E^2 \qquad (25.63)$$

where ϵ_0 is the permittivity of free space, c is the speed of light, and E is the value of the electric field. We can now obtain the intensity distribution by substituting E from equation 28.12 into equation 25.63. Thus,

$$I = \epsilon_0 c 4 E_0^2 \cos^2\left(\frac{\phi}{2}\right)\sin^2\left(kr_1 - \omega t + \frac{\phi}{2}\right) \qquad (28.13)$$

Because the frequency of visible light is so high (about 5×10^{14} cycles/s), the human eye cannot see the effect of each wave as it hits the screen but instead can see only their average value. We saw in chapter 24 that the average value of the $\sin^2\theta$ is

$$(\sin^2\theta)_{\text{avg}} = \tfrac{1}{2}$$

Hence, the average intensity on the screen becomes

$$I_{\text{avg}} = 4\epsilon_0 c E_0^2 \cos^2\left(\frac{\phi}{2}\right)\left(\frac{1}{2}\right)$$

or

$$I_{avg} = 2\epsilon_0 c E_0^2 \cos^2\left(\frac{\phi}{2}\right) \qquad (28.14)$$

This equation can be further simplified by letting

$$I_0 = 2\epsilon_0 c E_0^2 \qquad (28.15)$$

Thus, the intensity distribution of the light on the screen is

$$I_{avg} = I_0 \cos^2\left(\frac{\phi}{2}\right) \qquad (28.16)$$

Equation 28.16 says that the intensity of the light on the screen varies with the phase angle ϕ. But since ϕ is given by equation 28.9 as

$$\phi = \frac{2\pi d}{\lambda} \sin \theta$$

we see that *the intensity varies with the angle θ in figure 28.5.*

The location of a bright fringe on the screen can be determined by realizing that a bright fringe corresponds to maximum light intensity. The intensity I in equation 28.16 is a maximum whenever the cosine term is a maximum. This occurs whenever the angle $\phi/2$ is a multiple of π. That is, a bright fringe is obtained when

$$\frac{\phi}{2} = m\pi \qquad \text{Bright Fringe} \qquad (28.17)$$

where m is a whole number such as

$$m = 0, \pm 1, \pm 2, \pm 3, \ldots$$

For example, if we substitute equation 28.17 into equation 28.16, we get

$$I_{avg} = I_0 \cos^2(m\pi)$$

But $\cos(m\pi) = \pm 1$, and $\cos^2(m\pi) = 1$. Therefore,

$$I_{avg} = I_0$$

That is, the maximum intensity of light I_0 occurs for $\phi/2 = m\pi$. Substituting equation 28.9 into equation 28.17, gives

$$\frac{\phi}{2} = \frac{2\pi d}{2\lambda} \sin \theta = m\pi$$

Therefore,

$$\text{Bright Fringe} \qquad d \sin \theta = m\lambda \qquad \text{for } m = 0, \pm 1, \pm 2, \pm 3, \ldots \qquad (28.18)$$

Equation 28.18 is the condition for a bright fringe and is the same as equation 28.4, which was derived geometrically. Notice from figure 28.5 that $d \sin \theta$ is the difference in the path length between the wave from slit 1 and the wave from slit 2. *When this path difference is equal to a whole multiple of the wavelength ($m\lambda$), then the two waves are in phase at the point P, and there is constructive interference or a bright fringe.* The value $m = 0$ corresponds to the central maximum located at O in figure 28.5. The first bright fringe after the central maximum occurs for $m = 1$, the second fringe for $m = 2$, and so on. The location of the mth fringe on the screen is found from the geometry of figure 28.5 as

$$y_m = R \tan \theta$$

However, since R is so much larger than d in the experimental set-up, figure 28.5, the angle θ is a very small angle. When the angle θ is small, we can use the small angle approximation, and

$$\tan \theta \approx \sin \theta$$

Hence, the mth fringe is located at

$$y_m = R \sin \theta \qquad (28.19)$$

But from equation 28.18,

$$\sin \theta = \frac{m\lambda}{d} \qquad (28.20)$$

Substituting equation 28.20 into equation 28.19 *yields the location of the mth bright fringe on the screen at the position given by*

$$\text{Bright Fringe} \qquad y_m = \frac{Rm\lambda}{d} \qquad (28.21)$$

The dark fringes observed on the screen are at the positions of minimum light intensity. The intensity, equation 28.16, is a minimum whenever $\phi/2$ is equal to an odd multiple of $\pi/2$, because then the cosine of a multiple of $\pi/2$ is zero. Hence, the dark fringes occur when

$$\text{Dark Fringe} \qquad \frac{\phi}{2} = (2m - 1)\frac{\pi}{2} \qquad \text{for } m = 1, 2, 3, \ldots \qquad (28.22)$$

Note that the term $(2m - 1)$ is an odd number for $m = 1, 2, 3, \ldots$. Substituting equation 28.22 into the intensity equation 28.16 gives

$$I = I_0 \cos^2\left[(2m - 1)\frac{\pi}{2}\right]$$

But, $\cos(\pi/2) = 0$ and $\cos[(2m - 1)\pi/2] = 0$. *Therefore, the intensity I is equal to zero whenever the phase difference over two, $\phi/2$, is an odd multiple of $\pi/2$, and this gives a dark fringe.* Substituting the phase angle ϕ from equation 28.9 into equation 28.22, gives

$$\frac{\phi}{2} = \frac{2\pi d}{2\lambda} \sin \theta = (2m - 1)\frac{\pi}{2}$$

and the condition for a dark fringe becomes

$$\text{Dark Fringe} \qquad d \sin \theta = (2m - 1)\frac{\lambda}{2} \qquad \text{for } m = 1, 2, 3, \ldots \qquad (28.23)$$

As we can see in figure 28.5, $d \sin \theta$ corresponds to the path difference between the wave from slit 1 and the wave from slit 2. Equation 28.23 says that whenever this path difference is an odd multiple of half-wavelengths, the wave from slit 1 is 180° out of phase with the wave from slit 2, and destructive interference occurs at point P, thus producing zero light intensity and a dark fringe there. Notice that equation 28.23 is the same as equation 28.5 derived earlier in the geometrical solution.

We can also obtain the location of the dark fringe on the screen from the geometry of figure 28.5 as

$$y_m = R \tan \theta = R \sin \theta$$

$$\text{Location of Dark Fringe} \qquad y_m = \frac{R(2m - 1)}{d} \frac{\lambda}{2} \qquad (28.24)$$

The overall intensity distribution for the double-slit interference can be found by plotting the intensity I versus the angle $\phi/2$ for equation 28.16, as is shown

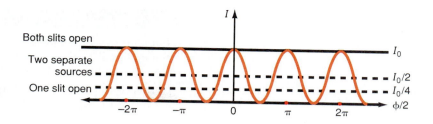

in figure 28.7. The maximum peaks correspond to the bright fringes, and the zeros of the intensity correspond to the dark fringes. Hence, it is easy to explain, by the wave theory of light, the bright and dark fringe pattern found on the screen.

If one of the slits in figure 28.5 were covered up, the interference pattern would disappear. For example, if slit 2 were covered up, then the only wave arriving at P would be wave 1, and the intensity associated with wave 1 would be

$$I_1 = \epsilon_0 c E_1^2 = \epsilon_0 c E_0^2 \sin^2(kr_1 - \omega t) \tag{28.25}$$

The average value of $\sin^2(kr_1 - \omega t)$ is again equal to $1/2$, so the average value of I_1 would be

$$I_1 = \frac{\epsilon_0 c E_0^2}{2} \tag{28.26}$$

which is a constant and does not vary with the angle θ. If we compare this intensity with the value of I_0 from equation 28.15, we see that

$$I_1 = \tfrac{1}{4} I_0$$

This intensity distribution is shown as the dashed line in figure 28.7. The screen would be uniformly illuminated.

If the two waves came from different sources they would not interfere with each other. To explain the reason for this, we need to describe the light that comes from a distributed light source. Light from a distributed light source consists of a very large number of atoms acting as independent light sources. The phase difference between the light waves from different atoms is random. Hence, the term containing the phase angle ϕ in equations 28.12 and 28.9 would not be constant in time but would change in a random way. Thus, there would not be any positions on the screen where the light would always be either interfering constructively or destructively to produce the bright or dark fringes.

If the phase difference between the two light waves of a source of light is a constant in time, we say the light source is *coherent*. If the phase difference varies with time in a random way, we say the source is *incoherent*. Young was able to produce coherent waves by using a single point source of light, which then fell on the two slits as shown in figure 28.4. If light comes from two different sources, the waves do not have a constant phase angle between them, and there is no interference. The intensity at the screen then is simply the sum of each individual intensity, that is,

$$I = I_1 + I_2$$
$$= \frac{\epsilon_0 c E_0^2}{2} + \frac{\epsilon_0 c E_0^2}{2}$$
$$= \epsilon_0 c E_0^2 = \frac{I_0}{2}$$

which is also shown in figure 28.7. That is, there is an even illumination of the entire screen, but there are no interference fringes.

Example 28.1

A double slit. In a double-slit experiment, light of 500.0-nm wavelength impinges on a double slit that has a separation of 0.350 mm. If the screen is placed 5.00 m from the double slit, find (a) the value of θ corresponding to the first four bright fringes, (b) the value of y locating the four bright fringes on the screen, (c) the value of θ corresponding to the first three dark fringes, and (d) the value of y locating the three dark fringes.

a. The angle θ corresponding to the bright fringes, found from equation 28.18, is

$$d \sin \theta = m\lambda$$

Therefore

$$\theta = \sin^{-1}\frac{m\lambda}{d}$$

The first fringe corresponds to the central maximum and occurs for $m = 0$. Hence,

For $m = 0$,

$$\theta_0 = 0$$

For $m = 1$,

$$\theta_1 = \sin^{-1}\frac{\lambda}{d}$$

$$= \sin^{-1}\left(\frac{500.0 \text{ nm}}{0.350 \text{ mm}}\right)\left(\frac{10^{-7} \text{ cm}}{1.00 \text{ nm}}\right)\left(\frac{10 \text{ mm}}{1 \text{ cm}}\right)$$

$$= \sin^{-1}(1.4286 \times 10^{-3})$$

$$= 0.0818°$$

For $m = 2$,

$$\theta_2 = \sin^{-1}\frac{2\lambda}{d}$$

$$= \sin^{-1}\frac{(2)(500.0 \text{ nm})}{(0.350 \text{ mm})}\left(\frac{10^{-6} \text{ mm}}{1 \text{ nm}}\right)$$

$$= 0.1640°$$

For $m = 3$,

$$\theta_3 = \sin^{-1}\frac{3\lambda}{d}$$

$$= \sin^{-1}\frac{(3)(500.0 \text{ nm})}{(0.350 \text{ mm})}\left(\frac{10^{-6} \text{ mm}}{1 \text{ nm}}\right)$$

$$= 0.246°$$

b. The value of y_m locating the bright fringes on the screen, found from equation 28.21, is

$$y_m = \frac{mR\lambda}{d}$$

For $m = 0$,

$$y_0 = 0$$

For $m = 1$,

$$y_1 = \frac{R\lambda}{d}$$

$$= \frac{(5.00 \text{ m})(500.0 \text{ nm})}{0.350 \text{ mm}} \left(\frac{10^{-6} \text{ mm}}{1 \text{ nm}}\right) \left(\frac{10^3 \text{ mm}}{1 \text{ m}}\right)$$

$$= 7.14 \text{ mm}$$

For $m = 2$,

$$y_2 = \frac{2R\lambda}{d} = 2y_1$$

$$= 14.3 \text{ mm}$$

For $m = 3$,

$$y_3 = \frac{3R\lambda}{d} = 3y_1$$

$$= 21.4 \text{ mm}$$

c. The value of θ corresponding to the dark fringes, found from equation 28.23, is

$$d \sin \theta = (2m - 1)\frac{\lambda}{2}$$

Hence,

$$\theta = \sin^{-1}\left[(2m - 1)\frac{\lambda}{2d}\right]$$

For $m = 1$,

$$\theta_1 = \sin^{-1}\left[(2 - 1)\frac{(500.0 \text{ nm})}{(2)(0.350 \text{ mm})}\left(\frac{10^{-6} \text{ mm}}{1 \text{ nm}}\right)\right]$$

$$= 0.0409°$$

For $m = 2$,

$$\theta_2 = \sin^{-1}\left[(4 - 1)\frac{(500.0 \text{ nm})}{(2)(0.350 \text{ mm})}\left(\frac{10^{-6} \text{ mm}}{1 \text{ nm}}\right)\right]$$

$$= 0.123°$$

For $m = 3$,

$$\theta_3 = \sin^{-1}\left[(6 - 1)\frac{(500.0 \text{ nm})}{(2)(0.350 \text{ mm})}\left(\frac{10^{-6} \text{ mm}}{1 \text{ nm}}\right)\right]$$

$$= 0.205°$$

d. The location y_m of the dark fringe on the screen, found from equation 28.24, is

$$y_m = (2m - 1)\frac{R\lambda}{2d}$$

For $m = 1$,

$$y_1 = (2 - 1)\frac{R\lambda}{2d} = \frac{R\lambda}{2d}$$

$$= \frac{(5.00 \text{ m})(500.0 \text{ nm})}{(2)(0.350 \text{ mm})}\left(\frac{10^{-6} \text{ mm}}{1 \text{ nm}}\right)\left(\frac{10^3 \text{ mm}}{1 \text{ m}}\right)$$

$$= 3.57 \text{ mm}$$

For $m = 2$,

$$y_2 = (4 - 1)\frac{R\lambda}{2d} = 3\left(\frac{R\lambda}{2d}\right) = 3y_1$$

$$= 10.7 \text{ mm}$$

For $m = 3$,

$$y_3 = (6 - 1)\frac{R\lambda}{2d} = 5\left(\frac{R\lambda}{2d}\right) = 5y_1$$

$$= 17.9 \text{ mm}$$

Note, in this example, the very small values of the angle θ, which justifies the assumption of the small angle approximation in the derivation.

Example 28.2

Using the double-slit experiment to determine the wavelength of light. A double-slit experiment is performed in order to determine the wavelength of the source of light. If $R = 5.00$ m, $d = 0.350$ mm, and the location of the first bright fringe on the screen is observed to be at 8.25 mm, find the value of the wavelength λ.

Solution

The wavelength, determined from equation 28.21, is

$$y_m = \frac{mR\lambda}{d}$$

$$\lambda = \frac{d\,y_m}{mR}$$

$$= \frac{(0.350 \text{ mm})(8.25 \text{ mm})}{1(5.00 \text{ m})}\left(\frac{1 \text{ m}}{10^3 \text{ mm}}\right)\left(\frac{1 \text{ nm}}{10^{-6} \text{ mm}}\right)$$

$$= 578 \text{ nm}$$

The double-slit experiment is, therefore, a simple experiment that can be used to determine the wavelength of a particular light source.

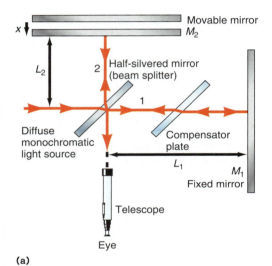

(a)

(b)

Figure 28.8

The Michelson interferometer.

28.3 The Interference of Light— The Michelson Interferometer

A very accurate device to measure the wavelength of light was designed by Albert Michelson (1852–1931). The device, now called the **Michelson interferometer,** is shown schematically in figure 28.8(a), and a typical laboratory interferometer is shown in figure 28.8(b). A diffuse source of monochromatic light impinges on a half-silvered mirror, called a beam-splitter. Because the beam-splitter mirror is not completely silvered, some of the incident light is transmitted through the beam-splitter and passes on to the fixed mirror M_1. This ray of light is reflected from mirror M_1 and returns to the beam-splitter where it is reflected into the telescope for the eye to observe. The rest of the original incident light falling on the beam-splitter is reflected from the half-silvered mirror and travels to the movable mirror M_2, where it is reflected back to the beam-splitter. Here it is transmitted through the beam-splitter and passes through the telescope where it can be observed by the eye. Thus, the incident ray is split into two separate rays, each of which follows a different

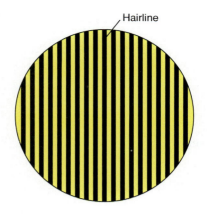

Figure 28.9

Interference fringes in the Michelson interferometer.

optical path, as seen in the figure. When these two rays come together at the eye they interfere with each other, thereby producing interference fringes. Note that ray 2 passes through the beam-splitter three times, whereas ray 1 only passes through the splitter once.

In order for both rays to pass through the same length of glass a *compensator plate* is placed between the beam-splitter and the fixed mirror M_1. The compensator plate is a piece of plane glass having the same index of refraction and the same thickness as the glass of the beam-splitter. Thus, ray 1 now goes through three thicknesses of glass, the same as ray 2. Hence, the only difference in the optical path between ray 1 and ray 2 depends on the difference in length between the path L_1 and the path L_2.

Because of the diffuse nature of the light source and the fact that, in practice, mirrors M_1 and M_2 are not exactly perpendicular to each other, the interference fringes appear as a series of straight lines when viewed through the telescope, as shown in figure 28.9. A hairline in the telescope is placed directly over the central bright fringe. Mirror M_2 in figure 28.8 is capable of being moved forward or backward by a very accurate micrometer screw. If M_2 is moved backward by a distance of $\lambda/4$, then ray 2 travels an additional distance of $\lambda/4$ on the way to the mirror and an additional distance of $\lambda/4$ after it is reflected from the mirror. Hence, in all, the path length of ray 2 changes by $\lambda/2$ from the path length of ray 1. Thus, ray 1 and ray 2 are now out of phase by a half-wavelength as they pass through the telescope and cause destructive interference. Therefore, a dark fringe appears where originally there was a bright fringe. If the mirror is moved an additional distance of $\lambda/4$, then there is another half-wavelength change in the optical path and the rays cause constructive interference or a bright fringe. Hence, a total movement of M_2 by a distance of $\lambda/2$ causes one bright fringe to move across the cross hair of the telescope. If the observer counts m fringes passing the cross hair of the telescope when the mirror is moved a distance x, as measured by the micrometer screw, then that distance x corresponds to a distance of m half-wavelengths or

$$x = m\frac{\lambda}{2} \tag{28.27}$$

If the wavelength λ of the light source is known then the interferometer, through the use of equation 28.27, can be used to measure a length x very accurately. On the other hand, if we wish to determine the wavelength λ of a light source, we can rearrange equation 28.27 to solve for λ when x is a measured reading of the micrometer screw. Thus,

$$\lambda = \frac{2x}{m} \tag{28.28}$$

Example 28.3

Using the Michelson interferometer. A Michelson interferometer is used to determine the D spectral line in sodium. If the movable mirror moves a distance of 0.2650 mm, as read from the micrometer screw, when 900.0 fringes are counted passing the telescope cross hair, find the wavelength of the D line.

The wavelength, found from equation 28.28, is

$$\lambda = \frac{2x}{m}$$

$$= \frac{2(0.2650 \text{ mm})}{900.0}$$

$$= (5.889 \times 10^{-4} \text{ mm})\left(\frac{1 \text{ nm}}{10^6 \text{ mm}}\right)$$

$$= 588.9 \text{ nm}$$

28.4 Interference—Thin Films

Interference fringes can also be caused by multiple reflections from a **thin film,** a very thin piece of transparent material. The interference can be caused by a phase change that occurs on reflection and/or a phase change that occurs because of the difference in the optical path of the two interfering light waves. Let us first consider what happens when a light wave is reflected from a boundary.

Reflection at a Boundary between a Less Dense Medium and a More Dense Medium

Recall from our study of waves on a string in chapter 12 that when the wave went from the less dense string to the more dense string, the reflected wave was inverted. That is, the reflected wave was 180° out of phase with the incident wave (see figure 12.10), while the transmitted wave did not change its phase. Although that result was obtained for waves on a string, it is a general characteristic of wave motion. *When a ray of light in air is incident on a piece of glass, the light wave is reflected at the boundary between the less optically dense air and the more optically dense glass. Hence, the reflected wave is found to be 180° out of phase with the incident wave, figure 28.10(a). The light wave transmitted through the glass does not change its phase with respect to the incident wave but its wavelength decreases in the more dense glass,* as shown in chapter 27, equation 27.7:

$$\lambda_2 = \frac{\lambda_1}{n_{21}}$$

Applying equation 27.7 to the current problem, the wavelength in the glass is

$$\lambda_g = \frac{\lambda_{air}}{n_{21}} = \frac{\lambda_{air}}{n_g/n_{air}}$$

$$= n_{air}\frac{\lambda_{air}}{n_g}$$

But since $n_{air} = 1$,

$$\lambda_g = \frac{\lambda_{air}}{n_g} \tag{28.29}$$

Thus, the wavelength of the transmitted wave in the glass is decreased.

Reflection at a Boundary between a More Dense Medium and a Less Dense Medium

It was shown in figure 12.11 that when a wave goes from a more dense medium to a less dense medium, the reflected wave is not inverted. That is the reflected wave is not out of phase with the incident wave. Although the result was established for

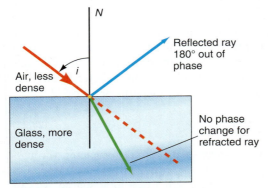

(a) Reflection from less dense to more dense

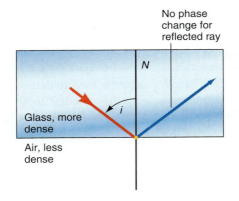

(b) Reflection from more dense to less dense

Figure 28.10

Reflection and phase changes of light at a boundary.

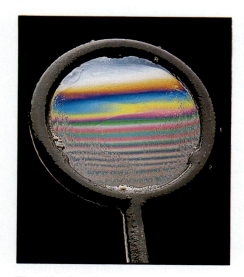

Thin film interference.

waves on a string it is a characteristic of wave motion in general. *Hence, when a light wave is reflected at a boundary between a more dense medium and a less dense medium, the reflected wave is not inverted but has the same phase as the incident wave.* Thus, in figure 28.10(b) the incident ray in the more dense glass is reflected at the boundary between the glass and the less optically dense air, and there is no phase change on reflection. *The transmitted wave is always in phase with the incident wave.* The effect of the phase change, or lack of it, in reflection becomes important in the interference of light waves from thin films.

Interference of Light from a Soap Bubble or Thin Film of Oil

Consider an incident ray of light in air that impinges on a thin film of water, such as a soap bubble found in a child's toy for making bubbles, and shown schematically in figure 28.11. We assume that the incident ray is almost normal to the thin film of water, although it is shown at an exaggerated angle in figure 28.11 so we can more easily see what is happening. As the incident ray hits the boundary between the air and the water, part of the incident light wave is reflected and part is transmitted into the film of water. The reflected ray is labeled as ray 1 in figure 28.11. Since this ray has been reflected from a boundary between a less optically dense medium, the air, and a more optically dense medium, the water, the reflected light wave undergoes a phase change of 180° on reflection. The part of the wave that is transmitted through the water, ray 2, undergoes a reflection at point A at the boundary between the water and the air, as shown in the figure. This reflection occurs at a boundary between a more optically dense medium, the water, and a less optically dense medium, the air, so there is no phase change on reflection for ray 2.

If the thickness of the film d were zero, then ray 1 would be 180° out of phase with ray 2, and then ray 1 and ray 2 would superimpose to give destructive interference. However, the thickness d is not zero, and ray 2 moves through the distance OAB before it emerges into the air, and this additional distance moved by ray 2 causes a path difference between ray 1 and ray 2. Because the incident ray is almost normal, the distance OAB is approximately twice the thickness of the film. That is, the path difference between ray 1 and ray 2 is

$$\text{Path difference} = \text{PD} = 2d$$

If this path difference is equal to a whole multiple of wavelengths, then ray 2 has the same phase at position B as ray 1 at position O, if ray 1 did not have a phase change of 180°. But ray 1 has experienced a phase change of 180° on reflection, so when ray 2 reaches point B it is 180° out of phase with ray 1 and causes destructive interference when viewed by the eye in figure 28.11. Hence, the condition for a dark fringe or destructive interference is

Dark Fringe   Path difference = PD = $2d = m\lambda_n$ $m = 1, 2, 3, \ldots$ **(28.30)**

Figure 28.11

Reflection from a thin layer of water.

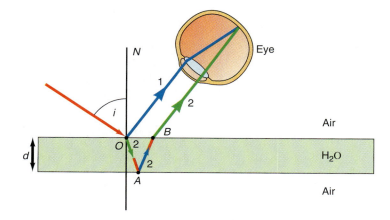

In equation 28.30 the subscript n on the wavelength indicates that this is the wavelength in the medium with the index of refraction n. Hence, λ_n represents the wavelength of light within the thin film of water. The wavelength of light within the water is related to the wavelength of light in air by equation 28.29 with a change in subscript from g to w representing the water medium rather than a glass medium. Hence,

$$\lambda_n = \lambda_w = \frac{\lambda_{air}}{n_w} \tag{28.31}$$

Substituting equation 28.31 back into equation 28.30, we get

$$\text{Dark Fringe} \qquad 2d = m\frac{\lambda_{air}}{n_w} \qquad m = 1, 2, 3, \ldots \tag{28.32}$$

Thus, the equation for destructive interference, or a dark fringe in the reflected light from a soap bubble, is

$$\text{Dark Fringe} \qquad 2n_w d = m\lambda_{air} \quad m = 1, 2, 3, \ldots \tag{28.33}$$

We use the index of refraction of water for the index of refraction of the soap bubble because it is made up largely of water. (If the interference were from a thin film of oil, then n would, of course, be the index of refraction of oil.)

If the path length of ray 2 in the water is equal to an odd multiple of $\lambda/2$, then ray 2 should be out of phase with ray 1. However, because ray 1 suffers a phase change of 180° by reflection, ray 1 is now in phase with ray 2. Therefore, the superposition of ray 1 and ray 2 now causes constructive interference or a bright fringe when viewed by the eye. Thus, the condition for a bright fringe is

$$PD = (2m - 1)\frac{\lambda_n}{2} \quad m = 1, 2, 3, \ldots$$

where the term $(2m - 1)$ always gives an odd number for any value of $m = 1, 2, 3, \ldots$. But we have already seen that the path difference for nearly normal incidence is $2d$. Thus,

$$\text{Bright Fringe} \qquad 2d = (2m - 1)\frac{\lambda_n}{2}$$

Substituting for λ_n from equation 28.31 *gives the condition for constructive interference or a bright fringe as*

$$\text{Bright Fringe} \qquad 2d = (2m - 1)\frac{\lambda_{air}}{2n_w} \tag{28.34}$$

Example 28.4

Minimum reflection from a thin film. A soap bubble is 220.0 nm thick. For what wavelength of the incident light will the intensity of the reflected light be zero?

Solution

The intensity of the reflected light is zero when destructive interference occurs between the wave reflected from the air-water boundary and the wave reflected from the water-air boundary. The condition for this dark fringe is found from equation 28.33. Solving that equation for the wavelength in air we get

$$\lambda_{air} = \frac{2n_w d}{m}$$

$$= \frac{2(1.33)(220.0 \text{ nm})}{1}$$

$$= 585 \text{ nm}$$

Thus, if a ray of light of wavelength 585 nm fell on the soap bubble there would be no reflected light and hence all the light would be transmitted through the soap bubble.

Example 28.5

Maximum reflection from a thin film. For the same soap bubble of 220.0-nm thickness, what wavelength of the incident light would give the maximum reflection?

Solution

The maximum reflection occurs when there is constructive interference between the reflected waves. The condition for this bright fringe is given by equation 28.34. Solving for the wavelength, we get

$$\lambda_{air} = \frac{4n_w d}{(2m - 1)}$$

For $m = 1$,

$$\lambda_{1air} = \frac{4(1.33)(220.0 \text{ nm})}{1}$$

$$= 1{,}170 \text{ nm}$$

But this wavelength falls out of the visible spectrum (380.0–720.0 nm) and could not be seen by the eye.

For $m = 2$,

$$\lambda_{2air} = \frac{\lambda_{1air}}{3} = \frac{1170 \text{ nm}}{3}$$

$$= 390 \text{ nm}$$

Therefore, a wavelength of 390 nm, a blue light just barely in the visible spectrum, gives a maximum intensity of reflected light. For values of $m = 3$ or greater, the wavelength occurs in the ultraviolet portion of the spectrum, which is of course, not visible.

If we shine white light, a mixture of all the wavelengths from 380.0 to 720.0 nm, the only bright reflection occurs for the blue wavelength of 390 nm. If we shine monochromatic light of any wavelength other than the 390 nm, we will not get a bright reflection.

In practice, the soap bubble is usually not of a constant thickness d. Therefore, many different wavelengths can be reflected for the different values of d, thus accounting for the beautiful array of colors usually associated with the soap bubble film.

Nonreflecting Glass

In many sophisticated optical systems, there may be as many as four to ten lenses in the optical system. The amount of light reflected from a single lens is actually quite small, of the order of 4%. However with four lenses the light lost by reflection is about 15%, whereas for 10 lenses it amounts to over 33%. Hence, for a multilens system, the amount of reflected light can become quite significant. Our analysis of the reflected light from the thin film of a soap bubble suggests that *a thin film of transparent material, placed on the surface of a lens, can be used to reduce the amount of reflected light from a lens.* Glass, so coated with a thin film, is called **nonreflecting glass.**

Figure 28.12

Nonreflecting glass.

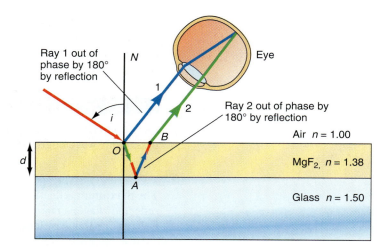

A thin film of magnesium fluoride (MgF_2) is, therefore, placed over a piece of glass as indicated in figure 28.12. The index of refraction of MgF_2 is 1.38, which is greater than the index of refraction of air, but less than the index of refraction of glass. When the incident ray impinges on the MgF_2 surface at the point O, part of the light is reflected as ray 1 and part is transmitted as ray 2. Ray 1 is 180° out of phase with the incident ray because of the reflection from a boundary between the less dense and more dense medium. When the transmitted ray, ray 2, hits the glass at point A of the figure, the reflected ray is also 180° out of phase with the incident ray because the reflection at the magnesium fluoride–glass boundary is a reflection at the boundary between the less dense medium, MgF_2 ($n = 1.38$) and the more dense glass ($n = 1.50$). Thus, if the thickness d of the magnesium fluoride were effectively zero, the two reflected waves would be in phase. However, d is not equal to zero and there will, therefore, be a difference in the optical paths between ray 1 and ray 2. *If this difference in the optical path is an odd multiple of a half-wavelength $\lambda_n/2$, ray 1 and ray 2 are out of phase. Their superposition produces destructive interference, and there is no reflected light.* Stated mathematically the condition for destructive interference is

$$\text{Path difference} = 2d = (2m - 1)\frac{\lambda_n}{2} \tag{28.35}$$

where the term $(2m - 1)$ always gives an odd number for $m = 1, 2, 3 \ldots$. The wavelength within the magnesium fluoride medium, λ_n, is given by equation 28.31 as

$$\lambda_n = \frac{\lambda_{air}}{n_{MgF_2}} \tag{28.36}$$

Substituting equation 28.36 into equation 28.35 gives the condition for zero reflection as

$$\text{Minimum reflection} \qquad 2d = (2m - 1)\frac{\lambda_{air}}{2n_{MgF_2}} \tag{28.37}$$

Note that equation 28.37 corresponds to destructive interference while equation 28.34 looks almost the same, yet equation 28.34 represents a constructive interference from a thin film of water. The difference between the two equations is based on the phase change at reflection. For the thin soap bubble film only the first reflected wave is out of phase by 180°. For the thin film on the glass lens, both reflected rays are 180° out of phase. Hence the problems are different and we should expect to get different results.

Example 28.6

Minimum reflection from nonreflecting glass. What thickness of magnesium fluoride should be deposited on a glass lens of $n_g = 1.50$, such that the greatest portion of incident white light should not be reflected?

Solution

Visible light lies in the range of 380.0 to 720.0 nm. It is desired that the thin film give minimum reflection at the center of the visible spectrum. That is, the center of the visible spectrum is equal to

$$\frac{380.0 \text{ nm} + 720.0 \text{ nm}}{2} = 550.0 \text{ nm}$$

Hence, a thickness d is picked that gives destructive interference for $\lambda = 550.0$ nm, the center of the visible spectrum. Solving equation 28.37 for the thickness d, gives

$$d = (2m - 1)\frac{\lambda_{air}}{4n_{MgF_2}}$$

For $m = 1$,

$$d = (2 - 1)\frac{550.0 \text{ nm}}{4(1.38)}$$

$$= 99.6 \text{ nm}$$

Therefore, a thickness of 99.6 nm of magnesium fluoride placed on a glass lens causes destructive interference for light of 550.0 nm and none of this light is reflected.

Example 28.7

Maximum reflection from nonreflecting glass. For the thickness of magnesium fluoride in example 28.6, is there any wavelength within the visible spectrum that will give maximum reflection of light, that is, give constructive interference?

Solution

For constructive interference the path difference between ray 1 and ray 2 must be a whole multiple of the wavelength λ of the light. Hence the path difference must be

$$\text{PD} = 2d = m\lambda_n \tag{28.38}$$

Substituting the value of λ_n from equation 28.36 into equation 28.38 yields for the condition of maximum reflection

Maximum reflection $$\qquad 2d = m\frac{\lambda_{air}}{n_{MgF_2}} \tag{28.39}$$

Solving for the wavelength that gives this maximum reflection we have

$$\lambda_{air} = \frac{2n_{MgF_2}d}{m}$$

$$= \frac{2(1.38)(99.6 \text{ nm})}{1}$$

$$= 275 \text{ nm}$$

That is, for the thin film thickness of 99.6 nm, maximum reflection occurs at 275 nm. But 275 nm lies outside the visible portion of the spectrum and cannot be seen.

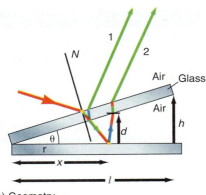

(a) Geometry

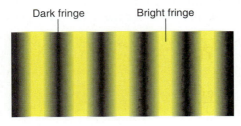

(b) Fringes

Figure 28.13

An air wedge.

At 550.0 nm complete constructive interference occurs. At wavelengths greater than 275 nm and less than 550.0 nm, superposition of rays 1 and 2 still occurs with some intermediate value of intensity. Thus, at the lower portion of the visible spectrum, 380.0 nm, some light is reflected. The closer we get to the 550.0 nm wavelength, the smaller is the reflection. There is also some reflection for wavelengths greater than 550.0 nm just as there were some in the lower region.

Interference from an Air Wedge

If two plates of glass are separated by a very small amount at one edge, as in figure 28.13(a), an interference pattern of dark and bright fringes occurs. Ray 1 is reflected from a glass-air interface (more dense to less dense) and no phase change occurs. Ray 2, on the other hand, is reflected from an air-glass interface (less dense to more dense) and suffers a 180° phase change on reflection. Thus, ray 1 and ray 2 are out of phase by 180°. However, there is an additional phase shift caused by the difference in path of $2d$ traversed by ray 2 as it passes through the air gap, between the upper and lower plate, after reflection. The sizes of h and d in figure 28.13(a) are greatly exaggerated so we more easily see the reflections and path differences. *Hence, the total phase difference between ray 1 and ray 2 is equal to the phase difference caused by the difference in optical path between the two rays plus the phase change that occurs to ray 2 on reflection. If the path difference is equal to a multiple of the wavelength λ, then the only phase difference is the 180° one caused by reflection and ray 1 is then out of phase with ray 2 and destructive interference occurs. Hence, a dark fringe occurs when*

$$\text{Dark Fringe} \qquad 2d = m\lambda \qquad m = 0, 1, 2, 3, \ldots \qquad \textbf{(28.40)}$$

We can find the location of this dark fringe from the geometry of figure 28.13(a):

$$\tan \theta = \frac{d}{x}$$

and also

$$\tan \theta = \frac{h}{l}$$

Therefore,

$$\frac{d}{x} = \frac{h}{l}$$

Solving for x,

$$x = \frac{dl}{h} \qquad \textbf{(28.41)}$$

The value of d for the mth fringe, found from equation 28.40, is

$$d = \frac{m\lambda}{2} \qquad \textbf{(28.42)}$$

Substituting equation 28.42 into equation 28.41 gives for the location of the mth fringe

$$\text{Location of Dark Fringe} \qquad x_m = \frac{m\lambda l}{2h} \qquad m = 0, 1, 2, 3, \ldots \qquad \textbf{(28.43)}$$

The separation between fringes is found by subtracting the location of the mth fringe from the $(m + 1)$th fringe. That is,

$$\Delta x = x_{m+1} - x_m$$

$$= \frac{(m + 1)\lambda l}{2h} - m\frac{\lambda l}{2h}$$

$$\Delta x = \frac{\lambda l}{2h} \tag{28.44}$$

Since λ, l, and h are all constant for the geometry of figure 28.13, the separation between dark fringes is uniform.

We should note that the special case of $m = 0$ in equation 28.43 corresponds to $d = 0$, or approximately zero, at $x = 0$, and is the location of a dark fringe. The reason for this is that the only phase change that occurs there, one of 180°, occurs from the reflection from the lower air-glass interface.

Between the dark fringes, where destructive interference occurs, there are bright fringes, where constructive interference occurs. The fringe pattern is shown in figure 28.13(b). We find the condition for the bright fringe from the geometry of figure 28.13(a), where the optical path difference must now be an odd multiple of a half a wavelength. That is, for a bright fringe we must have

$$\text{Bright Fringe} \qquad 2d = (2m - 1)\frac{\lambda}{2} \qquad m = 1, 2, 3, \ldots \tag{28.45}$$

And the location of the mth bright fringe, found from equations 28.41 and 28.45, is

$$\text{Location of Bright Fringe} \qquad x_m = (2m - 1)\frac{\lambda l}{4h} \qquad m = 1, 2, 3, \ldots \tag{28.46}$$

Example 28.8

The air wedge. A hair, 8.50×10^{-5} m in diameter, is placed between two 10.0-cm plates of glass ($n = 1.52$) at their edge. Light of 589.0-nm wavelength shines on the glasses and an interference pattern is formed. Find (a) the location of the fifth dark fringe, and (b) the location of the third bright fringe.

Solution

a. The location of the fifth dark fringe, found from equation 28.43, is

$$x_m = \frac{m\lambda l}{2h}$$

Because the first dark fringe occurs for $m = 0$, the fifth dark fringe occurs for $m = 4$. Therefore,

$$x_5 = \frac{4\lambda l}{2h}$$

$$= \frac{4(589.0 \times 10^{-9} \text{ m})(0.100 \text{ m})}{2(8.50 \times 10^{-5} \text{ m})}$$

$$= 1.39 \times 10^{-3} \text{ m} = 1.39 \text{ mm}$$

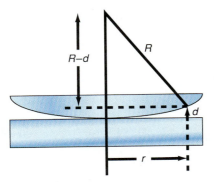

(a) Geometry

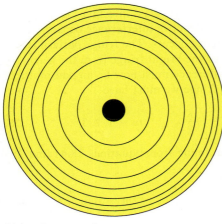

(b) Interference pattern

Figure 28.14

Newton's rings.

b. The location of the third bright fringe, found from equation 28.46, is

$$x_m = \frac{(2m - 1)\lambda l}{4h}$$

$$x_3 = \frac{[2(3) - 1](589.0 \times 10^{-9} \text{ m})(100 \text{ mm})}{4(8.50 \times 10^{-5} \text{ m})}$$

$$= 0.866 \text{ mm}$$

Newton's Rings—A Variable Air Wedge

An interesting interference pattern can be observed if light shines on a planoconvex lens that is placed on a flat piece of glass, as shown in figure 28.14(a). *The interference pattern consists of a family of concentric dark and bright circles, known as Newton's rings and shown in figure 28.14(b).*

The interference between the reflected light waves is the same as for the air wedge studied above, but because of the curvature of the lens, the angle of the air wedge is continuously varying. A dark fringe, or destructive interference, again occurs when the optical path difference is equal to a whole number of wavelengths, that is,

$$2d = m\lambda \tag{28.47}$$

where d, as shown in figure 28.14(a), is a variable. Since a fixed value of d occurs all around the lens, the fringes are circular. The radius of the fringe is related to the air gap d by the geometry of figure 28.14(a). Here R is the radius of curvature of the convex lens, while r is the location of the air gap d. From the figure we see that

$$R^2 = (R - d)^2 + r^2$$
$$= R^2 - 2Rd + d^2 + r^2$$

However, since $R \gg d$, $d^2 = 0$. Therefore,

$$R^2 = R^2 - 2Rd + r^2$$
$$0 = -2Rd + r^2$$

Solving for d, we get

$$d = \frac{r^2}{2R}$$

Substituting this value of d into equation 28.47, the condition for a dark fringe, we get

$$\frac{2(r^2)}{2R} = m\lambda$$

Solving for r_m, the radius of the mth fringe, we get

$$r_m = \sqrt{m\lambda R} \tag{28.48}$$

Example 28.9

Newton's rings. Light of 589.0-nm wavelength falls normally on a planoconvex lens whose radius of curvature is 10.0 m, and Newton's rings are observed. Find the radius of the 15th fringe.

Light and Optics

The radius of the 15th fringe, found from equation 28.48, is

$$r_m = \sqrt{m\lambda R}$$
$$= \sqrt{15(589.0 \times 10^{-9}\text{ m})(10.0\text{ m})}$$
$$= 9.40\text{ mm}$$

28.5 Diffraction from a Single Slit

As seen in the introduction to this chapter, diffraction is the bending of light waves around an obstacle. Let us now consider the diffraction of light from a single slit, where each side of the slit represents the obstacle that the light bends around.

There are two distinct ways to observe diffraction from a single slit. The first technique is called *Fresnel diffraction,* after its discoverer, Augustin Jean Fresnel (1788–1827), and is diagramed in figure 28.15. Monochromatic light from a point source impinges on the slit. The waves from the point source are spherical and the wave fronts are circular as they fall on the slit opening. The waves are diffracted at the slit opening and spread out into the shadow region and a diffraction pattern is found on the screen, figure 28.15(b). The pattern resembles the interference pattern of a double slit, with alternate dark and bright fringes, but with decreasing intensity. The screen is relatively close to the slit in this arrangement.

The second technique for observing the effect of diffraction from a single slit is due to Joseph von Fraunhofer (1787–1826) and is shown in figure 28.16(a). The source is very far away from the slit so that the incident light can be treated as a series of plane waves. The screen is so far away from the source that the light waves, after the diffraction, can again be assumed to be plane. Again a diffraction pattern occurs on the screen as in figure 28.15(b). This second technique is called **Fraunhofer diffraction** and can be approximated in the laboratory by the use of two converging lenses, as shown in figure 28.16(b). A converging lens is placed in front of the monochromatic light source such that the source is at the principal focus. Hence, light rays emanate from the lens as parallel rays, which means that the waves are, of course, plane light waves. The plane waves are diffracted by the slit and the light rays that head toward the slit are all parallel. A converging lens is placed a distance f in front of the screen and these parallel rays are then converged by the lens to the point P on the screen.

Figure 28.15

Fresnel diffraction.

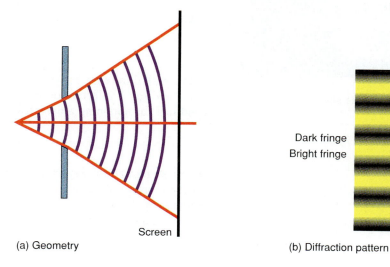

(a) Geometry

Screen

Dark fringe
Bright fringe

(b) Diffraction pattern

Figure 28.16

Fraunhofer diffraction.

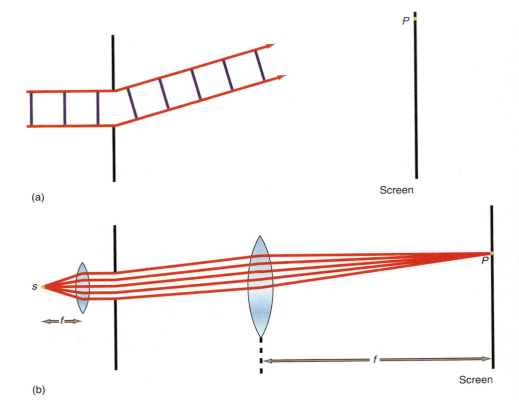

(a)

(b)

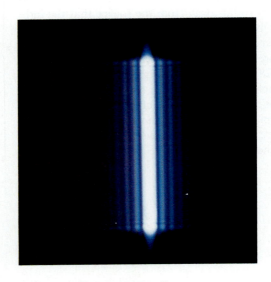

Diffraction from a single slit.

Our analysis of diffraction from the single slit will center on Fraunhofer diffraction because the plane waves are easier to handle than spherical waves. Figure 28.17(a) shows a single slit of width a. Points 1, 2, and 3 are points on the same wave front. Secondary Huygens' wavelets from each of these waves are in phase with each other before they arrive at the slit. These straight ahead waves are also in phase after the slit and are converged by the lens to the point P_0 to form the central bright fringe. Figure 28.17(b) shows rays 1, 2, 3, 4, and 5 associated with waves that are all in phase as the plane waves approach the slit. After the diffraction the rays are still parallel when they enter the lens, and are then converged to the point P. However, because the point P is no longer symmetrically located with respect to the slit, the light rays 1 through 5 travel different path lengths in arriving at the point P. We can see in figure 28.17(b) that the path difference between ray 1 and ray 3 is

$$\text{PD}_3 = \frac{a}{2} \sin \theta$$

If this path difference between ray 1 and ray 3 is exactly a half a wavelength, then ray 1 and ray 3 interfere destructively when they combine at point P. Hence, the condition for obtaining a dark fringe at P is that

$$\text{PD} = \frac{a}{2} \sin \theta = \frac{\lambda}{2} \tag{28.49}$$

Also note that all the rays between ray 1 and ray 2 are completely out of phase with all the corresponding rays between ray 3 and ray 4 and all interfere destructively at the point P.

The path difference between ray 2 and ray 1, from the figure, is

$$\text{PD}_2 = \frac{a}{4} \sin \theta$$

Figure 28.17

Single-slit diffraction.

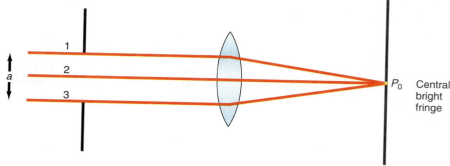

(a) Central bright fringe

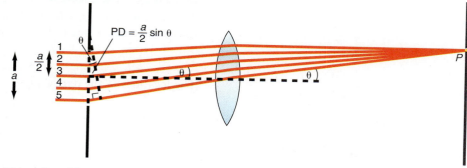

(b) Location of first dark fringe

while the path difference between ray 4 and ray 1 is

$$PD_4 = \frac{3a}{4} \sin \theta$$

Hence, the path difference between ray 2 and ray 4 is

$$PD_4 - PD_2 = \frac{3a}{4} \sin \theta - \frac{a}{4} \sin \theta = \frac{a}{2} \sin \theta$$

But as just seen in equation 28.49, $(a/2) \sin \theta = \lambda/2$, and therefore ray 2 and ray 4 are also 180° out of phase when they combine at point P, also producing a dark fringe. All the rays between ray 2 and ray 3 are also completely out of phase with all the corresponding rays between ray 4 and ray 5. Hence, all the rays from the single slit interfere destructively at the point P and form a dark fringe there. Therefore, the condition for the first dark fringe, obtained from equation 28.49, is

$$\text{First Dark Fringe} \qquad a \sin \theta = \lambda \qquad \qquad \textbf{(28.50)}$$

In general, there is more than one dark fringe and destructive interference occurs whenever the quantity $(a \sin \theta)$ is a whole multiple of the wavelength λ. Thus, the general condition for a dark fringe is

$$\text{Dark Fringe} \qquad \boxed{a \sin \theta = m\lambda} \qquad m = 1, 2, 3, \ldots \qquad \textbf{(28.51)}$$

Secondary bright fringes are found to exist approximately half way between the dark fringes. Although too long to derive here, the intensity distribution for the single-slit diffraction pattern is shown in figure 28.18 as a function of the angle θ. The intensity of the secondary maxima shown in the figure are exaggerated for clarity in the diagram. We should note a very interesting result of this intensity distribution. There is a central bright fringe occurring at $\theta = 0$, and secondary bright fringes occurring on both sides of the central maximum. However, the intensity of these secondary maxima diminish more rapidly than is shown in the figure.

Figure 28.18

Intensity distribution for single-slit diffraction.

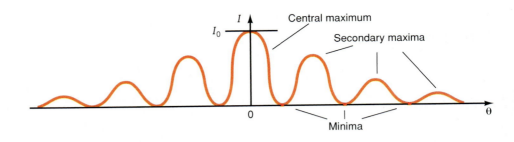

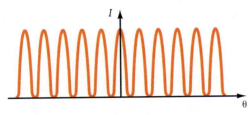

(a) Interference fringes

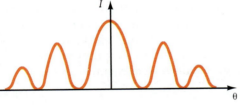

(b) Diffraction factor

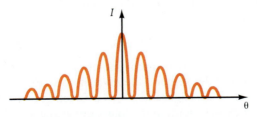

(c) Combined interference and diffraction pattern

Figure 28.19

Interference and diffraction combined.

Example 28.10

Diffraction from a single slit. Find the location for the first three dark fringes for diffraction from a single slit of width $a = 10\lambda$.

Solution

The locations of the dark fringes are found from equation 28.51 as

$$a \sin \theta = m\lambda$$

$$\sin \theta = \frac{m\lambda}{a}$$

$$\theta = \sin^{-1}\frac{m\lambda}{a} = \sin^{-1}\frac{m\lambda}{10\lambda}$$

$$= \sin^{-1}m(0.1)$$

For $m = 1$

$$\theta = \sin^{-1}(0.1) = 5.74°$$

For $m = 2$

$$\theta = \sin^{-1}(0.2) = 11.5°$$

For $m = 3$

$$\theta = \sin^{-1}(0.3) = 17.5°$$

It is interesting and informative to return to Young's double-slit experiment and note that each slit in that experiment is a single slit and must have a diffraction pattern associated with it. Thus, *the interference fringes of Young's double-slit experiment must be modified to show the effects of diffraction.* The combined effects of interference and diffraction are shown in figure 28.19. *The effect of the diffraction is to reduce the intensity of the interference fringes farther away from the central bright fringe.*

28.6 The Diffraction Grating

The interference of light from a double slit causes a characteristic interference pattern of bright and dark fringes. However, what is so special about two slits? Suppose three slits, or four slits, or even N slits, were used in the experiment. Would the characteristic interference pattern still be observed? The answer is yes. *Interference of light from any number of slits does give the characteristic pattern of bright and dark fringes.*

*Several parallel slits of equal width a, equally spaced a distance d apart, is called a **diffraction grating.*** Diffraction gratings are made by marking equally

Light and Optics

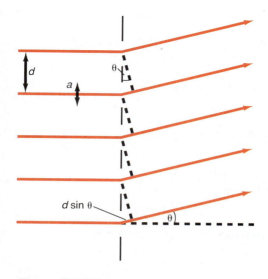

Figure 28.20

The diffraction grating.

Figure 28.21

Comparison of interference pattern from a double slit to a diffraction grating.

spaced grooves on a glass plate. The spaces between the grooves represent the slits. Typical gratings have anywhere between 400 slits per mm to 1200 slits per mm. A schematic of a diffraction grating is shown in figure 28.20. For light rays emerging from each slit at an angle θ, there is a path difference of $d \sin \theta$ between each ray and the one directly above it. If this path difference is an exact multiple of wavelengths then the rays from each slit would interfere constructively when they reach the screen and would therefore produce a bright fringe there. Thus, the condition for the formation of a bright fringe is

$$\text{Bright Fringe} \qquad d \sin \theta = m\lambda \qquad m = 0, 1, 2, \ldots \qquad (28.52)$$

Note that this is the same condition found for double-slit interference in equation 28.18. *The effect of the increased number of slits is to form a sharper and narrower interference pattern, as shown in figure 28.21.*

Example 28.11

The diffraction grating. Monochromatic light of 500.0-nm wavelength shines on a diffraction grating of 1200 lines per mm. Find the location of the first three bright fringes.

Solution

The distance d between slits is found by taking the reciprocal of the number of lines per mm, thus

$$d = \frac{1 \text{ mm}}{1200} = (8.33 \times 10^{-4} \text{ mm})\left(\frac{10^6 \text{ nm}}{1 \text{ mm}}\right) = 833.0 \text{ nm}$$

The location of the bright fringes, found from equation 28.52, is

$$d \sin \theta = m\lambda$$

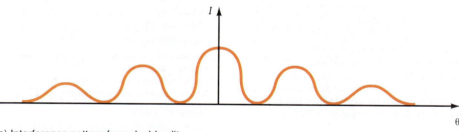

(a) Interference pattern from double slit

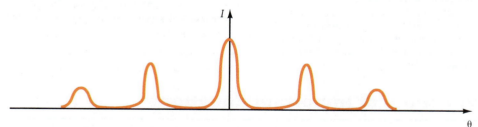

(b) Interference pattern from diffraction grating

and

$$\theta = \sin^{-1}\frac{m\lambda}{d}$$

The first bright fringe occurs for $m = 0$ and therefore

$$\theta = \sin^{-1} 0 = 0°$$

That is, the central maximum occurs at $\theta = 0°$, as expected. For $m = 1$, the first-order bright fringe is located at

$$\theta_1 = \sin^{-1}\frac{(1)(500.0\text{ nm})}{833.0\text{ nm}} = 36.9°$$

For $m = 2$,

$$\theta_2 = \sin^{-1}\frac{(2)(500.0\text{ nm})}{833.0\text{ nm}} = \sin^{-1} 1.20$$

But the sine of θ cannot exceed the value of 1. Yet for $m = 2$, the $\sin \theta = 1.20$, which is impossible. Therefore the second-order bright fringe (corresponding to $m = 2$) does not exist.

The Language of Physics

Diffraction
The bending of light around an obstacle, into the region that should be a shadow area (p. 807).

Monochromatic light
Light that consists of a single wavelength. In contrast, white light consists of light of very many wavelengths. A laser is a good source of monochromatic light (p. 809).

Young's double-slit experiment
An experiment that superimposes light from two different slits to form a series of bright and dark fringes upon a screen. The fringes can only be explained if light has the characteristics associated with waves (p. 809).

Michelson interferometer
An optical device that measures distances or wavelengths very accurately by superimposing light from two different paths to give a series of dark and bright fringes (p. 818).

Thin film
A very thin piece of transparent material. When monochromatic light shines on the film, interference effects can be observed due to a phase change that occurs on reflection and/or a phase change that occurs because of the difference in the optical path of the two interfering light waves (p. 820).

Nonreflecting glass
A piece of glass on which a thin film of transparent material has been placed to cause reflected rays to be out of phase and hence, interfere destructively (p. 823).

Newton's rings
An interference pattern that occurs when light shines on a planoconvex lens that is placed on a flat piece of glass. The pattern consists of a family of concentric dark and bright fringes (p. 828).

Fraunhofer diffraction
Diffraction from a slit in which the source is very far away from the slit, and the slit is very far away from the screen, so that the light waves can be assumed to be plane waves. When the effects of diffraction are combined with the interference of light from a double slit, the resulting intensity distribution is similar to the interference pattern except that the intensity of the fringes decreases farther away from the central bright fringe (p. 829).

Diffraction grating
Several parallel slits of equal width, equally spaced, that gives a characteristic intensity distribution of sharper and narrower fringes (p. 832).

Summary of Important Equations

Equation of plane light wave
$$E = E_0 \sin(kx - \omega t) \tag{25.38}$$

Phase difference in double slit
$$\phi = \frac{2\pi d}{\lambda} \sin \theta \tag{28.9}$$

Resultant wave in double slit
$$E = 2E_0 \cos\left(\frac{\phi}{2}\right)$$
$$\sin\left(kr_1 - \omega t + \frac{\phi}{2}\right) \tag{28.12}$$

Intensity of an electromagnetic wave
$$I_{avg} = \epsilon_0 c E^2 \tag{25.63}$$

Intensity distribution for a double slit
$$I = \epsilon_0 c 4E_0^2 \cos^2\left(\frac{\phi}{2}\right)$$
$$\sin^2\left(kr_1 - \omega t + \frac{\phi}{2}\right) \tag{28.13}$$

Average intensity distribution for a double slit
$$I_{avg} = I_0 \cos^2\left(\frac{\phi}{2}\right) \tag{28.16}$$

where
$$I_0 = 2\epsilon_0 c E_0^2 \tag{28.15}$$

Maximum intensity occurs for bright fringe for double slit
$$\frac{\phi}{2} = m\pi \tag{28.17}$$

Condition for bright fringe for double slit
$$d \sin \theta = m\lambda \qquad (28.18)$$

Location of mth bright fringe for double slit
$$y_m = \frac{Rm\lambda}{d} \qquad (28.21)$$

Minimum intensity for double slit
$$\frac{\phi}{2} = (2m - 1)\frac{\pi}{2} \qquad (28.22)$$

Condition for dark fringe for double slit
$$d \sin \theta = (2m - 1)\frac{\lambda}{2} \qquad (28.23)$$

Location of mth dark fringe for double slit
$$y_m = (2m - 1)\frac{R\lambda}{2d} \qquad (28.24)$$

Michelson interferometer
$$x = m\frac{\lambda}{2} \qquad (28.27)$$

$$\lambda = \frac{2x}{m} \qquad (28.28)$$

Decreased wavelength in a glass medium
$$\lambda_g = \frac{\lambda_{air}}{n_g} \qquad (28.29)$$

Destructive interference for thin film
$$2d = m\frac{\lambda_{air}}{n_w} \qquad (28.32)$$

Constructive interference for thin film
$$2d = (2m - 1)\frac{\lambda_{air}}{2n_w} \qquad (28.34)$$

Destructive interference for nonreflecting glass
$$2d = (2m - 1)\frac{\lambda_{air}}{2n_{MgF_2}} \qquad (28.37)$$

Constructive interference for nonreflecting glass
$$2d = \frac{m\lambda_{air}}{n_{MgF_2}} \qquad (28.39)$$

Condition for dark fringe for an air wedge
$$2d = m\lambda \qquad (28.40)$$

Location of dark fringe for an air wedge
$$x_m = \frac{m\lambda l}{2h} \qquad (28.43)$$

Separation between dark fringes
$$\Delta x = \frac{\lambda l}{2h} \qquad (28.44)$$

Condition for bright fringe for an air wedge
$$2d = (2m - 1)\frac{\lambda}{2} \qquad (28.45)$$

Location of bright fringe for an air wedge
$$x_m = (2m - 1)\frac{\lambda l}{4h} \qquad (28.46)$$

Radius of mth fringe for Newton's rings
$$r_m = \sqrt{m\lambda R} \qquad (28.48)$$

Condition for dark fringe for single slit diffraction
$$a \sin \theta = m\lambda \qquad (28.51)$$

Condition for bright fringe for diffraction grating
$$d \sin \theta = m\lambda \qquad (28.52)$$

Questions for Chapter 28

1. Why do we use monochromatic light rather than white light when doing experiments with interference and diffraction?
2. How can you hear a radio station that is on the other side of the mountain from you?
3. Two separate light sources, such as the headlights of your car, do not cause interference. Why not?

4. What effect does the medium have on interference fringes for (a) light and (b) sound?
5. When destructive interference occurs, what happens to the energy in the light waves?
6. What effect does changing the slit width and the slit separation have on the diffraction pattern?
7. Why don't you observe interference from a thick film, such as an ordinary piece of window glass?

8. Using an audio oscillator, can you set up a double-slit experiment with sound waves? What would the approximate size of the slits have to be? What would you hear on the far side of the double slit?
9. Why are diffraction gratings used more than prisms in spectroscopy?
†10. How is the resolving power of a telescope, the smallest resolution of an object that can be determined by the telescope, affected by diffraction?

Problems for Chapter 28

28.2 The Interference of Light— Young's Double-Slit Experiment

1. In a Young's double-slit experiment, a screen is placed 7.00 m behind a double slit of 0.200-mm separation. If light of 589.0 nm shines on the slit, find the value of the angle, θ, corresponding to the first 3 bright fringes.

2. In a Young's double-slit experiment, a screen is placed 5.00 m behind a double slit of 0.250-mm separation. If light of 589.0-nm wavelength shines on the slit, find the value of θ corresponding to the first three dark fringes.
3. In a Young's double-slit experiment, a screen is placed 7.00 m behind a double slit of 0.400-mm separation. If light of 589.0-nm wavelength shines on the slit, find the value of y on the screen corresponding to the first three bright fringes.

4. In a Young's double-slit experiment, a screen is placed 6.00 m behind a double slit of 0.387-mm separation. If light of 589.0-nm wavelength shines on the slit, find the value of y on the screen corresponding to the first three dark fringes.

†5. In a Young's double-slit experiment, a screen is placed 7.00 m behind a double slit of 0.200-mm separation. If light of 589.0-nm wavelength shines on the slit, find (a) the value of the angle θ corresponding to the first three bright fringes, (b) the value of θ corresponding to the first three dark fringes, and (c) the value of y on the screen corresponding to the first three bright and dark fringes.

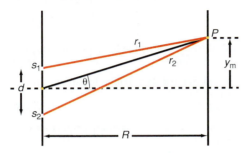

6. In a double-slit experiment the location of the second bright fringe on the screen is found to be at 12.5 cm. If the distance to the screen is 7.00 m and the separation between slits is 0.176 mm, find the wavelength of the light used.

7. Light of wavelength 600 nm is incident on a double slit whose separation is 0.0480 m. A bright fringe is observed at angular displacement $\theta = 0.0200°$. What is its order (i.e., the value of the integer m)?

8. Adjacent bright fringes from a double slit separated by 0.700 mm are measured to be 1.71 mm apart on a screen 1.00 m away. What is the wavelength of light used?

9. A Young's double-slit experiment is performed with a slit separation of 0.0230 m. A second-order bright fringe is observed at 0.200 mm from the center of the bright central fringe. If the wavelength of the light is 490 nm, find the distance between the double slit and the screen on which the fringes were observed.

10. Show that the distance between bright fringes in a double-slit experiment is equal to the distance between dark fringes.

11. What is the distance between fringes in a double-slit experiment when the slit separation is (a) doubled and (b) halved?

28.3 The Interference of Light—The Michelson Interferometer

12. A Michelson interferometer is used to determine the wavelength of a source of light. If the movable mirror moves a distance of 0.250 mm when 800 fringes move across the telescope cross hair, find the wavelength of the light source.

13. A length of 5.00 cm is to be measured very accurately. If a Michelson interferometer is to be used to measure this length, how many fringes of monochromatic light of 589.0-nm wavelengths must pass the cross hair of the telescope to correspond to this length?

14. How far must the movable mirror of the Michelson interferometer be moved in order to observe 900 fringes of light of 589.0-nm wavelength?

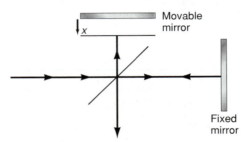

†15. A glass tube, 10.0 cm long, is used as one arm of the Michelson interferometer and the standard interference pattern is obtained using a wavelength of light of 589 nm. Water is now slowly introduced into the glass tube. When the tube is full, how many fringes will have passed through the cross hair of the eyepiece?

28.4 Interference—Thin Films

16. A light wave of 700 nm passes from air into glass. Find the wavelength of the light in the glass.

17. A light wave has a wavelength of 450 nm in water. Find its wavelength when it enters air.

18. A light wave of 589-nm wavelength in air passes from air into water and then into glass. Find the wavelength of the light in the water and the glass.

19. A wave of light will be reflected and transmitted at the interfaces of the air-glass-water-air surfaces, shown in the diagram. If the thickness of each surface is negligible, find the phase relations between waves 1, 2, 3, and 4.

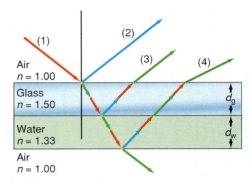

20. Repeat problem 19 but let the thickness of the water surface equal the thickness of the glass surface, which is equal to half of the incident wavelength of the light in air.

21. Repeat problem 19 but let the thickness of the water surface equal the thickness of the glass surface, which is equal to the incident wavelength of the light in air.

22. Repeat problem 19 but let the thickness of the water surface equal half of the incident wavelength of the light in air and the thickness of the glass surface be equal to the incident wavelength of the light in air.

23. A soap bubble is 200.0 nm thick. For what wavelength of light will (a) the intensity of the reflected light be zero and (b) the intensity of the reflected light be maximum?

24. A soap bubble is 300.0 nm thick. If white light shines on it, what color will it appear in reflected light?

25. The wavelength of the maximum reflected light from a soap bubble is 710 nm. Find the thickness of the soap bubble.

26. What thickness of an oil film ($n = 1.40$) will give complete destructive interference for the reflected light of 589.0-nm wavelength? For this thickness is there any wavelength that will be completely reflected?

27. A thin film of oil of 100-nm thickness lies on a surface. Maximum reflection from the oil occurs for a wavelength in air of 650 nm. Find the index of refraction of the oil.

28. We want to prevent any light of 700.0-nm wavelength from being reflected from a lens. What thickness of magnesium fluoride should be deposited on this lens? For this thickness is there any wavelength of light that gives maximum reflection?

29. If 4% of the incident light is lost by reflection at each lens of a multiple lens optical system, how much light is lost if the system contains six lenses?

30. An air wedge is formed by placing a sheet of paper between the edges of two glass plates 10.0 cm long. If 15 bright fringes are observed per cm, when light of 589.0-nm wavelength shines on the glasses, what is the thickness of the paper?

31. A hair is placed at the edge between two 10.0-cm glass plates of $n = 1.52$. Light of 400-nm wavelength shines on the glass and an interference pattern is formed. If the separation between the second and fourth dark fringe is measured to be 0.800 mm, find the diameter of the hair.

32. Repeat problem 31 but use fused quartz glass ($n = 1.46$) and the glass is immersed in glycerine ($n = 1.47$).

33. Show that the separation distance between bright fringes in an air wedge is equal to the separation distance between dark fringes.

34. Find the equation for the width of a fringe in the interference of light from an air wedge.

35. Newton's rings are used to determine the wavelength of light. If the radius of the fifth fringe is 7.00 mm and the radius of curvature of the lens is measured by a spherometer to be 10.0 m, find the wavelength of light.

36. The radius of the tenth fringe in Newton's rings is 10.0 mm. If the wavelength of light used is 589.0 nm, what is the radius of curvature of the lens?

28.5 Diffraction from a Single Slit

37. In the diffraction from a single slit of width, $a = 15\lambda$, find the angle θ at which the second dark fringe occurs.

38. The first dark fringe for a single slit is found at an angle θ of 5.74° for light of 589.0-nm wavelength. Find the value of the angle θ if the slit width is halved.

39. In a single-slit experiment, the first dark fringe occurs at $\theta = 0.010°$. If the slit is 0.035 m wide, find the wavelength of the light.

40. A source of light of $\lambda = 633$ nm shines on a single slit and a diffraction pattern is found on a wall 1.00 m away. The first dark fringe is found at $x = 5.00$ cm from the central maximum. Find the width of the slit.

41. In a single-slit experiment, a slit 0.0305 m wide receives light of wavelength 640 nm. Find the order of the dark fringe at the angular displacement $\theta = 0.600°$.

28.6 The Diffraction Grating

42. A diffraction grating has 800 lines per mm. Find the distance between each slit.

43. A diffraction grating has 800 lines per mm. If monochromatic light of 589.0-nm wavelength shines on the grating, where will the first three bright fringes be found?

44. How many bright fringes will occur on either side of the central maximum for a diffraction grating of 600 lines per mm if the wavelength of light is 589.0 nm?

45. The first bright fringe of a diffraction grating of 600 lines/mm, is found at an angle of 21.0°. Find the wavelength of light of the source.

46. A first dark fringe is observed at angular displacement $\theta = 15.0°$ due to light of frequency 6.00×10^{14} Hz passing through a diffraction grating. Find the separation of the slits in the grating.

47. Find the longest wavelength that can be observed in second order for a diffraction grating of 800 lines/mm.

48. Using light of 589.0-nm wavelength, a diffraction grating produces a first-order fringe at 44.9°. Find the number of lines/mm of this grating.

Additional Problems

†49. A double-slit experiment is performed with the light source, slits, and screen placed under water. If the separation of the double slit is 0.350 mm, the wavelength of light in air is 589.0 nm, and the screen is 2.00 m away from the slit, find the value of y on the screen corresponding to the third bright fringe.

†50. In a double-slit experiment the intensity of the light on the screen varies from zero at a dark fringe to I_0 at the center of a bright fringe. Find the width of a bright fringe assuming that a bright fringe exists for an intensity of $I \geq 0$.

†51. A piece of flint glass ($n = 1.57$) is being ground into a planoconvex lens. The focal length of the lens is to be 25.0 cm. The radius of curvature of the plane side of the glass is infinite. (a) Find the radius R_1 of the convex side. Newton's rings will be used to verify that the lens has been ground to the correct radius. The lens is placed on a flat piece of glass, and light of 589-nm wavelength illuminates the lens and the characteristic pattern of Newton's rings is obtained. (b) What must be the radius of the fifth fringe such that the lens has the correct focal length?

†52. When two objects are too close together the diffraction pattern of each object can overlap and hence the two objects cannot be seen distinctly. The smallest angular distance between two objects that can be seen distinctly with an optical system is called the resolution of the optical system. The smallest angular resolution is determined by the distance between the central maximum and the first intensity minimum (dark fringe) of the diffraction pattern of the single slit. (Then the two diffraction patterns will not overlap and each object is distinct.) For the single slit discussed in this chapter, the angular separation between the central bright fringe and the dark fringe was determined from $\sin \theta = \lambda/a$. For a circular aperture the result can be shown to be $\sin \theta = 1.22 \lambda/a$. What is the range of the angular resolution of a human eye if the pupil diameter can vary between 2 and 6 mm, and the wavelength of light used is 550 nm?

53. If the human eye can resolve two objects when the angle between them is 1 minute of arc, how far in advance can you detect two headlights of a car if they are separated by a distance of 1.32 m?

54. A diffraction grating of 1000 lines per cm is to be used for the entire spectrum of visible light. (a) Find the angle of diffraction for the first-order red light (720.0 nm) and first-order blue light (380.0 nm). (b) Find the angular separation. (c) If the screen is placed 1.00 m away, find the linear separation.

55. Calculate the angular separation between the first-order fringes of the two yellow lines in the sodium spectrum, having wavelengths of 589.0 nm and 589.6 nm, for a diffraction grating of 1200 lines/mm.

Interactive Tutorials

56. Young's double-slit intensity. In a double-slit experiment, light of wavelength $\lambda = 589.0$ nm impinges on a double slit that has a separation $d = 0.285$ mm. The screen is placed at a distance $R = 3.50$ m from the double slit. Find the intensity distribution on the screen as a function of the angle θ in figure 28.5.

57. A double slit. In a double-slit experiment, light of wavelength $\lambda = 589.0$ nm impinges on a double slit that has a separation $d = 0.285$ mm. The screen is placed at the distance $R = 3.50$ m from the double slit. Find (a) the value of θ corresponding to the $m = 3$ bright fringe; (b) the value of y locating the *bright* fringe on the screen, as seen in figure 28.5; (c) the value of θ corresponding to the $m = 3$ dark fringe; and (d) the value of y locating the *dark* fringe on the screen.

58. A soap bubble of thickness $d = 300$ nm is illuminated by white light. If the index of refraction of the soap film is $n = 1.28$, calculate (a) the first three wavelengths (lambda) of the incident light that will give a dark fringe and (b) the first three wavelengths of the incident light that will give a bright fringe.

59. A diffraction grating. Monochromatic light of wavelength $\lambda = 589.0$ nm shines on a diffraction grating that has 400 lines per mm. Find (a) the distance d between slits and (b) the value of θ corresponding to the $m = 3$ bright fringe.

Modern Physics

Color as a Study Aid
Chapters 29–30
The following color code is used in these chapters:

- Stationary frames S
- Moving frames S'
- Velocity vectors with respect to S frame
- Velocity vectors with respect to S' frame
- Distance with respect to S frame
- Distance with respect to S' frame
- World lines in S
- World lines in S'
- Light lines
- Light cone
- Hyperbolas of invariant interval
- Proper length
- Contracted length
- Proper time interval
- Time dilation interval

Chapter 31
The following color code is used in this chapter:

- Velocity vector
- Wave and wave envelope
- Matter wave
- Momentum vector
- Real particle
- Virtual particle
- Photon

Chapter 32
The following color code is used in this chapter:

- Angular momentum
- Component of angular momentum vector
- Matter wave
- Standing waves
- Energy levels
- Magnetic dipoles
- Magnetic fields
- Positive charges +
- Negative charges -

Chapter 33
The following color code is used in this chapter:

- Velocity vector
- Applied force vector
- Nucleus
- Nucleus fragments
- Neutrons
- Path of α particle
- α decay lines
- Path of β particle
- β decay lines
- Path of γ particle
- γ decay lines

Chapter 34
The following color code is used in this chapter (all positive charges are still red and negative charges are blue):

- Red quarks
- Green quarks
- Blue quarks

29 Special Relativity

A synchrotron at Fermilab.

And now, in our time, there has been unloosed a cataclysm which has swept away space, time, and matter hitherto regarded as the firmest pillars of natural science, but only to make place for a view of things of wider scope, and entailing a deeper vision. This revolution was promoted essentially by the thought of one man, Albert Einstein.

Hermann Weyl—Space-Time-Matter

29.1 Introduction to Relative Motion

Relativity has as its basis the observation of the motion of a body by two different observers in relative motion to each other. This observation, apparently innocuous when dealing with motions at low speeds has a revolutionary effect when the objects are moving at speeds near the velocity of light. At these high speeds, it becomes clear that the simple concepts of space and time studied in Newtonian physics no longer apply. Instead, *there becomes a fusion of space and time into one physical entity called spacetime. All physical events occur in the arena of spacetime. As we shall see, the normal Euclidean geometry, studied in high school, that applies to everyday objects in space does not apply to spacetime. That is, spacetime is non-Euclidean. The apparently strange effects of relativity, such as length contraction and time dilation, come as a result of this non-Euclidean geometry of spacetime.*

The earliest description of relative motion started with Aristotle who said that the earth was at absolute rest in the center of the universe and everything else moved relative to the earth. As a proof that the earth was at absolute rest, he reasoned that if you throw a rock straight upward it will fall back to the same place from which it was thrown. If the earth moved, then the rock would be displaced on landing by the amount that the earth moved. This is shown in figures 29.1(a) and 29.1(b).

Based on the prestige of Aristotle, the belief that the earth was at absolute rest was maintained until Galileo Galilee (1564–1642) pointed out the error in Aristotle's reasoning. Galileo suggested that if you throw a rock straight upward in a boat that is moving at constant velocity, then, as viewed from the boat, the rock goes straight up and straight down, as shown in figure 29.2(a). If the same projectile motion is observed from the shore, however, the rock is seen to be displaced to the right of the vertical path. The rock comes down to the same place on the boat only because the boat is also moving toward the right. Hence, to the observer on the boat, the rock went straight up and straight down and by Aristotle's reasoning the boat must be at rest. But as the observer on the shore will clearly state, the boat was not at rest but moving with a velocity **v.** Thus, Aristotle's argument is not valid. *The distinction between rest and motion at a constant velocity, is relative to the observer.* The observer on the boat says the boat is at rest while the observer on the shore says the boat is in motion. We then must ask, is there any way to distinguish between a state of rest and a state of motion at constant velocity?

Let us consider Newton's second law of motion studied in chapter 4,

$$\mathbf{F} = m\mathbf{a} \tag{4.9}$$

Figure 29.1

Aristotle's argument for the earth's being at rest.

(a) Earth at rest　　　　(b) Earth in motion

Displacement

Figure 29.2

Galileo's rebuttal of Aristotle's argument of absolute rest.

(a) Motion viewed from boat　　　(b) Motion viewed from shore

If the unbalanced external force acting on the body is zero, then the acceleration is also zero. But since $\mathbf{a} = \Delta\mathbf{v}/\Delta t$, this implies that there is no change in velocity of the body, and the velocity is constant. *We are capable of feeling forces and accelerations but we do not feel motion at constant velocity, and rest is the special case of zero constant velocity.* Recall from chapter 4, in section 4.5 concerning the weight of a person in an elevator, we saw that the scales read the same numerical value for the weight of the person (192 lb) when the elevator was either at rest or moving at a constant velocity. There is no way for the passenger to say he or she is at rest or moving at a constant velocity unless he or she can somehow look out of the elevator and see motion. When the elevator accelerated upward, on the other hand, the person experienced a force of 222 lb pushing upward on him. When the elevator accelerated downward, the person experienced a force of only 162 lb. Thus, accelerations are easily felt but not constant velocities. Only if the elevator accelerates can the passenger tell that he or she is in motion. While you sit there reading this sentence you are sitting on the earth, which is moving around the sun at about 30 km/s, yet you do not notice this motion.[1] When a person sits in a plane or a train moving at constant velocity, the motion is not sensed unless the person looks out the window. The person senses his or her motion only while the plane or train is accelerating.

　　　Since relative motion depends on the observer, there are many different ways to observe the same motion. For example, figure 29.3(a) shows body 1 at rest while body 2 moves to the right with a velocity **v.** But from the point of view of body 2, he can equally well say that it is he who is at rest and it is body 1 that is moving to the left with the velocity −**v,** figure 29.3(b). Or an arbitrary observer can be placed at rest between bodies 1 and 2, as shown in figure 29.3(c), and she will observe body 2 moving to the right with a velocity **v**/2 and body 1 moving to the left with a velocity

1. Actually the earth's motion around the sun constitutes an accelerated motion. The average centripetal acceleration is $a_c = v^2/r = (29.7 \times 10^3 \text{ m/s})^2/(1.5 \times 10^{11} \text{ m}) = 5.88 \times 10^{-3} \text{ m/s}^2 = 0.0059 \text{ m/s}^2$. This orbital acceleration is so small compared to the acceleration of gravity, 9.80 m/s², that we do not feel it and it can be ignored. Hence, we feel as though we were moving at constant velocity.

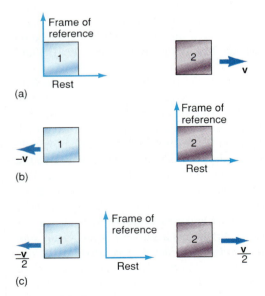

Frame of reference

1

Rest

2

v

(a)

1

−**v**

2

Rest

(b)

1

$-\dfrac{\mathbf{v}}{2}$

Frame of reference

Rest

2

$\dfrac{\mathbf{v}}{2}$

(c)

Figure 29.3

Relative motion.

of $-\mathbf{v}/2$. We can also conceive of the case of body 1 moving to the right with a velocity $\mathbf{v}$ and body 2 moving to the right with a velocity $2\mathbf{v}$, the relative velocities between the two bodies still being $\mathbf{v}$ to the right. Obviously an infinite number of such possible cases can be thought out. Therefore, we must conclude that, *if a body in motion at constant velocity is indistinguishable from a body at rest, then there is no reason why a state of rest should be called a state of rest, or a state of motion a state of motion.* Either body can be considered to be at rest while the other body is moving in the opposite direction with the speed v.

To describe the motion, we place a coordinate system at some point, either in the body or outside of it, and call this coordinate system a frame of reference. The motion of any body is then made with respect to this frame of reference. *A frame of reference that is either at rest or moving at a constant velocity is called an inertial frame of reference* or an **inertial coordinate system.** Newton's first law defines the inertial frame of reference. That is, when $\mathbf{F} = 0$, and the body is either at rest or moving uniformly in a straight line, then the body is in an inertial frame. There are an infinite number of inertial frames and Newton's second law, in the form $\mathbf{F} = m\mathbf{a}$, holds in all these inertial frames.

An example of a noninertial frame is an accelerated frame, and one is shown in figure 29.4. A rock is thrown straight up in a boat that is accelerating to the right. An observer on the shore sees the projectile motion as in figure 29.4(a). The observed motion of the projectile is the same as in figure 29.2(b), but now the observer on the shore sees the rock fall into the water behind the boat rather than back onto the same point on the boat from which the rock was launched. Because the boat has accelerated while the rock is in the air, the boat has a constantly increasing velocity while the horizontal component of the rock remains a constant. Thus the boat moves out from beneath the rock and when the rock returns to where the boat should be, the boat is no longer there. When the same motion is observed from the boat, the rock does not go straight up and straight down as in figure 29.2(a), but instead the rock appears to move backward toward the end of the boat as though there was a force pushing it backward. The boat observer sees the rock fall into the water behind the boat, figure 29.4(b). In this accelerated reference frame of the boat, there seems to be a force acting on the rock pushing it backward. Hence, Newton's second law, in the form $\mathbf{F} = m\mathbf{a}$, does not work on this accelerated boat. Instead a fictitious force must be introduced to account for the backward motion of the projectile.

For the moment, we will restrict ourselves to motion as observed from inertial frames of reference, the subject matter of the special or restricted theory of relativity. In chapter 30, we will discuss accelerated frames of reference, the subject matter of general relativity.

Figure 29.4

A linearly accelerated frame of reference.

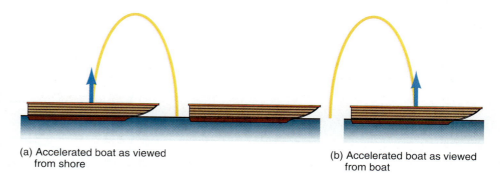

(a) Accelerated boat as viewed
　　from shore

(b) Accelerated boat as viewed
　　from boat

29.2 The Galilean Transformations of Classical Physics

The description of any type of motion in classical mechanics starts with an inertial coordinate system S, which is considered to be at rest. Let us consider the occurrence of some "event" that is observed in the S frame of reference, as shown in figure 29.5. The event might be the explosion of a firecracker or the lighting of a match, or just the location of a body at a particular instance of time. For simplicity, we will assume that the event occurs in the x,y plane. The event is located a distance r from the origin O of the S frame. The coordinates of the point in the S frame, are x and y. A second coordinate system S', moving at the constant velocity $\mathbf{v}$ in the positive x-direction, is also introduced. The same event can also be described in terms of this frame of reference. The event is located at a distance r' from the origin O' of the S' frame of reference and has coordinates x' and y', as shown in the figure. We assume that the two coordinate systems had their origins at the same place at the time, $t = 0$. At a later time t, the S' frame will have moved a distance, $d = vt$, along the x-axis. The x-component of the event in the S frame is related to the x'-component of the same event in the S' frame by

$$x = x' + vt \tag{29.1}$$

which can be easily seen in figure 29.5, and the y- and y'-components are seen to be

$$y = y' \tag{29.2}$$

Notice that because of the initial assumption, z and z' are also equal, that is

$$z = z' \tag{29.3}$$

It is also assumed, but usually never stated, that the time is the same in both frames of reference, that is,

$$t = t' \tag{29.4}$$

*These equations, that describe the event from either inertial coordinate system, are called the **Galilean transformations** of classical mechanics and they are summarized as*

$$x = x' + vt \tag{29.1}$$

$$y = y' \tag{29.2}$$

$$z = z' \tag{29.3}$$

$$t = t' \tag{29.4}$$

Figure 29.5

Inertial coordinate systems.

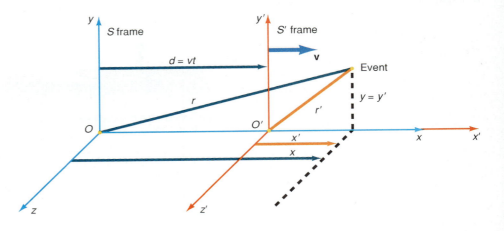

The inverse transformations from the S frame to the S' frame are

$$x' = x - vt \qquad (29.5)$$

$$y' = y \qquad (29.6)$$

$$z' = z \qquad (29.7)$$

$$t' = t \qquad (29.8)$$

Example 29.1

The Galilean transformation of distances. A student is sitting on a train 10.0 m from the rear of the car. The train is moving to the right at a speed of 4.00 m/s. If the rear of the car passes the end of the platform at $t = 0$, how far away from the platform is the student at 5.00 s?

Solution

The picture of the student, the train, and the platform is shown in figure 29.6. The platform represents the stationary S frame, whereas the train represents the moving S' frame. The location of the student, as observed from the platform, found from equation 29.1, is

$$x = x' + vt$$
$$= 10.0 \text{ m} + (4.00 \text{ m/s})(5.00 \text{ s})$$
$$= 30 \text{ m}$$

The speed of an object in either frame can also be easily found. If the body moves from x_1' to x_2' in the moving coordinate system in the time Δt, it moves from x_1 to x_2 in the stationary coordinate system in the same time Δt. Hence, the distance moved is

$$\Delta x = x_2 - x_1 = x_2' + vt_2 - x_1' - vt_1$$
$$= x_2' - x_1' + v(t_2 - t_1)$$

and

$$\Delta x = \Delta x' + v\Delta t \qquad (29.9)$$

If we divide both sides of equation 29.9 by Δt, we have

$$\frac{\Delta x}{\Delta t} = \frac{\Delta x'}{\Delta t} + v\frac{\Delta t}{\Delta t} \qquad (29.10)$$

Figure 29.6

An example of the Galilean transformations.

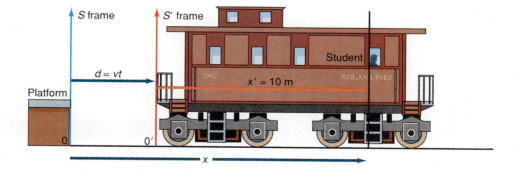

But $\Delta x / \Delta t = v_x$, the x-component of the velocity of the body in the stationary frame S, and $\Delta x' / \Delta t = v_x'$, the x-component of the velocity in the moving frame S'. Thus equation 29.10 becomes

$$v_x = v_x' + v \tag{29.11}$$

Equation 29.11 is a statement of the Galilean addition of velocities.

Example 29.2

The Galilean transformation of velocities. The student on the train of example 29.1, gets up and starts to walk. What is the student's speed relative to the platform if (a) the student walks toward the front of the train at a speed of 2.00 m/s and (b) the student walks toward the back of the train at a speed of 2.00 m/s?

Solution

a. The speed of the student relative to the stationary platform, found from equation 29.11, is

$$v_x = v_x' + v = 2.00 \text{ m/s} + 4.00 \text{ m/s}$$
$$= 6.00 \text{ m/s}$$

b. If the student walks toward the back of the train $\Delta x' = x_2' - x_1'$ is negative because x_1' is greater than x_2', and hence, v_x' is a negative quantity. Therefore,

$$v_x = v_x' + v$$
$$= -2.00 \text{ m/s} + 4.00 \text{ m/s}$$
$$= 2.00 \text{ m/s}$$

If there is more than one body in motion with respect to the stationary frame, the relative velocity between the two bodies is found by placing the S' frame on one of the bodies in motion. That is, if body A is moving with a velocity v_{AS} with respect to the stationary frame S, and body B is moving with a velocity v_{BS}, also with respect to the stationary frame S, the velocity of A as observed from B, v_{AB}, is simply

$$v_{AB} = v_{AS} - v_{BS} \tag{29.12}$$

as seen in figure 29.7(a).

If we place the moving frame of reference S' on body B, as in figure 29.7(b), then $v_{BS} = v$, the velocity of the S' frame. The velocity of the body A with respect to S, v_{AS}, is now set equal to v_x, the velocity of the body with respect to the S frame. The relative velocity of body A with respect to body B, v_{AB}, is now v_x', the velocity of the body with respect to the moving frame of reference S'. Hence the velocity v_x' of the moving body with respect to the moving frame is determined from equation 29.12 as

$$v_x' = v_x - v \tag{29.13}$$

Note that equation 29.13 is the inverse of equation 29.11.

Example 29.3

Relative velocity. A car is traveling at a velocity of 95.0 km/hr to the right, with respect to a telephone pole. A truck, which is behind the car, is also moving to the right at 65.0 km/hr with respect to the same telephone pole. Find the relative velocity of the car with respect to the truck.

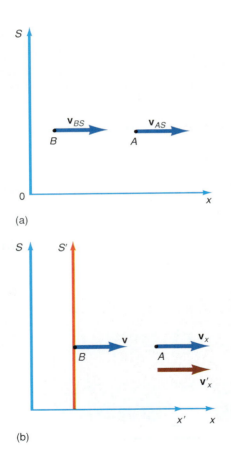

(a)

(b)

Figure 29.7

Relative velocities.

We represent the telephone pole as the stationary frame of reference S, while we place the moving frame of reference S' on the truck that is moving at a speed $v = 65.0$ km/hr. The auto is moving at the speed $v = 95.0$ km/hr with respect to S. The velocity of the auto with respect to the truck (or S' frame) is v_x' and is found from equation 29.13 as

$$v_x' = v_x - v$$
$$= 95.0 \text{ km/hr} - 65.0 \text{ km/hr}$$
$$= 30.0 \text{ km/hr}$$

The relative velocity is $+30.0$ km/hr. This means that the auto is pulling away or separating from the truck at the rate of 30.0 km/hr. If the auto were moving toward the S observer instead of away, then the auto's velocity with respect to S' would have been

$$v_x' = v_x - v = -95.0 \text{ km/hr} - 65.0 \text{ km/hr}$$
$$= -160.0 \text{ km/hr}$$

That is, the truck would then observe the auto approaching at a closing speed of -160 km/hr. Note that when the relative velocity v_x' is positive the two moving objects are separating, whereas when v_x' is negative the two objects are closing or coming toward each other.

To complete the velocity transformation equations, we use the fact that $y = y'$ and z and z', thereby giving us

$$v_y' = \frac{\Delta y'}{\Delta t} = \frac{\Delta y}{\Delta t} = v_y \qquad (29.14)$$

and

$$v_z' = \frac{\Delta z'}{\Delta t} = \frac{\Delta z}{\Delta t} = v_z \qquad (29.15)$$

The Galilean transformations of velocities can be summarized as:

$$v_x = v_x' + v \qquad (29.11)$$
$$v_x' = v_x - v \qquad (29.13)$$
$$v_y' = v_y \qquad (29.14)$$
$$v_z' = v_z \qquad (29.15)$$

29.3 The Invariance of the Mechanical Laws of Physics under a Galilean Transformation

Although the velocity of a moving object is different when observed from a stationary frame rather than a moving frame of reference, the acceleration of the body is the same in either reference frame. To see this, let us start with equation 29.13,

$$v_x' = v_x - v$$

The change in each term with time is

$$\frac{\Delta v_x'}{\Delta t} = \frac{\Delta v_x}{\Delta t} - \frac{\Delta v}{\Delta t} \qquad (29.16)$$

But v is the speed of the moving frame, which is a constant and does not change with time. Hence, $\Delta v/\Delta t = 0$. The term $\Delta v'_x/\Delta t = a'_x$ is the acceleration of the body with respect to the moving frame, whereas $\Delta v_x/\Delta t = a_x$ is the acceleration of the body with respect to the stationary frame. Therefore, equation 29.16 becomes

$$a'_x = a_x \qquad (29.17)$$

Equation 29.17 says that the acceleration of a moving body is invariant under a Galilean transformation. The word invariant when applied to a physical quantity means that the quantity remains a constant. We say that the acceleration is an **invariant quantity.** This means that either the moving or stationary observer would measure the same numerical value for the acceleration of the body.

If we multiply both sides of equation 29.17 by m, we get

$$ma'_x = ma_x$$

But the product of the mass and the acceleration is equal to the force F, by Newton's second law. Hence,

$$F' = F \qquad (29.18)$$

Thus, Newton's second law is also invariant to a Galilean transformation and applies to all inertial observers.

The laws of conservation of momentum and conservation of energy are also invariant under a Galilean transformation. We can see this for the case of the perfectly elastic collision illustrated in figure 29.8. We can write the law of conservation of momentum for the collision, as observed in the S frame, as

$$m_1v_1 + m_2v_2 = m_1V_1 + m_2V_2 \qquad (29.19)$$

Figure 29.8

A perfectly elastic collision as seen from two inertial frames.

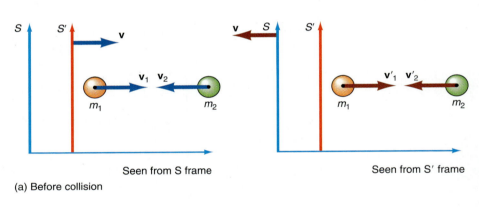

Seen from S frame   Seen from S' frame

(a) Before collision

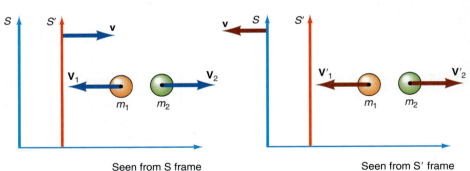

Seen from S frame   Seen from S' frame

(b) After collision

where v_1 is the velocity of ball 1 before the collision, v_2 is the velocity of ball 2 before the collision, V_1 is the velocity of ball 1 after the collision, and V_2 is the velocity of ball 2 after the collision. But the relation between the velocity in the S and S' frames, found from equation 29.11 and figure 29.8, is

$$\left.\begin{array}{c} v_1 = v_1' + v \\ v_2 = v_2' + v \\ V_1 = V_1' + v \\ V_2 = V_2' + v \end{array}\right\} \tag{29.20}$$

Substituting equations 29.20 into equation 29.19 for the law of conservation of momentum yields

$$m_1 v_1' + m_2 v_2' = m_1 V_1' + m_2 V_2' \tag{29.21}$$

Equation 29.21 is the law of conservation of momentum as observed from the moving S' frame. Note that it is of the same form as the law of conservation of momentum as observed from the S or stationary frame of reference. Thus, *the law of conservation of momentum is invariant to a Galilean transformation.*

The law of conservation of energy for the perfectly elastic collision of figure 29.8 as viewed from the S frame is

$$\tfrac{1}{2}m_1 v_1^2 + \tfrac{1}{2}m_2 v_2^2 = \tfrac{1}{2}m_1 V_1^2 + \tfrac{1}{2}m_2 V_2^2 \tag{29.22}$$

By replacing the velocities in equation 29.22 by their Galilean counterparts, equation 29.20, and after much algebra we find that

$$\tfrac{1}{2}m_1 v_1'^2 + \tfrac{1}{2}m_2 v_2'^2 = \tfrac{1}{2}m_1 V_1'^2 + \tfrac{1}{2}m_2 V_2'^2 \tag{29.23}$$

Equation 29.23 is the law of conservation of energy as observed by an observer in the moving S' frame of reference. Note again that the form of the equation is the same as in the stationary frame, and hence, *the law of conservation of energy is invariant to a Galilean transformation.* If we continued in this manner we would prove that all the laws of mechanics are invariant to a Galilean transformation.

29.4 Electromagnetism and the Ether

We have just seen that the laws of mechanics are invariant to a Galilean transformation. Are the laws of electromagnetism also invariant?

Consider a spherical electromagnetic wave propagating with a speed c with respect to a stationary frame of reference, as shown in figure 29.9. The speed of this electromagnetic wave is

$$c = \frac{r}{t}$$

where r is the distance from the source of the wave to the spherical wave front. We can rewrite this as

$$r = ct$$

or

$$r^2 = c^2 t^2$$

or

$$r^2 - c^2 t^2 = 0 \tag{29.24}$$

The radius r of the spherical wave is

$$r^2 = x^2 + y^2 + z^2$$

Figure 29.9

A spherical electromagnetic wave.

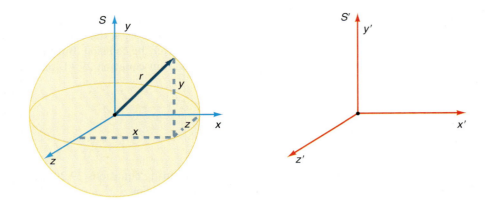

Substituting this into equation 29.24, gives

$$x^2 + y^2 + z^2 - c^2t^2 = 0 \qquad (29.25)$$

for the light wave as observed in the S frame of reference. Let us now assume that another observer, moving at the speed v in a moving frame of reference S' also observes this same light wave. The S' observer observes the coordinates x' and t', which are related to the x and t coordinates by the Galilean transformation equations as

$$x = x' + vt \qquad (29.1)$$

$$y = y' \qquad (29.2)$$

$$z = z' \qquad (29.3)$$

$$t = t' \qquad (29.4)$$

Substituting these Galilean transformations into equation 29.25 gives

$$(x' + vt)^2 + y'^2 + z'^2 - c^2t'^2 = 0$$

$$x'^2 + 2x'vt + v^2t^2 + y'^2 + z'^2 - c^2t'^2 = 0$$

or

$$x'^2 + y'^2 + z'^2 - c^2t'^2 = -2x'vt - v^2t^2 \qquad (29.26)$$

Notice that the form of the equation is not invariant to a Galilean transformation. That is, equation 29.26, the velocity of the light wave as observed in the S' frame, has a different form than equation 29.25, the velocity of light in the S frame. Something is very wrong either with the equations of electromagnetism or with the Galilean transformations. Einstein was so filled with the beauty of the unifying effects of Maxwell's equations of electromagnetism that he felt that there must be something wrong with the Galilean transformation and hence, a new transformation law was required.

A further difficulty associated with the electromagnetic waves of Maxwell was the medium in which these waves propagated. Recall from chapter 12 on wave motion, that a wave is a disturbance that propagates through a medium. When a rock, the disturbance, is dropped into a pond, a wave propagates through the water, the medium. Associated with a transverse wave on a string is the motion of the particles of the string executing simple harmonic motion perpendicular to the direction of the wave propagation. In this case, the medium is the particles of the string. A sound wave in air is a disturbance propagated through the medium air. In fact, when we say that a sound wave propagates through the air with a velocity of 330 m/s at 0 °C, we mean that the wave is moving at 330 m/s with respect to the air. A sound wave in water propagates through the water while a sound wave in a solid propagates through the solid. The one thing that all of these waves have in common is that they are all propagated through some medium. Classical physicists then naturally asked, "Through what medium does light propagate?" According to everything that was known in the field of physics in the nineteenth century, a wave must propagate through some medium. Therefore, it was reasonable to expect that a light wave, like any other wave, must propagate through some medium. This medium was called the luminiferous ether or just **ether** for short. It was assumed that this ether filled all of space, the inside of all material bodies, and was responsible for the transmission of all electromagnetic vibrations. Maxwell assumed that his electromagnetic waves propagated through this ether at the speed $c = 3 \times 10^8$ m/s.

An additional reason for the assumption of the existence of the ether was the phenomena of interference and diffraction of light that implied that light must be a wave. If light is an electromagnetic wave, then it is waving through the medium called ether.

There are, however, two disturbing characteristics of this ether. First, the ether had to have some very strange properties. The ether had to be very strong or rigid in order to support the extremely large speed of light. Recall from chapter 12 on wave motion that the speed of sound at 0 °C is 330 m/s in air, 1520 m/s in water, and 3420 m/s in iron. Thus, the more rigid the medium the higher the velocity of the wave. Similarly, we saw that for a transverse wave on a taut string the speed of propagation is

$$v = \sqrt{\frac{T}{m/l}} \qquad \text{(12.30)}$$

where T is the tension in the string. The greater the value of T, the greater the value of the speed of propagation. Greater tension in the string implies a more rigid string. Although a light wave is neither a sound wave nor a wave on a string, it is reasonable to assume that the ether, being a medium for propagation of an electromagnetic wave, should also be quite rigid in order to support the enormous speed of 3×10^8 m/s. Yet the earth moves through this rigid medium at an orbital speed of 3×10^4 m/s and its motion is not impeded one iota by this rigid medium. This is very strange indeed.

The second disturbing characteristic of this ether hypothesis is that if electromagnetic waves always move at a speed c with respect to the ether, then maybe the ether constitutes an absolute frame of reference that we have not been able to find up to now. *Newton postulated an absolute space and an absolute time in his Principia: "Absolute space, in its own nature without regard to anything external remains always similar and immovable."* And, *"Absolute, true, and mathematical time, of itself and from its own nature flows equally without regard to anything external."* Could the ether be the framework of absolute space? In order to settle these apparent inconsistencies, it became necessary to detect this medium, called the ether, and thus verify its very existence. Maxwell suggested a crucial experiment to detect this ether. The experiment was performed by A. A. Michelson and E. E. Morley and is described in section 29.5.

29.5　The Michelson-Morley Experiment

If there is a medium called the ether that pervades all of space then the earth must be moving through this ether as it moves in its orbital motion about the sun. From the point of view of an observer on the earth the ether must flow past the earth, that is, it must appear that the earth is afloat in an ether current. The ether current concept allows us to consider an analogy of a boat in a river current.

Consider a boat in a river, L meters wide, where the speed of the river current is some unknown quantity v, as shown in figure 29.10. The boat is capable of moving at a speed V with respect to the water. The captain would like to measure the river current v, using only his stopwatch and the speed of his boat with respect

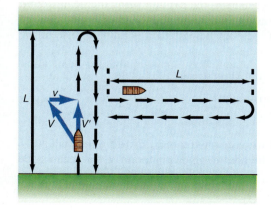

Figure 29.10

Current flowing in a river.

to the water. After some thought the captain proceeds as follows. He can measure the time it takes for his boat to go straight across the river and return. But if he heads straight across the river, the current pushes the boat downstream. Therefore, he heads the boat upstream at an angle such that one component of the boat's velocity with respect to the water is equal and opposite to the velocity of the current downstream. Hence, the boat moves directly across the river at a velocity $\mathbf{V}'$, as shown in the figure. The speed V' can be found from the application of the Pythagorean theorem to the velocity triangle of figure 29.10, namely

$$V^2 = V'^2 + v^2$$

Solving for V', we get

$$V' = \sqrt{V^2 - v^2}$$

Factoring out a V, we obtain, for the speed of the boat across the river,

$$V' = V\sqrt{1 - v^2/V^2} \tag{29.27}$$

We find the time to cross the river by dividing the distance traveled by the boat by the boat's speed, that is,

$$t_{across} = \frac{L}{V'}$$

The time to return is the same, that is,

$$t_{return} = \frac{L}{V'}$$

Hence, the total time to cross the river and return is

$$t_1 = t_{across} + t_{return} = \frac{L}{V'} + \frac{L}{V'} = \frac{2L}{V'}$$

Substituting V' from equation 29.27, the time becomes

$$t_1 = \frac{2L}{V\sqrt{1 - v^2/V^2}}$$

Hence, the time for the boat to cross the river and return is

$$t_1 = \frac{2L/V}{\sqrt{1 - v^2/V^2}} \tag{29.28}$$

The captain now tries another motion. He takes the boat out to the middle of the river and starts the boat downstream at the same speed V with respect to the water. After traveling a distance L downstream, the captain turns the boat around and travels the same distance L upstream to where he started from, as we can see in figure 29.10. The actual velocity of the boat downstream is found by use of the Galilean transformation as

$$V' = V + v \quad \text{Downstream}$$

while the actual velocity of the boat upstream is

$$V' = V - v \quad \text{Upstream}$$

We find the time for the boat to go downstream by dividing the distance L by the velocity V'. Thus the time for the boat to go downstream, a distance L, and to return is

$$t_2 = t_{downstream} + t_{upstream}$$

$$= \frac{L}{V + v} + \frac{L}{V - v}$$

Finding a common denominator and simplifying,

$$t_2 = \frac{L(V - v) + L(V + v)}{(V + v)(V - v)}$$

$$= \frac{LV - Lv + LV + Lv}{V^2 + vV - vV - v^2}$$

$$= \frac{2LV}{V^2 - v^2} = \frac{2LV/V^2}{V^2/V^2 - v^2/V^2}$$

$$t_2 = \frac{2L/V}{1 - v^2/V^2} \tag{29.29}$$

Hence, t_2 in equation 29.29 is the time for the boat to go downstream and return. Note from equations 29.28 and 29.29 that the two travel times are not equal.

The ratio of t_1, the time for the boat to cross the river and return, to t_2, the time for the boat to go downstream and return, found from equations 29.28 and 29.29, is

$$\frac{t_1}{t_2} = \frac{(2L/V)/\sqrt{1 - v^2/V^2}}{(2L/V)/(1 - v^2/V^2)}$$

$$= \frac{1 - v^2/V^2}{\sqrt{1 - v^2/V^2}}$$

$$\frac{t_1}{t_2} = \sqrt{1 - v^2/V^2} \tag{29.30}$$

Equation 29.30 says that if the speed v of the river current is known, then a relation between the times for the two different paths can be determined. On the other hand, if t_1 and t_2 are measured and the speed of the boat with respect to the water V is known, then the speed of the river current v can be determined. Thus, squaring equation 29.30,

$$\frac{t_1^2}{t_2^2} = 1 - \frac{v^2}{V^2}$$

$$\frac{v^2}{V^2} = 1 - \frac{t_1^2}{t_2^2}$$

or

$$v = V\sqrt{1 - \frac{t_1^2}{t_2^2}} \tag{29.31}$$

Thus, by knowing the times for the boat to travel the two paths the speed of the river current v can be determined.

Using the above analogy can help us to understand the experiment performed by Michelson and Morley to detect the ether current. The equipment used to measure the ether current was the Michelson interferometer described in chapter 28 and sketched in figure 29.11. The interferometer sits in a laboratory on the earth. Because the earth moves through the ether, the observer in the laboratory sees an ether current moving past him with a speed of approximately $v = 3.00 \times 10^4$ m/s, the orbital velocity of the earth about the sun. The motion of the light throughout the interferometer is the same as the motion of the boat in the river current. Light from the extended source is split by the half-silvered mirror. Half the light follows the path OM_1OE, which is perpendicular to the ether current. The rest follows the path OM_2OE, which is first in the direction of the ether current until it is reflected from mirror M_2, and is then in the direction that is opposite to the ether current. *The time for the light to cross the ether current is found from equation 29.28, but with V the speed of the boat replaced by c, the speed of light.* Thus,

$$t_1 = \frac{2L/c}{\sqrt{1 - v^2/c^2}}$$

Figure 29.11

The Michelson-Morley experiment.

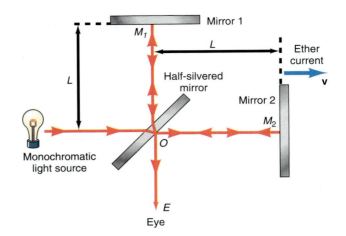

The time for the light to go downstream and upstream in the ether current is found from equation 29.29 but with V replaced by c. Thus,

$$t_2 = \frac{2L/c}{1 - v^2/c^2}$$

The time difference between the two optical paths because of the ether current is

$$\Delta t = t_2 - t_1$$

$$\Delta t = \frac{2L/c}{1 - v^2/c^2} - \frac{2L/c}{\sqrt{1 - v^2/c^2}} \qquad (29.32)$$

To simplify this equation, we use the *binomial theorem*. That is,

$$(1 - x)^n = 1 - nx + \frac{n(n-1)x^2}{2!} - \frac{n(n-1)(n-2)x^3}{3!} + \cdots \qquad (29.33)$$

This is a valid series expansion for $(1 - x)^n$ as long as x is less than 1. In this particular case,

$$x = \frac{v^2}{c^2} = \frac{(3.00 \times 10^4 \text{ m/s})^2}{(3.00 \times 10^8 \text{ m/s})^2} = 10^{-8}$$

which is much less than 1. In fact, since $x = 10^{-8}$, which is very small, it is possible to simplify the binomial theorem to

$$(1 - x)^n = 1 - nx \qquad (29.34)$$

That is, since $x = 10^{-8}$, $x^2 = 10^{-16}$, and $x^3 = 10^{-24}$, the terms in x^2 and x^3 are negligible when compared to the value of x, and can be set equal to zero. Therefore, we can write the denominator of the first term in equation 29.32 as

$$\frac{1}{1 - v^2/c^2} = \left(1 - \frac{v^2}{c^2}\right)^{-1} = 1 - (-1)\frac{v^2}{c^2} = 1 + \frac{v^2}{c^2} \qquad (29.35)$$

The denominator of the second term can be expressed as

$$\frac{1}{\sqrt{1 - v^2/c^2}} = (1 - v^2/c^2)^{-1/2} = 1 - \left(-\frac{1}{2}\right)\frac{v^2}{c^2}$$

$$\frac{1}{\sqrt{1 - v^2/c^2}} = 1 + \frac{1v^2}{2c^2} \qquad (29.36)$$

Substituting equations 29.35 and 29.36 into equation 29.32, yields

$$\Delta t = \frac{2L}{c}\left(1 + \frac{v^2}{c^2}\right) - \frac{2L}{c}\left(1 + \frac{1v^2}{2c^2}\right)$$

$$= \frac{2L}{c}\left(1 + \frac{v^2}{c^2} - 1 - \frac{1v^2}{2c^2}\right)$$

$$= \frac{2L}{c}\left(\frac{1v^2}{2c^2}\right)$$

The path difference d between rays OM_1OE and OM_2OE, corresponding to this time difference Δt, is

$$d = c\Delta t = c\left[\frac{2L}{c}\left(\frac{v^2}{2c^2}\right)\right]$$

or

$$d = L\frac{v^2}{c^2} \tag{29.37}$$

Equation 29.37 gives the path difference between the two light rays and would cause the rays of light to be out of phase with each other and should cause an interference fringe. However, as explained in chapter 28, the mirrors M_1 and M_2 of the Michelson interferometer are not quite perpendicular to each other and we always get interference fringes. However, if the interferometer is rotated through 90°, then the optical paths are interchanged. That is, the path that originally required a time t_1 for the light to pass through, now requires a time t_2 and vice versa. The new time difference between the paths, analogous to equation 29.32, becomes

$$\Delta t' = \frac{2L/c}{\sqrt{1 - v^2/c^2}} - \frac{2L/c}{1 - v^2/c^2}$$

Using the binomial theorem again, we get

$$\Delta t' = \frac{2L}{c}\left(1 + \frac{1v^2}{2c^2}\right) - \frac{2L}{c}\left(1 + \frac{v^2}{c^2}\right)$$

$$= \frac{2L}{c}\left(1 + \frac{1v^2}{2c^2} - 1 - \frac{v^2}{c^2}\right)$$

$$= \frac{2L}{c}\left(-\frac{1v^2}{2c^2}\right)$$

$$= -\frac{Lv^2}{cc^2}$$

The difference in path corresponding to this time difference is

$$d' = c\Delta t' = c\left(-\frac{Lv^2}{cc^2}\right)$$

or

$$d' = -\frac{Lv^2}{c^2}$$

By rotating the interferometer, the optical path has changed by

$$\Delta d = d - d' = \frac{Lv^2}{c^2} - \left(-\frac{Lv^2}{c^2}\right)$$

$$\Delta d = \frac{2Lv^2}{c^2} \tag{29.38}$$

This change in the optical paths corresponds to a shifting of the interference fringes. That is,

$$\Delta d = \Delta n \, \lambda$$

or

$$\Delta n = \frac{\Delta d}{\lambda} \qquad\qquad (29.39)$$

Using equations 29.38 for Δd, *the number of fringes, Δn, that should move across the screen when the interferometer is rotated is*

$$\Delta n = \frac{2Lv^2}{\lambda c^2} \qquad\qquad (29.40)$$

In the actual experimental set-up, the light path L was increased to 10.0 m by multiple reflections. The wavelength of light used was 500.0 nm. The ether current was assumed to be 3.00×10^4 m/s, the orbital speed of the earth around the sun. When all these values are placed into equation 29.40, the expected fringe shift is

$$\Delta n = \frac{2(10.0 \text{ m})(3.00 \times 10^4 \text{ m/s})^2}{(5.000 \times 10^{-7} \text{ m})(3.00 \times 10^8 \text{ m/s})^2}$$

$$= 0.400 \text{ fringes}$$

That is, if there is an ether that pervades all space, the earth must be moving through it. This ether current should cause a fringe shift of 0.400 fringes in the rotated interferometer, however, no fringe shift whatsoever was found. It should be noted that the interferometer was capable of reading a shift much smaller than the 0.400 fringe expected.

On the rare possibility that the earth was moving at the same speed as the ether, the experiment was repeated six months later when the motion of the earth was in the opposite direction. Again, no fringe shift was observed. *The ether cannot be detected. But if it cannot be detected there is no reason to even assume that it exists. Hence, the* **Michelson-Morley experiment's** *null result implies that the all pervading medium called the ether simply does not exist.* Therefore light, and all electromagnetic waves, are capable of propagating without the use of any medium. If there is no ether then the speed of the ether wind v is equal to zero. The null result of the experiment follows directly from equation 29.40 with $v = 0$.

The negative result also suggested a new physical principle. Even if there is no ether, when the light moves along path OM_2 the Galilean transformation equations with respect to the "fixed stars" still imply that the velocity of light along OM_2 should be $c + v$, where v is the earth's orbital velocity, with respect to the fixed stars, and c is the velocity of light with respect to the source on the interferometer. Similarly, it should be $c - v$ along path M_2O. But the negative result of the experiment requires the light to move at the same speed c whether the light was moving with the earth or against it. Hence, *the negative result implies that the speed of light in free space is the same everywhere regardless of the motion of the source or the observer.* This also implies that there is something wrong with the Galilean transformation, which gives us the $c + v$ and $c - v$ velocities. Thus, it would appear that a new transformation equation other than the Galilean transformation is necessary.

29.6 The Postulates of the Special Theory of Relativity

In 1905, Albert Einstein (1879–1955) formulated his **Special or Restricted Theory of Relativity** in terms of two postulates.

Postulate 1: The laws of physics have the same form in all frames of reference moving at a constant velocity with respect to one another. This first postulate is sometimes also stated in the more succinct form: The laws of physics are invariant to a transformation between all inertial frames.

Postulate 2: The speed of light in free space has the same value for all observers, regardless of their state of motion.

Postulate 1 is, in a sense, a consequence of the fact that all inertial frames are equivalent. If the laws of physics were different in different frames of reference, then we could tell from the form of the equation used which frame we were in. In particular, we could tell whether we were at rest or moving. But the difference between rest and motion at a constant velocity cannot be detected. Therefore, the laws of physics must be the same in all inertial frames.

Postulate 2 says that the velocity of light is always the same independent of the velocity of the source or of the observer. This can be taken as an experimental fact deduced from the Michelson-Morley experiment. However, Einstein, when asked years later if he had been aware of the results of the Michelson-Morley experiment, replied that he was not sure if he had been. Einstein came on the second postulate from a different viewpoint. According to his first postulate, the laws of physics must be the same for all inertial observers. If the velocity of light is different for different observers, then the observer could tell whether he was at rest or in motion at some constant velocity, simply by determining the velocity of light in his frame of reference. If the observed velocity of light c' were equal to c then the observer would be in the frame of reference that is at rest. If the observed velocity of light were $c' = c - v$, then the observer was in a frame of reference that was receding from the rest frame. Finally, if the observed velocity $c' = c + v$, then the observer would be in a frame of reference that was approaching the rest frame. Obviously these various values of c' would be a violation of the first postulate, since we could now define an absolute rest frame ($c' = c$), which would be different than all the other inertial frames.

The second postulate has revolutionary consequences. Recall that a velocity is equal to a distance in space divided by an interval of time. *In order for the velocity of light to remain a constant independent of the motion of the source or observer, space and time itself must change.* This is a revolutionary concept, indeed, because as already pointed out, Newton had assumed that space and time were absolute. A length of 1 m was considered to be a length of 1 m anywhere, and a time interval of 1 hr was considered to be a time interval of 1 hr anywhere. However, *if space and time change, then these concepts of absolute space and absolute time can no longer be part of the picture of the physical universe.*

The negative results of the Michelson-Morley experiment can also be explained by the second postulate. The velocity of light must always be c, never the $c + v$, $c - v$, or $\sqrt{c^2 - v^2}$ that were used in the original derivation. Thus, there would be no difference in time for either optical path of the interferometer and no fringe shift.

The Galilean equations for the transformation of velocity, which gave us the velocities of light as $c' = c + v$ and $c' = c - v$, must be replaced by some new transformation that always gives the velocity of light as c regardless of the velocity of the source or the observer. In section 29.7 we will derive such a transformation.

29.7 The Lorentz Transformation

Because the Galilean transformations violate the postulates of relativity, we must derive a new set of equations that relate the position and velocity of an object in one inertial frame to its position and velocity in another inertial frame. And we must derive the new transformation equations directly from the postulates of special relativity.

Since the Galilean transformations are correct when dealing with the motion of a body at low speeds, the new equations should reduce to the Galilean equations at low speeds. Therefore, the new transformation should have the form

$$x' = k(x - vt) \tag{29.41}$$

where x is the position of the body in the "rest" frame, t is the time of its observation, x' is the position of the body in the moving frame of reference, and finally k is some function or constant to be determined. For the classical case of low speeds, k should reduce to the value 1, and the new transformation equation would then reduce to the Galilean transformation, equation 29.5. This equation says that if the position x and velocity v of a body are measured in the stationary frame, then its position x' in the moving frame is determined by equation 29.41. Using the first postulate of relativity, this equation must have the same form in the frame of reference at rest. Therefore,

$$x = k(x' + vt') \tag{29.42}$$

where x' is the position of the body in the moving frame at the time t'. The sign of v has been changed to a positive quantity because, as shown in figure 29.3, a frame 2 moving to the right with a velocity v as observed from a frame 1 at rest, is equivalent to frame 2 at rest with frame 1 moving to the left with a velocity $-v$. This equation says that if the position x' and velocity v' of a body are measured in a moving frame, then its position x in the stationary frame is determined by equation 29.42. The position of the y- and z-coordinates are still the same, namely,

$$y' = y$$
$$z' = z \tag{29.43}$$

The time of the observation of the event in the moving frame is denoted by t'. We deliberately depart from our common experiences by arranging for the possibility of a different time t' for the event in the moving frame compared to the time t for the same event in the stationary frame. In fact, t' can be determined by substituting equation 29.41 into equation 29.42. That is,

$$x = k(x' + vt') = k[k(x - vt) + vt']$$
$$= k^2 x - k^2 vt + kvt'$$
$$kvt' = x - k^2 x + k^2 vt$$

and

$$t' = kt + \left(\frac{1 - k^2}{kv}\right)x \tag{29.44}$$

Thus, according to the results of the first postulate of relativity, the time t' in the moving coordinate system is not equal to the time t in the stationary coordinate system. The exact relation between these times is still unknown, however, because we still have to determine the value of k.

To determine k, we use the second postulate of relativity. Imagine a light wave emanating from a source that is located at the origin of the S and S' frame of reference, which momentarily coincide for $t = 0$ and $t' = 0$, figure 29.12. By

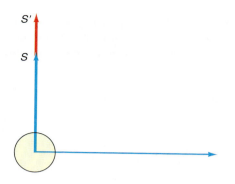

(a) Light wave emitted when frames coincide

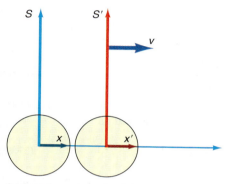

(b) Both frames see the light wave as eminating from its own origin

Figure 29.12

The same light wave observed from two inertial frames.

the second postulate both the stationary and moving observer must observe the same velocity c of the light wave. The distance the wave moves in the x-direction in the S frame is

$$x = ct \qquad (29.45)$$

whereas the distance the same wave moves in the x'-direction in the S' frame is

$$x' = ct' \qquad (29.46)$$

Substituting for x' from equation 29.41, and for t' from equation 29.44, into equation 29.46, yields

$$k(x - vt) = c\left[kt + \left(\frac{1 - k^2}{kv}\right)x\right]$$

Performing the following algebraic steps, we solve for x

$$kx - kvt = ckt + \frac{c(1 - k^2)}{kv}x$$

$$kx - \frac{c(1 - k^2)}{kv}x = ckt + kvt$$

$$x\left[k - \frac{c(1 - k^2)}{kv}\right] = ct\left(k + \frac{kv}{c}\right)$$

$$x = ct\left\{\frac{k + kv/c}{k - c[(1 - k^2)/kv]}\right\} \qquad (29.47)$$

But as already seen in equation 29.45, $x = ct$. Therefore, the term in braces in equation 29.47 must be equal to 1. Thus,

$$\frac{k + kv/c}{k - [c(1 - k^2)/kv]} = 1$$

$$\frac{k(1 + v/c)}{k\{1 - [c(1 - k^2)/k^2v]\}} = 1$$

$$1 + \frac{v}{c} = 1 - \frac{c}{v}\left(\frac{1}{k^2} - 1\right) = 1 - \frac{c}{vk^2} + \frac{c}{v}$$

$$1 + \frac{v}{c} - 1 - \frac{c}{v} = -\frac{c}{vk^2}$$

$$k^2\left(\frac{v}{c} - \frac{c}{v}\right) = -\frac{c}{v}$$

$$k^2 = \frac{-c/v}{v/c - c/v} = \frac{c/v}{c/v - v/c} = \frac{1}{1 - [(v/c)/(c/v)]} = \frac{1}{1 - v^2/c^2}$$

Thus, the function k becomes

$$k = \frac{1}{\sqrt{1 - v^2/c^2}} \qquad (29.48)$$

Substituting this value of k into equation 29.41 gives the first of the new transformation equations, namely

$$x' = \frac{x - vt}{\sqrt{1 - v^2/c^2}} \qquad (29.49)$$

Equation 29.49 gives the position x' of the body in the moving coordinate system in terms of its position x and velocity v at the time t in the stationary coordinate system. Before discussing its physical significance let us also substitute k into the time equation 29.44, that is,

$$t' = \frac{t}{\sqrt{1 - v^2/c^2}} + \left[\frac{1 - (1)^2/(\sqrt{1 - v^2/c^2})^2}{v/\sqrt{1 - v^2/c^2}} \right] x$$

Simplifying,

$$t' = \frac{t}{\sqrt{1 - v^2/c^2}} + \left(1 - \frac{1}{1 - v^2/c^2} \right) \left(\frac{x}{v} \right) \sqrt{1 - v^2/c^2}$$

$$= \frac{t}{\sqrt{1 - v^2/c^2}} + \frac{(1 - v^2/c^2 - 1)x}{(1 - v^2/c^2)v} \sqrt{1 - v^2/c^2}$$

$$= \frac{t}{\sqrt{1 - v^2/c^2}} - \frac{v^2/c^2}{\sqrt{1 - v^2/c^2}} \frac{x}{v}$$

and the second transformation equation becomes

$$t' = \frac{t - xv/c^2}{\sqrt{1 - v^2/c^2}} \tag{29.50}$$

These new transformation equations are called the **Lorentz transformations.**[2] The Lorentz transformation equations are summarized as

$$x' = \frac{x - vt}{\sqrt{1 - v^2/c^2}} \tag{29.49}$$

$$y' = y$$

$$z' = z$$

$$t' = \frac{t - xv/c^2}{\sqrt{1 - v^2/c^2}} \tag{29.50}$$

Now that we have obtained the new transformation equations, we must ask what they mean. First of all, note that the coordinate equation for the position does look like the Galilean transformation for position except for the term $\sqrt{1 - v^2/c^2}$ in the denominator of the x'-term. If the velocity v of the reference frame is small compared to c, then $v^2/c^2 \approx 0$, and hence,

$$x' = \frac{x - vt}{\sqrt{1 - v^2/c^2}} = \frac{x - vt}{\sqrt{1 - 0}} = x - vt$$

Similarly, for the time equation, if v is much less than c then $v^2/c^2 \approx 0$ and also $xv/c^2 \approx 0$. Therefore, the time equation becomes

$$t' = \frac{t - xv/c^2}{\sqrt{1 - v^2/c^2}} = \frac{t - 0}{\sqrt{1 - 0}} = t$$

Thus, the Lorentz transformation equations reduce to the classical Galilean transformation equations when the relative speed between the observers is small as compared to the speed of light. This reduction of a new theory to an old theory is called the correspondence principle and was first enunciated as a principle by Niels Bohr in 1923. It states that any new theory in physics must reduce to the well-established corresponding classical theory when the new theory is applied to the special situation in which the less general theory is known to be valid.

2. These equations were named for H. A. Lorentz because he derived them before Einstein's theory of special relativity. However, Lorentz derived these equations to explain the negative result of the Michelson-Morley experiment. They were essentially empirical equations because they could not be justified on general grounds as they were by Einstein.

Because of this reduction to the old theory, the consequences of special relativity are not apparent unless dealing with enormous speeds such as those comparable to the speed of light.

Example 29.4

The value of $1/\sqrt{1 - v^2/c^2}$ *for various values of* v. What is the value of $1/\sqrt{1 - v^2/c^2}$ for (a) $v = 1610$ km/hr $= 1000$ mph, (b) $v = 1610$ km/s $= 1000$ mi/s, and (c) $v = 0.8c$. Take $c = 3.00 \times 10^8$ m/s in SI units. It will be assumed in all the examples of relativity that the initial data are known to whatever number of significant figures necessary to demonstrate the principles of relativity in the calculations.

Solution

a. The speed $v = (1610$ km/hr$)(1$ hr$/3600$ s$) = 0.447$ km/s $= 447$ m/s. Hence,

$$\frac{1}{\sqrt{1 - v^2/c^2}} = \frac{1}{\sqrt{1 - (447 \text{ m/s})^2/(3.00 \times 10^8 \text{ m/s})^2}}$$

$$= \frac{1}{\sqrt{1 - 2.22 \times 10^{-12}}}$$

This can be further simplified by the binomial expansion as

$$(1 - x)^n = 1 - nx$$

and hence,

$$\frac{1}{\sqrt{1 - v^2/c^2}} = 1 - \left(\frac{-1}{2}\right)(2.22 \times 10^{-12}) = 1.00000000000111 = 1$$

That is, the value is so close to 1 that we cannot determine the difference.

b. The velocity $v = 1610$ km/s, gives a value of

$$\frac{1}{\sqrt{1 - v^2/c^2}} = \frac{1}{\sqrt{1 - (1.61 \times 10^6)^2/(3.00 \times 10^8)^2}}$$

$$= \frac{1}{\sqrt{1 - 2.88 \times 10^{-5}}}$$

$$= \frac{1}{\sqrt{0.99997}} = \frac{1}{0.99999}$$

$$= 1.00001$$

Now 1610 km/s is equal to 3,600,000 mph. Even though this is considered to be an enormous speed, far greater than anything people are now capable of moving at (for example, a satellite in a low earth orbit moves at about 18,000 mph, and the velocity of the earth around the sun is about 68,000 mph), the effect is still so small that it can still be considered to be negligible.

c. For a velocity of $0.8c$ the value becomes

$$\frac{1}{\sqrt{1 - v^2/c^2}} = \frac{1}{\sqrt{1 - (0.8c)^2/c^2}} = \frac{1}{\sqrt{1 - 0.64}}$$

$$= \frac{1}{\sqrt{0.36}} = \frac{1}{0.600}$$

$$= 1.67$$

Thus, at the speed of eight-tenths of the speed of light the factor becomes quite significant.

The Lorentz transformation equations point out that space and time are intimately connected. Notice that the position x' not only depends on the position x but also depends on the time t, whereas the time t' not only depends on the time t but also depends on the position x. *We can no longer consider such a thing as absolute time, because time now depends on the position of the observer. That is, all time must be considered relative. Thus, we can no longer consider space and time as separate entities. Instead there is a union or fusion of space and time into the single reality called spacetime.* That is, space by itself has no meaning; time by itself has no meaning; only spacetime exists. The coordinates of an event in four-dimensional spacetime are (x, y, z, t). We will say more about spacetime in chapter 30.

An interesting consequence of this result of special relativity is its effects on the fundamental quantities of physics. In chapter 1 we saw that the world could be described in terms of three fundamental quantities—space, time, and matter. It is now obvious that there are even fewer fundamental quantities. *Because space and time are fused into spacetime, the fundamental quantities are now only two, spacetime and matter.*

It is important to notice that the Lorentz transformation equations for special relativity put a limit on the maximum value of v that is attainable by a body, because if $v = c$,

$$x' = \frac{x - vt}{\sqrt{1 - v^2/c^2}} = \frac{x - vt}{\sqrt{1 - c^2/c^2}}$$

$$= \frac{x - vt}{0}$$

Since division by zero is undefined, we must take the limit as v approaches c. That is,

$$x' = \lim_{v \to c} \frac{x - vt}{\sqrt{1 - v^2/c^2}} = \infty$$

and similarly

$$t' = \lim_{v \to c} \frac{t - vx/c^2}{\sqrt{1 - v^2/c^2}} = \infty$$

That is, for $v = c$, the coordinates x' and t' are infinite, or at least undefinable. If $v > c$ then $v^2/c^2 > 1$ and $1 - v^2/c^2 < 1$. This means that the number under the square root sign is negative and the square root of a negative quantity is imaginary. Thus x' and t' become imaginary quantities. Hence, according to the theory of special relativity, no object can move at a speed equal to or greater than the speed of light.

Example 29.5

Lorentz transformation of coordinates. A man on the earth measures an event at a point 5.00 m from him at a time of 3.00 s. If a rocket ship flies over the man at a speed of 0.800c, what coordinates does the astronaut in the rocket ship attribute to this event?

Solution

The location of the event, as observed in the moving rocket ship, found from equation 29.49, is

$$x' = \frac{x - vt}{\sqrt{1 - v^2/c^2}}$$

$$= \frac{5.00 \text{ m} - (0.800)(3.00 \times 10^8 \text{ m/s})(3.00 \text{ s})}{\sqrt{1 - (0.800c)^2/c^2}}$$

$$= -1.20 \times 10^9 \text{ m}$$

This distance is quite large because the astronaut is moving at such high speed. The event occurs on the astronaut's clock at a time

$$t' = \frac{t - vx/c^2}{\sqrt{1 - v^2/c^2}}$$

$$= \frac{3.00 \text{ s} - (0.800)(3 \times 10^8 \text{ m/s})(5.00 \text{ m})/(3 \times 10^8 \text{ m/s})^2}{\sqrt{1 - (0.800c)^2/c^2}}$$

$$= 5.00 \text{ s}$$

The inverse Lorentz transformation equations from the moving system to the stationary system can be written down immediately by the use of the first postulate. That is, their form must be the same, but $-v$ is replaced by $+v$ and primes and unprimes are interchanged. Therefore, the inverse Lorentz transformation equations are

$$x = \frac{x' + vt'}{\sqrt{1 - v^2/c^2}} \qquad (29.51)$$

$$y = y' \qquad (29.52)$$

$$z = z' \qquad (29.53)$$

$$t = \frac{t' + vx'/c^2}{\sqrt{1 - v^2/c^2}} \qquad (29.54)$$

29.8 The Lorentz-Fitzgerald Contraction

Consider a rod at rest in a stationary coordinate system S on the earth, as in figure 29.13(a). What is the length of this rod when it is observed by an astronaut in the S' frame of reference, a rocket ship traveling at a speed v? One end of the rod is located at the point x_1, while the other end is located at the point x_2. *The length of this stationary rod, measured in the frame where it is at rest, is called its **proper length** and is denoted by L_0, where*

$$L_0 = x_2 - x_1 \qquad (29.55)$$

What is the length of this rod as observed in the rocket ship? The astronaut must measure the coordinates x_1' and x_2' for the ends of the rod at the same time t' in his frame S'.

The measurement of the length of any rod in a moving coordinate system must always be measured simultaneously in that coordinate system or else the ends of the rod will have moved during the measurement process and we will not be measuring the true length of the object. An often quoted example for the need of simultaneous measurements of length is the measurement of a fish in a tank. If the tail of the fish is measured first, and the head some time later, the fish has moved to the left and we have measured a much longer fish than the one in the tank, figure 29.14(a). If the head of the fish is measured first, and then the tail, the fish appears smaller than it is, figure 29.14(b). If, on the other hand, the head and tail are measured simultaneously we get the actual length of the fish, figure 29.14(c).

In a coordinate system where the rod or body is at rest, simultaneous measurements are not necessary because we can measure the ends at any time, since the rod is always there in that place and its ends never move. When the values of the coordinates of the end of the bar, x_1' and x_2', are measured at the time t', the values of x_1 and x_2 in the earth frame S are computed by the Lorentz transformation. Thus,

$$x_1 = \frac{x_1' + vt'}{\sqrt{1 - v^2/c^2}} \qquad (29.56)$$

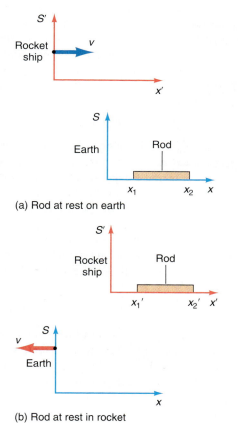

(a) Rod at rest on earth

(b) Rod at rest in rocket

Figure 29.13

The Lorentz-Fitzgerald contraction.

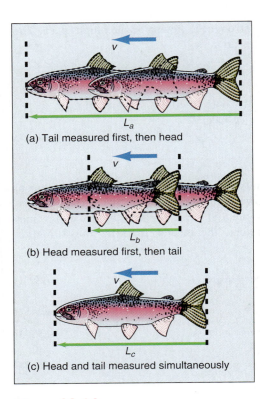

Figure 29.14

Measurement of the length of a moving fish.

(a) Tail measured first, then head

L_a

(b) Head measured first, then tail

L_b

(c) Head and tail measured simultaneously

L_c

while

$$x_2 = \frac{x_2' + vt'}{\sqrt{1 - v^2/c^2}} \qquad (29.57)$$

The length of the rod L_0, found from equations 29.55, 29.56, and 29.57, is

$$L_0 = x_2 - x_1 = \frac{x_2' + vt'}{\sqrt{1 - v^2/c^2}} - \frac{x_1' + vt'}{\sqrt{1 - v^2/c^2}}$$

$$= \frac{x_2' + vt' - x_1' - vt'}{\sqrt{1 - v^2/c^2}}$$

$$L_0 = \frac{x_2' - x_1'}{\sqrt{1 - v^2/c^2}} \qquad (29.58)$$

Let us designate L as the length of the rod as measured in the moving rocket frame S', that is,

$$L = x_2' - x_1' \qquad (29.59)$$

Then equation 29.58 becomes

$$L_0 = \frac{L}{\sqrt{1 - v^2/c^2}}$$

or

$$L = L_0\sqrt{1 - v^2/c^2} \qquad (29.60)$$

Because v is less than c, the quantity $\sqrt{1 - v^2/c^2} < 1$, which means that $L < L_0$. *That is, the rod at rest in the earth frame would be measured in the rocket frame to be smaller than it is in the earth frame. From the point of view of the astronaut in the rocket, the rocket is at rest and the rod in the earth frame is moving toward him at a velocity v. Hence, the astronaut considers the rod to be at rest in a moving frame, and he then concludes that a moving rod contracts, as given by equation 29.60.* That is, if the rod is a meterstick, then its proper length in the frame where it is at rest is $L_0 = 1.00$ m $= 100$ cm. If the rocket is moving at a speed of $0.8c$, then the length as observed from the rocket ship is

$$L = L_0\sqrt{1 - v^2/c^2} = 1.00 \text{ m}\sqrt{1 - (0.8c)^2/c^2} = 0.600 \text{ m}$$

Thus, the astronaut says that the meterstick is only 60.0 cm long. *This contraction of length is known as the **Lorentz-Fitzgerald contraction*** because it was derived earlier by Lorentz and Fitzgerald. However Lorentz and Fitzgerald attributed this effect to the ether. But since the ether does not exist, this effect cannot be attributed to it. It was Einstein's derivation of these same equations by the postulates of relativity that gave them real meaning.

This length contraction is a reciprocal effect. If there is a rod (a meterstick) at rest in the rocket S' frame, figure 29.13(b), then the astronaut measures the length of that rod by measuring the ends x_1' and x_2' at any time. The length of the rod as observed by the astronaut is

$$L_0 = x_2' - x_1' \qquad (29.61)$$

This length is now the proper length L_0 because the rod is at rest in the astronaut's frame of reference. The observer on earth (S frame) measures the coordinates of the ends of the rod, x_1 and x_2, simultaneously at the time t. The ends of the rod x_2' and x_1' are computed by the earth observer by the Lorentz transformations:

$$x_2' = \frac{x_2 - vt}{\sqrt{1 - v^2/c^2}}$$

$$x_1' = \frac{x_1 - vt}{\sqrt{1 - v^2/c^2}}$$

Thus, the length of the rod becomes

$$L_0 = x_2' - x_1' = \frac{x_2 - vt}{\sqrt{1 - v^2/c^2}} - \frac{x_1 - vt}{\sqrt{1 - v^2/c^2}}$$

$$= \frac{x_2 - vt - x_1 + vt}{\sqrt{1 - v^2/c^2}}$$

$$= \frac{x_2 - x_1}{\sqrt{1 - v^2/c^2}}$$

But

$$x_2 - x_1 = L$$

the length of the rod as observed by the earth man. Therefore,

$$L_0 = \frac{L}{\sqrt{1 - v^2/c^2}}$$

and

$$L = L_0\sqrt{1 - v^2/c^2} \qquad (29.62)$$

But this is the identical equation that was found before (equation 29.60). If L_0 is again the meterstick and the rocket ship is moving at the speed $0.800c$, then the length L as observed on earth is 60.0 cm. *Thus the length contraction effect is reciprocal.* When the meterstick is at rest on the earth the astronaut thinks it is only 60.0 cm long. When the meterstick is at rest in the rocket ship the earthbound observer thinks it is only 60.0 cm long. Thus each observer sees the other's length as contracted. This reciprocity is to be expected. If the two observers do not agree that the other's stick is contracted, they could use this information to tell which stick is at rest and which is in motion—a violation of the principle of relativity. One thing that is important to notice is that in equation 29.60, L_0 is the length of the rod at rest in the earth or S frame of reference, whereas in equation 29.62, L_0 is the length of the rod at rest in the moving rocket ship (S' frame). L_0 *is always the length of the rod in the frame of reference where it is at rest.* It does not matter if the frame of reference is at rest or moving so long as the rod is at rest in that frame. This is why L_0 is always called its proper length.

The Lorentz-Fitzgerald contraction can be summarized by saying that the *length of a rod in motion with respect to an observer is less than its length when measured by an observer who is at rest with respect to the rod.* This contraction occurs only in the direction of the relative motion. Let us consider the size of this contraction.

Example 29.6

Length contraction. What is the length of a meterstick when it is measured by an observer moving at (a) $v = 1610$ km/hr $= 1000$ mph, (b) $v = 1610$ km/s $= 1000$ miles/s, and (c) $v = 0.8c$. It is assumed in all these problems in relativity that the initial data are known to whatever number of significant figures necessary to demonstrate the principles of relativity in the calculations.

a. The speed $v = (1610 \text{ km/hr})(1 \text{ hr}/3600 \text{ s}) = 0.447 \text{ km/s} = 447 \text{ m/s}$. Take $c = 3.00 \times 10^8$ m/s in SI units. The length contraction, found from either equation 29.60 or equation 29.62, is

$$L = L_0 \sqrt{1 - \frac{v^2}{c^2}}$$

$$= (1.00 \text{ m}) \sqrt{1 - \frac{(447 \text{ m/s})^2}{(3.00 \times 10^8 \text{ m/s})^2}}$$

$$= (1.00 \text{ m}) \sqrt{1 - 2.22 \times 10^{-12}}$$

This can be further simplified by the binomial expansion as

$$(1 - x)^n = 1 - nx$$

and hence

$$\sqrt{1 - \frac{v^2}{c^2}} = 1 - \left(\frac{1}{2}\right)(2.22 \times 10^{-12}) = 1 - 0.00000000000111$$

$$= 0.99999999999888$$

and

$$L = 1.00 \text{ m}$$

Thus, at what might be considered the reasonably fast speed of 1000 mph, the contraction is so small that it is less than the width of one atom, and is negligible.

b. The contraction for a speed of 1610 km/s is

$$L = L_0 \sqrt{1 - \frac{v^2}{c^2}}$$

$$= (1.00 \text{ m}) \sqrt{1 - \frac{(1.610 \times 10^6)^2}{(3.00 \times 10^8)^2}}$$

$$= (1.00 \text{ m}) \sqrt{0.99997}$$

$$= 0.99997 \text{ m}$$

A speed of 1610 km/s is equivalent to a speed of 3,600,000 mph, which is an enormous speed, one man cannot even attain at this particular time. Yet the associated contraction is very small.

c. For a speed of $v = 0.8c$ the contraction is

$$L = L_0 \sqrt{1 - \frac{v^2}{c^2}}$$

$$= (1.00 \text{ m}) \sqrt{1 - \frac{(0.8c)^2}{c^2}} = (1.00 \text{ m}) \sqrt{0.360}$$

$$= 0.600 \text{ m}$$

At speeds approaching the speed of light the contraction is quite significant. Table 29.1 gives the Lorentz contraction for a range of values of speed approaching the speed of light. Notice that as v increases, the contraction becomes greater and greater, until at a speed of $0.999999c$ the meterstick has contracted to a thousandth of a meter or 1 mm. Therefore, the effects of relativity do not manifest themselves unless very great speeds are involved. This is why these effects had never been seen or even anticipated when Newton was formulating his laws of physics. However, in the present day it is possible to accelerate charged particles, such as electrons and protons, to speeds very near the speed of light, and the relativistic effects are observed with such particles.

Table 29.1

The Lorentz Contraction and Time Dilation

Speed	$L = L_0 \sqrt{1 - \dfrac{v^2}{c^2}}$	$\Delta t = \dfrac{\Delta t_0}{\sqrt{1 - v^2/c^2}}$
$0.1c$	$0.995L_0$	$1.01\Delta t_0$
$0.2c$	$0.980L_0$	$1.02\Delta t_0$
$0.4c$	$0.917L_0$	$1.09\Delta t_0$
$0.6c$	$0.800L_0$	$1.25\Delta t_0$
$0.8c$	$0.602L_0$	$1.66\Delta t_0$
$0.9c$	$0.437L_0$	$2.29\Delta t_0$
$0.99c$	$0.141L_0$	$7.08\Delta t_0$
$0.999c$	$0.045L_0$	$22.4\Delta t_0$
$0.9999c$	$0.014L_0$	$70.7\Delta t_0$
$0.99999c$	$0.005L_0$	$224\Delta t_0$
$0.999999c$	$0.001L_0$	$707\Delta t_0$

29.9 Time Dilation

Consider a clock at rest at the position x' in a moving coordinate system S' attached to a rocket ship. The astronaut sneezes ánd notes that he did so when his clock, located at x', reads a time t_1'. Shortly thereafter he sneezes again, and now notes that his clock indicates the time t_2'. The time interval between the two sneezes is

$$\Delta t' = t_2' - t_1' = \Delta t_0 \tag{29.63}$$

*This interval $\Delta t'$ is set equal to Δt_0, and is called the **proper time** because this is the time interval on a clock that is at rest relative to the observer.* The observer on earth in the S frame finds the time for the two sneezes to be

$$t_1 = \frac{t_1' + vx'/c_2}{\sqrt{1 - v^2/c^2}}$$

$$t_2 = \frac{t_2' + vx'/c_2}{\sqrt{1 - v^2/c^2}}$$

Thus, the time interval between the sneezes Δt, as observed by the earth man, becomes

$$\Delta t = t_2 - t_1 = \frac{t_2' + vx'/c^2 - t_1' - vx'/c^2}{\sqrt{1 - v^2/c^2}}$$

$$= \frac{t_2' - t_1'}{\sqrt{1 - v^2/c^2}}$$

But $t_2' - t_1' = \Delta t_0$, by equation 29.63, therefore,

$$\Delta t = \frac{\Delta t_0}{\sqrt{1 - v^2/c^2}} \tag{29.64}$$

Notice that because $v < c$, $v^2/c^2 < 1$ and thus $\sqrt{1 - v^2/c^2} < 1$. Therefore,

$$\Delta t > \Delta t_0 \tag{29.65}$$

*Equation 29.64 is the **time dilation** formula and equation 29.65 says that the clock on earth reads a longer time interval Δt than the clock in the rocket ship Δt_0. Or as is sometimes said, moving clocks slow down.* Thus, if the moving clock slows down, a smaller time duration is indicated on the moving clock than on a stationary clock. Hence, the astronaut ages at a slower rate than a person on earth. The amount of this slowing down of time is relatively small as seen in example 29.7.

Example 29.7

Time dilation. A clock on a rocket ship ticks off a time interval of 1 hr. What time elapses on earth if the rocket ship is moving at a speed of (a) 1610 km/hr = 1000 mph, (b) 1610 km/s = 1000 mi/s, and (c) 0.8c?

Solution

a. The time elapsed on earth, found from equation 29.64, is

$$\Delta t = \frac{\Delta t_0}{\sqrt{1 - v^2/c^2}}$$

$$= \frac{1 \text{ hr}}{\sqrt{1 - (447 \text{ m/s})^2/(3.00 \times 10^8 \text{ m/s})^2}}$$

$$= 1 \text{ hr}$$

The difference between the time interval on the astronaut's clock and the time interval on the earth man's clock is actually about 4 ns. This is such a small quantity that it is effectively zero and the difference between the two clocks can be considered to be zero for a speed of 1600 km/s = 1000 mph.

b. The time elapsed for a speed of 1610 km/s is

$$\Delta t = \frac{\Delta t_0}{\sqrt{1 - v^2/c^2}} = \frac{1 \text{ hr}}{\sqrt{1 - (1.61 \times 10^6)^2/(3.00 \times 10^8)^2}}$$

$$= 1.0000144 \text{ hr}$$

Even at the relatively large speed of 1610 km/s = 3,600,000 mph, the difference in the clocks is practically negligible, that is a difference of 0.05 s in a time interval of 1 hr.

c. The time elapsed for a speed of 0.8c is

$$\Delta t = \frac{\Delta t_0}{\sqrt{1 - v^2/c^2}} = \frac{1 \text{ hr}}{\sqrt{1 - (0.8c)^2/(c)^2}}$$

$$= 1.66 \text{ hr}$$

Therefore, at very high speeds the time dilation effect is quite significant. Table 29.1 shows the time dilation for various values of v. As we can see, the time dilation effect becomes quite pronounced for very large values of v.

It should be noted that the time dilation effect, like the Lorentz contraction, is also reciprocal. That is, a clock on the surface of the earth reads the proper time interval Δt_0 to an observer on the earth. An astronaut observing this earth clock assumes that he is at rest, but the earth is moving away from him at the velocity $-v$. Thus, he considers the earth clock to be the moving clock, and he finds that time on earth moves slower than the time on his rocket ship. This reciprocity of time dilation has led to the most famous paradox of relativity, called the twin paradox. (A paradox is an apparent contradiction.) The reciprocity of time dilation seems to be a contradiction when applied to the twins.

As an example, an astronaut leaves his twin sister on the earth as he travels, at a speed approaching the speed of light, to a distant star and then returns. According to the formula for time dilation, time has slowed down for the astronaut and when he returns to earth he should find his twin sister to be much older than he is. But by the first postulate of relativity, the laws of physics must be the same in all inertial coordinate systems. Therefore, the astronaut says that it is he who is the one at rest and the earth is moving away from him in the opposite direction. Thus, the astronaut says that it is the clock on earth that is moving and hence slowing down. He then concludes that his twin sister on earth will be younger than he is, when he returns. Both twins say that the other twin should be younger after the journey, and hence there seems to be a contradiction. How can we resolve this paradox?

With a little thought we can see that there is no contradiction here. The Lorentz transformations apply to inertial coordinate systems, that is, coordinates that are moving at a constant velocity with respect to each other. The twin on earth is in fact in an inertial coordinate system and can use the time dilation equation. The astronaut who returns home, however, is not in an inertial coordinate system. If the astronaut is originally moving at a velocity v, then in order for him to return home, he has to decelerate his spaceship to zero velocity and then accelerate to the velocity $-v$ to travel homeward. *During the deceleration and acceleration process the spaceship is not an inertial coordinate system, and we cannot justify using the time dilation formula that was derived on the basis of inertial coordinate systems.* Hence there is a very significant difference between the twin that stays home on the earth and the astronaut. Here again is that same conflict that occurs when we try to use an equation that was derived by using certain assumptions. When the assumptions hold, the equation is correct. When the assumptions do not hold, the equation no longer applies. In this example, the Lorentz transformation equations were derived on the assumption that two coordinate systems were moving with respect to each other at constant velocity. The astronaut is in an accelerated coordinate system when he turns around to come home. Hence, he is not in an inertial coordinate system and is not entitled to use the time dilation formula.[3] However, as correctly predicted by the earth twin, time has slowed down for the astronaut and when he returns to earth he should find his twin sister to be much older than he is.

We will consider a deeper insight into the slowing down of time in chapter 30 when we draw spacetime diagrams and discuss the general theory of relativity, and again in chapter 31 when we examine the gravitational red shift by the theory of the quanta.

29.10 Transformation of Velocities

We have seen that the Galilean transformation of velocities is incorrect when dealing with speeds at or near the speed of light. That is, velocities such as $V = c + v$ or $V = c - v$ are incorrect. Therefore, new transformation equations are needed for velocities. The necessary equations are found by the Lorentz equations. The components of the velocity of an object in a stationary coordinate system S are

$$V_x = \frac{\Delta x}{\Delta t} \tag{29.66}$$

$$V_y = \frac{\Delta y}{\Delta t} \tag{29.67}$$

$$V_z = \frac{\Delta z}{\Delta t} \tag{29.68}$$

---

3. This also points out a flaw in the derivation of the Lorentz transformation equations. Starting with inertial coordinate systems, if there is any time dilation caused by the acceleration of the coordinate system to the velocity v, it cannot be determined in this way.

whereas the components of the velocity of that same body, as observed in the moving coordinate system, S', are

$$V'_x = \frac{\Delta x'}{\Delta t'} \tag{29.69}$$

$$V'_y = \frac{\Delta y'}{\Delta t'} \tag{29.70}$$

$$V'_z = \frac{\Delta z'}{\Delta t'} \tag{29.71}$$

The transformation of the x-component of velocity is obtained as follows. The distance the object moves in the S' frame is

$$\Delta x' = x'_2 - x'_1$$

But x'_2 and x'_1, obtained from the Lorentz transformation equations 29.49, are

$$x'_2 = \frac{x_2 - vt_2}{\sqrt{1 - v^2/c^2}}$$

and

$$x'_1 = \frac{x_1 - vt_1}{\sqrt{1 - v^2/c^2}}$$

Therefore, $\Delta x'$ becomes

$$\Delta x' = x'_2 - x'_1 = \frac{x_2 - vt_2 - x_1 + vt_1}{\sqrt{1 - v^2/c^2}}$$

$$= \frac{x_2 - x_1 - v(t_2 - t_1)}{\sqrt{1 - v^2/c^2}}$$

But

$$x_2 - x_1 = \Delta x$$

$$t_2 - t_1 = \Delta t \tag{29.72}$$

Therefore,

$$\Delta x' = \frac{\Delta x - v\Delta t}{\sqrt{1 - v^2/c^2}} \tag{29.73}$$

The time interval $\Delta t'$ is

$$\Delta t' = t'_2 - t'_1$$

And t'_2 and t'_1, found from the Lorentz transformation equation 29.50, are

$$t'_2 = \frac{t_2 - x_2 v/c^2}{\sqrt{1 - v^2/c^2}}$$

and

$$t'_1 = \frac{t_1 - x_1 v/c^2}{\sqrt{1 - v^2/c^2}}$$

Thus,

$$\Delta t' = t'_2 - t'_1 = \frac{t_2 - x_2 v/c^2 - t_1 + x_1 v/c^2}{\sqrt{1 - v^2/c^2}}$$

$$= \frac{t_2 - t_1 - (x_2 - x_1)v/c^2}{\sqrt{1 - v^2/c^2}}$$

But $t_2 - t_1 = \Delta t$, and $x_2 - x_1 = \Delta x$, from equation 29.72. Therefore,

$$\Delta t' = \frac{\Delta t - \Delta x(v/c^2)}{\sqrt{1 - v^2/c^2}} \qquad (29.74)$$

The transformation for V'_x, found from equations 29.69, 29.73, and 29.74, is

$$V'_x = \frac{\Delta x'}{\Delta t'} = \frac{(\Delta x - v\Delta t)/\sqrt{1 - v^2/c^2}}{[\Delta t - \Delta x(v/c^2)]/\sqrt{1 - v^2/c^2}}$$

Cancelling out the square root term in both numerator and denominator, gives

$$V'_x = \frac{\Delta x - v\Delta t}{\Delta t - \Delta x(v/c^2)}$$

Dividing both numerator and denominator by Δt, gives

$$V'_x = \frac{\Delta x/\Delta t - v(\Delta t/\Delta t)}{\Delta t/\Delta t - (\Delta x/\Delta t)(v/c^2)}$$

But $\Delta x/\Delta t = V_x$ from equation 29.66, and $\Delta t/\Delta t = 1$. Hence,

$$V'_x = \frac{V_x - v}{1 - (v/c^2)V_x} \qquad (29.75)$$

Equation 29.75 is the Lorentz transformation for the x-component of velocity. Notice that if v is very small, compared to c, then the term $(v/c^2)V_x$ approaches zero, and this equation reduces to the Galilean transformation equation 29.13 as would be expected for low velocities.

The y-component of the velocity transformation is obtained similarly. Thus,

$$V'_y = \frac{\Delta y'}{\Delta t'} = \frac{\Delta y}{[\Delta t - (v/c^2)\Delta x]/\sqrt{1 - v^2/c^2}}$$

$$= \frac{\Delta y \sqrt{1 - v^2/c^2}}{\Delta t - (v/c^2)\Delta x}$$

Dividing numerator and denominator by Δt gives

$$V'_y = \frac{(\Delta y/\Delta t)\sqrt{1 - v^2/c^2}}{\Delta t/\Delta t - (v/c^2)\Delta x/\Delta t}$$

and therefore

$$V'_y = \frac{V_y \sqrt{1 - v^2/c^2}}{1 - (v/c^2)V_x} \qquad (29.76)$$

A similar analysis for the z-component of the velocity gives

$$V'_z = \frac{V_z \sqrt{1 - v^2/c^2}}{1 - (v/c^2)V_x} \qquad (29.77)$$

Note, that for v very much less than c, these equations reduce to the Galilean equations, $V'_y = V_y$ and $V'_z = V_z$, as expected.

The Lorentz velocity transformation equations are summarized as

$$V'_x = \frac{V_x - v}{1 - (v/c^2)V_x} \qquad (29.75)$$

$$V'_y = \frac{V_y \sqrt{1 - v^2/c^2}}{1 - (v/c^2)V_x} \qquad (29.76)$$

$$V'_z = \frac{V_z \sqrt{1 - v^2/c^2}}{1 - (v/c^2)V_x} \qquad (29.77)$$

The inverse transformations from the S' frame to the S frame can be written down immediately by changing primes for nonprimes and replacing $-v$ by $+v$. Thus,

$$V_x = \frac{V_x' + v}{1 + (v/c^2)V_x'} \tag{29.78}$$

$$V_y = \frac{V_y'\sqrt{1 - v^2/c^2}}{1 + (v/c^2)V_x'} \tag{29.79}$$

$$V_z = \frac{V_z'\sqrt{1 - v^2/c^2}}{1 + (v/c^2)V_x'} \tag{29.80}$$

Example 29.8

Galilean transformation of velocities versus the Lorentz transformation of velocities. Two rocket ships are approaching a space station, each at a speed of $0.9c$, with respect to the station, as shown in figure 29.15. What is their relative speed according to (a) the Galilean transformation and (b) the Lorentz transformation?

Solution

a. According to the space station observer, the space station is at rest and the two spaceships are closing on him, as shown in figure 29.15. The spaceship to the right is approaching at a speed $V_x = -0.9c$, in the space station coordinate system. The spaceship to the left is considered to be a moving coordinate system approaching with the speed $v = 0.9c$. The relative velocity according to the Galilean transformation, as observed in the moving spaceship to the left, is

$$V_x' = V_x - v$$

$$= -0.9c - 0.9c$$

$$= -1.8c$$

That is, the spaceship to the left sees the spaceship to the right approaching at a speed of $1.8c$. The minus sign means the velocity is toward the left in the S' frame of reference. Obviously this result is incorrect because the relative velocity is greater than c, which is impossible.

b. According to the Lorentz transformation the relative velocity of approach as observed by the S' spaceship, given by equation 29.75, is

$$V_x' = \frac{V_x - v}{1 - (v/c^2)V_x} = \frac{-0.9c - 0.9c}{1 - (0.9c/c^2)(-0.9c)} = \frac{-1.8c}{1 + 0.81c^2/c^2}$$

$$= \frac{-1.8c}{1.81} = -0.994c$$

Thus, the observer in the left-hand spaceship sees the right-hand spaceship approaching at the speed of $0.994c$. The minus sign means that the speed is toward the left in the diagram. Notice that the relative speed is less than c as it must be.

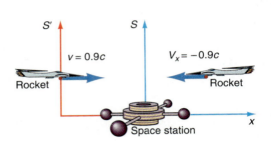

Figure 29.15

Galilean and Lorentz transformations of velocities.

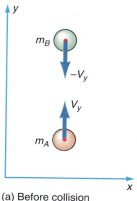

(a) Before collision

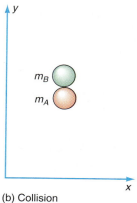

(b) Collision

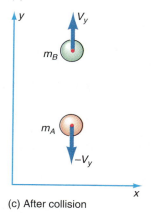

(c) After collision

Figure 29.16

A perfectly elastic collision in a stationary frame of reference.

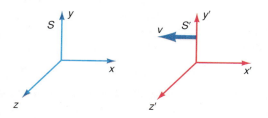

Figure 29.17

One observer is in rest frame and one in moving frame.

Example 29.9

Transformation of the speed of light. If a ray of light is emitted from a rocket ship moving at a speed v, what speed will be observed for that light on earth?

Solution

The speed of light from the rocket ship is $V'_x = c$. The speed observed on earth, found from equation 29.78, is

$$V_x = \frac{V'_x + v}{1 + (v/c^2)V'_x} = \frac{c + v}{1 + vc/c^2} = \frac{c + v}{1 + \left(\dfrac{v}{c}\right)}$$

$$= \frac{c + v}{(c + v)/c} = \left(\frac{c + v}{c + v}\right)c = c$$

Thus, all observers observe the same value for the speed of light.

29.11 The Law of Conservation of Momentum and Relativistic Mass

In section 29.3 we saw that momentum was conserved under a Galilean transformation. Does the law of conservation of momentum also hold in relativistic mechanics? Let us first consider the following perfectly elastic collision between two balls that are identical when observed in a stationary rest frame S in figure 29.16. The first ball m_A is thrown upward with the velocity V_y, whereas the second ball is thrown straight downward with the velocity $-V_y$. Thus the speed of each ball is the same. We assume that the original distance separating the two balls is small and the velocity V_y is relatively large, so that the effect of the acceleration of gravity can be ignored. Applying the law of conservation of momentum to the collision, we obtain

momentum before collision = momentum after collision

$$m_A V_y - m_B V_y = m_B V_y - m_A V_y$$

Simplifying,

$$2m_A V_y = 2m_B V_y$$

or

$$m_A V_y = m_B V_y \tag{29.81}$$

Equation 29.81 also indicates that $m_A = m_B$, as originally stated.

Let us now consider a similar perfectly elastic collision only now the ball B is thrown by a moving observer. The stationary observer is in the frame S and the moving observer is in the moving frame S', moving toward the left at the velocity $-v$, as shown in figure 29.17. In the stationary frame S, the observer throws a ball straight upward in the positive y-direction with the velocity V_y. The moving observer S' is on a truck moving to the left with the velocity $-v$. The moving observer throws an identical ball straight downward in the negative y-direction with the velocity $-u'$ in the moving frame of reference. In the stationary frame this velocity is observed as $-u$. We assume that both observers throw the ball with the same speed in their frames of reference. That is, the magnitude of the velocity V_y in the S frame is identical to the magnitude of the velocity u' in the S' frame. (As an example, let us assume that observer S throws the ball upward at a speed of 20 m/s and observer

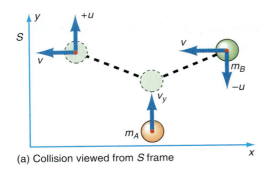

(a) Collision viewed from S frame

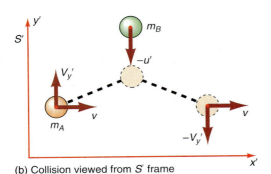

(b) Collision viewed from S' frame

Figure 29.18

A perfectly elastic collision viewed from different frames of reference.

S' throws his ball downward at a speed of 20 m/s.) The two balls are exactly alike in that they have the same mass and size when they are at rest before the experiment starts.

After some practice, the experimenters are able to throw the balls such that a collision occurs. As observed from the S frame of reference on the ground, the collision appears as in figure 29.18(a). The mass m_A goes straight up, collides in a perfectly elastic collision with mass m_B, and is reflected with the velocity $-V_y$, since no energy, and hence speed, was lost in the collision. Ball B has a velocity component $-u$ straight downward (as seen by the S observer) but it is also moving in the negative x-direction with the velocity $-v$, the velocity of the truck and hence the velocity of the S' frame.

The y-component of the law of conservation of momentum, as observed in the rest frame S, can be written as

$$\text{momentum before collision} = \text{momentum after collision}$$

$$m_A V_y - m_B u = -m_A V_y + m_B u$$

Simplifying,

$$2m_A V_y = 2m_B u$$

or

$$m_A V_y = m_B u \tag{29.82}$$

But the velocity u in the stationary frame S is related to the velocity u' in the moving frame of reference S' through the velocity transformation, equation 29.79, as

$$u = \frac{u'\sqrt{1 - v^2/c^2}}{1 + (v/c^2)V_x'}$$

However, $V_x' = 0$ in this experiment because m_B is thrown only in the y-direction. Hence, u becomes

$$u = u'\sqrt{1 - v^2/c^2} \tag{29.83}$$

Substituting u from equation 29.83 back into the law of conservation of momentum, equation 29.82, we obtain

$$m_A V_y = m_B u'\sqrt{1 - v^2/c^2}$$

But recall that the initial speed of each ball was the same in each reference frame, that is, $V_y = u'$. Hence,

$$m_A V_y = m_B V_y\sqrt{1 - v^2/c^2} \tag{29.84}$$

If we compare equation 29.84, for the conservation of momentum when one of the frames is in motion, with equation 29.81, for the conservation of momentum in a stationary frame, we see that the form of the equation is very different. Thus, in the form of equation 29.84, the law of conservation of momentum does not seem to hold. But the law of conservation of momentum is such a fundamental concept in physics that we certainly do not want to lose it in the description of relativistic mechanics. The law of conservation of momentum can be retained if we allow for the possibility that the moving mass changes its value because of that motion. That is, if both sides of equation 29.84 are divided by V_y we get

$$m_A = m_B\sqrt{1 - v^2/c^2} \tag{29.85}$$

Now m_A is the mass of the ball in the stationary frame and m_B is the mass of the ball in the moving frame. If we consider the very special case where V_y is zero in the S frame, then the mass m_A is at rest in the rest frame. We now let $m_A = m_0$, the mass when it is at rest, henceforth called the **proper mass or rest mass,** and we let $m_B = m$, the mass when it is in motion. Equation 29.85 then becomes

$$m_0 = m\sqrt{1 - v^2/c^2}$$

or, solving for m,

$$m = \frac{m_0}{\sqrt{1 - v^2/c^2}} \tag{29.86}$$

*Equation 29.86 defines the **relativistic mass** m in terms of its rest mass m_0. Because the term $\sqrt{1 - v^2/c^2}$ is always less than one, the relativistic mass m, the mass of a body in motion at the speed v, is always greater than m_0, the mass of the body when it is at rest.* The variation of mass with speed is again very small unless the speed is very great.

Example 29.10

The relativistic mass for various values of v. Find the mass m of a moving object when (a) $v = 1610$ km/hr $= 1000$ mph, (b) $v = 1610$ km/s $= 1000$ miles/s, (c) $v = 0.8c$, and (d) $v = c$.

Solution

The relativistic mass m is found in terms of its rest mass m_0 by equation 29.86.

a. For $v = 1610$ km/hr $= 447$ m/s, we obtain

$$m = \frac{m_0}{\sqrt{1 - v^2/c^2}}$$

$$= \frac{m_0}{\sqrt{1 - (447 \text{ m/s})^2/(3.00 \times 10^8 \text{ m/s})^2}}$$

$$= m_0$$

Thus, at this reasonably high speed there is no measurable difference in the mass of the body.

b. For $v = 1610$ km/s,

$$m = \frac{m_0}{\sqrt{1 - v^2/c^2}} = \frac{m_0}{\sqrt{1 - (1.610 \times 10^6)^2/(3.00 \times 10^8)^2}}$$

$$= \frac{m_0}{\sqrt{0.99997}} = \frac{m_0}{0.99999}$$

$$= 1.00001 m_0$$

Thus, for a speed of 1610 km/s $= 3,600,000$ mph, a speed so great that macroscopic objects cannot yet attain it, the relativistic increase in mass is still practically negligible.

c. For $v = 0.8c$,

$$m = \frac{m_0}{\sqrt{1 - v^2/c^2}} = \frac{m_0}{\sqrt{1 - (0.8c)^2/(c)^2}}$$

$$= 1.67 m_0$$

For the rather large velocity of $0.8c$, the increase in mass is very significant. We should note that it is almost routine today to accelerate elementary particles to speeds approaching the speed of light and in all such cases this variation of mass with speed is observed.

d. For $v = c$,

$$m = \frac{m_0}{\sqrt{1 - v^2/c^2}} = \frac{m_0}{\sqrt{1 - c^2/c^2}} = \frac{m_0}{0}$$

$$= \infty$$

Thus, as a particle approaches the speed of light c, the mass of the particle approaches infinity. Since an infinite force and infinite energy would be required to move an infinite mass it is obvious that a particle of a finite rest mass m_0 can never be accelerated to the speed of light.

The first of many experiments to verify the change in mass with speed was performed by A. H. Bucherer in 1909. Electrons were first accelerated by a large potential difference until they were moving at high speeds. They then entered a velocity selector, such as described in section 22.1. By varying the electric and magnetic field of the velocity selector, electrons with any desired velocity can be obtained by equation 22.10. These electrons were then sent through a uniform magnetic field B where they were deflected into a circular path, also described in section 22.1. The centripetal force was set equal to the magnetic force, and we obtained

$$\frac{mv^2}{r} = qvB \tag{22.8}$$

Simplifying,

$$mv = qBr \tag{29.87}$$

But now we must treat the mass in equation 29.87 as the relativistic mass in equation 29.86. Thus, equation 29.87 becomes

$$\frac{m_0 v}{\sqrt{1 - v^2/c^2}} = qBr \tag{29.88}$$

Because B, r, and v could be measured in the experiment, the ratio of the charge of the electron q to its rest mass m_0, found from equation 29.88, is

$$\frac{q}{m_0} = \frac{v}{Br\sqrt{1 - v^2/c^2}} \tag{29.89}$$

Bucherer's experiment confirmed equation 29.89 and hence the variation of mass with speed. Since 1909, thousands of experiments have been performed confirming the variation of mass with speed.

The variation of mass with speed truly points out the meaning of the concept of inertial mass as a measure of the resistance of matter to motion. As we can see with this relativistic mass, at higher and higher speeds there is a much greater resistance to motion and this is manifested as the increase in the mass of the body. The rest mass m_0 should probably be called the "quantity of matter" of a body since it is truly a measure of how much matter is present in the body, whereas the relativistic mass is the measure of the resistance of that quantity of matter to being put into motion.

*With this new definition of relativistic mass the **relativistic linear momentum** can now be defined as*

$$\mathbf{p} = m\mathbf{v} = \frac{m_0 \mathbf{v}}{\sqrt{1 - v^2/c^2}} \tag{29.90}$$

The law of conservation of momentum now holds for relativistic mechanics just as it did for Newtonian mechanics. In fact, we can rewrite equation 29.84 as

$$m_0 V_y = m V_y \sqrt{1 - v^2/c^2}$$

Substituting for m from equation 29.86,

$$m_0 V_y = \frac{m_0}{\sqrt{1 - v^2/c^2}} V_y \sqrt{1 - v^2/c^2}$$

Simplifying,

$$m_0 V_y = m_0 V_y \qquad (29.91)$$

Hence, using the concept of relativistic mass, the same form of the equation for the law of conservation of momentum 29.91 is obtained as for the Newtonian case in equation 29.81. Thus, momentum is always conserved if the relativistic mass is used and the law of conservation of momentum is preserved for relativistic mechanics.

Newton's second law is still valid for relativistic mechanics, but only in the form

$$F = \frac{\Delta p}{\Delta t} = \frac{\Delta(mv)}{\Delta t} = \frac{\Delta}{\Delta t}\left(\frac{m_0 v}{\sqrt{1 - v^2/c^2}}\right) \qquad (29.92)$$

29.12 The Law of Conservation of Mass-Energy

An important consequence of the variation of mass with speed can be found from equation 29.86 by rewriting it in the form

$$m = \frac{m_0}{\sqrt{1 - v^2/c^2}} = m_0(1 - v^2/c^2)^{-1/2} \qquad (29.93)$$

For relatively small velocities the term $(1 - v^2/c^2)^{-1/2}$ can be expanded by the binomial theorem, equation 29.34, as

$$(1 - x)^n = 1 - nx$$

$$\left(1 - \frac{v^2}{c^2}\right)^{-1/2} = 1 + \frac{1}{2}\frac{v^2}{c^2} \qquad (29.94)$$

Substituting equation 29.94 back into equation 29.93, we get

$$m = m_0\left(1 + \frac{1 v^2}{2 c^2}\right)$$

$$= m_0 + \frac{1}{2}\frac{m_0 v^2}{c^2}$$

Multiplying each term by c^2, gives

$$mc^2 = m_0 c^2 + \frac{1}{2} m_0 v^2$$

Rearranging terms,

$$\frac{1}{2} m_0 v^2 = mc^2 - m_0 c^2 \qquad (29.95)$$

The left-hand side of equation 29.95 is the kinetic energy of a slowly moving particle, so we can generalize 29.95 to

$$KE = mc^2 - m_0 c^2 \qquad (29.96)$$

We can also write the relativistic mass m as

$$m = m_0 + \Delta m \qquad (29.97)$$

That is, *the relativistic mass is equal to the rest mass plus the change in mass due to motion.* Substituting equation 29.97 back into equation 29.96, we have

$$KE = (m_0 + \Delta m)c^2 - m_0 c^2$$

or

$$KE = (\Delta m)c^2 \tag{29.98}$$

Thus, relativistically, the kinetic energy of a body is equal to the change in mass of the body caused by the motion times the velocity of light squared.

Notice that the left-hand side of either equation 29.96 or 29.98 represents an energy. Since the left-hand side of the equation is equal to the right-hand side of the equation, the right-hand side must also represent an energy. That is, *the product of a mass times the square of the speed of light must equal an energy. The total **relativistic energy** of a body is, therefore, defined as*

$$E = mc^2 \tag{29.99}$$

We can rewrite equation 29.96 as

$$mc^2 = KE + m_0c^2$$

In view of the definition in equation 29.99, the total energy of a body is

$$E = KE + m_0c^2 \tag{29.100}$$

When a particle is at rest, its kinetic energy KE is equal to zero. Therefore, the total energy of the particle when it is at rest must be equal to m_0c^2. *The **rest mass energy** of a particle can then be defined as*

$$E_0 = m_0c^2 \tag{29.101}$$

Substituting equation 29.101 back into equation 29.100, we get

$$E = KE + E_0 \tag{29.102}$$

Equation 29.102 states that the total energy of a body is equal to its kinetic energy plus its rest mass energy. The result of these equations is that energy can manifest itself as mass, and mass can manifest itself as energy. In a sense, mass can be thought of as being frozen energy.

Example 29.11

Energy in a 1-kg mass. How much energy is stored in a 1.00-kg mass?

Solution

The rest mass energy of a 1.00-kg mass, given by equation 29.101, is

$$E_0 = m_0c^2$$
$$= (1.00 \text{ kg})(3.00 \times 10^8 \text{ m/s})^2$$
$$= 9.00 \times 10^{16} \text{ J}$$

This is an enormous amount of energy to be sure. It is about a thousand times greater than the energy released from the atomic bomb dropped on Hiroshima, and could supply 2.85 gigawatts of power for a period of one year.

We should note that in the derivation of the mass-energy equivalence we assumed that the speeds were low so that the kinetic energy is equal to $\frac{1}{2}mv^2$. *At very high speeds where relativistic effects are important, the kinetic energy is no longer equal to $\frac{1}{2}mv^2$, but is still given by equation 29.96.*

Example 29.12

Relativistic and classical kinetic energy. A 1.00-kg object is accelerated to a speed of 0.4c. Find its kinetic energy (a) relativistically and (b) classically.

a.   The relativistic kinetic energy of the moving body, found from equation 29.96, is

$$KE = mc^2 - m_0c^2$$

$$= \frac{m_0c^2}{\sqrt{1 - v^2/c^2}} - m_0c^2$$

$$= \frac{1.00 \text{ kg}(3.0 \times 10^8 \text{ m/s})^2}{\sqrt{1 - (0.4c)^2/c^2}} - 1.00 \text{ kg}(3.00 \times 10^8 \text{ m/s})^2$$

$$= 8.20 \times 10^{15} \text{ J}$$

b.   The classical, and wrong, determination of the kinetic energy is

$$KE = \tfrac{1}{2}mv^2 = \tfrac{1}{2}(1.00 \text{ kg})[(0.4)(3.00 \times 10^8 \text{ m/s})]^2$$

$$= 7.20 \times 10^{15} \text{ J}$$

That is, if an experiment were performed to test these two results, the classical result would not agree with the experimental results, but the relativistic one would agree.

When dealing with charged elementary particles, the kinetic energy can be found as the work that you must do in order to accelerate the particle up to the speed v, and was given by equation 19.36 as

$$KE = \text{work done} = qV$$

Example 29.13

Kinetic energy of an electron. An electron is accelerated through a uniform potential difference of 2.00 × 10^6 V. What is its kinetic energy as it leaves the electric field?

The kinetic energy, found from equation 19.36, is

$$KE = qV$$

$$= (1.60 \times 10^{-19} \text{ C})(2.00 \times 10^6 \text{ V})$$

$$= 3.20 \times 10^{-13} \text{ J}$$

It is customary in relativity and modern physics to express energies in terms of electron volts, abbreviated eV. In section 19.9 we saw that the unit of energy called an electron volt is equal to the energy that an electron would acquire as it falls through a potential difference of 1 V. Hence,

$$KE = qV$$

$$1 \text{ eV} = (1.60 \times 10^{-19} \text{ C})(1.00 \text{ V})$$

$$1 \text{ eV} = 1.60 \times 10^{-19} \text{ J} \tag{29.103}$$

Thus, the electron volt is also a unit of energy.

Now we can express the KE in example 29.13 in electron volts as

$$KE = (3.20 \times 10^{-13} \text{ J})\left(\frac{1 \text{ eV}}{1.60 \times 10^{-19} \text{ J}}\right)$$

$$= 2.00 \times 10^6 \text{ eV}$$

For larger quantities of energy the following units of energy are used:

$$1 \text{ kilo electron volt} = 1 \text{ keV} = 10^3 \text{ eV}$$
$$1 \text{ mega electron volt} = 1 \text{ MeV} = 10^6 \text{ eV}$$
$$1 \text{ giga electron volt} = 1 \text{ GeV} = 10^9 \text{ eV}$$
$$1 \text{ tera electron volt} = 1 \text{ TeV} = 10^{12} \text{ eV}$$

Hence, the energy in example 29.13 can be expressed as

$$KE = 2.00 \text{ MeV}$$

By far, the greatest implication of equations 29.99 and 29.101 is that mass and energy must be considered as a manifestation of the same thing. *Thus, mass and energy are not independent quantities, just as we found that space and time are no longer independent quantities. Just as space and time are fused into space-time, we must now fuse the separate concepts of mass and energy into one concept called mass-energy. What was classically considered as two separate laws, namely the law of conservation of mass and the law of conservation of energy must now be considered as one single law—the law of conservation of mass-energy. That is, mass can be created or destroyed as long as an equal amount of energy vanishes or appears, respectively.*

Because mass and energy can be equated it is sometimes desirable to express the mass of a particle in terms of energy units. Let us start by *defining an atomic unit of mass, called the unified mass unit, and defined as one-twelfth of the mass of the carbon 12 atom.* In equation 15.47, the mass of a molecule was given by

$$m = \frac{M}{N_A}$$

where M was the molecular mass of the molecule and N_A was Avogadro's number. For a single atom the molecular mass is replaced by its atomic mass and the mass of a single atom is given by

$$m = \frac{\text{atomic mass}}{N_A}$$

Thus, we define the *unified mass unit,* u, as

$$1 \text{ u} = \frac{1}{12} m_C = \frac{1}{12} \frac{12 \text{ kg/kilomole}}{(6.0221367 \times 10^{26} \text{ molecules/kilomole})}$$

$$1 \text{ u} = 1.660540 \times 10^{-27} \text{ kg} \tag{29.104}$$

To express this mass unit in terms of energy, we use equation 29.101 as

$$E_0 = m_0 c^2$$

$$= (1 \text{ u})(c^2)$$

$$= \frac{(1.660540 \times 10^{-27} \text{ kg})(2.997925 \times 10^8 \text{ m/s})^2}{\left(1.60219 \times 10^{-19}\frac{\text{J}}{\text{eV}}\right)\left(\frac{10^6 \text{ eV}}{\text{MeV}}\right)}$$

$$= 931.493 \text{ MeV}$$

More significant figures have been used in this calculation than has been customary in this book. The additional accuracy is necessary because of the small quantities that are dealt with. Hence, a unified mass unit u has an energy equivalent of 931.493 MeV, that is,

$$1 \text{ u} = 931.493 \text{ MeV} \qquad (29.105)$$

The masses of some of the elementary particles in terms of unified mass units and MeVs are given as

$$\text{rest mass of proton} = m_p = 1.00726 \text{ u} = 938.256 \text{ MeV}$$

$$\text{rest mass of neutron} = m_n = 1.00865 \text{ u} = 939.550 \text{ MeV}$$

$$\text{rest mass of electron} = m_e = 0.00055 \text{ u} = 0.511006 \text{ MeV}$$

$$\text{rest mass of deuteron} = m_d = 2.01410 \text{ u} = 1875.580 \text{ MeV}$$

Example 29.14

The energy of the deuteron. Deuterium is an isotope of hydrogen whose nucleus, called a *deuteron,* consists of a proton and a neutron. Find the sum of the rest mass energies of the proton and the neutron, and compare it with the rest mass energy of the deuteron.

Solution

The sum of the rest mass energy of the proton and neutron is

$$m_p + m_n = 938.26 \text{ MeV} + 939.55 \text{ MeV} = 1877.81 \text{ MeV}$$

The actual rest mass of the deuteron is $m_d = 1875.58$ MeV. Thus, the sum of the masses of the individual proton and neutron is greater than the mass of the deuteron itself. The difference in mass is

$$\Delta m = (m_p + m_n) - m_d$$

$$= 1877.81 \text{ MeV} - 1875.58 \text{ MeV}$$

$$= 2.23 \text{ MeV}$$

That is, some mass Δm and hence energy is lost in combining the proton and the neutron. The lost energy that binds the proton and neutron together is called the binding energy of the system. This is the amount of energy that must be supplied to break up the deuteron.

A further and extremely important application of mass-energy conversions occurs in the fusion of light atoms into heavier atoms. The most famous of such fusion processes is the conversion of hydrogen to helium in the sun and in the hydrogen bomb. An extremely simplified version of the process can be obtained by considering the mass of helium as consisting of two protons, two neutrons, and two electrons. The atomic mass of helium, as determined by the rest masses of its constituents, is

$$m_{He} = 2m_p + 2m_n + 2m_e$$

$$= 2(938.256 \text{ MeV}) + 2(939.550 \text{ MeV}) + 2(0.511006 \text{ MeV})$$

$$= 1876.512 \text{ MeV} + 1879.100 \text{ MeV} + 1.0220 \text{ MeV}$$

$$= 3756.634 \text{ MeV}$$

If this value is compared to the atomic mass of helium from the table of elements we find

$$\text{Atomic mass of He} = (4.002603 \text{ u})\left(\frac{931.493 \text{ MeV}}{\text{u}}\right)$$

$$= 3728.397 \text{ MeV}$$

Hence, helium is lighter than the sum of its constituent parts. The difference in mass between helium and its constituent parts is

$$\Delta m = 3756.634 \text{ MeV} - 3728.397 \text{ MeV}$$

$$= 28.237 \text{ MeV}$$

Thus, 28.237 MeV of energy is given off for each atom of helium formed. For the formation of 1 mole of helium, there are 6.02×10^{23} atoms. Hence, the total energy released per mole of helium formed is

$$\frac{\text{Energy released}}{\text{mole}} = \left(28.237 \frac{\text{MeV}}{\text{atom}}\right)\left(6.02 \times 10^{23} \frac{\text{atoms}}{\text{mole}}\right)$$

$$= (1.70 \times 10^{25} \text{ MeV})\left(\frac{10^6 \text{ eV}}{\text{MeV}}\right)\left(\frac{1.60 \times 10^{-19} \text{ J}}{\text{eV}}\right)$$

$$= 2.73 \times 10^{12} \text{ J}$$

Hence, in the formation of 1 mole of helium, a mass of only 4 g, 2,730,000,000,000 J or 2,580,000,000 Btu of energy are released. This monumental amount of energy, which comes from the conversion of mass into energy, is continually being released by the sun. This fusion process is also the source of energy in the hydrogen bomb.

Example 29.15

A high-speed electron. An electron is accelerated from rest through a potential difference of 3.00×10^5 V. Find (a) the kinetic energy of the electron, (b) the total energy of the electron, (c) the speed of the electron, (d) the relativistic mass of the electron, and (e) the momentum of the electron.

Solution

a. The kinetic energy of the electron, found from equation 19.36, is

$$\text{KE} = \text{work done} = qV$$

$$\text{KE} = (1.60 \times 10^{-19} \text{ C})(3.00 \times 10^5 \text{ V})$$

$$\text{KE} = (4.80 \times 10^{-14} \text{ J})\left(\frac{1 \text{ eV}}{1.60 \times 10^{-19} \text{ J}}\right)$$

$$\text{KE} = (3.00 \times 10^5 \text{ eV})\left(\frac{1 \text{ MeV}}{10^6 \text{ eV}}\right) = 0.300 \text{ MeV}$$

b. The rest mass energy of the electron is

$$E_0 = (m_0 c^2)_{\text{electron}} = 0.511 \text{ MeV}$$

Thus, the total relativistic energy E, found from equation 29.100, is

$$E = \text{KE} + m_0 c^2$$

$$= 0.300 \text{ MeV} + 0.511 \text{ MeV}$$

$$= 0.811 \text{ MeV}$$

c. To determine the speed of the electron, equation 29.96 is rearranged as

$$KE = mc^2 - m_0c^2$$

$$= \frac{m_0c^2}{\sqrt{1 - v^2/c^2}} - m_0c^2 = \left(\frac{1}{\sqrt{1 - v^2/c^2}} - 1\right)m_0c^2$$

$$\frac{1}{\sqrt{1 - v^2/c^2}} - 1 = \frac{KE}{m_0c^2}$$

$$\frac{1}{\sqrt{1 - v^2/c^2}} = \frac{KE}{m_0c^2} + 1 = \frac{0.300 \text{ MeV}}{0.511 \text{ MeV}} + 1 = 1.587$$

$$\sqrt{1 - v^2/c^2} = \frac{1}{1.587} = 0.630$$

$$1 - \frac{v^2}{c^2} = (0.630)^2 = 0.397$$

$$\frac{v^2}{c^2} = 1 - 0.397 = 0.603$$

$$v = \sqrt{0.603c^2}$$

$$v = 0.776c$$

Hence, the speed of the electron is approximately seven-tenths the speed of light.

d. To determine the relativistic mass of the electron, we use equation 29.86:

$$m = \frac{m_0}{\sqrt{1 - v^2/c^2}}$$

$$= \frac{9.11 \times 10^{-31} \text{ kg}}{\sqrt{1 - (0.776c)^2/c^2}}$$

$$= 14.4 \times 10^{-31} \text{ kg}$$

The relativistic mass has increased by approximately 1.6 times the rest mass.

e. The momentum of the electron, found from equation 29.90, is

$$p = mv = \frac{m_0}{\sqrt{1 - v^2/c^2}}v$$

$$= (14.4 \times 10^{-31} \text{ kg})(0.776)(3.00 \times 10^8 \text{ m/s})$$

$$= 3.35 \times 10^{-22} \text{ kg m/s}$$

The Language of Physics

Relativity
The observation of the motion of a body by two different observers in relative motion to each other. At speeds approaching the speed of light, the length of a body contracts, its mass increases, and time slows down (p. 843).

Inertial coordinate system
A frame of reference that is either at rest or moving at a constant velocity (p. 845).

Galilean transformations
A set of classical equations that relate the motion of a body in one inertial coordinate system to that in a second inertial coordinate system. All the laws of classical mechanics are invariant under a Galilean transformation, but the laws of electromagnetism are not (p. 846).

Invariant quantity
A quantity that remains a constant whether it is observed from a system at rest or in motion (p. 850).

Ether
A medium that was assumed to pervade all space. This was the medium in which light was assumed to propagate (p. 853).

Michelson-Morley experiment
A crucial experiment that was performed to detect the presence of the ether. The results of the experiment indicated that if the ether exists it cannot be detected. The assumption is then made that if it cannot be detected, it does not exist. Hence, light does not need a medium to propagate through. The experiment also implied that the speed of light in free space is the same everywhere regardless of the motion of the source or the observer (p. 858).

Special or Restricted Theory of Relativity
Einstein stated his special theory of relativity in terms of two postulates.

Postulate 1: The laws of physics have the same form in all inertial frames of reference.

Postulate 2: The speed of light in free space has the same value for all observers, regardless of their state of motion.

In order for the speed of light to be the same for all observers, space and time itself must change. The special theory is restricted to inertial systems and does not apply to accelerated systems (p. 859).

Lorentz transformations
A new set of transformation equations to replace the Galilean transformations. These new equations are derived by the two postulates of special relativity. These equations show that space and time are intimately connected. The effects of relativity only manifests itself when objects are moving at speeds approaching the speed of light (p. 862).

Proper length
The length of an object that is measured in a frame where the object is at rest (p. 865).

Lorentz-Fitzgerald contraction
The length of a rod in motion as measured by an observer at rest is less than its proper length (p. 866).

Proper time
The time interval measured on a clock that is at rest relative to the observer (p. 869).

Time dilation
The time interval measured on a moving clock is less than the proper time. Hence, moving clocks slow down (p. 870).

Proper mass or rest mass
The mass of a body that is at rest in a frame of reference (p. 877).

Relativistic mass
The mass of a body that is in motion. The relativistic mass is always greater than the rest mass of the object (p. 877).

Relativistic linear momentum
The product of the relativistic mass of a body and its velocity (p. 878).

Relativistic energy
The product of the relativistic mass of a body and the square of the speed of light. This total energy is equal to the sum of the kinetic energy of the body and its rest mass energy (p. 880).

Rest mass energy
The product of the rest mass and the square of the speed of light. Hence, mass can manifest itself as energy, and energy can manifest itself as mass (p. 880).

The law of conservation of mass-energy
Mass can be created or destroyed as long as an equal amount of energy vanishes or appears, respectively (p. 882).

Summary of Important Equations

Galilean transformation of coordinates
$$x = x' + vt \quad (29.1)$$
$$y = y' \quad (29.2)$$
$$z = z' \quad (29.3)$$
$$t = t' \quad (29.4)$$

Galilean transformation of velocities
$$v_x = v'_x + v \quad (29.11)$$
$$v'_x = v_x - v \quad (29.13)$$
$$v'_y = v_y \quad (29.14)$$
$$v'_z = v_z \quad (29.15)$$

Lorentz transformation equations of coordinates
$$x' = \frac{x - vt}{\sqrt{1 - v^2/c^2}} \quad (29.49)$$
$$y' = y$$
$$z' = z$$
$$t' = \frac{t - xv/c^2}{\sqrt{1 - v^2/c^2}} \quad (29.50)$$

Inverse Lorentz transformation equations of coordinates
$$x = \frac{x' + vt'}{\sqrt{1 - v^2/c^2}} \quad (29.51)$$

$$y = y' \quad (29.52)$$
$$z = z' \quad (29.53)$$
$$t = \frac{t' + x'v/c^2}{\sqrt{1 - v^2/c^2}} \quad (29.54)$$

Length contraction
$$L = L_0\sqrt{1 - v^2/c^2} \quad (29.60)$$

Time dilation
$$\Delta t = \frac{\Delta t_0}{\sqrt{1 - v^2/c^2}} \quad (29.64)$$

Lorentz transformation of velocities
$$V'_x = \frac{V_x - v}{1 - (v/c^2)V_x} \quad (29.75)$$
$$V'_y = \frac{V_y\sqrt{1 - v^2/c^2}}{1 - (v/c^2)V_x} \quad (29.76)$$
$$V'_z = \frac{V_z\sqrt{1 - v^2/c^2}}{1 - (v/c^2)V_x} \quad (29.77)$$

Relativistic mass
$$m = \frac{m_0}{\sqrt{1 - v^2/c^2}} \quad (29.86)$$

Linear momentum
$$\mathbf{p} = m\mathbf{v} = \frac{m_0 \mathbf{v}}{\sqrt{1 - v^2/c^2}} \quad (29.90)$$

Newton's second law
$$F = \frac{\Delta p}{\Delta t} = \frac{\Delta(mv)}{\Delta t}$$
$$F = \frac{\Delta}{\Delta t}\left(\frac{(m_0 v)}{\sqrt{1 - v^2/c^2}}\right) \quad (29.92)$$

Relativistic kinetic energy
$$KE = mc^2 - m_0c^2 \quad (29.96)$$
$$KE = (\Delta m)c^2 \quad (29.98)$$

Total relativistic energy
$$E = mc^2 \quad (29.99)$$

Rest mass energy
$$E_0 = m_0c^2 \quad (29.101)$$

Law of conservation of relativistic energy
$$E = KE + E_0 \quad (29.102)$$

Electron volt
$$1 \text{ eV} = 1.60 \times 10^{-19} \text{ J} \quad (29.103)$$
$$u = 1.66 \times 10^{-27} \text{ kg} \quad (29.104)$$

Unified mass unit
$$u = 931.493 \text{ MeV} \quad (29.105)$$

1. If you are in an enclosed truck and cannot see outside, how can you tell if you are at rest, in motion at a constant velocity, speeding up, slowing down, turning to the right, or turning to the left?
†2. Does a length contract perpendicular to its direction of motion?
†3. Lorentz explained the negative result of the Michelson-Morley experiment by saying that the ether caused the length of the telescope in the direction of motion to be contracted by an amount given by $L = L_0\sqrt{1 - v^2/c^2}$. Would this give a satisfactory explanation of the Michelson-Morley experiment?

4. If the speed of light in our world was only 100 km/hr, describe some of the characteristics of this world.
†5. Does time dilation affect the physiological aspects of the human body, such as aging? How does the body know what time is?
6. Are length contraction and time dilation real or apparent?
7. An elementary particle called a neutrino moves at the speed of light. Must it have an infinite mass? Explain.
†8. It has been suggested that particles might exist that are moving at speeds greater than c. These particles, which have never been found, are called tachyons. Describe how such particles might exist and what their characteristics would have to be.

9. In the equation for the total relativistic energy of a body, could there be another term for the potential energy of a body? Does a compressed spring, which has potential energy, have more mass than a spring that is not compressed?
†10. When helium is formed, the difference in the mass of helium and the mass of its constituents is given off as energy. When the deuteron is formed, the difference in mass is also given off as energy. Could the formation of deuterium be used as a source of commercial energy?
11. If the speed of light were infinite, what would the Lorentz transformation equations reduce to?
†12. Can you apply the Lorentz transformations to a reference frame that is moving in a circle?

Problems for Chapter 29

29.1 Introduction to Relative Motion

1. A projectile is thrown straight upward at an initial velocity of 25.0 m/s from an open truck at the same instant that the truck starts to accelerate forward at 5.00 m/s². If the truck is 4.00 m long, how far behind the truck will the projectile land?
2. A projectile is thrown straight up at an initial velocity of 25.0 m/s from an open truck that is moving at a constant speed of 10.0 m/s. Where does the projectile land when (a) viewed from the ground (S frame) and (b) when viewed from the truck (S' frame)?
3. A truck moving east at a constant speed of 50.0 km/hr passes a traffic light where a car just starts to accelerate from rest at 2.00 m/s². At the end of 10.0 s, what is the velocity of the car with respect to (a) the traffic light and (b) with respect to the truck?
4. A woman is sitting on a bus 5.00 m from the end of the bus. If the bus is moving forward at a velocity of 7.00 m/s, how far away from the bus station is the woman after 10.0 s?

29.2 The Galilean Transformations of Classical Physics

5. The woman on the bus in problem 4 gets up and (a) walks toward the front of the bus at a velocity of 0.500 m/s. What is her velocity relative to the bus station? (b) The woman now walks toward the rear of the bus at a velocity of 0.500 m/s. What is her velocity relative to the bus station?

29.3 The Invariance of the Mechanical Laws of Physics under a Galilean Transformation

†6. Filling in the steps omitted in the derivation associated with figure 29.8, show that the law of conservation of momentum is invariant under a Galilean transformation.
†7. Show that the law of conservation of energy for a perfectly elastic collision is invariant under a Galilean transformation.

29.5 The Michelson-Morley Experiment

8. A boat travels at a speed V of 5.00 km/hr with respect to the water, as shown in figure 29.10. If it takes 90.0 s to cross the river and return and 95.0 s for the boat to go the same distance downstream and return, what is the speed of the river current?

29.7 The Lorentz Transformation

9. A woman on the earth observes a firecracker explode 10.0 m in front of her when her clock reads 5.00 s. An astronaut in a rocket ship who passes the woman on earth at $t = 0$, at a speed of $0.400c$ finds what coordinates for this event?
10. A clock in the moving coordinate system reads $t' = 0$ when the stationary clock reads $t = 0$. If the moving frame moves at a speed of $0.800c$, what time will the moving clock read when the stationary observer reads 15.0 hr on her clock?
†11. Use the Lorentz transformation to show that the equation for a light wave, equation 29.25, has the same form in a coordinate system moving at a constant velocity.

29.8 The Lorentz-Fitzgerald Contraction

12. The USS *Enterprise* approaches the planet Seti Alpha 5 at a speed of 0.800c. Captain Kirk observes an airplane runway on the planet to be 2.00 km long. The air controller on the planet says that the runway on the planet is how long?

13. The starship *Regulus* was measured to be 100 m long when in space dock. If it approaches a planet at a speed of 0.400c, how long does it appear to an observer on the planet?
14. How fast must a 15.0-ft car move in order to fit into a 1.00-ft garage? Could you park the car in this garage?

15. A comet is observed to be 130 km long as it moves past an observer at a speed of 0.700c. How long does the comet seem when it travels at a speed of 0.900c with respect to the observer?
16. A meterstick at rest makes an angle of 30.0° with the x-axis. Find the length of the meterstick and the angle it makes with the x'-axis for an observer moving parallel to the x-axis at a speed of 0.650c.

29.9 Time Dilation

17. A particle is observed to have a lifetime of 1.50×10^{-6} s when it is at rest in the laboratory. (a) What is its lifetime when it is moving at 0.800c? (b) How far will the particle move with respect to the moving frame of reference before it decays? (c) How far will the particle move with respect to the laboratory frame before it decays?

18. A stroboscope is flashing light signals at the rate of 2100 flashes/min. An observer in a rocket ship traveling toward the strobe light at 0.500c would see what flash rate?
19. A particle has a lifetime of 0.100 s when observed while it moves at a speed of 0.650c with respect to the laboratory. What is its lifetime in its rest frame?

29.10 Transformation of Velocities

20. A spaceship traveling at a speed of 0.600c relative to a planet launches a rocket backward at a speed of 0.500c. What is the velocity of the rocket as observed from the planet?
21. The three electrons are moving at the velocities shown in the diagram. Find the relative velocities between (a) electrons 1 and 2, (b) electrons 2 and 3, and (c) electrons 1 and 3.

29.11 The Law of Conservation of Momentum and Relativistic Mass

22. What is the mass of the following particles when traveling at a speed of 0.86c: (a) electron, (b) proton, and (c) neutron?
23. Find the speed of a particle at which the mass m is equal to (a) $0.100\ m_0$, (b) $1.00\ m_0$, (c) $10.0\ m_0$, (d) $100\ m_0$, and (e) $1000\ m_0$.
24. Determine the linear momentum of an electron moving at a speed of 0.990c.
25. How fast must a proton move so that its linear momentum is 8.08×10^{-19} kg m/s?
26. Compute the speed of a neutron whose total energy is 1.88×10^{-10} J.

29.12 The Law of Conservation of Mass-Energy

27. An isolated neutron is capable of decaying into a proton and an electron. How much energy is liberated in this process?
28. Since it takes 540 kcal to convert 1.00 kg of water to 1.00 kg of steam at 100 °C, what is the increase in mass of the steam?
29. What is the kinetic energy of a proton traveling at 0.800c?

30. Through what potential difference must an electron be accelerated if it is to attain a speed of 0.800c?
31. What is the total energy of a proton traveling at a speed of 2.50×10^8 m/s?
32. Calculate the speed of an electron whose kinetic energy is twice as large as its rest mass energy.

Additional Problems

33. If an ion-engine in a spacecraft can produce a continuous acceleration of 0.200 m/s², how long must the engine continue to accelerate if it is to reach the speed of 0.500c?
†34. The volume of a cube is V_0 in a frame of reference where it is at rest. Show that the volume observed in a moving frame of reference is given by

$$V = V_0 \sqrt{1 - v^2/c^2}$$

35. The distance to Alpha Centari, the closest star, is about 4.00 light years as measured from earth. What would this distance be as observed from a spaceship leaving earth at a speed of 0.500c? How long would it take to get there according to a clock on the spaceship and a clock on earth?
36. A muon is an elementary particle that is observed to have a lifetime of 2.00×10^{-6} s before decaying. It has a typical speed of 2.994×10^8 m/s. (a) How far can the muon travel before it decays? (b) These particles are observed high in our atmosphere, but with such a short lifetime how do they manage to get to the surface of the earth?
†37. Show that the formula for the density of a cube of material moving at a speed v is given by

$$\rho = \frac{\rho_0}{1 - v^2/c^2}$$

†38. A proton is accelerated to a speed of 0.500c. Find its (a) kinetic energy, (b) total energy, (c) relativistic mass, and (d) momentum.
†39. Show that the speed of a particle can be given by

$$v = c\sqrt{1 - (E_0/E)^2}$$

where E_0 is the rest mass energy of the particle and E is its total energy.
†40. An electron is accelerated from rest through a potential difference of 4.00×10^6 V. Find (a) the kinetic energy of the electron, (b) the total energy of the electron, (c) the velocity of the electron, (d) the relativistic mass, and (e) the momentum of the electron.

†**41.** From the solar constant, determine the total energy transmitted by the sun per second. How much mass is this equivalent to? If the mass of the sun is 1.99×10^{30} kg, approximately how long can the sun continue to radiate energy?

†**42.** A reference frame is accelerating away from a rest frame. Show that Newton's second law in the form $F = ma$ does not hold in the accelerated frame.

Interactive Tutorials

43. Length contraction. The length of a rod at rest is found to be $L_0 = 2.55$ m. Find the length L of the rod when observed by an observer in motion at a speed $v = 0.250c$.

44. Time dilation. A clock in a moving rocket ship reads a time duration $\Delta t_0 = 15.5$ hr. What time elapses, Δt, on earth if the rocket ship is moving at a speed $v = 0.355c$?

45. Relative velocities. Two spaceships are approaching a space station, as in figure 29.15. Spaceship 1 has a velocity of 0.55c to the left and spaceship 2 has a velocity of 0.75c to the right. Find the velocity of rocket ship 1 as observed by rocket ship 2.

46. Relativistic mass. A mass at rest has a value $m_0 = 2.55$ kg. Find the relativistic mass m when the object is moving at a speed $v = 0.355c$.

47. The length of a rod at rest is $L_0 = 1.00$ m and its mass is $m_0 = 1.00$ kg. Find the length L and mass m of the rod as its speed v in the axial direction increases from $0.00c$ to $0.90c$, where c is the speed of light ($c = 3.00 \times 10^8$ m/s). Plot the results.

48. An accelerated charged particle. An electron is accelerated from rest through a potential difference $V = 4.55 \times 10^5$ V. Find (a) the kinetic energy of the electron, (b) the rest mass energy of the electron, (c) the total relativistic energy of the electron, (d) the speed of the electron, (e) the relativistic mass of the electron, and (f) the momentum of the electron.

30

Spacetime and General Relativity

Artist's impression of massive black hole.

The views of space and time which I wish to lay before you have sprung from the soil of experimental physics, and therein lies their strength. They are radical. Henceforth, space by itself, and time by itself are doomed to fade away into mere shadows, and only a kind of union of the two will preserve an independent reality.

H. Minkowski—"Space and Time"

30.1 Spacetime Diagrams

Shortly after Einstein published his special theory of relativity, Hermann Minkowski (1864–1909), a former instructor of Einstein, set about to geometrize relativity. He said that time and space are inseparable. In his words, "Nobody has ever noticed a place except at a time, or a time except at a place. . . . A point of space at a point of time, that is, a system of values of x, y, z, t, I will call a world-point. The multiplicity of all thinkable x, y, z, t, systems of values we will christen the world."[1]

To simplify the discussion, we will consider only one space dimension, namely the x-coordinate. *Any occurrence in spacetime will be called an event*, and is represented in the **spacetime diagram** of figure 30.1(a). This event might be the explosion of a firecracker, let us say. The location of this event is the *world point*, and it has the coordinates x and t. (Many authors of more advanced relativity books interchange the coordinates, showing the time axis in the vertical direction to emphasize that this is a different graph than a conventional plot of distance versus time. However, we will use the conventional graphical format in this book because it is already familiar to the student and will therefore make spacetime concepts easier to understand.)

Figure 30.1(b) is a picture of a **world line** of a particle at rest at the position *x. The graph shows that even though the particle is at rest in space, it is still moving through time*. Its x-coordinate is a constant because it is not moving through space, but its time coordinate is continually increasing showing its motion through time. Figure 30.1(c) represents a rod at rest in spacetime. The top line represents the world line of the end of the rod at x_2, whereas the bottom line represents the world line of the opposite end of the rod at x_1. Notice that the stationary rod sweeps out an area in spacetime. Figure 30.1(d) shows the world line of particle A moving at a constant velocity v_A and the world line of particle B moving at the constant velocity v_B. The slope of a straight line on an x versus t graph represents the velocity of the particle. The greater the slope, the greater the velocity. Since particle A has the greater slope it has the greater velocity, that is, $v_A > v_B$. If the velocity of a particle changes with time, its world line is no longer a straight line, but becomes curved, as shown in figure 30.1(e). *Thus, the world line of an accelerated particle is curved in spacetime.* Figure 30.1(f) is the world line of a mass attached to a spring that is executing simple harmonic motion. Note that the world line is curved everywhere indicating that this is accelerated motion. Figure 30.1(g) is a two-space dimensional

1. "Space and Time," by H. Minkowski in *The Principle of Relativity,* Dover Publications.

Figure 30.1

Spacetime diagrams.

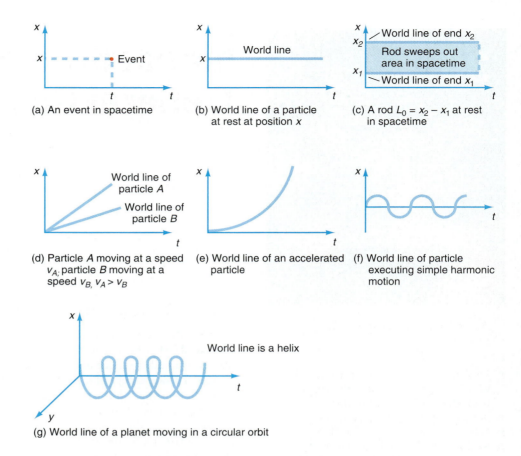

(a) An event in spacetime

(b) World line of a particle at rest at position x

(c) A rod $L_0 = x_2 - x_1$ at rest in spacetime

(d) Particle A moving at a speed v_A; particle B moving at a speed v_B, $v_A > v_B$

(e) World line of an accelerated particle

(f) World line of particle executing simple harmonic motion

(g) World line of a planet moving in a circular orbit

picture of a planet in its orbit about the sun. The motion of the planet is in the x,y plane but since the planet is also moving in time, its world line comes out of the plane and becomes a helix. Thus, when the planet moves from position x, goes once around the orbit, and returns to the same space point x, it is not at the same position in spacetime. It has moved forward through time.

A further convenient representation in spacetime diagrams is attained by changing the time axis to τ, where

$$\tau = ct \tag{30.1}$$

In this representation, τ is actually a length. (The product of a velocity times the time is equal to a length.) The length τ is the distance that light travels in a particular time. If t is measured in seconds, then τ becomes a light second, which is the distance that light travels in 1 s, namely,

$$\tau = ct = \left(3.00 \times 10^8 \, \frac{m}{s}\right)(1.00 \text{ s}) = 3.00 \times 10^8 \text{ m}$$

If t is measured in years, then τ becomes a light year, the distance that light travels in a period of time of 1 yr, namely,

$$\tau = ct = \left(3.00 \times 10^8 \, \frac{m}{s}\right)(1 \text{ yr})\left(\frac{365 \text{ days}}{1 \text{ yr}}\right)\left(\frac{24 \text{ hr}}{1 \text{ day}}\right)\left(\frac{3600 \text{ s}}{1 \text{ hr}}\right)$$

$$= 9.47 \times 10^{15} \text{ m} = 9.47 \times 10^{12} \text{ km}$$

The light year is a unit of distance routinely used in astronomy.

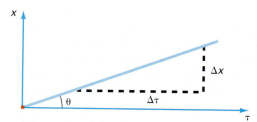

Figure 30.2

Changing the *t*-axis to a τ-axis.

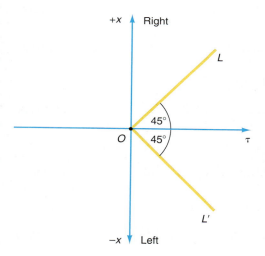

Figure 30.3

World lines of rays of light.

With this new notation, we draw the spacetime diagram as shown in figure 30.2. A straight line on this diagram can still represent a velocity. However, since a velocity is given as

$$v = \frac{\Delta x}{\Delta t}$$

and since $\tau = ct$,

$$c\Delta t = \Delta \tau$$

or

$$\Delta t = \frac{\Delta \tau}{c}$$

Thus, the velocity becomes

$$v = \frac{\Delta x}{\Delta t} = \frac{\Delta x}{\Delta \tau / c} = \frac{c\Delta x}{\Delta \tau}$$

but $\Delta x / \Delta \tau$ is the slope of the line and is given by

$$\frac{\Delta x}{\Delta \tau} = \text{slope of line} = \tan \theta$$

Then the velocity on such a diagram is given by

$$v = c \tan \theta \tag{30.2}$$

As a special case in such a diagram, if $\theta = 45°$, the tan 45° = 1 and equation 30.2 becomes

$$v = c$$

Thus, on a spacetime diagram of *x* versus τ, a straight line at an angle of 45° represents the world line of a light signal.

If a source of light at the origin emits a ray of light simultaneously toward the right and toward the left, we represent it on a spacetime diagram as shown in figure 30.3. Line *OL* is the world line of the light ray emitted toward the right, whereas *OL'* is the world line of the light ray emitted toward the left. Since the velocity of a particle must be less than *c*, the world line of any particle situated at *O* must have a slope less than 45° and is contained within the two light world lines *OL* and *OL'*. If the particle at *O* is at rest its world line is the τ-axis.

Example 30.1

The angle that a particle's world line makes as the particle moves through spacetime. If a particle moves to the right at a constant velocity of *c*/2, find the angle that its world line makes with the τ-axis.

Solution

Because the particle moves at a constant velocity through spacetime, its world line is a straight line. The angle that the world line makes with the τ-axis, found from equation 30.2, is

$$\theta = \tan^{-1}\frac{v}{c}$$

$$= \tan^{-1}\frac{c/2}{c} = \tan^{-1} 0.500$$

$$= 26.6° \tag{30.3}$$

Notice that the world line for this particle is contained between the lines *OL* and *OL'*.

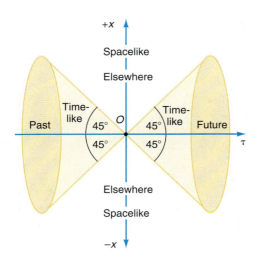

Figure 30.4

The light cone.

If we extend the diagram of figure 30.3 into two space dimensions, we obtain the **light cone** shown in figure 30.4. Straight lines passing through O and contained within the light cone are possible world lines of a particle or an observer at the origin O. Any world lines inside the left-hand cone come out of the observer's past, whereas any world line inside the right-hand cone goes into the observer's future. Only world lines within the cone can have a cause and effect relationship on the particle or observer at O. World lines that lie outside the cone can have no effect on the particle or observer at O and are world lines of some other particle or observer. Events that we actually "see," lie on the light cone because we see these events by light rays. *World lines within the cone are sometimes called timelike because they are accessible to us in time. Events outside the cone are called spacelike because they occur in another part of space that is not accessible to us and hence is called elsewhere.*

30.2 The Invariant Interval

From what has been said so far, it seems as if everything is relative. *In the varying world of spacetime is there anything that remains a constant?* Is there some one single thing that all observers, regardless of their state of motion, can agree on? In the field of physics, we are always looking for some characteristic constants of motion. Recall how in chapter 7 we studied the projectile motion of a particle in one dimension and saw that even though the projectile's position and velocity continually changed with time, there was one thing that always remained a constant, namely, the total energy of the projectile. In the same way we ask, isn't there a constant of the motion in spacetime? The answer is yes. *The constant value that all observers agree on, regardless of their state of motion, is called the **invariant interval**.*

Consider the Lorentz transformation equations from chapter 29,

$$\Delta x' = \frac{\Delta x - v\Delta t}{\sqrt{1 - v^2/c^2}} \tag{29.73}$$

$$\Delta t' = \frac{\Delta t - \Delta x \, (v/c^2)}{\sqrt{1 - v^2/c^2}} \tag{29.74}$$

Let us square each of these equations to get

$$(\Delta x')^2 = \frac{(\Delta x)^2 - 2v\Delta x\Delta t + v^2(\Delta t)^2}{1 - v^2/c^2} \tag{30.4}$$

and

$$(\Delta t')^2 = \frac{(\Delta t)^2 - (2v\Delta x\Delta t/c^2) + (v^2/c^4)(\Delta x)^2}{1 - v^2/c^2} \tag{30.5}$$

Let us multiply equation 30.5 by c^2 to get

$$c^2(\Delta t')^2 = \frac{c^2(\Delta t)^2 - 2v\Delta x\Delta t + (v^2/c^2)(\Delta x)^2}{1 - v^2/c^2} \tag{30.6}$$

Let us now subtract equation 30.4 from equation 30.6 to get

$$c^2(\Delta t')^2 - (\Delta x')^2 = \frac{c^2(\Delta t)^2 - 2v\Delta x\Delta t + (v^2/c^2)(\Delta x)^2}{1 - v^2/c^2} - \frac{(\Delta x)^2 - 2v\Delta x\Delta t + v^2(\Delta t)^2}{1 - v^2/c^2}$$

$$= \frac{c^2(\Delta t)^2 - v^2(\Delta t)^2 + (v^2/c^2)(\Delta x)^2 - (\Delta x)^2}{1 - v^2/c^2}$$

$$= \frac{(c^2 - v^2)(\Delta t)^2 - (1 - v^2/c^2)(\Delta x)^2}{1 - v^2/c^2}$$

$$c^2(\Delta t')^2 - (\Delta x')^2 = \frac{c^2(1 - v^2/c^2)(\Delta t)^2 - (1 - v^2/c^2)(\Delta x)^2}{1 - v^2/c^2}$$

Dividing each term on the right by $1 - v^2/c^2$ gives

$$c^2(\Delta t')^2 - (\Delta x')^2 = c^2(\Delta t)^2 - (\Delta x)^2 \qquad (30.7)$$

Equation 30.7 shows that the quantity $c^2(\Delta t)^2 - (\Delta x)^2$ as measured by the S observer is equal to the same quantity $c^2(\Delta t')^2 - (\Delta x')^2$ as measured by the S' observer. But how can this be? This can be true only if each side of equation 30.7 is equal to a constant. *Thus, the quantity $c^2(\Delta t)^2 - (\Delta x)^2$ is an invariant. That is, it is the same in all inertial systems. This quantity is called the invariant interval and is denoted by $(\Delta s)^2$.* Hence the invariant interval is given by

$$(\Delta s)^2 = c^2(\Delta t)^2 - (\Delta x)^2 \qquad (30.8)$$

The invariant interval is thus a constant in spacetime. All observers, regardless of their state of motion, agree on this value in spacetime. If the other two space dimensions are included, the invariant interval in four-dimensional spacetime becomes

$$(\Delta s)^2 = c^2(\Delta t)^2 - (\Delta x)^2 - (\Delta y)^2 - (\Delta z)^2 \qquad (30.9)$$

The invariant interval of spacetime is something of a strange quantity to us. In ordinary space, not spacetime, an invariant interval is given by the Pythagorean theorem as

$$(\Delta s)^2 = (\Delta x)^2 + (\Delta y)^2 = (\Delta x')^2 + (\Delta y')^2 \qquad (30.10)$$

as shown in figure 30.5, where Δs is the invariant, and is seen to be nothing more than the radius of the circle shown in figure 30.5 and given by equation 30.10. That is, equation 30.10 is of the form of the equation of a circle $r^2 = x^2 + y^2$. Even though Δx and $\Delta x'$ are different, and Δy and $\Delta y'$ are different, the quantity Δs is always the same positive quantity.

Now let us look at equation 30.8 for the invariant interval in spacetime. First, however, let $ct = \tau$ as we did previously in equation 30.1. Then we can express the invariant interval, equation 30.8, as

$$(\Delta s)^2 = (\Delta \tau)^2 - (\Delta x)^2 \qquad (30.11)$$

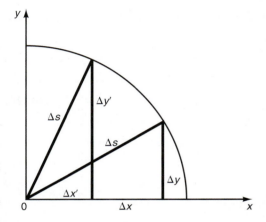

Figure 30.5

The invariant interval of space.

Because of the minus sign in front of $(\Delta x)^2$, the equation is not the equation of a circle $(x^2 + y^2 = r^2)$, but is rather the equation of a hyperbola, $x^2 - y^2 = $ constant.

The interval between two points in Euclidean geometry is represented by the hypotenuse of a right triangle and is given by the Pythagorean theorem: The square of the hypotenuse is equal to the *sum* of the squares of the other two sides of the triangle. However, *the square of the interval Δs in spacetime is not equal to the sum of the squares of the other two sides, but to their difference. Thus, the Pythagorean theorem of Euclidean geometry does not hold in spacetime. Therefore, spacetime is not Euclidean. This new type of geometry described by equation 30.11 is sometimes called flat-hyperbolic geometry.* However, since hyperbolic geometry is another name for the non-Euclidean geometry of the Russian mathematician, Nikolai Ivanovich Lobachevski (1793–1856), rather than calling spacetime hyperbolic, we say that spacetime is non-Euclidean. *Space by itself is Euclidean, but spacetime is not. The fact that spacetime is not Euclidean accounts for the apparently strange characteristics of length contraction and time dilation* as we will see shortly. The minus sign in equation 30.11 is the basis for all the differences between space and spacetime.

Also, because of that minus sign in equation 30.11, $(\Delta s)^2$ can be positive, negative, or zero. When $(\Delta \tau)^2 > (\Delta x)^2$, $(\Delta s)^2$ is positive. Because the time term predominates, the world line in spacetime is called timelike and is found in the future light cone. When $(\Delta x)^2 > (\Delta \tau)^2$, $(\Delta s)^2$ is negative. Because the space term predominates in this case, the world line is called spacelike. A spacelike world line lies outside the light cone in the region called elsewhere, figure 30.4. When

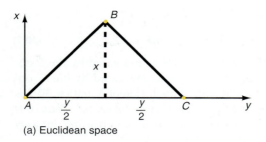

(a) Euclidean space

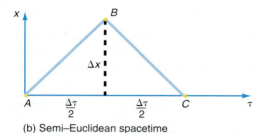

(b) Semi–Euclidean spacetime

Figure 30.6

Space versus spacetime.

$(\Delta x)^2 = (\Delta\tau)^2$, $(\Delta s)^2$ is equal to zero. In this case, $(\Delta x) = \Delta\tau = (c\Delta t)$. Hence, $\Delta x = c\Delta t$, or $\Delta x/\Delta t = c$. But $\Delta x/\Delta t$ is a velocity. For it to equal c, it must be the world line of something moving at the speed of light. Thus $(\Delta s)^2 = 0$ represents a light ray and the world line is called lightlike. Lightlike world lines make up the light cone.

Another characteristic of Euclidean space is that the straight line is the shortest distance between two points. Now we will see that *in non-Euclidean space-time, the straight line is the longest distance between two points.* Consider the distance traveled along the two space paths of figure 30.6(a). The distance traveled along path AB in Euclidean space is found from the Pythagorean theorem as

$$s_{AB} = \sqrt{\left(\frac{y}{2}\right)^2 + x^2}$$

And the distance along path BC is similarly

$$s_{BC} = \sqrt{\left(\frac{y}{2}\right)^2 + x^2}$$

The total distance traveled along path ABC is therefore

$$s_{ABC} = s_{AB} + s_{BC} = 2\sqrt{\left(\frac{y}{2}\right)^2 + x^2}$$

or

$$s_{ABC} = \sqrt{y^2 + (2x)^2} \tag{30.12}$$

The total distance traveled along path AC is

$$s_{AC} = \frac{y}{2} + \frac{y}{2} = y$$

But since

$$\sqrt{y^2 + (2x)^2} > y$$

the round-about path ABC is longer than the straight line path AC, as expected.

Example 30.2

Path length in Euclidean space. If $y = 8.00$ and $x = 3.00$ in figure 30.6(a), find the path lengths s_{ABC} and s_{AC}.

Solution

The length of the path along ABC, found from equation 30.12, is

$$s_{ABC} = \sqrt{y^2 + (2x)^2} = \sqrt{(8.00)^2 + (2(3.00))^2}$$
$$= 10.0$$

The length of path AC is simply

$$s_{AC} = y = 8.00$$

Thus, the straight line path in space is shorter than the round-about path.

Let us now look at the same problem in spacetime, as shown in figure 30.6(b). The distance traveled through spacetime along path AB is found by the invariant interval, equation 30.11, as

$$\Delta s_{AB} = \sqrt{\left(\frac{\Delta\tau}{2}\right)^2 - (\Delta x)^2}$$

Whereas the distance traveled through spacetime along path BC is

$$\Delta s_{BC} = \sqrt{\left(\frac{\Delta \tau}{2}\right)^2 - (\Delta x)^2}$$

The total distance traveled through spacetime along path ABC is thus,

$$\Delta s_{ABC} = \Delta s_{AB} + \Delta s_{BC}$$

$$= 2\sqrt{\left(\frac{\Delta \tau}{2}\right)^2 - (\Delta x)^2}$$

$$\Delta s_{ABC} = \sqrt{(\Delta \tau)^2 - (2\Delta x)^2} \tag{30.13}$$

Whereas the distance traveled through spacetime along the path AC is

$$\Delta s_{AC} = \frac{\Delta \tau}{2} + \frac{\Delta \tau}{2} = \Delta \tau$$

But comparing these two paths, ABC and AC, we see that

$$\sqrt{(\Delta \tau)^2 - (2\Delta x)^2} < \Delta \tau \tag{30.14}$$

Therefore, the distance through spacetime along the round-about path ABC is less than the straight line path AC through spacetime. Thus, the shortest distance between two points in spacetime is not the straight line. In fact the straight line is the longest distance between two points in spacetime. These apparently strange effects of relativity occur because spacetime is non-Euclidean. (It is that minus sign again!)

Example 30.3

Path length in non-Euclidean spacetime. If $\Delta \tau = 8.00$ and $\Delta x = 3.00$ in figure 30.6(b), find the path lengths Δs_{ABC} and Δs_{AC}.

Solution

The interval along path ABC, found from equation 30.13, is

$$\Delta s_{ABC} = \sqrt{(\Delta \tau)^2 - (2\Delta x)^2} = \sqrt{(8.00)^2 - (2(3.00))^2}$$

$$= 5.29$$

The interval along path AC is

$$\Delta s_{AC} = \Delta \tau = 8.00$$

Hence,

$$\Delta s_{ABC} < \Delta s_{AC}$$

and the straight line through spacetime is greater than the round-about line through spacetime.

The straight line AC in spacetime is the world line of an object or clock at rest at the origin of the coordinate system. The spacetime interval for a clock at rest ($\Delta x = 0$) is therefore

$$(\Delta s)^2 = (\Delta \tau)^2 - (\Delta x)^2 = (\Delta \tau_0)^2$$

or

$$\Delta s = \Delta \tau_0 \tag{30.15}$$

The subscript 0 has been used on τ to indicate that this is the time when the clock is at rest. *The time read by a clock at rest is called its proper time. But since this proper time is also equal to the spacetime interval, equation 30.15, and this space-time interval is an invariant, it follows that the interval measured along any timelike world line is equal to its proper time.* If a clock is carried along with a body from A to B, Δs_{AB} is the time that elapses on that clock as it moves from A to B, and Δs_{BC} is the time that elapses along path BC. Hence, from equation 30.14, *the time elapsed along path ABC is less than the time elapsed along path AC. Thus, if two clocks started out synchronized at A, they read different times when they come together at point C. It is therefore sometimes said that time, like distance, is a route-dependent quantity.* The path ABC represents an accelerated path. (Actually the acceleration occurs almost instantaneously at the point B.) Hence *the lapse of proper time for an accelerated observer is less than the proper time for an observer at rest. Thus, time must slow down during an acceleration,* a result that we will confirm in our study of general relativity.

In chapter 29 we discussed the twin paradox, whereby one twin became an astronaut and traveled into outer space while the second twin remained home on earth. The Lorentz time dilation equation showed that the traveling astronaut, on his return, would be younger than his stay-at-home twin. Figure 30.6(b) is essentially a spacetime diagram of the twin paradox. The world line through spacetime for the stay-at-home twin is shown as path AC, whereas the world line for the astronaut is given by path ABC. Path ABC through spacetime is curved because the astronaut went through an acceleration phase in order to turn around to return to earth. Hence, the astronaut can no longer be considered as an inertial observer. Since the stay-at-home twin's path AC is a straight line in spacetime, she is an inertial observer. As we have just seen in the last paragraph, the time elapsed along path ABC, the astronaut's path, is less than the time elapsed along path AC, the stay-at-home's path. Thus the astronaut does indeed return home younger than his stay-at-home twin.

Perhaps one of the most important characteristics of the invariant interval is that it allows us to draw a good geometrical picture of spacetime as it is seen by different observers. For example, a portion of spacetime for a stationary observer S is shown in figure 30.7. The x and τ coordinates of S are shown as the orthogonal axes. The light lines OL and OL' are drawn at angles of 45°. The interval, equation 30.11, is drawn for a series of values of x and τ and appear as the family of hyperbolas in the figure. (We might note that if spacetime were Euclidean the intervals would have been a family of concentric circles around the origin O instead of these hyperbolas.) The hyperbolas drawn about the τ-axis lie in the light cone future, while the hyperbolas drawn about the x-axis lie elsewhere. The interval has positive values within the light cone and negative values elsewhere.

A frame of reference S', moving at the velocity v, would have for the world line of its origin of coordinates, a straight line through spacetime inclined at an angle θ given by

$$\theta = \tan^{-1} \frac{v}{c} \tag{30.3}$$

For example, if S' is moving at a speed of $c/2$, $\theta = 26.6°$. This world line is drawn in figure 30.7. But the world line of the origin of coordinates ($x' = 0$) is the time axis τ' of the S' frame, and is thus so labeled in the diagram. *Where τ' intersects the family of hyperbolas at $\Delta s = 1, 2, 3, \ldots$, it establishes the time scale along the τ'-axis as $\tau' = 1, 2, 3, \ldots$.* (Recall that because $(\Delta s)^2 = (\Delta\tau')^2 - (\Delta x')^2$, and the origin of the coordinate system, $\Delta x' = 0$, hence $\Delta s = \Delta\tau'$.) *Note that the scale on the τ'-axis is not the same as the scale on the τ-axis.*

To draw the x'-axis on this graph, we note that the x'-axis represents all the points for which $t' = 0$. The Lorentz equation for t' was given in chapter 29 by equation 29.50 as

$$t' = \frac{t - xv/c^2}{\sqrt{1 - v^2/c^2}}$$

Figure 30.7

The invariant interval on a spacetime map.

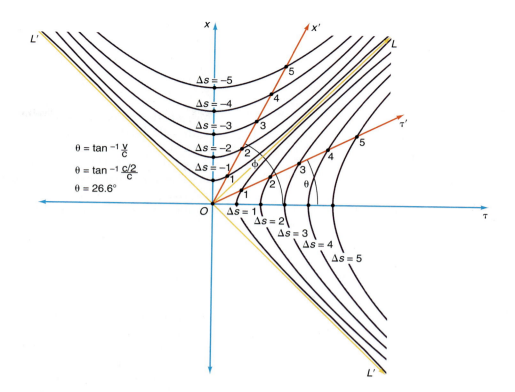

For $t' = 0$, we must have

$$t = \frac{xv}{c^2}$$

or

$$x = \frac{c^2}{v}t = \frac{c}{v}(ct)$$

$$x = \frac{c}{v}\tau \qquad (30.16)$$

Equation 30.16 is the equation of a straight line passing through the origin with the slope c/v. This line represents the x'-axis because it results from setting $t' = 0$ in the Lorentz equation. Because the slope of the τ'-axis was given by $\tan \theta = v/c$, the triangle of figure 30.8 can be drawn. Note that we can write the ratio of c/v, the slope of the x'-axis, as

$$\tan \phi = \frac{c}{v}$$

But from the figure $\theta + \phi = 90°$. Hence, the angle for the slope of the x'-axis must be

$$\phi = 90° - \theta \qquad (30.17)$$

In our example, $\theta = 26.6°$, thus $\phi = 63.4°$. The x'-axis is drawn in figure 30.7 at this angle. Note that the x'-axis makes an angle ϕ with the τ-axis, but an angle θ with the x-axis. The intersection of the x'-axis with the family of hyperbolas establishes the scale for the x'-axis. The interval is

$$(\Delta s)^2 = (\Delta \tau')^2 - (\Delta x')^2$$

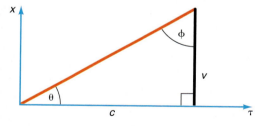

Figure 30.8

Determining the slope of the x'-axis.

But $\Delta\tau' = 0$ for the x'-axis, and Δs is a negative quantity elsewhere, hence

$$-(\Delta x')^2 = -(\Delta s)^2$$

and

$$\Delta x' = + \Delta s$$

Thus, where x' intersects the family of hyperbolas at $\Delta s = -1, -2, -3, \ldots$ the length scale along x' becomes $x' = 1, 2, 3, \ldots$. The scale on the x'-axis is now shown in the figure. Again note that the scale on the x'-axis is not the same as the scale on the x-axis. Having used the hyperbolas for the interval to establish the x'- and τ'-axes, and their scale, we can now dispense with them and the results of figure 30.7 are as shown in figure 30.9. Notice that the S' frame of reference is a skewed coordinate system, and the scales on S' are not the same as on S. Lines of constant values of x' are parallel to the τ'-axis, whereas lines of constant τ' are parallel to the x'-axis. The angle of the skewed coordinate system α is found from the figure to be

$$\alpha = 90° - 2\theta \tag{30.18}$$

The angle θ is found from equation 30.3. This S' frame is unique for a particular value of v. Another inertial observer moving at a different speed would have another skewed coordinate system. However, the angle θ and hence, the angle α, would be different, depending on the value of v.

The motion of the inertial observer S' seems to warp the simple orthogonal spacetime into a skewed spacetime. The length contraction and time dilation can easily be explained by this skewed spacetime. Figures 30.10 through 30.15 are a series of spacetime diagrams based on the invariant interval, showing length contraction, time dilation, and simultaneity.

Figure 30.10 represents a rod 4.00 units long at rest in a rocket ship S', moving at a speed of $c/2$. The world line of the top of the stick in S' is drawn parallel to the τ'-axis. (Any line parallel to the τ'-axis has one and only one value of x' and thus represents an object at rest in S'.)

If the world line is dashed backward to the x-axis, it intersects the x-axis at $x = 3.46$, which is the length of the rod L, as observed by the S frame observer. Thus, the rod at rest in the moving rocket frame appears contracted to the observer on earth, the S frame. The contraction of the moving rod is, of course, the Lorentz contraction. With the spacetime diagram it is easier to visualize.

Figure 30.9

Relation of S and S' frame of references.

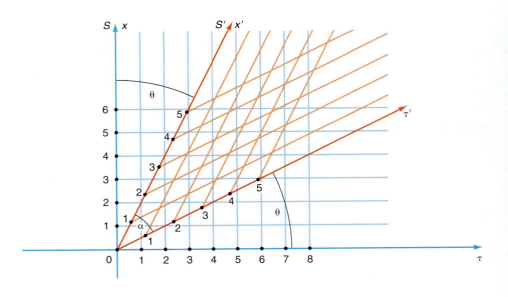

Figure 30.10

Length contraction, rod at rest in S' frame.

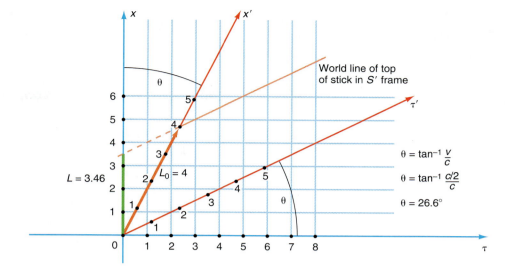

Figure 30.11

Length contraction, rod at rest in S frame.

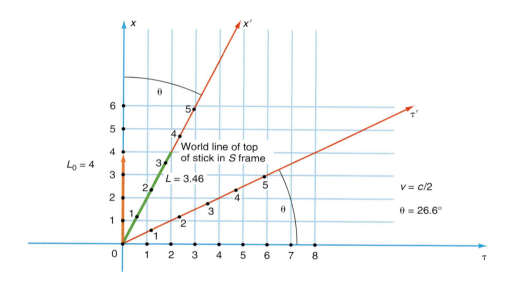

Figure 30.11 shows the same Lorentz contraction but as viewed from the S' frame. A rod 4.00 units long L_0 is at rest in the S frame, the earth. An observer in the rocket ship frame, the S' frame, considers himself to be at rest while the earth is moving away from him at a velocity $-v$. The astronaut sees the world line, which emanates from the top of the rod, as it intersects his coordinate system. The length of the rod that he sees is found by drawing the world line of the top of the rod in the S frame, as shown in the figure. This world line intersects the x'-axis at the position $x' = 3.46$. Hence, the rocket observer measures the rod on earth to be only 3.46 units long, the length L. Thus, the rocket ship observer sees the same length contraction. *The cause of these contractions is the non-Euclidity of spacetime.*

The effect of time dilation is also easily explained by the spacetime diagram, figure 30.12. A clock is at rest in a moving rocket ship at the position $x' = 2$. Its world line is drawn parallel to the τ'-axis, as shown. Between the occurrence of the events A and B a time elapses on the S' clock of $\Delta\tau' = 4.0 - 2.0 = 2.0$, as shown in the figure. This time interval, when observed by the S frame of the earthman, is found by dropping the dashed lines from the events A and B down to the τ-axis. (These lines are parallel to the x-axis, but because S is an orthogonal

Figure 30.12

Time dilation, clock at rest in S' frame.

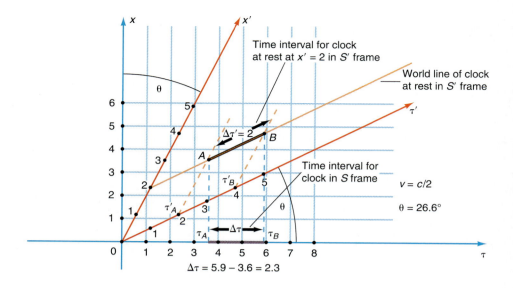

Figure 30.13

Time dilation, clock at rest in S frame.

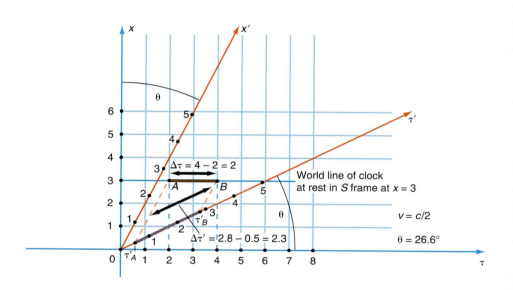

frame, they are also perpendicular to the τ-axis.) The time interval elapsed on earth is read from the graph as $\Delta\tau = 5.9 - 3.6 = 2.3$. A time lapse of 2 s on the rocket ship clock would appear as a lapse of 2.3 s on earth. Thus the moving clock in S' is running at a slower rate than a clock in S. Time has slowed down in the moving rocket ship. This is, of course, the Lorentz time dilation effect.

The inverse problem of time dilation is shown in figure 30.13. Here a clock is at rest on the earth, the S frame, at the position $x = 3$. The world line of the clock is drawn parallel to the τ-axis. The occurrence of two events, A and B, are noted and the time interval elapsed between these two events on earth is $\Delta\tau = 4.0 - 2.0 = 2.0$. The same events A and B are observed in the rocket ship, and the time of these events as observed on the rocket ship is found by drawing the dashed lines parallel to the x'-axis to where they intersect the τ'-axis. Thus, event A occurs at $\tau_A' = 0.5$, and event B occurs at $\tau_B' = 2.8$. The elapsed time on the rocket ship is thus

$$\Delta\tau' = \tau_B' - \tau_A' = 2.8 - 0.5 = 2.3$$

From the point of view of the rocket observer, he is at rest, and the earth is moving away from him at a velocity $-v$. Hence, he sees an elapsed time on the moving earth of 2 s while his own clock records a time interval of 2.3 s. He therefore concludes that time has slowed down on the moving earth.

Another explanation for this time dilation can be found in the concept of *simultaneity*. If we look back at figure 30.12 we see that the same event A occurs at the times $\tau_A = 3.6$ and $\tau_A' = 2.0$, whereas event B occurs at the times $\tau_B = 5.9$ and $\tau_B' = 4$. *The same event does not occur at the same time in the different coordinate systems.* Because the events occur at different times their time intervals should be expected to be different also. In fact, a more detailed picture of simultaneity can be found in figures 30.14 and 30.15.

Figure 30.14 shows two events A and B that occur simultaneously at the time $\tau' = 2$ on the moving rocket ship. However, the earth observer sees the two events occurring, not simultaneously, but rather at the two times $\tau_A = 3$ and $\tau_B = 4$. That is, the earth observer sees event A happen before event B. This same type of effect is shown in figure 30.15, where the two events A and B now occur simultaneously at $\tau = 4$ for the earth observer. However the rocket ship observer

Figure 30.14

Simultaneity, two events simultaneous in S' frame.

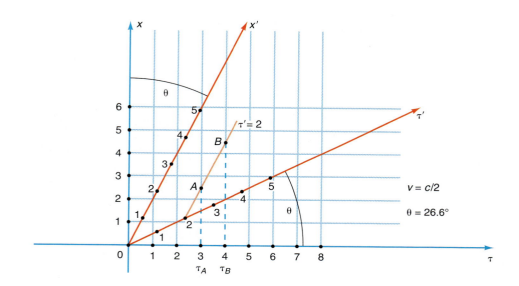

Figure 30.15

Simultaneity, two events simultaneous in S frame.

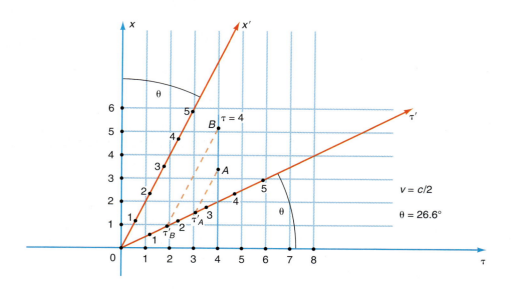

sees event B occurring at $\tau'_B = 1.6$ and event A at $\tau'_A = 2.7$. Thus, to the rocket ship observer events A and B are not simultaneous, but rather event B occurs before event A.

In summary, these spacetime diagrams are based on the invariant interval. Because the invariant interval is based on hyperbolas, spacetime is non-Euclidean. The S' frame of reference becomes a skewed coordinate system and the scales of the S' frame are not the same as the scales on the S frame.

Example 30.4

The skewing of the spacetime diagram with speed. Find the angles θ and α for a spacetime diagram if (a) $v = 1610$ km/hr $= 1000$ mph $= 477$ m/s, (b) $v = 1610$ km/s $= 1000$ miles/s, (c) $v = 0.8c$, (d) $v = 0.9c$, (e) $v = 0.99c$, (f) $v = 0.999c$, and (g) $v = c$.

Solution

a. The angle θ of the spacetime diagram, found from equation 30.3, is

$$\theta = \tan^{-1} \frac{v}{c} = \tan^{-1} \frac{477 \text{ m/s}}{3.00 \times 10^8 \text{ m/s}}$$

$$= (8.54 \times 10^{-5})°$$

The angle α, found from equation 30.18, is

$$\alpha = 90° - 2\theta = 90° - 2(8.54 \times 10^{-5})° = 90°$$

That is, for the reasonably large speed of 1000 mph, the angle θ is effectively zero and the angle $\alpha = 90°$. There is no skewing of the coordinate system and S and S' are orthogonal coordinate systems.

b. For $v = 1610$ km/s, the angle θ is

$$\theta = \tan^{-1} \frac{v}{c} = \tan^{-1} \frac{1.61 \times 10^6 \text{ m/s}}{3.00 \times 10^8 \text{ m/s}}$$

$$= 0.31°$$

The angle α is

$$\alpha = 90° - 2\theta = 90° - 2(0.31°) = 89.4°$$

That is, for $v = 1610$ km/s $= 3,600,000$ mph, the τ'- and x'-axes are just barely skewed.

c. For $v = 0.8c$,

$$\theta = \tan^{-1} \frac{v}{c} = \tan^{-1} \frac{0.8c}{c} = 38.7°$$

and

$$\alpha = 90° - 2\theta = 90° - 2(38.7) = 12.6°$$

For this large value of v, the axes are even more skewed than in figure 30.8.

d.–g. For these larger values of v, equations 30.3 and 30.18 give

$$v = 0.9c; \qquad \theta = 41.9°; \qquad \alpha = 6.2°$$
$$v = 0.99c; \qquad \theta = 44.7°; \qquad \alpha = 0.576°$$
$$v = 0.999c; \qquad \theta = 44.97°; \qquad \alpha = 0.057°$$
$$v = c; \qquad \theta = 45°; \qquad \alpha = 0$$

Hence, as v gets larger and larger the angle θ between the coordinate axes becomes larger and larger, eventually approaching 45°. The angle α gets smaller until at $v = c$, α has been reduced to zero and the entire S' frame of reference has been reduced to a line.

30.3 The General Theory of Relativity

We saw in the special theory of relativity that the laws of physics must be the same in all inertial reference systems. *But what is so special about an inertial reference system? The inertial reference frames are, in a sense, playing the same role as Newton's absolute space.* That is, absolute space has been abolished only to replace it by absolute inertial reference frames. Shouldn't the laws of physics be the same in all coordinate systems, whether inertial or noninertial? The inertial frame should not be such a privileged frame. But clearly, accelerations can be easily detected, whereas constant velocities cannot. How can this very obvious difference be reconciled? That is, we must show that even all accelerated motions are relative. How can this be done?

Let us consider the very simple case of a mass m on the floor of a rocket ship that is at rest in a uniform gravitational field on the surface of the earth, as depicted in figure 30.16(a). The force acting on the mass is its weight w, which we write as

$$F = w = mg \qquad (30.19)$$

Let us now consider the case of the same rocket ship in interstellar space far removed from all gravitational fields. Let the rocket ship now accelerate upward,

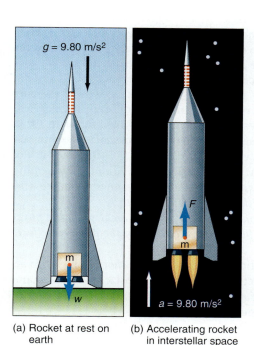
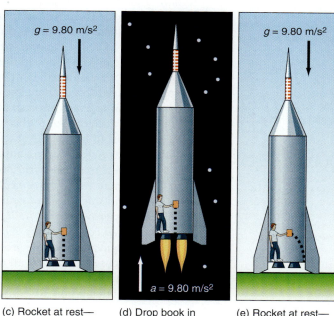
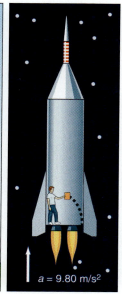

(a) Rocket at rest on earth

(b) Accelerating rocket in interstellar space

(c) Rocket at rest— book falls to floor

(d) Drop book in accelerated rocket ship

(e) Rocket at rest— thrown book follows parabolic trajectory

(f) Thrown book follows parabolic trajectory in accelerated rocket

Figure 30.16

An accelerated frame of reference is equivalent to an inertial frame of reference plus gravity.

The astronauts on board are in an accelerated frame of reference.

as in figure 30.16(b), with an acceleration a that is numerically equal to the acceleration due to gravity g, that is, $a = g = 9.80$ m/s². The mass m that is sitting on the floor of the rocket now experiences the force, given by Newton's second law as

$$F = ma = mg = w \qquad (30.20)$$

That is, the mass m sitting on the floor of the accelerated rocket experiences the same force as the mass m sitting on the floor of the rocket ship when it is at rest in the uniform gravitational field of the earth. Therefore, there seems to be some relation between accelerations and gravity.

Let us experiment a little further in the rocket ship at rest by holding a book out in front of us and then dropping it, as in figure 30.16(c). The book falls to the floor and if we measured the acceleration we would, of course, find it to be the acceleration due to gravity, $g = 9.80$ m/s². Now let us take the same book in the accelerated rocket ship and again drop it, as in figure 30.16(d). An inertial observer outside the rocket would see the book stay in one place but would see the floor accelerating upward toward the book at the rate of $a = 9.80$ m/s². The astronaut in the accelerated rocket ship sees the book fall to the floor with the acceleration of 9.80 m/s² just as the astronaut at rest on the earth observed.

The astronaut in the rocket at rest on the earth now throws the book across the room of the rocket ship. He observes that the book follows the familiar parabolic trajectory of the projectile that we studied in chapter 3 and that is again shown in figure 30.16(e). Similarly, the astronaut in the accelerated rocket also throws the book across the room. An outside inertial observer would observe the book moving across the room in a straight line and would also see the floor accelerating upward toward the book. The accelerated astronaut would simply see the book following the familiar parabolic trajectory it followed on earth, figure 30.16(f).

Hence, the same results are obtained in the accelerated rocket ship as are found in the rocket ship at rest in the gravitational field of the earth. Thus, *the effects of gravity can be either created or eliminated by the proper choice of coordinate systems.* Our experimental considerations suggest that *the accelerated frame of reference is equivalent to an inertial frame of reference in which gravity is present.* Einstein, thus found a way to make accelerations relative. He stated his results in what he called the ***equivalence principle.*** Calling the inertial system containing gravity the K system and the accelerated frame of reference the K' system, Einstein said, "we assume that we may just as well regard the system K as being in a space free from gravitational field if we then regard K as uniformly accelerated. This assumption of exact physical equivalence makes it impossible for us to speak of the absolute acceleration of the system, just as the usual (special) theory of relativity forbids us to talk of the absolute velocity of a system. . . . But this view of ours will not have any deeper significance unless *the systems K and K' are equivalent with respect to all physical processes, that is, unless the laws of nature with respect to K are in entire agreement with those with respect to K'.*"[2]

Einstein's principle of equivalence is stated as: on a local scale the physical effects of a gravitational field are indistinguishable from the physical effects of an accelerated coordinate system.

The equivalence of the gravitational field and acceleration "fields" also accounts for the observation that all objects, regardless of their size, fall at the same rate in a gravitational field. If we write m_g for the mass that experiences the gravitational force in equation 30.19 and figure 30.16(a), then

$$F = w = m_g g$$

2. "On the Influence of Gravitation on the Propagation of Light," from A. Einstein, *Annalen der Physik* 35, 1911, in *The Principle of Relativity,* Dover Publishing Co.

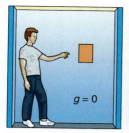
(a) Book remains at rest

(c) Book remains at rest

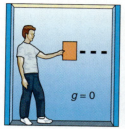
(b) Thrown book moves in straight line at constant velocity

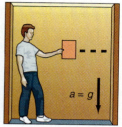
(d) Thrown book moves in straight line at constant velocity

(a and b)
Truly inertial system in interstellar space

(c and d)
Inside a freely falling elevator

Figure 30.17

A freely falling frame of reference is locally the same as an inertial frame of reference.

And if we write m_i for the inertial mass that resists the motion of the rocket in figure 30.16(b) and equation 30.20, then

$$F = m_i a = m_i g$$

Since we have already seen that the two forces are equal, by the equivalence principle, it follows that

$$m_g = m_i \qquad (30.21)$$

That is, the gravitational mass is in fact equal to the inertial mass. Thus, the equivalence principle implies the equality of inertial and gravitational mass and this is the reason why all objects of any size fall at the same rate in a gravitational field.

As a final example of the equivalence of a gravitational field and an acceleration let us consider an observer in a closed room, such as a nonrotating space station in interstellar space, far removed from all gravitating matter. This space station is truly an inertial coordinate system. Let the observer place a book in front of him and then release it, as shown in figure 30.17(a). Since there are no forces present, not even gravity, the book stays suspended in space, at rest, exactly where the observer placed it. If the observer then took the book and threw it across the room, he would observe the book moving in a straight line at constant velocity, as shown in figure 30.17(b).

Let us now consider an elevator on earth where the supporting cables have broken and the elevator goes into free-fall. An observer inside the freely falling elevator places a book in front of himself and then releases it. The book appears to that freely falling observer to be at rest exactly where the observer placed it, figure 30.17(c). (Of course, an observer outside the freely falling elevator would observe both the man and the book in free-fall but with no relative motion with respect to each other.) If the freely falling observer now takes the book and throws it across the elevator room he would observe that the book travels in a straight line at constant velocity, figure 30.17(d).

Because an inertial frame is defined by Newton's first law as a frame in which a body at rest, remains at rest, and a body in motion at some constant velocity continues in motion at that same constant velocity, we must conclude from the illustration of figure 30.17 that the freely falling frame of reference acts exactly as an inertial coordinate system to anyone inside of it. *Thus, the acceleration due to gravity has been transformed away by accelerating the coordinate system by the same amount as the acceleration due to gravity.* If the elevator were completely closed, the observer could not tell whether he was in a freely falling elevator or in a space station in interstellar space.

The equivalence principle allows us to treat an accelerated frame of reference as equivalent to an inertial frame of reference with gravity present, figure 30.16, or to consider an inertial frame as equivalent to an accelerated frame in which gravity is absent, figure 30.17. By placing all frames of reference on the same footing, Einstein was then able to *postulate* **the general theory of relativity,** namely, *the laws of physics are the same in all frames of reference.*

A complete analysis of the general theory of relativity requires the use of very advanced mathematics, called tensor analysis. However, many of the results of the general theory can be explained in terms of the equivalence principle, and this is the path that we will follow in the rest of this chapter.

From his general theory of relativity, Einstein was quick to see its relation to gravitation when he said, "It will be seen from these reflections that in pursuing the General Theory of Relativity we shall be led to a theory of gravitation, since we are able to produce a gravitational field merely by changing the system of coordinates. *It will also be obvious that the principle of the constancy of the velocity of light in vacuo must be modified.*"[3]

---

3. "The Foundation of the General Theory of Relativity," from A. Einstein, *Annalen der Physik* 49, 1916, in *The Principle of Relativity,* Dover Publishing Co.

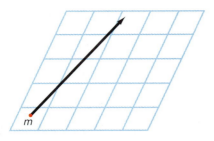

(a) Flat spacetime

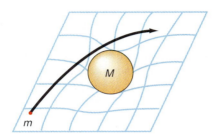

(b) Warping of spacetime by matter

Figure 30.18

Flat and curved spacetime.

Although the general theory was developed by Einstein to cover the cases of accelerated reference frames, it soon became obvious to him that the general theory had something quite significant to say about gravitation. Since the world line of an accelerated particle in spacetime is curved, then by the principle of equivalence, a particle moving under the effect of gravity must also have a curved world line in spacetime. *Hence, the mass that is responsible for causing the gravitational field, must warp spacetime to make the world lines of spacetime curved.* This is sometimes expressed as, matter warps spacetime and spacetime tells matter how to move.

A familiar example of the visualization of curved or **warped spacetime** is the rubber sheet analogy. A flat rubber sheet with a rectangular grid painted on it is stretched, as shown in figure 30.18(a). By Newton's first law, a free particle, a small rolling ball *m* moves in a straight line as shown. A bowling ball is then placed on the rubber sheet distorting or warping the rubber sheet, as shown in figure 30.18(b). When the small ball *m* is rolled on the sheet it no longer moves in a straight line path but it now curves around the bowling ball *M*, as shown. *Thus, gravity is no longer to be thought of as a force in the Newtonian tradition but it is rather a consequence of the warping or curvature of spacetime caused by mass.* The amount of warping is a function of the mass.

The four experimental confirmations of the general theory of relativity are

1. The bending of light in a gravitational field.
2. The advance of the perihelion of the planet Mercury.
3. The gravitational red shift of spectral lines.
4. The Shapiro experiment, which shows the slowing down of the speed of light near a large mass.

Let us now look at each of these confirmations.

30.4 The Bending of Light in a Gravitational Field

Let us consider a ray of light that shines through a window in an elevator at rest, as shown in figure 30.19(a). The ray of light follows a straight line path and hits the opposite wall of the elevator at the point *P*. Let us now repeat the experiment, but let the elevator accelerate upward very rapidly, as shown in figure 30.19(b). The ray of light enters the window as before, but before it can cross the room to the opposite wall the elevator is displaced upward because of the acceleration. Instead of the ray of light hitting the wall at the point *P*, it hits at some lower point *Q* because of the upward acceleration of the elevator. To an observer in the elevator, the ray of light follows the parabolic path, as shown in figure 30.19(c). Thus, *in the accelerated coordinate system of the elevator, light does not travel in a straight line, but instead follows a curved path. But by the principle of equivalence the accelerated elevator can be replaced by a gravitational field. Therefore light should be*

Figure 30.19

The bending of light in an accelerated elevator.

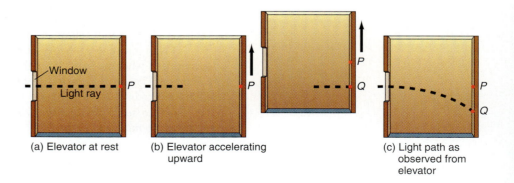

(a) Elevator at rest

(b) Elevator accelerating upward

(c) Light path as observed from elevator

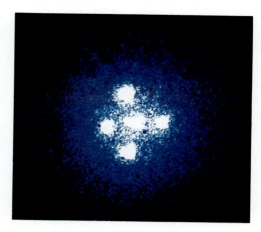

The four images that make up the "Einstein Cross" are actually images of the same quasar. This phenomenon occurs when the light from a background object passes close to a massive foreground object—in this case, the galaxy in the middle image.

bent from a straight line path in the presence of a gravitational field. The gravitational field of the earth is relatively small and the bending cannot be measured on earth. However, the gravitational field of the sun is much larger and Einstein predicted in 1916 that rays of light that pass close to the sun should be bent by the gravitational field of the sun.

Another way of considering this bending of light is to say that light has energy and energy can be equated to mass, thus the light-mass should be attracted to the sun. Finally, we can think of this bending of light in terms of the curvature of spacetime caused by the mass of the sun. Light follows the shortest path, called a *geodesic,* and is thus bent by the curvature of spacetime.

Regardless of which conceptual picture we pick, Einstein predicted that a ray of light should be deflected by the sun by the angle of 1.75 seconds of arc. In order to observe this deflection it was necessary to measure the angular deviation between two stars when they are far removed from the sun, and then measure the deflection again when they are close to the sun (see figure 30.20). Of course when they are close to the sun, there is too much light from the sun to be able to see the stars. Hence, to test out Einstein's prediction it was necessary to measure the separation during a total eclipse of the sun. Sir Arthur Eddington led an expedition to the west coast of Africa for the solar eclipse of May 29, 1919, and measured the deflection. On November 6, 1919, the confirmation of Einstein's prediction of the bending of light was announced to the world.

More modern techniques used today measure radio waves from the two quasars, 3c273 and 3c279 in the constellation of Virgo. A quasar is a quasi-stellar object, a star, that emits very large quantities of radio waves. Because the sun is very dim in the emission of radio waves, radio astronomers do not have to wait for an eclipse to measure the angular separation but can measure it at any time. On October 8, 1972, when the quasars were close to the sun, radio astronomers measured the angular separation between 3c273 and 3c279 in radio waves and found that the change in the angular separation caused by the bending of the radio waves around the sun was 1.73 seconds of arc, in agreement with the general theory of relativity.

Figure 30.20

Bending of light by the sun.

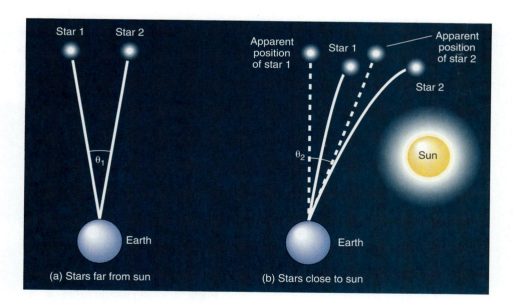

(a) Stars far from sun

(b) Stars close to sun

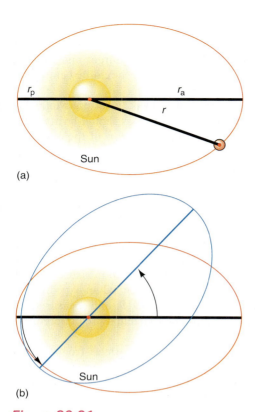

(a)

(b)

Figure 30.21

Advance of the perihelion of the planet
Mercury.

Figure 30.22

A clock in a gravitational field.

30.5 The Advance of the Perihelion of the Planet Mercury

According to Newton's laws of motion and his law of universal gravitation, each planet revolves around the sun in an elliptic orbit, as shown in figure 30.21. The closest approach of the planet to the sun is called its *perihelion distance* r_p, whereas its furthest distance is called its *aphelion distance* r_a. If there were only one planet in the solar system, the elliptical orbit would stay exactly as it is in figure 30.21(a). However, there are other planets in the solar system and each of these planets exert forces on every other planet. Because the masses of each of these planets is small compared to the mass of the sun, their gravitational effects are also relatively small. These extra gravitational forces cause a perturbation of the elliptical orbit. In particular, they cause the elliptical orbit to rotate in its plane, as shown in figure 30.21(b). The total precession of the perihelion of the planet Mercury is 574 seconds of arc in a century. The perturbation of all the other planets can only explain 531 seconds of arc by the Newtonian theory of gravitation, leaving a discrepancy of 43 seconds of arc per century of the advance of the perihelion of Mercury. Einstein, using the full power of his tensor equations, predicted an advance of the perihelion by 43 seconds of arc per century in agreement with the known observational discrepancy.

30.6 The Gravitational Red Shift

Let us consider the two clocks A and B located at the top and bottom of the rocket, respectively, in figure 30.22(a). The rocket is in interstellar space where we assume that all gravitational fields, if any, are effectively zero. The rocket is accelerating uniformly, as shown. Located in this interstellar space is a clock C, which is at rest. At the instant that the top of the rocket accelerates past clock C, clock A passes clock C at the speed v_A. Clock A, the moving clock, when observed from clock C, the stationary clock, shows an elapsed time Δt_A, given by the time dilation equation 29.64 as

$$\Delta t_C = \frac{\Delta t_A}{\sqrt{1 - v_A^2/c^2}} \tag{30.22}$$

Thus, the moving clock A runs slow compared to the stationary clock C.

A few moments later, clock B passes clock C at the speed v_B, as in figure 30.22(b). The speed v_B is greater than v_A because of the acceleration of the rocket.

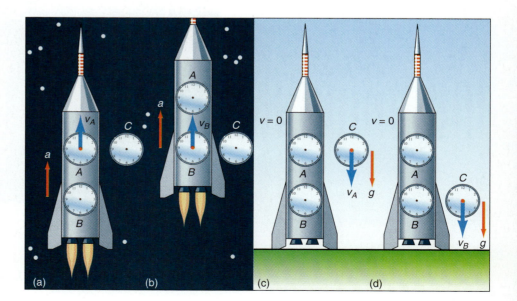

(a)   (b)   (c)   (d)

Let us read the same time interval Δt_C on clock C when clock B passes as we did for clock A so the two clocks can be compared. The difference in the time interval between the two clocks, B and C, is again given by the time dilation equation 29.64 as

$$\Delta t_C = \frac{\Delta t_B}{\sqrt{1 - v_B^2/c^2}} \tag{30.23}$$

Because the time interval Δt_C was set up to be the same in both equations 30.22 and 30.23, the two equations can be equated to give a relation between clocks A and B. Thus,

$$\frac{\Delta t_A}{(1 - v_A^2/c^2)^{1/2}} = \frac{\Delta t_B}{(1 - v_B^2/c^2)^{1/2}}$$

Rearranging terms, we get

$$\frac{\Delta t_A}{\Delta t_B} = \frac{(1 - v_A^2/c^2)^{1/2}}{(1 - v_B^2/c^2)^{1/2}}$$

$$\frac{\Delta t_A}{\Delta t_B} = (1 - v_A^2/c^2)^{1/2}(1 - v_B^2/c^2)^{-1/2} \tag{30.24}$$

But the two terms on the right-hand side of equation 30.24 can be expanded by the binomial theorem, equation 29.33, as

$$(1 - x)^n = 1 - nx$$

$$(1 - v_A^2/c^2)^{1/2} = 1 - \left(\frac{1}{2}\right)\frac{v_A^2}{c^2} = 1 - \frac{v_A^2}{2c^2}$$

and

$$(1 - v_B^2/c^2)^{-1/2} = 1 - \left(\frac{-1}{2}\right)\frac{v_B^2}{c^2} = 1 + \frac{v_B^2}{2c^2}$$

where again the assumption is made that v is small enough compared to c, to allow us to neglect the terms x^2 and higher in the expansion. Thus, equation 30.24 becomes

$$\frac{\Delta t_A}{\Delta t_B} = \left(1 - \frac{v_A^2}{2c^2}\right)\left(1 + \frac{v_B^2}{2c^2}\right)$$

$$= 1 + \frac{v_B^2}{2c^2} - \frac{v_A^2}{2c^2} - \frac{1}{4}\frac{v_B^2 v_A^2}{c^4}$$

The last term is set equal to zero on the same assumption that the speeds v are much less than c. Finally, rearranging terms,

$$\frac{\Delta t_A}{\Delta t_B} = 1 + \left(\frac{v_B^2}{2} - \frac{v_A^2}{2}\right)\frac{1}{c^2} \tag{30.25}$$

But by Einstein's principle of equivalence, we can equally well say that the rocket is at rest in the gravitational field of the earth, whereas the clock C is accelerating toward the earth in free-fall. When the clock C passes clock A it has the instantaneous velocity v_A, figure 30.22(c), and when it passes clock B it has the instantaneous velocity v_B, figure 30.22(b). We can obtain the velocities v_A and v_B by the law of conservation of energy, that is,

$$\tfrac{1}{2}mv^2 + \text{PE} = E_0 = \text{Constant} = \text{Total energy} \tag{30.26}$$

The total energy per unit mass, found by dividing equation 30.26 by m, is

$$\frac{v^2}{2} + \frac{\text{PE}}{m} = \frac{E_0}{m}$$

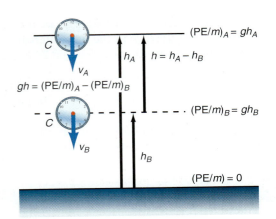

Figure 30.23

Freely falling clock C.

The conservation of energy per unit mass when clock C is next to clock A, obtained with the aid of figure 30.23, is

$$\frac{v_A^2}{2} + \frac{mgh_A}{m} = \frac{E_0}{m}$$

or

$$\frac{v_A^2}{2} + gh_A = \frac{E_0}{m} \tag{30.27}$$

Similarly, when the clock C is next to clock B, the conservation of energy per unit mass becomes

$$\frac{v_B^2}{2} + gh_B = \frac{E_0}{m} \tag{30.28}$$

Subtracting equation 30.27 from equation 30.28, gives

$$\frac{v_B^2}{2} + gh_B - \frac{v_A^2}{2} - gh_A = \frac{E_0}{m} - \frac{E_0}{m} = 0$$

Hence,

$$\frac{v_B^2}{2} - \frac{v_A^2}{2} = gh_A - gh_B = gh \tag{30.29}$$

where h is the distance between A and B, and gh is the gravitational potential energy per unit mass, which is sometimes called the *gravitational potential*. Substituting equation 30.29 back into equation 30.25, gives

$$\frac{\Delta t_A}{\Delta t_B} = 1 + \frac{gh}{c^2} \tag{30.30}$$

For a clearer interpretation of equation 30.30, let us change the notation slightly. Because clock B is closer to the surface of the earth where there is a stronger gravitational field than there is at a height h above the surface where the gravitational field is weaker, we will let

$$\Delta t_B = \Delta t_g$$

and

$$\Delta t_A = \Delta t_f$$

where Δt_g is the elapsed time on a clock in a strong gravitational field and Δt_f is the elapsed time on a clock in a weaker gravitational field. If we are far enough away from the gravitational mass, we can say that Δt_f is the elapsed time in a gravitational-field-free space. With this new notation equation 30.30 becomes

$$\frac{\Delta t_f}{\Delta t_g} = 1 + \frac{gh}{c^2}$$

or

$$\Delta t_f = \Delta t_g\left(1 + \frac{gh}{c^2}\right) \tag{30.31}$$

Since $(1 + gh/c^2) > 0$, *the elapsed time on the clock in the gravitational-field-free space Δt_f is greater than the elapsed time on a clock in a gravitational field Δt_g. Thus, the time elapsed on a clock in a gravitational field is less than the time elapsed on a clock in a gravity-free space. Hence, a clock in a gravitational field runs slower than a clock in a field-free space.*

We can find the effect of the slowing down of a clock in a gravitational field by placing an excited atom in a gravitational field, and then observing a spectral line from that atom far away from the gravitational field. The speed of the light from that spectral line is, of course, given by

$$c = \lambda \nu = \frac{\lambda}{T} \tag{30.32}$$

where λ is the wavelength of the spectral line, ν is its frequency, and T is the period or time interval associated with that frequency. Hence, if the time interval $\Delta t = T$ changes, then the wavelength of that light must also change. Solving for the period or time interval from equation 30.32, we get

$$T = \frac{\lambda}{c} \tag{30.33}$$

Substituting T from 30.33 for Δt in equation 30.31, we get

$$T_f = T_g\left(1 + \frac{gh}{c^2}\right)$$
$$\frac{\lambda_f}{c} = \frac{\lambda_g}{c}\left(1 + \frac{gh}{c^2}\right) \tag{30.34}$$

$$\lambda_f = \lambda_g\left(1 + \frac{gh}{c^2}\right) \tag{30.35}$$

where λ_g is the wavelength of the emitted spectral line in the gravitational field and λ_f is the wavelength of the observed spectral line in gravity-free space, or at least farther from where the atom is located in the gravitational field. Because the term $(1 + gh/c^2)$ is a positive number, it follows that

$$\lambda_f > \lambda_g \tag{30.36}$$

That is, *the wavelength observed in the gravity-free space is greater than the wavelength emitted from the atom in the gravitational field.* Recall from chapter 25 that the visible portion of the electromagnetic spectrum runs from violet light at around 380.0 nm to red light at 720.0 nm. Thus, red light is associated with longer wavelengths. *Hence, since $\lambda_f > \lambda_g$, the wavelength of the spectral line increases toward the red end of the spectrum, and the entire process of the slowing down of clocks in a gravitational field is referred to as the **gravitational red shift.***

A similar analysis in terms of frequency can be obtained from equations 30.32, 30.34, and the binomial theorem equation 29.34, to yield

$$\nu_f = \nu_g\left(1 - \frac{gh}{c^2}\right) \tag{30.37}$$

Where now the frequency observed in the gravitational-free space is less than the frequency emitted in the gravitational field because the term $\left(1 - \dfrac{gh}{c^2}\right)$ is less than one. The change in frequency per unit frequency emitted, found from equation 30.37, is

$$\nu_f - \nu_g = -\frac{gh}{c^2}\nu_g$$

$$\frac{\nu_g - \nu_f}{\nu_g} = \frac{gh}{c^2}$$

$$\frac{\Delta\nu}{\nu_g} = \frac{gh}{c^2} \tag{30.38}$$

The gravitational red shift was confirmed on the earth by an experiment by R. V. Pound and G. A. Rebka at Harvard University in 1959 using a technique called the *Mossbauer effect*. Gamma rays were emitted from radioactive cobalt in the basement of the Jefferson Physical Laboratory at Harvard University. These gamma rays traveled 22.5 m, through holes in the floors, up to the top floor. The difference between the emitted and absorbed frequency of the gamma ray was found to agree with equation 30.38.

Example 30.5

Gravitational frequency shift. Find the change in frequency per unit frequency for a γ-ray traveling from the basement, where there is a large gravitational field, to the roof of the building, which is 22.5 m higher, where the gravitational field is weaker.

Solution

The change in frequency per unit frequency, found from equation 30.38, is

$$\frac{\Delta \nu}{\nu_g} = \frac{gh}{c^2}$$

$$= \left(9.80 \, \frac{m}{s^2}\right)\left[\frac{22.5 \, m}{(3 \times 10^8 \, m/s)^2}\right]$$

$$= 2.50 \times 10^{-15}$$

The experiment was repeated by Pound and J. L. Snider in 1965, with another confirmation. Since then the experiment has been repeated many times, giving an accuracy to the gravitational red shift to within 1%.

Further confirmation of the gravitational red shift came from an experiment by Joseph Hafele and Richard Keating. Carrying four atomic clocks, previously synchronized with a reference clock in Washington, D.C., Hafele and Keating flew around the world in 1971. On their return they compared their airborne clocks to the clock on the ground and found the time differences associated with the time dilation effect and the gravitational effect exactly as predicted. Further tests with atomic clocks in airplanes and rockets have added to the confirmation of the gravitational red shift.

30.7 The Shapiro Experiment

Einstein's theory of general relativity not only predicts the slowing down of clocks in a gravitational field but it also predicts a contraction of the length of a rod in a gravitational field. The shrinking of rods and slowing down of clocks in a gravitational field can also be represented as a curvature of spacetime caused by mass. The slowing down of clocks and gravitational length contraction result in a reduction in the speed of light near a large massive body such as the sun. I. I. Shapiro performed an experiment in 1970 where he measured the time it takes for a radar signal (a light wave) to bounce off the planet Venus and return to earth at a time when Venus is close to the sun. The slowing down of light as it passes the sun causes the radar signal to be delayed by about 240×10^{-6} s. Shapiro's results agree with Einstein's theory to an accuracy of about 3%.

As an additional confirmation the delay in the travel time of radio signals to the spacecraft *Mariner 6* and *Mariner 7* showed the same kind of results.

An Essay on the Application of Physics

The Black Hole

Have you ever wondered, while watching those science fiction movies, why the astronauts were afraid of a black hole? They certainly make them seem very sinister. Are they really that dangerous? What is a black hole? How is it formed? What are its characteristics? What would happen if you went into one? Is it possible to go space traveling through a black hole?

The simplest way to describe the black hole is to start with a classical analogue. Suppose we wished to launch a rocket from the earth to a far distant place in outer space. How fast must the rocket travel to escape the gravitational pull of the earth? When we launch the rocket it has a velocity v, and hence, a kinetic energy. As the rocket proceeds into space, its velocity decreases but its potential energy increases. In chapter 7 we found that the potential energy of an object near the surface of the earth is PE $= mgh$, where g is the acceleration due to gravity at the surface of the earth. When an object is moved a large distance from the earth, g is no longer a constant and a new equation must be developed for the potential energy. The potential energy of an object when it is a distance r away from the center of the earth is found by the calculus to be

$$PE = -\frac{GM_e m}{r}$$

where G is the universal gravitational constant, M_e is the mass of the earth, and m is the mass of the object. Let us now apply this potential energy term to a rocket that is trying to escape from the gravitational pull of the earth. The total energy of the rocket at any time is equal to the sum of its potential energy and its kinetic energy, that is,

$$E = \text{KE} + \text{PE} = \frac{1}{2}mv^2 - \left[GM_e m\left(\frac{1}{r}\right)\right] \quad \textbf{(30H.1)}$$

When the rocket is fired from the surface of the earth, $r = R$, at an escape velocity v_e its total energy will be

$$E = \frac{1}{2}mv_e^2 - \left[GM_e m\left(\frac{1}{R}\right)\right]$$

By the law of conservation of energy, the total energy of the rocket remains a constant. Hence, we can equate the total energy at the surface of the earth to the total energy when the rocket is far removed from the earth. That is,

$$\frac{1}{2}mv_e^2 - \left[GM_e m\left(\frac{1}{R}\right)\right] = \frac{1}{2}mv^2$$
$$- \left[GM_e m\left(\frac{1}{r}\right)\right] \quad \textbf{(30H.2)}$$

When the rocket escapes the pull of the earth it has effectively traveled to infinity, that is, $r = \infty$, and its velocity at that time is reduced to zero, that is, $v = 0$. Hence, equation 30H.2 reduces to

$$\frac{1}{2}mv_e^2 - \left[GM_e m\left(\frac{1}{R}\right)\right]$$
$$= 0 - \left[GM_e m\left(\frac{1}{\infty}\right)\right] = 0$$
$$\frac{1}{2}mv_e^2 = \frac{GM_e m}{R}$$
$$v_e^2 = \frac{2GM_e}{R}$$
$$v_e = \sqrt{\frac{2GM_e}{R}} \quad \textbf{(30H.3)}$$

Equation 30H.3 is called the *escape velocity of the earth*. This is the velocity that an object must have if it is to escape the gravitational field of the earth. Now it was first observed by a British amateur astronomer, the Rev. John Michell, in 1783, and then 15 years later by Marquis Pierre de Laplace, that if light were a particle, as originally proposed by Sir Isaac Newton, then there was a limit to the size the earth could be and still have light escape from it. That is, if we solve equation 30H.3 for R, and replace the velocity of escape v_e by the velocity of light c, we get

$$R_S = \frac{2GM_e}{c^2} \quad \textbf{(30H.4)}$$

For reasons that will be explained later, this value of R is called the *Schwarzschild radius*, and is designated as R_S. Solving equation 30H.4 for the Schwarzschild radius of the earth we get 8.85×10^{-3} m, which means that if the earth were contracted to a sphere of radius smaller than 8.85×10^{-3} m, then the escape velocity from the earth would be greater than the velocity of light. That is, nothing, not even light could escape from the earth if it were this small. The earth would then be called a black hole because we could not see anything coming from it.

The reason for the name, black hole, comes from the idea that if we look at an object in space, such as a star, we see light coming from that star. If the star became a black hole, no light could come from that star. Hence, when we look into space we would no longer see a bright star at that location, but rather nothing but the blackness of space. There seems to be a hole in space where the star used to be and therefore we say that there is a black hole there.

Solving equation 30H.4 for the Schwarzschild radius of the sun, by replacing the mass of the earth by the mass of the sun, we get 2.95×10^3 m. Thus, if the sun were to contract to a radius below 2.95×10^3 m the gravitational force would become so great that no light could escape from the sun, and the sun would become a black hole.

Up to this point the arguments have been strictly classical. Since Einstein's theory of general relativity is a theory of gravitation, what does it say about black holes?

As we have seen, Einstein's theory of general relativity says that mass warps spacetime and we saw this in the rubber sheet

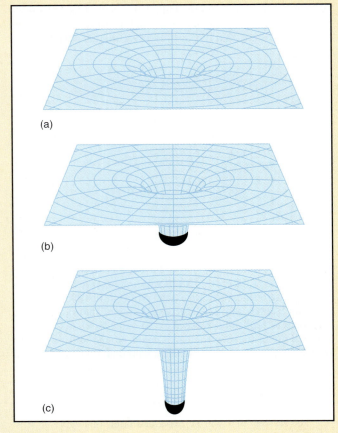

Figure 1

The warping of spacetime.

r	$\Delta r/\Delta t$
$R_S/10$	$-9c$
$R_S/5$	$-4c$
$R_S/2$	$-c$
R_S	0
$2R_S$	$0.5c$
$10R_S$	$0.9c$
$100R_S$	$0.99c$
$1000R_S$	$0.999c$

We saw there that if $\Delta s = 0$, then $\Delta x/\Delta t = c$, the velocity of light, and it is a constant, hence $\Delta s = 0$ represents the world line of a ray of light. Using the same analogy for the radial portion of the Schwarzschild solution we have

$$(\Delta s)^2 = \frac{(\Delta r)^2}{1 - 2GM/rc^2} - \left(1 - \frac{2GM}{rc^2}\right)c^2(\Delta t)^2$$

As we have just seen, $\Delta s = 0$ represents the world line of a ray of light. Applying this to the Schwarzschild solution we get

$$\frac{(\Delta r)^2}{1 - 2GM/rc^2} = \left(1 - \frac{2GM}{rc^2}\right)c^2(\Delta t)^2$$

$$\frac{(\Delta r)^2}{(\Delta t)^2} = \left(1 - \frac{2GM}{rc^2}\right)^2 c^2$$

$$\frac{\Delta r}{\Delta t} = \left(1 - \frac{2GM}{rc^2}\right)c \qquad \textbf{(30H.6)}$$

Notice that if $r = 2GM/c^2$, then $\Delta r/\Delta t = 0$. This means that the velocity of light $\Delta r/\Delta t$ is then zero, and no light is able to leave the gravitating body. But notice that this quantity is exactly what we already called the Schwarzschild radius. The Schwarzschild radius is also called the *event horizon of the black hole.* We can generalize equation 30H.6 to the form

$$\frac{\Delta r}{\Delta t} = \left(1 - \frac{R_S}{r}\right)c \qquad \textbf{(30H.7)}$$

The solution of equation 30H.7 for various values of r is shown in table 30H.1. Notice that the velocity of light is not a constant near the black hole, but in a distance of only 1000 times the radius of the black hole, the velocity of light approaches the constant value c. Note that the constancy of the velocity of light is not a postulate of general relativity as it is for special relativity. Also note that as we get far away from the black hole, $r \gg R_S$, we enter the region of flat spacetime and the velocity of light has the constant value c of special relativity. However, within the event horizon, equation 30H.7 and table 30H.1 show that the velocity of light can be greater than c.

analogy in figure 30.18. The greater the mass of the gravitating body the greater the warping of spacetime. Figure 1(a) shows the warping of spacetime by a star. Figure 1(b) shows the warping for a much more massive star. As the radius of the star becomes much smaller, the warping becomes more pronounced as the star approaches the size of a black hole, figure 1(c).

Shortly after Einstein stated his principle of general relativity, K. Schwarzschild solved Einstein's equations for the gravitational field of a point mass. For the radial portion of the solution he obtained

$$(\Delta s)^2 = \frac{(\Delta r)^2}{1 - 2GM/rc^2} - (1 - 2GM/rc^2)c^2(\Delta t)^2 \qquad \textbf{(30H.5)}$$

Equation 30H.5 is called the *radial portion of the Schwarzschild metric* and is the radial portion of the invariant interval of spacetime curved by the presence of a point mass. The invariant interval found previously in equation 30.9 is the metric for a flat spacetime, that is, one in which there is no mass to warp spacetime. That is, for flat spacetime

$$(\Delta s)^2 = c^2(\Delta t)^2 - (\Delta x)^2 - (\Delta y)^2 - (\Delta z)^2 \qquad \textbf{(30.9)}$$

and in only one space dimension by

$$(\Delta s)^2 = c^2(\Delta t)^2 - (\Delta x)^2 \qquad \textbf{(30.8)}$$

The argument up to now may seem somewhat academic, in that we have described some of the characteristics of black holes, but do they really exist in nature? That is, is it possible for any objects in the universe to become black holes? The answer is yes. In the ordinary evolution of very massive stars, black holes can be formed. A star is essentially a gigantic nuclear reactor converting hydrogen to helium in a process called *nuclear fusion*. Think of the star as millions of hydrogen bombs going off at the same time, thereby producing enormous quantities of energy and enormous forces outward from the star. There is an equilibrium between the gravitational forces inward and the forces outward caused by the exploding gases. Eventually, when all the nuclear fuel is used, there is no longer an equilibrium condition. The gravitational force causes the gas to become very compact. If the star is large enough, it is compressed below its Schwarzschild radius and a black hole is formed. For an evolving star to condense into a black hole it must be approximately 25 times the mass of the sun. When the star condenses to a black hole it does not stop at the event horizon but continues to reduce in size until it becomes a singularity, a point mass. That is, the entire mass of the star has condensed to the size of a point.

There is experimental evidence that a black hole has been found as a companion of the star Cygnus X-1 and more are looked for every day.

Since time slows down in a gravitational field, the effect becomes much more pronounced in the vicinity of the black hole. If a person were to fall into the black hole he would eventually be crushed due to the enormous gravitational forces. Time would slow down for him as he approached the event horizon. At the event horizon, time would stand still for him.

The Schwarzschild black hole is an example of a nonrotating massive body. However, just as the sun and planets rotate about their axes, a more general solution of a black hole should also be concerned with the rotation of the massive body. The solution to the rotating black hole is called a *Kerr black hole*, after Roy Kerr, a New Zealand mathematician. The rotating black hole[4] (essentially an accelerating black hole) drags spacetime around with it, forming a second event horizon, thus leaving a space between the first event horizon and the second event horizon. It has been speculated that it may be possible to enter the first event horizon, but not the second, and exit somewhere else in either another universe or in this universe in another place and/or time.

It has also been speculated that there might also exist white holes in space. That is, mass is drawn into a black hole, but would be spewed out of a white hole. In fact some physicists have speculated that a black hole in one universe is a white hole in another universe.

4. See interactive tutorial problem 15.

The Language of Physics

Spacetime diagram
A graph of a particle's space and time coordinates. The time coordinate is usually expressed as τ, which is equal to the product of the speed of light and the time (p. 891).

World line
A line in a spacetime diagram that shows the motion of a particle through spacetime. A world line of a particle at rest or moving at a constant velocity is a straight line in spacetime. The world line of a light ray makes an angle of 45° with the τ-axis in spacetime. The world line of an accelerated particle is a curve in spacetime (p. 891).

Light cone
A cone that is drawn in spacetime showing the relation between the past and the future of a particle in spacetime. World lines within the cone are called timelike because they are accessible to us in time. Events outside the cone are called spacelike because they occur in another part of space that is not accessible to us and hence is called elsewhere (p. 894).

Invariant interval
A constant value in spacetime that all observers agree on, regardless of their state of motion. The equation of the invariant interval is in the form of a hyperbola in spacetime. Because of the hyperbolic form of the invariant interval, Euclidean geometry does not hold in spacetime. The reason for length contraction and time dilation is the fact that spacetime is non-Euclidean. The longest distance in spacetime is the straight line (p. 894).

Equivalence principle
On a local scale, the physical effects of a gravitational field are indistinguishable from the physical effects of an accelerated coordinate system. Hence, an accelerated frame of reference is equivalent to an inertial frame of reference in which gravity is present, and an inertial frame is equivalent to an accelerated frame in which gravity is absent (p. 906).

The general theory of relativity
The laws of physics are the same in all frames of reference (note that there is no statement about the constancy of the velocity of light as in the special theory of relativity) (p. 907).

Warped spacetime
Matter causes spacetime to be warped so that the world lines of particles in spacetime are curved. Hence, matter warps spacetime and spacetime tells matter how to move. Gravity is a consequence of the warping of spacetime by matter (p. 908).

Gravitational red shift
Time elapsed on a clock in a gravitational field is less than the time elapsed on a clock in a gravity-free space. This effect of the slowing down of a clock in a gravitational field is manifested by observing a spectral line from an excited atom in a gravitational field. The wavelength of the spectral line of that atom is shifted toward the red end of the electromagnetic spectrum (p. 913).

Summary of Important Equations

Tau in spacetime
$$\tau = ct \tag{30.1}$$

Velocity in a spacetime diagram
$$v = c \tan \theta \tag{30.2}$$

The square of the invariant interval
$$(\Delta s)^2 = c^2(\Delta t)^2 - (\Delta x)^2 \tag{30.8}$$
$$(\Delta s)^2 = c^2(\Delta t)^2 - (\Delta x)^2 - (\Delta y)^2 - (\Delta z)^2 \tag{30.9}$$
$$(\Delta s)^2 = (\Delta \tau)^2 - (\Delta x)^2 \tag{30.11}$$

Slowing down of a clock in a gravitational field
$$\Delta t_{\text{f}} = \Delta t_{\text{g}}\left(1 + \frac{gh}{c^2}\right) \tag{30.31}$$

Gravitational red shift of wavelength

$$\lambda_f = \lambda_g\left(1 + \frac{gh}{c^2}\right) \qquad (30.35)$$

Gravitational red shift of frequency

$$\nu_f = \nu_g\left(1 - \frac{gh}{c^2}\right) \qquad (30.37)$$

Change in frequency per unit frequency

$$\frac{\Delta\nu}{\nu_g} = \frac{gh}{c^2} \qquad (30.38)$$

Questions for Chapter 30

1. Discuss the concept of spacetime. How is it like space and how is it different?
2. How many light cones are there in your classroom?
3. Why can't a person communicate with another person who is elsewhere?

†4. Discuss the twin paradox on the basis of figure 30.6(b).
5. Using figure 30.7, discuss why the scales in the S' system are not the same as the scales in the S system.
†6. Considering some of the characteristics of spacetime, that is, it can be warped, and so forth, could spacetime be the elusive ether?

7. What does it mean to say that spacetime is warped?
8. Describe length contraction by a spacetime diagram.
9. Describe time dilation by a spacetime diagram.
10. Discuss simultaneity with the aid of a spacetime diagram.

Problems for Chapter 30

30.1 Spacetime Diagrams

1. Draw the world line in spacetime for a particle moving in (a) an elliptical orbit, (b) a parabolic orbit, and (c) a hyperbolic orbit.

30.2 The Invariant Interval

2. Find the angle that the world line of a particle moving at a speed of $c/4$ makes with the τ-axis in spacetime.
3. The world line of a particle is a straight line making an angle of 30° below the τ-axis. Determine the speed of the particle.
4. The world line of a particle is a straight line of length 150 m. Find the value of Δx if $\Delta\tau = 200$ m.

5. (a) On a sheet of graph paper draw the hyperbolas representing the invariant interval of spacetime as shown in figure 30.7. (b) Draw the S'-axes on this diagram for a particle moving at a speed of $c/4$.
6. Using the graph of problem 5, draw a rod 1.50 units long at rest in the S frame of reference. (a) From the graph determine the length of the rod in the S' frame of reference. (b) Determine the length of the rod using the Lorentz contraction equation.
7. Using the graph of problem 5, draw a rod 1.50 units long at rest in the S' frame of reference. (a) From the graph determine the length of the rod in the S frame of reference. (b) Determine the length of the rod using the Lorentz contraction equation.

30.6 The Gravitational Red Shift

8. One twin lives on the ground floor of a very tall apartment building, whereas the second twin lives 200 ft above the ground floor. What is the difference in their age after 50 years?
9. The lifetime of a subatomic particle is 6.25×10^{-7} s on the earth's surface. Find its lifetime at a height of 500 km above the earth's surface.
10. An atom on the surface of Jupiter ($g = 23.1$ m/s²) emits a ray of light of wavelength 528.0 nm. What wavelength would be observed at a height of 10,000 m above the surface of Jupiter?

Additional Problems

†11. Using the principle of equivalence, show that the difference in time between a clock at rest and an accelerated clock should be given by

$$\Delta t_R = \Delta t_A \left(1 + \frac{ax}{c^2} \right)$$

where Δt_R is the time elapsed on a clock at rest, Δt_A is the time elapsed on the accelerated clock, a is the acceleration of the clock, and x is the distance that the clock moves during the acceleration.

†12. A particle is moving in a circle of 1.00-m radius and undergoes a centripetal acceleration of 9.80 m/s². Using the results of problem 11, determine how many revolutions the particle must go through in order to show a 10% variation in time.

13. The pendulum of a grandfather clock has a period of 0.500 s on the surface of the earth. Find its period at an altitude of 200 km. *Hint:* Note that the change in the period is due to two effects. The acceleration due to gravity is smaller at this height even in classical physics, since

$$g = \frac{GM}{(R + h)^2}$$

To solve this problem, use the fact that the average acceleration is

$$g = \frac{GM}{R(R + h)}, \text{ and assume that}$$

$$\Delta t_f = \Delta t_g \left(1 + \frac{gh}{c^2} \right).$$

14. Compute the fractional change in frequency of a spectral line that occurs between atomic emission on the earth's surface and that at a height of 325 km.

Interactive Tutorials

15. A rotating black hole. Assume the sun were to collapse to a black hole as described in the "Have you ever wondered . . . ?" section.
(a) Calculate the radius of the black hole, which is called the Schwarzschild radius R_S. Since the sun is also rotating, angular momentum must be conserved. Therefore as the sun collapses the angular velocity of the sun must increase, and hence the tangential velocity of a point on the surface of the sun must also increase. (b) Find the radius of the sun during the collapse such that the tangential velocity of a point on the equator is equal to the velocity of light c. Compare this radius to the Schwarzschild radius. Some characteristics of the sun are radius, $r_0 = 6.96 \times 10^8$ m, mass of sun $M = 1.99 \times 10^{30}$ kg, and the angular velocity of the sun $\omega_0 = 2.86 \times 10^{-6}$ rad/s.

16. Gravitational red shift. An atom on the surface of the earth emits a ray of light of wavelength $\lambda_g = 528.0$ nm, straight upward. (a) What wavelength λ_f would be observed at a height $y = 10,000$ m? (b) What frequency ν_f would be observed at this height? (c) What change in time would this correspond to?

Quantum Physics

Particle tracks.

Newton himself was better aware of the weakness inherent in his intellectual edifice than the generations which followed him. This fact has always aroused my admiration.

Albert Einstein

31.1 The Particle Nature of Waves

Up to now in our study of physics, we considered (1) the motion of particles and their interaction with other particles and their environment and (2) the nature, representation, and motion of waves. We considered particles as little hard balls of matter while a wave was a disturbance that was spread out through a medium. There was certainly a significant difference between the two concepts, and one of the most striking of these is illustrated in figure 31.1. In figure 31.1(a), two particles collide, bounce off each other, and then continue in a new direction. In figure 31.1(b), two waves collide, but they do not bounce off each other. They add together by the principle of superposition, and then each continues in its original direction as if the waves never interacted with each other.

Another difference between a particle and a wave is that the total energy of the particle is concentrated in the localized mass of the particle. In a wave, on the other hand, the energy is spread out throughout the entire wave. Thus, there is a very significant difference between a particle and a wave.

We have seen that light is an electromagnetic wave. The processes of interference, diffraction, and polarization are characteristic of wave phenomena and

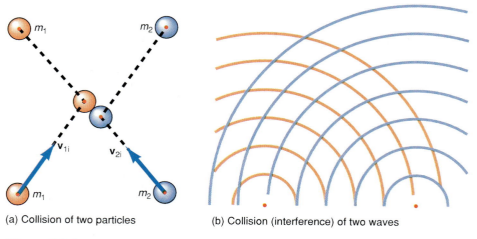

(a) Collision of two particles

(b) Collision (interference) of two waves

Figure 31.1

Characteristics of particles and waves.

921

have been studied and verified in the laboratory many times over. Yet there has appeared with time, some apparent contradictions to the wave nature of light. We will discuss the following three of these physical phenomena:

1. Blackbody radiation.
2. The photoelectric effect.
3. Compton scattering.

31.2 Blackbody Radiation

In chapter 16, we saw that all bodies emit and absorb radiation. (Recall that radiation is heat transfer by electromagnetic waves.) The Stefan-Boltzmann law showed that the amount of energy radiated is proportional to the fourth power of the temperature, but did not say how the heat radiated was a function of the wavelength of the radiation. Because the radiation consists of electromagnetic waves, we would expect that the energy should be distributed evenly among all possible wavelengths. However, as shown in figure 16.10, the energy distribution is not even but varies according to wavelength and frequency. All attempts to account for the energy distribution by classical means failed.

Let us consider for a moment how a body can radiate energy. We saw in chapter 25 that an oscillating electric charge generates an electromagnetic wave. A body can be considered to be composed of a large number of atoms in a lattice structure as shown in figure 31.2(a). For a metallic material the positively ionized atom is located at the lattice site and the outermost electron of the atom moves throughout the lattice as part of the electron gas. Each atom of the lattice is in a state of equilibrium under the action of all the forces from all its neighboring atoms. The atom is free to vibrate about this equilibrium position. A mechanical analogue to the lattice structure is shown in figure 31.2(b) as a series of masses connected by springs. Each mass can oscillate about its equilibrium position. To simplify the picture further, let us consider a single ionized atom with a charge q and let it oscillate in simple harmonic motion, as shown in figure 31.2(c). The oscillating charge generates an electromagnetic wave that is emitted by the body. Each ionized atom is an oscillator and each has its own fixed frequency and emits radiation of this frequency. Because the body is made up of millions of these oscillating charges, the body always emits radiation of all these different frequencies, and hence the emission spectrum should be continuous. The intensity of the radiation depends on the amplitude of the oscillation. A typical radiated wave is given by equation 25.38 as

$$E = E_0 \sin(kx - \omega t)$$

Figure 31.2

A solid body emits electromagnetic radiation.

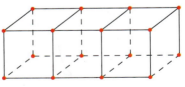

(a) Lattice structure

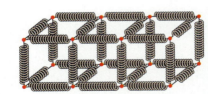

(b) Equivalent mass-spring structure

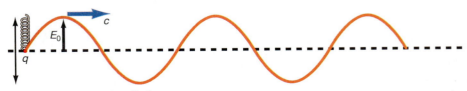

(c) Simplified picture of oscillation

Modern Physics

where

$$k = \frac{2\pi}{\lambda} \qquad \text{(12.9)}$$

and

$$\omega = 2\pi\nu \qquad \text{(12.12)}$$

Thus, the frequency of the oscillating charge is the frequency of the electromagnetic wave. The amplitude of the wave E_0 depends on the amplitude of the simple harmonic motion of the oscillating charge. When the body is heated, the heat energy causes the ionized atoms to vibrate with greater amplitude about their equilibrium position. The energy density of the emitted waves is given by equation 25.61 as

$$u = \epsilon_0 E^2$$

or

$$u = \epsilon_0 E_0^2 \sin^2(kx - \omega t) \qquad \text{(31.1)}$$

Thus, when the amplitude of the oscillation E_0 increases, more energy is emitted. When the hot body is left to itself it loses energy to the environment by this radiation process and the amplitude of the oscillation decreases. The amplitude of the oscillation determines the energy of the electromagnetic wave. Because of the extremely large number of ionized atoms in the lattice structure that can participate in the oscillations, all modes of vibration of the lattice structure are possible and hence all possible frequencies are present. Thus, the classical picture of blackbody radiation permits all frequencies and energies for the electromagnetic waves. However, this classical picture does not agree with experiment.

Max Planck (1858–1947), a German physicist, tried to "fit" the experimental results to the theory. However, he found that he had to break with tradition and propose a new and revolutionary concept. *Planck assumed that the atomic oscillators cannot take on all possible energies, but could only oscillate with certain discrete amounts of energy given by*

$$E = nh\nu \qquad \text{(31.2)}$$

where h is a constant, now called *Planck's constant,* and has the value

$$h = 6.625 \times 10^{-34} \text{ J s}$$

In equation 31.2 ν is the frequency of the oscillator and n is an integer, a number, now called a *quantum number.* The energies of the vibrating atom are now said to be quantized, or limited to only those values given by equation 31.2. Hence, the atom can have energies $h\nu$, $2h\nu$, $3h\nu$, and so on, but never an energy such as 2.5 $h\nu$. This concept of quantization is at complete variance with classical electromagnetic theory. In the classical theory, as the oscillating charge radiates energy it loses energy and the amplitude of the oscillation decreases continuously. If the energy of the oscillator is quantized, the amplitude cannot decrease continuously and hence the oscillating charge cannot radiate while it is in this quantum state. If the oscillator now drops down in energy one quantum state, the difference in energy between the two states is now available to be radiated away. Hence, *the assumption of discrete energy states entails that the radiation process can only occur when the oscillator jumps from one quantized energy state to another quantized energy state.* As an example, if the oscillating charge is in the quantum state 4 it has an energy

$$E_4 = 4h\nu$$

When the oscillator drops to the quantum state 3 it has the energy

$$E_3 = 3h\nu$$

When the oscillator drops from the 4 state to the 3 state it can emit the energy

$$\Delta E = E_4 - E_3 = 4h\nu - 3h\nu = h\nu$$

Thus, the amount of energy radiated is always in small bundles of energy of amount $h\nu$. *This little bundle of radiated electromagnetic energy was called a quantum of energy. Much later, this bundle of electromagnetic energy came to be called a* **photon.**

Although this quantum hypothesis led to the correct formulation of black-body radiation, it had some serious unanswered questions. Why should the energy of the oscillator be quantized? If the energy from the blackbody is emitted as a little bundle of energy how does it get to be spread out into Maxwell's electromagnetic wave? How does the energy, which is spread out in the wave, get compressed back into the little quantum of energy so it can be absorbed by an atomic oscillator? These and other questions were very unsettling to Planck and the physics community in general. Although Planck started what would be eventually called *quantum mechanics,* and won the Nobel Prize for his work, he spent many years trying to disprove his own theory.

Example 31.1

Applying the quantum condition to a vibrating spring. A weightless spring has a spring constant k of 29.4 N/m. A mass of 300 g is attached to the spring and is then displaced 5.00 cm. When the mass is released, find (a) the total energy of the mass, (b) the frequency of the vibration, (c) the quantum number n associated with this energy, and (d) the energy change when the oscillator changes its quantum state by one value, that is, for $n = 1$.

Solution

a. The total energy of the vibrating spring comes from its potential energy, which it obtained when work was done to stretch the spring to give an amplitude A of 5.00 cm. The energy, found from equation 11.24 with $x = A$, is

$$E_{total} = PE = \tfrac{1}{2}kA^2$$

$$E = \tfrac{1}{2}(29.4 \text{ N/m})(0.0500 \text{ m})^2$$

$$= 3.68 \times 10^{-2} \text{ J}$$

b. The frequency ν of the vibration, found from equation 11.21, is

$$\nu = \frac{1}{2\pi} \sqrt{\frac{k}{m}}$$

$$= \frac{1}{2\pi} \sqrt{\frac{29.4 \text{ N/m}}{0.300 \text{ kg}}}$$

$$= 1.58 \text{ Hz}$$

c. The quantum number n associated with this energy, found from equation 31.2, is

$$E = nh\nu$$

$$n = \frac{E}{h\nu} = \frac{3.68 \times 10^{-2} \text{ J}}{6.625 \times 10^{-34} \text{ J s} \times 1.58 \text{ s}^{-1}}$$

$$= 3.52 \times 10^{31}$$

This is an enormously large number. Therefore, the effect of a quantum of energy is very small unless the vibrating system itself is very small, as in the case of the vibration of an atom.

924

d. The energy change associated with the oscillator changing one energy state, found from equation 31.2, is

$$E = nh\nu = h\nu$$

$$= (6.625 \times 10^{-34} \text{ J s})(1.58 \text{ s}^{-1})$$

$$= 1.05 \times 10^{-33} \text{ J}$$

This change in energy is so small that for all intents and purposes, the energy of a vibrating spring-mass system is continuous.

Example 31.2

The energy of a photon of light. An atomic oscillator emits radiation of 700.0-nm wavelength. How much energy is associated with a photon of light of this wavelength?

Solution

The energy of the photon, given by equation 31.2, is

$$E = h\nu$$

but since the frequency ν can be written as c/λ, the energy of the photon can also be written as

$$E = h\nu = \frac{hc}{\lambda}$$

$$= \frac{(6.625 \times 10^{-34} \text{ J s})(3.00 \times 10^8 \text{ m/s})}{700.0 \text{ nm}} \left(\frac{1 \text{ nm}}{10^{-9} \text{ m}} \right)$$

$$= 2.84 \times 10^{-19} \text{ J}$$

Thus, the photon of light is indeed a small bundle of energy.

31.3 The Photoelectric Effect

When Heinrich Hertz performed his experiments in 1887 to prove the existence of electromagnetic waves, he accidently found that when light fell on a metallic surface, the surface emitted electrical charges. *This effect, whereby light falling on a metallic surface produces electrical charges, is called the* **photoelectric effect.** The photoelectric effect was the first proof that light consists of small particles called photons. Thus, the initial work that showed light to be a wave would also show that light must also be a particle.

Further experiments by Philipp Lenard in 1900 confirmed that these electrical charges were electrons. These electrons were called *photoelectrons.* The photoelectric effect can best be described by an experiment, the schematic diagram of which is shown in figure 31.3. The switch S is thrown to make the anode of the phototube positive and the cathode negative. Monochromatic light (light of a single frequency ν) of intensity I_1, is allowed to shine on the cathode of the phototube, causing electrons to be emitted.

The positive anode attracts these electrons, and they flow to the anode and then through the connecting circuit. The ammeter in the circuit measures this current. Starting with a positive potential V, the current is observed for decreasing values of V. When the potential V is reduced to zero, the switch S is reversed to make the anode negative and the cathode positive. The negative anode now repels the photoelectrons as they approach the anode. If this potential is made more and

Figure 31.3

Schematic diagram for the photoelectric effect.

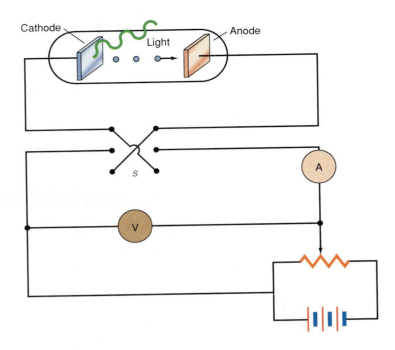

Figure 31.4

Current i as a function of voltage V for the photoelectric effect.

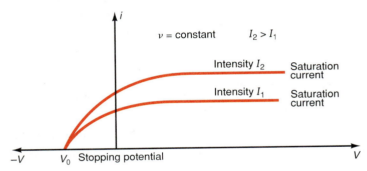

more negative, however, a point is eventually reached when the kinetic energy of the electrons is not great enough to overcome the negative stopping potential, and no more electrons reach the anode. The current i, therefore, becomes zero. A plot of the current i in the circuit, as a function of the potential between the plates, is shown in figure 31.4. If we increase the intensity of the light to I_2 and repeat the experiment, we obtain the second curve shown in the figure.

An analysis of figure 31.4 shows that when the value of V is high and positive, the current i is a constant. This occurs because all the photoelectrons formed at the cathode are reaching the anode. By increasing the intensity I, we obtain a higher constant value of current, because more photoelectrons are being emitted per unit time. This shows that the number of electrons emitted (the current) is proportional to the intensity of the incident light, that is,

$$i \propto I$$

Notice that when the potential is reduced to zero, there is still a current in the tube. Even though there is no electric field to draw them to the anode, many of the photoelectrons still reach the anode because of the initial kinetic energy they possess when they leave the cathode. As the switch S in figure 31.3 is reversed, the potential V between the plates becomes negative and tends to repel the photoelectrons. As the retarding potential V is made more negative, the current i (in figure 31.4) decreases, indicating that fewer and fewer photoelectrons are reaching the anode. When

V is reduced to V_0, there is no current at all in the circuit; V_0 is called the *stopping potential*. Note that it is the same value regardless of the intensity. (Both curves intersect at V_0.) Hence the stopping potential is independent of the intensity of light, or stated another way, the stopping potential is not a function of the intensity of light. Stated mathematically this becomes,

$$V_0 \neq V_0(I) \qquad (31.3)$$

The retarding potential is related to the kinetic energy of the photoelectrons. For the electron to reach the anode, its kinetic energy must be equal to the potential energy between the plates. (A mechanical analogy might be helpful at this point. If we wish to throw a ball up to a height h, where it will have the potential energy PE $= mgh$, we must throw the ball with an initial velocity v_0 such that the initial kinetic energy of the ball KE $= \frac{1}{2}mv_0^2$, is equal to the final potential energy of the ball.) Hence, the kinetic energy of the electron must be

$$\text{KE of electron} = \text{PE between the plates}$$

or

$$\text{KE} = eV \qquad (31.4)$$

where e is the charge on the electron and V is the potential between the plates.

The retarding potential acts on electrons that have less kinetic energy than that given by equation 31.4. When $V = V_0$, the stopping potential, even the most energetic electrons (those with maximum kinetic energy) do not reach the anode. Therefore,

$$\text{KE}_{\text{max}} = eV_0 \qquad (31.5)$$

As equations 31.3 and 31.5 show, the maximum kinetic energy of the photoelectrons is not a function of the intensity of the incident light, that is,

$$\text{KE}_{\text{max}} \neq \text{KE}_{\text{max}}(I)$$

It is also found experimentally that there is essentially no time lag between the time the light shines on the cathode and the time the photoelectrons are emitted.

If we keep the intensity constant and perform the experiment with different frequencies of light, we obtain the curves shown in figure 31.5. As the graph in figure 31.5 shows, the saturation current (the maximum current) is the same for any frequency of light, as long as the intensity is constant. But the stopping potential is different for each frequency of the incident light. Since the stopping potential is proportional to the maximum kinetic energy of the photoelectrons by equation 31.5, the maximum kinetic energy of the photoelectrons should be proportional to the frequency of the incident light. The maximum kinetic energy of the photoelectrons is plotted as a function of frequency in the graph of figure 31.6.

The first thing to observe is that the maximum kinetic energy of the photoelectrons is proportional to the frequency of the incident light. That is,

$$\text{KE}_{\text{max}} \propto \nu$$

The second thing to observe is that there is a cutoff frequency ν_0 below which there is no photoelectronic emission. That is, no photoelectric effect occurs unless the incident light has a frequency higher than the threshold frequency ν_0. For most metals, ν_0 lies in the ultraviolet region of the spectrum, but for the alkali metals it lies in the visible region.

Failure of the Classical Theory of Electromagnetism to Explain the Photoelectric Effect

The classical theory of electromagnetism was initially used to try to explain the results of the photoelectric effect. The results of the experiment are compared with

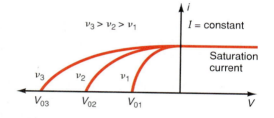

Figure 31.5

Current i as a function of voltage V for different light frequencies.

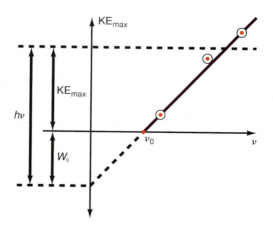

Figure 31.6

Maximum kinetic energy (KE_{max}) as a function of frequency ν for the photoelectric effect.

Table 31.1
The Photoelectric Effect

Experimental Results	Theoretical Predictions of Classical Electromagnetism	Agreement
$i \propto I$	$i \propto I$	Yes
Cutoff frequency ν_0	There should not be a cutoff frequency	No
No time lag for emission of electrons	There should be a time lag	No
$KE_{max} \propto \nu$	KE_{max} not $\propto \nu$	No
$KE_{max} \neq KE_{max}(I)$	$KE_{max} \propto I$	No

the predictions of classical electromagnetic theory in table 31.1. The only agreement between theory and experiment is the fact that the photocurrent is proportional to the intensity of the incident light. According to classical theory, there should be no minimum threshold frequency ν_0 for emission of photoelectrons. This prediction does not agree with the experimental results.

According to classical electromagnetic theory, energy is distributed equally throughout the entire electric wave front. When the wave hits the electron on the cathode, the electron should be able to absorb only the small fraction of the energy of the total wave that is hitting the electron. Therefore, there should be a time delay to let the electron absorb enough energy for it to be emitted. Experimentally, it is found that emission occurs immediately on illumination; there is no time delay for emission.

Finally, classical electromagnetic theory predicts that a very intense light of very low frequency will cause more emission than a high-frequency light of very low intensity. Again the theory fails to agree with the experimental result. Therefore, classical electromagnetic theory cannot explain the photoelectric effect.

Einstein's Theory of the Photoelectric Effect

In the same year that Einstein published his special theory of relativity, 1905, he also proposed a new and revolutionary solution for the problem of the photoelectric effect. Using the concept of the quantization of energy as proposed by Planck for the solution to the blackbody radiation problem, Einstein assumed that the energy of the electromagnetic wave was not spread out equally along the wave front, but that it was concentrated into Planck's little bundles or quanta of energy. Planck had assumed that the atomic radiators were quantized, but he still believed that the energy became spread out across the wave as the wave propagated. Einstein, on the other hand, assumed that as the wave progressed, the energy did not spread out with the wave front, but stayed in the little bundle or quanta of energy that would later become known as the photon. The photon thus contained the energy

$$E = h\nu \tag{31.6}$$

Einstein assumed that this concentrated bundle of radiant energy struck an electron on the metallic surface. The electron then absorbed this entire quantum of energy ($E = h\nu$). A portion of this energy is used by the electron to break away from the solid, and the rest shows up as the kinetic energy of the electron. That is,

(incident absorbed energy) − (energy to break away from solid)
= (maximum KE of electron) (31.7)

We call the energy for the electron to break away from the solid the *work function of the solid* and denote it by W_0. We can state equation 31.7 mathematically as

$$E - W_0 = KE_{max} \tag{31.8}$$

or

$$h\nu - W_0 = KE_{max} \tag{31.9}$$

We find the final maximum kinetic energy of the photoelectrons from equation 31.9 as

$$KE_{max} = h\nu - W_0 \tag{31.10}$$

Equation 31.10 is known as Einstein's photoelectric equation.

Notice from figure 31.6, when the KE_{max} of the photoelectrons is equal to zero, the frequency ν is equal to the cutoff frequency ν_0. Hence, equation 31.10 becomes

$$0 = h\nu_0 - W_0$$

Thus, we can also write the work function of the metal as

$$W_0 = h\nu_0 \tag{31.11}$$

Hence, we can also write Einstein's photoelectric equation as

$$KE_{max} = h\nu - h\nu_0 \tag{31.12}$$

For light frequencies equal to or less than ν_0, there is not enough energy in the incident wave to remove the electron from the solid, and hence there is no photoelectric effect. This explains why there is a threshold frequency below which there is no photoelectric effect.

When Einstein proposed his theory of the photoelectric effect, there were not enough quantitative data available to prove the theory. In 1914, R. A. Millikan performed experiments (essentially the experiment described here) that confirmed Einstein's theory of the photoelectric effect.

Einstein's theory accounts for the absence of a time lag for photoelectronic emission. As soon as the electron on the metal surface is hit by a photon, the electron absorbs enough energy to be emitted immediately. Einstein's equation also correctly predicts the fact that the maximum kinetic energy of the photoelectron is dependent on the frequency of the incident light. Thus, Einstein's equation completely predicts the experimental results.

Einstein's theory of the photoelectric effect is outstanding because it was the first application of quantum concepts. *Light should be considered as having not only a wave character, but also a particle character. (The photon is the light particle.)*

For his explanation of the photoelectric effect, Einstein won the Nobel Prize in physics in 1921. As mentioned earlier, Einstein's paper on the photoelectric effect was also published in 1905 around the same time as his paper on special relativity. Thus, he was obviously thinking about both concepts at the same time. It is no wonder then that he was not too upset with dismissing the concept of the ether for the propagation of electromagnetic waves. Because he could now picture light as a particle, a photon, he no longer needed a medium for these waves to propagate in.

Example 31.3

The photoelectric effect. Yellow light of 577.0-nm wavelength is incident on a cesium surface. It is found that no photoelectrons flow in the circuit when the cathode-anode voltage drops below 0.250 V. Find (a) the frequency of the incident photon,

(b) the initial energy of the photon, (c) the maximum kinetic energy of the photo-electron, (d) the work function of cesium, (e) the threshold frequency, and (f) the corresponding threshold wavelength.

Solution

a. The frequency of the photon is found from

$$\nu = \frac{c}{\lambda} = \left(\frac{3.00 \times 10^8 \text{ m/s}}{577.0 \text{ nm}}\right)\left(\frac{1 \text{ nm}}{10^{-9} \text{ m}}\right)$$

$$= 5.20 \times 10^{14} \text{ Hz}$$

b. The energy of the incident photon, found from equation 31.6, is

$$E = h\nu = (6.625 \times 10^{-34} \text{ J s})(5.20 \times 10^{14} \text{ s}^{-1})$$

$$= 3.45 \times 10^{-19} \text{ J}$$

c. The maximum kinetic energy of the photoelectron, found from equation 31.5, is

$$\text{KE}_{max} = eV_0$$

$$= (1.60 \times 10^{-19} \text{ C})(0.250 \text{ V})$$

$$= 4.00 \times 10^{-20} \text{ J}$$

d. The work function of cesium is found by rearranging Einstein's photoelectric equation, 31.8, as

$$W_0 = E - \text{KE}_{max}$$

$$= 3.45 \times 10^{-19} \text{ J} - 4.00 \times 10^{-20} \text{ J}$$

$$= 3.05 \times 10^{-19} \text{ J}$$

$$= 1.91 \text{ eV}$$

e. The threshold frequency is found by solving equation 31.11 for ν_0. Thus,

$$\nu_0 = \frac{W_0}{h} = \frac{3.05 \times 10^{-19} \text{ J}}{6.625 \times 10^{-34} \text{ J s}}$$

$$= 4.60 \times 10^{14} \text{ Hz}$$

f. The wavelength of light associated with the threshold frequency is found from

$$\lambda_0 = \frac{c}{\nu_0} = \left(\frac{3.00 \times 10^8 \text{ m/s}}{4.60 \times 10^{14} \text{ s}^{-1}}\right)\left(\frac{1 \text{ nm}}{10^{-9} \text{ m}}\right)$$

$$= 653 \text{ nm}$$

This wavelength lies in the red portion of the visible spectrum.

31.4 The Properties of the Photon

According to classical physics light must be a wave. But the results of the photoelectric effect require light to be a particle, a photon. What then is light? Is it a wave or is it a particle?

If light is a particle then it must have some of the characteristics of particles, that is, it should possess mass, energy, and momentum. Let us first consider the mass of the photon. The relativistic mass of a particle was given by equation 29.86 as

$$m = \frac{m_0}{\sqrt{1 - v^2/c^2}}$$

But the photon is a particle of light and must therefore move at the speed of light c. Hence, its mass becomes

$$m = \frac{m_0}{\sqrt{1 - c^2/c^2}} = \frac{m_0}{0} \qquad (31.13)$$

But division by zero is undefined. The only way out of this problem is to *define the rest mass of a photon as being zero,* that is,

$$\text{Photon} \qquad m_0 = 0 \qquad (31.14)$$

At first this may seem a contradiction, but since the photon always moves at the speed c, it is never at rest, and therefore does not need a rest mass. With $m_0 = 0$, equation 31.13 becomes $0/0$, which is an indeterminate form. Although the mass of the photon still cannot be defined by equation 29.86 it can be defined from equation 29.99, namely

$$E = mc^2$$

Hence,

$$m = \frac{E}{c^2} \qquad (31.15)$$

The energy of the photon was given by

$$\text{Energy of Photon} \qquad E = h\nu \qquad (31.6)$$

Therefore, the mass of the photon can be found by substituting equation 31.6 into equation 31.15, that is,

$$\text{Mass of Photon} \qquad m = \frac{E}{c^2} = \frac{h\nu}{c^2} \qquad (31.16)$$

Example 31.4

The mass of a photon. Find the mass of a photon of light that has a wavelength of (a) 380.0 nm and (b) 720.0 nm.

Solution

a. For $\lambda = 380.0$ nm, the frequency of the photon is found from

$$\nu = \frac{c}{\lambda} = \left(\frac{3.00 \times 10^8 \text{ m/s}}{380.0 \text{ nm}}\right)\left(\frac{1 \text{ nm}}{10^{-9} \text{ m}}\right)$$

$$= 7.89 \times 10^{14} \text{ Hz}$$

Now we can find the mass from equation 31.16 as

$$m = \frac{E}{c^2} = \frac{h\nu}{c^2}$$

$$= \frac{(6.625 \times 10^{-34} \text{ J s})(7.89 \times 10^{14} \text{ s}^{-1})}{(3.00 \times 10^8 \text{ m/s})^2}\left[\frac{(\text{kg m/s}^2)\text{m}}{\text{J}}\right]$$

$$= 5.81 \times 10^{-36} \text{ kg}$$

b. For $\lambda = 720.0$ nm, the frequency is

$$\nu = \frac{c}{\lambda} = \left(\frac{3.00 \times 10^8 \text{ m/s}}{720.0 \text{ nm}} \right) \left(\frac{1 \text{ nm}}{10^{-9} \text{ m}} \right)$$

$$= 4.17 \times 10^{14} \text{ 1/s}$$

and the mass is

$$m = \frac{h\nu}{c^2} = \frac{(6.625 \times 10^{-34} \text{ J s})(4.17 \times 10^{14} \text{ 1/s})}{(3.00 \times 10^8 \text{ m/s})^2}$$

$$= 3.07 \times 10^{-36} \text{ kg}$$

As we can see from these examples, the mass of the photon for visible light is very small.

The momentum of the photon can be found as follows. Starting with the relativistic mass

$$m = \frac{m_0}{\sqrt{1 - v^2/c^2}} \qquad (29.86)$$

we square both sides of the equation and obtain

$$m^2 \left(1 - \frac{v^2}{c^2} \right) = m_0^2$$

$$m^2 - \frac{m^2 v^2}{c^2} = m_0^2 \qquad (31.17)$$

Multiplying both sides of equation 31.17 by c^4, we obtain

$$m^2 c^4 - m^2 v^2 c^2 = m_0^2 c^4$$

But $m^2 c^4 = E^2$, $m_0^2 c^4 = E_0^2$, and $m^2 v^2 = p^2$, thus,

$$E^2 - p^2 c^2 = E_0^2 \qquad (31.18)$$

Hence, we find the momentum of any particle from equation 31.18 as

$$p = \frac{\sqrt{E^2 - E_0^2}}{c} \qquad (31.19)$$

For the special case of a particle of zero rest mass, $E_0 = m_0 c^2 = 0$, and *the momentum of a photon*, found from equation 31.19, is

$$\textit{Momentum of Photon} \qquad p = \frac{E}{c} \qquad (31.20)$$

Using equation 31.6, we can write the momentum of a photon in terms of its frequency as

$$p = \frac{E}{c} = \frac{h\nu}{c}$$

Since $\nu/c = 1/\lambda$, this is also written as

$$\textit{Momentum of Photon} \qquad p = \frac{E}{c} = \frac{h\nu}{c} = \frac{h}{\lambda} \qquad (31.21)$$

Example 31.5

The momentum of a photon. Find the momentum of visible light for (a) $\lambda = 380.0$ nm and (b) $\lambda = 720.0$ nm.

Solution

a. The momentum of the photon, found from equation 31.21, is

$$p = \frac{h}{\lambda} = \left(\frac{6.625 \times 10^{-34} \text{ J s}}{380.0 \text{ nm}} \right) \left(\frac{1 \text{ nm}}{10^{-9} \text{ m}} \right) \left[\frac{(\text{kg m/s}^2)\text{m}}{\text{J}} \right]$$

$$= 1.74 \times 10^{-27} \text{ kg m/s}$$

b. The momentum of the second photon is found similarly

$$p = \frac{h}{\lambda} = \frac{6.625 \times 10^{-34} \text{ J s}}{720.0 \text{ nm}}$$

$$= 9.20 \times 10^{-28} \text{ kg m/s}$$

According to this quantum theory of light, light spreads out from a source in small bundles of energy called quanta or photons. *Although the photon is treated as a particle, its properties of mass, energy, and momentum are described in terms of frequency or wavelength, strictly a wave concept.*

Thus, we say that *light has a dual nature. It can act as a wave or it can act as a particle, but never both at the same time.* To answer the question posed at the beginning of this section, is light a wave or a particle, the answer is that light is both a wave and a particle. This dual nature of light is stated in the ***principle of complementarity:*** *The wave theory of light and the quantum theory of light complement each other. In a specific event, light exhibits either a wave nature or a particle nature, but never both at the same time.*

When the wavelength of an electromagnetic wave is long, its frequency and hence its photon energy ($E = h\nu$) are small and we are usually concerned with the wave characteristics of the electromagnetic wave. For example, radio and television waves have relatively long wavelengths and they are usually treated as waves. When the wavelength of the electromagnetic wave is small, its frequency and hence its photon energy are large. The electromagnetic wave is then usually considered as a particle. For example, X rays have very small wavelengths and are usually treated as particles. However, this does not mean that X rays cannot also act as waves. In fact they do. When X rays are scattered from a crystal, they behave like waves, exhibiting the usual diffraction patterns associated with waves. The important thing is that light can act either as a wave or a particle, but never both at the same time.

Let us *summarize the characteristics of the photon:*

Rest Mass	$m_0 = 0$	(31.14)
Energy	$E = h\nu$	(31.6)
Mass	$m = \dfrac{E}{c^2} = \dfrac{h\nu}{c^2}$	(31.16)
Momentum	$p = \dfrac{E}{c} = \dfrac{h\nu}{c} = \dfrac{h}{\lambda}$	(31.21)

Although the two examples considered were for photons of visible light, do not forget that the photon is a particle in the entire electromagnetic spectrum.

Example 31.6

The mass of an X ray and a gamma ray. Find the mass of a photon for (a) an X ray of 100.0-nm wavelength and (b) for a gamma ray of 0.0500 nm.

a. The mass of an X-ray photon, found from equation 31.16, is

$$m = \frac{h\nu}{c^2} = \frac{h}{c\lambda}$$

$$= \frac{6.625 \times 10^{-34} \text{ J s}}{(3.00 \times 10^8 \text{ m/s})(100.0 \text{ nm})}\left(\frac{1 \text{ nm}}{10^{-9} \text{ m}}\right)$$

$$= 2.20 \times 10^{-35} \text{ kg}$$

b. The mass of the gamma ray is

$$m = \frac{h\nu}{c^2} = \frac{h}{c\lambda}$$

$$= \frac{6.625 \times 10^{-34} \text{ J s}}{(3.00 \times 10^8 \text{ m/s})(0.0500 \text{ nm})}\left(\frac{1 \text{ nm}}{10^{-9} \text{ m}}\right)$$

$$= 4.42 \times 10^{-32} \text{ kg}$$

Comparing the mass of a photon for red light, violet light, X rays, and gamma rays we see

$$m_{\text{red}} = 3.07 \times 10^{-36} \text{ kg}$$

$$m_{\text{violet}} = 5.81 \times 10^{-36} \text{ kg}$$

$$m_{\text{X ray}} = 22.0 \times 10^{-36} \text{ kg}$$

$$m_{\text{gamma ray}} = 44,200 \times 10^{-36} \text{ kg}$$

Thus, as the frequency of the electromagnetic spectrum increases (wavelength decreases), the mass of the photon increases.

31.5 The Compton Effect

If light sometimes behaves like a particle, the photon, why not consider the collision of a photon with a free electron from the same point of view as the collision of two billiard balls? Such a collision between a photon and a free electron is called Compton scattering, or the **Compton effect,** in honor of Arthur Holly Compton (1892–1962). In order to get a massive photon for the collision, X rays are used. (Recall that X rays have a high frequency ν, and therefore the energy of the X ray, $E = h\nu$, is large, and thus its mass, $m = E/c^2$, is also large.) In order to get a free electron, a target made of carbon is used. The outer electrons of the carbon atom are very loosely bound, so compared with the initial energy of the photon, the electron looks like a free electron. Thus, the collision between the photon and the electron can be pictured as shown in figure 31.7. We assume that the electron is initially at rest and that the incident photon has an energy ($E = h\nu$) and a momentum ($p = E/c$). After the collision, the electron is found to be scattered at an angle θ from the original direction of the photon. *Because the electron has moved after the collision, some energy must have been imparted to it. But where could this energy come from? It must have come from the incident photon. But if that is true, then the*

Figure 31.7

Compton scattering.

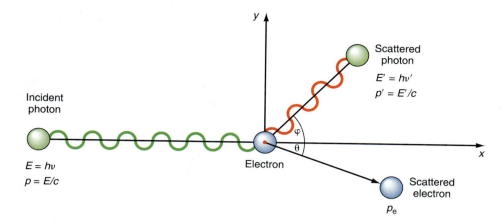

scattered photon must have less energy than the incident photon, and therefore, its wavelength should also have changed. Let us call the energy of the scattered photon E', where

$$E' = h\nu'$$

and hence, its final momentum is

$$p' = \frac{E'}{c} = \frac{h\nu'}{c}$$

Because momentum is conserved in all collisions, the law of conservation of momentum is applied to the collision of figure 31.7. First however, notice that the collision is two dimensional. Because the vector momentum is conserved, the x-component of the momentum and the y-component of momentum must also be conserved. The law of conservation for the x-component of momentum can be written as

$$p_\mathbf{p} + 0 = p_\mathbf{p}' \cos \phi + p_\mathbf{e} \cos \theta$$

and for the y-component,

$$0 + 0 = p_\mathbf{p}' \sin \phi - p_\mathbf{e} \sin \theta$$

where $p_\mathbf{p}$ is the momentum of the incident photon, $p_\mathbf{p}'$ the momentum of the scattered photon, and $p_\mathbf{e}$ the momentum of the scattered electron. Substituting the values for the energy and momentum of the photon, these equations become

$$\frac{h\nu}{c} = \frac{h\nu'}{c} \cos \phi + p_\mathbf{e} \cos \theta$$

$$0 = \frac{h\nu'}{c} \sin \phi - p_\mathbf{e} \sin \theta$$

There are more unknowns (ν', θ, ϕ, $p_\mathbf{e}$) than we can handle at this moment, so let us eliminate θ from these two equations by rearranging, squaring, and adding them. That is,

$$p_\mathbf{e} \cos \theta = \frac{h\nu}{c} - \frac{h\nu'}{c} \cos \phi$$

$$p_\mathbf{e} \sin \theta = \frac{h\nu'}{c} \sin \phi$$

$$p_\mathbf{e}^2 \cos^2\theta = \frac{(h\nu)^2}{c^2} - \frac{2h\nu h\nu' \cos \phi}{c^2} + \frac{(h\nu')^2}{c^2} \cos^2\phi$$

$$p_\mathbf{e}^2 \sin^2\theta = \frac{(h\nu')^2}{c^2} \sin^2\phi$$

$$p_\mathbf{e}^2 (\sin^2\theta + \cos^2\theta) = \frac{(h\nu)^2}{c^2} + \frac{(h\nu')^2}{c^2}(\sin^2\phi + \cos^2\phi) - \frac{2(h\nu)(h\nu')}{c^2} \cos \phi$$

But since $\sin^2\theta + \cos^2\theta = 1$, we get

$$p_e^2 = \frac{(h\nu)^2}{c^2} + \frac{(h\nu')^2}{c^2} - \frac{2(h\nu)(h\nu')}{c^2}\cos\phi \qquad (31.22)$$

The angle θ has thus been eliminated from the equation. Let us now look for a way to eliminate the term p_e, the momentum of the electron. If we square equation 31.19, we can solve for p_e^2 and obtain

$$p_e^2 = \frac{E_e^2 - E_{0e}^2}{c^2} \qquad (31.23)$$

But the total energy of the electron E_e, given by equation 29.102, is

$$E_e = \mathrm{KE}_e + E_{0e}$$

where KE_e is the kinetic energy of the electron and E_{0e} is its rest mass. Substituting equation 29.102 back into equation 31.23, gives, for the momentum of the electron,

$$p_e^2 = \frac{(\mathrm{KE}_e + E_{0e})^2 - E_{0e}^2}{c^2}$$

$$= \frac{\mathrm{KE}_e^2 + 2E_{0e}\,\mathrm{KE}_e + E_{0e}^2 - E_{0e}^2}{c^2}$$

$$p_e^2 = \frac{\mathrm{KE}_e^2 + 2E_{0e}\,\mathrm{KE}_e}{c^2} \qquad (31.24)$$

But if the law of conservation of energy is applied to the collision of figure 31.7, we get

$$E = E' + \mathrm{KE}_e$$

$$h\nu = h\nu' + \mathrm{KE}_e \qquad (31.25)$$

where E is the total energy of the system, E' is the energy of the scattered photon and KE_e is the kinetic energy imparted to the electron during the collision. Thus, the kinetic energy of the electron, found from equation 31.25, is

$$\mathrm{KE}_e = h\nu - h\nu' \qquad (31.26)$$

Substituting the value of the kinetic energy from equation 31.26 and $E_{0e} = m_0 c^2$, the rest energy of the electron, back into equation 31.24, we get, for the momentum of the electron,

$$p_e^2 = \frac{(h\nu - h\nu')^2 + 2m_0 c^2(h\nu - h\nu')}{c^2}$$

$$p_e^2 = \frac{(h\nu)^2}{c^2} + \frac{(h\nu')^2}{c^2} - \frac{2h\nu h\nu'}{c^2} + 2m_0(h\nu - h\nu') \qquad (31.27)$$

Since we now have two separate equations for the momentum of the electron, equations 31.22 and 31.27, we can equate them to eliminate p_e. Therefore,

$$\frac{(h\nu)^2}{c^2} + \frac{(h\nu')^2}{c^2} - \frac{2h\nu h\nu'}{c^2} + 2m_0(h\nu - h\nu') = \frac{(h\nu)^2}{c^2} + \frac{(h\nu')^2}{c^2} - \frac{2(h\nu)(h\nu')}{c^2}\cos\phi$$

Simplifying,

$$2m_0(h\nu - h\nu') = \frac{2h\nu h\nu'}{c^2} - \frac{2(h\nu)(h\nu')}{c^2}\cos\phi$$

$$h\nu - h\nu' = \frac{h\nu h\nu'}{m_0 c^2}(1 - \cos\phi)$$

$$\frac{\nu - \nu'}{\nu\nu'} = \frac{h}{m_0 c^2}(1 - \cos\phi)$$

However, since $\nu = c/\lambda$ this becomes

$$\frac{c/\lambda - c/\lambda'}{(c/\lambda)(c/\lambda')} = \frac{h}{m_0 c^2}(1 - \cos \phi)$$

$$\lambda \lambda'\left(\frac{1}{\lambda} - \frac{1}{\lambda'}\right) = \frac{h}{m_0 c}(1 - \cos \phi)$$

$$\lambda' - \lambda = \frac{h}{m_0 c}(1 - \cos \phi) \qquad (31.28)$$

Equation 31.28 is called the Compton scattering formula. It gives the change in wavelength of the scattered photon as a function of the scattering angle ϕ. The quantity,

$$\frac{h}{m_0 c} = 2.426 \times 10^{-12} \text{ m} = 0.002426 \text{ nm}$$

which has the dimensions of a length, is called the *Compton wavelength*.

Thus, in a collision between an energetic photon and an electron, the scattered light shows a different wavelength than the wavelength of the incident light. In 1923, A. H. Compton confirmed the modified wavelength of the scattered photon and received the Nobel Prize in 1927 for his work.

Example 31.7

Compton scattering. A 90.0-KeV X-ray photon is fired at a carbon target and Compton scattering occurs. Find the wavelength of the incident photon and the wavelength of the scattered photon for scattering angles of (a) 30.0° and (b) 60.0°.

Solution

The frequency of the incident photon is found from $E = h\nu$ as

$$\nu = \frac{E}{h} = \left(\frac{90.0 \times 10^3 \text{ eV}}{6.625 \times 10^{-34} \text{ J s}}\right)\left(\frac{1.60 \times 10^{-19} \text{ J}}{1 \text{ eV}}\right)$$

$$= 2.17 \times 10^{19} \text{ Hz}$$

The wavelength of the incident photon is found from

$$\lambda = \frac{c}{\nu} = \left(\frac{3.00 \times 10^8 \text{ m/s}}{2.17 \times 10^{19} \text{ 1/s}}\right)\left(\frac{1 \text{ nm}}{10^{-9} \text{ m}}\right)$$

$$= 0.0138 \text{ nm}$$

The modified wavelength is found from the Compton scattering formula, equation 31.28, as

$$\lambda' = \lambda + \frac{h}{m_0 c}(1 - \cos \phi)$$

a.
$$\lambda' = 0.0138 \text{ nm} + (0.002426 \text{ nm})(1 - \cos 30.0°)$$

$$= 0.0141 \text{ nm}$$

b.
$$\lambda' = 0.0138 \text{ nm} + (0.002426 \text{ nm})(1 - \cos 60.0°)$$

$$= 0.0150 \text{ nm}$$

In an actual experiment both the incident and modified wavelengths are found in the scattered photons. The incident wavelength is found in the scattered photons because some of the incident photons are scattered by the atom. In this case, the rest mass of the electron m_0 must be replaced in equation 31.28 by the mass M of the entire atom. Because M is so much greater than m_0, the Compton wavelength

h/MC is so small that the change in wavelength for these photons is too small to be observed. Thus, these incident photons are scattered with the same wavelength.

31.6 The Wave Nature of Particles

We have seen that light displays a dual nature; it acts as a wave and it acts as a particle. Assuming symmetry in nature, the French physicist Louis de Broglie (1892–1987) proposed, in his 1924 doctoral dissertation, that particles should also possess a wave characteristic. Because the momentum of a photon was shown to be

$$p = \frac{h}{\lambda} \tag{31.21}$$

de Broglie assumed that the wavelength of the wave associated with a particle of momentum p, should be given by

$$\lambda = \frac{h}{p} \tag{31.29}$$

*Equation 31.29 is called the **de Broglie relation**. Thus, de Broglie assumed that the same wave-particle duality associated with electromagnetic waves should also apply to particles.* Hence, an electron can be considered to be a particle and it can also be considered to be a wave. Instead of solving the problem of the wave-particle duality of electromagnetic waves, de Broglie extended it to include matter as well.

Example 31.8

The wavelength of a particle. Calculate the wavelength of (a) a 0.140-kg baseball moving at a speed of 44.0 m/s, (b) a proton moving at the same speed, and (c) an electron moving at the same speed.

Solution

a. A baseball has an associated wavelength given by equation 31.29 as

$$\lambda = \frac{h}{p} = \frac{h}{mv} = \frac{6.625 \times 10^{-34} \text{ J s}}{(0.140 \text{ kg})(44.0 \text{ m/s})}$$
$$= 1.08 \times 10^{-34} \text{ m}$$

Such a small wavelength cannot be measured and therefore baseballs always appear as particles.

b. The wavelength of the proton, found from equation 31.29, is

$$\lambda = \frac{h}{p} = \frac{h}{mv}$$
$$= \frac{6.625 \times 10^{-34} \text{ J s}}{(1.67 \times 10^{-27} \text{ kg})(44.0 \text{ m/s})}\left(\frac{1 \text{ nm}}{10^{-9} \text{ m}}\right)$$
$$= 9.02 \text{ nm}$$

Although this wavelength is small (it is in the X-ray region of the electromagnetic spectrum), it can be detected.

c. The wavelength of the electron is found from

$$\lambda = \frac{h}{p} = \frac{h}{mv}$$
$$= \frac{6.625 \times 10^{-34} \text{ J s}}{(9.11 \times 10^{-31} \text{ kg})(44.0 \text{ m/s})}\left(\frac{1 \text{ nm}}{10^{-9} \text{ m}}\right)$$
$$= 1.65 \times 10^4 \text{ nm}$$

which is a very large wavelength and can be easily detected.

Note from example 31.8, that because Planck's constant h is so small, the wave nature of a particle does not manifest itself unless the mass m of the particle is also very small (of the order of an atom or smaller). This is why the wave nature of particles is not part of our everyday experience.

de Broglie's hypothesis was almost immediately confirmed when in 1927 C. J. Davisson and L. H. Germer performed an experiment that showed that electrons could be diffracted by a crystal. G. P. Thomson performed an independent experiment at the same time by scattering electrons from very thin metal foils and obtained the standard diffraction patterns that are usually associated with waves. Since that time diffraction patterns have been observed with protons, neutrons, hydrogen atoms, and helium atoms, thereby giving substantial evidence for the wave nature of particles.

For his work on the dual nature of particles, de Broglie received the 1929 Nobel Prize in physics. Davisson and Thomson shared the Nobel Prize in 1937 for their experimental confirmation of the wave nature of particles.

31.7 The Wave Representation of a Particle

We have just seen that a particle can be represented by a wave. The wave associated with a photon was an electromagnetic wave. *But what kind of wave is associated with a particle?* It is certainly not an electromagnetic wave. de Broglie called the wave a *pilot wave* because he believed that it steered the particle during its motion. The waves have also been called *matter waves* to show that they are associated with matter. Today, the wave is simply referred to as the *wave function* and is represented by Ψ.

Because this wave function refers to the motion of a particle we say that *the value of the wave function Ψ is related to the probability of finding the particle at a specific place and time.* The probability P that something can be somewhere at a certain time, can have any value between 0 and 1. If the probability $P = 0$, then there is an absolute certainty that the particle is absent. If the probability $P = 1$, then there is an absolute certainty that the particle is present. If the probability P lies somewhere between 0 and 1, then that value is the probability of finding the particle there. That is, if the probability $P = 0.20$, there is a 20% probability of finding the particle at the specified place and time.

Because the amplitude of any wave varies between positive and negative values, the wave function Ψ cannot by itself represent the probability of finding the particle at a particular time and place. However, the quantity Ψ^2 is always positive and is called the probability density. *The probability density Ψ^2 is the probability of finding the particle at the position (x, y, z) at the time t.* The new science of wave mechanics, or as it was eventually called, quantum mechanics, has to do with determining the wave function Ψ for any particle or system of particles.

How can a particle be represented by a wave? Recall from chapter 12, that a wave moving to the right is defined by the function

$$y = A \sin(kx - \omega t) \qquad (12.13)$$

where the wave number k is

$$k = \frac{2\pi}{\lambda} \qquad (12.9)$$

and the angular frequency ω is given by

$$\omega = 2\pi f \qquad (12.12)$$

or since $f = \nu$, in our new notation,

$$\omega = 2\pi\nu$$

Also recall from equation 12.14 that the velocity of the wave is given by

$$v = \frac{\omega}{k}$$

We will therefore begin, in our analysis of matter waves, by trying to define the wave function as

$$\Psi = A \sin(kx - \omega t) \tag{31.30}$$

A plot of this wave function for $t = 0$ is shown in figure 31.8(a). The first thing to observe in this picture is that the wave is too spread out to be able to represent a particle. Remember the particle must be found somewhere within the wave. Because the wave extends out to infinity the particle could be anywhere.

Because one of the characteristics of waves is that they obey the superposition principle, perhaps a wave representation can be found by adding different waves together. As an example, let us add two waves of slightly different wave numbers and slightly different angular frequencies. That is, consider the two waves

$$\Psi_1 = A \sin(k_1 x - \omega_1 t)$$

$$\Psi_2 = A \sin(k_2 x - \omega_2 t)$$

where

$$k_2 = k_1 + \Delta k$$

and

$$\omega_2 = \omega_1 + \Delta \omega$$

Figure 31.8

Representation of a particle as a wave.

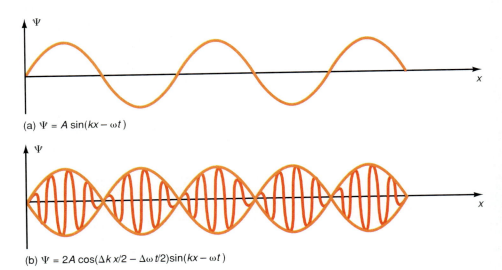

(a) $\Psi = A \sin(kx - \omega t)$

(b) $\Psi = 2A \cos(\Delta k\, x/2 - \Delta \omega\, t/2)\sin(kx - \omega t)$

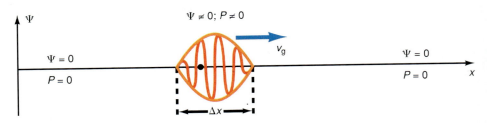

(c) Wave packet representing a particle $\quad \Psi = \sum\limits_{i=1}^{\infty} A \sin(k_i x - \omega_i t)$

The addition of these two waves gives

$$\Psi = \Psi_1 + \Psi_2$$

$$= A \sin(k_1 x - \omega_1 t) + A \sin(k_2 x - \omega_2 t)$$

The addition of two sine waves is shown in appendix B, and in equation 12.48, to be

$$\sin B + \sin C = 2 \sin\left(\frac{B + C}{2}\right)\cos\left(\frac{B - C}{2}\right)$$

Letting

$$B = k_1 x - \omega_1 t$$

and

$$C = k_2 x - \omega_2 t$$

we find

$$\Psi = 2A \sin\left(\frac{k_1 x - \omega_1 t + k_2 x - \omega_2 t}{2}\right)\cos\left(\frac{k_1 x - \omega_1 t - k_2 x + \omega_2 t}{2}\right)$$

$$= 2A \sin\left[\frac{k_1 x - \omega_1 t + (k_1 + \Delta k)x - (\omega_1 + \Delta\omega)t}{2}\right]$$

$$\cos\left[\frac{k_1 x - \omega_1 t - (k_1 + \Delta k)x + (\omega_1 + \Delta\omega)t}{2}\right]$$

$$= 2A \sin\left[\frac{2kx + (\Delta k)x - 2\omega t - (\Delta\omega)t}{2}\right]\cos\left(-\frac{\Delta k}{2}x + \frac{\Delta\omega}{2}t\right)$$

We have dropped the subscript 1 on k and ω to establish the general case. Now as an approximation

$$2kx + (\Delta k)x \approx 2kx$$

and

$$-2\omega t - (\Delta\omega)t \approx -2\omega t$$

Therefore,

$$\Psi = 2A \sin(kx - \omega t)\cos\left(-\frac{\Delta k}{2}x + \frac{\Delta\omega}{2}t\right)$$

One of the properties of the cosine function is that $\cos(-\theta) = \cos\theta$. Using this relation the wave function becomes

$$\Psi = 2A \cos\left(\frac{\Delta k}{2}x - \frac{\Delta\omega}{2}t\right)\sin(kx - \omega t) \tag{31.31}$$

A plot of equation 31.31 is shown in figure 31.8(b). The amplitude of this wave is modulated and is given by the first part of equation 31.31 as

$$A_{\mathbf{m}} = 2A \cos\left(\frac{\Delta k}{2}x - \frac{\Delta\omega}{2}t\right) \tag{31.32}$$

This wave superposition gives us a closer representation of a particle. Each modulated portion of the wave represents a group of waves and any one group can represent a particle. The velocity of the group of waves represents the velocity of the particle.

Equation 31.31 and figure 31.8(b) approaches a wave representation of the particle. If an infinite number of waves, each differing slightly in wave number and angular frequency, were added together we would get the wave function

$$\Psi = \sum_{i=1}^{\infty} A \sin(k_i x - \omega_i t) \tag{31.33}$$

which is shown in figure 31.8(c) and is called a *wave packet. This wave packet can indeed represent the motion of a particle. Because the wave function Ψ is zero everywhere except within the packet, the probability of finding the particle is zero everywhere except within the packet.* The wave packet localizes the particle to be within the region Δx shown in figure 31.8(c), and the wave packet moves with the group velocity of the waves and this is the velocity of the particle. The fundamental object of wave mechanics or quantum mechanics is to find the wave function Ψ associated with a particle or a system of particles.

31.8 The Heisenberg Uncertainty Principle

One of the characteristics of the dual nature of matter is a fundamental limitation in the accuracy of the measurement of the position and momentum of a particle. This can be seen in a very simplified way by looking at the modulated wave of figure 31.8(b) and reproduced in figure 31.9. A particle is shown located in the first group of the modulated wave. Since the particle lies somewhere within the wave packet its exact position is uncertain. The amount of the uncertainty in its position is no greater than Δx, the width of the entire wave packet or wave group. The wavelength of the modulated amplitude λ_m is shown in figure 31.9 and we can see that a wave group is only half that distance. Thus, the uncertainty in the location of the particle is given by

$$\Delta x = \frac{\lambda_m}{2} \tag{31.34}$$

The uncertainty in the momentum can be found by solving the de Broglie relation, equation 31.29, for momentum as

$$p = \frac{h}{\lambda} \tag{31.35}$$

and the fact that the wavelength is given in terms of the wave number by rearranging equation 12.9 into

$$\lambda = \frac{2\pi}{k} \tag{31.36}$$

Figure 31.9

Limitations on position and momentum of a particle.

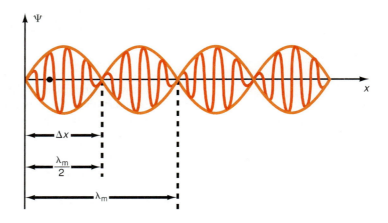

Figure 31.9

Limitations on position and momentum of a particle.

Substituting equation 31.36 into equation 31.35 gives

$$p = \frac{h}{\lambda} = \frac{h}{2\pi/k} = \frac{h}{2\pi}k \qquad (31.37)$$

The uncertainty in the momentum, found from equation 31.37, is

$$\Delta p = \frac{h}{2\pi}\Delta k \qquad (31.38)$$

Because the wave packet is made up of many waves, there is a Δk associated with it. This means that in representing a particle as a wave, there is automatically an uncertainty in the wave number, k, which we now see implies an uncertainty in the momentum of that particle. For the special case considered in figure 31.9, the wave number of the modulated wave Δk_m is found from

$$A_m = 2A\cos(k_m x - \omega_m t)$$

and from equation 31.32 as

$$k_m = \frac{\Delta k}{2} \qquad (31.39)$$

But from the definition of a wave number

$$k_m = \frac{2\pi}{\lambda_m} \qquad (31.40)$$

Substituting equation 31.40 into equation 31.39 gives, for Δk,

$$\Delta k = 2k_m = 2\left(\frac{2\pi}{\lambda_m}\right) \qquad (31.41)$$

Substituting the uncertainty for Δk, equation 31.41, into the uncertainty for Δp, equation 31.38, gives

$$\Delta p = \frac{h}{2\pi}\Delta k = \frac{h}{2\pi}2\left(\frac{2\pi}{\lambda_m}\right) = \frac{h}{\lambda_m/2} \qquad (31.42)$$

The uncertainty between the position and momentum of the particle is obtained by substituting equation 31.34 for $\lambda_m/2$ into equation 31.42 to get

$$\Delta p = \frac{h}{\Delta x}$$

or

$$\Delta p\Delta x = h \qquad (31.43)$$

Because Δp and Δx are the smallest uncertainties that a particle can have, their values are usually greater than this, so their product is usually greater than the value of h. To show this, equation 31.43 is usually written with an inequality sign also, that is,

$$\Delta p\Delta x \geq h$$

The analysis of the wave packet was greatly simplified by using the modulated wave of figure 31.8(b). A more sophisticated analysis applied to the more reasonable wave packet of figure 31.8(c) yields the relation

$$\Delta p\Delta x \geq \hbar \qquad (31.44)$$

where the symbol $\hbar$, called h bar, is

$$\hbar = \frac{h}{2\pi} = 1.05 \times 10^{-34} \text{ J s} \qquad (31.45)$$

Figure 31.10
Wave packets of different size.

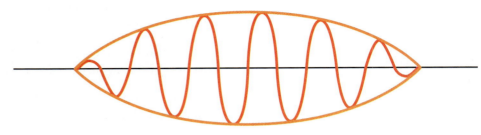

(a) Good Δx estimate, poor λ estimate

(b) Good λ estimate, poor Δx estimate

*Equation 31.44 is called the **Heisenberg uncertainty principle**. It says that the position and momentum of a particle cannot both be measured simultaneously with perfect accuracy. There is always a fundamental uncertainty associated with any measurement. This uncertainty is not associated with the measuring instrument. It is a consequence of the wave-particle duality of matter.*

As an example of the application of equation 31.44, if the position of a particle is known exactly, then $\Delta x = 0$ and Δp would have to be infinite in order for the product $\Delta x \Delta p$ to be greater than $\hbar$. If Δp is infinite, the value of the momentum of the particle is completely unknown. A wave packet associated with a very accurate value of position is shown in figure 31.10(a). Although this wave packet gives a very small value of Δx, it gives an exceedingly poor representation of the wavelength. Because the uncertainty in the wavelength λ is large, the uncertainty in the wave number is also large. Since the uncertainty in the wave number is related to the uncertainty in the momentum of the particle by equation 31.38, there is also a large uncertainty in the momentum of the particle. Thus, a good Δx estimate always gives a poor Δp estimate.

If the momentum of a particle is known exactly, then $\Delta p = 0$, and this implies that Δx must approach infinity. That is, if the momentum of a particle is known exactly, the particle could be located anywhere. A wave packet approximating this case is shown in figure 31.10(b). Because the wave packet is spread out over a large area it is easy to get a good estimate of the de Broglie wavelength, and hence a good estimate of the momentum of the particle. On the other hand, since the wave packet is so spread out, it is very difficult to locate the particle inside the wave packet. Thus a good Δp estimate always gives a poor Δx estimate.

Example 31.9

The uncertainty in the velocity of a baseball. A 0.140-kg baseball is moving along the x-axis. At a particular instant of time it is located at the position $x = 0.500$ m with an uncertainty in the measurement of $\Delta x = 0.001$ m. How accurately can the velocity of the baseball be determined?

The uncertainty in the momentum is found by the Heisenberg uncertainty principle, equation 31.44, as

$$\Delta p \geq \frac{\hbar}{\Delta x}$$

$$\geq \frac{1.05 \times 10^{-34} \text{ J s}}{0.001 \text{ m}}$$

$$\geq 1.05 \times 10^{-31} \text{ kg m/s}$$

Since $p = mv$, the uncertainty in the velocity is

$$\Delta v \geq \frac{\Delta p}{m} = \frac{1.05 \times 10^{-31} \text{ kg m/s}}{0.140 \text{ kg}}$$

$$\geq 7.50 \times 10^{-31} \text{ m/s}$$

The error in Δp and Δv caused by the uncertainty principle is so small for macroscopic bodies moving around in the everyday world that it can be neglected.

Example 31.10

The uncertainty in the velocity of an electron confined to a box the size of the nucleus. We want to confine an electron, $m_e = 9.11 \times 10^{-31}$ kg, to a box, 1.00×10^{-14} m long (approximately the size of a nucleus). What would the speed of the electron be if it were so confined?

Solution

Because the electron can be located anywhere within the box, the worst case of locating the electron is for the uncertainty of the location of the electron to be equal to the size of the box itself. That is, $\Delta x = 1.00 \times 10^{-14}$ m.

We also assume that the uncertainty in the velocity is so bad that it is equal to the velocity of the electron itself. The uncertainty in the speed, found from the Heisenberg uncertainty principle, is

$$\Delta p \geq \frac{\hbar}{\Delta x}$$

$$m\Delta v \geq \frac{\hbar}{\Delta x}$$

$$\Delta v \geq \frac{\hbar}{m\Delta x} \tag{31.46}$$

$$\geq \frac{1.05 \times 10^{-34} \text{ J s}}{(9.11 \times 10^{-31} \text{ kg})(1.00 \times 10^{-14} \text{ m})}\left[\frac{(\text{kg m/s}^2) \text{ m}}{1 \text{ J}}\right]$$

$$\geq 1.15 \times 10^{10} \text{ m/s}$$

Hence, for the electron to be confined in a box about the size of the nucleus, its speed would have to be greater than the speed of light. Because this is impossible, we must conclude that an electron can never be found inside of a nucleus.

Example 31.11

The uncertainty in the velocity of an electron confined to a box the size of an atom.
An electron is placed in a box about the size of an atom, that is, $\Delta x = 1.00 \times 10^{-12}$
m. What is the velocity of the electron?

Solution

We again assume that the velocity of the electron is of the same order as the uncertainty in the velocity, then from equation 31.46, we have

$$\Delta v \geq \frac{\hbar}{m\Delta x}$$

$$\geq \frac{1.05 \times 10^{-34} \text{ J s}}{(9.11 \times 10^{-31} \text{ kg})(1.00 \times 10^{-12} \text{ m})}$$

$$\geq 1.15 \times 10^{8} \text{ m/s}$$

Because this velocity is less than the velocity of light, an electron can exist in an atom. Notice from these examples that the uncertainty principle is only important on the microscopic level.

Another way to observe the effect of the uncertainty principle from a more physical viewpoint is to see what happens when we "see" a particle in order to locate its position. Figure 31.11(a) shows how we locate a moving baseball. The process is basically a collision between the photon of light and the baseball. The photon hits the baseball and then bounces off (is reflected) and proceeds to our eye. We then can say that we saw the baseball at a particular location. Because the mass of the photon is so small compared to the mass of the baseball, the photon bounces off the

Figure 31.11

The Heisenberg uncertainty principle.

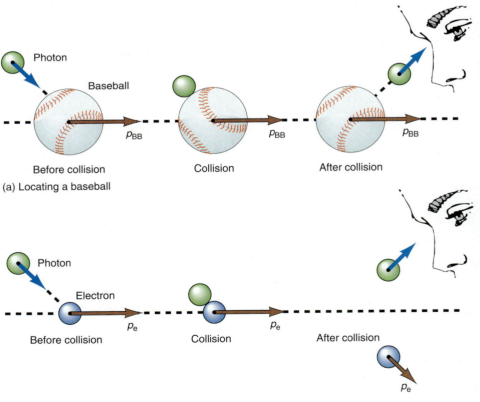

Photon

Baseball

p_{BB} p_{BB} p_{BB}

Before collision   Collision   After collision

(a) Locating a baseball

Photon

Electron

p_e p_e

Before collision   Collision   After collision

p_e

(b) Locating an electron

baseball without disturbing the momentum of the baseball. Thus, in the process of "locating" the baseball, we have done nothing to disturb its momentum.

Now let us look at the problem of "seeing" an electron, figure 31.11(b). The process of "seeing" again implies a collision between the photon of light and the object we wish to see; in this case, the electron. However, the momentum of the photon is now of the same order of magnitude as the momentum of the electron. Hence, as the photon hits the electron, the electron's momentum is changed just as in the Compton effect. Thus, we have located the electron by "seeing" it, but in the process of "seeing" it, we have disturbed or changed its momentum. *Hence, in the process of determining its position, we have caused an uncertainty in its momentum. The uncertainty occurs because the mass of the photon is of the same order of magnitude as the mass of the electron. Thus, the uncertainty always occurs when dealing with microscopic objects.*

The classical picture of being able to predict the exact position and velocity of a particle by Newton's second law and the kinematic equations obviously does not hold in the microscopic region of atoms because of the uncertainty principle. The exact positions and velocities are replaced by a probabilistic determination of position and velocity. That is, we now speak of the probability of finding a particle at a particular position, and the probability that its velocity is a particular value.

On the macroscopic level, the mass of the photon is totally insignificant with respect to the mass of the macroscopic body we wish to see and there is, therefore, no intrinsic uncertainty in measuring the position and velocity of the particle. This is why we are not concerned with the uncertainty principle in classical mechanics.

31.9 Different Forms of the Uncertainty Principle

The limitation on simultaneous measurements is limited not only to the position and momentum of a particle but also to its angular position and angular momentum, and also to its energy and the time in which the measurement of the energy is made.

The angular momentum of a particle was defined in equation 9.48 as

$$L = I\omega$$

We can also write the angular momentum for a particle in a different form by recalling that the moment of inertia I for a particle is

$$I = mr^2 \qquad\qquad (9.17)$$

Also recall that the angular velocity ω is related to the linear velocity by rearranging equation 9.2 into the form

$$\omega = \frac{v}{r}$$

Substituting this and equation 9.17 into equation 9.48 gives

$$L = I\omega = (mr^2)\left(\frac{v}{r}\right)$$

$$L = rmv = rp \qquad\qquad (31.47)$$

Thus, we can also write the angular momentum of a particle as the product of the radius of the circle and the linear momentum of the particle. (In this form the angular momentum is sometimes referred to as the *moment of momentum*.) The generalization of the angular momentum of a particle is found as

$$L = rp \sin \theta \qquad\qquad (31.48)$$

Notice that equation 31.47 is the special case of a particle moving in a circle of radius r, and the momentum is perpendicular to the radius. Hence, $\theta = 90°$, and $\sin 90° = 1$. With this new definition of angular momentum, we can easily see the effect of the uncertainty principle on a particle in rotational motion.

Calling x the displacement of a particle along the arc of the circle, when the particle moves through the angle θ, we have

$$x = r\theta$$

The uncertainty Δx in terms of the uncertainty $\Delta\theta$ in angle, becomes

$$\Delta x = r\Delta\theta$$

Substituting this uncertainty into Heisenberg's uncertainty relation, we get

$$\Delta x\Delta p \geq \hbar$$

$$r\Delta\theta\Delta p \geq \hbar$$

$$(\Delta\theta)(r\Delta p) \geq \hbar \qquad (31.49)$$

But equation 31.47, which gave us the angular momentum of the particle, also gives us the uncertainty in this angular momentum as

$$\Delta L = r\Delta p \qquad (31.50)$$

But this is exactly one of the terms in equation 31.49. Therefore, substituting equation 31.50 into equation 31.49 gives the Heisenberg uncertainty principle for rotational motion as

$$\Delta\theta\,\Delta L \geq \hbar \qquad (31.51)$$

Heisenberg's uncertainty principle in this form says that the product of the uncertainty in the angular position and the uncertainty in the angular momentum of the particle is always equal to or greater than the value $\hbar$. Thus, if the angular position of a particle is known exactly, $\Delta\theta = 0$, then the uncertainty in the angular momentum is infinite. On the other hand, if the angular momentum is known exactly, $\Delta L = 0$, then we have no idea where the particle is located in the circle.

The relationship between the uncertainty in the energy of a particle and the uncertainty in the time of its measurement is found as follows. Because the velocity of a particle is given by $v = \Delta x/\Delta t$, the distance that the particle moves during the measurement process is

$$\Delta x = v\Delta t \qquad (31.52)$$

The momentum of the particle is given by the de Broglie relation as

$$p = \frac{h}{\lambda} = \frac{h\nu}{v} = \frac{E}{v} \qquad (31.53)$$

because $1/\lambda = \nu/v$ and $h\nu = E$. The uncertainty of momentum in terms of the uncertainty in its energy, found from equation 31.53, is

$$\Delta p = \frac{\Delta E}{v} \qquad (31.54)$$

Substituting equations 31.52 and 31.54 into the Heisenberg uncertainty relation, gives

$$\Delta x\Delta p \geq \hbar$$

$$(v\Delta t)\left(\frac{\Delta E}{v}\right) \geq \hbar$$

or

$$\Delta E\Delta t \geq \hbar \qquad (31.55)$$

Equation 31.55 says that the product of the uncertainty in the measurement of the energy of a particle and the uncertainty in the time of the measurement of the particle is always equal to or greater than $\hbar$. Thus, in order to measure the energy of a particle exactly, $\Delta E = 0$, it would take an infinite time for the measurement. To measure the particle at an exact instant of time, $\Delta t = 0$, we will have no idea of the energy of that particle (ΔE would be infinite).

Example 31.12

The uncertainty in the energy of an electron in an excited state. The lifetime of an electron in an excited state is about 10^{-8} s. (This is the time it takes for the electron to stay in the excited state before it jumps back to the ground state.) What is its uncertainty in energy during this time?

Solution

The energy uncertainty, found from equation 31.55, is

$$\Delta E \Delta t \geq \hbar$$

$$\Delta E \geq \frac{\hbar}{\Delta t}$$

$$\geq \frac{1.05 \times 10^{-34} \text{ J s}}{1.00 \times 10^{-8} \text{ s}}$$

$$\geq 1.05 \times 10^{-26} \text{ J}$$

31.10 The Heisenberg Uncertainty Principle and Virtual Particles

It is a truly amazing result of the uncertainty principle that it is possible to violate the law of conservation of energy by borrowing an amount of energy ΔE, just as long as it is paid back before the time Δt, required by the uncertainty principle, equation 31.55, has elapsed. That is, the energy ΔE can be borrowed if it is paid back before the time

$$\Delta t = \frac{\hbar}{\Delta E} \tag{31.56}$$

This borrowed energy can be used to create particles. The borrowed energy ΔE is converted to a mass Δm, given by Einstein's mass-energy relation

$$\Delta E = (\Delta m)c^2$$

The payback time thus becomes

$$\Delta t = \frac{\hbar}{(\Delta m)c^2} \tag{31.57}$$

These particles must have very short lifetimes because the energy must be repaid before the elapsed time Δt. These ghostlike particles are called **virtual particles.** Around any real particle there exists a host of these virtual particles. We can visualize virtual particles with the help of figure 31.12. The real particle is shown in figure 31.12(a). In the short period of time Δt, another particle, the virtual particle, materializes as in figure 31.12(b). Before the time Δt is over, the virtual particle returns to the original particle, repaying its energy, and leaving only the real particle, figure 31.12(c). The original particle continues to fluctuate into the two particles.

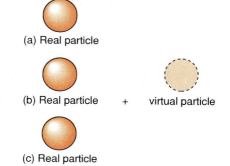

(a) Real particle

(b) Real particle   +   virtual particle

(c) Real particle

Figure 31.12

The virtual particle.

We can determine approximately how far the virtual particle moves away from the real particle by assuming that the maximum speed at which it could possibly move is the speed of light. The distance that the virtual particle can move and then return is then found from

$$d = c\frac{\Delta t}{2} \tag{31.58}$$

As an example, suppose the real particle is a proton. Let us assume that we borrow enough energy from the proton to create a particle called the pi-meson (*pion* for short). The mass of the pion is about 2.48×10^{-28} kg. How long can this virtual pion live? From equation 31.57, we have

$$\Delta t = \frac{\hbar}{(\Delta m)c^2}$$

$$= \frac{1.05 \times 10^{-34} \text{ J s}}{(2.48 \times 10^{-28} \text{ kg})(3.00 \times 10^8 \text{ m/s})^2}$$

$$= 4.70 \times 10^{-24} \text{ s}$$

The approximate distance that the pion can move in this time and return, found from equation 31.58, is

$$d = c\frac{\Delta t}{2}$$

$$= (3.00 \times 10^8 \text{ m/s})\left(\frac{4.70 \times 10^{-24} \text{ s}}{2}\right)$$

$$= 0.705 \times 10^{-15} \text{ m}$$

This distance is, of course, only approximate because the pion does not move at the speed of light. However, the calculation does give us the order of magnitude of the distance. What is interesting is that the radius of the nucleus of hydrogen is 1.41×10^{-15} m and for uranium it is 8.69×10^{-15} m. Thus, the distance that a virtual particle can move is of the order of the size of the nucleus.

If there are two real protons relatively close together as in the nucleus of an atom as shown in figure 31.13(a), then one proton can emit a virtual pion that

Figure 31.13

The exchange of a virtual pion.

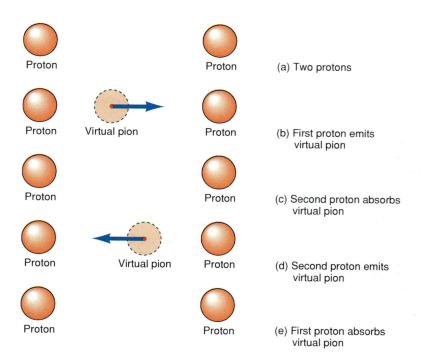

Proton   Proton   (a) Two protons

Proton   Virtual pion   Proton   (b) First proton emits
virtual pion

Proton   Proton   (c) Second proton absorbs
virtual pion

Proton   Virtual pion   Proton   (d) Second proton emits
virtual pion

Proton   Proton   (e) First proton absorbs
virtual pion

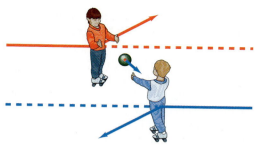

(a) Force of repulsion by exchange of bowling ball

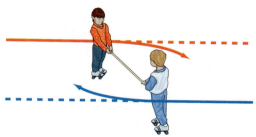

(b) Force of attraction by both boys holding onto a meter stick as they pass each other

Figure 31.14

Classical analogy of exchange force.

can travel to the second proton, figure 31.13(b). The second proton can absorb the virtual pion, figure 31.13(c). The second pion then emits a virtual pion that can travel to the first pion, figure 31.13(d). The first proton then absorbs the virtual pion, figure 31.13(e). Thus, the protons can exchange virtual pions with one another. *In 1934, the Japanese physicist, Hideki Yukawa (1907–1981), proposed that if two protons exchanged virtual mesons, the result of the exchange would be a very strong attractive force between the protons.* The exchange of virtual mesons between neutrons would also cause a strong attractive force between the neutrons. This exchange force must be a very short-ranged force because it is not observed anywhere outside of the nucleus. The predicted pi-meson was found in cosmic rays by Cecil F. Powell in 1947. Yukawa won the Nobel Prize in physics in 1949, and Powell in 1950.

The concept of a force caused by the exchange of particles is a quantum mechanical concept that is not found in classical physics. The best way to try to describe it classically is to imagine two boys approaching each other on roller skates, as shown in figure 31.14(a). Each boy is moving in a straight line as they approach each other. When the boys are relatively close, the first boy throws a bowling ball to the second boy. By the law of conservation of momentum the first boy recoils after he throws the ball, whereas the second boy recoils after he catches it. The net effect of throwing the bowling ball is to deviate the boys from their straight line motion as though a force of repulsion acted on the two boys. In this way we can say that the exchange of the bowling ball caused a repulsive force between the two boys.

A force of attraction can be similarly analyzed. Suppose, again, that the two boys are approaching each other on roller skates in a straight line motion. When the boys are relatively close the first boy holds out a meterstick for the second boy to grab, figure 31.14(b). As both boys hold on to the meterstick as they pass, they exert a force on each other through the meterstick. The force pulls each boy toward the other boy and deviates the straight line motion into the curved motion toward each other. When the first boy lets go of the meterstick, the attractive force disappears and the boys move in a new straight line motion. Thus, the exchange of the meterstick acted like an attractive force.

The exchange of the virtual pions between the protons in the nucleus cause a very large attractive force that is able to overcome the electrostatic force of repulsion between the protons. The virtual pions can be thought of as a *nuclear glue* that holds the nucleus together. *The tremendous importance of the concept of borrowing energy to form virtual particles, a concept that comes from the Heisenberg uncertainty principle, allows us to think of all forces as being caused by the exchange of virtual particles. Thus, the electrical force can be thought of as caused by the exchange of virtual photons and the gravitational force by the exchange of virtual gravitons (a particle not yet discovered).*

31.11 The Gravitational Red Shift by the Theory of Quanta

The relation for the gravitational red shift was derived in chapter 30 by observing how a clock slows down in a gravitational field. A remarkably simple derivation of this red shift can be obtained by treating light as a particle.

Let an atom at the surface of the earth emit a photon of light of frequency ν_g. This photon has the energy

$$E_g = h\nu_g \tag{31.59}$$

The subscript g is to remind us that this is a photon in the gravitational field. Let us assume that the light source was pointing upward so that the photon travels upward against the gravitational field of the earth until it arrives at a height y above the surface, as shown in figure 31.15. (We have used y for the height instead of h, as used previously, so as not to confuse the height with Planck's constant h.) As the

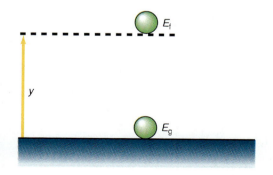

Figure 31.15

A photon in a gravitational field.

photon rises it must do work against the gravitational field. When the photon arrives at the height y, its energy E_f must be diminished by the work it had to do to get there. Thus

$$E_f = E_g - W \qquad (31.60)$$

Because the gravitational field is weaker at the height y than at the surface, the subscript f has been used on E to indicate that this is the energy in the weaker field or even in a field-free space. The work done by the photon in climbing to the height y is the same as the potential energy of the photon at the height y. Therefore,

$$W = PE = mgy \qquad (31.61)$$

Substituting equation 31.61 and the values of the energies back into equation 31.60, gives

$$h\nu_f = h\nu_g - mgy \qquad (31.62)$$

But the mass of the emitted photon is

$$m = \frac{E_g}{c^2} = \frac{h\nu_g}{c^2}$$

Placing this value of the mass back into equation 31.62, gives

$$h\nu_f = h\nu_g - \frac{h\nu_g}{c^2}gy$$

or

$$\nu_f = \nu_g\left(1 - \frac{gy}{c^2}\right) \qquad (31.63)$$

Equation 31.63 says that the frequency of a photon associated with a spectral line that is observed away from the gravitational field is less than the frequency of the spectral line emitted by the atom in the gravitational field itself. Since the frequency ν is related to the wavelength λ by $c = \lambda\nu$, the observed wavelength in the field-free space λ_f is longer than the wavelength emitted by the atom in the gravitational field λ_g. Therefore, the observed wavelength is shifted toward the red end of the spectrum. Note the equation 31.63 is the same as equation 30.37. *The slowing down of a clock in a gravitational field follows directly from equation (31.63)* by noting that the frequency ν is related to the period of time T by $\nu = 1/T$. Hence

$$\frac{1}{T_f} = \frac{1}{T_g}\left(1 - \frac{gy}{c^2}\right)$$

$$T_f = \frac{T_g}{1 - gy/c^2}$$

$$T_f = T_g\left(1 - \frac{gy}{c^2}\right)^{-1}$$

But by the binomial theorem,

$$\left(1 - \frac{gy}{c^2}\right)^{-1} = 1 + \frac{gy}{c^2}$$

Thus,

$$T_f = T_g\left(1 + \frac{gy}{c^2}\right) \qquad (31.64)$$

Equation 31.64 is identical to equation 30.34. Finally calling the period of time T an elapsed time, Δt, we have

$$\Delta t_f = \Delta t_g \left(1 + \frac{gy}{c^2} \right) \tag{31.65}$$

which is identical to equation 30.31, which shows the slowing down of a clock in a gravitational field.

31.12 An Accelerated Clock

An extremely interesting consequence of the gravitational red shift can be formulated by invoking Einstein's principle of equivalence discussed in chapter 30. Calling the inertial system containing gravity the K system and the accelerated frame of reference the K' system, Einstein stated, "we assume that we may just as well regard the system K as being in a space free from a gravitational field if we then regard K as uniformly accelerated." *Einstein's principle of equivalence was thus stated as: on a local scale the physical effects of a gravitational field are indistinguishable from the physical effects of an accelerated coordinate system.* "Hence the systems K and K' are equivalent with respect to all physical processes, that is, the laws of nature with respect to K are in entire agreement with those with respect to K'." Einstein then *postulated his theory of general relativity, as: The laws of physics are the same in all frames of reference.*

Since a clock slows down in a gravitational field, equation 31.65, using the equivalence principle, an accelerated clock should also slow down. Replacing the acceleration due to gravity g by the acceleration of the clock a, equation 31.65 becomes

$$\Delta t_f = \Delta t_a \left(1 + \frac{ay}{c^2} \right) \tag{31.66}$$

Note that the subscript g on Δt_g in equation 31.65 has now been replaced by the subscript a, giving Δt_a, to indicate that this is the time elapsed on the accelerated clock. Notice from equation 31.66 that

$$\Delta t_f > \Delta t_a$$

indicating that time slows down on the accelerated clock. *That is, an accelerated clock runs more slowly than a clock at rest.* In section 29.8 we saw, using the Lorentz transformation equations, that a clock at rest in a moving coordinate system slows down, and called the result the Lorentz time dilation. However, nothing was said at that time to show how the coordinate system attained its velocity. *Except for zero velocity, all bodies or reference systems must be accelerated to attain a velocity. Thus, there should be a relation between the Lorentz time dilation and the slowing down of an accelerated clock.* Let us change our notation slightly and call Δt_f the time Δt in a stationary coordinate system and Δt_a the time interval on a clock that is at rest in a coordinate system that is accelerating to the velocity v. Assuming that the acceleration is constant, we can use the kinematic equation

$$v^2 = v_0^2 + 2ay$$

Further assuming that the initial velocity v_0 is equal to zero and solving for the quantity ay we obtain

$$ay = \frac{v^2}{2} \tag{31.67}$$

Substituting equation 31.67 into equation 31.66, yields

$$\Delta t = \Delta t_a \left(1 + \frac{v^2}{2c^2} \right) \tag{31.68}$$

Using the binomial theorem in reverse

$$1 - nx = (1 - x)^n$$

with $x = v^2/c^2$ and $n = -1/2$, we get

$$\left(1 + \frac{v^2}{2c^2}\right) = \left[1 - \left(-\frac{1}{2}\right)\frac{v^2}{c^2}\right] = \left(1 - \frac{v_2}{c^2}\right)^{-1/2} = \frac{1}{\sqrt{1 - v^2/c^2}} \quad \text{(31.69)}$$

Equation 31.68 becomes

$$\Delta t = \frac{\Delta t_a}{\sqrt{1 - v^2/c^2}} \quad \text{(31.70)}$$

But this is exactly the time dilation formula, equation 29.64, found by the Lorentz transformation. Thus the Lorentz time dilation is a special case of the slowing down of an accelerated clock. This is a very important result. *Therefore, it is more reasonable to take the slowing down of a clock in a gravitational field, and thus by the principle of equivalence, the slowing down of an accelerated clock as the more basic physical principle. The Lorentz transformation for time dilation can then be derived as a special case of a clock that is accelerated from rest to the velocity v.*

Just as the slowing down of a clock in a gravitational field can be attributed to the warping of spacetime by the mass, it is reasonable to assume that the slowing down of the accelerated clock can also be thought of as the warping of spacetime by the increased mass, due to the increase in the velocity of the accelerating mass.

The Lorentz length contraction can also be derived from this model by the following considerations. Consider the emission of a light wave in a gravitational field. We will designate the wavelength of the emitted light by λ_g, and the period of the light by T_g. The velocity of the light emitted in the gravitational field is given by

$$c_g = \frac{\lambda_g}{T_g} \quad \text{(31.71)}$$

We will designate the velocity of light in a region far removed from the gravitational field as c_f for the velocity in a field-free region. The velocity of light in the field-free region is given by

$$c_f = \frac{\lambda_f}{T_f} \quad \text{(31.72)}$$

where λ_f is the wavelength of light, and T_f is the period of the light as observed in the field-free region. If the gravitating mass is not too large, then we can make the reasonable assumption that the velocity of light is the same in the gravitational field region and the field-free region, that is, $c_g = c_f$. We can then equate equation 31.71 to equation 31.72 to obtain

$$\frac{\lambda_g}{T_g} = \frac{\lambda_f}{T_f}$$

Solving for the wavelength of light in the field-free region, we get

$$\lambda_f = \frac{T_f}{T_g}\lambda_g$$

Substituting the value of T_f from equation 31.64 into this we get

$$\lambda_f = \frac{T_g}{T_g}\left(1 + \frac{gy}{c^2}\right)\lambda_g$$

$$= \left(1 + \frac{gy}{c^2}\right)\lambda_g \quad \text{(31.73)}$$

Equation 31.73 gives the wavelength of light λ_f in the gravitational-field-free region. By the principle of equivalence, the wavelength of light emitted from an accelerated observer, accelerating with the constant acceleration a through a distance y is obtained from equation 31.73 as

$$\lambda_0 = \left(1 + \frac{ay}{c^2}\right)\lambda_a \qquad (31.74)$$

where λ_0 is the wavelength of light that is observed in the region that is not accelerating, that is, the wavelength observed by an observer who is at rest. This result can be related to the velocity v that the accelerated observer attained during the constant acceleration by the kinematic equation

$$v^2 = v_0^2 + 2ay$$

Further assuming that the initial velocity v_0 is equal to zero and solving for the quantity ay we obtain

$$ay = \frac{v^2}{2}$$

Substituting this result into equation 31.74 we obtain

$$\lambda_0 = \left(1 + \frac{v^2}{2c^2}\right)\lambda_a \qquad (31.75)$$

Using the binomial theorem in reverse as in equation 31.69,

$$\left(1 + \frac{v^2}{2c^2}\right) = \frac{1}{\sqrt{1 - v^2/c^2}}$$

equation 31.75 becomes

$$\lambda_0 = \frac{\lambda_a}{\sqrt{1 - v^2/c^2}}$$

Solving for λ_a we get

$$\lambda_a = \sqrt{1 - v^2/c^2}\,\lambda_0 \qquad (31.76)$$

But λ is a length, in particular λ_a is a length that is observed by the observer who has accelerated from 0 up to the velocity v and is usually referred to as L, whereas λ_0 is a length that is observed by an observer who is at rest relative to the measurement and is usually referred to as L_0. Hence, we can write equation 31.76 as

$$L = L_0\sqrt{1 - v^2/c^2} \qquad (31.77)$$

But equation 31.77 is the Lorentz contraction of special relativity. *Hence, the Lorentz contraction is a special case of contraction of a length in a gravitational field, and by the principle of equivalence, a rod L_0 that is accelerated to the velocity v is contracted to the length L.* (That is, if a rod of length L_0 is at rest in a stationary spaceship, and the spaceship accelerates up to the velocity v, then the stationary observer on the earth would observe the contracted length L.) *Hence, the acceleration of the rod is the basic physical principle underlying the length contraction.*

Thus, both the time dilation and length contraction of special relativity should be attributed to the warping of spacetime by the accelerating mass.

The warping of spacetime by the accelerating mass can be likened to the Doppler effect for sound. Recall from section 12.8 that if a source of a sound wave is stationary, the sound wave propagates outward in concentric circles. When the sound source is moving, the waves are no longer circular but tend to bunch up in advance of the moving source. Since light does not require a medium for propagation, the Doppler effect for light is very much different. However, we can speculate that the warping of spacetime by the accelerating mass is comparable to the bunching up of sound waves in air. In fact, if we return to equation 31.63, for the gravitational red shift, and again, using the principle of equivalence, let $g = a$, and dropping the subscript f, this becomes

$$\nu = \nu_a\left(1 - \frac{ay}{c^2}\right) \tag{31.78}$$

Using the kinematic equation for constant acceleration, $ay = v^2/2$. Hence equation 31.78 becomes

$$\nu = \nu_a\left(1 - \frac{v^2}{2c^2}\right) \tag{31.79}$$

Again using the binomial theorem

$$\left(1 - \frac{v^2}{2c^2}\right) = \sqrt{1 - v^2/c^2}$$

Equation 31.78 becomes

$$\nu = \nu_a\sqrt{1 - v^2/c^2} \tag{31.80}$$

Equation 31.80 is called the transverse Doppler effect. It is a strictly relativistic result and has no counterpart in classical physics. The frequency ν_a is the frequency of light emitted by a light source that is at rest in a coordinate system that is accelerating past a stationary observer, whereas ν is the frequency of light observed by the stationary observer. *Notice that the transverse Doppler effect comes directly from the gravitational red shift by using the equivalence principle.* Thus the transverse Doppler effect should be looked on as a frequency shift caused by accelerating a light source to the velocity v.

It is important to notice here that this entire derivation started with the gravitational red shift by the theory of the quanta, then the equivalence principle was used to obtain the results for an accelerating system. The Lorentz time dilation and length contraction came out of this derivation as a special case. Thus, the Lorentz equations should be thought of as kinematic equations, whereas the gravitational and acceleration results should be thought of as a dynamical result.

Time dilation and length contraction have always been thought of as only depending upon the velocity of the moving body and not upon its acceleration. As an example, in Wolfgang Rindler's book *Essential Relativity,*[1] he quotes results of experiments at the CERN laboratory where muons were accelerated. He states "that accelerations up to 10^{19} g (!) do not contribute to the muon time dilation." The only time dilation that could be found came from the Lorentz time dilation formula. They could not find the effect of the acceleration because they had it all the time. The Lorentz time dilation formula itself is a result of the acceleration. Remember, it is impossible to get a nonzero velocity without an acceleration.

In our study in chapter 30 we discussed how a very large collapsing star could become a black hole. Pursuing the equivalence principle further, if gravitational mass can warp spacetime into a black hole, can the singularity that would occur if a body could be accelerated to the velocity c, be considered as an accelerating black hole, and if so what implications would this have?

1. Springer-Verlag, New York, 1979, Revised 2nd edition, p. 44.

The Language of Physics

Photon
A small bundle of electromagnetic energy that acts as a particle of light. The photon has zero rest mass and its energy and momentum are determined in terms of the wavelength and frequency of the light wave (p. 924).

Photoelectric effect
Light falling on a metallic surface produces electrical charges. The photoelectric effect cannot be explained by classical electromagnetic theory. Einstein used the quantum theory to successfully explain this effect and won the Nobel Prize in physics. He said that a photon of light collides with an electron and imparts enough energy to it to remove it from its position in the metal (p. 925).

Principle of complementarity
The wave theory of light and the quantum theory of light complement each other. In a specific case, light exhibits either a wave nature or a particle nature, but never both at the same time (p. 933).

Compton effect
Compton bombarded electrons with photons and found that the scattered photon has a different wavelength than the incident light. The photon lost energy to the electron in the collision (p. 934).

de Broglie relation
de Broglie assumed that the same wave-particle duality associated with electromagnetic waves should also apply to particles. Thus, particles should also act as waves. The wave was first called a pilot wave, and then a matter wave. Today, it is simply called the wave function (p. 938).

Heisenberg uncertainty principle
The position and momentum of a particle cannot both be measured simultaneously with perfect accuracy. There is always a fundamental uncertainty associated with any measurement. This uncertainty is not associated with the measuring instrument. It is a consequence of the wave-particle duality of matter (p. 944).

Virtual particles
Ghostlike particles that exist around true particles. They exist by borrowing energy from the true particle, and converting this energy into mass. The energy must, however, be paid back before the time Δt, determined by the uncertainty principle, elapses. The virtual particles supply the force necessary to keep protons and neutrons together in the nucleus (p. 949).

Summary of Important Equations

Planck's relation
$$E = nh\nu \tag{31.2}$$

Einstein's photoelectric equation
$$KE_{max} = h\nu - W_0 \tag{31.10}$$

The work function
$$W_0 = h\nu_0 \tag{31.11}$$

Properties of the photon
Rest mass
$$m_0 = 0 \tag{31.14}$$

Energy
$$E = h\nu \tag{31.6}$$

Relativistic mass
$$m = \frac{E}{c^2} = \frac{h\nu}{c^2} \tag{31.16}$$

Momentum
$$p = \frac{E}{c} = \frac{h\nu}{c} = \frac{h}{\lambda} \tag{31.21}$$

Momentum of any particle
$$p = \frac{\sqrt{E^2 - E_0^2}}{c} \tag{31.19}$$

Compton scattering formula
$$\lambda' - \lambda = \frac{h}{m_0 c}(1 - \cos\phi) \tag{31.28}$$

de Broglie relation
$$\lambda = \frac{h}{p} \tag{31.29}$$

The uncertainty principle
$$\Delta p\,\Delta x \geq \hbar \tag{31.44}$$
$$\Delta\theta\,\Delta L \geq \hbar \tag{31.51}$$
$$\Delta E\,\Delta t \geq \hbar \tag{31.55}$$

Angular momentum of a particle
$$L = rp\sin\theta \tag{31.48}$$
$$L = rp = rm\upsilon \tag{31.47}$$

Payback time for a virtual particle
$$\Delta t = \frac{\hbar}{(\Delta m)c^2} \tag{31.57}$$

Gravitational red shift
$$\nu_f = \nu_g\left(1 - \frac{gy}{c^2}\right) \tag{31.63}$$

$$T_f = T_g\left(1 + \frac{gy}{c^2}\right) \tag{31.64}$$

Slowing down of a clock in a gravitational field
$$\Delta t_f = \Delta t_g\left(1 + \frac{gy}{c^2}\right) \tag{31.65}$$

Slowing down of an accelerated clock
$$\Delta t_f = \Delta t_a\left(1 + \frac{ay}{c^2}\right) \tag{31.66}$$

$$\Delta t = \frac{\Delta t_a}{\sqrt{1 - \upsilon^2/c^2}} \tag{31.70}$$

Length contraction in a gravitational field
$$\lambda_f = \left(1 + \frac{gy}{c^2}\right)\lambda_g \tag{31.73}$$

Length contraction in an acceleration
$$\lambda_0 = \left(1 + \frac{ay}{c^2}\right)\lambda_a \tag{31.74}$$
$$L = L_0\sqrt{1 - \upsilon^2/c^2} \tag{31.77}$$

Questions for Chapter 31

†1. How would the world appear if Planck's constant h were very large? Describe some common occurrences and how they would be affected by the quantization of energy.

2. When light shines on a surface, is momentum transferred to the surface?

3. Could photons be used to power a spaceship through interplanetary space?

4. Should the concept of the cessation of all molecular motion at absolute zero be modified in view of the uncertainty principle?

5. Which photon, red, green, or blue, carries the most (a) energy and (b) momentum?

6. Discuss the entire wave-particle duality. That is, is light a wave or a particle, and is an electron a particle or a wave?

†7. Discuss the concept of determinism in terms of the uncertainty principle.

†8. Why isn't the photoelectric effect observed in all metals?

9. Ultraviolet light has a higher frequency than infrared light. What does this say about the energy of each type of light?

†10. Why can red light be used in a photographic dark room when developing pictures, but a blue or white light cannot?

Problems for Chapter 31

31.2 Blackbody Radiation

1. A weightless spring has a spring constant of 18.5 N/m. A 500-g mass is attached to the spring. It is then displaced 10.0 cm and released. Find (a) the total energy of the mass, (b) the frequency of the vibration, (c) the quantum number n associated with this energy, and (d) the energy change when the oscillator changes its quantum state by one value.

2. Find the energy of a photon of light of 400.0-nm wavelength.

3. A radio station broadcasts at 92.4 MHz. What is the energy of a photon of this electromagnetic wave?

31.3 The Photoelectric Effect

4. The work function of a material is 4.52 eV. What is the threshold wavelength for photoelectronic emission?

5. The threshold wavelength for photoelectronic emission for a particular material is 518 nm. Find the work function for this material.

†6. Light of 546.0-nm wavelength is incident on a cesium surface that has a work function of 1.91 eV. Find (a) the frequency of the incident light, (b) the energy of the incident photon, (c) the maximum kinetic energy of the photoelectron, (d) the stopping potential, and (e) the threshold wavelength.

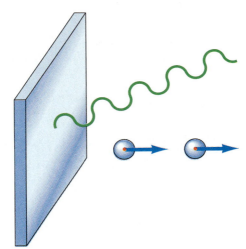

31.4 The Properties of the Photon

7. A photon has an energy of 5.00 eV. What is its frequency and wavelength?

8. Find the mass of a photon of light of 500.0-nm wavelength.

9. Find the momentum of a photon of light of 500.0-nm wavelength.

10. Find the wavelength of a photon whose energy is 500 MeV.

11. What is the energy of a 650 nm photon?

31.5 The Compton Effect

12. An 80.0-KeV X ray is fired at a carbon target and Compton scattering occurs. Find the wavelength of the incident photon and the wavelength of the scattered photon for an angle of 40.0°.

13. If an incident photon has a wavelength of 0.0140 nm, and is found to be scattered at an angle of 50.0° in Compton scattering, find the energy of the recoiling electron.

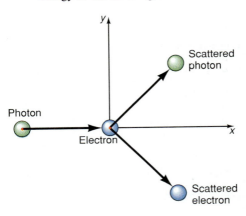

14. In a Compton scattering experiment, 400-nm photons are scattered by the target, yielding 769-nm photons. What is the angle at which the 769-nm photons are scattered?

31.6 The Wave Nature of Particles

15. Find the wavelength of a 4.60×10^{-2} kg golf ball moving at a speed of 60.0 m/s.
16. Find the wavelength of a proton moving at 10.0% of the speed of light.
17. Find the wavelength of an electron moving at 10.0% of the speed of light.
18. Find the wavelength of a 5.00-KeV electron.
19. Find the wavelength of an oxygen molecule at room temperature.
20. What is the frequency of the matter wave representing an electron moving at a speed of $2c/3$?
21. (a) Find the total energy of a proton moving at a speed of $c/2$. (b) Compute the wavelength of this proton.

31.8 The Heisenberg Uncertainty Principle

22. A 4.6×10^{-2} kg golf ball is in motion along the x-axis. If it is located at the position $x = 1.00$ m, with an uncertainty of 0.005 m, find the uncertainty in the determination of the momentum and velocity of the golf ball.
23. Find the minimum uncertainty in the determination of the momentum and speed of a 1300-kg car if the position of the car is to be known to a value of 10 nm.
24. The uncertainty in the position of a proton is 100 nm. Find the uncertainty in the kinetic energy of the proton.

31.9 Different Forms of the Uncertainty Principle

†25. The lifetime of an electron in an excited state of an atom is 10^{-8} s. From the uncertainty in the energy of the electron, determine the width of the spectral line centered about 550 nm.

Additional Problems

26. Approximately 5.00% of a 100-W incandescent lamp falls in the visible portion of the electromagnetic spectrum. How many photons of light are emitted from the bulb per second, assuming that the wavelength of the average photon is 550 nm?

Interactive Tutorials

27. The photoelectric effect. Light of wavelength $\lambda = 577.0$ nm is incident on a cesium surface. Photoelectrons are observed to flow when the applied voltage $V_0 = 0.250$ V. Find (a) the frequency ν of the incident photon, (b) the initial energy E of the incident photon, (c) the maximum kinetic energy KE_{max} of the photoelectrons, (d) the work function W_0 of cesium, (e) the threshold frequency ν_0, and (f) the corresponding threshold wavelength λ_0.

28. Light of wavelength $\lambda = 460$ nm is incident on a cesium surface. The work function of cesium is $W_0 = 3.42 \times 10^{-19}$ J. Find (a) the frequency ν of the incident photon, (b) the initial energy E of the incident photon, (c) the maximum kinetic energy KE_{max} of the emitted photoelectrons, (d) the maximum speed υ of the electron, (e) the threshold frequency ν_0, and (f) the corresponding longest wavelength λ_0 that will eject electrons from the metal.

29. Properties of a photon. A photon of light has a wavelength $\lambda = 420.0$ nm, find (a) the frequency ν of the photon, (b) the energy E of the photon, (c) the mass m of the photon, and (d) the momentum p of the photon.

30. The Compton effect. An x-ray photon of energy $E = 90.0$ KeV is fired at a carbon target and Compton scattering occurs at an angle $\phi = 30.0°$. Find (a) the frequency ν of the incident photon, (b) the wavelength λ of the incident photon, and (c) the wavelength λ' of the scattered photon.

31. Using the concept of wave particle duality, calculate the wavelength λ of a golf ball whose mass $m = 4.60 \times 10^{-2}$ kg and is traveling at a speed $\upsilon = 60.0$ m/s.

Atomic Physics

An atom consists of electrons that whirl about the nucleus. The nucleus consists of protons and neutrons, which in turn consist of quarks.

Nature resolves everything into it component atoms and never reduces anything to nothing.

Lucretius (95 BC - 55 BC)

32.1 The History of the Atom

As mentioned previously in section 18.2, the earliest attempt to find simplicity in matter occurred in the fifth century B.C., when the Greek philosophers Leucippus and Democritus stated that matter is composed of very small particles called atoms. The Greek word for *atom* means "that which is indivisible." The concept of an atom of matter was to lie dormant for hundreds of years until 1803 when John Dalton, an English chemist, introduced his atomic theory of matter in which he proposed that to every known chemical element there corresponds an atom of matter. Today there are known to be 105 chemical elements. All other substances in the world are combinations of these elements.

The Greek philosophers' statement about atoms was based on speculation, whereas Dalton's theory was based on experimental evidence. Dalton's world of the atom was a simple and orderly place until some new experimental results appeared on the scene. M. Faraday performed experiments in electrolysis by passing electrical charges through a chemical solution of sodium chloride, NaCl. His laws of electrolysis showed that one unit of electricity was associated with one atom of a substance. He assumed that this charge was carried by the atom. A study of the conduction of electricity through rarefied gases led the English physicist J. J. Thomson, in 1898, to the verification of an independent existence of very small negatively charged particles. Even before this, in 1891, the Irish physicist George Stoney had made the hypothesis that there is a natural unit of electricity and he called it an electron. In 1896, Henri Becquerel discovered radiation from the atoms of a uranium salt.

The results of these experiments led to some inevitable questions. Where did these negatively charged particles, the electrons, come from? Where did the radiations from the uranium atom come from? The only place they could conceivably come from was from inside the atom. But if they came from within the atom, the atom could no longer be considered indivisible. That is, the "indivisible" atom must have some internal structure. Also, because the atom is observed to be electrically neutral, there must be some positive charge within it in order to neutralize the effect of the negative electrons. It had also been determined experimentally that these negative electrons were thousands of times lighter than the entire atom. Therefore, whatever positive charges existed within the atom, they must contain most of the mass of the atom.

J. J. Thomson proposed the first picture of the atom in 1898. He assumed that atoms are uniform spheres of positively charged matter in which the negatively charged electrons are embedded, figure 32.1.

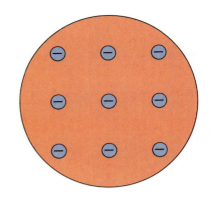

Figure 32.1

The plum pudding model of the atom.

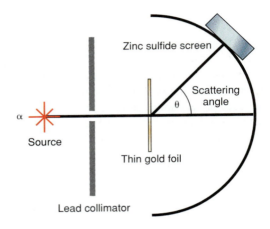

Figure 32.2

Rutherford scattering.

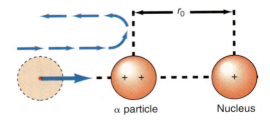

Figure 32.3

Approximate nuclear dimensions by a head-on collision.

This model of the atom was called the plum pudding model because it resembled the raisins embedded in a plum pudding. That is, the electrons were like the raisins, and the pudding was the positively charged matter. But how are these electrons distributed within the atom? It became obvious that the only way to say exactly what is within the atom is to take a "look" inside the atom. But how do you "look" inside an atom?

In 1911, Hans Geiger (1882–1945) and Ernest Marsden, following a suggestion by Ernest Rutherford (1871–1937), bombarded atoms with alpha particles to "see" what was inside of the atom. The alpha particles, also written α particles, were high-energy particles emitted by radioactive substances that were discovered by Rutherford in 1899. These α particles were found to have a mass approximately four times the mass of a hydrogen atom and carried a positive charge of two units. (Today we know that the α particle is the nucleus of the helium atom.) Rutherford's idea was that the direction of motion of the α particle should be changed or deflected by the electrical charges within the atom. In this way, we can "see" within the atom. The experimental arrangement is illustrated in figure 32.2. A polonium-214 source emits α particles of 7.68 MeV. A lead screen with a slit (the lead collimator) allows only those α particles that pass through the slit to fall on a very thin gold foil. It is expected that most of these α particles should go straight through the gold atoms and arrive at the zinc sulfide (ZnS) screen. When the α particle hits the ZnS screen a small flash of light is given off (called a *scintillation*). It is expected that some of these α particles should be slightly deflected by some of the positive charge of the atomic pudding. The ZnS screen can be rotated through any angle around the gold foil and can therefore observe any deflected α particles. *This process of deflection of the α particle is called Rutherford scattering.* The expected scattering from the distributed positive charge should be quite small and this was what at first was observed. Then Rutherford suggested that Geiger and Marsden should look for some scattering through large angles. To everyone's surprise, some α particles were found to come straight backward ($\theta \approx 180°$). This back-scattering was such a shock to Rutherford that as he described it, "It was almost as incredible as if you fired a 15-inch shell at a piece of paper and it bounced off and came right back to you."

The only way to explain this back-scattering is to assert that the positive charge is not distributed over the entire atom but instead it must be concentrated in a very small volume. Thus, the experimental results of large-angle scattering are not consistent with the plum pudding model of an atom. Rutherford, therefore, proposed a new model of the atom, the *nuclear atom*. In this model, the atom consists of a very small, dense, positively charged nucleus. Because the negatively charged electrons would be attracted to the positive nucleus and would crash into it if they were at rest, it was necessary to assume that the electrons were in motion orbiting around the nucleus somewhat in the manner of the planets orbiting about the sun in the solar system. In 1919, Rutherford found this positive particle of the nucleus and named it the *proton*. The proton has a positive charge equal in magnitude to the charge on the electron (i.e., 1.60×10^{-19} C). The atom must contain the same number of protons as electrons in order to account for the fact that the atom is electrically neutral. The mass of the proton is 1.67×10^{-27} kg, which is 1836 times more massive than the electron.

A simple analysis allows us to determine the approximate dimensions of the nucleus. Consider a head-on collision between the α particle and the nucleus, as shown in figure 32.3. Because both particles are positive, there is an electrostatic force of repulsion between them. The α particle, as it approaches the nucleus, slows down because of the repulsion. Eventually it comes to a stop close to the nucleus, shown as the distance r_0 in the figure. The repulsive force now causes the α particle to accelerate away from the nucleus giving the back-scattering of 180°. When the α particle left the source, it had a kinetic energy of 7.68 MeV. When it momentarily came to rest at the position r_0, its velocity was zero and hence its kinetic energy was

Modern Physics

Transmission Electron Micrograph of the atomic lattice of a thin gold crystal.

also zero there. Where did all the energy go? The whereabouts of this energy can be determined by referring back to figure 19.19, which showed the potential hill of a positive point charge, such as a nucleus. As the α particle approaches the nucleus, it climbs up the potential hill, losing its kinetic energy but gaining potential energy. The potential energy that the α particle gains is found by rearranging equation 19.19 to the form

$$PE = q_{alpha}V$$

where q_{alpha} is the charge on the α particle. The potential V of the positive nucleus is given by equation 19.31, the equation for the potential of a point charge, as

$$V = \frac{kq_n}{r}$$

where q_n is the charge of the nucleus and k is the constant in Coulomb's law. Thus, the potential energy that the α particle gains as it climbs up the potential hill is

$$PE = \frac{kq_{alpha}q_n}{r}$$

When the α particle momentarily comes to rest at the distance r_0 from the nucleus, all its kinetic energy has been converted to potential energy. Equating the kinetic energy of the α particle to its potential energy in the field of the nucleus, we get

$$KE = PE = \frac{kq_{alpha}q_n}{r_0} \tag{32.1}$$

Because the kinetic energy of the α particle is known, equation 32.1 can be solved for r_0, the approximate radius of the nucleus, to give

$$r_0 = \frac{kq_{alpha}q_n}{KE} \tag{32.2}$$

It had previously been determined that the charge on the α particle was twice the charge of the electron, that is,

$$q_{alpha} = 2e \tag{32.3}$$

To determine the charge of the nucleus, more detailed scattering experiments were performed with different foils, and it was found that the positive charge on the nucleus was approximately

$$q_n = \frac{Ae}{2} \tag{32.4}$$

where A is the mass number of the foil material. *The mass number A is the nearest whole number to the atomic mass of an element.* For example, the atomic mass of nitrogen (N) is 14.0067, its mass number A is 14. A new number, *the atomic number Z is defined from equation 32.4 as*

$$Z = \frac{A}{2} \tag{32.5}$$

The atomic number of nitrogen is thus $14/2 = 7$. The atomic number Z represents the number of positive charges in the nucleus, that is, the number of protons in the nucleus. Because an atom is neutral, Z also represents the total number of electrons in the atom. Using this notation the positive charge of the nucleus q_n is given by

$$q_n = Ze \tag{32.6}$$

Substituting equations 32.3 and 32.6 into equation 32.2 gives, for the value of r_0,

$$r_0 = \frac{k(2e)(Ze)}{KE}$$

or

$$r_0 = \frac{2kZe^2}{KE} \qquad (32.7)$$

Example 32.1

The radius of the gold nucleus found by scattering. Find the maximum radius of a gold nucleus that is bombarded by 7.68-MeV α particles.

Solution

The maximum radius of the gold nucleus is found from equation 32.7 with the atomic number Z for gold equal to 79,

$$r_0 = \frac{2kZe^2}{KE}$$

$$= \left[\frac{2(9.00 \times 10^9 \text{ Nm}^2/\text{C}^2)(79)(1.60 \times 10^{-19} \text{ C})^2}{7.68 \text{ MeV}} \right] \left(\frac{1 \text{ MeV}}{10^6 \text{ eV}} \right) \left(\frac{1 \text{ eV}}{1.60 \times^{-19} \text{ J}} \right)$$

$$= 2.96 \times 10^{-14} \text{ m}$$

Thus, the approximate radius of a gold nucleus is 2.96×10^{-14} m. The actual radius is somewhat less than this value. Although this calculation is only an approximation, it gives us the order of magnitude of the radius of the nucleus.

It was already known from experimental data, that the radius of the atom r_a is of the order

$$r_a = 10^{-10} \text{ m}$$

Comparing the relative size of the atom r_a to the size of the nucleus r_n, we get

$$\frac{r_a}{r_n} = \frac{10^{-10} \text{ m}}{10^{-14} \text{ m}} = 10^4$$

or

$$r_a = 10^4 \, r_n = 10{,}000 \, r_n \qquad (32.8)$$

That is, the radius of the atom is about 10,000 times greater than the radius of the nucleus.

More detailed scattering experiments have since led to the following approximate formula for the size of the nucleus

$$R = R_0 A^{1/3} \qquad (32.9)$$

where A is again the mass number of the element and R_0 is a constant equal to 1.20×10^{-15} m.

Example 32.2

A more accurate value for the radius of a gold nucleus. Find the radius of the gold nucleus using equation 32.9.

The mass number A for gold is found by looking up the atomic mass of gold in the periodic table of the elements (figure 32.18 or appendix E). The atomic mass is 197.0. The mass number A for gold is the nearest whole number to the atomic mass, so $A = 197$. Now we find the radius of the gold nucleus from equation 32.9, as

$$R = R_0 A^{1/3}$$
$$= (1.20 \times 10^{-15} \text{ m})(197)^{1/3}$$
$$= 0.70 \times 10^{-14} \text{ m}$$

Notice that the radius obtained in this way is less than the value obtained by equation 32.7, which is, of course, to be expected.

Because a nucleus contains Z protons, its mass should be Zm_p. Since the proton mass is so much greater than the electron mass, the mass of the atom should be very nearly equal to the mass of the nucleus. However, the atomic mass of an element was more than twice the mass of the Z protons. This led Rutherford in 1920 to predict the existence of another particle within the nucleus having about the same mass as the proton. Because this particle had no electric charge, Rutherford called it a neutron, for the neutral particle. In 1932, James Chadwick found this neutral particle, the neutron. The mass of the neutron was found to be $m_n = 1.6749 \times 10^{-27}$ kg, which is very close to the mass of the proton, $m_p = 1.6726 \times 10^{-27}$ kg.

With the finding of the neutron, the discrepancy in the mass of the nucleus was solved. *It can now be stated that an atom consists of Z electrons that orbit about a positive nucleus that contains Z protons and $(A - Z)$ neutrons.*

Example 32.3

The number of electrons, protons, and neutrons in a gold atom. How many electrons, protons, and neutrons are there in a gold atom?

Solution

From the table of the elements, we see that the atomic number Z for gold is 79. Hence, there are 79 protons in the nucleus of the gold atom surrounded by 79 orbiting electrons. The mass number A for gold is found from the table of the elements to be 197. Hence, the number of neutrons in the nucleus of a gold atom is

$$A - Z = 197 - 79 = 118 \text{ neutrons}$$

A very interesting characteristic of nuclear material is that all nuclei have the same density. For example, to find the density of nuclear matter, we use the definition of density defined in chapter 13,

$$\rho = \frac{m}{V} \qquad (13.1)$$

Because the greatest portion of the matter of an atom resides in the nucleus, we can take for the mass of the nucleus

$$m = \frac{\text{Atomic mass}}{N_A} \qquad (32.10)$$

where $N_A = 6.022 \times 10^{26}$ atoms/kmole $= 6.022 \times 10^{23}$ atoms/mole is Avogadro's number. Since the atomic mass, which has units of kg/kmole, is numerically very close to the mass number A, a dimensionless quantity, we can write the mass as

$$m = \frac{A(\text{kg/kmole})}{N_A} \tag{32.11}$$

We find the volume of the nucleus from the assumption that the atom is spherical, and hence

$$V = \frac{4\pi r^3}{3} \tag{32.12}$$

Substituting the value for the radius of the nucleus found in equation 32.9 into equation 32.12, we get, for the volume of the nucleus,

$$V = \tfrac{4}{3}\pi (R_0 A^{1/3})^3$$

or

$$V = \tfrac{4}{3}\pi R_0^3 A \tag{32.13}$$

Substituting the mass from equation 32.11 and the volume from equation 32.13 back into equation 13.1 for the density, we get

$$\rho = \frac{m}{V} = \frac{A(\text{kg/kmole})/N_A}{\tfrac{4}{3}\pi R_0^3 A} \tag{32.14}$$

or

$$\rho = \frac{3(\text{kg/kmole})}{4\pi N_A R_0^3} \tag{32.15}$$

$$= \frac{3(\text{kg/kmole})}{4\pi (6.022 \times 10^{26}\ \text{atoms/kmole})(1.2 \times 10^{-15}\ \text{m})^3}$$

$$= 2.29 \times 10^{17}\ \text{kg/m}^3$$

Because the mass number A cancelled out of equation 32.14, the density is the same for all nuclei. To get a "feel" for the magnitude of this nuclear density, note that a density of $2.29 \times 10^{17}\ \text{kg/m}^3$ is roughly equivalent to a density of a billion tons of matter per cubic inch, an enormously large number in terms of our usual experiences.

Now that the **Rutherford model of the atom** has been developed, let us look at some of its dynamical aspects. Let us consider the dynamics of the hydrogen atom. Because this new model of the atom is very similar to the planetary system of our solar system, we would expect the dynamics of the atom to be very similar to the dynamics of the planetary system. Recall from chapter 6 that we assumed that a planet moved in a circular orbit and the necessary centripetal force for that circular motion was supplied by the gravitational force. In a similar analysis, let us now assume that the negative electron moves in a circular orbit about the positive nucleus. The Coulomb attractive force between the electron and the proton supplies the necessary centripetal force to keep the electron in its orbit. Therefore, equating the centripetal force F_c to the electric force F_e, we obtain

$$F_c = F_e \tag{32.16}$$

$$\frac{mv^2}{r} = \frac{k(e)(e)}{r^2} \tag{32.17}$$

Solving for the speed of the electron in its circular orbit, we get

$$v = \sqrt{\frac{ke^2}{mr}} \tag{32.18}$$

Thus, for a particular orbital radius r, there corresponds a particular velocity of the electron. This is, of course, the same kind of a relation found for the planetary case.

The total energy of the electron is equal to the sum of its kinetic and potential energy. Thus,

$$E = KE + PE$$

$$E = \frac{1}{2} mv^2 + \left(-\frac{ke^2}{r} \right) \tag{32.19}$$

where the negative potential energy of the electron follows from the definition of potential energy. (The zero of potential energy is taken at infinity, and since the electron can do work, as it approaches the positive nucleus, it loses some of its electric potential energy. Because it started with zero potential energy at infinity, its potential energy becomes more negative as it approaches the nucleus.) Substituting the speed of the electron for the circular orbit found in equation 32.18 into equation 32.19, we have, for the total energy,

$$E = \frac{1}{2} m \left(\frac{ke^2}{mr} \right) - \frac{ke^2}{r}$$

$$= \frac{ke^2}{2r} - \frac{ke^2}{r}$$

$$E = -\frac{ke^2}{2r} \tag{32.20}$$

The total energy of the electron is negative indicating that the electron is bound to the atom. Equation 32.20 says that the total energy of an electron in the Rutherford atom is not quantized—that is, the electron could be in any orbit of radius r and would have an energy consistent with that value of r.

From chemical analysis, it is known that it takes 13.6 eV of energy to ionize a hydrogen atom. This means that it takes 13.6 eV of energy to remove an electron from the hydrogen atom to infinity, where it would then have zero kinetic energy. Looking at this from the reverse process, it means that we need -13.6 eV of energy to bind the electron to the atom. This energy is called the *binding energy of the electron*. Knowing this binding energy permits us to calculate the orbital radius of the electron. Solving equation 32.20 for the orbital radius gives

$$r = -\frac{ke^2}{2E} \tag{32.21}$$

$$= -\left[\frac{(9.00 \times 10^9 \text{ Nm}^2/\text{C}^2)(1.60 \times 10^{-19} \text{ C})^2}{2(-13.6 \text{ eV})} \right] \left(\frac{1 \text{ eV}}{1.60 \times 10^{-19} \text{ J}} \right)$$

$$= 5.29 \times 10^{-11} \text{ m}$$

which is certainly the right order of magnitude.

Because the electron is in a circular orbit, it is undergoing accelerated motion. From the laws of classical electromagnetic theory, an accelerated electric charge should radiate electromagnetic waves. The frequency of these electromagnetic waves should correspond to the frequency of the accelerating electron. The frequency ν of the moving electron is related to its angular velocity ω by

$$\nu = \frac{\omega}{2\pi}$$

But the linear velocity v of the electron is related to its angular velocity by

$$v = \omega r$$

or

$$\omega = \frac{v}{r}$$

The frequency becomes

$$\nu = \frac{\omega}{2\pi} = \frac{v}{2\pi r} \tag{32.22}$$

Substituting for the speed v from equation 32.18, we get

$$\nu = \frac{v}{2\pi r} = \frac{1}{2\pi r}\sqrt{\frac{ke^2}{mr}}$$

$$\nu = \frac{1}{2\pi}\sqrt{\frac{ke^2}{mr^3}} \tag{32.23}$$

The frequency of the orbiting electron, and hence the frequency of the electromagnetic wave radiated, should be given by equation 32.23. Assuming the value of r computed above, the frequency becomes

$$\nu = \frac{1}{2\pi}\sqrt{\frac{ke^2}{mr^3}}$$

$$= \frac{1}{2\pi}\sqrt{\frac{(9.00 \times 10^9 \text{ Nm}^2/\text{C}^2)(1.60 \times 10^{-19} \text{ C})^2}{(9.11 \times 10^{-31} \text{ kg})(5.29 \times 10^{-11} \text{ m})^3}}$$

$$= 6.56 \times 10^{15} \text{ Hz}$$

This frequency corresponds to a wavelength of

$$\lambda = \frac{c}{\nu} = \frac{3.00 \times 10^8 \text{ m/s}}{6.56 \times 10^{15} \text{ 1/s}}$$

$$= 4.57 \times 10^{-8} \text{ m}\left(\frac{1 \text{ nm}}{10^{-9} \text{ m}}\right)$$

$$= 45.7 \text{ nm}$$

The problem with this wavelength is that it is in the extreme ultraviolet portion of the electromagnetic spectrum, whereas some spectral lines of the hydrogen atom are known to be in the visible portion of the spectrum. An even greater discrepancy associated with Rutherford's model of the atom is that if the orbiting electron radiates electromagnetic waves, it must lose energy. If it loses energy by radiation, its orbital radius must decrease. For example, if the electron is initially in the state given by equation 32.20 as

$$E_i = -\frac{ke^2}{2r_i} \tag{32.24}$$

After it radiates energy, it will have the smaller final energy E_f, given by

$$E_f = -\frac{ke^2}{2r_f}$$

The energy lost by radiation is

$$E_f - E_i = -\frac{ke^2}{2r_f} - \left(-\frac{ke^2}{2r_i}\right)$$

$$= \frac{ke^2}{2}\left(\frac{1}{r_i} - \frac{1}{r_f}\right)$$

But since the final state has less energy than the initial state

$$E_f - E_i < 0$$

this implies that the quantity

$$\frac{ke^2}{2}\left(\frac{1}{r_i} - \frac{1}{r_f}\right) < 0$$

The only way for this quantity to be less than zero is for

$$\frac{1}{r_i} < \frac{1}{r_f}$$

which requires that

$$r_f < r_i$$

That is, the final orbital radius is less than the initial radius. Hence, when the electron radiates energy, its orbital radius must decrease. But as the electron keeps orbiting, it keeps losing energy and its orbital radius keeps decreasing. As r decreases, the frequency of the electromagnetic waves increases according to equation 32.23 and its wavelength decreases continuously. Hence, the radiation from the atom should be continuous. Experimentally, however, it is found that the radiation from the atom is not continuous but is discrete. As the radius of the orbit keeps decreasing, the electron spirals into the nucleus and the atom should collapse. Because the entire world is made of atoms, it too should collapse. Since it does not, there is something very wrong with the dynamics of the Rutherford model of the atom.

32.2 The Bohr Theory of the Atom

The greatest difficulty with the Rutherford planetary model is that the accelerated electron should radiate a continuous spectrum of electromagnetic waves, thereby losing energy, and should thus, spiral into the nucleus. There is certainly merit in the planetary model, but it is not completely accurate. As we have seen, the search for truth in nature follows the path of successive approximations. Each approximation gets us closer to the truth, but we are still not there yet. How can this radiation problem of the atomic model be solved?

Niels Bohr (1885–1962), a young Danish physicist, who worked with J. J. Thomson, and then Rutherford, felt that the new success of the quantum theory by Planck and Einstein must be the direction to take in understanding the atom, that is, the atom must be quantized. But how?

In the **Bohr theory of the atom,** Bohr took the ingenious step of restricting the electron orbits to those for which the angular momentum is quantized. That is, *Bohr postulated that the electron could only be found in those orbits for which the angular momentum, L, is given by*

$$L = mvr = n\hbar \qquad (32.25)$$

where n is called the principal quantum number and takes on the values 1, 2, 3, 4. . . . The value $\hbar$ thus becomes a fundamental unit of angular momentum. *The consequence of this postulate is that the electron, which can now be looked on as a matter wave by the de Broglie hypothesis, can be represented in its orbit as a standing wave.* Consider the standing wave in figure 32.4(a). As seen in chapter 12 for the vibrating string, the nodes of a standing wave remain nodes for all time. The string cannot move up or down at that point, and, hence, cannot transmit any energy past that point. Thus, the standing wave does not move along the string, but is instead stationary or standing.

For the vibrating string fixed at both ends in chapter 12, we found that the only waves that can stand for such a configuration are those for which the length of the string is equal to a multiple of a half wavelength, that is, for $l = n\lambda/2$. If the vibrating string is bent into a circle, figure 32.4(b), (perhaps this should be called a vibrating wire to justify bending it into a circle), the traveling waves are

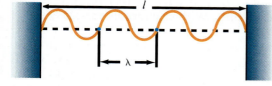

(a) Standing wave on a string

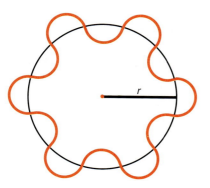

(b) Standing wave bent into a circular path

Figure 32.4

Standing wave of an electron in its orbit.

not reflected at a fixed boundary because there is now no fixed boundary. The waves keep passing around the circle. *The only waves that can stand in this circular configuration are those for which the length of the wire is a whole number of wavelengths.* Thus,

$$l = n\lambda \qquad (32.26)$$

but the length of the wire is the circumference of the circle and is equal to $2\pi r$. Hence,

$$l = 2\pi r = n\lambda \qquad (32.27)$$

The wavelength of the matter wave is given by the de Broglie relation as

$$\lambda = \frac{h}{p} = \frac{h}{mv} \qquad (32.28)$$

Substituting equation 32.28 into equation 32.27, gives

$$2\pi r = \frac{nh}{mv}$$

or

$$mvr = \frac{nh}{2\pi} = n\hbar \qquad (32.29)$$

But this is precisely the Bohr postulate for the allowed electron orbits, previously defined in equation 32.25. *Hence, Bohr's postulate of the quantization of the orbital angular momentum is equivalent to a standing matter wave on the electron orbit. But because standing waves do not change with time and thus, do not transmit energy, these matter waves representing the electron should not radiate electromagnetic waves. Thus, an electron in this prescribed orbit does not radiate energy and hence it does not spiral into the nucleus.* This state wherein an electron does not radiate energy is called a *stationary state.* With electrons in stationary states the atom is now stable.

The quantization of the orbital angular momentum displays itself as a quantization of the orbital radius, the orbital velocity, and the total energy of the electron. As an example, let us consider the dynamics of the Bohr model of the atom. Because it is basically still a planetary model, equations 32.16 and 32.17 still apply. However, equation 32.18 for the orbital speed is no longer applicable. Instead, we use equation 32.29 to obtain the speed of the electron as

$$v = \frac{n\hbar}{mr} \qquad (32.30)$$

Substituting this value of v into equation 32.17, we get

$$\frac{mv^2}{r} = \frac{m}{r}\left(\frac{n\hbar}{mr}\right)^2 = \frac{ke^2}{r^2}$$

$$\frac{mn^2\hbar^2}{rm^2r^2} = \frac{ke^2}{r^2}$$

$$\frac{n^2\hbar^2}{rm} = ke^2$$

Solving for r, we get

$$r_n = \frac{\hbar^2}{kme^2}n^2 \qquad (32.31)$$

Because of the n on the right-hand side of equation 32.31, the electron orbits are quantized. The subscript n has been placed on r to remind us that there is one value of r corresponding to each value of n. From the derivation, we see that r must be quantized in order to have standing or stationary waves.

Example 32.4

The radius of a Bohr orbit. Find the radius of the first Bohr orbit.

Solution

The radius of the first Bohr orbit is found from equation 32.31 with $n = 1$. Thus,

$$r_1 = \frac{\hbar^2}{kme^2}$$

$$= \frac{(1.0546 \times 10^{-34} \text{ J s})^2}{(8.9878 \times 10^9 \text{ Nm}^2/\text{C}^2)(9.1091 \times 10^{-31} \text{ kg})(1.6022 \times 10^{-19} \text{ C})^2}$$

$$= 5.2919 \times 10^{-11} \text{ m} \left(\frac{1 \text{ nm}}{10^{-9} \text{ m}} \right)$$

$$= 0.0529 \text{ nm}$$

This is the radius of the electron orbit for the $n = 1$ state (called the *ground state*) and is called the *Bohr radius*.

Now we can also write the radius of the nth orbit, equation 32.31, as

$$r_n = r_1 n^2 \tag{32.32}$$

Thus, the only allowed orbits are those for $r_n = r_1, 4r_1, 9r_1, 16r_1, \ldots$

The speed of the electron in its orbit can be rewritten by substituting equation 32.31 back into equation 32.30, yielding

$$v_n = \frac{n\hbar}{mr_n} = \frac{n\hbar}{m (\hbar^2/kme^2)n^2}$$

or

$$v_n = \frac{ke^2}{n\hbar} \tag{32.33}$$

Because of the n on the right-hand side of equation 32.33, the speed of the electron is also quantized.

Example 32.5

The speed of an electron in a Bohr orbit. Find the speed of the electron in the first Bohr orbit.

Solution

The speed, found from equation 32.33 with $n = 1$, is

$$v_1 = \frac{ke^2}{\hbar}$$

$$= \left(9.00 \times 10^9 \frac{\text{N m}^2}{\text{C}^2} \right) \frac{(1.60 \times 10^{-19} \text{ C})^2}{1.05 \times 10^{-34} \text{ J s}}$$

$$= 2.19 \times 10^6 \text{ m/s}$$

The speed of the electron in higher orbits is obtained from equation 32.33 as

$$v_n = \frac{v_1}{n} \tag{32.34}$$

We will now see that the quantizing of the orbital radius and speed leads to the quantizing of the electron's energy. The total energy of the electron still follows from equation 32.19, but now $v = v_n$ and $r = r_n$. Thus,

$$E = \frac{1}{2} mv_n^2 - \frac{ke^2}{r_n}$$

Substituting for r_n and v_n from equations 32.31 and 32.33, respectively, leads to

$$E = \frac{m}{2} \left(\frac{ke^2}{n\hbar} \right)^2 - \frac{ke^2}{(\hbar^2 n^2)/kme^2}$$

$$= \frac{mk^2 e^4}{2n^2 \hbar^2} - \frac{mk^2 e^4}{n^2 \hbar^2}$$

$$E_n = -\frac{mk^2 e^4}{2n^2 \hbar^2} \tag{32.35}$$

Because of the appearance of the quantum number n in equation 32.35, the electron's energy is seen to be quantized. To emphasize this fact, we have placed the subscript n on E.

Example 32.6

The energy of an electron in a Bohr orbit. Find the energy of the electron in the first Bohr orbit.

Solution

The energy is found from equation 32.35 with $n = 1$ as follows

$$E_1 = -\frac{mk^2 e^4}{2\hbar^2} \tag{32.36}$$

$$= -\left[\frac{(9.1092 \times 10^{-31}\ \text{kg})(8.9878 \times 10^9\ \text{Nm}^2/\text{C}^2)^2(1.6022 \times 10^{-19}\text{C})^4}{2(1.054 \times 10^{-34}\ \text{J s})^2} \right] \times$$

$$\left(\frac{1\ \text{eV}}{1.6022 \times 10^{-19}\ \text{J}} \right)$$

$$= -13.6\ \text{eV}$$

Thus, the energy of the electron in the first Bohr orbit is $-13.6\ \text{eV}$. If this electron were to be removed from the atom, it would take 13.6 eV of energy. But the energy necessary to remove an electron from an atom is called the *ionization energy,* and it was previously known that the ionization energy of hydrogen was indeed 13.6 eV. Thus, the Bohr model of the atom seems to be on the right track in its attempt to represent the hydrogen atom.

When the electron is in the first Bohr orbit, it is said to be in the *ground state.* When it is in a higher orbit, it is said to be in an *excited state.* The energy of the electron in an excited state is given by equation 32.35, and in conjunction with equation 32.36, we can also write it as

$$E_n = -\frac{E_1}{n^2} \tag{32.37}$$

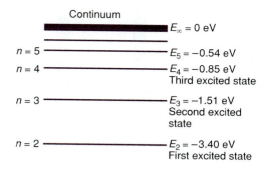

$E_n > 0$

Continuum

$E_\infty = 0$ eV

$n = 5$ ——————— $E_5 = -0.54$ eV
$n = 4$ ——————— $E_4 = -0.85$ eV
Third excited state

$n = 3$ ——————— $E_3 = -1.51$ eV
Second excited state

$n = 2$ ——————— $E_2 = -3.40$ eV
First excited state

$n = 1$ ——————— $E_1 = -13.6$ eV
Ground state

Figure 32.5

An energy-level diagram for the hydrogen atom.

with $E_1 = 13.6$ eV. These different energy levels for the different states of the electron are drawn, in what is called an *energy-level diagram,* in figure 32.5. Note that as *n* gets larger the energy states get closer together until the difference between one energy state and another is so small that there are no longer any observable quantization effects. The energy spectrum is then considered continuous just as it is in classical physics. For positive values of energy ($E > 0$) the electron is no longer bound to the atom and is free to go anywhere.

32.3 The Bohr Theory and Atomic Spectra

Whenever sufficient energy is added to an electron it jumps to an excited state. The electron only stays in that excited state for a very short time (10^{-8} s). *Bohr next postulated that when the electron jumps from its initial higher energy state, E_i, to a final lower energy state, E_f, a photon of light is emitted in accordance with Einstein's relation*

$$h\nu = E_i - E_f \qquad (32.38)$$

Using equation 32.37, we can write this as

$$h\nu = -\frac{E_1}{n_i^2} - \left(-\frac{E_1}{n_f^2}\right)$$

The frequency of the emitted photon is thus

$$\nu = \frac{E_1}{h}\left(\frac{1}{n_f^2} - \frac{1}{n_i^2}\right) \qquad (32.39)$$

The wavelength of the emitted photon, found from $\nu = c/\lambda$, is

$$\nu = \frac{c}{\lambda} = \frac{E_1}{h}\left(\frac{1}{n_f^2} - \frac{1}{n_i^2}\right)$$

or

$$\frac{1}{\lambda} = \frac{E_1}{hc}\left(\frac{1}{n_f^2} - \frac{1}{n_i^2}\right) \qquad (32.40)$$

Computing the value of E_1/hc gives

$$\frac{E_1}{hc} = \left[\frac{13.6 \text{ eV}}{(6.6262 \times 10^{-34} \text{ J s})(2.9979 \times 10^8 \text{ m/s})}\right]\left(\frac{1.6022 \times 10^{-19} \text{ J}}{1 \text{ eV}}\right)$$

$$= \left(1.0969 \times 10^7 \frac{1}{\text{m}}\right)\left(\frac{10^{-9} \text{ m}}{1 \text{ nm}}\right)$$

$$= 1.097 \times 10^{-2} \text{ (nm)}^{-1}$$

As seen in chapter 27, when white light is passed through a prism, it is broken up into a *continuous spectrum* of color from red through violet. On the other hand, if a gas such as hydrogen is placed in a tube under very low pressure and an electrical field is applied between two electrodes of the tube, the energy gained by the electrons from the field causes the electrons of the hydrogen atoms to jump to higher energy states. The gas glows with a characteristic color as the electrons fall back to the lower energy states. If the light from this spectral tube is passed through a prism or a diffraction grating, a line spectrum such as shown in figure 32.6 is found. That is, instead of the continuous spectrum of all the colors of the rainbow, only a few discrete colors are found with the wavelengths indicated. The discrete spectra of hydrogen were known as far back as 1885 when Johann Jakob Balmer (1825–1898), a Swiss mathematician and physicist, devised the mathematical formula

$$\lambda = (364.56 \text{ nm})\frac{n^2}{n^2 - 4}$$

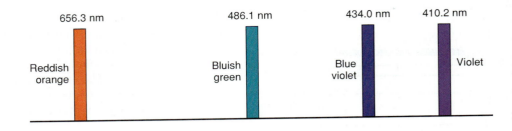

Figure 32.6

Line spectrum of hydrogen.

to describe the wavelength of the hydrogen spectrum. In 1896, the Swedish spectroscopist J. R. Rydberg (1853–1919) found the empirical formula

$$\frac{1}{\lambda} = R\left(\frac{1}{2^2} - \frac{1}{n^2}\right)$$ (32.41)

for $n = 3, 4, 5, \ldots$, and where the constant R, now called the *Rydberg constant*, was given by $R = 1.097 \times 10^{-2}$ (nm)$^{-1}$. Comparing equation 32.41 with 32.40, we see that the Rydberg constant R is equal to the quantity E_1/hc and, if $n_f = 2$, the two equations are identical. Equation 32.41 was a purely empirical result from experiment with no indication as to why the spectral lines should be ordered in this way, whereas equation 32.40 is a direct result of the Bohr model of the hydrogen atom.

Example 32.7

Spectral lines with the Bohr model. Using the Bohr model of the hydrogen atom, determine the wavelength of the spectral lines associated with the transitions from the (a) $n_i = 3$ to $n_f = 2$ state, (b) $n_i = 4$ to $n_f = 2$ state, (c) $n_i = 5$ to $n_f = 2$ state, and (d) $n_i = 6$ to $n_f = 2$ state.

Solution

The wavelength of the spectral line is found from equation 32.40.

a. $n_i = 3$; $n_f = 2$:

$$\frac{1}{\lambda} = \frac{E_1}{hc}\left(\frac{1}{n_f^2} - \frac{1}{n_i^2}\right)$$

$$= [1.097 \times 10^{-2}\ (nm)^{-1}]\left(\frac{1}{2^2} - \frac{1}{3^2}\right)$$

$$= 1.5236 \times 10^{-3}\ (nm)^{-1}$$

$$\lambda = 656.3\ nm$$

b. $n_i = 4$; $n_f = 2$:

$$\frac{1}{\lambda} = [1.097 \times 10^{-2}\ (nm)^{-1}]\left(\frac{1}{2^2} - \frac{1}{4^2}\right)$$

$$= 2.0569 \times 10^{-3}\ (nm)^{-1}$$

$$\lambda = 486.1\ nm$$

c. $n_i = 5$; $n_f = 2$:

$$\frac{1}{\lambda} = [1.097 \times 10^{-2}\ (nm)^{-1}]\left(\frac{1}{2^2} - \frac{1}{5^2}\right)$$

$$= 2.3037 \times 10^{-3}\ (nm)^{-1}$$

$$\lambda = 434.0\ nm$$

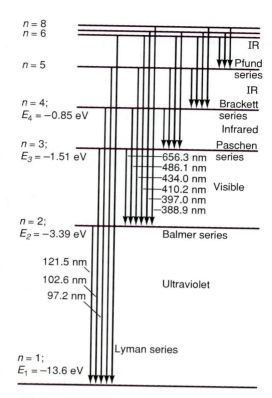

Figure 32.7

Associated spectrum for the energy-level diagram of hydrogen.

d. $n_i = 6$; $n_f = 2$:

$$\frac{1}{\lambda} = [1.097 \times 10^{-2} \text{ (nm)}^{-1}]\left(\frac{1}{2^2} - \frac{1}{6^2}\right)$$

$$= 2.4370 \times 10^{-3} \text{ (nm)}^{-1}$$

$$\lambda = 410.2 \text{ nm}$$

Thus, the Bohr theory of the hydrogen atom agrees with the experimental values of the wavelengths of the spectral lines shown in figure 32.6. In fact, the Bohr formula is more complete than that given by the Balmer formula. That is, the Balmer series is associated with transitions to the $n = 2$ state. According to the Bohr formula, there should be spectral lines associated with transitions to the $n_f = 1, 2, 3, 4, \ldots$, states. With such a prediction, it was not long before experimental physicists found these spectral lines. The reason they had not been observed before is because they were not in the visible portion of the spectrum. Lyman found the series associated with transitions to the ground state $n_f = 1$ in the ultraviolet portion of the spectrum. Paschen found the series associated with the transitions to the $n_f = 3$ state in the infrared portion of the spectrum. Brackett and Pfund found the series associated with the transitions to the $n_f = 4$ and $n_f = 5$ states, respectively, also in the infrared spectrum. The energy-level diagram and the associated spectral lines are shown in figure 32.7. Hence, the Bohr theory had great success in predicting the properties of the hydrogen atom.

Obviously, the Bohr theory of the atom was on the right track in explaining the nature and characteristics of the atom. However, as has been seen over and over again, physics arrives at a true picture of nature only through a series of successive approximations. The Bohr theory would be no different. As great as it was, it had its limitations. Why should the electron orbit be circular? The most general case would be elliptical. Arnold Sommerfeld (1868–1951) modified the Bohr theory to take into account elliptical orbits. The result increased the number of quantum numbers from one to two. With the advent of more refined spectroscopic equipment, it was found that some spectral lines actually consisted of two or more spectral lines. The Bohr theory could not account for this. Perhaps the greatest difficulty with the Bohr theory was its total inability to account for the spectrum of multielectron atoms. Also, some spectral lines were found to be more intense than others. Again, the Bohr theory could not explain why. Thus, the Bohr theory contains a great deal about the true nature of the atom, but it is not the complete picture. It is just a part of the successive approximations to a true picture of nature.

32.4 The Quantum Mechanical Model of the Hydrogen Atom

Obviously the Bohr theory of the hydrogen atom, although not wrong, was also not quite right. A new approach to the nature of the hydrogen atom was necessary. When de Broglie introduced his matter waves in 1924 to describe particles, it became necessary to develop a technique to find these matter waves mathematically. Erwin Schrödinger (1887–1961), an Austrian physicist, developed a new equation to describe these matter waves. This new equation is called the *Schrödinger wave equation*. (The Schrödinger wave equation is to quantum mechanics what Newton's second law is to classical mechanics. In fact, Newton's second law can be derived as a special case of the Schrödinger wave equation.) The solution of the wave equation is the wave function Ψ. For the **quantum mechanical model of the hydrogen atom,** the Schrödinger wave equation is applied to the hydrogen atom. It was found that it was necessary to have three quantum numbers to describe the electron in the hydrogen atom in this model,[1] whereas the Bohr theory had required only one. The

---

1. Actually the quantum mechanical model of the hydrogen atom requires four quantum numbers for its description. The fourth quantum number m_s, called the spin magnetic quantum number, is associated with the spin of the electron. The concept of the spin of the electron is introduced in section 32.7 and its effects are described there.

three quantum numbers are (1) the principal quantum number n, which is the same as that used in the Bohr theory; (2) the orbital quantum number l; and (3) the magnetic quantum number m_l. These quantum numbers are not completely independent; n can take on any value given by

$$n = 1, 2, 3, \ldots \tag{32.42}$$

Whereas l, the orbital quantum number, can only take on the values

$$l = 0, 1, 2, \ldots, (n - 1) \tag{32.43}$$

Thus, l is limited to values up to $n - 1$. The magnetic quantum number m_l can take on only the values given by

$$m_l = 0, \pm 1, \pm 2, \ldots, \pm l \tag{32.44}$$

Hence, m_l is limited to values up to $\pm l$. Let us now see a physical interpretation for each of these quantum numbers.

The Principal Quantum Number n

The principal quantum number n plays the same role in the quantum mechanical model of the hydrogen atom as it did in the Bohr theory in that it quantizes the possible energy of the electron in a particular orbit. The solution of the Schrödinger wave equation for the allowed energy values is

$$E_n = -\frac{k^2 e^4 m}{2\hbar^2} \frac{1}{n^2} \tag{32.45}$$

which we see has the same energy values as given by the Bohr theory in equation 32.35.

The Orbital Quantum Number l

In the Bohr theory, the angular momentum of the electron was quantized according to the relation $L = n\hbar$. The solution of the Schrödinger wave equation gives, for the angular momentum, the relation

$$L = \sqrt{l(l + 1)}\, \hbar \tag{32.46}$$

where $l = 0, 1, 2, \ldots n - 1$.

Example 32.8

The angular momentum of an electron in a quantum mechanical model of the atom. Determine the angular momentum of an electron in the hydrogen atom for the orbital quantum numbers of (a) $l = 0$, (b) $l = 1$, (c) $l = 2$, and (d) $l = 3$.

Solution

The angular momentum of the electron is quantized according to equation 32.46 as

$$L = \sqrt{l(l + 1)}\, \hbar$$

a. $l = 0$;

$$L = \sqrt{0(0 + 1)}\, \hbar$$
$$= 0$$

Thus for the $l = 0$ state, the angular momentum of the electron is zero. This is a very different case than anything found in classical physics. For an orbiting electron there must be some angular momentum, and yet for the $l = 0$ state, we get $L = 0$. Thus, the model of the atom with the electron orbiting the nucleus must now be considered questionable. We still speak of orbits, but they are apparently not the same simple concepts used in classical physics.

b. $l = 1; \quad L = \sqrt{1(1 + 1)}\ \hbar = \sqrt{2}\ \hbar = 1.414\hbar$

c. $l = 2; \quad L = \sqrt{2(2 + 1)}\ \hbar = \sqrt{6}\ \hbar = 2.449\hbar$

d. $l = 3; \quad L = \sqrt{3(3 + 1)}\ \hbar = \sqrt{12}\ \hbar = 3.464\hbar$

Note that in the Bohr theory the angular momentum was a whole number times $\hbar$. Here in the quantum mechanical treatment the angular momentum is no longer a whole multiple of $\hbar$.

The different angular momentum quantum states are usually designated in terms of the spectroscopic notation shown in table 32.1. The different states of the electron in the hydrogen atom are now described in terms of this spectroscopic notation in table 32.2. For the ground state, $n = 1$. However, because l can only take on values up to $n - 1$, l must be zero. Thus, the only state for $n = 1$ is the $1s$ or ground state of the electron. When $n = 2$, l can take on the values 0 and 1. Hence, there can be only a $2s$ and $2p$ state associated with $n = 2$. For $n = 3$, l can take on the values $l = 0, 1, 2$, and hence the electron can take on the states $3s$, $3p$, and $3d$. In this way, for various values of n, the states in table 32.2 are obtained.

The Magnetic Quantum Number m_l

Recall that angular momentum is a vector quantity and thus has a direction as well as a magnitude. We have just seen that the magnitude of the angular momentum

Table 32.1

Spectroscopic Notation for Angular Momentum

Orbital Quantum Number l	Angular Momentum	State	Spectroscopic Name
0	0	s	Sharp
1	$\sqrt{2}\ \hbar$	p	Principal
2	$\sqrt{6}\ \hbar$	d	Diffuse
3	$2\sqrt{3}\ \hbar$	f	Fundamental
4	$\sqrt{20}\ \hbar$	g	
5	$\sqrt{30}\ \hbar$	h	

Table 32.2

Atomic States in the Hydrogen Atom

	$l = 0$	$l = 1$	$l = 2$	$l = 3$	$l = 4$	$l = 5$
$n = 1$	$1s$					
$n = 2$	$2s$	$2p$				
$n = 3$	$3s$	$3p$	$3d$			
$n = 4$	$4s$	$4p$	$4d$	$4f$		
$n = 5$	$5s$	$5p$	$5d$	$5f$	$5g$	
$n = 6$	$6s$	$6p$	$6d$	$6f$	$6g$	$6h$

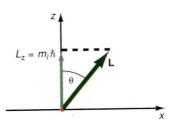

Figure 32.8

Direction of the angular momentum vector.

is quantized. The result of the Schrödinger equation applied to the hydrogen atom shows that the direction of the angular momentum vector must also be quantized. The magnetic quantum number m_l specifies the direction of $\mathbf{L}$ by requiring the z-component of $\mathbf{L}$ to be quantized according to the relation

$$L_z = m_l \hbar \qquad (32.47)$$

Quantization of the z-component of angular momentum specifies the direction of the angular momentum vector. As an example, let $\mathbf{L}$ be the angular momentum vector of the electron shown in figure 32.8. The z-component of $\mathbf{L}$ is found from the diagram to be

$$L_z = L \cos \theta \qquad (32.48)$$

Substituting the value of L_z from equation 32.48 and the value of L from equation 32.46, gives

$$m_l \hbar = \sqrt{l(l+1)} \, \hbar \cos \theta$$

Solving for the angle θ that determines the direction of $\mathbf{L}$, we get

$$\theta = \cos^{-1} \frac{m_l}{\sqrt{l(l+1)}} \qquad (32.49)$$

Thus, for particular values of the quantum numbers l and m_l, the angle θ specifying the direction of $\mathbf{L}$ is determined.

Example 32.9

The direction of the angular momentum vector. For an electron in the state determined by $n = 4$ and $l = 3$, determine the magnitude and the direction of the possible angular momentum vectors.

Solution

For $n = 4$ and $l = 3$, the electron is in the $4f$ state. The magnitude of the angular momentum vector, found from equation 32.46, is

$$L = \sqrt{l(l+1)} \, \hbar$$
$$= \sqrt{3(3+1)} \, \hbar$$
$$= 2\sqrt{3} \, \hbar$$

The possible values of m_l, found from equation 32.44, are

$$m_l = 0, \pm 1, \pm 2, \pm 3$$

The angle θ, that the angular momentum vector makes with the z-axis, found from equation 32.49, is

$$\theta = \cos^{-1} \frac{m_l}{\sqrt{l(l+1)}}$$

$m_l = 0;$ $\qquad \theta_0 = \cos^{-1} 0 = 90°$

$m_l = \pm 1;$ $\qquad \theta_1 = \cos^{-1} \dfrac{+1}{2\sqrt{3}} = \pm 73.2°$

$m_l = \pm 2;$ $\qquad \theta_2 = \cos^{-1} \dfrac{+2}{2\sqrt{3}} = \pm 54.7°$

$m_l = \pm 3;$ $\qquad \theta_3 = \cos^{-1} \dfrac{+3}{2\sqrt{3}} = \pm 30.0°$

The various orientations of the angular momentum vector are shown in figure 32.9.

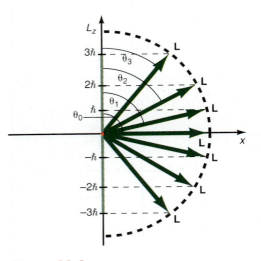

Figure 32.9

Quantization of the direction of the angular momentum vector.

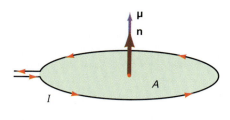

(a)

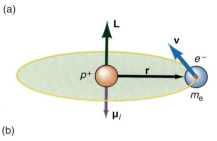

(b)

Figure 32.10

Orbital magnetic dipole moment.

32.5 The Magnetic Moment of the Hydrogen Atom

Using the picture of an atom as an electron orbiting about a nucleus, we see that the orbiting electron looks like the current loop already studied in chapter 22. Because a current loop has a magnetic dipole moment, it is logical to assume that the orbiting electron must also have a magnetic dipole moment associated with it. Figure 32.10(a) shows a current loop, while figure 32.10(b) shows the electron in its orbit. The usual notation for a magnetic dipole moment is the Greek letter μ (mu). In chapter 22, the letter m was used for the dipole moment so as not to confuse the dipole moment with the permeability constant, also denoted by the symbol μ. In this chapter, we will change the notation because permeabilities are not used here. Thus, μ now is the designation for the magnetic dipole moment. The magnetic dipole moment for the current loop in figure 32.10(a), with the aid of the equation 22.40, becomes

$$\mu = IA\mathbf{n}$$

Recall that I is the current in the loop, A is the area of the loop, and $\mathbf{n}$ is a unit vector that is normal to the current loop. The orbiting electron of figure 32.10(b) constitutes a current given by

$$I = \frac{-e}{T} \tag{32.50}$$

where $-e$ is the negative electronic charge and T is the time it takes for the electron to go once around its orbit. But the time to go once around its orbit is its period and, as seen previously, the period T is equal to the reciprocal of its frequency ν. That is, $T = 1/\nu$. Hence, equation 32.50 becomes

$$I = -e\nu \tag{32.51}$$

where ν is the number of times the electron circles its orbit in one second.

Thus, the orbiting electron looks like a current loop, with a current given by equation 32.51. We attribute to this current loop a magnetic dipole moment and call it the *orbital magnetic dipole moment,* designated by $\boldsymbol{\mu}_l$, and now given by

$$\boldsymbol{\mu}_l = IA\mathbf{n} = -e\nu A\mathbf{n}$$

Assuming the orbit to be circular, $A = \pi r^2$ and hence,

$$\boldsymbol{\mu}_l = -e\nu\pi r^2\mathbf{n} \tag{32.52}$$

Note that $\boldsymbol{\mu}_l$ is in the opposite direction of $\boldsymbol{\mu}$ because the electron is negative. We will return to this equation shortly.

The angular momentum of the electron is given by equation 31.48 as

$$L = rp \sin \theta$$

Since $p = m_e v$ we can write

$$L = rm_e v \sin \theta$$

where m_e is the mass of the electron. The vector $\mathbf{L}$ is opposite to the direction of the orbital magnetic dipole moment $\boldsymbol{\mu}_l$, as shown in figure 32.10(b). Because the orbital radius r is perpendicular to the orbital velocity $\mathbf{v}$, we can write the magnitude of the angular momentum of the electron as

$$L = rm_e v \sin 90° = rm_e v$$

Using the same unit vector $\mathbf{n}$ to show the direction perpendicular to the orbit in figure 32.10(a), we can express the angular momentum of the orbiting electron as

$$\mathbf{L} = m_e r v \mathbf{n} \tag{32.53}$$

The speed of the electron in its orbit is just the distance s it travels along its arc divided by the time, that is,

$$v = \frac{s}{T} = \frac{2\pi r}{T} = 2\pi \nu r$$

Substituting this into equation 32.53, yields

$$\mathbf{L} = m_e r(2\pi \nu r)\mathbf{n} = 2\pi m_e \nu r^2 \mathbf{n} \tag{32.54}$$

Dividing equation 32.54 by $2m_e$, we get

$$\frac{\mathbf{L}}{2m_e} = \pi \nu r^2 \mathbf{n} \tag{32.55}$$

Returning to equation 32.52 and dividing by e, we get

$$-\frac{\boldsymbol{\mu}_l}{e} = \pi \nu r^2 \mathbf{n} \tag{32.56}$$

Comparing the right-hand sides of equations 32.55 and 32.56, we see that they are identical and can therefore be equated to each other giving

$$-\frac{\boldsymbol{\mu}_l}{e} = \frac{\mathbf{L}}{2m_e}$$

Solving for $\boldsymbol{\mu}_l$, we have

$$\boldsymbol{\mu}_l = -\frac{e}{2m_e}\mathbf{L} \tag{32.57}$$

Equation 32.57 is the orbital magnetic dipole moment of the electron in the hydrogen atom. That is, the orbiting electron has a magnetic dipole associated with it, and, as seen from equation 32.57, it is related to the angular momentum $\mathbf{L}$ of the electron. The quantity $e/2m_e$ is sometimes called the *gyromagnetic ratio*.

The magnitude of the orbital magnetic dipole moment is found from equation 32.57, with the value of L determined from equation 32.46. Hence,

$$\mu_l = \frac{eL}{2m_e} = \frac{e}{2m_e}\sqrt{l(l+1)}\,\hbar$$

$$\mu_l = \frac{e\hbar}{2m_e}\sqrt{l(l+1)} \tag{32.58}$$

The quantity $e\hbar/2m_e$ is considered to be the smallest unit of magnetism, that is, an atomic magnet, and is called the *Bohr magneton*. Its value is

$$\frac{e\hbar}{2m_e} = \frac{(1.6021 \times 10^{-19} \text{ C})(1.0546 \times 10^{-34} \text{ J s})}{2(9.1091 \times 10^{-31} \text{ kg})}$$

$$= 9.274 \times 10^{-29} \text{ A m}^2 = 9.274 \times 10^{-24} \text{ J/T}$$

Example 32.10

The orbital magnetic dipole moment. Find the orbital magnetic dipole moment of an electron in the hydrogen atom when it is in (a) an s state, (b) a p state, and (c) a d state.

Solution

The magnitude of μ_l, found from equation 32.58, is

$$\mu_l = \frac{e\hbar}{2m_e}\sqrt{l(l+1)}$$

a. For an s state, $l = 0$,

$$\mu_l = 0$$

b. For a p state, $l = 1$,

$$\mu_l = \frac{e\hbar}{2m_e}\sqrt{1(1+1)}$$

$$= \left(9.274 \times 10^{-24}\frac{J}{T}\right)\sqrt{2}$$

$$= 1.31 \times 10^{-23} \text{ J/T}$$

c. For a d state, $l = 2$,

$$\mu_l = \frac{e\hbar}{2m_e}\sqrt{2(2+1)}$$

$$= \left(9.274 \times 10^{-24}\frac{J}{T}\right)\sqrt{6}$$

$$= 2.26 \times 10^{-23} \text{ J/T}$$

The Potential Energy of a Magnetic Dipole in an External Magnetic Field

In chapter 22 we saw that when a magnetic dipole μ is placed in an external magnetic field **B,** it experiences a torque given by

$$\tau = \mu B \sin \theta$$

This torque acts to rotate the dipole until it is aligned with the external magnetic field. Because the orbiting electron constitutes a magnetic dipole, if the hydrogen atom is placed in an external magnetic field, the orbital magnetic dipole of the atom rotates in the external field until it is aligned with it. Of course, since μ_l is 180° opposite to **L,** the angular momentum of the atom, aligning the dipole in the field is equivalent to aligning the angular momentum vector of the atom. (Actually **L** is antiparallel to **B.**)

Because the natural position of μ_l is parallel to the field, as shown in figure 32.11(a), work must be done to rotate μ_l in the external magnetic field. When work was done in lifting a rock in a gravitational field, the rock then possessed potential energy. In the same way, work done in rotating the dipole in the magnetic field shows up as potential energy of the dipole figure 32.11(b). That is, the electron now possesses an additional potential energy associated with the work done in rotating μ_l. It can be shown that the potential energy of the dipole in an external magnetic field **B** is

$$PE = -\mu_l B \cos \theta \qquad (32.59)$$

Example 32.11

The potential energy of an orbital magnetic dipole moment. Find the potential energy of the orbital magnetic dipole in an external field when (a) it is antiparallel to **B** (i.e., $\theta = 180°$), (b) it is perpendicular to **B** (i.e., $\theta = 90°$), and (c) it is aligned with **B** (i.e., $\theta = 0$).

Solution

The potential energy of the dipole, found from equation 32.59, is

a. $PE = -\mu_l B \cos 180°$
$\quad = +\mu_l B$

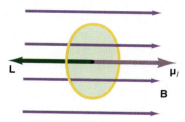

(a)

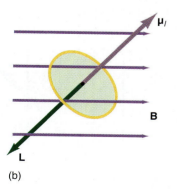

(b)

Figure 32.11

Orbital magnetic dipole in an external magnetic field **B.**

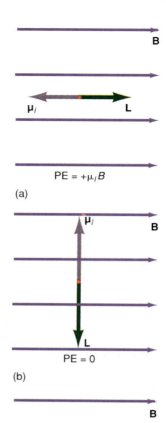

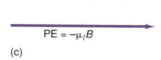

(a)

(b)

(c)

Figure 32.12

Potential energy of a dipole in an external magnetic field.

b. $PE = -\mu_l B \cos 90°$
$= 0$

c. $PE = -\mu_l B \cos 0°$
$= -\mu_l B$

Thus, the dipole has its highest potential energy when it is antiparallel (180°), decreases to zero when it is perpendicular (90°), and decreases to its lowest potential energy, a negative value, when it is aligned with the magnetic field, $\theta = 0°$. This is shown in figure 32.12. So, just as the rock falls from a position of high potential energy to the ground where it has its lowest potential energy, the dipole, if given a slight push to get it started, rotates from its highest potential energy (antiparallel) to its lowest potential energy (parallel).

32.6 The Zeeman Effect

The fact that there is a potential energy associated with a magnetic dipole placed in an external magnetic field has an important consequence on the energy of a particular atomic state, because the energy of a particular quantum state can change because of the acquired potential energy of the dipole. This acquired potential energy manifests itself as a splitting of a single energy state into multiple energy states, with a consequent splitting of the spectral lines associated with the transitions from these multiple energy states to lower energy states. The entire process is called the **Zeeman effect** after the Dutch physicist Pieter Zeeman (1865–1943) who first observed the splitting of spectral lines into several components when the atom was placed in an external magnetic field. Let us now analyze the phenomenon.

Let us begin by orienting an ordinary magnetic dipole, as shown in figure 32.13(a). A uniform magnetic field **B** is then turned on, as shown in figure 32.13(b). A torque acts on the dipole and the dipole becomes aligned with the field as expected. If the orbital magnetic dipole is oriented in the same way, figure 32.13(c), and then the magnetic field **B** is turned on, a torque acts to align $\boldsymbol{\mu}_l$. But the quantum conditions say that the angular momentum vector **L** can only be oriented such that its z-component L_z must be equal to $m_l \hbar$, equation 32.47. Hence, the dipole cannot rotate completely to align itself with **B,** but stops rotating at a position, such as in figure 32.13(d), where $L_z = m_l \hbar$. The orbital magnetic dipole has the potential energy given by equation 32.59 when stopped in this position.

This is strictly a quantum mechanical phenomenon, not found in classical physics. Its analogue in classical physics would be dropping a rock in the gravitational field, where the rock falls a certain distance and then comes to a stop some distance above the surface of the earth. This is an effect never observed classically.

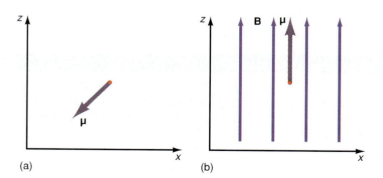

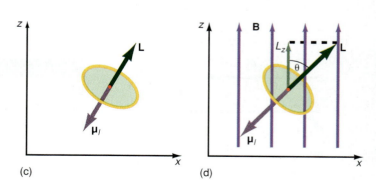

(a)   (b)   (c)   (d)

Figure 32.13

Orientation of the magnetic dipole in an external magnetic field.

is   The potential energy of the orbital magnetic dipole, given by equation 32.59, is

$$PE = -\mu_l B \cos \theta$$

But the orbital magnetic dipole moment was found in equation 32.57 as

$$\boldsymbol{\mu}_l = -\frac{e}{2m_e} \mathbf{L}$$

Substituting equation 32.57 into equation 32.59, gives

$$PE = +\frac{e}{2m_e} LB \cos \theta \qquad (32.60)$$

But,

$$LB \cos \theta = B(L \cos \theta) \qquad (32.61)$$

And, as seen from figure 32.13(d),

$$L \cos \theta = L_z$$

Substituting this into equation 32.60, gives

$$PE = \frac{e}{2m_e} L_z B$$

Finally, substituting for $L_z = m_l \hbar$, we get

$$PE = m_l \frac{e\hbar}{2m_e} B \qquad (32.62)$$

where $m_l = 0, \pm 1, \pm 2, \ldots, \pm l$. *Equation 32.62 represents the additional energy that an electron in the hydrogen atom can possess when it is placed in an external magnetic field. Hence, the energy of a particular atomic state depends on m_l as well as n.* Also, note that for s states, $l = 0$, and hence $m_l = 0$. Therefore, there is no potential energy for the dipole when it is in an s state.

Perhaps the best way to explain the Zeeman effect is by an example. Suppose an electron is in the $2p$ state, as shown in figure 32.14(a), with no applied external magnetic field. In the $2p$ state, the electron possesses an energy given by equation 32.45 as

$$E_2 = -\frac{k^2 e^4 m}{2\hbar^2} \frac{1}{2^2} = \frac{E_1}{4} \qquad (32.63)$$

When the electron drops to the $1s$ state it has the energy

$$E_1 = -\frac{k^2 e^4 m}{2\hbar^2} = -13.6 \text{ eV}$$

Figure 32.14
The Zeeman effect.

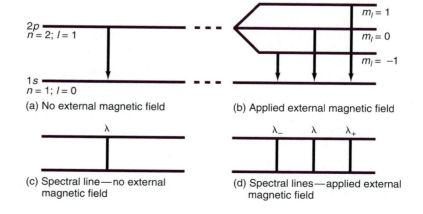

(a) No external magnetic field

(b) Applied external magnetic field

(c) Spectral line—no external magnetic field

(d) Spectral lines—applied external magnetic field

and has emitted a photon of energy

$$hv = E_2 - E_1 = \frac{E_1}{4} - E_1 = -\frac{3E_1}{4}$$

with a frequency of

$$v_0 = -\frac{3E_1}{4h} \tag{32.64}$$

and a wavelength of

$$\lambda = \frac{c}{v_0} = \frac{ch}{\frac{3}{4}E_1} \tag{32.65}$$

Example 32.12

An electron drops from the 2p state to the 1s state. Find (a) the energy of an electron in the 2p state, (b) the energy lost by the electron as it drops from the 2p state to the 1s state, (c) the frequency of the emitted photon, and (d) the wavelength of the emitted photon.

Solution

a. The energy of the electron in the 2p state, given by equation 32.63, is

$$E_2 = \frac{E_1}{4} = -\frac{13.6 \text{ eV}}{4} = -3.40 \text{ eV}$$

b. The energy lost by the electron when it drops from the 2p state to the 1s state is found from

$$\Delta E = E_2 - E_1 = -3.40 \text{ eV} - (-13.6 \text{ eV}) = 10.2 \text{ eV}$$

c. The frequency of the emitted photon, found from equation 32.64, is

$$v_0 = -\frac{3E_1}{4h}$$

$$= \left[\frac{-3(-13.6 \text{ eV})}{4(6.63 \times 10^{-34} \text{ J s})}\right]\left(\frac{1.60 \times 10^{-19} \text{ J}}{1 \text{ eV}}\right)$$

$$= 2.46 \times 10^{15} \text{ Hz}$$

d. The wavelength of the emitted photon, found from equation 32.65, is

$$\lambda_0 = \frac{c}{v_0} = \frac{3.00 \times 10^8 \text{ m/s}}{2.46 \times 10^{15} \text{ 1/s}} = 1.22 \times 10^{-7} \text{ m}$$

When the magnetic field is turned on, the electron in the 2p state acquires the potential energy of the dipole and now has the energy

$$E_2' = E_2 + \text{PE} = E_2 + m_l\frac{(e\hbar)B}{2m_e} \tag{32.66}$$

But for a p state, $l = 1$ and m_l then becomes equal to 1, 0, and −1. Thus, there are now three energy levels associated with E_2' because there are now three values of m_l. Therefore,

$$E_{2+}' = E_2 + \frac{(e\hbar)B}{2m_e} \tag{32.67}$$

$$E_2' = E_2 \tag{32.68}$$

$$E_{2-}' = E_2 - \frac{(e\hbar)B}{2m_e} \tag{32.69}$$

These energy states are shown in figure 32.14(b). Thus, the application of the magnetic field has split the single $2p$ state into 3 states. Since there are now three energy states, an electron can be in any one of them and hence, there are now three possible transitions to the ground state, where before there was only one. Corresponding to each of these three transitions are three spectral lines, as shown in figure 32.14(d). *Thus, the application of the magnetic field splits the single spectral line of figure 32.14(c) into the three spectral lines of figure 32.14(d).* The emitted photon associated with the transition from the $m_l = +1$ state is

$$h_+ = E'_{2+} - E_1 = E_2 + \frac{e\hbar B}{2m_e} - E_1$$

$$= \frac{E_1}{4} - E_1 + \frac{e\hbar B}{2m_e} = -\frac{3E_1}{4} + \frac{e\hbar B}{2m_e}$$

Using equation 32.64 the frequency of this spectral line becomes

$$\nu_+ = \nu_0 + \frac{e\hbar B}{2m_e h}$$

The wavelength of the spectral line is given by

$$\lambda_+ = \frac{c}{\nu_+} = \frac{c}{\nu_0 + e\hbar B/2m_e h} \tag{32.70}$$

Comparing equation 32.70 with equation 32.65, we see that λ_+ is slightly smaller than the original wavelength λ_0.

The transition from the $m_l = 0$ state is the same as the original transition from the $2p$ state because the electron has no potential energy associated with the magnetic dipole for $m_l = 0$. Thus, the spectral line is of the same wavelength λ_0 observed in the nonsplit spectral line.

The transition from the $m_l = -1$ state to the $1s$ state emits a photon of energy,

$$h\nu_- = E'_2 - E_1 = E_2 - \frac{e\hbar B}{2m_e} - E_1$$

$$= \frac{E_1}{4} - E_1 - \frac{e\hbar B}{2m_e} = -\frac{3E_1}{4} - \frac{e\hbar B}{2m_e}$$

Using equation 32.64, the frequency of this spectral line becomes

$$\nu_- = \nu_0 - \frac{e\hbar B}{2m_e h}$$

The wavelength of the spectral line is

$$\lambda_- = \frac{c}{\nu_-} = \frac{c}{\nu_0 - e\hbar B/2m_e h} \tag{32.71}$$

Comparing equation 32.71 with equation 32.65, we see that the wavelength λ_- is slightly larger than the original wavelength λ_0.

It turns out that all transitions from split states are not necessarily observed. Certain transitions are forbidden, and the allowed transitions are given by a set of selection rules on the allowed values of the quantum numbers. *Allowed transitions are possible only for changes in states where*

Selection
Rules
$$\Delta l = \pm 1$$
$$\Delta m_l = 0, \pm 1 \tag{32.72}$$

Note that these selection rules were obeyed in the preceding example.

The selection rule requiring l to change by ± 1 means that the emitted photon must carry away angular momentum equal to the difference between the angular momentum of the atom's initial and final states.

32.7 Electron Spin

The final correction to the model of the atom assumes that the electron is not quite a point charge, but its charge is distributed over a sphere. As early as 1921 A. H. Compton suggested that the electron might be a spinning particle. The Dutch-American physicists, Samuel Goudsmit (1902–) and George Uhlenbeck (1900–) inferred, in 1925, that the electron did spin about its own axis and because of this spin had an additional angular momentum, **S**, associated with this spin. Thus, this semiclassical model of the atom has an electron orbiting the nucleus, just as the earth orbits the sun, and the electron spinning on its own axis, just as the earth does about its axis. The model of the electron as a rotating charged sphere gives rise to an equivalent current loop and hence a magnetic dipole moment μ_s associated with this spinning electron.

Associated with the orbital angular momentum **L** was the orbital quantum number l. Similarly, associated with the spin angular momentum **S** is the spin quantum number s. Whereas l could take on the values $l = 0, 1, 2, \ldots , n - 1$, s can only take on the value

$$s = \tfrac{1}{2} \tag{32.73}$$

Similar to the magnitude of the orbital angular momentum given in equation 32.46, the magnitude of the spin angular momentum is given by

$$S = \sqrt{s(s + 1)}\, \hbar \tag{32.74}$$

Because s can only take on the value $\tfrac{1}{2}$, the magnitude of the spin angular momentum S can only be

$$S = \sqrt{\tfrac{1}{2}(\tfrac{1}{2} + 1)}\, \hbar = \tfrac{1}{2}\sqrt{3}\, \hbar \tag{32.75}$$

Just as the direction of the orbital angular momentum vector was quantized according to equation 32.47, the z-component of the spin angular momentum is quantized to

$$S_z = m_s \hbar \tag{32.76}$$

where m_s is called the spin magnetic quantum number. Just as the angular momentum vector **L** could have $2l + 1$ directions, that is, $m_l = 0, \pm 1, \pm l$, the spin angular momentum vector **S** can have $2s + 1$ directions, that is, $2s + 1 = 2(\tfrac{1}{2}) + 1 = 2$ directions specified by $m_s = \tfrac{1}{2}$ and $m_s = -\tfrac{1}{2}$. Thus, the z-component of the spin angular momentum can only be

$$S_z = \pm \frac{\hbar}{2} \tag{32.77}$$

The only two possible spin angular momentum orientations are shown in figure 32.15. When $m_s = +\tfrac{1}{2}$, the electron is usually designated as spin-up, while $m_s = -\tfrac{1}{2}$ is referred to as spin-down. The state of any electron in an atom is now specified by the four quantum numbers n, l, m_l, and m_s.

We should note that the semiclassical picture of the spinning electron is not quite correct. Using the picture of a spinning sphere, we can find its angular momentum about its own axis from the study of rotational motion in chapter 9 as

$$L = I\omega = \left(\frac{2}{5}mr^2\right)\left(\frac{v}{r}\right)$$

$$= \frac{2}{5}mrv$$

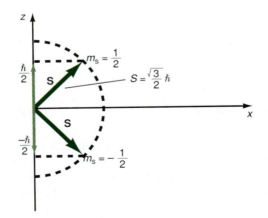

Figure 32.15

Orientations of the spin angular momentum vector.

where r is the radius of the electron, which is of the order of 10^{-15} m. Because the spin angular momentum has only the one value given by equation 32.75, for these angular momenta to be equal, $\mathbf{L} = \mathbf{S}$, or

$$\frac{2}{5}mrv = \frac{1}{2}\sqrt{3}\,\hbar$$

The speed of a point on the surface of the spinning electron would have to be

$$v = \frac{5\sqrt{3}\,\hbar}{4mr}$$

$$= \frac{5\sqrt{3}\,(1.0546 \times 10^{-34}\text{ J s})}{4(9.11 \times 10^{-31}\text{ kg})(1.00 \times 10^{-15}\text{ m})}$$

$$= 2.5 \times 10^{11}\text{ m/s}$$

But this is a velocity greater than the velocity of light, which cannot be. Hence, the classical picture of the charged rotating sphere cannot be correct. However, in 1928, Paul A. M. Dirac (1902–1984) joined together the special theory of relativity and quantum mechanics, and from this merger of the two theories found that the electron must indeed have an intrinsic angular momentum that is the same as that given by the semiclassical spin angular momentum. This angular momentum is purely a quantum mechanical effect and although of the same magnitude as the spin angular momentum, it has no classical analogue. However, because the value of the angular momentum is the same, it is still customary to speak of the spin of the electron.

Just as the orbital angular momentum has an orbital magnetic dipole moment $\boldsymbol{\mu}_l$ given by equation 32.57, the spin angular momentum has a spin magnetic dipole moment given by

$$\boldsymbol{\mu}_s = -2.0024\left(\frac{e}{2m_e}\right)\mathbf{S} \qquad (32.78)$$

and the spin magnetic dipole moment is shown in figure 32.16. In what follows, we will round off the value 2.0024 in equation 32.78 to the value 2.

When the orbital magnetic dipole $\boldsymbol{\mu}_l$ was placed in a magnetic field, a torque acted on $\boldsymbol{\mu}_l$ trying to align it with the magnetic field. The space quantization of $\mathbf{L}$ made it impossible for $\mathbf{L}$ to become aligned, and hence, the electron had the potential energy given by equation 32.62. In the same way, the spin magnetic dipole moment $\boldsymbol{\mu}_s$ should try to align itself in any magnetic field and because of the space quantization of the spin angular momentum, the electron should have the potential energy

$$PE = -\mu_s B \cos\theta \qquad (32.79)$$

Substituting equation 32.78 into equation 32.79, gives

$$PE = +2\frac{e}{2m_e}SB \cos\theta \qquad (32.80)$$

If there is an applied magnetic field $\mathbf{B}$, the electron acquires the additional potential energy given by equation 32.80. However, if there is no applied magnetic field, this potential energy term is still present, because from the frame of reference of the electron, the proton is in orbit about the electron, figures 32.17(a) and 32.17(b). The revolving proton constitutes a current loop and produces a magnetic field $\mathbf{B}_{so}$ at the location of the electron. This magnetic field interacts with the spin magnetic dipole moment $\boldsymbol{\mu}_s$ as given by equation 32.80. *The interaction of the spin magnetic dipole with the magnetic field $\mathbf{B}_{so}$ produced by the orbiting proton is called the spin-orbit interaction.*

We can now write equation 32.80 as

$$PE = 2\frac{e}{2m_e}SB_{so} \cos\theta$$

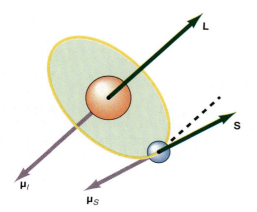

Figure 32.16

Orbital and spin angular momentum vectors and their associated dipole moments.

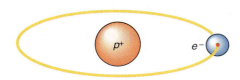

(a) Frame of reference of proton. Electron orbits about proton.

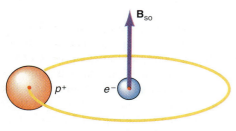

(b) Frame of reference of electron. Proton appears to orbit about electron.

Figure 32.17

The spin-orbit interaction.

But $S \cos \theta = S_z$, the z-component of the spin angular momentum vector. Hence,

$$PE = 2\frac{e}{2m_e}S_z B_{so}$$

But $S_z = \pm \hbar/2$ from equation 32.77, thus

$$PE = 2\frac{e}{2m_e}\frac{\pm \hbar}{2}B_{so}$$

or the acquired potential energy of an electron caused by the spin-orbit interaction is

$$PE = \pm \frac{e\hbar}{2m_e}B_{so} \qquad (32.81)$$

Hence, every quantum state, except s states, splits into two energy states, one corresponding to the electron with its spin-up

$$E_1 = E_0 + \frac{e\hbar}{2m_e}B_{so} \qquad (32.82)$$

and one corresponding to the electron with its spin-down

$$E_1 = E_0 - \frac{e\hbar}{2m_e}B_{so} \qquad (32.83)$$

The splitting of the energy state causes a splitting of the spectral line associated with each energy state into two component lines. *This spin-orbit splitting of spectral lines is sometimes called the fine structure of the spectral lines.* The variation in wavelength is quite small, of the order of 0.2 nm, which is, however, measurable.

There is no splitting of s states because for an s state, $l = 0$, which implies that there is no angular momentum. Thus, in the s state, the electron does not orbit about the proton in any classical sense. This implies that the proton cannot orbit about the electron and hence, cannot create the magnetic field, B_{so}, at the location of the electron. Therefore, if $B_{so} = 0$ in equation 32.81, then the potential energy term must also equal zero, and there can be no splitting of such a state.

32.8 The Pauli Exclusion Principle and the Periodic Table of the Elements

An electron in the hydrogen atom can now be completely specified by the four quantum members:

$n =$ the principal quantum number
$l =$ the orbital quantum number
$m_l =$ the orbital magnetic quantum number
$m_s =$ the spin magnetic quantum number

To obtain the remaining chemical elements a building up process occurs. That is, protons and neutrons are added to the nucleus, and electrons are added to the orbits to form the rest of the chemical elements. As an example, the chemical element helium is formed by adding one proton and two neutrons to the nucleus, and one orbital electron to give a total of two electrons, two protons, and two neutrons. The next chemical element, lithium, contains three protons, four neutrons, and three electrons. Beryllium has four protons, five neutrons, and four electrons. In this fashion of adding electrons, protons, and neutrons, the entire table of chemical elements can be generated. But where are these additional electrons located in the atom? Can they all be found in the same orbit? The answer is no, and was stated in the form of the Pauli exclusion principle by the Austrian physicist Wolfgang Pauli (1900–1958) in 1925. *The **Pauli exclusion principle** states that no two electrons in an*

Table 32.3

Atomic Shells

Quantum number n	1	2	3	4	5
Shell	K	L	M	N	O

atom can exist in the same quantum state. Because the state of any electron is specified by the quantum numbers n, l, m_l, and m_s, *the exclusion principle states that no two electrons can have the same set of the four quantum numbers.*

Electrons with the same value of n are said to be in the same orbital shell, and the shell designation is shown in table 32.3. Electrons that have the same value of l in a shell are said to occupy the same subshell. Electrons fill up a shell by starting at the lowest energy.

The building up of the chemical elements is shown in table 32.4. Thus, the first electron is found in the K shell with quantum numbers $(100\frac{1}{2})$ and this is the configuration of the hydrogen atom. When the next electron is placed in the atom,

Table 32.4

Electron States in Terms of the Quantum Numbers

		n	l	m_l	m_s		Numbers of States for Value of l	Total # of States for Value of n
$n = 1$	$l = 0$	1	0	0	1/2	1s↑	2	2
		1	0	0	−1/2	1s↓		
$n = 2$	$l = 0$	2	0	0	1/2	2s↑	2	8
		2	0	0	−1/2	2s↓		
	$l = 1$	2	1	1	1/2			
		2	1	1	−1/2			
		2	1	0	1/2		6	
		2	1	0	−1/2			
		2	1	−1	1/2			
		2	1	−1	−1/2			
$n = 3$	$l = 0$	3	0	0	1/2	3s↑	2	18
		3	0	0	−1/2	3s↓		
	$l = 1$	3	1	1	1/2			
		3	1	1	−1/2		6	
		3	1	0	1/2			
		3	1	0	−1/2			
		3	1	−1	1/2			
		3	1	−1	−1/2			
	$l = 2$	3	2	2	1/2			
		3	2	2	−1/2			
		3	2	1	1/2			
		3	2	1	−1/2		10	
		3	2	0	1/2			
		3	2	0	−1/2			
		3	2	−1	1/2			
		3	2	−1	−1/2			
		3	2	−2	1/2			
		3	2	−2	−1/2			

989

Table 32.5

Electron Configuration for the First Few Chemical Elements

Chemical Element	Electron Configuration
H	$1s^1$
He	$1s^2$
Li	$1s^2 2s^1$
Be	$1s^2 2s^2$
B	$1s^2 2s^2$ $2p^1$
C	$1s^2 2s^2$ $2p^2$
N	$1s^2 2s^2$ $2p^3$
O	$1s^2 2s^2$ $2p^4$
F	$1s^2 2s^2$ $2p^5$
Ne	$1s^2 2s^2$ $2p^6$
Na	$1s^2 2s^2$ $2p^6$ $3s^1$
Mg	$2p^6$ $3s^2$
Al	$3s^2$ $3p^1$
Si	$3s^2$ $3p^2$
P	$3s^2$ $3p^3$
S	$3s^2$ $3p^4$
Cl	$3s^2$ $3p^5$
Ar	$3s^2$ $3p^6$

it cannot have the quantum numbers of the first electron, so it must now be the electron given by the quantum numbers $(100-\frac{1}{2})$, that is, this second electron must have its spin-down. The atom with two electrons is the helium atom, and, as we now see, its two electrons are found in the K shell, one with spin-up the other with spin-down. The addition of the third electron cannot go into the K shell because all of the quantum numbers associated with $n = 1$ are already used up. Hence, a third electron must go into the $n = 2$ state or L shell. The rest of the table shows how the process of building up the set of quantum numbers continues. The notation $1s\uparrow$ means that the electron is in the $1s$ state with its spin-up. The notation $1s\downarrow$ means that the electron is in the $1s$ state with its spin-down.

The electron configuration is stated symbolically in the form

$$n(l)^{\#}$$

where n is the principal quantum number, l is the orbital quantum number expressed in the spectroscopic notation, and # stands for the number of electrons in that subshell. Hence, the electron configuration for the hydrogen atom would be $1s^1$, and for the helium atom, $1s^2$. The electron configuration for the first few chemical elements is shown in table 32.5. Note that the difference between one chemical element and the next is the addition of one more proton and electron.

The complete electron configurations for the ground states of all the chemical elements is shown somewhat differently in table 32.6. Thus, the entire set of chemical elements can be built-up in this way.

The way that an element reacts chemically depends on the number of electrons in the outer shell. Hence, all the chemical elements can be grouped into a table that shows how these elements react. Such a table, called the Periodic Table of the Elements and first formulated by the Russian chemist, Dmitri Mendeleev (1834–1907) around 1869, is shown in figure 32.18. Notice that there are vertical columns called groups and the chemical elements within each group have very similar properties. As an example, Group I contains elements that have only one electron in their

Period	Group I	Group II											Group III	Group IV	Group V	Group VI	Group VII	Group VIII
1	1 H 1.00																	2 He 4.00
2	3 Li 6.94	4 Be 9.01											5 B 10.81	6 C 12.00	7 N 14.01	8 O 15.99	9 F 19.00	10 Ne 20.18
3	11 Na 22.99	12 Mg 24.31											13 Al 26.98	14 Si 28.09	15 P 30.98	16 S 32.06	17 Cl 35.46	18 Ar 39.95
4	19 K 39.10	20 Ca 40.08	21 Se 44.96	22 Ti 47.90	23 V 50.94	24 Cr 52.00	25 Mn 54.94	26 Fe 55.85	27 Co 58.93	28 Ni 58.71	29 Cu 63.54	30 Zn 65.37	31 Ga 69.72	32 Ge 72.59	33 As 74.92	34 Se 78.96	35 Br 79.91	36 Kr 83.8
5	37 Rb 85.47	38 Sr 87.66	39 Y 88.91	40 Zr 91.22	41 Nb 92.91	42 Mo 95.94	43 Tc (98.91)	44 Ru 101.1	45 Rh 102.91	46 Pd 106.4	47 Ag 107.87	48 Cd 112.40	49 In 114.82	50 Sn 118.69	51 Sb 121.75	52 Te 127.60	53 I 126.90	54 Xe 131.30
6	55 Cs 132.91	56 Ba 137.34	57-71 *	72 Hf 178.49	73 Ta 180.95	74 W 183.85	75 Re 186.2	76 Os 190.2	77 Ir 192.2	78 Pt 195.09	79 Au 197.0	80 Hg 200.59	81 Tl 204.37	82 Pb 207.19	83 Bi 208.98	84 Po (210)	85 At (218)	86 Rn 222
7	87 Fr (223)	88 Ra 226.05	89-103 **															
	*Rare earths		57 La 138.91	58 Ce 140.12	59 Pr 140.91	60 Nd 144.24	61 Pm (145)	62 Sm 150.35	63 Eu 152.0	64 Gd 157.25	65 Tb 158.92	66 Dy 162.50	67 Ho 164.92	68 Er 167.26	69 Tm 168.93	70 Yb 173.04	71 Lu 174.97	
	**Actinides		89 Ac 227	90 Th 232.04	91 Pa 231	92 U 238.03	93 Np (237)	94 Pu (242)	95 Am (243)	96 Cm (247)	97 Bk (249)	98 Cf (251)	99 Es (254)	100 Fm (253)	101 Md (256)	102 No (254)	103 Lr (257)	

Figure 32.18

Periodic table of the elements.

Table 32.6
Electron Configurations for the Ground States of the Elements

		K	L		M			N				O				P			Q
		1s	2s	2p	3s	3p	3d	4s	4p	4d	4f	5s	5p	5d	5f	6s	6p	6d	7f
1	H	1																	
2	He	2																	
3	Li	2	1																
4	Be	2	2																
5	B	2	2	1															
6	C	2	2	2															
7	N	2	2	3															
8	O	2	2	4															
9	F	2	2	5															
10	Ne	2	2	6															
11	Na	2	2	6	1														
12	Mg	2	2	6	2														
13	Al	2	2	6	2	1													
14	Si	2	2	6	2	2													
15	P	2	2	6	2	3													
16	S	2	2	6	2	4													
17	Cl	2	2	6	2	5													
18	Ar	2	2	6	2	6													
19	K	2	2	6	2	6		1											
20	Ca	2	2	6	2	6		2											
21	Sc	2	2	6	2	6	1	2											
22	Ti	2	2	6	2	6	2	2											
23	V	2	2	6	2	6	3	2											
24	Cr	2	2	6	2	6	5	1											
25	Mn	2	2	6	2	6	5	2											
26	Fe	2	2	6	2	6	6	2											
27	Co	2	2	6	2	6	7	2											
28	Ni	2	2	6	2	6	8	2											
29	Cu	2	2	6	2	6	10	1											
30	Zn	2	2	6	2	6	10	2											
31	Ga	2	2	6	2	6	10	2	1										
32	Ge	2	2	6	2	6	10	2	2										
33	As	2	2	6	2	6	10	2	3										
34	Se	2	2	6	2	6	10	2	4										
35	Br	2	2	6	2	6	10	2	5										
36	Kr	2	2	6	2	6	10	2	6										
37	Rb	2	2	6	2	6	10	2	6			1							
38	Sr	2	2	6	2	6	10	2	6			2							
39	Y	2	2	6	2	6	10	2	6	1		2							
40	Zr	2	2	6	2	6	10	2	6	2		2							
41	Nb	2	2	6	2	6	10	2	6	4		1							
42	Mo	2	2	6	2	6	10	2	6	5		1							
43	Tc	2	2	6	2	6	10	2	6	5		2							
44	Ru	2	2	6	2	6	10	2	6	7		1							
45	Rh	2	2	6	2	6	10	2	6	8		1							
46	Pd	2	2	6	2	6	10	2	6	10									
47	Ag	2	2	6	2	6	10	2	6	10		1							
48	Cd	2	2	6	2	6	10	2	6	10		2							
49	In	2	2	6	2	6	10	2	6	10		2	1						
50	Sn	2	2	6	2	6	10	2	6	10		2	2						
51	Sb	2	2	6	2	6	10	2	6	10		2	3						
52	Te	2	2	6	2	6	10	2	6	10		2	4						
53	I	2	2	6	2	6	10	2	6	10		2	5						

Table 32.6
Continued

		K	L		M					N			O				P		Q
		1s	2s	2p	3s	3p	3d	4s	4p	4d	4f	5s	5p	5d	5f	6s	6p	6d	7f
54	Xe	2	2	6	2	6	10	2	6	10		2	6						
55	Cs	2	2	6	2	6	10	2	6	10		2	6			1			
56	Ba	2	2	6	2	6	10	2	6	10		2	6			2			
57	La	2	2	6	2	6	10	2	6	10		2	6	1		2			
58	Ce	2	2	6	2	6	10	2	6	10	2	2	6			2			
59	Pr	2	2	6	2	6	10	2	6	10	3	2	6			2			
60	Nd	2	2	6	2	6	10	2	6	10	4	2	6			2			
61	Pm	2	2	6	2	6	10	2	6	10	5	2	6			2			
62	Sm	2	2	6	2	6	10	2	6	10	6	2	6			2			
63	Eu	2	2	6	2	6	10	2	6	10	7	2	6			2			
64	Gd	2	2	6	2	6	10	2	6	10	7	2	6	1		2			
65	Tb	2	2	6	2	6	10	2	6	10	9	2	6			2			
66	Dy	2	2	6	2	6	10	2	6	10	10	2	6			2			
67	Ho	2	2	6	2	6	10	2	6	10	11	2	6			2			
68	Er	2	2	6	2	6	10	2	6	10	12	2	6			2			
69	Tm	2	2	6	2	6	10	2	6	10	13	2	6			2			
70	Yb	2	2	6	2	6	10	2	6	10	14	2	6			2			
71	Lu	2	2	6	2	6	10	2	6	10	14	2	6	1		2			
72	Hf	2	2	6	2	6	10	2	6	10	14	2	6	2		2			
73	Ta	2	2	6	2	6	10	2	6	10	14	2	6	3		2			
74	W	2	2	6	2	6	10	2	6	10	14	2	6	4		2			
75	Re	2	2	6	2	6	10	2	6	10	14	2	6	5		2			
76	Os	2	2	6	2	6	10	2	6	10	14	2	6	6		2			
77	Ir	2	2	6	2	6	10	2	6	10	14	2	6	7		2			
78	Pt	2	2	6	2	6	10	2	6	10	14	2	6	9		1			
79	Au	2	2	6	2	6	10	2	6	10	14	2	6	10		1			
80	Hg	2	2	6	2	6	10	2	6	10	14	2	6	10		2			
81	Tl	2	2	6	2	6	10	2	6	10	14	2	6	10		2	1		
82	Pb	2	2	6	2	6	10	2	6	10	14	2	6	10		2	2		
83	Bi	2	2	6	2	6	10	2	6	10	14	2	6	10		2	3		
84	Po	2	2	6	2	6	10	2	6	10	14	2	6	10		2	4		
85	At	2	2	6	2	6	10	2	6	10	14	2	6	10		2	5		
86	Rn	2	2	6	2	6	10	2	6	10	14	2	6	10		2	6		
87	Fr	2	2	6	2	6	10	2	6	10	14	2	6	10		2	6		1
88	Ra	2	2	6	2	6	10	2	6	10	14	2	6	10		2	6		2
89	Ac	2	2	6	2	6	10	2	6	10	14	2	6	10		2	6	1	2
90	Th	2	2	6	2	6	10	2	6	10	14	2	6	10		2	6	2	2
91	Pa	2	2	6	2	6	10	2	6	10	14	2	6	10	2	2	6	1	2
92	U	2	2	6	2	6	10	2	6	10	14	2	6	10	3	2	6	1	2
93	Np	2	2	6	2	6	10	2	6	10	14	2	6	10	4	2	6	1	2
94	Pu	2	2	6	2	6	10	2	6	10	14	2	6	10	5	2	6	1	2
95	Am	2	2	6	2	6	10	2	6	10	14	2	6	10	6	2	6	1	2
96	Cm	2	2	6	2	6	10	2	6	10	14	2	6	10	7	2	6	1	2
97	Bk	2	2	6	2	6	10	2	6	10	14	2	6	10	8	2	6	1	2
98	Cf	2	2	6	2	6	10	2	6	10	14	2	6	10	10	2	6		2
99	Es	2	2	6	2	6	10	2	6	10	14	2	6	10	11	2	6		2
100	Fm	2	2	6	2	6	10	2	6	10	14	2	6	10	12	2	6		2
101	Md	2	2	6	2	6	10	2	6	10	14	2	6	10	13	2	6		2
102	No	2	2	6	2	6	10	2	6	10	14	2	6	10	14	2	6		2
103	Lr	2	2	6	2	6	10	2	6	10	14	2	6	10	14	2	6	1	2
104	Rf																		
105	Ha																		

outermost shell and these chemicals react very strongly. The horizontal rows are called periods, and progressing from one column to another in a particular row, the chemical element contains one more electron. Thus in column I, there is one outer electron; in column II, there are two; in column III, there are three; and so on until we get to column VIII, where there are eight electrons in a closed shell. The chemical properties within a period change gradually as the additional electron is added. However, the first element of the period is very active chemically, whereas the last element of a period contains the inert gases. These gases are inert because the outer electron shell is closed and there is no affinity to either gain or lose electrons, and, hence, these elements do not react chemically with any of the other elements. Thus, there is a drastic chemical difference between the elements in Group I and Group VIII. The chemical properties of any element is a function of the number of electrons in the outer shell.

As mentioned earlier, an electron always falls into the state of lowest energy, and this can be seen in table 32.6. For the early elements, each lower quantum state is filled before a higher one starts. However, starting with the element potassium (K), a change occurs in the sequence of quantum numbers. Instead of the 19th electron going into a 3d state, it goes into the 4s state, as shown in table 32.6. The reason for this is that these elements of higher atomic number start to have an energy dependence on the quantum number l, because the higher orbits are partially shielded from the nuclear charge by the inner electrons of low values of l. Thus, as l increases, the energy of the state also increases. Hence, a 4s state is actually at a lower energy than a 3d state. Therefore, the 19th electron goes into the $4s^1$ state; the 20th electron goes into the $4s^2$ state; and the 21st electron goes into the 3d state, which is lower than the 4p state. This is seen in both table 32.6 and figure 32.18. Additional electrons now start to fill up the 3d shell as shown. The order in which electron subshells are filled in atoms is given by 1s, 2s, 2p, 3s, 3p, 4s, 3d, 4p, 5s, 4d, 5p, 6s, 4f, 5d, 6p, 7s, 6d. Table 32.7 is a reproduction of table 32.2 and it can be used to generate the sequence in which electrons fill the orbital subshells by following the diagonal lines traced on the table.

One of the characteristics of a closed shell is that the total orbital angular momentum **L** is zero and the total spin angular momentum **S** is also zero. To see this, let us consider the electrons in a closed 2p subshell. The angular momentum

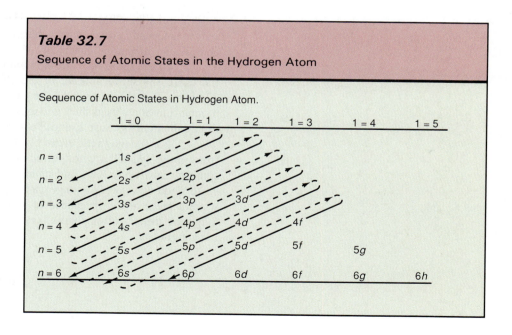

Table 32.7

Sequence of Atomic States in the Hydrogen Atom

Figure 32.19

The total angular momentum of a closed
shell is zero.

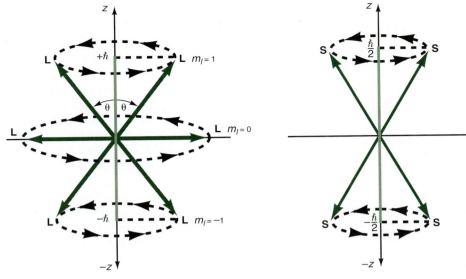

(a) Orbital angular momentum   (b) Spin angular momentum

vectors associated with the (211), (210), and (21–1) quantum numbers are shown
in figure 32.19(a). The angular momentum vector associated with the state (211)
should be fixed in its direction in space by the requirement that $L_z = \hbar$, as shown
in the figure. However, by the *Heisenberg uncertainty principle,* the direction of **L**
cannot be so precisely stated. Hence, the angular momentum vector can precess
around the z-axis as shown. Thus, the value of θ is fixed, but **L** precesses around z,
always at the same angle θ. Sometimes, **L** is toward the right, sometimes toward
the left, sometimes toward the back, and sometimes toward the front. Its mean position is, therefore, in the positive z-direction. The angular momentum vector associated with (21–1) precesses in the same way about the negative z-axis and its
mean position is in the negative z-direction. Since the magnitude of **L** is the same
for both vectors, the average or mean value of L for the two states adds up to zero.
The mean value of **L** for the (210) state is zero, because sometimes it is toward the
right and sometimes toward the left, and so on. Hence, for any closed subshell the
total angular momentum **L** is zero. Because the angular momentum vector for the
s state is already equal to zero, because $l = 0$, the orbital angular momentum of a
completely filled shell is zero.

    In the same way, the spin angular momentum **S** also adds up to zero when
there are the same number of electrons with spin-up as with spin-down, figure
32.19(b). But this is exactly the case of a closed shell, so the total spin angular
momentum of a closed shell is also zero. An atom with a closed shell also has a zero
dipole moment because **L** and **S** are both zero.

    For the magnetic elements iron (Fe), cobalt (Co), and nickel (Ni), the elec-
trons in the 3d shell are not paired off according to spin. Iron has five electrons with
spin-up, cobalt has four, and nickel has three. Thus, the spin angular momentum
vectors add up very easily to give a rather large spin magnetic dipole moment. Hence,
when a piece of iron is placed in an external magnetic field, all these very strong
magnetic dipoles align themselves with the field, thereby producing the ordinary
bar magnet.

An Essay on the Application of Physics

Is This World Real or Just an Illusion?

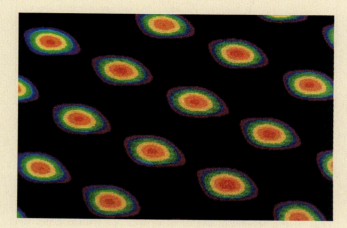

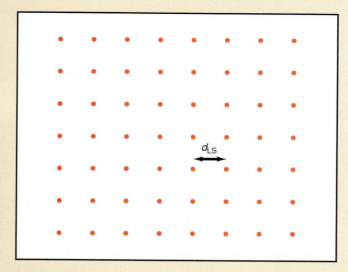

Figure 1

The lattice structure.

Have you ever wondered if this solid world that we see around us is really an illusion? Philosophers have argued this question for centuries. To see for ourselves all we have to do is slam our fist down on the table. Ouch! That table is real, I can tell because my hand hurts where I hit the table. That table is solid and is no illusion.

Let us look a little bit more carefully at the solid table. It certainly looks solid. If we were to take a very powerful microscope and look at the smooth table we would see that the table is made up of a lattice structure, the simplest lattice structure is shown in figure 1. Each one of those dots represents an atom of the material. They are arranged in a symmetric net. The distance separating each atom in the lattice structure is actually quite small, a typical distance is about 5.6×10^{-10} m. The diameter of an atom is about 1.1×10^{-10} m. Hence, the ratio of the separation between atoms in the lattice structure d_{LS} to the diameter of the atom d_A is

$$\frac{d_{LS}}{d_A} = \frac{5.6 \times 10^{-10} \text{ m}}{1.1 \times 10^{-10} \text{ m}} = 5.09 \qquad \textbf{(32H.1)}$$

or the distance separating the atoms in the lattice structure is

$$d_{LS} = 5.09 \, d_A \qquad \textbf{(32H.2)}$$

Equation 32H.2 says that each nearest atom is about 5.09 atomic diameters distant.

If we now look at the problem in three dimensions, the lattice structure looks like a box with the atom at each corner of the box, each separated by the distance d_{LS}. The box is called a *unit cell,* and is shown in figure 2. The volume of the box is given by

$$V_{box} = (d_{LS})^3$$
$$= (5.6 \times 10^{-10} \text{ m})^3 = 1.76 \times 10^{-28} \text{ m}^3 \qquad \textbf{(32H.3)}$$

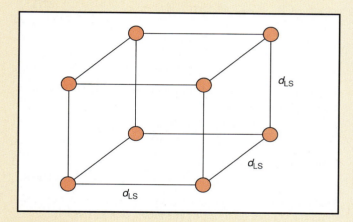

Figure 2

The unit cell.

But part of the volume of each atom is shared with the surrounding boxes. Figure 3 shows four of these boxes where they join. Consider the atom as the sphere located at the bottom corner of box 1 and protruding into boxes 2, 3, and 4. Each box shown contains $\frac{1}{4}$ of the volume of the sphere. There are four more boxes in front of the four boxes shown here. Hence, each box contains $\frac{1}{8}$ of the volume of the sphere. Therefore, each box contains the atomic volume of

$$\left(8 \, \frac{\text{atoms}}{\text{box}}\right)\left(\frac{1}{8}\right)\left(\frac{\text{atomic volume}}{\text{atom}}\right)$$
$$= \frac{1 \text{ atomic volume}}{\text{box}} \qquad \textbf{(32H.4)}$$

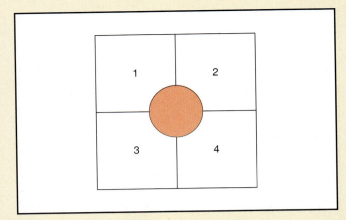

Figure 3

Determining the number of atoms in a unit cell.

That is, the unit cell or box contains the equivalent of one atom. The volume occupied by the atom in the box is

$$V_{atomic} = \frac{4}{3}\pi r^3 \qquad \text{(32H.5)}$$

$$= \frac{4}{3}\pi(0.55 \times 10^{-10} \text{ m})^3$$

$$= 6.95 \times 10^{-31} \text{ m}^3$$

The ratio of the volume of the box to the volume of the atom contained in the box is

$$\frac{V_{box}}{V_{atomic}} = \frac{1.76 \times 10^{-28} \text{ m}^3}{6.95 \times 10^{-31} \text{ m}^3} = 2.53 \times 10^2$$

Hence, the volume of the box is

$$V_{box} = 253 \, V_{atomic} \qquad \text{(32H.6)}$$

That is, the volume of one unit cell of the lattice structure is 253 times the volume occupied by the atom. Hence, the solid table that you see before you, constructed from that lattice structure, is made up of a great deal of empty space.

The atoms making up the lattice structure are also composed of almost all empty space. For example, the simplest atom, hydrogen, shown in figure 4, has a diameter d_A of about 1.1×10^{-10} m. The diameter of the nucleus is about 1×10^{-14} m. Because the mass of the electron is so small compared to the mass of the proton and neutron, the nucleus contains about 99.9% of the mass of the atom. The size of the electron is so small that its volume can be neglected compared to the volume of the nucleus and the atom. The ratio of the volume of the atom to the volume of the nucleus is about

$$\frac{V_A}{V_N} = \frac{\frac{4}{3}\pi r_A^3}{\frac{4}{3}\pi r_N^3} = \frac{r_A^3}{r_N^3}$$

$$= \frac{(0.55 \times 10^{-10} \text{ m})^3}{(0.5 \times 10^{-14} \text{ m})^3} = 1.1 \times 10^{12} \qquad \text{(32H.7)}$$

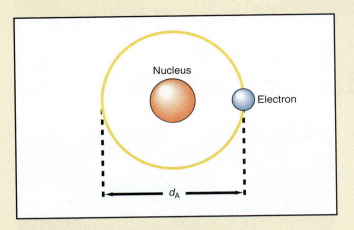

Figure 4

The size of an atom.

Hence, the volume of the atom is

$$V_A = 1.1 \times 10^{12} \, V_N = 1,100,000,000,000 \, V_N \qquad \text{(32H.8)}$$

That is, the volume of the atom is over one trillion times the volume of the nucleus. Hence, the atoms that make up that lattice structure are also composed of almost all empty space. When we combine the volume of the box (unit cell) with respect to the volume of each atom in the box, equation 32H.6, with the volume of the atom with respect to the nucleus, equation 32H.8, we get

$$V_{box} = 253 \, V_{atomic} = 253(1.1 \times 10^{12} \, V_N) = 278 \times 10^{12} \, V_N$$

or *the volume of the box (unit cell) is 278 trillion times the volume of the nucleus. Therefore, the solid is made up almost entirely of empty space.*

But if a solid consists almost entirely of empty space, then why can't you put your hand through the solid? You can't place your hand through the solid because there are electrical and atomic forces that hold the atom and lattice structure together, and your hand cannot penetrate that force field.

You can't put your hand through a block of ice either, but by heating the ice you give energy to the water molecules that make up the ice, and that energy is enough to pull the molecules away from the lattice structure, thereby melting the ice. You can now put your hand in the water, even though you could not put it through the ice. If you heat the water further, the water evaporates into the air and becomes invisible. You can walk through the air containing the water vapor as though it weren't even there.

So, is the world real or only an illusion? The world is certainly real because it is made up of all those atoms and molecules. But is it an illusion? In the sense described here, yes it is. But it is truly a magnificent illusion. For this solid world that we live in is composed almost entirely of empty space. It is a beautiful stage on which we all act out our lives.

The Language of Physics

Rutherford model of the atom
A planetary model of the atom wherein the negative electron orbits about the positive nucleus in a circular orbit. The orbiting electron is an accelerated charge and should radiate energy. As the electron radiates energy it loses energy and should spiral into the nucleus. Therefore, the Rutherford model of the atom is not correct (p. 966).

Bohr theory of the atom
A revised Rutherford model, wherein the electron can be found only in an orbit for which the angular momentum is quantized in multiples of $\hbar$. The consequence of the quantization postulate is that the electron can be considered as a standing matter wave in the electron orbit. Because standing waves do not transmit energy, the electron does not radiate energy while in its orbit and does not spiral into the nucleus. The Bohr model is thus stable. Bohr then postulated that when the electron jumps from a higher energy orbit to a lower energy orbit, a photon of light is emitted.

Thus, the spectral lines of the hydrogen atom should be discrete, agreeing with experimental results. However, the Bohr theory could not explain the spectra from multielectron atoms and it is not, therefore, a completely accurate model of the atom (p. 969).

Quantum mechanical model of the atom
This model arises from the application of the Schrödinger equation to the atom. The model says that the following four quantum numbers are necessary to describe the electron in the atom: (1) the principal quantum number n, which quantizes the energy of the electron; (2) the orbital quantum number l, which quantizes the magnitude of the orbital angular momentum of the electron; (3) the magnetic quantum number m_l, which quantizes the direction of the orbital angular momentum of the electron; and (4) the spin quantum number s, which quantizes the spin angular momentum of the electron (p. 975).

Zeeman effect
When an atom is placed in an external magnetic field a torque acts on the orbital magnetic dipole moment of the atom giving it a potential energy. The energy of the electron depends on the magnetic quantum number as well as the principal quantum number. For a particular value of n, there are multiple values of the energy. Hence, instead of a single spectral line associated with a transition from the nth state to the ground state, there are many spectral lines depending on the value of m_l. Thus, a single spectral line has been split into several spectral lines (p. 982).

Pauli exclusion principle
No two electrons in an atom can exist in the same quantum state. Hence, no two electrons can have the same quantum numbers (p. 988).

Summary of Important Equations

Atomic number
$$Z = \frac{A}{2} \tag{32.5}$$

Distance of closest approach to nucleus
$$r_0 = \frac{2kZe^2}{\text{KE}} \tag{32.7}$$

Relative size of atom
$$r_a = 10{,}000 \; r_n \tag{32.8}$$

Radius of nucleus
$$R = R_0 A^{1/3} \tag{32.9}$$

Bohr Theory of the Hydrogen Atom

Angular momentum is quantized
$$L = mvr = n\hbar \tag{32.25}$$

Orbital radius
$$r_n = \frac{\hbar^2}{kme^2} \, n^2 \tag{32.31}$$
$$r_n = r_1 n^2 \tag{32.32}$$

Orbital velocity
$$v_n = \frac{ke^2}{n\hbar} \tag{32.33}$$
$$v_n = \frac{v_1}{n} \tag{32.34}$$

Electron energy
$$E_n = -\frac{mk^2 e^4}{2n^2\hbar^2} \tag{32.35}$$
$$E_n = -\frac{E_1}{n^2} \tag{32.37}$$

Einstein's relation
$$h\nu = E_i - E_f \tag{32.38}$$

Frequency of emitted photon
$$\nu = \frac{E_1}{h}\left(\frac{1}{n_f^2} - \frac{1}{n_i^2}\right) \tag{32.39}$$

Wavelength of emitted photon
$$\frac{1}{\lambda} = \frac{E_1}{hc}\left(\frac{1}{n_f^2} - \frac{1}{n_i^2}\right) \tag{32.40}$$

Quantum Mechanical Theory of the Hydrogen Atom

Principal quantum number
$$n = 1, 2, 3, \ldots \tag{32.42}$$

Orbital quantum number
$$l = 0, 1, 2, \ldots, (n-1) \tag{32.43}$$

Magnetic quantum number
$$m_l = 0, \pm 1, \pm 2,$$
$$\ldots, \pm l \tag{32.44}$$

Electron energy
$$E_n = -\frac{k^2 e^4 m}{2\hbar^2}\frac{1}{n^2} \tag{32.45}$$

Angular momentum
$$L = \sqrt{l(l+1)}\,\hbar \tag{32.46}$$

z-component of angular momentum
$$L_z = m_l \hbar \tag{32.47}$$

Direction of **L**
$$\theta = \cos^{-1}\frac{m_l}{\sqrt{l(l+1)}} \tag{32.49}$$

Orbital magnetic dipole moment

$$\boldsymbol{\mu}_l = -\frac{e}{2m_e}\mathbf{L} \qquad (32.57)$$

$$\mu_l = \frac{eL}{2m_e}$$

$$= \frac{e}{2m_e}\sqrt{l(l+1)}\,\hbar \qquad (32.58)$$

Potential energy of a dipole in an external magnetic field

$$PE = -\mu_l B \cos\theta \qquad (32.59)$$

$$PE = m_l \frac{e\hbar}{2m_e}B \qquad (32.62)$$

Zeeman Effect

Splitting of energy state

$$E' = E + m_l \frac{(e\hbar)B}{2m_e} \qquad (32.66)$$

Splitting of spectral lines

$$\lambda_+ = \frac{c}{\nu_+}$$

$$\lambda_+ = \frac{c}{\nu_0 + e\hbar B/2m_e h} \qquad (32.70)$$

in an external

$$\lambda = \frac{c}{\nu_0} = \frac{ch}{\frac{3}{4}E_1} \qquad (32.65)$$

magnetic field

$$\lambda_- = \frac{c}{\nu_-}$$

$$\lambda_- = \frac{c}{\nu_0 - e\hbar B/2m_e h} \qquad (32.71)$$

Selection rules for transitions

$$\Delta l = \pm 1$$

$$\Delta m_l = 0, \pm 1 \qquad (32.72)$$

Spin quantum number

$$s = \tfrac{1}{2} \qquad (32.73)$$

Spin angular momentum

$$S = \sqrt{s(s+1)}\,\hbar \qquad (32.74)$$

$$S = \tfrac{1}{2}\sqrt{3}\,\hbar \qquad (32.75)$$

z-component of spin

$$S_z = m_s \hbar \qquad (32.76)$$

$$S_z = \pm\frac{\hbar}{2} \qquad (32.77)$$

Spin magnetic dipole moment

$$\boldsymbol{\mu}_s = -2.0024\left(\frac{e}{2m_e}\right)\mathbf{S} \qquad (32.78)$$

Potential energy

$$PE = -\mu_s B \cos\theta \qquad (32.79)$$

of electron

$$PE = +2\frac{e}{2m_e}SB \cos\theta \qquad (32.80)$$

due to spin

$$PE = \pm\frac{e\hbar}{2m_e}B_{so} \qquad (32.81)$$

Spin-orbit splitting

$$E_1 = E_0 + \frac{e\hbar}{2m_e}B_{so} \qquad (32.82)$$

of energy state

$$E_1 = E_0 - \frac{e\hbar}{2m_e}B_{so} \qquad (32.83)$$

Questions for Chapter 32

1. Discuss the differences among (a) the plum pudding model of the atom, (b) the Rutherford model of the atom, (c) the Bohr theory of the atom, and (d) the quantum mechanical theory of the atom.

†2. Discuss the effect of the uncertainty principle and the Bohr theory of electron orbits.

†3. How can you use spectral lines to determine the chemical composition of a substance?

4. If you send white light through a prism and then send it through a tube of hot hydrogen gas, what would you expect the spectrum to look like when it emerges from the hydrogen gas?

†5. In most chemical reactions, why are the outer electrons the ones that get involved in the reaction? Is it possible to get the inner electrons of an atom involved?

6. When an atom emits a photon of light what does this do to the angular momentum of the atom?

7. Explain how the Bohr theory can be used to explain the spectra from singly ionized atoms.

†8. Discuss the process of absorption of light by matter in terms of the atomic structure of the absorbing medium.

†9. Rutherford used the principle of scattering to "see" inside the atom. Is it possible to use the principle of scattering to "see" inside a proton and a neutron?

†10. How can you determine the chemical composition of a star?

†11. How can you determine if a star is approaching or receding from you? How can you determine if it is a small star or a very massive star?

Problems for Chapter 32

Section 32.1 The History of the Atom

1. Find the potential energy of two α particles when they are brought together to a distance of 1.20×10^{-15} m, the approximate size of a nucleus.
2. Find the potential energy of an electron and a proton when they are brought together to a distance of 5.29×10^{-11} m to form a hydrogen atom.
3. A silver nucleus is bombarded with 8-MeV α particles. Find (a) the maximum radius of the silver nucleus and (b) the more probable radius of the silver nucleus.
4. Estimate the radius of a nucleus of $^{238}_{92}$U.
5. How many electrons, protons, and neutrons are there in a silver atom?
6. How many electrons, protons, and neutrons are there in a uranium atom?
7. Find the difference in the orbital radius of an electron in the Rutherford atom if the electron is initially in an orbit of 5.29×10^{-11} m radius and the atom radiates 2.00 eV of energy.

Section 32.2 The Bohr Theory of the Atom

†8. An electron is in the third Bohr orbit. Find (a) the radius, (b) the speed, (c) the energy, and (d) the angular momentum of the electron in this orbit.
9. The orbital electron of a hydrogen atom moves with a speed of 5.459×10^5 m/s. (a) Determine the value of the quantum number n associated with this electron. (b) Find the radius of this orbit. (c) Find the energy of the electron in this orbit.

†10. An electron in the third Bohr orbit drops to the ground state. Find the angular momentum of the electron in (a) the third Bohr orbit, and (b) the ground state. (c) Find the change in the angular momentum of the electron. (d) Where did the angular momentum go?
11. At what temperature will the average thermal speed of a free electron equal the speed of an electron in its second Bohr orbit?
12. The lifetime of an electron in an excited state of an atom is about 10^{-8} s. How many orbits will an electron in the 2p state execute before falling back to the ground state?
13. Show that the ratio of the speed of an electron in the first Bohr orbit to the speed of light is equal to 1/137. This ratio is called the *fine-structure constant*.
14. Find the radius of the first Bohr orbit of an electron in a singly ionized helium atom.
15. Find the angular momentum of an electron in the third Bohr orbit and the second Bohr orbit. How much angular momentum is lost when the electron drops from the third orbit to the second orbit?

Section 32.3 The Bohr Theory and Atomic Spectra

16. An electron in the third Bohr orbit drops to the second Bohr orbit. Find (a) the energy of the photon emitted, (b) its frequency, and (c) its wavelength.
17. An electron in the third Bohr orbit drops to the ground state. Find (a) the energy of the photon emitted, (b) its frequency, and (c) its wavelength.
18. Calculate the wavelength of the first two lines of the Paschen series.

Section 32.4 The Quantum Mechanical Model of the Hydrogen Atom

19. Find the angular momentum of the electron in the quantum mechanical model of the hydrogen atom when it is in the 2p state. How much angular momentum is lost when the electron drops to the 1s state?

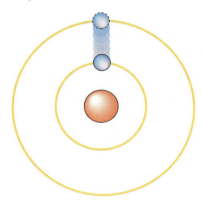

20. Find the angular momentum of the electron in the quantum mechanical model of the hydrogen atom when it is in the 3d state. (a) How much angular momentum is lost when the electron drops to the 1s state? (b) How much energy is lost when the electron drops to the 1s state?
21. An orbital electron is in the 5d state. (a) Find the energy of the electron in this state. (b) Find the orbital angular momentum of the electron. (c) Compute all possible values of the z-component of the orbital angular momentum of the electron. (d) Determine the largest value of θ, the angle between the orbital angular momentum vector and the z-axis.

22. Find the angle θ that the angular momentum vector makes with the z-axis when the electron is in the $3d$ state.

23. Find (a) the z-component of the angular momentum of an electron when it is in the $2p$ state and (b) the angle that **L** makes with the z-axis.

Section 32.5 The Magnetic Moment of the Hydrogen Atom

24. Find the orbital magnetic dipole moment of an electron in the hydrogen atom when it is in the $4f$ state.

25. Find the torque acting on the magnetic dipole of an electron in the hydrogen atom when it is in the $3d$ state and is in an external magnetic field of 2.50×10^{-3} T.

26. Find the additional potential energy of an electron in a $2p$ state when the atom is placed in an external magnetic field of 2.00 T.

†27. Find (a) the total energy of an electron in the three $2p$ states when it is placed in an external magnetic field of 2.00 T, (b) the energy of the photons emitted when the electrons fall to the ground state, (c) the frequencies of the spectral lines associated with these transitions, and (d) the wavelengths of their spectral lines.

Section 32.7 Electron Spin

28. Calculate the magnitude of the spin magnetic dipole moment of the electron in a hydrogen atom.

29. Find the potential energy associated with the spin magnetic dipole moment of the electron in a hydrogen atom when it is placed in an external magnetic field of 2.50×10^{-3} T.

Section 32.8 The Pauli Exclusion Principle and the Periodic Table of the Elements

30. How many electrons are necessary to fill the N shell of an atom?

31. Write the electron configuration for the chemical element potassium.

32. Write the electron configuration for the chemical element iron.

33. Enumerate the quantum states (n, l, m_l, m_s) of each of the orbital electrons in the element $^{40}_{20}\text{Ca}$.

34. Find the velocity of an electron in a 5.29×10^{-11} m radius orbit by (a) the Rutherford model and (b) the Bohr model of the hydrogen atom.

35. In what quantum state must an orbital electron be such that its orbital angular momentum is 4.719×10^{-34} J s (i.e., find l, the orbital angular momentum quantum number).

†36. Determine the angular momentum of the moon about the earth. (a) Use Bohr's postulate of quantization of angular momentum and determine the quantum number associated with this orbit. (b) If the quantum number n increases by 1, what is the new angular momentum of the moon? (c) What is the change in the orbital radius of the moon for this change in the quantum number? (d) Is it reasonable to neglect quantization of angular momentum for classical orbits?

37. From the frame of reference of the electron in the hydrogen atom, the proton is in an orbit about the electron and constitutes a current loop. Determine the magnitude of the magnetic field produced by the proton when the electron is in the $2p$ state.

†38. Using the results of problem 37, (a) determine the additional potential energy of an electron caused by the spin-orbit interaction. (b) Find the change in energy when electrons drop from the $2p$ state back to the ground state. (c) Find the frequencies of the emitted photons. (d) Find the wavelengths of the emitted photons.

⌨ 39. Bohr theory of the atom. An electron is in a Bohr orbit with a principal quantum number $n_i = 3$, and then jumps to a final orbit for the final value $n_f = 1$, find (a) the radius of the nth orbit, (b) the speed of the electron in the nth orbit, (c) the energy of the electron in the initial n_i orbit, (d) the energy of the electron in the final n_f orbit, (e) the energy given up by the electron as it jumps to the lower orbit, (f) the frequency, and (g) the wavelength of the spectral line associated with the transition from the initial $n_i = 3$ state to the final $n_f = 1$ state.

33 | **N**uclear Physics

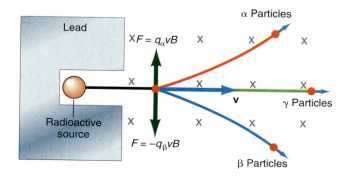

Nuclear magnetic resonance (NMR) image of a human head.

Some 15 years ago the radiation of uranium was discovered by Henri Becquerel and two years later the study of this phenomenon was extended to toher substances, first by me, and then by Pierre Curie and myself. This study rapidly led us to the discovery of new elements, the radiation of which, while being analogous with that of uranium, was far more intense. All the elements emitting such radiation I have termed rardioactive, and the new property of matter revealed in this emission has thus received the name radioactivity.

Marie Curie, 1911

33.1 Introduction

In 1896, Henri Becquerel (1852–1908) found that an ore containing uranium emits an invisible radiation that can penetrate paper and expose a photographic plate. After Becquerel's discovery, Marie (1867–1934) and Pierre (1859–1906) Curie discovered two new radioactive elements that they called polonium and radium. The Curies performed many experiments on these new elements and found that their radioactivity was unaffected by any physical or chemical process. As seen in chapter 32, chemical effects are caused by the interaction with atomic electrons. The reason for the lack of chemical changes affecting the radioactivity implied that radioactivity has nothing to do with the orbital electrons. Hence, the radioactivity must come from within the nucleus.

Rutherford investigated this invisible radiation from the atomic nucleus by letting it move in a magnetic field that is perpendicular to the paper, as shown in figure 33.1. Some of the particles were bent upward, some downward, while others went straight through the magnetic field without being bent at all. The particles that were bent upward were called alpha particles, α; those bent downward, beta particles, β; and those that were not deviated, gamma particles, γ.

We saw in chapter 22, equation 22.1, that the magnitude of the force that acts on a particle of charge q moving at a velocity $\mathbf{v}$ in a magnetic field $\mathbf{B}$ is given by

$$F = qvB \sin \theta \qquad (22.1)$$

(Recall from chapter 22, that the direction of the magnetic force is found by the right-hand rule. Place your right hand in the direction of the velocity vector $\mathbf{v}$. On rotating your hand toward the magnetic field $\mathbf{B}$, your thumb points in the direction

Figure 33.1

Radioactive particles.

of the force **F** acting on the particle.) If the charge of the α particle is positive, then the force acts upward in figure 33.1, and the α particle should be deflected upward. Because the α particle is observed to move upward, its charge must indeed be positive. Its magnitude was found to be twice that of the electronic charge. (Later the α particle was found to be the nucleus of the helium atom.)

Because the β particle is deflected downward in the magnetic field, it must have a negative charge. (The β particles were found to be high energy electrons.) The fact that the γ particle was not deflected in the magnetic field indicated that the γ particle contained no electric charge. The γ particles have since been found to be very energetic photons.

The energies of the α, β, and γ particles are of the order of 0.1 MeV up to 10 MeV, whereas energies of the orbital electrons are of the order of electron volts. Also, the α particles were found to be barely able to penetrate a piece of paper, whereas β particles could penetrate a few millimeters of aluminum, and the γ rays could penetrate several centimeters of lead. Hence, these high energies were further evidence to support the idea that these energetic particles must be coming from the nucleus itself.

33.2 Nuclear Structure

After quantum mechanics successfully explained the properties of the atom, the next questions asked were, What is the nature and structure of the nucleus? How are the protons and neutrons arranged in the nucleus? Why doesn't the nucleus blow itself apart by the repulsive force of the protons? If β particles that come out of a nucleus are electrons, are there electrons in the nucleus? We will discuss these questions shortly.

As seen in chapter 31, the nucleus is composed of protons and neutrons. *These protons and neutrons are collectively called* nucleons. *The number of protons in the nucleus is given by the* **atomic number Z,** *whereas the* **mass number A** *is equal to the number of protons plus neutrons in the nucleus. The number of neutrons in a nucleus is given by the* **neutron number N,** *which is just the difference between the mass number and the atomic number, that is,*

$$N = A - Z \tag{33.1}$$

A nucleus is represented symbolically in the form

$$^A_Z X \tag{33.2}$$

with the mass number A displayed as a superscript and the atomic number Z displayed as a subscript and where X is the nucleus of the chemical element that is given by the atomic number Z. As an example, the notation

$$^{12}_6 C$$

represents the nucleus of the carbon atom that has an atomic number of 6 indicating that it has 6 protons, while the 12 is the mass number indicating that there are 12 nucleons in the nucleus. The number of neutrons, given by equation 33.1, is

$$N = A - Z$$
$$= 12 - 6$$
$$= 6$$

Every chemical element is found to have isotopes. *An* **isotope** *of a chemical element has the same number of protons as the element but a different number of neutrons than the element.* Hence, an isotope of a chemical element has the same atomic number Z but a different mass number A and a different neutron number N. Since the chemical properties of an element are determined by the number of orbiting electrons, an isotope also has the same number of electrons and hence reacts chemically in the same way as the parent element. Its only observable difference chemically is its different atomic mass, which comes from the excess or deficiency of neutrons in the nucleus.

An example of an isotope is the carbon isotope

$$^{14}_{6}C$$

which has the same 6 protons as the parent element but now has 14 nucleons, indicating that there are now $14 - 6 = 8$ neutrons. The simplest element, hydrogen, has two isotopes, so there are three types of hydrogen:

$^{1}_{1}H$—Normal hydrogen contains 1 proton and 0 neutrons
$^{2}_{1}H$—Deuterium contains 1 proton and 1 neutron
$^{3}_{1}H$—Tritium contains 1 proton and 2 neutrons

Most elements have two or more stable isotopes. Hence, any chemical sample usually contains isotopes. The **atomic mass** of an element is really an average of the masses of the different isotopes. The abundance of isotopes of a particular element is usually quite small. For example, deuterium has an abundance of only 0.015%. Hence, the actual atomic mass is very close to the mass number A. There are a few exceptions to this, however, one being the chemical element chlorine. As seen from the table of the elements, the atomic mass of chlorine is 35.5, rounded to three significant figures. Contained in that chlorine sample is $^{35}_{17}Cl$ and $^{37}_{17}Cl$. The abundance of $^{35}_{17}Cl$ is 75.5%, whereas the abundance of $^{37}_{17}Cl$ is 24.5%. The atomic mass of chlorine is the average of these two forms of chlorine, weighted by the amount of each present in a sample. Thus, the atomic mass of chlorine is

$$\text{Atomic mass} = 35(0.755) + 37(0.245) = 35.5$$

In general, the atomic mass of any element is

$$\text{Atomic mass} = A_1(\% \text{ Abundance}) + A_2(\% \text{ Abundance}) + A_3(\% \text{ Abundance}) + \cdots$$

where A_1, A_2, and A_3, is the mass number of a particular isotope.

In chapters 29 and 32, the masses of the proton and neutron are given as

$$m_p = 1.6726 \times 10^{-27} \text{ kg} = 1.00726 \text{ u} = 938.256 \text{ MeV}$$

$$m_n = 1.6749 \times 10^{-27} \text{ kg} = 1.00865 \text{ u} = 939.550 \text{ MeV}$$

The protons in a nucleus are charged positively, thus Coulomb's law mandates a force of repulsion between these protons and the nucleus should blow itself apart. Because the nucleus does not blow itself apart, we conclude that there must be another force within the nucleus holding these protons together. *This nuclear force is called the* **strong nuclear force** *or the strong interaction. The strong force acts not only on protons but also on neutrons and is thus the force that binds the nucleus together.* The strong force has a very short range. That is, it acts within a distance of approximately 10^{-14} m, the order of the size of the nucleus. Outside the nucleus, there is no trace whatsoever of this force. The strong nuclear force is the strongest force known.

If we plot the number of neutrons in a nucleus N against the number of protons in that same nucleus Z for several nuclei, we obtain a graph similar to the one in figure 33.2. For light nuclei the number of neutrons is approximately equal to the number of protons, as seen by the line labeled $N = Z$. As the atomic number Z increases there are more neutrons in the nucleus than there are protons. Recall that the electrostatic repulsive force acts only between the protons, while the strong nuclear force of attraction acts between the protons and the neutrons. Hence, the additional neutrons increase the attractive force without increasing the repulsive electric force and, thereby, add to the stability of the nucleus. Whenever the nuclear force of attraction is greater than the electrostatic force of repulsion, the nucleus is

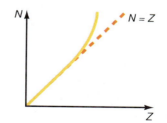

Figure 33.2

Plot of N versus Z for atomic nuclei.

stable. *Whenever the nuclear force is less than the electrostatic force, the nucleus breaks up or decays, and emits radioactive particles.* The chemical elements with atomic numbers Z greater than 83 have unstable nuclei and decay.

The internal structure of the nucleus is determined in much the same way as in Rutherford scattering. The nucleus is bombarded by high energy electrons (several hundred mega electron volts), that penetrate the nucleus and react electrically with the protons within the nucleus. The results of such scattering experiments seem to indicate that the protons and neutrons are distributed rather evenly throughout the nucleus, and the nucleus itself is generally spherical or ellipsoidal in shape.

It is usually assumed that the whole is always equal to the sum of its parts. This is not so in the nucleus. *The results of experiments on the masses of different nuclei shows that the mass of the nucleus is always less than the total mass of all the protons and neutrons making up the nucleus.* In the nucleus, the missing mass is called the **mass defect,** Δm, given by

$$\Delta m = Zm_p + (A - Z)m_n - m_{nucleus} \tag{33.3}$$

Because Z is the total number of protons, and m_p is the mass of a proton, Zm_p is the total mass of all the protons. As shown in equation 33.1, $A - Z$ is the total number of neutrons, and since m_n is the mass of a single neutron, $(A - Z)m_n$ is the total mass of all the neutrons. The term $m_{nucleus}$ is the experimentally measured mass of the entire nucleus. Hence, equation 33.3 represents the difference in mass between the sum of the masses of its constituents and the mass of the nucleus itself.

The missing mass is converted to energy in the formation of the nucleus. This energy is found from Einstein's mass-energy relation,

$$E = (\Delta m)c^2 \tag{33.4}$$

and is called the **binding energy** (BE) of the nucleus. From equations 33.3 and 33.4, the binding energy of a nucleus is

$$\text{BE} = (\Delta m)c^2 = Zm_pc^2 + (A - Z)m_nc^2 - m_{nucleus}c^2 \tag{33.5}$$

Example 33.1

The mass defect and the binding energy of the deuteron. Find the mass defect and the binding energy of the deuteron nucleus. The experimental mass of the deuteron is 3.3435×10^{-27} kg.

Solution

The mass defect for the deuteron is found from equation 33.3, with

$$\Delta m = m_p + m_n - m_D$$
$$= 1.6726 \times 10^{-27} \text{ kg} + 1.6749 \times 10^{-27} \text{ kg} - 3.3435 \times 10^{-27} \text{ kg}$$
$$= 3.9754 \times 10^{-30} \text{ kg}$$

The binding energy of the deuteron, found from equation 33.4, is

$$\text{BE} = (\Delta m)c^2$$
$$= (3.9754 \times 10^{-30} \text{ kg})(2.9979 \times 10^8 \text{ m/s})^2$$
$$= (3.5729 \times 10^{-13} \text{ J})\left(\frac{1 \text{ eV}}{1.60218 \times 10^{-19} \text{ J}}\right)\left(\frac{1 \text{ MeV}}{10^6 \text{ eV}}\right)$$
$$= 2.23 \text{ MeV}$$

Therefore, the bound constituents have less energy than when they are free. That is, the binding energy comes from the mass that is lost in the process of formation. Conversely, an amount of energy equal to the binding energy is the amount of energy that must be supplied to a nucleus if the nucleus is to be broken up into protons and neutrons. Thus, the binding energy of a nucleus is similar to the ionization energy of an electron in the atom.

33.3 Radioactive Decay Law

The spontaneous emission of radiation from the nucleus of an atom is called radioactivity. Radioactivity is the result of the decay or disintegration of unstable nuclei. Radioactivity occurs naturally from all the chemical elements with atomic numbers greater than 83, and can occur naturally from some of the isotopes of the chemical elements below atomic number 83. Some can also occur artificially from nearly all of the chemical elements.

The rate of radioactive emission is measured by the *radioactive decay law.* The number of nuclei ΔN that disintegrate during a particular time interval Δt is directly proportional to the number of nuclei N present. That is

$$\frac{\Delta N}{\Delta t} \propto -N$$

To make an equality of this, we introduce the constant of proportionality λ, called the *decay constant* or *disintegration constant,* and we obtain

$$\frac{\Delta N}{\Delta t} = -\lambda N \tag{33.6}$$

The minus sign in equation 33.6 is necessary because the final number of nuclei N_f is always less than the initial number of nuclei N_i; hence $\Delta N = N_f - N_i$ is always a negative quantity because there is always less radioactive nuclei with time. The decay constant λ is a function of the particular isotope of the chemical element. A large value of λ indicates a large decay rate, whereas a small value of λ indicates a small decay rate. The quantity $-\Delta N/\Delta t$ in equation 33.6 is the rate at which the nuclei decay with time and it is also called the **activity** and designated by the symbol A. Hence,

$$A = -\frac{\Delta N}{\Delta t} = \lambda N \tag{33.7}$$

We must be careful in what follows not to confuse the symbol A for activity with the same symbol A for mass number. It should always be clear in the particular context used.

The SI unit of activity is the becquerel where 1 becquerel (Bq) is equal to one decay per second. That is,

$$1 \text{ Bq} = 1 \text{ decay/s}$$

An older unit of activity, the curie, abbreviated Ci, is equivalent to

$$1 \text{ Ci} = 3.7 \times 10^{10} \text{ Bq}$$

Smaller units of activity are the millicurie (10^{-3} curie = mCi) and the microcurie (10^{-6} curie = μCi).

The total number of nuclei present at any instant of time is found from equation 33.6, with the help of the calculus, to be

$$N = N_0 e^{-\lambda t} \tag{33.8}$$

and is the radioactive decay law. Here N_0 is the number of nuclei present at the time $t = 0$, which is the time that the observations of the nuclei is started. A plot of the radioactive decay law, equation 33.8, is shown in figure 33.3. The curve represents

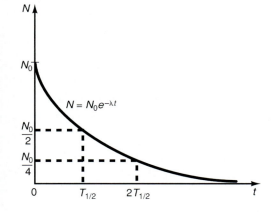

Figure 33.3

The radioactive decay law.

the number of radioactive nuclei still present at any time t. A very interesting quantity is found by looking for the time it takes for half of the original nuclei to decay, so that only half of the original nuclei are still present. Half the original nuclei is $N_0/2$ and is shown in the figure. A horizontal line for the value of $N_0/2$ is drawn until it intersects the curve $N = N_0 e^{-\lambda t}$. A vertical line is dropped from this point to the t-axis. The value of the time read on the t-axis is the time it takes for half the original nuclei to decay. Hence, this time read from the t-axis is called the half-life of the radioactive nuclei and is denoted by $T_{1/2}$. *The **half-life** of a radioactive substance is thus the time it takes for half the original radioactive nuclei to decay.*

Example 33.2

The number of radioactive nuclei for several half-lives. One mole of a radioactive substance starts to decay. How many radioactive nuclei will be left after $t =$ (a) $T_{1/2}$, (b) $2T_{1/2}$, (c) $3T_{1/2}$, (d) $4T_{1/2}$, and (e) $nT_{1/2}$ half-lives?

Solution

Since one mole of any substance contains 6.022×10^{23} atoms/mole (Avogadro's number), $N_0 = 6.022 \times 10^{23}$ nuclei.

a. At the end of one half-life, there will be

$$\frac{N_0}{2} = \frac{6.022 \times 10^{23} \text{ nuclei}}{2} = 3.011 \times 10^{23} \text{ nuclei}$$

b. At the end of another half-life, $t = 2T_{1/2}$, half of those present at the time $t = T_{1/2}$ will be lost, or

$$N = \frac{1}{2} \frac{N_0}{2} = \frac{N_0}{4} = \frac{6.022 \times 10^{23}}{4} \text{ nuclei}$$

$$= 1.506 \times 10^{23} \text{ nuclei}$$

c. At $t = 3T_{1/2}$, the number of radioactive nuclei remaining is

$$N = \frac{1}{2} \frac{N_0}{4} = \frac{N_0}{8} = 0.753 \times 10^{23} \text{ nuclei}$$

d. At $t = 4T_{1/2}$, the number of radioactive nuclei remaining is

$$N = \frac{1}{2} \frac{N_0}{8} = \frac{N_0}{16} = 0.376 \times 10^{23} \text{ nuclei}$$

e. At a period of time equal to n half-lives, we can see from the above examples that the number of nuclei remaining is

$$N = \frac{N_0}{2^n} \quad \text{for } t = nT_{1/2} \tag{33.9}$$

An important relationship between the half-life and the decay constant can be found by noting that when $t = T_{1/2}$, $N = N_0/2$. If these values are placed into the decay law in equation 33.8, we get

$$\frac{N_0}{2} = N_0 e^{-\lambda T_{1/2}}$$

or

$$\tfrac{1}{2} = e^{-\lambda T_{1/2}}$$

Taking the natural logarithm of both sides of this equation, we get

$$\ln \tfrac{1}{2} = \ln e^{-\lambda T_{1/2}} \qquad (33.10)$$

But the natural logarithm ln is the inverse of the exponential function e, and when applied successively, as in the right-hand side of equation 33.10, they cancel each other leaving only the function. Hence, equation 33.10 becomes

$$\ln \tfrac{1}{2} = -\lambda T_{1/2}$$

Taking the natural logarithm ln of 1/2 on the electronic calculator gives -0.693. Thus,

$$-0.693 = -\lambda T_{1/2}$$

Solving for the decay constant, we get

$$\lambda = \frac{0.693}{T_{1/2}} \qquad (33.11)$$

Thus, if we know the half-life $T_{1/2}$ of a radioactive nuclide, we can find its decay constant λ from equation 33.11. Conversely, if we know λ, then we can find the half-life from equation 33.11.

Example 33.3

Finding the decay constant and the activity for $^{90}_{38}Sr$. The half-life of strontium-90, $^{90}_{38}Sr$, is 28.8 yr. Find (a) its decay constant and (b) its activity for 1 g of the material.

Solution

a. The decay constant, found from equation 33.11, is

$$\lambda = \frac{0.693}{T_{1/2}}$$

$$= \left(\frac{0.693}{28.8 \text{ yr}}\right)\left(\frac{1 \text{ y}}{365 \text{ days}}\right)\left(\frac{1 \text{ day}}{24 \text{ hr}}\right)\left(\frac{1 \text{ hr}}{3600 \text{ s}}\right)$$

$$= 7.63 \times 10^{-10}/\text{s}$$

b. Before the activity can be determined, the number of nuclei present must be known. The atomic mass of $^{90}_{38}Sr$ is 89.907746. Thus, 1 mole of it has a mass of approximately 89.91 g. The mass of 1 mole contains Avogadro's number or 6.022×10^{23} molecules. We find the number of molecules in 1 g of the material from the ratio

$$\frac{N_0}{N_A} = \frac{1 \text{ g}}{89.91 \text{ g}}$$

or the number of nuclei in 1 g of strontium-90 is

$$N_0 = \frac{1 \text{ g } (N_A)}{89.91 \text{ g}} = \frac{1(6.022 \times 10^{23})}{89.91} = 6.70 \times 10^{21} \text{ nuclei}$$

We can now find the activity from equation 33.7 as

$$A_0 = \lambda N_0$$

$$= \left(7.63 \times 10^{-10} \frac{1}{\text{s}}\right)(6.70 \times 10^{21} \text{ nuclei})$$

$$= 5.11 \times 10^{12} \text{ nuclear disintegrations/s}$$

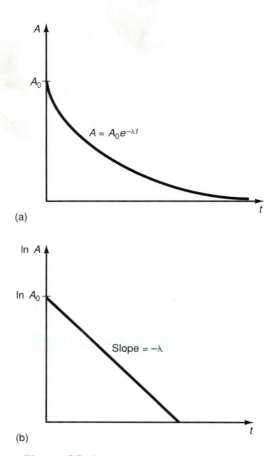

(a)

(b)

Figure 33.4

Radioactive activity.

Note that the activity (that is, the number of disintegrations per second) is not a constant because it depends on N, which is decreasing with time by equation 33.8. In fact, if equation 33.8 is substituted into equation 33.7 for the activity, we get

$$A = \lambda N = \lambda N_0 e^{-\lambda t}$$

Letting

$$\lambda N_0 = A_0$$

the rate at which the nuclei are decaying at the time $t = 0$, we obtain for the activity

$$A = A_0 e^{-\lambda t} \tag{33.12}$$

Recalling that the activity is the number of disintegrations per second, we see that the rate of decay is not a constant but decreases exponentially. A plot of the activity as a function of time is shown in figure 33.4(a). Notice the similarity of this diagram with figure 33.3. For the time t, equal to a half-life, the activity is

$$A = A_0 e^{-\lambda T_{1/2}}$$

Substituting λ from equation 33.11, gives

$$A = A_0 e^{-[0.693/T_{1/2}]\,T_{1/2}} = A_0 e^{-0.693}$$

Using the electronic calculator, we obtain $e^{-0.693} = 0.500 = 1/2$. Hence,

$$A = \frac{A_0}{2} \quad \text{for } t = T_{1/2} \tag{33.13}$$

That is, the rate of decay is cut in half for a time period of one half-life.

<hr>

Example 33.4

The number of radioactive nuclei and their rate of decay. Find the number of radioactive nuclei and their rate of decay for $t = T_{1/2}$ in the 1.00-g sample in example 33.3.

Solution

The number of nuclei left at the end of one half-life, found from equation 33.9, is

$$N = \frac{N_0}{2}$$

$$= \frac{6.70 \times 10^{21}}{2} \text{ nuclei}$$

$$= 3.35 \times 10^{21} \text{ nuclei}$$

While the rate of decay at the end of one half-life is found from equation 33.13 as

$$A = \frac{A_0}{2}$$

$$= \frac{5.11 \times 10^{12}}{2} \text{ decays/s}$$

$$= 2.55 \times 10^{12} \text{ decays/s}$$

Thus, at the end of 28 years, the number of strontium-90 radioactive nuclei have been cut in half and the rate at which they decay is also cut in half. That is, *there are less radioactive nuclei present at the end of the half-life, but the rate at which they decay also decreases.*

Example 33.5

The decay constant and activity for $^{91}_{38}Sr$. The half-life of $^{91}_{38}$Sr is 9.70 hr. Find (a) its decay constant and (b) its activity for 1.00 g of the material.

Solution

a. This problem is very similar to example 33.3 except that this isotope of strontium has a very short half-life. The decay constant, found from equation 33.11, is

$$\lambda = \frac{0.693}{T_{1/2}}$$

$$= \left(\frac{0.693}{9.70 \text{ hr}}\right)\left(\frac{1 \text{ hr}}{3600 \text{ s}}\right)$$

$$= 1.99 \times 10^{-5}/\text{s}$$

b. The number of nuclei in a 1-g sample is found as before but the atomic mass of $^{91}_{38}$Sr is 90.90. Hence,

$$N_0 = \frac{1 \text{ g}}{90.90 \text{ g}}(N_A)$$

$$= \frac{1}{90.90}(6.022 \times 10^{23} \text{ nuclei})$$

$$= 6.63 \times 10^{21} \text{ nuclei}$$

The activity, again found from equation 33.7, is

$$A_0 = \lambda N_0$$

$$= (1.99 \times 10^{-5}/\text{s})(6.63 \times 10^{21} \text{ nuclei})$$

$$= 1.32 \times 10^{17} \text{ disintegrations/s}$$

Comparing this example to example 33.3, we see *that for a smaller half-life, we get a larger decay constant λ, and hence a greater activity, or decays per second.*

The decay constant λ can be found experimentally using the following technique. First, let us return to equation 33.12 and take the natural logarithm of both sides of the equation, that is,

$$\ln A = \ln(A_0 e^{-\lambda t})$$

Now from the rules of manipulating logarithms, the logarithms of a product is equal to the sum of the logarithms of each term. Therefore,

$$\ln A = \ln A_0 + \ln e^{-\lambda t}$$

But as mentioned before, the natural log and the exponential are inverses of each other, and hence

$$\ln e^{-\lambda t} = -\lambda t$$

Thus,

$$\ln A = \ln A_0 - \lambda t$$

Rearranging, this becomes

$$\ln A = -\lambda t + \ln A_0 \qquad (33.14)$$

If we now go into the laboratory and count the number of disintegrations per unit time A, at different times t, we can plot the $\ln A$ on the y-axis versus t on the

x-axis, and obtain the straight line shown in figure 33.4(b). The slope of the line is $-\lambda$, and thus, λ can be determined experimentally. Once we know λ, we determine the half-life from equation 33.11.

We should also note that sometimes it is convenient to use the mean life or average life T_{avg} of a sample. The mean or average life is defined as the average lifetime of all the particles in a given sample of the material. It turns out to be just the reciprocal of the decay constant, that is,

$$T_{avg} = \frac{1}{\lambda} \qquad (33.15)$$

33.4 Forms of Radioactivity

Up to now, only the number of decaying nuclei has been discussed, without specifying the details of the disintegrations. Nuclei can decay by:

1. Alpha decay, α
2. Beta decay, β^-
3. Beta decay, β^+, positron emission
4. Electron capture
5. Gamma decay, γ

Now let us discuss each of these in more detail.

Alpha Decay

When a nucleus has too many protons compared to the number of neutrons, the electrostatic force of repulsion starts to dominate the nuclear force of attraction. When this occurs, the nucleus is unstable and emits an α particle in radioactive decay. The nucleus thus loses two protons and two neutrons. Hence, its atomic number Z, which represents the number of protons in the nucleus, decreases by 2, while its mass number A, which is equal to the number of protons and neutrons in the nucleus, decreases by 4. Before the decay, the nucleus is called the *"parent" nucleus;* after the decay the nucleus is referred to as the *"daughter" nucleus.* Hence, *we represent an **alpha decay** symbolically as*

$$^A_Z X \rightarrow \,^{A-4}_{Z-2} X + \,^4_2 He \qquad (33.16)$$

where $^A_Z X$ is the parent nucleus, which decays into the daughter nucleus $^{A-4}_{Z-2} X$, and $^4_2 He$ is the α particle, which is the helium nucleus. *Notice that the atomic number Z has decreased by two units. This means that in alpha decay, one chemical element of atomic number Z has been transmuted into a new chemical element of atomic number $Z - 2$.* The dream of the ancient alchemists was to transmute the chemical elements, in particular to turn the baser metals into gold. This result was never attained because they were working with chemical reactions, which as has been seen, depends on the electronic structure of the atom and not its nucleus.

An example of a naturally occurring alpha decay can be found in uranium-238, which decays by α particle emission with a half-life of 4.51×10^9 yr. We find its daughter nucleus by using equation 33.16. Hence,

$$^{238}_{92} U \rightarrow \,^{234}_{90} X + \,^4_2 He \qquad (33.17)$$

Notice that the atomic number Z has dropped from 92 to 90. Consulting the table of the elements, we see that the chemical element with $Z = 90$ is thorium. Hence, uranium has been transmuted to thorium by the emission of an α particle. Equation 33.17 is now written as

$$^{238}_{92} U \rightarrow \,^{234}_{90} Th + \,^4_2 He \qquad (33.18)$$

Also note from the periodic table that the mass number A for thorium is 232, whereas in equation 33.18, the mass number is 234. This means that an isotope of thorium has been formed. (As a matter of fact $^{234}_{90}$Th is an unstable isotope and it also decays, only this time by beta emission. We will say more about this later.)

Beta Decay, β^-

In beta decay, an electron is observed to leave the nucleus. However, as seen in chapter 31, an electron cannot be contained in a nucleus because of the Heisenberg uncertainty principle. Hence, the electron must be created within the nucleus at the moment of its emission. In fact, it has been found that a neutron within the nucleus decays into a proton and an electron, plus another particle called an *antineutrino*. The antineutrino is designated by the greek letter nu, ν, with a bar over the ν, that is $\bar{\nu}$. The antineutrino is the antiparticle of the neutrino ν. The neutron decay is written as

$$^{1}_{0}n \rightarrow {}^{1}_{1}p + {}^{0}_{-1}e + \bar{\nu} \qquad (33.19)$$

The notation $^{0}_{-1}$ is used to designate the electron or β particle. It has a mass number A of 0 because it has no nucleons, and an atomic number of -1 to signify that it is a negative particle. The proton is written as $^{1}_{1}p$ because it has a mass number and atomic number of one. Hence, in beta decay the nucleus loses a neutron but gains a proton, while the β particle, the electron, and the antineutrino are emitted from the nucleus. Thus, the atomic number Z increases by 1 in the decay because the nucleus gained a proton, but the mass number A stays the same because even though 1 neutron is lost, we have gained 1 proton.

It is perhaps appropriate to mention an interesting historical point here. The original assumption about neutron decay shown in equation 33.19 did not contain the antineutrino particle. The original decay seemed to violate the principle of conservation of energy. However, Wolfgang Pauli proposed the existence of a particle to account for the missing energy. Since the particle had to be neutral because of the law of conservation of electrical charge, the new particle was called a "neutrino" by the Italian-American physicist Enrico Fermi (1902–1954), for the "little" neutral particle. The antineutrino is the antiparticle of the neutrino, it is a particle of the same mass (zero rest mass) but has a spin component opposite to that of the neutrino. The neutrino was found experimentally in 1956. It is such an elusive particle that some move right through the earth without ever hitting anything.

*A **beta decay**, β^- can be written symbolically as*

$$^{A}_{Z}X \rightarrow {}^{A}_{Z+1}X + {}^{0}_{-1}e + \bar{\nu} \qquad (33.20)$$

Note that in beta decay, Z increases to $Z + 1$. *Hence, a chemical element of atomic number Z is transmuted into another chemical element of atomic number $Z + 1$.*

As an example, the isotope $^{234}_{90}$Th is unstable and decays by beta emission with a half-life of 24 days. Its decay can be represented with the use of equation 33.20 as

$$^{234}_{90}Th \rightarrow {}^{234}_{91}X + {}^{0}_{-1}e + \bar{\nu}$$

Looking up the periodic table of the elements, we see that the chemical element corresponding to the atomic number 91 is protactinium (Pa). Hence, the element thorium has been transmuted to the element protactinium. Also note from the periodic table that the mass number A for protactinium should be 231. Since we have a mass number of 234, this is an isotope of protactinium. (One that is also unstable and decays again.) The beta decay of thorium is now written as

$$^{234}_{90}Th \rightarrow {}^{234}_{91}Pa + {}^{0}_{-1}e + \bar{\nu}$$

Example 33.6

Beta decay, β^-. The element $^{234}_{91}\text{Pa}$ is unstable and decays by beta emission with a half-life of 6.66 hr. Find the nuclear reaction and the daughter nuclei.

Solution

Because $^{234}_{91}\text{Pa}$ decays by beta emission, it follows the form of equation 33.20. Hence,

$$^{234}_{91}\text{Pa} \rightarrow {}^{234}_{92}X + {}^{0}_{-1}\text{e} + \bar{\nu}$$

But from the table of elements, $Z = 92$ is the atomic number of uranium. Hence, the daughter nuclei is $^{234}_{92}\text{U}$, and the entire reaction is written as

$$^{234}_{91}\text{Pa} \rightarrow {}^{234}_{92}\text{U} + {}^{0}_{-1}\text{e} + \bar{\nu}$$

Beta Decay, β^+ Positron Emission

In this type of decay, a positron is emitted from the nucleus. A *positron* is the antiparticle of the electron. It has all the characteristics of the electron except it carries a positive charge. Because there are no positrons in the nucleus, a positron must be created immediately before emission. Positron emission is the result of the decay of a proton into a neutron, a positron, and a neutrino ν, and is written symbolically as

$$^{1}_{1}\text{p} \rightarrow {}^{1}_{0}\text{n} + {}^{0}_{+1}\text{e} + \nu \tag{33.21}$$

The positron $^{0}_{+1}\text{e}$ is emitted with the neutrino ν. The neutron stays behind in the nucleus. *Hence, in a **beta decay**, β^+ the atomic number Z decreases by one because of the loss of the proton. The mass number A stays the same because even though a proton is lost, a neutron is created to keep the same number of nucleons. Hence, a β^+ decay can be written symbolically as*

$$^{A}_{Z}X \rightarrow {}_{Z-1}^{A}X + {}^{0}_{+1}\text{e} + \nu \tag{33.22}$$

As an example, the isotope of aluminum $^{26}_{13}\text{Al}$ is unstable and decays by β^+ emission with a half-life of 7.40×10^5 yr. The reaction is written with the help of equation 33.22 as

$$^{26}_{13}\text{Al} \rightarrow {}^{26}_{12}X + {}^{0}_{+1}\text{e} + \nu$$

Looking at the periodic table of the elements, we find that the atomic number 12 corresponds to the chemical element magnesium Mg. Hence,

$$^{26}_{13}\text{Al} \rightarrow {}^{26}_{12}\text{Mg} + {}^{0}_{+1}\text{e} + \nu$$

Because the mass number A of magnesium is 24, we see that this transmutation created an isotope of magnesium.

It is important to note that the decay of the proton, equation 33.21, can only occur within the nucleus. A free proton cannot decay into a neutron because the mass of the proton is less than the mass of the neutron.

Electron Capture

Occasionally an orbital electron gets too close to the nucleus and gets absorbed by the nucleus. Since the electron cannot remain as an electron within the nucleus, it combines with a proton and in the process creates a neutron and a neutrino. We represent this as

$$^{0}_{-1}\text{e} + {}^{1}_{1}\text{p} \rightarrow {}^{1}_{0}\text{n} + \nu \tag{33.23}$$

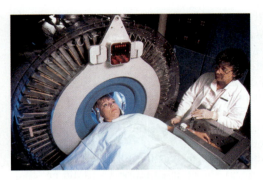

A patient undergoing a PET (Positron Emission Tomography) scan.

The net result of this process decreases the number of protons in the nucleus by one hence changing Z to $Z - 1$, while keeping the number of nucleons A constant. Hence, *this decay can be written as*

$$_{-1}^{0}e + {}_{Z}^{A}X \rightarrow {}_{Z-1}^{A}X + \nu \qquad (33.24)$$

When the electron that is close to the nucleus is captured by the nucleus, it leaves a vacancy in the electron orbit. An electron from a higher energy orbit falls into this vacancy. The difference in the energy of the electron in the higher orbit from the energy in the lower orbit is emitted as a photon in the X-ray portion of the spectrum.

As an example of electron capture, we consider the isotope of mercury $_{80}^{197}$Hg that decays by electron capture with a half-life of 65 hr. This decay can be represented, with the help of equation 33.24, as

$$_{-1}^{0}e + {}_{80}^{197}Hg \rightarrow {}_{79}^{197}X + \nu$$

Consulting the table of elements, we find that the atomic number $Z = 79$ represents the chemical element gold, Au. Hence,

$$_{-1}^{0}e + {}_{80}^{197}Hg \rightarrow {}_{79}^{197}Au + \nu$$

Thus, the dreams of the ancient alchemists have been fulfilled. An isotope of mercury has been transmuted into the element gold. Also note that mass number A of gold is 197. Hence, the transmutation has given the stable element gold.

Gamma Decay

A nucleus undergoing a decay is sometimes left in an excited state. Just as an electron in an excited state of an atom emits a photon and drops down to the ground state, a proton or neutron can be in an excited state in the nucleus. When the nucleon drops back to its ground state, it also emits a photon. Because the energy given off is so large, the frequency of the photon is in the gamma ray portion of the electromagnetic spectrum. Hence, the excited nucleus returns to its ground state and a gamma ray is emitted. Thus *gamma decay is represented symbolically as*

$$_{Z}^{A}X^* \rightarrow {}_{Z}^{A}X + \gamma \qquad (33.25)$$

Where the * on the nucleus indicates an excited state. *In this type of decay, neither the atomic number Z nor the mass number A changes. Hence, gamma decay does not transmute any of the chemical elements.*

33.5 Radioactive Series

As indicated earlier, elements with atomic numbers Z greater than 83 are unstable and decay naturally. Most of these unstable elements have very short lifetimes and decay rather quickly. Hence, they are not easily found in nature. The exceptions to this are the elements thorium-232, uranium-238, and the uranium isotope 235. The element $_{90}^{232}$Th has a half-life of 1.39×10^{10} yr, $_{92}^{238}$U has a half-life of 4.50×10^9 yr, and $_{92}^{235}$U has a half-life of 7.10×10^8 yr. Moreover, these elements decay into a series of daughters, granddaughters, great granddaughters, and so on.

As an example, the series decay $_{90}^{232}$Th is shown in figure 33.5, which is a plot of the neutron number N versus the atomic number Z. Because $_{90}^{232}$Th has a Z value of 90 and an N value of $232 - 90 = 142$, $_{90}^{232}$Th is plotted with the coordinates $N = 142$ and $Z = 90$. First $_{90}^{232}$Th decays by alpha emission with a half-life of 1.39×10^{10} yr. As seen in section 33.4, equation 33.16, an alpha decay changes the atomic number Z to $Z - 2$, and decreases the mass number A by 4 to $A - 4$. Thus, $_{90}^{232}$Th decays as

$$_{90}^{232}Th \rightarrow {}_{88}^{228}X + {}_{2}^{4}He$$

Figure 33.5

Thorium $^{232}_{90}$Th decay series.

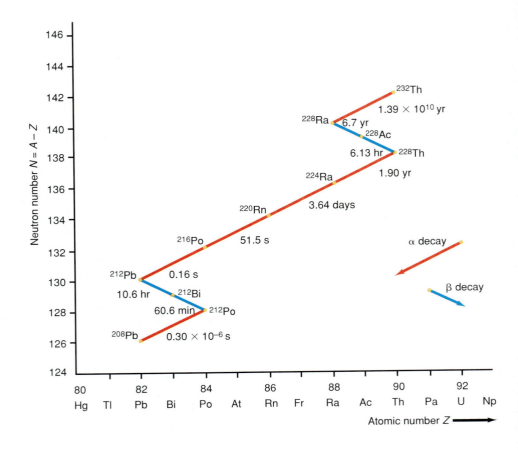

But atomic number 88 corresponds to the chemical element radium (Ra). Hence,

$$^{232}_{90}\text{Th} \rightarrow \,^{228}_{88}\text{Ra} + \,^{4}_{2}\text{He}$$

The neutron number N for $^{228}_{88}$Ra is $228 - 88 = 140$. Thus, $^{228}_{88}$Ra is found in the diagram with coordinates, $N = 140$ and $Z = 88$. The original neutron number is given by

$$N_0 = A - Z$$

But in alpha emission, A goes to $A - 4$ and Z goes to $Z - 2$, equation 33.16. Hence, the new neutron number is given by

$$N_1 = (A - 4) - (Z - 2)$$
$$= A - Z - 2$$
$$N_1 = N_0 - 2 \quad \text{(alpha decay)} \tag{33.26}$$

Thus, for all alpha emissions, the neutron number decreases by 2. Hence, in the diagram, for every alpha emission the element has both N and Z decreased by 2.

Radium-228 is also unstable and decays by beta emission with a half-life of 6.7 yr. As shown in equation 33.20, the value of the atomic number Z increases to $Z + 1$, while the mass number A remains the same. The neutron number for beta emission becomes

$$N_1 = A - (Z + 1) = A - Z - 1$$
$$N_1 = N_0 - 1 \quad \text{(beta}^- \text{ decay)} \tag{33.27}$$

Thus, $^{228}_{88}$Ra becomes actinium, $^{228}_{89}$Ac, with coordinates $N = 139$ and $Z = 89$.

Therefore, in the series diagram, alpha emission appears as a line sloping down toward the left, with both N and Z decreasing by 2 units. Beta emission, on the other hand, appears as a line sloping downward to the right with N decreasing by 1 and Z increasing by 1. The entire decay of the family is shown in figure 33.5: thorium-232 decays by α emission to radium-228, which then decays by β^- emission to actinium-228, which then decays by β^- to thorium-228, which then decays by α emission to radium-224, which then decays by α emission to radon-220, which then decays by α emission to polonium-216, which then decays by α emission to lead-212, which then decays by β^- to bismuth-212, which then decays by β^- to polonium-212, which finally decays by α emission to the stable lead-208. The half-life for each decay is shown in the diagram.

The radioactive chain is called a *series*. The decay series for uranium-238 is shown in figure 33.6. It starts with $^{238}_{92}U$ and ends in the stable isotope of lead-206. Figure 33.7 shows the decay series for uranium-235. As we can see, the series ends with the stable chemical element lead-207. Figure 33.8 shows the neptunium series that ends in the stable chemical element bismuth-209. Neptunium is called a *transuranic element* because it lies beyond uranium in the periodic table. Uranium with an atomic number $Z = 92$ is the highest chemical element found in nature. Elements with Z greater than 92 have been made by man. Many different isotopes of these new elements can also be created.

As an example of the creation of a transuranic element, bombarding $^{238}_{92}U$ with neutrons creates neptunium by the reaction

$$^{238}_{92}U + {}^{1}_{0}n \rightarrow {}^{239}_{93}Np + {}^{0}_{-1}e + \bar{\nu} \tag{33.28}$$

That is, $^{238}_{92}U$ absorbs the neutron and then goes through a beta decay by emitting an electron. The atomic number is increased by one, from $Z = 92$ to $Z = 93$, thus creating an isotope of a new chemical element, which is called neptunium.

Figure 33.6

Uranium $^{238}_{92}U$ decay series.

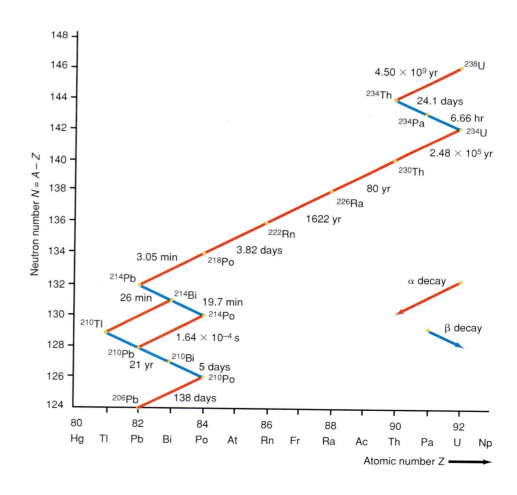

Figure 33.7

Uranium $^{235}_{92}$U decay series.

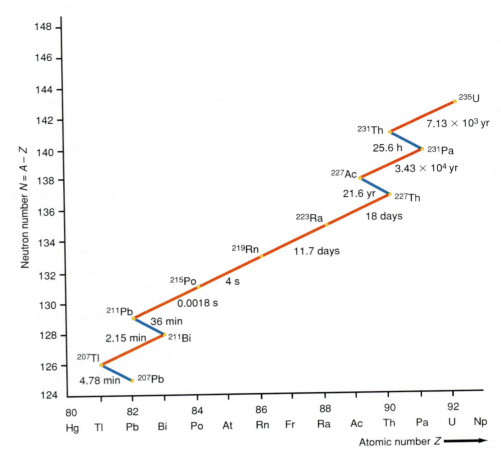

Figure 33.8

The neptunium series.

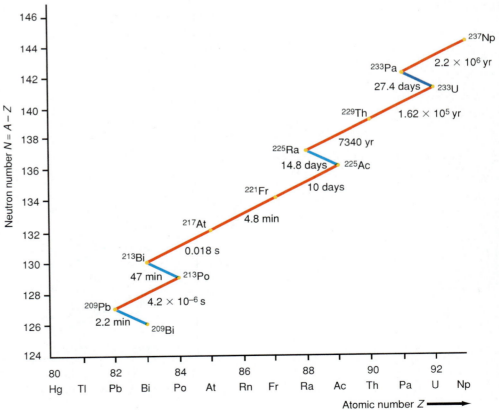

Neptunium-239 is itself unstable and decays by beta emission creating still another chemical element called plutonium, according to the reaction

$$^{239}_{93}\text{Np} \rightarrow\ ^{239}_{94}\text{Pu} +\ ^{0}_{-1}e + \bar{\nu} \tag{33.29}$$

The next transuranic element to be created was americium, which was created by the series of processes given by

$$^{239}_{94}\text{Pu} +\ ^{1}_{0}n \rightarrow\ ^{240}_{94}\text{Pu} + \gamma$$

$$^{240}_{94}\text{Pu} +\ ^{1}_{0}n \rightarrow\ ^{241}_{94}\text{Pu} + \gamma$$

$$^{241}_{94}\text{Pu} \rightarrow\ ^{241}_{95}\text{Am} +\ ^{0}_{-1}e + \bar{\nu}$$

That is, plutonium is bombarded with neutrons until the isotope $^{241}_{94}\text{Pu}$ is created, which then beta decays producing the isotope of the new chemical element americium. Bombarding elements with various other particles and elements, created still more elements. As examples of the creation of some other new elements, we have

(curium) $^{239}_{94}\text{Pu} +\ ^{4}_{2}\text{He} \rightarrow\ ^{242}_{96}\text{Cm} +\ ^{1}_{0}n$
(berkelium) $^{241}_{94}\text{Am} +\ ^{4}_{2}\text{He} \rightarrow\ ^{243}_{97}\text{Bk} +\ ^{0}_{-1}e + \bar{\nu} + 2\ ^{1}_{0}n$
(californium) $^{242}_{96}\text{Cm} +\ ^{4}_{2}\text{He} \rightarrow\ ^{245}_{98}\text{Cf} +\ ^{1}_{0}n$
(nobelium) $^{246}_{96}\text{Cm} +\ ^{12}_{6}\text{C} \rightarrow\ ^{254}_{102}\text{No} + 4\ ^{1}_{0}n$
(lawrencium) $^{252}_{98}\text{Cf} +\ ^{10}_{5}\text{B} \rightarrow\ ^{257}_{103}\text{Lr} + 5\ ^{1}_{0}n$

The neptunium decay series was later found to actually start with plutonium, $^{241}_{94}\text{Pu}$, which decays by beta emission to americium, $^{241}_{95}\text{Am}$, which then decays by alpha emission to neptunium, $^{237}_{93}\text{Np}$.

Because of the very long lifetimes of the parent element of these series, most of the members of the series are found naturally. An equilibrium condition is established within the series with as many isotopes decaying as are being formed. Artificial isotopes are those that are made by man. They most probably also existed in nature at the time of the creation of the earth. But because of their relatively short lifetimes and lack of a continuing source, they have all decayed away. Thus, there is nothing essentially different between radioisotopes found in nature and those made by man. In addition to the four natural radioactive series there are a host of other series from the decay of artificial isotopes. Such series are similar to the natural series and are called *collateral series*.

33.6 Energy in Nuclear Reactions

Let us generalize the nuclear reactions discussed so far by considering the reaction shown in figure 33.9. The initial reactants are a particle x of mass m_x moving at a velocity $\mathbf{v}_x$ toward a target element X of mass M_X, which is at rest. After the nuclear reaction, a particle y of mass m_y leaves with a velocity $\mathbf{v}_y$ while the product nucleus Y of mass M_Y moves at a velocity $\mathbf{V}_Y$. We can write the nuclear reaction in a general format as

$$x + X = y + Y \tag{33.30}$$

where x and X are the reactants and y and Y are the products of the reaction. Applying the law of conservation of energy to this reaction, we get

$$m_x c^2 + \text{KE}_x + M_X c^2 = m_y c^2 + \text{KE}_y + M_Y c^2 + \text{KE}_Y$$

Rearranging,

$$(m_x + M_X)c^2 - (m_y + M_Y)c^2 = \text{KE}_y + \text{KE}_Y - \text{KE}_x$$

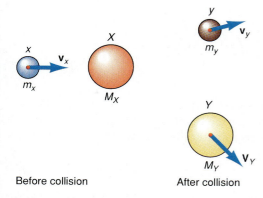

Before collision   After collision

Figure 33.9

Nuclear reaction as a collision.

The **Q value of a nuclear reaction** is now defined as the energy available in a reaction caused by the difference in mass between the reactants and the products. Thus,

$$Q = (m_x + M_X)c^2 - (m_y + M_Y)c^2 \qquad (33.31)$$

or

$$Q = [(\text{Input mass}) - (\text{Output mass})]c^2 \qquad (33.32)$$

or

$$Q = E_{in} - E_{out} \qquad (33.33)$$

That is, the Q value is the difference between the energy put into a nuclear reaction E_{in} and the energy that comes out E_{out}.

If $m_x + M_X$ is greater than $m_y + M_Y$, then Q is greater than zero ($Q > 0$). That is, the input mass energy is greater than the output mass energy. Thus, mass is lost in the nuclear reaction and an amount of energy Q is released in the process. *A nuclear reaction in which energy is released is called an **exoergic reaction** (sometimes called an exothermic reaction).*

Example 33.7

Energy released in a nuclear reaction. How much energy is released or absorbed in the following reaction?

$$^{219}_{86}\text{Rn} \rightarrow {}^{215}_{84}\text{Po} + {}^{4}_{2}\text{He}$$

Solution

The mass of radon-219 is 219.009523 unified mass units (u) while the mass of polonium, $^{215}_{84}\text{Po}$, is 214.999469 u. The mass of the α particle is 4.002603 u. The total output mass is

$$\begin{array}{r} 214.999469 \text{ u} \\ +4.002603 \text{ u} \\ \hline 219.002072 \text{ u} \end{array}$$

Hence, the difference in mass between the input mass and the output mass is

$$219.009523 \text{ u} - 219.002072 \text{ u} = 0.007451 \text{ u}$$

Converting this to an energy

$$Q = (0.007451 \text{ u})\left(\frac{931.49 \text{ MeV}}{\text{u}}\right)$$

$$= 6.94 \text{ MeV}$$

Since Q is greater than zero, energy is released in this reaction, and the reaction is exoergic. We might note that $^{219}_{86}\text{Rn}$ is one of the isotopes of the $^{235}_{92}\text{U}$ decay series. Because all of the isotopes of this chain decay naturally, Q is positive for such natural decays.

If in a nuclear reaction, $m_x + M_X$ is less than $m_y + M_Y$, then the Q value is negative ($Q < 0$). In such a reaction, mass is created if an amount of energy Q is added to the system. This energy is usually added by way of the kinetic energy of the reacting particle and nuclei. *A nuclear reaction in which energy is added to the system is called an **endoergic reaction** (sometimes called an endothermic reaction).*

A nuclear reaction proceeds naturally in the direction of minimum energy. Thus, in the decay of a natural radioactive nuclide, the nucleus emits a particle in order to reach a lower equilibrium energy state. The excess energy is given off in the process. Endoergic reactions, on the other hand, do not occur naturally in the physical world because the energy of the reactants is less than the required energy for the products to be created. Thus, endoergic reactions cannot take place unless energy is added to the system. The energy is added by accelerating the particle to very high speeds in an accelerator. When the particle hits the target, this additional kinetic energy is the energy necessary to make the reaction proceed. It is sometimes necessary to have additional kinetic energy to overcome the Coulomb barrier.

Example 33.8

Find the Q value of a nuclear reaction. The first artificial transmutation of an element was performed by Rutherford in 1919 when he bombarded nitrogen with alpha particles according to the reaction

$$^{14}_{7}N + ^{4}_{2}He \rightarrow ^{17}_{8}O + ^{1}_{1}p$$

Find the Q value associated with this reaction.

Solution

The Q value, found from equation 33.31, is

$$Q = (m_x + M_X)c^2 - (m_y + M_Y)c^2$$

where

$$m_x = m(^{4}_{2}He) = 4.002603 \text{ u}$$
$$\underline{M_X = m(^{14}_{7}N) = 14.003242 \text{ u}}$$
$$(m_x + M_X) = 18.005845 \text{ u}$$

and

$$m_y = m_p = 1.007825 \text{ u}$$
$$\underline{M_Y = m(^{17}_{8}O) = 16.999133 \text{ u}}$$
$$(m_y + M_Y) = 18.006958 \text{ u}$$

Hence,

$$Q = 18.005845 \text{ u} - 18.006958 \text{ u}$$

$$= -(0.001113 \text{ u})\left(\frac{931.49 \text{ MeV}}{1 \text{ u}}\right)$$

$$= -1.04 \text{ MeV}$$

Therefore, Q is negative, and this much energy must be supplied to start the reaction. The initial α particle used by Rutherford had energies of about 5.5 MeV, well above the amount of energy needed.

We should note that the amount of energy necessary to break up the nucleus, its Q value, is the same as the binding energy of the nucleus, BE, discussed in section 33.2. We can now write a nuclear reaction in the form

$$x + X \rightarrow y + Y + Q \tag{33.34}$$

When $Q > 0$, Q is the amount of energy released in a reaction. When $Q < 0$, Q is the amount of energy that must be added to the system in order for the reaction to proceed.

33.7 Nuclear Fission

In 1934, Enrico Fermi, beginning at the bottom of the periodic table, fired neutrons at each chemical element in order to create isotopes of the elements. He systematically worked his way up the periodic table until he came to the last known element, at that time, uranium. He assumed that bombarding uranium with neutrons would make it unstable. He then felt that if the unstable uranium nucleus went through a beta decay, the atomic number would increase from 92 to 93 and he would have created a new element. (He was the first to coin the word *transuranic*.) However after the bombardment of uranium, he could not figure out what the products of the reaction were.

From 1935 through 1938, the experiments were repeated in Germany by Otto Hahn and Lise Meitner. The German chemist, Ida Noddack, analyzed the products of the reaction and said that *it appeared as if the uranium atom had been split into two lighter elements*. Lise Meitner and her nephew, Otto Frisch, considered these results and concluded that indeed the atom had been split into two lighter elements. The splitting of an atom resembled the splitting of one living cell into two cells of equal size. This biological process is called fission. Otto Frisch then used this biological term, fission, to describe the splitting of an atom. Hence, **nuclear fission** is *the process of splitting a heavy atom into two lighter atoms*. The isotope of uranium that undergoes fission is $^{235}_{92}\text{U}$. The process can be described in general as

$$_{0}^{1}\text{n} + _{92}^{235}\text{U} \rightarrow y + Y + _{0}^{1}\text{n} + Q \tag{33.35}$$

The fission process does not always produce the same fragments, however. It was found that the product or fragment nuclei, y and Y, varied between the elements $Z = 36$ to $Z = 60$. Some typical fission reactions are:

$$_{0}^{1}\text{n} + _{92}^{235}\text{U} \rightarrow _{56}^{141}\text{Ba} + _{36}^{92}\text{Kr} + 3_{0}^{1}\text{n} + Q \tag{33.36}$$

$$_{0}^{1}\text{n} + _{92}^{235}\text{U} \rightarrow _{56}^{144}\text{Ba} + _{36}^{89}\text{Kr} + 3_{0}^{1}\text{n} + Q \tag{33.37}$$

$$_{0}^{1}\text{n} + _{92}^{235}\text{U} \rightarrow _{54}^{140}\text{Xe} + _{38}^{94}\text{Sr} + 2_{0}^{1}\text{n} + Q \tag{33.38}$$

$$_{0}^{1}\text{n} + _{92}^{235}\text{U} \rightarrow _{50}^{132}\text{Sn} + _{42}^{101}\text{Mo} + 3_{0}^{1}\text{n} + Q \tag{33.39}$$

In all cases, the masses of the product nuclei are less than the masses of the reactants, indicating that the Q value is greater than zero. The reaction is, therefore, exoergic and energy is given off in the process.

Example 33.9

The Q value of a nuclear fission reaction. Find the Q value associated with the nuclear fission process given by equation 33.36.

Solution

The mass of the reactants are

$$m_{\text{n}} = 1.008665 \text{ u}$$
$$m(_{92}^{235}\text{U}) = 235.043933 \text{ u}$$
$$m_{\text{n}} + m(_{92}^{235}\text{U}) = 236.052598 \text{ u}$$

The mass of the products are

$$3m_{\text{n}} = 3.025995 \text{ u}$$
$$m(_{56}^{141}\text{Ba}) = 140.913740 \text{ u}$$
$$m(_{36}^{92}\text{Kr}) = 91.925765 \text{ u}$$
$$m_{\text{Ba}} + m_{\text{Kr}} = 235.865500 \text{ u}$$

The mass lost in the process is

$$\Delta m = 236.052598 \text{ u} - 235.865500 \text{ u}$$

$$= +0.187098 \text{ u}$$

The Q value is obtained by multiplying Δm by the conversion factor 931.49 MeV $= 1$ u.

$$Q = (0.187098 \text{ u})\left(\frac{931.49 \text{ MeV}}{\text{u}}\right)$$

$$= 174 \text{ MeV}$$

Hence, the splitting of only one nucleus of $^{235}_{92}$U gives off an enormous quantity of energy. The actual energy in the fission process turns out to be even greater than this because the fragments themselves are radioactive and give off an additional 15 to 20 MeV of energy as they decay. Hence, in the entire fission process of $^{235}_{92}$U, some 200 MeV of energy are given off per nucleus.

Example 33.10

The energy of fission of uranium. If 1 kg of $^{235}_{92}$U were to go through the fission process, how much energy would be released?

Solution

Because the mass of any quantity is equal to the mass of one atom times the total number of atoms, that is,

$$m = m_{\text{atom}}N$$

the number of atoms is

$$N = \frac{m}{m_{\text{atom}}}$$

$$= \left(\frac{1 \text{ kg}}{235.04 \text{ u}}\right)\left(\frac{1 \text{ u}}{1.66 \times 10^{-27} \text{ kg}}\right)$$

$$= 2.56 \times 10^{24} \text{ atoms}$$

But the number of nuclei is exactly equal to the number of atoms, hence, there are 2.56×10^{24} uranium nuclei in 1 kg of uranium-235. Assuming a total energy release of 200 MeV per nucleus, the total energy released is

$$E = \left(\frac{200 \text{ MeV}}{\text{nuclei}}\right)(2.56 \times 10^{24} \text{ nuclei})$$

$$= 5.12 \times 10^{26} \text{ MeV} = 8.19 \times 10^{13} \text{ J}$$

which is an absolutely immense amount of energy. It is comparable to the amount of energy released from the explosion of 20,000 tons of TNT.

A theoretical model of nuclear fission, developed by Niels Bohr and John A. Wheeler, and called the *liquid-drop model,* is sketched in figure 33.10. When the bombarding neutron is captured by the uranium nucleus, the nucleus becomes unstable, vibrates, and becomes deformed as in figure 33.10(c). In the deformed state, the nuclear force is not as great as usual because the nucleus is spread so far apart. The Coulomb force of repulsion is, however, just as strong as always and acts to split the drop (nucleus) into fragments, figure 33.10(d). Thus, the uranium nucleus is split into fragment nuclei accompanied by extra neutrons and a large amount of energy.

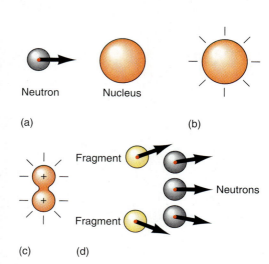

Neutron   Nucleus

(a)   (b)

Fragment

Neutrons

Fragment

(c)   (d)

Figure 33.10
The liquid-drop model of nuclear fission.

All in all, there are about 90 different daughter nuclei formed in the fission process. The initial neutrons that are used to bombard the uranium are called *slow neutrons* because they have very small kinetic energies and, hence, low velocities and, therefore, move slowly. The slow neutrons have a large probability of capture by the uranium-235 nucleus because they are moving so slowly. There are about two or three neutrons released per each fission.

A historical anecdote relating to nuclear fission might be interesting to mention here. In 1906, at McGill University in Montreal, Canada, Lord Rutherford said: "If it were ever found possible to control at will, the rate of disintegration of the radioactive elements, an enormous amount of energy could be obtained from a small quantity of matter."[1] With age, Rutherford was to change that vision to, "The energy produced by breaking down the atom is a very poor kind of thing. Anyone who expects a source of power from the transformation of these atoms is talking moonshine." That statement was a challenge to Leo Szilard (1898–1964), a Hungarian physicist working for Rutherford. Szilard thought, "What if you found an element in which nuclei throw off energy? What if you could make it happen at will? What if this element's atoms threw off two new neutrons to strike two more nuclei. Two twos are four, four fours are sixteen—in a flash, the number would be astronomical. Moonshine? All you need do is to find the right element!"[2]

A by-product of fission is that it produces the same particles that initiated the fission in the first place, namely neutrons. If more neutrons are produced than started the reaction, the result is a multiplication. If the excess product neutrons can initiate more fission, more neutrons are produced to produce more fission, and so on and on. The result is a chain reaction, as shown in figure 33.11. The multiplication of neutrons is given by a multiplication factor, k. If $k < 1$, the reaction gives less neutrons than initiated the reaction, and the chain dies out. If $k > 1$ the reaction gives too many neutrons and the reaction escalates and runs wild. If $k = 1$, just the right number of neutrons are produced to keep the process going at a constant rate.

1. *Uranium, The Deadly Element,* by Lennard Bikel, p. 43.

2. Ibid, p. 72.

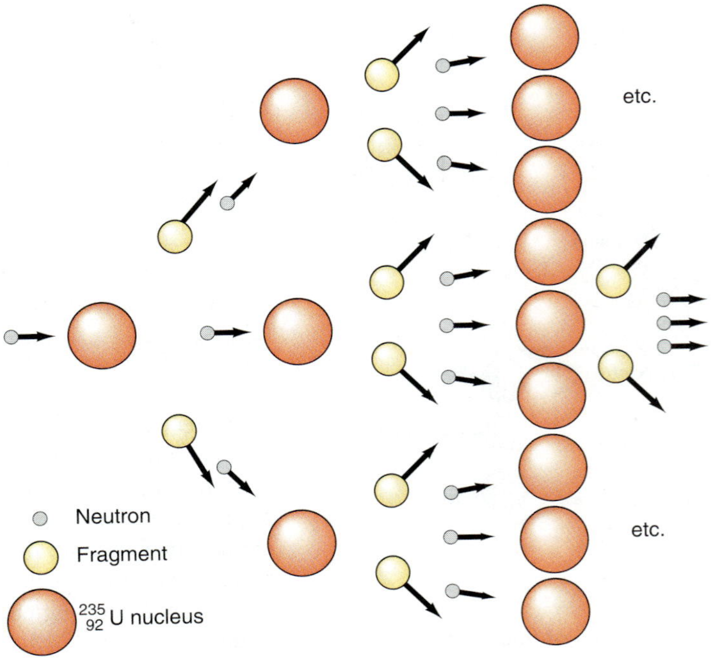

Figure 33.11

The chain reaction.

etc.

etc.

○ Neutron

○ Fragment

○ $^{235}_{92}$U nucleus

Natural uranium contains 99.3% of $^{238}_{92}U$ and only 0.7% of $^{235}_{92}U$, and cannot chain react. To get a chain reaction, the percentage of $^{235}_{92}U$ must be increased. Weapons grade uranium contains about 50% of $^{235}_{92}U$, whereas *nuclear reactor grade uranium contains only about 3.6% of $^{235}_{92}U$, which is much too small to produce a nuclear explosion.*

To finish our little story, Leo Szilard filed an application in the London Patents Office on June 28, 1934. It was the world's first registration of a nuclear process chain reaction by neutron bombardment. Szilard was afraid his chain reaction idea would fall into Nazi hands, so he assigned his patent to the British Admiralty. In general, the heavier the nucleus that is split, the greater the energy given off. Szilard made the mistake of proposing to split the lighter elements instead of the heavier ones.

The Atomic Bomb

The possibility of a chain reaction in uranium with its extremely large energy release, led some of the nuclear scientists to conceive of making a bomb—an atomic bomb. The Second World War was raging in Europe and the scientists were afraid that Hitler might develop such a bomb. Such a bomb in his hands, it was felt, would mean the end of the civilized world. For our own protection, it was imperative that we should develop such a bomb as quickly as possible. Leo Szilard and Edward Teller (later to become the Father of the hydrogen bomb), both Hungarian physicists who were refugees from Hitler's Europe, approached Albert Einstein and had him draft a letter to President Roosevelt on the possibility of making an atomic bomb. The letter was given to Dr. Alexander Sachs who personally delivered it to President Roosevelt on October 11, 1939. Ironically, the final decision to go ahead with the development of the A-bomb was made on December 6, 1941, under the name of the Manhattan Project.

In order to make an atomic bomb, enough uranium-235 had to be assembled to make the chain reaction. The amount of mass of uranium-235 needed to start the chain reaction was called the *critical mass*. The uranium-235 had to consist of two pieces, both below the critical mass. When one piece, in the form of a bullet, was fired into the second piece, the critical mass was obtained and the chain reaction would lead to a violent explosion. This was the type of bomb called the "Thin Man" that was detonated at Hiroshima on August 5, 1945. The difficulty with a uranium bomb was that it was relatively difficult to separate $^{235}_{92}U$ from $^{238}_{92}U$.

As already seen in equation 33.28, bombarding $^{238}_{92}U$ with neutrons produces the element neptunium, $^{239}_{93}Np$, which decays into plutonium, $^{239}_{94}Pu$, equation 33.29. It turns out that plutonium-239 is even more fissionable than uranium-235, so a much smaller mass of it is necessary for its critical mass. By making a nuclear reactor, which we will discuss in a moment, a very large, relatively cheap supply of plutonium was made available. So the Manhattan Project proceeded to make another type of atomic bomb—a plutonium bomb. The plutonium bomb was made in the form of a sphere, with pieces of plutonium, each below the critical mass, at the edge of the sphere, as shown in figure 33.12. For ignition, a series of chemical explosions fired the plutonium pieces all toward the center of the sphere at the same time. When all these pieces of plutonium came together they constituted the critical mass of plutonium, and the chain reaction was initiated and the bomb exploded.

The first test of an atomic device, was a test of the plutonium bomb on July 16, 1945, at a site called "Trinity" in the New Mexico desert. The first plutonium bomb, called "Fat Boy" was dropped on Nagasaki on August 9, 1945.

Fission Nuclear Reactors

The first nuclear reactor was built by Enrico Fermi on the squash court under the west stands of Stagg Field at the University of Chicago. It was started in October of 1942 and began operating on December 2, 1942. This was the first controlled use of nuclear fission.

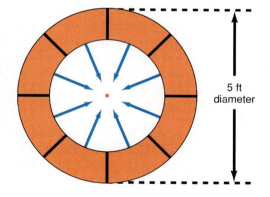

Figure 33.12

Triggering the plutonium bomb.

5 ft diameter

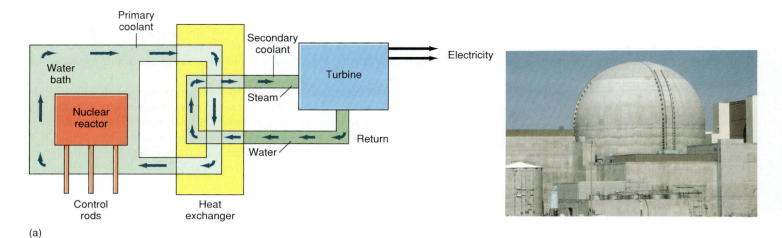

Figure 33.13

A typical nuclear reactor.

A typical nuclear reactor is sketched in figure 33.13. The reactor itself contains uranium, $^{238}_{92}U$, enriched with 3.6% of $^{235}_{92}U$. Neutrons are given off by a reaction such as equation 33.36. The neutrons given off have a rather high kinetic energy and are called *fast neutrons* because of the high speed associated with the large kinetic energy. These fast neutrons are moving too fast to initiate more fission reactions and must be slowed down. One such way is to enclose the entire reactor in a water bath under high pressure. Such a reactor is called a pressurized water reactor (PWR). The neutrons now collide with the water molecules and are slowed down so that they can be used in the fission process. The water is called the *moderator,* because it moderates or slows down the neutrons. The slow neutrons now proceed to split more $^{235}_{92}U$ nuclei until a chain reaction is obtained. The chain reaction is not allowed to run wild as in an atomic bomb but is controlled by a series of rods, usually made of cadmium, that are inserted into the reactor. Cadmium is an element that is capable of absorbing a large number of neutrons without becoming unstable or radioactive. Hence, when the cadmium control rods are inserted into the reactor they absorb neutrons to cut down on the number of neutrons that are available for the fission process. In this way, the fission reaction is controlled. The water moderator also acts as a coolant. The tremendous heat generated by the fission process heats up the water, which is then pumped to a heat exchanger. The hot moderator water is at a very high temperature and pressure, and the boiling point of water increases with pressure. Thus, the moderator water could be at a couple of hundred degrees Celsius without boiling. When this water enters the heat exchanger, it heats up the secondary water coolant. Because of the high temperatures of the primary coolant, the secondary coolant at relatively normal pressure is immediately converted to steam. This steam is then passed to a turbine, which drives an electric generator, thereby producing electricity.

The energy that comes from a reactor is quite large. As shown in example 33.10, there are approximately 200 MeV of energy given off in the splitting of only one $^{235}_{92}U$ nucleus. Typical energies given off in chemical reactions are only of the order of 3 or 4 eV. Hence, *fission of $^{235}_{92}U$ yields approximately 2.5 million times as much energy as found in the combination of the same mass of carbon (such as in coal or gasoline).*

The one drawback to a fission reactor is the nuclear waste material. As shown in equations 33.36 through 33.39, fission fragments such as $^{141}_{56}Ba$, $^{92}_{36}Kr$, $^{144}_{56}Ba$, $^{89}_{36}Kr$, $^{140}_{54}Xe$, $^{94}_{38}Sr$, $^{132}_{50}Sn$, and $^{101}_{42}Mo$ are some of the possible products of the reaction. These isotopes are unstable and decay into other radioactive nuclei. Eventually all these dangerous radioactive waste nuclei must be discarded. Some of them have relatively long half-lives and will, therefore, be around for a long time. They cannot be dumped into oceans or left in any place where they

The enormous energy generated by stars like our Sun is a result of nuclear fusion.

A tokamak apparatus, designed to generate the magnetic field necessary for containing a fusion reaction.

will contaminate the environment, such as through the soil or the air. They must not be allowed to get into the drinking water. The best place so far found to store these wastes is in the bottom of old salt mines, which are very dry and are thousands of feet below the surface of the earth. Here they can sit and decay without polluting the environment.

One unfounded fear of many people is that a nuclear reactor may explode like an atomic bomb and kill all the people in its neighborhood. A nuclear reactor does not contain enough $^{235}_{92}U$ to explode as an atomic bomb. What is more, the cadmium control rod's normal position is in the reactor. They must be pulled out to get and keep the reactor in operation. Any failure of any mechanism of the reactor causes the control rods to fall back into the reactor, thereby, stopping the chain reaction and shutting down the reactor.

Another type of a fission nuclear reactor is the *breeder reactor*. A breeder reactor uses uranium $^{238}_{92}U$, or thorium $^{232}_{90}Th$, as the nuclear fuel and uses fast high-energy neutrons instead of the slow ones used in the PWR. The fast neutrons react with the $^{238}_{92}U$, according to equation 33.28, and form neptunium, $^{239}_{93}Np$. The neptunium, $^{239}_{93}Np$, decays according to equation 33.29, and produces plutonium, $^{239}_{94}Pu$. The plutonium is highly fissionable and it too can supply energy in the reactor. The net result of forming plutonium in the reactor is to create more fissionable material than is used. Hence, the name breeder reactor; it "breeds" nuclear fuel. Of course, the breeder reactor can also generate electricity while it is creating more fuel. Breeder reactors are used to create plutonium for nuclear weapons.

33.8 Nuclear Fusion

It has been long observed that the sun emits tremendous quantities of energy for an enormous quantity of time. There was much speculation as to the source of this energy. In 1938, Hans Bethe (1906–) suggested that the fusion of hydrogen nuclei into helium nuclei was responsible for the tremendous energy released. *Nuclear fusion is a process in which lighter nuclei are joined together to produce a heavier nucleus and a good deal of energy.* Bethe proposed that the energy was released in the sun in what he called the proton-proton cycle. The first part of the cycle consists of two protons combining to form an unstable isotope of helium.

$$^{1}_{1}p + ^{1}_{1}p \rightarrow ^{2}_{2}He$$

But one of these combined protons in the nucleus of the unstable isotope immediately decays by equation 33.21 as

$$^{1}_{1}p \rightarrow ^{1}_{0}n + ^{0}_{+1}e + \nu$$

The neutron now combines with the first proton to form the deuteron, and we can write the entire reaction as

$$^{1}_{1}p + ^{1}_{1}p \rightarrow ^{2}_{1}H + ^{0}_{+1}e + \nu \qquad (33.40)$$

Note here that the decay of a proton or a neutron in the nucleus is caused by the weak nuclear force, which we will describe in more detail in chapter 34. The deuteron formed in equation 33.40 now combines with another proton to form the isotope of helium, $^{3}_{2}He$, according to the reaction

$$^{2}_{1}H + ^{1}_{1}p \rightarrow ^{3}_{2}He + \gamma \qquad (33.41)$$

The process represented by equations 33.40 and 33.41 must occur twice to form two $^{3}_{2}He$ nuclei, which then react according to the equation

$$^{3}_{2}He + ^{3}_{2}He \rightarrow ^{4}_{2}He + 2^{1}_{1}p \qquad (33.42)$$

The energy released in an explosion resulting from the nuclear fusion of even small amounts of material is enormous.

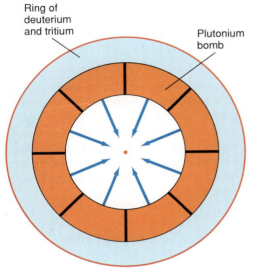

Ring of deuterium and tritium

Plutonium bomb

Figure 33.14

A hydrogen bomb.

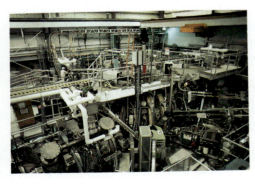

Thermonuclear fusion reactor.

Thus, the stable element helium has been formed from the fusion of the nuclei of the hydrogen atom. We can write the entire proton-proton cycle in the short-hand version as

$$4\,^1_1p \rightarrow\,^4_2He + 2\,^0_{+1}e + 2\gamma + 2\nu + Q \qquad (33.43)$$

The net Q value, or energy released in the process, is about 26 MeV.

The Hydrogen Bomb

In July of 1942, Robert Oppenheimer (1904–1967), reporting on the work of Edward Teller, Enrico Fermi, and Hans Bethe, noted that the extremely high temperature of an atomic bomb could be used to trigger a fusion reaction in deuterium, thus producing a fusion bomb or a hydrogen bomb. The reaction between deuterium and tritium, both isotopes of hydrogen, is given by

$$^2_1H +\,^3_1H \rightarrow\,^4_2He +\,^1_0n + 17.6\ MeV$$

Deuterium is relatively abundant in ocean water but tritium is relatively scarce. However, tritium can be generated in a nuclear reactor by surrounding the core with lithium. The neutron from the reactor causes the reaction

$$^7_3Li +\,^1_0n \rightarrow\,^4_2He +\,^3_1H +\,^1_0n$$

Thus, all the tritium desired can be relatively easily created.

The hydrogen bomb is effectively a bomb within a bomb, as illustrated in figure 33.14. A conventional atomic bomb made of plutonium is ignited. The tremendous heat given off by the A-bomb supplies the high temperature to start the fusion process of the deuterium-tritium mixture. The size of an A-bomb is limited by the critical mass of plutonium. We cannot assemble an amount of plutonium greater than the critical mass without it exploding. We can, however, assemble as much deuterium and tritium as we please. It will never go off unless supplied with the extremely high temperature necessary for fusion.

The first H-bomb was detonated on October 31, 1952. It completely eliminated the island of Eniwetok in the Marshall Islands. The Soviet Union quickly followed suit by exploding their H-Bomb on August 12, 1953. The Soviets used lithium in place of tritium in their fusion reaction, because it is cheaper and more easily available.

The Fusion Reactor

One of the difficulties of a fission reactor is the radioactive fragments or waste that is a by-product of the reaction. *The fusion process has no by-product that is radioactive.* That is, the only result of fusion, is helium, which is an inert gas and is not radioactive. The proton-proton cycle in the sun is too slow to take place in a reactor. Hence, the fusion cycle in a fusion reactor is given by

$$^2_1H +\,^2_1H \rightarrow\,^3_1H +\,^1_1p$$
$$^2_1H +\,^3_1H \rightarrow\,^4_2He +\,^1_0n$$

The difficulty in the design of a fusion reactor has to do with the extremely high temperatures associated with the fusion process, that is, millions of kelvins. (Remember the surface temperature of the sun is about 6000 K, and the core, thousands of times higher.) At these high temperatures, all materials that could be made to contain the reaction would melt. The task of building a fusion reactor is not, however, impossible, just difficult. At the high temperatures of fusion, electrons and nuclei are completely separated from each other in what is called a *plasma,* an ionized fluid. Because of the electric charges of the fluid, the fusion reaction can be contained within magnetic fields. Experimental fusion reactors have been built on a very limited scale using magnetic confinement with some slight success. A great deal of work still has to be done to perfect the fusion reactor. This work must be done, because the fusion reactor promises to be a source of enormous energy produced by a very cheap fuel, effectively water, with no radioactive contaminants as a by-product.

33.9 Nucleosynthesis

It is a fact of life that we all take for granted the things that are around us. On this planet earth, the materials we see are made out of molecules and atoms. We saw in the discussion of the periodic table of the elements in chapter 32 how each element differs from each other by the number of electrons, protons, and neutrons contained within each atom. *But what is the origin of all these elements? How were they originally formed? The elements found on the earth and throughout the universe were originally synthesized by the process of fusion within the stars, a process called* **nucleosynthesis.** The proton-proton cycle formed helium from hydrogen. As a continuation of that cycle, $_2^3\text{He}$ can fuse with $_2^4\text{He}$ to produce beryllium according to the equation

$$_2^3\text{He} + _2^4\text{He} \rightarrow _4^7\text{Be} + \gamma$$

Beryllium can now capture an electron to form lithium according to the relation

$$_4^7\text{Be} + _{-1}^0\text{e} \rightarrow _3^7\text{Li} + \nu$$

As most of the hydrogen is used, the helium nuclei can now fuse to form the following nuclei

$$_2^4\text{He} + _2^4\text{He} \rightarrow _4^8\text{Be} + \gamma$$
$$_4^8\text{Be} + _2^4\text{He} \rightarrow _6^{12}\text{O*} \rightarrow _6^{12}\text{C} + 2\gamma$$
$$_6^{12}\text{C} + _2^4\text{He} \rightarrow _8^{16}\text{O} + \gamma$$
$$_8^{16}\text{O} + _2^4\text{He} \rightarrow _{10}^{20}\text{Ne} + \gamma$$

If the star continues burning, additional elements are formed, such as

$$_6^{12}\text{C} + _1^1\text{p} \rightarrow _7^{13}\text{N} + \gamma$$
$$_7^{13}\text{N} \rightarrow _6^{13}\text{C} + _{+1}^0\text{e} + \nu$$
$$_6^{13}\text{C} + _1^1\text{p} \rightarrow _7^{14}\text{N} + \gamma$$
$$_7^{14}\text{N} + _1^1\text{p} \rightarrow _8^{15}\text{O} + \gamma$$
$$_8^{15}\text{O} \rightarrow _7^{15}\text{N} + _{+1}^0\text{e} + \nu$$
$$_7^{15}\text{N} + _1^1\text{p} \rightarrow _8^{16}\text{O} + \gamma$$
$$_8^{16}\text{O} + _1^1\text{p} \rightarrow _9^{17}\text{F} + \gamma$$

As the nuclear fusion process continues, all the chemical elements and their isotopes up to about an atomic number of 56 or so, are created within the stars. If the star is large enough it eventually explodes as a supernova, spewing its contents into interstellar space. It is believed that the high temperatures in the explosion cause the formation of the higher chemical elements above iron. Some of the dust from these clouds is gradually pulled together by gravity to form still new stars. *Hence, stars are factories for the creation of the chemical elements.* If some of these fragments of the supernova are caught up in the gravitational field of another star, they could, with the correct initial velocity, go into orbit around the new star. The captured fragments of the star would slowly condense and become a planet with a complete set of elements as are now found on the planet earth.

An Essay on the Application of Physics

Radioactive Dating

A fossil of a seed fern. How can you tell how old it is?

Have you ever wondered how scientists are able to determine the age of very old objects? The technique used to determine their age is called **radioactive dating** and it is based upon the amount of unstable isotopes still contained in them. Perhaps the most famous of these techniques is *carbon dating*. Cosmic rays, which are high-energy protons and neutrons from outer space, impinge on the earth's upper atmosphere, and cause nuclear reactions with the nitrogen present there. The result of these nuclear reactions is to create an unstable isotope of carbon, namely, $^{14}_{6}C$, which has a half-life of 5770 yr. It is assumed that the total amount of this isotope remains constant with time because of an equilibrium between the amount being formed at any time and the amount decaying at any time. This isotope of carbon combines chemically with the oxygen, O_2, in the atmosphere to form carbon dioxide, CO_2. Most of the carbon dioxide in the atmosphere is, of course, formed from ordinary carbon, $^{12}_{6}C$. Because the chemical properties depend on the orbital electrons and not the nucleus, $^{14}_{6}C$ reacts chemically the same as $^{12}_{6}C$. Hence, we cannot determine chemically whether the carbon dioxide is made from carbon $^{12}_{6}C$ or carbon $^{14}_{6}C$.

The green plants in the environment convert water, H_2O, and carbon dioxide, CO_2, into carbohydrates by the process of photosynthesis. Hence, the radioactive isotope $^{14}_{6}C$ becomes a part of every living plant. Animals and humans eat these plants while also exhaling carbon dioxide. Thus plants, animals, and humans are found to contain the radioactive isotope $^{14}_{6}C$. The ratio of the carbon isotope $^{14}_{6}C$ to ordinary carbon $^{12}_{6}C$ is a constant in the atmosphere and all living things. The ratio is of course quite small, approximately 1.3×10^{-12}. That is, the amount of carbon $^{14}_{6}C$ is equal to 0.0000000000013 times the amount of ordinary carbon. Whenever any living thing dies, the radioactive isotope $^{14}_{6}C$ is no longer replenished and decreases by beta decay according to the reaction

$$^{14}_{6}C \rightarrow {}^{14}_{7}N + {}_{-1}^{0}e + \bar{\nu} \qquad \text{(33H.1)}$$

Thus, the ratio of $^{14}_{6}C/^{12}_{6}C$ is no longer a constant, but starts to decay with time. Thus, by knowing the present ratio of $^{14}_{6}C/^{12}_{6}C$, the age of the particular object can be determined. In practice, the amount of $^{14}_{6}C$ nuclei is relatively difficult to measure, whereas its activity, the number of disintegrations per unit time, is not. Using equation 33.12 for the activity of a radioactive nucleus, we get

$$\frac{A}{A_0} = e^{-\lambda t}$$

Taking the natural logarithms of both sides of the equation, we get

$$\ln\left(\frac{A}{A_0}\right) = -\lambda t$$

Solving for the time t, we get

$$t = \frac{-\ln(A/A_0)}{\lambda} \qquad \text{(33H.2)}$$

Thus, if the activity A_0 of a present living thing is known, and the activity A of the object we wish to date is measured, we can solve equation 33H.2 for its age.

Example 33H.1

Carbon dating. A piece of wood believed to be from an ancient Egyptian tomb is tested in the laboratory for its carbon-14 activity. It is found that the old wood has an activity of 10.0 disintegrations/min, whereas a new piece of wood has an activity of 15.0 disintegrations/min. Find the age of the wood.

Solution

First, we find the decay constant of $^{14}_{6}C$ from equation 33.11 as

$$\lambda = \frac{0.693}{T_{1/2}}$$

$$= \frac{(0.693)(1 \text{ yr})}{(5770 \text{ yr})(3.1535 \times 10^7 \text{ s})}$$

$$= 3.86 \times 10^{-12}/\text{s}$$

We find the age of the wood from equation 33H.2 with $A = 10.0$ disintegrations/min and $A_0 = 15$ disintegrations/min. Thus,

$$t = \frac{-\ln(A/A_0)}{\lambda}$$

$$= \frac{\ln(10.0/15.0)}{3.86 \times 10^{-12}/\text{s}}$$

$$= (+1.05 \times 10^{11} \text{ s})\left(\frac{1 \text{ yr}}{3.1535 \times 10^7 \text{ s}}\right)$$

$$= 3330 \text{ yr}$$

Hence, the wood must be 3330 yr old.

Similar dating techniques are used in geology to determine the age of rocks. As an example, the uranium atom $^{238}_{92}U$ decays through a series of steps and ends up as the stable isotope of lead, $^{206}_{82}Pb$. The ratio of the abundance of $^{238}_{92}U$ to $^{206}_{82}Pb$ can be used to determine the age of a rock.

The Language of Physics

Atomic number Z
The number of protons or electrons in an atom (p. 1004).

Mass number A
The number of protons plus neutrons in the nucleus (p. 1004).

Neutron number N
The number of neutrons in the nucleus. It is equal to the difference between the mass number and the atomic number (p. 1004).

Isotope
An isotope of a chemical element has the same number of protons as the element but a different number of neutrons. An isotope reacts chemically in the same way as the parent element. Its observable difference is its different atomic mass, which comes from the excess or deficiency of neutrons in the nucleus (p. 1004).

Atomic mass
The mass of a chemical element that is listed in the periodic table of the elements. That atomic mass is an average of the masses of its different isotopes (p. 1005).

Strong nuclear force
The force that binds protons and neutrons together in the nucleus. Whenever the nuclear force is less than the electrostatic force, the nucleus breaks up or decays, and emits radioactive particles (p. 1005).

Mass defect
The difference in mass between the sum of the masses of the constituents of a nucleus and the mass of the nucleus (p. 1006).

Binding energy
The energy that binds the nucleus together. It is the mass defect expressed as an energy (p. 1006).

Radioactivity
The spontaneous disintegration of the nuclei of an atom with the emission of α, β, or γ particles (p. 1007).

Activity
The rate at which nuclei decay with time (p. 1007).

Half-life
The time it takes for half the original radioactive nuclei to decay (p. 1008).

Alpha decay
A disintegration of an atomic nucleus whereby an α particle is emitted. The original element of atomic number Z is transmuted into a new chemical element of atomic number $Z - 2$ (p. 1012).

Beta decay, β^-
A nuclear decay whereby a neutron within the nucleus decays into a proton, an electron, and an antineutrino. The proton stays in the nucleus, but the electron and antineutrino are emitted. Thus, the atomic number Z increases by 1, but the mass number A stays the same. Hence, a chemical element Z is transmuted into the element $Z + 1$ (p. 1013).

Beta decay, β^+
A nuclear decay whereby a proton within the nucleus decays into a neutron, a positron, and a neutrino. The positron and neutrino are emitted but the neutron stays behind in the nucleus. The atomic number Z of the element decreases by one because of the loss of the proton. Hence, an element of atomic number Z is converted into the element $Z - 1$ (p. 1014).

Q value of a nuclear reaction
The energy available in a reaction caused by the difference in mass between the reactants and the products (p. 1020).

Exoergic reaction
A nuclear reaction in which energy is released. It is sometimes called an exothermic reaction (p. 1020).

Endoergic reaction
A nuclear reaction in which energy must be added to the system to make the reaction proceed. It is sometimes called an endothermic reaction (p. 1020).

Nuclear fission
The process of splitting a heavy atom into two lighter atoms (p. 1022).

Nuclear fusion
The process in which lighter nuclei are joined together to produce a heavier nucleus with a large amount of energy released (p. 1027).

Nucleosynthesis
The formation of the nuclei of all the chemical elements by the process of fusion within the stars (p. 1029).

Radioactive dating
A technique in which the age of very old objects can be determined by the amount of unstable isotopes still contained in them (p. 1030).

Summary of Important Equations

Neutron number
$$N = A - Z \tag{33.1}$$

Representation of a nucleus
$$^A_Z X \tag{33.2}$$

Mass defect
$$\Delta m = Z m_p + (A - Z)m_n - m_{\text{nucleus}} \tag{33.3}$$

Binding energy
$$BE = (\Delta m)c^2 \tag{33.5}$$

Rate of nuclear decay
$$\frac{\Delta N}{\Delta t} = -\lambda N \tag{33.6}$$

Activity
$$A = -\frac{\Delta N}{\Delta t} = \lambda N \tag{33.7}$$

Radioactive decay law
$$N = N_0 e^{-\lambda t} \tag{33.8}$$

Decay constant
$$\lambda = \frac{0.693}{T_{1/2}} \tag{33.11}$$

Alpha decay
$$^A_Z X \rightarrow {}^{A-4}_{Z-2} X + {}^4_2 He \tag{33.16}$$

Neutron decay
$$^1_0 n \rightarrow {}^1_1 p + {}^{\ 0}_{-1} e + \bar{\nu} \tag{33.19}$$

Beta$^-$ decay
$$^A_Z X \rightarrow {}^{\ \ A}_{Z+1} X + {}^{\ 0}_{-1} e + \bar{\nu} \tag{33.20}$$

Proton decay
$$^1_1 p \rightarrow {}^1_0 n + {}^0_{+1} e + \nu \tag{33.21}$$

Beta$^+$ decay
$$^A_Z X \rightarrow {}^{\ \ A}_{Z-1} X + {}^0_{+1} e + \nu \tag{33.22}$$

Electron capture
$$^{\ 0}_{-1} e + {}^1_1 p \rightarrow {}^1_0 n + \nu \tag{33.23}$$

Electron capture
$$^{\ 0}_{-1} e + {}^A_Z X \rightarrow {}^{\ \ A}_{Z-1} X + \nu \tag{33.24}$$

Gamma decay
$$^A_Z X^* \rightarrow\, ^A_Z X + \gamma \qquad (33.25)$$

Q value of a nuclear reaction
$$Q = (m_x + M_X)c^2$$
$$\quad - (m_y + M_Y)c^2 \qquad (33.31)$$
$$Q = [(\text{Input mass})$$
$$\quad - (\text{Output mass})]c^2 \qquad (33.32)$$

$$Q = E_{\text{in}} - E_{\text{out}} \qquad (33.33)$$
General form of equation for nuclear reaction
$$x + X = y + Y + Q \qquad (33.34)$$

Nuclear fission of $^{235}_{92}U$
$$^1_0 n + \,^{235}_{92}U \rightarrow y + Y + \,^1_0 n + Q \qquad (33.35)$$

Proton-proton cycle of nuclear fusion
$$^1_1 p + \,^1_1 p \rightarrow \,^2_1 H$$
$$\quad + \,^0_{+1}e + \nu \qquad (33.40)$$
$$^2_1 H + \,^1_1 p \rightarrow \,^3_2 He + \gamma \qquad (33.41)$$
$$^3_2 He + \,^3_2 He \rightarrow \,^4_2 He + 2\,^1_1 p \qquad (33.42)$$

Radioactive age
$$t = \frac{-\ln(A/A_0)}{\lambda} \qquad (33H.2)$$

Questions for Chapter 33

1. What are isotopes? What do they have in common and what are their differences?
2. What is the difference between fast neutrons and slow neutrons, and how do they have an effect on nuclear reactions?
3. What do we mean by the term critical mass?

4. Discuss the advantages and disadvantages of nuclear power compared to the use of fossil-fuel-generated power.
†5. What is a radioactive tracer and how is it used in medicine?
6. Explain the difference between nuclear fission and nuclear fusion.
7. Should an atomic bomb really be called a nuclear bomb?

8. How is the half-life of a radioactive substance related to its activity?
†9. Was the Chernobyl Nuclear Reactor explosion in the Soviet Union a nuclear explosion? Does the fact that the reactor was a breeder reactor, rather than a commercial electricity generator, have anything to do with the severity of the disaster?

Problems for Chapter 33

Section 33.2 Nuclear Structure

1. Find the atomic number, the mass number, and the neutron number for (a) $^{58}_{29}Cu$, (b) $^{24}_{11}Na$, (c) $^{210}_{84}Po$, (d) $^{45}_{20}Ca$, and (e) $^{206}_{82}Pb$.
2. Determine the number of protons and neutrons in one atom of (a) $^{87}_{37}Rb$, (b) $^{40}_{19}K$, (c) $^{137}_{55}Cs$, (d) $^{60}_{27}Co$, and (e) $^{131}_{53}I$.
3. Find the number of protons in 1 g of $^{40}_{19}K$.
4. $^{63}_{29}Cu$ has an atomic mass of 62.929595 u and an abundance of 69.09%, whereas $^{65}_{29}Cu$ has an atomic mass of 64.927786 u and an abundance of 30.91%. Find the atomic mass of the element copper.
5. $^{107}_{47}Ag$ has an atomic mass of 106.905095 u and an abundance of 51.83%, whereas $^{109}_{47}Ag$ has an atomic mass of 108.904754 u and an abundance of 48.17%. Find the atomic mass of the element silver.
6. Find the mass defect and the binding energy for the helium nucleus if the atomic mass of the helium nucleus is 4.0026 u.
7. Find the mass defect and the binding energy for tritium if the atomic mass of tritium is 3.016049 u.
8. How much energy would be released if six hydrogen atoms and six neutrons were combined to form $^{12}_6 C$?

Section 33.3 Radioactive Decay Law

9. $^{63}_{28}Ni$ has a half-life of 92 yr. Find its decay constant.
10. $^{235}_{92}U$ has a half-life of 7.038×10^8 yr. Find its decay constant.
11. An unknown sample has a decay constant of 2.83×10^{-6} 1/s. Find the half-life of the sample.
12. The decay constant of $^{14}_6 C$ is $\lambda = 3.86 \times 10^{-12}$ s^{-1}. If there are 7.35×10^{90} atoms of carbon fourteen at $t = 0$, how many of them will decay in a time of $t = 2.00 \times 10^{12}$ s?
13. A sample contains 0.200 moles of $^{65}_{30}Zn$. If $^{65}_{30}Zn$ has a decay constant of 3.27×10^{-8} /s, find the number of $^{65}_{30}Zn$ nuclei present at the end of 1 day.
14. One gram of $^{87}_{36}Kr$ has a half-life of 78.0 min. How many of these nuclei are still present at the end of 15.0 min?
15. $^{60}_{27}Co$ has a half-life of 5.27 yr. How long will it take for 90.0% of the original sample to disintegrate?
16. $^{90}_{38}Sr$ has a half-life of 28.8 yr. How long will it take for it to decay to 10.0% of its original value?
17. A dose of 1.85×10^6 Bq of radioactive iodine, $^{131}_{53}I$, is used in the treatment of a disorder of the thyroid gland. If its half-life is 8 days, find the activity after (a) 8 days, (b) 16 days, and (c) 32 days.

18. In a given sample of radioactive material, the number of original nuclei drops from 6.00×10^{50} to 1.50×10^{50} in 4.50 s. Find (a) the half-life and (b) the mean lifetime (τ_{avg}) of the material.

Section 33.4 Forms of Radioactivity

19. $^{220}_{86}Rn$ decays by alpha emission. What isotope is formed?
20. $^{230}_{90}Th$ decays by alpha emission. What isotope is formed?
21. If ^{223}U decays twice by alpha emission, what is the resulting isotope?
22. $^{214}_{84}Po$ decays by β^- decay. What isotope is formed?
23. $^{210}_{82}Pb$ decays by β^- decay. What isotope is formed?
24. $^{33}_{17}Cl$ decays by β^+ decay. What isotope is formed?
25. $^{49}_{24}Cr$ decays by β^+ decay. What isotope is formed?
26. $^{41}_{20}Ca$ decays by electron capture. What isotope is formed?
27. $^{52}_{25}Mn$ decays by electron capture. What isotope is formed?

28. How much energy is released or absorbed in the following reaction?

$$^{216}_{84}\text{Po} \rightarrow \,^{212}_{82}\text{Pb} + \,^{4}_{2}\text{He}$$

The atomic mass of $^{216}_{84}\text{Po}$ is 216.0019 u, $^{4}_{2}\text{He}$ is 4.002603 u, and $^{212}_{82}\text{Pb}$ is 211.9919 u.

29. Determine the energy associated with the reactions

$$^{1}_{0}\text{n} + \,^{1}_{1}\text{p} \rightarrow \,^{0}_{-1}\text{e} + \bar{\nu}$$
$$^{1}_{1}\text{p} \rightarrow \,^{1}_{0}\text{n} + \,^{0}_{+1}\text{e} + \nu$$

30. Find the Q value associated with the reaction

$$^{1}_{1}\text{H} + \,^{14}_{7}\text{N} \rightarrow \,^{15}_{8}\text{O} + \nu$$

The atomic mass of $^{14}_{7}\text{N}$ is 14.003074 and $^{15}_{8}\text{O}$ is 15.003072 u.

31. Find the Q value associated with the reaction

$$^{14}_{6}\text{C} \rightarrow \,^{14}_{7}\text{N} + \,^{0}_{-1}\text{e} + \bar{\nu} + Q$$

32. Find the Q value associated with the nuclear fission reaction

$$^{1}_{0}\text{n} + \,^{235}_{92}\text{U} \rightarrow \,^{132}_{50}\text{Sn} + \,^{101}_{42}\text{Mo}$$
$$+ \, 3^{1}_{0}\text{n} + Q$$

The atomic mass of $^{235}_{92}\text{U}$ is 235.043933 u, $^{132}_{50}\text{Sn}$ is 49.917756 u, and $^{101}_{42}\text{Mo}$ is 41.910346 u.

33. Find the Q value of the fusion reaction

$$^{2}_{1}\text{H} + \,^{3}_{1}\text{H} \rightarrow \,^{4}_{2}\text{He} + \,^{1}_{0}\text{n} + Q$$

Additional Problems

†34. A 5.00-g sample of $^{60}_{27}\text{Co}$ has a half-life of 5.27 yr. Find (a) the decay constant, (b) the activity of the material when $t = 0$, (c) the activity when $t = 1.00$ yr, and (d) the number of nuclei present after 1.00 yr.

†35. A 5.00-g sample of $^{230}_{90}\text{Th}$ has a half-life of 80.0 yr, and a 5.00-g sample of $^{222}_{86}\text{Rn}$ has a half-life of 3.82 days. For each sample find (a) the decay constant, (b) the activity of the material when $t = 0$, (c) the activity when $t = 100$ days, (d) the number of nuclei present after 100 days. (e) Comparing the activities and the number of radioactive nuclei remaining at 100 days for the two samples, what can you conclude?

36. If ^{231}Pa decays first by beta decay, and then by alpha emission, what is the resulting isotope?

37. A bone from an animal is found in a very old cave. It is tested in the laboratory and it is found that it has a carbon-14 activity of 13.0 disintegrations per minute. A similar bone from a new animal is tested and found to have an activity of 25.0 disintegrations per minute. What is the age of the bone?

38. A wooden statue is observed to have a carbon fourteen activity of 7.0 disintegrations per minute. How old is the statue? (New wood was found to have an activity of 15.0 disintegrations/min.)

Interactive Tutorials

◻39. Radioactive decay. A mass of 8.55 g of the isotope $^{90}\text{Sr}_{38}$ has a half-life $T_{1/2} = 28.8$ yr. Find (a) the decay constant λ, (b) the number of nuclei N_0 present at the start, (c) the activity A_0 at the start, (d) the number of nuclei N present for $t = T_{1/2}$, (e) the rate of decay of the nuclei at $t = T_{1/2}$, (f) the number of nuclei present for any time t, and (g) the activity at any time t.

34 Elementary Particle Physics and The Unification of the Forces

The Milky Way.

Three quarks for Muster Mark!
Sure he hasn't got much of a bark
and sure any he has it's all the mark.

James Joyce, Finnegans Wake

34.1 Introduction

Man has always searched for simplicity in nature. Recall that the ancient Greeks tried to describe the entire physical world in terms of the four quantities of earth, air, fire, and water. These, of course, have been replaced with the fundamental quantities of length, mass, charge, and time in order to describe the physical world of space, matter, and time. We have seen that space and time are not independent quantities, but rather are a manifestation of the single quantity—spacetime—and that mass and energy are interchangeable, so that energy could even be treated as one of the fundamental quantities. We also found that energy is quantized and therefore, matter should also be quantized. What is the smallest quantum of matter? That is, what are the fundamental or elementary building blocks of matter? What are the forces that act on these fundamental particles? Is it possible to combine these forces of nature into one unified force that is responsible for all the observed interactions? We shall attempt to answer these questions in this chapter.

34.2 Particles and Antiparticles

As mentioned in chapter 18, the Greek philosophers Leucippus and Democritus suggested that matter is composed of fundamental or elementary particles called atoms. The idea was placed on a scientific foundation with the publication, by John Dalton, of *A New System of Chemical Philosophy* in 1808, in which he listed about 20 chemical elements, each made up of an atom. By 1896 there were about 60 known elements. It became obvious that there must be a way to arrange these different atoms in an orderly way in order to make sense of what was quickly becoming chaos. In 1869 the Russian chemist, Dimitri Mendeleev, developed the periodic table of the elements based on the chemical properties of the elements. Order was brought to the chaos of the large diversity of elements. In fact, new chemical elements were predicted on the basis of the blank spaces found in the periodic table. Later with the discovery of the internal structure of the atom, the atom could no longer be considered as elementary.

By 1932, only four elementary particles were known; the electron, the proton, the neutron, and the photon. Things looked simple again. But this simplicity was not to last. Other particles were soon discovered in cosmic rays. Cosmic rays are particles from outer space that impinge on the top of the atmosphere. Some of them make it to the surface of the earth, whereas others decay into still other particles before they reach the surface. Other new particles were found in the large accelerating machines made by man. Today, there are hundreds of such particles. Except

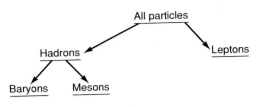

Figure 34.1

First classification of the elementary particles.

Table 34.1

List of Some of the Elementary Particles

Leptons	electron,	e^-
	muon,	μ^-
	tauon,	τ^-
	neutrinos,	ν_e, ν_μ, ν_τ
Hadrons		
Baryons	proton,	p
	neutron,	n
	delta,	Δ
	lambda,	λ
	Sigma,	Σ
	Hyperon,	Λ
	Omega	Ω
Mesons	pi,	π
	eta,	η
	rho,	ρ
	omega,	Ω
	delta,	δ
	phi	ϕ

for the electron, proton, and neutron, most of these elementary particles decay very quickly. We are again in the position of trying to make order out of the chaos of so many particles.

The first attempt at order is the classification of particles according to the scheme shown in figure 34.1. All the elementary particles can be grouped into particles called hadrons or leptons.

Leptons

The Leptons are particles that are not affected by the strong nuclear force. They are very small in terms of size, in that they are less than 10^{-19} m in diameter. They all have spin $1/2$ in units of $\hbar$. There are a total of six leptons: the electron, e^-; the muon, μ^-; and the tauon, τ^-, each with an associated neutrino. They can be grouped in the form

$$\begin{array}{ccc} (\nu_e) & (\nu_\mu) & (\nu_\tau) \\ (e^-) & (\mu^-) & (\tau^-) \end{array} \qquad (34.1)$$

There are thus three neutrinos: the neutrino associated with the electron, ν_e; the neutrino associated with the muon, ν_μ; and the neutrino associated with the tauon, ν_τ. The muon is very much like an electron but it is much heavier. It has a mass about 200 times greater than the electron. It is not stable like the electron but decays in about 10^{-6} s.

Originally the word lepton, which comes from the Greek word *leptos* meaning small or light in weight, signified that these particles were light. However, in 1975 the τ lepton was discovered and it has twice the mass of the proton. That is, the τ lepton is a heavy lepton, certainly a misnomer.

Leptons are truly elementary in that they apparently have no structure. That is, they are not composed of something still smaller. Leptons participate in the weak nuclear force, while the charged leptons, e^-, μ^-, τ^-, also participate in the electromagnetic interaction.

The muon was originally thought to be Yukawa's meson (discussed in chapter 31) that mediated the strong nuclear force, and hence it was called a μ^- meson. This is now known to be a misnomer, since the muon is not a meson but a lepton.

Hadrons

Hadrons are particles that are affected by the strong nuclear force. There are hundreds of known hadrons. Hadrons have an internal structure, composed of what appears to be truly elementary particles called quarks. The hadrons can be further broken down into two subgroups, the baryons and the mesons.

1. **Baryons.** Baryons are heavy particles that, *when they decay, contain at least one proton or neutron in the decay products.* The baryons have half-integral spin, that is, $\frac{1}{2}\hbar$, $\frac{3}{2}\hbar$, and so on. We will see in a moment that all *baryons are particles that are composed of three quarks.*

2. **Mesons.** Originally, mesons were particles of intermediate-sized mass between the electron and the proton. However many massive mesons have since been found, so the original definition is no longer appropriate. *A meson is now defined as any particle whose decay products do not include a baryon.* We will see that *mesons are particles that are composed of a quark-antiquark pair.* All mesons have integral spin, that is, 0, $1\hbar$, $2\hbar$, $3\hbar$, and so on. The mass of the meson increases with its spin. A list of some of the elementary particles is shown in table 34.1.

In 1928, Paul Dirac merged special relativity with the quantum theory to give a relativistic theory of the electron. A surprising result of that merger was that his equations predicted two energy states for each electron. One is associated with

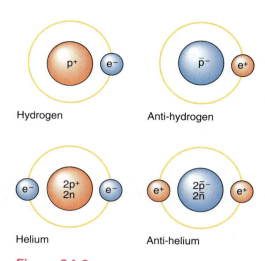

Hydrogen   Anti-hydrogen

Helium   Anti-helium

Figure 34.2

Matter and antimatter.

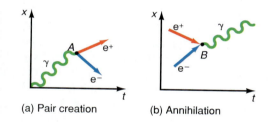

(a) Pair creation   (b) Annihilation

Figure 34.3

Creation and annihilation of particles.

the electron, whereas the other is associated with a particle, like the electron in every way, except that it carries a positive charge. This new particle was called the *antielectron* or the *positron*. This was the first prediction of the existence of antimatter. The positron was found in 1932.

For every particle in nature there is associated an **antiparticle.** *The antiparticle of the proton is the antiproton.* It has all the characteristics of the proton except that it carries a negative charge. Some purely neutral particles such as the photon and the π^0 meson are their own antiparticles. Antiparticles are written with a bar over the symbol for the particle. Hence, $\bar{p}$ is an antiproton and $\bar{n}$ is an antineutron.

Matter consists of electrons, protons and neutrons, whereas **antimatter** *consists of antielectrons (positrons), antiprotons, and antineutrons.* Figure 34.2 shows atoms of matter and antimatter. The same electric forces that hold matter together, hold antimatter together. (Note that the positive and negative signs are changed in antimatter.) The antihelium nucleus has already been made in high-energy accelerators.

Whenever particles and antiparticles come together they annihilate each other and only energy is left. For example, when an electron comes in contact with a positron they annihilate according to the reaction

$$e^- + e^+ \rightarrow 2\gamma \tag{34.2}$$

where the 2γ's are photons of electromagnetic energy. (Two gamma rays are necessary in order to conserve energy and momentum.) This energy can also be used to create other particles. Conversely, particles can be created by converting the energy in the photon to a particle-antiparticle pair such as

$$\gamma \rightarrow e^- + e^+ \tag{34.3}$$

Creation or annihilation can be shown on a spacetime diagram, called a *Feynman diagram,* after the American physicist Richard Feynman (1918–1988), such as in figure 34.3. Figure 34.3(a) shows the creation of an electron-positron pair. A photon γ moves through spacetime until it reaches the spacetime point A, where the energy of the photon is converted into the electron-positron pair. Figure 34.3(b) shows an electron and positron colliding at the spacetime point B where they annihilate each other and only the photon γ now moves through spacetime. (In order to conserve momentum and energy in the creation process, the presence of a relatively heavy nucleus is required.)

34.3 The Four Forces of Nature

In the study of nature, four forces that act on the particles of matter are known. They are:

1. *The Gravitational Force.* The gravitational force is the oldest known force. It holds us to the surface of the earth and holds the entire universe together. It is a long-range force, varying as $1/r^2$. Compared to the other forces of nature it is by far the weakest force of all.

2. *The Electromagnetic Force.* The electromagnetic force was the second force known. In fact, it was originally two forces, the electric force and the magnetic force, until the first unification of the forces tied them together as a single electromagnetic force. The electromagnetic force holds atoms, molecules, solids, and liquids together. Like gravity, it is a long-range force varying as $1/r^2$.

3. *The Weak Nuclear Force.* The weak nuclear force manifests itself not so much in holding matter together, but in allowing it to disintegrate, such as in the decay of the neutron and the proton. The weak force is responsible for the fusion process occurring in the sun by allowing a proton to decay into a

An aerial view of the particle accelerator at the Fermi National Accelerator Laboratory, where protons and antiprotons are accelerated and collided in an effort to analyze subnuclear particles.

Analyzing the particle tracks from a bubble chamber.

neutron such as given in equation 33.21. The proton-proton cycle then continues until helium is formed and large quantities of energy are given off. The nucleosynthesis of the chemical elements also occurred because of the weak force. Unlike the gravitational and electromagnetic force, the weak nuclear force is a very short range force.

4. *The Strong Nuclear Force.* The strong nuclear force is responsible for holding the nucleus together. It is the strongest of all the forces but is a very short range force. That is, its effects occur within a distance of about 10^{-15} m, the diameter of the nucleus. At distances greater than this, there is no evidence whatsoever for its very existence. The strong nuclear force acts only on the hadrons.

Why should there be four forces in nature? Einstein, after unifying space and time into spacetime, tried to unify the gravitational force and the electromagnetic force into a single force. Although he spent a lifetime trying, he did not succeed. The hope of a unification of the forces has not died, however. In fact, we will see shortly that the electromagnetic force and the weak nuclear force have already been unified theoretically into the electroweak force by Glashow, Weinberg, and Salam, and experimentally confirmed by Rubbia. A grand unification between the electroweak and the strong force has been proposed. Finally an attempt to unify all the four forces into one superforce is presently underway.

34.4 Quarks

In the attempt to make order out of the very large number of elementary particles, Murray Gell-Mann and George Zweig in 1964, independently proposed that the hadrons were not elementary particles but rather were made of still more elementary particles. Gell-Mann called these particles, **quarks.** He initially assumed there were only three such quarks, but with time the number has increased to six. The six quarks are shown in table 34.2. *The names of the quarks are: up, down, strange, charmed, bottom, and top.* (Sometimes, the bottom quark is called "beauty" and the top quark, "truth.") One of the characteristics of these quarks is that they have fractional electric charges. That is, the up, charmed, and top quark has 2/3 of the charge found on the proton, whereas the down, strange, and bottom quark has 1/3 of the charge found on the electron. They all have spin 1/2, in units of $\hbar$. Each quark has an antiquark, which is the same as the original quark except it has an opposite charge. The antiquark is written with a bar over the letter, that is $\bar{q}$.

We will now see that all of the hadrons are made up of quarks. The baryons are made up of three quarks:

$$\text{Baryon} = qqq \tag{34.4}$$

Table 34.2
The Quarks

Name (Flavor)	Symbol	Charge	Spin
up	u	2/3	1/2
down	d	−1/3	1/2
strange	s	−1/3	1/2
charmed	c	2/3	1/2
bottom	b	−1/3	1/2
top	t	2/3	1/2

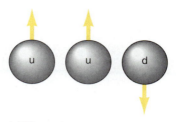

(a) The proton, p

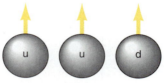

(b) The delta plus, Δ^+

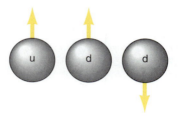

(c) The neutron, n

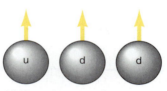

(d) The delta zero, Δ^0

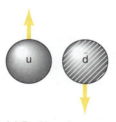

(e) Positive pi-meson, π^+

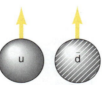

(f) Positive rho-meson, ρ^+

Figure 34.4

Some quark configurations of baryons and mesons.

While the mesons are made up of a quark-antiquark pair:

$$\text{Meson} = q\bar{q} \qquad (34.5)$$

As an example of the formation of a baryon from quarks, consider the proton. The proton consists of two up quarks and one down quark, as shown in figure 34.4(a). The electric charge of the proton is found by adding the charges of the constitutive quarks. That is, since the u quark has a charge of 2/3, and the d quark has a charge of $-1/3$, the charge of the proton is

$$\tfrac{2}{3} + \tfrac{2}{3} - \tfrac{1}{3} = 1$$

which is exactly as expected. Now the proton should have a spin of 1/2 in units of $\hbar$. In figure 34.4(a), we see the two up quarks as having their spin up by the direction of the arrow on the quark. The down quark has its arrow pointing down to signify that its spin is down. Because each quark has spin 1/2, the spin of the proton is found by adding the spins of the quarks as

$$\tfrac{1}{2} + \tfrac{1}{2} - \tfrac{1}{2} = \tfrac{1}{2}$$

We should note that the names up and down for the quarks are just that, a name, and have nothing to do with the direction of the spin of the quark. For example, the delta plus Δ^+ baryon is made from the same three quarks as the proton, but their spins are all aligned in the same direction, as shown in figure 34.4(b). Thus, the spin of the Δ^+ particle is

$$\tfrac{1}{2} + \tfrac{1}{2} + \tfrac{1}{2} = \tfrac{3}{2}$$

That is, the Δ^+ particle has a spin of 3/2 $\hbar$. Since it takes more energy to align the spins in the same direction, when quark spins are aligned, they have more energy. This manifests itself as an increased mass by Einstein's equivalence of mass and energy ($E = mc^2$). Thus, we see that the mass of the Δ^+ particle has a larger mass than the proton. Hence, in the formation of particles from quarks, we not only have to know the types of quarks making up the particle but we must also know the direction of their spin.

Figure 34.4(c) shows that a neutron is made up of one up quark and two down quarks. The total electric charge is

$$\tfrac{2}{3} - \tfrac{1}{3} - \tfrac{1}{3} = 0$$

While its spin is

$$\tfrac{1}{2} + \tfrac{1}{2} - \tfrac{1}{2} = \tfrac{1}{2}$$

Again note that the delta zero Δ^0 particle is made up of the same three quarks, figure 34.4(d), but their spins are all aligned.

As an example of the formation of a meson from quarks, consider the pi plus π^+ meson in figure 34.4(e). It consists of an up quark and an antidown quark. Its charge is found as

$$\tfrac{2}{3} + [-(-\tfrac{1}{3})] = \tfrac{2}{3} + \tfrac{1}{3} = 1$$

That is, the d quark has a charge of $-1/3$, so its antiquark $\bar{d}$ has the same charge but of opposite sign $+1/3$. The spin of the π^+ is

$$\tfrac{1}{2} - \tfrac{1}{2} = 0$$

Thus, the π^+ meson has a charge of $+1$ and a spin of zero.

If the spins of these same two quarks are aligned, as in figure 34.4(f), the meson is the positive rho-meson ρ^+, with electric charge of $+1$ and spin of 1.

The quark structure of some of the baryons is shown in table 34.3, whereas table 34.4 shows the quark structure for some mesons.

Table 34.3
Quark Structure of Some of the Baryons

Name	Symbol	Structure	Charge (units of e)	Spin (units of $\hbar$)	Mass (GeV)
Proton	p	u u d	1	1/2	0.938
Neutron	n	u d d	0	1/2	0.940
Delta plus plus	Δ^{++}	u u u	2	3/2	1.232
Delta plus	Δ^{+}	u u d	1	3/2	
Delta zero	Δ^{0}	u d d	0	3/2	
Delta minus	Δ^{-}	d d d	−1	3/2	
Lambda zero	Λ^{0}	u d s	0	1/2	1.116
Positive sigma	Σ^{*+}	u u s	1	3/2	1.385
Positive sigma	Σ^{+}	u u s	1	1/2	1.189
Neutral sigma	Σ^{*0}	u d s	0	3/2	1.385
Neutral sigma	Σ^{0}	u d s	0	1/2	1.192
Negative sigma	Σ^{*-}	d d s	−1	3/2	1.385
Negative sigma	Σ^{-}	d d s	−1	1/2	1.197
Negative xi	Ξ^{-}	s d s	−1	1/2	1.321
Neutral xi	Ξ^{0}	s u s	0	1/2	1.315
Omega minus	Ω^{-}	s s s	−1	3/2	1.672
Charmed lambda	Λ_c^{+}	u d c	+1	1/2	2.281

Table 34.4
Quark Structure of Some Mesons

Name	Symbol	Structure	Charge (units of e)	Spin (units of $\hbar$)	Mass (GeV)
Positive pion	π^{+}	$\bar{d}$ u	+1	0	0.140
Positive rho	ρ^{+}	$\bar{d}$ u	+1	1	0.770
Negative pion	π^{-}	$\bar{u}$ d	−1	0	0.140
Negative rho	ρ^{-}	$\bar{u}$ d	−1	1	0.770
Pi zero	π^{0}	50%($\bar{u}$ u) + 50%($\bar{d}$ d)	0	0	0.135
Positive kaon	K^{+}	u $\bar{s}$	+1	0	0.494
Neutral kaon	K^{0}	$\bar{s}$ d	0	0	0.498
Negative kaon	K^{-}	$\bar{u}$ s	−1	0	0.494
J/Psi (charmonium)	J/Ψ	c $\bar{d}$	0	1	3.097
Charmed eta	η_c	c $\bar{c}$	0	0	2.980
Neutral D	D^{0}	$\bar{u}$ c	0	0	1.863
Neutral D	D^{*0}	$\bar{u}$ c	0	1	
Positive D	D^{+}	$\bar{d}$ c	1	0	1.868
Zero B-meson	B^{0}	$\bar{d}$ b	0		5.26
Negative B-meson	B^{-}	$\bar{u}$ b	−1		5.26
Upsilon	υ	$\bar{b}$ b	0	1	9.46
Phi-meson	Φ	s $\bar{s}$	0	1	1.020
F-meson	F^{+}	c $\bar{s}$	0	1	2.040

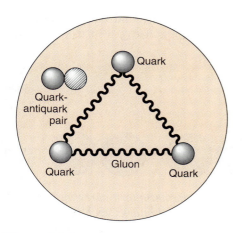

Figure 34.5

Structure of the proton.

(Source: Data from D. H. Perkins, "Inside the Proton" in *The Nature of Matter*, Oxford University Press, 1981.)

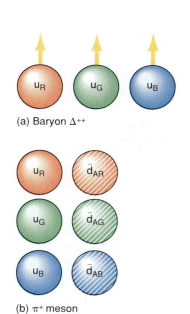

(a) Baryon Δ^{++}

(b) π^+ meson

Figure 34.6

Colored quarks.

Particles that contain the strange quark are called strange particles. The reason for this name is because these particles took so much longer to decay than the other elementary particles, that it was considered strange.

If a proton or neutron consists of quarks, we would like to "see" them. Just as Rutherford "saw" inside the atom by bombarding it with alpha particles, we can "see" inside a proton by bombarding it with electrons or neutrinos. In 1969, at the Stanford Linear Accelerator Center (SLAC), protons were bombarded by high-energy electrons. *It was found that some of these electrons were scattered at very large angles, just as in Rutherford scattering, indicating that there are small constituents within the proton.* Figure 34.5 shows the picture of a proton as observed by scattering experiments. The scattering appears to come from particles with charges of $+2/3$ and $-1/3$ of the electronic charge. (Recall that the up quark has a charge of $+2/3$, whereas the down quark has a charge of $-1/3$.) *There is thus, experimental evidence for the quark structure of the proton.* Similar experiments have also been performed on neutrons with the same success. The scattering also confirmed the existence of some quark-antiquark pairs within the proton. Recall that quark-antiquark pairs are the constituents of mesons. The experiments also showed the existence of other particles within the nucleons, called *gluons*. The gluons are the exchange particles between the quarks that act to hold the quarks together. They are the nuclear glue.

The one difficulty with the quark model at this point is that there seems to be a violation of the Pauli exclusion principle. Recall from chapter 32, section 32.8, that the Pauli exclusion principle stated that no two electrons can have the same quantum numbers at the same time. The Pauli exclusion principle is actually more general than that, in that it applies not only to electrons, but to any particles that have half-integral spin, such as $1/2$, $3/2$, $5/2$, and so on. *Particles that have half-integral spin are called fermions.* Because quarks have spin $1/2$, they also must obey the Pauli exclusion principle. But the Δ^{++} particle is composed of three up quarks all with the same spin, and the Ω^- particle has three strange quarks all with the same spin. *Thus, there must be an additional characteristic of each quark, that is different for each quark, so that the Pauli exclusion principle will not be violated. This new attribute of the quark is called "color."*

Quarks come in three colors: red, green, and blue. We should note that these colors are just names and have no relation to the real colors that we see everyday with our eyes. The words are arbitrary. As an example, they could just as easily have been called A, B, and C. We can think of color in the same way as electric charges. Electric charges come in two varieties, positive and negative. Color charges come in three varieties: red, green, and blue. *Thus, there are three types of up quarks; a red-up quark u_R, a green-up quark u_G, and a blue-up quark u_B.* Hence the delta plus-plus particle Δ^{++} can be represented as in figure 34.6(a). In this way there is no violation of the Pauli exclusion principle since each up quark is different.

All baryons are composed of red, green, and blue quarks. Just as the primary colors red, green, and blue add up to white, the combination of a red, green, and blue quark is said to make up the color white. All baryons are, therefore, said to be white, or colorless. Just as a quark has an antiquark, each color of quark has an anticolor. Hence, a red-up quark has an up antiquark that carries the color antired, and is called an antired-up quark. The varieties of quarks are called flavors, such as up, down, strange, and so on. Hence, each flavor of quark comes in three colors to give a total of six flavors times three colors equals 18 quarks. Associated with the 18 quarks are 18 antiquarks. *Mesons, like baryons, must also be white or colorless. Hence, one colored quark of a meson must always be associated with an anticolor, since a color plus its anticolor gives white.* Thus, possible formations of a π^+ meson are shown in figure 34.6(b). That is, a red-up quark u_R combines with an antidown quark that carries the color antired $\bar{d}_{AR}$ to form the white π^+ meson. (The anticolor quark is shown with the hatched lines in figure 34.6.) Similarly the

π^+ meson can be made out of green and antigreen $u_G\bar{d}_{AG}$ and blue and antiblue quarks $u_B\bar{d}_{AB}$ and a linear combination of them, such as $u_R\bar{d}_{AR} + u_G\bar{d}_{AG} + u_B\bar{d}_{AB}$. We can rewrite equations 34.4 and 34.5 as

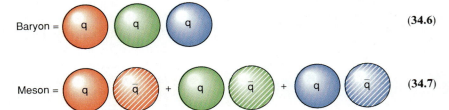

(34.6)

(34.7)

The force between a quark carrying a color and its antiquark carrying anticolor is always attractive. Similarly the force between three quarks each of a different color is also attractive. All other combinations of colors gives a repulsive force. We will say more about colored quarks when we discuss the strong nuclear force in section 34.8.

34.5 The Electromagnetic Force

The electromagnetic force has been discussed in some detail in chapters 18 through 25. To summarize the results from there, Coulomb's law gave the electric force between charged particles, and the electric field was the mediator of that force. The relation between electricity and magnetism was first discovered by Ampère when he found that a current flowing in a wire produced a magnetic field. Faraday found that a changing magnetic field caused an electric current. James Clerk Maxwell synthesized all of electricity with all of magnetism into his famous equations of electromagnetism. That is, the separate force of electricity and the force of magnetism were unified into one electromagnetic force.

The merger of electromagnetic theory with quantum mechanics has led to what is now called **quantum electrodynamics,** *which is abbreviated QED. In QED the electric force is transmitted by the exchange of a virtual photon.* That is, the force between two electrons can be visualized as in figure 34.7.

Recall from chapter 31 that the Heisenberg uncertainty relation allows for the creation of a virtual particle as long as the energy associated with the mass of the virtual particle is repaid in a time interval Δt that satisfies equation 31.56. In figure 34.7, two electrons approach each other. The first electron emits a virtual photon and recoils as shown. When the second electron absorbs that photon it also recoils as shown, leading to the result that the exchange of the photon caused a force of repulsion between the two electrons. As pointed out in chapter 31, this exchange force is strictly a quantum mechanical phenomena with no real classical analogue. So it is perhaps a little more difficult to visualize that the exchange of a photon between an electron and a proton produces an attractive force between them. *The exchanged photon is the mediator or transmitter of the force. All of the forces of nature can be represented by an exchanged particle.*

Because the rest mass of a photon is equal to zero, the range of the electric force is infinite. This can be shown with the help of a few equations from chapter 31. The payback time for the uncertainty principle was

$$\Delta t = \frac{\hbar}{\Delta E}$$

(31.56)

While the energy ΔE was related to the mass Δm of the virtual particle by

$$\Delta E = (\Delta m)c^2$$

Substituting this into equation 31.56, gave for the payback time

$$\Delta t = \frac{\hbar}{(\Delta m)c^2}$$

(31.57)

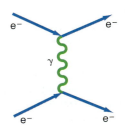

Figure 34.7

The electric force as an exchange of a virtual photon.

The distance a virtual particle could move and still return during that time Δt, was given as

$$d = c\frac{\Delta t}{2} \qquad (31.58)$$

This distance is called the *range* of the virtual particle. Substituting equation 31.57 into 31.58 gives for the range

$$d = c\frac{\hbar}{2(\Delta m)c^2}$$

$$d = \frac{\hbar}{2c}\frac{1}{\Delta m} \qquad (34.8)$$

For a photon, the rest mass Δm is equal to zero. So as the denominator of a fraction approaches zero, the fraction approaches infinity. Hence, the range d of the particle goes to infinity. Thus, the electric force should extend to infinity, which, of course, it does.

34.6 The Weak Nuclear Force

The weak nuclear force is best known for the part it plays in radioactive decay. Recall from chapter 33 on nuclear physics that the initial step in beta β^- decay is for a neutron in the nucleus to decay according to the relation

$$n \rightarrow p + e^- + \bar{\nu}_e \qquad (34.9)$$

Whereas the proton inside the nucleus decays as

$$p \rightarrow n + e^+ + \nu_e \qquad (34.10)$$

and is the initial step in the beta β^+ decay. Finally, the radioactive disintegration caused by the capture of an electron by the nucleus (electron capture), is initiated by the reaction

$$e^- + p \rightarrow n + \nu_e \qquad (34.11)$$

These three reactions are just some of the reactions that are mediated by the weak nuclear force.

The weak nuclear force does not exert the traditional push or pull type of force known in classical physics. Rather, it is responsible for the transmutation of the subatomic particles. The weak nuclear force is independent of electric charge and acts between leptons and hadrons and also between hadrons and hadrons. The range of the weak nuclear force is very small, only about 10^{-17} m. The decay time is relatively large in that the weak decay occurs in about 10^{-10} seconds, whereas decays associated with the strong interaction occur in approximately 10^{-23} seconds.

The weak nuclear force is the weakest force after gravity. A product of weak interactions is the neutrino. The neutrinos are very light particles. Some say they have zero rest mass while others consider them to be very small, with an upper limit of about 10–30 eV for the ν_e neutrino. The neutrino is not affected by the strong or electromagnetic forces, only by the weak force. Its interaction is so weak that it can pass through the earth or the sun without ever interacting with anything.

34.7 The Electroweak Force

Steven Weinberg, Abdus Salam, and Sheldon Glashow proposed a unification of the electromagnetic force with the weak nuclear force and received the Nobel Prize for their work in 1979. This force is called the **electroweak force.** Just as a virtual photon mediates the electromagnetic force between charged particles, it became obvious that there should also be some particle to mediate the weak nuclear force. The new

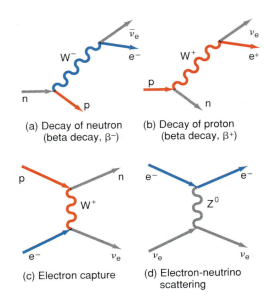

(a) Decay of neutron
(beta decay, β⁻)

(b) Decay of proton
(beta decay, β⁺)

(c) Electron capture

(d) Electron-neutrino
scattering

Figure 34.8

Examples of the electroweak force.

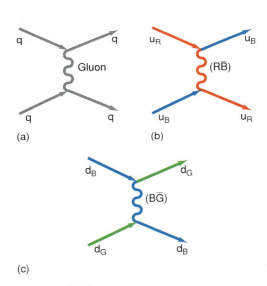

(a)

(b)

(c)

Figure 34.9

Exchange of gluons between quarks.

electroweak force is mediated by four particles: the photon and three intermediate *vector bosons* called W^+, W^-, and Z^0. The photon mediates the electromagnetic force, whereas the vector bosons mediate the weak nuclear force. In terms of the exchange particles, the decay of a neutron, equation 34.9, is shown in figure 34.8(a). *A neutron decays by emitting a W^- particle, thereby converting the neutron into a proton. The W^- particle subsequently decays within 10^{-26} s into an electron and an antineutrino.* The decay of the proton in a radioactive nucleus, equation 34.10, is shown in figure 34.8(b). *The proton emits the positive intermediate vector boson, W^+, and is converted into a neutron. The W^+ subsequently decays into a positron and a neutrino.* An electron capture, equation 34.11, is shown in figure 34.8(c) as a collision between a proton and an electron. The proton emits a W^+ and is converted into a neutron. The W^+ then combines with the electron forming a neutrino. The Z^0 particle is observed in electron-neutrino scattering, as shown in figure 34.8(d).

The vector bosons, W^+ and W^-, were found experimentally in proton-antiproton collisions at high energies, at the European Center for Nuclear Research (CERN), in January 1983, by a team headed by Carlo Rubbia of Harvard University. The Z^0 was found a little later in May 1983. The mass of the $W^{\pm}$ was around 80 GeV, while the mass of the Z^0 was about 90 GeV. Referring to equation 34.8, we see that for such a large mass, Δm in that equation gives a very short range d for the weak force, as found experimentally.

At very high energies, around 100 GeV, the electromagnetic force and the weak nuclear force merge into one electroweak force that acts equally between all particles: hadrons and leptons, charged and uncharged.

34.8 The Strong Nuclear Force

As mentioned previously, the **strong nuclear force** is responsible for holding the protons and neutrons together in the nucleus. The strong nuclear force must indeed be very strong to overcome the enormous electrical force of repulsion between the protons. Yukawa proposed that an exchange of mesons between the nucleons was the source of the nuclear force. But the nucleons are themselves made up of quarks. What holds these quarks together?

*In quantum electrodynamics (QED), the electric force was caused by the exchange of virtual photons. One of the latest theories in elementary particle physics is called **quantum chromodynamics (QCD)** and the force holding quarks together is caused by the exchange of a new particle, called a "gluon."* That is, a gluon is the nuclear glue that holds quarks together in a nucleon. Figure 34.9(a) shows the force between quarks as the exchange of a virtual gluon. Gluons, like quarks, come in colors and anticolors. *A gluon interacting with a quark changes the color of a quark.*

As an example, figure 34.9(b) shows a red-up quark u_R emitting a red-antiblue gluon $(R\overline{B})$. The up quark loses its red color and becomes blue. That is, *in taking away an anticolor, the color itself must remain.* Hence, taking away an antiblue from the up quark, the color blue must remain. When the first blue-up quark receives the red-antiblue gluon $(R\overline{B})$, the blue of the up quark combines with the antiblue of the gluon cancelling out the color blue. (A color and its anticolor always gives white.) The red color of the gluon is now absorbed by the up quark turning it into a red-up quark. *Thus, in the process of exchanging the gluon, the quarks changed color.* Figure 34.9(c) shows a blue-down quark emitting a blue-antigreen gluon $(B\overline{G})$, changing the blue-down quark into a green-down quark. When the first green-down quark absorbs the $(B\overline{G})$ gluon, the color green cancels and the down quark becomes a blue-down quark.

All told, there are eight different gluons and each gluon has a mass. Each gluon always carries one color and one anticolor. Occasionally a gluon can transform to a quark-antiquark pair.

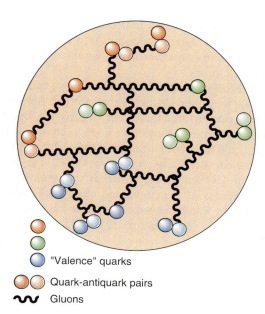

"Valence" quarks

Quark-antiquark pairs

Gluons

Figure 34.10

More detailed structure of the proton.

(Source: Data from D. H. Perkins, "Inside the Proton" in *The Nature of Matter,* Oxford University Press, 1981.)

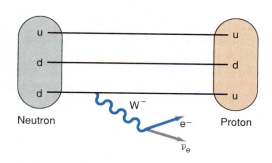

Figure 34.11

The decay of the neutron.

At energies greater than that used for the scattering shown in figure 34.5, scattering from protons reveals even more detail, as shown in figure 34.10. The three valence quarks are shown as before, but now there is also observed a large number of quark-antiquark pairs. Recall that a quark-antiquark pair constitutes a meson. *Hence, the proton is seething with virtual mesons.* Also observed are the gluons. To answer the traditional questions concerning what holds the protons together in the nucleus, we can say that the strong force is the result of the color forces between the quarks within the nucleons. At relatively large separation distances within the nucleus, the quark-antiquark pair (meson), which is created by the gluons, is exchanged between the nucleons. At shorter distances within the nucleus, the strong force can be explained either as an exchange between the quarks of one proton and the quarks of another proton, or perhaps as a direct exchange of the gluons themselves, which give rise to the quark-quark force within the nucleon. *Thus, the strong force originates with the quarks, and the force binding the protons and neutrons together in the nucleus is the manifestation of the force between the quarks.*

If quarks are the constituents of all the hadrons, why have they never been isolated? The quark-quark force is something like an elastic force given by Hooke's law, $F = -kx$. For small values of the separation distance x, the force between the quarks is small and the quarks are relatively free to move around within the particle. However, if we try to separate the quarks through a large separation distance x, then the force becomes very large, so large, in fact, that the quarks cannot be separated at all. *This condition is called the confinement of quarks. Thus, quarks are never seen in an isolated state because they cannot escape from the particle in which they are the constituents.*

But is there any evidence for the existence of quarks? The answer is yes. Experiments were performed in the new PETRA storage ring at DESY (Deutsches Electronen-Synchronton) in Hamburg, Germany, in 1978. Electrons and positrons, each at an energy of 20 GeV, were fired at each other in a head-on collision. The annihilation of the electron and its antiparticle, the positron, produce a large amount of energy; it is from this energy that the quarks are produced. The experimenters found a series of "quark jets," which were the decay products of the quarks, exactly as predicted. (A quark jet is a number of hadrons flying off from the interaction in roughly the same direction.) These quark jets were an indirect proof of the existence of quarks. Similar experiments have been performed at CERN and other accelerators.

As far as can be determined presently, quarks and gluons are not made up of still smaller particles; that is, they appear to be truly elementary. However, there are some speculative theories that suggest that quarks are made up of even smaller particles called preons. There is no evidence at this time, however, for the existence of preons.

34.9 Grand Unified Theories (GUT)

If it is possible to merge the electric force with the weak nuclear force, into a unified electroweak force, why not merge the electroweak force with the strong nuclear force? In 1973 Sheldon Glashow and Howard Georgi did exactly that, when they published a theory merging the electroweak with the strong force. This new theory was the first of many to be called the **grand unified theory,** or GUT.

The first part of this merger showed how the weak nuclear force was related to the strong nuclear force. Let us consider the decay of the neutron shown in equation 34.9:

$$n \rightarrow p + e^- + \bar{\nu}_e$$

We can now visualize this decay according to the diagram in figure 34.11. According to the quark theory, a neutron is composed of one up quark and two down quarks. One of the down quarks of the neutron emits the W$^-$ boson and is changed into an

Table 34.5

Family of Particles that Mediate the Unified Force

Particle	Number	Force Mediated
Photon	1	Electromagnetic
Vector bosons (W^+, W^-, Z^0)	3	Weak
Gluons	8	Strong
X particles	12	Strong-electroweak

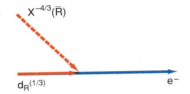

Figure 34.12

Changing a quark to an electron.

up quark, transforming the neutron into a proton. (Recall, that the proton consists of two up quarks and one down quark.) The W^- boson then decays into an electron and an antineutrino. Thus, *the weak force changes the flavor of a quark, whereas the strong force changes only the color of a quark.*

Above 10^{15} GeV of energy, called the *grand unification energy*, we can no longer tell the difference between the strong, weak, and electromagnetic forces. Above this energy there is only one unified interaction or force that occurs. Of course, this energy is so large that it is greater than anything we could ever hope to create experimentally. As we shall see, however, it could have been attained in the early stages of the creation of the universe—the so-called "Big Bang."

The strong nuclear force operates between quarks, whereas the weak nuclear force operates between quarks and leptons. If the strong and weak forces are to be combined, then the quarks and leptons should be aspects of one more fundamental quantity. That is, the grand unified force should be able to transform quarks into leptons and vice versa.

In the grand unified theories, there are 24 particles that mediate the unified force and they are listed in table 34.5. In grand unified theories, the forces are unified because the forces arise through the exchange of the same family of particles. As seen before, the photon mediates the electromagnetic force; the vector bosons mediate the weak force; the gluons mediate the strong force; and there are now 12 new particles called X particles (sometimes progenitor and/or lepto-quark particles) that mediate the unified force. It is these X particles that are capable of converting hadrons into leptons by changing quarks to leptons. The X particles come in four different electrical charges, $\pm 1/3$ and $\pm 4/3$. Thus, the X particles can be written as $X^{1/3}$, $X^{-1/3}$, $X^{4/3}$, and $X^{-4/3}$. Each of these X particles also comes in the three colors red, blue, and green, thereby giving the total of 12 X particles. The X particles can change a quark into a lepton, as shown in figure 34.12. An X particle carrying an electrical charge of $-4/3$, and a color charge of antired combines with a red-down quark, which carries an electrical charge of $1/3$. The colors red and antired cancel to give white, while the electrical charge becomes $1/3 - 4/3 = -3/3 = -1$ and an electron is created out of a quark. This type of process is not readily seen in our everyday life because the mass of the virtual X particle must be of the order of 10^{15} GeV, which is an extremely large energy. A similar analysis shows that an isolated proton should also decay. The lifetime, however, is predicted to be 10^{32} yr. Experiments are being performed to look for the predicted decays. However, at the present time no such decay of an isolated proton has been found. An isolated proton seems to be a very stable particle, indicating that either more experiments are needed, or the GUT model needs some modifications.

34.10 The Gravitational Force and Quantum Gravity

As has been seen throughout this book, physics is a science of successive approximations to the truth hidden in nature. Newton found that celestial gravity was of the same form as terrestrial gravity and unified them into his law of universal gravitation. However, it turned out that it was not quite so universal. Einstein started the change in his special theory of relativity, which governed systems moving with respect to each other at constant velocity. As he generalized this theory to systems that were accelerated with respect to each other, he found the equivalence between accelerated systems and gravity. The next step of course was to show that matter warped spacetime and gravitation was a manifestation of that warped spacetime. Thus, general relativity became a law of gravitation, and it was found that Newton's law of gravitation was only a special case of Einstein's theory of general relativity.

We have also seen that the quantum theory is one of the great new theories of modern physics, which seems to say that nature is quantized. There are quanta of energy, mass, angular momentum, charge, and the like. But general relativity, in its present format, is essentially independent of the quantum theory. It is, in this sense, still classical physics. It, too, must be only an approximation to the truth hidden in nature. *A more general theory should fuse quantum mechanics with general relativity—that is, we need a quantum theory of gravity.*

In order to combine quantum theory with general relativity (hereafter called Einstein's theory of gravitation), we have to determine where these two theories merge. Remember the quantum theory deals with very small quantities, because of the smallness of Planck's constant h, whereas Einstein's theory of gravitation deals with very large scale phenomena, or at least with very large masses that can significantly warp spacetime.

One of the important characteristics of the quantum theory is the wave-particle duality; waves can act as particles and particles can act as waves. And as has also been seen, waves can exist in the electromagnetic field. Let us, for the moment, compare electromagnetic fields with gravitational fields. On a large scale the electric field appears smooth. It is only when we go down to the microscopic level that we see that the electric field is not smooth at all, but rather is quite bumpy, because the energy of the electric field is not spread out in space but is, instead, stored in little bundles of electromagnetic energy, called photons. *Similarly, from the quantum theory we should expect that on a microscopic level the gravitational field should also be quantized into little particles, which we will call the quanta of the gravitational field—the* **gravitons.**

But what is a gravitational field but the warping of spacetime? Hence, a quantum of gravitation must be a quantum of spacetime itself. Thus, the graviton would appear to be a quantum of spacetime. Therefore, on a microscopic level, spacetime itself is probably not smooth but probably has a graininess or bumpiness to it. At this time, no one knows for sure what happens to spacetime on this microscopic level, but it has been conjectured that spacetime may look something like a foam that contains "wormholes."

At what point do the quantum theory and Einstein's theory of gravitation merge? The answer is to be found in Heisenberg's uncertainty principle.

$$\Delta E \Delta t \geq \hbar \tag{31.55}$$

For the electric field, small quantities of energy ΔE of the electric field are turned into small quanta of energy, the photons. In a similar manner, small quantities of energy ΔE of the gravitational field should be turned into little bundles or quantums of gravity, the gravitons. Since the range of a force is determined by the mass of the exchanged particle, and the range of the gravitational force is known to be infinite, it follows that the rest mass of the graviton must be zero. Hence, a quantum fluctuation should appear as a gravitational wave moving at the speed of light c. Therefore, if we consider a fluctuation of the gravitational field that spreads out spherically, the small time for it to move a distance r is

$$\Delta t = \frac{r}{c} \tag{34.12}$$

To obtain an order of magnitude for the energy, we drop the greater than sign in the uncertainty principle and on substituting equation 34.12 into equation 31.55 we get, for the energy of the fluctuation,

$$\Delta E \Delta t = \Delta E \left(\frac{r}{c} \right) = \hbar$$

and

$$\Delta E = \frac{\hbar c}{r} \tag{34.13}$$

The value of r in equation 34.13, wherein the quantum effects become important, is unknown at this point; in fact, it is one of the things that we wish to find. So further information is needed. Let us consider the amount of energy required to pull this little graviton or bundle of energy apart against its own gravity. The work to pull the graviton apart is equal to the energy necessary to assemble that mass by bringing small portions of it together from infinity. Let us first consider the problem for the electric field, and then use the analogy for the gravitational field. We saw in chapter 19 that the electric potential for a small spherical charge is

$$V = \frac{kq}{r} \tag{19.31}$$

But the electric potential V was defined as the potential energy per unit charge, that is,

$$V = \frac{PE}{q} \tag{19.19}$$

So if a second charge q is brought from infinity to the position r, the potential energy of the system of two charges is

$$PE = qV = \frac{kq^2}{r}$$

In a similar vein, a gravitational potential Φ could have been derived using the same general technique used to derive equation 19.31. The result for the gravitational potential would be

$$\Phi = \frac{GM}{r} \tag{34.14}$$

where G, of course, is the gravitational constant, M is the mass, and r is the distance from the mass to the point where we wish to determine the gravitational potential. The gravitational potential of a spherical mass is defined, similar to the electric potential of equation 19.19, as the gravitational potential energy per unit mass. That is,

$$\Phi = \frac{PE}{M} \tag{34.15}$$

Hence, if another mass M is brought from infinity to the position r, the potential energy of the system of two equal masses is

$$PE = M\Phi = \frac{GM^2}{r} \tag{34.16}$$

This value of the potential energy, PE to assemble the two masses, is the same energy that would be necessary to pull the two masses apart. Applying the same reasoning to the assembly of the masses that constitutes the graviton, the potential energy given by equation 34.16 is equal to the energy that would be necessary to pull the graviton apart. This energy can be equated to the energy of the graviton found from the uncertainty principle. Thus,

$$PE = \Delta E$$

Substituting for the PE from equation 34.16 and the energy ΔE from the uncertainty principle, equation 34.13, we get

$$\frac{GM^2}{r} = \frac{\hbar c}{r} \tag{34.17}$$

But the mass of the graviton M can be related to the energy of the graviton by Einstein's mass-energy relation as

$$\Delta E = Mc^2$$

or

$$M = \frac{\Delta E}{c^2} \qquad \text{(34.18)}$$

Substituting equation 34.18 into equation 34.17 gives

$$\frac{G(\Delta E)^2}{r(c^2)^2} = \frac{\hbar c}{r}$$

Solving for ΔE, we get

$$\Delta E = \sqrt{\frac{\hbar c^5}{G}} \qquad \text{(34.19)}$$

Equation 34.19 represents the energy of the graviton.

Example 34.1

The energy of the graviton. Find the energy of the graviton.

Solution

The energy of the graviton, found from equation 34.19, is

$$\Delta E = \sqrt{\frac{\hbar c^5}{G}}$$

$$= \sqrt{\frac{(1.05 \times 10^{-34} \text{ J s})(3.00 \times 10^8 \text{ m/s})^5}{6.67 \times 10^{-11} \text{ (N m}^2\text{)}/\text{kg}^2}}$$

$$= 1.96 \times 10^9 \text{ J}$$

This can also be expressed in terms of electron volts as

$$\Delta E = (1.96 \times 10^9 \text{ J})\left(\frac{1 \text{ eV}}{1.60 \times 10^{-19} \text{ J}}\right)\left(\frac{1 \text{ GeV}}{10^9 \text{ eV}}\right)$$

$$= 1.20 \times 10^{19} \text{ GeV}$$

This is the energy of a graviton; it is called the *Planck energy*.

From the point of view of particle physics, then, the graviton looks like a particle of mass 10^{19} GeV$/c^2$. This is an enormous mass and energy when compared to the masses and energies of all the other elementary particles. However, for any elementary particles of this size or larger, both quantum theory and gravitation must be taken into account. Recall that in all the other interactions of the elementary particles, gravity was ignored. From the point of view of ordinary gravity, this energy is associated with a mass of 2×10^{-5} g, a very small mass.

The distance in which this quantum fluctuation occurs can now be found by equating ΔE from equation 34.13 to ΔE from equation 34.19, that is,

$$\Delta E = \frac{\hbar c}{r} = \Delta E = \sqrt{\frac{\hbar c^5}{G}}$$

Solving for r we get

$$r = \frac{\hbar c}{\sqrt{\hbar c^5/G}}$$

$$r = \sqrt{\frac{\hbar G}{c^3}} \qquad \text{(34.20)}$$

Equation 34.20 is the distance or length where quantum gravity becomes significant. This distance turns out to be the same distance that Max Planck found when he was trying to establish some fundamental units from the fundamental constants of nature, and is called the *Planck length L_P*. Hence, the Planck length is

$$L_P = \sqrt{\frac{\hbar G}{c^3}} \qquad (34.21)$$

Example 34.2

The Planck length. Determine the size of the Planck length.

Solution

The Planck length, determined from equation 34.21, is

$$L_P = \sqrt{\frac{\hbar G}{c^3}}$$

$$= \sqrt{\frac{(1.05 \times 10^{-34} \text{ J s})[6.67 \times 10^{-11} \text{ (N m}^2)/\text{kg}^2]}{(3.00 \times 10^8 \text{ m/s})^3}}$$

$$= 1.61 \times 10^{-35} \text{ m} = 1.61 \times 10^{-33} \text{ cm}$$

Thus, quantum fluctuations of spacetime start to occur at distances of the order of 1.61×10^{-33} cm. We can now find the interval of time, within which this quantum fluctuation of spacetime occurs, from equation 34.12 as

$$\Delta t = \frac{r}{c} = \frac{L_P}{c}$$

This time unit is called the *Planck time T_P* and is

$$T_P = \frac{L_P}{c}$$

$$= \frac{1.61 \times 10^{-35} \text{ m}}{3.00 \times 10^8 \text{ m/s}} \qquad (34.22)$$

$$= 5.37 \times 10^{-44} \text{ s}$$

Thus, intervals of space and time given by the Planck length and the Planck time are the regions in which quantum gravity must be considered. This distance and time are extremely small. Recall that the size of the electron is about 10^{-19} m. Thus, quantum gravity occurs on a scale much smaller than that of an atom, a nucleus, or even an electron. There is relatively little known about quantum gravity at this time, but research is underway to find more answers dealing with the ultimate structure of spacetime itself.

34.11 The Superforce—Unification of All the Forces

An attempt to unify all the forces into one single force—a kind of **superforce**—continues today. One of the techniques followed is called *supersymmetry*, where the main symmetry element is spin. (Recall that all particles have spin.) Those particles that obey the Pauli exclusion principle have half-integral spin, that is, spin $\hbar/2$, $3\hbar/2$, and so on. Those particles that obey the Pauli exclusion principle are called *fermions*. All the quarks and leptons are fermions. Particles that have integral spin, $\hbar$, $2\hbar$, and so on, do not obey the Pauli exclusion principle. These particles are called *bosons*. All the mediating particles, such as the photon, $W^\pm$, Z^0, gluons, and the like, are bosons. *Hence, fermions are associated with particles of matter, whereas bosons are associated with the forces of nature, through an exchange of bosons. The new theories of supersymmetry attempt to unite bosons and fermions.*

A further addition to supersymmetry unites gravity with the electroweak-strong or GUT force into the superforce that is also called super gravity. Super gravity requires not only the existence of the graviton but also a new particle, the "gravitino," which has spin 3/2. However, this unification exists only at the extremely high energy of 10^{19} GeV, an energy that cannot be produced in a laboratory. However, in the initial formation or creation of the universe, a theory referred to as the Big Bang, such energies did exist.

The latest attempt to unify all the forces is found in the *superstring theory.* The superstring theory assumes that the ultimate building blocks of nature consist of very small vibrating strings. As we saw in chapter 12 on wave motion, a string is capable of vibrating in several different modes. The superstring theory assumes that each mode of vibration of a superstring can represent a particle or a force. Because there are an infinite number of possible modes of vibration, the superstring can represent an infinite number of possible particles. The graviton, which is responsible for the gravitational interaction, is caused by the lowest vibratory mode of a circular string. (Superstrings come in two types: open strings, which have ends, and closed strings, which are circular.) The photon corresponds to the lowest mode of vibration of the open string. Higher modes of vibrations represent different particles, such as quarks, gluons, protons, neutrons, and the like. In fact, the gluon is considered to be a string that is connected to a quark at each end. In this theory, no particle is more fundamental than any other, each is just a different mode of vibration of the superstrings. The superstrings interact with other superstrings by breaking and reforming. The four forces are considered just different manifestations of the one unifying force of the superstring. The superstring theory assumes that the universe originally existed in ten dimensions, but broke into two pieces—one of the pieces being our four-dimensional universe. Like the theories of supersymmetry and super gravity, the energies needed to test this theory experimentally are too large to be produced in any laboratory.

A simple picture of the unifications is shown in table 34.6. A great deal more work is necessary to complete this final unification.

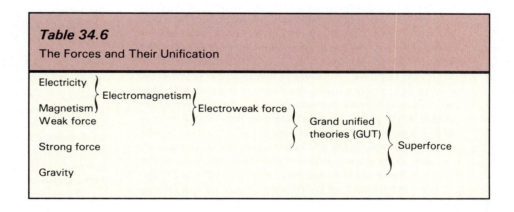

Table 34.6
The Forces and Their Unification

An Essay on the Application of Physics

The Big Bang Theory and the Creation of the Universe

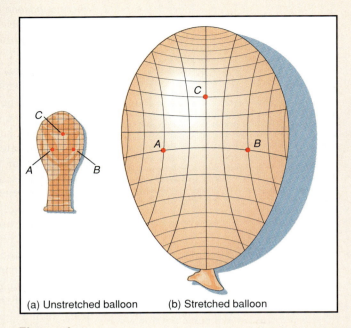

(a) Unstretched balloon   (b) Stretched balloon

Figure 1

An analogy to the expanding universe.

Have you ever wondered how the world was created? In every civilization throughout time and throughout the world, there has always been an account of the creation of the world. Such discussions have always belonged to religion and philosophy. It might seem strange that astronomers, astrophysicists, and physicists have now become involved in the discussion of the creation of the universe. Of course, if we think about it, it is not strange at all. *Since physics is a study of the entire physical world; it is only natural that physics should try to say something about the world's birth.*

The story starts in 1923 when the American astronomer, Edwin Hubble, using the Doppler effect for light, observed that all the galactic clusters, outside our own, in the sky were receding away from the earth. When we studied the Doppler effect for sound in chapter 12, we saw that when a train recedes from us its frequency decreases. A decrease in the frequency means that there is an increase in the wavelength. Similarly, a Doppler effect for light waves can be derived. The equations are different than those derived in chapter 12 because, in the special theory of relativity, the velocity of light is independent of the source. However, the effect is the same. That is, a receding source that emits light at a frequency ν, is observed by the stationary observer to have a frequency ν', where ν' is less than ν. Thus, since the frequency decreases, the wavelength increases. Because long waves are associated with the red end of the visible spectrum, all the observed wavelengths are shifted toward the red end of the spectrum. The effect is called the *cosmological red shift,* to distinguish it from the gravitational red shift discussed in chapter 30. *Hubble found that the light from the distant galaxies were all red shifted indicating that the distant galaxies were receding from us.*

It can, therefore, be concluded that if all the galaxies are receding from us, the universe itself must be expanding. Hubble was able to determine the rate at which the universe is expanding. If the universe is expanding now, then in some time in the past it must have been closer together. If we look far enough back in time, we should be able to find when the expansion began. (Imagine taking a movie picture of an explosion showing all the fragments flying out from the position of the explosion. If the movie is run backward, all the fragments would be seen moving backward toward the source of the explosion.)

The best estimate for the creation of the universe, is that the universe began as a great bundle of energy that exploded outward about 15 billion years ago. This great explosion has been called the **Big Bang.** *It was not an explosion of matter into an already existing space and time, rather it was the very creation of space and time, or spacetime, and matter themselves.*

As the universe expanded from this explosion, all objects became farther and farther apart. A good analogy to the expan-

sion of spacetime is the expansion of a toy balloon. A rectangular coordinate system is drawn on an unstretched balloon, as shown in figure 1(a), locating three arbitrary points, *A, B,* and *C*. The balloon is then blown up. As the balloon expands the distance between points *A* and *B, A* and *C,* and *B* and *C* increases. So no matter where you were on the surface of the balloon you would find all other points moving away from you. This is similar to the distant galaxies moving away from the earth. To complete the analogy to the expanding universe, we note that the simple flat rectangular grid in which Euclidean geometry holds now become a curved surface in which Euclidean geometry no longer holds.

If everything in the universe is spread out and expanding, the early stages of the universe must have been very compressed. To get all these masses of stars of the present universe back into a small compressed state, that compressed state must have been a state of tremendous energy and exceedingly high density and temperature. Matter and energy would be transforming back and forth through Einstein's mass-energy formula, $E = mc^2$. Work done by particle physicists at very high energies allows us to speculate what the universe must have looked like at these very high energies at the beginning of the universe.

The early history of the universe is sketched in figure 2. The Big Bang is shown occurring at time $t = 0$, which is approximately 15 billion years ago.

1. *From the Big Bang to 10^{-43} s*

   Between the creation and the Planck time, 0 to 10^{-43} s, the energy of the universe was enormous, dropping to about 10^{19} GeV at the Planck time. The temperature was greater than 10^{33} K. Relatively little is known about this era, but the extremely high energy would cause all the

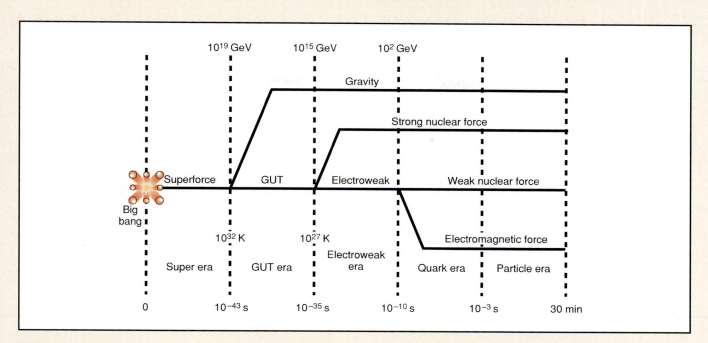

Figure 2

Creation of the four forces from the superforce.

forces to merge into one superforce. That is, gravity, the strong force, the weak force, and the electromagnetic force would all be replaced by one single superforce. This is the era being researched by present physicists in the supersymmetry and super gravity theories. There is only one particle, a super particle, that decays into bosons and fermions, and continually converts fermions to bosons and vice versa, so that there is no real distinction between them.

2. *From 10^{-43} s to 10^{-35} s*
As the universe expands, the temperature drops and the universe cools to about 10^{32} K. The energy drops below 10^{19} GeV and the gravitational force breaks away from the superforce as a separate force, leaving the grand unified force of GUT as a separate force. Now two forces exist in nature. We are now in the GUT era, that era governed by the grand unified theories. The X particle and its antiparticle $\overline{X}$ are in abundance. The X particles decay into quarks and leptons, whereas the $\overline{X}$ particles decay into antiquarks and antileptons. However, the decay rate of X and $\overline{X}$ are not the same and more particles than antiparticles are formed. This will eventually lead to the existence of more matter than antimatter in the universe. The X particles continually convert quarks into leptons and vice versa. There are plenty of quarks, electrons, neutrinos, photons, gluons, X particles, and their antiparticles present, but they have effectively lost their individuality.

3. *From 10^{-35} s to 10^{-10} s*
As further expansion of the universe continues, the temperature drops to 10^{27} K and the energy drops to 10^{15}

GeV. At this low energy all the X particles disappear, and quarks and leptons start to have an individual identity of their own. No longer can they be converted into each other. The lower energy causes the strong nuclear force to break away leaving the electroweak force as the only unified force left. There are now three forces of nature: gravity, strong nuclear, and the electroweak. There are quarks, leptons, photons, neutrinos, $W^{\pm}$ and Z^0, and gluon particles present. It is still too hot for the quarks to combine.

4. *From 10^{-10} s to 10^{-3} s*
As the universe continues to expand, it cools down to an energy of 10^2 GeV. The $W^{\pm}$ and Z^0 particles disappear because there is not enough energy to form them anymore. The weak nuclear force breaks away from the electroweak force, leaving the electromagnetic force. There are now present the four familiar forces of nature: gravity, strong nuclear, weak nuclear, and electromagnetic. Quarks now combine to form baryons, qqq, and mesons, q$\overline{q}$. The familiar protons and neutrons are now formed. Because of the abundance of quarks over antiquarks, there will also be an excess of protons and neutrons over antiprotons and antineutrons.

5. *From 10^{-3} s to 30 min*
The universe has now expanded and cooled to the point where protons and neutrons can combine to form the nucleus of deuterium. The deuterium nuclei combine to form helium as described in section 33.9 on fusion. There are about 77% hydrogen nuclei, and 23% helium nuclei

present at this time and this ratio will continue about the same to the present day. There are no atoms formed yet because the temperature is still too high. What is present is called a *plasma*.

6. *From 30 min to 1 Billion Years*
   Further expansion and cooling now allows the hydrogen and helium nuclei to capture electrons and the first chemical elements are born. Large clouds of hydrogen and helium are formed.

7. *From 1 Billion Years to 10 Billion Years*
   The large rotating clouds of hydrogen and helium matter begin to concentrate due to the gravitational force. As the radius of the cloud decreases, the angular velocity of the cloud increases in order to conserve angular momentum. (Similar to the spinning ice skater discussed in section 9.7.) These condensing, rotating masses are the beginning of galaxies.

   Within the galaxies, gravitation causes more and more matter to be compressed into spherical objects, the beginning of stars. More and more matter gets compressed until the increased pressure of that matter

causes a high enough temperature to initiate the fusion process of converting hydrogen to helium and the first stars are formed. Through the fusion process, more and more chemical elements are formed. The higher chemical elements are formed by neutron absorption until all the chemical elements are formed.

These first massive stars did not live very long and died in an explosion—a supernova—spewing the matter of all these heavier elements out into space. The fragments of these early stars would become the nuclei of new stars and planets.

8. *From 10 Billion Years to the Present*
   The remnants of dead stars along with hydrogen and helium gases again formed new clouds, which were again compressed by gravity until our own star, the sun, and the planets were formed. All the matter on earth is the left over ashes of those early stars. Thus, even we ourselves are made up of the ashes of these early stars. As somebody once said, there is a little bit of star dust in each of us.

The Language of Physics

Leptons
Particles that are not affected by the strong nuclear force (p. 1036).

Hadrons
Particles that are affected by the strong nuclear force (p. 1036).

Baryons
A group of hadrons that have half-integral spin and are composed of three quarks (p. 1036).

Mesons
A group of hadrons that have integral spin, that are composed of quark-antiquark pairs (p. 1036).

Antiparticles
To each elementary particle in nature there corresponds another particle that has the characteristics of the original particle but opposite charge. Some neutral particles have antiparticles that have opposite spin, whereas the photon is its own antiparticle. The antiparticle of the proton is the antiproton. The antiparticle of the electron is the antielectron or positron. If a particle collides with its antiparticle both are annihilated with the emission of radiation or other particles. Conversely, photons can be converted to particles and antiparticles (p. 1037).

Antimatter
Matter consists of protons, neutrons, and electrons, whereas antimatter consists of antiprotons, antineutrons, and antielectrons (p. 1037).

Quarks
Elementary particles that are the building blocks of matter. There are six quarks and six antiquarks. The six quarks are: up, down, strange, charmed, bottom, and top. Each quark and antiquark also comes in three colors, red, green, and blue. Each color quark also has an anticolor quark. Baryons are composed of red, green, and blue quarks and mesons are made up of a linear combination of colored quark-antiquark pairs (p. 1038).

Quantum electrodynamics (QED)
The merger of electromagnetic theory with quantum mechanics. In QED, the electric force is transmitted by the exchange of a virtual photon (p. 1042).

Weak nuclear force
The weak nuclear force does not exert the traditional push or pull type of force known in classical physics. Rather, it is responsible for the transmutation of the subatomic particles. The weak force is independent of electric charge and acts between leptons and hadrons and also between hadrons and hadrons. The weak force is the weakest force after gravity (p. 1043).

Electroweak force
A unification of the electromagnetic force with the weak nuclear force. The force is mediated by four particles: the photon and three intermediate vector bosons called W^+, W^-, and Z^0 (p. 1043).

The strong nuclear force
The force that holds the nucleons together in the nucleus. The force is the result of the color forces between the quarks within the nucleons. At relatively large separation distances within the nucleus, the quark-antiquark pair (meson), which is created by the gluons, is exchanged between the nucleons. At shorter distances within the nucleus, the strong force can be explained either as an exchange between the quarks of one proton and the quarks of another proton, or perhaps as a direct exchange of the gluons themselves, which give rise to the quark-quark force within the nucleon (p. 1044).

Quantum chromodynamics (QCD)
In QCD, the force holding quarks together is caused by the exchange of a new particle, called a gluon. A gluon interacting with a quark changes the color of a quark (p. 1044).

Grand unified theory

A theory that merges the electroweak force with the strong nuclear force. This force should be able to transform quarks into leptons and vice versa. The theory predicts the existence of 12 new particles, called X particles that are capable of converting hadrons into leptons by changing quarks to leptons. This theory also predicts that an isolated proton should decay. However, no such decays have ever been found, so the theory may have to be modified (p. 1045).

Gravitons

The quanta of the gravitational field. Since gravitation is a warping of spacetime, the graviton must be a quantum of spacetime (p. 1047).

Superforce

An attempt to unify all the forces under a single force. The theories go under the names of supersymmetry, super gravity, and superstrings (p. 1050).

The Big Bang theory

The theory of the creation of the universe that says that the universe began as a great bundle of energy that exploded outward about 15 billion years ago. It was not an explosion of matter into an already existing space and time, rather it was the very creation of spacetime and matter (p. 1052).

Questions for Chapter 34

†1. Discuss the statement, "A graviton is a quantum of gravity. But gravity is a result of the warping of spacetime. Therefore, the graviton should be a quantum of spacetime. But just as a quantum of the electromagnetic field, the photon, has energy, the graviton should also have energy. In fact, we can estimate the energy of a graviton. Therefore, is spacetime another aspect of energy? Is there only one fundamental quantity, energy?"

†2. Does antimatter occur naturally in the universe? How could you detect it? Where might it be located?

3. When an electron and positron annihilate, why are there two photons formed instead of just one?

4. Murray Gell-Mann first introduced three quarks to simplify the number of truly elementary particles present in nature. Now there are six quarks and six antiquarks, and each can come in three colors and three anticolors. Are we losing some of the simplicity? Discuss.

5. Discuss the experimental evidence for the existence of structure within the proton and the neutron.

6. How did the Pauli exclusion principle necessitate the introduction of colors into the quark model?

†7. If the universe is expanding from the Big Bang, will the gravitational force of attraction of all the masses in the universe eventually cause a slowing of the expansion, a complete stop to the expansion, and finally a contraction of the entire universe?

†8. Just as there are electromagnetic waves associated with a disturbance in the electromagnetic field, should there be gravitational waves associated with a disturbance in a gravitational field? How might such gravitational waves be detected?

†9. Einstein's picture of gravitational attraction is a warping of spacetime by matter. This has been pictured as the rubber sheet analogy in chapter 30. What might antimatter do to spacetime? Would it warp spacetime in the same way or might it warp spacetime to cause a gravitational repulsion? Would this be antigravity? Would the antiparticle of the graviton then be an antigraviton? Instead of a black hole, would there be a white hill?

10. Discuss the similarities and differences between the photon and the neutrino.

Problems for Chapter 34

Section 34.2 Particles and Antiparticles

1. How much energy is released when an electron and a positron annihilate? What is the frequency and wavelength of the two photons that are created?

2. How much energy is released when a proton and antiproton annihilate?

3. How much energy is released if 1.00 kg of matter annihilates with 1.00 kg of antimatter? Find the wavelength and frequency of the resulting two photons.

4. A photon "disintegrates," creating an electron-positron pair. If the frequency of the photon is 5.00×10^{24} Hz, determine the linear momentum and the energy of each product particle.

Section 34.4 Quarks

5. If the three quarks shown in the diagram combine to form a baryon, find the charge and spin of the resulting particle.

6. If the three quarks shown in the diagram combine to form a baryon, find the charge and spin of the resulting particle.

7. If the three quarks shown in the diagram combine to form a baryon, find the charge and spin of the resulting particle.

8. Find the charge and spin of the baryon that consists of the three quarks shown in the diagram.

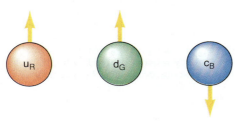

9. If the two quarks shown in the diagram combine to form a meson, find the charge and spin of the resulting particle.

10. If the two quarks shown in the diagram combine to form a meson, find the charge and spin of the resulting particle.

11. Find the charge and spin of the meson that consists of the two quarks shown in the diagram.

12. Which of the combinations of particles in the diagram are possible and which are not. If the combination is not possible, state the reason.

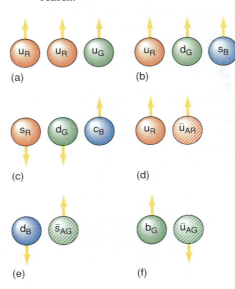

(a)   (b)

(c)   (d)

(e)   (f)

13. Why are the two particles in the diagram impossible?

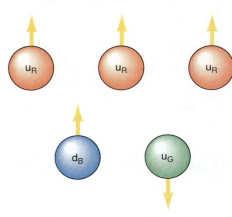

14. A baryon is composed of three quarks. It can be made from a total of six possible quarks, each in three possible colors, and each with either a spin-up or spin-down. From this information, how many possible baryons can be made?

15. A meson is composed of a quark-antiquark pair. It can be made from a total of six possible quarks, each in three possible colors, and each with either a spin-up or spin-down, and six possible antiquarks each in three possible colors, and each with either a spin-up or spin-down. Neglecting linear combinations of these quarks, how many possible mesons can be made?

16. From problems 14 and 15 determine the total number of possible hadrons, ignoring possible mesons made from linear combinations of quarks and antiquarks. Could you make a "periodic table" from this number? Discuss the attempt to attain simplicity in nature.

17. Determine all possible quark combinations that could form a baryon of charge $+1$ and spin $\frac{1}{2}$.

Epilogue

In the first chapter of the book we said that physics had its birthplace in mankind's quest for knowledge and truth. It started with the earliest man as he came out of his cave. We have surely progressed a great deal since that long ago time, as is evidenced by such topics covered in this book as atomic and nuclear physics, quantum physics, special and general relativity, and the unification of all the forces. However, in terms of what still lies ahead for us in this universe, we have barely taken one step out of the cave.

Appendix A
Conversion Factors

Length

1 meter (m) = 100 cm = 39.4 in. = 3.28 ft
    = 6.21×10^{-4} mi
1 centimeter (cm) = 10^{-2} m = 10 mm
    = 0.394 in.
1 inch (in.) = 2.54 cm = 0.0254 m = 0.083 ft
1 foot (ft) = 0.305 m = 30.5 cm = 12 in.
1 mile (mi) = 1610 m = 1.61 km = 5280 ft
1 kilometer (km) = 1000 m = 0.621 mi
1 nautical mile = 1.15 mi = 6076 ft = 1852 m
1 nanometer (nm) = 10^{-9} m = 10^{-7} cm
1 micron (μ) = 10^{-6} m = 1 μm = 10^{-4} cm
1 angstrom (Å) = 10^{-10} m = 10^{-8} cm
1 mil = 10^{-3} in.
1 yard (yd) = 0.9144 m = 91.44 cm

Area

1 m^2 = 10^4 cm^2 = 1.55×10^3 in.^{2}
1 cm^2 = 10^{-4} m^2 = 0.155 in.^{2}
1 in.^{2} = 6.45 cm^2 = 1.27×10^6 circular mils
    = 6.45×10^{-4} m^2
1 ft^2 = 144 in.^{2} = 929 cm^2 = 9.29×10^{-2} m^2
1 km^2 = 10^6 m^2
1 yd^2 = 0.836 m^2

Volume

1 m^3 = 35.3 ft^3 = 6.1×10^4 in.^{3} = 10^3 L
1 ft^3 = 2.83×10^{-2} m^3 = 1.73×10^3 in.^{3}
    = 28.3 L = 7.48 gal
1 U.S. gallon = 231 in.^{3} = 0.134 ft^3 = 3.79
    $\times 10^{-3}$ m^3
1 in.^{3} = 1.64×10^{-5} m^3

Mass

1 kg = 1000 g = 6.85×10^{-2} slugs
1 slug = 14.59 kg

Weight

1 lb = 4.45 N
1 N = 0.225 lb

Time

1 year (yr) = 365.24 day = 8.76×10^3 hr
    = 5.26×10^5 min = 3.16×10^7 s
1 day = 1.44×10^3 min = 8.64×10^4 s

Density

1 kg/m^3 = 1×10^{-3} g/cm^3 = 1.94×10^{-3} slug/ft^3
1 g/cm^3 = 1000 kg/m^3 = 1.94 slug/ft^3

Velocity

1 m/s = 3.28 ft/s = 2.24 mi/hr = 3.60 km/hr
    = 1.94 knot
1 ft/s = 0.305 m/s = 0.682 mi/hr
    = 1.10 km/hr
88 ft/s = 60 mph
1 mi/hr = 1.47 ft/s = 1.61 km/hr
    = 0.869 knot = 0.447 m/s
1 km/hr = 0.278 m/s = 0.621 mph = 0.91 ft/s

Acceleration

1 m/s^2 = 3.281 ft/s^2 = 3.60 km/hr/s
    = 2.24 mph/s
1 ft/s^2 = 0.3048 m/s^2
1 mph/s = 1.467 ft/s^2 = 0.447 m/s^2

Angles

1 radian (rad) = 57.30° = 57°18′ = 0.159 rev
1 degree (°) = 0.01745 rad
360° = 2π radians
1 rev/min (rpm) = 0.1047 rad/s
1 rad/s = 9.55 rev/min

Force

1 newton (N) = 10^5 dynes = 0.225 lb
    = 3.60 oz
1 pound (lb) = 4.45 N = 16 ounces (oz)

Pressure

1 N/m^2 = 1.00 pascal (Pa) = 2.09×10^{-2}
    lb/ft^2
    = 1.45×10^{-4} lb/in.^{2}
    = 9.87×10^{-6} atm = 7.50×10^{-4} cm
    of Hg
    = 4.01×10^{-3} in. of H$_2$O = 10^{-5} bar
    = 10^{-2} millibars (mb)
1 lb/in.^{2} = 144 lb/ft^2 = 6.90×10^3 N/m^2 =
    5.17 cm of Hg
    = 27.68 in. of H$_2$O
1 atmosphere (atm) = 1.013×10^5 N/m^2 =
    1013 mb
    = 14.7 lb/in.^{2}
    = 2.12×10^3 lb/ft^2 = 760
    torr = 76 cm of Hg
    = 406.8 in. H$_2$O

Energy, Work, Thermal Energy

1 joule (J) = 0.738 ft lb = 2.39×10^{-4} kcal =
    6.24×10^{18} eV
    = 9.481×10^{-4} Btu = 10^7 ergs =
    0.239 cal = 1 N m
1 kilocalorie (kcal) = 4185 J = 3.97 Btu
    = 3077 ft lb
1 foot pound (ft lb) = 1.36 J = 1.29×10^{-3} Btu
    = 3.25×10^{-4} kcal
1 electron volt (eV) = 1.60×10^{-19} J = 1.18
    $\times 10^{-19}$ ft lb
1 kilowatt hour (kWh) = 3.6×10^6 J =
    3413 Btu = 860 kcal
    = 1.34 hp hr
1 calorie (cal) = 4.185 J
1 British thermal unit (Btu) = 0.252 kcal
    = 778 ft lb
    = 1.05×10^3 J

Power

1 watt (W) = 1 J/s = 0.738 ft lb/s =
    1.34×10^{-3} hp
    = 2.39×10^{-4} kcal/s
1 horsepower (hp) = 550 ft lb/s = 746 W =
    2545 Btu/hr
    = 0.1782 kcal/s
1 kilowatt (kW) = 1000 W = 1.34 hp
1 refrigeration ton = 12,000 Btu/hr

Electricity and Magnetism

1 volt (V) = 1 joule/coulomb = J/C
1 ampere (A) = 1 C/s = 6.3×10^{18} electrons/s
1 ohm (Ω) = 1 volt/ampere
1 farad = 1 coulomb/volt
Magnetic Field Intensity
1 tesla (T) = $\dfrac{1 \text{ N}}{\text{A m}}$ = 10^4 gauss =
    1 weber/meter^2
Magnetic Flux = 1 weber = 1 T m^2
Inductance = 1 henry = $\dfrac{1 \text{ J}}{\text{A}^2}$ = $\dfrac{\text{V s}}{\text{A}}$

Appendix B
Useful Mathematical Formulas

Geometry

$C = 2\pi r$ Circumference of circle

$A = \pi r^2$ Area of circle

$A = 4\pi r^2$ Area of sphere

$V = \frac{4}{3}\pi r^3$ Volume of sphere

$V = \pi r^2 h$ Volume of cylinder

Algebra

$x^n x^m = x^{(n+m)}$ $(x^n)^m = x^{nm}$

$\dfrac{x^n}{x^m} = x^{(n-m)}$ $x^{1/m} = (x)^{1/m}$

$\dfrac{1}{1/x} = x$ $x^{n/m} = (x^n)^{1/m} = (x^{1/m})^n$

$$(x + y)^2 = x^2 + 2xy + y^2$$
$$(x - y)^2 = x^2 - 2xy + y^2$$
$$(x + y)^3 = x^3 + 3x^2y + 3xy^2 + y^3$$
$$(x - y)^3 = x^3 - 3x^2y + 3xy^2 - y^3$$
$$(x + y)(x - y) = x^2 - y^2$$

Quadratic Equation

if $ax^2 + bx + c = 0$

then $x = \dfrac{-b \pm \sqrt{b^2 - 4ac}}{2a}$

Binomial Expansion

$$(1 + x)^n = 1 + nx + \frac{n(n-1)x^2}{2!} + \frac{n(n-1)(n-2)x^3}{3!} + \ \cdots$$

Powers of Ten and Scientific Notation

Very large or very small numbers can be written in a simple way by expressing them as powers of 10. For example, 100,000 may be written as 10^5. That is,

$$100,000 = (10)(10)(10)(10)(10) = 1 \times 10^5$$

Table B.1
Powers of 10

$10^{-4} = \dfrac{1}{10^4} = \dfrac{1}{10,000} = 0.0001$

$10^{-3} = \dfrac{1}{10^3} = \dfrac{1}{1000} = 0.001$

$10^{-2} = \dfrac{1}{10^2} = \dfrac{1}{100} = 0.01$

$10^{-1} = \dfrac{1}{10} = 0.1$

$10^0 = 1$

$10^1 = 10$

$10^2 = (10)(10) = 100$

$10^3 = (10)(10)(10) = 1000$

$10^4 = (10)(10)(10)(10) = 10,000$

$10^5 = (10)(10)(10)(10)(10) = 100,000$

$10^6 = (10)(10)(10)(10)(10)(10) = 1,000,000$

Table B.1 is a list of some powers of 10. Notice that $10^0 = 1$. *In numbers with positive exponents, the exponent is equal to the number of zeros following the 1. In the cases of negative exponents, the exponent is equal to the number of places the decimal point is moved to the left of the 1.* That is,

$$1 \times 10^{-1} = 0.1$$

The operations using powers of 10 follow the same rules of exponents as in algebra. For example, in algebra

$$x^n x^m = x^{(n+m)}$$

whereas for powers of 10 it becomes

$$10^n 10^m = 10^{(n+m)}$$

For example,

$$10^2 10^3 = 10^{2+3} = 10^5$$

The division of powers of ten becomes

$$\frac{10^n}{10^m} = 10^{(n-m)}$$

For example,

$$\frac{10^5}{10^3} = 10^{5-3} = 10^2$$

The reciprocal of a power of ten becomes

$$\frac{1}{10^m} = 10^{-m}$$

For example,

$$\frac{1}{10^2} = 10^{-2}$$

Finally, we have

$$(10^n)^m = 10^{nm}$$

and

$$10^{n/m} = (10^n)^{1/m} = (10^{1/m})^n$$

with the examples

$$(10^2)^3 = 10^6$$

$$10^{4/2} = (10^4)^{1/2} = 10^2$$

Scientific Notation

Very large or very small numbers can be written in a very convenient way with the use of powers of ten. For example, the number 583,000 is equal to $5.83 \times 100,000$. However, since $100,000 = 10^5$, we have

$$583,000 = 5.83 \times 10^5$$

Writing numbers in this notation is called expressing the number in scientific notation. The expression of any number in scientific notation is based on the fact that all numbers can be expressed as a number between 1 and 10 multiplied by a power of ten. As examples of some numbers expressed in scientific notation, we have

$$583,000 = 5.83 \times 10^5$$

$$1430 = 1.43 \times 10^3$$

$$0.025 = 2.5 \times 10^{-2}$$

$$0.00045 = 4.5 \times 10^{-4}$$

For a discussion of significant figures in your calculations see your laboratory text.

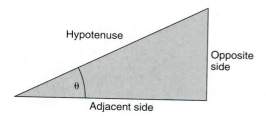

Figure B.1

A right triangle for defining the trigonometric functions.

Trigonometry

1. *Definitions.* The trigonometric functions are defined as ratios of the sides of a right triangle. Using figure B.1, they are defined as

$$\sin \theta = \frac{\text{opposite side}}{\text{hypotenuse}}$$

$$\cos \theta = \frac{\text{adjacent side}}{\text{hypotenuse}}$$

$$\tan \theta = \frac{\text{opposite side}}{\text{adjacent side}}$$

$$\sec \theta = \frac{1}{\cos \theta}$$

$$\csc \theta = \frac{1}{\sin \theta}$$

$$\cot \theta = \frac{1}{\tan \theta}$$

$$\tan \theta = \frac{\sin \theta}{\cos \theta} \qquad \cot \theta = \frac{\cos \theta}{\sin \theta}$$

2. *Some Important Trigonometric Identities*

$$\sin^2\theta + \cos^2\theta = 1$$

$$\sec^2\theta = 1 + \tan^2\theta$$

$$\csc^2\theta = 1 + \cot^2\theta$$

3. *Sums and Differences of the Trigonometric Functions*

$$\sin(A + B) = \sin A \cos B + \cos A \sin B$$

$$\sin(A - B) = \sin A \cos B - \cos A \sin B$$

$$\cos(A + B) = \cos A \cos B - \sin A \sin B$$

$$\tan(A + B) = \frac{\tan A + \tan B}{1 - \tan A \tan B}$$

$$\tan(A - B) = \frac{\tan A + \tan B}{1 + \tan A \tan B}$$

4. *Double Angles and Half Angles*

$$\sin 2\theta = 2 \sin \theta \cos \theta$$

$$\cos 2\theta = 1 - 2 \sin^2\theta$$

$$\sin\left(\frac{\theta}{2}\right) = \pm \sqrt{\frac{1 - \cos \theta}{2}}$$

$$\cos\left(\frac{\theta}{2}\right) = \pm \sqrt{\frac{1 + \cos \theta}{2}}$$

5. *Addition and Subtraction of the Trigonometric Functions*

$$\sin A + \sin B = 2 \sin\left(\frac{A + B}{2}\right) \cos\left(\frac{A - B}{2}\right)$$

$$\sin A - \sin B = 2 \cos\left(\frac{A + B}{2}\right) \sin\left(\frac{A - B}{2}\right)$$

$$\cos A + \cos B = 2 \cos\left(\frac{A + B}{2}\right) \cos\left(\frac{A - B}{2}\right)$$

$$\cos A - \cos B = 2 \sin\left(\frac{A + B}{2}\right) \sin\left(\frac{A - B}{2}\right)$$

6. *Trigonometric Functions for Negative Angles*

$$\cos(-\theta) = \cos \theta$$

$$\sin(-\theta) = -\sin \theta$$

$$\tan(-\theta) = -\tan \theta$$

Figure B.2 should be used for the following formulas:

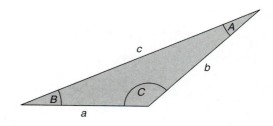

Figure B.2

An obtuse triangle.

7. Law of Sines

$$\frac{a}{\sin A} = \frac{b}{\sin B} = \frac{c}{\sin C}$$

8. Law of Cosines

$$c^2 = a^2 + b^2 - 2ab \cos C$$

9. Special Case of Law of Cosines ($C = 90°$)

$$c^2 = a^2 + b^2 \qquad \text{Pythagorean theorem}$$

Appendix C
Proportionalities

In any science, especially in physics, we constantly come on quantities that are proportional to other quantities. *A proportion is a relation among different quantities.* There are two main types of proportions: (1) direct proportions and (2) inverse proportions.

Direct Proportions

Two physical quantities are said to be directly proportional to each other if the ratio of these quantities is always equal to a constant. As an example, consider the tabulated values of x and y in table C.1. Taking the ratio of y to x for each of these values, we get the constant value 2, as shown in the third row of the table; that is,

$$\frac{y}{x} = 2 = \text{constant} \qquad \text{(C.1)}$$

Thus, the ratio of y to x is always a constant, and by the definition, y is directly proportional to x. If we solve equation C.1 for y, we get

$$y = 2x \qquad \text{(C.2)}$$

Equation C.2 says that y is directly proportional to x. Thus, as x increases, so does the value of y. In particular, y is always twice as large as x. The number 2, the value of the constant ratio y/x, is called the *constant of proportionality.*

In general, if y is directly proportional to x, it can be written in the form

$$y = kx \qquad \text{(C.3)}$$

where k is called the constant of proportionality and can have any constant value. Thus, in table C.2, the constant of proportionality k would be equal to the value 3.

Table C.2				
x	1	2	3	4	5
y	3	6	9	12	15
y/x	3	3	3	3	3

In summary, the general statement for a direct proportion is usually written in the form

$$y \propto x \qquad \text{(C.4)}$$

where the symbol $\propto$ is a shorthand notation for the words "is proportional to." The main characteristic of a direct proportion is that as one variable increases, so does the other; or if one variable decreases, so does the other. When it is desirable to express this proportionality in terms of an equation, we introduce k, the constant of proportionality, and express the proportionality as the equation

$$y = kx \qquad \text{(C.5)}$$

Equation C.4 says that as x increases, y also increases. Equation C.5, on the other hand, is more general and says that as x increases, y also increases, and the amount of the increase depends on the value of k.

If we plot the data of table C.1, we get the graph shown in figure C.1. Notice that when this data is plotted the result is a straight line that passes through the origin. This is a result of all

Table C.1							
x	2	3	4	5	6	7	8
y	4	6	8	10	12	14	16
y/x	4/2 = 2	6/3 = 2	8/4 = 2	10/5 = 2	12/6 = 2	14/7 = 2	16/8 = 2

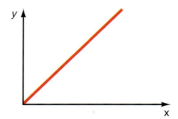

Figure C.1

Graph of a direct proportion.

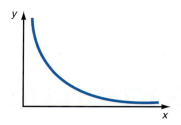

Figure C.2

Graph of *y* versus *x* showing an inverse proportionality.

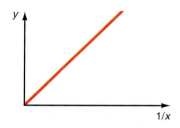

Figure C.3

Graph of *y* versus 1/*x* showing an inverse proportionality.

direct proportions. *When the direct proportion is expressed as an equation, such as that given by equation C.5, the equation is the equation of a straight line that passes through the origin.* The slope of the line is equal to the constant of proportionality *k*. Thus, *as a corollary, we can say that if the relation between two variables is a straight line going through the origin of a graph, then those two variables are directly proportional to each other.*

Inverse Proportion

An inverse proportion is one in which there is an inverse relationship between the variables. Thus,

$$y \propto \frac{1}{x} \tag{C.6}$$

says that y is inversely proportional to x. This means that if x increases, y must decrease, or if x decreases, y must increase. To make an equality of C.6, we introduce a constant of proportionality *k* and obtain

$$y = \frac{k}{x} \tag{C.7}$$

This is sometimes written in the equivalent form

$$yx = k \tag{C.8}$$

and sometimes this form is used for the defining form for an inverse relationship.

As an example of an inverse proportionality, consider the data of table C.3. Note that as *x* increases, *y* decreases, but the product of *yx* remains a constant.

A graph of the data of table C.3, which is an inverse relationship, is shown in figure C.2 and is a graph of a rectangular hyperbola. *Thus, as a corollary, if the graph of y versus x is a rectangular hyperbola, this implies that y is inversely proportional to x.* Sometimes it is more convenient to plot *y* versus the reciprocal of *x* to show the inverse proportion, because the plot of *y* versus 1/*x* is a straight line, as shown in figure C.3. *Thus, if the graph of y versus 1/x is a straight line, then y is inversely proportional to x. The slope of the straight line is equal to the proportionality constant k.*

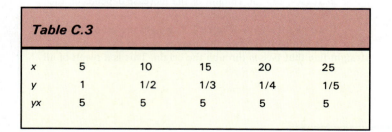

Table C.3				
x	5	10	15	20	25
y	1	1/2	1/3	1/4	1/5
yx	5	5	5	5	5

Appendix D
Physical Constants

Speed of light, $c = 2.998 \times 10^8$ m/s $= 1.86 \times 10^5$ mi/s

Gravitational constant, $G = 6.67 \times 10^{-11}$ N m^2/kg^2

Standard acceleration of gravity, $g = 9.80$ m/s$^2 = 32.0$ ft/s^2

Heat of fusion of water, $L_f = 80$ kcal/kg $= 3.33 \times 10^5$ J/kg

Heat of vaporization of water, $L_v = 540$ kcal/kg
$\qquad = 2.26 \times 10^6$ J/kg

Mass of earth $= 5.98 \times 10^{24}$ kg

Mean radius of earth, $r_e = 6.37 \times 10^6$ m $= 3960$ mi

Angular velocity of earth $\omega_e = 7.27 \times 10^{-5}$ rad/s

Mean earth-sun distance, $r_{es} = 1.49 \times 10^8$ km $= 9.29 \times 10^7$ mi

Speed of sound in air, $v_a = 331$ m/s $= 1090$ ft/s

Speed of sound in water, $v_w = 1460$ m/s $= 4790$ ft/s

Density of dry air (STP) $= 1.29$ kg/m^3

Universal gas constant, $R = 8.314$ J/(mole K)

Mechanical equivalent of heat $= 4185$ J/kcal

Permittivity of free space, $\epsilon_0 = 8.85 \times 10^{-12}$ C^2/(N m^2)

Permeability of free space, $\mu_0 = 4\pi \times 10^{-7}$ (T m)/A

Electronic charge, $e = 1.6021 \times 10^{-19}$ C

Electron rest mass, $m_e = 9.1090 \times 10^{-31}$ kg

Mass of proton, $m_p = 1.6726 \times 10^{-27}$ kg

Mass of neutron, $m_n = 1.6749 \times 10^{-27}$ kg

Avogadro's number, $N_A = 6.022 \times 10^{23}$ molecules/mole
$\qquad = 6.022 \times 10^{26}$ molecules/kmole

Atmospheric pressure, $P_{atm} = 1.013 \times 10^5$ N/m^2

Mass of moon, $M_M = 7.343 \times 10^{22}$ kg

Mass of sun, $M_S = 1.987 \times 10^{30}$ kg

Planck's constant, $h = 6.63 \times 10^{-34}$ J s $= 4.14 \times 10^{-15}$ eV s

Boltzmann's constant, $k = 1.38 \times 10^{-23}$ J/molecules K
$\qquad = 8.63 \times 10^{-5}$ eV/molecules K

Ratio of electronic charge to mass, $e/m_e = 1.756 \times 10^{11}$ C/kg

Constant in Coulomb's law, $k = 1/4\pi\epsilon_0 = 9.00 \times 10^9$ N m^2/C^2

Appendix E
Table of the Elements

Atomic Number	Element	Symbol	Atomic Mass
1	Hydrogen	H	1.008
2	Helium	He	4.003
3	Lithium	Li	6.941
4	Beryllium	Be	9.012
5	Boron	B	10.81
6	Carbon	C	12.00
7	Nitrogen	N	14.01
8	Oxygen	O	15.99
9	Fluorine	F	19.00
10	Neon	Ne	20.18
11	Sodium	Na	22.99
12	Magnesium	Mg	24.31
13	Aluminum	Al	26.98
14	Silicon	Si	28.09
15	Phosphorus	P	30.98
16	Sulfur	S	32.06
17	Chlorine	Cl	35.46
18	Argon	Ar	39.95
19	Potassium	K	39.10
20	Calcium	Ca	40.08
21	Scandium	Sc	44.96
22	Titanium	Ti	47.90
23	Vanadium	V	50.94
24	Chromium	Cr	52.00
25	Manganese	Mn	54.94
26	Iron	Fe	55.85
27	Cobalt	Co	58.93
28	Nickel	Ni	58.71

Atomic Number	Element	Symbol	Atomic Mass
29	Copper	Cu	63.54
30	Zinc	Zn	65.37
31	Gallium	Ga	69.72
32	Germanium	Ge	72.59
33	Arsenic	As	74.92
34	Selenium	Se	78.96
35	Bromine	Br	79.91
36	Krypton	Kr	83.80
37	Rubidium	Rb	85.47
38	Strontium	Sr	87.62
39	Yttrium	Y	88.91
40	Zirconium	Zr	91.22
41	Niobium	Nb	92.91
42	Molybdenum	Mo	95.94
43	Technetium	Tc	98.91
44	Ruthenium	Ru	101.1
45	Rhodium	Rh	102.9
46	Palladium	Pd	106.4
47	Silver	Ag	107.9
48	Cadmium	Cd	112.4
49	Indium	In	114.8
50	Tin	Sn	118.7
51	Antimony	Sb	121.8
52	Tellurium	Te	127.6
53	Iodine	I	126.9
54	Xenon	Xe	131.3
55	Cesium	Cs	132.9
56	Barium	Ba	137.3

Atomic Number	Element	Symbol	Atomic Mass		Atomic Number	Element	Symbol	Atomic Mass
57	Lanthanum	La	138.9		81	Thallium	Tl	204.4
58	Cerium	Ce	140.1		82	Lead	Pb	207.2
59	Praseodymium	Pr	140.9		83	Bismuth	Bi	209.0
60	Neodymium	Nd	144.2		84	Polonium	Po	210
61	Promethium	Pm	145		85	Astatine	At	218
62	Samarium	Sm	150.4		86	Radon	Rn	222
63	Europium	Eu	152.0		87	Francium	Fr	223
64	Gadolinium	Gd	157.3		88	Radium	Ra	226.0
65	Terbium	Tb	158.9		89	Actinium	Ac	227.0
66	Dysprosium	Dy	162.5		90	Thorium	Th	232.0
67	Holmium	Ho	164.9		91	Protactinium	Pa	231.0
68	Erbium	Er	167.3		92	Uranium	U	238.0
69	Thulium	Tm	168.9		93	Neptunium	Np	237
70	Ytterbium	Yb	173.0		94	Plutonium	Pu	244
71	Lutetium	Lu	175.0		95	Americium	Am	243
72	Hafnium	Hf	178.5		96	Curium	Cm	247
73	Tantalum	Ta	181.0		97	Berkelium	Bk	247
74	Tungsten	W	183.9		98	Californium	Cf	251
75	Rhenium	Re	186.2		99	Einsteinium	Es	254
76	Osmium	Os	190.2		100	Fermium	Fm	257
77	Iridium	Ir	192.2		101	Mendelevium	Md	258
78	Platinum	Pt	195.1		102	Nobelium	No	259
79	Gold	Au	197.0		103	Lawrencium	Lr	260
80	Mercury	Hg	200.6					

Appendix F
Vector Multiplication

F.1 Introduction

Some instructors of college physics like to use the multiplication of vectors in their courses while others do not. Therefore, the multiplication of vectors is covered in this appendix, so those wishing to use it may do so. Section F.2 gives a general introduction to the multiplication of vectors, whereas each topic in the book that might utilize the multiplication of vectors is also treated here as a separate section. The section in the book where that topic is first introduced is listed in parentheses behind the section title.

F.2 The Multiplication of Vectors

Vectors were introduced in chapter 2 and we saw how vectors could be added and subtracted. That description of vectors was somewhat incomplete in that the possibility of multiplying vectors together was never discussed. That situation will be remedied here because there are many physical laws that are expressed as the product of vectors. There are three types of multiplication involving vectors:

1. Multiplication of a vector by a scalar.
2. The scalar product (the multiplication of two vectors, the product of which is a scalar).
3. The vector product (the multiplication of two vectors, the product of which is a vector).

Let us look at each of these products in detail.

The Multiplication of a Vector by a Scalar

Consider the vector **a** in figure F.1(a). This vector has both magnitude and direction. If **a** is multiplied by a scalar, then the product $k\mathbf{a}$ is a new vector, say **b**, where

$$\mathbf{b} = k\mathbf{a} \tag{F.1}$$

If k is a positive number greater than one, then the product represents a vector in the same direction as **a** but elongated by a factor k, as shown in figure F.1(b). If k is less than one, but greater than zero, the new vector is shorter than **a**, figure F.1(c), and if k is negative, the new vector is in the opposite direction of **a**, figure F.1(d).

The Scalar Product or Dot Product

*The scalar product of two vectors **a** and **b**, figure F.2, is defined as*

$$\mathbf{a} \cdot \mathbf{b} = ab \cos \theta \tag{F.2}$$

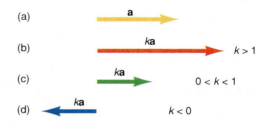

Figure F.1

Multiplication of a vector by a scalar.

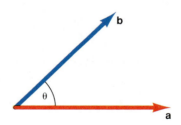

Figure F.2

The scalar product.

where θ is the angle between the two vectors when they are drawn tail-to-tail, and a and b are the magnitudes of the two vectors. The symbol for this type of multiplication is the dot "·" between the vectors **a** and **b.** Hence, *the scalar product is also called the dot product.* The definition, equation F.2, can be used to represent many different physical concepts and phenomena.

As we can observe in figure F.2, $b \cos \theta$ is the component of the vector **b** in the **a** direction. Thus, *the scalar product is the component of the vector **b** in the direction of the vector **a**, multiplied by the magnitude of the vector **a** itself.* We can also say that the quantity, $a \cos \theta$, is the component of the vector **a** in the **b** direction and hence the *scalar product is also the component of the vector **a** in the **b** direction, multiplied by the magnitude of the vector **b** itself.* Either description is correct.

The inclusion of the cos θ in the definition of the scalar product gives rise to some interesting special cases. If the vectors **a** and **b** are parallel to each other, then the angle θ between **a** and **b** is equal to zero. In that case, the scalar product becomes

$$\mathbf{a} \cdot \mathbf{b} = ab \cos 0°$$

But the cos 0° = 1, therefore

$$\mathbf{a} \cdot \mathbf{b} = ab \quad \text{(for } \mathbf{a} \parallel \mathbf{b}\text{)} \qquad \text{(F.3)}$$

(As another special case, $\mathbf{a} \cdot \mathbf{a} = a^2$, and $a = \sqrt{\mathbf{a} \cdot \mathbf{a}}$ defines the magnitude of any vector.) On the other hand, if **a** is perpendicular to **b**, then the angle between the two vectors is 90°, and the scalar product becomes

$$\mathbf{a} \cdot \mathbf{b} = ab \cos 90°$$

But the cos 90° = 0, and hence

$$\mathbf{a} \cdot \mathbf{b} = 0 \quad \text{(for } a \perp b\text{)} \qquad \text{(F.4)}$$

For example, if the vector **a** has a magnitude of 2 units and vector **b** a magnitude of 3 units, then their dot product can take on any value between −6 and +6 depending on the angle θ between the two vectors. For scalars, on the other hand, 2 times 3 is always equal to 6 and nothing else.

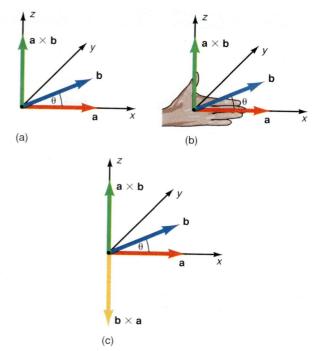

Figure F.3

The cross product.

One of the characteristics of this vector product is that the order of the multiplication is extremely important, for as we can see from figure F.3(c),

$$\mathbf{a} \times \mathbf{b} = -\mathbf{b} \times \mathbf{a} \qquad \text{(F.6)}$$

The magnitude of $\mathbf{a} \times \mathbf{b}$ is the same as the magnitude of $\mathbf{b} \times \mathbf{a}$ from the defining equation F.5, but their directions are opposite to each other as seen in figure F.3(c). This result is often stated as: vectors are noncommutative under vector multiplication.

If the vectors **a** and **b** are parallel to each other then $\theta = 0°$ and

$$|\mathbf{a} \times \mathbf{b}| = ab \sin 0°$$

But sin 0° = 0, therefore,

$$|\mathbf{a} \times \mathbf{b}| = 0 \quad \text{(for } a \parallel b\text{)} \qquad \text{(F.7)}$$

If the vector **a** is perpendicular to the vector **b**, then $\theta = 90°$, and

$$|\mathbf{a} \times \mathbf{b}| = ab \sin 90°$$

But the sin 90° = 1, therefore

$$|\mathbf{a} \times \mathbf{b}| = ab \quad \text{(for } a \perp b\text{)} \qquad \text{(F.8)}$$

Note that these two cases are the opposite of what was found for the dot product. This is because the dot product contains the cos θ term, whereas the cross product contains the sin θ. The cos θ and sin θ are complementary functions and they are out of phase with each other by 90°.

Example F.1

The scalar product. The vector **a** has a magnitude of 5.00 units and the vector **b** a magnitude of 7.00 units. If the angle between the vectors is 53.0°, find their scalar product.

Solution

The scalar product of the two vectors, given by equation F.2, is

$$\mathbf{a} \cdot \mathbf{b} = ab \cos \theta$$
$$= (5.00)(7.00)\cos 53.0°$$
$$= 21.1$$

The Vector Product or Cross Product
*The magnitude of the vector product of two vectors **a** and **b** is defined to be*

$$|\mathbf{a} \times \mathbf{b}| = ab \sin \theta \qquad \text{(F.5)}$$

where θ is the angle between the two vectors when they are drawn tail-to-tail, as shown in figure F.3(a). The symbol for the multiplication is designated by the cross sign "×" between the two vectors, and hence the name, *cross product*. The result of the cross product of two vectors, **a** and **b**, is another vector, $\mathbf{a} \times \mathbf{b}$, which is perpendicular to the plane made by the vectors **a** and **b**. *The direction of $\mathbf{a} \times \mathbf{b}$ is found by taking your right hand with the fingers in the same direction as the vector **a**, your palm facing toward the vector **b**, and then rotating your right hand through the angle between **a** and **b**. Your thumb will then point in the direction of $\mathbf{a} \times \mathbf{b}$ as seen in figure F.3(b). This rule is called the right-hand rule and it is imperative that the right hand be used.* The new vector is perpendicular both to the vector **a** and the vector **b**, and hence to the plane generated by **a** and **b**.

Example F.2

The vector product. The vector **a** has a magnitude of 5.00 units and the vector **b** of 7.00 units. If the angle between the vectors is 53.0°, find their cross product.

Solution

The magnitude of their cross product, found from equation F.5, is

$$|\mathbf{a} \times \mathbf{b}| = ab \sin \theta$$
$$= (5.00)(7.00)\sin 53.0°$$
$$= 28.0$$

The direction of $\mathbf{a} \times \mathbf{b}$ is as shown in figure F.3(a).

The area of a surface can be represented by the cross product of the two vectors that generate the area. As an example, consider the two vectors, **a** and **b**, in figure F.4. If vectors **a** and **b** are moved parallel to themselves they are still the same vectors, since they still have the same magnitude and direction. But in the process of moving them parallel to themselves they have generated a parallelogram. Recall from elementary geometry that the area of a parallelogram is equal to the product of its base times its altitude, that is,

$$A = (\text{base})(\text{altitude})$$

The base of the parallelogram is given by the magnitude of the vector **a**, and, we can see from figure F.4, the altitude is given by

$$\text{altitude} = b \sin \theta$$

Therefore, the area of the parallelogram is

$$A = ab \sin \theta$$

But

$$ab \sin \theta = |\mathbf{a} \times \mathbf{b}|$$

Therefore, the area of a parallelogram generated by sides **a** and **b** is

$$A = |\mathbf{a} \times \mathbf{b}| = ab \sin \theta \qquad \text{(F.9)}$$

Because $\mathbf{a} \times \mathbf{b}$ is a vector perpendicular to the surface generated by **a** and **b**, the area of the surface can be represented as the vector **A**, where

$$\mathbf{A} = \mathbf{a} \times \mathbf{b} \qquad \text{(F.10)}$$

Thus, **A** is perpendicular to the surface. An area can be represented as a vector in many physical applications.

F.3 Torque (Section 5.2)

An important application of the cross product is in the definition of the concept of torque, which is defined in section 5.2. A force **F** acting at the point *A*, figure F.5, a distance *r* from the axis of rotation, 0, creates a torque given by

$$\tau = r_\perp F = rF_\perp = rF \sin \theta \qquad \text{(5.21)}$$

Notice, from equation 5.21, *that the torque acting about an axis should be defined as the cross product of r and F,* that is,

$$\tau = \mathbf{r} \times \mathbf{F} \qquad \text{(F.11)}$$

with magnitude

$$\tau = |\tau| = |\mathbf{r} \times \mathbf{F}| = rF \sin \theta \qquad \text{(F.12)}$$

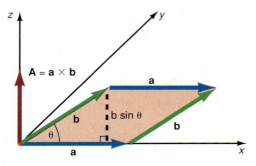

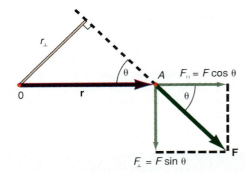

Figure F.4

The area of a surface.

Figure F.5

Torque.

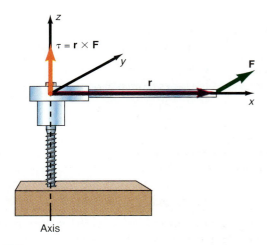

Figure F.6

Torque and the cross product.

The torque vector lies along the axis of rotation. It is perpendicular to the page and points into the page in figure F.5.

As an example of the vector concept of torque, consider the screw with right-handed threads shown in figure F.6. A screwdriver head is attached to a ratchet wrench and is inserted into the slot at the top of the screw. A force **F** is exerted at the end of the ratchet as shown. If **r** is the vector distance from the axis of the screw to the point of application of the force, then a torque is created given by

$$\tau = \mathbf{r} \times \mathbf{F}$$

The torque vector τ lies along the axis of rotation and points upward, in the figure, in the direction that the screw moves while the torque is applied. A force in the direction shown causes the screw to rotate in a counterclockwise direction, which causes the screw to move out from the surface in an upward direction. If the direction of the force is reversed by 180°, the screw rotates in a clockwise direction, the torque vector points downward along the axis of the screw, and the screw moves in that direction into the wood.

F.4 Work (Section 7.2)

One of the most important examples of the dot product is found in the concept of work. As defined in section 7.2, *the work done in moving a block a distance x by a force **F**, figure F.7, is determined by multiplying the component of the force in the direction of the displacement, by the displacement itself.* The component of the force in the direction of the displacement, seen from figure F.7, is

$$F_x = \mathbf{F} \cos \theta$$

From the definition, the work done becomes

$$W = F_x x = (F \cos \theta)x = Fx \cos \theta$$

But $Fx \cos \theta$ is the dot product of $\mathbf{F} \cdot \mathbf{x}$. Therefore, work should be defined as the scalar or dot product of $\mathbf{F}$ and $\mathbf{x}$, that is,

$$W = \mathbf{F} \cdot \mathbf{x} = Fx \cos \theta \qquad \text{(F.13)}$$

It is now obvious why the scalar product was defined the way it was in equation F.2, because it does indeed represent a physical concept.

F.5 The Force on an Electric Charge in a Magnetic Field (Section 22.1)

If an electric charge is fired with a velocity **v** into a magnetic field **B**, it is observed that a force acts on the charge. The force, however, acts at an angle of 90° to the plane determined by the velocity vector **v** and the magnetic field **B**, as shown in figure F.8. In fact, *the force acting on the moving charge is given by the cross product of **v** and **B** as*

$$\mathbf{F} = q\mathbf{v} \times \mathbf{B} \qquad \text{(F.14)}$$

Recall that in determining the direction of the cross product, point the fingers of your right hand in the direction of the first vector, in this case **v**. Your palm should be facing the second vector, in this case **B**. Rotate your right hand from **v** to **B** and your thumb will point in the direction of the cross product **v** × **B**, in this case upward. Hence the direction of the force is also upward.

The magnitude of the force is determined from the definition of the cross product, equation F.5, and is given by

$$F = qvB \sin \theta \qquad \text{(F.15)}$$

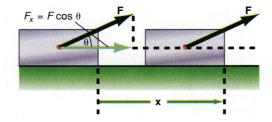

Figure F.7
The concept of work.

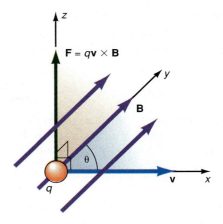

Figure F.8
Force acting on a charge in a magnetic field.

The angle θ is the angle between the velocity vector **v** and the magnetic field vector **B**.

F.6 Force on a Current-Carrying Conductor in an External Magnetic Field (Section 22.2)

A wire carrying a current I in any external magnetic field **B**, will experience a force given by

$$\mathbf{F} = I\boldsymbol{\ell} \times \mathbf{B} \qquad \text{(F.16)}$$

The force is given by a cross product term, and the direction of the force is found from $\boldsymbol{\ell} \times \mathbf{B}$. With $\boldsymbol{\ell}$ in the direction of the current and **B** pointing into the page in figure F.9, the direction of $\boldsymbol{\ell} \times \mathbf{B}$ is found by rotating the vector $\boldsymbol{\ell}$ toward the vector **B** in the cross product, the thumb points upward, indicating that the direction of the force on the wire is also upward. If the direction of the current flow is reversed, $\boldsymbol{\ell}$ is reversed and $\boldsymbol{\ell} \times \mathbf{B}$ then points downward. *The magnitude of the force, determined from equation F.16, is*

$$F = I\ell B \sin \theta \qquad \text{(F.17)}$$

where θ is the angle between $\boldsymbol{\ell}$ and **B**.

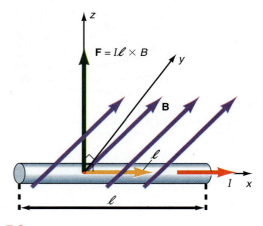

Figure F.9
Force on a current-carrying wire in an external magnetic field.

F.7 The Biot-Savart Law (Section 22.4)

The Biot-Savart law relates the amount of magnetic field ΔB produced by a small element $\Delta \ell$ of a wire carrying a current I and is given by

$$\Delta \mathbf{B} = \frac{\mu_0 I}{4\pi} \frac{\Delta \boldsymbol{\ell} \times \mathbf{r}}{r^3} \qquad \text{(F.18)}$$

and is shown in figure F.10(a). The cross product term $\Delta \boldsymbol{\ell} \times \mathbf{r}$ immediately determines the direction of $\Delta \mathbf{B}$, and is shown in figure F.10(a). The magnitude of $\Delta \mathbf{B}$, found from equation F.18, is

$$\Delta B = \frac{\mu_0 I}{4\pi} \frac{\Delta \ell \, r \sin \theta}{r^3} \qquad \text{(F.19)}$$

The Biot-Savart law says that the small current element $\Delta \boldsymbol{\ell}$ produces a small amount of magnetic field $\Delta \mathbf{B}$ at the point P. But the entire length of wire can be cut up into many $\Delta \boldsymbol{\ell}$'s, and each contributes to the total magnetic field at P. This is shown in figure F.10(b). Therefore, *the total magnetic field at point P is the vector sum of all the $\Delta \mathbf{B}$'s associated with each current element,* that is,

$$\mathbf{B} = \Sigma \Delta \mathbf{B}$$

F.8 Ampère's Circuital Law (Section 22.6)

Ampère's circuital law is also used to determine the magnetic field for different current distributions. *Ampère's law states that along any arbitrary path encircling a total current I, the sum of the scalar product of the magnetic field $\mathbf{B}$ with the element of length $\Delta \boldsymbol{\ell}$ of the path, is equal to the permeability μ_0, times the total current I enclosed by the path.* That is,

$$\Sigma \, \mathbf{B} \cdot \Delta \boldsymbol{\ell} = \mu_0 I \qquad \text{(F.20)}$$

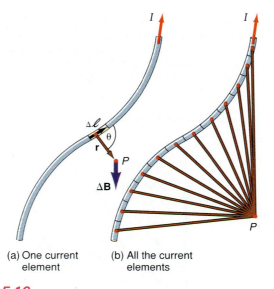

(a) One current element

(b) All the current elements

Figure F.10

The magnetic field produced by a current element.

F.9 Torque on a Current Loop in an External Magnetic Field (Section 22.8)

If a rectangular coil of wire, carrying a current I, is placed in a uniform magnetic field **B**, figure F.11(a), each segment of the coil represents a current-carrying wire in an external magnetic field and thus experiences a force on it, given by equation F.16,

$$\mathbf{F} = I\boldsymbol{\ell} \times \mathbf{B}$$

The net result of the current flowing in a coil in an external magnetic field is to produce two equal and opposite forces acting on the coil. But since the forces do not lie along the same line of action, the forces cause a torque to act on the coil as is readily observable in figure F.11(b). The torque acting on the coil, given by equation F.11, is

$$\boldsymbol{\tau} = \mathbf{r} \times \mathbf{F}$$

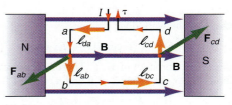

(a) Side view

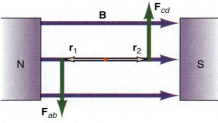

(b) Top view

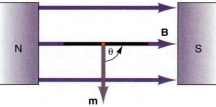

(c) Top view—**m** perpendicular to **B**

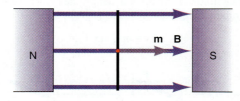

(d) Top view—**m** parallel to **B**

Figure F.11

Torque on a current-carrying loop in an external magnetic field.

As seen in section 22.8, this reduces to

$$\tau = IAB \sin \theta \qquad \text{(F.21)}$$

Thus, the torque on the coil in a magnetic field depends on the current I in the coil, the area A of the coil, the intensity of the magnetic field $\mathbf{B}$, and the angle θ between $\mathbf{r}$ and $\mathbf{F}$. Since the magnetic dipole moment $\mathbf{m}$ of a current loop is defined as $m = IA$, we can also write the torque acting on the coil, equation F.21, in terms of the magnetic dipole moment of the coil as

$$\tau = mB \sin \theta \qquad \text{(F.22)}$$

or in the vector notation

$$\tau = \mathbf{m} \times \mathbf{B} \qquad \text{(F.23)}$$

Equation F.23 shows that the torque acting on a coil in a magnetic field is equal to the cross product of the magnetic dipole moment of the coil and the magnetic field $\mathbf{B}$. Note that the angle θ is the angle between the magnetic dipole moment vector $\mathbf{m}$ and the magnetic field vector $\mathbf{B}$ in figure F.11(c). The magnetic field causes a torque to act on the coil until the magnetic dipole moment of the coil is aligned with the magnetic field, figure F.11(d).

F.10 Magnetic Flux (Section 23.2)

Flux is a quantitative measure of the number of lines of a vector field that passes perpendicularly through a surface. The magnetic flux Φ_M can be defined as a quantitative measure of the number of magnetic field lines $\mathbf{B}$ passing normally through a particular surface area $\mathbf{A}$. That is,

$$\Phi_M = \mathbf{B} \cdot \mathbf{A} \qquad \text{(F.24)}$$

as shown in figure F.12(a). The area is represented as a vector $\mathbf{A}$ that has the magnitude of the area of the surface and a direction that is perpendicular to the surface. From the definition of the scalar product we can also write equation F.24 as

$$\Phi_M = BA \cos \theta \qquad \text{(F.25)}$$

where θ is the angle between $\mathbf{B}$ and $\mathbf{A}$.

The vector $\mathbf{B}$, at the point P of figure F.12(a), can be resolved into the components $B_\perp$, the component perpendicular to the area, and $B_\parallel$, the parallel component. The perpendicular component is

$$B_\perp = B \cos \theta$$

whereas the parallel component is given by

$$B_\parallel = B \sin \theta$$

The parallel component $B_\parallel$ lies in the surface itself and therefore does not pass through the surface, whereas the perpendicular component $B_\perp$ completely passes through the surface at the point P. The product of the perpendicular component and the area

$$B_\perp A = (B \cos \theta)A = AB \cos \theta = \mathbf{A} \cdot \mathbf{B} = \Phi_M$$

is therefore a quantitative measure of the number of lines of $\mathbf{B}$ passing normally through the entire area $\mathbf{A}$. The number of lines represents the strength of the field. If the angle θ in equation F.25 is zero, then $\mathbf{B}$ is parallel to the vector $\mathbf{A}$ and all the lines of $\mathbf{B}$ pass normally through the area A, as seen in figure F.12(b). If the angle θ in equation F.25 is 90°, then $\mathbf{B}$ is perpendicular to the area vector $\mathbf{A}$, and none of the lines of $\mathbf{B}$ pass through the surface A, as seen in figure F.12(c).

F.11 Induced Electric Field (Section 23.3)

A metal wire is pulled at a velocity $\mathbf{v}$ to the right through a uniform magnetic field $\mathbf{B}$ directed into the paper, as shown in figure F.13. As the wire is moved to the right, any charge q within the wire experiences the force

$$\mathbf{F} = q\mathbf{v} \times \mathbf{B}$$

as was shown in section F.5. If both sides of the equation are divided by $\mathbf{q}$, we have

$$\frac{\mathbf{F}}{q} = \mathbf{v} \times \mathbf{B}$$

But an electrostatic field was originally defined as

$$\mathbf{E} = \frac{\mathbf{F}}{q} \qquad \text{(F.26)}$$

where $\mathbf{F}$ is the force acting on a test charge q placed at rest in the electrostatic field. It is therefore reasonable to now *define an induced electric field E by equation F.27 as*

$$\mathbf{E} = \frac{\mathbf{F}}{q} = \mathbf{v} \times \mathbf{B} \qquad \text{(F.27)}$$

Figure F.12
The magnetic flux.

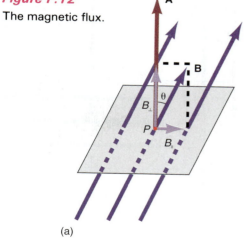

(a)

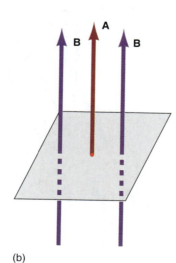

(b)

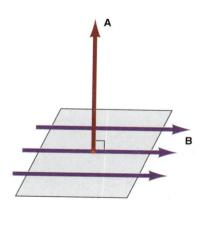

(c)

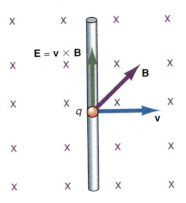

Figure F.13

An induced electric field.

This is a different type of field than the electrostatic field. *The induced electric field exists only when the charge is in motion at a velocity v. When $v = 0$, the induced electric field is also zero* as can be seen by equation F.27. The direction of the induced electric field in figure F.13, and hence the direction that a positive charge q within the wire moves, is found from the definition of the cross product $\mathbf{v} \times \mathbf{B}$. The direction is found by pointing the fingers of your right hand in the direction of the velocity vector **v**. Your palm should be facing the magnetic field vector **B**. Rotate your right hand from **v** to **B** and your thumb will point in the direction of the vector $\mathbf{v} \times \mathbf{B}$, and hence in the direction of the electric field vector, in this case upward. The magnitude of the induced electric field, found from equation F.27, is

$$E = vB \sin \theta \qquad \text{(F.28)}$$

F.12 Faraday's Law of Induction (Section 23.3)

Faraday's law of electromagnetic induction states that whenever the magnetic flux changes with time, there is an induced emf. Faraday's law is written as

$$\mathscr{E} = -\frac{\Delta \Phi_M}{\Delta t} \qquad \text{(F.29)}$$

The magnetic flux Φ_M is defined as

$$\Phi_M = \mathbf{B} \cdot \mathbf{A}$$

In general the change in the magnetic flux caused by a change in area $\Delta \mathbf{A}$ and a change in the magnetic field $\Delta \mathbf{B}$, is

$$\Delta \Phi_M = \mathbf{B} \cdot \Delta \mathbf{A} + \mathbf{A} \cdot \Delta \mathbf{B}$$

and we can also write Faraday's law, equation F.29, as

$$\mathscr{E} = -\frac{\Delta \Phi_M}{\Delta t} = -\mathbf{B} \cdot \frac{\Delta \mathbf{A}}{\Delta t} - \mathbf{A} \cdot \frac{\Delta \mathbf{B}}{\Delta t} \qquad \text{(F.30)}$$

Faraday's law, in this form, says that an emf can be induced either by changing the area of a loop of wire with time, while the magnetic field remains constant, or by keeping the area of the loop of wire constant and changing the magnetic field with time. Because the flux is given by a scalar product, it is also possible to keep the magnitudes B and A constant, but vary the angle between them. This also causes a change in the magnetic flux and hence an induced emf.

F.13 Electric Flux (Section 25.2)

The electric flux is defined as

$$\Phi_E = \mathbf{E} \cdot \mathbf{A} = EA \cos \theta \qquad \text{(F.31)}$$

and is a quantitative measure of the number of lines of **E** that pass normally through the surface area **A**. Figure F.14(a) shows an electric field **E** passing through a portion of a surface of area **A**. The area of the surface is represented by a vector **A**, whose magnitude is the area A of the surface, and whose direction is perpendicular to the surface. The number of lines represents the strength of the field. The vector **E**, at the point P of figure F.14(a), can be resolved into the components $E_\perp$, the component perpendicular to the area, and $E_\parallel$, the parallel component. The parallel component $E_\parallel$ lies in the surface itself and therefore does not pass through the surface, whereas the perpendicular component $E_\perp$ completely passes through the surface at the point P. The product of the perpendicular component and the area

$$E_\perp A = (E \cos \theta)A = EA \cos \theta = \mathbf{E} \cdot \mathbf{A} = \Phi_E \qquad \text{(F.32)}$$

is therefore a quantitative measure of the number of lines of **E** passing normally through the entire area, A. If the angle θ in

Figure F.14

Electric flux.

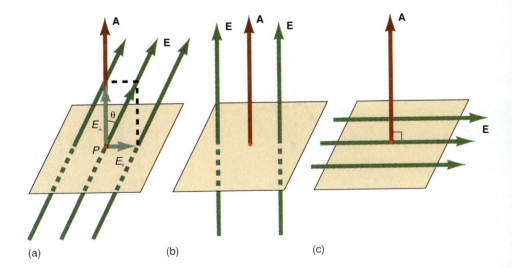

(a)   (b)   (c)

equation F.31 is zero, then **E** is parallel to the vector **A** and all the lines of **E** pass normally through the area A, as seen in figure F.14(b). If the angle θ in equation F.31 is 90° and then **E** is perpendicular to the area vector **A**, and none of the lines of **E** pass through the surface A, as seen in figure F.14(c).

F.14 Gauss's Law for Electricity (Section 25.2)

Gauss's law for electricity says that the electric flux Φ_E through a surface surrounding the point charge q is a measure of the amount of charge q contained within the Gaussian surface. It is stated mathematically as

$$\Phi_E = \Sigma\, \mathbf{E} \cdot \Delta\mathbf{A} = \frac{q}{\epsilon_0} \tag{F.33}$$

where q is now the net charge contained within the Gaussian surface. Here $\mathbf{E} \cdot \Delta\mathbf{A}$ is the amount of electric flux passing through a small area ΔA of the surface, and $\Sigma\, \mathbf{E} \cdot \Delta\mathbf{A}$ is the sum of that flux passing through the entire surface. When Φ_E is a positive quantity, the Gaussian surface surrounds a source of positive charge, and electric flux diverges out of the surface. Whenever Φ_E is negative,

the Gaussian surface surrounds a negative charge distribution and the electric flux converges into the Gaussian surface. If there is no enclosed charge, $q = 0$, and hence the flux $\Phi_E = 0$. In this case, whatever electric flux enters one part of a Gaussian surface, the same amount must leave somewhere else.

F.15 Gauss's Law for Magnetism (Section 25.3)

Gauss's law for magnetism takes the same form as Gauss's law for electricity. However, the electric flux is a measure of the electric charge q enclosed within the Gaussian surface. The magnetic flux should be a measure of the amount of magnetic pole enclosed within the Gaussian surface. But, isolated magnetic poles do not exist. Thus, the expected term for the enclosed magnetic pole on the right side of Gauss's law must be equal to zero. *Thus, Gauss's law for magnetism becomes*

$$\Phi_M = \Sigma\, \mathbf{B} \cdot \Delta\mathbf{A} = 0 \tag{F.34}$$

The net magnetic flux through any Gaussian surface anywhere in a magnetic field is zero because there are no isolated magnetic poles.

Appendix G
Answers to Odd-Numbered Problems

Chapter 1

1. 16.4 cm³
3. 88 ft/s
5. 86,400 s; 2.59×10^6 s; 3.15×10^7 s
7. 55.9 mph
9. 3280 ft
11. 1090 ft/s; 736 mph
13. 13.4 m²
15. **a.** 2.37×10^9 s   **b.** 3.95×10^7 min; 2.76×10^9 pulses/lifetime
17. 6.70×10^8 mph; 2.99×10^8 m/s
19. 9.14 m; 91.4 m
21. 380 m; 0.236 mi; 1.49×10^4 in.; 3.80×10^5 mm
23. 1.275×10^4 km
25. **a.** 5.89×10^5 pm   **b.** 5.89×10^{-4} mm
    c. 5.89×10^{-5} cm
    d. 5.89×10^{-7} m; 4.31×10^4 waves
27. 2.13 m
29. 169 m
31. 1×10^{-14} m³ to 1×10^{-12} m³; 6.10×10^{-10} in.³ to 6.10×10^{-8} in.³
33. 8.6×10^{-7} mm; 3.39×10^{-8} in.
35. **a.** 2.0×10^5 pm   **b.** 200 nm   **c.** 0.2 μm
    d. 0.0002 mm   **e.** 2×10^{-5} cm
37. 0.00994 slugs
39. 0.159 m³
41. 3.58×10^4 ft
43. 5.10×10^{14} m²; 5.49×10^{15} ft²; 1.08×10^{21} m³; 3.83×10^{22} ft³; 5.51×10^3 kg/m³
45. $v = 24.84$ mi/hr
47. 4.25 min/month; 51.74 min/yr

Chapter 2

1. 57.4 lb; 81.9 lb
3. 6.47 lb; 24.2 lb
5. 262 ft
7. $v_x = 198$ km/hr; $v_y = 27.8$ km/hr
9. $w_\| = 6070$ N; $w_\perp = 6510$ N
11. 292 m/s; 956 m/s
13. 8.09 mi at 63.7° north of east
15. 174 mph at 9.26° south of east
17. 85.2 N; 148° from + x - axis
19. **a.** $F_h = 39.9$ lb   **b.** $F_v = 30.1$ lb   **c.** -119.9 lb
21. **a.** 100 ft; 50° N of E   **b.** 25 ft; 50° N of E
    c. 50 ft; 50° S of W   **d.** 250 ft; 50° S of W
    e. 250 ft; 50° N of E   **f.** 150 ft; 50° S of W

23. 4.85 m/s; $\theta = 48.7°$
25. 22.7 lb at 85.6° above the +x-axis
27. **a.** The velocity of the aircraft is 169 km/hr in the direction 12.1° south of west.   **b.** The pilot must point the plane in a direction 10.2° N of W.   **c.** 2.47 hr
29. **a.** 9.43 mph   **b.** 15 min   **c.** 1.25 mi   **d.** 38.7° S of E
33. 5.71 N; $\theta = 82°$; $\theta = 82°$ below + x-axis
37. 228.3 miles east of its starting point A, and 28.3 mi south of its starting point or 40.0 mi to the south east of the city at B
39. $T_y = 123$ N; $T_x = 158$ N
41. 21.3°
43. 14.4 m; $-27.8°$

Chapter 3

1. **a.** 55.4 mph   **b.** 81.3 ft/s   **c.** 89.1 km/hr   **d.** 24.8 m/s
3. 1720 m
5. 4.33 min
7. 4.59 m/s
9. **a.** 3 m/s²;   **b.** 0 m/s²;   **c.** 2 m/s²;   **d.** -4 m/s²
11. 0.300 m/s²
13. 1.86 ft/s²; 475 ft
15. 54.5 mph
17. $t = 10$ s: $x = 147$ ft; $v = 29.3$ ft/s
    $t = 15$ s: $x = 330$ ft; $v = 44$ ft/s
    $t = 20$ s: $x = 586$ ft; $v = 58.6$ ft/s
    $t = 25$ s: $x = 916$ ft; $v = 73.3$ ft/s
19. 5.62×10^{17} cm/s²
21. -20.2 ft/s; 6.29 s
23. 200 ft
25. 59.4 ft; 193 ft
27. $a = 0.579$ m/s²; $v = 13.9$ m/s
29. 650 m
31. -17.1 m/s
33. 121 ft
35. 2.5 s
37. 0.500 s
39. 35.1 m; 75.9 m; 77.5 m; 39.9 m
41. 4.51 s
43. **a.** -31.8 m/s;   **b.** -1.71 s
45. 158 m
47. 5430 m
49. 9.55°
51. **a.** 281 ft   **b.** 8.38 s   **c.** 943 ft
53. **a.** 122 ft/s²   **b.** 37.3 m/s²   **c.** 3.81g
55. **a.** 12.6 s   **b.** 556 ft   **c.** 60.1 mph
57. **a.** -1.1 m/s²   **b.** 29.1 m   **c.** 14.54 s   **d.** -8.00 m/s

59. 10.7 s for train 2 to overtake train 1; 218 m

61. 10.5 s; 3.03 m/s²

63. $x_1 + x_2 = 466$ m; the trains will stop in time

65. 193 m

67. 19.8 m/s $= v_{20}$ the initial speed of the ball thrown upward

71. **a.** 3.97 s **b.** 117 ft **c.** 62.9 ft **d.** 70 ft/s at 65° below
$+x$-axis

73. **a.** 127 m **b.** 7.44 s **c.** 143 m **d.** 53.5 m/s at 68.9° below
$+x$-axis

75. $\dfrac{g}{2v_{0x}} x^2 - \dfrac{v_{0y}}{v_{0x}} x - y - y_0 = 0$, which is the standard form of the
equation of a parabola, $A_x{}^2 + D_x + E_y + F = 0$. Hence the
trajectory of the projectile is a parabola.

Chapter 4

1. 6.25 slugs

3. 75.0 N

5. −1500 lb

7. −1.82 × 10⁴ lb opposite to direction of motion; −909 lb

9. 130 lb

11. **a.** 9220 N **b.** 6780 N

13. 417 N

15. 2.27 × 10⁵ lb

17. 61.1 m

19. **a.** 0.500 m/s² **b.** 2.50 m/s **c.** 6.25 m **d.** 3.5 m/s

21. **a.** 4.9 m/s² **b.** 9.9 m/s

23. **a.** 1020 N **b.** 882 N **c.** 747 N **d.** 882 N **e.** 0

25. **a.** 0.0989 m **b.** 0.165 s

27. **a.** 0.891 m/s² **b.** 2.67 m/s **c.** 4.01 m to the right

29. $F = 2.4$ N; $T_2 = 1.2$ N; $T_1 = 0.400$ N

31. 403 ft

33. 100 N; 80 N

35. **a.** 23.2 lb **b.** 15.1 lb **c.** Method b is better; it requires less force to
move the block.

37. 0.0085

39. **a.** 10.8 m/s **b.** 6.05 m/s

41. **a.** 10.6 N **b.** 0.312

43. **a.** 79.3 ft/s **b.** 983 ft from the bottom of the slope

45. **a.** $a = 20.0$ m/s² **b.** $v = 11.0$ m/s

47. .600

49. 35.3 N

51. 800 g

53. acceleration $a = 0.887$ m/s²; tension $= 65.1$ N

55. $T_C = 25.6$ N; $T_B = 64.0$ N; $T_A = 128$ N

57. 7.54 N

59. **a.** 4.69 m/s² **b.** 0.586 m to the left of its original position

61. **a.** $\dfrac{w_A \sin\theta - w_B \sin\phi}{m_A + m_B}$

63. **a.** 2.07 m/s² **b.** 1.14 N

65. 3.83 m/s²; $T_A = T_B = 35.8$ N; $T_C = 15.5$ N

67. 1.25 m/s²; $T_A = T_B = 51.3$ N; $T_C = 42.9$ N

69. **a.** 3.27 m/s² **b.** 0.333

71. $v = \dfrac{mg}{k}$. This means, that after a relatively long time, the velocity of
the object becomes constant. This is referred to as the terminal
velocity.

Chapter 5

1. 0.902 N at 101° from the $+x$-axis

3. 2.87 N

5. 133 N

7. $T_1 = 2134$ lb; $T_2 = 2130$ lb

9. 20.0 N

11. $T_1 = 489$ N; $T_2 = 553$ N

13. **a.** 196 N **b.** 392 N

15. 25.0 lb ft

17. 17.9 Nm

19. $\tau = 5.00$ mN

21. $F_2 = 120$ lb; $F_1 = 80.0$ lb

23. 725 N; 525 N

25. 45.0 lb ft counterclockwise

27. 3.68 m from end that the son is carrying

29. $x_{cg} = 1.00$ ft

31. 3.43″, 8.58″

33. $F_B = 5000$ N; $F_A = 5900$ N

35. 30.0 cm from the origin

37. 0.0154 m to the right of center of large plate

39. $T = 1710$ N; $H = 1310$ N; $V = 101$ N

41. 1.12 m

43. $T = 897$ N;
Hinge forces: horizontal $= 777$ N, vertical $= 752$ N

45. $F_{wall} = 10.1$ N; $F_{Hfloor} = 10.1$ N; $F_{Vfloor} = 120$ N

47. no

49. 10.5 ft up the ladder

51. $F_M = 518$ lb; $F_J = 702$ lb

53. 187 lb; 304 lb

55. $T_A = 1170$ N; $F_T = 1500$ N

57. 1.09 kg

59. 34.3 lb

61. **a.** 18.8 lb ft **b.** 18.8 lb ft **c.** 226 lb

63. 0.625 m

65. 51.3°

67. $F_{wall} = 12.1$ lb;
$f_{wall\ friction} = 4.84$ lb;
vert. component of floor $= 25.2$ lb;
horiz. component of floor $= 12.1$ lb

Chapter 6

1. **a.** 2π rad **b.** $\dfrac{3\pi}{2}$ rad **c.** π rad **d.** $\dfrac{\pi}{2}$ rad
e. $\dfrac{1}{3}\pi$ rad **f.** $\dfrac{1\pi}{6}$ rad **g.** 2π rad

3. 174 ft

5. 284 m/s²

7. $a_c = 1.76$ m/s²; $F_c = 2630$ N

9. 0.540 lb

11. $a = 15.5$ ft/s²; $\mu_s = 0.484$

13. 34.6°

15. **a.** .711 **b.** 35.4°

17. 11.14°

19. 22.6°

21. **a.** 3.34 × 10⁻⁹ N **b.** 3.34 × 10⁻⁹ N **c.** $a_{5\ kg} = 6.68 \times 10^{-10}$ m/s²; $a_{10\ kg} = 3.34 \times 10^{-10}$ m/s²

23. no; $F = 2.78 \times 10^{-8}$ N

25. **a.** 9.81 m/s² **b.** 2.45 m/s² **c.** 0.0981 m/s² **d.** 2.70 × 10⁻³ m/s²

27. $g_{mars} = 3.87$ m/s²; $w_{mars} = 71.1$ lb

29. 270 N on the sun

31. $v = 29.7$ km/s; $t = 367$ days

33. $v = 3530$ m/s; $t = 1.69$ hr

35. $r = 8.50 \times 10^6$ m; $H = 2130$ km; $v = 6850$ m/s

37. **a.** 3.37 N **b.** 2.39 N **c.** 0

39. $a_c = 65.2$ ft/s²; $F_c = 367$ lb; $F_N = 547$ lb

41. 179 ft/s

45. 1.67 m/s

47. 1.42 × 10⁻⁷ N along the diagonal (45° above the horizontal)

49. 7900 m/s; 1.41 hr; 0

51. 3.52 × 10²² N; 1.98 × 10²⁰ N; the moon

53. **a.** 5.899 × 10⁻³ m/s² **b.** 3.433 × 10⁻⁵ m/s²

55. 3.46 × 10⁸ m

57. $\dfrac{GmM}{r} = \dfrac{mv^2}{r}$;
$v = \dfrac{2\pi r}{T}$
Substitute and simplify
$T^2 = \dfrac{4\pi r^3}{GM}$ (Kepler's third!)

59. 2.98×10^4 m/s = 66,700 mph

61. a. 4.66×10^6 m **b.** 1.14×10^7 m **c.** 3.812×10^8 m
   d. 1.963×10^8 m **e.** 7.69×10^3 m/s **f.** 8240 m/s
   g. 550 m/s **h.** 1610 m/s **i.** 246 m/s **j.** -1360 m/s

Chapter 7

1. 7.50×10^3 ft lb
3. 613 J
5. 3.90×10^9 J
7. a. 196 J **b.** 196 J **c.** The force in part **a** is one-half the size of the force in part **b.**
9. 1.35×10^4 J
11. a. 300 ft lb/s **b.** 3000 ft lb
13. 1600 hp
15. 137 J; 206 J
17. 294 J
19. 2.69×10^{33} J
21. a. 4.17×10^4 J **b.** 1.67×10^5 J **c.** 6.67×10^5 J
23. $KE = 7.57 \times 10^5$ ft lb; $PE = 3.00 \times 10^6$ ft lb
25. 4.05×10^3 J; -2.03×10^5 N
27. a. 36.0 J **b.** 36.0 J **c.** 6.00 m/s **d.** 6.00 m/s²; 6.00 m/s
29. 625 ft
31. 1.04 m/s
33. a. 0.0835 J **b.** 0.0835 J **c.** 0.817 m/s
35. 1.08 m
37. a. 1.23 J **b.** 0 **c.** -1.23 J **d.** 0.616 N **e.** 0.768
39. 2.25×10^3 ft lb
41. 105.4 J
43. a. 98.0 J **b.** 6.26 m/s **c.** 6.67 m
45. (H7.7) = MA = $\dfrac{F_{out}}{F_{in}} = \dfrac{r_{in}}{r_{out}}$
47. 2.06 J
49. 118,000 W
51. a. 52.7 J **b.** $+38.9$ J **c.** 18.3 N
53. a. 4.35 trips, or a little more than 2 oscillations **b.** 1.05 m; the object stops 1.05 m to the right of point B
55. $.0188 = \mu$
57. $E_{total} = 14.7$ J; $PE_1 = 9.80$ J; $PE_2 = 1.96$ J; $v_1 = v_2 = 2.90$ m/s; $KE_1 = 2.10$ J; $KE_2 = 0.841$ J
59. a. 1680 m/s **b.** 3590 m/s **c.** 4.25×10^4 m/s
61. 10.4 m
63. 0.256 m
65. a. 1.88 J **b.** 0.787 J **c.** 0.189 J **d.** 0.751 J
   e. 0.661 m/s **f.** $KE_1 = 0.109$ J; $KE_2 = 0.0437$ J

Chapter 8

1. 8.07×10^3 slug ft/s
3. -8.07×10^3 slug ft/s; -2.69×10^4 lb; -8.07×10^3 lb
5. 2.4 s
7. -0.600 m/s
9. 1.50×10^{-3} m/s
11. 0.333 ft/s
13. 1.20 ft/s
15. -30 N
17. a. 12.4 Ns **b.** 12.4 kg m/s **c.** 49.6 m/s
19. a. 125.0 Ns **b.** 3.13×10^4 N
21. $v_{1f} = -0.417$ m/s; $v_{2f} = 0.183$ m/s
23. a. 0.600 kg **b.** $\dfrac{1}{2}v_{1i}$
25. a. -19.9 cm/s $= V_{1f}$ (to the left); 8.4 cm/s (to the right) **b.** 8.97×10^{-3} J **c.** 3.85×10^{-3} J **d.** -57.1%
27. $V_f = -23.8$ km/hr
29. 1.71 ft/s
31. a. -0.083 m/s **b.** 0.0216 J **c.** 1.55×10^{-3} J **d.** -2.01×10^{-2} J; the energy is dissipated as heat
33. 41.8 km/hr at 78.7° N of E
35. a. $\Delta p = -4.82$ kg m/s **b.** Magnitude = 4.82 kg m/s; Direction = into the wall $\rightarrow$ to the right

37. a. 8.00 kg m/s **b.** 0.050 kg m/s **c.** 160 J
   d. 6.25×10^{-3} J **e.** 6.25×10^{-3}
39. a. 0.748 **b.** 0.259 J
41. a. 1.99 m/s **b.** 0.202 m
43. $9.6 \rightarrow 10$ bullets needed
45. 42 cm to the right
47. 2.75 ft/s (to the left)
49. $KE = \dfrac{1}{2}mv^2$; $p = mv$

   $KE = \dfrac{1}{2}m\left(\dfrac{p}{m}\right)^2 = \dfrac{p^2}{2m}$
51. 25 cm/s
53. Yes. See Solutions Guide.
55. a. 0.556 ft/s **b.** 1.25 ft/s **c.** 5.00 ft/s
57. 7.24 mm
59. a. 293 m **b.** 1.65m
61. 2.33 cm
63. $V_{Ci} = 89.9$ m/s; $V_{Ti} = 4.50$ m/s

Chapter 9

1. a. 3.49 rad/s **b.** 4.71 rad/s **c.** 8.17 rad/s
3. a. 105 rad/s **b.** 7.85 m/s
5. -62.8 rad/s²; 180 rev
7. 2.36×10^{-3} J
9. 0.0417 kg m²
11. a. 3.79 kg m² **b.** 0
13. 6.86 rad/s²
15. 2.80 m/s²
17. a. 1.23 m/s² **b.** 1.92 m/s
19. 157 rad/s
21. a. 5.11 m/s **b.** 4.43 m/s
23. a. 5.00 rad/s **b.** 0.188 J **c.** 7.50 J **d.** 7.69 J
25. 9.82 J
27. a. 7.27×10^{-5} rad/s **b.** 9.69×10^{37} kg m²
   c. 2.56×10^{29} J **d.** 2.66×10^{33} J; linear KE is approximately 10,000 times greater than rotational KE **e.** 7.05×10^{33} kg m²/s
29. a. 5.00×10^{-4} kg m² **b.** 0.200 m N
   c. 400 rad/s² **d.** 800 rad/s **e.** 800 rad **f.** 160 J
   g. 0.400 kg m²/s
31. a. 7.272×10^{-5} rad/s **b.** 1.263×10^{-12} rad/s **c.** yes
33. a. 13.3 rad/s² **b.** 133 rad/s **c.** 829 J **d.** 831 J
35. a. 2.5 m/s **b.** 2.50 m/s **c.** 55.6 rad/s
37. a. 0.300 m N **b.** 9.88×10^{-3} kg m² **c.** 30.4 rad/s²
   d. 122 rad/s
39. a. 5.88 J **b.** 6.26 m/s
41. 1.74 rad/s
43. a. 4.02 m/s² **b.** 2.65 m/s **c.** 0.659 s **d.** 1.52 m
   e. 6.10 m from the base of the incline

Chapter 10

1. 1.76×10^9 N/m²
3. 1.33×10^{-3}
5. 0.0191 in.
7. a. 1.36×10^8 N/m² **b.** 2.91×10^{-3} m **c.** 1.94×10^{-3}
9. 1.08×10^3 kg
11. 1.55×10^4 N
13. 2.84 mm
15. 5.34×10^4 N
17. 123.0 N/m
19. a. 294 N/m **b.** 2.67 cm
21. 3.81 N
23. 1.48×10^{-5} rad
25. 2.64×10^{-5} rad; the copper cylinder deforms 1.78 times more
27. 0.467×10^{10} N/m²
29. -0.0338 m³
31. 1.0002334
33. 0.32 mm; 5.01 m
35. 1082 masses

37. 1.67×10^{-4} m
39. 6.60×10^4 N
41. **a.** 33.7 N   **b.** 64.3 N   **c.** 3.2×10^{-4} m
43. 4.29×10^{-4} m
45. **a.** 5.33 m   **b.** 2.00 m   **c.** 3.33 m

Chapter 11

1. 1.95 Hz
3. 0.05 kg
5. **a.** 0   **b.** 3.14 m/s
7. 0.808 m/s; 6.53 m/s²
9. **a.** 1.58 Hz   **b.** 0.633 s   **c.** -1.11 m/s; -9.86 m/s²
11. 0.0563 J
13. 0.0338 J; 3.75×10^{-3} J; 0.0301 J
15. 1.74 s; 0.575 Hz
17. .276 Hz
19. **a.** 2.46 s   **b.** an infinitely long period of oscillation
21. 9.852 m/s²
23. **a.** 3.33 N/m   **b.** 7.70 s
25. **a.** .156 m/s²   **b.** 0   **c.** .078 m/s²
27. 0.709 m
29. $T_{10} = 2.0109178$ s, 0.190%; $T_{30} = 2.041969$ s, 1.71%;
   $T_{50} = 2.105713$ s, 4.68%
31. 0.231 m
33. 4.59 km
35. **a.** 444 J; 0.330 J; 0.443 J   **b.** 0.444 J   **c.** 2.03 m/s
37. 45.4 N/m
39. $A_1 = 0.0949$ m; $A_2 = 0.090$ m; $A_4 = 0.081$ m;
   $A_6 = 0.0729$ m; $A_8 = 0.0656$ m
41. 0.256 m
43. 1.35 m/s
45. See Solutions Guide.
47. 0.848 m/s upward
49. $-\dfrac{C\theta}{T}$

Chapter 12

1. **a.** 0.050 s   **b.** 5.00×10^{-5} s
3. 5.72 m
5. **a.** $f = 400$ Hz   **b.** $k = 3.14$ m⁻¹   **c.** $w = 2513$ rad/s
7. **a.** $k = 25.1$ m⁻¹   **b.** $w = 1450$ rad/s   **c.** $y = (0.0185$ m$) \sin[(25.1$ m⁻¹$)x - (1450$ Hz$)t$
9. 6.48 N
11. **a.** 302 m/s   **b.** 1.01 cm
13. **a.** 20.0 m/s   **b.** 800 N   **c.** $\lambda_2 = 1.20$ m; $f = 16.7$ Hz
15. 4.51 m
17. $f_3 = 1980$ Hz; $\lambda_3 = 40$ cm
   $f_5 = 3300$ Hz; $\lambda_5 = 24$ cm
   $f_7 = 4610$ Hz; $\lambda_7 = 17.1$ cm
19. **a.** 482 Hz   **b.** 394 Hz
21. **a.** 3160 N   **b.** $f_2 = 880$ Hz; $f_3 = 1320$ Hz; $f_4 = 1760$ Hz
   c. $\lambda_{fundamental} = 120$ cm; $\lambda_2 = 60$ cm; $\lambda_3 = 40$ cm; $\lambda_4 = 30$ cm
23. 4080 m
25. 1080 m
27. 0.3277 m; 0.2920 m; 0.2601 m
29. **a.** 542 Hz   **b.** 317 Hz
31. 357 Hz
33. 337 Hz
35. 0.0134 s
37. 1110 m
39. 64.8 dB
41. 2.14×10^{-9} m
45. **a.** 650 N   **b.** $f_2 = 880$ Hz; $f_3 = 1320$ Hz; $f_4 = 1760$ Hz;
   $f_5 = 2200$ Hz
   c. $\lambda_1 = 1.20$ m; $\lambda_2 = 0.600$ m; $\lambda_3 = 0.400$ m; $\lambda_4 = 0.300$ m;
   $\lambda_5 = 0.240$ m

Chapter 13

1. 0.707 gm/cm³
3. **a.** 193 kg   **b.** 1890 N
5. 5510 kg/m³
7. No, the crown is not pure gold.
9. 1.12×10^4 kg/m³
11. 5.27×10^{18} kg
13. **a.** 14.9 lb/in.²   **b.** 14.2 lb/in.²
15. 1.91×10^8 Pa
17. 3.433×10^5 Pa
19. 7.60×10^4 N
21. **a.** 10.3 m   **b.** 11.8 m   **c.** 12.8 m   **d.** 8.2 m
23. 1.50×10^5 Pa; 21.3 m
25. 2.52×10^5 N
27. .414 m or 41.4 cm
29. 9.63 N; 8.41 N
31. **a.** 19.0 N   **b.** 23.9 N   **c.** 259 N
33. 7.77 cm
35. 4.50 ft/s; 0.383 lb/s or 6.14×10^{-3} ft³/s
37. 5.87 ft/s
39. 9.13×10^4 Pa; 1.6 m/s
41. $v_1 = 1.28$ m/s; $v_2 = 5.12$ m/s; mass flow $= 1.61$ kg/s; volume flow $= 1.61 \times 10^{-3}$ m³/s
43. **a.** 1.993×10^5 Pa   **b.** 1.50×10^5 Pa   **c.** 4.51×10^7 N
45. 1.96×10^5 Pa
47. 8.55 m/s². It can be seen that the diameter of the ball does not play a role in the computation. This can only be true if the water is assumed to be nonviscous. Viscosity is an "internal friction" of the fluid that would decrease the acceleration of the ball.
49. 92%
51. 2.65 cm
53. 0.300 m
55. 1.29×10^3 Pa
57. $\Delta p = 5.22 \times 10^3$ Pa; $F = 7.84 \times 10^4$ N

Chapter 14

1. **a.** 37.0 °C   **b.** 37.6 °C   **c.** 36.4 °C
3. -40.0 °F
5. **a.** -30.6 °C   **b.** -10.8 °C   **c.** 12.8 °C
   d. 32.2 °C   **e.** 82.2 °C
7. $K = \dfrac{5}{9}(t \,°F + 459.4°)$
9. 3.31 kcal
11. 400 kcal
13. 0.0622 °C
15. 195 kcal
17. 21.8 °C
19. 0.0911 kcal/kg °C; copper
21. 16.6 °C
23. 5.25 kcal
25. 27.5 g
27. 7.30 kcal
29. 13.5 g
31. 161 g
33. 0.721 hr
35. 635 Btu
37. 0.358 kcal/s; 5.12×10^3 Btu/hr
39. Assume 50% of all the kinetic energy becomes thermal energy. Then the change in temperature is equal to 236 °C.
41. 35.2 N
43. 588 kcal
45. 0.836 gal
47. 2.28×10^4 kcal
49. 6.75 m³
51. 9.01 kg
53. 1.17×10^6 Btu
55. 2.88×10^5 Btu; 1.20×10^4 Btu/hr; 3520 W
57. **a.** 1.50×10^9 m³   **b.** 2.60×10^7 kg   **c.** 7.27×10^6 kg
   d. 594 kcal/kg   **e.** 1.11×10^{10} kcal

f. Thunderstorms may cover areas the size of several states. Our dimensions for the box of air may be on the order of 1000 km by 1500 km by 1 km. This volume contains $(1 \times 10^6$ m) $(1.5 \times 10^6$ m)$(1 \times 10^3$ m) or 1×10^{15} m³ of air or about 1×10^6 times greater than our sample box in this problem. Now imagine the amount of energy released in a real thunderstorm, given everything else the same.

Chapter 15

1. 2.006 m
3. 125 °C; -124.8 °C
5. 0.79904 m
7. $\Delta L = 0.288$ in.; $F = 4.32 \times 10^5$ lb
9. 50.36 ft²
11. 1261 cm²
13. 0.5964 cm³
15. $\Delta V = \beta V_0 \Delta T$
$V = V_0 + \Delta V = V_0(1 + \beta \Delta T)$
$m = \rho V; m = \rho_0 V_0$ (mass remains unchanged)
$\rho = \dfrac{\rho_0 V_0}{V} = \dfrac{\rho_0 V_0}{V_0(1 + \beta \Delta T)}$
$\rho = \dfrac{\rho_0}{1 + \beta \Delta T}$
17. 96.2 atm
19. 2.45×10^{21} molecules
21. $2.90 \times 10^5 \; \dfrac{N}{m^2}$
23. .0167 kg
25. 2.00 atm
27. 5.09 L
29. 66.3 cm³
31. 250 m/s
33. 7.31×10^{-23} g/molecule
35. 903 °C
37. 4.25×10^7 J
39. 76.5 cm Hg
41. See Solutions Guide.
43. **a.** 2.8×10^{-5} kg m²   **b.** -3.49×10^{-4} J
45. 1.44×10^{-5} kg m²/s assuming constant $\omega = 11.4$ rad/s;
$\Delta L = 0$ if ω is allowed to vary (conservation of angular momentum)
47. 1.401×10^{-4} s increase
49. **a.** 100 °C   **b.** 2.38 cm³
51. **a.** 1.01×10^4 K   **b.** 2.01×10^4 K   **c.** 1.41×10^5 K
   d. 1.61×10^5 K   **e.** 2.22×10^5 K   **f.** 9.06×10^4 K
53. $v_1 = \sqrt{\dfrac{3kT_1}{m}}$
$v_2 = \sqrt{\dfrac{3kT_2}{m}}$
$\dfrac{v_2}{v_1} = \sqrt{\dfrac{\dfrac{3kT_2}{m}}{\dfrac{3kT_1}{m}}} = \sqrt{\dfrac{T_2}{T_1}}$
$v_2 = \sqrt{\dfrac{T_2}{T_1}} v_1$
55. $v_1 = \sqrt{\dfrac{3kT}{m_1}}$
$v_2 = \sqrt{\dfrac{3kT}{m_2}}$
$\dfrac{v_2}{v_1} = \sqrt{\dfrac{\dfrac{3kT}{m_2}}{\dfrac{3kT}{m_1}}} = \sqrt{\dfrac{m_1}{m_2}}$
$v_2 = \sqrt{\dfrac{m_1}{m_2}} v_1$

Chapter 16

1. 0.900 kcal/kg
3. 1.27×10^4 Btu/hr
5. 35,000 Btu/hr
7. 122.3 °C
9. 2.62×10^8 J
11. 5.11×10^4 kcal
13. 1.55 kcal/hr
15. 1.48×10^5 Btu
17. **a.** 73.6 kg   **b.** 5.89×10^3 kcal   **c.** 55.1 hr to melt
19. 20%
21. **a.** 13.8 hr ft² °F/Btu   **b.** 20.7 hr ft² °F/Btu
23. 7.4×10^4 Btu
25. **a.** 1.75×10^6 Btu   **b.** 9.68×10^3 Btu transferred from the attic into the house over a period of 2 hr
27. 1.13×10^6 J
29. 5.07×10^{21} J
31. 3.28×10^7 J
33. **a.** 13.1 J/s   **b.** 89.2 J/s
35. 6.27×10^3 J/s; 9.89×10^{-6} m
37. 6.13×10^{-6} m
39. 4140 K
41. 4.27×10^8 J
43. **a.** 4.68×10^4 J/s   **b.** 2.34×10^4 J/s
   c. 4.07×10^3 J/s   **d.** 2.85×10^4 J/s
45. **a.** 31.9 °C   **b.** 18.6 °C
47. $Q = 256 \, Q_0$
49. 33.6 g/min enters the container
51. 1.87×10^6 kcal/hr
53. For large values of r_1 and r_2, the product $r_1 r_2$ is approximately equal to r^2 and $r_2 - r_1$ is the thickness of the material (d).
$\dfrac{\Delta Q}{\Delta t} = \dfrac{4\pi k \Delta T}{(r_2 - r_1)/r_1 r_2} = \dfrac{4\pi r^2 k \Delta T}{r_2 - r_1} = \dfrac{4\pi r^2 k \Delta T}{d}$
$4\pi r^2$ is the area of the surface of the sphere.
$\dfrac{\Delta Q}{\Delta t} = \dfrac{kA\Delta t}{d}$, which is the same relation as that for a wall of surface area A.
55. 7.16×10^3 W
57. 17.1 kcal/min
59. $\dfrac{P_2}{P_1} = \dfrac{\sigma A(T + \Delta T)^4}{\sigma A T^4} = \dfrac{T^4(1 + \Delta T/T)^4}{T^4} = \left(1 + \dfrac{\Delta T}{T}\right)^4$
$\dfrac{P_2}{P_1} = \left(1 + \dfrac{0.9K}{310K}\right)^4 = 1.012$
$P_2 = 1.012 P_1 = 1.00 \, P_1 + 0.012 P_1$
Power increased by 1.2%.

Chapter 17

1. 9.86×10^4 J
3. 1.52×10^5 J
5. **a.** 3.00×10^5 J   **b.** 0   **c.** -7.50×10^4 J   **d.** 0   **e.** 2.25×10^5 J
7. 4220 cal
9. **a.** 596 cal   **b.** 994 cal
11. 749 J
13. -300 J
15. Since the result is positive, the internal energy of the system is increased by 300 J.
17. 0.266 m³; 0.133 m³
19. 250 J
21. 1.20 J
23. **a.** -8.02 K   **b.** -8.02 °C   **c.** -14.4 °F
25. 55 J
27. 1.85×10^5 Pa
29. 7290 W
31. 40%
33. 49.2 J
35. 3.29 kcal/K
37. $+0.0359$ kcal/K
39. 3.88×10^{-4} kcal/K
43. 0.031 m³
45. 17.9 °C; 8.82×10^4 Btu/hr

Chapter 18

1. 1.89×10^{13} electrons
3. 1.97×10^{-1} N; repulsive, away from q_2
5. 8.75×10^{-3} N
7. 2.19×10^6 m/s
9. 4.80 N attractive; 1.60 N repulsive
11. 4.80×10^{-4} N
13. On q_1, 0.719 N in $-x$-direction; on q_2, 4.36 N in $+x$-direction; on q_3, 5.08 N in $+x$-direction
15. On 2.00 μC charge, 2.00 N in $+x$-direction
    On -4.00 μC charge, 6.40 N in $+x$-direction
    On 6.00 μC charge, 8.40 N in $-x$-direction
17. 1.21 m to the left of q_1 and along the extension of the straight line between q_1 and q_2
19. 10.8 N, outward along diagonal line between q_2 and q_3
21. 2.24×10^{-3} N, perpendicular to line l and in the directional sense from q_1 to q_2
23. 6.10 N toward and perpendicular to line connecting q_1 and q_2
25. 4.31×10^{-8} C
27. 1.00×10^{-1} N
29. 0.137 N, 327.2° relative to line q_1 to q_3
31. 525 m/s² away from q_1
33. (0.414 m, 1.17 m)
35. 2.10×10^{-14} in the negative x direction
37. 36.0 N m
39. 4.00×10^{-2} N. Note that since $d \gg a$, this force is identical, to three significant figures, with that found when the ring is treated as a point charge.

Chapter 19

1. 6.75×10^{-3} N/C
3. **a.** 1.26×10^5 N/C away from q_2 **b.** -9.00×10^4 N/C toward q_2
5. $x = 22.2$ cm
7. $E = 9.53 \times 10^4$ N/C; $\theta = 341°$
9. 2.25×10^{-7} C m
11. 6.43×10^5 N/C at 26.5° relative to the diagonal from the 2.00 μC to the -6.00 μC charge
13. 4.68×10^4 N/C, 222.9° relative to direction q_1 to q_2
15. 2.40×10^{-11} J; 2.40×10^{-11} J
17. **a.** 2.75×10^3 V; 6.88×10^2 V   **b.** 2.06×10^3 V
19. 2.40×10^3 J
21. **a.** 9.00×10^4 V   **b.** -1.80×10^4 V
23. 18.0 cm; zero work
25. 2.04×10^5 V
27. **a.** -8.75×10^3 V   **b.** -1.53×10^{-2} J
29. **a.** 1.00×10^5 V/m   **b.** 1.60×10^{-14} N   **c.** 1.76×10^{16} m/s²
31. **a.** 2.40×10^{-17} J   **b.** 2.40×10^{-17} J
33. For an electron, $v = 1.19 \times 10^6$ m/s;
    For a proton, $v = 2.77 \times 10^4$ m/s
35. **a.** 0.108 N/C in $+ x$-direction   **b.** 54.0 mV
    c. 2.16×10^{-7} N in $+ x$-direction
37. zero
39. $V = 3.63 \times 10^{-13}$ volts
41. 1.12 J
43. **a.** $V_A = 1.80 \times 10^4$ V; $V_B = 2.25 \times 10^4$ V
    b. 4.50×10^3 V   **c.** 5.94×10^{-3} J   **d.** -5.94×10^{-3} J
45. $V_1 = 270$ mV; $V_2 = 135$ mV; $V_3 = 108$ mV; $V_4 = 96.4$ mV;
    $V_5 = 93.1$ mV; $V_6 = 90.0$ mV; $V_7 = 87.1$ mV; $V_8 = 84.4$ mV; $V_9 = 77.1$ mV; $V_{10} = 67.5$ mV; $V_{11} = 54.0$ mV. $E_{30} = +540$ mV/m; $E_{30} = + 338$ mV/m; $E_{30} = +309$ mV/m; $E_{30} = +300$ mV/m; $E_{30} = +300$ mV/m; $E = 300$ mV/m

Chapter 20

1. 3.12×10^{19} electrons/s
3. 4.40×10^{-13} A
5. 48.0 Ω
7. 5.28 Ω
9. The resistance halves.

11. **a.** 1.41×10^{-4} Ω **b.** 5.64×10^{-8} Ω
13. 34.1 Ω
15. 147 °C
17. 539 Ω
19. 0.500 A; 1.08×10^6 J
21. 20.0 MW; 20.0 MJ/s
23. 600 Ω
25. 54.6 Ω
27. 43.3 Ω
29. **a.** 700 Ω **b.** 17.1 mA   **c.** 8.57 V   **d.** 3.43 V   **e.** $I_1 = 17.1$ mA; $I_2 = 11.4$ mA; $I_3 = 5.72$ mA   **f.** 0.205 W   **g.** $P_1 = 0.147$ W; $P_2 = 3.90 \times 10^{-2}$ W; $P_3 = 1.96 \times 10^{-2}$ W
31. **a.** 0.923 Ω **b.** 1.92 Ω **c.** 6.24 A   **d.** 6.24 V
    e. 5.76 V   **f.** $I_1 = 6.24$ A; $I_2 = 2.88$ A; $I_3 = 1.92$ A; $I_4 = 1.44$ A   **g.** $P_1 = 38.9$ W; $P_2 = 16.6$ W; $P_3 = 11.1$ W; $P_4 = 8.29$ W
33. **a.** 252 Ω **b.** $I_1 = I_4 = 47.6$ mA; $I_2 = 31.0$ mA; $I_3 = 16.6$ mA; $I_5 = 28.6$ mA; $I_6 = 19.0$ mA
35. 7.14 Ω
37. 9.00 V; 6.60 V
39. 13.3 V
41. 6.01×10^{-3} Ω
43. 200 Ω
45. $R = 32.7$ Ω; $I_1 = 0.117$ A; $I_3 = 6.67 \times 10^2$ A; $I_2 = 0.117$ A; $I_4 = 6.67 \times 10^{-2}$ A
47. $I_3 = 0.27$ A; $V_3 = 8.1$ V
49. $I_2 = 41.7$ mA; $I_3 = 0.275$ A; $I_1 = 0.233$ A
51. 16.5 V

53.

Resistance	Current (Amps)	Voltage (volts)
R_1	1.67	8.33
R_2	1.67	25.0
R_3	0.667	13.3
R_4	0.667	13.3
R_5	0.667	6.67
R_6	0.334	6.67
R_7	1.00	20.0

55. **a.** 19.2 Ω **b.** $I_1 = 0.624$ A; $I_2 = 0.288$ A; $I_3 = 0.192$ A; $I_4 = 0.144$ A
57. 837 s
59. $V_{out} = \left(\dfrac{I_{out}}{I_{in}}\right) V_{in}$
61. **a.** $V_{AC} = 2.05$ V   **b.** $V_{AD} = 2.43$ V   **c.** $V_{CD} = 0.385$ V
    d. $V_{CB} = 3.96$ V   **e.** $V_{DB} = 3.55$ V

Chapter 21

1. 48.0 μC
3. half
5. 2.66 pF
7. $C = 0.300$ pF; $d = 0.295$ m
9. 2.41 pF
11. 37.1 pF
13. $w = 6.80 \times 10^{-2}$ J; $q = 1.13 \times 10^{-3}$ C
15. 1.14 μF
17. 14.0 μF
19. **a.** 300 μC   **b.** $q_1 = 50.0$ μC; $q_2 = 250$ μC   **c.** 1.50×10^{-2} J
    d. $W_1 = 4.17 \times 10^{-4}$ J; $W_2 = 2.08 \times 10^{-3}$ J
21. **a.** 4.29 μF   **b.** 25.7 μC   **c.** $V_1 = 5.14$ V;
    $V_2 = V_3 = 0.857$ V   **d.** $q_1 = 25.7$ μC; $q_2 = 8.57$ μC; $q_3 = 17.1$ μC
    e. $W_1 = 66.0$ μJ; $W_2 = 3.67$ μJ; $W_3 = 7.34$ μJ
23. **a.** 6.90 μF   **b.** 33.1 μC   **c.** 2.07 V
25. $C = 8.43$ μF; $q_1 = 38.9$ μC; $q_3 = 62.3$ μC; $q_2 = 63.2$ μC; $q_4 = 37.9$ μC
27. 2.60×10^{-11} C²/N m²
29. 197 μF
31. 4.50×10^3 V
33. 3.20×10^{-4} J
35. **a.** 1.20×10^4 V/m   **b.** 2.54×10^4 V/m

37. $C = \dfrac{4\pi\epsilon_0 r_a r_b}{r_b - r_a} \approx \dfrac{\epsilon_0 A_s}{d}$ for r_a, r_b large since $4\pi r_a r_b \approx 4\pi \bar{r}^2 = A_s$ and $r_b - r_a = d$.

39. $709\ \mu F$; -4.51×10^5 C

41. a. $5.00 \times 10^{-2}\ \mu F$, series **b.** $125\ \mu F$, parallel

43. a. $C = 2.51\ \mu F$ **b.** $V_3 = V_5 = V_6 = V_{356} = 0.445$ V **c.** $q_1 = 30.1\ \mu C$ from part b

45. a. $I = 6.00 \times 10^{-2}$ A; $I = 0$ with C_1 or C_2 **b.** $V = 12.0$ V across R; $V_1 = V_2 = 12.0$ V across the capacitors
c. $q_1 = 24.0\ \mu C$; $q_2 = 60.0\ \mu C$

47. a. $C_0 = 17.7$ pF **b.** $C_d = 95.6$ pF **c.** $q_0 = 106$ pC;
$q_d = 573$ pC **d.** $E = 3.00 \times 10^3$ V/m
e. $W_0 = 3.19 \times 10^{-10}$ J; $W_d = 1.72 \times 10^{-9}$ J

49. a. $q_1 = 300\ \mu C$; $C_2 = 16.2\ \mu F$; $q_2 = 1620\ \mu C$
b. 2.00×10^5 V/m **c.** $W_1 = 1.50 \times 10^{-2}$ J; $W_2 = 8.10 \times 10^{-2}$ J

51. a. $C_0 = \dfrac{\epsilon_0 A}{d_0}$ **b.** $C = \dfrac{\epsilon_0 A}{d_0} = C_0$

53. a. $C = \left(\dfrac{\kappa_2 \kappa_1}{\kappa_2 + \kappa_1}\right) C_0$ **b.** $C = \kappa C_0$

55. a. 5.65×10^{-17} F **b.** 5.65×10^{-18} C

Chapter 22

1. 4.47×10^{-17} N perpendicular to **v** and **B**
3. a. 6.01×10^{-15} N **b.** 6.60×10^{15} m/s²
5. 4.09×10^{-5} m/s
7. 5.33×10^{-4} m
9. 9.63×10^{-4} m
11. 0.200 T
13. 1.57×10^{-3} N
15. 3.50×10^{-3} T
17. 6.28×10^{-4} T
19. 11.9
21. $4.00\ \mu T$
23. 5.00×10^4 A
25. a. 0 **b.** $26.7\ \mu T$
27. $50.0\ \mu T$; $53.1°$ counterclockwise from line $\overline{AI_2}$
29. 7.96×10^3 turns
31. 1.44×10^{-4} N/m to the left
33. 0.157 A m²
35. a. 0.150 A m² **b.** 2.12 A
37. $\tau = (1.07 \times 10^{-4}$ N m$)\sin\theta$; $\tau_{max} = 1.07 \times 10^{-4}$ N m
39. a. 2.40×10^{-19} N **b.** 1.44×10^8 m/s² in direction $-\mathbf{v} \times \mathbf{B}$, perpendicular to the velocity and the magnetic field **c.** 1.67×10^{-7} s **d.** 3.67×10^{-3} m
41. $B = 113\ \mu T$ directed along the diagonal of the square and toward wire 3.
43. 4.40×10^{-10} N
45. $F = 0$
47. 12.5 T
49. a. $1.20\ \Omega$ **b.** 22.7 m **c.** 181 **d.** 8.12×10^{-3} T
e. Choose a cardboard core of smaller diameter and shorter length. The increase in n increases B. Use a larger diameter wire. The required length of wire will be greater, resulting in more turns and hence a larger n.
51. a. $\ell = 379$ m **b.** $\ell = 0.121$ m **c.** $A = 4.57 \times 10^{-2}$ m² **d.** $m = 0.457$ A m², at maximum current
e. $\tau_{max} = 1.14 \times 10^{-2}$ N m **f.** $\phi = 7.27 \times 10^{-3}$ N m/rad
53. 1.31 N m; $\tau_{max}\omega$; 137 W
55. a. $m = \dfrac{qBR}{v}$ **b.** $1.13R$

Chapter 23

1. a. 1.20×10^{-3} Wb **b.** 0 **c.** 1.04×10^{-3} Wb
3. -0.450 V
5. a. 1.77 mV **b.** $70.8\ \mu A$
7. a. $\Delta\mathbf{A}$ is perpendicular to the loop plane and opposite **B**.
b. $180°$

c. $\mathscr{E} = B\dfrac{\Delta A}{\Delta t}$
d. counterclockwise, looking in the direction of **B**
9. a. 15.0 mV **b.** 30.0 mA **c.** counterclockwise
11. 300 T/s
13. 43.2 mV; no
15. 17.5 mV
17. 0.239 T
19. $I = \dfrac{\Delta Q}{\Delta t} = \dfrac{\mathscr{E}}{R} = \dfrac{1}{R}\left(-N\dfrac{\Delta\Phi_M}{\Delta t}\right)$; $\Delta Q = -\dfrac{N\Delta\Phi_M}{R}$
21. clockwise, as viewed from the bar magnet side
23. 255 rad/s
25. 51.7 V
27. 4.29×10^4 rad/s
29. 1.77 cm
31. 4.00 mH
33. -0.750 V
35. -2.00×10^4 A/s
37. 6.25×10^{-2} J
39. 0.800 H
41. a. the current in R is downward **b.** the current in R is upward
43. 3.14 mA; right to left
45. 6.70×10^{-10} A
47. 8.84 mV
49. 4.00 H
51. a. $\varepsilon = 1.23$ mV **b.** $\Phi_M = 2.46 \times 10^{-5}$ Wb **c.** $\varepsilon = 4.92 \times 10^{-4}$ V $\sin\left(20.0\dfrac{\text{rad}}{\text{s}}t\right)$ **d.** $I_{max} = 16.4\ \mu A$
55. $\varepsilon = \left(\dfrac{1}{2}gB\ell \sin 2\theta\right)t$

Chapter 24

1. 70.7 V
3. 7.07 A
5. $1.26 \times 10^4\ \Omega$
7. 44.5 A
9. $265\ \Omega$
11. a. 1.59 Hz **b.** 15.9 Hz **c.** 159 Hz; low frequencies
13. a. $0.754\ \Omega$ **b.** 146 A
15. $1.03 \times 10^3\ \Omega$
17. a. $-17.1°$ **b.** $107°$ **c.** $-72.9°$
19. 0.780 H
21. $Z = 280\ \Omega$; $\phi = 0.370°$
23. 1.07 H
25. 4.49×10^{-7} H
27. a. 6.75×10^5 H **b.** $1.20 \times 10^4\ \Omega = 12.0$ kΩ **c.** 10.4 pF
29. 2.85×10^{-18} F
31. $\pm 45.0\ \Omega$
33. a. $1.88 \times 10^3\ \Omega$ **b.** 0.110 A **c.** 58.5 mA
d. 0.125 A **e.** $-28.0°$ **f.** $880\ \Omega$
35. a. $663\ \Omega$ **b.** 0.220 A **c.** 0.166 A **d.** 0.276 A
e. $37.0°$ **f.** $399\ \Omega$
37. 1.20 A
39. a. 66.7 turns **b.** 75.0 V; 22.5 V
41. $P_1 = 1.00 \times 10^5$ W; $P_2 = 1.00 \times 10^3$ W
43. a. $1.76 \times 10^3\ \Omega$ **b.** $497\ \Omega$ **c.** $1.55 \times 10^3\ \Omega$
d. 0.142 A **e.** $V_R = 128$ V; $V_L = 250$ V; $V_C = 70.5$ V
f. 220 V, the applied voltage **g.** $54.5°$ **h.** 21.3 Hz
45. a. $884\ \Omega$ **b.** $2.01 \times 10^3\ \Omega$ **c.** 54.8 mA
d. 98.7 V **e.** 48.4 V **f.** 110 V, the applied voltage
g. $-26.1°$ **h.** 0.898
47. $\varepsilon = \varepsilon_1 + \varepsilon_2 + \varepsilon_3$
$$-L\dfrac{\Delta i}{\Delta t} = -L_1\dfrac{\Delta i}{\Delta t} - L_2\dfrac{\Delta i}{\Delta t} - L_3\dfrac{\Delta i}{\Delta t}$$
49. The current through R, C, and L is 83.3 mA. $i_1 = 32.0$ mA (current in C_1); $i_2 = 51.3$ mA (current in C_2)

51. $i = 88.0$ mA = current through R_1, R_2, C_1, and C_2.
$i_{L1} = 38.5$ mA through L_1; $i_{L2} = 49.1$ mA through L_2

53. $V_{AB} = \dfrac{\varepsilon}{\left(1 - \dfrac{1}{4\pi^2 LCf^2}\right)}$

V_{AB} increases as f increases, and approaches the applied voltage ε as f reaches high values. For low frequencies, V_{AB} is small.

Chapter 25

1. 2.26×10^5 N m²/C
3. 3.39×10^8 N/C
5. 7.53×10^{14} N/C s
9. 60.0 mA
13. $E = (0.85$ V/m$)\sin[(1.88$ m$^{-1})x - (5.65 \times 10^8$ s$^{-1})t]$
15. 8.78 min
17. 4.17×10^{14} to 7.89×10^{14} Hz
19. 6.00×10^{12} to 4.16×10^{14} Hz
21. 3.21 m
23. 23.8 MJ/m²
25. a. 199 W/m² **b.** 49.7 W/m² **c.** 22.1 W/m²
   d. 12.4 W/m² **e.** 7.96 W/m²
27. a. 274 V/m; 9.13×10^{-7} T **b.** 137 V/m; 4.57×10^{-7} T
   c. 91.2 V/m; 3.04×10^{-7} T **d.** 68.4 V/m; 2.28×10^{-7} T
   e. 54.8 V/m; 1.83×10^{-7} T
29. 2.74 V/m; 9.13×10^{-9} T
31. 1.73×10^{-2} V/m
33. 2.86×10^5 W/m²
35. 2.82×10^7 V/m
37. 3.21 W
39. a. 1.38×10^{-6} W = 1.38 μW **b.** 7.62×10^{-2} V/m
41. a. 60.2 MHz **b.** 60.2 MHz **c.** 4.98 m
43. $\sqrt{\kappa}$

Chapter 26

1. a. 7.89×10^{14} Hz **b.** 4.17×10^{14} Hz
3. 2.22×10^4
5. $q = 20.0$ cm; $h_i = 10.0$ cm
7. a mirror of length 2′10″, or one half the height of the student
9. 12.3 cm
11. a. 12.0 cm **b.** 13.3 cm **c.** 20.0 cm **d.** $q =$ is at infinity
   e. $q = -10.0$ cm
13. a. 1.00 cm **b.** 1.67 cm **c.** 5.00 cm **d.** is infinite **e.** 10.0 cm
15. $+7.5$ cm
17. 30.0 cm
19. Case I, $p = 25.0$ cm; Case II, $p = 18.8$ cm
21. 75.0 cm
23. 60.0 cm
25. -2.00 cm
27. a. -8.75 cm **b.** -8.00 cm **c.** -6.67 cm
   d. -5.00 cm **e.** -3.33 cm
29. 15.5 cm
37. Case I, $R = 60.0$ cm; Case II, $R = 20.0$ cm
39. $q = -60.0$ cm; M $= -3.00$; image is virtual and inverted
41. Rays 1, 2, and 3 pass very close to the focal point while rays 4 and 5, which are further removed from the axis of the mirror, come nowhere near the focal point. A spherical mirror forms a good image if we use only a small portion of the spherical surface.

Chapter 27

1. 30.3°
3. 40.3°
5. a. 2.25×10^8 m/s **b.** 2.04×10^8 m/s **c.** 1.24×10^8 m/s
7. 773 nm
9. a. 18.6° **b.** 1.91×10^8 m/s **c.** 376 nm
11. 90.0 cm
13. 61.0°
15. 1.50
17. $+21.0$ cm; converging

19. $+46.2$ cm; converging
21. a. 60.0 cm **b.** -3.00 **c.** 9.00 cm **d.** The image is real since q is positive. **e.** Since M is negative, the image is inverted.
23. 40.0 cm
25. 11.1 cm
27. 10.0 cm
29. 1.67 cm
31. a. The image is real and located 70.0 cm from the object.
   b. 10.0 cm
33. -60.0 cm
35. $f_c = +.650$ cm
37. $v_2 = 1.17 \times 10^8$ m/s
39. a. $r = 67.5°$ **b.** $\sin r = 1.06$. There is no refracted ray; we have total internal reflection.
41. 43.4 cm; 16.6 cm
43. $P = +2.33$ diopters
45. The wavelength is 380 nm.

Chapter 28

1. 0.169°; 0.338°; 0.507°
3. 1.03 cm; 2.06 cm; 3.09 cm
5. a. 0.169°; 0.338°; 0.507° **b.** 0.084°; 0.253°; 0.421° **c.** 2.06 cm; 4.12 cm; 6.18 cm; 1.03 cm; 3.09 cm; 5.15 cm
7. m = 28
9. R = 4.69 m
11. a. halved **b.** doubled
13. 169,779
15. 1.50×10^5
17. 613 nm
19. Wave 1 is out of phase with wave 2 but in phase with waves 3 and 4.
21. Wave 1 is out of phase with wave 2, in phase with wave 3, and not completely out of phase with wave 4.
23. a. 532 nm **b.** nothing in the visible spectrum
25. 133 nm; 400 nm
27. 1.63
29. $0.738I_0$
31. 5.00×10^{-5} m
33. $\lambda I/2h$; $\lambda I/2h$
35. 9.80×10^{-7} m
37. 7.66°
39. $\lambda = 6.13 \times 10^{-6}$ m
41. m = 499
43. 28.1°; 70.4°; no third bright spot
45. 5.98×10^{-7} m
47. 6.25×10^{-7} m
49. 7.42 mm
51. a. 14.3 cm **b.** 0.649 mm
53. 4.54×10^3 m
55. $\Delta\theta = 0.1°$

Chapter 29

1. 65.0 m; 61.0 m
3. a. 20.0 m/s **b.** 6.1 m/s
5. a. 7.50 m/s **b.** 6.50 m/s
9. 6.55×10^8 m; 5.46 s
13. 91.7 m
15. $L_0 = 182$ km; $L_2 = 79.3$ km
17. a. 2.50×10^{-6} s **b.** 360 m **c.** 600 m
19. $T_0 = 0.07605$
21. a. $0.263c$ **b.** $0.806c$ **c.** $0.882c$
23. a. impossible **b.** mass at rest
   c. $0.995c$ **d.** $0.99995c$ **e.** $0.9999995c$
25. V $= 2.55 \times 10^8\ \dfrac{\text{m}}{\text{s}}$
27. 0.78 MeV
29. 1.00×10^{-10} J
31. 2.71×10^{-10} J
33. 7.50×10^8 s
35. 3.46 light years; $t_{space} = 6.92$ yr; $t_{earth} = 8.00$ yr

37. $\dfrac{\rho_0}{1-v^2/c^2}$

39. $v = c\sqrt{1-\left(\dfrac{E_0}{E}\right)^2}$

41. $E = 3.82 \times 10^{26}$ J; $m = 4.24 \times 10^9$ kg; $T = 1.48 \times 10^{13}$ yr

Chapter 30

3. $v = 0.577c$

7. b. 1.45 m

9. $\Delta t_f = (6.25 \times 10^{-7})(1 + 5.44 \times 10^{-11})$s

13. $\Delta t_f = (0.500)(1 + 2.11 \times 10^{-11})$s

Chapter 31

1. a. 9.25×10^{-2} J **b.** 0.968 Hz **c.** 1.44×10^{32}
   d. imperceptible

3. 6.14×10^{-26} J

5. 2.40 eV

7. 1.20×10^{15} Hz; 250 nm

9. 1.33×10^{-27} kg m/s

11. 3.06×10^{-19} J

13. 8.59×10^{-16} J

15. 2.41×10^{-34} m

17. 2.43×10^{-11} m

19. 3.69×10^{-11} m

21. a. $E = 1.74 \times 10^{-10}$ J **b.** $\lambda = 2.30 \times 10^{-15}$ m

23. 5.11×10^{-29} m/s

25. 1.59×10^{-14} m

Chapter 32

1. 7.68×10^{-13} J

3. a. 1.69×10^{-14} m **b.** 5.71×10^{-15} m

5. # electrons $= Z = 47$; # protons $= Z = 47$;
   # neutrons $= A - Z = 61$

7. $E_1 = -13.6$ eV; $r_2 = 4.62 \times 10^{-11}$

9. a $n = 4$ **b.** $r4 = 0.846$ nm **c.** $E4 = -0.85$eV

11. 5.33×10^4 K

15. 3.15×10^{-34} J s; 2.10×10^{-34} J s; 1.05×10^{-34} J s

17. a. 1.95×10^{-18} J **b.** 2.94×10^{15} Hz **c.** 102 nm

19. 1.48×10^{-34} J s; 1.48×10^{-34} J s

21. a. $E_5 = -0.544$eV **b.** $L = 2.57 \times 10^{-34}$ Js **c.** $L_z = 0, \pm 1.05 \times 10^{-34}$ Js, $\pm 2.10 \times 10^{-34}$ Js **d.** $\theta = 90°$

23. a. $L_{z1} = 1.05 \times 10^{-34}$ J s; $L_{z0} = 0$; $L_{z,-1} = -1.05 \times 10^{-34}$ J s **b.** $\theta_1 = 45°$; $\theta_0 = 90°$; $\theta_{-1} = 135°$

25. $\tau_1 = 3.26 \times 10^{-26}$ N m; $\tau_2 = 5.16 \times 10^{-26}$ N m; $\tau_3 = 5.65 \times 10^{-26}$ N m; $\tau_4 = 5.16 \times 10^{-26}$ N m; $\tau_5 = 3.26 \times 10^{-26}$ N m

27. a. $(-5.48 \times 10^{-19} - 1.84 \times 10^{-23})$J **b.** One energy photon will be $+1.64 \times 10^{-18}$ J. The other two photons will have an energy of $\pm 1.84 \times 10^{-23}$ J. **c.** 2.47×10^{15} Hz. The other two photons will have frequencies with an additional frequency of $\pm 2.77 \times 10^{10}$ Hz. **d.** 1.21×10^{-7} m

29. $\pm 2.31 \times 10^{-26}$ J

31. $4s^1$

33.

$n = 1$	$l = 0$	$m_e = 0$	$m_s = \pm\frac{1}{2}$
$n = 2$	$l = 0$	$m_e = 0$	$m_s = \pm\frac{1}{2}$
	$l = 1$	$m_e = 0, \pm 1$	$m_s = \pm\frac{1}{2}$
$n = 3$	$l = 0$	$m_e = 0$	$m_s = \pm\frac{1}{2}$
	$l = 1$	$m_e = 0, \pm 1$	$m_s = \pm\frac{1}{2}$
$n = 4$	$l = 0$	$m_e = 0$	$m_s = \pm\frac{1}{2}$

35. $l = 4$. The orbit must be $n \geq 5$.

37. 7.11×10^{-27} J/T

39. a. $r_n = r_1 n^2$ **b.** $v_n = \dfrac{v_1}{n}$ **c.** $E_3 = -1.51$ eV
   d. $E_1 = -13.6$ eV **e.** $\Delta E = 12.1$ eV
   f. $f = 2.92 \times 10^{15}$ Hz **g.** $\lambda = 1.03 \times 10^{-7}$ m $= 103$ nm

Chapter 33

1. a. 29; 58; 29 **b.** 11; 24; 13 **c.** 84; 210; 126 **d.** 20; 45; 25
   e. 82; 206; 124

3. 2.86×10^{23}

5. 107.868331 u

7. 0.00857 u; 7.98 MeV

9. 3.44×10^{-10} s^{-1}

11. 4.86 days

13. The half-life is about 245 days; very few decay in one day.

15. 17.6 yr

17. a. 9.25×10^5 Bg **b.** 4.63×10^5 Bg **c.** 2.31×10^5 Bg

19. $^{216}_{84}$Po

21. $^{225}_{88}$Ra

23. $^{210}_{82}$Pb $\rightarrow$ $^{0}_{-1}$e $+ \bar{v} + ^{210}_{83}$Bi

25. $^{49}_{24}$Cr $\rightarrow$ $^{0}_{+1}$e $+ v + ^{49}_{23}$V

27. $^{52}_{25}$Mn $\rightarrow$ $^{0}_{-1}$e $+ v + ^{52}_{24}$Cr

29. 1.88×10^3 MeV; -1.81×10^3 MeV

31. 156 KeV

33. 17.6 MeV

35. a. $\lambda_{Th} = 2.37 \times 10^{-5}$ days^{-1}; $\lambda_{Rn} = 0.181$ days^{-1} **b.** $A_{Th} = 3.59 \times 10^{12}$ Bq; $A_{Rn} = 2.85 \times 10^{16}$ Bq **c.** $A_{Th} = 3.58 \times 10^{12}$ Bq; $A_{Rn} = 3.93 \times 10^8$ Bq **d.** $N_{Th} = 1.305 \times 10^{22}$ nuclei; $N_{Rn} = 1.88 \times 10^{14}$ nuclei **e.** The activity of Rn has decreased considerably, whereas the activity of Th has hardly changed. The same is true of the number of radioactive nuclei. When the radioactive Rn has for all practical purposes disintegrated, the Th will still last for many years.

37. 5350 yr

Chapter 34

1. $1.64 \times 10^{3-13}$ J; 2.47×10^{20} Hz; 1.21×10^{-12} m

3. 1.80×10^{17} J; 1.36×10^{50} Hz; 2.21×10^{-42} m

5. charge is 0; spin is $+\dfrac{1}{2}$

7. $q_T = +$e; $s_T = +\dfrac{1}{2}$

9. $q = 0, s = 0$

11. spin is 1; charge is 0

13. The three quarks all have the same quantum numbers which violates the Pauli exclusion principle. The two quarks give a fractional charge and the net color is not white.

15. 432

17. There are no 2 quark combinations; because spin cannot be made to equal ½.
   The following are 3 quark combinations; the spins cannot be all aligned and the three colors, RGB, must be present.

uud	ctd	b̲b̲d̲
uus	cts	b̲b̲s̲
uub	ctb	b s d
ucd	ttd	
ucs	tts	
ucb	ttb̲	
utd	d̲d̲d̲	
uts	d̲d̲s̲	
utb	d̲d̲b̲	
ccd	s̲ s̲ s̲	
ccs	s s d	
ccb	s s b	
	bbb	

Bibliography

College Physics

There are so many good books on college physics that it is impossible to list them all. Let me mention a few standards and/or favorites.

Brancazio, Peter J. *The Nature of Physics.* New York: Macmillan Publishing Co., 1975. An absolutely delightful textbook on college physics. It is a little more descriptive than most books.

Holton, G., and Roller, D. *Foundations of Modern Physical Science.* Reading, Mass.: Addison-Wesley Pub. Co., 1965.

Lindsay, Robert B. *Basic Concepts of Physics.* New York: Van Nostrand Reinhold Co., 1971.

Sears, F. W., Zemansky, M. W., and Young, H. *College Physics.* Reading, Mass.: Addison-Wesley Pub. Co. This book could probably be called the "bible" of college physics texts. It has been around for about thirty years.

Stevenson, R., and Moore, R. B. *Theory of Physics.* Philadelphia, Pa.: W. B. Saunders Co., 1967.

University Physics

University physics covers essentially the same topics as college physics, but it uses calculus and is therefore at a higher level. For anyone wishing a reference to this more advanced level, the "bible" of university physics texts is:

Halliday, David, Resnick, Robert. *Physics.* New York: John Wiley and Sons, 1977.

Modern Physics

Although modern physics textbooks are usually at a calculus level and therefore above the level of a college physics textbook, I would like to list a favorite of mine that the more advanced student might wish to pursue.

Beiser, Arthur. *Perspectives of Modern Physics.* New York: McGraw-Hill Book Co., 1969. This is a delightful text on modern physics. I have used it in my modern physics classes since it was first published and its flavor can be seen throughout the modern physics topics in this text.

Popular Physics Books

There are a large number of books on the market today that treat some of the latest topics in the field of physics. These books are very meaningful and very accessible to the student who is taking or has completed a course in college physics. The list is divided into two groups—quantum physics and relativity—although there is an overlap in many of the books, as might be expected.

Quantum Physics

Atkins, P. W. *The Creation.* San Francisco: W. H. Freeman & Co., 1981.

Barrow, J. D., Silk, J. *The Left Hand of Creation, The Origin and Evolution of the Expanding Universe.* New York: Basic Books, Inc., Publishers, 1983.

Bickel, Lennard. *The Deadly Element, The Story of Uranium.* New York: Stein and Day Publishers, 1979.

Chaisson, Eric. *Cosmic Dawn, The Origins of Matter and Life.* Boston: Little Brown and Co., 1981.

Chester, Michael. *Particles, An Introduction to Particle Physics.* New York: Mentor Books, New American Library, 1978.

Davies, P. C. W. *The Forces of Nature.* Cambridge: Cambridge University Press, 1979.

Davies, Paul. *Other Worlds, A Portrait of Nature in Rebellion, Space, Superspace and the Quantum Universe.* New York: Simon & Schuster, 1980.

Davies, Paul. *SuperForce, The Search for a Grand Unified Theory of Nature.* New York: Simon and Schuster, 1984. All of Davies' books are very enjoyable.

Davies, Paul. *The Runaway Universe.* New York: Harper & Row, Publishers, 1978.

Davies, Paul, and Gribbin, John. *The Matter Myth, Dramatic Discoveries That Challenge Our Understanding of Physical Reality.* New York: Touchstone Book, Simon & Schuster, 1992.

Feinberg, Gerald. *What is the World Made Of? Atoms, Leptons, Quarks, and Other Tantalizing Particles.* Garden City, N.Y.: Anchor Press/Doubleday, 1978.

Ferris, Timothy. *The Red Limit, The Search for the Edge of the Universe,* 2nd ed. New York: Quill, 1983.

Fritzsch, Harald. *Quarks, The Stuff of Matter.* New York: Basic Books, Inc., 1983.

Fritzsch, Harald. *The Creation of Matter, The Universe from Beginning to End.* New York: Basic Books, Inc., 1984.

Gribbin, John. *In Search of Schrödinger's Cat, Quantum Physics and Reality.* New York: Bantam Books, 1984.

Gribbin, John. *In Search of the Big Bang, Quantum Physics and Cosmology.* Toronto: Bantam Books, 1986.

Jastrow, Robert. *God and the Astronomers.* New York: W. W. Norton & Co., 1978.

Kaku, Michio, Trainer, Jennifer. *Beyond Einstein, The Cosmic Quest for the Theory of the Universe.* New York: Bantam Books, 1987. An easy-to-read book that discusses some of the ideas of the superstring theory.

Mulvey, J. H., ed. *The Nature of Matter.* Oxford: Clarendon Press, 1981.

Polkinghorne, J. C. *The Particle Play, An Account of the Ultimate Constituents of Matter.* San Francisco: W. H. Freeman and Co., 1979.

Schechter, Bruce. *The Path of No Resistance, The Story of the Revolution in Superconductivity.* New York: Simon and Schuster, 1989.

Sutton, Christine. *The Particle Connection.* New York: Simon and Schuster, 1984.

Trefil, James S. *From Atoms to Quarks, An Introduction to the Strange World of Particle Physics.* New York: Charles Scribner's Sons, 1980. Trefil is an excellent science writer and all his books are very enjoyable.

Trefil, James S. *The Moment of Creation, Big Bang Physics From Before the First Millisecond to the Present Universe.* New York: Charles Scribner's Sons, 1983.

Relativity

Bergmann, Peter G. *The Riddle of Gravitation.* New York: Charles Scribner's Sons, 1968.

Bernstein, Jeremy. *Einstein.* New York: Penguin Books, 1982.

Bondi, Hermann. *Relativity and Common Sense, A New Approach to Einstein.* New York: Dover Publications, 1964.

Born, Max. *Einstein's Theory of Relativity.* New York: Dover Publications, 1965. Excellent text.

Davies, P. C. W. *The Edge of Infinity, Where the Universe came from and how it will end.* New York: Simon and Schuster, 1981.

Davies, P. C. W. *The Search for Gravity Waves.* New York: Cambridge University Press, 1980.

Davies, P. C. W. *Space and Time in the Modern Universe.* New York: Cambridge University Press, 1978.

Eddington, Sir Arthur. *Space, Time, and Gravitation, An outline of the General Relativity Theory.* New York: Harper Torchbooks, 1959. A delightful little book.

Einstein, Albert. *Relativity, The Special and General Theory.* New York: Crown Publishers, Inc., 1961.

Ellis, George F. R., and Williams, Ruth M. *Flat and Curved Space-Times.* Oxford: Clarendon Press, 1988.

Gardner, Martin. *The Relativity Explosion.* New York: Vintage Books, 1976.

Gibilisco, Stan. *Understanding Einstein's Theories of Relativity, Man's New Perspective on the Cosmos.* New York: Dover Publications, 1983.

Gribbin, John. *TimeWarps, Is time travel possible? An exploration into today's best scientific knowledge about the nature of time.* New York: Delacorte Press, 1979.

Gribbin, John. *Unveiling the Edge of Time, Black Holes, White Holes, Wormholes.* New York: Harmony Books, 1992.

Halpern, Paul. *Time Journeys, A Search for Cosmic Destiny and Meaning.* New York: McGraw-Hill, 1990.

Halpern, Paul. *Cosmic Wormholes, The Search for Interstellar Shortcuts.* New York: Dutton Books, 1992.

Harpaz, Amos. *Relativity Theory, Concepts and Basic Principles.* Boston: Jones and Bartlett Publishers, 1992. A truly excellent book.

Hawking, Stephen W. *A Brief History of Time, From the Big Bang to Black Holes.* New York: Bantam Books, 1988.

Hoffmann, Banesh. *Relativity and its Roots.* New York: Scientific American Books, 1983.

Kaufmann, William J. *Black Holes and Warped Spacetime.* San Francisco: W. H. Freeman and Company, 1979.

Kaufmann, William J. *The Cosmic Frontiers of General Relativity.* Boston: Little Brown and Company, 1977.

Luminet, Jean-Pierre. *Black Holes.* Cambridge: Cambridge University Press, 1992.

Macvey, John W. *Time Travel, A Guide to Journeys in the Fourth Dimension.* Chelsea, Michigan: Scarborough House Publishers, 1990.

Marder, L. *Time and the Space Traveller.* Philadelphia: University of Pennsylvania Press, 1974.

Mermin, N. David. *Space and Time in Special Relativity.* New York: McGraw-Hill Book Company, 1968.

Morris, Richard. *Time's Arrows, Scientific Attitudes Toward Time.* New York: Simon and Schuster, 1984.

Narlikar, Jayant V. *The Lighter Side of Gravity.* San Francisco: W. H. Freeman and Company, 1982.

Nicolson, Iain. *Gravity, Black Holes and the Universe.* New York: Halsted Press, 1981.

Novikov, Igor. *Black Holes and the Universe.* Cambridge: Cambridge University Press, 1990.

Reichenbach, Hans. *The Philosophy of Space and Time.* New York: Dover Publications, 1958.

Sexl, Roman, and Sexl, Hannelore. *White Dwarfs, Black Holes, An Introduction to Relativistic Astrophysics.* New York: Academic Press, 1979. A fascinating little book. The discussion of the falling clock is the one used in this book.

Shipman, Harry L. *Black Holes, Quasars, and the Universe,* 2nd ed. Boston: Houghton Mifflin Company, 1980.

Taylor, John G. *Black Holes.* New York: Avon Books, 1973.

Wald, Robert M. *Space, Time, and Gravity, The Theory of the Big Bang and Black Holes.* Chicago: The University of Chicago Press, 1977.

Wheeler, John Archibald. *A Journey into Gravity and Spacetime.* New York: Scientific American Library, 1990. A beautiful popular account of relativity written by one of the masters.

Will, Clifford M. *Was Einstein Right? Putting General Relativity to the Test.* New York: Basic Books, Inc., 1986.

redits

Photographs

Chapter 1

Opener: © NASA/Rainbow; **1.1:** Field Museum of Natural History, #110a2513c; **1.5B:** NASA; **1.6:** National Institute of Standards & Technology; **1.7B:** Courtesy of Central Scientific Company; **1.8:** © National Institute of Standards & Technology, Boulder Labs, US Department of Commerce; **1.9:** NASA

Chapter 2

Opener: © Rhoda Sidney/The Image Works; **2.18A:** © Eric Millette/The Picture Cube

Chapter 3

Opener: © Yoav Levy/Phototake; **3.1:** © The Granger Collection; **3.14B:** © R. P. Kingston/The Picture Cube; **3.16:** © NASA; **3.18A & B:** © Mark Antman/The Image Works; **p. 62:** © Yoav Levy/Phototake; **3.24:** © Sheila Sharidan/Monkmeyer Press; **p. 70:** © Joe Sohm/The Image Works

Chapter 4

Opener: © Julian Baum/Science Photo Library/Photo Researchers, Inc.; **4.1A:** © The Granger Collection; **4.3:** © M. E. Warren/Photo Researchers, Inc.; **4.15B:** © Wm. C. Brown Communications, Inc./Photo by Toni Michaels; **4.20:** © Jaye R. Phillips/The Picture Cube; **p. 112A–C:** © Tom Doody/The Picture Cube

Chapter 5

Opener: Dan McCoy/Rainbow; **5.6B:** © Michal Heron/Monkmeyer Press; **5.20B:** © Bruce Iverson; **5.21B:** © Wm. C. Brown Communications, Inc./Photo by Toni Michaels; **5.23C:** © Bob Daemmrich/The Image Works; **p. 147:** © Will McIntyre/Photo Researchers, Inc.

Chapter 6

Opener: NASA; **6.12C:** © Gloria Karlson/The Picture Cube; **6.14C:** © Charles Feil/Stock Boston; **6.15B:** © Galen Rowell/Peter Arnold, Inc; **6.17A:** © Visual Images West/The Image Works; **6.18B:** © Jerome Yeats/Science Photo Library/Photo Researchers, Inc.; **pp. 176, 181:** © NASA

Chapter 7

Opener: © Robert Mathena/Fundamental Photographs; **7.8:** © David Ball/The Picture Cube; **7.10:** © Tony Savino/The Image Works; **p. 207:** © Farrell Grehan/Photo Researchers, Inc.; **p. 209:** From *Secrets of the Great Pyramid* by Peter Tomkins. Copyright © 1971 by Peter Tomkins. Reprinted by permission of Harper Collins Publishers, Inc.

Chapter 8

Opener: © Richard Megna/Fundamental Photographs; **8.1:** © Richard Megna/Fundamental Photographs; **8.4:** © J. L. Barken/The Picture Cube; **8.7A:** © Tim Davis/Photo Researchers; **8.7B:** © Y. Arthus-Bertrand/Gerard Vandystadt/Photo Researchers, Inc.; **8.8B:** © C. E. Miller/Peter Arnold, Inc.; **8.11B:** © Richard Megna/Fundamental Photographs; **8.12B:** © Dan Burns/Monkmeyer Press; **p. 235:** © Terry Barner/Unicorn Stock Photos

Chapter 9

Opener: © Richard Megna/Fundamental Photographs; **9.2B:** © Bob Daemmrich/The Image Works; **p. 243:** © Richard Megna/Fundamental Photographs; **9.7:** © Frank Pedrick/The Image Works; **9.13:** NASA; **9.14:** © Dave Schaffer/Monkmeyer Press; **9.15(Right):** © Arthur Grace/Stock Boston; **p. 268:** © D. Wray/The Image Works; **p. 275:** © Gerard Lacz/NHPA

Chapter 10

Opener: © Fundamental Photographs; **10.1A:** © D.O.E./Science Source/Photo Researchers, Inc.

Chapter 11

Opener: © Paul Silverman/Fundamental Photographs; **11.7D:** © Mark Antman/The Image Works

Chapter 12

Opener: © Richard Megna/Fundamental Photographs; **12.1G, p. 336:** © Richard Megna/Fundamental Photographs; **12.16B:** © Toni Michaels; **12.19D:** © Eric Roth/The Picture Cube; **12.22B:** © Martha Cooper/Peter Arnold, Inc.

Chapter 13

Opener: © Bruce Coleman; **13.4C:** © Larry Miller/Photo Researchers, Inc.; **p. 377:** © Tony Freeman/Photo Edit; **13.7B:** © Herb Snitzer/The Picture Cube; **p. 381:** © Renee Lynn/Photo Researchers, Inc.; **13.13C:** © Jerry Wachter/Photo Researchers, Inc.; **13.14B:** © Dick Davis/Photo Researchers, Inc.; **p. 394:** © Blair Seitz/Photo Researchers, Inc.

Chapter 14

Opener: © Mark Antman/The Image Works

Chapter 15

Opener: © Bruce Iverson; **p. 427:** © Mark Burnett/Photo Researchers, Inc.; **p. 435:** © Tom Branch/Science Source/Photo Researchers, Inc.; **p. 445:** © Toni Michaels

Index

Springs
    applying quantum condition to vibrating,
        924–25
    compression and rarefaction of, 321
    conservation of energy and vibrating, 307–9
    Hook's law on elasticity for, 287–88
    in parallel, 312–13
    potential energy of, **306–7**
    in series, 313–14
    simple harmonic motion in vibrating, 297–99
Standard rays, 779
Standing waves, 339–46
    of electron in orbit, *969*
    formation of, *339*
    frequency, resonance, and, 343–45
    nodes and antinodes of, 341–42
    on string, *341, **342***
    vibration modes, 342, *343*
Stanford Linear Accelerator Center (SLAC),
        1041
Static equilibrium, 121, 123
Static friction, 102–3
    coefficients of, 103(table), 111
Statics, **121**
    fluid, 367
Stationary state, electron in, 970
Stationary wave. *See* Standing waves
Statistical mechanics, 501
Stefan, Joseph, 465
Stefan-Boltzmann constant, 503
Stefan-Boltzmann law, **465**
Stoney, George, 961
Stopping potential, 927
Strain
    elasticity and, **285**, *286*
    shearing, 289
Streamline flow, 382
Stress
    elasticity and, **285**, *286*
    shearing, 289
Strings
    creation of pulse on, *331–42*
    reflection and transmission of wave on, in two
        different media, 333–36
    reflection of wave on not firmly fixed, *332*
    reflection of wave on rigidly fixed, 332, *333*
    speed of transverse wave on, 329–31
    standing waves on, *341, **342***
    vibration modes on, *343*
Strong nuclear force, 520, **1005**, 1038, **1044–45**
Sublimation, latent heat of, 414–**15**
Sun, light rays bent by, *909*
Superconductivity, 571
Superforce, **1050–51**
Superposition, principle of, **336–39**, 921, 940
Superposition of electric fields, 537–42
    calculating electric field of two positive
        charges, 538–41
    electric field of two like charges, 540
    electric field of two unlike charges (electric
        dipole), 541–42
    mathematical statement and definition of,
        538
    of potentials, 551–52

Superstring theory, 1051
Supersymmetry, 1050–51
Surface area, pressure and, 369–70
Surface charge density, 719
System
    closed, and conservation of energy, 198
    conservative, **205**
    defined, **189**
System of units, 16(table)–17. *See also*
        International System (SI) of Units
Szilard, Leo, 1025

T

Tangent function, **26**
    determining side of triangle using, 27
Tangential velocity, 242–43
Telescope, 796–98
Television waves, 736–37
Teller, Edward, 1025
Temperature, **403**–8. *See also* Heat; Heat
        transfer
    of gas, calculating, 436
    isothermal process at constant, 482, 492
    planetary, 472
    scales, *404,* 405
    scales, conversions of, 406–8
    SI system of noting, 406
    speed of sound as function of, 347
    thought experiment on, *9,* 403, *404*
    variation of resistance with, 569–71
Temperature gradient, **453**
Tensor analysis, 907
Tesla, Nikola, 632
Tesla (measurement unit), **632**
Thales of Miletus, 6
Theory as step in scientific method, 6, 7
Thermal energy. *See* Heat
Thermal equilibrium, **408**
Thermal expansion, **425**
    area expansion of solids as, 427–28
    coefficients of, 426(table)
    linear expansion of solids as, 425–27
    volume expansion of gases, 430–32
    volume expansion of solids and liquids as,
        429–30
Thermal resistance *R,* **463**
Thermodynamic process, heat absorbed or
        liberated in, **484,** 485
Thermodynamics, **479–512**
    Carnot cycle, 496–97
    entropy, 499–501
    entropy, statistical interpretation of, 501–5
    first law of, 485–92
    first law of, special cases of, 492–93
    gasoline engine and, 493–95
    heat added to and removed from
        thermodynamic system, 483–85
    ideal heat engine, 495–96
    meteorology and, 505–8
    second law of, 498–99
    work concept applied to thermodynamic
        system, 479–83
Thermometer, 403, **404,** 405

Thin film, interference of light from, **820**–29
    air wedge, 826–28
    density of medium and, 820–21
    Newton's rights, 828–29
    nonreflecting glass, 823–26
    soap bubble or film of oil, 821–23
Thin lens(es), **776–77**
    in contact, 789–92
Third law of motion, Newton's, **81**–82
    law of conservation of momentum as
        consequence of, 217–18
Third law of rotational motion, Newton's, **251**
Thompson, Benjamin, 408
Thompson, G. P., 939
Thompson, William, 8, 405
Thomson, J. J., 519
    plum pudding model of atom by, 961, *962*
Thorium, radioactive decay series of, 1015, *1016*
Thought experiment on temperature, *9*
Time, 9, 10. *See also* Spacetime
    displacement as function of, 47–48
    flow of, in direction of increase in entropy,
        505
    projectile motion and dimension of, 57–58,
        63–64
    proper, 869
    SI vs. BES system of noting, 16(table)
    standard of, 14–15
    velocity as function of, 46
Time dilation, 843, 869–71
    defined, **870**
    Lorentz contraction and, 869(table)
    nonEuclidian nature of spacetime and, 895,
        901, *902*
    proper time and, **869**
    slowing of accelerated clock and, 953–56
    twin paradox and, 870–71, 898
Tornado, atmospheric pressure in, 374–75
Torque, 126–29
    clockwise and counterclockwise, 130, 143
    in current loop in external magnetic field,
        646–52
    defined, **127**
    in induced emf in rotating loop of wire in
        magnetic field, 669–74
    second condition of equilibrium and, 129–31
Torr (measurement unit), 371
Torricelli, Evangelista, 371
Total internal reflection, 773–74
Traction, equilibrium and medical, 147–48
Traffic
    kinematics and congested, 70–72
    noise from, 360–61
Trajectory of moving projectile, **56**, 57, *66, 68,*
    69
Transformer, **705–8**
Translational equilibrium, 126, 133
Translational kinematics. *See also* Kinematics
    angular momentum and, 260
    combined rotational motion and, treated by
        law of conservation of energy, 264–66
    combined rotational motion and, treated by
        Newton's second law, 254–57
    compared to rotational motion, 242(table)